#1

General data and fundamental constants

Quantity	Symbol	Value	Power of ten	Units
Speed of light	c	2.997 924 58*	10^8	$m\,s^{-1}$
Elementary charge	e	1.602 177	10^{-19}	C
Faraday constant	$F = N_A e$	9.6485	10^4	$C\,mol^{-1}$
Boltzmann constant	k	1.380 66	10^{-23}	$J\,K^{-1}$
Gas constant	$R = N_A k$	8.314 51		$J\,K^{-1}\,mol^{-1}$
		8.314 51	10^{-2}	$L\,bar\,K^{-1}\,mol^{-1}$
		8.205 78	10^{-2}	$L\,atm\,K^{-1}\,mol^{-1}$
		6.2364	10	$L\,Torr\,K^{-1}\,mol^{-1}$
Planck constant	h	6.626 08	10^{-34}	$J\,s$
	$\hbar = h/2\pi$	1.054 57	10^{-34}	$J\,s$
Avogadro constant	N_A	6.022 14	10^{23}	mol^{-1}
Atomic mass unit	u	1.660 54	10^{-27}	kg
Mass				
electron	m_e	9.109 39	10^{-31}	kg
proton	m_p	1.672 62	10^{-27}	kg
neutron	m_n	1.674 93	10^{-27}	kg
Vacuum permittivity	$\varepsilon_0 = 1/c^2\mu_0$	8.854 19	10^{-12}	$J^{-1}\,C^2\,m^{-1}$
	$4\pi\varepsilon_0$	1.112 65	10^{-10}	$J^{-1}\,C^2\,m^{-1}$
Vacuum permeability	μ_0	4π	10^{-7}	$J\,s^2\,C^{-2}\,m^{-1}$ $(=T^2\,J^{-1}\,m^3)$
Magneton				
Bohr	$\mu_B = e\hbar/2m_e$	9.274 02	10^{-24}	$J\,T^{-1}$
nuclear	$\mu_N = e\hbar/2m_p$	5.050 79	10^{-27}	$J\,T^{-1}$
g value	g_e	2.002 32		
Bohr radius	$a_0 = 4\pi\varepsilon_0\hbar^2/m_e e^2$	5.291 77	10^{-11}	m
Fine-structure constant	$\alpha = \mu_0 e^2 c/2h$	7.297 35	10^{-3}	
Rydberg constant	$R_\infty = m_e e^4/8h^3 c\varepsilon_0^2$	1.097 37	10^5	cm^{-1}
Standard acceleration of free fall	g	9.806 65*		$m\,s^{-2}$
Gravitational constant	G	6.672 59	10^{-11}	$N\,m^2\,kg^{-2}$

* Exact value.

Physical Chemistry

PHYSICAL CHEMISTRY

FIFTH EDITION

P. W. Atkins

*University Lecturer and Fellow
of Lincoln College, Oxford*

W. H. FREEMAN AND COMPANY
NEW YORK

The cover as illustrated by Ian Worpole, is based on Peter Atkins' representation of the amplitude of the antibonding orbital formed from the overlap of two H1s orbitals.

Library of Congress Cataloging-in-Publication Data

Atkins, P. W. (Peter William), 1940-
    Physical Chemistry/Peter Atkins. – 5th ed.
      p.   cm.
    Includes index.
    ISBN O-7167-2402-2
    1. Chemistry, Physical and theoretical. I. Title
QD4531.2.A88 1994b
541.3–dc20

Printed in the United States of America

1 2 3 4 5 6 7 8 9 0 KP 9 9 8 7 6 5 4 3

This edition has been authorized by the Oxford University Press for sale in the USA and Canada only and not for export therefrom.

Preface

A fifth edition of a text has all the advantages of maturity: it represents the accumulated wisdom of all readers who have taken the trouble to suggest variations in former editions and ideas on organization and content. At the same time it stands in danger of succumbing, perhaps unwittingly, to one of the problems of maturity: an unwillingness to change. Although the advantages of maturity will be perceived throughout the text, I hope the danger has been avoided. From their familiarity with successive editions, users will be aware that in each case I have provided not a slight tinkering with the text but a wholesale rewrite. This edition is no exception, for yet again I have considered every word. Moreover, to quash the charge of sameness, I have introduced a number of innovations in this text which I hope will improve its pedagogical value.

The most obvious new feature of the text is the use of a second colour. It has been considered for some time now that students who are introduced to chemistry through the glitter of freshman texts and their equivalents in various parts of the world, are suddenly confronted with grey again when they move on to more advanced courses. The books look dull after encounters with the expensive texts of introductory chemistry. Partly to offset this attitude and to make the text look more lively and enticing, I have redrawn all the art with the systematic use of a second colour. Broadly speaking, I have used the second colour for emphasis, or for an abstract aspect of a diagram (for example, showing the symmetry axis of an object). Only rarely is it used just for decoration. The use of colour entailed redrawing nearly 1000 illustrations, and I have taken the opportunity to do so in a more consistent, modern style using the electronic aids that are now available.

The second major innovation is the use of *Justifications*. I explored the use of this device in my *Elements of physical chemistry*, where it proved popular, and have made extensive use of *Justifications* in this text. Briefly, the format of a *Justification* is that a conclusion (typically a mathematical formula) is stated in the text, and the derivation then follows the statement of the result. There are several advantages in this approach. One is that a result is immediately presented, and not lost in a thicket of other equations. Another is that a *Justification* captures the manner in which lectures are often given, where the *result* is presented and discussed, and the detailed derivation is left to private study. Thirdly, *Justifications* allow the text to be used at different levels: some students will find that they need not grapple with some derivations, or find that they can return to them later when their ideas have matured. The *Justifications* help to make the subject more digestible and accessible without damaging the authoritative completeness of the text. Some background material which I consider to be important but too intrusive to appear in the body of the text is present as *Further information* at the end of the volume. (The *Further reading* sections are now collected there too.)

In keeping with the general aim to make the text more accessible I have introduced several more changes. Thus, in the Examples that occur throughout the text, I have introduced a *Method* section. The most difficult part of answering a problem is knowing where to start; the *Methods* suggest how to collect one's thoughts, what expressions to use, where to find the data, and generally how to think strategically about the problem. Keen-eyed, long-memoried readers will remember that I used *Methods* in the third edition, but dropped them from the fourth; these *Methods*, though, are more extensive and more structured than in the third edition.

Another feature is the introduction of *Molecular interpretations*, which are new to this edition. These sections, which are found in Part 1, stemmed from the views that have been expressed concerning the question of whether to start from classical thermodynamics or from atomic properties. I have always maintained that instructors can easily use Part 2 of this text before Part 1 (indeed, few instructors consider themselves bound by the order of a text at this level), but I wanted to find a way of introducing atomic concepts (which have a strong intuitive appeal and an undeniable instructive value) without damaging what I perceive to be the intellectual self-sufficiency of thermodynamics. So, interspersed into the text of Part 1 are optional *Molecular interpretations*, which show (as their name suggests) the interpretation of a bulk result in terms of atomic and molecular properties. I believe that this approach enriches and deepens thermodynamics without damaging its integrity.

I have introduced a number of devices for helping readers to keep the subject in mind. First, I have reinstated an entirely new version of Chapter 0, which gives an overview of the whole subject and introduces some fundamental concepts by way of review. Chapter 0, now called the *Introduction*, gives the big picture. Secondly, I have moved the *Key ideas* to the ends of their respective chapters. I consider that they act better as a check list after the topics have been introduced than as a somewhat daunting list of terms at the head of a chapter. In their place at the start of the chapter, I have put a paragraph of introduction, an abstract of the chapter, which maps out the sequence of ideas that the reader will encounter, and highlights the key issues.

There are a considerable number of changes within the text. For example, I have responded to the advice that suggested that the kinetic theory of gases came far too late. It now comes in Chapter 1, where it augments the phenomenological aspects of gases. I have shifted several sections on the properties of surfaces from their original location in Chapter 7 to a chapter that now deals with all aspects of surfaces, both liquid and solid (Chapter 28). Some chapters have been completely rewritten. An example is Chapter 15 on group theory: it has been urged on me that it is inappropriate to attempt to give a formal introduction to group theory in a single chapter. Now I give a highly pragmatic introduction—essentially confined to showing how character tables are used—but a lot of the background is still present in several *Justifications*. The material on transport properties has been combined with the material on ionic motion and diffusion into a single chapter; this reorganization brings out more clearly the unity of the ideas underlying molecular migrations of various kinds.

While rewriting a text keeps the narrative young (at the risk of ageing the author), I know that it is important to try to ensure accuracy. In this edition I have called on the good offices of a number of colleagues who have scrutinized the text at several stages of its production. The typescript was read by over 50 people in part or in its entirety, and I have corrected all the errors and infelicities of expression they found. We took particular care at proof stage. Galley proofs were read several times by Michael Clugston (the whole) and Dixie Goss and Stephen Davis (sharing the whole), and we collectively tried to perfect the proofs at that stage. The entire set of page proofs was then read again, with a fresh eye, by Carmen Giunta, who happily found few points to correct at that stage. Charles Trapp, who has been largely responsible for compiling the *Solutions Manual* for this edition, also provided many helpful suggestions and corrections. We are very conscious of the need to perfect a text, and have put a considerable effort into cleansing this edition of irritating typographical slips, cross-references, and the like.

To these readers I owe my very special thanks. I also owe a great debt to all those colleagues, users, students, and translators around the world who sent me the distillation of their wisdom. If a book is a monument to anyone, then it is a monument to its well-wishers who have done so much to ensure that it is exactly what is needed for modern courses in physical chemistry.

As always, I wish to record my thanks to my publishers for the understanding and unstinting assistance they have provided at all stages of the massive and complex task of producing a book such as this.

Oxford, 1993   P.W.A.

Acknowledgements

I wish to record publicly my deep appreciation of the help given to me by countless people around the world. Those who contributed to earlier editions have already been thanked, and their influence and contribution is the foundation for this edition. Many people continue to write to me, and I am always grateful and consider carefully what they suggest. Those who have contributed explicitly to the preparation of this edition are:

Professor J. Acrivos, San Jose State University, San Jose, California

Dr D. L. Andrews, University of East Anglia, Norwich

Dr R. D. Armstrong, University of Newcastle upon Tyne

Professor N. M. Atherton, University of Sheffield

Professor F. Baglin, University of Nevada at Reno, Nevada

Professor D. L. Baulch, University of Leeds

Professor C. Bergo, East Stroudsburg University, East Stroudsberg, Pennsylvania

Professor R. Binning, Rice University, Houston, Texas

Dr I. M. Campbell, University of Leeds

Dr J. H. Carpenter, University of Newcastle upon Tyne

Dr P. A. Christensen, University of Newcastle upon Tyne

Dr W. Clegg, University of Newcastle upon Tyne

Dr A. A. Clifford, University of Leeds

Dr M. J. Clugston, Tonbridge School

Dr D. B. Cook, University of Sheffield

Dr I. L. Cooper, University of Newcastle upon Tyne

Professor T. Darrah Thomas, Oregon State University, Corvallis, Oregon

Professor S. Davis, George Mason University, Fairfax, Virginia

Professor J. de la Vega, Villanova University, Villanova, Pennsylvania

Dr R. Devonshire, University of Sheffield

Dr M. B. Ewing, University College, London

Professor A. V. Fratini, University of Dayton, Dayton, Ohio

Professor C. Giunta, Le Moyne College, Syracuse, New York

Professor D. Goss, Hunter College, New York

Professor A. Hamnett, University of Newcastle upon Tyne

Professor H. Harris, University of Missouri at St Louis, Missouri

Dr J. Henderson, University of Leeds

Professor S. Hunnicutt, University of Dayton, Dayton, Ohio

Dr G. Jackson, University of Sheffield

Dr P. Jones, University of Newcastle upon Tyne

Professor N. Kestner, Louisiana State University, Baton Rouge, Louisiana

Professor H. Kutz, University of Tennessee at Chattanooga, Tennessee

Dr P. G. Laye, University of Leeds

Dr T. H. Lilley, University of Sheffield

Dr W. MacFarlane, University of Newcastle upon Tyne

Dr A. O. S. Maczek, University of Sheffield

Dr M. McCoustra, University of East Anglia, Norwich

Professor M. Mueller, Rose-Hulman Institute of Technology, Terre Haute, Indianapolis

Dr J. E. Parkin, University College, London

Professor M. J. Pilling, University of Leeds

Mr R. Pilling, University of Nottingham

Dr S. L. Price, University College of London

Dr W. T. Raynes, University of Sheffield

Professor B. Robinson, University of East Anglia

Dr S. K. Scott, University of Leeds

Dr P. W. Seakins, University of Leeds

Dr A. J. Smith, University of Sheffield

Dr D. Smithies, University of Leeds

Dr N. Taylor, University of Leeds

Dr T. Thirunamachandran, University College, London

Dr K. M. Thomas, University of Newcastle upon Tyne

Dr S. H. Walmsley, University College of London

Dr J. B. Whittaker, University of Leeds

Professor D. E. Williams, University College, London

Summary of contents

Contents

Contents

0

Introduction: orientation and background

Physical chemistry establishes and develops the principles that are used to explain and interpret the observations made in the other branches of chemistry. The subject is characterized by three main approaches: the discussion of bulk properties in terms of thermodynamics, the use of spectroscopy to explore the behaviour of individual atoms and molecules, and the analysis of the rates and mechanisms of chemical change. This opening chapter surveys these approaches and puts them in a general context. It also reviews some basic ideas on which the entire text hinges. All the new material will be dealt with in greater detail later in the text.

SOME BASIC IDEAS

Certain concepts are common to the whole of the subject (and to other branches of science too), and here we review some of them.

The structure of science

The observations that physical chemistry (like other branches of science) organizes and explains are summarized by scientific laws. A **scientific law** is a summary of experience. Thus, we shall encounter the laws of thermodynamics, which are summaries of the relations between bulk properties and particularly observations on the transfer of energy. We shall also encounter the laws of quantum mechanics, which summarize observations on the behaviour of individual particles. The first step in accounting for a law is to propose a **hypothesis**, which is essentially a guess at an explanation in terms of more fundamental concepts. Dalton's atomic hypothesis, which was proposed to account for the laws of chemical composition in terms of atoms, is an example. When the hypothesis has become established, perhaps as a result of the success of further experiments it has inspired or by a more elaborate formulation (often in terms of mathematics) that puts it into the context of broader aspects of science, it is promoted to the status of a **theory**. We shall encounter a number of theories in this text: among them will be the theories of atomic structure and the chemical bond.

A characteristic of physical chemistry (like other branches of physical science) is that, to develop theories, it adopts **models** of the system it is seeking to describe. A model is a simplified version of the system that focuses on the essentials of the problem. That is, a model seeks to identify the heart of the problem and ignores possible complications that are considered to be of only secondary importance. Once a successful model has been constructed (and tested against known observations and any experiments it inspires), it can be made more sophisticated so as to incorporate some of the complications the original model ignored. Thus, models provide the framework for discussions, and reality is captured rather like a building is completed, decorated, and furnished. We shall encounter a number of such models. One example is the **kinetic model** of gases, in which a gas is regarded as a collection of particles in ceaseless, chaotic motion. Another example is the **nuclear model** of an atom, and in particular a hydrogen atom, which is used as a basis for the discussion of the structures of all atoms. A third very important type of example is that of a **perfect gas**, which is an idealized model of the gaseous state of matter in which the pressure is related to the volume, temperature, and amount of substance by the relation

$$pV = nRT$$

where R is a constant called the **gas constant**. This expression, which is the starting point of the discussion of **real gases** (that is, actual gaseous chemical species) is one of the most important equations in physical chemistry, and a wide variety of thermodynamic expressions are based on it. It is introduced in Chapter 1.

It is commonly the case that the form of equations developed on the basis of a simple model is maintained in the elaboration of the model. The advantage of such a procedure is that the appearance of many equations is then preserved, and the equations remain familiar. An example of such a modification is the replacement of a concentration term in certain thermodynamic expressions (such as an equilibrium constant) by an effective concentration called an **activity**. To make this a practically useful procedure it is then necessary to find the relation between this effective concentration and the true concentration. That step is often done by proposing a more detailed model of the behaviour of the system that takes into account the interactions between molecules.

The amount of substance: moles

Another preliminary idea we need is that in chemistry we are normally concerned with enormous numbers of atoms. The number of atoms in 1 g of matter is of the order of 10^{22} to 10^{23}, which is larger than the number of stars in the visible universe. To express these large numbers it is conventional in chemistry to refer to the **amount of substance** n and to express that amount in the unit called the **mole**. The formal definition of a mole is that it is the amount of substance that contains as many objects (atoms, molecules, formula units, ions, or other specified entities) as there are atoms in exactly 12 g of carbon-12. This number is found experimentally to be approximately 6.02×10^{23}. Therefore, if a sample contains N specified entities, then the amount of substance it contains is $n = N/N_A$, where N_A is **Avogadro's constant**, $N_A = 6.02 \times 10^{23} \, \text{mol}^{-1}$. Note that N_A is a quantity with units, not a pure number. Likewise, if the amount of a substance is n (for example, 2.0 mol O_2), then the number of elementary objects is nN_A (in this example, 1.2×10^{24} O_2 molecules).

A distinction is made in chemistry between **extensive properties** and **intensive properties**. The former depend on the size (the extent) of the sample; the latter are independent of it. Examples of extensive properties are mass and volume. Examples of intensive properties are temperature, density, and pressure. A **molar property**, such as the mass per mole of specified entities (in practice, the value of an extensive property X of the sample divided by the amount of substance the sample contains, $X_m = X/n$), is also intensive because the property and the amount are both proportional to the size of the sample so that the ratio is independent of its extent. With one exception, molar quantities are denoted by the subscript m. An example is the **molar volume** V_m, the volume per unit amount of substance (the volume per mole). The one exception to the notation is the **molar mass** M, the mass of sample divided by the amount of substance it contains. If the sample consists of atoms, then the molar mass is the mass per mole of atoms; if it consists of molecules, then the molar mass is the mass per mole of molecules. If the sample is an ionic solid (such as NaCl), then the molar mass is the mass per mole of formula units. The names 'atomic weight' and 'molecular weight' are still widely used in place of molar mass, but we shall not use them in this text.

We shall often need to consider the amounts of matter present in solutions of various kinds. The term **molar concentration** of a solute in a solution refers to the amount of substance of the solute divided by the volume of the solution. Molar concentration is usually expressed in moles per litre (mol L^{-1} or mol dm^{-3}; 1 L is exactly the same as 1 dm^3). A solution in which the molar concentration of the solute is 1 mol L^{-1} is prepared by dissolving 1 mol of the solute in sufficient solvent to prepare 1 L of solution. Such a solution is widely called a '1 molar' solution and denoted 1 M. However, the symbol M is not a part of the *Système International* of units (SI units), and it is best to use it adjectivally. That is, we can speak of a 1 M NaCl(aq) solution, but we should refer to the molar concentration of NaCl in that solution as 1 mol L^{-1}. The term **molality** refers to the amount of substance of solute divided by the mass of solvent used to prepare the solution. Its units are typically moles per kilogram (mol kg^{-1}). Note that, whereas the molar concentration of a solute varies with temperature (because the volume of the solution changes), the molality is independent of temperature.

It is often helpful to be able to picture solutions on a molecular scale (indeed, the development of an appreciation of events on an atomic scale

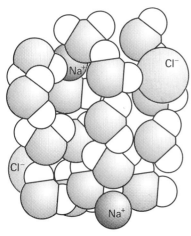

0.1 A schematic indication of the relative sizes of ions and molecules and the average separation of ions in a 1 M NaCl aqueous solution. There are typically about three H_2O molecules between ions. Cations tend to be found near anions, and vice versa. Cations are hydrated by weak bonding with the O atoms of neighbouring H_2O molecules; anions are hydrated by weak bonding through the H atoms.

is a valuable talent in physical chemistry). In a 1 M NaCl(aq) solution, the average separation between oppositely charged ions is about 1 nm, which is enough to accommodate about three H_2O molecules (Fig. 0.1). A dilute solution typically means a solution containing no more than about 0.01 mol L^{-1} of solute. In such solutions, the ions are separated by about 10 H_2O molecules.

Energy

The central concept of all explanations in physical chemistry, as in so many other branches of physical science, is that of **energy**. A formal definition of this quantity will be given in Part 1; here we shall make use of the somewhat bald definition of energy as *the capacity to do work*. We shall often make use of the universal law of nature that *energy is conserved*; that is, energy can be neither created nor destroyed. Therefore, although energy can be transferred from one part of the universe to another (as when water in a beaker is heated by electricity generated in a power station), the total quantity of energy available is a constant.

Much of physical chemistry involves the monitoring of transfers of energy and then interpreting the observation on the basis that the total energy is conserved. The great edifice of thermodynamics, which occupies Part 1 of this text and helps to co-ordinate and interrelate many of the bulk properties of matter, is constructed on the observation that energy is conserved but may be transferred from one region to another or may be converted from one form into another. The conservation of energy is also utilized in the interpretation of spectroscopy, which occupies the bulk of Part 2. In spectroscopy, energy is transferred between a molecule and the electromagnetic field (in the form of radiation); the latter is monitored and then interpreted in terms of the geometrical or electronic structure of the molecule. Finally, the conservation of energy is crucial to the process of change in chemistry, the topic of Part 3, for the rate of a reaction is largely determined by the rate at which excess energy accumulates in a molecule or an individual bond. That energy cannot simply be created; it must be gathered from the immediate surroundings, and the time it takes to accumulate where it is needed governs the rate at which chemical change can occur.

There are essentially three kinds of energy of direct significance in chemistry: kinetic energy, potential energy, and the energy of the electromagnetic field. The **kinetic energy** T of a body is the energy it possesses as a result of its motion. For a body of mass m travelling at a speed v, the kinetic energy is $\frac{1}{2}mv^2$, so a heavy body travelling rapidly has a high kinetic energy. A stationary body has zero kinetic energy.

The **potential energy** V of a body is the energy it possesses as a result of its position. The zero of potential energy is arbitrary. For example, the potential energy of a body is often set to zero at the surface of the Earth; the zero of potential energy for two charged particles is often taken to be at infinite distance apart. No universal expression for the potential energy can be given because it depends on the mode of interaction (the type of field) that the body experiences. However, there are two common types of interaction that give rise to simple expressions for the potential energy.

One is the potential energy of a body of mass m in the gravitational field of the Earth (a gravitational field acts on the mass of a body). If the body is at a height h above the surface of the Earth, then its potential energy is mgh, where g is a constant called the acceleration of free fall, $g = 9.81\,\mathrm{m\,s^{-2}}$, and $V = 0$ at $h = 0$ (the arbitrary zero mentioned previously). Of greater importance in chemistry is the potential energy of a charged body in the vicinity of another charged body (an electric field acts on the charge carried by a body). If a particle (a point-like body) of charge q_1 is at a distance r in a vacuum from another particle of charge q_2, then their potential energy is given by the expression

$$V = \frac{q_1 q_2}{4\pi\varepsilon_0 r}$$

The constant ε_0 is the **vacuum permittivity**, a fundamental constant with the value $8.85 \times 10^{-12}\,\mathrm{C^2\,J^{-1}\,m^{-1}}$. (Precise values of the quantities we introduce here and throughout the text will be found listed in the tables inside the front cover of the book; the imprecise values given here are often worth remembering because they help to establish the orders of magnitudes of quantities.) Note that, as remarked previously, $V = 0$ at infinite separation. This very important relation is called the **Coulomb potential energy**. It is important in chemistry because there we deal frequently with the interactions between electrons, nuclei, and ions, all of which are described by this potential energy.

Electromagnetic radiation

The energy of the electromagnetic field will be important to us because we shall often be concerned with the absorption and emission of radiation. Indeed, not only does spectroscopy depend crucially on the electromagnetic field, but some species acquire the energy they need to react from radiation. The latter subject is the domain of **photochemistry**. An electromagnetic field is a disturbance that spreads through empty space, the **vacuum**, at a fixed speed called the **speed of light** c, which is about $3 \times 10^8\,\mathrm{m\,s^{-1}}$.

An electromagnetic field consists of two components, an **electric field** that acts on charged particles (whether stationary or moving) and a **magnetic field** that acts only on moving charged particles. Each field produces a force that can accelerate the particle. An electromagnetic field is generated when charges move. That is the principle of a radio transmitter, where electrons are moved backwards and forwards in an antenna and hence generate an electromagnetic disturbance that propagates through space. An electromagnetic field can also induce motion in charged particles, which is what happens when the electromagnetic field of a distant transmitter moves past the antenna of a receiving set. The electromagnetic field propagates as a wave of disturbance, and is characterized by a **wavelength** λ, the distance between the neighbouring peaks of the wave, and its frequency ν, the number of times the wave returns to its original displacement per unit time as it passes a fixed point (Fig. 0.2). According to the original theories of the electromagnetic field, as established by the nineteenth-century physicist J. C. Maxwell, the

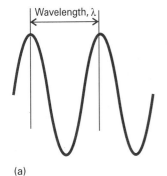

(a)

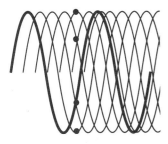

(b)

0.2 (a) The wavelength λ of a wave is the peak-to-peak distance. (b) The wave is shown travelling to the right at a speed c; at a given location, the instantaneous amplitude of the wave changes through a complete cycle (the four dots show half a cycle) as it passes a given point, and the frequency ν is the number of cycles per second that occur at a given point. Wavelength and frequency are related by $\lambda\nu = c$.

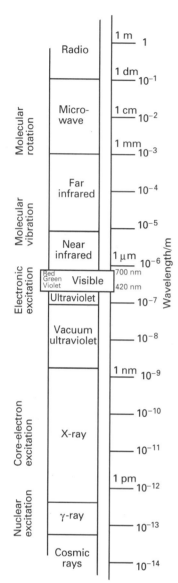

0.3 The regions of the electromagnetic spectrum and the types of excitation that give rise to each region.

energy of the electromagnetic field is determined by the overall **amplitude** of the wave, the maximum magnitude of the disturbance (as distinct from the *instantaneous* amplitude, the displacement at a given point in space at a given instant). We do not need the exact relation because twentieth-century views have modified this conclusion.

The classification of electromagnetic radiation into different regions according to its frequency and wavelength is summarized in Fig. 0.3. We shall see that different frequencies of radiation are absorbed or emitted by different types of electronic and molecular motions, and these associations are indicated in the illustration too.

Energy units

All three types of energy, kinetic, potential, and electromagnetic, have two features in common. The most important feature is that they are freely interconvertible from one form to the other. Thus, the potential energy of a ball is converted into kinetic energy when it is released. If the ball happens to be charged, then it generates electromagnetic radiation as it accelerates. The *total* energy, however, remains constant despite these transformations. The second feature is that the quantity of energy may be expressed in the same units for each type. In SI units, the unit of energy is the **joule** (J), which is defined as $1\,J = 1\,kg\,m^2\,s^{-2}$. A joule is quite a small unit of energy: for instance, each beat of the human heart consumes about $1\,J$. The unit is named after the nineteenth-century scientist J. P. Joule who helped to establish the role of energy in science. The **molar energy** is the energy per unit amount of substance and is normally expressed in joules per mole ($J\,mol^{-1}$) or a multiple of this unit (most commonly kilojoules per mole, $kJ\,mol^{-1}$). For example, if the average kinetic energy of a molecule in a gas is $6 \times 10^{-21}\,J$ (a typical value at room temperature), then the molar energy is this quantity multiplied by Avogadro's constant, or $4\,kJ\,mol^{-1}$. Many chemical reactions involve energies of the order of $10^2\,kJ$ per mole of molecules that react; for example, when natural gas (methane) burns to carbon dioxide, the reaction releases $890\,kJ$ per mole of CH_4 molecules.

Although the joule is the preferred unit of energy, it is sometimes convenient to employ other units. One of the most useful alternative units in chemistry is the **electronvolt** (eV), which is defined as the kinetic energy acquired when an electron is accelerated through a potential difference of $1\,V$. The energy that a 1.5-V battery in a cassette player provides for each electron that travels from one terminal to the other is $1.5\,eV$. The conversion is $1\,eV = 1.6 \times 10^{-19}\,J$. (Once again, for more precise values, see inside the front cover.) Many processes in chemistry involve energies of a few electronvolts. For example, to remove an electron from a sodium atom requires about $5\,eV$. Energies are also sometimes expressed as a **wavenumber** by dividing the energy by hc, where h is Planck's constant. The significance of a wavenumber and of Planck's constant will be explained shortly: all we need at this stage is to note that $1.0\,cm^{-1}$ is equivalent to $2.0 \times 10^{-23}\,J$, where the reciprocal centimetre, cm^{-1}, is the favoured unit for wavenumber.

EQUILIBRIUM

With some basic concepts established, we shall now consider the structure of physical chemistry as it is set out in this text. We shall trace the subject through the three main divisions that we have adopted, namely 'Equilibrium', 'Structure', and 'Change'. 'Equilibrium' is largely the domain of chemical thermodynamics, 'Structure' the domain of quantum mechanics and spectroscopy, and 'Change' the domain of chemical reaction. No concept is an island, however, and aspects of all three branches pervade the entire subject.

First, we consider 'Equilibrium' and the role of thermodynamics in chemistry. The role of energy was thought to be thoroughly understood towards the end of the nineteenth century. Considerable progress had been made in understanding the behaviour of energy when it is transformed from one form to another. These observations were originally motivated by a desire to improve the efficiencies of steam engines, in which the energy released when a fuel burns is transformed into motive power. It was found that the essential features (as distinct from the individual vagaries of particular engines) of such engines could be summarized by two laws, which in due course became the First and Second Laws of thermodynamics. The **First Law of thermodynamics** expresses the conservation of energy. The **Second Law of thermodynamics** expresses the observation about the *direction* of natural change. (Both laws will be explained fully in the early chapters of Part 1.) The introduction of the laws of thermodynamics through an analysis of steam engines may seem remote from the concerns of chemistry. However, it turns out that steam engines epitomize the processes that occur in chemistry. Thus, the conversion of heat to work, a process in which random molecular motion (heat) is converted to orderly motion (the motion characteristic of work, as in the motion of the atoms of a piston and a wheel), is analogous to the generation of a complicated molecule (an enzyme, for instance) from a chaotic collection of smaller molecules by using the energy provided by an appropriate source.

The deep connection of steam engines and chemistry was developed by two giants of nineteenth-century science, Josiah Willard Gibbs and Ludwig Boltzmann. The latter's contribution we shall consider later. Gibbs's contribution was to import a whole corpus of information about the transformation of energy in engines and apply them to discuss the physical and chemical transformations of matter. Part 1 of this text is essentially a monument to Gibbs, and is built on the foundations he laid.

When thinking about thermodynamics, there are two features to keep in mind which help to explain a wide range of phenomena. First, there is the conservation of energy (the essential content of the First Law). This subject is developed in Chapters 2 and 3 after an initial introduction to the properties of the simplest state of matter, the gaseous state, and an introduction to the kinetic model. The principal application of the First Law in chemistry is to the subject of **thermochemistry**, the study of the heat produced or required in the course of chemical reactions. The second feature to keep in mind is the extent to which energy spreads in a disordered manner. The crucial term here is *disordered*, for we shall see that all natural chemical processes are accompanied by a net increase in the disorder of the world, and that by looking for the direction in which disorder will increase we can identify whether a process has a tendency to occur. The extent of the dispersal of energy is measured by the property called the **entropy**, so considerations of the direction of natural change are based on the characteristics of entropy. Entropy is the subject of Chapters 3 and 4, which treat the Second Law of thermodynamics.

The quantity and extent of dispersal of the energy need to be kept in mind. To make this easier, a property called the **Gibbs energy** is introduced. This quantity combines the essential features of the First and Second Laws of thermodynamics in a manner that makes it exceptionally useful in chemistry. As we shall see in Chapters 5 to 10, the Gibbs energy is the root of most of the applications that we shall describe, and can be used to describe physical and chemical transformations and the compositions of systems that have reached equilibrium. At root though, beneath the mathematics that is used to develop the properties of the Gibbs energy, lie the simple concepts of the conservation of energy and the extent of its chaotic dispersal.

STRUCTURE

The second great branch of chemistry is its concern with the structures of atoms, molecules, and the physical states of matter (solids, liquids, and gases). In Part 2, therefore, we turn the spotlight on to the description of the properties of individual atoms and molecules, and build up a picture of how those properties are investigated and contribute to the properties of matter in bulk.

The quantization of energy

The great revolution in physics that occurred in the opening decades of the twentieth century and that introduced **quantum mechanics** is of crucial importance to chemistry. Chemistry is concerned with the behaviour of subatomic particles, particularly electrons, and for their description quantum mechanics is essential. The essential feature of quantum mechanics that distinguishes it from the **classical mechanics** of Isaac Newton and his immediate successors, is that particles have a wave-like character. That is, instead of particles and waves being distinct entities, particles have some of the properties of waves and waves have some of the properties of particles. This blurring of the distinction between particles and waves will be discussed fully in Part 2, and is epitomized by the **de Broglie relation** between the momentum of a particle (a particle-like property) and the wavelength of the particle (a wave-like property):

$$\lambda = \frac{h}{p}$$

where h is **Planck's constant**, a fundamental constant of nature with the value 6.6×10^{-34} J s. The de Broglie relation shows that the greater the

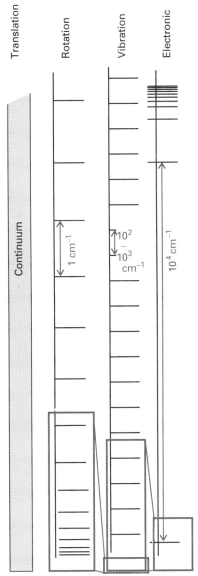

0.4 A representation of the quantization of the energy of different types of motion. Free translational motion in an infinite region is not quantized, and the permitted energy levels form a continuum. Rotation is quantized, and the separation increases as the state of excitation increases. The separation between levels depends on the moment of inertia of the molecule. Vibrational motion is quantized, and note the change in scale between the ladders. The separation of levels depends on the masses of atoms in the molecules and the rigidities of the bonds linking them. Electronic energy levels are quantized, and the separations are typically very large (of the order of 1 eV).

linear momentum p of an object (where p is the product of mass and velocity), the shorter the wavelength of the object. The important consequence of this **wave–particle duality** is that, as we shall see in Chapters 11 and 12, the energy of a body cannot be varied arbitrarily. That is, energy is **quantized**, which means that it is confined to certain discrete values that are characteristic of the system. In other words, a system can exist only in certain discrete **energy levels**.

An example will help to make this quantization of energy clear. According to classical physics, a pendulum can be made to swing with any energy. A small impulse sets it swinging with its natural frequency ν that is determined by its length and the force of gravity, but the amplitude of the motion is small. A bigger impulse sets the pendulum swinging with the same frequency, but the amplitude is greater because more energy has been transferred. An impulse of any reasonable magnitude can set the pendulum in motion and, by governing the impulse, any amplitude of swing can be achieved. According to quantum mechanics, however, the pendulum can possess one of a precisely defined, discrete set of energies. Specifically, the energy of the pendulum must have one of the values $\frac{1}{2}h\nu$, $\frac{3}{2}h\nu$, $\frac{5}{2}h\nu$, and so on. In a sense, quantum mechanics permits only some amplitudes of swing, and it is impossible to set the pendulum swinging with any intermediate amplitude. Furthemore, because the minimum energy is $\frac{1}{2}h\nu$, according to quantum mechanics a pendulum can never be completely still.

The reason why it took until the beginning of the twentieth century to discover the discrete character of the possible energies of a pendulum, and of many other mechanical systems, is the smallness of Planck's constant. Thus, a pendulum of natural frequency 1 Hz can accept energy in steps of about 6.6×10^{-34} J, which are so small that for all practical purposes it seems that energy can be transferred continuously. Only for oscillators with very high natural frequencies does the quantum $h\nu$ become large enough to give rise to observable differences from classical mechanics. A particle on a spring oscillates with simple harmonic motion, and its natural frequency increases as its mass decreases ($\nu \propto 1/\sqrt{m}$). When the mass is as small as that of atoms vibrating in molecules, the size of the quantum needed to excite the vibration to its next level is so great that significant quantum effects can be expected.

All kinds of motion other than completely free translational motion are quantized, and Fig. 0.4 depicts the permitted energy values in typical systems. Translational motion is quantized if the particle is confined to a finite region of space, but for containers of macroscopic dimensions the separation between levels is so small that for all practical purposes the motion is unquantized and the levels form a continuum. The separation between adjacent energy levels is typically smallest for rotational motion, larger for vibrational motion, and greatest for the energy of electrons in atoms and molecules. The separations of energy levels for a small molecule are about 10^{-23} J for rotational motion (which corresponds to 0.01 kJ mol^{-1}), 10^{-20} J for vibrational translational motion (10 kJ mol^{-1}), and 10^{-18} J for electronic excitation (10^3 kJ mol^{-1}). The features that determine the allowed energies of each type of motion (the size and mass of molecules for rotational motion, the masses of atoms and the rigidities of bonds for vibrational motion, and the allowed distributions of electrons for electronic excitation) are discussed in the opening chapters of Part 2.

The detection of energy levels: spectroscopy

The quantization of energy is the basis of the usefulness of **spectroscopy** for identifying species and exploring their structures. In spectroscopy molecules absorb or emit electromagnetic radiation. From the conservation of energy, the energy of the radiation emitted (or absorbed) must be equal to the difference in energy between the initial and final states of the molecule. Because energy levels are determined by the constitution of the molecule, an analysis of the characteristic energies of the emitted or absorbed radiation can be interpreted in terms of the constitution of the molecule, including bond lengths and angles and the strengths and rigidities of bonds.

The great usefulness of spectroscopy, however, stems from a quantum mechanical feature of electromagnetic radiation. We have seen that, according to classical physics (in particular Maxwell's theory of electromagnetic radiation), the energy of radiation is determined by the overall amplitude of the waves. Quantum theory provides a totally different interpretation by combining the concepts of particles with those of waves. Specifically, without abandoning the wave-like description of electromagnetic radiation, it introduces the concept of particle-like packets of electromagnetic energy called **photons**, each one of which travels at the speed of light. The **intensity** of the radiation is determined by the number of photons in the ray: an intense ray consists of a large number of photons; a feeble ray consists of only a few photons. (A human eye can respond to a single photon; a 100-W lamp generates about 10^{19} photons each second, but even so takes several hours to generate a mole of them.) The special feature of photons that concerns us immediately, however, is that the **energy** of each one is determined by the frequency of the radiation to which it corresponds. The relation is very simple: if the frequency of the radiation is v, then the energy of one photon of that radiation is hv, where h is Planck's constant. This relation implies that photons of microwave radiation (long-wavelength, low-frequency radiation) have lower energy than photons of visible light, and that the energies of photons of visible light increase as the light is changed from red (long-wavelength) to violet (short-wavelength).

The importance of photons is that they enable us to explore the energy levels of molecules by monitoring the frequencies of the radiation they emit and absorb. Thus, when a transition occurs between energy levels, the difference in energy is carried away as a photon. Because the energy of the photon is equal to the difference in energy of the molecular states, it has a characteristic frequency. In fact, if the difference in energy between the initial and final molecular states is ΔE, then from the conservation of energy it follows that the frequency of the corresponding radiation is given by the **Bohr frequency condition**

$$\Delta E = hv$$

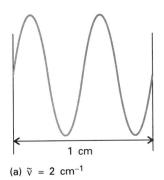

(a) $\tilde{\nu} = 2\ \text{cm}^{-1}$

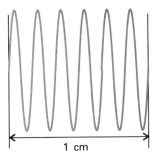

(b) $\tilde{\nu} = 6\ \text{cm}^{-1}$

0.5 The wavenumber $\tilde{\nu}$ is the number of wavelengths per unit length (commonly per centimetre). On the scale shown, the radiation corresponds to wavenumbers of (a) $2\ \text{cm}^{-1}$ and (b) $6\ \text{cm}^{-1}$. Note that the shorter the wavelength, the greater the wavenumber. The frequency is proportional to the wavenumber: $\nu = \tilde{\nu}c$.

This simple relation is the basis of all the forms of spectroscopy that we shall encounter in Chapters 16 to 18. As we shall see, **microwave spectroscopy**, which employs electromagnetic radiation of wavelengths of the order of millimetres and centimetres, is used to explore the rotational energy levels of molecules and hence their bond lengths and bond angles. **Infrared spectroscopy**, which makes use of radiation of the order of $1\ \mu\text{m}$, gives information on the rigidities of bonds. Visible and ultraviolet spectroscopy, which makes use of radiation with wavelengths of the order of $10^2\ \text{nm}$, gives information about the electronic energy levels of molecules. One very useful spectroscopic probe of molecules makes use of radiofrequency radiation, with a frequency of the order of $10^2\ \text{MHz}$ and a wavelength of the order of $300\ \text{cm}$. It may seem extraordinary that such long wavelengths can be used to determine the structures of molecules, but **nuclear magnetic resonance**, one of the most useful of all spectroscopic techniques, achieves this feat, as will be explained in Chapter 18.

In spectroscopy, it is common to specify the radiation in terms of a closely related quantity, the **wavenumber** $\tilde{\nu}$ of the radiation. The formal definition of wavenumber is $\tilde{\nu} = \nu/c$. It follows from the definition of wavelength that in a vacuum (where the radiation travels at a speed c) $\tilde{\nu} = 1/\lambda$. Therefore, as the wavelength increases, the wavenumber decreases. More specifically, the wavenumber can be envisaged as the number of complete wavelengths of radiation that fit into unit length (conventionally $1\ \text{cm}$). This interpretation is illustrated in Fig. 0.5. Visible light has a wavenumber of the order of $10^4\ \text{cm}^{-1}$, corresponding to 10^4 complete wavelengths per centimetre.

The reason for expressing energies as a wavenumber (through $\tilde{\nu} = E/hc$) should now be clear. Because the energy of a photon emitted by a molecule is equal to the difference in energies of the states involved in the transition, the wavenumber of the radiation is equal to the difference in the energies expressed as wavenumbers. Thus, reporting energies of states in wavenumbers is very appropriate when treating spectroscopic transitions.

Atomic and molecular structure

A major component of physical chemistry is its ability to provide explanations of the properties of atoms and the structures of molecules. These are the topics described in Chapters 13 and 14, respectively. The quantum mechanical ideas needed for these discussions are introduced in Chapters 11 and 12 together with the **Schrödinger equation**, the fundamental equation of quantum mechanics.

The description of the structures of atoms is an excellent example of how models are used in chemistry and how parameters are introduced to summarize properties. It is also a good example of the application of quantum mechanics to systems that are of central importance to chemistry. Thus, the Schrödinger equation for the hydrogen atom can be solved exactly, and serves to introduce and illustrate a number of quantum mechanical ideas, such as orbitals, quantum numbers, and angular momentum. However, to go beyond this simple one-electron atom requires sophisticated numerical analysis. That level of detail is inappropriate to much of chemistry, which generally finds it appropriate to use qualitative ideas that have been introduced as a result of empirical observations. One task for physical chemistry is to justify the use of these parameters and to relate them to more fundamental properties.

The hydrogen atom is taken as a model of many-electron atoms, and a simple scheme based on its structure is constructed that will account for the known electronic structures of the atoms (this is the **building-up principle**). That approach turns out to be very fruitful, because it serves to account for the layout of the periodic table (which is shown inside the back cover of the book), the most important unifying device in chemistry. It also accounts for the trends in the properties of atoms, particularly their sizes and the energies involved in removing or adding electrons to form ions.

A similar approach is taken for molecular structure in Chapter 14. There we see how the simple ideas first put forward by Gilbert Lewis, starting in about 1916, on the structure of the chemical bond can be modelled in terms of orbitals. In this connection, it will be important to keep in mind the distinction introduced in elementary chemistry courses between ionic bonding and covalent bonding.

An **ionic bond** is formed when one or more electrons are transferred from one atom to another, and the resulting ions cohere by a direct Coulombic interaction. A typical example is the crystal structure of sodium chloride, an array of Na^+ cations and Cl^- anions. As in this example, ionic bonding is non-directional and results in extensive aggregates of ions and the formation of an ionic solid. Such solids are generally hard and brittle and have high melting points. A **covalent bond** is formed when two atoms share a pair of electrons. The simplest example is the bond between two hydrogen atoms in H_2. When the electrons are shared equally, the bond is classified as **non-polar** (as in H_2 and Cl_2). When the electron pair is shared unequally, the bond is **polar** (as in HCl). Whether or not a molecule as a whole is polar (and possesses an **electric dipole moment**) depends on the extent to which the polar bonds cancel each other when the symmetry of the molecule is taken into account. Thus, benzene is non-polar because the diametrically opposite C—H bonds cancel one another. Covalently bonded molecules are usually discrete units, like N_2 or H_2O, but nevertheless may be very large. Human DNA, for example, is a covalently bonded molecule that is several metres long.

This discussion suggests that the symmetry of a molecule is important in determining its physical properties. Indeed, we shall see that symmetry considerations are a very powerful approach to the prediction and rationalization of molecular properties, particularly when developed systematically through the approach called **group theory** (Chapter 15). We shall see that symmetry considerations help not only in the assessment of molecular properties, but are essential to the modern description of the distribution of electrons in molecules (molecular orbital theory makes heavy use of symmetry considerations for deciding which atomic orbitals to combine into molecular orbitals) and to spectroscopy for deciding which transitions can occur.

Symmetry considerations are also helpful in the discussion of solids, for atoms, ions, and molecules stack together in symmetrical arrays when they form crystals. In Chapter 21 we shall come to understand some of the patterns that crystals form through the insight that diffraction techniques provide. We shall also see in Chapter 22 the origin of the forces that are responsible for holding molecules together in aggregates even though their ability to form chemical bonds is complete. This is the domain of **intermolecular forces**, which affect not only the structure of solids but also the properties of liquids and of real gases. Much less is known about liquids than solids or gases (or, at least, the description of their properties cannot be discussed in simple terms or parametrized so fruitfully as the other states of matter), and the lengths of the discussion will reflect this still primitive field of physical chemistry.

The scattering of radiation

Electromagnetic radiation can interact with molecules by being **scattered**, or deflected from its initial direction. If the energy remains unchanged in this process, the scattering is called **elastic**. If the frequency is changed when scattering occurs, the process is said to be **inelastic**. One of the most useful applications of elastic scattering is the technique of **X-ray diffraction**, in which X-rays scattered by individual atoms in a crystalline sample interfere with one another to give a characteristic pattern of intensities. This pattern can be interpreted in terms of the location of atoms in the molecule, and gives unparalleled information about molecular structure (Fig. 0.6). Almost the whole of the progress in molecular biology that has characterized the second half of the twentieth century has stemmed from the application of this technique to biochemically significant molecules. The basis of the technique is described in Chapter 21. Note that, whereas spectroscopy probes energy levels, from which molecular structure is inferred, scattering experiments explore electron distributions directly.

The populations of energy levels

A major question tackled by physical chemistry is the relationship between the energy levels characteristic of individual atoms and molecules and the properties of bulk samples. This connection is the domain of **statistical thermodynamics**, which is described in Chapters 19 and 20. The great importance of this subject is that it provides a point of unification between the material of thermodynamics treated in Part 1 and the structural and spectroscopic aspects treated in Part 2. Thus, through it, it is possible to relate such characteristic chemical quantities as equilibrium constants to the spectroscopically determined energy levels of molecules.

At this stage it will prove helpful to be aware of two principal conclusions from statistical thermodynamics, for they help to forge a link between thermodynamic quantities and molecular properties. (Indeed, many of the *Molecular interpretations* provided in the early chapters of the text draw on the material described here.) One conclusion concerns the numbers of molecules that occupy each energy level; the other is the mean energy associated with each type of motion.

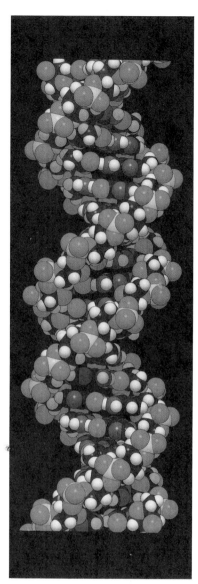

0.6 X-ray diffraction has led to highly detailed structural information on the arrangement of atoms and ions in crystals and molecules. The information obtained is the basis of computer graphic representations of molecular structure, such as this image of a fragment of DNA.

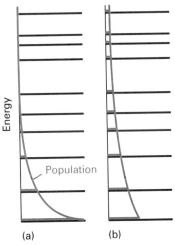

Energy

Population

(a)   (b)

0.7 The Boltzmann distribution predicts that the population of a state decreases exponentially with the energy of the state. (a) At low temperatures, only the lowest states are significantly populated; (b) at high temperatures, there is significant population in high-energy states as well as in low-energy states. At infinite temperature, all states are equally populated.

First, consider the number of molecules in each of the available energy levels of sample at a given temperature. The continuous thermal agitation that the molecules experience ensures that they are distributed over all the available energy levels. One particular molecule may be in one energy state at one instant, and then be knocked into another state a moment later. Although we cannot keep track of the energy state of a single molecule, we can speak of the *average* numbers of molecules in each state, and these average numbers are constant in time so long as the temperature remains the same. The average number of molecules in a state is called the **population** of the state.

Only the lowest energy level is occupied at $T = 0$. (Here and throughout this text, the symbol T will denote an absolute temperature that begins at 0 for the lowest attainable temperature, at $-273°C$.) Raising the temperature excites some molecules into higher energy levels, and more and more levels become accessible as the temperature is raised further (Fig. 0.7). Nevertheless, whatever the temperature, there is always a higher population in a state of low energy than one of high energy.

The formula for calculating the population of states of various energies is called the **Boltzmann distribution** and was derived by Ludwig Boltzmann towards the end of the nineteenth century. This formula gives the ratio of the numbers of particles in states with energies E_i and E_j as

$$\frac{N_i}{N_j} = e^{-(E_i - E_j)/kT}$$

The fundamental constant k is **Boltzmann's constant**, $k = 1.38 \times 10^{-23} \, J \, K^{-1}$. Boltzmann's constant is replaced by the gas constant, $R = N_A k$, if the energies in the Boltzmann distribution are replaced by molar energies.

An important detail concerning the interpretation of the Boltzmann distribution is that it refers to the populations of *states*: it may happen that several different states have the same energy. The technical term for this equivalence of energies is **degeneracy**. Thus, a pendulum free to swing in two dimensions may have degenerate states (for example, swinging in the x direction with a certain amplitude, or swinging in the y direction with the same amplitude). Because the Boltzmann distribution refers to states, all the members of a degenerate set of states will have the same population. We shall see a consequence of this feature shortly.

Consider first the distribution of atoms over their available electronic levels. A typical energy separation between the ground state and the first excited state of a molecule is about 3 eV, which corresponds to $300 \, kJ \, mol^{-1}$. In a sample at 25°C (298 K) the ratio of the populations of the two states is about e^{-121}, or about 10^{-53}. Therefore, essentially every atom in a sample is in its ground state. The population of the upper state rises to about 1 per cent of the population of the ground state only when the temperature reaches $10^4 °C$. The separation of vibrational energy levels is very much less than that of electronic energy levels (about 0.1 eV, corresponding to $10 \, kJ \, mol^{-1}$), but nevertheless at room temperature only the lowest energy level is significantly populated. Only about 1 in 10^4 molecules are not in their ground state. Rotational energy

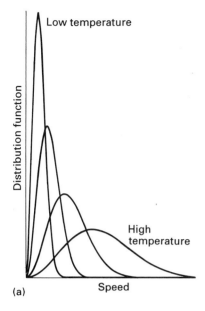

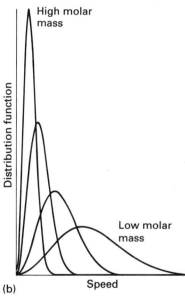

0.8 The essential content of the Maxwell distribution of molecular speeds is summarized by these two diagrams (which have the same shape but a different significance). (a) The distribution of speeds for a given species of molecule at different temperatures. (b) The distribution of speeds for molecules with different masses at the same temperature.

levels are much more closely spaced than vibrational energy levels (typically, about 100 to 1000 times closer), and even at room temperatures we can expect many rotational states to be occupied. Therefore, when considering the contribution of rotational motion to the properties of a sample, we need to take into account the fact that molecules occupy a wide range of different states, with some rotating rapidly and others slowly.

The Boltzmann distribution takes a special form when we consider the free translational motion of non-interacting gas molecules. Different energies now correspond to different speeds, so the Boltzmann formula can be used to predict the proportions of molecules having a specific speed at a particular temperature. The distribution of speeds is called the **Maxwell distribution** and has the features summarized in Fig. 0.8. The bulge in the distribution represents the facts that translational energy levels are highly degenerate and that, although the populations of individual states decrease with increasing energy (and hence speed), there are many more states of a given energy at high energies. Notice how the tail toward high speeds is longer at high temperatures than at low, which indicates that at high temperatures many molecules in a sample have speeds much higher than average. The speed corresponding to the maximum in the graph is the most probable speed, the speed most likely to be found for a molecule selected at random. Figure 0.8 also shows how the distribution varies with mass for some components of air at 25°C. The lighter particles move, on average, much faster than the heavier ones.

The important features of the Boltzmann distribution to bear in mind are that the distribution of populations is an exponential function of energy and temperature, and that more states are significantly populated if they are close together in comparison with kT (like rotational and translational states), than if they are far apart (like vibrational and electronic states). Moreover, more states are occupied at high temperatures than at low temperatures. Figure 0.9 summarizes the form of the Boltzmann distribution for some typical sets of energy levels.

Equipartition

Now we consider the average energy of each type of molecular motion. The electronic and vibrational contributions are dealt with very simply because, as we have seen, only the corresponding ground states are significantly occupied at room temperature. (Excited states will be occupied at higher temperatures, but we shall not consider that complication here.) When many energy levels are populated (as is the case for translational and rotational motion), the mean energy can be calculated by using a very useful and simple conclusion of statistical thermodynamics which states that the mean energy of each quadratic contribution to the total energy of a molecule is the same (hence 'equipartition', because the energy is partitioned equally over the available modes) and equal to $\frac{1}{2}kT$. By 'quadratic contribution' is meant a contribution to the energy that is proportional to the square of a velocity or a position.

As an illustration, a particle of mass m that is free to travel parallel to the x-axis has a kinetic energy equal to $\frac{1}{2}mv_x^2$, and so, according to the

Rotation   Vibration   Electronic

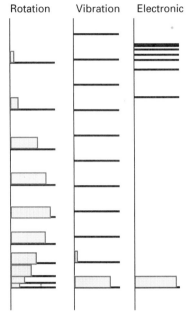

0.9 The Boltzmann distribution for three types of motion at a single temperature. There is a change in scale between the three stacks of levels (recall Fig. 0.4). Only the ground electronic state is populated at room temperature in most systems, and the bulk of the molecules are also in their ground vibrational state. Many rotational states are populated at room temperature as the energy levels are so close. The peculiar shape of the distribution over rotational states arises from the fact that each energy level actually corresponds to a number of degenerate states in which the molecule is rotating at the same speed but in different orientations. Each of these states is populated according to the Boltzmann distribution, and the shape of the distribution reflects the total population of each level.

equipartition theorem, the mean energy of a large number of molecules that are at a temperature T is $\frac{1}{2}kT$. For a gas of molecules free to move in three dimensions, like an actual gas in a container, there are three contributions to the kinetic energy of the same kind, so the average kinetic energy of a molecule in such a gas is $\frac{3}{2}kT$. The total energy per mole of such a sample is this average energy multiplied by Avogadro's constant, or $\frac{3}{2}RT$ (because $R = N_A k$). The theorem also implies that the average energy of rotation is the sum of $\frac{1}{2}kT$ for each axis about which it can rotate. Therefore, the mean rotational energy of a CH_4 molecule (which can rotate around three perpendicular axes) is $\frac{3}{2}kT$. As we have indicated, it is unsafe to apply the equipartition theorem to vibrations and electronic motion because too few states are occupied for the theorem to be valid.

CHANGE

The third great branch of physical chemistry is its treatment of chemical change. Part 3 builds up a detailed understanding of how chemical reactions occur and how we obtain our knowledge of them. Many of the topics treated here draw on the material earlier in the book, for chemical reactions involve both bulk properties and a detailed under-standing of processes occurring at a molecular level.

Chapter 24 introduces the simplest aspects of change, those associated with the motion of molecules from one part of a system to another. This topic provides another good example of model building, because the kinetic theory of gases introduced in Chapter 1 can be used to build a quantitative theory of the transport of properties through gases: such properties include the transport of matter by **diffusion** and **effusion**, **thermal conduction**, and **viscosity**. Another class of transport properties is the transport of electric charge, **electrical conduction**, by ions in a solution. We shall see how this transport process can be described and related to the microscopic nature of solutions. In fact, motion in fluids of all kinds can be expressed in a unified way by setting up one of the most important types of equation for the description of matter, the **diffusion equation**. This equation (which resembles the Schrödinger equation in appearance) greatly unifies the treatment of processes in fluids, including chemical reactions. We shall see that diffusion is yet another example of the natural tendency of matter to spread chaotically, and the diffusion equation provides a relation between the rate of dispersal in a particular region and the unevenness of the concentration there. We shall see that the equation expresses the intuitively plausible result that wrinkles in a distribution tend to spread and the distribution becomes uniform (Fig 0.10).

After the treatment of these physical processes, it is natural to turn to a consideration of chemical change. Indeed, this subject is so important that in large part it occupies all the remaining chapters of the text. Chemistry, after all, is largely the science of change. Initially (in Chapter 25) we examine how information is obtained about the rates of chemical reactions. This is the field of **chemical kinetics**. The results of these investigations allow us to relate the rates of reactions to the concentrations of the reactants and products by establishing the differential equations

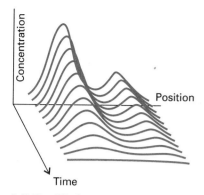

0.10 The diffusion equation relates the rate of spreading of a non-uniform concentration of molecules to the spatial variation of the concentration. When it is solved, it shows how a non-uniform distribution spreads and becomes uniform.

known as **rate laws**. These rate laws, which are constructed empirically by investigating the relation between reaction rates and the concentrations of species, are useful in a variety of ways. First, they are a valuable clue (and in earlier times often the only clue) to the **mechanism** of a reaction, the molecular steps by which it takes place. However, rate laws are very interesting in their own right, as we shall see in Chapter 26, for their solutions often predict highly unusual behaviour, such as oscillation and irregularity in concentrations. Such considerations have recently been the subject of considerable interest, and we shall see just a little of the conclusions that have been drawn.

Physical chemistry, though, has developed more detailed approaches to the mechanisms of chemical reactions than can be obtained from a consideration of rate laws alone. Rate laws are the analogue of the bulk properties treated by thermodynamics. The approach corresponding to a molecular interpretation of reaction rates (Chapter 27) seeks to relate the rate of a chemical reaction to the structures of the participating molecules. We shall see that the relation can be approached in three steps. The first is an elaboration of the kinetic theory of gases that treats a reaction as the outcome of an energetic collision; this is the domain of **collision theory**. The second, **activated complex theory**, draws on the statistical thermodynamic ideas introduced in Part 2, and seeks to forge a link between the population of energy levels and the rates of changes that they can undergo. The third, **molecular reaction dynamics**, seeks to predict reaction rates in terms of solutions of the Schrödinger equation: in essence, it regards a chemical reaction as a very floppy molecule in which the atoms happen to have one arrangement (the 'reactants') initially, but then move into a new arrangement (the 'products').

The work up to this point has concerned chemical reactions taking part throughout a bulk medium (typically a gas or a solution). The reactions of species at surfaces provide an interesting and important set of problems, because they include the action of **catalysts** and the processes that occur at electrodes. These two special topics are treated in Chapters 28 and 29, respectively. In fact, surfaces play a special role in chemistry, and the opportunity is taken to include in the discussion their general features, including the surfaces of liquids.

CONCLUSION

This introduction has displayed the broad sweep of physical chemistry and its treatment of equilibrium, structure, and change. The central unifying feature is the role of energy, in its conservation, its dispersal in disorder, and the rate at which it can be accumulated to allow chemical reaction. The language of physical chemistry is essentially that of thermodynamics and quantum mechanics, which embrace bulk and individual molecular properties, respectively, and their union in statistical thermodynamics. The techniques are primarily those of monitoring changes in energy in thermodynamics and spectroscopy. The achievements of physical chemistry are the establishment of models of behaviour and their elaboration by reference to experiment. The outcome is deep insight into the properties of matter and the changes it can undergo.

PART ONE

Equilibrium

Part 1 of the text develops the concepts that are needed for the discussion of equilibria in chemistry. Equilibria include physical change, such as fusion and vaporization, and chemical change, including electrochemistry. The discussion is in terms of thermodynamics, and particularly in terms of enthalpy and entropy. We shall see that a unified view of equilibrium and the direction of spontaneous change can be obtained in terms of the chemical potential of substances. The chapters of Part 1 deal with the properties of matter in bulk; those of Part 2 will show how these properties stem from the behaviour of individual atoms.

1

The properties of gases

This chapter establishes the properties of gases that will be used throughout the text. It begins with an account of an idealized version of a gas, a perfect gas, and shows how its equation of state—perhaps the single most important equation in physical chemistry—may be assembled experimentally. We shall then see how this experimentally determined relation between the properties of the gas can be explained in terms of the kinetic theory, in which the gas is modelled as a collection of mass points in continuous chaotic motion. The calculation illustrates one of the uses of physical chemistry, which is to take a conceptual model and develop it into an experimentally testable theory. We shall then move on to a consideration of real gases and see how their properties differ from those of a perfect gas. Once again, it is possible to construct a model of the departures of real gases from perfect gases in terms of forces acting between molecules, and we shall see how to construct, interpret and use the van der Waals equation of state, which is an approximate equation of state for real gases.

The simplest state of matter is a **gas**, a form of matter that fills any container it occupies. Initially we shall consider only pure gases, but later in the chapter we shall see that the same ideas and equations apply to mixtures of gases too.

THE PERFECT GAS

We shall find it helpful to picture a gas as a collection of molecules (or atoms) in continuous random, chaotic motion, with speeds that increase as the temperature is raised. A gas differs from a liquid (in which the molecules are also moving chaotically) in that the molecules of a gas are widely separated from each other, except during collisions, and move largely independently of each other. We shall use this qualitative model to interpret the experimental observations we are about to describe, and later in the chapter shall show that the model can be developed into a quantitative theory.

1.1 The states of gases

The space occupied by a sample of gas is called its **volume** V. The number of molecules present in the sample is expressed as the **amount of substance** (more colloquially, the number of moles) n. The condition of a sample of gas (and of any other substance) also requires the specification of the **pressure** p and the **temperature** T. The first part of this section will be taken up with explaining these last two properties.

The physical **state** of a sample of a substance is defined by its physical properties, and two samples of a substance that have the same physical properties are in the same state. The state of a pure gas, for example, is specified by giving the values of its volume, amount of substance, pressure, temperature, density, and so on. Although this definition suggests that to define a state we need to specify a large number of properties, it is found experimentally that not all the properties of a substance are independent. Thus, it has been established that it is sufficient to give the values of only a few of the properties of a substance (typically three or four), for then all the other properties are fixed. We shall see, for example, that the state of a pure gas is specified by *three* variables (for example, the volume, amount of substance, and the temperature) and that the pressure then has a value that depends on the values of the other three. In other words, a pure gas is described by an **equation of state** that expresses one of the four variables in terms of the other three. For example, one of the most famous equations of state is the one that describes a gas at low pressures:

$$p = \frac{nRT}{V} \tag{1}$$

where R is a constant common to all gases. It follows from this equation of state that, if we know the values of n, V, and T for a sample of gas, then we can state its pressure too. Much of the rest of this chapter will describe some of the equations of state of gases that have been obtained experimentally or proposed theoretically.

Pressure

Pressure is defined as force per unit area. The greater the force acting on a given surface, the greater the pressure. The origin of the force exerted by a gas is the incessant battering of the molecules on the walls of the container. The collisions are so numerous that they exert an effectively steady force, which is experienced as a steady pressure.

If two gases are in separate containers that share a common movable wall (Fig. 1.1), then the gas that has the higher pressure will force the wall to move and compress the low-pressure gas. The pressure of the former will fall and that of the latter will rise, and there will come a stage when the two pressures are equal and the wall has no further tendency to move. This condition of equality of pressure is a state of **mechanical**

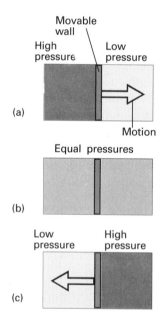

1.1 (a), (c) When a region of high pressure is separated from a region of low pressure by a movable wall, the wall will be pushed into one region or the other. (b) However, if the two pressures are identical, the wall will not move. The latter condition is one of mechanical equilibrium between the two regions.

equilibrium between the two gases. The numerical value of the pressure of a gas is therefore an indication of whether a container that contains the gas will be in mechanical equilibrium with another gas that shares a movable wall with it.

The SI unit of pressure, the **pascal** (Pa), is defined as 1 newton per square metre:

$$1\,\text{Pa} = 1\,\text{N}\,\text{m}^{-2}$$

However, several other units are still widely used. These units include the **bar**, the **atmosphere** (atm), and the **Torr**:

$$1\,\text{bar} = 10^5\,\text{Pa} = 100\,\text{kPa}$$

$$1\,\text{atm} = 101.325\,\text{kPa}$$

$$1\,\text{atm} = 760\,\text{Torr}$$

All three relations are exact. A pressure of $10^5\,\text{Pa}$ (1 bar) is the standard pressure for reporting data, and we shall denote it $p^{\ominus}$:

$$p^{\ominus} = 10^5\,\text{Pa}$$

The Torr is almost exactly equal to the **millimetre of mercury** (mmHg), which is defined so that 1 mmHg is the pressure exerted by a column of mercury 1 mm high. For most purposes, the units mmHg and Torr can be used interchangeably.

Example 1.1 *Calculating pressure*

Suppose Isaac Newton weighed 65 kg. Calculate the pressure he exerted on the ground when wearing (a) boots with soles of total area 250 cm² in contact with the ground, (b) ice skates, of total area 2.0 cm².

Method. We know that pressure is force per unit area, so the calculation depends on being able to calculate the force that Newton exerts on the ground, and then to divide it by the area over which the force is exerted. To calculate the force, we need to know (from elementary physics) that the downward force an object of mass m exerts on the surface of the Earth is $F = mg$ where g is the acceleration of free fall, $9.81\,\text{m}\,\text{s}^{-2}$. We need to note that $1\,\text{N} = 1\,\text{kg}\,\text{m}\,\text{s}^{-2}$ and $1\,\text{Pa} = 1\,\text{N}\,\text{m}^{-2}$.

Answer. The force exerted by Newton is

$$F = 65\,\text{kg} \times 9.81\,\text{m}\,\text{s}^{-2} = 6.4 \times 10^2\,\text{N}$$

The force is the same whatever his footwear. The pressure he exerts in each case is $p = F/A$, where A is the area over which the force acts. Hence:

(a) $p = \dfrac{6.4 \times 10^2\,\text{N}}{2.50 \times 10^{-2}\,\text{m}^2} = 2.6 \times 10^4\,\text{Pa}$,   or 26 kPa

(b) $p = \dfrac{6.4 \times 10^2\,\text{N}}{2.0 \times 10^{-4}\,\text{m}^2} = 3.2 \times 10^6\,\text{Pa}$,   or 3.2 MPa

Comment. A pressure of 26 kPa corresponds to 0.26 atm and a pressure of 3.2 MPa corresponds to 31 atm.

Exercise E1.1. Calculate the pressure exerted by a mass of 1.0 kg pressing through the point of a pin of area $1.0 \times 10^{-2} \, \text{mm}^2$ on the surface of the Earth.   [$9.8 \times 10^2 \, \text{MPa}$, $9.7 \times 10^3 \, \text{atm}$]

The pressure exerted by the atmosphere is measured with a **barometer**. The original version of a barometer (which was invented by Torricelli, a student of Galileo) was an inverted tube of mercury sealed at the upper end. When the column of mercury is in mechanical equilibrium with the atmosphere, the pressure at its base is equal to that exerted by the atmosphere. It follows that the height of the mercury column is proportional to the external pressure.

Example 1.2 *Calculating the pressure exerted by a column of liquid*

Calculate the pressure at the base of a column of liquid of density ρ and height h at the surface of the Earth.

Method. We know that to calculate pressure we need to calculate the force exerted over a given area, and then to divide that force by the area. To calculate the force exerted by an object (solid or liquid) we need to know its mass and then multiply that mass by the acceleration of free fall g. To calculate the mass of a liquid of given density, we need to multiply the density by the volume of the liquid (mass = density × volume). So, the first step is to calculate the volume of a column of liquid.

Answer. Suppose the column has cross-sectional area A; then its volume is Ah. The mass of this column of liquid of density ρ is

$$m = \rho \times Ah$$

The force the column of this mass exerts at its base is

$$F = mg = \rho Ah \times g$$

The pressure is this force divided by the area on which it acts, which is the cross-sectional area (A) of the column:

$$p = \frac{\text{force}}{\text{area}} = \frac{\rho g Ah}{A} = \rho g h$$

Comment. Note that the pressure is independent of the shape or cross-sectional area of the column. The mass of the column increases as the area, but so does the area on which the force acts; thus the two cancel.

Exercise E1.2. Calculate the pressure at the base of a column of length l held at an angle θ to the vertical.   [$p = \rho g l \cos \theta$]

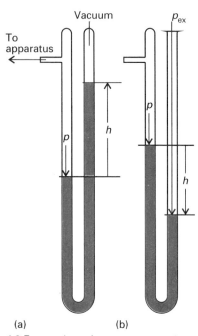

1.2 Two versions of a manometer used to measure the pressure of a sample of gas. (a) The height difference h of the two columns in the sealed-tube manometer is directly proportional to the pressure of the sample, and $p = \rho g h$ where ρ is the density of the liquid. (b) The difference in heights of the columns in the open-tube manometer is proportional to the difference between the pressure of the sample and that of the atmosphere p_{ex}. In the example shown, the pressure of the sample is lower than that of the atmosphere.

The pressure of a sample of gas in a container is measured with a **manometer** (Fig. 1.2). In its simplest form, a manometer is a U-tube filled with a liquid of low volatility (such as silicone oil). When the system has reached mechanical equilibrium, the pressure of the gaseous sample p balances the pressure exerted by the column of liquid, which is equal to

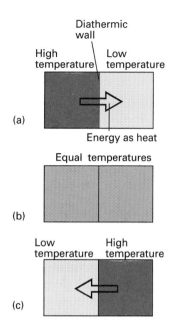

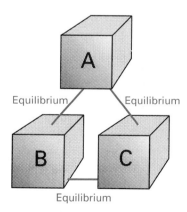

1.3 (a), (c) Energy flows as heat from a region at a higher temperature to one at a lower temperature if the two are in contact through a diathermic wall. (b) However, if the two regions have identical temperatures, there is no net transfer of energy as heat even though the two regions are separated by a diathermic wall. The latter condition corresponds to the two regions being at thermal equilibrium.

1.4 The experience summarized by the Zeroth Law of thermodynamics is that, if an object A is in thermal equilibrium with B and B is in thermal equilibrium with C, then C is in thermal equilibrium with A.

ρgh, if the column is of height h, plus the external pressure p_{ex}, if one tube is open to the atmosphere:

$$p = p_{ex} + \rho gh \qquad (2)$$

It follows that the pressure of the sample can be obtained by measuring the height of the column and noting the external pressure. More sophisticated techniques are used at lower pressures. Methods that avoid the complication of having to account for the vapour from the manometer fluid are also available. These superior methods include monitoring the deflection of a diaphragm, either mechanically or electrically, or monitoring the change in a pressure-sensitive electrical property.

Temperature

The concept of temperature springs from the observation that a change in physical state (for example, a change of volume) can occur when two objects are in contact with one another (as when a red-hot metal is plunged into water). In Section 2.1 we shall see that the change in state can be interpreted as arising from a flow of energy as heat from one object to another. The **temperature** is the property that tells us the *direction* of the flow of energy, and, if energy flows from A to B, then we say that A has a higher temperature than B (Fig. 1.3). It will prove useful to distinguish between two types of wall that can separate the objects. A wall is **diathermic** if a change of state is observed when two objects at different temperatures are brought into contact (the word *dia* is from the Greek for 'through'). A metal container has diathermic walls. A wall is **adiabatic** if no change occurs even though the two objects have different temperatures. A Dewar vessel is a good example of a container with adiabatic walls.

If no change of state occurs when two objects A and B are in contact through diathermic walls, we say that A and B are in **thermal equilibrium** with one another. Suppose an object A (which we can think of as a block of iron) is in thermal equilibrium with an object B (a block of copper), and that B is also in thermal equilibrium with another object C (a flask of water). Then it has been found experimentally that A and C will also be in thermal equilibrium when they are put in contact (Fig. 1.4). This observation is summarized by the following statement.

> **The Zeroth Law of thermodynamics.** If A is in thermal equilibrium with B, and B is in thermal equilibrium with C, then C is also in thermal equilibrium with A.

In science, a **law** is a summary of experience, so the Zeroth Law summarizes the observed behaviour of bodies that are in contact with each other through diathermic walls.

The importance of the Zeroth Law is that it is the fundamental principle that allows us to build a **thermometer,** a device that displays a change in temperature as a change in some physical property (such as the length of a column of mercury). Thus, suppose that B is a glass capillary containing mercury. Then, when A is in contact with B, the mercury column in the latter has a certain length. According to the Zeroth Law, if the mercury column in B has the same length when it is placed in contact with another object C, then we can predict that no change of state of A and C will occur when they are in contact whatever their composition. Moreover, we can use the length of the mercury column as a measure of the temperatures of A and C.

The relation of the numerical value of the temperature to the property selected for monitoring it is arbitrary. In the early days of thermometry (and still in laboratory practice today) temperatures were related to the length of a column of liquid, and the difference in lengths shown when the thermometer was first in contact with melting ice and then with boiling water was divided into 100 steps called 'degrees', the lowest point being labelled 0. That led to the **Celsius scale** of temperature. Celsius temperatures are denoted θ and expressed in °C. However, because different liquids expand to different extents, and do not always expand uniformly over a given range, thermometers constructed from different materials showed different numerical values of temperature between their fixed points. Thus, while A, B, C, . . . may all have been ascribed the temperature 28.7°C when a mercury-in-glass thermometer was used, they may all have been ascribed 28.8°C when an alcohol-in-glass thermometer was used. The variation of quoted temperatures is even greater when electrical measurements, such as the resistance of a wire, are included. The pressure of a gas, however, can be used to construct a **perfect-gas temperature scale**, which is independent of the identity of the gas, as we shall explain shortly. The perfect-gas scale turns out to be identical to the **thermodynamic temperature scale** that we shall introduce in Section 4.6, so we shall use the latter term from now on to avoid a proliferation of names. On the thermodynamic temperature scale, temperatures are denoted T and are normally reported in **kelvins**, K. Thermodynamic and Celsius temperatures are related by the exact expression

$$T/K = \theta/°C + 273.15 \qquad (3)$$

That is, 0°C corresponds to 273 K.

Example 1.3 *Converting temperatures from one scale to another*

Express 25°C as a temperature in kelvins.

Method. This example is numerically trivial but is included to illustrate the convention for writing relations, as in eqn 3. We have used the procedure called 'quantity calculus' in which a physical quantity (such as the temperature θ) is the product of a numerical value (25) and a unit (°C). That is, 25°C stands for $25 \times$ °C. Therefore, $\theta/°C$ stands for the physical quantity ($25 \times$ °C) divided by the unit (°C). Both sides of eqn 3 are therefore pure numbers, and the expression may be manipulated like any arithmetical expression between numbers.

Answer. Because $\theta/°C = 25$ it follows from eqn 3 that

$$T/K = 25 + 273.15 = 298$$

Multiplication of both sides by the unit K then gives

$$T = 298\,\text{K}$$

Comment. Quantity calculus is a very powerful way of keeping track of units, and we shall use it whenever appropriate. The essence of quantity calculus is to treat units as algebraic quantities that can be multiplied and divided like numbers. The numerical values of all physical quantities should be accompanied by their units at all stages of a calculation.

Exercise E1.3. Use the result obtained in Example 1.2 (that $p = \rho g h$) to calculate (using quantity calculus) the pressure at the base of a 1.0 m high column of water of density $1.0\,\text{g cm}^{-3}$. [9.8 kPa]

1.2 The gas laws

The equation of state of a low-pressure gas was established by combining a series of experimental results obtained starting in the seventeenth century. We shall introduce these laws in this section, and then show how they can be combined into a single equation of state.

The individual gas laws

Robert Boyle, acting on the suggestion of a correspondent John Townley, showed in 1661 that, to a good approximation, the pressure and volume of a fixed amount of gas at constant temperature are related by

Boyle's law: $pV = \text{constant}$ (4)°

The variation of the volume of a sample of gas as the pressure is changed is depicted in Fig. 1.5. Each of the curves in the illustration corresponds to a single temperature and hence is called an **isotherm** (from the Greek words for 'equal heat'). According to Boyle's law, the isotherms of gases are hyperbolas (graphs of y against x with $xy = \text{constant}$). We now know that Boyle's law is valid only at low pressures, and that real gases have hyperbolic isotherms only in the limit of $p \to 0$. Boyle's law is therefore an example of a **limiting law**, a law that is strictly true only in the limit of low pressure. Equations that are valid in this limiting sense will be signalled by a superscript ° on the equation number, as in eqn 4.

The molecular explanation of Boyle's law can be traced to the fact that, if the volume of a sample is reduced by half, then there will be twice as many molecules per unit volume. As a result, twice as many molecules strike the walls in a given period of time so that the average force exerted on the walls is doubled. Hence, when the volume is halved, the pressure of the gas is doubled and $p \times V$ is a constant. The reason why Boyle's law applies to all gases regardless of their chemical identity (as long as the pressure is low) is that at low pressures the molecules are so far apart on average that they exert no influence on one another and travel independently.

The French scientist Jacques Charles established the next important empirical (based on observation) property of gases. He studied the effect of temperature on a sample of gas that was subjected to constant pressure. He found that the volume varied linearly with the temperature

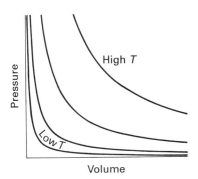

1.5 The pressure–volume dependence of a fixed amount of perfect gas at different temperatures. Each curve is a hyperbola ($p \propto 1/V$) and is called an *isotherm*.

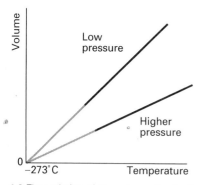

1.6 The variation of the volume of a fixed amount of gas with the temperature constant at constant pressure. Note that in each case the curves extrapolate to zero volume at −273°C.

and that, whatever the identity of the gas so long as it was at low pressure,

$$V = \text{constant} \times (\theta + 273°C) \text{ (at constant pressure)}$$

(Recall that we use θ to denote temperatures on the Celsius scale.) The linear variation of volume with temperature summarized by this expression is illustrated in Fig. 1.6. The lines are examples of **isobars**, lines showing the variation of properties at constant pressure. Charles's law suggests that the volume of any gas should extrapolate to zero at $\theta = -273°C$ and therefore that $-273°C$ is a natural zero of a fundamental temperature scale. As we have already indicated, a scale with 0 set at $-273°C$ is equivalent to the thermodynamic temperature scale devised by Kelvin, so Charles's law can be written

Charles's law: $V = \text{constant} \times T$ (at constant pressure)　　(5)°

The molecular explanation of Charles's law lies in the fact that raising the temperature of a gas increases the average speed of its molecules. The molecules collide with the walls more frequently and do so with greater impact. Therefore they drive back the walls, and the gas occupies a greater volume than it did initially. A problem we shall leave until later is why the change in volume should depend so simply on the temperature.

The final piece of experimental information we need is that, at a given pressure and temperature, the molar volumes ($V_m = V/n$, the volume per mole of molecules) of all gases are approximately the same. This equality becomes increasingly exact as the pressure of the gas is reduced. For example, the molar volume of carbon dioxide at 0°C and 1 atm pressure is $22.26 \, \text{L mol}^{-1}$ and that of argon under the same conditions is $22.09 \, \text{L mol}^{-1}$. Because the molar volume of a gas is almost the same whatever the chemical identity of the gas (under the same conditions of pressure and temperature), it follows that the total volume of a sample of gas is proportional to the amount (in moles) present, and the constant of proportionality is independent of the identity of the gas:

$$V = \text{constant} \times n \text{ (at constant pressure and temperature)}$$　(6)°

This conclusion is a modern form of **Avogadro's principle**, that equal volumes of gases at the same pressure and temperature contain the same numbers of molecules.

The combined gas law

The three experimental observations we have summarized, namely:

1. Boyle's law: $pV = \text{constant}$ (for n and T constant)
2. Charles's law: $V = \text{constant} \times T$ (for n and p constant)
3. Avogadro's principle: $V = \text{constant} \times n$ (for p and T constant)

can be combined into a single expression:

$$pV = \text{constant} \times nT$$

The constant, which experimentally is found to be the same for all gases,

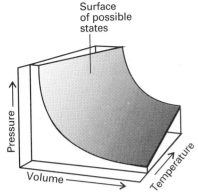

1.7 A region of the *p, V, T*-surface of a fixed amount of perfect gas. The points forming the surface represent the only states of the gas that can exist.

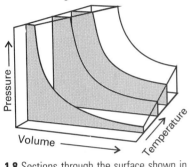

1.8 Sections through the surface shown in Fig. 1.7 at constant temperature give the isotherms shown in Fig. 1.5.

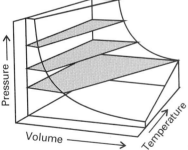

1.9 Sections through the surface shown in Fig. 1.7 at constant pressure give the straight lines shown in Fig. 1.6.

Table 1.1 The gas constant in various units

R
$8.31451 \, \text{J K}^{-1} \, \text{mol}^{-1}$
$8.20578 \times 10^{-2} \, \text{L atm K}^{-1} \, \text{mol}^{-1}$
$8.31451 \times 10^{-2} \, \text{L bar K}^{-1} \, \text{mol}^{-1}$
$8.31451 \, \text{Pa m}^3 \, \text{K}^{-1} \, \text{mol}^{-1}$
$62.364 \, \text{L Torr K}^{-1} \, \text{mol}^{-1}$
$1.98722 \, \text{cal K}^{-1} \, \text{mol}^{-1}$

is denoted R and is called the **gas constant**. The combined expression

$$pV = nRT \tag{7}°$$

is called the **perfect gas equation**. It is the *approximate* equation of state of any gas, and becomes increasingly exact as the pressure of the gas approaches zero. A gas that obeys eqn 7 exactly under all conditions is called a **perfect gas** (or ideal gas). A **real gas**, an actual gas, behaves more like a perfect gas the lower its pressure and is described exactly by eqn 7 in the limit of $p \to 0$.

The surface in Fig. 1.7 is a plot of the pressure of a fixed amount of perfect gas against its volume and thermodynamic temperature as given by eqn 7. The surface depicts the only possible states of a perfect gas: the gas cannot exist in states that do not correspond to points on the surface. The isotherms in Fig. 1.5 and the isobars in Fig. 1.6 correspond to the sections through the surface shown in Figs 1.8 and 1.9, respectively.

The value of the gas constant R can be obtained by evaluating pV/nT for a gas in the limit of zero pressure (to guarantee that it is behaving perfectly). However, a more accurate value can be obtained by measuring the speed of sound in a low-pressure gas and extrapolating its value to zero pressure. The currently accepted best value is

$$R = 8.31451 \, \text{J K}^{-1} \, \text{mol}^{-1}$$

The values of R in other units, which are useful in practice, are given in Table 1.1.

The perfect gas equation is of the greatest importance in physical chemistry because it is used to derive a wide range of relations that are used throughout thermodynamics. However, it is also of considerable practical utility for calculating the properties of a gas under a variety of conditions.

Example 1.4 *Using the perfect gas equation*

In an industrial process, nitrogen is heated to 500 K in a vessel of constant volume. If it enters the vessel at a pressure of 100 atm and a temperature of 300 K, what pressure would it exert at the working temperature if it behaved as a perfect gas?

Method. When confronted with a problem on perfect gases, the solution can usually be found by rearranging eqn 7 into a form that gives the unknown quantity in terms of the data. Some problems (this is an example) do not give all the physical properties (we are not told the amount or the volume), but this deficiency can be overcome by writing the equation as

$$\frac{pV}{nT} = R$$

for both sets of conditions, and noting that, as R is a constant, then

$$\frac{p_1 V_1}{n_1 T_1} = \frac{p_2 V_2}{n_2 T_2}$$

The constant quantities (amount and volume in this example) cancel, and the data may then be substituted into the resulting expression.

Answer. Cancellation of the amounts ($n_1 = n_2$) and volumes ($V_1 = V_2$) on each side of this expression results in

$$\frac{p_1}{T_1} = \frac{p_2}{T_2}$$

which can be rearranged into

$$p_2 = \frac{T_2}{T_1} \times p_1$$

Substitution of the data then gives

$$p_2 = \frac{500\,\text{K}}{300\,\text{K}} \times 100\,\text{atm} = 167\,\text{atm}$$

Comment. Experiment shows that the pressure is actually 183 atm under these conditions; thus the assumption that the gas is perfect leads to a 10 per cent error.

Exercise E1.4. What temperature would result in the same sample exerting a pressure of 300 atm?

[900 K]

Two sets of conditions are currently used as 'standard' values for reporting data. One is **standard temperature and pressure** (STP), which corresponds to 0°C and 1 atm. The other is **standard ambient temperature and pressure** (SATP), which corresponds to 25°C (more precisely, to 298.15 K) and 10^5 Pa (that is, to $p^{\ominus}$). The molar volumes of a perfect gas under these conditions are obtained by substitution of the values into

$$V_{\text{m}} = \frac{V}{n} = \frac{(nRT/p)}{n} = \frac{RT}{p} \qquad (8)°$$

Then

STP: $V_{\text{m}} = 22.414\,\text{L mol}^{-1}$

SATP: $V_{\text{m}} = 24.789\,\text{L mol}^{-1}$

We shall denote molar volume at SATP by $V_{\text{m}}^{\ominus}$.

Dalton's law

Now we consider the properties of mixtures of gases, such as the atmosphere, and see that very similar equations apply to them as apply to a single substance. This should not be surprising: the same limiting law $pV = nRT$ applies to every pure gas, so we can expect it to apply to a mixture of gases too. In this chapter we limit ourselves to gases that do not react when they mix.

The kind of question we need to answer when dealing with gaseous mixtures is the contribution that each component gas makes to the total pressure of the sample. In the nineteenth century John Dalton[1] made observations that provide the answer and summarized them in a law.

> **Dalton's law:** The pressure exerted by a mixture of perfect gases is the sum of the pressures exerted by the individual gases occupying the same volume alone.

The pressure that a gas would exert if it occupied the container alone and if it behaved perfectly is called the **partial pressure** of the gas. (Shortly we shall give a more general definition, which is applicable to real gases too.) That is, if a certain amount of H_2 exerts 0.25 atm when present alone in a container and an amount of N_2 exerts 0.80 atm when present alone in the same container at the same temperature, then the total pressure when both are present is the sum of these two partial pressures, or 1.05 atm. More generally, if the partial pressure of a perfect gas A is p_A, that of a perfect gas B is p_B, and so on, then the total pressure when all the gases occupy the same container at the same temperature is

$$p = p_A + p_B + \ldots = \sum_J p_J \qquad (9a)°$$

where, for each substance,

$$p_J = \frac{n_J RT}{V} \qquad (9b)°$$

The partial pressure p_A is the force per unit area exerted on a wall by the molecules of gas A, p_B is likewise the force per unit area exerted by the molecules B, and so on. Therefore, Dalton's law implies that the total force per unit area exerted by the gas molecules is the sum of the forces that each species in the mixture exerts per unit area. Dalton's law is valid if the forces exerted by the molecules of one gas are unaffected by the presence of another gas, and is therefore a description of perfect gases.

Example 1.5 *Using Dalton's law*

A container of volume 10.0 L holds 1.00 mol N_2 and 3.00 mol H_2 at 298 K. What is the total pressure if each component behaves as a perfect gas?

Method. The data are the volume, amount, and temperature of each gas, so there is enough information to use the perfect gas equation to calculate the partial pressure of each component. If an amount n_A of a perfect gas A occupies a container of volume V at a temperature T, then its pressure is $p_A = n_A(RT/V)$. If, instead, an amount n_B of another perfect gas B occupies the container, then its pressure is $p_B = n_B(RT/V)$. The total pressure when both gases occupy the same container simultaneously is

$$p = p_A + p_B = (n_A + n_B)\frac{RT}{V}$$

---

1 John Dalton (1766–1844) is the Dalton of the atomic hypothesis, and also the Dalton of daltonism, or colour-blindness (from which he suffered and which he described). He was described as 'an indifferent experimenter, and singularly wanting in the language and power of illustration'. He paid another price: 'Into society he rarely went, and amusement he had none, with the exception of a game of bowls on Thursday afternoons.'

Answer. Under the stated conditions

$$\frac{RT}{V} = \frac{(8.206 \times 10^{-2}\,\text{L atm K}^{-1}\,\text{mol}^{-1}) \times (298\,\text{K})}{10.0\,\text{L}} = 2.45\,\text{atm mol}^{-1}$$

It follows that

$$p(N_2) = 1.00\,\text{mol} \times 2.45\,\text{atm mol}^{-1} = 2.45\,\text{atm}$$

$$p(H_2) = 3.00\,\text{mol} \times 2.45\,\text{atm mol}^{-1} = 7.35\,\text{atm}$$

and the total pressure is

$$p = 2.45\,\text{atm} + 7.35\,\text{atm} = 9.80\,\text{atm}$$

Exercise E1.5. Calculate the total pressure when 1.00 mol N_2 and 2.00 mol O_2 are added to the same container (with the nitrogen and hydrogen still inside) at 298 K.   [17.1 atm]

Mole fractions and partial pressures

We can take a step closer to the discussion of mixtures of real gases by introducing the **mole fraction**, x_J, of each component J. The mole fraction of J in a mixture is the amount of J molecules present (n_J, the number of moles of J) expressed as a fraction of the total amount of molecules (n) in the sample:

$$x_J = \frac{n_J}{n}, \qquad \text{with } n = n_A + n_B + \ldots \tag{10}$$

When no J molecules are present, $x_J = 0$; when only J molecules are present, $x_J = 1$. A mixture of 1.0 mol N_2 and 3.0 mol H_2, and therefore of 4.0 mol molecules in all, consists of mole fractions 0.25 of N_2 and 0.75 of H_2. It follows from the definition of x_J that, whatever the composition of the mixture,

$$x_A + x_B + \ldots = 1 \tag{11}$$

Next, we define the partial pressure p_J of a gas in a mixture (*any* gas, not just a perfect gas), as

$$p_J = x_J p \tag{12}$$

where p is the total pressure of the mixture. It follows from eqn 11 that the sum of the partial pressures is equal to the total pressure (for real and perfect gases):

$$p_A + p_B + \ldots = (x_A + x_B + \ldots)p = p$$

Figure 1.10 shows how the partial pressures of a binary (two-component) mixture contribute to the total pressure as the mole fraction of one component increases from 0 to 1.

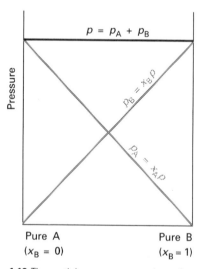

1.10 The partial pressures p_A and p_B of a binary mixture of (real or perfect) gases of total pressure p as the composition changes from pure A to pure B. The sum of the partial pressures is equal to the total pressure. If the gases are perfect, then the partial pressure is also the pressure that each gas would exert if it were present alone in the container.

Example 1.6 *Calculating partial pressures*

The mass percentage composition of dry air at sea level is approximately: N_2, 75.5; O_2, 23.2; Ar, 1.3. What is the partial pressure of each component when the total pressure is 1.00 atm?

Method. Partial pressures are defined by eqn 12, so that is the expression we examine to decide what data are required. To use the equation we need the mole fractions of the components. To calculate mole fractions, which are defined by eqn 10, we need the amounts of each substance and must use the fact that the amount of molecules of molar mass M in a sample of mass m is m/M. The mole fractions are independent of the total mass of the sample, so we can choose it to be 100 g (which makes the conversion from mass percentages very easy).

Answer. The amounts of each type of molecule present in 100 g of air are

$$n(N_2) = \frac{(100 \text{ g}) \times 0.755}{28.02 \text{ g mol}^{-1}} = 2.69 \text{ mol}$$

$$n(O_2) = \frac{(100 \text{ g}) \times 0.232}{32.00 \text{ g mol}^{-1}} = 0.725 \text{ mol}$$

$$n(Ar) = \frac{(100 \text{ g}) \times 0.013}{39.95 \text{ g mol}^{-1}} = 0.033 \text{ mol}$$

Because, overall, $n = 3.45$ mol, the mole fractions and partial pressures (obtained by multiplying the mole fraction by the total pressure, 1.00 atm) are as follows:

	N_2	O_2	Ar
Mole fraction:	0.780	0.210	0.0096
Partial pressure/atm:	0.780	0.210	0.0096

Comment. We have not had to assume that the gases are perfect: partial pressures are *defined* as $p_J = x_J p$ for any gas.

Exercise E1.6. When carbon dioxide is taken into account the mass percentages are 75.52 (N_2), 23.15 (O_2), 1.28 (Ar), and 0.046 (CO_2). What are the partial pressures when the total pressure is 0.900 atm?

[0.703, 0.189, 0.0084, 0.00027 atm]

To confirm that the definition of partial pressure in eqn 12 is consistent with the interpretation of partial pressure as the pressure that a perfect gas would exert if it were alone in the container we write $p = nRT/V$ and $x_J = n_J/n$ in eqn 12 and obtain

$$p_J = \frac{n_J}{n} \times \frac{nRT}{V} = \frac{n_J RT}{V}$$

in accord with eqn 9b. It is important to remember that the partial pressure is defined by eqn 12 for *any* gas but is equal to the pressure that it would exert alone only in the case of a *perfect* gas.

1.3 The kinetic theory of gases

We have seen that properties of a perfect gas can be rationalized qualitatively in terms of a model in which the molecules of the gas are in continuous chaotic motion. We shall now see how this model can be expressed quantitatively in terms of the **kinetic theory of gases**, in which it is assumed that the only contribution to the energy of the gas is from the kinetic energies of the molecules (that is, the potential energy of the interactions between molecules makes a negligible contribution to the total energy of the gas).

The kinetic theory of gases is based on three assumptions:

1. The gas consists of molecules of mass m and diameter d in ceaseless random motion.
2. The size of the molecules is negligible (in the sense that their diameters are much smaller than the average distance travelled between collisions).
3. The molecules do not interact, except that they make perfectly elastic collisions when the separation of their centres is equal to d.

An **elastic collision** is one in which the translational kinetic energy of a molecule is the same before and after a collision: no energy is transferred to its internal modes of motion.

The force of collision

The new physical concept we need for the kinetic theory of gases is that of the **momentum** of a molecule, the product of its mass and velocity. Momentum is important because it determines the impact that a molecule makes when it collides with a wall, and therefore the contribution of that molecule to the pressure of the gas. In brief, the greater the momentum of a molecule, the greater the force it exerts on a wall during a collision. The quantitative connection between the momentum of a molecule and the force it exerts during a collision is provided by one of Newton's laws of motion, which states that the force is equal to the rate of change of momentum. Therefore, to change the momenta of a large number of molecules in a brief interval requires a large force. The strategy of the calculation is therefore to calculate the rate of change of momentum of the molecules in the gas, and to interpret it as the force exerted by the molecules on the walls of the container. The tactics of the calculation involve three steps:

1. The calculation of the change in momentum that occurs when one molecule strikes a wall.
2. The calculation of the total number of collisions with a wall of given area in a given interval.
3. The conversion of the total change in momentum to a force per unit area.

The details of the calculation are set out in the *Justification* that follows; the result is that the pressure is related to the volume V, amount n, and molar mass M of the molecules by the expression

$$pV = \tfrac{1}{3}nMc^2 \tag{13}°$$

The quantity c is the **root mean square speed** of the molecules:

$$c = (\langle v_x^2 \rangle + \langle v_y^2 \rangle + \langle v_z^2 \rangle)^{\frac{1}{2}} \tag{14}$$

where v_x, etc. is the x component of velocity and $\langle \ldots \rangle$ denotes a mean value.

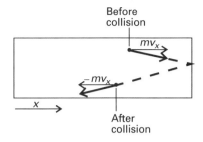

1.11 The pressure of a gas arises from the impact of its molecules on the walls. In an elastic collision of a molecule with a wall perpendicular to the x-axis, the x-component of velocity is reversed but the y- and z-components are unchanged.

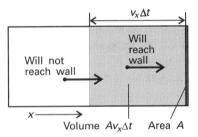

1.12 A molecule will reach the wall on the right within an interval Δt if it is within a distance $v_x \Delta t$ of the wall and travelling to the right.

JUSTIFICATION

Consider the arrangement in Fig. 1.11. When a particle of mass m that is travelling with a velocity v_x parallel to the x axis collides with the wall on the right and is deflected back along its path, its momentum changes from mv_x before the collision to $-mv_x$ after the collision (when it is travelling in the opposite direction). The momentum therefore changes by $2mv_x$ on each collision (the y and z components are unchanged). Many molecules collide with the wall in a time interval Δt, and the total change of momentum is the product of the change in momentum of each molecule multiplied by the number of molecules that reach the wall during the interval.

Next we calculate the number of molecules that collide with the wall in the interval Δt. Because a molecule with velocity v_x can travel a distance $v_x \Delta t$ along the x axis in an interval Δt, all the molecules within a distance $v_x \Delta t$ of the wall will strike it if they are travelling towards it (Fig. 1.12). It follows that, if the wall has area A, then all the particles in a volume $A \times v_x \Delta t$ will reach the wall (if they are travelling towards it). If the number of particles per unit volume is nN_A/V, where n is the amount of molecules and N_A is Avogadro's constant, then the number in the volume $Av_x \Delta t$ is $(nN_A/V) \times Av_x \Delta t$.

On average, at any instant half the particles are moving to the right, and half are moving to the left. Therefore, the average number of collisions with the wall during the interval Δt is $\frac{1}{2}nN_A Av_x \Delta t/V$. The total momentum change in that interval is the product of this number and the change $2mv_x$:

$$\text{Momentum change} = \frac{nN_A Av_x \Delta t}{2V} \times 2mv_x = \frac{nN_A mAv_x^2 \Delta t}{V}$$

The product of the molecular mass m and Avogadro's constant is the molar mass M of the molecules, so

$$\text{Momentum change} = \frac{nMv_x^2 A \Delta t}{V}$$

Next, to identify the force, we calculate the rate of change of momentum, which is this change of momentum divided by the interval Δt during which it occurs:

$$\text{Rate of change of momentum} = \frac{nMv_x^2 A}{V}$$

This rate of change of momentum is equal to the force (by Newton's second law of motion). It follows that the pressure, the force per unit area, is

$$\text{Pressure} = \frac{nMv_x^2}{V}$$

Not all the molecules travel with the same velocity, so the detected pressure p is the average (denoted $\langle\ldots\rangle$) of the quantity just calculated:

$$p = \frac{nM\langle v_x^2\rangle}{V}, \qquad \text{or } pV = nM\langle v_x^2\rangle$$

This expression already resembles the perfect gas equation of state.

Because the molecules are moving randomly (and there is no net flow in a particular direction), the average speed along x is the same as that in the y and z directions. It follows from eqn 14 that

$$c = \surd(3\langle v_x^2\rangle), \text{ implying that } \langle v_x^2\rangle = \tfrac{1}{3}c^2$$

Therefore,

$$pV = \tfrac{1}{3}nMc^2$$

as in eqn 13.

Equation 13 is one of the key results of kinetic theory. We see that if the root mean square speed of the molecules depends only on the temperature, then at constant temperature

$$pV = \text{constant}$$

which is the content of Boyle's law.

Molecular speeds

If eqn 13 is to be *precisely* the equation of state of a perfect gas, then the right-hand side must be equal to nRT. For this to be true, it follows that the root mean square speed of the molecules in a gas at a temperature T must be given by the expression

$$c = \left(\frac{3RT}{M}\right)^{\frac{1}{2}} \tag{15}°$$

We can conclude that *the root mean square speed of the molecules of a gas is proportional to the square root of the temperature and inversely proportional to the square root of the molar mass.* That is, the higher the temperature, the faster the molecules travel and, at a given temperature, heavy molecules travel more slowly than light molecules.

Equation 15 can be used to calculate the root mean square speeds of molecules of any molar mass. For example, because the molar mass of CO_2 is 44.01 g mol^{-1}, we find that at 298 K

$$c = \left(\frac{3 \times (8.3145 \text{ J K}^{-1} \text{ mol}^{-1}) \times (298 \text{ K})}{44.01 \times 10^{-3} \text{ kg mol}^{-1}}\right)^{\frac{1}{2}} = 411 \text{ m s}^{-1}$$

(To obtain this result, we have used $1 \text{ J} = 1 \text{ kg m}^2 \text{ s}^{-2}$.) It is no accident that the root mean square speed is not very different from the speed of sound in air, which is about 340 m s^{-1} at sea level. Sound waves are pressure waves and, for them to propagate, the molecules of the gas must move to form regions of high and low pressure. Therefore, we should expect the speed of sound to be similar to the average speed of the molecules of the gas.

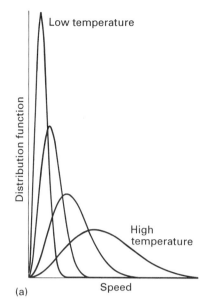

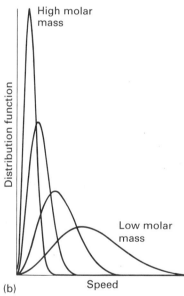

(a)

(b)

1.13 The distribution of molecular speeds with (a) temperature and (b) molar mass. Note that the most probable speed (corresponding to the peak of the distribution) increases with temperature and with decreasing molar mass and, simultaneously, the distribution becomes broader.

Equation 15 is an expression for the mean square speed of molecules. However, in an actual gas the speeds of individual molecules span a wide range, and the collisions in the gas continually redistribute the speeds among the molecules. Before a collision a particular molecule may be travelling rapidly, but after a collision it may be accelerated to a very high velocity, only to be slowed again by the next collision. The fraction of molecules that have speeds in the range s to $s + ds$ is proportional to the width of the range, and is written $f(s)\,ds$, where f, which varies with the speed s, is called the **distribution** of speeds. The precise form of f was derived by James Clerk Maxwell (see *Further information 1*) and is

$$f(s) = 4\pi\left(\frac{M}{2\pi RT}\right)^{\frac{3}{2}} s^2 e^{-Ms^2/2RT} \qquad (16)°$$

This expression is called the **Maxwell distribution of speeds**. Figure 1.13, which summarizes the main features of the Maxwell distribution, shows that the distribution of speeds broadens as the temperature increases. Lighter molecules also have a broader distribution of speeds than heavy molecules.

Example 1.7 *Calculating the mean speed of molecules in a gas*

What is the mean speed $\bar{c}$ of N_2 molecules in air at 25°C?

Method. We are asked to calculate the *mean* speed, not the root mean square speed. A mean value is calculated by multiplying each speed by the fraction of molecules that have that speed and then adding all the products together. When the speed varies continuously, the sum is replaced by an integral. To employ this approach here, we note that the fraction of molecules with a speed in the range s to $s + ds$ is $f(s)\,ds$, so the product of this fraction and the speed is $sf(s)\,ds$. The mean speed $\bar{c}$ is obtained by evaluating the integral

$$\bar{c} = \int_0^\infty sf(s)\,ds$$

with f given in eqn 16.

Answer. The integral required is

$$\bar{c} = \int_0^\infty sf(s)\,ds = 4\pi\left(\frac{M}{2\pi RT}\right)^{\frac{3}{2}}\int_0^\infty s^3 e^{-Ms^2/2RT}\,ds$$

$$= 4\pi\left(\frac{M}{2\pi RT}\right)^{\frac{3}{2}} \times \frac{1}{2}\left(\frac{2RT}{M}\right)^2 = \left(\frac{8RT}{\pi M}\right)^{\frac{1}{2}}$$

To evaluate the integral we have used the standard result that

$$\int_0^\infty x^3 e^{-ax^2}\,dx = \frac{1}{2a^2}$$

Substitution of the data then gives

$$\bar{c} = \left(\frac{8\times(8.3145\,\text{J K}^{-1}\,\text{mol}^{-1})\times(298\,\text{K})}{\pi\times(28.02\times10^{-3}\,\text{kg mol}^{-1})}\right)^{\frac{1}{2}} = 475\,\text{m s}^{-1}$$

Exercise E1.7. Evaluate the root mean square speed of the molecules by integration. You will need the integral

$$\int_0^\infty x^4 e^{-ax^2} = \frac{3}{8}\left(\frac{\pi}{a^5}\right)^{\frac{1}{2}}$$

$$[c = (3RT/M)^{\frac{1}{2}}, 515 \text{ m s}^{-1}]$$

The Maxwell distribution has been verified experimentally. For example, molecular speeds can be measured directly with a velocity selector like the one shown in Fig. 1.14. The spinning disks have slots that permit the passage of only those molecules moving through them at the appropriate speed, and the number of molecules can be determined by collecting them at a detector.

Intermolecular collisions

The kinetic theory enables us to make more quantitative our picture of the events that take place in a gas. In particular, it enables us to calculate the frequency with which molecular collisions occur and the distance a molecule travels on average between collisions. (It also provides a basis for calculating how rapidly physical properties are transported through a gas, Chapter 24, and the rates of chemical reactions, Chapter 27.)

We count a 'hit' whenever the centres of two molecules come within a distance d of each other where d, the **collision diameter**, is of the order of the actual diameters of the molecules (for impenetrable hard spheres d is the diameter). The simplest approach to calculating the frequency of such collisions is to freeze the positions of all the molecules except one. Then we note what happens as that mobile molecule travels through the gas with a mean speed $\bar{c}$ for a time Δt. In doing so it sweeps out a 'collision tube' of cross-sectional area $\sigma = \pi d^2$ and length $\bar{c}\,\Delta t$, and therefore of volume $\sigma\bar{c}\,\Delta t$ (Fig. 1.15). The area σ is called the **collision cross-section** of the molecules. Some typical collision cross-sections are given in Table 1.2 (they are obtained by the techniques described in Section 22.5). As we show in the *Justification* that follows, kinetic theory can be used to deduce that the **collision frequency** z, the number of collisions made by one molecule per unit time, when there are N molecules in a volume V is

$$z = \sqrt{2}\,\sigma\bar{c}\times\frac{N}{V} \qquad\qquad (\mathbf{17a})^\circ$$

1.14 A velocity selector. The molecules are produced in the source (which may be an oven with a small hole in one wall), and travel in a beam towards the rotating disks. Only if the speed of a molecule is such as to carry it through each slot that rotates into its path will it reach the detector. Thus, the number of slow molecules can be counted by rotating the disks slowly, and the number of fast molecules counted by rotating the disks rapidly.

Detector

Source

Selector

1.15 In an interval Δt a molecule of diameter d sweeps out a tube of radius d and length $\bar{c}\,\Delta t$. As it does so it encounters other molecules with centres that lie within the tube, and each such encounter counts as one collision. In practice, the tube is not straight, but changes direction at each collision. Nevertheless, the volume swept out is the same, and this straightened version of the tube can be used as a basis of the calculation.

Table 1.2* Collision cross-sections

	σ/nm^2
Benzene, C_6H_6	0.88
Carbon dioxide, CO_2	0.52
Helium, He	0.21
Nitrogen, N_2	0.43

* More values are given in the Data section at the end of this volume.

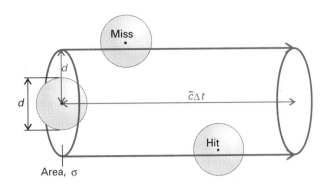

Alternatively, in terms of the pressure,

$$z = \sqrt{2}\,\sigma\bar{c} \times \frac{p}{kT} \qquad\qquad \textbf{(17b)}^{\circ}$$

In this expression k is **Boltzmann's constant**,

$$k = \frac{R}{N_A} = 1.380\,66 \times 10^{-23}\,\mathrm{J\,K^{-1}}$$

JUSTIFICATION

The number of stationary molecules with centres inside the collision tube is given by the volume of the tube multiplied by the number density N/V, and is $(N/V)\sigma\bar{c}\,\Delta t$. The number of hits scored in the interval Δt is equal to this number, so the number of collisions per unit time is $(N/V)\sigma\bar{c}$. However, the molecules are not stationary, so we should use the *relative* speed of the colliding molecules:

$$\bar{c}_{rel} = \sqrt{2} \times \bar{c}$$

where $\bar{c}$ was calculated in Example 1.7 and is

$$\bar{c} = \left(\frac{8RT}{\pi M}\right)^{\frac{1}{2}}$$

Therefore,

$$z = \sqrt{2}\,\sigma\bar{c} \times \frac{N}{V}$$

The expression in terms of the pressure of the gas is obtained by using the perfect gas equation to write

$$\frac{N}{V} = \frac{nN_A}{V} = \frac{pN_A}{RT} = \frac{p}{kT}$$

Equation 17a shows that the collision frequency increases with increasing temperature in a sample held at constant volume because the mean speed increases with temperature. On the other hand, eqn 17b shows that, at constant temperature, the collision frequency is proportional to the pressure. Such a proportionality is plausible, for the greater the pressure, the greater the number of molecules in the sample, and the rate at which they encounter one another is greater even though their average speed is constant. For an N_2 molecule in a sample at 1 atm and 25°C (as in air at sea level on a warm day), the collision frequency works out as about $5 \times 10^9 \, s^{-1}$, so in 1 s a given molecule collides about 5×10^9 times. We are beginning to appreciate the *time scale* of events in gases.

Once we have the collision frequency, we can calculate the **mean free path** λ, the average distance a molecule travels between collisions. If a molecule moving with a mean speed $\bar{c}$ collides with a frequency z, it spends a time $1/z$ in free flight between collisions, and therefore travels a distance $(1/z)\bar{c}$. Therefore, the mean free path is

$$\lambda = \frac{\bar{c}}{z} \tag{18}$$

Substitution of the expression for z derived in eqn 17 gives

$$\lambda = \frac{kT}{\sqrt{2}\,\sigma p} \tag{19}°$$

Doubling the pressure reduces the mean free path by half. A typical mean free path in nitrogen gas at 1 atm is 70 nm, or about 10^3 molecular diameters. Although the temperature appears in eqn 19, in a sample of constant volume, the pressure is proportional to T, so T/p remains constant when the temperature is increased. Therefore, the mean free path is independent of the temperature in a sample of gas in a container of fixed volume. The distance between collisions is determined by the *number* of molecules present in the given volume, not by the speed at which they travel.

In summary, a typical gas (N_2 or O_2) at 1 atm and 25°C can be thought of as a collection of molecules travelling with a mean speed of about $350 \, m \, s^{-1}$ (close to the speed of sound). Each molecule makes a collision within about 1 ns, and between collisions it travels about 10^3 molecular diameters. The kinetic theory of gases is valid (and the gas behaves nearly perfectly) if the diameter of the molecules is much smaller than the mean free path $(d \ll \lambda)$, for then the molecules spend most of their time far from each other.

REAL GASES

Real gases do not obey the perfect gas equation exactly. The deviations from the law are particularly important at high pressures and low temperatures, especially when the gas is on the point of condensing to a liquid.

1.4 Molecular interactions

Real gases show deviations from the perfect gas equation because molecules interact with each other. Repulsive forces between molecules assist expansion and attractive forces assist compression.

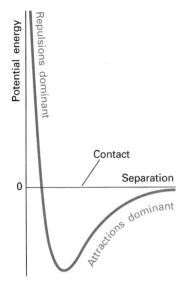

1.16 The variation of the potential energy of two molecules with their separation. High positive potential energy (at very small separations) indicates that the interactions between them are strongly repulsive at these distances. At intermediate separations, where the potential energy is negative, the attractive interactions dominate. At large separations (on the right) the potential energy is zero and there is no interaction between the molecules.

Repulsive forces between neutral molecules are significant only when the molecules are almost in contact: they are short-range interactions, even on a scale measured in molecular diameters (Fig. 1.16). Because they are short-range interactions, repulsions can be expected to be important only when the molecules are close together on average. This is the case at high pressure when a large number of molecules occupy a small volume. On the other hand, attractive intermolecular forces have a relatively long range and are effective over several molecular diameters. They are important when the molecules are fairly close together but not necessarily touching (at the intermediate separations in Fig. 1.16). Attractive forces are ineffective when the molecules are far apart (well to the right in Fig. 1.16).

It follows that at low pressure, when the molecules occupy a large volume, the molecules are so far apart for most of the time that the intermolecular forces play no significant role and the gas behaves perfectly. At moderate pressure, when the molecules are on average only a few molecular diameters apart, the attractive forces dominate the repulsive forces. In this case, the gas can be expected to be more compressible than a perfect gas because the forces are helping to draw the molecules together. At high pressure, when the molecules are on average close together, the repulsive forces dominate and the gas can be expected to be less compressible because now the forces help to drive the molecules apart.

The compression factor

The way in which the behaviour of real gases reflects this distance dependence of the forces can be demonstrated by plotting the **compression factor** Z against pressure, where Z is defined as

$$Z = \frac{pV_m}{RT} \tag{20}$$

Because, for a perfect gas, $Z = 1$ under all conditions, deviation of Z from 1 is a measure of imperfection.

Some experimental data on Z are plotted in Fig. 1.17. At very low pressures, all the gases shown have $Z \approx 1$ and are behaving nearly perfectly. At high pressures, all the gases have $Z > 1$, signifying that they are more difficult to compress than a perfect gas (the product pV_m is greater than RT). Repulsive forces are now dominant. At intermediate pressures, most of the gases have $Z < 1$, indicating that the attractive forces are dominant and favour compression.

Virial coefficients

Figure 1.18 shows some experimental isotherms for carbon dioxide. At large molar volumes and high temperatures the real and perfect isotherms do not differ greatly. The small differences suggest that the perfect gas law is correct at low pressures and is in fact the first term in an expression of the form

$$pV_m = RT(1 + B'p + C'p^2 + \ldots) \tag{21a}$$

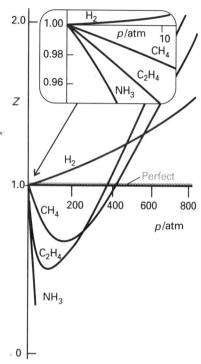

1.17 The variation of the compression factor $Z = pV_m/RT$ with pressure for several gases at 0°C. A perfect gas has $Z = 1$ at all pressures. Notice that, although the curves approach 1 as $p \to 0$, they do so with different slopes.

In many applications, a more convenient expansion is

$$pV_m = RT\left(1 + \frac{B}{V_m} + \frac{C}{V_m^2} + \ldots\right) \qquad \text{(21b)}$$

These expressions are two versions of the **virial equation of state** (the name comes from the Latin word for 'force'). The coefficients B, C, ..., which depend on the temperature, are the second, third, ... **virial coefficients** (Table 1.3); the first virial coefficient is 1. The third virial coefficient C is usually less important than the second B in the sense that at typical molar volumes $C/V_m^2 \ll B/V_m$. The virial equation is an example of a common procedure in physical chemistry, in which a simple law (in this case $pV = nRT$) is treated as the first term in a series in powers of a variable (in this case p or V_m).

The virial equation can be used to demonstrate the important point that, although the equation of state of a real gas may coincide with the perfect gas law as $p \to 0$, all its *properties* do not necessarily coincide with those of a perfect gas in that limit. Consider, for example, the value of dZ/dp, the slope of the graph of compression factor against pressure. For a perfect gas $dZ/dp = 0$ (because $Z = 1$ at all pressures), but for a real gas

$$\frac{dZ}{dp} = B' + 2pC' + \ldots \to B' \quad \text{as} \quad p \to 0$$

However, B' is not necessarily zero and the *slope* of Z with respect to p does not approach 0 (the perfect gas value). Because several properties depend on derivatives (as we shall see), the properties of real gases do not always coincide with the perfect gas values at low pressures. By a similar argument based on eqn 21b, the variation of Z with molar volume at large molar volumes (the equivalent of low pressures) is

$$\frac{dZ}{d(1/V_m)} \to B \quad \text{as} \quad V_m \to \infty \text{ (corresponding to } p \to 0)$$

Because the virial coefficients depend on the temperature, there may be a temperature at which $Z \to 1$ with *zero* slope at low pressure or high molar volume (Fig. 1.19). At this temperature, which is called the **Boyle temperature** T_B, the properties of the real gas do coincide with those of a perfect gas as $p \to 0$. According to the relation above, Z has zero slope as $p \to 0$ if $B = 0$, so we can conclude that at the Boyle temperature $B = 0$. It then follows from eqn 21b that $pV_m \approx RT_B$ over a more extended range of pressures than at other temperatures because the first term after 1 (that is, B/V_m) in the virial equation is zero and C/V_m^2 and higher terms are negligibly small. For helium $T_B = 22.64$ K and for air $T_B = 346.8$ K; some more values are given in Table 1.4.

Condensation

Now consider what happens when the volume of a sample of gas initially in the state marked A in Fig. 1.18 is decreased at constant temperature (by pushing in a piston). Near A, the pressure of the gas rises in approximate agreement with Boyle's law. Serious deviations from that law begin to appear when the volume has been reduced to B.

Table 1.3* Second virial coefficients, $B/(\text{cm}^3\,\text{mol}^{-1})$

	Temperature	
	273 K	600 K
Ar	−21.7	11.9
CO_2	−149.7	−12.4
N_2	−10.5	21.7
Xe	−153.7	−19.6

* More values are given in the Data section.

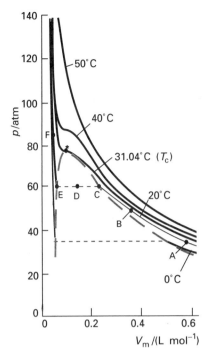

1.18 Experimental isotherms of carbon dioxide at several temperatures. The 'critical isotherm', the isotherm at the critical temperature, is at 31.04°C. The critical point is marked with a star.

Table 1.4* Critical constants of gases

	p_c/atm	V_c/(cm^3 mol^{-1})	T_c/K	Z_c	T_B/K
Ar	48.0	75.3	150.7	0.292	411.5
CO$_2$	72.9	94.0	304.2	0.274	714.8
He	2.26	57.8	5.2	0.305	22.6
O$_2$	50.14	78.0	154.8	0.308	405.9

* More values are given in the Data section.

At C (which corresponds to about 60 atm for carbon dioxide), all similarity to perfect behaviour is lost, for suddenly the piston slides in without any further rise in pressure: this stage is represented by the horizontal line CDE. Examination of the contents of the vessel shows that just to the left of C a liquid appears, and there are two phases separated by a sharply defined surface. As the volume is decreased from C through D to E, the amount of liquid increases. There is no additional resistance to the piston because the gas can respond by condensing. The pressure corresponding to the line CDE, when both liquid and vapour are present in equilibrium, is called the **vapour pressure** of the liquid at the temperature of the experiment.

At E, the sample is entirely liquid and the piston rests on its surface. Any further reduction of volume requires the exertion of considerable pressure, as is indicated by the sharply rising line to the left of E. Even a small reduction of volume from E to F requires a great increase in pressure.

Critical constants

The isotherm at the temperature T_c (304.19 K or 31.04°C for CO$_2$) plays a special role in the theory of the states of matter. An isotherm slightly below T_c behaves as we have already described: at a certain pressure, a liquid condenses from the gas and is distinguishable from it by the presence of a visible surface. If, however, the compression takes place at T_c itself, a surface separating two phases does not appear and the volumes at each end of the horizontal part of the isotherm have merged to a single point, the **critical point** of the gas. The temperature, pressure, and molar volume at the critical point are called the **critical temperature** T_c, **critical pressure** p_c, and **critical molar volume** V_c of the substance. Collectively, p_c, V_c, and T_c are the **critical constants** of a substance (Table 1.4).

At and above T_c the sample has a single phase that occupies the entire volume of the container. Such a phase is, by definition, a gas. Hence, *the liquid phase of a substance does not form above the critical temperature.* The critical temperature of oxygen, for instance, signifies that it is impossible to produce liquid oxygen by compression alone if its temperature is greater than 155 K: to liquefy it—to obtain a fluid phase that does not occupy the entire volume—the temperature must first be lowered to below 155 K, and then the gas compressed isothermally. The single phase that fills the entire volume at $T > T_c$ may be much denser than we

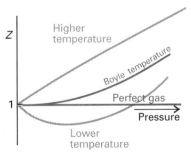

1.19 The compression factor approaches 1 at low pressures, but does so with different slopes. For a perfect gas, the slope is zero, but real gases may have either positive or negative slopes, and the slope may vary with temperature. At the Boyle temperature, the slope is zero and the gas behaves perfectly over a wider range of conditions than at other temperatures.

normally consider typical of gases, and the name **supercritical fluid** is preferred.

1.5 The van der Waals equation

Conclusions can be drawn from the virial equations of state only by inserting specific values of the coefficients. It is often useful to have a broader, if less precise, view of all gases. Therefore, we introduce the approximate equation of state suggested by Johannes van der Waals in 1873. It is an excellent example of an equation that can be obtained by thinking scientifically about a mathematically complicated but physically simple problem. Van der Waals himself proposed it on the basis of experimental evidence available to him in conjunction with rigorous thermodynamic arguments.

Constructing the equation

The repulsive interactions between molecules are taken into account by supposing that they cause the molecules to behave as small but impenetrable spheres. The non-zero volume of the molecules implies that instead of moving in a volume V they are restricted to a smaller volume $V - nb$, where nb is approximately the total volume taken up by the molecules themselves. This argument suggests that the perfect gas law $p = nRT/V$ should be replaced by

$$p = \frac{nRT}{V - nb}$$

The pressure depends on both the frequency of collisions with the walls and the force of each collision. Both the frequency of the collisions and their force are reduced by the attractive forces, which act with a strength proportional to the molar concentration n/V of molecules in the sample. Therefore, because both the frequency and the force of the collisions are reduced by the attractive forces, the pressure is reduced in proportion to the square of this concentration. If the reduction of pressure is written as $-a(n/V)^2$, where a is a positive constant characteristic of each gas, the combined effect of the repulsive and attractive forces is the **van der Waals equation of state**:

$$p = \frac{nRT}{V - nb} - a\left(\frac{n}{V}\right)^2 \tag{22a}$$

This equation is often written in terms of the molar volume $V_{\mathrm{m}} = V/n$ as

$$p = \frac{RT}{V_{\mathrm{m}} - b} - \frac{a}{V_{\mathrm{m}}^2} \tag{22b}$$

The values of the **van der Waals coefficients** a and b, which are characteristic of each gas but independent of the temperature, are listed in Table 1.5.

Table 1.5* van der Waals coefficients

	$a/(\mathrm{atm\,L^2\,mol^{-2}})$	$b/(10^{-2}\,\mathrm{L\,mol^{-1}})$
Ar	1.345	3.22
CO_2	3.592	4.267
He	0.034	2.37
N_2	1.390	3.913

* More values are given in the Data section.

Example 1.8 *Using the van der Waals equation to estimate a molar volume*

Estimate the molar volume of CO_2 at 500 K and 100 atm by treating it as a van der Waals gas.

Answer. We rearrange eqn 22b into an equation for V_m:

$$V_m^3 - \left(b + \frac{RT}{p}\right)V_m^2 + \left(\frac{a}{p}\right)V_m - \frac{ab}{p} = 0$$

According to Table 1.5, $a = 3.592\,L^2\,atm\,mol^{-2}$ and $b = 4.267 \times 10^{-2}\,L\,mol^{-1}$. Therefore, because

$$\frac{RT}{p} = \frac{(8.206 \times 10^{-2}\,L\,atm\,K^{-1}\,mol^{-1}) \times (500\,K)}{100\,atm} = 0.410\,L\,mol^{-1}$$

the coefficients in the equation for V_m are

$b + RT/p = 0.453\,L\,mol^{-1}$

$a/p = 3.59 \times 10^{-2}\,(L\,mol^{-1})^2$

$ab/p = 1.53 \times 10^{-3}\,(L\,mol^{-1})^3$

Then, on writing $x = V_m/(L\,mol^{-1})$, we must solve

$$x^3 - 0.453x^2 + (3.59 \times 10^{-2})x - (1.53 \times 10^{-3}) = 0$$

The most elementary way of solving this cubic equation is by starting from the perfect gas value $x = 0.410$ and then looking for the value of x that actually solves the equation. (A better way is to use a computer or programmable calculator.) We find $x = 0.366$, which implies that $V_m = 0.366\,L\,mol^{-1}$.

Comment. Cubic equations can be solved analytically. The formula is given in Section 3.8.2 of *Handbook of mathematical functions*, M. Abramowitz and I. Stegun, Dover (1965), a rich source of this kind of information.

Exercise E1.8. Calculate the molar volume of argon at 100°C and 100 atm on the assumption that it is a van der Waals gas.   [$0.298\,L\,mol^{-1}$]

We have built the van der Waals equation by using vague arguments about the volumes of molecules and the effects of forces. It can be derived in other ways, but the present method has the advantage that it shows how to derive the form of an equation out of general ideas. The derivation also has the advantage of keeping imprecise the significance of the coefficients a and b: they are much better regarded as empirical parameters than as precisely defined molecular properties.

The reliability of the equation

We now examine to what extent the van der Waals equation predicts the behaviour of real gases. It is too optimistic to expect a single, simple expression to be the true equation of state of all substances, and accurate work on gases must resort to the virial equation, use tabulated values of the coefficients at various temperatures, and analyse the systems numerically. The advantage of the van der Waals equation is that it is analytical and allows us to draw some general conclusions about real gases. When the equation fails we must use one of the other equations of state that have been proposed (some are listed in Table 1.6), invent a new one, or go back to the virial equation.

Table 1.6 Selected equations of state

Equation	Reduced form	Critical constants		
		p_c	V_c	T_c
Perfect gas $\quad p = \dfrac{RT}{V_m}$	$p_r = \dfrac{8T_r}{3V_r - 1} - \dfrac{3}{V_r^2}$			
Van der Waals $\quad p = \dfrac{RT}{V_m - b} - \dfrac{a}{V_m^2}$	$p_r = \dfrac{8T_r}{3V_r - 1} - \dfrac{3}{T_r V_r^2}$	$\dfrac{a}{27b^2}$	$3b$	$\dfrac{8a}{27bR}$
Berthelot $\quad p = \dfrac{RT}{V_m - b} - \dfrac{a}{TV_m^2}$	$p_r = \dfrac{e^2 T_r e^{-2/T_r V_r}}{2V_r - 1}$	$\dfrac{1}{12}\left(\dfrac{2aR}{3b^3}\right)^{\frac{1}{2}}$	$3b$	$\dfrac{2}{3}\left(\dfrac{2a}{3bR}\right)^{\frac{1}{2}}$
Dieterici $\quad p = \dfrac{RT\,e^{-a/RTV_m}}{V_m - b}$		$\dfrac{a}{4e^2 b^2}$	$2b$	$\dfrac{a}{4Rb}$
Beattie–Bridgman $\quad p = \dfrac{(1-\gamma)RT(V_m + \beta) - \alpha}{V_m^2}$ with $\alpha = a_0\left(1 + \dfrac{a}{V_m}\right)$ $\qquad \beta = b_0\left(1 - \dfrac{b}{V_m}\right)$ $\qquad \gamma = \dfrac{c_0}{V_m T^3}$				
Virial (Kammerlingh Onnes) $\quad p = \dfrac{RT}{V_m}\left\{1 + \dfrac{B(T)}{V_m} + \dfrac{C(T)}{V_m^2} + \ldots\right\}$				

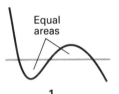

Equal areas

1

That having been said, we can begin to judge the reliability of the equation by comparing the isotherms it predicts with the experimental isotherms in Fig. 1.18. Some calculated isotherms are shown in Fig. 1.20 and, apart from the oscillations below the critical temperature, they do resemble experimental isotherms quite well. The oscillations, the **van der Waals loops**, are unrealistic because they suggest that under some conditions an increase of pressure results in an increase of volume. Therefore they are replaced by horizontal lines drawn so the loops define equal areas above and below the lines: this procedure is called the **Maxwell construction (1)**. The van der Waals coefficients are found by fitting the calculated curves to the experimental curves, and the values for some gases are listed in Table 1.5.

The features of the equation

The principal features of the van der Waals equation can be summarized as follows.

(1) *Perfect gas isotherms are obtained at high temperatures and large molar volumes.* When the temperature is high, RT may be so large that the first term in eqn 22b greatly exceeds the second. Furthermore, if the molar volume is large (in the sense $V_m \gg b$), then the denominator $V_m - b \approx V_m$. Under these conditions, the equation reduces to $p = RT/V_m$, the perfect gas equation.

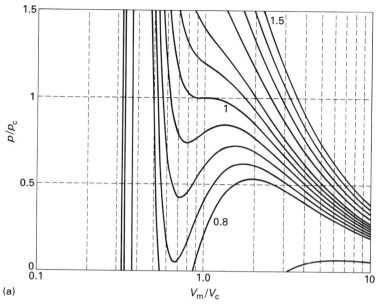

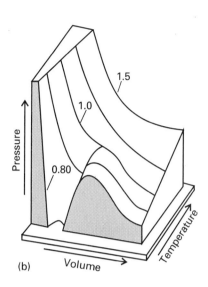

(a) V_m/V_c

(b) Volume

1.20 The van der Waals isotherms at several values of T/T_c. (a) A selection of individual isotherms; compare these curves with those in Fig. 1.5. (b) The shape of the surface; compare this surface with that shown in Fig. 1.7.

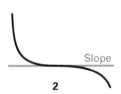

Slope

2

(2) *Liquids and gases coexist when cohesive and dispersing effects are in balance.* The van der Waals loops occur when both terms in eqn 22b have similar magnitudes. The first term arises from the kinetic energy of the molecules and their repulsive interactions; the second represents the effect of the attractive interactions.

(3) *The critical constants are related to the van der Waals coefficients.* For $T < T_c$, the calculated isotherms oscillate, and each one passes through a minimum followed by a maximum. These extrema converge as $T \to T_c$ and coincide at $T = T_c$; at the critical point the curve has a flat inflection (**2**). From the properties of curves, we know that an inflection of this type occurs when both the first and second derivatives are zero. Hence, we can find the critical constants by calculating these derivatives and setting them equal to zero:

$$\frac{dp}{dV_m} = -\frac{RT}{(V_m - b)^2} + \frac{2a}{V_m^3} = 0$$

$$\frac{d^2p}{dV_m^2} = \frac{2RT}{(V_m - b)^3} - \frac{6a}{V_m^4} = 0$$

at the critical point. The solution of these two equations is

$$V_c = 3b \qquad p_c = \frac{a}{27b^2} \qquad T_c = \frac{8a}{27Rb} \tag{23}$$

These relations can be tested by noting that the **critical compression factor** Z_c is predicted to be equal to

$$Z_c = \frac{p_c V_c}{RT_c} = \frac{3}{8} \tag{24}$$

for all gases. We see from Table 1.4 that, although $Z_c < \frac{3}{8}$ (or 0.375), it is approximately constant (at 0.3) and the discrepancy is reasonably small.

1.6 The principle of corresponding states

An important technique in science in general for comparing the properties of objects is to choose a related fundamental property of the same kind and to set up a relative scale on that basis. We have seen that the critical constants are characteristic properties of gases, so it may be that a scale can be set up by using them as yardsticks. We therefore introduce the **reduced variables** of a gas by dividing the actual variable by the corresponding critical constant:

$$p_r = \frac{p}{p_c} \qquad V_r = \frac{V_m}{V_c} \qquad T_r = \frac{T}{T_c} \tag{25}$$

If the reduced pressure of a gas is given, then we can easily calculate its actual pressure by using

$$p = p_r p_c$$

and likewise for the volume and temperature. Van der Waals, who first tried this procedure, hoped that gases confined to the same reduced volume (V_r) at the same reduced temperature (T_r) would exert the same reduced pressure (p_r). The hope was largely fulfilled. Figure 1.21 shows the dependence of the compression factor Z on the reduced pressure for a variety of gases at various reduced temperatures. The success of the procedure is strikingly clear: compare this graph with Fig. 1.17, where similar data are plotted without using reduced variables. The observation that real gases at the same reduced volume and reduced temperature exert the same reduced pressure is called the **principle of corresponding states**. It is only an approximation and works best for gases composed of spherical molecules. It fails, sometimes badly, when the molecules are non-spherical or polar.

The van der Waals equation sheds some light on the principle. First, we express eqn 22b in terms of the reduced variables, which gives

$$p_r p_c = \frac{R T_r T_c}{V_r V_c - b} - \frac{a}{V_r^2 V_c^2}$$

Then we express the critical constants in terms of a and b by using eqn

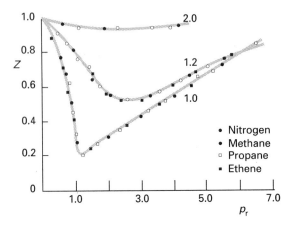

1.21 The compression factors of four of the gases shown in Fig. 1.17 plotted using reduced variables. The use of reduced variables organizes the data on to single curves for each reduced temperature.

23:

$$\frac{ap_r}{27b^2} = \frac{8aT_r}{27b(3bV_r - b)} - \frac{a}{9b^2V_r^2}$$

which can be reorganized into

$$p_r = \frac{8T_r}{3V_r - 1} - \frac{3}{V_r^2} \tag{26}$$

This equation has the same form as the original, but the coefficients a and b, which differ from gas to gas, have disappeared. It follows that if the isotherms are plotted in terms of the reduced variables (as we did in fact in Fig. 1.20 without drawing attention to the fact), then the same curves are obtained whatever the gas. This is precisely the content of the principle of corresponding states, so the van der Waals equation is compatible with it.

Looking for too much significance in this apparent triumph is mistaken, because other equations of state also accommodate the principle (Table 1.6). In fact, all we need are two parameters playing the roles of a and b, for then the equation can always be manipulated into reduced form. The observation that real gases obey the principle approximately amounts to saying that the effects of the attractive and repulsive interactions can each be approximated in terms of a single parameter. The importance of the principle is then not so much its theoretical interpretation but the way that it enables the properties of a range of gases to be coordinated on to a single diagram (e.g. Fig. 1.21 instead of Fig. 1.17).

CHECK LIST OF KEY IDEAS

1 The concept of **state** and **equation of state** (Section 1.1).

2 The definition of **pressure**, its measurement, and its units (Section 1.1).

3 Introduction of the concept of **temperature** in terms of **thermal equilibrium** and the summary of experience called the **Zeroth Law of thermodynamics** (Section 1.1).

4 The empirical behaviour of gases at low pressures as expressed by **Boyle's law** (eqn 4) and **Charles's law** (eqn 5), and **Avogadro's principle**.

5 The introduction of the **perfect gas equation** (Section 1.2, eqn 7) as a **limiting law**.

6 The extension of the description of gases to include mixtures of gases in terms of **Dalton's** law for perfect gases (eqn 9) and **partial pressures** in general (eqn 12).

7 The formulation of the **kinetic theory of gases** (Section 1.3) and the derivation of the expression for the pressure of a perfect gas (eqn 13).

8 The justification of the expression for the temperature dependence of the **root mean square speed** of molecules in a gas (eqn 15) and the properties of the **Maxwell distribution of speeds** (eqn 16).

9 The derivation of expressions for the **collision frequency** (eqn 17) and the **mean free path** (eqn 18) and their physical interpretation.

10 The properties of **real gases** (Section 1.4) expressed in terms of the **isotherms** and the **compression factor** and summarized by the **virial equation of state** (eqn 21).

11 The physical significance of the **critical constants** of a gas (Section 1.4).

12 The formulation of the **van der Waals equation** (eqn 22) as an approximate equation of state of real gases and some of the conclusions that can be drawn from it (Section 1.5).

13 The unification of the description of real gases using the **principle of corresponding states** (Section 1.6) and an illustration of the principle using the van der Waals equation.

EXERCISES

1.1. A sample of air occupies 1.0 L at 25°C and 1.00 atm. What pressure is needed to compress it to 100 cm^3 at this temperature?

1.2. (a) Could 131 g of xenon gas in a vessel of volume 1.0 L exert a pressure of 20 atm at 25°C if it behaved as a perfect gas? If not, what pressure would it exert? (b) What pressure would it exert if it behaved as a van der Waals gas?

1.3. A perfect gas undergoes isothermal compression, which reduces its volume by 2.20 L. The final pressure and volume of the gas are 3.78×10^3 Torr and 4.65 L, respectively. Calculate the original pressure of the gas in (a) Torr, (b) atm.

1.4. To what temperature must a 1.0-L sample of a perfect gas be cooled from room temperature in order to reduce its volume to 100 cm^3?

1.5. A car tyre (i.e. an automobile tire) was inflated to a pressure of 24 lb in^{-2} (1 atm = 14.7 lb in^{-2}) on a winter's day when the temperature was -5°C. What pressure will be found, assuming no leaks have occurred and that the volume is constant, on a subsequent summer's day when the temperature is 35°C? What complications should be taken into account in practice?

1.6. A sample of 255 mg of neon occupies 3.00 L at 122 K. Use the perfect gas law to calculate the pressure of the gas.

1.7. A homeowner uses 4.00×10^3 m^3 of natural gas in a year to heat a home. Assume that natural gas is all methane, CH$_4$, and that methane is a perfect gas for the conditions of this problem, which are 1.00 atm and 20°C. What is the mass of gas used?

1.8. In an attempt to determine an accurate value of the gas constant R, a student heated a 20.000 L container filled with 0.25132 g of helium gas to 500°C and measured the pressure as 206.402 cm of water in a manometer at 25°C. Calculate the value of R from these data. (The density of water at 25°C is 0.99707 g cm^{-3}.)

1.9. The following data have been obtained for oxygen gas at 0°C. Calculate the best value of the gas constant R from them and the best value of the molar mass of O$_2$.

p/atm	0.750 000	0.500 000	0.250 000
V_m/L mol^{-1}	29.9649	44.8090	89.6384
ρ/(g L^{-1})	1.071 44	0.714 110	0.356 975

1.10. At 500°C and 699 Torr, the density of sulfur vapour is 3.71 g L^{-1}. What is the molecular formula of sulfur under these conditions?

1.11. Calculate the mass of water vapour present in a room of volume 400 m^3 that contains air at 27°C on a day when the relative humidity is 60 per cent. The vapour pressure of water at 27°C is 26.74 Torr.

1.12. Given that the density of air at 740 Torr and 27°C is 1.146 g L^{-1}, calculate the mole fraction and partial pressure of nitrogen and oxygen assuming that (a) air consists only of these two gases, (b) air also contains 1.00 mole per cent Ar.

1.13. A gas mixture consists of 320 mg of methane, 175 mg of argon, and 225 mg of neon. The partial pressure of neon at 300 K is 66.5 Torr. Calculate (a) the volume and (b) the total pressure of the mixture.

1.14. The density of a gaseous compound was found to be 1.23 g L^{-1} at 330 K and 150 Torr. What is the molar mass of the compound?

1.15. In an experiment to measure the molar mass of a gas, 250 cm^3 of the gas was confined in a glass vessel. The pressure was 152 Torr at 298 K and, after correcting for buoyancy effects, the mass of the gas was 33.5 mg. What is the molar mass of the gas?

1.16. The density of water is 0.99707 g cm^{-3} at 25°C and its molar mass is 18.016 g mol^{-1}. What is the pressure that the atmosphere must have in order to support a column of water 10.0 m high in a cylindrical tube of diameter 1.00 cm?

1.17. The densities of air at -85°C, 0°C, and 100°C are 1.877 g L^{-1}, 1.294 g L^{-1}, and 0.946 g L^{-1}, respectively. From these data and assuming that air obeys Charles's law, determine a value for the absolute zero of temperature in degrees Celsius.

1.18. A certain sample of a gas has a volume of 20.00 L at 0°C and 1.000 atm pressure. A plot of the experimental data of its volume against the Celsius temperature θ at constant p gives a straight line of slope $0.0741 \, L \, °C^{-1}$. From these data alone (without making use of the perfect gas law), determine the absolute zero of temperature in degrees Celsius.

1.19. Recent communication with the inhabitants of Neptune has revealed that they have a Celsius-type temperature scale, but based on the melting point (0°N) and boiling point (100°N) of their most common substance, hydrogen. Further communications have revealed that the Neptunians know about perfect gas behaviour and that they find, in the limit of zero pressure, that the value of pV is $28.0 \, L \, atm$ at 0°N and $40.0 \, L \, atm$ at 100°N. What is the value of the absolute zero for temperature on their temperature scale?

1.20. Determine the ratios of (a) the mean speeds, (b) the mean translational kinetic energies of H_2 molecules and Hg atoms at 20°C.

1.21. A 1.0-L glass bulb contains $1.0 \times 10^{23} \, H_2$ molecules. If the pressure exerted by the gas is 100 kPa, what is (a) the temperature of the gas, (b) the root mean square speed of the molecules? (c) Would the temperature be different if they were O_2 molecules?

1.22. The best laboratory vacuum pump can generate a vacuum of about 1 nTorr. At 25°C and assuming that air consists of N_2 molecules with a collision diameter of 395 pm, calculate (a) the mean speed of the molecules, (b) the mean free path, (c) the collision frequency in the gas.

1.23. Calculate the mean speed of (a) He atoms, (b) CH_4 molecules at (i) 77 K, (ii) 298 K, (iii) 1000 K.

1.24. At what pressure does the mean free path of argon at 25°C become comparable to the size of a 1.0 L vessel that contains it? Take $\sigma = 0.36 \, nm^2$.

1.25. At what pressure does the mean free path of argon at 25°C become comparable to the diameters of the atoms themselves?

1.26. At an altitude of 20 km the temperature is 217 K and the pressure 0.05 atm. What is the mean free path of N_2 molecules? ($\sigma = 0.43 \, nm^2$.)

1.27. How many collisions does a single Ar atom make in 1.0 s when the temperature is 25°C and the pressure is (a) 10 atm, (b) 1.0 atm, (c) 1.0 μatm?

1.28. How many collisions per second does an N_2 molecule make at an altitude of 20 km? (Take $\sigma = 0.43 \, nm^2$.)

1.29. Calculate the mean free path of diatomic molecules in air using $\sigma = 0.43 \, nm^2$ at 25°C and (a) 10 atm, (b) 1 atm, (c) 10^{-6} atm.

1.30. Use the Maxwell distribution of speeds to estimate the fraction of N_2 molecules at 500 K that have speeds in the range 290 to 300 m s^{-1}.

1.31. How does the mean free path in a sample of a gas vary with temperature in a constant-volume container?

1.32. Calculate the pressure exerted by 1.0 mol C_2H_6 behaving as (a) a perfect gas, (b) a van der Waals gas when it is confined under the following conditions: (i) at 273.15 K in 22.414 L, (ii) at 1000 K in 100 cm^3. Use the data in Table 1.5.

1.33. Estimate the critical constants of a gas with van der Waals parameters $a = 0.751 \, atm \, L^2 \, mol^{-2}$ and $b = 0.0226 \, L \, mol^{-1}$.

1.34. A gas at 250 K and 15 atm has a molar volume 12 per cent smaller than that calculated from the perfect gas law. Calculate (a) the compression factor under these conditions and (b) the molar volume of the gas. Which are dominating in the sample, the attractive or the repulsive forces?

1.35. In an industrial process, nitrogen is heated to 500 K at a constant volume of 1.000 m^3. The gas enters the container at 300 K and 100 atm pressure. The mass of the gas is 92.4 kg. Use the van der Waals equation to determine the approximate pressure of the gas at its working temperature of 500 K. For nitrogen, $a = 1.39 \, L^2 \, atm \, mol^{-2}$, $b = 0.0391 \, L \, mol^{-1}$.

1.36. Cylinders of compressed gas are typically filled to a pressure of 200 bar. For oxygen, what would be the molar volume at this pressure and 25°C based on (a) the perfect gas equation, (b) the van der Waals equation. For oxygen, $a = 1.360 \, L^2 \, atm \, mol^{-2}$, $b = 3.183 \times 10^{-2} \, L \, mol^{-1}$.

1.37. The density of water vapour at 327.6 atm and 776.4 K is 133.2 g L^{-1}. Determine the molar volume V_m of water and the compression factor Z from these data. Calculate Z from the van der Waals equation with $a = 5.464 \, L^2 \, atm \, mol^{-2}$ and $b = 0.03049 \, L \, mol^{-1}$.

1.38. Suppose that 10.0 mol $C_2H_6(g)$ is confined to a volume of 4.860 L at 27°C. Predict the pressure exerted by the ethane from (a) the perfect gas and (b) the van der Waals equations of state. Calculate the compression factor based on these calculations. For ethane, $a = 5.489 \, L^2 \, atm \, mol^{-2}$, $b = 0.06380 \, L \, mol^{-1}$.

1.39. At 300 K and 20 atm, the compression factor of a gas is 0.86. Calculate (a) the volume occupied by 8.2 mmol of the gas under these conditions and (b) an approximate value of the second virial coefficient B at 300 K.

1.40. A vessel of volume 22.4 L contains 2.0 mol H_2 and 1.0 mol N_2 at 273.15 K. Calculate (a) the mole fractions of each component, (b) their partial pressures, and (c) their total pressure.

1.41. The critical constants of methane are $p_c = 45.6$ atm, $V_c = 98.7$ cm^3 mol^{-1}, and $T_c = 190.6$ K. Calculate the van der Waals coefficients of the gas and estimate the radius of the molecules.

1.42. Use the van der Waals coefficients for chlorine to calculate approximate values of (a) the Boyle temperature of chlorine and (b) the radius of a Cl_2 molecule regarded as a sphere.

1.43. Suggest the pressure and temperature at which 1.0 mol of (a) NH_3, (b) Xe, (c) He will be in states that correspond to 1.0 mol H_2 at 1.0 atm and 25°C.

1.44. A certain gas obeys the van der Waals equation with $a = 0.50$ m^6 Pa mol^{-2}. Its molar volume is found to be 5.00×10^{-4} m^3 mol^{-1} at 273 K and 3.0 MPa. From this information calculate the van der Waals constant b. What is the compression factor for this gas at the prevailing temperature and pressure?

PROBLEMS
Numerical problems

1.1. A diving bell has an air space of 3.0 m^3 when on the deck of a boat. What is the volume of the air space when the bell has been lowered to a depth of 50 m? Take the mean density of sea water to be 1.025 g cm^{-3} and assume that the temperature is the same as on the surface.

1.2. What pressure difference must be generated across the length of a 15-cm vertical drinking straw in order to drink a water-like liquid of density 1.0 g cm^{-3}?

1.3. A meteorological balloon had a radius of 1.0 m when released at sea level at 20°C and expanded to a radius of 3.0 m when it had risen to its maximum altitude where the temperature was -20°C. What is the pressure inside the balloon at that altitude?

1.4. Deduce the relation between the pressure p and density ρ of a perfect gas of molar mass M. Confirm graphically, using the following data on dimethyl ether at 25°C, that perfect behaviour is reached at low pressures and find the molar mass of the gas.

p/Torr	91.74	188.98	277.3	452.8	639.3	760.0
ρ/(g L^{-1})	0.232	0.489	0.733	1.25	1.87	2.30

1.5. Charles's law is sometimes expressed in the form $V = V_0(1 + \alpha\theta)$ where θ is the Celsius temperature, α is a constant, and V_0 is the volume of the sample at 0°C. The following values for α have been reported for nitrogen at 0°C.

p/Torr	749.7	599.6	333.1	98.6
$10^3\alpha$/°C^{-1}	3.6717	3.6697	3.6665	3.6643

From these data calculate the best value for the absolute zero of temperature on the Celsius scale.

1.6. Investigate some of the technicalities of ballooning using the perfect gas law. Suppose your balloon has a radius of 3.0 m and that it is spherical. (a) What amount of H_2 (in moles) is needed to inflate it to a pressure of 1.0 atm in an ambient temperature of 25°C at sea level? (b) What mass can the balloon lift at sea level, where the density of air is 1.22 kg m^{-3}? (c) What would be the payload if He were used instead of H_2?

1.7. The molar mass of a newly synthesized fluorocarbon was measured in a gas microbalance. This device consists of a glass bulb forming one end of a beam, the whole surrounded by a closed container. The beam is pivoted, and the balance point is attained by raising the pressure of gas in the container, so increasing the buoyancy of the enclosed bulb. In one experiment, the balance point was reached when the fluorocarbon pressure was 327.10 Torr; for the same setting of the pivot, a balance was reached when CHF_3 ($M = 70.014$ g mol^{-1}) was introduced at 423.22 Torr. A repeat of the experiment with a different setting of the pivot required a pressure of 293.22 Torr of the fluorocarbon and 427.22 Torr of the CHF_3. What is the molar mass of the fluorocarbon? Suggest a molecular formula.

1.8. A constant-volume perfect gas thermometer indicates a pressure of 50.2 Torr at the triple point temperature of water (273.16 K). (a) What change of pressure indicates a change of 1 K at this temperature? (b) What pressure indicates a temperature of 100.00°C? (c) What change of pressure indicates a change of 1 K at the latter temperature?

1.9. A vessel of volume 22.4 L contains 2.0 mol H_2 and 1.0 mol N_2 at 273.15 K initially. All the H_2 reacted with sufficient N_2 to form NH_3. Calculate the partial pressures and the total pressure of the final mixture.

1.10. In an experiment to measure the speed of molecules by a rotating slotted-disk experiment, the apparatus consisted of five coaxial 5.0 cm diameter disks separated by 1.0 cm, the slots in their rims being displaced by 2.0° between neighbours. The relative intensities I of the detected beam of Kr atoms for two different temperatures and at a series of rotation rates were as follows:

v/Hz	20	40	80	100	120
I(40 K)	0.846	0.513	0.069	0.015	0.002
I(100 K)	0.592	0.485	0.217	0.119	0.057

Find the distributions of molecular velocities $f(v_x)$ at these temperatures, and check that they conform to the theoretical prediction for a one-dimensional system.

1.11. Cars were timed by police radar as they passed in both directions below a bridge. Their velocities (m.p.h., numbers of cars in brackets) to the east and west were as follows: 50 E (40), 55 E (62), 60 E (53), 65 E (12), 70 E (2); 50 W (38), 55 W (59), 60 W (50), 65 W (10), 70 W (2). What are (a) the mean velocity, (b) the mean speed, (c) the root mean square speed?

1.12. A population consists of people of the following heights (in feet and inches, numbers of individuals in parentheses): 5′5″ (1), 5′6″ (2), 5′7″ (4), 5′8″ (7), 5′9″ (10), 5′10″ (15), 5′11″ (9), 6′0″ (4), 6′1″ (0), 6′2″ (1). What is (a) the mean height, (b) the root mean square height of the population.

1.13. Calculate the escape velocity (the minimum initial velocity that will take an object to infinity) from the surface of a planet of radius R. What is the value for (a) the Earth, $R = 6.37 \times 10^6$ m, $g = 9.81$ m s^{-2}, (b) Mars, $R = 3.38 \times 10^6$ m, $m_{Mars}/m_{Earth} = 0.108$? At what temperatures do H_2, He, and O_2 molecules have mean speeds equal to their escape speeds? What proportion of the molecules have enough speed to escape when the temperature is (a) 240 K, (b) 1500 K? Calculations of this kind are very important in considering the composition of planetary atmospheres.

1.14. Calculate the molar volume of Cl_2 at 350 K and 2.30 atm using (a) the perfect gas law and (b) the van der Waals equation. Use the answer to part (a) to calculate a first approximation to the correction term for attraction and then use successive approximations to obtain a numerical answer for part (b).

1.15. At 273 K measurements on argon gave $B = -21.7$ cm^3 mol^{-1} and $C = 1200$ cm^6 mol^{-2}, where B and C are the second and third virial coefficients in the expansion of Z in powers of $1/V_m$. Assuming that the perfect gas law holds sufficiently well for the estimation of the second and third terms of the expansion, calculate the compression factor of argon at 100 atm and 273 K. From your result, estimate the molar volume of argon under these conditions.

1.16. Calculate the volume occupied by 1.00 mol N_2 using the van der Waals equation in the form of a virial expansion at (a) its critical temperature, (b) its Boyle temperature, and (c) its inversion temperature (see Table 3.2). Assume 10.0 atm pressure throughout. At what temperature is the gas most perfect? Use the following data: $T_c = 126.3$ K, $a = 1.390$ L^2 atm mol^{-2}, $b = 0.0391$ L mol^{-1}.

1.17. The density of water vapour at 327.6 atm and 776.4 K is 1.332×10^2 g L^{-1}. Given that for water $T_c = 647.4$ K, $p_c = 218.3$ atm, $a = 5.464$ L^2 atm mol^{-2}, $b = 0.03049$ L mol^{-1}, and $M = 18.02$ g mol^{-1}, calculate (a) the molar volume. Then calculate the compression factor (b) from the data, (c) from the virial expansion of the van der Waals equation.

1.18. The critical volume and critical pressure of a certain gas are 160 cm^3 mol^{-1} and 40 atm respectively. Estimate the critical temperature by assuming that the gas obeys the Berthelot equation of state (see Table 1.6). Estimate the radii of the gas molecules on the assumption that they are spheres.

1.19. Estimate the coefficients a and b in the Dieterici equation of state (see Table 1.6) from the critical constants of xenon. Calculate the pressure exerted by 1.0 mol Xe when it is confined to 1.0 L at 25°C.

Theoretical problems

1.20. The Maxwell distribution was derived from arguments about probability, but it can also be derived from the Boltzmann distribution (see the Introduction). Do so.

1.21. Start from the Maxwell distribution and derive an expression for the most probable speed of a gas of molecules at a temperature T. Go on to demonstrate the validity of the equipartition conclusion (see Introduction) that the average translational kinetic energy of molecules free to move in three dimensions is $\frac{3}{2}kT$.

1.22. Consider molecules that are confined to move in a plane (a two-dimensional gas). Calculate the distribution of speeds and determine the mean speed of the molecules at a temperature T.

1.23. A specially constructed velocity selector accepts a beam of molecules from an oven at a temperature T but blocks the passage of molecules with a speed greater than the mean. What is the mean speed of the emerging beam, relative to the initial value, treated as a one-dimensional problem?

1.24. What is the proportion of gas molecules having (a) more than, (b) less than the root mean square speed? (c) What are the proportions having speeds greater and smaller than the mean speed?

1.25. Calculate the fractions of molecules in a gas that have a speed in a range Δs at the speed nc^* relative to those in the same range at c^* (the most probable speed) itself? This calculation can be used to estimate the fraction of very energetic molecules (which is important for reactions). Evaluate the ratio for $n = 3$ and $n = 4$.

1.26. Show that the van der Waals equation leads to values of $Z < 1$ and $Z > 1$, and identify the conditions for which these values are obtained.

1.27. Express the van der Waals equation of state as a virial expansion in powers of $1/V_m$ and obtain expressions for B and C in terms of the parameters a and b. The expansion you will need is

$$\frac{1}{1-x} = 1 + x + x^2 + \ldots$$

Measurements on argon gave $B = -21.7 \text{ cm}^3 \text{ mol}^{-1}$ and $C = 1200 \text{ cm}^6 \text{ mol}^{-2}$ for the virial coefficients at 273 K. What are the values of a and b in the corresponding van der Waals equation of state?

1.28. A scientist proposed the following equation of state:

$$p = \frac{RT}{V_m} - \frac{B}{V_m^2} + \frac{C}{V_m^3}$$

Show that the equation leads to critical behaviour. Find the critical constants of the gas in terms of B and C and an expression for the critical compression factor.

1.29. Equations 21a and 21b are expansions in p and $1/V_m$, respectively. Find the relation between B, C and B', C'.

1.30. The second virial coefficient B' can be obtained from measurements of the density ρ of a gas at a series of pressures. Show that the graph of p/ρ against p should be a straight line with slope proportional to B'. Use the data on dimethyl ether in Problem 1.4 to find the values of B' and B at 25°C.

1.31. The equation of state of a certain gas is given by $p = RT/V_m + (a + bT)/V_m^2$, where a and b are constants. Find $(\partial V/\partial T)_p$.

1.32. The following equations of state are occasionally used for approximate calculations on gases: (gas A) $pV_m = RT(1 + b/V_m)$, (gas B) $p(V_m - b) = RT$. Assuming that there were gases that actually obeyed these equations of state, would it be possible to liquefy either gas A or B? Would they have a critical temperature? Explain your answer.

1.33. Derive an expression for the compression factor of a gas that obeys the equation of state $p(V - nb) = nRT$, where b and R are constants. If the pressure and temperature are such that $V_m = 10b$, what is the numerical value of the compression factor?

1.34. The barometric formula

$$p = p_0 e^{-Mgh/RT}$$

relates the pressure of a gas of molar mass M at an altitude h to its pressure p_0 at sea level. Derive this relation by showing that the change in pressure dp for an infinitesimal change in altitude dh where the density ρ is $dp = -\rho g\, dh$. Remember that ρ depends on the pressure. Evaluate the pressure difference between the top and bottom of (a) a laboratory vessel of height 15 cm, and (b) the World Trade Center in New York, 1350 ft. Ignore temperature variations.

2

The First Law: the concepts

This chapter introduces some of the basic concepts of thermodynamics. It concentrates on the conservation of energy—the experimental observation that it can be neither created nor destroyed—and shows how the principle of conservation of energy can be used to assess the energy changes that accompany physical and chemical processes. Thermodynamics is a subtle subject, and it is essential to take into account all the energy transactions that may take place between a system of interest and its immediate surroundings. Much of this chapter is based on a careful development of the means by which a system can exchange energy with its surroundings in terms of the work it may do or the heat that it may produce. The target concept of the chapter is enthalpy, which is a very useful book-keeping property for keeping track of the heat output (or requirements) of physical processes and chemical reactions under conditions of constant pressure (which is typical of many laboratory processes in vessels open to the atmosphere). Enthalpy will figure in discussions throughout the rest of the text.

The release of energy can be used to provide heat when a fuel burns in a furnace, to produce mechanical work when a fuel burns in an engine, and to produce electrical work when a chemical reaction pumps electrons through a circuit. In chemistry, we encounter reactions that can be harnessed to provide heat and work, reactions that liberate energy that is squandered (often to the detriment of the environment) but which give products we require, and reactions that constitute the processes of life. **Thermodynamics,** the study of the transformations of energy, enables us to discuss all these matters quantitatively.

THE BASIC CONCEPTS

For the purposes of physical chemistry, the universe is divided into two parts, the system and its surroundings. A **system** is the part of the world in which we have a special interest. It may be a reaction vessel, an

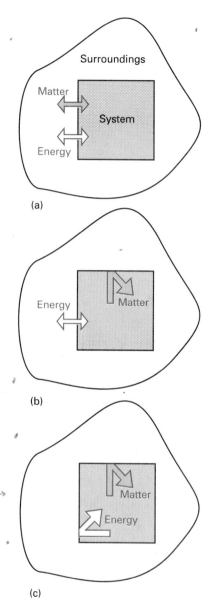

2.1 (a) An open system can exchange matter and energy with its surroundings. (b) A closed system can exchange energy with its surroundings, but it cannot exchange matter or undergo a change of composition. (c) An isolated system can exchange neither energy nor matter with its surroundings.

engine, an electrochemical cell, a biological cell, and so on. The **surroundings** are where we make our observations. The two parts are separated by a boundary, and to specify the system and its surroundings we need to specify the boundary between them. The type of system is determined by the characteristics of the boundary that divides it from the surroundings (Fig. 2.1). When matter can be transferred through the boundary between the system and its surroundings the system is classified as **open**; otherwise it is **closed**. Both open and closed systems can exchange energy with their surroundings. For example, a closed system can expand and thereby raise a weight in the surroundings, and it may also transfer energy to them if they are at a lower temperature. An **isolated system** is a closed system that has neither mechanical nor thermal contact with its surroundings.

2.1 Work, heat, and energy

Work, heat, and energy are the basic concepts of thermodynamics, and of these concepts the most fundamental is work. As we shall see, all measurements of heat and changes in energy can be expressed in terms of measurements of work.

Thermodynamics is concerned with energy changes that accompany a **process**. A process may be a simple change of state (such as expansion or cooling), a change in physical state (such as melting or freezing), or a complex chemical change in which new substances are produced from old. A process that does **work** is one that could be used to bring about a change in the height of a weight somewhere in the surroundings. An example of doing work is the expansion of a gas that pushes out a piston and raises a weight. A chemical reaction that drives an electric current through a resistance is also an example of a process that does work because the same current could be driven through a motor and used to raise a weight.

We shall say that work is done *by* the system if, as a result of a process taking place in the system, a weight has been raised in the surroundings. Work has been done *on* the system if, as a result of a process in the system, a weight is lowered. When we need to measure the work we use its definition as *opposing force × distance*, which we develop below.

By the **energy** of a system we mean its capacity to do work. When work is done on an otherwise isolated system (for instance, by compressing a gas or winding a spring), its capacity to do work is increased and thus the energy of the system is increased. When the system does work (when the piston moves out or the spring unwinds), its energy is reduced because it can do less work than before.

Experiments have shown that the energy of a system (its capacity to do work) may be changed by means other than work itself. When the energy of a system changes as a result of a temperature difference between it and its surroundings we say that energy has been transferred as **heat**. When a beaker of water (the system) stands on a hot plate, the capacity of the system to do work increases. Heated water, for example, can be used to expand gas through a greater volume than cold water can, so its energy is

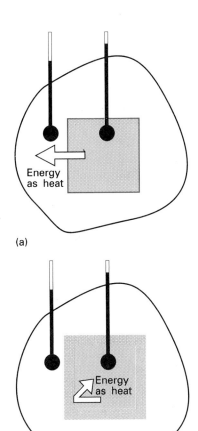

2.2 (a) A diathermic system is one that allows energy to escape as heat through its boundary if there is a difference in temperature between the system and its surroundings. (b) An adiabatic system is one that does not permit the passage of energy as heat through its boundary even if there is a temperature difference between the system and its surroundings.

greater; because the increase has occurred as a result of a temperature difference, the increase in energy has been transferred to the system as heat.[1]

Not all boundaries permit the transfer of energy even though there is a temperature difference between the system and its surroundings. It should be recalled from Section 1.1 that walls that do permit energy transfer as heat (such as steel and glass) are called diathermic and walls that do not permit energy transfer as heat are called adiabatic (Fig. 2.2).

A process that releases energy as heat is called **exothermic**. All combustion reactions are exothermic. Processes that absorb energy as heat are called **endothermic**. An example of an endothermic process is the vaporization of water. An endothermic process in a diathermic container that is maintained at constant temperature by being immersed in a water bath results in energy flowing into the system as heat. An exothermic process in a similar diathermic container results in a release of energy as heat into the surroundings. When an endothermic process takes place in an adiabatic container, it results in a lowering of temperature of the system; an exothermic process results in a rise of temperature. These features are summarized in Fig. 2.3.

MOLECULAR INTERPRETATION 2.1

In molecular terms, *heat is the transfer of energy that makes use of chaotic molecular motion*. The chaotic motion of molecules is called **thermal motion**. The thermal motion of the molecules in the hot surroundings stimulates the molecules in the cooler system to move more vigorously and, as a result, the energy of the system increases. When a system heats its surroundings, molecules of the system stimulate the thermal motion of the molecules in the surroundings.

In molecular terms, *work is the transfer of energy that makes use of organized motion*. When a weight is raised or lowered, its atoms move in an organized way. When a system does work it causes atoms or electrons in its surroundings to move in an organized way. Likewise, when work is done on a system, molecules in the surroundings are used to transfer energy to it in an organized way, as the atoms in a weight are lowered or a current of electrons is passed.

The distinction between work and heat is made *in the surroundings*. When work is done on a system, the energy leaves the surroundings in an orderly way, but it does not always arrive in the system that way. Thus, the work done by an electric current on a heater may end up as thermal motion in the system. The fact that a falling weight may stimulate thermal motion in the system is irrelevant to the distinction between heat and work: *work is identified as energy transfer making use of the organized motion of atoms in the surroundings*, and *heat is identified as energy transfer*

1 If the heater is regarded as a part of the system, then the transfer of energy is a result of doing *work* on the system. In this case, there is no temperature difference between the system and its surroundings, and the external source has driven the current through the system, a form of work. Electric heaters are devices for converting work into an increase in temperature.

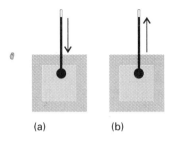

(a)   (b)

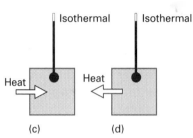

(c)   (d)

2.3 (a) When an endothermic process occurs in an adiabatic system, the temperature falls; (b) if the process is exothermic, then the temperature rises. (c) When an endothermic process occurs in a diathermic container, energy enters as heat from the surroundings, and the system remains at the same temperature. (d) If the process is exothermic, then the energy leaves as heat, and the process is isothermal.

making use of thermal motion in the surroundings (see Fig. 2.4). In the adiabatic compression of a gas, for instance, work is done as the particles of the compressing weight descend in an orderly way, but the effect of the incoming piston is to accelerate the gas molecules to higher average speeds. Because collisions between molecules quickly randomize their directions, the orderly motion of the atoms of the weight is in effect stimulating thermal motion in the gas. We observe the falling weight, the orderly descent of its atoms, and report that work is being done even though it is stimulating thermal motion.

2.2 The First Law

In thermodynamics, the total energy of a system is called its **internal energy**, U. We denote by ΔU the change in internal energy when a system changes from an initial state i with internal energy U_i to a final state f of internal energy U_f:

$$\Delta U = U_f - U_i \tag{1}$$

The internal energy is a **state function** in the sense that its value depends only on the current state of the system and is independent of how that state has been prepared. In other words, it is a function of the properties that determine the current state of the system:

$$U = U(n, p, \ldots)$$

A 1-L sample of hydrogen at 500 K and 100 kPa, for instance, has the same internal energy however the sample has been prepared. The internal energy is an extensive property, a property that is proportional to the size (extent) of the system (see Introduction, p. 3). The molar internal energy, the internal energy divided by the amount of substance, is an intensive property.

Internal energy, heat, and work are all measured in the same units, the **joule** (J), which is defined as

$$1\,J = 1\,kg\,m^2\,s^{-2}$$

For example, a mass of 2 kg travelling at 1 m s⁻¹ has a kinetic energy of

$$E_{kinetic} = \tfrac{1}{2}mv^2 = \tfrac{1}{2} \times (2\,kg) \times (1\,m\,s^{-1})^2 = 1\,kg\,m^2\,s^{-2} = 1\,J$$

Each beat of the human heart uses about 1 J of energy. We shall normally express changes in internal energy in kilojoules, where

$$1\,kJ = 10^3\,J$$

Thus, just to keep the heart pumping for a day requires the expenditure of about 100 kJ of energy. Changes in molar internal energy are expressed in kilojoules per mole (kJ mol⁻¹).

The conservation of energy

Although thermodynamics may appear to be a highly mathematical subject, it is firmly based in experimental observations. For instance, it

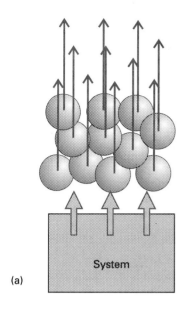

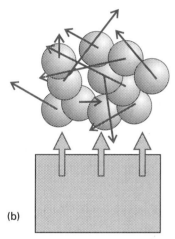

2.4 The distinction between work and heat is made in the surroundings. (a) Work is the transfer of energy that changes the motion of atoms in the surroundings in a uniform manner. (b) Heat is the transfer of energy that changes the motion of atoms in the surroundings in a chaotic manner.

has been found experimentally that the internal energy of a system may be changed either by doing work on the system or by heating it. Whereas we may know how the energy transfer has occurred (because we can see if a weight has been raised or lowered in the surroundings, indicating transfer of energy by doing work, or if ice has melted in the surroundings, indicating transfer of energy as heat), the system is blind to the mode employed. *Heat and work are equivalent ways of changing a system's internal energy.* A system is like a bank: it accepts deposits in either currency, but stores its reserves as internal energy. If we write w for the work done *on* a system, q for the energy transferred as heat *to* a system, and ΔU for the resulting change in internal energy, then it follows that

$$\Delta U = q + w \tag{2}$$

This equation states that the *change in internal energy of a closed system is equal to the energy that passes through its boundary as heat or work.* If 10 kJ of work is done on the system, then we write $w = +10$ kJ and conclude that $\Delta U = +10$ kJ. If 10 kJ of work is done by the system, we write $w = -10$ kJ and it follows that $\Delta U = -10$ kJ. If 5 kJ of energy is supplied to the system as heat, then we write $q = +5$ kJ and the internal energy of the system increases by 5 kJ: $\Delta U = +5$ kJ. If 5 kJ of energy escapes from the system as heat, we write $q = -5$ kJ and conclude that the internal energy has decreased by 5 kJ: $\Delta U = -5$ kJ.

Example 2.1 *Calculating a change in internal energy*

A certain electric motor produced 15 kJ of energy each second as mechanical work and lost 2 kJ as heat to the surroundings. What was the change in the internal energy of the motor and its power supply each second?

Method. The net change in energy is calculated from eqn 2: work done by the system and energy lost as heat are negative (because they correspond to a reduction in internal energy).

Answer. Because energy was lost from the system as work, we write $w = -15$ kJ. Energy was also lost as heat, so $q = -2$ kJ. The total change in internal energy each second is therefore

$$\Delta U = -2 \text{ kJ} - 15 \text{ kJ} = -17 \text{ kJ}$$

Comment. If we had decided to call the motor alone the system, then its internal energy would not have changed: the motor lost 17 kJ to the surroundings but the power supply did 17 kJ of work on the motor. The net loser is the power supply, the motor being merely a device that converts one form of energy into another.

Exercise E2.1. When a spring was wound, 100 J of work was done on it, but 15 J escaped to the surroundings as heat. What is the change in internal energy of the spring?   [+85 J]

It is also found experimentally that if a system is isolated from its surroundings, then *no change in internal energy takes place.* We cannot use a system to do work, leave it isolated for a month, and then come back expecting to find it restored to its original state and ready to do the same work again. The evidence for this property is that no perpetual motion machine (a machine that does work without consuming fuel or some other source of energy) has ever been built. The **First Law of thermodynamics** is a statement of this observation:

First Law. The internal energy of an isolated system is constant.

MOLECULAR INTERPRETATION 2.2

The internal energy is the total energy of the molecules composing the system. If a molecule can possess the energies E_1, E_2, and so on, then the average energy of a molecule is:

$$E = P_1 E_1 + P_2 E_2 + \ldots$$

where P_1, P_2, etc. are the probabilities that the molecule has the energy E_1, E_2, etc., so the internal energy is NE, where N is the total number of molecules in the sample.

The probabilities P_i are given by the Boltzmann distribution (see the 'Introduction' to this volume), which is illustrated in Fig. 2.5. We see that, although very high energy levels may be occupied at room temperature, the probability that they are occupied is very low (Fig. 2.5a). However, as the temperature of the system is increased, the exponential tail of the distribution stretches to higher energies (Fig. 2.5b). Correspondingly, the internal energy—the weighted average of the energy levels—contains a greater contribution from the higher energy levels, and its value increases.

For a perfect monatomic gas, the only contribution to the internal energy (other than the energy stored in the atomic structure itself) is translational motion of the atoms. Because the mean kinetic energy of the molecules is $\frac{1}{2}mc^2$, the molar energy is $N_A \times \frac{1}{2}mc^2$, or $\frac{1}{2}Mc^2$, where M is the molar mass of the molecules. Substitution of the expression for c^2

2.5 The Boltzmann distribution for two temperatures. (a) At low temperatures, the bulk of the population is in the lowest energy levels. (b) At high temperatures, the higher energy levels are more highly populated at the expense of the lower levels.

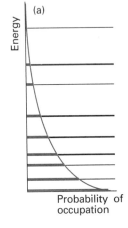
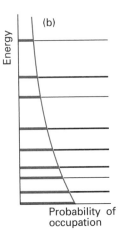

given in eqn 1.15 results in the following expression for the molar internal energy of a perfect gas at a temperature T:

$$U_m = U_m(0) + \tfrac{3}{2}RT$$

where $U_m(0)$ is the molar internal energy at $T = 0$, when all translational motion has ceased and the sole contribution to the internal energy arises from the internal structure of the molecules. This equation shows that the internal energy of a perfect gas increases in proportion to the temperature because the only contribution to the energy is the kinetic energy of the molecules, which is proportional to c^2, and c^2 is proportional to the temperature.

The formal statement of the First Law

The expression of the First Law that we have given is adequate for most purposes in thermodynamics. However, there are several unsatisfactory features about it, such as how we define and measure 'heat'. This section gives a more sophisticated version of the law, and shows how eqn 2 can be put on a firmer foundation. As the remainder of the text does not depend on this material, it is possible to omit it and go immediately to the following section ('Work and heat', p. 63).

We shall begin by pretending that we do not know what we mean by 'energy'. We shall pretend that we know only what is meant by work, because we can observe a weight being raised or lowered in the surroundings. We also know how to measure work by noting the height through which the weight is raised. Throughout this section, work will be the fundamental, measurable quantity, and we shall define energy, heat, and the First Law in terms of it alone. We shall employ terms that have been established by the Zeroth Law of thermodynamics (Section 1.1), namely, state and temperature and the concepts of adiabatic and diathermic walls.

In an adiabatic system of a particular composition (such as 1 kg of water), it is known experimentally that the same increase in temperature is brought about by the same quantity of any kind of work we do on the system. Thus, if 1 kJ of mechanical work is done on the system (by stirring it with rotating paddles, for instance), or 1 kJ of electrical work is done (by passing an electric current through a heater), and so on, then the same rise in temperature is produced. The following statement of the First Law of thermodynamics is a summary of a large number of observations of this kind:

> The work needed to change an adiabatic system from one specified state to another specified state is the same however the work is done.

This form of the law looks completely different from the form we gave before, but we shall now see how it implies that law.

Suppose we do work w_{ad} on an adiabatic system to change it from an initial state i to a final state f. To be definite, we might be changing the temperature of 1 kg of water in an adiabatic container from 20°C to 30°C at constant pressure by doing work on it. The work may be of any kind (mechanical or electrical) and may take the system through different intermediate states (different temperatures and pressures, for instance). We might (in ignorance of the First Law) think that we need to label w_{ad} with the path and to write w_{ad}(mechanical) or w_{ad}(electrical). However, the First Law tells us that w_{ad} is the same for all paths and depends only on the initial and final states. This conclusion is analogous to climbing a mountain: the *height* we must climb between any two points is independent of the path we take. In mountain-climbing we can attach a number, the altitude A, to each point on the mountain and express the height h of the climb as a difference in altitudes:

$$h = A_f - A_i = \Delta A$$

That is, in mountain climbing, *the observation that h is independent of the path taken implies the existence of the state function A.* The First Law has exactly the same implication. The fact that w_{ad} is independent of the path implies that to each state of the system we can attach a value of a quantity—we call it the 'internal energy' U—and express the work as a difference in internal energies:

$$w_{ad} = U_f - U_i = \Delta U \tag{3}$$

This equation also shows that we can measure the change in the internal energy of a system by measuring the work needed to bring about the change in an adiabatic system. That is, if we can devise an adiabatic path between the two states of interest, and can measure the work needed to drive the system between the two states along that path, then the work required is equal to the change in internal energy between the two states.

The mechanical definition of heat

The First Law in the form it has been expressed in this section does not mention heat. However, we shall now show that not only does the law imply the existence of heat but that it also provides a fundamental definition of heat in terms of work.

Suppose we strip away the thermal insulation around the system and make it diathermic. The system is now in thermal contact with its surroundings as we drive it from the same initial state to the same final state (such as changing the temperature of 1 kg of water from 20°C to 30°C at constant pressure). The change in internal energy is the same as before, because U is a state function, but we might find that the work we must do is not the same as before. Thus, whereas we might have needed to do 42 kJ of work when the system was in an adiabatic container, to achieve the same change we might now have to do 50 kJ of work. The difference between the work done in the two cases is defined as the heat absorbed by the system in the process:

$$q = w_{ad} - w \tag{4}$$

In the present case, we would conclude that

$$q = 42 \text{ kJ} - 50 \text{ kJ} = -8 \text{ kJ}$$

and report that 8 kJ of energy had left the system as heat. We see that *we now have a purely mechanical definition of heat in terms of work*. We know how to measure work in terms of the height through which a weight falls, so we now also have a method for measuring heat in terms of work.

Finally, we can express eqn 4 in a more familiar way. Because we already know that ΔU is (by definition) equal to w_{ad}, the expression for the energy transferred to the system as heat is

$$q = \Delta U - w$$

However, this expression is equivalent to eqn 2, the mathematical form of the First Law that we saw earlier.

WORK AND HEAT

The way can now be opened to powerful methods of calculation by switching attention to infinitesimal changes of state (such as infinitesimal changes in temperature) and infinitesimal changes in the internal energy dU. Then, if the work done on a system is dw and the energy supplied to it as heat is dq, in place of $\Delta U = q + w$ we have

$$dU = dq + dw \tag{5}$$

To use eqn 5 we must be able to relate dq and dw to events taking place in the surroundings. We begin by discussing **expansion work**, the work arising from a change in volume. This type of work includes the work done by a gas as it expands and drives back the atmosphere. Many chemical reactions result in the generation or consumption of gases, and the thermodynamic characteristics of a reaction, such as the heat it generates, depend on the work done during the reaction.

2.3 Expansion work

In thermodynamics we are often concerned with the work done on or by a system as it expands. This work can be calculated by considering the arrangement shown in Fig. 2.6 in which one wall of a system is a massless, frictionless, rigid, perfectly fitting piston of area A. If the external pressure is p_{ex}, the force on the outer face of the piston is $F = p_{ex}A$; this force is equivalent to a constant weight of magnitude $p_{ex}A$ pressing down on the system.

We shall suppose that the motion of the piston is **quasistatic**, or so slow compared with any processes that spread energy and matter through the surroundings that no regions of non-uniform temperature or pressure are generated. Another way of expressing the quasistatic character of the process is to say that the surroundings must remain in **internal equilibrium**. That is, there should be no flow of energy or matter from one region of the surroundings to another region of the surroundings when the piston halts. Quasistatic motion ensures that the surroundings are in internal equilibrium.

The calculation of work starts from the definition used in physics which states that the work required to move an object a distance dz against an opposing force F is

$$dw = -F\,dz \tag{6}$$

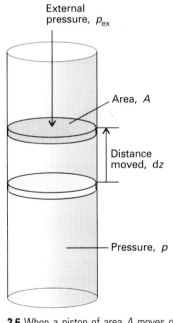

External pressure, p_{ex}

Area, A

Distance moved, dz

Pressure, p

2.6 When a piston of area A moves out through a distance dz, it sweeps out a volume $dV = A\,dz$. The external pressure p_{ex} is equivalent to a weight pressing on the piston, and the force opposing expansion is $F = p_{ex}A$.

Table 2.1 Varieties of work*

Type of work	dw	Comments	Units†
Expansion	$-p_{ex}\,dV$	p_{ex} is the external pressure dV is the change in volume	Pa m^3
Surface expansion	$\gamma\,d\sigma$	γ is the surface tension dσ is the change in area	N m^{-1} m^2
Extension	$f\,dl$	f is the tension dl is the change of length	N m
Electrical	$\phi\,dq$	ϕ is the electric potential dq is the change in charge	V C

* In general, the work done on a system can be expressed in the form d$w = -F\,dz$, where
  F is a 'generalized force' and dz is a 'generalized displacement'.
† For work in joules (J). Note that 1 N m = 1 J and 1 V C = 1 J.

The negative sign tells us that, when the system moves an object against an opposing force, the internal energy of the system doing the work will decrease. When a system expands quasistatically through a distance dz against an external pressure p_{ex}, it raises the effective weight $p_{ex}A$ through a distance dz, so the work done is d$w = -p_{ex}A \times dz$. But $A\,dz$ is the volume swept out in the course of the expansion, which we write as dV. Therefore, the work done when the system expands through dV against a constant pressure p_{ex} is

$$dw = -p_{ex}\,dV \tag{7}$$

If the system is compressed instead, then a weight of magnitude $p_{ex}A$ is lowered in the surroundings, so work of magnitude $p_{ex}A \times |dz|$ is now done on the system. ($|dz|$ is the absolute value of dz.) It is important to note that it is still the external pressure that determines the magnitude of the work even though it is the system that is opposing the insertion of the piston. The external pressure occurs in the expression because the act of compression corresponds to the lowering of a weight of magnitude $p_{ex}A$ in the surroundings. In either case, therefore, the work done on the system is given by eqn 7, but in a compression dV is negative (a reduction of volume) so dw is positive. Work is done on the system by compression and, so long as no other energy changes take place, its internal energy increases.

Other types of work (for example, electrical work) have analogous expressions, with each one the product of an intensive property (the pressure, for instance) and an extensive property (the change in volume). Some are collected in Table 2.1. For the present we continue with the work associated with changing the volume, the expansion work, and see what we can extract from eqn 7.

Free expansion

Free expansion occurs when $p_{ex} = 0$ and there is no opposing force. No work is done, and d$w = 0$ for each stage of the expansion. Hence, overall:

$$\text{Free expansion: } w = 0 \tag{8}$$

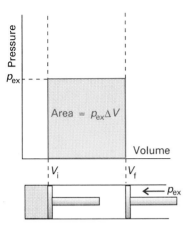

2.7 The work done by a gas when it expands against a constant external pressure p_{ex} is equal to the shaded area in this example of an indicator diagram.

Expansion against constant pressure

Because the external pressure p_{ex} is constant throughout the expansion (for example, the piston may be pressed on by the atmosphere, which exerts the same pressure throughout the expansion), the work done as the system passes quasistatically through each successive infinitesimal displacement dV is $dw = -p_{ex}dV$. The total work done in the expansion from V_i to V_f is the sum (integral) of all these equal contributions:

$$w = -\int_{V_i}^{V_f} p_{ex}\,dV = -p_{ex}\int_{V_i}^{V_f} dV = -p_{ex}(V_f - V_i)$$

Therefore, writing the change in volume as $\Delta V = V_f - V_i$,

$$\text{Expansion against constant pressure: } w = -p_{ex}\Delta V \qquad (9)$$

This result is illustrated graphically in Fig. 2.7: the magnitude of w, which is denoted $|w|$, is equal to the area beneath the horizontal line at $p = p_{ex}$ lying between the initial and final volumes. A p,V-graph used to compute expansion work is called an **indicator diagram**. (James Watt first used one to indicate aspects of the operation of his steam engine.)

Reversible expansion

A **reversible change** in thermodynamics is a change that can be reversed by an *infinitesimal* modification of a variable. The key word 'infinitesimal' sharpens the everyday meaning of the word 'reversible' as something that can change direction. We say that a system is in **equilibrium** with its surroundings if an infinitesimal change in the conditions in opposite directions results in opposite changes in its state. One example of reversibility that we have encountered already is the thermal equilibrium of two systems with the same temperature. The transfer of energy as heat between the two is reversible because if the temperature of either system is lowered infinitesimally, then energy flows into the system with the lower temperature. If the temperature of either system at thermal equilibrium is raised infinitesimally, then energy flows out of the hotter system.

Suppose a gas is confined by a piston and that the external pressure p_{ex} is set equal to the pressure of the confined gas p. Such a system is in mechanical equilibrium with its surroundings (as illustrated in Section 1.1) because an infinitesimal change in the external pressure in either direction causes changes in volume in opposite directions. If the external pressure is reduced infinitesimally, then the gas expands slightly. If the external pressure is increased infinitesimally, then the gas contracts slightly. In either case the change is reversible in the thermodynamic sense. If, on the other hand, the external pressure differs measurably from the internal pressure, then changing p_{ex} infinitesimally will not decrease it below the pressure of the gas, and so will not change the direction of the process. Such a system is not in mechanical equilibrium with its surroundings and the expansion is thermodynamically irreversible.

To achieve reversible expansion we must match p_{ex} to p at each state:

$$dw = -p_{ex}\,dV = -p\,dV \qquad (10)_r$$

(Equations valid only for reversible processes are labelled with a subscript r.) Although the pressure inside the system appears in this expression for the work, it does so only because p_{ex} has been set equal to p to ensure reversibility. The total work of reversible expansion is therefore

$$w = -\int_{V_i}^{V_f} p \, dV \tag{11}_r$$

The integral can be evaluated once we know how the pressure of the confined gas depends on its volume. Equation 11 is the link with the material covered in Chapter 1, for, if we know the equation of state of the gas, then we can express p in terms of V and evaluate the integral.

Isothermal reversible expansion

We shall illustrate the use of an equation of state to evaluate the work by considering the isothermal, reversible expansion of a perfect gas. The expansion is made isothermal by keeping the system in thermal contact with its surroundings (which may be a constant-temperature bath). Because the equation of state is $pV = nRT$, we know that at each stage $p = nRT/V$, with V the volume at that stage of the expansion. The temperature T is constant in an isothermal expansion, and so (together with n and R) may be taken outside the integral. It follows that the work of reversible isothermal expansion of a perfect gas from V_i to V_f at a temperature T is

$$w = -nRT \int_{V_i}^{V_f} \frac{dV}{V} = -nRT \ln\left(\frac{V_f}{V_i}\right) \tag{12}_r^\circ$$

When the final volume is greater than the initial volume, as in an expansion, the logarithm in eqn 12 is positive and hence $w < 0$. In this case, the system has done work on the surroundings and the internal energy of the system has decreased as a result of the work it has done. The equations also show that more work is done for a given change of volume when the temperature is increased. The greater pressure of the confined gas then needs a higher opposing pressure to ensure reversibility.

The result of the calculation can be expressed as an indicator diagram, for the magnitude of the work done is equal to the area under the isotherm $p = nRT/V$ (Fig. 2.8). Superimposed on the diagram is the rectangular area obtained for irreversible expansion against a constant external pressure fixed at the same final value as that reached in the reversible expansion. More work is obtained when the expansion is reversible (the area is greater) because matching the external pressure to the internal pressure at each stage of the process ensures that none of the system's pushing power is wasted. We cannot obtain more work than for the reversible process because increasing the external pressure even infinitesimally results in contraction. We may infer from this discussion that, because some pushing power is wasted when $p > p_{ex}$, *the maximum work available from a system operating between specified initial and final states and passing along a specified path is obtained when it is operating reversibly.*

We have introduced the connection between reversibility and maximum work for the special case of a perfect gas undergoing expansion.

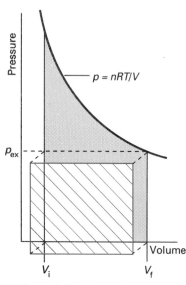

2.8 The work done by a perfect gas when it expands reversibly and isothermally is equal to the area under the isotherm $p = nRT/V$. The work done during the irreversible expansion against the same final pressure is equal to the rectangular area shown hatched. Note that the reversible work is greater than the irreversible work.

Later (in Section 4.6) we shall see that it applies to all substances and to all kinds of work.

Example 2.2 *Calculating the work of gas production*

Calculate the work done when 50 g of iron reacts with hydrochloric acid in: (a) a closed vessel of fixed volume; (b) an open beaker at 25°C.

Method. We need to judge the magnitude of the volume change, and then to decide how the process occurs. If there is no change in volume, then there is no expansion work however the process takes place. If the system expands against a constant external pressure, then the work it does can be calculated from eqn 9. A general feature of processes in which a condensed phase changes into a gas is that the volume of the former may be neglected relative to that of the gas it forms.

Answer. In (a) the volume cannot change, so no work is done and $w = 0$. In (b) the gas drives back the atmosphere and therefore $w = -p_{ex} \Delta V$. We can neglect the initial volume because the final volume (after the production of gas) is so much larger and $\Delta V = V_f - V_i \approx V_f = nRT/p_{ex}$ where n is the amount of H_2 produced. Therefore,

$$w = -p_{ex} \times \frac{nRT}{p_{ex}} = -nRT$$

Because the reaction is

$$Fe(s) + 2HCl(aq) \rightarrow FeCl_2(aq) + H_2(g)$$

we know that 1 mol H_2 is generated when 1 mol Fe is consumed, and n can be taken as the amount of Fe atoms that react. Because the molar mass of Fe is 55.85 g mol^{-1}, it follows that

$$w = \frac{-50 \text{ g}}{55.85 \text{ g mol}^{-1}} \times (8.3145 \text{ J K}^{-1} \text{ mol}^{-1}) \times (298.15 \text{ K})$$

$$= -2.2 \text{ kJ}$$

The system, the reaction mixture, does 2.2 kJ of work driving back the atmosphere.

Comment. Note that (for this perfect gas system) the external pressure does not affect the final result: the lower the pressure, the larger the volume occupied by the gas, so the effects cancel.

Exercise E2.2. Calculate the expansion work done when 50 g of water is electrolysed under constant pressure at 25°C. $\qquad$ [-10 kJ]

2.4 Heat and enthalpy

In general, the change in internal energy of a system is

$$dU = dq + dw_e + dw_{exp}$$

where dw_e is work in addition (e for 'extra') to the expansion work, dw_{exp}. For instance, dw_e might be the electrical work of driving a current

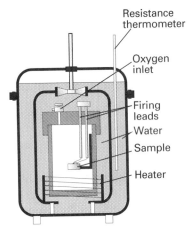

2.9 A constant-volume bomb calorimeter. The 'bomb' is the central vessel, which is massive enough to withstand high pressures. The calorimeter (for which the heat capacity must be known) is the entire assembly shown here. In order to ensure adiabaticity, the calorimeter may be immersed in a water bath with a temperature continuously readjusted to that of the calorimeter at each stage of the combustion.

through a circuit. A system kept at constant volume can do no expansion work, and so $dw_{exp} = 0$. If the system is also incapable of doing any other kind of work (if it is not, for instance, an electrochemical cell connected to an electric motor), then $dw_e = 0$ too. Under these circumstances:

$$dU = dq \quad \text{(at constant volume, no additional work)} \qquad \textbf{(13a)}$$

We express this relation by writing $dU = dq_V$. For a measurable change,

$$\Delta U = q_V \qquad \textbf{(13b)}$$

It follows that by measuring the energy supplied to a constant-volume system as heat ($q > 0$) or obtained from it as heat ($q < 0$) when it undergoes a change of state, we are in fact measuring the change in its internal energy.

Calorimetry

The most common device for measuring ΔU is the **adiabatic bomb calorimeter** (Fig. 2.9). The process we wish to study—which may be a chemical reaction—is initiated inside a constant-volume container. The bomb is immersed in a stirred water bath, and the whole device is the calorimeter. The calorimeter is also immersed in an outer water bath. The water in the calorimeter and in the outer bath are both monitored and adjusted to the same temperature. This arrangement ensures that there is no net loss of heat from the calorimeter to the surroundings (the bath) and hence that the calorimeter is adiabatic.

The change in temperature ΔT of the calorimeter is proportional to the heat that the reaction releases or absorbs. Therefore, by measuring ΔT we can determine q_V and hence find ΔU. The conversion of ΔT to q_V is best achieved by calibrating the calorimeter by using a process of known energy output and determining the **calorimeter constant**, the constant C in the relation

$$q = C \times \Delta T \qquad \textbf{(14)}$$

The calorimeter constant may be measured electrically by passing a current I from a source of known potential V through a heater for a known period of time t:

$$q = IVt \qquad \textbf{(15)}$$

For instance, if we pass a current of 10.0 A from a 12-V supply for 300 s,

$$q = (10.0\,\text{A}) \times (12\,\text{V}) \times (300\,\text{s}) = 3.6 \times 10^4\,\text{A V s} = 36\,\text{kJ}$$

(because $1\,\text{A V s} = 1\,\text{J}$). If the observed rise in temperature is 5.5 K, then the calorimeter constant is

$$C = \frac{q}{\Delta T} = \frac{36\,\text{kJ}}{5.5\,\text{K}} = 6.5\,\text{kJ K}^{-1}$$

Alternatively, C may be determined by burning a known mass of substance (benzoic acid is often used) that has a known heat output. With C known, it is simple to interpret an observed temperature rise as a release of heat.

Labels on figure:
Resistance thermometer
Oxygen inlet
Firing leads
Water
Sample
Heater

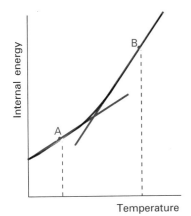

2.10 The internal energy of a system increases as the temperature is raised; this graph shows its variation as the system is heated at constant volume. The slope of the graph at any temperature (as shown by the tangents at A and B) is the heat capacity at constant volume at that temperature. Note that, for the system illustrated, the heat capacity is greater at B than at A.

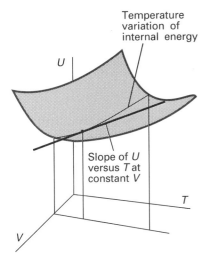

2.11 The internal energy of a system varies with volume and temperature, perhaps as shown here by the surface. The variation of the internal energy with temperature at one particular constant volume is illustrated by the curve drawn parallel to *T*. The slope of this curve at any point is the partial derivative $(\partial U/\partial T)_V$.

Heat capacity

The internal energy of a substance increases when its temperature is raised. The extent of the increase depends on the conditions under which the heating takes place, and for the present we shall suppose that the sample is confined to a constant volume. For example, it may be a gas in a container of fixed volume. If the internal energy is plotted against temperature, then a curve like that in Fig. 2.10 is obtained. The *slope* of the curve at any temperature is called the **heat capacity** of the system at that temperature. The heat capacity at constant volume is denoted C_V and is defined formally as

$$C_V = \left(\frac{\partial U}{\partial T}\right)_V \tag{16}$$

The notation is that of a partial derivative. A partial derivative is a slope calculated with all except one variable held constant. In this case, the internal energy varies with the temperature and the volume of the sample, but we are interested only in the variation of U with the temperature, the volume being held constant (Fig. 2.11).

Heat capacities are extensive properties: 100 g of water, for instance, has 100 times the heat capacity of 1 g of water (and therefore requires 100 times the heat to bring about the same rise in temperature). The **molar heat capacity at constant volume** $C_{V,\mathrm{m}}$ is the heat capacity per mole of material, and is an intensive property (all molar quantities are intensive). Typical values of molar heat capacities of gases are close to $25\,\mathrm{J\,K^{-1}\,mol^{-1}}$. For certain applications it is useful to know the **specific heat capacity** (more informally, the 'specific heat') of a substance, which is the heat capacity per unit mass, usually per gram of material. The specific heat capacity of water at room temperature is $4\,\mathrm{J\,K^{-1}\,g^{-1}}$. In general, heat capacities depend on the temperature (Fig. 2.12) and approaches zero at very low temperatures. However, over small ranges of temperature at and above room temperature, the variation is quite small and for approximate calculations heat capacities can be treated as almost independent of temperature.

The heat capacity can be used to relate a change in internal energy to a change in temperature of a constant-volume system. It follows from eqn 16 (and, more informally, from the graph in Fig. 2.13) that

$$dU = C_V\,dT \quad \text{at constant volume} \tag{17a}$$

That is, an infinitesimal change in temperature brings about an infinitesimal change in internal energy, and the constant of proportionality is the heat capacity at constant volume. If the heat capacity is independent of temperature over the range of temperatures of interest, then a measurable change of temperature ΔT brings about a measurable increase in internal energy ΔU, where

$$\Delta U = C_V\,\Delta T \tag{17b}$$

Because a change in internal energy can be identified with the heat

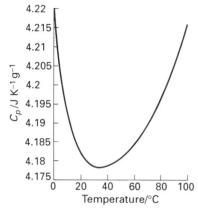

2.12 In general, heat capacities vary with temperature. This graph shows the specific heat capacity at constant pressure C_p of liquid water (the heat capacity per gram). Note that, although the curve suggests a strong variation with temperature, the scale on the left shows that in fact the variation is quite small over the entire range shown (from 4.18 to 4.22 J K^{-1} g^{-1}). The specific heat capacity of ice at 0°C is 2.1 J K^{-1} g^{-1}, and that of water vapour (at 25°C) is 1.9 J K^{-1} g^{-1}.

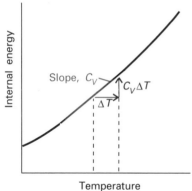

2.13 When the temperature of a constant-volume system is increased by ΔT, the internal energy increases by $C_V \Delta T$, where C_V is the constant-volume heat capacity at the temperature of interest. The relation $\Delta U = C_V \Delta T$ is strictly valid only for infinitesimal increases in temperature, but, if, as is shown here, the heat capacity is almost constant over the temperature range of interest, then the approximate expression may be used.

supplied at constant volume (eqn 13b), the last equation can be written

$$q_V = C_V \Delta T \tag{17c}$$

This relation provides a simple way of measuring the heat capacity of a sample: a measured quantity of heat is supplied to the sample (electrically, for example), and the resulting increase in temperature is monitored. The ratio of the heat supplied to the temperature rise it causes is the heat capacity of the sample.

A large heat capacity implies that, for a given quantity of heat, there will be only a *small* increase in temperature (the sample has a large capacity for heat). An infinite heat capacity implies that there will be no increase in temperature however much heat is supplied. At a phase transition (such as at the boiling point of water), the temperature of a substance does not rise as heat is supplied (the energy is used to drive the endothermic phase transition, in this case to vaporize the water, rather than to increase its temperature), so, at the temperature of a phase transition, the heat capacity of a sample is infinite.

Enthalpy

The change in internal energy is not equal to the heat supplied when the system is free to change its volume. Under these circumstances some of the energy supplied as heat to the system is returned to the surroundings as expansion work (Fig. 2.14), so dU is less than dq. However, we shall now show that in this case the heat supplied is equal to the change in another thermodynamic property of the system, the **enthalpy** H. The enthalpy is defined as

$$H = U + pV \tag{18}$$

where p is the pressure of the system and V is its volume. (The term pV is a part of the definition of H for any system, and the fact that the same term appears in the perfect gas equation of state is only a coincidence.) Because U, p, and V all depend solely on the current state of the system, the enthalpy is a state function. As is true of any state function, the change in enthalpy between any pair of initial and final states is independent of the path between them.

A very important interpretation of the enthalpy is that *a change in enthalpy is equal to the heat supplied at constant pressure to a system* (so long as the system does no additional work):

$$dH = dq \text{ (at constant pressure, no additional work)} \tag{19a}$$

For a measurable change,

$$\Delta H = q_p \tag{19b}$$

JUSTIFICATION

For a general change in the state of the system, U changes to $U + dU$, p changes to $p + dp$, and V changes to $V + dV$, so H changes from $U + pV$

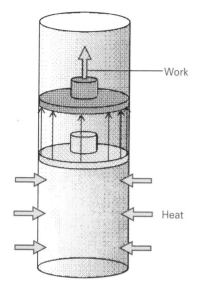

2.14 When a system is subjected to constant pressure and is free to change its volume, then some of the energy supplied as heat may escape back into the surroundings as work. In such a case, the change in internal energy is smaller than the energy supplied as heat.

to

$$H + dH = (U + dU) + (p + dp)(V + dV)$$
$$= U + dU + pV + p\,dV + V\,dp + dp\,dV$$

The last term is the product of two infinitesimally small quantities, and can be neglected. As a result, after recognizing $U + pV = H$ on the right, we find that H changes to

$$H + dH = H + dU + p\,dV + V\,dp$$

and hence that

$$dH = dU + p\,dV + V\,dp$$

If we now substitute

$$dU = dq + dw$$

into this expression, we get

$$dH = dq + dw + p\,dV + V\,dp$$

If the system is in mechanical equilibrium with its surroundings at a pressure p and does only expansion work, we can write $dw = -p\,dV$ and obtain

$$dH = dq + V\,dp$$

Now we impose the condition that the heating occurs at constant pressure by writing $dp = 0$. Then

$$dH = dq \quad \text{(at constant pressure, no additional work)}$$

as in eqn 19a.

The result expressed in eqn 19 states that, when a system is subjected to a constant pressure and only expansion work can occur, the change in enthalpy is equal to the energy supplied as heat. For example, if we supply 36 kJ of energy through an electric heater immersed in an open beaker of water, then the enthalpy of the water increases by 36 kJ and we write $\Delta H = +36$ kJ.

The measurement of an enthalpy change

An enthalpy change can be measured calorimetrically by monitoring the temperature change that accompanies a physical or chemical change occurring at constant pressure. For a combustion reaction an **adiabatic flame calorimeter** (Fig. 2.15) may be used to measure ΔT when a given amount of substance burns in a supply of oxygen. Another route to ΔH is to measure the internal energy change using a bomb calorimeter, and then to convert ΔU to ΔH. Because solids and liquids have small molar volumes, for them pV_m is so small that the molar enthalpy and molar internal energy are almost identical. Consequently, if a process involves only solids or liquids, the values of ΔH and ΔU are almost identical. Physically, such processes are accompanied by a very small change in

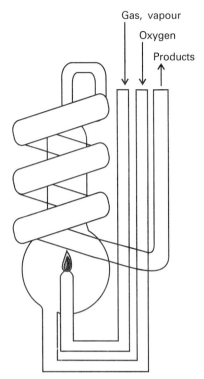

Gas, vapour

Oxygen

Products

2.15 A constant-pressure flame calorimeter consists of this element immersed in a stirred water bath. Combustion occurs as a known amount of reactant is passed through to fuel the flame, and the rise of temperature is monitored.

volume, the system does negligible work on the surroundings when the process occurs, so the energy supplied as heat stays entirely within the system.

Example 2.3 *Relating ΔH and ΔU (1)*

The internal energy change when 1.0 mol $CaCO_3$ in the form of calcite converts to aragonite is $+0.21$ kJ. Calculate the difference between the enthalpy change and the change in internal energy when the pressure is 1.0 bar given that the densities of the solids are 2.71 and 2.93 g cm^{-3}, respectively.

Method. The starting point for the calculation is the relation between the enthalpy of a substance and its internal energy (eqn 18). The difference between the two quantities can be expressed in terms of the pressure and the difference of their molar volumes, and the latter can be calculated from their molar masses M and their densities ρ by using $M = \rho V_m$.

Answer. The change in enthalpy when transformation occurs is

$$\Delta H = H(\text{aragonite}) - H(\text{calcite})$$
$$= \{U(a) + pV(a)\} - \{U(c) + pV(c)\}$$
$$= \Delta U + p\{V(a) - V(c)\} = \Delta U + p\,\Delta V$$

The volume of 1.0 mol $CaCO_3$ (100 g) as aragonite (a) is 34 cm^3, and that of 1.0 mol $CaCO_3$ as calcite (c) is 37 cm^3. Therefore,

$$p\,\Delta V = (1.0 \times 10^5\,\text{Pa}) \times (34 - 37) \times 10^{-6}\,\text{m}^3 = -0.3\,\text{J}$$

(because 1 Pa m^3 = 1 J). Hence,

$$\Delta H - \Delta U = -0.3\,\text{J}$$

which is only 0.1 per cent of the value of ΔU.

Comment. It is usually justifiable to ignore the difference between the enthalpy and internal energy of condensed phases, except at very high pressures, when pV is no longer negligible.

Exercise E2.3. Calculate the difference between ΔH and ΔU when 1.0 mol of grey tin (density 5.75 g cm^{-3}) changes to white tin (density 7.31 g cm^{-3}) under 10.0 bar pressure. At 298 K, $\Delta H = +2.1$ kJ.

$$[\Delta H - \Delta U = -4.4\,\text{J}]$$

The enthalpy of a perfect gas is related to its internal energy by substituting the perfect gas equation of state into the definition of H:

$$H = U + pV = U + nRT$$

This relation implies that the change of enthalpy in a reaction that produces or consumes gas is

$$\Delta H = \Delta U + \Delta n_g RT \tag{20}°$$

where Δn_g is the change in the amount of gas molecules in the reaction.

For example, in the reaction

$$2H_2(g) + O_2(g) \rightarrow 2H_2O(l) \qquad \Delta n_g = -3 \, mol$$

because 3 mol of gas-phase molecules are replaced by 2 mol of liquid-phase molecules, and at 298 K the enthalpy and internal energy changes taking place in the system are related by

$$\Delta H - \Delta U = (-3 \, mol) \times RT \approx -7.5 \, kJ$$

Note that the difference is in *kilo*joules, not joules as in Example 2.3.

Example 2.4 *Relating ΔH and ΔU (2)*

The enthalpy change accompanying the formation of 1.00 mol $NH_3(g)$ from its elements at 298 K is -46.1 kJ. Estimate the change in internal energy.

Method. To answer questions like this it is necessary to compute Δn_g from the stoichiometric numbers in the chemical equation and then to use eqn 20.

Answer. The chemical equation is

$$\tfrac{3}{2}H_2(g) + \tfrac{1}{2}N_2(g) \rightarrow NH_3(g)$$

The change in the amount of gas-phase molecules is

$$\Delta n_g = 1.00 \, mol - 1.50 \, mol - 0.50 \, mol = -1.00 \, mol$$

Because $RT = 2.48$ kJ mol^{-1} at 298 K,

$$\Delta U = \Delta H - (-1.00 \, mol) \times RT = -43.6 \, kJ$$

Comment. The difference is 5 per cent of the enthalpy change, a much more significant difference than for the conversion of one condensed phase into another. One of the most serious sources of error in this procedure is the assumption that the gases are perfect.

Exercise E2.4. Calculate ΔU for the combustion of 1.000 mol of propene given that $\Delta H = -2058$ kJ. $\qquad\qquad$ [-2052 kJ]

Example 2.5 *Calculating a change in enthalpy*

Water is heated to boiling under a pressure of 1.0 atm. When an electric current of 0.50 A from a 12-V supply is passed for 300 s through a resistance in thermal contact with it, it is found that 0.798 g of water is vaporized. Calculate the molar internal energy and enthalpy changes at the boiling point (373.15 K).

Method. Because the vaporization occurs at constant pressure, the enthalpy change is equal to the heat supplied by the heater. Therefore, the strategy is to calculate the heat supplied (from $q = IVt$), express that as an enthalpy change, and then convert the result to a molar enthalpy change by division by the amount of H_2O molecules vaporized. To convert from enthalpy change to internal energy change, we assume that the vapor is a perfect gas and use eqn 20.

Answer. The enthalpy change is

$$\Delta H = q_p = (0.50 \text{ A}) \times (12 \text{ V}) \times (300 \text{ s}) = +1.8 \text{ kJ}$$

Because 0.798 g of water is 0.0443 mol H_2O, the molar enthalpy of vaporization is

$$\Delta H_m = + \frac{1.8 \text{ kJ}}{0.0443 \text{ mol}} = +41 \text{ kJ mol}^{-1}$$

In the process

$$H_2O(l) \rightarrow H_2O(g)$$

the change in the amount of gas molecules is $\Delta n_g = +1$ mol, so

$$\Delta U_m = \Delta H_m - RT = +38 \text{ kJ mol}^{-1}$$

Comment. The plus sign is added to positive quantities to emphasize that they represent an increase in internal energy or enthalpy. Notice that the internal energy change is smaller than the enthalpy change because energy has been used to drive back the surrounding atmosphere to make room for the vapour.

Exercise E2.5. The molar enthalpy of vaporization of benzene at its boiling point (353.25 K) is 30.8 kJ mol^{-1}. What is the molar internal energy change? For how long would the same 12-V source need to supply a 0.50-A current in order to vaporize a 10-g sample?

$$[+27.9 \text{ kJ mol}^{-1}, 660 \text{ s}]$$

Enthalpy and internal energy changes may also be measured by non-calorimetric methods (see Chapters 8 and 10).

The variation of enthalpy with temperature

The enthalpy of a substance increases as its temperature is raised. The relation between the increase in enthalpy and the increase in temperature depends on the conditions (for example, constant pressure or constant volume). The most important condition is constant pressure, and the slope of a graph of enthalpy against temperature at constant pressure (Fig. 2.16) is called the **heat capacity at constant pressure**, C_p. More formally:

$$C_p = \left(\frac{\partial H}{\partial T} \right)_p \tag{21}$$

The heat capacity at constant pressure is the analogue of the heat capacity at constant volume, and is an extensive property. The **molar heat capacity at constant pressure**, $C_{p,m}$, is the heat capacity per mole of material; it is an intensive property.

The heat capacity at constant pressure is used to relate the change in enthalpy to a change in temperature. For infinitesimal changes of

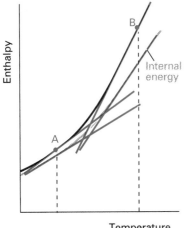

2.16 The slope of a graph of the enthalpy of a system subjected to a constant pressure plotted against temperature is the constant-pressure heat capacity. The slope of the graph may change with temperature, in which case the heat capacity varies with temperature. Thus, the heat capacities at A and B are different. For gases, the slope of the graph of enthalpy versus temperature is steeper than that of the graph of internal energy versus temperature, and $C_{p,m}$ is larger than $C_{V,m}$.

temperature,

$$dH = C_p \, dT \quad \text{(at constant pressure)} \tag{22a}$$

If the heat capacity is constant over the range of temperatures of interest, then for a measurable increase in temperature

$$\Delta H = C_p \, \Delta T \quad \text{(at constant pressure)} \tag{22b}$$

Because an increase in enthalpy can be equated with the heat supplied at constant pressure, the practical form of the latter equation is

$$q_p = C_p \, \Delta T \tag{23}$$

This expression shows us how to measure the heat capacity of a sample: a measured quantity of heat is supplied under conditions of constant pressure (as in a sample exposed to the atmosphere and free to expand), and the temperature rise is monitored.

The variation of heat capacity with temperature can sometimes be ignored if the temperature range is small; this approximation is highly accurate for a monatomic perfect gas (one of the noble gases). However, when it is necessary to take the variation into account, a convenient approximate empirical expression is

$$C_{p,m} = a + bT + \frac{c}{T^2} \tag{24}$$

The empirical parameters a, b, and c are independent of temperature. Some typical values are given in Table 2.2.

Table 2.2* Temperature variation of molar heat capacities, $C_{p,m}/(\text{J K}^{-1}\,\text{mol}^{-1}) = a + bT + c/T^2$

	a	$b/(10^{-3}\,\text{K}^{-1})$	$c/(10^5\,\text{K}^2)$
C(s, graphite)	16.86	4.77	−8.54
CO_2(g)	44.22	8.79	−8.62
H_2O(l)	75.29	0	0
N_2(g)	28.58	3.77	−0.50

* More values are given in the Data section at the end of this volume.

Example 2.6 *Evaluating an increase in enthalpy with temperature*

What is the change in molar enthalpy of N_2 when it is heated from 25°C to 100°C? Use the heat capacity information in Table 2.2.

Method. The heat capacity of N_2 changes with temperature, so we cannot use eqn 22b (which assumes that the heat capacity of the substance is constant). Therefore, we must use eqn 22a, substitute eqn 24 for the temperature dependence of the heat capacity, and integrate the resulting expression from 25°C to 100°C.

Answer. For convenience, we shall denote the two temperatures T_1 (298 K) and T_2 (373 K). The integrals we require are

$$\int_{H(T_1)}^{H(T_2)} dH = \int_{T_1}^{T_2} \left(a + bT + \frac{c}{T^2} \right) dT$$

which evaluate to

$$H(T_2) - H(T_1) = a(T_2 - T_1) + \tfrac{1}{2}b(T_2^2 - T_1^2) - c\left(\frac{1}{T_2} - \frac{1}{T_1} \right)$$

Substitution of the numerical data results in

$$H(373\,\text{K}) = H(298\,\text{K}) + 2.20\,\text{kJ mol}^{-1}$$

If we had assumed a constant heat capacity of $29.14\,\text{J K}^{-1}\,\text{mol}^{-1}$ (the value given by eqn 24 at 25°C), we would have found that the two enthalpies differed by 2.19 kJ mol^{-1}.

Exercise E2.6. At very low temperatures the heat capacity of a solid is proportional to T^3, and we can write $C_V = aT^3$. What is the change in enthalpy of such a substance when it is heated from 0 to a temperature T (with T close to 0)?

$$[\Delta H = \tfrac{1}{4}aT^4]$$

The relation between heat capacities

Most systems expand when heated at constant pressure. Such systems do work on the surroundings and some of the energy supplied to them as heat escapes back to the surroundings. As a result, the temperature of the system rises less than when the heating occurs at constant volume. A smaller increase in temperature implies a larger heat capacity, so we conclude that in most cases *the heat capacity at constant pressure of a system is larger than its heat capacity at constant volume.*

There is a simple relation between the two heat capacities of a perfect gas:

$$C_p - C_V = nR \qquad\qquad\qquad (25)°$$

(This relation will be derived in Section 3.3.) It follows that the molar heat capacity of a perfect gas is about 8 J K^{-1} mol^{-1} larger at constant pressure than at constant volume. Because the heat capacity at constant volume of a monatomic gas is about 25 J K^{-1} mol^{-1}, the difference is highly significant and must be taken into account.

THERMOCHEMISTRY

The study of the heat produced or required by chemical reactions is called **thermochemistry**. Thermochemistry is a branch of thermodynamics because a reaction vessel and its contents form a system, and chemical reactions result in the exchange of energy between the system and the surroundings. Thus we can use calorimetry to measure the heat produced or absorbed by a reaction, and can identify q with a change in internal energy (if the reaction occurs at constant volume) or in enthalpy (if the reaction occurs at constant pressure). Conversely, if we know the ΔU or ΔH for a reaction, then we can predict the heat the reaction can produce.

We have already remarked that a process that releases heat is classified as exothermic and one that absorbs heat is classified as endothermic. Because the release of heat signifies a decrease in the enthalpy of a system (at constant pressure), we can now see that *an exothermic process is a process for which* $\Delta H < 0$. Conversely, because the absorption of heat results in an increase in enthalpy, *an endothermic process is a process for which* $\Delta H > 0$.

2.5 Standard enthalpy changes

Changes in enthalpy are normally reported for processes taking place under a set of standard conditions. In most of our discussions we shall

Table 2.3* Standard enthalpies of fusion and vaporization at the transition temperature, $\Delta_{trs}H^{\ominus}/(\text{kJ mol}^{-1})$

	T_f/K	Fusion	T_b/K	Vaporization
Ar	83.81	1.188	87.29	6.506
C_6H_6	278.61	10.59	353.2	30.8
H_2O	273.15	6.008	373.15	40.656
				44.016 at 298 K
He	3.5	0.021	4.22	0.084

* More values are given in the Data section.

consider the **standard enthalpy change** $\Delta H^{\ominus}$, the change in enthalpy for a process in which the initial and final species are in their **standard states:**[2]

> The standard state of a substance at a specified temperature is its pure form at 1 bar.

For example, the standard state of liquid ethanol at 298 K is pure liquid ethanol at 298 K and 1 bar; the standard state of solid iron at 500 K is pure iron at 500 K and 1 bar. The standard enthalpy change for a reaction or a physical process is the difference between the products in their standard states and the reactants in their standard states, all at the same specified temperature.

As an example of a standard enthalpy change, the **standard enthalpy of vaporization**[3] $\Delta_{vap}H^{\ominus}$, is the enthalpy change per mole when a pure liquid at 1 bar vaporizes to a gas at 1 bar, as in

$$H_2O(l) \rightarrow H_2O(g) \qquad \Delta_{vap}H^{\ominus} = +40.66 \text{ kJ mol}^{-1} \text{ at } 373 \text{ K}$$

As implied by the example, standard enthalpies may be reported for any temperature. However, the recommended temperature for reporting thermodynamic data is 298.15 K (corresponding to 25°C). Because this temperature is frequently mentioned, it is sometimes given the special symbol $\mathcal{T}$, standing for 'conventional temperature'. Unless otherwise mentioned, all thermodynamic data in this text will refer to this conventional temperature.

Enthalpies of physical change

The standard enthalpy change that accompanies a change of physical state is called the **standard enthalpy of transition** and is denoted $\Delta_{trs}H^{\ominus}$ (Table 2.3). The standard enthalpy of vaporization, $\Delta_{vap}H^{\ominus}$, is

2 The choice of 1 bar (1 bar = 10^5 Pa) as the standard pressure (in place of 1 atm) for reporting thermodynamic data is recent. At the level of accuracy of most data, especially those relating to liquids and solids, the differences in values are usually negligible except in work of the highest precision. A meticulous analysis of the consequences of the change has been given by R. D. Freeman in *Bulletin of Chemical Thermodynamics*, **25,** 523 (1982). Modern tables of data now widely, but not universally, adopt the new standard.

3 The IUPAC recommendation, which we shall follow, is to attach the subscript to Δ, as in $\Delta_{vap}H$. In casual use, however, the subscript is still widely attached to ΔH, as in ΔH_{vap}. The latter usage will no doubt continue, despite IUPAC's disapproval, for a considerable time.

one example. Another is the **standard enthalpy of fusion** $\Delta_{fus}H^{\ominus}$, the enthalpy change accompanying the conversion of a solid to a liquid, as in

$$H_2O(s) \rightarrow H_2O(l) \qquad \Delta_{fus}H^{\ominus}(273\ K) = +6.01\ kJ\ mol^{-1}$$

As in this case, it is sometimes convenient to know the standard enthalpy change at the transition temperature as well as at the conventional temperature. A third example is the **standard enthalpy of sublimation** $\Delta_{sub}H^{\ominus}$, the standard enthalpy for a process in which a solid is converted directly into a vapour. An example is

$$C(s,\ graphite) \rightarrow C(g) \qquad \Delta_{sub}H^{\ominus}(298\ K) = +716.68\ kJ\ mol^{-1}$$

Because enthalpy is a state function, a change in enthalpy is independent of the path between the two states. This feature is of great importance in thermochemistry, for it implies that the same value of $\Delta H^{\ominus}$ will be obtained however the change is brought about (so long as the initial and final states are the same). For example, we can picture the conversion of a solid to a vapour either as occurring by sublimation

$$H_2O(s) \rightarrow H_2O(g) \qquad \Delta_{sub}H^{\ominus}$$

or as occurring in two steps, first fusion (melting) and then vaporization of the resulting liquid:

$$H_2O(s) \rightarrow H_2O(l) \qquad \Delta_{fus}H^{\ominus}$$
$$H_2O(l) \rightarrow H_2O(g) \qquad \Delta_{vap}H^{\ominus}$$
$$\text{Overall: } H_2O(s) \rightarrow H_2O(g) \qquad \Delta_{fus}H^{\ominus} + \Delta_{vap}H^{\ominus}$$

Because the overall result of the indirect path is the same as that of the direct path, the overall enthalpy change is the same in each case (**1**), and we can conclude that (for processes occurring at the same temperature)

$$\Delta_{sub}H^{\ominus} = \Delta_{fus}H^{\ominus} + \Delta_{vap}H^{\ominus}$$

An immediate conclusion is that, because all enthalpies of fusion are positive, the enthalpy of sublimation of a substance is greater than its enthalpy of vaporization (at a given temperature).

Another consequence of H being a state function is that the standard enthalpy changes of a forward process and its reverse must differ only in sign (**2**):

$$\Delta H^{\ominus}(A \rightarrow B) = -\Delta H^{\ominus}(B \rightarrow A)$$

For instance, because the enthalpy of vaporization of water is $+44\ kJ\ mol^{-1}$ at 298 K, its enthalpy of condensation at that temperature is $-44\ kJ\ mol^{-1}$.

The **standard enthalpy of solution** $\Delta_{sol}H^{\ominus}$ of a substance is the standard enthalpy change when it dissolves in a specified quantity of solvent. The **limiting enthalpy of solution** is the standard enthalpy change when the substance dissolves in an infinite amount of solvent and the interactions between the ions (or solute molecules if it is a non-electrolyte) are negligible. For HCl:

$$HCl(g) \rightarrow HCl(aq) \qquad \Delta_{sol}H^{\ominus} = -75.14\ kJ\ mol^{-1}$$

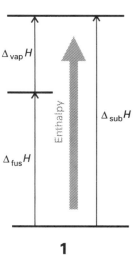

1

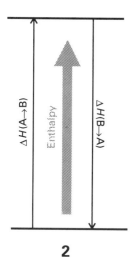

2

Table 2.4* Limiting enthalpies of solution, $\Delta_{sol}H^{\ominus}$/(kJ mol^{-1}), at 298 K

NaF(s)	+1.90
KF(s)	−17.74
NH$_4$NO$_3$(s)	+25.69
NaCl(s)	+3.89
KCl(s)	+17.22
H$_2$SO$_4$(l)	−95.28

* More values are given in the Data section.

and so 75 kJ of heat is released when 1.0 mol of HCl dissolves to produce an infinitely dilute solution. Some values for other substances are given in Table 2.4.

The different types of enthalpies encountered in thermochemistry are summarized in Table 2.5.

Table 2.5 Enthalpies of transition

Transition	Process	Symbol†
Transition	Phase $\alpha \rightarrow$ Phase β	$\Delta_{trs}H$
Fusion	s $\rightarrow$ l	$\Delta_{fus}H$
Vaporization	l $\rightarrow$ g	$\Delta_{vap}H$
Sublimation	s $\rightarrow$ g	$\Delta_{sub}H$
Mixing of fluids	Pure $\rightarrow$ mixture	$\Delta_{mix}H$
Solution	Solute $\rightarrow$ solution	$\Delta_{sol}H$
Hydration	$X^{\pm}(g) \rightarrow X(aq)$	$\Delta_{hyd}H$
Atomization	Species(s, l, g) $\rightarrow$ atoms(g)	$\Delta_{at}H$
Ionization	$X(g) \rightarrow X^+(g) + e^-(g)$	$\Delta_{ion}H$
Electron gain	$X(g) + e^-(g) \rightarrow X^-(g)$	$\Delta_{eg}H$
Reaction	Reactants $\rightarrow$ products	$\Delta_{r}H$
Combustion	Compound(s, l, g) + O_2(g) $\rightarrow CO_2$(g), H_2O(l, g)	$\Delta_{c}H$
Formation	Elements $\rightarrow$ compound	$\Delta_{f}H$
Activation	Reactants $\rightarrow$ activated complex	$\Delta^{\ddagger}H$

† IUPAC recommendations. In common usage, the transition subscript is attached to ΔH, as in ΔH_{trs}.

Enthalpies of ionization

Two very important enthalpy changes relating to the properties of atoms are those accompanying the formation of cations and anions of atoms and molecules in the gas phase. The **enthalpy of ionization** $\Delta_{ion}H^{\ominus}$ is the standard enthalpy change per mole for the removal of an electron from a species in the gas phase:

$$Na(g) \rightarrow Na^+(g) + e^-(g)$$

Because 1 mol of gaseous reactants give 2 mol of gaseous products (so $\Delta n_g = +1$ mol), the **ionization energy,** E_i (the molar internal energy of ionization), and the enthalpy of ionization differ by about RT:

$$\Delta_{ion}H^{\ominus} = E_i + RT \tag{26}$$

Table 2.6* First and second ionization energies, E_i/(kJ mol^{-1})

H	1312	
He	2372	5251
Mg	738	1451
Na	496	4563

* More values are given in the Data section.

Because RT is only about 2.5 kJ mol^{-1} at room temperature and ionization energies are typically more than 100 times larger, in approximate work it is normally safe to ignore the differences between $\Delta_{ion}H^{\ominus}$ and E_i. The ionization energies of the elements are given in Table 2.6.

A cation may itself be ionized, in which case the change in internal energy is called the **second ionization energy,** E_{i2}, of the element. The second ionization energy (and the corresponding enthalpy change) is always larger than the first, in part because more energy is required to remove an electron from a positively charged species than from a neutral

Table 2.7* Electron affinities, $E_{ea}/(\text{kJ mol}^{-1})$

Cl	349		
F	328		
H	73		
O	141	O^-	-844

* More values are given in the Data section.

Table 2.8* Bond dissociation enthalpies, $\Delta H^{\ominus}(\text{A–B})/(\text{kJ mol}^{-1})$, at 298 K

H–CH$_3$	435
H–Cl	431
H–H	436
H–OH	492
H–O	428
H$_3$C–CH$_3$	368

* More values are given in the Data section.

Table 2.9* Mean bond enthalpies, $B(\text{A–B})/(\text{kJ mol}^{-1})$

	H	C	N	O
H	436			
C	412	348		
		612		
		838		
N	388	305	163	
		613	409	
			944	
O	463	360	157	146
				497

Successive values are for single, double, and triple bonds, respectively.
* More values are given in the Data section.

Table 2.10* Standard enthalpies of atomization, $\Delta_{at}H^{\ominus}/(\text{kJ mol}^{-1})$, at 298 K

C(s, graphite)	$+716.7$
Cu(s)	$+338.3$
K(s)	$+89.2$
Na(s)	$+107.3$

* More values are given in the Data section.

one. Ionization energies and, from them, enthalpies of ionization are widely obtained from spectroscopic measurements, as we shall see in Chapter 13.

The standard enthalpy change accompanying electron attachment to an atom, ion, or molecule in the gas phase is the **electron gain enthalpy**, $\Delta_{eg}H^{\ominus}$:

$$\text{Cl(g)} + e^-(g) \to \text{Cl}^-(g) \qquad \Delta_{eg}H^{\ominus} = -351.2 \text{ kJ mol}^{-1}$$

The negative of the corresponding internal energy change is widely called the **electron affinity** E_{ea} of the element (Table 2.7). With this change of sign, *a positive electron affinity corresponds to exothermic electron gain*. For example, the attachment of one electron to an O atom is exothermic (Table 2.7), so oxygen has a positive electron affinity. However, the attachment of a second electron is strongly endothermic, so O^- has a negative electron affinity. As in the case of cation formation, the internal energy change and the enthalpy change differ by RT:

$$\Delta_{eg}H^{\ominus} = -E_{ea} - RT \tag{27}$$

Enthalpies of bond formation and dissociation

The **bond dissociation enthalpy**, $\Delta H^{\ominus}(\text{A–B})$, is the standard reaction enthalpy for the breaking of the A–B bond:

$$\text{A–B(g)} \to \text{A(g)} + \text{B(g)} \qquad \Delta H^{\ominus}(\text{A–B})$$

A and B may be atoms or groups of atoms, as in

$$\text{CH}_3\text{OH(g)} \to \text{CH}_3(g) + \text{OH(g)} \quad \Delta H^{\ominus}(\text{CH}_3\text{–OH}) = +380 \text{ kJ mol}^{-1}$$

Some experimental values are listed in Table 2.8.

The dissociation enthalpy of a given bond depends on the structure of the rest of the molecule: for H$_2$O, for instance, $\Delta H^{\ominus}(\text{HO–H}) = +492 \text{ kJ mol}^{-1}$, whereas for the OH fragment $\Delta H^{\ominus}(\text{O–H}) = +428 \text{ kJ mol}^{-1}$. Removing the first and second H atoms results in different enthalpy changes because the electronic structure of the molecule adjusts after the first atom is removed. The **mean bond enthalpy**, $B(\text{A–B})$, is the bond dissociation enthalpy of the A–B bond averaged over a series of related compounds (Table 2.9). For instance, the O–H bond enthalpy is calculated using the data for H$_2$O and similar compounds, such as $\Delta H^{\ominus}(\text{CH}_3\text{O–H}) = +437 \text{ kJ mol}^{-1}$ in methanol. Mean bond enthalpies are useful because they let us make estimates of enthalpy changes in reactions where data might not be available (see Example 2.7). However, they should be used only if more accurate data are not available, and conclusions based on them should be regarded with great caution.

The standard enthalpy change that accompanies the separation of all the atoms in a substance (which may be an element or a compound) is called the **enthalpy of atomization** $\Delta_{at}H^{\ominus}$. For example, the enthalpy of atomization of gaseous H$_2$O is the sum of the HO–H and H–O bond dissociation enthalpies, which from Table 2.8 is $+920 \text{ kJ mol}^{-1}$. Values for other substances are given in Table 2.10. The enthalpy of atomization is the same as the enthalpy of sublimation for an elemental solid that evaporates to a monatomic gas, as in the process $\text{Na(s)} \to \text{Na(g)}$.

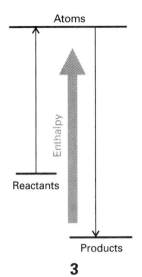

Atoms

Enthalpy

Reactants

Products

3

Example 2.7 *Using mean bond enthalpy data*

Use mean bond enthalpy and enthalpy of atomization data to estimate the standard enthalpy change accompanying the reaction

$$C(s, graphite) + 2H_2(g) + \tfrac{1}{2}O_2(g) \rightarrow CH_3OH(l)$$

Method. The aim is to express the enthalpy required in terms of the enthalpies given. The overall procedure is to disassemble the reactant molecules into atoms and then to use those atoms to form the product, the reverse of its atomization (**3**). The enthalpy change in the first stage is therefore the enthalpy of atomization of the reactants, and the second stage is the negative of the energy of atomization of the products. The gaseous product must be condensed to form the stated product. Use data from Tables 2.9 and 2.10.

Answer. The enthalpy change for the reaction

$$C(s, graphite) + 2H_2(g) + \tfrac{1}{2}O_2(g) \rightarrow C(g) + 4H(g) + O(g)$$

in which 1 mol C is consumed is

$$\Delta H^{\ominus} = (1\ mol) \times \Delta_{sub}H^{\ominus}(C, s) + (2\ mol) \times \Delta H^{\ominus}(H{-}H)$$
$$+ (\tfrac{1}{2}\ mol) \times \Delta H^{\ominus}(O{=}O) = +1837\ kJ$$

The enthalpy change when these atoms form $CH_3OH(g)$ in the reaction

$$C(g) + 4H(g) + O(g) \rightarrow CH_3OH(g)$$

is

$$\Delta H^{\ominus} = -\{(3\ mol) \times B(C{-}H) + (1\ mol) \times B(C{-}O)$$
$$+ (1\ mol) \times B(O{-}H)\} = -2061\ kJ$$

The enthalpy of the overall reaction is therefore the sum of the two enthalpy changes, $-223\ kJ$. Because the enthalpy of vaporization of $CH_3OH(l)$ is $+37.99\ kJ\ mol^{-1}$, the enthalpy of condensation of 1 mol CH_3OH is $-37.99\ kJ$. The overall enthalpy of the reaction is therefore $-261\ kJ$.

Comment. The measured value is $-239\ kJ\ mol^{-1}$. The discrepancy stems from the use of mean bond enthalpies in the calculation of the enthalpy of construction of CH_3OH.

Exercise E2.7. Estimate the standard enthalpy change of the reaction in which 1 mol $C_2H_5OH(l)$ is formed from its elements.   [$-287\ kJ$]

Enthalpies of chemical change

Now we consider enthalpy changes that accompany chemical reactions. Broadly speaking, the **standard reaction enthalpy**, $\Delta_r H^{\ominus}$, is the change in enthalpy when reactants in their standard states change to products in their standard states, as in

$$CH_4(g) + 2O_2(g) \rightarrow CO_2(g) + 2H_2O(l)$$
$$\Delta_r H^{\ominus}(298\ K) = -890\ kJ\ mol^{-1}$$

This standard value refers to the reaction in which 1 mol CH_4 in the form of pure methane gas at 1 bar reacts completely with 2 mol O_2 in the form of pure oxygen gas at 1 bar to produce 1 mol CO_2 as pure carbon dioxide at 1 bar and 2 mol H_2O as pure liquid water at 1 bar, all the substances being at 298 K. The combination of a chemical equation and a standard reaction enthalpy is called a **thermochemical equation**. A standard reaction enthalpy refers to the overall process

>   Pure, unmixed reactants in their standard states
>
> $\rightarrow$ Pure, separated products in their standard states

However, except in the case of ionic reactions in solution, the enthalpy changes accompanying mixing and separation are insignificant in comparison with the contribution from the reaction itself.

We said 'broadly speaking' above, because the precise specification of the standard reaction enthalpy is more specific about the significance of the 'per mole' that appears in the value of $\Delta_r H^\ominus$. To establish this precise definition, consider the reaction

$$2A + B \rightarrow 3C + D$$

written in the symbolic form

$$0 = 3C + D - 2A - B$$

by subtracting the reactants from both sides (and replacing the arrow by an equals sign). This equation has the form

$$0 = \sum_J v_J J \tag{28}$$

where J denotes substances and the v_J are the corresponding **stoichiometric numbers** in the chemical equation. These numbers have the values

$$v_A = -2 \qquad v_B = -1 \qquad v_C = +3 \qquad v_D = +1$$

Note that, in the convention we shall adopt, the stoichiometric numbers of the products are positive and those of the reactants are negative.

The standard reaction enthalpy for the specimen reaction above is defined as the difference between the standard molar enthalpies of the products and reactants weighted by the stoichiometric numbers that appear in the chemical equation:

$$\Delta_r H^\ominus = \{3 \times H^\ominus(C) + H^\ominus(D)\} - \{2 \times H^\ominus(A) + H^\ominus(B)\}$$

where $H^\ominus(J)$ is the standard *molar* enthalpy of species J at the temperature of interest. This expression is a special case of the general definition

$$\Delta_r H^\ominus = \sum_J v_J H^\ominus(J) \tag{29}$$

as may be verified by substituting the values of the stoichiometric numbers given above. For example, if the reaction is

$$N_2(g) + 3H_2(g) \rightarrow 2NH_3(g)$$

Table 2.11* Standard enthalpies of formation and combustion of organic compounds at 298 K

	$\Delta_f H^{\ominus}/(\text{kJ mol}^{-1})$	$\Delta_c H^{\ominus}/(\text{kJ mol}^{-1})$
Benzene, $C_6H_6(l)$	+49.0	−3268
Ethane, $C_2H_6(g)$	−84.7	−1560
Glucose, $C_6H_{12}O_6(s)$	−1274	−2808
Methane, $CH_4(g)$	−74.8	−890
Methanol, $CH_3OH(l)$	−238.7	−726

* More values are given in the Data section.

then the stoichiometric numbers are

$$v(N_2) = 1 \qquad v(H_2) = -3 \qquad v(NH_3) = +2$$

and the standard reaction enthalpy is

$$\Delta_r H^{\ominus} = 2 \times H^{\ominus}(NH_3) - \{H^{\ominus}(N_2) + 3 \times H^{\ominus}(H_2)\}$$

There are some reaction enthalpies that have special names and a particular significance. The **standard enthalpy of combustion** $\Delta_c H^{\ominus}$ is the standard reaction enthalpy for the complete oxidation of an organic compound to CO_2 and H_2O if the compound contains C, H, and O, and to N_2 if N is also present. The thermochemical reaction to which the enthalpy of combustion refers is always written with the stoichiometric number of the compound equal to −1. An example is the combustion of glucose:

$$C_6H_{12}O_6(s) + 6O_2(g) \rightarrow 6CO_2(g) + 6H_2O(l)$$
$$\Delta_c H^{\ominus} = -2808 \text{ kJ mol}^{-1}$$

The value quoted shows that 2808 kJ of heat is released when 1 mol $C_6H_{12}O_6$ burns under standard conditions (at 298 K). Some further values are listed in Table 2.11.

The **standard enthalpy of hydrogenation** is the standard reaction enthalpy for the hydrogenation of an unsaturated organic compound (with the compound ascribed the stoichiometric number −1 in the chemical equation). Two especially important cases are the hydrogenations of ethene and benzene:

$$CH_2{=}CH_2(g) + H_2(g) \rightarrow CH_3CH_3(g) \qquad \Delta_r H^{\ominus} = -137 \text{ kJ mol}^{-1}$$

$$\bigcirc + 3H_2(g) \rightarrow \bigcirc \qquad \Delta_r H^{\ominus} = -205 \text{ kJ mol}^{-1}$$

The interest in these two values lies in the observation that the second is not three times the first, as might be expected on the basis that benzene contains three double bonds. The value for benzene is less than $3 \times (-137 \text{ kJ mol}^{-1})$ by 206 kJ mol^{-1}, which therefore represents a thermochemical stabilization of benzene (so benzene lies closer in energy than expected to the fully hydrogenated form). This stabilization is explained in Section 14.9.

Hess's law

Standard enthalpies of individual reactions can be combined to obtain the enthalpy of another reaction. This application of the First Law is widely called **Hess's law**.

> **Hess's law**: The standard enthalpy of an overall reaction is the sum of the standard enthalpies of the individual reactions into which a reaction may be divided.

The individual steps need not be realizable in practice: they may be hypothetical reactions, the only requirement being that they should balance. The thermodynamic basis of the law is the path independence of the value of $\Delta_r H^{\ominus}$ and the implication that we may take the specified reactants, pass through any (possibly hypothetical) set of reactions to the specified products, and overall obtain the same change of enthalpy.

Example 2.8 *Using Hess's law*

The standard reaction enthalpy for the hydrogenation of propene,

$$CH_2{=}CHCH_3(g) + H_2(g) \rightarrow CH_3CH_2CH_3(g)$$

is $-124 \, kJ \, mol^{-1}$. The standard reaction enthalpy for the combustion of propane,

$$CH_3CH_2CH_3(g) + 5O_2(g) \rightarrow 3CO_2(g) + 4H_2O(l)$$

is $-2220 \, kJ \, mol^{-1}$. Calculate the standard reaction enthalpy for the combustion of propene.

Method. Add and subtract the reactions given, together with any others needed, so as to reproduce the reaction required. Then add and subtract the reaction enthalpies in the same way. Additional data are in Table 2.11.

Answer. The combustion reaction we require is

$$C_3H_6(g) + \tfrac{9}{2}O_2(g) \rightarrow 3CO_2(g) + 3H_2O(l)$$

This reaction can be recreated from the following sum:

	$\Delta_r H^{\ominus}/kJ \, mol^{-1}$
$C_3H_6(g) + H_2(g) \rightarrow C_3H_8(g)$	-124
$C_3H_8(g) + 5O_2(g) \rightarrow 3CO_2(g) + 4H_2O(l)$	-2220
$H_2O(l) \rightarrow H_2(g) + \tfrac{1}{2}O_2(l)$	$+286$
$C_3H_6(g) + \tfrac{9}{2}O_2(g) \rightarrow 3CO_2(g) + 3H_2O(l)$	-2058

Comment. The skill to develop is the ability to assemble a given thermochemical equation from others.

Exercise E2.8. Calculate the enthalpy of hydrogenation of benzene from its enthalpy of combustion and the enthalpy of combustion of cyclohexane. $[-205 \, kJ \, mol^{-1}]$

2.6 Enthalpies of formation

Thermochemical data are often reported in terms of the enthalpy of a compound relative to its elements under standard conditions:

The **standard enthalpy of formation** $\Delta_f H^\ominus$ of a substance is the standard reaction enthalpy for the formation of the compound from its elements in their reference states.

The **reference state** of an element is its most stable state at the specified temperature and 1 bar. For example, at 298 K the reference state of nitrogen is a gas of N_2 molecules, that of mercury is liquid mercury, that of carbon is graphite, and that of tin is the white (metallic) form. There is one exception to this general prescription of reference states: the reference state of phosphorus is taken to be white phosphorus despite this allotrope not being the most stable form but simply the most reproducible form of the element. Standard enthalpies of formation are expressed as enthalpies per mole of the compound (that is, the thermochemical equation for the formation is written with the stoichiometric number of the compound equal to +1). The standard enthalpy of formation of liquid benzene at 298 K, for example, refers to the reaction

$$6C(s, \text{graphite}) + 3H_2(g) \rightarrow C_6H_6(l)$$

and is $+49.0\ \text{kJ mol}^{-1}$. The standard enthalpies of formation of elements in their reference states are zero at all temperatures because they are the enthalpies of such 'null' reactions as

$$N_2(g) \rightarrow N_2(g)$$

Some standard enthalpies of formation are listed in Tables 2.11 and 2.12.

Table 2.12* Standard enthalpies of formation of inorganic compounds, $\Delta_f H^\ominus/(\text{kJ mol}^{-1})$, at 298 K

$H_2O(l)$	−285.8	$H_2O_2(l)$	−187.8
$NH_3(g)$	−46.1	$N_2H_4(l)$	+50.6
$NO_2(g)$	+33.2	$N_2O_4(g)$	+9.2
$NaCl(s)$	−411.2	$KCl(s)$	−436.8

* More values are given in the Data section.

The Born–Haber cycle

The enthalpy of formation of a solid compound may be analysed into several contributions. For solid sodium chloride, for instance, the overall reaction $Na(s) + \frac{1}{2}Cl_2(g) \rightarrow NaCl(s)$ can be regarded as the outcome of five steps.

1.	Sublimation of $Na(s)$	$Na(s) \rightarrow Na(g)$
2.	Ionization of $Na(g)$	$Na(g) \rightarrow Na^+(g) + e^-(g)$
3.	Dissociation of $\frac{1}{2}Cl_2(g)$	$\frac{1}{2}Cl_2(g) \rightarrow Cl(g)$
4.	Electron gain by $Cl(g)$	$Cl(g) + e^-(g) \rightarrow Cl^-(g)$
5.	Formation of NaCl from $Na^+(g)$ and $Cl^-(g)$	$Na^+(g) + Cl^-(g) \rightarrow NaCl(s)$

Table 2.13* Lattice enthalpies, $\Delta H_L^\ominus/(\text{kJ mol}^{-1})$, at 298 K

NaF	926
NaCl	787
KCl	717
MgO	3850
MgS	3406

* More values are given in the Data section.

The enthalpy change for Step 5 is the negative of the **lattice enthalpy**, $\Delta H_L^\ominus$. In general, the lattice enthalpy is the standard reaction enthalpy for the formation of a gas of ions from the crystalline solid:

$$MX(s) \rightarrow M^+(g) + X^-(g)$$

All lattice enthalpies are positive (Table 2.13).

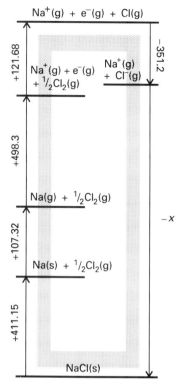

$Na^+(g) + e^-(g) + Cl(g)$

$+121.68$

$Na^+(g) + e^-(g) + \frac{1}{2}Cl_2(g)$

$Na^+(g) + Cl^-(g)$

-351.2

$+498.3$

$Na(g) + \frac{1}{2}Cl_2(g)$

$-x$

$+107.32$

$Na(s) + \frac{1}{2}Cl_2(g)$

$+411.15$

$NaCl(s)$

2.17 A Born–Haber cycle for the determination of the lattice enthalpy. The sum of the enthalpy changes round the cycle is zero. In other words, the distance up on the left must be equal to the distance up on the right. From this equality, x may be determined.

The sequence of steps is depicted in Fig. 2.17. When the step

$$NaCl(s) \rightarrow Na(s) + \tfrac{1}{2}Cl_2(g)$$

which is the reverse of the formation reaction for NaCl from its elements, is included, we obtain a **thermochemical cycle**, a closed path of processes, known as a **Born–Haber cycle**. The importance of a cycle is that *the sum of enthalpy changes around a cycle is zero* (because the initial and final states are the same). It follows that, if all but one of the enthalpy changes in a cycle are known, then the unknown enthalpy change may be found.

Example 2.9 *Using a Born–Haber cycle*

Calculate the lattice enthalpy of sodium chloride using the Born–Haber cycle in Fig. 2.17.

Method. We need to find the value of x in Fig. 2.17 given that the sum of enthalpy changes around the cycle is zero. The data may be assembled from tables and then the value of x determined algebraically.

Answer. From the tables of data

$$\tfrac{1}{2}\Delta H^{\ominus}(Cl\text{—}Cl) = +121.68 \text{ kJ mol}^{-1}$$
$$\Delta_{sub}H^{\ominus}(Na, s) = +107.32 \text{ kJ mol}^{-1}$$
$$\Delta_{ion}H^{\ominus}(Na, g) = +498.3 \text{ kJ mol}^{-1}$$
$$\Delta_{eg}H^{\ominus}(Cl, g) = -351.2 \text{ kJ mol}^{-1}$$
$$\Delta_f H^{\ominus}(NaCl, s) = -411.15 \text{ kJ mol}^{-1} \text{ (the negative of this value is used in the cycle)}$$

The sum of the enthalpy changes round the cycle is

$$(411.5 + 121.68 + 107.32 + 498.3 - 351.2 - x) \text{ kJ} = 0$$

This expression solves to $x = 787.2$. Because the lattice enthalpy is positive (it corresponds to the reverse of the final step in the cycle), $\Delta H_L^{\ominus} = +787.2 \text{ kJ mol}^{-1}$.

Exercise E2.9. Calculate the lattice enthalpy of calcium bromide.

$$[2148 \text{ kJ mol}^{-1}]$$

The reaction enthalpy in terms of enthalpies of formation

Conceptually, we can regard a reaction as proceeding by decomposing the reactants into their elements and then forming those elements into the products. The value of $\Delta_r H^{\ominus}$ for the overall reaction is the sum of these 'unforming' and forming enthalpies. Because 'unforming' is the reverse of forming, the enthalpy of an unforming step is the negative of the enthalpy of formation (**4**). Hence, in the enthalpies of formation of substances we have enough information to calculate the enthalpy of any reaction. For example, the reaction enthalpy of

$$2HN_3(l) + 2NO(g) \rightarrow H_2O_2(l) + 4N_2(g)$$

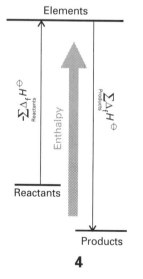

Elements

$-\sum \Delta_f H^{\ominus}_{\text{Reactants}}$

$\sum \Delta_f H^{\ominus}_{\text{Products}}$

Enthalpy

Reactants

Products

4

is the sum of the following contributions:

$$2HN_3(l) \rightarrow 3N_2(g) + H_2(g) \qquad -2\Delta_f H^{\ominus}(HN_3, l)$$
$$2NO(g) \rightarrow N_2(g) + O_2(g) \qquad -2\Delta_f H^{\ominus}(NO, g)$$
$$H_2(g) + O_2(g) \rightarrow H_2O_2(l) \qquad \Delta_f H^{\ominus}(H_2O_2, l)$$

and, from the data in Table 2.12, is $-956.5 \text{ kJ mol}^{-1}$.

The calculation of a reaction enthalpy from enthalpies of formation can be expressed by writing the chemical equation in terms of eqn 28, identifying the stoichiometric coefficients, and then using the expression

$$\Delta_r H^{\ominus} = \sum_J v_J \Delta_f H^{\ominus}(J) \tag{30}$$

Example 2.10 *Calculating a standard reaction enthalpy*

Express the standard reaction enthalpy of

$$2HN_3(l) + 2NO(g) \rightarrow H_2O_2(l) + 4N_2(g)$$

in terms of the standard enthalpies of formation of the components.

Method. We express the reaction in the form of eqn 28, identify the stoichiometric numbers, and then use eqn 30 to express $\Delta_r H^{\ominus}$ in terms of the $\Delta_f H^{\ominus}$.

Answer. The reaction is written

$$0 = -2HN_3(l) - 2NO(g) + H_2O_2(l) + 4N_2(g)$$

and we identify the following stoichiometric numbers

$$v(HN_3) = -2 \qquad v(NO) = -2 \qquad v(H_2O_2) = +1 \qquad v(N_2) = +4$$

The standard reaction enthalpy is therefore

$$\Delta_r H^{\ominus} = -2\Delta_f H^{\ominus}(HN_3) - 2\Delta_f H^{\ominus}(NO) + \Delta_f H^{\ominus}(H_2O_2) + 4\Delta_f H^{\ominus}(N_2)$$

When numerical values are inserted we use $\Delta_f H^{\ominus}(N_2, g) = 0$.

Exercise E2.10. Express the standard reaction enthalpy of

$$2C_3H_6(g) + 9O_2(g) \rightarrow 6CO_2(g) + 6H_2O(l)$$

in terms of enthalpies of formation.

$$[\Delta_r H^{\ominus} = 6\Delta_f H^{\ominus}(CO_2) + 6\Delta_f H^{\ominus}(H_2O) - 2\Delta_f H^{\ominus}(C_3H_6) - 9\Delta_f H^{\ominus}(O_2)]$$

The enthalpies of formation of substances in solution

Two examples of the enthalpy of formation of a substance in solution are

$$\tfrac{1}{2}H_2(g) + \tfrac{1}{2}Cl_2(g) \rightarrow HCl(aq) \qquad \Delta_f H^{\ominus}(HCl, aq) = -167 \text{ kJ mol}^{-1}$$
$$Na(s) + \tfrac{1}{2}Cl_2(g) \rightarrow NaCl(aq) \qquad \Delta_f H^{\ominus}(NaCl, aq) = -407 \text{ kJ mol}^{-1}$$

We can analyse the enthalpy of formation in solution using a cycle like that used for the analysis of the formation of the solid. The only difference is that in place of the lattice formation step (step 5) we use

5'. Hydration of ions: $Na^+(g) + Cl^-(g) \rightarrow$

$$Na^+(aq) + Cl^-(aq) = NaCl(aq)$$

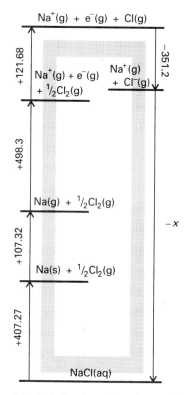

2.18 A similar thermodynamic cycle for the determination of the enthalpy of hydration of Na^+ and Cl^- ions. The distance up on the left is equal to the distance up on the right.

In this case, the standard enthalpy change is the **standard enthalpy of hydration** $\Delta_{hyd}H^{\ominus}$, of the ions. Because all the other enthalpy changes in the cycle are known, this quantity may be determined: the sum of the enthalpy changes around the cycle in Fig. 2.18 is zero, so we can conclude that the enthalpy of hydration of NaCl is $-783.4\,kJ\,mol^{-1}$. Other values may be obtained in this way, and some are listed in Table 2.14. The values there refer to the formation of the solution at zero concentration (infinite dilution) and do not take into account the interactions between the ions that are important in more concentrated solutions.[4]

The enthalpies of formation of individual ions in solution

The enthalpy of formation of a fully ionized compound in solution (that is, of a strong electrolyte) can be regarded as the sum of the enthalpies of formation of solutions of the constituent ions. Thus, we can think of the enthalpy of formation of the strong electrolyte HCl(aq) as the sum of the enthalpies of formation of $H^+(aq)$ and $Cl^-(aq)$.

The problem that confronts us is how to apportion the overall value between the two kinds of ions. It is conventional to set the standard enthalpy of formation of $H^+(aq)$ equal to zero:

$$\tfrac{1}{2}H_2(g) \rightarrow H^+(g) \qquad \Delta_f H^{\ominus}(H^+, aq) = 0 \text{ at all temperatures}$$

It follows that the standard enthalpy of formation of $Cl^-(aq)$ is $-167.45\,kJ\,mol^{-1}$. Combining that value with the observed enthalpy of formation of NaCl(aq) then leads to a value for $Na^+(aq)$, and so on. This procedure leads to the values in Table 2.15.

Just as the enthalpy of hydration of NaCl can be analysed in terms of a thermodynamic cycle, so individual hydration enthalpies of the ions may also be analysed into physically meaningful contributions. However, we cannot adopt an arbitrary convention about the hydration enthalpy of the proton because it would conflict with the choice already made about its enthalpy of formation: we have selected zero for the step $\tfrac{1}{2}H_2(g) \rightarrow H^+(aq)$, and we cannot also select zero for $H^+(g) \rightarrow H^+(aq)$. We are therefore forced to estimate the actual value of the enthalpy change in the step by calculating the energy of interaction of a proton and the surrounding water molecules. There is some agreement (from mass spectrometry and the energy of proton attachment to small clusters of water molecules) that

$$H^+(g) \rightarrow H^+(aq) \qquad \Delta_{hyd}H^{\ominus} \approx -1090\,kJ\,mol^{-1}$$

With this value accepted, the value for $Cl^-(g) \rightarrow Cl^-(aq)$ can be obtained from data on HCl, and then the value for Cl^- combined with the value of $\Delta_{hyd}H^{\ominus}$ for NaCl as obtained above, to arrive at the value of about $-400\,kJ\,mol^{-1}$ for $Na^+(g) \rightarrow Na^+(aq)$. The individual ion hydration enthalpies in Table 2.16 have been obtained in this way. The data show that small, highly charged ions have the most negative (exothermic) hydration enthalpies: such ions attract the solvent strongly.

Table 2.14* Standard molar enthalpies of hydration at infinite dilution, $\Delta_{hyd}H^{\ominus}/(kJ\,mol^{-1})$, at 298 K

	Li^+	Na^+	K^+
F^-	-1026	-911	-828
Cl^-	-884	-783	-685
Br^-	-856	-742	-658

The values at the intersection of the column headed M^+ and the row labelled X^- is for the process $M^+(g) + X^-(g) \rightarrow M^+(aq) + X^-(aq)$.
* More values are given in the Data section.

4 We explain what is meant by the standard state of a solution in Section 7.7.

Table 2.15* Limiting enthalpies of formation of ions in aqueous solution, $\Delta_f H^\ominus/(\text{kJ mol}^{-1})$, at 298 K

Cations

H^+	0
Na^+	-240.1
Cu^{2+}	$+64.8$
Al^{3+}	-531

Anions

OH^-	-230.0
Cl^-	-167.2
SO_4^{2-}	-909.3
PO_4^{3-}	-1277.4

* More values are given in the Data section.

Table 2.16* Standard ion hydration enthalpies, $\Delta_{hyd} H^\ominus/(\text{kJ mol}^{-1})$

Li^+	-520	F^-	-506
Na^+	-405	Cl^-	-364
K^+	-321	Br^-	-337

* More values are given in the Data section.

2.7 The temperature dependence of reaction enthalpies

The standard enthalpies of many important reactions have been measured at different temperatures, and for serious work these accurate data must be used. However, in the absence of this information, standard reaction enthalpies at different temperatures may be estimated from heat capacities and the reaction enthalpy at some other temperature.

It follows from eqn 22 that, when a substance is heated from T_1 to T_2, its enthalpy changes from $H(T_1)$ to

$$H(T_2) = H(T_1) + \int_{T_1}^{T_2} C_p \, dT$$

(We have assumed that no phase transition takes place in the temperature range of interest.) Because this equation applies to each substance in the reaction, the standard reaction enthalpy changes from $\Delta_r H^\ominus(T_1)$ to

$$\Delta_r H^\ominus(T_2) = \Delta_r H^\ominus(T_1) + \int_{T_1}^{T_2} \Delta_r C_p \, dT \tag{31}$$

where

$$\Delta_r C_p = \sum_J \nu_J C_{p,m}(J) \tag{32}$$

The quantity $\Delta_r C_p$ is essentially the difference of the molar heat capacities of products and reactants weighted by the stoichiometric numbers that appear in the chemical equation. Equation 31 is known as **Kirchhoff's law**. It is normally a good approximation to assume that $\Delta_r C_p$ is independent of the temperature, at least over reasonably limited ranges, as illustrated in the following example. However, in some cases the temperature dependence of heat capacities is taken into account by using eqn 24.

Example 2.11 *Using Kirchhoff's law*

The standard enthalpy of formation of gaseous H_2O at 25°C is $-241.82 \text{ kJ mol}^{-1}$. Estimate its value at 100°C given the following values of the molar heat capacities at constant pressure: $H_2O(g)$, $33.58 \text{ J K}^{-1} \text{ mol}^{-1}$; $H_2(g)$, $28.84 \text{ J K}^{-1} \text{ mol}^{-1}$; $O_2(g)$, $29.37 \text{ J K}^{-1} \text{ mol}^{-1}$. Assume that the heat capacities are independent of temperature.

Method. If $\Delta_r C_p$ is independent of temperature in the range T_1 to T_2, then the integral in eqn 31 evaluates to $\Delta_r C_p \times (T_2 - T_1)$. Therefore,

$$\Delta_r H^\ominus(T_2) = \Delta_r H^\ominus(T_1) + \Delta_r C_p \times (T_2 - T_1)$$

To proceed, write the chemical equation, identify the stoichiometric numbers, and calculate $\Delta_r C_p$ from the data.

Answer. The reaction is

$$H_2(g) + \tfrac{1}{2}O_2(g) \rightarrow H_2O(g),$$

or equivalently $0 = H_2O(g) - H_2(g) - \tfrac{1}{2}O_2(g)$

Hence

$$\Delta_r C_p = C_{p,m}(H_2O, g) - C_{p,m}(H_2, g) - \tfrac{1}{2}C_{p,m}(O_2, g)$$
$$= -9.94 \, J \, K^{-1} \, mol^{-1}$$

It then follows that

$$\Delta_f H^{\ominus}(373 \, K) = -241.82 \, kJ \, mol^{-1} + (75 \, K) \times (-9.94 \, J \, K^{-1} \, mol^{-1})$$
$$= -242.6 \, kJ \, mol^{-1}$$

Exercise E2.11. Estimate the standard enthalpy of formation of $NH_3(g)$ at 400 K from the data in Table 2.12. $[-48.4 \, kJ \, mol^{-1}]$

CHECK LIST OF KEY IDEAS

1   The introduction of the concepts of a **system** and its **surroundings** and **work, heat,** and **internal energy** (Section 2.1).

2   The classification of processes as **exothermic** and **endothermic** (Section 2.1).

3   The statement of the **First Law of thermodynamics** in a simple but practical form (Section 2.2), and a second formulation of the law that shows how the concepts of heat and energy are based on and are measured in terms of work (Section 2.2).

4   The introduction of the internal energy as a **state function** (Section 2.2).

5   The derivation of the expression for **expansion work** (eqn 7) and its application to expansion against constant external pressure (eqn 9) and **reversible expansion** (eqn 12).

6   An introduction to the principles of **calorimetry** (Section 2.4).

7   The definition and significance of the **heat capacity at constant volume** (eqn 16).

8   The introduction of the state function **enthalpy** (eqn 18) and its relation to the energy transferred as heat at constant pressure (eqn 19).

9   The definition and significance of the **heat capacity at constant pressure** (eqn 21) and its relation to the heat capacity at constant volume (eqn 25).

10   The definition of the **standard state** of a substance and of the **standard enthalpy change** for a physical transformation and a chemical reaction (Section 2.5).

11   An introduction to a number of different types of enthalpy change, including the **enthalpy of phase transition**, of **solution**, of **ionization**, and of **dissociation** (Section 2.5).

12   The statement of **Hess's law** as a special case of the enthalpy being a state function, and the manipulation of **thermochemical equations** (Section 2.5).

13   The concept of a **thermochemical cycle** and the particular case of a **Born–Haber cycle** for the discussion of **lattice enthalpy** (Section 2.6).

14   The expression of standard reaction enthalpies in terms of **standard enthalpies of formation,** including a method of expressing chemical equations that will be used throughout the text (eqns 28 and 29).

15   The variation of the reaction enthalpy with temperature and **Kirchhoff's law** (eqn 31).

EXERCISES

Assume all gases are perfect unless stated otherwise. To two significant figures, 1.0 atm is the same as 1.0 bar.

Unless otherwise stated, thermochemical data are for 298 K.

2.1. Calculate the work done to raise a mass of 1.0 kg through 10 m on the surface of (a) the Earth ($g = 9.81 \text{ m s}^{-2}$) and (b) the moon ($g = 1.60 \text{ m s}^{-2}$).

2.2. Calculate the work needed for a 65-kg person to climb through 4.0 m on the surface of the Earth.

2.3. A chemical reaction takes place in a container of cross-sectional area 100 cm², the container has a loosely fitted piston at one end. As a result of the reaction, the piston is pushed out through 10 cm against an external pressure of 1.0 atm. Calculate the work done by the system.

2.4. A sample consisting of 1.00 mol Ar is expanded isothermally at 0°C from 22.4 to 44.8 L (a) reversibly, (b) against a constant external pressure equal to the final pressure of the gas, and (c) freely (against zero external pressure). For the three processes calculate q, w, ΔU, and ΔH.

2.5. A 1.00-mol sample of monatomic perfect gas, for which $C_{V,m} = \frac{3}{2}R$, initially at $p_1 = 1.00$ atm and $T_1 = 300$ K, is heated reversibly to 400 K at constant volume. Calculate the final pressure, ΔU, q, and w.

2.6. A sample of 4.50 g of methane occupies 12.7 L at 310 K. (a) Calculate the work done when the gas expands isothermally against a constant external pressure of 200 Torr until its volume has increased by 3.3 L. (b) Calculate the work that would be done if the same expansion occurred reversibly.

2.7. In the isothermal reversible compression of 52.0 mmol of a perfect gas at 260 K, the volume of the gas is reduced to one-third its initial value. Calculate w for this process.

2.8. A sample of 1.00 mol $H_2O(g)$ is condensed isothermally and reversibly to liquid water at 100°C. The standard enthalpy of vaporization of water at 100°C is $+40.656 \text{ kJ mol}^{-1}$. Find w, q, ΔU, and ΔH for this process.

2.9. A strip of magnesium of mass 15 g is dropped into a beaker of dilute hydrochloric acid. Calculate the work done by the system as a result of the reaction. The atmospheric pressure is 1.0 atm and the temperature 25°C.

2.10. Calculate the heat required to melt 750 kg of sodium metal at 371 K.

2.11. The value of $C_{p,m}$ for a sample of a perfect gas was found to vary with temperature according to the expression $C_{p,m}/\text{J K}^{-1}\text{mol}^{-1} = 20.17 + 0.3665(T/\text{K})$. Calculate q, w, ΔU, and ΔH for 1.00 mol of the gas when the temperature is raised from 25°C to 200°C (a) at constant pressure, (b) at constant volume.

2.12. Calculate the standard enthalpy of formation of butane at 25°C from its standard enthalpy of combustion.

2.13. When 229 J of energy is supplied as heat to 3.0 mol Ar(g) at constant pressure, the temperature of the sample increases by 2.55 K. Calculate the molar heat capacities at constant volume and constant pressure of the gas.

2.14. A 25-g sample of a liquid is cooled from 290 K to 275 K at constant pressure by the extraction of 1.2 kJ of energy as heat. Calculate q and ΔH and estimate the heat capacity of the sample.

2.15. When 3.0 mol O_2 is heated at a constant pressure of 3.25 atm, its temperature increases from 260 K to 285 K. Given that the molar heat capacity of O_2 at constant pressure is $29.4 \text{ J K}^{-1} \text{mol}^{-1}$, calculate q, ΔH, and ΔU.

2.16. A certain liquid has $\Delta_{vap}H^{\ominus} = +26.0 \text{ kJ mol}^{-1}$. Calculate q, w, ΔH, and ΔU when 0.50 mol is vaporized at 250 K and 750 Torr.

2.17. The standard enthalpy of formation of ethylbenzene is $-12.5 \text{ kJ mol}^{-1}$. Calculate its standard enthalpy of combustion.

2.18. Calculate the standard enthalpy of hydrogenation of 1-hexene to hexane given that the standard enthalpy of combustion of 1-hexene is $-4003 \text{ kJ mol}^{-1}$.

2.19. The standard enthalpy of combustion of cyclopropane is $-2091 \text{ kJ mol}^{-1}$ at 25°C. From this information and enthalpy of formation data for $CO_2(g)$ and $H_2O(l)$, calculate the enthalpy of formation of cyclopropane. The standard enthalpy of formation of propene is $+20.42 \text{ kJ mol}^{-1}$. Calculate the enthalpy of isomerization of cyclopropane to propene.

2.20. From the following data, determine $\Delta_f H^{\ominus}$ for diborane, $B_2H_6(g)$, at 298 K:

(1) $B_2H_6(g) + 3O_2(g) \rightarrow B_2O_3(s) + 3H_2O(g)$
$\Delta_r H^{\ominus} = -1941 \text{ kJ mol}^{-1}$

(2) $2B(s) + \frac{3}{2}O_2(g) \rightarrow B_2O_3(s)$ $\Delta_r H^{\ominus} = -2368 \text{ kJ mol}^{-1}$

(3) $H_2(g) + \frac{1}{2}O_2(g) \rightarrow H_2O(g)$ $\Delta_r H^{\ominus} = -241.8 \text{ kJ mol}^{-1}$

2.21. Calculate the standard internal energy of formation of liquid methyl acetate from its standard enthalpy of formation, which is -442 kJ mol^{-1}.

2.22. The temperature of a bomb calorimeter rose by 1.617 K when a current of 3.20 A was passed for 27.0 s from a 12.0-V source. Calculate the heat capacity of the calorimeter.

2.23. When 120 mg of naphthalene, $C_{10}H_8(s)$, was burned in a bomb calorimeter the temperature rose by 3.05 K. Calculate the heat capacity of the calorimeter. By how much will the temperature rise when 100 mg of phenol, $C_6H_5OH(s)$, is burned in the calorimeter under the same conditions?

2.24. When 0.3212 g of glucose was burned in a bomb calorimeter of calorimeter constant 641 J K⁻¹ the temperature rose by 7.793 K. Calculate (a) the standard molar internal energy of combustion, (b) the standard enthalpy of combustion, and (c) the standard enthalpy of formation of glucose.

2.25. Calculate the standard enthalpy of solution of AgCl(s) in water from the enthalpies of formation of the solid and the aqueous ions.

2.26. The standard enthalpy of decomposition of the yellow complex $H_3N-SO_2(s)$ into $NH_3(g)$ and $SO_2(g)$ is $+40$ kJ mol^{-1}. Calculate the standard enthalpy of formation of $H_3N-SO_2(s)$.

2.27. Given that the standard enthalpy of combustion of graphite is -393.51 kJ mol^{-1} and that of diamond is -395.412 kJ mol^{-1}, calculate the enthalpy of the graphite → diamond transition.

2.28. The mass of a typical sugar cube is 1.5 g. Calculate the energy released as heat when a cube is burned in air. To what height could you climb on the energy a cube provides assuming 25 per cent of the energy is available for work?

2.29. The standard enthalpy of a combustion of propane gas is -2220 kJ mol^{-1} and the standard enthalpy of vaporization of the liquid is $+15$ kJ mol^{-1}. Calculate (a) the standard enthalpy and (b) the standard internal energy of combustion of the liquid.

2.30. Classify as endothermic or exothermic the following reactions:

(a) $CH_4(g) + 2O_2(g) \rightarrow CO_2(g) + 2H_2O(l)$
$\Delta_r H^\ominus = -890$ kJ mol^{-1}

(b) $2C(s) + H_2(g) \rightarrow C_2H_2(g)$
$\Delta_r H^\ominus = +227$ kJ mol^{-1}

(c) $NaCl(s) \rightarrow NaCl(aq)$ $\Delta_r H^\ominus = +3.9$ kJ mol^{-1}

2.31. Express the reactions in Exercise 2.30 in the form $0 = \sum_J \nu_J J$, and identify the stoichiometric numbers.

2.32. Use standard enthalpies of formation to calculate the standard enthalpies of the following reactions:

(a) $2NO_2(g) \rightarrow N_2O_4(g)$
(b) $NH_3(g) + HCl(g) \rightarrow NH_4Cl(s)$
(c) Cyclopropane(g) → propene(g)
(d) $HCl(aq) + NaOH(aq) \rightarrow NaCl(aq) + H_2O(l)$

2.33. Given the reactions (1) and (2) below, determine (a) $\Delta_r H^\ominus$ and $\Delta_r U^\ominus$ for reaction (3), (b) $\Delta_f H^\ominus$ for both HCl(g) and $H_2O(g)$, all at 298 K. Assume all gases are perfect.

(1) $H_2(g) + Cl_2(g) \rightarrow 2HCl(g)$
$\Delta_r H^\ominus = -184.62$ kJ mol^{-1}

(2) $2H_2(g) + O_2(g) \rightarrow 2H_2O(g)$
$\Delta_r H^\ominus = -483.64$ kJ mol^{-1}

(3) $4HCl(g) + O_2(g) \rightarrow 2Cl_2(g) + 2H_2O(g)$

2.34. For the reaction $C_2H_5OH(l) + 3O_2(g) \rightarrow 2CO_2(g) + 3H_2O(g)$, $\Delta_r U^\ominus = -1373$ kJ mol^{-1} at 298 K. Calculate $\Delta_r H^\ominus$.

2.35. Calculate the standard enthalpies of formation of (a) $KClO_3(s)$ from the enthalpy of formation of KCl, (b) $NaHCO_3(s)$ from the enthalpies of formation of CO_2 and NaOH, and (c) NOCl(g) from the enthalpy of formation of NO given in Table 2.12, together with the following information:

$2KClO_3(s) \rightarrow 2KCl(s) + 3O_2(g)$ $\Delta_r H^\ominus = -89.4$ kJ mol^{-1}
$NaOH(s) + CO_2(g) \rightarrow$

$\qquad\qquad NaHCO_3(s)$ $\Delta_r H^\ominus = -127.5$ kJ mol^{-1}

$2NOCl(g) \rightarrow 2NO(g) + Cl_2(g)$ $\Delta_r H^\ominus = +76.5$ kJ mol^{-1}

2.36. Use the information in Table 2.12 to predict the standard reaction enthalpy of $2NO_2(g) \rightarrow N_2O_4(g)$ at 100°C from its value at 25°C.

2.37. From the data in Table 2.12, calculate $\Delta_r H^\ominus$ and $\Delta_r U^\ominus$ at (a) 298 K, (b) 378 K for the reaction $C(graphite) + H_2O(g) \rightarrow CO(g) + H_2(g)$. Assume all heat capacities to be constant over the temperature range of interest.

2.38. Calculate $\Delta_r H^\ominus$ and $\Delta_r U^\ominus$ at 298 K and $\Delta_r H^\ominus$ at 348 K for the hydrogenation of ethyne (acetylene) to ethene (ethylene) from the enthalpy of combustion and heat capacity data in Table 2.11. Assume the heat capacities to be constant over the temperature range involved.

2.39. Set up a thermodynamic cycle for determining the enthalpy of hydration of Mg^{2+} ions using the following data: enthalpy of sublimation of Mg(s), $+167.2$ kJ mol^{-1}; first and second ionization energies of Mg(g), 7.646 eV and 15.035 eV; dissociation enthalpy of $Cl_2(g)$, $+241.6$ kJ mol^{-1}; electron affinity of Cl(g), $+3.78$ eV; enthalpy of solution of $MgCl_2(s)$, -150.5 kJ mol^{-1}; enthalpy of hydration of Cl$^-$(g), -383.7 kJ mol^{-1}.

PROBLEMS

Assume all gases are perfect unless stated otherwise. To two significant figures, 1.0 atm is the same as 1.0 bar.

Unless otherwise stated, thermochemical data are for 298 K.

Numerical problems

2.1. Calculate the heat needed to heat the air in a house from 20°C to 25°C. Assume that the house contains 600 m³ of air, which should be taken to be a perfect diatomic gas. The density of air is 1.21 kg m⁻³ at 20°C. Calculate ΔU and ΔH for the heating of the air.

2.2. An average human produces about 10 MJ of heat each day through metabolic activity. If a human body were an isolated system of mass 65 kg with the heat capacity of water, what temperature rise would the body experience? Human bodies are actually open systems, and the main mechanism of heat loss is through the evaporation of water. What mass of water should be evaporated each day to maintain constant temperature?

2.3. Consider a perfect gas contained in a cylinder and separated by a frictionless adiabatic piston into two sections, A and B; section B is in contact with a water bath that maintains it at constant temperature. Initially $T_A = T_B = 300$ K, $V_A = V_B = 2.00$ L, and $n_A = n_B = 2.00$ mol. Heat is supplied to Section A and the piston moves to the right reversibly until the final volume of Section B is 1.00 L. Calculate (a) the work done by the gas in Section A, (b) ΔU for the gas in Section B, (c) q for the gas in B, (d) ΔU for the gas in A, and (e) q for the gas in A. Assume $C_{V,m} = 20.0$ J K⁻¹ mol⁻¹.

2.4. A sample consisting of 1 mol of a monatomic perfect gas (for which $C_{V,m} = \frac{3}{2}R$) is taken through the cycle shown in Fig. 2.19. (a) Determine the temperatures at 1, 2, and 3. (b) Calculate q, w, ΔU, and ΔH for each step and for the overall cycle. If a numerical answer cannot be obtained, then write +, −, or ? as appropriate.

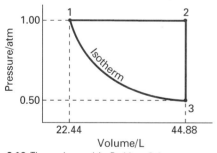

2.19 The cycle used in Problem 2.4.

2.5. A 5.0-g block of solid carbon dioxide is allowed to evaporate in a vessel of volume 100 cm³ maintained at 20°C. Calculate the work done when the system expands (a) isothermally against a pressure of 1.0 atm, and (b) isothermally and reversibly to the same volume as in (a).

2.6. A sample consisting of 1.0 mol CaCO₃(s) was heated to 800°C, when it decomposed. The heating was carried out in a container fitted with a piston which was initially resting on the solid. Calculate the work done during complete decomposition at 1.0 atm. What work would be done if instead of having a piston the container was open to the atmosphere?

2.7. A sample consisting of 2.0 mol CO₂ occupies a fixed volume of 15.0 L at 300 K. When it is supplied with 2.35 kJ of energy as heat its temperature increases to 341 K. Assume that CO₂ is described by the van der Waals equation of state, and calculate w, ΔU, and ΔH.

2.8. A sample of 70 mmol Kr(g) expands reversibly and isothermally at 373 K from 5.25 cm³ to 6.29 cm³, and the internal energy of the sample is known to increase by 83.5 J. Use the virial equation of state up to the second coefficient $B = -28.7$ cm³ mol⁻¹ to calculate w, q, and ΔH for this change of state.

2.9. A piston exerting a pressure of 1.0 atm rests on the surface of water at 100°C. The pressure is reduced infinitesimally, and as a result 10 g of water evaporates and absorbs 22.2 kJ of heat. Calculate w, ΔU, ΔH, and ΔH_m.

2.10. A new fluorocarbon of molar mass 102 g mol⁻¹ was placed in an electrically heated vessel. When the pressure was 650 Torr, the liquid boiled at 78°C. After the boiling point had been reached, it was found that a current of 0.232 A from a 12.0-V supply passed for 650 s vaporized 1.871 g of the sample. Calculate the molar enthalpy and molar internal energy of vaporization.

2.11. An object is cooled by the evaporation of liquid methane at its normal boiling point (112 K). What volume of methane gas at 1.00 atm pressure must be formed from the liquid methane in order to remove 32.5 kJ of energy as heat from the object?

2.12. The molar heat capacity of ethane is represented in the temperature range 298 K to 400 K by the empirical expression $C_{p,m}/(\text{J K}^{-1}\text{mol}^{-1}) = 14.73 + 0.1272(T/\text{K})$. The corresponding expressions for C(s) and H₂(g) are given in Table 2.2. Calculate the standard enthalpy of formation of ethane at 350 K from its value at 298 K.

2.13. Several characteristics of a hydrocarbon must be considered when it is being considered as a fuel. Among them are the specific enthalpy, the heat evolved per unit mass, for the advantage of a high molar enthalpy of combustion may be eliminated if a large mass of fuel is to be transported. Use the data in Tables 2.11 and 2.12 to calculate the following information on butane, pentane, and octane: (a) the heat output per mole, (b) the heat output per gram.

2.14. Geophysical conditions are sometimes so extreme that quantities neglected in normal laboratory experiments take on an overriding importance. For example, consider the formation of diamond under geophysically typical conditions. The density of graphite is 2.27 g cm⁻³ and that of diamond is 3.52 g cm⁻³ at a certain temperature and 500 kbar. By how much does ΔU differ from ΔH for the graphite $\rightarrow$ diamond transition under these conditions?

2.15. A sample of the sugar D-ribose ($C_5H_{10}O_5$) of mass 0.727 g was placed in a bomb calorimeter and then ignited in the presence of excess oxygen. The temperature rose by 0.910 K. In a separate experiment in the same calorimeter, the combustion of 0.825 g of benzoic acid, for which the internal energy of combustion is -3251 kJ mol^{-1}, gave a temperature rise of 1.940 K. Calculate the internal energy of combustion of D-ribose and its enthalpy of formation.

2.16. The standard enthalpy of formation of the metallocene bis(benzene)chromium was investigated in a calorimeter. It was found that for the reaction

$$Cr(C_6H_6)_2(s) \rightarrow Cr(s) + 2C_6H_6(g)$$

$\Delta_r U^{\ominus}(583\ K) = +8.0$ kJ mol^{-1}. Find the corresponding reaction enthalpy and estimate the standard enthalpy of formation of the compound at 583 K. The molar heat capacity of benzene is 140 J K^{-1} mol^{-1} in its liquid range and 28 J K^{-1} mol^{-1} as a gas.

2.17. The standard enthalpy of combustion of sucrose is -5645 kJ mol^{-1}. What is the advantage (in kJ mol^{-1} of energy released as heat) of complete aerobic oxidation compared with anaerobic hydrolysis of sucrose to lactic acid?

Theoretical problems

2.18. In a machine of a particular design, the opposing force acting on a mass m varies as $F \sin(\pi x/a)$. Calculate the work needed to move the mass (a) from $x = 0$ to $x = a$ and (b) from $x = 0$ to $x = 2a$.

2.19. Calculate the work done during the isothermal reversible expansion of a gas that satisfies the virial equation of state, eqn 1.21b. Evaluate (a) the work for 1.0 mol Ar at 273 K (for data, see Table 1.3) and (b) the same amount of a perfect gas. Let the expansion be from 500 to 1000 cm^3 in each case.

2.20. With reference to Fig. 2.20 and assuming perfect gas behaviour, calculate: (a) the amount of gas molecules (in moles) in this system and its volume in states B and C, (b) the work done on the gas along the paths ACB and

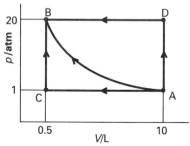

2.20 The cycle used in Problem 2.20.

ADB, (c) The work done on the gas along the isotherm AB, (d) q and ΔU for each of the three paths. Take $C_{V,m} = \frac{3}{2}R$.

2.21. Calculate the work done during the isothermal reversible expansion of a van der Waals gas. Account physically for the way in which the coefficients a and b appear in the final expression. Plot on the same graph the indicator diagrams for the isothermal reversible expansion of (a) a perfect gas, (b) a van der Waals gas in which $a = 0$ and $b = 5.11 \times 10^{-2}$ mol^{-1}, and (c) $a = 4.2$ L^2 atm mol^{-2} and $b = 0$. (Use other values too if you have a computer available.) The values selected exaggerate the imperfections but give rise to significant effects on the indicator diagrams. Take $V_i = 1.0$ L, $V_f = 2.0$ L, $n = 1.0$ mol, and $T = 298$ K.

2.22. When a system is taken from state A to state B along the path ACB in Fig. 2.21, 80 J of heat flows into the system and the system does 30 J of work. (a) How much heat flows into the system along path ADB if the work done is 10 J? (b) When the system is returned from state B to A along the curved path, the work done on the system is 20 J. Does the system absorb or liberate heat, and how much? (c) If $U_D - U_A = +40$ J, find the heat absorbed in the processes AD and DB.

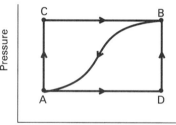

2.21 The cycle used in Problem 2.22.

2.23. Express the work of isothermal reversible expansion of a van der Waals gas in reduced variables and find a definition of reduced work that makes the overall expression independent of the identity of the gas. Calculate the work of isothermal reversible expansion along the critical isotherm from V_c to $x \times V_c$.

2.24. Use (a) the perfect gas equation and (b) the Dieterici equation of state (Table 1.6) to evaluate $(\partial p/\partial T)_V$ and $(\partial p/\partial V)_T$. Go on to confirm that $\partial^2 p/\partial V\,\partial T = \partial^2 p/\partial T\,\partial V$.

2.25. The molar heat capacity of a substance can often be expressed as in eqn 24. Deduce an expression for the enthalpy of the reaction $0 = \sum_J \nu_J J$ at T' in terms of its enthalpy at T and the three coefficients a, b, and c that occur in the expression for $C_{p,m}$. Estimate the error involved in ignoring the temperature variation of C_p for the formation of $H_2O(l)$ at 100°C.

3

The First Law: the machinery

In this chapter we begin to unfold some of the power of thermodynamics by showing how it is used to establish relations between different properties of a system. The procedure we use is based on the experimental fact that the internal energy and the enthalpy are state functions, and we derive a number of relations between observables by exploring the mathematical consequences of this fact. We shall see that the response of the internal energy and enthalpy to changes in temperature under a variety of conditions can be related not only to one another but also to properties such as the compressibility and thermal expansivity of the system. In particular, we shall see that one very useful aspect of thermodynamics is that a property can be measured indirectly by measuring others and then combining their values. The relations we derive also enable us to discuss the liquefaction of gases and to establish a quantitative relation between the heat capacities of a substance at constant pressure and constant volume. Other results that can be derived include the characteristics of the adiabatic, reversible expansion of a perfect gas. Although these characteristics have little direct application in chemistry, they are crucial to the establishment of the properties of the entropy of a system, so we need them in Chapter 4.

Properties that are independent of how a sample is prepared are called **state functions**. Such properties can be regarded as functions of variables, such as pressure and temperature, that define the current state of the system. The internal energy, enthalpy, and heat capacity are examples of state functions, for they depend on the current state of the system and are independent of its previous history. Properties that relate to the preparation of the state are called **path functions**. Examples of path functions are the work that is done in preparing a state and the energy transferred as heat. We do not speak of a system in a particular state as *possessing* work or heat. In each case, the energy transferred as work or heat relates to the path being taken, not the current state itself.

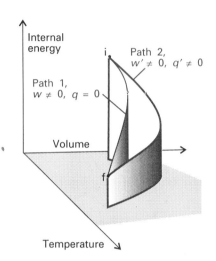

3.1 As the volume and temperature of a system are changed (as indicated by the paths in the T,V-plane), the internal energy changes (vertical axis). An adiabatic and a non-adiabatic path are shown (path 1 and path 2, respectively): they correspond to different values of q and w but to the same value of ΔU.

STATE FUNCTIONS AND EXACT DIFFERENTIALS

The fact that a property such as the internal energy or the enthalpy is a state function is of considerable power. As we shall see, we can use the mathematical properties of state functions to draw far-reaching conclusions about the relations between physical properties and establish connections that are completely unexpected. In practice, it turns out to be possible to combine measurements of different properties to obtain the value of a property we require.

This section will start the task of deducing the consequences of a property being a state function. We can begin to see the importance of the distinction between state functions and path functions by considering the First Law.

3.1 State functions

Consider a system undergoing the changes depicted in Fig. 3.1. The initial state of the system is i and in this state the internal energy is U_i. Work is done on the system to compress it adiabatically to a state f. In this state the system has an internal energy U_f and the work done on the system as it changes along path 1 from i to f is w. Notice our use of language: U is a property of the state; w is a property of the path. Now consider another process, path 2, in which the initial and final states are the same but in which the compression is not adiabatic. The internal energies of both the initial and the final states are the same as before (because U is a state function). However, in the second path an energy q' enters the system as heat and the work w' is not the same as w. The work and the heat are path functions.

Exact and inexact differentials

If a system is taken along a path (e.g. by heating it), U changes from U_i to U_f, and the overall change is the sum of all the infinitesimal changes along the path:

$$\Delta U = \int_i^f dU = U_f - U_i$$

The value of ΔU depends on the initial and final states of the system but is independent of the path between them. This path independence of the integral is expressed by saying that dU is an *exact* differential. In general, an **exact differential** is an infinitesimal quantity which, when integrated, gives a result that is independent of the path between the initial and final states of the system.

When a system is heated, the total energy transferred as heat is the sum of all individual contributions at each point of the path:

$$q = \int_{i,path}^f dq$$

Notice the difference between this and the preceding equation. First, we do not write Δq, because q is not a state function and the energy supplied as heat cannot be expressed as $q_f - q_i$. Second, it is necessary to specify the path of integration because q depends on the path selected (for example, an adiabatic path has $q = 0$, whereas a non-adiabatic path between the same two states would have $q \neq 0$). This path dependence is expressed by saying that dq is an inexact differential. In general, an **inexact differential** is an infinitesimal quantity that, when integrated, gives a result that depends on the path between the initial and final states. Often dq is written $đq$ to emphasize that it is inexact.

The work done on a system to change it from one state to another depends on the path travelled between the two specified states; for example, it is different when the change takes place adiabatically from when it takes place diathermically. It follows that dw is an inexact differential. It is often written $đw$.

Example 3.1 *Calculating work, heat, and internal energy*

Consider a perfect gas inside a cylinder fitted with a piston. Let the initial state be T, V_i and the final state be T, V_f. The change of state can be brought about in many ways, of which the two simplest are the following: path 1, in which there is free, irreversible expansion against zero external pressure, and path 2, in which there is reversible, isothermal expansion. Calculate w, q, and ΔU for each process. It is necessary to know that the internal energy of a perfect gas is independent of its volume (a result we prove later).

Method. To find a starting point for a calculation in thermodynamics, it is often a good idea to go back to first principles, and to look for a way of expressing the quantity you are asked to calculate in terms of other quantities that are easier to calculate. From the information given, $\Delta U = 0$ for the specified process, and we know from Chapter 2 that, for any change, $\Delta U = q + w$. The question hangs on being able to combine the two expressions. We derived a number of expressions for the work done in a variety of processes in Section 2.3, and here we need to select the appropriate ones.

Answer. Because $\Delta U = 0$ for both paths and $\Delta U = q + w$, in each case $q = -w$. The work of irreversible expansion is zero (because $p_{ex} = 0$, Section 2.3), so in path 1 $w = 0$ and $q = 0$. For path 2, the work is given by eqn 2.12, so $w = -nRT \ln(V_f/V_i)$ and $q = nRT \ln(V_f/V_i)$.

Exercise E3.1. Calculate the values of q, w, and ΔU for an irreversible isothermal expansion of a perfect gas against a constant non-zero external pressure.
$$[q = +p_{ex} \Delta V, \ w = -p_{ex} \Delta V, \ \Delta U = 0]$$

Changes in internal energy

We shall now begin to unfold the consequences of dU being an exact differential by noting that, for a closed system of constant composition (the only type of system considered in this chapter), U is a function of volume and temperature.[1] When V changes to $V + dV$ at constant

---

1 U could be regarded as a function of V, T, and p; but because there is an equation of state, it is possible to express p in terms of V and T so p is not an independent variable. We could choose p, T or p, V as independent variables, but V, T fit our purpose.

temperature, U changes to

$$U' = U + \left(\frac{\partial U}{\partial V}\right)_T dV$$

The coefficient $(\partial U/\partial V)_T$, the slope of a graph of U against V at constant temperature, is the partial derivative of U with respect to V. If instead, T changes to $T + dT$ at constant volume, then the internal energy changes to

$$U' = U + \left(\frac{\partial U}{\partial T}\right)_V dT$$

Now suppose that both V and T change infinitesimally. The new internal energy, neglecting second-order infinitesimals (those proportional to $dV\,dT$), is

$$U' = U + \left(\frac{\partial U}{\partial V}\right)_T dV + \left(\frac{\partial U}{\partial T}\right)_V dT$$

The internal energy U' differs from U by the infinitesimal amount dU. Therefore, from the last equation we obtain the very important result that

$$dU = \left(\frac{\partial U}{\partial V}\right)_T dV + \left(\frac{\partial U}{\partial T}\right)_V dT \tag{1}$$

The interpretation of this equation is that, *in a closed system of constant composition, any infinitesimal change in the internal energy is proportional to the infinitesimal changes of volume and temperature,* the coefficients of proportionality being the partial derivatives.

In every case, a partial derivative is the slope of a graph of the property of interest against *one* of the variables on which it depends (recall Fig. 2.11), all the other variables being held constant. In many cases the slopes have a straightforward physical interpretation, and thermodynamics gets shapeless and difficult only when that meaning is not kept in sight. In the present case, we have already met $(\partial U/\partial T)_V$ in Section 2.4, where we saw that it is the heat capacity at constant volume:

$$C_V = \left(\frac{\partial U}{\partial T}\right)_V \tag{2}$$

Hence

$$dU = \left(\frac{\partial U}{\partial V}\right)_T dV + C_V\, dT$$

The other coefficient, $(\partial U/\partial V)_T$ plays a major role in thermodynamics because it is a measure of the variation of the internal energy of a substance as the volume it occupies is changed at constant temperature. We shall denote it π_T (because it has the same dimensions as pressure):

$$\pi_T = \left(\frac{\partial U}{\partial V}\right)_T \tag{3}$$

Then

$$dU = \pi_T\, dV + C_V\, dT \tag{4}$$

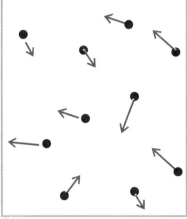

(a)

(b)

3.2 (a) A perfect gas at a particular temperature consists of a collection of molecules that have a certain kinetic energy and zero potential energy. (b) When the volume available to the molecules is increased isothermally, the mean kinetic energy remains the same (as symbolized by the arrows being the same lengths as in part (a)), and the potential energy remains zero. There is therefore no change in the internal energy of the gas.

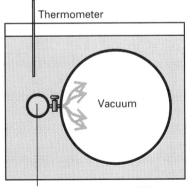

Thermometer

Vacuum

High-pressure gas

Water bath

3.3 A schematic diagram of the apparatus used by Joule in an attempt to measure the change in internal energy when a gas expands isothermally. The heat absorbed by the gas is proportional to the change in temperature of the bath.

If the internal energy increases ($dU > 0$) as the volume of the sample is expanded isothermally ($dV > 0$), which is the case when there are attractive forces between the particles, a graph of internal energy against volume slopes upwards and $\pi_T > 0$. When there are no interactions between the molecules, the internal energy is independent of their separation and hence independent of the volume the sample occupies (Fig. 3.2); hence $\pi_T = 0$ for a perfect gas. The statement $\pi_T = 0$ (that is, the independence of the internal energy and the volume occupied by the sample) can be taken to be the definition of a perfect gas, for later we shall see that it implies the equation of state $pV = nRT$.

The Joule experiment

James Joule thought that he could measure π_T by observing the change in temperature of a gas when it is allowed to expand into a vacuum. He used two metal vessels immersed in a water bath (Fig. 3.3). One was filled with air at about 22 atm and the other was evacuated. He then tried to measure the change in temperature of the water of the bath when a stopcock was opened and the air expanded into a vacuum. He observed no change in temperature.

The thermodynamic implications of the experiment are as follows. No work was done in the expansion into a vacuum, so $w = 0$. No heat entered or left the system (the gas) because the temperature of the bath did not change, so $q = 0$. Consequently, within the accuracy of the experiment, $\Delta U = 0$. It follows that U does not change much when a gas expands isothermally.

Joule's experiment was crude. In particular, the heat capacity of the apparatus was so large that the temperature change that gases do in fact cause was too small to measure. His experiment was on a par with Boyle's: he extracted an essential limiting property of a gas, a property of a perfect gas, without detecting the small deviations characteristic of real gases.

Example 3.2 *Estimating a change in internal energy*

It can be estimated from the van der Waals equation for ammonia that, for a sample of NH_3, $\pi_T = 840$ Pa at 300 K, and it is known experimentally that $C_{V,m} = 27.32$ J K^{-1} mol^{-1}. What is the change of internal energy of 1.00 mol NH_3 when it is heated through 2.0 K and compressed through 100 cm^3?

Method. Infinitesimal changes in volume and temperature result in the infinitesimal change in internal energy that is given by eqn 4. Because the changes in the present problem are small, we may use an approximate form of this expression and write

$$\Delta U \approx \pi_T \, \Delta V + C_V \, \Delta T$$

In the numerical part of the calculation we can use the fact that 1 Pa $= 1$ J m^{-3}.

Answer. From the approximate expression and with $\pi_T = 840 \, \mathrm{J \, m^{-3}}$,

$$\Delta U \approx (840 \, \mathrm{J \, m^{-3}}) \times (-100 \times 10^{-6} \, \mathrm{m^3})$$
$$+ (1.00 \, \mathrm{mol} \times 27.32 \, \mathrm{J \, K^{-1} \, mol^{-1}}) \times (2.0 \, \mathrm{K})$$
$$\approx -0.084 \, \mathrm{J} + 55 \, \mathrm{J} = 55 \, \mathrm{J}$$

Comment. Note that the change is dominated by the effect of the temperature. If ammonia behaved as a perfect gas, π_T would be zero and the volume change would have no effect on U.

Exercise E3.2. Confirm that π_T has the same dimensions as pressure, and express the value for ammonia in atmospheres. $[8.3 \times 10^{-3} \, \mathrm{atm}]$

Changes in internal energy at constant pressure

Partial derivatives have many useful properties and some that we shall draw on frequently are reviewed in *Further information 2*. Skillful use of them can often turn some unfamiliar quantity into a quantity that can be recognized, interpreted, or measured.

As an example, suppose we want to find out how the internal energy varies with temperature when the *pressure* of the system is kept constant. We shall now show that it is possible to use the relations nos 1–4 in *Further information 2* to extract an expression for $(\partial U/\partial T)_p$ from eqn 4. It follows from eqn 4 and relation no. 1 that

$$\left(\frac{\partial U}{\partial T}\right)_p = \pi_T \left(\frac{\partial V}{\partial T}\right)_p + C_V$$

It is usually sensible in thermodynamics to inspect the output of a manipulation like this to see if it contains any recognizable physical quantity. The differential coefficient on the right in this expression is the slope of the graph of volume against temperature (at constant pressure). This property is normally tabulated as the **expansion coefficient** α of a substance, which is defined as

$$\alpha = \frac{1}{V}\left(\frac{\partial V}{\partial T}\right)_p \tag{5}$$

Example 3.3 *Using the expansion coefficient of a gas*

Calculate the volume change that occurs when $50 \, \mathrm{cm^3}$ of neon, treated as a perfect gas, is heated through $5.0 \, \mathrm{K}$ at $298 \, \mathrm{K}$.

Method. To calculate the change in volume we need the value of α for a perfect gas. To calculate its value from the definition in eqn 5 we need to know how V varies with temperature for then we can evaluate the differential coefficient. This variation is given by the perfect gas equation of state.

Answer. From eqn 5 and $pV = nRT$ we can write

$$\alpha = \frac{1}{V}\left(\frac{\partial}{\partial T}\frac{nRT}{p}\right)_p = \frac{nR}{pV} = \frac{1}{T}$$

The increase in volume when the temperature is raised by the small amount ΔT is therefore

$$\Delta V \approx \left(\frac{\partial V}{\partial T}\right)_p \times \Delta T = \alpha V \Delta T = V \times \frac{\Delta T}{T}$$

Substitution of the data gives

$$\Delta V \approx 50 \text{ cm}^3 \times \frac{5.0 \text{ K}}{298 \text{ K}} = 0.84 \text{ cm}^3$$

Exercise E3.3. For copper, $\alpha = 5.01 \times 10^{-5} \text{ K}^{-1}$; calculate the change in volume that occurs when a copper block of volume 50 cm^3 is heated through 5.0 K.   [12 mm^3]

Introduction of the general definition of α into the equation for $(\partial U/\partial T)_p$ gives

$$\left(\frac{\partial U}{\partial T}\right)_p = \alpha \pi_T V + C_V \qquad (6)$$

This equation is entirely general (so long as the system is closed and its composition is constant). It expresses the dependence of the internal energy on the temperature at constant pressure in terms of C_V, which can be measured in one experiment, in terms of α, which can be measured in another (Table 3.1), and in terms of the quantity $\pi_T = (\partial U/\partial V)_T$. For a perfect gas, $\pi_T = 0$, so

$$\text{for a perfect gas: } \left(\frac{\partial U}{\partial T}\right)_p = C_V \qquad (7)°$$

That is, the constant-volume heat capacity of a perfect gas is equal to the slope of a graph of the internal energy against temperature at constant pressure as well as (by definition) to the slope at constant volume. Note that eqn 2 is a *definition* of heat capacity for all substances; only for perfect gases can we also write eqn 7.

Table 3.1* Expansion coefficients (α) and isothermal compressibilities (κ_T)

Substance	$\alpha/$ (10^{-4} K^{-1})	$\kappa_T/$ $(10^{-6} \text{ atm}^{-1})$
Benzene	12.4	92.1
Diamond	0.030	0.187
Lead	0.861	2.21
Water	2.1	49.6

* More values are given in the data section at the end of this volume.

3.2 The temperature dependence of the enthalpy

We can carry out a similar set of operations on the enthalpy H where

$$H = U + pV$$

The quantities U, p, and V are all state functions; therefore H is also a state function, and hence dH is an exact differential. If we regard H as a function of p and T, then by the same argument as for U (but with p in place of V) we find that, for a closed system of constant composition,

$$dH = \left(\frac{\partial H}{\partial p}\right)_T dp + \left(\frac{\partial H}{\partial T}\right)_p dT$$

The second coefficient is the definition of the constant-pressure heat capacity,

$$C_p = \left(\frac{\partial H}{\partial T}\right)_p$$

Therefore

$$dH = \left(\frac{\partial H}{\partial p}\right)_T dp + C_p \, dT \tag{8}$$

The variation of the enthalpy at constant volume

So far we know how U varies with temperature at constant pressure and constant volume (the latter is the constant-volume heat capacity). We also know how H varies with temperature at constant pressure (this is the constant-pressure heat capacity). The only missing temperature variation is that of H at constant volume, $(\partial H/\partial T)_V$. We can obtain this quantity from the last equation by repeating the development that we have just carried through for U. Indeed, it is often helpful to keep in mind when manipulating U and H that V and p play analogous roles, respectively, in the two functions. We shall show in the *Justification* that follows that the slope of a graph of enthalpy against temperature at constant volume is given by

$$\left(\frac{\partial H}{\partial T}\right)_V = \left(1 - \frac{\alpha\mu}{\kappa_T}\right)C_p \tag{9}$$

where the **isothermal compressibility** κ_T is defined as

$$\kappa_T = -\frac{1}{V}\left(\frac{\partial V}{\partial p}\right)_T \tag{10}$$

and the **Joule–Thomson coefficient** μ is defined as

$$\mu = \left(\frac{\partial T}{\partial p}\right)_H \tag{11}$$

Equation 9 applies to any substance. Because all the quantities that appear in it can be measured in suitable experiments, we now know how H varies with T when the volume of the sample is held constant.

JUSTIFICATION

First, we divide eqn 8 through by dT and impose constant volume:

$$\left(\frac{\partial H}{\partial T}\right)_V = \left(\frac{\partial H}{\partial p}\right)_T\left(\frac{\partial p}{\partial T}\right)_V + C_p$$

The third differential coefficient looks like something we ought to recognize, and is perhaps related to $(\partial V/\partial T)_p$, the expansion coefficient. Relation no. 3 shuffles p, V, and T around inside partial differentials, and acting on $(\partial p/\partial T)_V$ it produces

$$\left(\frac{\partial p}{\partial T}\right)_V = -\frac{1}{\left(\frac{\partial T}{\partial V}\right)_p\left(\frac{\partial V}{\partial p}\right)_T}$$

Unfortunately, $(\partial T/\partial V)_p$ occurs instead of $(\partial V/\partial T)_p$, but relation no. 2 inverts partial differentials and leads to

$$\left(\frac{\partial p}{\partial T}\right)_V = -\frac{\left(\frac{\partial V}{\partial T}\right)_p}{\left(\frac{\partial V}{\partial p}\right)_T} = \frac{\alpha}{\kappa_T}$$

Next, we change $(\partial H/\partial p)_T$ into something recognizable. Using relation no. 3 we find

$$\left(\frac{\partial H}{\partial p}\right)_T = -\frac{1}{\left(\frac{\partial p}{\partial T}\right)_H \left(\frac{\partial T}{\partial H}\right)_p}$$

A double use of the inverter, relation no. 2, then gives

$$\left(\frac{\partial H}{\partial p}\right)_T = -\left(\frac{\partial T}{\partial p}\right)_H \left(\frac{\partial H}{\partial T}\right)_p$$

and we can recognize both the constant-pressure heat capacity C_p and the Joule–Thomson coefficient μ. Hence,

$$\left(\frac{\partial H}{\partial p}\right)_T = -\mu C_p$$

It follows from these conclusions that the first equation of this *Justification* can be written

$$\left(\frac{\partial H}{\partial T}\right)_V = \left(1 - \frac{\alpha\mu}{\kappa_T}\right)C_p$$

which is eqn 9.

The isothermal compressibility

First, we shall make a few remarks about the isothermal compressibility κ_T. The negative sign in its definition ensures that κ_T is positive, because an increase of pressure, implying a positive dp, brings about a reduction of volume, a negative dV. The isothermal compressibility is proportional to the slope of the graph of volume plotted against pressure at constant temperature (that is, it is proportional to the slope of an isotherm). Some values of κ_T are listed in Table 3.1. Its value for a perfect gas is obtained by substitution of the perfect gas equation of state into eqn 10, which gives

$$\kappa_T = \frac{1}{p} \tag{12}$$

This expression shows that the higher the pressure of the gas, the lower its compressibility.

Example 3.4. *Using the isothermal compressibility*

The isothermal compressibility of water at 20°C and 1 atm is $4.94 \times 10^{-6} \, atm^{-1}$. What change of volume occurs when a sample of volume 50 cm^3 is subjected to an additional 1000 atm?

Method. We know from the definition of compressibility that, for an infinitesimal change of pressure, the volume changes by $dV = (\partial V/\partial p)_T \, dp = -\kappa_T V \, dp$. Therefore, for a finite change in pressure, we need to integrate both sides. When confronted by an integration, it is often a useful first approximation (for substances other than gases) to suppose that the integrand is a constant over the range of integration.

Answer. The integral we need to evaluate is

$$\Delta V = -\int \kappa_T V \, dp$$

If we suppose that κ_T and V are constant over the range of pressures, we can write

$$\Delta V = -\kappa_T V \, \Delta p$$

Substitution of the data into the last expression then gives

$$\Delta V = -(4.94 \times 10^{-6} \, atm^{-1}) \times (50 \, cm^3) \times (1000 \, atm) = -0.25 \, cm^3$$

Comment. Because the compression results in a decrease in volume of only 0.5 per cent, the assumption of constant V and κ_T is probably acceptable as a first approximation. Note that very high pressures are needed to bring about significant changes of volume.

Exercise E3.4. A sample of copper of volume 50 cm^3 is subjected to an additional pressure of 100 atm and a temperature increase of 5.0 K. Estimate the total change in volume.   [8.8 mm^3]

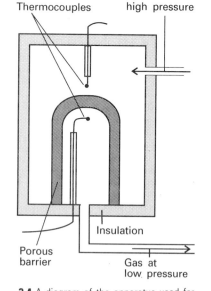

3.4 A diagram of the apparatus used for measuring the Joule–Thomson effect. The gas expands through the porous barrier, which acts as a throttle, and the whole apparatus is thermally insulated. As explained in the text, this arrangement corresponds to an isenthalpic expansion (expansion at constant enthalpy). Whether the expansion results in a heating or a cooling of the gas depends on the conditions.

The Joule–Thomson effect

As we shall see, the analysis of the Joule–Thomson coefficient is central to the technological problems associated with the liquefaction of gases. We need to be able to interpret it physically and we need to be able to measure it.

The cunning required to impose the constraint of constant enthalpy on a change of state was supplied by James Joule and William Thomson (later Lord Kelvin). They let a gas expand through a throttle from one constant pressure to another, and monitored the difference of temperature that arose from the expansion (Fig. 3.4). The whole apparatus was insulated so that the process was adiabatic. They observed a lower temperature on the low-pressure side, the difference being proportional to the pressure difference they maintained. This cooling by adiabatic expansion is now called the **Joule–Thomson effect**.

The thermodynamic analysis of the experiment takes as the system a sample of fixed amount of gas. Because all changes to the gas occur adiabatically, $q = 0$. To calculate the work done as the gas passes through the throttle, we consider the passage of a fixed amount of gas from the high-pressure side, where the pressure is p_i, the temperature T_i, and the gas occupies a volume V_i. The gas emerges on the low-pressure side, where the same amount of gas has a pressure p_f and a temperature T_f and occupies a volume V_f. The gas on the left (Fig. 3.5) is compressed isothermally by the upstream gas acting as a piston. The relevant pressure is p_i and the volume changes from V_i to 0; therefore, the work done on the gas is $-p_i(0 - V_i)$, or $p_i V_i$. The gas expands isothermally on the right of the throttle (but possibly at a different constant temperature) against the pressure p_f provided by the downstream gas acting as a piston to be driven out. The volume changes from 0 to V_f, so the work done on the gas in this stage is $-p_f(V_f - 0)$, or $-p_f V_f$. The total work done on the gas is the sum of these two quantities, or $p_i V_i - p_f V_f$. It follows that the change of internal energy of the gas as it moves from one side of the throttle to the other is

$$U_f - U_i = w = p_i V_i - p_f V_f$$

Reorganization of this expression gives

$$U_f + p_f V_f = U_i + p_i V_i, \qquad \text{or} \qquad H_f = H_i$$

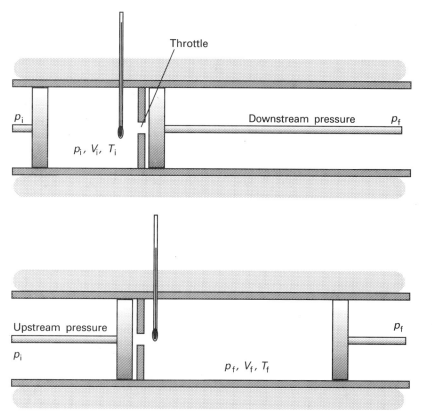

3.5 A diagram representing the thermodynamic basis of Joule–Thomson expansion. The pistons represent the upstream and downstream gases, which maintain constant pressures either side of the throttle. The transition from the upper diagram to the lower, which represents the passage of a given amount of gas through the throttle, occurs without change of enthalpy.

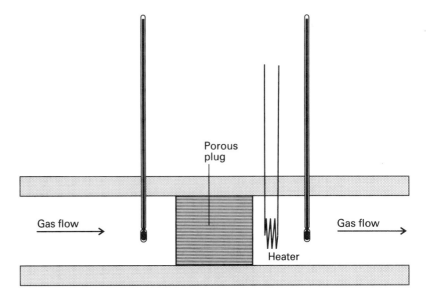

3.6 A schematic diagram of the apparatus used for measuring the isothermal Joule–Thomson coefficient. The electrical heating required to offset the cooling arising from expansion is interpreted as ΔH and used to calculate $(\partial H/\partial p)_T$, which is then converted to μ as explained in the text.

Therefore, the expansion occurs without change of enthalpy: it is an **isenthalpic** process.

The property measured in the experiment is the temperature change per unit change of pressure, $\Delta T/\Delta p$. Adding the constraint of constant enthalpy and taking the limit of small Δp implies that the thermodynamic quantity measured is $(\partial T/\partial p)_H$, which is the Joule–Thomson coefficient μ. In other words, the physical interpretation of μ is that *it is the change in temperature per unit pressure change when a gas expands under adiabatic conditions.*

The modern method of measuring μ is indirect, and involves measuring the **isothermal Joule–Thomson coefficient**, the quantity $(\partial H/\partial p)_T$. The two coefficients are related by the equation derived in the *Justification* of eqn 9:

$$\mu = -\frac{1}{C_p}\left(\frac{\partial H}{\partial p}\right)_T \tag{13}$$

The gas is pumped continuously at a steady pressure through a heat exchanger (which brings it to the required temperature), and then through a throttle inside a thermally insulated container. The steep pressure drop is measured, and the cooling effect is exactly offset by an electric heater placed immediately after the throttle (Fig. 3.6). The energy provided by the heater is monitored. Because the heat can be identified with the value of ΔH for the gas, and the pressure change Δp is known, the value of $(\partial H/\partial p)_T$ can be obtained from the limiting value of $\Delta H/\Delta p$ as $\Delta p \to 0$, and then converted to μ. Some values obtained in this way are listed in Table 3.2.

Real gases have non-zero Joule–Thomson coefficients and, depending on the identity of the gas, the pressure, and the temperature, the sign of the coefficient may be either positive or negative (Fig. 3.7). A positive sign implies that $\mathrm{d}T$ is negative when $\mathrm{d}p$ is negative, in which case the gas cools on expansion. Gases that show a heating effect ($\mu < 0$) at one

Table 3.2* Inversion temperatures (T_i), normal freezing (T_f) and boiling (T_b) points, and Joule–Thomson coefficients (μ) at 1 atm and 298 K

	T_i/K	T_f/K	T_b/K	μ/ (K atm^{-1})
Ar	723	83.8	87.3	
CO_2	1500		194.7	1.11
He	40		4.2	−0.060
N_2	621	63.3	77.4	0.25

* More values are given in the Data section.

3.7 (a) The sign of the Joule–Thomson coefficient μ depends on the conditions. Inside the boundary, the shaded area, μ is positive and outside it is negative. The temperature corresponding to the boundary at a given pressure is the 'inversion temperature' of the gas at that pressure. For a given pressure, the temperature must be below a certain value if cooling is required but, if it becomes too low, the boundary is crossed again and heating occurs. Reduction of pressure under adiabatic conditions moves the system along one of the isenthalps, or curves of constant enthalpy. The inversion temperature curve runs through the points of the isenthalps where their slope changes from negative to positive. (b) The inversion temperatures for three real gases.

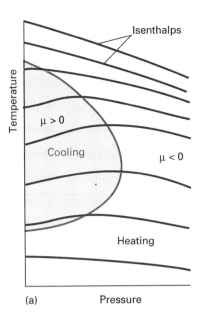

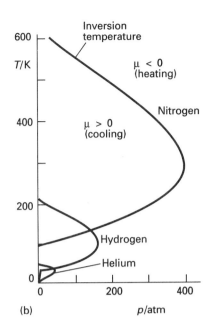

temperature show a cooling effect ($\mu > 0$) when the temperature is below their **inversion temperature** T_I (see Table 3.2).

The technological importance of the Joule–Thomson effect lies in its application to the cooling and liquefaction of gases. The **Linde refrigerator** works on the principle that, below its inversion temperature, a gas cools on expansion. It is then essential to know the conditions under which μ is positive, as was illustrated in Fig. 3.7. The principle of the refrigerator is illustrated in Fig. 3.8: after recirculation of the cooling gas, and for a big enough pressure drop across the throttle, the temperature falls below the condensation temperature, and the liquid forms. Note the importance of working beneath the inversion temperature, and therefore the need to cool some gases by other means initially: using helium at room temperature would turn the refrigerator into an expensive oven.

For a perfect gas, $\mu = 0$; hence, the temperature of a perfect gas is unchanged by Joule–Thomson expansion. This characteristic points clearly to the involvement of intermolecular forces in determining the size of the effect. However, the Joule–Thomson coefficient of a real gas does not necessarily approach zero as the pressure is reduced even though the equation of state of the gas approaches that of a perfect gas. The coefficient is an example of a property mentioned in Section 1.4 that depends on derivatives and not on p, V, and T themselves.

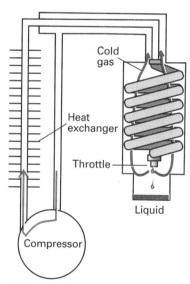

3.8 The principle of the Linde refrigerator is shown in this diagram. The gas is recirculated and, so long as it is beneath its inversion temperature, it cools on expansion through the throttle. The cooled gas cools the high-pressure gas, which cools still further as it expands. Eventually liquefied gas drips from the throttle.

3.3 The relation between C_v and C_p

The constant-pressure heat capacity C_p differs from the constant-volume heat capacity C_V by the work needed to change the volume of the system when the pressure is held constant. This work arises in two ways. One is the work of driving back the atmosphere; the other is the work of stretching the bonds in the material, including any weak intermolecular interactions. In the case of a perfect gas, the second makes no contribution. We shall now derive a general relation between the two heat capacities, and show that it reduces to the perfect gas result in the absence of intermolecular forces.

The relation for a perfect gas

First, we carry through the calculation for a perfect gas. In this special case, we can use eqn 7 to express both heat capacities in terms of derivatives at constant pressure:

$$C_p - C_V = \left(\frac{\partial H}{\partial T}\right)_p - \left(\frac{\partial U}{\partial T}\right)_p$$

Then we introduce

$$H = U + pV = U + nRT$$

into the first term, which results in

$$C_p - C_V = \left(\frac{\partial U}{\partial T}\right)_p + nR - \left(\frac{\partial U}{\partial T}\right)_p = nR \tag{14}°$$

This is the result we quoted in Section 2.4 (eqn 2.25).

The general case

We shall now demonstrate that the general relation between the two heat capacities for *any* substance is

$$C_p - C_V = \frac{\alpha^2 TV}{\kappa_T} \tag{15}$$

This formula is a **thermodynamic expression**, which means that it applies to any substance (i.e. that it is 'universally true'). It reduces to eqn 14 for a perfect gas when we set $\alpha = 1/T$ and $\kappa_T = 1/p$.

JUSTIFICATION

A useful rule when doing a problem in thermodynamics is to go back to first principles. In the present problem we do this twice, first by expressing C_p and C_V in terms of their definitions:

$$C_p - C_V = \left(\frac{\partial H}{\partial T}\right)_p - \left(\frac{\partial U}{\partial T}\right)_V$$

and then by inserting the definition $H = U + pV$:

$$C_p - C_V = \left(\frac{\partial U}{\partial T}\right)_p + \left(\frac{\partial (pV)}{\partial T}\right)_p - \left(\frac{\partial U}{\partial T}\right)_V$$

We have already calculated the difference of the first and third terms on the right (eqn 6):

$$\left(\frac{\partial U}{\partial T}\right)_p - \left(\frac{\partial U}{\partial T}\right)_V = \alpha \pi_T V$$

The factor αV gives the change in volume when the temperature is raised, and $\pi_T = (\partial U/\partial V)_T$ converts this change in volume into a change in internal energy. We can simplify the remaining term by noting that, because p is constant,

$$\left(\frac{\partial(pV)}{\partial T}\right)_p = p\left(\frac{\partial V}{\partial T}\right)_p = \alpha pV$$

The middle term of this expression identifies it as the contribution to the work of pushing back the atmosphere: $(\partial V/\partial T)_p$ is the change of volume caused by a change of temperature, and multiplication by p converts the expansion into work.

Collecting the two contributions gives

$$C_p - C_V = \alpha(p + \pi_T)V$$

The first term on the right, αpV, is a measure of the work needed to push back the atmosphere; the second term on the right, $\alpha \pi_T V$, is the work required to separate the molecules composing the system.

At this point we can go further by using the result we prove in Chapter 5 that

$$\pi_T = T\left(\frac{\partial p}{\partial T}\right)_V - p$$

When this expression is inserted in the last equation we obtain

$$C_p - C_V = \alpha TV\left(\frac{\partial p}{\partial T}\right)_V$$

The same coefficient as appears here was encountered in the *Justification* of eqn 9, where we saw that it is equal to α/κ_T. Therefore, we can conclude that

$$C_p - C_V = \frac{\alpha^2 TV}{\kappa_T}$$

which is eqn 15.

Because thermal expansivities α of liquids and solids are small, it is tempting to deduce from eqn 15 that, for them, $C_p \approx C_V$. But this is not always so, because the compressibility κ_T might also be small and so α^2/κ_T might be large. That is, although only a little work need be done to push back the atmosphere, a great deal of work may have to be done to pull atoms apart from one another as the solid expands. As an illustration, for water at 25°C, eqn 15 gives $C_{p,\text{m}} = 75.3\,\text{J K}^{-1}\,\text{mol}^{-1}$ compared with $C_{V,\text{m}} = 74.8\,\text{J K}^{-1}\,\text{mol}^{-1}$. In some cases, the two heat capacities differ by as much as 30 per cent.

Example 3.5 *Evaluating the difference between C_p and C_V*

Estimate the difference between $C_{p,\text{m}}$ and $C_{V,\text{m}}$ for tetrachloromethane (CCl$_4$) at 25°C, for which $C_{p,\text{m}} = 132\,\text{J K}^{-1}\,\text{mol}^{-1}$. At this temperature, its density ρ is $1.59\,\text{g cm}^{-3}$, its expansion coefficient is $1.24 \times 10^{-3}\,\text{K}^{-1}$, and its isothermal compressibility is $9.05 \times 10^{-5}\,\text{atm}^{-1}$.

Method. The result involves substitution of the data into eqn 15 after it has been modified into an expression in terms of the density ρ of the liquid of molar mass M by writing $M = \rho V_m$:

$$C_{p,m} - C_{V,m} = \frac{\alpha^2 TM}{\rho \kappa_T}$$

The compressibility converts to

$$\kappa_T = \frac{9.05 \times 10^{-5}\,\text{atm}^{-1}}{1.013 \times 10^5\,\text{Pa atm}^{-1}} = 8.93 \times 10^{-10}\,\text{Pa}^{-1}$$

Answer. Substitution of the data gives

$$C_{p,m} - C_{V,m} = \frac{(1.24 \times 10^{-3}\,\text{K}^{-1})^2 \times (298\,\text{K}) \times (153.82\,\text{g mol}^{-1})}{(1.59\,\text{g cm}^{-3}) \times (8.93 \times 10^{-10}\,\text{Pa}^{-1})}$$

$$= 4.96 \times 10^7\,\text{Pa cm}^3\,\text{K}^{-1}\,\text{mol}^{-1}$$

$$= 4.96 \times 10\,\text{Pa m}^3\,\text{K}^{-1}\,\text{mol}^{-1} = 49.6\,\text{J K}^{-1}\,\text{mol}^{-1}$$

because $1\,\text{Pa m}^3 = 1\,\text{N m} = 1\,\text{J}$.

Comment. The difference in heat capacities is 38 per cent of $C_{p,m}$ itself.

Exercise E3.5. Repeat the calculation for benzene, for which $\rho = 0.88$ g cm^{-3}, $\alpha = 1.24 \times 10^{-3}\,\text{K}^{-1}$, and $\kappa_T = 9.21 \times 10^{-5}\,\text{atm}^{-1}$.

$$[45\,\text{J K}^{-1}\,\text{mol}^{-1}]$$

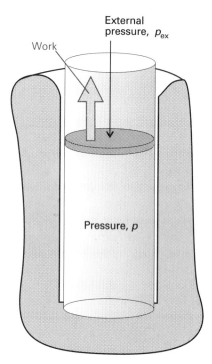

3.9 The arrangement to achieve adiabatic expansion work. The insulation prevents an exchange of heat with the surroundings.

WORK OF ADIABATIC EXPANSION

We are now equipped to deal with the work of *adiabatic* expansion, in which the system does work while it is thermally insulated from the outside world (Fig. 3.9). The approach we use illustrates a useful rule in thermodynamics, that, *when calculating a property, it may be possible to find a state function related to it and to calculate the change in that function by the most convenient path.* In the present case, because the change is adiabatic, $dq = 0$ at each stage of the expansion. Consequently, $dU = dw$. Therefore, instead of calculating the work done during the expansion, we can calculate the change in internal energy between the same initial and final states:

$$w = \int_i^f dU$$

The remainder of this section deals with a perfect gas with a heat capacity that is independent of temperature. The restriction to perfect gases allows us to relate dU to the change in volume in a simple way, because we can use eqn 4 with $\pi_T = 0$:

$$w = \int_i^f C_V\,dT$$

For a gas with a heat capacity that is independent of temperature,

$$w = C_V \int_i^f dT = C_V \times (T_f - T_i) = C_V \Delta T \tag{16}°$$

That is, the work done during an adiabatic expansion of a perfect gas is proportional to the temperature difference between the initial and final states.

3.4 Special cases

A general conclusion from eqn 16 is that, if $w < 0$ (so the system has done work), then $\Delta T < 0$. This conclusion should not be surprising: if the system does work then, because no heat can enter, the internal energy must fall. However, because the internal energy of a perfect gas is unaffected by changes of volume alone, a reduction of internal energy on expansion must mean that the temperature has fallen too.

The most important type of adiabatic expansion (and the only kind that we need for later) is reversible expansion, in which the external pressure is matched to the internal pressure throughout the process (in the arrangement illustrated in Fig. 3.9, p_{ex} would be continuously adjusted to be equal to p). We shall now show that for the adiabatic reversible expansion of a perfect gas the work of expansion can be calculated from eqn 16 by using the following relation between the initial and final temperatures:

$$V_f T_f^c = V_i T_i^c \tag{17}_r^\circ$$

where $c = C_{V,m}/R$.

JUSTIFICATION

Consider a stage of expansion at which the pressure (inside and out) is p; then when the volume changes by dV the work done is $dw = -p\,dV$. Because $dq = 0$, it follows that $dU = -p\,dV$. However, by the argument above, for a perfect gas $dU = C_V\,dT$. These two terms must be equal (dU is an exact differential), so at each stage

$$C_V\,dT = -p\,dV$$

The pressure of the gas changes as the volume changes, but because a perfect gas obeys $pV = nRT$ we can write

$$C_V \frac{dT}{T} = -nR \frac{dV}{V}$$

The temperature falls from T_i to T_f as the volume increases from V_i to V_f. Because C_V can be taken to be independent of temperature (this is true for monatomic perfect gases, and approximately true for others), it may be treated as a constant. The integrations we need are therefore

$$C_V \int_{T_i}^{T_f} \frac{dT}{T} = -nR \int_{V_i}^{V_f} \frac{dV}{V}$$

and so we conclude that

$$C_V \ln\left(\frac{T_f}{T_i}\right) = -nR \ln\left(\frac{V_f}{V_i}\right)$$

On writing $c = C_V/nR = C_{V,\mathrm{m}}/R$ and doing a little rearranging, we get

$$\ln\left(\frac{T_\mathrm{f}}{T_\mathrm{i}}\right)^c = \ln\left(\frac{V_\mathrm{i}}{V_\mathrm{f}}\right)$$

and hence conclude that the initial and final temperatures are related to the initial and final volumes by

$$V_\mathrm{f} T_\mathrm{f}^c = V_\mathrm{i} T_\mathrm{i}^c$$

which is eqn 17.

Example 3.6 *Calculating the work of adiabatic expansion*

A sample of argon at 1.0 atm pressure and 25°C expands reversibly and adiabatically from 0.50 L to 1.00 L. Calculate its final temperature, the work done during the expansion, and the change in internal energy. The molar heat capacity of argon at constant volume is $12.48\,\mathrm{J\,K^{-1}\,mol^{-1}}$.

Method. The final temperature can be calculated from eqn 17 after it has been rearranged to

$$T_\mathrm{f} = \left(\frac{V_\mathrm{i}}{V_\mathrm{f}}\right)^{1/c} \times T_\mathrm{i}$$

The result depends only on the molar heat capacity of the gas (through c) so it is independent of the amount of argon present. The work, however, does depend on the amount of argon in the system because, according to eqn 16, w is proportional to C_V and the latter is an extensive property ($C_V = nC_{V,\mathrm{m}}$). The amount can be obtained from the perfect gas equation.

Answer. From the data, $c = 1.501$, so

$$T_\mathrm{f} = \left(\frac{0.50\,\mathrm{L}}{1.0\,\mathrm{L}}\right)^{1/1.501} \times 298\,\mathrm{K} = 188\,\mathrm{K}$$

For the amount of Ar present we use

$$n = \frac{pV}{RT} = \frac{(1.0\,\mathrm{atm}) \times (0.50\,\mathrm{L})}{(8.206 \times 10^{-2}\,\mathrm{L\,atm\,K^{-1}\,mol^{-1}}) \times (298\,\mathrm{K})} = 0.020\,\mathrm{mol}$$

It follows that, because $w = nC_{V,\mathrm{m}}\Delta T$,

$$w = (0.020\,\mathrm{mol}) \times (12.48\,\mathrm{J\,K^{-1}\,mol^{-1}}) \times (188\,\mathrm{K} - 298\,\mathrm{K}) = -27\,\mathrm{J}$$

Because $q = 0$ for an adiabatic change, it follows that $\Delta U = -27\,\mathrm{J}$.

Exercise E3.6. Calculate the final temperature, the work done, and the change of internal energy when ammonia is used in a reversible adiabatic expansion from 0.50 L to 2.00 L, the other initial conditions being the same.   [196 K, −57 J, −57 J]

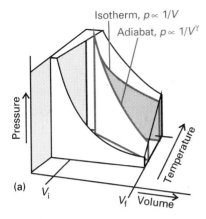

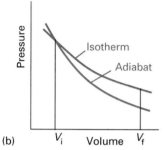

3.10 The isotherms and adiabats corresponding to the expansion of a perfect gas are paths on the p, V, T-surface, as shown in (a). (b) The projections of these paths on to the p, V-plane show that the pressure decreases more rapidly along an adiabat than an isotherm and therefore that the work corresponding to the process (the area under the curves) is less for adiabatic reversible expansion than for isothermal reversible expansion between the same initial and final volumes.

3.5 Perfect gas adiabats

It is now easy to find the variation of the pressure of a perfect gas when it undergoes a reversible adiabatic change. From the perfect gas equation in the form

$$\frac{p_i V_i}{p_f V_f} = \frac{T_i}{T_f}$$

and from eqn 17 we have

$$\frac{T_i}{T_f} = \left(\frac{V_f}{V_i}\right)^{1/c}$$

Combining the two gives

$$p_i V_i^\gamma = p_f V_f^\gamma \qquad (18)_r^\circ$$

where

$$\gamma = \frac{R}{C_{V,m}} + 1 = \frac{R + C_{V,m}}{C_{V,m}}$$

For a perfect gas $C_{V,m} + R = C_{p,m}$, so γ is the ratio of molar heat capacities:

$$\gamma = \frac{C_{p,m}}{C_{V,m}} \qquad (19)$$

Equation 18 is often expressed in the form

$$pV^\gamma = \text{constant} \qquad (20)_r^\circ$$

For all gases $\gamma > 1$ (for a perfect monatomic gas, for example, $\gamma = \frac{5}{3}$, a result we establish in Section 20.4). As a result, an **adiabat**, a graph showing the dependence of the pressure on the volume (Fig. 3.10), falls more steeply ($p \propto 1/V^\gamma$) than the corresponding isotherm ($p \propto 1/V$). The physical reason for the difference is that, in an isothermal expansion, energy flows into the system as heat and maintains the temperature, so the pressure does not fall as much as in an adiabatic expansion.

Example 3.7 *Calculating the pressure change in an adiabatic expansion*

A sample of argon (for which $\gamma = \frac{5}{3}$) at 1.00 atm expands reversibly and adiabatically to twice its initial volume. Calculate its final pressure.

Method. We use eqn 18 rearranged into

$$p_f = \left(\frac{V_i}{V_f}\right)^\gamma \times p_i$$

Note that we do not need to know the amount of gas present (because γ is independent of the amount).

Answer. Substitution of the data gives

$$p_f = \left(\tfrac{1}{2}\right)^{\frac{5}{3}} \times 1.00 \text{ atm} = 0.31 \text{ atm}$$

Comment. If the initial temperature is 298 K, then the final temperature will be 188 K (from eqn 17).

Exercise E3.7. Calculate the final pressure when neon at 1.0 atm is compressed reversibly and adiabatically to 75 per cent of its initial volume.   [1.6 atm]

Table 3.3 Work done on expansion*

Type of work	w	q	ΔU	ΔT
Expansion against $p = 0$				
Isothermal	0	$0°$	$0°$	0
Adiabatic	0	0	0	0
Expansion against constant pressure				
Isothermal	$-p_{ex}\Delta V$	$p_{ex}\Delta V°$	$0°$	0
Adiabatic	$-p_{ex}\Delta V$	0	$-p_{ex}\Delta V$	$-\dfrac{p_{ex}\Delta V°}{C_V}$
Reversible expansion or compression				
Isothermal	$-nRT\ln\left(\dfrac{V_f}{V_i}\right)°$	$nRT\ln\left(\dfrac{V_f}{V_i}\right)°$	$0°$	0
Adiabatic	$C_V\Delta T°$	0	$C_V\Delta T°$	$\left\{\left(\dfrac{V_i}{V_f}\right)^{1/c}-1\right\}T_i°$

* The entries marked ° are for a perfect gas; the rest apply to any substance. $c = C_{V,m}/R$.

Table 3.3 summarizes the behaviour of a system when it undergoes a variety of different types of change.

CHECK LIST OF KEY IDEAS

1 Introduction of the concepts of **exact and inexact differentials** and their significance in thermodynamics.

2 The **Joule experiment** to measure the variation of the internal energy of a gas with volume (Section 3.1).

3 The derivation of an expression for the **variation of internal energy with temperature** when the pressure is held constant (eqn 6).

4 The derivation of an expression for the **variation of enthalpy with temperature** when the volume is held constant (eqn 9).

5 The **Joule–Thomson experiment** to measure the change in temperature of a gas when it expands adiabatically (Section 3.2).

6 The relation between the heat capacities at constant volume and constant pressure for a perfect gas (eqn 14) and for any substance (eqn 15).

7 The derivation of an expression for the **change in temperature** when a perfect gas expands adiabatically either reversibly (eqn 17) or in general (Table 3.3).

8 The derivation of perfect-gas **adiabats** showing how the pressure and volume of a perfect gas are related when it undergoes reversible adiabatic expansion (eqn 20).

EXERCISES

Assume that all gases are perfect and that all data refer to 298 K unless stated otherwise.

3.1. Show that the following functions have exact differentials: (a) $x^2y + 3y^2$, (b) $x \cos xy$, (c) $t(t + e^s) + s$.

3.2. Let $z = ax^2y^3$. Find dz.

3.3. (a) What is the total differential of $z = x^2 + 2y^2 - 2xy + 2x - 4y - 8$? (b) Show that $\partial^2z/\partial y\,\partial x = \partial^2z/\partial x\,\partial y$ for this function.

3.4. Let $z = xy - y + \ln x + 2$. Find dz and show that it is exact.

3.5. Express $(\partial C_V/\partial V)_T$ as a second-derivative of U and find its relation to $(\partial U/\partial V)_T$. From this relation show that $(\partial C_V/\partial V)_T = 0$ for a perfect gas.

3.6. By direct differentiation of $H = U + pV$, obtain a relation between $(\partial H/\partial U)_p$ and $(\partial U/\partial V)_p$. Confirm the result by expressing $(\partial H/\partial U)_p$ as the ratio of two derivatives with respect to volume and then using the definition of enthalpy.

3.7. Write an expression for dV given that V is a function of p and T. Deduce an expression for $d\ln V$ in terms of the expansion coefficient and the isothermal compressibility.

3.8. The internal energy of a perfect monatomic gas relative to its value at $T = 0$ is $\frac{3}{2}nRT$. Calculate $(\partial U/\partial V)_T$ and $(\partial H/\partial V)_T$ for the gas.

3.9. The coefficient of thermal expansion α is defined in eqn 5 and the isothermal compressibility κ_T is defined in eqn 10. (a) Starting from the expression for the total differential dV in terms of T and p, show that $(\partial p/\partial T)_V = \alpha/\kappa_T$. (b) Evaluate α and κ_T for a perfect gas.

3.10. When a certain freon used in refrigeration was expanded from an initial pressure of 32 atm and 0°C to a final pressure of 1.00 atm, the temperature fell by 22 K. Calculate the Joule–Thomson coefficient μ at 0°C, assuming it remains constant over this temperature range.

3.11. For a van der Waals gas, $(\partial U/\partial V)_T = a/V_{\mathrm{m}}^2$. Calculate ΔU_{m} for an isothermal expansion of nitrogen gas from an initial volume of 1.00 L to 24.8 L at 298 K. What are the values of q and w?

3.12. The volume of 1.00 g of a certain liquid varies with temperature as

$$V = V'\{0.75 + 3.9 \times 10^{-4}(T/\mathrm{K}) + 1.48 \times 10^{-6}(T/\mathrm{K})^2\}$$

where V' is its volume at 300 K. Given that its density at 300 K is $0.875\,\mathrm{g\,cm}^{-3}$, calculate its expansion coefficient at 320 K.

3.13. The isothermal compressibility of copper at 293 K is $7.35 \times 10^{-7}\,\mathrm{atm}^{-1}$. Calculate the pressure that must be applied in order to increase its density by 0.08 per cent.

3.14. Given that $\mu = 0.25\,\mathrm{K\,atm}^{-1}$ for nitrogen, calculate the value of its isothermal Joule–Thomson coefficient. Calculate the energy that must be supplied as heat to maintain constant temperature when $15.0\,\mathrm{mol\,N_2}$ flows through a throttle in an isothermal Joule–Thomson experiment and the pressure drop is 75 atm.

3.15. A sample of $4.0\,\mathrm{mol\,O_2}$ is originally confined in 20 L at 270 K and then undergoes adiabatic expansion against a constant pressure of 600 Torr until the volume has tripled. Calculate q, w, ΔT, ΔU, and ΔH. (The final pressure of the gas is not necessarily 600 Torr.)

3.16. A sample of 3.0 mol of gas at 200 K and 2.00 atm is compressed reversibly and adiabatically until the temperature reaches 250 K. Given that its molar constant-volume heat capacity is $27.5\,\mathrm{J\,K^{-1}\,mol^{-1}}$, calculate q, w, ΔU, ΔH, and the final pressure and volume.

3.17. A sample of 1.0 mol of perfect gas with $C_p = 20.8\,\mathrm{J\,K^{-1}}$ is initially at 3.25 atm and 310 K. It undergoes reversible adiabatic expansion until its pressure reaches 2.50 atm. Calculate the final volume and temperature and the work done.

3.18. Estimate the changes in volume that occur when 1.0-cm^3 blocks of (a) mercury and (b) diamond are heated through 5 K at room temperature.

3.19. To design a particular kind of refrigerator we need to know the temperature drop brought about by adiabatic expansion of the refrigerant gas. For one type of freon, $\mu = 1.2\,\mathrm{K\,atm}^{-1}$. What pressure difference is needed to produce a temperature drop of 5.0 K?

3.20. Consider a system consisting of $2.0\,\mathrm{mol\,CO_2}$ (assumed to be a perfect gas) at 25°C confined to a cylinder of cross-section $10\,\mathrm{cm}^2$ at 10 atm. The gas is allowed to expand adiabatically and reversibly against a constant pressure of 1.0 atm. Calculate w, q, ΔU, ΔH, and ΔT when the piston has moved 200 cm.

3.21 A sample consisting of 65.0 g of xenon is confined in a container at 2.00 atm and 298 K and then allowed to expand adiabatically (a) reversibly to 1.00 atm, (b) against a constant pressure of 1.00 atm. Calculate the final temperature in each case.

PROBLEMS

Assume that all gases are perfect and that all data refer to 298 K unless stated otherwise.

Numerical problems

3.1. The isothermal compressibility of lead is 2.3×10^{-6} atm^{-1}. Express this value in Pa^{-1}. A cube of lead of side 10 cm at 25°C was to be inserted in the keel of an underwater exploration TV camera, and its designers needed to know the stresses in the equipment. Calculate the change of volume of the cube at a depth of 1000 m (disregarding the effects of temperature). Take the mean density of sea water as 1.03 g cm^{-3}. Given that the expansion coefficient of lead is 8.61×10^{-5} K^{-1} and that the temperature where the camera operates is −5°C, calculate the volume of the block taking the temperature into account too.

3.2. Calculate the change in (a) the molar internal energy and (b) the molar enthalpy of water when its temperature is raised by 10 K. Account for the difference between the two quantities.

3.3. The constant-volume heat capacity of a gas can be measured by observing the decrease in temperature when it expands adiabatically and reversibly. If the decrease in pressure is also measured we can use it to infer the value of γ (the ratio of heat capacities, C_p/C_V) and hence, by combining the two values, deduce the constant-pressure heat capacity. A fluorocarbon gas was allowed to expand reversibly and adiabatically to twice its volume, the temperature fell from 298.15 K to 248.44 K and its pressure fell from 1522.2 Torr to 613.85 Torr. Evaluate C_p.

3.4. A sample consisting of 1.00 mol of a van der Waals gas is compressed from 20.0 L to 10.0 L at 300 K. In the process, 20.2 kJ of work is done on the gas. Given that $\mu = \{(2a/RT) - b\}/C_{p,m}$, with $C_{p,m} = 38.4$ J K^{-1} mol^{-1}, $a = 3.60$ L^2 atm mol^{-2}, and $b = 0.044$ L mol^{-1}, calculate ΔH for the process.

3.5. 1.000 mol of a perfect monatomic gas, for which $C_{p,m} = \frac{5}{2}R$, is put through the cycle shown in Fig. 3.11.

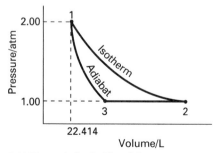

3.11 The cycle for Problem 3.5.

(a) Calculate the volumes of the states 2 and 3 and the temperatures of states 1, 2, and 3. (b) Calculate ΔU and ΔH for each step, taking steps $1 \rightarrow 2$ and $3 \rightarrow 4$ as reversible. Arrange your results in a table.

3.6. A sample of 1.00 mol perfect gas at 1.00 atm and 298 K with $C_{p,m} = \frac{7}{2}R$ is put through the following cycle: (a) constant-volume heating to twice its initial temperature; (b) reversible, adiabatic expansion back to its initial temperature; (c) reversible isothermal compression back to 1.00 atm. Calculate q, w, ΔU, and ΔH for each step and overall.

3.7. Estimate γ (the ratio of heat capacities) for xenon at 100°C and 1.00 atm on the assumption that it is a van der Waals gas.

Theoretical problems

3.8. Determine whether or not $dz = xy\,dx + xy\,dy$ is exact by integrating it around the closed curve formed by the paths $y = x$ and $y = x^2$ between the points $(0, 0)$ and the $(1, 1)$.

3.9. Decide whether $dq = (RT/p)\,dp - R\,dT$ is exact. Then determine whether multiplication of dq by $1/T$ is exact. Comment on the significance of your result.

3.10. Obtain the total differential of the function $w = xy + yz + xz$. Demonstrate that dw is exact by integration between the points $(0, 0, 0)$ and $(1, 1, 1)$ along the two different paths: (1) $z = y = x$ and (2) $z = y = x^2$.

3.11. Derive the relation $C_V = -(\partial U/\partial V)_T(\partial V/\partial T)_U$ from the expression for the total differential of $U(T, V)$.

3.12. Starting from the expression for the total differential of $H(T, p)$, express $(\partial H/\partial p)_T$ in terms of C_p and the Joule–Thomson coefficient μ.

3.13. Starting from the expression $C_p - C_V = T(\partial p/\partial T)_V \times (\partial V/\partial T)_p$, use the appropriate relations between partial derivatives to show that

$$C_p - C_V = -\frac{T\left(\dfrac{\partial V}{\partial T}\right)_p^2}{\left(\dfrac{\partial V}{\partial p}\right)_T}$$

Evaluate $C_p - C_V$ for a perfect gas.

3.14. From an analysis of Joule's free expansion experiment, demonstrate that it is possible to calculate the change in internal energy of a perfect gas for any process by knowing only C_V and ΔT.

3.15. By the consideration of a suitable cycle involving a perfect gas, demonstrate that dq is an inexact differential and that, therefore, heat is not a state function.

3.16. Use the fact that $(\partial U/\partial V)_T = a/V_m^2$ for a van der Waals gas to show that $\mu C_{p,m} \approx (2a/RT) - b$ by using the definition of μ and appropriate relations between partial derivatives. (*Hint:* Use the approximation $pV_m \approx RT$ when it is justifiable to do so.)

3.17. Obtain the expression for the total differential dp for a van der Waals gas in terms of dT and dV. Also obtain $(\partial V/\partial T)_p$. Demonstrate that dp is an exact differential by integrating it from (T_1, V_1) to (T_2, V_2) along the two different paths, namely, (1) $(T_1, V_1) \rightarrow (T_2, V_1) \rightarrow (T_2, V_2)$ and (2) $(T_1, V_1) \rightarrow (T_1, V_2) \rightarrow (T_2, V_2)$.

3.18 Take nitrogen to be a van der Waals gas with $a = 1.390 \text{ L}^2 \text{ atm mol}^{-2}$ and $b = 0.03913 \text{ L mol}^{-1}$, and calculate ΔH_m when the pressure on the gas is decreased from 500 atm to 1.00 atm at 300 K. For a van der Waals gas, $\mu \approx \{(2a/RT) - b\}/C_{p,m}$. Assume $C_{p,m} = \frac{7}{2}R$.

3.19. The pressure of a given amount of a van der Waals gas depends on T and V. Find an expression for dp in terms of dT and dV.

3.20. Rearrange the van der Waals equation of state to give an expression for T as a function of p and V (with n constant). Calculate $(\partial T/\partial p)_V$ and confirm that $(\partial T/\partial p)_V = 1/(\partial p/\partial T)_V$. Go on to confirm Euler's chain relation.

3.21. Calculate the isothermal compressibility and the expansion coefficient of a van der Waals gas. Show, using Euler's chain relation, that

$$\kappa_T R = \alpha(V_m - b)$$

3.22. Given that

$$\mu C_p = T\left(\frac{\partial V}{\partial T}\right)_p - V$$

derive an expression for μ in terms of the van der Waals parameters a and b, and express it in terms of reduced variables. Evaluate μ at 25°C and 1.0 atm, when its molar volume is 24.6 L mol^{-1}. Use the expression obtained to derive a formula for the inversion temperature of a van der Waals gas in terms of reduced variables and evaluate it for xenon.

3.23. The thermodynamic equation of state $(\partial U/\partial V)_T = T(\partial p/\partial T)_V - p$ was quoted in the chapter. Derive its partner

$$\left(\frac{\partial H}{\partial p}\right)_T = -T\left(\frac{\partial V}{\partial T}\right)_p + V$$

from it and the general relations between partial differentials.

3.24. Show that, for a van der Waals gas,

$$C_{p,m} - C_{V,m} = \lambda R \qquad \text{with} \quad \frac{1}{\lambda} = 1 - \frac{(3V_r - 1)^2}{4V_r^3 T_r}$$

and evaluate the difference for xenon at 25°C and 10.0 atm.

3.25. Show that the value of ΔH for the adiabatic expansion of a perfect gas may be calculated by integration of $dH = V \, dp$, and evaluate the integral for the reversible adiabatic expansion of a perfect gas.

3.26. The speed of sound in a gas of molar mass M is related to the ratio of heat capacities γ by

$$c = \left(\frac{\gamma RT}{M}\right)^{\frac{1}{2}}$$

Show that $c = (\gamma p/\rho)^{\frac{1}{2}}$, where ρ is the density of the gas. Calculate the speed of sound in argon at 25°C.

4

The Second Law: the concepts

The purpose of this chapter is to explain the origin of the driving force of physical and chemical change. We shall identify the driving force by examining two simple processes and show that a quantitative measure of the driving force—the entropy—can be formulated. The chapter explains how the change in entropy can be calculated for some fundamental processes and how the entropies of substances can be measured experimentally. Two major applications of entropy are then described. One application is to the efficiencies of heat engines and refrigerators. Although this application may seem to be remote from chemical applications, it illustrates a type of argument that appears later when we consider the more complex case of chemical reactions; these considerations also let us define a fundamental scale of temperature. The second application of entropy is to chemical reactions, where we see how to use tabulated data to decide whether any chemical reaction has a spontaneous tendency to occur.

The chapter also introduces a major subsidiary thermodynamic property, the free energy. We shall see that the introduction of free energy enables the spontaneity of a process to be expressed solely in terms of the properties of a system (instead of having to consider entropy changes in the system and its surroundings). The free energy also enables us to predict the maximum work that a process may achieve.

Some things happen naturally; some things don't. A gas expands to fill the available volume, a hot body cools to the temperature of its surroundings, and a chemical reaction runs in one direction rather than another. Some aspect of the world determines the **spontaneous** direction of change, the direction of change that does not require work to be done to bring it about. We can confine a gas to a smaller volume, we can cool an object with a refrigerator, and we can force some reactions to go in reverse (as in the electrolysis of water). However, none of these happens spontaneously; each one must be brought about by doing work.

The recognition of two classes of process, those that are spontaneous

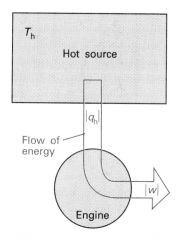

4.1 The Kelvin statement of the Second Law denies the possibility of the process illustrated here, in which heat is changed completely into work, there being no other change. The process is not in conflict with the First Law because energy is conserved.

and those that are not, is summarized by the Second Law of thermodynamics. This law may be expressed in a variety of equivalent ways. One statement was formulated by Kelvin:

> **Second Law of thermodynamics:** No process is possible in which the sole result is the absorption of heat from a reservoir and its complete conversion into work.

For example, it has proved impossible to construct an engine like that shown in Fig. 4.1 in which heat is drawn from a hot reservoir and *completely* converted into work. All real engines (as we shall see) have both a hot source and a cold sink, and some heat is always discarded into the cold sink and not converted into work. The Kelvin statement is a generalization of another everyday observation—a ball at rest on a surface has never been observed to leap spontaneously upwards. The leap of the ball is equivalent to the conversion of heat from the surface into work (the rise of the weight representing the ball). The reverse processes, the conversion of work completely into heat and the conversion of the bouncing of a ball into heat, are both spontaneous.

We shall begin by identifying the characteristic property of a system that determines the direction of spontaneous change. To be definite in what follows, we shall consider an isolated system inside which the events are taking place (Fig. 4.2). We shall divide this isolated system into two parts. One part of the isolated system is the system of interest, such as a reaction mixture. The other is the immediate surroundings of the system of interest, which we shall usually take to be an infinitely large thermal reservoir (for example, a water bath).

THE DIRECTION OF SPONTANEOUS CHANGE

What determines the direction of spontaneous change? It is not the total energy of the isolated system. The First Law of thermodynamics states that energy is conserved in any process, and we cannot disregard that law now and say that everything tends towards a state of lower energy: the total energy of an isolated system is constant.

Is it perhaps the energy of the system of interest that tends towards a minimum? Two arguments show that this cannot be so. First, a perfect gas expands spontaneously into a vacuum, yet its internal energy remains constant as it does so. Secondly, if the energy of a system does happen to decrease during a spontaneous change, then the energy of its surroundings must increase by the same amount (by the First Law). The increase in energy of the surroundings is just as spontaneous a process as the decrease in energy of the system.

When a change occurs, the total energy of an isolated system remains constant but it is parcelled out in different ways. Can it be, therefore, that the direction of change is related to the *distribution* of energy? We shall see that this idea is the key and that *spontaneous changes are always accompanied by a dispersal of energy into a more disordered form.*

4.1 The dispersal of energy

The role of the distribution of energy can be illustrated by thinking about a ball (the system of interest) bouncing on a floor (the surroundings).

4.2 The global isolated system that we consider in this chapter, and its division into the system of interest and that system's surroundings.

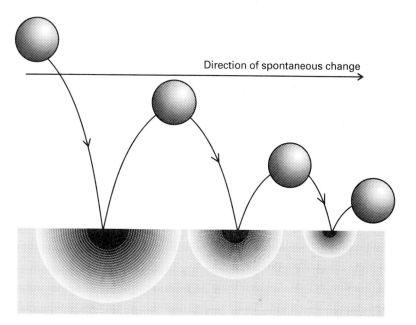

Direction of spontaneous change

4.3 The direction of spontaneous change for a ball bouncing on a floor. On each bounce some of its energy is degraded into the thermal motion of the atoms of the floor, and that energy disperses. The reverse has never been observed.

The ball does not rise as high after each bounce because there are inelastic losses in the materials of the ball and floor (that is, the conversion of kinetic energy of the ball's overall motion into the energy of thermal motion). The direction of spontaneous change is towards a state in which the ball is at rest with all its energy degraded into the thermal motion of the atoms of the virtually infinite floor (Fig. 4.3).

A ball resting on a warm floor has never been observed to start bouncing. For bouncing to begin, something rather special would need to happen. In the first place, some of the thermal motion of the atoms in the floor would have to accumulate in a single, small object, the ball. This accumulation requires a spontaneous localization of energy from the myriad of vibrations of the atoms of the floor into the much smaller number of atoms that constitute the ball (Fig. 4.4). Furthermore, whereas the thermal motion is disorderly, for the ball to move upwards

4.4 The molecular interpretation of the irreversibility expressed by the Second Law. (a) A ball resting on a warm surface; the atoms are undergoing thermal motion (chaotic vibration, in this instance) as indicated by the arrows. (b) For the ball to fly upwards, some of the random vibrational motion would have to change into coordinated, directed motion. Such a conversion is highly improbable.

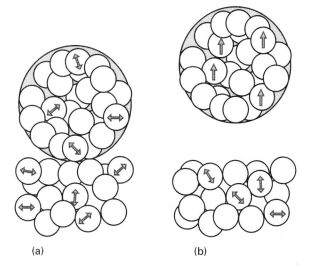

(a)   (b)

its atoms must all move in the same direction. The localization of *random* motion as *orderly* motion is so unlikely that we can dismiss it as virtually impossible.

We have found the signpost of spontaneous change: *we look for the direction of change that leads to the greater chaotic dispersal of the total energy of the isolated system.* This principle accounts for the direction of change of the bouncing ball, because its energy is dissipated as thermal motion of the atoms of the floor. The reverse process is not spontaneous because it is highly improbable that the chaotic distribution of energy will become organized into localized, uniform motion. A gas does not spontaneously contract, because to do so the chaotic motion of its molecules would have to take them all into the same region of the container; the opposite change, spontaneous expansion, is a natural consequence of increasing chaos. An object does not spontaneously become warmer than its surroundings because it is highly improbable that the jostling of randomly vibrating atoms in the surroundings will lead to the accumulation of excess thermal motion in the object. The opposite change, the spreading of the object's energy into the surroundings as thermal motion, is a natural consequence of chaos.

It may seem very puzzling that collapse into disorder can account for the formation of such ordered substances as crystals as a result of a chemical reaction or the evaporation of a solvent, or the formation of the highly ordered macromolecular structures characteristic of proteins. Nevertheless, in due course we shall see that highly ordered structures can emerge as energy and matter disperse in chaos and that collapse into disorder accounts for change in all its forms.

4.2 Entropy

The First Law of thermodynamics led to the introduction of the internal energy U. The internal energy is a state function that lets us assess whether a change is permissible: only those changes may occur for which the internal energy of an isolated system remains constant. The law that is used to identify the signpost of spontaneous change, the Second Law of thermodynamics, may also be expressed in terms of another state function, the **entropy** S. We shall see that the entropy (which we shall define shortly) lets us assess whether one state is accessible from another by a spontaneous change. The First Law used the internal energy to identify *permissible* changes (those that conserve energy); the Second Law uses the entropy to identify the *spontaneous* changes among those permissible changes:

> **Second Law**: The entropy of an isolated system increases in the course of a spontaneous change:
>
> $$\Delta S_{tot} > 0 \tag{1}$$
>
> where S_{tot} is the total entropy of the isolated system that contains the system of interest.

Thermodynamically irreversible processes (like cooling to the temperature of the surroundings and the free expansion of gases) are spontaneous processes, and hence must be accompanied by an increase in entropy. We can express this conclusion by saying that *irreversible processes generate entropy*. On the other hand, reversible processes are finely balanced changes in which the system is in equilibrium with its surroundings at every stage. Each infinitesimal step along a reversible path occurs without dispersing energy chaotically and hence without increasing the entropy: *reversible processes do not generate entropy*. At most, reversible processes transfer entropy from one part of an isolated system to another.

MOLECULAR INTERPRETATION 4.1

The definition of entropy in terms of the properties of atoms and molecules will be presented in detail in Chapter 19, but it is helpful to be aware of it from the outset because it helps us to visualize the thermodynamic definition. This **statistical definition** of entropy allows us to calculate the degree of disorder in a system by using a formula proposed by Ludwig Boltzmann in 1896:

$$S = k \ln W$$

where k is Boltzmann's constant: $k = 1.381 \times 10^{-23} \, \text{J K}^{-1}$ (with $R = N_A k$, Section 1.3). The quantity W is the number of different ways in which the energy of the system can be achieved by rearranging the atoms or molecules among the states available to them.

As an illustration of how Boltzmann's equation may be used, in a solid composed of N HCl molecules at $T = 0$, the only state available to each molecule is the state of lowest energy (see Fig. 4.5a). Then $W = 1$ and $S = 0$ (because $\ln 1 = 0$). This perfectly ordered system has zero entropy.

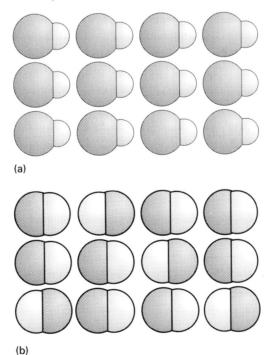

(a)

(b)

4.5 (a) According to the Boltzmann formula, the entropy of a perfectly ordered solid (such as solid HCl, shown here schematically) is zero since $W = 1$. (b) The entropy of a disorderly solid (such as solid CO) in which each diatomic molecule can adopt either of two orientations at $T = 0$ is greater than 0 (specifically $Nk \ln 2$).

Now consider a sample of solid carbon monoxide that consists of N CO molecules. A special feature of solid carbon monoxide is that, largely because its molecular dipole moment is so small, neighbouring molecules have virtually the same energy irrespective of whether they lie head to head or head to tail. It follows that even at $T = 0$ each CO molecule can adopt either of *two* orientations with equal probability (Fig. 4.5b). The crystal of CO has a greater disorder than a crystal of HCl, because we cannot predict in advance whether a given molecule will be in one orientation or the other, so we expect its entropy to be greater than zero. Because each molecule can lie in two orientations in the crystal, the total number of states (orientations) accessible to the solid is

$$W = 2 \times 2 \times \ldots = 2^N$$

The entropy of a sample of 1.00 mol CO (so $N = 6.02 \times 10^{23}$) is therefore

$$S = k \ln 2^N = Nk \ln 2 = (6.02 \times 10^{23}) \times (1.381 \times 10^{-23} \, \text{J K}^{-1}) \times \ln 2$$
$$= 5.76 \, \text{J K}^{-1}$$

Notice that, unlike U and H, the entropy has an absolute value; we shall see more of this feature later (Section 4.5).

The entropy of solid HCl increases as the temperature is increased, because then the molecules have enough energy to be found in different relative orientations in the crystal. Indeed, the molar entropy of solid HCl just before its melting point at 159 K is 64 J K^{-1} mol^{-1}. Once the solid melts, many more states become accessible to the molecules because they can now move from place to place as well as change their relative orientations, and the entropy of liquid hydrogen chloride at 159 K is 77 J K^{-1} mol^{-1}.

The thermodynamic definition of entropy

The thermodynamic definition of entropy concentrates on the change in entropy dS that occurs as a result of a physical or chemical change (in general, as a result of a 'process'). The definition is motivated by the idea that a change in the extent to which energy is dispersed chaotically can be derived by noting the quantity of energy that is transferred as heat during a process. As we have remarked, the transfer of energy as heat makes use of the chaotic motion of atoms in the surroundings. The transfer of energy as work, which makes use of the *uniform* motion of atoms in the surroundings, does not change their degree of disorder and so does not change their entropy.

We begin by defining the entropy change in the surroundings dS'. (In this chapter, the prime will usually denote the surroundings, and the absence of a prime will imply the system; so $dS_{\text{tot}} = dS + dS'$.) We represent the surroundings by a large thermal reservoir (a water bath in practice) that remains at the temperature T'. Suppose a falling weight is coupled to the reservoir (for example, by driving a generator connected to a heater, as in Fig. 4.6) and that, when the weight falls, a quantity of

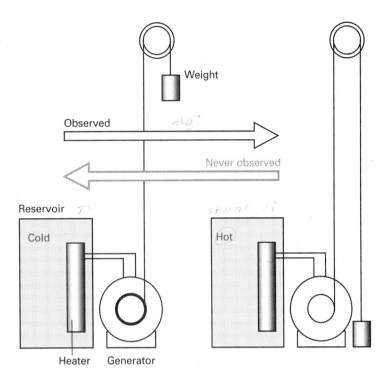

4.6 The fundamental spontaneous process is represented by a falling weight and a thermal reservoir. The potential energy of the weight is transferred to the thermal motion of the reservoir, never vice versa spontaneously. Note that a reservoir is a sink of infinite extent, and its temperature remains constant however much energy is transferred as heat.

heat dq' is transferred to the reservoir. The greater the quantity of heat transferred to the reservoir, the greater the thermal motion that is stimulated in it, and hence the greater the dispersal of energy that occurs. This argument suggests that we write

$$dS' \propto dq'$$

Because energy has a spontaneous tendency to flow as heat from a hot body to a cold body, it follows that a given quantity of energy stored at high temperature has a lower entropy than the same quantity of energy stored at a lower temperature. Therefore, if an energy dq' is transferred to a body at a low temperature, then the change in entropy will be larger than if it had been transferred at a high temperature. The simplest way of taking this dependence on temperature into account is to write

$$dS' = \frac{dq'}{T'} \tag{2a}$$

where T' is the temperature of the surroundings. For a finite change (with the reservoir at a constant temperature)

$$\Delta S' = \frac{q'}{T'} \tag{2b}$$

Large changes in entropy occur when a lot of thermal motion is generated at low temperature.

MOLECULAR INTERPRETATION 4.2

The molecules in a system at high temperature are highly disorganized, either in terms of their locations or in terms of the occupation of their available translational, rotational, and vibrational energy states. A small additional transfer of energy will result in a relatively small additional disorder, much as sneezing in a busy street may be barely noticed. In contrast, the molecules in a system at low temperature have access to far fewer energy states (at $T = 0$, only the lowest state is accessible), and the transfer of the same quantity of energy as heat will have a pronounced effect on the degree of disorder, much as sneezing in a quiet library can be very disruptive. Hence, the change in entropy when a given quantity of heat is transferred will be greater when it is transferred to a cold body than when it is transferred to a hot body. This argument suggests that the change in entropy should be inversely proportional to the temperature at which the transfer takes place, as in eqn 2.

Note that, according to the expression for $\Delta S'$, when the heat transferred is expressed in joules and the temperature is in kelvins, the units of entropy are joules per kelvin ($J\,K^{-1}$), in accord with the statistical definition in *Molecular interpretation 4.1*. Molar entropies, entropies per unit amount of material, are expressed in joules per kelvin per mole ($J\,K^{-1}\,mol^{-1}$), the same units as those of the gas constant R and molar heat capacities.

Equation 2 makes it very simple to calculate the changes in entropy of the surroundings that accompany any process. For instance, for any adiabatic change,

$$\Delta S' = 0 \qquad \text{when } q' = 0$$

This expression is true however the change takes place, reversibly or irreversibly, so long as no local hot spots are formed in the surroundings. That is, it is true so long as the surroundings remain in internal equilibrium. If hot spots do form, then the localized energy may subsequently disperse spontaneously and hence generate entropy.

When a chemical reaction takes place in a system in thermal equilibrium with its surroundings (so that $T' = T$), with enthalpy change ΔH, the heat that enters the surroundings at constant pressure is $q' = -\Delta H$. It follows that the entropy change of the surroundings is

$$\Delta S' = -\frac{\Delta H}{T} \tag{3}$$

A strongly exothermic reaction (for which $\Delta H < 0$) generates a large amount of entropy in the surroundings, particularly if it occurs at low temperature. An endothermic reaction *reduces* the entropy of the surroundings as energy flows out of them and enters the system. We shall see later that the relation between entropy changes in the surroundings and the enthalpy of reaction plays a vital role in determining the direction of spontaneous chemical changes.

Example 4.1 *Calculating the entropy change in the surroundings*

Calculate the entropy change in the surroundings when $1.00\,mol\,H_2O(l)$ is formed from its elements under standard conditions at $298.15\,K$.

Method. To calculate the entropy change *in the surroundings* from eqn 3 we need to know the enthalpy change accompanying the reaction. That is readily obtained from the standard enthalpy of formation of water, as explained in Section 2.6; data are in Table 2.12 in the Data section at the end of this volume.

Answer. The reaction is

$$H_2(g) + \tfrac{1}{2}O_2(g) \rightarrow H_2O(l)$$ $\Delta S\degree$ $69.91 - 130.684 - \frac{205.138}{2} =$

The change in enthalpy when 1.00 mol H_2O is formed is

$$\Delta H^{\ominus} = (1.00\ \text{mol}) \times \Delta_f H^{\ominus}(H_2O, l) = -286\ \text{kJ}$$

Therefore,

$$\Delta S' = -\frac{(-286 \times 10^3\ \text{J})}{298.15\ \text{K}} = +959\ \text{J K}^{-1}$$

Comment. This strongly exothermic reaction results in an increase in the entropy of the surroundings as heat is released into them. We shall see later that the entropy of the system undergoes a considerable decrease as a compact liquid is formed from two gases.

Exercise E4.1. Calculate the entropy change in the surroundings when 1.00 mol $N_2O_4(g)$ is formed from 2.00 mol $NO_2(g)$ under standard conditions at 25°C. $[-192\ \text{J K}^{-1}]$

The entropy change in the system

Now we adapt the definition of the entropy change in the surroundings to find an expression for the entropy change of the system itself. The strategy we adopt is to use the surroundings to restore the system to its initial state *reversibly*, with no further generation of entropy. Then we inspect the surroundings to see how much entropy has been transferred in the process. We shall set the temperature of the surroundings equal to that of the system (that is, $T' = T$), so that they are in thermal equilibrium and the transfer of heat is reversible.

Let the original change in the entropy of the system when the process of interest occurs be dS (this is the change we want to measure). The process need not be reversible, but we suppose that we can find a path that joins the same initial and final states and which is reversible. For example, the change may be the isothermal, irreversible expansion of a gas through a volume dV, in which case the reversible path will be the reversible isothermal expansion of the gas through the same change in volume. The same change of entropy of the system is obtained in each case because (as we prove later) S is a state function, but the energy absorbed as heat, dq in general and dq_{rev} for the reversible path, might be different.

Now suppose that the system is restored reversibly to its initial state. Its entropy changes by $-dS$ (because the entropy is a state function, and its value must return to what it was originally if the initial state is restored). The energy we must supply as heat is also the negative of the change in the forward step, and hence is equal to $-dq_{rev}$. Because this energy comes from the surroundings, they undergo a change of energy $dq' = dq_{rev}$ so their entropy changes by $dS' = dq_{rev}/T$. However, the total entropy change of the global, isolated system during the restoration is zero (because it is carried out reversibly). Therefore:

$$-dS + \frac{dq_{rev}}{T} = 0$$

from which it follows that

$$dS = \frac{dq_{rev}}{T} \tag{4a}$$

For a measurable change, the entropy change is the sum (integral) of the infinitesimal changes:

$$\Delta S = \int_i^f \frac{dq_{rev}}{T} \tag{4b}$$

That is, the entropy change of a system when it changes between two specified states can be determined by finding the heat necessary to take it along a *reversible* path between the same two states.

Example 4.2 *Calculating the entropy change during the isothermal expansion of a perfect gas*

Calculate the entropy change of a sample of perfect gas when it expands isothermally from a volume V_i to a volume V_f.

Method. Because entropy is a state function, we can use eqn 4b to calculate the change in entropy even though we are not told whether the expansion is reversible or irreversible: the change in entropy of the system is the same in either case. The equation instructs us to find the heat absorbed for a *reversible* path between the stated initial and final states. A simplification is that the expansion is isothermal; so, once we know the heat supplied during the process, we can divide it by the constant temperature at which the heat transfer takes place. Energy transfers during reversible expansion were calculated in Section 2.3, particularly eqn 2.12.

Answer. Because the temperature is constant,

$$\Delta S = \frac{1}{T} \int_i^f dq_{rev} = \frac{q_{rev}}{T}$$

The calculations in Section 2.3 led to the result that, for the isothermal reversible expansion of a gas,

$$\Delta U = 0, \qquad q_{rev} = -w_{rev}, \qquad q_{rev} = nRT \ln\left(\frac{V_f}{V_i}\right)$$

Therefore, the entropy change accompanying this change of state is

$$\Delta S = nR \ln\left(\frac{V_f}{V_i}\right)$$

Comment. As an illustration of this formula, when 1.00 mol of any perfect gas doubles its volume at any temperature

$$\Delta S = (1.00 \text{ mol}) \times (8.3145 \text{ J K}^{-1} \text{ mol}^{-1}) \times \ln 2 = +5.76 \text{ J K}^{-1}$$

Exercise E4.2. Calculate the change in entropy when the pressure of a perfect gas is changed isothermally from p_i to p_f. $[\Delta S = nR \ln (p_i/p_f)]$

MOLECULAR INTERPRETATION 4.3

The calculation in Example 4.2 shows that the entropy of a perfect gas changes by

$$\Delta S = nR \ln \left(\frac{V_f}{V_i}\right)$$

when it expands isothermally from V_i to V_f. The increase in the entropy of a gas with the volume of the sample stems from the greater number of states that become available to the gas molecules as the volume is increased.

A very simple approach to the calculation of the entropy change accompanying the isothermal expansion of a gas is to suppose that a molecule occupies a small volume v and, therefore, that the number of places that the molecule may be in if the volume of the container is V is V/v. Because there are N molecules in the sample, the total number of ways of distributing the molecules is $W = (V/v)^N$. Therefore, when the volume of the system is changed from V_i to V_f, the change in entropy is

$$\Delta S = k \ln \left(\frac{V_f}{v}\right)^N - k \ln \left(\frac{V_i}{v}\right)^N$$

$$= Nk \ln \left(\frac{V_f}{v}\right) - Nk \ln \left(\frac{V_i}{v}\right) = Nk \ln \left(\frac{V_f}{V_i}\right)$$

This expression is exactly the same as that derived thermodynamically in Example 4.2 because $N = nN_A$ and $R = N_A k$. In accord with the conclusion in Example 4.2 we see that, when the volume is doubled isothermally, the change in entropy is $Nk \ln 2$.

The entropy as a state function

The entropy of a system is a state function. To prove this assertion, we need to show that the integral of dS as expressed in eqn 4b is independent of path. To do so, it is sufficient to prove that the integral of eqn 4a around an arbitrary cycle is zero, for that guarantees that the entropy is the same at the initial and final states of the system regardless of the path taken between them (Fig. 4.7). That is, we need to show that

$$\oint \frac{dq_{rev}}{T} = 0 \tag{5}$$

where the symbol $\oint$ denotes integration around a closed path.

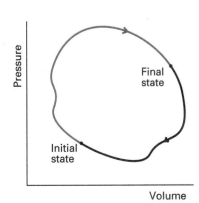

4.7 In a thermodynamic cycle, the overall change in a state function (from the initial state to the final state and then back to the initial state again) is zero.

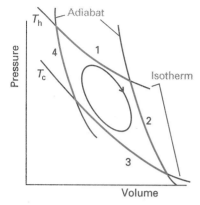

4.8 The basic structure of a Carnot cycle. In step 1, there is isothermal reversible expansion at the temperature T_h. Step 2 is a reversible adiabatic expansion in which the temperature falls from T_h to T_c. In step 3 there is an isothermal reversible compression at T_c, and that is followed by an adiabatic reversible compression, which restores the system to its initial state.

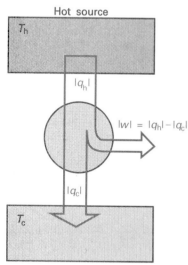

4.9 The thermodynamic basis of heat engines. When a quantity of energy $|q_h|$ is withdrawn from a hot reservoir the entropy decreases by $|q_h|/T_h$, and when $|q_c|$ is added to a cold reservoir its entropy is increased by $|q_c|/T_c$. So long as $T_c < T_h$, the overall entropy change may still be positive even if less energy is returned to the cold reservoir than was removed from the hot. The difference $|q_h| - |q_c|$ may be withdrawn as work, the overall process being spontaneous.

To prove eqn 5 we first consider the special **Carnot cycle** shown in Fig. 4.8. A Carnot cycle, which is named after the French engineer Sadi Carnot, consists of four reversible stages:

Stage 1. Reversible isothermal expansion at T_h; the entropy change of the system is $+|q_h|/T_h$, where $|q_h|$ is the heat taken from the hot source.

Stage 2. Reversible adiabatic expansion. No heat leaves the system, so the change in its entropy is zero. In the course of this expansion, the temperature falls from T_h to T_c, the temperature of the cold sink.

Stage 3. Reversible isothermal compression at T_c. The heat $|q_c|$, is released to the cold sink, so the change in entropy of the system is $-|q_c|/T_c$.

Stage 4. Reversible adiabatic compression. No heat enters the system, so the change in entropy is zero. The temperature rises from T_c to T_h.

The total change in entropy around the cycle is

$$\oint dS = \frac{|q_h|}{T_h} - \frac{|q_c|}{T_c} \tag{6}$$

We shall now build an argument to prove that the right-hand side of this expression is zero. First, we shall show that all reversible engines have the same efficiency whatever their working substance: this step will provide a relation between q_h and q_c. Then we shall calculate the efficiency of an engine in which the working material is a perfect gas, which will provide a relation between the heat transactions and the temperatures of the hot and cold reservoirs.

By the **efficiency** ε of an engine we shall mean

$$\varepsilon = \frac{\text{work performed}}{\text{heat absorbed}} = \frac{|w|}{|q_h|} \tag{7}$$

The definition implies that, the greater the work output for a given supply of heat from the hot reservoir, then the greater the efficiency of the engine. The definition can be expressed in terms of the heat transactions alone, because (as shown in Fig. 4.9) the work supplied by the engine is the difference between the heat supplied by the hot reservoir and returned to the cold reservoir:

$$\varepsilon = \frac{|q_h| - |q_c|}{|q_h|} = 1 - \frac{|q_c|}{|q_h|} \tag{8}$$

Although the remainder of the discussion will consider a perfect gas as the working substance in the Carnot cycle, it follows from the Second Law of thermodynamics (see the *Justification* below) that *all reversible engines have the same efficiency regardless of their construction*. Therefore, the conclusions we draw are relevant to a system of any composition.

JUSTIFICATION

Suppose that two reversible engines are coupled together and run between the same two reservoirs (Fig. 4.10). The working substances

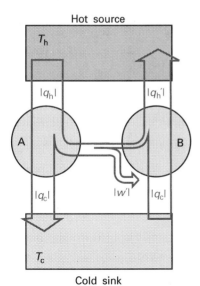

4.10 The demonstration of the equivalence of the efficiencies of all reversible engines working between the same thermal reservoirs is based on the flow of energy represented in this diagram (see text).

and details of construction of the two engines are entirely arbitrary, the only requirement being that they are reversible. Initially we suppose that engine A is more efficient than engine B, so we can choose a setting of its controls that causes engine B to withdraw the heat $|q_c|$ from the cold reservoir and to deposit a certain quantity of heat into the hot reservoir. However, because engine A is more efficient than engine B, not all the work it produces is needed for this process, and the difference can be used for other purposes. The net result is that the cold reservoir is unchanged, work has been produced, and the hot reservoir has lost a certain amount of energy. This outcome is consistent with the First Law of thermodynamics (no energy has been created or destroyed; some heat has been converted into work), but it is contrary to the Kelvin statement of the Second Law, because some heat has been converted directly into work. (In molecular terms, the disordered thermal motion of the hot reservoir has been converted into ordered motion characteristic of work.) Because the conclusion is contrary to experience, the initial assumption that engines A and B can have different efficiencies must be false. Therefore, the Second Law implies that all engines working reversibly between reservoirs of the same temperatures have the same efficiencies.

Now we shall show that the efficiency of a reversible engine in which the working substance is a perfect gas is given by the expression

$$\varepsilon_{rev} = 1 - \frac{T_c}{T_h} \qquad (9)_r$$

This expression is the **Carnot efficiency** of a heat engine. Although it will be derived for a perfect gas, because all reversible engines have the same efficiency it applies to any substance so long as the engine is operating reversibly.

JUSTIFICATION

The work done and the heat transferred in each stage of a Carnot cycle in which the working substance is a perfect gas can be taken from Table 3.3, and the resulting expressions are given in Fig. 4.11. The work done on the system on going round the cycle is the sum of the work done in each stage:

$$w = -nRT_h \ln\left(\frac{V_B}{V_A}\right) + C_V(T_c - T_h) - nRT_c \ln\left(\frac{V_D}{V_C}\right) + C_V(T_h - T_c)$$

$$= -nRT_h \ln\left(\frac{V_B}{V_A}\right) - nRT_c \ln\left(\frac{V_D}{V_C}\right)$$

(The adiabatic stages cancel.) The energy supplied to the system as heat by the hot source is

$$q_h = nRT_h \ln\left(\frac{V_B}{V_A}\right)$$

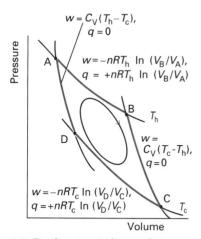

4.11 The Carnot cycle for a perfect gas. All stages are thermodynamically reversible. A to B is an isothermal expansion at a temperature T_h, B to C an adiabatic expansion that lowers the temperature to T_c. C to D is an isothermal compression at T_c. D to A completes the cycle with an adiabatic compression that raises the temperature to T_h again. The work done at each stage and the heat absorbed are shown.

It is a straightforward piece of algebra to simplify the resulting expression for $|w|/|q_h|$ by using the relations between temperature and volume for reversible adiabatic processes (Section 3.5):

$$V_A T_h^c = V_D T_c^c \qquad \text{and} \qquad V_C T_c^c = V_B T_h^c$$

Multiplication of the first expression by the second gives

$$V_A V_C T_h^c T_c^c = V_B V_D T_h^c T_c^c$$

which simplifies to

$$\frac{V_A}{V_B} = \frac{V_D}{V_C}$$

Consequently, the work done on the system is

$$w = -nRT_h \ln\left(\frac{V_B}{V_A}\right) - nRT_c \ln\left(\frac{V_A}{V_B}\right) = -nR(T_h - T_c) \ln\left(\frac{V_B}{V_A}\right)$$

Division of this expression by the expression for q_h then results in

$$\varepsilon_{rev} = \left(\frac{|w|}{|q_h|}\right)_{rev} = \frac{T_h - T_c}{T_h}$$

which rearranges to eqn 9.

It now follows, by equating eqns 8 and 9, that

$$\frac{|q_c|}{|q_h|} = \frac{T_c}{T_h}, \quad \text{or} \quad \frac{|q_c|}{T_c} = \frac{|q_h|}{T_h}$$

Then substitution of this relation into eqn 6 gives

$$\oint dS = 0$$

as we set out to prove.

Finally, we note that any reversible cycle can be approximated as a collection of Carnot cycles (Fig. 4.12). This approximation becomes exact as the individual cycles are allowed to become infinitesimal. The entropy change around each individual cycle is zero (as demonstrated above), so the sum of entropy changes for all the cycles is zero. However, in the interior of the overall cycle, the entropy change along any path is cancelled by the entropy change along the path it shares with the neighbouring cycle. Therefore, all the entropy changes cancel except for those along the perimeter of the overall cycle. That is,

$$\sum_{all} \frac{q_{rev}}{T} = \sum_{perimeter} \frac{q_{rev}}{T} = 0$$

In the limit of infinitesimal cycles, the non-cancelling edges of the Carnot cycles match the overall cycle exactly, and the sum becomes an integral. It then follows that

$$\oint \frac{dq_{rev}}{T} = 0$$

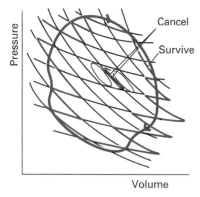

4.12 A general cycle can be divided into Carnot cycles. The match is exact in the limit of infinitesimally small cycles. Paths cancel in the interior of the collection, and only the perimeter, an increasingly good approximation to the true cycle as the number of cycles increases, survives. Because the entropy change around every individual cycle is zero, the integral of the entropy around the perimeter is zero as well.

This is the result we wanted to prove, for it shows that dS is an exact differential and therefore that S is a state function.

4.3 The entropy of irreversible change

Consider a system in thermal and mechanical contact with its surroundings at the same temperature. The system and the surroundings are not necessarily in mechanical equilibrium (for instance, a gas might have a greater pressure than its surroundings). Any change of state is accompanied by a change in entropy dS of the system and dS' of the surroundings. In general, the total entropy of the system and its surroundings will increase when a process occurs in the system because the process might be irreversible:

$$dS + dS' \geq 0, \qquad \text{or } dS \geq -dS'$$

(The equality applies if the process is reversible.) Because $dS' = -dq/T$, where dq is the heat supplied to the system during the process, it follows that for any change

$$dS \geq \frac{dq}{T} \tag{10}$$

This expression is the **Clausius inequality**. We shall now use this result to show that the entropy does indeed increase for two of the cases mentioned in the introduction to this chapter, the free expansion of a perfect gas and the cooling of a hot substance.

Spontaneous expansion

First, suppose a system undergoes an irreversible adiabatic change. Then $dq = 0$ and, by the Clausius inequality, $dS > 0$. That is, for this type of spontaneous change the entropy of the system has increased. (We consider the entropy of the surroundings shortly.) Now consider irreversible isothermal expansion. We saw in Chapter 3 that, when a perfect gas expands isothermally, its internal energy remains constant. Therefore, according to the First Law,

$$dU = dq + dw = 0, \qquad \text{so } dq = -dw$$

If the gas expands freely into a vacuum, it does no work and $dw = 0$, which implies that $dq = 0$ too. Therefore, according to the Clausius inequality, $dS > 0$.

Now we consider the surroundings. In both cases $dq = 0$, and no heat is transferred into the surroundings. Because eqn 2 gives us the change in entropy of the surroundings however the change comes about (so long as the surroundings remain in internal equilibrium), for both types of change $dS' = 0$. The overall entropy change is the sum of the changes in the system and its surroundings. Because $dS > 0$ and $dS' = 0$ for both, we conclude that for irreversible adiabatic expansion and for free isothermal expansion of a perfect gas $dS_{tot} > 0$. Hence, the processes are spontaneous.

Spontaneous cooling

Consider a transfer of energy as heat dq from one system—the hot source—at a temperature T_h to another system—the cold sink—at a

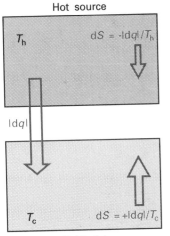

Hot source

T_h $dS = -|dq|/T_h$

$|dq|$

T_c $dS = +|dq|/T_c$

Cold sink

4.13 When energy leaves a hot reservoir as heat, the entropy of the reservoir decreases. When the same quantity of energy enters a cooler reservoir, the entropy increases by a larger amount. Hence, overall there is an increase in entropy and the process is spontaneous. Relative changes in entropy are indicated by the sizes of the arrows.

temperature T_c (Fig. 4.13). When $|dq|$ leaves the hot source, the entropy of the source changes by $-|dq|/T_h$ (a decrease). When $|dq|$ enters the cold sink its entropy changes by $+|dq|/T_c$ (an increase). The overall change in entropy is therefore

$$dS_{tot} = \frac{|dq|}{T_c} - \frac{|dq|}{T_h} = \left(\frac{1}{T_c} - \frac{1}{T_h}\right) \times |dq|$$

which is positive (because $T_h \geq T_c$). Hence, cooling (the transfer of heat from hot to cold) is spontaneous, as we know from experience. When the temperatures of the two systems are equal, $dS_{tot} = 0$: the two systems are then at thermal equilibrium.

4.4 Entropy changes accompanying specific processes

We shall now see how to calculate the entropy change that occurs as a result of a variety of simple processes.

The entropy of phase transition at the transition temperature

Because a change in the degree of molecular order occurs when a substance freezes or boils, we should expect the transition to be accompanied by a change in entropy. For example, when a substance vaporizes, a compact condensed phase changes into a widely dispersed gas, and we can expect the entropy of the substance to increase considerably. The entropy of a solid substance increases when it melts to a liquid, and it also increases when the liquid phase turns into a gas.

Consider a system and its surroundings at the transition temperature T_t, the temperature at which two phases are in equilibrium at a pressure of 1 atm. This temperature is 0°C (273 K) for ice in equilibrium with liquid water at 1 atm, and 100°C (373 K) for water in equilibrium with its vapour at 1 atm. At the transition temperature any transfer of heat between the system and its surroundings is reversible because the two phases in the system are in equilibrium. Because at constant pressure $q = \Delta_{trs}H$, the change in entropy of the system is

$$\Delta S = \frac{\Delta_{trs}H}{T_t} \tag{11}$$

If the phase transition is exothermic ($\Delta_{trs}H < 0$, as in freezing or condensing), then the entropy change is negative. This decrease in entropy is consistent with the system becoming more ordered when a solid forms from a liquid. If the transition is endothermic ($\Delta_{trs}H > 0$, as in melting), then the entropy change is positive, which is consistent with the system becoming more disordered. Melting and vaporizing are endothermic processes, so both are accompanied by an increase in the system's entropy. This increase is consistent with liquids being more disordered than solids, and gases more disordered than liquids. Some experimental molar entropies of transition are listed in Table 4.1.

In Table 4.2 we list in more detail the standard molar enthalpies of vaporization of several liquids at their boiling points. An interesting feature of the data is that a wide range of liquids give approximately the same standard molar entropy of vaporization (about $85 \, J \, K^{-1} \, mol^{-1}$): this empirical observation is called **Trouton's rule**.

Table 4.1* Standard entropies (and temperatures) of phase transitions at 1 atm, $\Delta_{trs}S^{\ominus}/(J\,K^{-1}\,mol^{-1})$

	Fusion (at T_f)	Vaporization (at T_b)
Argon, Ar	14.2 (at 83.8 K)	74.5 (at 87.3 K)
Benzene, C_6H_6	38.0 (at 279 K)	87.2 (at 353 K)
Water, H_2O	22.0 (at 273.15 K)	109.0 (at 373.15 K)
Helium, He	4.8 (at 1.8 K and 30 bar)	19.9 (at 4.22 K)

* More values are given in the Data section at the end of this volume.

Table 4.2* The standard molar entropies of vaporization of liquids

	$\Delta_{vap}H^{\ominus}/(kJ\,mol^{-1})$	$\theta_b/°C$	$\Delta_{vap}S^{\ominus}/(J\,K^{-1}\,mol^{-1})$
Benzene	+30.8	80.1	+87.2
Carbon tetrachloride	+30.00	76.1	+85.9
Cyclohexane	+30.1	80.7	+85.1
Hydrogen sulfide	+18.7	−60.4	+87.9
Methane	+8.18	−161.5	+73.2
Water	+40.7	100.0	+109.1

* More values are given in the Data section.

MOLECULAR INTERPRETATION 4.4

The explanation of Trouton's rule is that a comparable amount of disorder is generated when any liquid evaporates and becomes a gas. Liquids that show significant deviations from Trouton's rule do so on account of the molecules in the liquid being arranged in a partially orderly manner. In such cases, a greater change of disorder occurs when the liquid evaporates. An example is water, where the large entropy of vaporization reflects the presence of structure arising from hydrogen bonding in the liquid. Hydrogen bonds tend to organize the molecules in the liquid so that they are less random than, for example, the molecules in liquid hydrogen sulfide (which is not hydrogen bonded).

Methane has an unusually low standard entropy of vaporization. A part of the reason is that the entropy of the gas itself is slightly low ($186\,J\,K^{-1}\,mol^{-1}$; the entropy of N_2 under the same conditions is $192\,J\,K^{-1}\,mol^{-1}$). As we shall see in Chapter 19, molecules with low moments of inertia are difficult to excite into rotation; as a result, only a few rotational states are accessible at room temperature, and the disorder associated with the population of rotational states is low. Hydrogen gas has an even lower entropy than CH_4 ($131\,J\,K^{-1}\,mol^{-1}$) because its rotational disorder is low, and also because relatively few translational states are available as hydrogen molecules are so light.

Example 4.3 *Using Trouton's rule*

Predict the standard molar enthalpy of vaporization of bromine given that it boils at 59.2°C.

Method. We need to judge whether there is the likelihood of anomalous structural organization in the liquid phase or some anomaly in the gas phase. If there is not, it is permissible to use Trouton's rule in the form

$$\Delta_{vap}H^{\ominus} = T_b \times 85 \, J \, K^{-1} \, mol^{-1}$$

Answer. There is no hydrogen bonding in liquid bromine and Br_2 is a heavy molecule that is unlikely to display unusual behaviour in the gas phase, so it would seem safe to use Trouton's rule. Substitution of the data then gives

$$\Delta_{vap}H^{\ominus} = (332.4 \, K) \times 85 \, J \, K^{-1} \, mol^{-1} = +28 \, kJ \, mol^{-1}$$

Comment. The experimental value is $+29.45 \, kJ \, mol^{-1}$.

Exercise E4.3. Predict the enthalpy of vaporization of ethane from its boiling point, $-88.6°C$. $[+16 \, kJ \, mol^{-1}]$

The expansion of a perfect gas

We established in Example 4.2 that the change in entropy of a perfect gas that expands isothermally from V_i to V_f is

$$\Delta S = nR \ln\left(\frac{V_f}{V_i}\right) \tag{12}°$$

Because S is a state function and independent of the path, this expression applies whether the change of state occurs reversibly or irreversibly. If the change is reversible, then the entropy change in the surroundings (which are in thermal and mechanical equilibrium with the system) must be such as to give $\Delta S_{tot} = 0$. Therefore, in this case, the change in entropy of the surroundings is

$$\Delta S' = -nR \ln\left(\frac{V_f}{V_i}\right)$$

However, if the expansion occurs freely and irreversibly, if no work is done $(w = 0)$, and if the temperature remains constant (implying $\Delta U = 0$), then there is no energy transferred between the system and its surroundings as heat $(q = 0)$. Consequently, the entropy of the surroundings does not change. The entropy of the system changes by the same amount as before (eqn 12, because entropy is a state function), and so the total change in entropy is

$$\Delta S_{tot} = nR \ln\left(\frac{V_f}{V_i}\right)$$

The variation of entropy with temperature

Equation 4b can be used to calculate the entropy of a system at a temperature T_f from a knowledge of its entropy at a temperature T_i and the heat supplied to change its temperature from one value to the other:

$$S(T_f) = S(T_i) + \int_i^f \frac{dq_{rev}}{T}$$

We shall be particularly interested in the entropy change when the system is subjected to constant pressure (such as from the atmosphere) during the heating. Then from the definition of heat capacity (Section 2.4),

$$dq_{rev} = C_p \, dT$$

so long as the system is doing no non-expansion work. Consequently, at constant pressure,

$$S(T_f) = S(T_i) + \int_{T_i}^{T_f} \frac{C_p \, dT}{T} \tag{13a}$$

Similarly, at constant volume,

$$S(T_f) = S(T_i) + \int_{T_i}^{T_f} \frac{C_V \, dT}{T} \tag{13b}$$

These expressions let us find the entropy of a substance at any temperature provided the heat capacity has been measured in the range of interest.

When C_p and C_V are independent of temperature in the temperature range of interest, we obtain

$$S(T_f) = S(T_i) + C_p \int_{T_i}^{T_f} \frac{dT}{T} = S(T_i) + C_p \ln \left(\frac{T_f}{T_i} \right) \tag{14a}$$

when heating occurs at constant pressure, and

$$S(T_f) = S(T_i) + C_V \int_{T_i}^{T_f} \frac{dT}{T} = S(T_i) + C_V \ln \left(\frac{T_f}{T_i} \right) \tag{14b}$$

when heating occurs at constant volume.

Example 4.4 *Calculating the entropy change*

Calculate the entropy change when argon at 25°C and 1.00 atm in a container of volume 500 cm³ is allowed to expand to 1000 cm³ and is simultaneously heated to 100°C.

Method. Because S is a state function, we are free to choose the most convenient path from the initial state. One such path is reversible isothermal expansion to the final volume followed by reversible heating at constant volume to the final temperature. The entropy change in the first step is given by eqn 12 and that in the second step by eqn 14b. In each case we need to know n, the amount of gas, and can calculate it from the perfect gas equation and the data for the initial state. The heat capacity at constant volume can be obtained from the value of $C_{p,m}$ in Table 2.12 and the relation $C_{p,m} - C_{V,m} = R$.

Answer. The amount of Ar present (from $n = pV/RT$) is 0.0204 mol. The entropy change in the first step (expansion from 500 cm³ to 1000 cm³ at 298 K) is

$$\Delta S = nR \ln 2.00 = +0.118 \, J \, K^{-1}$$

The entropy change in the second step, heating from 298 K to 373 K at

(a)

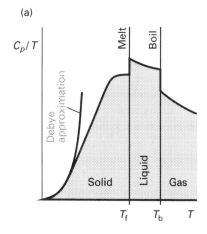

(b)

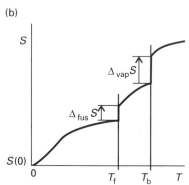

4.14 The calculation of a Third-Law entropy from heat capacity data. (a) The variation of C_p/T with the temperature for a sample. (b) The entropy, which is equal to the area beneath the upper curve up to the corresponding temperature, plus the entropy of each phase transition passed.

constant volume, is

$$\Delta S = (0.0204\,\text{mol}) \times (12.47\,\text{J K}^{-1}\,\text{mol}^{-1}) \times \ln\left(\frac{373\,\text{K}}{298\,\text{K}}\right) = +0.057\,\text{J K}^{-1}$$

The overall entropy change is the sum of these two changes: $\Delta S = +0.175\,\text{J K}^{-1}$.

Exercise E4.4. Calculate the entropy change when the same initial sample is compressed to $50.0\,\text{cm}^3$ and cooled to $-25°\text{C}$. $[-0.44\,\text{J K}^{-1}]$

The measurement of entropy

The entropy of a system at a temperature T can be related to its entropy at $T = 0$ by measuring its heat capacity C_p at different temperatures and evaluating the integral in eqn 13a. The entropy of transition $(\Delta_{\text{trs}}H/T_{\text{t}})$ must be added for each phase transition between $T = 0$ and the temperature of interest. For example, if a substance melts at T_{f} and boils at T_{b}, then its entropy above its boiling temperature is given by

$$S(T) = S(0) + \int_0^{T_{\text{f}}} \frac{C_p(\text{s})}{T}\,\text{d}T + \frac{\Delta_{\text{fus}}H}{T_{\text{f}}} + \int_{T_{\text{f}}}^{T_{\text{b}}} \frac{C_p(\text{l})}{T}\,\text{d}T$$
$$+ \frac{\Delta_{\text{vap}}H}{T_{\text{b}}} + \int_{T_{\text{b}}}^{T} \frac{C_p(\text{g})}{T}\,\text{d}T$$

All the properties required, except $S(0)$, can be measured calorimetrically, and the integrals can be evaluated either graphically or, as is now more usual, by numerical integration on a computer. The procedure is illustrated in Fig. 4.14: the area under the curve of C_p/T against T is the integral required. Because $\text{d}T/T = \text{d}\ln T$, an alternative procedure is to evaluate the area under a plot of C_p against $\ln T$. Examples of this procedure are given in the problems at the end of the chapter.

One problem with the measurement of entropy is the difficulty of measuring heat capacities near $T = 0$. There are good theoretical grounds for assuming that the heat capacity is proportional to T^3 when T is low (see Section 11.1), and this dependence is the basis of the **Debye extrapolation**. In this method, C_p is measured down to as low a temperature as possible, and a curve of the form aT^3 is fitted to the data. That fit determines the value of a, and the expression $C_p = aT^3$ is assumed valid down to $T = 0$.

Example 4.5 *Calculating the entropy at low temperatures*

The molar heat capacity of a certain solid at $10\,\text{K}$ is $0.43\,\text{J K}^{-1}\,\text{mol}^{-1}$. What is its molar entropy at that temperature?

Method. Because the temperature is so low, we can assume that the heat capacity varies with temperature as aT^3, in which case we can use eqn 13 to calculate the entropy at a temperature T in terms of the entropy at $T = 0$ and the constant a. When the integration is carried out, it turns out that the result can be expressed in terms of the heat capacity at the temperature T, so the data can be used directly to calculate the entropy.

Answer. The integration we require is

$$S(T) = S(0) + a \int_0^T T^2 \, dT = S(0) + \tfrac{1}{3}aT^3$$

However, because aT^3 is the heat capacity at the temperature T,

$$S_m(T) = S_m(0) + \tfrac{1}{3}C_{p,m}(T)$$

from which it follows that

$$S_m(10\,K) = S_m(0) + 0.14\,J\,K^{-1}\,mol^{-1}$$

Comment. Because the heat capacity is very small at low temperatures, only small errors arise from the Debye extrapolation.

Exercise E4.5. For metals there is also a contribution to the heat capacity from the electrons which is linearly proportional to T when the temperature is low. Find its contribution to the entropy at low temperatures. $[S(T) = S(0) + C_p(T)]$

As an example of the determination of an entropy, the standard molar entropy of nitrogen gas at 25°C has been calculated from the following data:

	$S_m^{\ominus}/(J\,K^{-1}\,mol^{-1})$
Debye extrapolation (0 to 10 K)	1.92
Integration, eqn 13a, from 10 K to 35.61 K	25.25
Phase transition (at 35.61 K)	6.43
Integration, eqn 13a, from 35.61 K to 63.14 K	23.38
Phase transition (melting, 63.14 K)	11.42
Integration, eqn 13a, from 63.14 K to 77.32 K	11.41
Phase transition (vaporization, 77.32 K)	72.13
Constant C_p, eqn 14 from 77.32 K to 298.15 K	39.20
Correction for gas imperfection[1]	0.92
Total:	192.06

Hence,

$$S_m^{\ominus}(298.15\,K) = S_m(0) + 192.06\,J\,K^{-1}\,mol^{-1}$$

We deal with the value of $S(0)$ next.

4.5 The Third Law of thermodynamics

At $T = 0$ all thermal motion has been quenched and in a perfect crystal all the particles are in a regular, uniform array. The absence of both spatial disorder and thermal motion suggests that such materials also have zero entropy. This conclusion is consistent with the Boltzmann formula, for if $W = 1$ (only one way of arranging the molecules), then $S = 0$.

1 Gas imperfections are discussed in Section 22.6.

The Nernst heat theorem

The thermodynamic observation that turns out to be consistent with the view that the entropy of regular arrays of molecules is zero at absolute zero is known as the **Nernst heat theorem**:

> The entropy change accompanying any physical or chemical transformation approaches zero as the temperature approaches zero:
>
> $\Delta S \rightarrow 0$ as $T \rightarrow 0$

For example, the entropy of the transition between orthorhombic sulfur $S(\alpha)$ and monoclinic sulfur $S(\beta)$ can be measured by determining its enthalpy ($-402\,\mathrm{J\,mol^{-1}}$) at the transition temperature (369 K):

$$\Delta S_\mathrm{m} = S_\mathrm{m}(\alpha) - S_\mathrm{m}(\beta) = \frac{(-402\,\mathrm{J\,mol^{-1}})}{369\,\mathrm{K}} = -1.09\,\mathrm{J\,K^{-1}\,mol^{-1}}$$

The two individual entropies can also be determined by measuring the heat capacities from $T = 0$ up to $T = 369\,\mathrm{K}$. It is found that

$$S_\mathrm{m}(\alpha) = S_\mathrm{m}(\alpha, 0) + 37\,\mathrm{J\,K^{-1}\,mol^{-1}}$$
$$S_\mathrm{m}(\beta) = S_\mathrm{m}(\beta, 0) + 38\,\mathrm{J\,K^{-1}\,mol^{-1}}$$

implying that at the transition temperature

$$\Delta S_\mathrm{m} = S_\mathrm{m}(\alpha, 0) - S_\mathrm{m}(\beta, 0) - 1\,\mathrm{J\,K^{-1}\,mol^{-1}}$$

On comparing this value with the one above, we conclude that

$$S_\mathrm{m}(\alpha, 0) - S_\mathrm{m}(\beta, 0) = 0$$

in accord with the theorem.

It follows from the Nernst theorem that, if we arbitrarily ascribe the value zero to the entropies of elements in their perfect crystalline form at $T = 0$, then all perfect crystalline compounds also have zero entropy at $T = 0$ (because the change in entropy that accompanies the formation of the compounds, like the entropy of all transformations at that temperature, is zero). Hence, all perfect crystals may be taken to have zero entropy at $T = 0$. This conclusion is summarized by the 'Third Law' of thermodynamics:

> **Third Law:** If the entropy of every element in its most stable state at $T = 0$ is taken as zero, then every substance has a positive entropy which at $T = 0$ may become zero, and which does become zero for all perfect crystalline substances, including compounds.

Note that a non-crystalline perfect state, such as the superfluid state of He, is included by the opening phrase. It should be noted that the Third Law does not state that entropies *are* zero at $T = 0$: it merely implies that all perfect materials have the same entropy at that temperature. As far as thermodynamics is concerned, choosing this common value as zero is then a matter of convenience.

Third-Law entropies

The choice $S(0) = 0$ for perfect crystals will be made from now on. Entropies reported on the basis of this choice are called **Third-Law**

Table 4.3* Standard Third-Law entropies at 298 K

	$S_m^{\ominus}/$ (J K^{-1} mol^{-1})
Solids:	
Graphite, C(s)	5.7
Diamond, C(s)	2.4
Sucrose, $C_{12}H_{22}O_{11}$(s)	360.2
Iodine, I_2(s)	116.1
Liquids:	
Benzene, C_6H_6(l)	173.3
Water, H_2O(l)	69.9
Mercury, Hg(l)	76.0
Gases:	
Methane, CH_4(g)	186.3
Carbon dioxide, CO_2(g)	213.7
Hydrogen, H_2(g)	130.7
Helium, He(g)	126.2
Ammonia, NH_3(g)	192.3

* More values are given in the Data section.

entropies (and often just 'entropies'). When the substance is in its standard state at the temperature T, the standard (Third-Law) entropy is denoted $S^{\ominus}(T)$. A list of values at 298 K is given in Table 4.3.

The **standard reaction entropy** $\Delta_r S^{\ominus}$ is defined, like the standard reaction enthalpy, as the difference between the entropies of the pure, separated products and the pure, separated reactants, all substances being in their standard states at the specified temperature:

$$\Delta_r S^{\ominus} = \sum_J \nu_J S_m^{\ominus}(J) \tag{15}$$

(For a review of the notation employed in this definition see Sections 2.5 and 2.6.)

Example 4.6 *Calculating a standard reaction entropy*

Calculate the standard entropy of $H_2(g) + \frac{1}{2}O_2(g) \rightarrow H_2O(l)$ at 25°C.

Method. This is a straightforward exercise in the use of eqn 15. Standard entropies are listed in Table 4.3 (a longer list is given in Table 2.12 of the Data section at the end of the volume).

Answer. Equation 15 takes the form

$$\Delta_r S^{\ominus} = S_m^{\ominus}(H_2O, l) - \{S_m^{\ominus}(H_2, g) + \frac{1}{2} \times S_m^{\ominus}(O_2, g)\}$$
$$= (69.9 - 130.7 - \tfrac{1}{2} \times 205.0)\, J\, K^{-1}\, mol^{-1} = -163.3\, J\, K^{-1}\, mol^{-1}$$

Comment. The large decrease in entropy is largely due to the formation of a compact liquid from two gases.

Exercise E4.6. Calculate the standard reaction entropy for the combustion of 1 mol CH_4(g) to carbon dioxide and liquid water at 25°C.

$$[-243\, J\, K^{-1}\, mol^{-1}]$$

THE EFFICIENCIES OF THERMAL PROCESSES

As we have seen, thermodynamics grew out of the study of ways of improving the efficiencies of **heat engines**, which are devices for converting heat into work and which include steam engines, internal combustion engines, and jet engines. While these early applications of thermodynamics are apparently far removed from chemistry, we shall see that analogous considerations are relevant to discussions of chemical processes, such as the use of chemical reactions in electrochemical cells to produce electric currents and the biosynthesis of proteins. They are also relevant to considerations of how it is possible to achieve low temperatures.

4.6 The efficiencies of heat engines

The process used to produce work in a heat engine must be spontaneous (an engine is worthless if it has to be driven!). Hence, the flow of energy as heat from the hot source to the cold sink must be accompanied by an overall increase in entropy. We can now, therefore, establish the equivalence of Kelvin's statement of the Second Law and its statement in

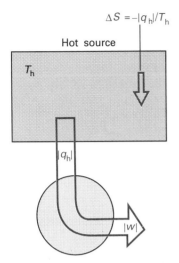

$$\Delta S = -|q_h|/T_h$$

Hot source

T_h

$|q_h|$

$|w|$

4.15 The change in entropy when heat is converted completely into work. There is only a decrease in entropy as heat is withdrawn from the hot source. Hence, the overall process is not spontaneous, in accord with Kelvin's statement of the Second Law.

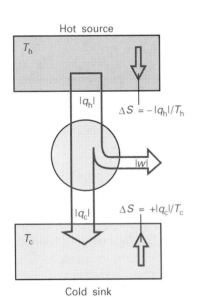

Hot source

T_h

$|q_h|$

$$\Delta S = -|q_h|/T_h$$

$|w|$

$|q_c|$

$$\Delta S = +|q_c|/T_c$$

T_c

Cold sink

4.16 When heat leaves a hot source, the source undergoes a decrease in entropy. To achieve a matching increase in entropy, less energy need be transferred as heat to a cold sink. The balance of energy may be extracted as work. This diagram illustrates the energy flow.

terms of entropy. Thus, if the hypothetical engine extracts heat $|q|$ from a hot sink of temperature T_h and converts it entirely into work (Fig. 4.15), then the change in entropy is

$$\Delta S = -\frac{|q|}{T_h}$$

This change in entropy is negative, and so the process is non-spontaneous, in accord with the Kelvin statement.

Engines in general

When a cold sink is present, some heat can be discarded into it, and the overall change in entropy need not be negative. Because the removal of $|q_h|$ from the hot reservoir changes its entropy by $-|q_h|/T_h$ and a transfer of $|q_c|$ to the cold reservoir increases its entropy by $|q_c|/T_c$, the overall entropy change in general is

$$\Delta S = -\frac{|q_h|}{T_h} + \frac{|q_c|}{T_c}$$

It follows from this expression that ΔS is not negative if

$$|q_c| \geq \frac{T_c}{T_h} \times |q_h|$$

Therefore, we are free to use some kind of device, a heat engine, to draw off the difference $|q_h| - |q_c|$ as work, yet still have a spontaneous process (Fig. 4.16). It then follows that the maximum work the engine can do is

$$|w_{max}| = |q_h| - |q_{c,min}| = \left(1 - \frac{T_c}{T_h}\right) \times |q_h|$$

We can conclude from the definition of the Carnot efficiency that the maximum possible efficiency of the engine is

$$\varepsilon_{rev} = \frac{|w_{max}|}{|q_h|} = 1 - \frac{T_c}{T_h} \qquad (16)_r$$

as we established earlier by a different argument.

Selecting the most efficient conditions

Equation 16 is of major economic significance, for it shows that an engine cannot convert heat into work with 100 per cent efficiency (unless $T_c = 0$). The limit implied by the Carnot efficiency applies to all engines, whatever their construction and working substance. It is an upper limit to their conversion efficiency, and technological deficiencies (such as friction) reduce the actual efficiency of real engines. Its positive feature is that it indicates how high conversion efficiencies may be achieved:

$$\varepsilon_{rev} \rightarrow 1 \text{ as } T_c \rightarrow 0 \text{ or } T_h \rightarrow \infty$$

Lowering the temperature of the cold sink (typically a lake, river, or the atmosphere) is not a practicable option, so engines and turbines are designed to run at high temperatures. A typical generating station using superheated steam at about 550°C and a cold sink at about 100°C has a thermodynamic efficiency of only 55 per cent. The remaining 45 per cent of heat taken from the hot source is discarded into the surroundings to ensure that enough entropy is generated to make the overall process spontaneous. An internal combustion engine operates between about 3200 K (the high temperature being brought about by the combustion of fuel) and 1400 K (the temperature in the exhaust manifold). Its thermodynamic efficiency is therefore only 56 per cent, but other losses reduce this ideal value to around 25 per cent in practice.

The thermodynamic temperature scale

The analysis of the efficiency of heat engines leads to a definition of a fundamental temperature scale that is independent of the thermometric substance.

Suppose we have an engine that is working reversibly between a hot source at a temperature T_h and a cold sink at temperature T, then we know that

$$\frac{|q_c|}{|q_h|} = \frac{T}{T_h}$$

This expression enabled Kelvin to define the **thermodynamic temperature scale** in terms of the ratio of the heat withdrawn from the hot source and the heat supplied to the cold sink, both of which can be measured. The zero of the thermodynamic temperature scale is defined as the value of T at which the Carnot efficiency becomes equal to 1 and the work output is equal to the heat supplied:

$$\varepsilon_{rev} = 1 \text{ at } T = 0$$

The size of the unit of the temperature scale is defined by choosing as a single fixed point the triple point of water and defining the temperature of the triple point T_3 as 273.16 K exactly (with this choice, the kelvin is almost exactly equal to a Celsius degree). Then, if the heat engine has a hot source at this temperature, the temperature of the cold sink—the object with the temperature we want to measure—can be found by measuring q_h and q_c for an engine that is working reversibly between the two reservoirs and using

$$T = \frac{|q_c|}{|q_h|} \times T_3 \tag{17}_r$$

This result is independent of the working substance.

An additional point is that, as we saw in Chapter 2, heat transferred can, in principle, be measured mechanically (in terms of the raising of a weight). Therefore, it is possible, in principle at least, to use the distance moved by a weight to measure temperature. Kelvin's definition of the temperature scale puts the measurement of temperature on to a purely mechanical basis.

4.7 The energetics of refrigeration

Work must be done to transfer heat from a system to its warmer surroundings and to go against the natural tendency of change, the spontaneous flow from hot to cold. Cooling against a temperature

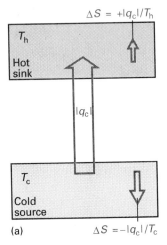

$\Delta S = +|q_c|/T_h$

T_h
Hot sink

$|q_c|$

T_c
Cold source

(a) $\Delta S = -|q_c|/T_c$

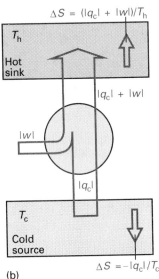

$\Delta S = (|q_c| + |w|)/T_h$

T_h
Hot sink

$|q_c| + |w|$

$|w|$

$|q_c|$

T_c
Cold source

(b) $\Delta S = -|q_c|/T_c$

4.17 (a) The flow of energy as heat from a cold source to a hot body is not spontaneous. As shown here, the entropy increase of the hot sink is smaller than the entropy decrease of the cold source, so there is a net decrease in entropy. (b) The process becomes feasible if work is provided to augment the flow of energy. Then the increase in entropy of the hot sink can be made to cancel the entropy decrease of the cold source.

gradient is achieved using a **refrigerator**, a heat engine operating in reverse, in which work is done to transfer energy from a cold source to a hot sink.

The work to achieve low temperatures

The ideas behind refrigeration and the calculation of the minimum amount of work that must be done to achieve a stated amount of cooling are illustrated in Fig. 4.17. When an energy $|q_c|$ is removed from a cool source, the entropy of the source is lowered. When the same energy is released into a hot reservoir as heat, the entropy of the sink increases, and the overall change is

$$\Delta S = -\frac{|q_c|}{T_c} + \frac{|q_c|}{T_h} \leq 0$$

The process is not spontaneous because not enough entropy is generated in the hot sink to overcome the loss from the cold source (Fig. 4.17a). To generate more entropy, the energy transferred as heat to the hot reservoir must exceed that taken from the cold reservoir. This additional entropy can be achieved by adding to the stream of energy by doing work (Fig. 4.17b). Our task is to find the minimum quantity of work that has to be added in order for the process to take place. The outcome is expressed as the **coefficient of performance** c:

$$c = \frac{|q_c|}{|w|} \tag{18}$$

where $|q_c|$ is the energy to be removed from the cold source and w the work required to bring that removal about. The less the work required, the more efficient the operation, and the greater the coefficient of performance.

If $|q_c|$ is withdrawn from the cold source and work $|w|$ is done, then energy that must be dissipated into the hot sink (such as the surrounding room) as heat is $|q_h| = |q_c| + |w|$. Therefore,

$$\frac{1}{c} = \frac{|q_h| - |q_c|}{|q_c|} = \frac{|q_h|}{|q_c|} - 1$$

The refrigerator is at its most efficient when it is working reversibly because then $|w|$ is a minimum. The best coefficient of performance, c_{rev}, is therefore given by

$$\frac{1}{c_{rev}} = \frac{|q_{h,rev}|}{|q_{c,rev}|} - 1$$

The next step is to express the ratio of qs in terms of the temperatures of the hot and cold parts of the system. To do so, we make use of the fact that the entropy of the cold source changes by $-|q_{c,rev}|/T_c$ and that of the hot sink changes by $+|q_{h,rev}|/T_h$, but (because the process is occurring reversibly) there is no net entropy production. In other words, just enough work is being done to ensure that overall the process occurs without reduction of entropy. Therefore

$$\Delta S = \frac{|q_{h,rev}|}{T_h} - \frac{|q_{c,rev}|}{T_c} = 0$$

which solves to

$$\frac{|q_{h,rev}|}{|q_{c,rev}|} = \frac{T_h}{T_c}$$

It follows that the coefficient of performance of a perfect refrigerator working reversibly between the temperatures T_c and T_h is

$$c_{rev} = \frac{T_c}{T_h - T_c} \qquad (19)_r$$

The ideal value makes no reference to the type of refrigerator or the working substance: it depends only on the temperatures of the hot and cold reservoirs. Practical refrigerators (which do not work reversibly) have coefficients of performance lower than c_{rev}.

The work to maintain low temperatures

A related problem is the work required to maintain a low temperature once it has been reached. No thermal insulation is perfect, so there is always a flow of heat from the warm surroundings back into the sample. According to Newton's law of cooling, the rate of this re-entry of energy as heat is proportional to the temperature difference $T_h - T_c$. To maintain the low temperature, heat must be removed from the cold object at the same rate as it leaks in. If the rate at which heat leaks in is

$$\frac{dq_c}{dt} = A(T_h - T_c)$$

where A is a constant that depends on the size of the sample and the thermal conductivity of its insulation, then the minimum power P, the minimum rate at which work must be done to maintain the temperature difference, is

$$P = \frac{d|w|}{dt} = \frac{1}{c_{rev}} \times \frac{d|q_c|}{dt} = \frac{1}{c_{rev}} \times A(T_h - T_c)$$

$$= A \times \frac{(T_h - T_c)^2}{T_c} \qquad (20)_r$$

We see that the power increases as the square of the temperature difference we are trying to maintain. For this reason, air-conditioners are much more expensive to run on hot days than on mild days. Notice too that the power depends inversely on the temperature of the cold object: very high powers must be dissipated when the temperature is very low. As an illustration of the consequences of eqn 20, about 7×10^2 times as much power must be expended to maintain the same object at 1 mK with surroundings at 1.0 K as must be expended at 0°C with surroundings at 20°C. Therefore, if 150 W is needed at 0°C, then 100 kW is needed at 1 mK. In practice, the experimenter would use a much smaller sample at 1 mK and would take great care with its insulation.

Adiabatic demagnetization

The world record low temperature stands at about $20\,nK$ ($2 \times 10^{-8}\,K$). Gases may be cooled by Joule–Thomson expansion below their inversion temperatures, and temperatures as low as about $4\,K$ (the boiling point of helium) may be reached without great difficulty. Temperatures lower than $4\,K$ can be reached by evaporating liquid helium by pumping rapidly through large-diameter pipes: as the helium evaporates it withdraws energy from the object being cooled. Temperatures as low as about $1\,K$ can be reached, but below this temperature the volatility of helium is too low for the process to be effective; moreover, the superfluid phase begins to interfere with the cooling process by creeping around the apparatus.

The method used to reach very low temperatures is **adiabatic demagnetization**, a cooling technique that utilizes the magnetic properties of a substance. In Part 2 we shall see that magnetic properties arise because electrons behave as tiny magnets and that a **paramagnetic substance** is one that has unpaired electrons. In normal circumstances the unpaired electrons of a paramagnetic material are orientated at random. However, in a magnetic field more of the electron magnets are aligned along the field than against it. In thermodynamic terms, the application of a magnetic field lowers the entropy of the paramagnetic system by the ordering it induces (Fig. 4.18). Therefore, a sample has different entropy versus temperature curves, depending on the magnetic field it experiences. At a given temperature, the entropy is lower when the field is on than when it is off.

A sample of paramagnetic material, such as a *d*- or *f*-metal complex, is cooled to about $1\,K$ in the way already described. Gadolinium(III) sulfate octahydrate, $Gd_2(SO_4)_3 \cdot 8H_2O$, has been used because each gadolinium ion carries several electrons, but is separated from its neighbours by a sheath of solvating water molecules. The sample is then magnetized by the application of a strong magnetic field. This magnetization occurs while the sample is surrounded by helium gas, which provides a thermal contact with a cold reservoir. The magnetization is therefore isothermal, and heat leaves the sample as the electron magnets adopt lower energy by aligning with the applied field. This stage is represented by the line AB in Fig. 4.18.

Thermal contact between sample and surroundings is now broken by pumping away the gas; then the magnetic field is reduced slowly to zero. This reversible step is adiabatic ($q_{rev} = 0$) and consequently the entropy of the sample remains constant. Hence, the state of the sample changes from B to C in Fig. 4.18. When the field has reached zero, the sample is the same as it was initially, except that it now has a lower entropy. Therefore, it now also has a lower temperature: adiabatic demagnetization has cooled the sample.

Even lower temperatures can be reached if, instead of electron magnetic moments, nuclear magnetic moments are used. This process of **adiabatic nuclear demagnetization** works on the same principle as the electronic method, and has been used to establish the world record (in copper).

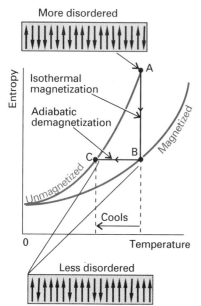

4.18 The technique of adiabatic demagnetization is used to attain very low temperatures. The upper curve shows the variation of the entropy of a paramagnetic system in the absence of an applied field. the lower curve shows the variation in entropy when a field has been applied and has made the electron magnets more orderly. The isothermal magnetization step is from A to B; the adiabatic demagnetization step (at constant entropy) is from B to C.

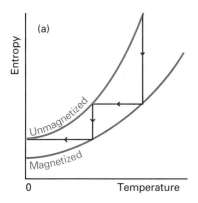

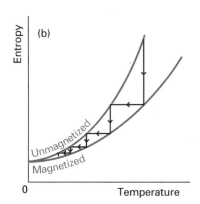

4.19 The connection between the Nernst theorem and the unattainability of absolute zero is illustrated in this pair of diagrams. In (a) we assume that the theorem is false, and that the entropies do not coincide as $T \to 0$: we see that $T = 0$ can be reached in a finite number of steps. In (b) we see that the steps down in temperature become progressively smaller if, as the Nernst theorem asserts, the entropies do coincide as $T \to 0$. A finite number of steps does not now take the system to $T = 0$.

From the Third Law of thermodynamics, we know that the entropies of substances coincide as $T \to 0$. The convergence of their values implies that adiabatic demagnetization—and, in fact, any process—cannot be used to cool an object to absolute zero in a finite number of steps. To see why this is so, suppose the law were false and that the entropy varied as shown in Fig. 4.19a. The last step indicated could cool the system to $T = 0$. However, the Third Law asserts that the curves actually coincide at $T = 0$, as shown in Fig. 4.19b, so no finite sequence of steps can cool the system to $T = 0$. By this method (as by any method) absolute zero is unattainable. This conclusion is summarized by an alternative version of the Third Law of thermodynamics:

Third Law: It is impossible to reach $T = 0$ in a finite number of steps.

No one has ever succeeded in cooling any system to $T = 0$, and the Third Law generalizes this experience into a principle.

Heat pumps

Refrigeration is closely related to the operation of a **heat pump**, in which work is done to pump heat from a cold reservoir (such as a river or the surrounding land) into a hot sink (such as a house). A heat pump is like a refrigerator operating in reverse. Because work is supplied to the energy stream and delivered to the hot sink (the house), the total heat delivered when the pump is working reversibly is

$$|q_h| = |q_c| + |w_{rev}| = (c_{rev} + 1)\,|w_{rev}|$$

If indoors the temperature is $T_h = 300\,\text{K}$ and outside $T_c = 290\,\text{K}$, then $c_{rev} = 29$; consequently, for each 1 kJ of energy used to drive the pump ($w = 1\,\text{kJ}$), 30 kJ of heat will be supplied to the house. A 1 kW pump would therefore be a 30-kW heater if it worked reversibly. Commercial heat pumps have $c \approx 5$, which still amounts to a good yield in terms of the energy used to drive the equipment.

CONCENTRATING ON THE SYSTEM

Entropy is the basic concept for discussing the direction of natural change, but to use it we have to analyse changes in both the system and its surroundings. We have seen that it is always very simple to calculate the entropy change in the surroundings, and we shall now see that it is possible to devise a simple method for taking that contribution into account automatically. This approach focuses our attention on the system and simplifies discussions. Moreover, it is the foundation of all the applications of chemical thermodynamics that follow.

4.8 The Helmholtz and Gibbs energies

Consider a system in thermal equilibrium with its surroundings at a temperature T. When a change in the system occurs and there is a transfer of energy as heat between the system and the surroundings, the

Clausius inequality, eqn 10, reads

$$dS - \frac{dq}{T} \geq 0$$

This inequality can be developed in two ways according to the conditions (of constant volume or constant pressure) under which the process occurs.

First, consider heat transfer at constant volume. Then, in the absence of non-expansion work, we can write $dq_V = dU$; consequently

$$dS - \frac{dU}{T} \geq 0 \tag{21}$$

The importance of the inequality in this form is that *it expresses the criterion for spontaneous change solely in terms of the state functions of the system*. The inequality is easily rearranged to

$$T\,dS \geq dU \quad \text{(constant } V, \text{ no non-expansion work)} \tag{22a}$$

At either constant internal energy ($dU = 0$) or constant entropy ($dS = 0$), respectively, this expression becomes

$$dS_{U,V} \geq 0 \qquad dU_{S,V} \leq 0 \tag{22b}$$

These are the criteria for natural changes in terms of properties relating to the system. The first states that, in a system at constant volume and constant internal energy (such as an isolated system), the entropy increases in a spontaneous change. That statement is essentially the content of the Second Law. The second inequality is less obvious, for it says that, if the entropy and volume of the system are constant, then the internal energy must decrease in a spontaneous change. We should interpret this requirement as implying that, if the entropy of the *system* is unchanged, then there must be an increase in entropy of the surroundings, which can be achieved only if the energy of the system decreases as energy flows out as heat.

When heat is transferred at constant pressure and there is no work other than expansion work, we can write $dq_p = dH$ and obtain

$$T\,dS \geq dH \quad \text{(constant } p, \text{ no non-expansion work)} \tag{23a}$$

At either constant enthalpy or constant entropy this inequality becomes, respectively,

$$dS_{H,p} \geq 0 \qquad dH_{S,p} \leq 0 \tag{23b}$$

The interpretations of these inequalities are similar to those of eqn 22. The entropy of the system must increase if its enthalpy remains constant (for there can be no change in entropy of the surroundings). Alternatively, the enthalpy must decrease if the entropy of the system is constant, for then it is essential to have an increase in entropy of the surroundings.

Because eqns 22 and 23 have the forms $dU - T\,dS \leq 0$ and $dH - T\,dS \leq 0$, respectively, they can be expressed more simply by introducing two more thermodynamic functions:

Helmholtz energy: $A = U - TS$ **(24a)**

Gibbs energy: $G = H - TS$ **(24b)**

The Helmholtz energy is also called the Helmholtz function or Helmholtz free energy. Similarly, the Gibbs energy is also called the Gibbs function or the Gibbs free energy. The Gibbs energy is far more widely encountered in chemistry than the Helmholtz energy and, after a few remarks about the latter, almost the whole of the following discussion will be in terms of G. Indeed, G is so widely used that it is normally referred to as simply the 'free energy'.

All the symbols in eqn 24 refer to the system. When the state of the system changes at constant temperature, the two functions change as follows:

$$dA = dU - T\,dS \qquad dG = dH - T\,dS \qquad \textbf{(25a)}$$

if the change is infinitesimal, and

$$\Delta A = \Delta U - T\,\Delta S \qquad \Delta G = \Delta H - T\,\Delta S \qquad \textbf{(25b)}$$

if it is measurable. When we introduce eqn 22 into eqns 23 and 24, we obtain the criteria of spontaneous change as

$$dA_{T,V} \leq 0 \qquad dG_{T,p} \leq 0 \qquad \textbf{(26)}$$

These inequalities are the most important conclusions from thermodynamics for chemistry. They are developed in subsequent sections and chapters.

Some remarks on the Helmholtz energy

A change in a system at constant temperature and volume is spontaneous if $dA_{T,V} \leq 0$. That is, a change under these conditions is spontaneous if it corresponds to a *decrease* in the Helmholtz energy. Such systems move spontaneously towards states of lower A if a path is available; the criterion of equilibrium, when neither the forward nor reverse process has a tendency to occur, is $dA_{T,V} = 0$.

The expression $dA = dU - T\,dS$ and $dA < 0$ is sometimes interpreted as follows. A negative value of dA is favoured by a negative value of dU and a positive value of $T\,dS$. This observation suggests that the tendency of a system to move to lower A is due to its tendency to move towards states of lower internal energy and higher entropy. However, this interpretation is false (even though it is a good rule of thumb for remembering the expression for dA) because the tendency to lower A is solely a tendency towards states of greater overall entropy. *Systems change spontaneously if in doing so the total entropy of the system and its surroundings increases, not because they tend to lower internal energy.* The form of dA may give the impression that systems favour lower energy, but that is misleading: dS is the entropy change of the system, $-dU/T$ is the entropy change of the surroundings, and their total tends to a maximum.

Maximum work

It turns out that A carries a greater significance than being simply a signpost of spontaneous change: the change in the Helmholtz energy is equal to the maximum work the system can do:

$$w_{max} = \Delta A \tag{27}$$

As a result, A is sometimes called the **maximum work function**, or the **work function** (*Arbeit* is the German word for work; hence the symbol A). The proof is in two stages.

First we prove that *a system does maximum work when it is working reversibly*. (This conclusion was demonstrated in Section 2.3 for the expansion of a perfect gas; now we prove its universal validity.) We combine the Clausius inequality $dS \geq dq/T$ in the form $T\,dS \geq dq$ with the First Law, $dU = dq + dw$, and obtain

$$dU \leq T\,dS + dw$$

(dU is smaller than the term on the right because we are replacing dq by $T\,dS$, which in general is larger). This expression rearranges to

$$dw \geq dU - T\,dS$$

It follows that the most negative value of dw, and therefore the *maximum* value of $|dw|$, the maximum energy that can be obtained from the system as work, is given by

$$dw_{max} = dU - T\,dS$$

and that this work is done only when the path is traversed reversibly (because then the equality applies). Because at constant temperature $dA = dU - T\,dS$, we conclude that $dw_{max} = dA$. Equation 27 then follows for a measurable change when the Helmholtz energy changes by ΔA.

We can write eqn 27 in the form

$$w_{max} = \Delta U - T\,\Delta S \tag{28}$$

This expression shows that in some cases, depending on the sign of $T\,\Delta S$, not all the change in internal energy may be available for doing work. If the change occurs with a decrease in entropy (of the system) so that $T\,\Delta S < 0$, then the right-hand side of this equation is not as negative as ΔU itself and the maximum work is less than ΔU. For the change to be spontaneous, some of the energy must escape as heat so as to generate enough entropy in the surroundings to overcome the reduction in entropy in the system (Fig. 4.20a). In this case Nature is demanding a tax on the internal energy as it is converted into work. This is the origin of the alternative name Helmholtz *free* energy for A, because ΔA is that part of the change in internal energy that we are free to use to do work.

MOLECULAR INTERPRETATION 4.5

Further insight into the relation between the work that a system can do and the Helmholtz energy is obtained by recalling that work is energy transferred to the surroundings as the uniform motion of atoms. The expression $A = U - TS$ can be interpreted as showing that A is the total internal energy of the system U less a contribution that is stored chaotically (the quantity TS). Because chaotically stored energy cannot

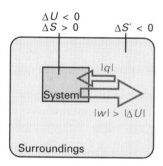

(a)

(b)

4.20 In a system not isolated from its surroundings, the work done may be different from the change in internal energy. Moreover, the process is spontaneous if overall the entropy of the global, isolated system increases. In (a) the entropy of the system decreases, so that of the surroundings must increase in order for the process to be spontaneous, which means that energy must pass from the system to the surroundings as heat. Therefore, less work than ΔU can be obtained. In (b) the entropy of the system increases and hence we can afford to lose some entropy of the surroundings; that is, some of their energy may be lost as heat to the system. This energy can be returned to them as work. Hence the work done can exceed ΔU.

be used to achieve uniform motion in the surroundings, only the part of U that is not stored chaotically, the quantity $U - TS$, is available for conversion into work.

If the change occurs with an increase of entropy of the system (in which case $T \Delta S > 0$), then the right-hand side of the equation is more negative than ΔU, so the maximum work that can be obtained from the system is greater than ΔU. The explanation of this apparent paradox is that the system is not isolated and energy may flow in as heat as work is done. Because the entropy of the system increases, we can afford a reduction of the entropy of the surroundings yet still have, overall, a spontaneous process. Therefore, some heat (no more than the value of $T \Delta S$) may leave the surroundings and contribute to the work the change is generating (Fig. 4.20b). Nature is now providing a tax refund.

Example 4.7 *Calculating the maximum available work*

When 1.000 mol glucose is oxidized to carbon dioxide and water at 25°C according to the equation

$$C_6H_{12}O_6(s) + 6O_2(g) \rightarrow 6CO_2(g) + 6H_2O(l)$$

calorimetric measurements give $\Delta U = -2808$ kJ and $\Delta S = +182.4$ J K^{-1} at 25°C and standard conditions. How much of this energy change can be extracted as (a) heat at constant pressure, (b) work?

Method. We know that the heat released at constant pressure is equal to the value of ΔH, so we need to relate ΔH to ΔU, which is given. To do so we suppose that all the gases involved are perfect, and use eqn 2.20 (this is easy, because $\Delta n_g = 0$). For the work available from the process we should use eqn 28.

Answer. (a) Because $\Delta n_g = 0$, we know that $\Delta H = \Delta U = -2808$ kJ. Therefore, at constant pressure, the energy available as heat is 2808 kJ. (b) Because $T = 298$ K, the value of ΔA is

$$\Delta A = \Delta U - T \Delta S = -2862 \text{ kJ}$$

Therefore, the combustion of 1.000 mol $C_6H_{12}O_6$ can be used to produce up to 2862 kJ of work.

Comment. The maximum work available is greater than the change in internal energy on account of the positive entropy of reaction (which is partly due to the generation of a large number of small molecules from one big one). The system can therefore draw in energy from the surroundings (so reducing their entropy) and make it available for doing work.

Exercise E4.7. Repeat the calculation for the combustion of 1.000 mol $CH_4(g)$ under the same conditions using enthalpy data from Table 2.11 and $\Delta S = -140.3$ J K^{-1}.   [$|q_p| = 890$ kJ, $|w_{max}| = 845$ kJ]

Some remarks on the Gibbs energy

As we have indicated, the Gibbs energy (the 'free energy') is more common in chemistry than the Helmholtz energy because, at least in laboratory chemistry, we are usually more interested in changes occurring at constant pressure than at constant volume. The criterion $dG_{T,p} \leq 0$ carries over into chemistry as the observation that, *at constant temperature and pressure, chemical reactions are spontaneous in the direction of decreasing free energy*. Therefore, if we want to know whether a reaction is spontaneous, the pressure and temperature being constant, we assess the change in the Gibbs energy that accompanies it. If G decreases as the reaction proceeds, then the reaction has a spontaneous tendency to convert the reactants into products. If G increases, then the reverse reaction is spontaneous.

The spontaneity of endothermic reactions

An illustration of the role of G is provided by the existence of spontaneous endothermic reactions. In such reactions, H increases, the system rises spontaneously to states of higher enthalpy, and $dH > 0$. Because the reaction is spontaneous we know that $dG < 0$ despite $dH > 0$; it follows that the entropy of the system increases so much that $T \, dS$ is strongly positive and outweighs dH in $dG = dH - T \, dS$. Endothermic reactions are therefore driven by the increase in entropy of the system, and this entropy change overcomes the reduction of entropy brought about in the surroundings by the inflow of heat into the system $(dS' = -dH/T)$.

Example 4.8 *Calculating the change in Gibbs energy accompanying a reaction*

Calculate the change in Gibbs energy when $1.00 \, mol \, N_2O_4(g)$ forms $2.00 \, mol \, NO_2(g)$ under standard conditions at 25°C in the reaction $N_2O_4(g) \rightarrow 2NO_2(g)$. The entropy change accompanying the reaction is $+4.8 \, J \, K^{-1}$.

Method. To calculate the change in G we need to know the changes in enthalpy and entropy, and then to combine them by using eqn 25b ($\Delta G = \Delta H - T \, \Delta S$). The standard reaction enthalpy is $\Delta_r H^{\ominus} = +57.2 \, kJ \, mol^{-1}$, so the enthalpy change for the reaction of $1.00 \, mol \, NO$ is $+57.2 \, kJ$. Therefore, at 25°C,

$$\Delta G = +57.2 \, kJ - (298.15 \, K) \times (4.8 \, J \, K^{-1}) = +55.8 \, kJ$$

Comment. The reaction as specified is not spontaneous at 25°C, but we see that, as the temperature increases, the negative entropy change of the system (which is multiplied by the temperature) becomes more important and eventually may dominate the positive enthalpy change of the endothermic reaction. We discuss this question in detail in Chapter 9 when we shall see that partial disssociation occurs even at 25°C.

Exercise E4.8. Is the oxidation of iron to $Fe_2O_3(s)$ at 298 K spontaneous given that the formation of 1.00 mol $Fe_2O_3(s)$ is accompanied by an entropy change of $-272\,J\,K^{-1}$?   [$\Delta G = -741$ kJ; yes]

Maximum non-expansion work

The analogue of the maximum work interpretation of ΔA, and the origin of the name free energy, can be found for ΔG. Because $H = U + pV$, in a general change

$$dH = dq + dw + d(pV)$$

When the change is reversible, $dw = dw_{rev}$ and $dq = dq_{rev} = T\,dS$, so

$$dG = T\,dS + dw_{rev} + d(pV) - T\,dS = dw_{rev} + d(pV)$$

The work consists of expansion work, which for a reversible change is given by $-p\,dV$, and possibly some other kind of work (for instance, the electrical work of pushing electrons through a circuit or of raising a column of liquid); this non-expansion work we denote dw_e. Therefore, with $d(pV) = p\,dV + V\,dp$,

$$dG = (-p\,dV + dw_{e,rev}) + p\,dV + V\,dp = dw_{e,rev} + V\,dp$$

If the change occurs at constant pressure (as well as constant temperature), the last term disappears, and $dG = dw_{e,rev}$. Therefore,

$$dw_{e,rev} = dG \qquad (T, p \text{ constant})$$

However, because the process is reversible, the work done must now have its maximum value, and so for a measurable change we can conclude that

$$w_{e,max} = \Delta G \qquad (T, p \text{ constant}) \tag{29}$$

That is, *the maximum non-expansion work we can obtain from a process at constant pressure and temperature is given by the value of ΔG for the process.* This expression is particularly useful for assessing the electrical work that may be produced by electrochemical cells, and we shall see many applications of it.

Example 4.9 *Calculating the maximum non-expansion work of a reaction*

How much energy is available for sustaining muscular and nervous activity from the combustion of 1.000 mol of glucose molecules under standard conditions at 37°C (blood temperature)? The standard entropy of reaction is $+182.4\,J\,K^{-1}$.

Method. The non-expansion work available from the reaction is equal to the Gibbs energy of reaction. To calculate this quantity, it is legitimate to ignore the temperature dependence of the reaction enthalpy and obtain ΔH from Table 2.11; then substitute the data into $\Delta G = \Delta H - T\,\Delta S$.

Answer. Because the standard enthalpy of reaction is -2808 kJ, it follows that the standard reaction Gibbs energy is

$$\Delta G = -2808\,\text{kJ} - (310\,\text{K}) \times (182.4\,J\,K^{-1}) = -2864\,\text{kJ}$$

Therefore, $w_{e,max} = -2864 \, kJ$ for the combustion of 1 mol glucose molecules, and the reaction can be used to do up to 2864 kJ of non-expansion work.

Comment. A 70-kg person would need to do 2.1 kJ of work to climb vertically through 3 m: therefore, at least 0.13 g of glucose is needed to complete the task (and in practice significantly more).

Exercise E4.9. How much non-expansion work can be obtained from the combustion of 1.00 mol $CH_4(g)$ under standard conditions at 298 K? Use $\Delta S = -140 \, J \, K^{-1}$.   [849 kJ]

4.9 Standard molar Gibbs energies

Standard entropies and enthalpies of reaction can be combined to obtain the **standard Gibbs energy of reaction** $\Delta_r G^{\ominus}$:

$$\Delta_r G^{\ominus} = \Delta_r H^{\ominus} - T \, \Delta_r S^{\ominus} \qquad (30)$$

The standard Gibbs energy of reaction (the 'standard reaction free energy') is the difference in standard Gibbs energies of the products and reactants in their standard states at the temperature specified for the reaction as written. As in the case of reaction enthalpies, it is convenient to define the standard Gibbs energies of formation $\Delta_f G^{\ominus}$:

> The **standard Gibbs energy of formation** is the standard reaction Gibbs energy for the formation of a compound from its elements in their reference states.

Standard Gibbs energies of formation of the elements are zero, because their formation is a 'null' reaction, as in

$$Cl_2(g) \rightarrow Cl_2(g) \qquad \Delta_r G^{\ominus} = 0$$

A selection of values for compounds is given in Table 4.4. From the values there, it is a simple matter to obtain the standard Gibbs energy of reaction by taking the appropriate combination:

$$\Delta_r G^{\ominus} = \sum_J \nu_J \Delta_f G^{\ominus}(J) \qquad (31)$$

Table 4.4* Standard Gibbs energies of formation at 298 K

	$\Delta_f G^{\ominus}/(kJ \, mol^{-1})$
Diamond, C(s)	+2.9
Benzene, C_6H_6(l)	+124.3
Methane, CH_4(g)	−50.7
Carbon dioxide, CO_2(g)	−394.4
Water, H_2O(l)	−237.1
Ammonia, NH_3(g)	−16.5
Sodium chloride, NaCl(s)	−384.1

* More values are given in the Data section.

Example 4.10 *Calculating a standard reaction Gibbs energy*

Calculate the standard reaction Gibbs energy for $CO(g) + \frac{1}{2}O_2(g) \rightarrow CO_2(g)$ at 25°C.

Method. We use eqn 31 with data from Table 4.4 after identifying the stoichiometric numbers.

Answer.

$$\Delta_r G^{\ominus} = \Delta_f G^{\ominus}(CO_2, g) - \{\Delta_f G^{\ominus}(CO, g) + \frac{1}{2} \times \Delta_f G^{\ominus}(O_2, g)\}$$
$$= \{-394.4 - (-137.2) - \frac{1}{2}(0)\} \, kJ \, mol^{-1} = -257.2 \, kJ \, mol^{-1}$$

Comment. The quantity $\Delta_r G^{\ominus}$ is the change of Gibbs energy that occurs when the pure, separated gases are converted completely into the pure product. In due course we shall have to deal with systems in which there are mixtures of reactants and products.

Exercise E4.10. Calculate the standard reaction Gibbs energy for the combustion of $CH_4(g)$ at 25°C. $[-818 \text{ kJ mol}^{-1}]$

Calorimetry (for ΔH directly and for S via heat capacities) is only one of the ways of determining the values of Gibbs energies. They may also be obtained from equilibrium constants (Chapter 9) and electrochemical measurements (Chapter 10), and they may be calculated using data from spectroscopic observations (Chapter 20). The information in Table 2.12 of the Data section, however, together with the machinery we shall now construct, is all we need in order to draw far-reaching conclusions about reactions and other processes of interest in chemistry.

CHECK LIST OF KEY IDEAS

1 The **Kelvin statement** of the **Second Law** in terms of heat engines.

2 The definition of a **spontaneous change** and its interpretation in terms of the tendency of an isolated system to achieve greater disorder.

3 The statement of the Second Law of thermodynamics in terms of the **entropy** (Section 4.2).

4 The **thermodynamic definition** of entropy in terms of the energy transferred as heat to the surroundings (eqn 2) and the adaptation of this definition to the calculation of the entropy change of the system (eqn 4).

5 The proof that the entropy is a state function (Section 4.2) by using a **Carnot cycle**.

6 The statement of the **Clausius inequality** for the change of entropy (eqn 10).

7 The confirmation that the entropy of an isolated system does increase when a gas expands irreversibly and an object cools spontaneously (Section 4.3).

8 The calculation of the entropy change of a substance when it undergoes a phase transition (eqn 11).

9 The calculation of the entropy change when a perfect gas expands isothermally (eqn 12).

10 The entropy of a substance at one temperature in terms of its entropy at another temperature (eqn 13).

11 The measurement of the entropy of a substance calorimetrically (Section 4.4).

12 The **Nernst heat theorem** for the entropy change accompanying a transformation at $T = 0$ (Section 4.5).

13 The **Third Law** of thermodynamics, **Third-Law entropies** of substances, and the definition of the **standard reaction entropy** (eqn 15).

14 The calculation of the **efficiency** of heat engines from considerations of entropy (Section 4.6).

15 The use of the thermodynamic efficiency to establish the **thermodynamic temperature scale** (Section 4.6).

16 The best **coefficient of performance** of a refrigerator (eqn 19) and the power needed to sustain low temperatures (eqn 20).

17 The principle of **adiabatic demagnetization** for reaching low temperatures and the operation of **heat pumps** (Section 4.7).

18 The **criteria for spontaneous change** in terms of the entropy, the internal energy, and the enthalpy (eqns 22 and 23).

19 The definition of the **Helmholtz energy** and the **Gibbs energy** (eqn 24) and the criteria for spontaneous change (eqn 26).

20 The relation between the Helmholtz energy and the **maximum work** done in a process (eqn 27).

21 The relation between the Gibbs energy and the **maximum non-expansion work** (eqn 29).

22 The definition of the **standard reaction Gibbs energies of formation** of substances and their use to calculate $\Delta_r G^{\ominus}$ (eqn 31).

EXERCISES

Assume that all gases are perfect and that data refer to 298 K unless otherwise stated.

4.1. Calculate the change in entropy when 25 kJ of energy is transferred reversibly and isothermally as heat to a large block of iron at (a) 0°C, (b) 100°C.

4.2. Calculate the molar entropy of a constant-volume sample of neon at 500 K given that it is 146.22 J K^{-1} mol^{-1} at 298 K.

4.3. A sample consisting of 1.00 mol of a monatomic perfect gas with $C_{V,m} = \frac{3}{2}R$ is heated from 100°C to 300°C at constant pressure. Calculate ΔS (for the system).

4.4. Calculate ΔS (for the system) when 3.00 mol of a monatomic perfect gas, for which $C_{p,m} = \frac{5}{2}R$, is heated and compressed from 25°C and 1.00 atm to 125°C and 5.00 atm. How do you rationalize the sign of ΔS?

4.5. A sample of 3.00 mol of a diatomic perfect gas at 200 K is compressed reversibly and adiabatically until its temperature reaches 250 K. Given that $C_{V,m} = 27.5$ J K^{-1} mol^{-1}, calculate q, w, ΔU, ΔH, and ΔS.

4.6. Calculate the increase in entropy when 1.00 mol of a monatomic perfect gas with $C_{p,m} = \frac{5}{2}R$, is heated from 300 K to 600 K and simultaneously expanded from 30.0 L to 50.0 L.

4.7. A system undergoes a process in which the entropy change is $+2.41$ J K^{-1}. During the process, 1.00 kJ of heat is added to the system at 500 K. Is the process thermodynamically reversible? Explain your reasoning.

4.8. A 1.75-kg sample of aluminium is cooled at constant pressure from 300 K to 265 K. Calculate the amount of energy that must be removed as heat and the change in entropy of the sample.

4.9. A 25-g sample of methane gas at 250 K and 18.5 atm expands isothermally until its pressure is 2.5 atm. Calculate the change in entropy of the gas.

4.10. A sample of perfect gas that initially occupies 15.0 L at 250 K and 1.00 atm is compressed isothermally. To what volume must the gas be decreased in order to reduce its entropy by 5.0 J K^{-1}?

4.11. Calculate the change in entropy when 50 g of water at 80°C is poured into 100 g of water at 10°C in an insulated vessel given that $C_{p,m} = 75.5$ J K^{-1} mol^{-1}.

4.12. Calculate ΔH and ΔS_{tot} when two 10.0-kg copper bricks, one at 100°C and the other at 0°C, are placed in contact in an isolated container. The specific heat capacity of copper is 0.385 J K^{-1} g^{-1} and is constant over the temperature range involved.

4.13. Consider a system consisting of 2.0 mol $CO_2(g)$, initially at 25°C and 10 atm and confined to a cylinder of 10 cm^2 cross-section. It is allowed to expand adiabatically against an external pressure of 1.0 atm until the piston has moved outwards through 20 cm. Assume that carbon dioxide may be considered a perfect gas with $C_{V,m} = 28.8$ J K^{-1} mol^{-1} and calculate (a) q, (b) w, (c) ΔU, (d) ΔT, (e) ΔS.

4.14. The enthalpy of vaporization of chloroform ($CHCl_3$) is 29.4 kJ mol^{-1} at its normal boiling point of 334.88 K. Calculate the entropy of vaporization of chloroform at this temperature. What is the entropy change in the surroundings?

4.15. Calculate the standard reaction entropy at 298 K of

(a) $2CH_3CHO(g) + O_2(g) \rightarrow 2CH_3COOH(l)$

(b) $2AgCl(s) + Br_2(l) \rightarrow 2AgBr(s) + Cl_2(g)$

(c) $Hg(l) + Cl_2(g) \rightarrow HgCl_2(s)$

(d) $Zn(s) + Cu^{2+}(aq) \rightarrow Zn^{2+}(aq) + Cu(s)$

(e) $C_{12}H_{22}O_{11}(s) + 12O_2(g) \rightarrow 12CO_2(g) + 11H_2O(l)$

4.16. Combine the reaction entropies calculated in Exercise 4.15 with the reaction enthalpies and calculate the standard reaction Gibbs energies of the reactions at 298 K.

4.17. Use standard Gibbs energies of formation to calculate the standard reaction Gibbs energies at 298 K of the reactions in Exercise 4.15.

4.18. Calculate the standard Gibbs energy of the reaction $4HCl(g) + O_2(g) \rightarrow 2Cl_2(g) + 2H_2O(l)$ at 298 K from the standard entropies and enthalpies of formation given in Table 2.12.

4.19. The standard enthalpy of combustion of solid phenol (C_6H_5OH) is -3054 kJ mol^{-1} at 298 K and its standard molar entropy is 144.0 J K^{-1} mol^{-1}. Calculate the standard Gibbs energy of formation of phenol at 298 K.

4.20. Calculate the change in the entropies of the system and the surroundings, and the total change in entropy, when a 14-g sample of nitrogen gas at 298 K and 1.00 bar doubles its volume in (a) an isothermal reversible expansion, (b) an isothermal irreversible expansion against $p_{ex} = 0$, and (c) an adiabatic reversible expansion.

4.21. Calculate the change in entropy when a perfect gas is compressed to half its volume and simultaneously heated to twice its initial temperature.

4.22. Calculate the maximum non-expansion work per mole that may be obtained from a fuel cell in which the chemical reaction is the combustion of methane at 298 K.

4.23. Calculate the Carnot efficiency of a primitive steam engine operating on steam at 100°C and discharging at 60°C. Repeat the calculation for a modern steam turbine that operates with steam at 300°C and discharges at 80°C.

4.24. A heat engine operates between 1000 K and 500 K. (a) What is the maximum efficiency of the engine? (b) Calculate the maximum work that can be done by each 1.0 kJ of heat supplied by the hot source. (c) How much heat is discharged into the cold sink in a reversible process for each 1.0 kJ supplied by the hot source?

4.25. The enthalpy of the graphite → diamond phase transition, which under 100 kbar occurs at 2000 K, is

+1.9 kJ mol^{-1}. Calculate the entropy change of the transition.

4.26. How much work must be done in order to cool the air in an otherwise empty room of dimensions 5.0 m × 5.0 m × 3.0 m from 30°C to 22°C when the ambient temperature is (a) 20°C, (b) 30°C? (Assume that c_{rev} is constant over the range and that $C_{p,m} = 29$ J K^{-1} mol^{-1} for the air of mean density 1.2 mg cm^{-3} and molar mass 29 g mol^{-1}.)

4.27. A refrigerator operating reversibly extracts 45 kJ of energy as heat from a cold source and delivers 67 kJ to a hot sink at 300 K. Calculate the temperature of the source.

4.28. A refrigerator operates reversibly between 80 K and 200 K. Calculate the work required to remove 2.10 kJ of energy from the cold source to the hot sink.

4.29. Calculate the coefficient of performance of a perfect refrigerator operating in a room at 20°C when the interior is at (a) 0°C, (b) −10°C.

4.30. Calculate the minimum work needed to freeze 250 g of water originally at 0°C standing in a room at 20°C. What would be the minimum time required in a refrigerator operating ideally at 100 W?

PROBLEMS

Assume that all gases are perfect and that data refer to 298 K unless otherwise stated.

Numerical problems

4.1. Calculate the difference in molar entropy (a) between liquid water and ice at −5°C, (b) between liquid water and its vapour at 95°C and 1.00 atm. The differences in heat capacities on melting and on vaporization are 37.3 J K^{-1} mol^{-1} and −41.9 J K^{-1} mol^{-1}, respectively. Distinguish between the entropy changes of the sample, the surroundings, and the total system, and discuss the spontaneity of the transitions at the two temperatures.

4.2. The heat capacity of chloroform (trichloromethane, CHCl$_3$) in the range 240 K to 330 K is given by $C_{p,m}/(\text{J K}^{-1}\,\text{mol}^{-1}) = 91.47 + 7.5 \times 10^{-2}(T/\text{K})$. In a particular experiment, 1.00 mol CHCl$_3$ is heated from 273 K to 300 K. Calculate the change in molar entropy of the sample.

4.3. A block of copper of mass 2.00 kg ($C_{p,s} = 0.385$ J K^{-1} g^{-1}) and temperature 0°C is introduced into an insulated container in which there is 1.00 mol H$_2$O(g) at 100°C and 1.00 atm. (a) Assuming all the steam is condensed to water, what will be the final temperature of the system, the heat transferred from water to copper, and

the entropy change of the water, copper, and the total system? (b) In fact, some water vapour is present at equilibrium. From the vapour pressure of water at the temperature calculated in (a), and assuming that the heat capacities of both gaseous and liquid water are constant and given by their values at that temperature, obtain an improved value of the final temperature, the heat transferred, and the various entropies. (*Hint.* You will need to make plausible approximations.)

4.4. Consider a perfect gas contained in a cylinder and separated by a frictionless adiabatic piston into two sections A and B. All changes in B are isothermal; that is, a thermostat surrounds B to keep its temperature constant. There is 2.00 mol of the gas in each section. Initially $T_A = T_B = 300$ K, $V_A = V_B = 2.00$ L. Heat is added to Section A and the piston moves to the right reversibly until the final volume of Section B is 1.00 L. Calculate (a) ΔS_A and ΔS_B, (b) ΔA_A and ΔA_B, (c) ΔG_A and ΔG_B, (d) ΔS of the total system and its surroundings. If numerical values cannot be obtained, indicate whether the values should be positive, negative, or zero or are indeterminate from the information given. (Assume $C_{V,m} = 20$ J K^{-1} mol^{-1}.)

4.5. A Carnot cycle uses 1.00 mol of a monatomic perfect gas as the working substance from an initial state of 10.0 atm and 600 K. It expands isothermally to a pressure of of 1.00 atm (step 1), and then adiabatically to a temperature of 300 K (step 2). This expansion is followed by an isothermal compression (step 3), and then an adiabatic compression (step 4) back to the initial state. Determine the values of q, w, ΔU, ΔH, ΔS, and $\Delta S'$ for each step in the cycle and the cycle as a whole. (Express your answer as a table of values.)

4.6. A sample of 1.00 mol of a monatomic perfect gas for which $C_{V,m} = \frac{3}{2}R$ is taken through the cycle shown in Fig. 4.21. (a) Determine the temperatures at each state. (b) Give the values of q, w, ΔU, ΔH, ΔS, ΔS_{tot}, and ΔG for each stage and for the cycle as a whole. (Express your answer as a table of values.)

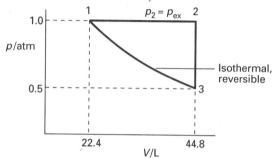

4.7. 1.00 mol of perfect gas at 27°C is expanded isothermally from an initial pressure of 3.00 atm to a final pressure of 1.00 atm in two ways: (1) reversibly, and (2) against constant external pressure of 1.00 atm. Calculate the final temperature and q, w, ΔH, ΔS, $\Delta S'$, and ΔS_{tot} for each path.

4.8. A sample of 1.00 mol of a perfect gas at 27°C and 1.00 atm pressure is expanded adiabatically in two ways: (1) reversibly to 0.50 atm, and (2) against a constant external pressure of 0.50 atm. Calculate the final temperature and q, w, ΔH, ΔS, $\Delta S'$, and ΔS_{tot} for each path. Take $C_{V,m} = \frac{3}{2}R$.

4.9. A sample of 1.00 mol of a monatomic perfect gas with $C_{V,m} = \frac{3}{2}R$, initially at 298 K and 10 L, is expanded, with the surroundings maintained at 298 K, to a final volume of 20 L, in three ways: (1) isothermally and reversibly; (2) isothermally against a constant external pressure of 0.50 atm; (3) adiabatically against a constant external pressure of 0.50 atm. Calculate ΔS, $\Delta S'$, ΔH, ΔT, ΔA, and ΔG for each path. If a numerical answer cannot be obtained from the data, write $+$, $-$, or ? as appropriate.

4.10. The standard molar entropy of $NH_3(g)$ is 192.45 J K^{-1} mol^{-1} at 298 K, and its heat capacity is given by eqn 2.24 with the coefficients given in Table 2.2. Calculate the standard molar entropy at (a) 100°C and (b) 500°C.

4.11. A block of copper of mass 500 g and initially at 293 K is in thermal contact with an electric heater of

resistance 1.00 kΩ and negligible mass. A current of 1.00 A is passed for 15.0 s. Calculate the change in entropy of the copper, taking $C_{p,m} = 24.4$ J K^{-1} mol^{-1}. The experiment is then repeated with the copper immersed in a stream of water that maintains its temperature at 293 K. Calculate the change in entropy of the copper and the water in this case.

4.12. Calculate the standard Helmholtz energy of formation, $\Delta_f A^{\ominus}$, of $CH_3OH(l)$ at 298 K from its standard Gibbs energy of formation and the assumption that H_2 and O_2 are perfect gases.

4.13. Calculate the change in entropy when 200 g of (a) water at 0°C, (b) ice at 0°C is added to 200 g of water at 90°C in an insulated container.

4.14. Calculate (a) the maximum work and (b) the maximum non-expansion work that can be obtained from the freezing of supercooled water at $-5°C$ and 1.0 atm. The densities of water and ice are 0.999 and 0.917 g cm^{-3} respectively at $-5°C$.

4.15. The molar heat capacity of lead varies with temperature as follows:

T/K	10	15	20	25	30	50
$C_{p,m}/$J K^{-1} mol^{-1}	2.8	7.0	10.8	14.1	16.5	21.4
T/K	70	100	150	200	250	298
$C_{p,m}/$J K^{-1} mol^{-1}	23.3	24.5	25.3	25.8	26.2	26.6

Calculate the standard Third-Law entropy of lead at (a) 0°C and (b) 25°C.

4.16. Suppose that an internal combustion engine runs on octane, for which the enthalpy of combustion is -5512 kJ mol^{-1} and take the mass of 1 gallon of fuel as 3 kg. What is the maximum height, neglecting all forms of friction, to which a 1000-kg car can be driven on 1 gallon of fuel given that the engine cylinder temperature is 2000°C and the exit temperature is 800°C?

4.17. From standard enthalpies of formation, standard entropies, and standard heat capacities available from tables in the Data section, calculate the standard enthalpies and entropies at 298 K and 398 K for the reaction $CO_2(g) + H_2(g) \rightarrow CO(g) + H_2O(g)$. Assume that the heat capacities are constant over the temperature range involved.

4.18. The standard reaction Gibbs energy of

$$K_4[Fe(CN)_6] \cdot 3H_2O(s)$$
$$\rightarrow 4K^+(aq) + [Fe(CN)_6]^{4-}(aq) + 3H_2O(l)$$

is $+26.120$ kJ mol^{-1} (I. R. Malcolm, L. A. K. Staveley, and R. D. Worswick, *J. Chem. Soc. Faraday Trans.* **I**, 1532 (1973)). The enthalpy of solution of the trihydrate is $+55.000$ kJ mol^{-1}. Calculate the standard molar entropy of solution of the hexacyanoferrate(II) ion in water given that the standard molar entropy of the solid trihydrate is 599.7 J K^{-1} mol^{-1} and that of the K$^+$ ion in water is 102.5 J K^{-1} mol^{-1}.

4.19. The heat capacity of anhydrous potassium hexacyanoferrate(II) varies with temperature as follows:

T/K	$C_{p,m}/JK^{-1}$ mol^{-1}	T/K	$C_{p,m}/JK^{-1}$ mol^{-1}
10	2.09	90	165.3
20	14.43	100	179.6
30	36.44	110	192.8
40	62.55	150	237.6
50	87.03	160	247.3
60	111.0	170	256.5
70	131.4	180	265.1
80	149.4	190	273.0
		200	280.3

Calculate the molar enthalpy relative to its value at $T = 0$ and the Third-Law entropy at each of these temperatures.

4.20. The compound 1,3,5-trichloro-2,4,6-trifluorobenzene is an intermediate in the conversion of hexachlorobenzene to hexafluorobenzene, and its thermodynamic properties have been examined by measuring its heat capacity over a wide temperature range (R. L. Andon and J. F. Martin, *J. Chem. Soc. Faraday Trans.* **I**, 871 (1973)). Some of the data are as follows:

T/K	14.14	16.33	20.03	31.15	44.08	64.81
$C_{p,m}/JK^{-1}mol^{-1}$	9.492	12.70	18.18	32.54	46.86	66.36
T/K	100.90	140.86	183.59	225.10	262.99	298.06
$C_{p,m}/JK^{-1}mol^{-1}$	95.05	121.3	144.4	163.7	180.2	196.4

Calculate the molar enthalpy relative to its value at $T = 0$ and the Third-Law entropy of the compound at these temperatures.

4.21. Calculate the minimum work needed to reduce the temperature of a 1.0-g block of copper from 1.10 K to 0.10 K, the surroundings being at 1.20 K. Proceed by supposing that the heat capacity remains constant at $39 \, \mu JK^{-1}mol^{-1}$ and that the coefficient of performance can be evaluated at the mean temperature of the block. Then go on to do a more realistic calculation in which $C_{p,m} = AT^3 + BT$, with $A = 48.2 \, \mu JK^{-4}mol^{-1}$ and $B = 688 \, \mu JK^{-2}mol^{-1}$, and taking into account the variation of the coefficient of performance with temperature.

Theoretical problems

4.22. Show that the integral of dq_{rev}/T round a Carnot cycle is zero. Then show that, if the isothermal reversible expansion stage is replaced by an isothermal irreversible expansion stage, the integral is negative.

4.23. Prove that two reversible adiabatic paths can never cross. Assume that the energy of the system under consideration is a function of temperature only. (*Hint.* Suppose that two such paths can intersect, and complete a cycle with the two paths plus one isothermal path. Consider the changes accompanying each stage of the cycle and show that they conflict with the Kelvin statement of the Second Law.)

4.24. Represent the Carnot cycle on a temperature-entropy diagram and show that the area enclosed by the cycle is equal to the work done.

4.25. Find an expression for the change in entropy when two blocks of the same substance and of equal mass, one at the temperature T_h and the other at T_c, are brought into thermal contact and allowed to reach equilibrium. Evaluate the change for two 500-g blocks of copper with $C_{p,m} = 24.4 \, JK^{-1}mol^{-1}$, taking $T_h = 500 \, K$ and $T_c = 250 \, K$.

4.26. A gaseous sample consisting of 1.00 mol molecules is described by the equation of state $pV_m = RT(1 + Bp)$. Initially at 373 K, it undergoes Joule-Thomson expansion from 100 atm to 1.00 atm. Given that $C_{p,m} = \frac{5}{2}R$, $\mu = 0.21 \, K \, atm^{-1}$, $B = -0.525 \, (K/T) \, atm^{-1}$, and that these are constant over the temperature range involved, calculate ΔT for the gas.

4.27. The cycle involved in the operation of an internal combustion engine is called the Otto cycle. Air can be considered to be the working substance and can be assumed to be a perfect gas. The cycle consists of the following steps: (1) reversible adiabatic compression from A to B; (2) reversible constant-volume pressure increase from B to C due to the combustion of a small amount of fuel; (3) reversible adiabatic expansion from C to D; and (4) reversible and constant-volume pressure decrease back to state A. Determine the change in entropy (of the system and of the surroundings) for each step of the cycle and determine an expression for the efficiency of the cycle, assuming that the heat is supplied in step 2. Evaluate the efficiency for a compression ratio of 10:1. Assume that in state A $V = 4.00 \, L$, $p = 1.00 \, atm$, and $T = 300 \, K$, that $V_A = 10 \, V_B$, $p_C/p_B = 5$, and that $C_{p,m} = \frac{7}{2}R$.

4.28. Prove that the perfect-gas temperature scale and the thermodynamic temperature scale based on the Second Law of thermodynamics differ from each other by at most a constant numerical factor.

4.29. The definitions of the enthalpy, Gibbs energy, and Helmholtz energy have all been of the form $g = f + yz$. Show that the addition of the product yz is a general way of converting a function of x and y to a function of x and z in the sense that if $df = a \, dx - z \, dy$, then $dg = a \, dx + y \, dz$.

4.30. Calculate the minimum work needed to cool an object of finite size from T_i to T_f using a refrigerator in a room at temperature T_h. (*Hint.* Write $|dw| = |dq|/c_{rev}$, and express dq in terms of dT by using the heat capacity, which is assumed to be constant. Then integrate the expression.) Find the work needed to freeze 250 g of water initially at 20°C, the temperature of the room. What work would be required if the water was initially at 25°C (and the room still at 20°C)?

5

The Second Law: the machinery

In the first part of this chapter we see how the fact that a property is a state function can be used to derive relations between properties that might not be thought to be related. Then, in the major part of the chapter, we establish several important properties of the Gibbs energy. We see how to use these properties to derive expressions for the variation of G with temperature and pressure. These expressions will prove useful later when we need to discuss the effect of temperature on equilibrium constants.

The material in this chapter also serves as an introduction to two aspects of matter. First, we see how the discussion of the pressure dependence of G leads to the introduction of a property, the chemical potential, which will be at the centre of discussions in the remaining chapters of this part of the text. Secondly, we see how to break away from perfect gases and formulate expressions that are valid for real systems.

In this chapter we begin to derive some of the rich consequences that stem from combining the First and Second Laws of thermodynamics. In particular, we explore the Gibbs energy—the free energy—and show how it varies with temperature, pressure, and composition. As we have remarked, the Gibbs energy is the concept of thermodynamics that is of central importance to chemistry, and in this chapter it begins to move to the centre of the stage. We shall also meet the 'chemical potential', the quantity on which almost all the most important applications of thermodynamics to chemistry are based.

COMBINING THE FIRST AND SECOND LAWS

We have seen that the First Law of thermodynamics may be written

$$dU = dq + dw$$

For a reversible change in a closed system (one in which there is no change of composition) and in the absence of any non-expansion work,

$$\mathrm{d}w_{\mathrm{rev}} = -p\,\mathrm{d}V \qquad \mathrm{d}q_{\mathrm{rev}} = T\,\mathrm{d}S$$

Therefore,

$$\mathrm{d}U = T\,\mathrm{d}S - p\,\mathrm{d}V \tag{1}$$

However, because $\mathrm{d}U$ is an exact differential, its value is independent of path, so the same value of $\mathrm{d}U$ is obtained whether the change is brought about irreversibly or reversibly. Therefore, eqn 1 applies to any change—reversible or irreversible—of a closed system that does no non-expansion work. We shall call this combination of the First and Second Laws the **fundamental equation**.

The fact that the fundamental equation applies to both reversible and irreversible changes may be puzzling at first sight. The reason is that only in the case of a reversible change may $T\,\mathrm{d}S$ be identified with $\mathrm{d}q$ and $-p\,\mathrm{d}V$ with $\mathrm{d}w$. When the change is irreversible, $|T\,\mathrm{d}S| > |\mathrm{d}q|$ (the Clausius inequality) and $|p\,\mathrm{d}V| > |\mathrm{d}w|$. The sum of $\mathrm{d}w$ and $\mathrm{d}q$ remains equal to the sum of $T\,\mathrm{d}S$ and $-p\,\mathrm{d}V$, provided the composition is constant.

5.1 Properties of the internal energy

Equation 1 shows that the internal energy of a closed system changes in a simple way when S and V are changed ($\mathrm{d}U \propto \mathrm{d}S$ and $\mathrm{d}U \propto \mathrm{d}V$). These simple proportionalities suggest that U should be regarded as a function of S and V. We could regard U as a function of other variables, such as S and p or T and V, because they are all interrelated, but the simplicity of the fundamental equation suggests that $U(S, V)$ is the best choice.

The mathematical consequence of U being a function of S and V is that a change $\mathrm{d}U$ can be expressed in terms of the changes $\mathrm{d}S$ and $\mathrm{d}V$ by

$$\mathrm{d}U = \left(\frac{\partial U}{\partial S}\right)_V \mathrm{d}S + \left(\frac{\partial U}{\partial V}\right)_S \mathrm{d}V \tag{2}$$

This expression states that U changes by an amount that is proportional to the change in S and to the change in V, the two coefficients being the slopes of the graphs of U against S and V, respectively. When this expression is compared to the thermodynamic relation, eqn 1, we see that, for systems of constant composition,

$$\left(\frac{\partial U}{\partial S}\right)_V = T \qquad \left(\frac{\partial U}{\partial V}\right)_S = -p \tag{3}$$

The first of these two equations is a purely thermodynamic definition of temperature as the ratio of the changes in the internal energy and entropy of a constant-volume closed system. We are beginning to generate relations between the properties of a system and to discover the power of thermodynamics for establishing relations.

The Maxwell relations

Because the fundamental equation is an expression for an exact differential, the coefficients of $\mathrm{d}S$ and $\mathrm{d}V$ must pass the test of relation no. 4 in *Further Information 2* (the test for exact differentials). As shown in that discussion, $df = g\,dx + h\,dy$ is exact if

$$\left(\frac{\partial g}{\partial y}\right)_x = \left(\frac{\partial h}{\partial x}\right)_y$$

Therefore, the expression $dU = T\,dS - p\,dV$ is exact if

$$\left(\frac{\partial T}{\partial V}\right)_S = -\left(\frac{\partial p}{\partial S}\right)_V \tag{4}$$

We have generated a relation between quantities that, at first sight, would not seem to be related.

The equation just derived is an example of a **Maxwell relation**. However, apart from being unexpected, it does not look particularly interesting. Nevertheless, it does suggest that there may be other similar relations that are more useful. Indeed, the fact that H, G, and A are state functions can be used to derive three more Maxwell relations. The argument to obtain them runs in the same way in each case: because H, G, and A are state functions, the expression for dH, dG, and dA satisfy relation no. 4 in *Further information 2*. All four relations are listed in Table 5.1. In the next section we derive one of them but, as no new principles are involved, we shall not derive them all.

The variation of internal energy with volume

The coefficient

$$\pi_T = \left(\frac{\partial U}{\partial V}\right)_T$$

played a central role in the manipulation of the First Law, and in Section 3.3 we used the relation

$$\pi_T = T\left(\frac{\partial p}{\partial T}\right)_V - p \tag{5}$$

which was to be proved in this chapter. We are now ready to derive it from the relations we have just established.

We can obtain the coefficient π_T from eqn 2 by dividing both sides by dV, imposing the constraint of constant temperature, and then introducing the two relations in eqn 3:

$$\left(\frac{\partial U}{\partial V}\right)_T = \left(\frac{\partial U}{\partial S}\right)_V\left(\frac{\partial S}{\partial V}\right)_T + \left(\frac{\partial U}{\partial V}\right)_S = T\left(\frac{\partial S}{\partial V}\right)_T - p$$

This equation is already beginning to look like the expression we want. One of the Maxwell relations does the job of turning $(\partial S/\partial V)_T$ into something else:

$$\left(\frac{\partial S}{\partial V}\right)_T = \left(\frac{\partial p}{\partial T}\right)_V$$

The substitution of this relation completes the proof of eqn 5.

Equation 5 is called a **thermodynamic equation of state** because it

Table 5.1 The Maxwell relations

$$\left(\frac{\partial T}{\partial V}\right)_S = -\left(\frac{\partial p}{\partial S}\right)_V$$

$$\left(\frac{\partial T}{\partial p}\right)_S = \left(\frac{\partial V}{\partial S}\right)_p$$

$$\left(\frac{\partial p}{\partial T}\right)_V = \left(\frac{\partial S}{\partial V}\right)_T$$

$$\left(\frac{\partial V}{\partial T}\right)_p = -\left(\frac{\partial S}{\partial p}\right)_T$$

expresses a quantity in terms of the two variables T and p and applies to any material and any phase.

Example 5.1 *Deriving a thermodynamic relation*

Show thermodynamically that $\pi_T = 0$ for a perfect gas, and compute its value for a van der Waals gas.

Method. Proving a result 'thermodynamically' means basing it entirely on general thermodynamic relations and equations of state, without drawing on molecular arguments (such as the existence of intermolecular forces). We know that, for a perfect gas, $p = nRT/V$, so this relation should be used in eqn 5. Similarly, the van der Waals equation is given in Table 1.6 and for the second part of the question it should be used in eqn 5.

Answer. Because $(\partial p/\partial T)_V = nR/V$ for a perfect gas (by differentiation of the equation of state), eqn 5 becomes

$$\pi_T = T\left(\frac{nR}{V}\right) - p = 0$$

The equation of state of a van der Waals gas is

$$p = \frac{nRT}{V - nb} - a\frac{n^2}{V^2}$$

Therefore, because a and b are independent of temperature,

$$\left(\frac{\partial p}{\partial T}\right)_V = \frac{nR}{V - nb}$$

we can write

$$\pi_T = \frac{nRT}{V - nb} - \frac{nRT}{V - nb} + a\frac{n^2}{V^2} = a\frac{n^2}{V^2} = \frac{a}{V_m^2}$$

Comment. This calculation shows that the internal energy of a van der Waals gas increases when it expands isothermally, and that the increase is related to the parameter that models the attractive interactions between the particles: a larger molar volume implies a weaker mean attraction.

Exercise E5.1. Calculate π_T for a gas that obeys the virial equation of state. $[\pi_T = RT^2(\partial B/\partial T)_V/V_m + \ldots]$

5.2 Properties of the Gibbs energy

The same arguments that were applied to the fundamental equation for U may be applied to the Gibbs energy $G = H - TS$. When the system changes its state, G may change because H, T, and S change. For infinitesimal changes in each property (and neglecting second-order infinitesimals),

$$dG = dH - T\,dS - S\,dT$$

Because $H = U + pV$, we know that

$$dH = dU + p\,dV + V\,dp$$

For a closed system doing no non-expansion work, dU can be replaced by the fundamental equation $dU = T\,dS - p\,dV$. The result of these steps is

$$dG = (T\,dS - p\,dV) + p\,dV + V\,dp - T\,dS - S\,dT$$

That is, for a closed system in the absence of non-expansion work,

$$dG = V\,dp - S\,dT \tag{6}$$

This expression, which shows that a change in G is proportional to changes in p and T, suggests that G may be regarded as a function of p and T. It confirms that G is an important quantity in chemistry because the pressure and temperature are usually the variables under our control. Hence, *G carries around the combined consequences of the First and Second Laws in a way that makes it particularly suitable for chemical applications.*

The same argument that led to eqn 3, when applied to the exact differential dG, now gives

$$\left(\frac{\partial G}{\partial T}\right)_p = -S \qquad \left(\frac{\partial G}{\partial p}\right)_T = V \tag{7}$$

JUSTIFICATION

If we regard G as a function of p and T, then a change in its value can be expressed as

$$dG = \left(\frac{\partial G}{\partial T}\right)_p dT + \left(\frac{\partial G}{\partial p}\right)_T dp$$

However, dG is also given by eqn 6:

$$dG = V\,dp - S\,dT$$

Because dG is an exact differential, the coefficients of dT and dp must be equal in the two cases, so

$$\left(\frac{\partial G}{\partial T}\right)_p = -S \qquad \left(\frac{\partial G}{\partial p}\right)_T = V$$

as given in eqn 7.

The relations in eqn 7 show how the Gibbs energy varies with temperature and pressure. Because S is positive, it follows that G decreases when the temperature is raised at constant pressure and composition. Moreover, the relation shows that G decreases most sharply when the entropy of the system is large. Therefore, the Gibbs energy of the gaseous phase of a substance, which has a high molar entropy, is more sensitive to temperature than its liquid and solid phases (Fig. 5.1). Because V is positive, G always increases when the pressure of the system is increased at constant temperature (and composition). Because the molar volumes of gases are large, G is more sensitive to changes of pressure for the gaseous phase of a substance than for its liquid and solid phases (Fig. 5.2).

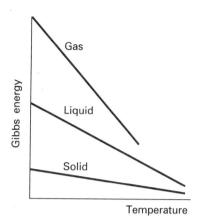

5.1 The variation of the Gibbs energy with the temperature is determined by the entropy. Because the entropy of the gaseous phase of a substance is greater than that of the liquid phase, and the entropy of the solid phase is smallest, the Gibbs energy changes most steeply for the gas phase, followed by the liquid phase, and then by the solid phase of the substance.

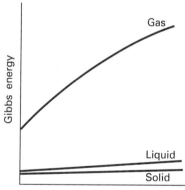

5.2 The variation of the Gibbs energy with the pressure is determined by the volume of the sample. Because the volume of the gaseous phase of a substance is greater than that of the liquid phase, and the volume of the solid phase is smallest (for almost all substances), the Gibbs energy changes most steeply for the gas phase, followed by the liquid phase, and then by the volumes of the solid and liquid phases of a substance are similar, their Gibbs energies vary by similar amounts as the pressure is changed.

Example 5.2 *Calculating the effect of pressure on the Gibbs energy*

Calculate the change in the molar Gibbs energy of (a) liquid water treated as an incompressible fluid and (b) water vapour treated as a perfect gas when the pressure is increased isothermally from 1.0 bar to 2.0 bar at 298 K.

Method. In each case the change in molar Gibbs energy can be obtained by integration of eqn 6 with the temperature held constant (that is, setting $dT = 0$):

$$\Delta G_m = \int V_m \, dp$$

For an incompressible fluid, the molar volume is independent of the pressure, so V_m can be treated as a constant. For a perfect gas, the molar volume varies with pressure as $V_m = RT/p$, so this expression must be used in the integrand and the integration performed treating RT as a constant.

Answer. For the incompressible liquid, V_m is constant at $18.0 \, \text{cm}^3 \, \text{mol}^{-1}$, so

$$\Delta G_m = V_m \, \Delta p = (18.0 \times 10^{-6} \, \text{m}^3 \, \text{mol}^{-1}) \times (1.0 \times 10^5 \, \text{Pa})$$
$$= +1.8 \, \text{J} \, \text{mol}^{-1}$$

(because $1 \, \text{Pa} \, \text{m}^3 = 1 \, \text{N} \, \text{m} = 1 \, \text{J}$). For the perfect gas:

$$\Delta G_m = RT \int_{p_i}^{p_f} \frac{dp}{p} = RT \ln\left(\frac{p_f}{p_i}\right)$$
$$= (2.48 \, \text{kJ} \, \text{mol}^{-1}) \times \ln 2.0 = +1.7 \, \text{kJ} \, \text{mol}^{-1}.$$

Comment. Note that G increases in both cases, and that the increase for a gas is 1000 times greater than for the liquid.

Exercise E5.2. Calculate the change in G_m for ice at $-10°C$, when it has density $0.917 \, \text{g} \, \text{cm}^{-3}$, when the pressure is increased from 1.0 bar to 2.0 bar. $[+2.0 \, \text{J} \, \text{mol}^{-1}]$

When we examine eqn 6 with the test for an exact differential using relation no. 4 of *Further information 2*, and set $g = V$ and $h = -S$, we find that

$$\left(\frac{\partial V}{\partial T}\right)_p = -\left(\frac{\partial S}{\partial p}\right)_T \tag{8}$$

This expression is another of the Maxwell relations in Table 5.1.

The temperature dependence of the Gibbs energy

In due course we shall see that the equilibrium composition of a system depends on its Gibbs energy and that to discuss the response of the composition to temperature it is necessary to know how G varies with temperature. Equation 7 is the starting point; although it expresses the variation of G in terms of the entropy, it can be expressed in terms of the enthalpy by using the definition of G to write

$$S = \frac{H - G}{T}$$

Then

$$\left(\frac{\partial G}{\partial T}\right)_p = \frac{G - H}{T} \tag{9}$$

We shall see later that the equilibrium constant of a reaction is related to G/T rather than to G itself,[1] and it turns out that the variation of this quantity with temperature is simpler than the temperature variation of G itself. In fact, it is easy to deduce from the last equation (see the following *Justification*) that

$$\left(\frac{\partial}{\partial T}\left(\frac{G}{T}\right)\right)_p = -\frac{H}{T^2} \tag{10}$$

This expression is called the **Gibbs–Helmholtz equation**. ($G–H$ is a helpful way of remembering what this equation relates.) It shows that, if the enthalpy of the system is known, then the temperature dependence of G/T is also known.

JUSTIFICATION

First, we write eqn 9 as

$$\left(\frac{\partial G}{\partial T}\right)_p - \frac{G}{T} = -\frac{H}{T}$$

The expression on the left is simplified by noting that

$$\left(\frac{\partial G}{\partial T}\right)_p - \frac{G}{T} = T\left(\frac{\partial}{\partial T}\left(\frac{G}{T}\right)\right)_p$$

This result is proved using the rules for differentiating a product:

$$\left(\frac{\partial}{\partial T}\left(\frac{G}{T}\right)\right)_p = \frac{1}{T}\left(\frac{\partial G}{\partial T}\right)_p + G\left(\frac{\partial}{\partial T}\left(\frac{1}{T}\right)\right)_p = \frac{1}{T}\left(\frac{\partial G}{\partial T}\right)_p - \frac{G}{T^2}$$

$$= \frac{1}{T}\left\{\left(\frac{\partial G}{\partial T}\right)_p - \frac{G}{T}\right\}$$

It follows that

$$T\left(\frac{\partial}{\partial T}\left(\frac{G}{T}\right)\right)_p = -\frac{H}{T}$$

The Gibbs–Helmholtz equation is then obtained by dividing both sides by T.

---

1 In Chapter 9 we derive the result that the equilibrium constant for a reaction is related to its standard Gibbs energy by the relation $\Delta_r G^\ominus / T = -R \ln K$.

Example 5.3 *Manipulating the Gibbs–Helmholtz equation*

Show that

$$\left(\frac{\partial(G/T)}{\partial(1/T)}\right)_p = H$$

Method. This example is an exercise in manipulating partial differentials. The desired expression resembles the Gibbs–Helmholtz equation, so eqn 10 is a good starting point. To obtain the desired result, we need to convert the variable of differentiation from T to $1/T$, which can be done by standard techniques of manipulating derivatives.

Answer. The left-hand side of eqn 10 can be written

$$\left(\frac{\partial(G/T)}{\partial T}\right)_p = \left(\frac{\partial(G/T)}{\partial(1/T)}\right)_p \left(\frac{\partial(1/T)}{\partial T}\right)_p$$

$$= \left(\frac{\partial(G/T)}{\partial(1/T)}\right)_p \times \left(-\frac{1}{T^2}\right)$$

because

$$\frac{\mathrm{d}(1/T)}{\mathrm{d}T} = \frac{\mathrm{d}(T^{-1})}{\mathrm{d}T} = -\frac{1}{T^2}$$

Substitution of this result into eqn 10 and multiplication of both sides by $-T^2$ gives the expression required.

Comment. The result shows that, if H is independent of temperature over a limited range, then a plot of G/T against $1/T$ should be a straight line of slope H. We shall see the usefulness of this result in Chapter 9.

Exercise E5.3. Find the equation for the temperature dependence of A that corresponds to that just derived for G. $[(\partial(A/T)/\partial(1/T))_V = U]$

The Gibbs–Helmholtz equation is most useful when it is applied to changes, including changes of physical state and chemical reactions at constant pressure. Then, with $\Delta G = G_f - G_i$ for the change of Gibbs energy between the final and initial states, because the equation applies to both G_f and G_i, we can write

$$\left(\frac{\partial}{\partial T}\left(\frac{\Delta G}{T}\right)\right)_p = -\frac{\Delta H}{T^2} \tag{11}$$

The pressure dependence of the Gibbs energy

Now we turn to a consideration of the variation of the free energy with pressure. To find the Gibbs energy at one pressure in terms of its value at another pressure, the temperature being constant, we integrate eqn 6 with $\mathrm{d}T = 0$:

$$G(p_f) = G(p_i) + \int_{p_i}^{p_f} V \, \mathrm{d}p \tag{12}$$

For a liquid or solid, the volume changes only slightly as the pressure

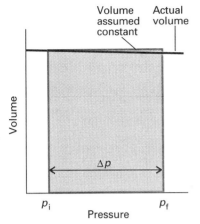

5.3 The difference in Gibbs energy of a solid or liquid at two pressures is equal to the rectangular area shown. We have assumed that the variation of volume with pressure is negligible.

changes (Fig. 5.3), so V may be treated as a constant and taken outside the integral. Then, for molar quantities,

$$G_{m,f} = G_{m,i} + (p_f - p_i)V_m$$
$$= G_{m,i} + V_m \Delta p \tag{13}$$

Under normal laboratory conditions $V_m \Delta p$ is very small, as we saw in Example 5.2, and may be neglected. Hence, we may usually suppose that the Gibbs energies of solids and liquids are independent of pressure. However, if we are interested in geophysical problems then, because pressures in the Earth's interior are huge, their effect on the Gibbs energy cannot be ignored and we have to use eqn 13. If the pressures are so great that there are substantial volume changes, we must use the complete expression, eqn 12.

Example 5.4 *Calculating the effect of pressure and temperature on ΔG*

The pressure deep inside the Earth is probably greater than 3×10^3 kbar, and the temperature there is around 4×10^3 °C. Estimate the change in ΔG on going from crust to core for a process in which $\Delta V = +1.0$ cm^3 mol^{-1} and $\Delta S = +2.1$ J K^{-1} mol^{-1}.

Method. Because $(\partial \Delta G/\partial p)_T = \Delta V$ and $(\partial \Delta G/\partial T)_p = -\Delta S$, we can use $d(\Delta G) = (\Delta V)\,dp$ and $d(\Delta G) = -(\Delta S)\,dT$. For changes in pressure and temperature (δp and δT, respectively) that are small enough to leave the values of ΔV and ΔS unchanged, these relations are approximately

$$\delta(\Delta G) \approx \Delta V \times \delta p \qquad \delta(\Delta G) = -\Delta S \times \delta T$$

and hence we estimate the total change as the sum of the two contributions.

Answer. The total change is

$$\Delta G(\text{core}) - \Delta G(\text{crust}) = \Delta V \times \{p(\text{core}) - p(\text{crust})\}$$
$$- \Delta S \times \{T(\text{core}) - T(\text{crust})\}$$
$$= 3 \times 10^2 \text{ kJ mol}^{-1} - 8 \text{ kJ mol}^{-1}$$
$$= 3 \times 10^2 \text{ kJ mol}^{-1}$$

Comment. The effect of pressure dominates, and results in a very large modification of ΔG for the process. This is the thermodynamic reason why materials change their forms at great depths in the Earth's interior.

Exercise E5.4. Calculate the difference in molar Gibbs energy between the top and bottom of a column of mercury in a barometer. The density of mercury is 13.6 g cm^{-3}. $\qquad$ [+1.5 J mol^{-1}]

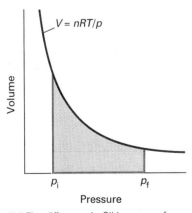

5.4 The difference in Gibbs energy for a perfect gas at two pressures is equal to the area shown below the perfect-gas isotherm.

The molar volumes of gases are large, so the correction term may be large even if the pressure difference is small. Furthermore, because the volume depends strongly on the pressure, we cannot treat it as a constant in the integral in eqn 12 (Fig. 5.4). For a perfect gas we substitute

$V = nRT/p$ into the integral, and find

$$G(p_f) = G(p_i) + nRT \int_{p_i}^{p_f} \frac{dp}{p} = G(p_i) + nRT \ln\left(\frac{p_f}{p_i}\right) \qquad (14)°$$

This expression shows that when the pressure is increased 10-fold at room temperature, the molar Gibbs energy increases by about $6 \, \text{kJ mol}^{-1}$.

THE CHEMICAL POTENTIAL

Now we switch attention from the Gibbs energy itself to a quantity, the chemical potential, which is closely related and which will play the central role in all subsequent discussions of equilibrium, including (in Chapter 9) chemical equilibrium. First, we introduce the chemical potential of a pure substance and, in particular, the chemical potential of a perfect gas. At this stage its introduction will seem to be no more than a change of notation. However, the definition prepares the ground for the introduction (in Section 5.4) of the chemical potential of a substance in a mixture (including a reaction mixture), which is a powerful and general concept.

5.3 The chemical potential of a pure substance

The **chemical potential** μ of a pure substance is defined as

$$\mu = \left(\frac{\partial G}{\partial n}\right)_{p,T} \qquad (15)$$

That is, *the chemical potential shows how the Gibbs energy of a system changes when the substance is added to it*: if $\mu > 0$, the Gibbs energy increases as n is increased. For a pure substance the Gibbs energy is simply $G = n \times G_m$, so

$$\mu = \left(\frac{\partial(nG_m)}{\partial n}\right)_{p,T} = G_m$$

and the chemical potential is the same as the molar Gibbs energy.

We have seen that G_m for a solid or liquid is only weakly dependent on the pressure (eqn 13), so the same is true of the chemical potential. However, the molar Gibbs energy of a gas is strongly dependent on the pressure, and we can adapt eqn 14 to show how the chemical potential of a perfect gas varies. To do so, we note that a perfect gas is in its standard state when its pressure is $p^\ominus$ (that is, 1 bar) and its Gibbs energy is then $G^\ominus$. According to eqn 14, at any other pressure p, its Gibbs energy G is

$$G = G^\ominus + nRT \ln\left(\frac{p}{p^\ominus}\right) \qquad (16)°$$

It then follows from eqn 15 that, for a perfect gas,

$$\mu = \mu^\ominus + RT \ln\left(\frac{p}{p^\ominus}\right) \qquad (17)°$$

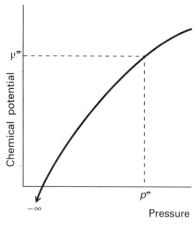

5.5 The chemical potential of a perfect gas varies as ln p and the standard state is reached when $p = p^\ominus$. Note that as $p \to 0$, μ becomes negatively infinite.

The variation of the chemical potential with the pressure given by this relation is illustrated in Fig. 5.5. We can already suspect that μ will have an important role to play in discussions of chemical equilibria, because it carries the information about how G changes as the amount of substance changes, and hence how G changes as the composition of a system changes.

Example 5.5 *Calculating the change in chemical potential*

Calculate the change in chemical potential when water vaporizes at 1 bar and 25°C.

Method. For a pure substance, the change in chemical potential is the same as the change in molar Gibbs energy. Therefore, because both the liquid and the vapour are in their standard states, we can use the data in Table 2.12 for the transformation $H_2O(l) \rightarrow H_2O(g)$.

Answer. The change in standard molar Gibbs energy is

$$\Delta G_m^{\ominus} = \Delta_f G^{\ominus}(H_2O, g) - \Delta_f G^{\ominus}(H_2O, l)$$
$$= (-228.57 \text{ kJ mol}^{-1}) - (-237.13 \text{ kJ mol}^{-1})$$
$$= +8.56 \text{ kJ mol}^{-1}$$

That is, the chemical potential of water vapour at 1 bar and 25°C is 8.56 kJ mol^{-1} greater than that of liquid water under the same conditions.

Comment. The greater chemical potential of the vapour is consistent with the chemical intuition of water vapour as being chemically more active than the liquid. We shall make this idea precise later.

Exercise E5.5. What is the change in chemical potential of any pure liquid when it vaporizes at its boiling point? [0]

5.4 The chemical potential of a substance in a mixture

In an open system (one in which the composition may vary) the Gibbs energy depends on the composition as well as the pressure and the temperature. Thus G may change when p, T, and the composition change and, for a binary system of components 1 and 2,

$$dG = \left(\frac{\partial G}{\partial p}\right)_{T,n_1,n_2} dp + \left(\frac{\partial G}{\partial T}\right)_{p,n_1,n_2} dT$$
$$+ \left(\frac{\partial G}{\partial n_1}\right)_{p,T,n_2} dn_1 + \left(\frac{\partial G}{\partial n_2}\right)_{p,T,n_1} dn_2 \tag{18}$$

This fearsome expression simply says that G may change because any of the four variables that define the state may change.

Clearly, the first job is to simplify the notation. As a first step, consider how G changes when the composition is constant but the pressure and temperature change infinitesimally:

$$dG = \left(\frac{\partial G}{\partial p}\right)_{T,n_1,n_2} dp + \left(\frac{\partial G}{\partial T}\right)_{p,n_1,n_2} dT$$

We already know that $dG = V \, dp - S \, dT$ under the same conditions.

Therefore, because dG is an exact differential, we may identify the coefficients:

$$\left(\frac{\partial G}{\partial p}\right)_{T,n_1,n_2} = V \qquad \left(\frac{\partial G}{\partial T}\right)_{p,n_1,n_2} = -S$$

These relations are the same as in eqn 7 but are dressed more elaborately with the constant composition stated explicitly.

The remaining differential coefficients should be recognized as slightly more elaborate versions of the chemical potentials of a single substance. Specifically, the coefficients with respect to the composition are, by definition, the chemical potentials of substances in a mixture:

$$\left(\frac{\partial G}{\partial n_1}\right)_{p,T,n_2} = \mu_1 \qquad \left(\frac{\partial G}{\partial n_2}\right)_{p,T,n_1} = \mu_2$$

That is, the chemical potential μ_1 expresses how G changes as substance 1 is added to the system (the pressure, temperature, and amount of substance 2 being constant); μ_2 does the same for the addition of substance 2. The chemical potentials depend on the composition of the mixture. For instance, adding 10^{-3} mol CH_3OH (almost an infinitesimal amount of methanol) to 1 L of a 20 per cent by mass methanol/water mixture leads to a change in the overall G which is different from the change brought about by the addition of the same amount of CH_3OH to an 80 per cent mixture.

In general, we define the **chemical potential of a substance J** as

$$\mu_J = \left(\frac{\partial G}{\partial n_J}\right)_{p,T,n'} \tag{19}$$

where the subscript n' signifies that the amounts of all the other components (those other than J) are constant. The introduction of these conclusions into eqn 18 (and allowing for the possibility of more components) leads to

$$\begin{aligned} dG &= V\,dp - S\,dT + \mu_1\,dn_1 + \mu_2\,dn_2 + \ldots \\ &= V\,dp - S\,dT + \sum_J \mu_J\,dn_J \end{aligned} \tag{20}$$

This is the **fundamental equation of chemical thermodynamics**. Its implications and consequences are explored and developed in the next five chapters.

At constant pressure and temperature eqn 20 simplifies to

$$dG = \mu_1\,dn_1 + \mu_2\,dn_2 + \ldots = \sum_J \mu_J\,dn_J$$

We saw in Section 4.8 that, under the same conditions, $dG = dw_{e,max}$. Therefore,

$$dw_{e,max} = \sum_J \mu_J\,dn_J \qquad \text{(at constant } p, T) \tag{21}$$

That is, non-expansion work can arise from the changing composition of a system that is not at internal equilibrium. For instance, in an electrochemical cell, the chemical reaction is arranged to take place at two distinct sites (at the two electrodes). The cell is not at internal equilibrium, and the electrical work it performs can be traced to its changing composition as products are formed from reactants.

5.5 The wider significance of μ

The chemical potential does more than show how G varies with composition. Because

$$G = U + pV - TS$$

a general infinitesimal change in U for a system of variable composition can be written

$$\begin{aligned}
\mathrm{d}U &= -p\,\mathrm{d}V - V\,\mathrm{d}p + S\,\mathrm{d}T + T\,\mathrm{d}S + \mathrm{d}G \\
&= -p\,\mathrm{d}V - V\,\mathrm{d}p + S\,\mathrm{d}T + T\,\mathrm{d}S \\
&\quad + (V\,\mathrm{d}p - S\,\mathrm{d}T + \mu_1\,\mathrm{d}n_1 + \mu_2\,\mathrm{d}n_2 + \ldots) \\
&= -p\,\mathrm{d}V + T\,\mathrm{d}S + (\mu_1\,\mathrm{d}n_1 + \mu_2\,\mathrm{d}n_2 + \ldots)
\end{aligned}$$

This is the generalization of eqn 1 (that $\mathrm{d}U = T\,\mathrm{d}S - p\,\mathrm{d}V$) to systems in which the composition may change. It follows that, at constant volume and entropy,

$$\mathrm{d}U = \mu_1\,\mathrm{d}n_1 + \mu_2\,\mathrm{d}n_2 + \ldots = \sum_J \mu_J\,\mathrm{d}n_J$$

and hence that

$$\mu_J = \left(\frac{\partial U}{\partial n_J}\right)_{S,V,n'} \tag{22a}$$

Therefore, not only does the chemical potential show how G changes when the composition changes, it also shows how the internal energy changes (but under a different set of conditions). In the same way it is easy to deduce that

$$\mu_J = \left(\frac{\partial H}{\partial n_J}\right)_{S,p,n'} \tag{22b}$$

$$\mu_J = \left(\frac{\partial A}{\partial n_J}\right)_{V,T,n'} \tag{22c}$$

Thus we see that the μ_J show how all the extensive thermodynamic properties U, H, A, and G depend on the composition. This is why the chemical potential is so central to chemistry.

REAL GASES: THE FUGACITY

At various stages in the development of physical chemistry it is necessary to switch from a consideration of idealized systems to real systems. An example of the procedure was encountered in Chapter 1 when we turned our attention from perfect gases to real gases. In many cases it is desirable to make the transition by preserving the form of the expressions that have been derived for the idealized system, so that equations are preserved as much as possible, and deviations from the idealized behaviour can be expressed most simply. We shall illustrate such a

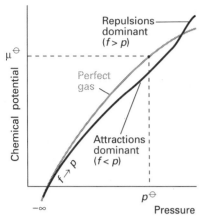

5.6 The chemical potential of a real gas. As $p \to 0$, μ coincides with the value for a perfect gas (shown by the pale line). When attractive forces are dominant (at intermediate pressures), the chemical potential is less than that of a perfect gas and the molecules have a lower 'escaping tendency'. At high pressures, when repulsive forces are dominant, the chemical potential of a real gas is greater than that of a perfect gas. Then the 'escaping tendency' is increased.

procedure in this section, by considering how the expressions that have been derived for perfect gases, particularly eqn 17 for the chemical potential of a perfect gas, are adapted to describe real gases.

The pressure dependence of the chemical potential of a real gas might resemble that shown in Fig. 5.6. To adapt eqn 17 to this case, we replace the true pressure p by an *effective* pressure, called the **fugacity** f, and write

$$\mu = \mu^{\ominus} + RT \ln\left(\frac{f}{p^{\ominus}}\right) \qquad (23)$$

The name 'fugacity' comes from the Latin for 'fleetness' in the sense of 'escaping tendency'; fugacity has the same dimensions as pressure. In later chapters we derive thermodynamically exact expressions in terms of chemical potentials, and therefore in terms of fugacities. For example, although from elementary chemistry we know that the equilibrium constant for a reaction such as

$$H_2(g) + Br_2(g) \rightleftharpoons 2HBr(g)$$

would be written

$$K = \frac{p_{HBr}^2}{p_{H_2} p_{Br_2}}$$

where p_J is the partial pressure of substance J, this expression is only an approximation. The thermodynamically exact expression is

$$K = \frac{f_{HBr}^2}{f_{H_2} f_{Br_2}}$$

where f_J is the fugacity of species J. Although the latter expression is exact, it is useful only if we know how to interpret the fugacities in terms of the partial pressures. This is the task we deal with in the remainder of this section.

5.6 Standard states of real gases

A perfect gas is in its standard state when its pressure is $p^{\ominus}$ (that is, 1 bar): the pressure arises solely from the kinetic energy of the molecules and there are no intermolecular forces to take into account. We aim to recapture this 'kinetic energy only' definition for a real gas by picturing it as a hypothetical state in which all the intermolecular forces have been extinguished:

> The **standard state of a real gas** is a hypothetical state in which the gas is at a pressure $p^{\ominus}$ and behaving perfectly.

The advantage of this definition is that it ensures that the standard state of a real gas has the simple properties of a perfect gas. If we had defined the standard state as the one for which $f = p^{\ominus}$, then the standard states of different gases would have had relatively complex properties. The choice of a hypothetical standard state literally standardizes the interactions between the particles by setting them to zero.[2] Then differences of standard chemical potential of different gases arise solely from the internal structure and properties of the molecules, not from the way they interact with each other.

5.7 The relation between fugacity and pressure

We shall write the fugacity as

$$f = \phi p \tag{24}$$

where ϕ is the dimensionless **fugacity coefficient**. In general, ϕ depends on the identity of the gas, the pressure, and the temperature. Then

$$\mu = \mu^{\ominus} + RT \ln\left(\frac{p}{p^{\ominus}}\right) + RT \ln \phi$$

As $\mu^{\ominus}$ refers to a hypothetical 'kinetic energy only' gas, and the term $\ln(p/p^{\ominus})$ is the same as for a perfect gas, the term $RT \ln \phi$ must express the entire effect of all the intermolecular forces.

Because all gases become perfect as the pressure approaches zero (so $f \to p$ as $p \to 0$), we know that

$$\phi \to 1 \qquad \text{as } p \to 0$$

We shall now show that, at a general pressure p, the fugacity coefficient of a gas is given by the expression

$$\ln \phi = \int_0^p \left(\frac{Z-1}{p}\right) \mathrm{d}p \tag{25}$$

where Z is the compression factor of the gas ($Z = pV_\mathrm{m}/RT$; this quantity was introduced in Section 1.4). Equation 25 is an explicit expression for the fugacity coefficient at any pressure p and therefore, through eqn 24, for the fugacity of the gas at that pressure.

JUSTIFICATION

Equation 12 is true for all gases, whether real or perfect. Expressing it in terms of molar quantities and then using eqn 23 gives

$$\int_{p'}^p V_\mathrm{m} \, \mathrm{d}p = \mu - \mu' = RT \ln\left(\frac{f}{f'}\right)$$

In this expression, f is the fugacity when the pressure is p and f' is the fugacity when the pressure is p'. If the gas were perfect we could write

$$\int_{p'}^p V_\mathrm{pg,m} \, \mathrm{d}p = \mu_\mathrm{pg} - \mu'_\mathrm{pg} = RT \ln\left(\frac{p}{p'}\right)$$

where the subscript pg denotes quantities relating to a perfect gas. The

2 The alternative choice, of selecting as the standard state the gas at zero pressure, at which it certainly behaves perfectly, runs into difficulty because (by eqn 17) $\mu \to -\infty$ as $p \to 0$.

difference of the two equations is

$$\int_{p'}^{p} (V_m - V_{pg,m})\, dp = RT\left\{\ln\left(\frac{f}{f'}\right) - \ln\left(\frac{p}{p'}\right)\right\}$$

which can be rearranged to

$$\ln\left(\frac{f}{p} \times \frac{p'}{f'}\right) = \frac{1}{RT}\int_{p'}^{p} (V_m - V_{pg,m})\, dp$$

When $p' \to 0$, the gas behaves perfectly and f' becomes equal to the pressure p'. Therefore, $f'/p' \to 1$ as $p' \to 0$. If we take this limit (which means setting $f'/p' = 1$ on the left and $p' = 0$ on the right), then the last equation becomes

$$\ln\left(\frac{f}{p}\right) = \frac{1}{RT}\int_{0}^{p} (V_m - V_{pg,m})\, dp$$

Then, with $\phi = f/p$,

$$\ln \phi = \frac{1}{RT}\int_{0}^{p} (V_m - V_{pg,m})\, dp$$

For a perfect gas $V_{pg,m} = RT/p$. For a real gas, $V_m = RTZ/p$, where Z is the compression factor. Therefore

$$\ln \phi = \int_{0}^{p} \left(\frac{Z-1}{p}\right) dp$$

which is eqn 25.

To evaluate ϕ from eqn 25, we need experimental data on the compression factor from very low pressures up to the pressure of interest. Some information of this kind is available in numerical tables, in which case the integral may be evaluated numerically. Sometimes an algebraic expression is available for Z (for instance, from one of the equations of state, Table 1.6) and it may be possible to evaluate the integral analytically. Thus, if we know the virial coefficients for the gas we can obtain the fugacity using

$$\ln \phi = B'p + \tfrac{1}{2}C'p^2 + \ldots$$

This expression was obtained by explicit evaluation of eqn 25.

Example 5.6 *Calculating a fugacity*

Suppose that the attractive interactions between gas particles can be neglected and find an expression for the fugacity of a van der Waals gas in terms of the pressure. Estimate its value for ammonia at 10.00 atm and 298.15 K.

Method. The starting point for the calculation is eqn 25. To evaluate the integral, we need an analytical expression for Z, which can be obtained from the equation of state. We saw in Section 1.5 that the van der Waals coefficient a represents the attraction between molecules, so it may be set equal to zero in this calculation.

Answer. When we neglect a in the van der Waals equation, that equation becomes

$$p = \frac{RT}{V_m - b}$$

and hence

$$Z = 1 + \frac{bp}{RT}$$

The integral required is therefore

$$\int_0^p \left(\frac{Z-1}{p}\right) dp = \int_0^p \left(\frac{b}{RT}\right) dp = \frac{bp}{RT}$$

Consequently, from eqns 24 and 25, the fugacity at the pressure p is

$$f = p \times e^{bp/RT}$$

From Table 1.5, $b = 3.707 \times 10^{-2} \, \text{L mol}^{-1}$, so $bp/RT = 0.015$, giving

$$f = (10.00 \, \text{atm}) \times e^{0.015} = 10.15 \, \text{atm}$$

Comment. The effect of the repulsive term (as represented by the coefficient b in the van der Waals equation) is to increase the fugacity above the pressure, and so the effective pressure of the gas—its 'escaping tendency'—is greater than if it were perfect.

Exercise E5.6. Find an expression for the fugacity coefficient when the attractive interaction is dominant in a van der Waals gas, and the pressure is low enough to make the approximation $4ap/(RT)^2 \ll 1$. Evaluate the fugacity for ammonia, as above.

$$[\ln \phi = -ap/(RT)^2, \, 9.33 \, \text{atm}]$$

It is clear from Fig. 1.17 that, for most gases, $Z < 1$ up to moderate pressures, but that $Z > 1$ at higher pressures. If $Z < 1$ throughout the range of integration, then the integrand in eqn 25 is negative and $\phi < 1$. This implies that $f < p$ (the molecules tend to stick together) and that the chemical potential of the gas is less than that of a perfect gas (Fig. 5.6). At higher pressures, the range over which $Z > 1$ may dominate the range over which $Z < 1$. The integral is then positive, $\phi > 1$, and $f > p$ (the repulsive interactions are dominant and tend to drive the particles apart). Now the chemical potential of the gas is greater than that of the perfect gas at the same pressure.

Figure 5.7, which has been calculated using the full van der Waals equation of state, shows how the fugacity depends on the pressure in terms of their reduced variables (Section 1.6). Because the critical constants are available in Table 1.5, the graphs can be used for quick estimates of the fugacities of a wide range of gases. Table 5.2 gives some explicit values for nitrogen.

Table 5.2* The fugacity of nitrogen at 273 K

p/atm	f/atm
1	0.99955
10	9.9560
100	97.03
1000	1839

* More values are given in the Data section.

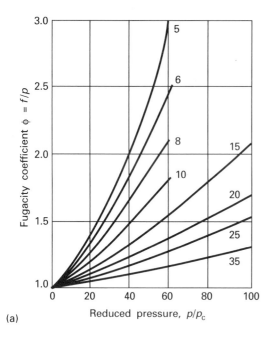

(a)

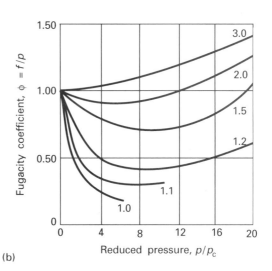

(b)

5.7 The fugacity coefficient of a van der Waals gas plotted using the reduced variables of the gas. The curves are labelled with the reduced temperature $T_r = T/T_c$.

Example 5.7 *Estimating the fugacity of a gas*

Estimate the fugacity of nitrogen at 500 atm and 0°C.

Method. To use the graphs in Fig. 5.7, we need to determine the reduced pressure and temperature of the gas, so we should begin by calculating $p_r = p/p_c$ and $T_r = T/T_c$, where p_c and T_c are the critical pressure and temperature, respectively.

Answer. Because the critical pressure and temperature of nitrogen are 33.5 atm and 126.2 K, the reduced pressure and temperature of the sample are

$$p_r = \frac{500 \text{ atm}}{33.5 \text{ atm}} = 14.9$$

$$T_r = \frac{273 \text{ K}}{126.2 \text{ K}} = 2.16$$

These values correspond to approximately $\phi = 1.15$ in Fig. 5.7b, so the fugacity of nitrogen is approximately

$$f = 1.15 \times 500 \text{ atm} = 575 \text{ atm}$$

under the stated conditions.

Comment. Because $\phi > 1$, the repulsive contributions are dominant in nitrogen at 500 atm and 0°C.

Exercise E5.7. Estimate the fugacity of carbon dioxide at 90°C and 580 atm.   [230 atm]

CHECK LIST OF KEY IDEAS

1 The formulation of the **fundamental equation** (eqn 1) expressing a change in internal energy in terms of changes in the volume and the entropy.

2 The derivation of the **Maxwell relations** (Section 5.1) between differential coefficients.

3 The derivation of a **thermodynamic equation of state** (eqn 5).

4 The variation of the Gibbs energy with the temperature and the pressure (eqn 6) and the derivation of the **Gibbs–Helmholtz equation** (eqn 10).

5 The pressure dependence of the Gibbs energy of a solid and liquid (eqn 13) and of a perfect gas (eqn 14).

6 The introduction of the **chemical potential** (eqn 15) and the variation with pressure of the chemical potential of a perfect gas (eqn 17).

7 The chemical potential of a substance in a mixture (eqn 19) and the derivation of the **fundamental equation of chemical thermodynamics** (eqn 20).

8 The relation of the chemical potential to the composition dependence of thermodynamic functions (Section 5.5).

9 The definition of the **fugacity** and the **standard state of a real gas** (Sections 5.6 and 5.7) and the calculation of the **fugacity coefficient** from compressibility data (eqn 25).

EXERCISES

Assume all gases are perfect and that the temperature is 298 K unless stated otherwise.

5.1. Express $(\partial S/\partial V)_T = (\partial p/\partial T)_V$ and $(\partial S/\partial p)_T = -(\partial V/\partial T)_p$ in terms of α and κ_T (see eqns 3.5 and 3.10).

5.2. Suppose that 3.0 mmol of N_2(g) occupies 36 cm³ at 300 K and expands to 60 cm³. Calculate ΔG for the process.

5.3. The change in the Gibbs energy of a certain constant-pressure process was found to fit the expression $\Delta G/J = -85.40 + 36.5(T/K)$. Calculate the value of ΔS for the process.

5.4. When the pressure on a 35-g sample of a liquid was increased isothermally from 1 atm to 3000 atm, the Gibbs energy increased by 12 kJ. Calculate the density of the liquid.

5.5. When 2.00 mol of a gas at 330 K and 3.50 atm is subjected to isothermal compression, its entropy decreases by 25.0 J K⁻¹. Calculate the final pressure of the gas and ΔG for the compression.

5.6. Calculate the change in chemical potential of a perfect gas that is compressed isothermally from 1.8 atm to 29.5 atm at 40°C.

5.7. The fugacity coefficient of a certain gas at 200 K and 50 bar is 0.72. Calculate the difference of its chemical potential from that of a perfect gas in the same state.

5.8. At 373 K, the second virial coefficient B of xenon is -81.7 cm³ mol⁻¹. Calculate the value of B' and hence estimate the fugacity coefficient of xenon at 50 atm and 373 K.

5.9. Estimate the change in the Gibbs energy of 1.0 L of benzene when the pressure acting on it is increased from 1.0 atm to 100 atm.

5.10. Calculate the change in the molar Gibbs energy of hydrogen gas when it is compressed isothermally from 1.0 atm to 100.0 atm at 298 K.

5.11. The molar Helmholtz energy of a certain gas is given by:

$$A = -\frac{a}{V_m} - RT \ln(V_m - b) + f(T)$$

where a and b are constants and $f(T)$ is a function of temperature only. Obtain the equation of state of the gas.

5.12. The molar Gibbs energy of a certain gas is given by:

$$G = RT \ln p + A' + B'p + \tfrac{1}{2}C'p^2 + \tfrac{1}{3}D'p^3$$

where A', B', C', and D' are constants. Obtain the equation of state of the gas.

5.13. Evaluate $(\partial S/\partial V)_T$ for a van der Waals gas. For an isothermal expansion will ΔS be greater for a perfect gas or a van der Waals gas? Explain your conclusion.

PROBLEMS

Numerical problems

5.1. Calculate $\Delta_r G^{\ominus}$ (375 K) for the reaction

$$2CO(g) + O_2(g) \rightarrow 2CO_2(g)$$

from the value of $\Delta_r G^{\ominus}$ (298 K), $\Delta_r H^{\ominus}$ (298 K), and the Gibbs–Helmholtz equation.

5.2. Estimate the standard reaction Gibbs energy of

$$N_2(g) + 3H_2(g) \rightarrow 2NH_3(g)$$

at (a) 500 K, (b) 1000 K from their values at 298 K.

5.3. At 298 K the standard enthalpy of combustion of sucrose is $-5645 \text{ kJ mol}^{-1}$ and the standard Gibbs energy of the reaction is $-5797 \text{ kJ mol}^{-1}$. Estimate the additional non-expansion work that may be obtained by raising the temperature to blood temperature, 37°C.

5.4. At 200 K, the compression factor of oxygen varies with pressure as shown below. Evaluate the fugacity of oxygen at this temperature and 100 atm.

p/atm	1.0000	4.00000	7.00000	10.0000	40.00	70.00	100.0
Z	0.9971	0.98796	0.97880	0.96956	0.8734	0.7764	0.6871

Theoretical problems

5.5. Show that $C_p = T(\partial S/\partial T)_p$ and $C_V = T(\partial S/\partial T)_V$.

5.6. Two of the four Maxwell relations were derived in the text, but two were not. Complete their derivation by showing that $(\partial S/\partial V)_T = (\partial p/\partial T)_V$ and $(\partial T/\partial p)_S = (\partial V/\partial S)_p$. Use the Maxwell relations to express the derivatives $(\partial S/\partial V)_T$ and $(\partial V/\partial S)_p$ in terms of the expansion coefficient α and the isothermal compressibility κ_T.

5.7. Use the Maxwell relations and Euler's chain relation to express $(\partial p/\partial S)_V$ in terms of the heat capacities, the expansion coefficient, and the isothermal compressibility.

5.8. Use the Maxwell relations to show that the entropy of a perfect gas depends on the volume as $S \propto R \ln V$.

5.9. Derive the thermodynamic equation of state

$$\left(\frac{\partial H}{\partial p}\right)_T = V - T\left(\frac{\partial V}{\partial T}\right)_p$$

Derive an expression for $(\partial H/\partial p)_T$ for (a) a perfect gas and (b) a van der Waals gas. In the latter case, estimate its value for 1.0 mol Ar(g) at 298 K and 10 atm. By how much does the enthalpy of the argon change when the pressure is increased isothermally to 11 atm?

5.10. Prove the following two relations:

(a) $\left(\dfrac{\partial S}{\partial p}\right)_V \left(\dfrac{\partial T}{\partial V}\right)_p - \left(\dfrac{\partial T}{\partial p}\right)_V \left(\dfrac{\partial S}{\partial V}\right)_p = -1$

(b) $\left(\dfrac{\partial H}{\partial V}\right)_T = -V^2 \left(\dfrac{\partial p}{\partial T}\right)_V \left(\dfrac{\partial}{\partial V}\left(\dfrac{T}{V}\right)\right)_p$

5.11. Show that if $B(T)$ is the second virial coefficient of a gas, and $\Delta B = B(T'') - B(T')$, $\Delta T = T'' - T'$, and T is the mean of T'' and T', then

$$\pi_T = \frac{RT^2}{V_m^2} \frac{\Delta B}{\Delta T}$$

Estimate π_T for argon at 275 K given that $B(250 \text{ K}) = -28.0 \text{ cm}^3 \text{ mol}^{-1}$ and $B(300 \text{ K}) = -15.6 \text{ cm}^3 \text{ mol}^{-1}$ at (a) 1.0 atm, (b) 10.0 atm.

5.12. (a) Prove that the heat capacities C_V and C_p of a perfect gas are independent of both volume and pressure. May they depend on the temperature? (b) Deduce an expression for the dependence of C_V on volume of a gas that is described by the equation of state $pV_m/RT = 1 + B/V_m$.

5.13. The Joule coefficient μ_J is defined as $\mu_J = (\partial T/\partial V)_U$. Show that

$$\mu_J C_V = p - \frac{\alpha T}{\kappa_T}$$

5.14. Evaluate π_T for a Dieterici gas (Table 1.6). Justify the form of the expression obtained.

5.15. Instead of assuming that the volume of a condensed phase is constant when pressure is applied, assume only that the compressibility is constant. Show that, when the pressure is changed isothermally by Δp, G changes to

$$G' = G + V_m \Delta p(1 - \tfrac{1}{2}\kappa_T \Delta p)$$

Assess the error in assuming that a solid is incompressible by applying this expression to the compression of copper when $\Delta p = 500$ atm. (For copper at 25°C, $\kappa_T = 0.8 \times 10^{-6} \text{ atm}^{-1}$ and $\rho = 8.93 \text{ g cm}^{-3}$.)

5.16. Derive an expression for the reaction Gibbs energy $\Delta_r G^{\ominus}$ at a temperature T' in terms of its value $\Delta_r G^{\ominus}$ at T by using the Gibbs–Helmholtz equation and (a) assuming that $\Delta_r H^{\ominus}$ does not vary with temperature, (b) assuming instead that $\Delta_r C_p$ does not vary with temperature and using Kirchhoff's law.

5.17. The adiabatic compressibility, κ_S, is defined like κ_T but at constant entropy. Show that for a perfect gas $p\gamma\kappa_S = 1$ (where γ is the ratio of heat capacities).

5.18. Show that, if S is regarded as a function of T and V, then

$$T \, dS = C_V \, dT + T\left(\frac{\partial p}{\partial T}\right)_V dV$$

Calculate the energy that must be transferred as heat to a van der Waals gas that expands reversibly and isothermally from V_i to V_f.

5.19. Suppose that S is regarded as a function of p and T. Show that

$$T\, dS = C_p\, dT - \alpha TV\, dp$$

Hence, show that the energy transferred as heat when the pressure on an incompressible liquid or solid is increased by Δp is equal to $-\alpha TV\, \Delta p$. Evaluate q when the pressure acting on $100\,\mathrm{cm}^3$ of mercury at $0°C$ is increased by $1.0\,\mathrm{kbar}$. ($\alpha = 1.82 \times 10^{-4}\,\mathrm{K}^{-1}$.)

5.20. The volume of a newly synthesized polymer was found to depend exponentially on the pressure as $V = V_0 e^{-p/p^*}$ where p is the excess pressure and p^* is a constant. Deduce an expression for the Gibbs energy of the polymer as a function of excess pressure. What is the natural direction of change of the compressed material when the pressure is relaxed?

5.21. Find an expression for the fugacity coefficient of a gas that obeys the equation of state

$$\frac{pV_m}{RT} = 1 + \frac{B}{V_m} + \frac{C}{V_m^2}$$

Use the resulting expression to estimate the fugacity of argon at $1.00\,\mathrm{atm}$ at $100\,\mathrm{K}$ using $B = -21.13\,\mathrm{cm}^3\,\mathrm{mol}^{-1}$ and $C = 1054\,\mathrm{cm}^6\,\mathrm{mol}^{-2}$.

5.22. Derive an expression for the fugacity coefficient of a gas that obeys the equation of state

$$\frac{pV_m}{RT} = 1 + \frac{BT}{V_m}$$

and plot ϕ against $4pB/R$.

Physical transformations of pure substances

*The simplest applications of thermodynamics to chemically significant systems
are to the discussion of the phase transitions that pure substances undergo. In
each case the process involves a single substance that undergoes a physical
change. We shall see that a phase diagram is a map of the ranges of pressure
and temperature at which each phase of a substance is the most stable (that is,
has the lowest Gibbs energy). First, the interpretation of empirically determined
phase diagrams is illustrated for four important materials. Then we turn to a
consideration of the factors that determine the positions and shapes of the
boundaries between the regions on a phase diagram. The practical importance
of the expressions we derive is that they show how the vapour pressure of a
substance varies with temperature and how the melting point varies with
pressure. Finally, we shall see that the transitions between phases can be
classified by noting how various thermodynamic functions change when the
transition occurs.*

Boiling, freezing, and the conversion of graphite to diamond are all
examples of changes of phase without change of chemical composition.
In this chapter we describe such processes thermodynamically using as
the guiding principle the tendency of systems at constant temperature
and pressure to minimize their Gibbs energy. Because we are dealing
with pure substances, the molar Gibbs energy of the system is the same
as the chemical potential μ, so the tendency to change is in the direction
of decreasing chemical potential. Once again we see how the properties
of the chemical potential mirror its name: because a spontaneous change
is accompanied by a decrease in Gibbs energy, a pure substance with a
high chemical potential has a spontaneous tendency to move to a state
with lower chemical potential.

It must never be forgotten that the alternative ways of expressing the
direction of spontaneous change all stem from the tendency of the system

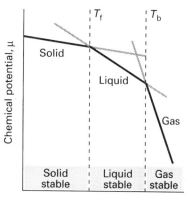

6.1 The schematic temperature dependence of the chemical potential of the solid, liquid, and gas phases of a substance (in practice, the lines are curved). The phase with the lowest chemical potential at a specified temperature is the most stable one at that temperature. The transition temperatures, the melting and boiling temperatures, are the temperatures at which the chemical potentials of two phases are equal.

and its surroundings to change in the direction of greater *total* entropy. Although a system may take on greater order and its entropy decrease, such as when a liquid freezes to a solid, that transition will be accompanied by an increase in entropy of the surroundings as a result of the heat released into them. As we shall see, below a certain temperature, the increase in entropy of the surroundings may be greater than the decrease in entropy of the system, and the transition is then spontaneous even though it brings about an increase in the order of the system.

A **phase** of a substance is a form of matter that is uniform throughout in chemical composition and physical state. (The word phase comes from the Greek word for appearance.) Thus, we speak of solid, liquid, and gas phases of a substance, and of its various solid phases (such as white phosphorus and black phosphorus). A **phase transition**, the spontaneous conversion of one phase to another phase, occurs at a characteristic temperature for a given pressure. Thus, at 1 atm, ice is the stable phase of water below 0°C, but above 0°C the liquid is more stable. This difference indicates that, below 0°C, the chemical potential of ice is lower than that of liquid water, $\mu(s) < \mu(l)$ (Fig. 6.1), and that, above 0°C, $\mu(l) < \mu(s)$. The **transition temperature** is the temperature at which the two chemical potentials coincide and $\mu(s) = \mu(l)$.

When considering phase transitions it is always important to distinguish between the thermodynamics of the transition and the rate at which it occurs. A transition that is predicted from thermodynamics to be spontaneous may occur too slowly to be significant in practice. For instance, at normal temperatures and pressures the chemical potential of graphite is lower than that of diamond, so there is a thermodynamic tendency for diamond to convert to graphite. However, for this transition to take place, the C atoms must change their locations, which is an immeasurably slow process in a solid except at high temperatures. The rate of attainment of equilibrium is a *kinetic* problem and is outside the range of thermodynamics. In gases and liquids the mobilities of the molecules allow phase transitions to occur rapidly, but in solids thermodynamic instability may be frozen in. Thermodynamically unstable phases that persist because the transition is kinetically hindered are called **metastable phases**. Diamond is a metastable phase of carbon under normal conditions. Metastable phases are sometimes said to be **kinetically stable** as distinct from thermodynamically stable.

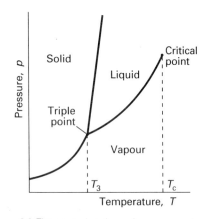

6.2 The general regions of pressure and temperature where solid, liquid, or gas is stable (i.e. has lowest chemical potential) are shown on this phase diagram. For example, the solid phase is the most stable phase at low temperatures and high pressures. In the following paragraphs we locate the precise boundaries between the regions.

PHASE DIAGRAMS

The **phase diagram** of a substance shows the regions of pressure and temperature at which its various phases are thermodynamically stable (Fig. 6.2). The boundaries between regions, the **phase boundaries**, show the values of p and T at which two phases coexist in equilibrium. As we shall see, the solid–liquid phase boundary is a plot of the freezing point at various pressures, the liquid–vapour boundary is a plot of the vapour pressure of the liquid against temperature, and the solid–vapour boundary is a plot of the sublimation vapour pressure against temperature.

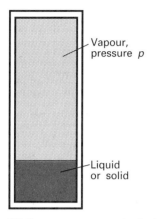

6.3 The vapour pressure of a liquid or solid is the pressure exerted by the vapour in equilibrium with the condensed phase.

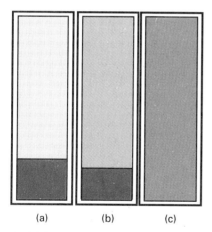

(a)   (b)   (c)

6.4 (a) A liquid in equilibrium with its vapour. (b) When a liquid is heated in a sealed container, the density of the vapour phase increases and that of the liquid decreases slightly. (The decrease in *quantity* of liquid is a result of vaporization.) (c) There comes a stage at which the two densities are equal and the interface between the fluids disappears. This occurs at the critical temperature. The container needs to be strong: the critical temperature of water is 374°C and the vapour pressure is then 218 atm.

6.1 Phase boundaries

Consider a sample of a pure substance in a closed vessel of constant volume. The pressure of a vapour in equilibrium with its condensed phase at a specified temperature is called the **vapour pressure** of the substance at that temperature (Fig. 6.3). Hence, as anticipated above, the phase boundaries between the liquid and the vapour and between the solid and the vapour show how the vapour pressures of the two condensed phases vary with temperature. The vapour pressure of a substance increases with temperature because, at higher temperatures, the molecules can escape more readily from the attractive interactions that bind them to their neighbours in the condensed phase.

Critical points and boiling points

The behaviour of a liquid heated in an open vessel differs from that of a liquid in a sealed vessel. In an open vessel, the liquid vaporizes from its surface as it is heated. At the temperature at which its vapour pressure would be equal to the external pressure, vaporization can occur throughout the bulk of the liquid and the vapour can expand freely into the surroundings. The condition of free vaporization throughout the liquid is called **boiling**. The temperature at which the vapour pressure of a liquid is equal to the external pressure is called the **boiling temperature** at that pressure. Note that a liquid does not suddenly start to form a vapour at its boiling temperature, for even at lower temperatures there is an equilibrium between the liquid and its vapour: at the boiling point the vapour pressure is great enough to drive back the atmosphere and vaporization can occur freely. For the special case of an external pressure of 1 atm, the boiling temperature is called the **normal boiling point** T_b. With the replacement of 1 atm by 1 bar as standard pressure, there is some advantage in modifying the definition so that the transition temperature refers to that pressure; the term **standard boiling point** is then used. Because 1 bar is slightly less than 1 atm (1.00 bar = 0.987 atm), the standard boiling point of a liquid is slightly lower than its normal boiling point. The normal boiling point of water is 100.0°C; its standard boiling point is 99.6°C.

When a liquid is heated in a sealed vessel, boiling does not occur. Instead, the temperature, vapour pressure, and the density of the vapour rise continuously (Fig. 6.4). At the same time, the density of the liquid decreases as a result of its expansion. There comes a stage at which the density of the vapour is equal to that of the remaining liquid and the surface between the two phases disappears. The temperature at which the surface disappears is the **critical temperature** T_c (which we first encountered in Section 1.4). The corresponding vapour pressure is the **critical pressure** p_c. At and above this temperature a single uniform phase fills the container and an interface no longer exists. That is, above the critical temperature the liquid phase of the substance does not exist.

Melting points and triple points

The temperature at which, under a specified pressure, liquid and solid coexist in equilibrium is called the **melting temperature**. Because a substance melts at exactly the same temperature as that at which it freezes, the melting temperature of a substance is the same as its **freezing temperature**. The freezing temperature when the pressure is 1 atm is called the **normal freezing point** T_f and that when the pressure is 1 bar is called the **standard freezing point**. The normal and standard freezing points are negligibly different for most purposes. The normal freezing point is also called the normal melting point.

There is a set of conditions under which three different phases (typically solid, liquid, and vapour) all simultaneously coexist in equilibrium. It is represented by the **triple point**, where the three phase boundaries coincide (this point is marked in Fig. 6.2; the temperature at the triple point is denoted T_3). The location of the triple point of a pure substance is outside our control: it occurs at a single definite pressure and temperature characteristic of the substance. The triple point of water lies at 273.16 K and 611 Pa (6.11 mbar, 4.58 Torr), and the three phases of water (that is, frozen water, liquid water, and water vapour) coexist in equilibrium at no other combination of pressure and temperature. This invariance of the triple point is the basis of its use in the definition of the thermodynamic temperature scale (Section 4.6).

As can be seen from Fig. 6.2, the triple point marks the lowest pressure at which a liquid phase of a substance can exist. If (as is common) the slope of the solid–liquid phase boundary is as shown in the diagram, then the triple point also marks the lowest temperature at which the liquid can exist; the critical temperature is the upper limit.

6.2 Phase diagrams of single substances

We shall now show how these general features appear in the phase diagrams of pure substances. In Chapter 8 we shall consider the more elaborate phase diagrams that summarize the phase equilibria of mixtures and for which composition is an additional variable.

Water

Figure 6.5 is the phase diagram for water. The liquid–vapour line summarizes how the vapour pressure of liquid water varies with temperature. It also summarizes how the boiling temperature varies with pressure. The solid–liquid line shows how the melting temperature varies with the pressure and indicates that enormous pressures are needed to bring about significant changes. Notice that the line has a negative slope up to 2 kbar, which means that the melting temperature falls as the pressure is raised. The reason for this unusual behaviour can be traced to the decrease in volume that occurs on melting, and hence it being more favourable for the solid to transform into the liquid as the pressure is raised. The decrease in volume is a result of the very open molecular structure of ice: the H_2O molecules are held apart (as well as together) by the hydrogen bonds between them, but the structure partially collapses on melting and the liquid is denser than the solid.

The motion of glaciers may be a consequence of the decrease in melting temperature with pressure: glacial ice melts where it is pressed against the sharp edges of stones and rocks and the glacier inches

6.5 The experimental phase diagram for water showing the different solid phases. Note the change of vertical scale at 2 bar.

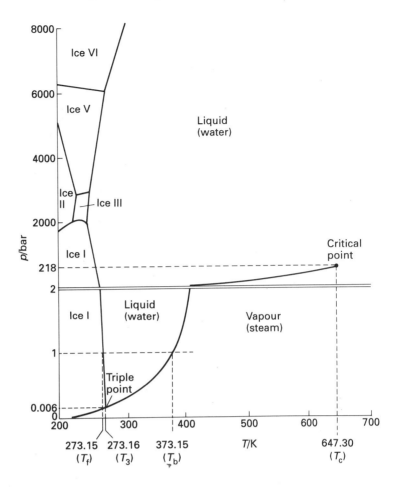

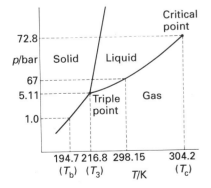

6.6 The experimental phase diagram for carbon dioxide. Note that, as the triple point lies at pressures well above atmospheric, liquid carbon dioxide does not exist under normal conditions (a pressure of at least 5.1 bar must be applied).

forwards. However, for many substances surface melting also occurs below the normal melting point, and the explanation of glacier motion (and ice skating) may be more subtle. The reduction in chemical potential of the water below that of the ice may stem from differences in the energy of interaction between water and ice and the rock surface.

At high pressures, different structural forms of ice come into stability as the bonds between molecules are modified by the stress. Some of these phases (which are called ice-II, III, V, VI, and VII)[1] melt at high temperatures. Ice-VII, for instance, melts at 100°C, but exists only above 25 kbar. Note that five more triple points occur in the diagram other than the one where vapour, liquid, and ice-I coexist. Each one occurs at a definite pressure and temperature that cannot be changed.

Carbon dioxide

The phase diagram for carbon dioxide is shown in Fig. 6.6. The features to notice include the positive slope of the solid–liquid boundary (the direction of this line is characteristic of most substances), which indicates that the melting temperature of solid carbon dioxide rises as the pressure is increased. Notice also that, as the triple point lies above 1 atm, the

1 Ice-IV was an illusion, like the once-fashionable alternative liquid phase called 'polywater'.

liquid cannot exist at normal atmospheric pressures whatever the temperature, and the solid sublimes when left in the open (hence the name 'dry ice'). To obtain the liquid, it is necessary to exert a pressure of at least 5.11 atm. Cylinders of carbon dioxide generally contain the liquid or compressed gas; at a temperature of 25°C that implies a vapour pressure of 67 atm if both gas and liquid are present in equilibrium. When the gas squirts through the throttle it cools by the Joule–Thomson effect, so that when it emerges into a region where the pressure is only 1 atm it condenses into a finely divided snow-like solid.

Supercritical carbon dioxide (that is, compressed carbon dioxide heated to above its critical temperature) is used in **supercritical fluid chromatography** (SFC), a form of chromatography in which the supercritical fluid is used as the mobile phase. The technique can be used to separate lipids and phospholipids and to separate fuel oil into alkanes, alkenes, and arenes. A more mundane application of supercritical carbon dioxide is in the decaffeination of coffee, in which green coffee beans are extracted by the fluid.

Carbon

Figure 6.7 is the phase diagram for carbon. It is ill defined because the various phases come into stability at high temperatures and pressures, and gathering the data is very difficult. The third solid phase of carbon, the fullerene phase (which is a solid composed of C_{60} molecules) might appear somewhere on the phase diagram, but the location of its region of thermodynamic stability has not yet been determined. It is not yet known whether the fullerene phase is thermodynamically more stable than other allotropes of carbon: it might be metastable, and hence not appear in the phase diagram.

As an indication of the difficulty of collecting data on carbon, it should be noted that liquid carbon can be obtained at 1 bar, but only at temperatures above 4000 K; to make the liquid at 2000 K requires a pressure of more than about 550 kbar. Making diamonds is a minor problem in comparison, because the diamond phase becomes stable at 400 kbar at 1000 K. Small diamonds are synthesized and are widely used in industry, but the phase diagram does not reveal the full problem. The rate of conversion is an important factor, and pure graphite changes into

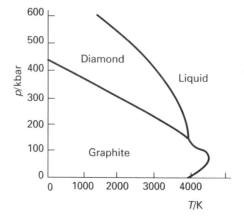

6.7 The phase diagram for carbon. There are very large uncertainties about the precise form of this phase diagram because the data are so difficult to obtain.

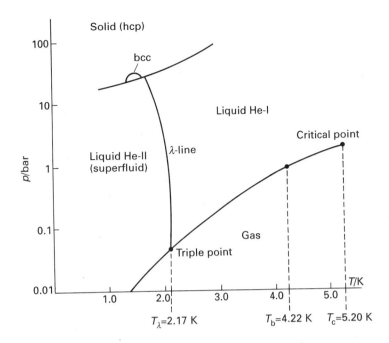

6.8 The phase diagram for helium (^{4}He). The λ-line marks the conditions under which the two liquid phases are in equilibrium. Helium-II is the superfluid phase. Note that a pressure of over 20 bar must be exerted before solid helium can be obtained. The labels hcp and bcc denote different solid phases in which the atoms pack together differently: hcp denotes hexagonal closed packing and bcc denotes body-centred cubic (see Section 21.1 for a description of these structures).

diamond at a useful rate only when the temperature is close to 4000 K, but then the apparatus tends to disappear first. Therefore catalysts are added in commercial syntheses; then the conversion proceeds at more attainable conditions. The contamination by the metal catalysts such as molten nickel (which also acts as a solvent for the carbon) enables commercial and natural diamonds to be distinguished from one another. This high-pressure procedure has been partially replaced by one that makes use of the decomposition of methane and the preferential deposition of the diamond phase.[2]

Helium

The phase diagram of helium is shown in Fig. 6.8. Helium behaves unusually at low temperatures. For instance, the solid and gas phases are never in equilibrium however low the temperature: the He atoms are so light that they vibrate with a large-amplitude motion even at very low temperatures and the solid simply shakes itself apart. Solid helium can be obtained, but only by holding the atoms together by applying pressure.

When considering helium at low temperatures it is necessary to distinguish between the isotopes ^{3}He and ^{4}He, because quantum mechanical effects become important and these two isotopes differ not only in mass but also in the spin of their nuclei.[3] As a result of quantum

2 The mechanism is obscure. It appears that both diamond and graphite phases are formed, but that the radicals in the gas, which include H atoms and CH radicals, react more rapidly with the graphite phase and remove it, leaving the diamond phase. This process is an example of a non-equilibrium, kinetically controlled reaction.

3 Nuclear spin is a quantum mechanical property that will be encountered in more detail in Part 2. At this stage it can be visualized as a spinning motion of the nucleus. A ^{4}He nucleus has zero spin; a ^{3}He nucleus has non-zero spin. The difference in spin leads to markedly different properties at temperatures that are low enough for quantum mechanics to be relevant to the description of bulk samples.

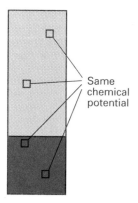

Same
chemical
potential

6.9 When two or more phases are in equilibrium, the chemical potential of a substance (and, in a mixture, a component) is the same in each phase and is the same at all points in each phase.

mechanical effects, pure helium-4 has a liquid–liquid phase transition at its **λ-line** (the reason for this name is explained in Section 6.5). The liquid phase marked He-I behaves like a normal liquid. The other phase, He-II, is a **superfluid** and is so called because it flows without viscosity. The phase diagram of helium-3 differs from the phase diagram of helium-4, but it also possesses a superfluid phase. Helium-3 is unusual in that the entropy of the liquid is lower than that of the solid and melting is exothermic.

PHASE STABILITY AND PHASE TRANSITIONS

We shall now see how thermodynamic considerations can account for the features of the phase diagrams we have just described. To do so we shall base our discussion on the following consequence of the Second Law:

> At equilibrium, the chemical potential of a substance is the same throughout a sample, regardless of how many phases are present.

When the liquid and solid phases of a substance are in equilibrium, the chemical potential of the substance is the same throughout the liquid and throughout the solid, and is the same in the solid as in the liquid (Fig. 6.9).

JUSTIFICATION

The proof of the uniformity of chemical potentials is as follows. Consider a system in which the chemical potential of a substance is μ_1 at one location and μ_2 at another location. (The locations may be in the same or in different phases.) When an amount dn of the substance is transferred from one location to the other, the Gibbs energy of the system changes by $-\mu_1 \, dn$ when material is removed from location 1 and by $+\mu_2 \, dn$ when that material is added to location 2. The overall change is therefore $dG = (\mu_2 - \mu_1) \, dn$. If the chemical potential at location 1 is higher than that at location 2, then the transfer is accompanied by a decrease in G and thus has a spontaneous tendency to occur. Only if $\mu_1 = \mu_2$ is there no change in G and only then is the system at equilibrium.

MOLECULAR INTERPRETATION 6.1

The tendency towards equality of chemical potential is a disguised form of the tendency to greater total entropy expressed by the Second Law. Suppose that in location 1 the molecules experience less favourable attractive forces than in location 2. Then the entropy of the *surroundings* will increase if molecules migrate from location 1 to location 2 because heat will be released into the surroundings and increase their disorder. However, to judge whether a process is spontaneous, we need to consider the total entropy change, and there may be a difference in molecular disorder between the two locations. If the disorder of the molecules in location 1 is greater than in location 2, then the entropy of the system will *decrease* when molecules migrate from location 1 to location 2. For example, location 1 might be a gas phase and location 2 might be a liquid phase. The transfer of molecules from location 1 (gas) to location 2 (liquid) will be spontaneous if the increase in entropy of the surroundings exceeds the decrease in entropy of the system. The opposite transfer, from location 2 (liquid) to location 1 (gas), will be spontaneous if the increase in disorder of the system is greater than the decrease in disorder of the surroundings. The change will be spontaneous in neither direction if the changes in entropies of the system and its surrounding cancel, which is expressed by setting the chemical potentials of the two phases equal to one another, for the chemical potential takes into account the contributions of the enthalpy change to the entropy of the surroundings and the change in entropy of the molecules themselves.

6.3 The dependence of stability on the conditions

We shall denote the chemical potentials of the solid, liquid, and gas phases of a substance by $\mu(s)$, $\mu(l)$, and $\mu(g)$, respectively. At a given pressure, a phase is thermodynamically stable over the range of temperatures at which it has a lower chemical potential than any other phase.

A solid phase has the lowest chemical potential and, so long as the pressure is not too low, is the most stable at low temperatures. However, the chemical potentials of phases change with temperature in different ways and, as the temperature is raised, the chemical potential of another phase (perhaps another solid phase, or a liquid, or a gas) will fall below that of the solid. When that happens, a phase transition occurs if it is kinetically feasible to do so.

The temperature dependence of phase stability

The temperature dependence of the Gibbs energy is expressed in terms of the entropy of the system by eqn 5.7 (that $(\partial G/\partial T)_p = -S$). Because the chemical potential of a pure substance is equal to the *molar* Gibbs energy of that substance, it follows that the temperature dependence of the chemical potential is expressed in terms of the molar entropy of the substance by

$$\left(\frac{\partial \mu}{\partial T}\right)_p = -S_{\mathrm{m}} \tag{1}$$

This relation shows that, *as the temperature is raised, the chemical potential of a pure substance decreases* (because $S_{\mathrm{m}} > 0$ always, so the slope of a graph of μ against T is negative). It also implies that *the slope of a plot of μ against temperature is steeper for gases than for liquids*, because $S_{\mathrm{m}}(g) > S_{\mathrm{m}}(l)$, *and steeper for a liquid than the corresponding solid*, because $S_{\mathrm{m}}(l) > S_{\mathrm{m}}(s)$ almost always. The steep negative slope of $\mu(l)$ results in its falling below $\mu(s)$ when the temperature is high enough, and then the liquid becomes the stable phase: the solid melts. The chemical potential of the gas phase plunges steeply downwards as the temperature is raised (because the molar entropy of the vapour is so high), and there comes a temperature at which it lies lowest. Then the gas is the stable phase and the liquid vaporizes. When we bring about a phase transition what we are

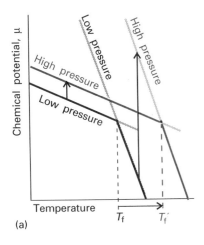

(a)

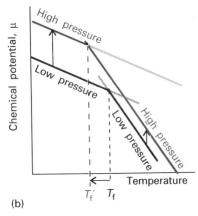

(b)

6.10 The pressure dependence of the chemical potential of a substance depends on the molar volume of the phase. The lines show schematically the effect of increasing pressure on the chemical potentials of the solid and liquid phases (in practice, the lines are curved), and the corresponding effects on the freezing temperatures. (a) In this case the molar volume of the solid is less than that of the liquid and $\mu(\text{s})$ increases less than $\mu(\text{l})$. As a result the freezing temperature rises. (b) Here the molar volume is greater for the solid than the liquid (as for water), so $\mu(\text{s})$ increases more strongly than $\mu(\text{l})$, and the freezing temperature is lowered.

actually doing is modifying the relative values of the chemical potentials of the phases, and the easiest way of doing that is by changing the temperature of the sample.

The response of melting to applied pressure

Most substances melt at a higher temperature when subjected to pressure. It is as though the pressure is preventing the formation of the less dense liquid phase. Exceptions to this behaviour include water, for which the liquid is denser than the solid, and the application of pressure to water encourages the formation of the liquid phase. That is, water freezes at a lower temperature when it is under pressure.

We can rationalize the response of melting temperatures to pressure in terms of the different pressure dependence of the chemical potentials of the solid and liquid phases, and the displacement of the temperature at which their graphs cross in Fig. 6.10. The variation of the chemical potential with pressure is expressed by

$$\left(\frac{\partial \mu}{\partial p}\right)_T = V_m \tag{2}$$

This equation shows that the slope of a graph of chemical potential against pressure is proportional to the molar volume of the substance. An increase in pressure raises the chemical potential of any pure substance (because $V_m > 0$). In most cases, $V_m(\text{l}) > V_m(\text{s})$ and the equation predicts that an increase in pressure increases the chemical potential of the liquid more than that of the solid. As shown in Fig. 6.10a, the effect is to raise the melting temperature slightly. For water, $V_m(\text{l}) < V_m(\text{s})$, and an increase in pressure increases the chemical potential of the solid more than that of the liquid. In this case, the melting temperature is lowered slightly (Fig. 6.10b).

Example 6.1 *Assessing the effect of pressure on the chemical potential*

Calculate the effect on the chemical potentials of increasing the pressure from 1.00 bar to 2.00 bar on ice and water at 0°C. The density of ice is 0.917 g cm^{-3} and that of liquid water is 0.999 g cm^{-3} under these conditions.

Method. From eqn 5.13 we know that the change in chemical potential of an incompressible substance is

$$\Delta \mu = V_m \Delta p$$

Therefore, to answer the question, we need to know the molar volumes of the two phases of water. These are obtained from the density ρ and the molar mass M by using

$$V_m = \frac{M}{\rho}$$

Answer. It follows from the expression just given that the molar volumes of ice and water (of molar mass $18.02\ \text{g mol}^{-1}$) are $19.7\ \text{cm}^3\ \text{mol}^{-1}$ and $18.0\ \text{cm}^3\ \text{mol}^{-1}$. Therefore

$$\Delta\mu(\text{ice}) = (1.97\times10^{-5}\ \text{m}^3\ \text{mol}^{-1})\times(1.0\times10^5\ \text{Pa})$$
$$= +1.97\ \text{J mol}^{-1}$$
$$\Delta\mu(\text{water}) = (1.80\times10^{-5}\ \text{m}^3\ \text{mol}^{-1})\times(1.0\times10^5\ \text{Pa})$$
$$= +1.80\ \text{J mol}^{-1}$$

Comment. The chemical potential of ice rises more sharply than that of water, so that if they are initially in equilibrium at 1 bar there will be a tendency for the ice to melt at 2 bar.

Exercise E6.1. Calculate the effect of an increase in pressure of 1.00 bar on a liquid and a solid of molar mass $44.0\ \text{g mol}^{-1}$ that are in equilibrium with densities $2.35\ \text{g cm}^{-3}$ and $2.50\ \text{g cm}^{-3}$, respectively.

$$[\Delta\mu(\text{l}) = +1.87\ \text{J mol}^{-1},\ \Delta\mu(\text{s}) = +1.76\ \text{J mol}^{-1},\ \text{solid forms}]$$

The effect of applied pressure on vapour pressure

We have considered the effect of applying pressure simultaneously to the solid and liquid phases of a substance. Now we consider the effect of applying pressure solely to the condensed phase. When pressure is applied to a condensed phase, its vapour pressure rises: in effect, molecules are squeezed out of the phase and escape as a gas. Compression can be achieved mechanically or by subjecting the condensed phase to the applied pressure of an inert gas (Fig. 6.11); in the latter case the vapour pressure is the *partial* pressure of the vapour in equilibrium with the condensed phase. One complication (which we ignore here) is that, if the condensed phase is a liquid, the pressurizing gas might dissolve and change its properties. Another complication is that the gas-phase molecules might attract molecules out of the liquid by the process of **gas solvation**, the attachment of molecules to gas-phase species.

The quantitative relation between the vapour pressure p when a pressure ΔP is applied and the vapour pressure p^* of the liquid in the absence of an additional pressure is

$$p = p^* e^{V_m\Delta P/RT} \qquad (3)^\circ$$

This equation shows how the vapour pressure increases when the pressure acting on the condensed phase is increased.

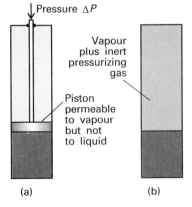

↓ Pressure ΔP

Vapour plus inert pressurizing gas

Piston permeable to vapour but not to liquid

(a)   (b)

6.11 Pressure may be applied to a condensed phase either (a) by compressing the condensed phase or (b) by subjecting it to an inert pressurizing gas. When pressure is applied, the vapour pressure of the condensed phase increases.

JUSTIFICATION

We calculate the vapour pressure of a pressurized liquid by using the fact that at equilibrium the chemical potentials of the liquid and its vapour are equal: $\mu(\text{l}) = \mu(\text{g})$. It follows that for any change that preserves equilibrium, the resulting change in $\mu(\text{l})$ must be equal to the change in $\mu(\text{g})$, and we can write $d\mu(g) = d\mu(l)$. When the pressure P on the liquid is increased by dP, the chemical potential of the liquid changes by

$$d\mu(l) = V_m(l)\,dP$$

The chemical potential of the vapour changes by

$$d\mu(g) = V_m(g)\,dp$$

where dp is the change in its pressure p, the vapour pressure we are trying to find. If we treat the vapour as a perfect gas, then the molar volume can be replaced by $V_m(g) = RT/p$, and we obtain

$$d\mu(g) = RT\,\frac{dp}{p}$$

Equating the changes in chemical potentials of the vapour and the liquid gives

$$RT\,\frac{dp}{p} = V_m(l)\,dP$$

This expression can be integrated once we know the limits of integration.

When there is no additional pressure acting on the liquid, P (the pressure experienced by the liquid) is equal to the normal vapour pressure p^*; so when $P = p^*$, $p = p^*$ too. When there is an additional pressure ΔP on the liquid, with the result that $P = p + \Delta P$, the vapour pressure is p (the value we want to find). The effect of pressure on the vapour pressure is so small that it is a good approximation to replace the p in $p + \Delta P$ by p^* itself, and to set the upper limit of the integral to $p^* + \Delta P$. The integrations required are therefore as follows:

$$RT\int_{p^*}^{p}\frac{dp}{p} = \int_{p^*}^{p^*+\Delta P} V_m(l)\,dP$$

We now assume that the molar volume of the liquid is the same throughout the small range of pressures involved. Then both integrations are straightforward, and lead to

$$RT\ln\!\left(\frac{p}{p^*}\right) = V_m(l)\,\Delta P$$

which rearranges to eqn 3.

Example 6.2 *Estimating the effect of presure on the vapour pressure*

Derive an expression from eqn 3 that is valid for small changes in p and calculate the percentage increase of the vapour pressure of water for an increase in pressure of 10 bar at 25°C.

Method. The question centres on the approximation of the right-hand side of eqn 3 when the exponent is small. For the approximation we note that the exponential function e^x is equal to the expansion $1 + x + x^2/2! + \ldots$, so, if $x \ll 1$, a good approximation is $e^x \approx 1 + x$.

Answer. If $V_m \Delta P/RT \ll 1$, the exponential function on the right of eqn 3 may be approximated by $1 + V_m \Delta P/RT$:

$$p \approx p^* \left(1 + \frac{V_m(l) \, \Delta P}{RT} \right)$$

This expression rearranges to

$$\frac{p - p^*}{p^*} \approx \frac{V_m \, \Delta P}{RT}$$

For water (which has density 0.997 g cm^{-3} at 25°C and therefore molar volume $18.1 \text{ cm}^3 \text{ mol}^{-1}$),

$$\frac{V_m \, \Delta P}{RT} = \frac{(1.81 \times 10^{-5} \text{ m}^3 \text{ mol}^{-1}) \times (1.0 \times 10^6 \text{ Pa})}{(8.3145 \text{ J K}^{-1} \text{ mol}^{-1}) \times (298 \text{ K})} = 7.3 \times 10^{-3}$$

Because $V_m \Delta P/RT \ll 1$, the approximate formula can be used, and we obtain

$$\frac{p - p^*}{p^*} \times 100 \text{ per cent} \approx 0.73 \text{ per cent}$$

Exercise E6.2. Calculate the percentage effect of an increase in pressure of 100 bar on the vapour pressure of benzene at 25°C, which has density 0.879 g cm^{-3}.

[43 per cent; because the change is so large, use eqn 3]

6.4 The location of phase boundaries

We can find the precise locations of the phase boundaries—the pressures and temperatures at which two phases can coexist—by making use of the fact that when two phases are in equilibrium their chemical potentials must be equal. Therefore, where the phases α and β are in equilibrium,

$$\mu_\alpha(p, T) = \mu_\beta(p, T)$$

By solving this equation for p in terms of T we shall get an equation for the phase boundary.

The slopes of the phase boundaries

It turns out to be simplest to discuss the phase boundaries in terms of their slopes, so we begin by finding an equation for dp/dT.

Let p and T be changed infinitesimally, but in such a way that the two phases α and β remain in equilibrium. The chemical potentials of the phases are initially equal (the two phases are in equilibrium). They remain equal when the conditions are changed to another point on the phase boundary, where the two phases continue to be in equilibrium. Therefore, the changes in the chemical potentials of the two phases must be equal and we can write $d\mu_\alpha = d\mu_\beta$. Because, from eqn 5.6, we know that

$$d\mu = -S_m \, dT + V_m \, dp$$

for each phase, it follows that

$$-S_{\alpha,\mathrm{m}}\,\mathrm{d}T + V_{\alpha,\mathrm{m}}\,\mathrm{d}p = -S_{\beta,\mathrm{m}}\,\mathrm{d}T + V_{\beta,\mathrm{m}}\,\mathrm{d}p$$

where $S_{\alpha,\mathrm{m}}$ and $S_{\beta,\mathrm{m}}$ are the molar entropies of the phases, and $V_{\alpha,\mathrm{m}}$ and $V_{\beta,\mathrm{m}}$ are their molar volumes. Hence

$$(V_{\beta,\mathrm{m}} - V_{\alpha,\mathrm{m}})\,\mathrm{d}p = (S_{\beta,\mathrm{m}} - S_{\alpha,\mathrm{m}})\,\mathrm{d}T$$

which rearranges into the **Clapeyron equation**:

$$\frac{\mathrm{d}p}{\mathrm{d}T} = \frac{\Delta S_{\mathrm{m}}}{\Delta V_{\mathrm{m}}} \tag{4}$$

In this expression $\Delta S_{\mathrm{m}} = S_{\beta,\mathrm{m}} - S_{\alpha,\mathrm{m}}$ and $\Delta V_{\mathrm{m}} = V_{\beta,\mathrm{m}} - V_{\alpha,\mathrm{m}}$ are the changes in molar entropy and molar volume when the transition occurs. This important result for the slope of the phase boundary at any point is exact; it applies to any phase equilibrium of any pure substance.

The solid–liquid boundary

Melting (fusion) is accompanied by a molar enthalpy change $\Delta_{\mathrm{fus}}H$ and occurs at a temperature T. The molar entropy of melting at T is therefore $\Delta_{\mathrm{fus}}H/T$, and the Clapeyron equation becomes

$$\frac{\mathrm{d}p}{\mathrm{d}T} = \frac{\Delta_{\mathrm{fus}}H}{T\,\Delta_{\mathrm{fus}}V} \tag{5}$$

where $\Delta_{\mathrm{fus}}V$ is the change in molar volume that occurs on melting. The enthalpy of melting is positive (the only exception is helium-3) and the volume change is usually positive and always small. This means that the slope $\mathrm{d}p/\mathrm{d}T$ is steep and usually positive. The curve itself can be obtained by integrating $\mathrm{d}p/\mathrm{d}T$ assuming that $\Delta_{\mathrm{fus}}H$ and $\Delta_{\mathrm{fus}}V$ barely change with temperature and pressure, and thus can be treated as constant in the integration. If the melting temperature is T^* when the pressure is p^*, and is T when the pressure is p, then the integration required is

$$\int_{p^*}^{p} \mathrm{d}p = \frac{\Delta_{\mathrm{fus}}H}{\Delta_{\mathrm{fus}}V} \int_{T^\star}^{T} \frac{\mathrm{d}T}{T}$$

Therefore, the approximate equation of the solid–liquid boundary is

$$p = p^* + \frac{\Delta_{\mathrm{fus}}H}{\Delta_{\mathrm{fus}}V} \ln\left(\frac{T}{T^*}\right) \tag{6}$$

This equation was originally obtained by yet another Thomson—James, the brother of William, Lord Kelvin. When T is close to T^*, the logarithm can be approximated by using

$$\ln\left(\frac{T}{T^*}\right) = \ln\left(1 + \frac{T - T^*}{T^*}\right) \approx \frac{T - T^*}{T^*}$$

because $\ln(1 + x) \approx x$ when x is small; therefore,

$$p = p^* + \frac{\Delta_{\mathrm{fus}}H}{T^*\,\Delta_{\mathrm{fus}}V} \times (T - T^*) \tag{7}$$

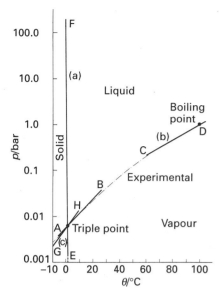

6.12 The phase boundaries for water as calculated in Examples 6.3, 6.5, and 6.6. Note that the vertical axis is logarithmic so that it has the effect of squashing down the upper parts of the diagram (compare the lower part of Fig. 6.5 which is on a linear scale).

This expression is the equation of a steep straight line when p is plotted against T (Fig. 6.12).

Example 6.3 *Constructing a solid–liquid phase boundary*

Although phase diagrams are constructed *experimentally*, it is instructive to see how their features are related to the thermodynamic properties of the substance through expressions like eqns 6 and 7. Construct the ice–liquid phase boundary for water at temperatures between $-1°C$ and $0°C$. What is the melting temperature of ice under a pressure of 1.5 kbar?

Method. The construction makes use of eqn 7. To use this equation, we need to select the point (p^*, T^*) and then to plot the straight line that passes through it. A sensible point to select (because it lies within the range we are asked to plot) is the standard freezing point (1 bar, 273 K). For the second part of the question, we need to rearrange eqn 6 into an expression for T. (Equation 7 could be used instead, but it is just as easy to use the more accurate equation.)

Answer. From Table 2.3 we have $\Delta_{fus}H = +6.008 \text{ kJ mol}^{-1}$ and, from Example 6.1, $\Delta_{fus}V = -1.7 \text{ cm}^3 \text{ mol}^{-1}$. Hence,

$$p/\text{bar} = 1.0 - (3.53 \times 10^4) \times \ln(T/273.15 \text{ K})$$

This formula gives the following values:

$\theta/°C$	-1.0	-0.8	-0.6	-0.4	-0.2	0.0
p/bar	130	105	79	53	27	1.0

These points are plotted in Fig. 6.12 as the line EF. For the second part, we rearrange the formula into

$$T = (273.15 \text{ K}) \times \exp\left(\frac{1.0 - p/\text{bar}}{3.53 \times 10^4}\right)$$

Then, with $p = 1.5$ kbar, $T = 262$ K or $-11°C$.

Comment. Notice the decrease in melting temperature with increasing pressure: water is denser than ice, so ice responds to pressure by tending to melt.

Exercise E6.3. Estimate the value of the difference between the standard melting point of ice and its normal melting point.   [About $+0.1$ mK]

The liquid–vapour boundary

The molar entropy of vaporization at a temperature T is equal to $\Delta_{vap}H/T$; the Clapeyron equation for the liquid–vapour boundary is therefore

$$\frac{dp}{dT} = \frac{\Delta_{vap}H}{T\,\Delta_{vap}V} \tag{8}$$

The enthalpy of vaporization is positive; $\Delta_{vap}V$ is large and positive.

Therefore dp/dT is positive, but it is much smaller than for the solid–liquid boundary.

Example 6.4 *Estimating the effect of pressure on the boiling point*

Estimate the typical size of the effect of increasing pressure on the boiling point of a liquid.

Method. To use eqn 8 we need to estimate the right-hand side. The term $\Delta_{vap}H/T$, with T the normal boiling point, should be recognized as Trouton's constant (Section 4.4). Because the molar volume of a gas is so much greater than the molar volume of a liquid, we can write

$$\Delta_{vap}V = V_m(g) - V_m(l) \approx V_m(g)$$

and take for $V_m(g)$ the molar volume of a perfect gas (at low pressures, at least).

Answer. According to Trouton's law,

$$\Delta_{vap}H \approx 85\,\mathrm{J\,K^{-1}\,mol^{-1}} \times T_b$$

The molar volume of a perfect gas is about $25\,\mathrm{L\,mol^{-1}}$ at 1 atm and near but above room temperature. Therefore,

$$\frac{dp}{dT} = \frac{85\,\mathrm{J\,K^{-1}\,mol^{-1}}}{25 \times 10^{-3}\,\mathrm{m^3\,mol^{-1}}} = 3.4 \times 10^3\,\mathrm{Pa\,K^{-1}} = 0.034\,\mathrm{atm\,K^{-1}}$$

This value corresponds to a dT/dp slope of $30\,\mathrm{K\,atm^{-1}}$; hence a change of pressure of $+0.1$ atm can be expected to change a boiling temperature by about $+3$ K.

Exercise E6.4. Estimate dT/dp for water at its normal boiling point using the information in Table 4.2 and $V_m(g) \approx RT/p$.   [$28\,\mathrm{K\,atm^{-1}}$]

As in the example, because the molar volume of a gas is so much greater than the molar volume of a liquid, we can write $\Delta_{vap}V \approx V_m(g)$. Moreover, if the gas behaves perfectly, $V_m(g) = RT/p$. These two approximations turn the exact Clapeyron equation into the approximate **Clausius–Clapeyron equation**:

$$\frac{d\ln p}{dT} \approx \frac{\Delta_{vap}H}{RT^2} \qquad\qquad (9)°$$

for the variation of vapour pressure with temperature. (We have used $dx/x = d\ln x$.) If we also assume that the enthalpy of vaporization is independent of temperature, this equation integrates to

$$p = p^* e^{-\chi} \quad \text{with} \quad \chi = \frac{\Delta_{vap}H}{R}\left(\frac{1}{T} - \frac{1}{T^*}\right) \qquad (10)°$$

where p^* is the vapour pressure when the temperature is T^* and p the vapour pressure when the temperature is T. Equation 10 is the curve plotted as the liquid–vapour boundary in Fig. 6.12 (see Example 6.5).

The line does not extend beyond the critical temperature T_c because above this temperature the liquid does not exist.

Example 6.5 *Constructing a vapour pressure curve*

Construct the vapour pressure curve for water between $-5°C$ and $100°C$.

Method. To use eqn 10, we need to select the fixed point (p^*, T^*). A good strategy is to select known values that are close to the range of temperatures being plotted, for the derivation of eqn 10 has assumed that the enthalpy of vaporization is a constant. At the lower end of the temperature range we can select the triple point (6.11 mbar, 273.16 K) and at the upper end of the range we can select the normal boiling point (1.01 bar, 373.15 K). Use $\Delta_{vap}H = +45.05 \, kJ \, mol^{-1}$ at 273 K and temperatures nearby, and $+40.66 \, kJ \, mol^{-1}$ at 373 K and temperatures nearby. Note that the assumption that the enthalpy of vaporization is constant in each range is an approximation.

Answer. Evaluation of eqn 10 for several temperatures gives the following values:

$\theta/°C$	-5	0	5	10	20	30	70	80	90	100
p/atm	0.004	0.006	0.009	0.012	0.024	0.044	0.32	0.48	0.70	1.0

The curve is plotted as AB and CD in Fig. 6.12, and compared there with the experimental vapour pressure.

Comment. The negative curvature of the curve as plotted comes from the use of a logarithmic scale. Note that we can use eqn 10 in conjunction with experimental vapour pressure data to determine enthalpies of vaporization.

Exercise E6.5. Calculate the standard boiling point of water from its normal boiling point.   [99.6°C]

The solid–vapour boundary

The only difference between this case and the last is the replacement of the enthalpy of vaporization by the enthalpy of sublimation $\Delta_{sub}H$. The approximations that led to the Clausius–Clapeyron equation give the following expressions for the temperature dependence of the sublimation vapour pressure:

$$\frac{d \ln p}{dT} \approx \frac{\Delta_{sub}H}{RT^2}$$

$$p = p^* e^{-\chi'} \quad \text{with} \quad \chi' = \frac{\Delta_{sub}H}{R}\left(\frac{1}{T} - \frac{1}{T^*}\right) \tag{11}°$$

Because the enthalpy of sublimation is greater than the enthalpy of vaporization, the equation predicts a steeper slope for the sublimation curve than for the vaporization curve near where they meet (Fig. 6.12).

Example 6.6 *Constructing a solid–vapour phase boundary*

Construct the ice–vapour phase boundary over the range −10°C to +5°C, using the information that, at 273 K, $\Delta_{vap}H = +45.05\ kJ\ mol^{-1}$ and $\Delta_{fus}H = +6.01\ kJ\ mol^{-1}$.

Method. To use eqn 11, we need to know the enthalpy of sublimation. From the First Law and the fact that H is a state function we can write (for data at the same temperature)

$$\Delta_{sub}H = \Delta_{fus}H + \Delta_{vap}H$$

The fixed point (p^*, T^*) can be taken as the triple point (6.11 mbar, 273.16 K), as in Example 6.5, for this point lies in the range of interest.

Answer. The enthalpy of sublimation is

$$\Delta_{sub}H = 6.01\ kJ\ mol^{-1} + 45.05\ kJ\ mol^{-1} = +51.06\ kJ\ mol^{-1}$$

Then substitution of this value into eqn 11 gives

$\theta/°C$	−10	−5	0	5
p/bar	0.003	0.004	0.006	0.009

These points are plotted as GH in Fig. 6.12 and compared with the experimental curve.

Comment. The small discrepancy between the experimental and calculated curves is a result of assuming that the enthalpy of sublimation is a constant.

Exercise E6.6. Construct the phase diagram for carbon dioxide using the data given in the margin.   [compare with Fig. 6.6]

Density of solid: 1.53 g cm^{-3}
Density of liquid: 0.78 g cm^{-3}
$\Delta_{sub}H$: +25.2 kJ mol^{-1}
$\Delta_{fus}H$: +8.3 kJ mol^{-1}
Triple point: −57°C, 5.11 bar
Critical constants: 31°C, 72.8 bar

6.5 The Ehrenfest classification of phase transitions

There are many different types of phase transition, including the common examples of fusion and vaporization and the less common examples of solid–solid, conducting–superconducting, and fluid–superfluid transitions. The question arises as to whether these transitions have a common character or whether they fall into different classes. We shall now see that it is possible to use thermodynamic properties of substances and, in particular, the behaviour of the chemical potential to classify phase transitions into different types. The classification scheme was originally proposed by Paul Ehrenfest, and is known as the **Ehrenfest classification**.

Many familiar phase transitions, like fusion and vaporization, are accompanied by changes of enthalpy and volume. These changes have implications for the slopes of the chemical potentials of the phases at either side of the phase transition. Thus, at the transition from a phase α

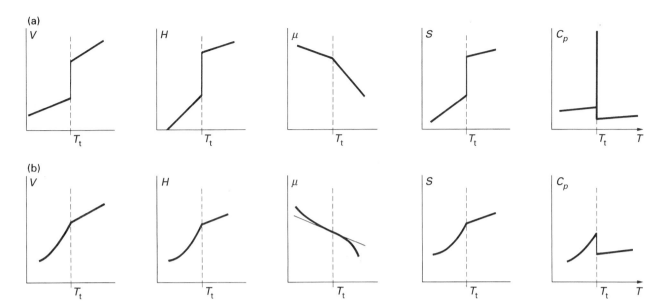

6.13 The changes in thermodynamic properties accompanying (a) first-order and (b) second-order phase transitions.

to another phase β,

$$\left(\frac{\partial \mu_\beta}{\partial p}\right)_T - \left(\frac{\partial \mu_\alpha}{\partial p}\right)_T = V_{\beta,m} - V_{\alpha,m} = \Delta_{trs}V$$

$$\left(\frac{\partial \mu_\beta}{\partial T}\right)_p - \left(\frac{\partial \mu_\alpha}{\partial T}\right)_p = -S_{\beta,m} + S_{\alpha,m} = -\Delta_{trs}S = -\frac{\Delta_{trs}H}{T}$$

(12)

Because $\Delta_{trs}V$ and $\Delta_{trs}S$ are non-zero for melting and vaporization for such transitions, it follows that the slopes of the chemical potential plotted against either pressure or temperature are different on either side of the transition (Fig. 6.13a). In other words, the first derivatives of the chemical potentials with respect to pressure and temperature are discontinuous at the transition. A transition for which the first derivative with respect to temperature is discontinuous is classified as a **first-order phase transition**.

The heat capacity C_p of a substance is the slope of the enthalpy with respect to temperature. At a first-order phase transition, H changes by a finite amount for an infinitesimal change of temperature (as shown in Fig. 6.13a). Therefore, at the transition the slope of H and thus the heat capacity are infinite. The physical reason is that heating drives the transition rather than raising the temperature. For example, boiling water stays at the same temperature even though heat is being supplied. It follows that *a first-order phase transition is also characterized by an infinite heat capacity at the transition temperature*.

A **second-order phase transition** in the Ehrenfest sense is one in which the first derivative of μ with respect to temperature is continuous but its second derivative with respect to temperature is discontinuous. A continuous slope of μ (a graph with the same slope on either side of the transition) implies that the volume and entropy (and hence the enthalpy) do not change at the transition (Fig. 6.13b). The heat capacity is

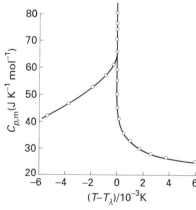

6.14 The λ-curve for helium, where the heat capacity rises to infinity. The shape of this curve is the origin of the name λ-transition.

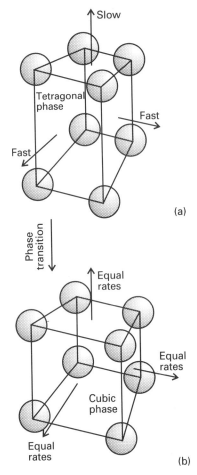

6.15 One version of a second-order phase transition in which a tetragonal phase expands more rapidly in two directions than in a third, and hence becomes a cubic phase, which expands uniformly in three directions as the temperature is raised. There is no rearrangement of atoms at the transition temperature and hence no enthalpy of transition.

discontinuous at the transition but does not become infinite there. An example of a second-order transition is the conducting–superconducting transition in metals at low temperatures. The term **λ-transition** is applied to a phase transition that is not first-order and in which the heat capacity becomes infinite at the transition temperature. Typically, the heat capacity of a system that shows such a transition begins to increase well before the transition (compare Figs. 6.13 and 6.14), and the shape of the heat capacity curve resembles the Greek letter *lambda*. This type of transition includes order–disorder transitions in alloys, the onset of ferromagnetism, and the fluid–superfluid transition of liquid helium.

MOLECULAR INTERPRETATION 6.2

One type of second-order transition is that associated with a change in symmetry of the crystal structure of a solid. Thus, suppose the arrangement of atoms in a solid is like that represented in Fig. 6.15a, with one dimension (technically, of the unit cell) longer than the other two, which are equal. Such a crystal structure is classified as tetragonal (see Section 21.1). Moreover, suppose the two shorter dimensions increase more than the long dimension when the temperature is raised. There may come a stage when the three dimensions become equal. At that point the crystal has cubic symmetry (Fig. 6.15b), and at higher temperatures it will expand equally in all three directions (because there is no longer any distinction between them). The tetragonal → cubic phase transition has occurred but, as it has not involved a discontinuity in the interaction energy between the atoms or the volume they occupy, the transition is not first-order.

The order–disorder transition in β-brass (CuZn) is an example of a λ-transition. The low-temperature phase is an orderly array of alternating Cu and Zn atoms. The high-temperature phase is a random array of the atoms (Fig. 6.16). At $T = 0$ the order is perfect, but islands of disorder appear as the temperature is raised. The islands form because the transition is co-operative in the sense that, once two atoms have exchanged locations, it is easier for their neighbours to exchange their

6.16 An order–disorder transition. (a) At $T = 0$, there is perfect order, with different kinds of atoms occupying alternate sites. (b) As the temperature is increased, atoms exchange locations and islands of each kind of atom form in regions of the solid. Some of the original order survives. (c) At and above the transition temperature the islands occur at random throughout the sample.

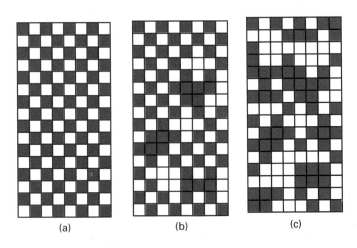

locations. The islands grow in extent, and merge throughout the crystal at the transition temperature (which occurs at 742 K). The heat capacity increases as the transition temperature is approached because the co-operative nature of the transition means that it is increasingly easy for the heat supplied to drive the phase transition rather than to be stored as thermal motion.

CHECK LIST OF KEY IDEAS

1 The use of a **phase diagram** to depict the temperatures and pressures at which different phases are stable (Section 6.1) and the significance of the **phase boundaries**.

2 The significance of the **melting point**, the **boiling point**, the **critical temperature**, and the **triple point** (Section 6.1).

3 The interpretation of representative phase diagrams of four substances (Section 6.2).

4 The expression of phase equilibria in terms of the chemical potential of the phases and the response of the phase equilibrium to the temperature (eqn 1) and the pressure (eqn 2).

5 The calculation of the vapour pressure when an additional pressure is applied to a liquid (eqn 3).

6 The calculation of the phase boundaries, using the **Clapeyron equation** (eqn 4) for the solid–liquid phase boundary and the approximate **Clausius–Clapeyron equation** for the liquid–vapour (eqn 9) and solid–vapour phase boundaries.

7 The **Ehrenfest classification** of phase transitions as **first-order** or **second-order** and the meaning of the term **λ-transition** (Section 6.5).

EXERCISES

6.1. The vapour pressure of dichloromethane at 24.1°C is 400 Torr and its enthalpy of vaporization is 28.7 kJ mol^{-1}. Estimate the temperature at which its vapour pressure is 500 Torr.

6.2. The molar volume of a certain solid is 161.0 cm^3 mol^{-1} at 1.00 atm and 350.75 K, its melting temperature. The molar volume of the liquid at this temperature and pressure is 163.3 cm^3 mol^{-1}. At 100 atm the melting temperature changes to 351.26 K. Calculate the molar enthalpy and entropy of fusion of the solid.

6.3. The vapour pressure of a liquid in the temperature range 200 K to 260 K was found to fit the expression

$$\ln (p/\text{Torr}) = 16.255 - \frac{2501.8}{T/\text{K}}$$

Calculate the enthalpy of vaporization of the liquid.

6.4. The vapour pressure of benzene between 10°C and 30°C fits the expression

$$\log (p/\text{Torr}) = 7.960 - \frac{1780}{T/\text{K}}$$

Calculate (a) the enthalpy of vaporization and (b) the normal boiling point of benzene.

6.5. When benzene freezes at 5.5°C its density changes from 0.879 g cm^{-3} to 0.891 g cm^{-3}. Its enthalpy of fusion is 10.59 kJ mol^{-1}. Estimate the freezing point of benzene at 1000 atm.

6.6. In July in Los Angeles, the incident sunlight at ground level has a power density of 1.2 kW m^{-2} at noon. A swimming pool of area 50 m^2 is directly exposed to the sun. What is the maximum rate of loss of water?

6.7. An open vessel containing (a) water, (b) benzene, (c) mercury stands in a laboratory measuring 5 m × 5 m × 3 m at 25°C. What mass of each substance will be found in the air if there is no ventilation? (The vapour pressures are (a) 24 Torr, (b) 98 Torr, (c) 1.7 mTorr.)

6.8. On a cold, dry morning after a frost, the temperature was −5°C and the partial pressure of water in the atmosphere fell to 2 Torr. Will the frost sublime? What partial pressure of water would ensure that the frost remained?

6.9. Refer to Fig. 6.5 and describe the changes that would be observed when water vapour at 1.0 bar and 400 K is cooled at constant pressure to 260 K. Suggest the appearance of a plot of temperature against time.

6.10. Refer to Fig. 6.5 again and describe the changes that would be observed when cooling takes place at the pressure of the triple point.

6.11. Use the phase diagram in Fig. 6.6 to state what would be observed when a sample of carbon dioxide, initially at 1.0 bar and 298 K, is subjected to the following cycle: (a) isobaric (constant-pressure) heating to 320 K, (b) isothermal compression to 100 bar, (c) isobaric cooling to 210 K, (d) isothermal decompression to 1.0 bar, isobaric heating to 298 K.

6.12. Naphthalene, $C_{10}H_8$, melts at 80.2°C. If the vapour pressure of the liquid is 10 Torr at 85.8°C and 40 Torr at 119.3°C, use the Clausius–Clapeyron equation to calculate (a) the enthalpy of vaporization, (b) the normal boiling point, and (c) the entropy of vaporization at the boiling point.

6.13. The boiling point of hexane is 69.0°C. Estimate (a) its molar enthalpy of vaporization and (b) its vapour pressure at 25°C and 60°C.

6.14. Calculate the melting point of ice under a pressure of 50 bar. Assume that the density of ice under these conditions is approximately $0.92\ g\ cm^{-3}$ and that of water is $1.00\ g\ cm^{-3}$.

6.15. What fraction of the enthalpy of vaporization of water is spent on expanding the water vapour?

6.16. The temperature dependence of the vapour pressure of solid sulfur dioxide can be approximately represented by the relation $\log(p/\text{Torr}) = 10.5916 - 1871.2/(T/\text{K})$ and that of liquid sulfur dioxide by $\log(p/\text{Torr}) = 8.3186 - 1425.7/(T/\text{K})$. Estimate the temperature and pressure of the triple point of sulfur dioxide.

PROBLEMS

Numerical problems

6.1. Prior to the discovery that Freon-12 (CF_2Cl_2) was harmful to the Earth's ozone layer, it was frequently used as the dispersing agent in spray cans for hair spray, etc. Its enthalpy of vaporization at its normal boiling point of −29.2°C is $20.25\ kJ\ mol^{-1}$. Estimate the pressure that a can of hair spray using Freon-12 had to withstand at 40°C, the temperature of a can that has been standing in sunlight. Assume that $\Delta_{vap}H$ is a constant over the temperature range involved and equal to its value at −29.2°C.

6.2. The enthalpy of vaporization of a certain liquid is found to be $14.4\ kJ\ mol^{-1}$ at 180 K, its normal boiling point. The molar volumes of the liquid and the vapour at the boiling point are $115\ cm^3\ mol^{-1}$ and $14.5\ L\ mol^{-1}$, respectively. Estimate dp/dT from the Clapeyron equation and estimate the percentage error in its value if the Clausius–Clapeyron equation is used instead.

6.3. Calculate the difference in slope of the chemical potential against temperature on either side of (a) the normal freezing point of water and (b) the normal boiling point of water. By how much does the chemical potential of water supercooled to −5.0°C exceed that of ice at that temperature?

6.4. Calculate the difference in slope of the chemical potential against pressure on either side of (a) the normal freezing point of water and (b) the normal boiling point of water. The densities of ice and water at 0°C are $0.917\ g\ cm^{-3}$ and $1.000\ g\ cm^{-3}$, and those of water and water vapour at 100°C are $0.958\ g\ cm^{-3}$ and $0.598\ g\ L^{-1}$.

By how much does the chemical potential of water vapour exceed that of liquid water at 1.2 atm and 100°C?

6.5. The enthalpy of fusion of mercury is $2.292\ kJ\ mol^{-1}$ and its normal freezing point is 234.3 K with a change in molar volume of $+0.517\ cm^3\ mol^{-1}$ on melting. At what temperature will the bottom of a 10-m high column of mercury (of density $13.6\ g\ cm^{-3}$) be expected to freeze?

6.6. 50.0 L of dry air was slowly bubbled through a thermally insulated beaker containing 250 g of water initially at 25°C. Calculate the final temperature. (The vapour pressure of water is approximately constant at 23.8 Torr throughout, and its heat capacity is $75.5\ J\ K^{-1}\ mol^{-1}$. Assume that the air is not heated or cooled and that water vapour is a perfect gas.)

6.7. The vapour pressure, p of nitric acid varies with temperature as follows:

θ/°C	0	20	40	50	70	80	90	100
p/Torr	14.4	47.9	133	208	467	670	937	1282

What is (a) the normal boiling point and (b) the enthalpy of vaporization of nitric acid?

6.8 The vapour pressure of the ketone carvone ($M = 150.2\ g\ mol^{-1}$), a component of oil of spearmint, is as follows:

θ/°C	57.4	100.4	133.0	157.3	203.5	227.5
p/Torr	1.00	10.0	40.0	100	400	760

What is (a) the normal boiling point and (b) the enthalpy of vaporization of carvone?

6.9. Construct the phase diagram for benzene near its triple point at 36 Torr and 5.50°C using the following data: $\Delta_{fus}H = +10.6 \text{ kJ mol}^{-1}$, $\Delta_{vap}H = +30.8 \text{ kJ mol}^{-1}$, $\rho(s) = 0.891 \text{ g cm}^{-3}$, $\rho(l) = 0.879 \text{ g cm}^{-3}$.

Theoretical problems

6.10. Show that, for a transition between two incompressible solid phases, ΔG is independent of the pressure.

6.11. In the 'gas saturation method' for the measurement of vapour pressure, a volume V of gas (as measured at a temperature T and a pressure P) is bubbled slowly through the liquid that is maintained at the temperature T and a mass loss m is measured. Show that the vapour pressure p of the liquid is related to its molar mass M by

$$p = \frac{AmP}{1 + Am} \quad \text{where } A = \frac{RT}{MPV}$$

The vapour pressure of geraniol ($M = 154.2 \text{ g mol}^{-1}$), which is a component of oil of roses, was measured at 110°C. It was found that when 5.00 L of N_2 at 760 Torr was passed slowly through the heated liquid, the loss of mass was 0.32 g. Calculate the vapour pressure of geraniol.

6.12. Combine the barometric formula (stated in Problem 1.34) for the dependence of the pressure on altitude with the Clausius–Clapeyron equation, and predict how the boiling temperature of a liquid depends on the altitude and the ambient temperature. Take the mean ambient temperature as 20°C and predict the boiling temperature of water at 3000 m.

6.13. Figures 6.1 and 6.10 give schematic representations of how the chemical potentials of the solid, liquid, and gaseous phases of a substance vary with temperature. All have a negative slope, but it is unlikely that they are truly straight lines as indicated in the illustrations. Derive an expression for the curvatures (specifically, the second derivatives with respect to temperature) of these lines. Is there a restriction on the curvature of these lines? Which state of matter shows the greatest curvature?

6.14. The Clapeyron equation does not apply to second-order phase transitions, but there are two analogous equations, the Ehrenfest equations, that do. They are:

$$\frac{dp}{dT} = \frac{\alpha_2 - \alpha_1}{\kappa_{T,2} - \kappa_{T,2}} \qquad \frac{dp}{dT} = \frac{C_{p,m2} - C_{p,m1}}{TV_m(\alpha_2 - \alpha_1)}$$

where α is the expansion coefficient, κ_T the isothermal compressibility, and the subscripts 1 and 2 refer to two different phases. Derive these two equations. Why does the Clapeyron equation not apply to second-order transitions?

6.15. For a first-order phase transitions, to which the Clapeyron equation does apply, prove the relation

$$C_S = C_p - \alpha V \left(\frac{\Delta_{trs}H}{\Delta_{trs}V} \right)$$

where $C_S = (dq/dT)_S$ is the heat capacity along the coexistence curve of two phases.

7

The properties of simple mixtures

The chapter begins by developing the concept of chemical potential to show that it is a particular case of a class of properties called partial molar quantities. Then it explores how the chemical potential of a substance is used to describe the physical properties of a mixture in which it occurs. The underlying principle to keep in mind, as in Chapter 6, is that, at equilibrium, the chemical potential of a species is the same in every phase.

We shall see, by making use of the experimental observations known as Raoult's and Henry's laws, that the chemical potential of a substance can be expressed in terms of its mole fraction in a mixture. With this result established, it is quite easy to calculate the effect of a solute on certain thermodynamic properties of a solution. These properties include the lowering of vapour pressure of the solvent, the elevation of its boiling point, the depression of its freezing point, and the origin of osmotic pressure. Finally, we shall see how the chemical potential of a substance in a real mixture can be expressed in terms of a property known as the activity. We shall see how the activity may be measured and shall conclude with a brief discussion of how the standard states of solutes and solvents are defined.

We now leave single substances and the limited but important changes that they undergo and examine mixtures. Here we shall consider only mixtures of substances that do not react together and shall leave the description of the mixtures of substances that do react together to Chapter 9. At this stage we shall deal mainly with binary mixtures (mixtures of two components, A and B). We shall therefore often be able to simplify equations by making use of the relation $x_A + x_B = 1$.

Another restriction of this chapter is that we shall consider mainly **non-electrolyte solutions,** in which the solute is not present as ions. We shall delay until Chapter 10 the special problems of **electrolyte solutions,** in which the solute is ionized and the ions generally interact strongly with each other.

THE THERMODYNAMIC DESCRIPTION OF MIXTURES

In this and the following chapters we need a set of concepts that enable us to apply thermodynamics to mixtures of variable composition. We have already seen that the partial pressure, which is the contribution of one component to the total pressure, is used to discuss the properties of mixtures of gases. For a more general description of the thermodynamics of mixtures we need to introduce other analogous 'partial' properties.

7.1 Partial molar quantities

The easiest partial molar property to visualize is the partial molar volume, the contribution that a component makes to the total volume of a sample.

Partial molar volume

Imagine a huge volume of pure water at 25°C. When a further 1 mol H_2O is added, the volume increases by 18 cm³ and we can report that 18 cm³ mol⁻¹ is the molar volume of pure water. However, when we add 1 mol H_2O to a huge volume of pure ethanol, the volume increases by only 14 cm³. The reason for the different increase in volume is that the volume occupied by a given number of water molecules depends on the identity of the molecules that surround them. In the latter case there is so much ethanol present that each H_2O molecule is surrounded by ethanol molecules, and the packing of the molecules results in the H_2O molecules occupying a total volume of only 14 cm³. The quantity 14 cm³ mol⁻¹ is the partial molar volume of water in pure ethanol. In general, the **partial molar volume** of a substance A in a mixture is the change in volume on the addition of 1 mol A to a large excess of the mixture.

The partial molar volumes of the components of a binary mixture of A and B vary with composition because the environment of each type of molecule changes as the composition changes from pure A to pure B. It is this changing molecular environment, and the consequential modification of the forces acting between molecules, that results in the variation of the thermodynamic properties of a mixture as its composition is changed. The partial molar volumes of water and ethanol across the full composition range at 25°C are shown in Fig. 7.1.

The partial molar volume V_J of a substance J at some general composition is defined formally as follows:

$$V_J = \left(\frac{\partial V}{\partial n_J}\right)_{p,T,n'} \tag{1}$$

where n_J is the amount (the number of moles) of J and the subscript n' signifies that the amounts of all other substances present are constant. The partial molar volume is the slope of the graph of the total volume as the amount of J is changed, the pressure, temperature, and amount of the other components being constant (Fig. 7.2). Its value depends on the composition, as we saw for water and ethanol. The definition implies that, when the composition of the mixture is changed by the addition of

7.1 The partial molar volumes of water and ethanol at 25°C. Note the different scales (water on the left, ethanol on the right).

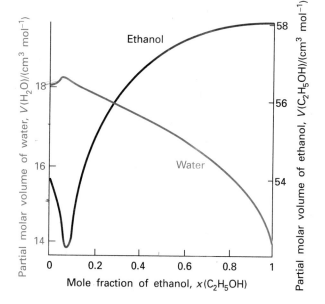

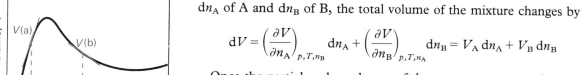

dn_A of A and dn_B of B, the total volume of the mixture changes by

$$dV = \left(\frac{\partial V}{\partial n_A}\right)_{p,T,n_B} dn_A + \left(\frac{\partial V}{\partial n_B}\right)_{p,T,n_A} dn_B = V_A\, dn_A + V_B\, dn_B \qquad (2)$$

Once the partial molar volumes of the two components of a mixture at the composition (and temperature) of interest are known, we can state the total volume V of the mixture by using

$$V = n_A V_A + n_B V_B \qquad (3)$$

JUSTIFICATION

Consider a very large sample of the mixture of the specified composition. Then, when an amount n_A of A is added to the mixture, the composition remains virtually unchanged, the partial molar volume V_A is constant, and the volume of the sample changes by $n_A V_A$. When n_B of B is added, the volume changes by $n_B V_B$ for the same reason. The total change of volume is therefore $n_A V_A + n_B V_B$. The sample now occupies a larger volume, but the proportions of the components are still the same. At this stage, scoop out of the enlarged volume a sample containing n_A of A and n_B of B. Its volume is $n_A V_A + n_B V_B$. Because V is a state function, the same sample could have been prepared simply by mixing the appropriate amounts of A and B. This justifies eqn 3.

7.2 The partial molar volume of a substance is the slope of the variation of the total volume of the sample plotted against the composition. In general, partial molar quantities vary with the composition, as shown by the different slopes at the compositions a and b. Note that the partial molar volume at b is negative: the overall volume of the sample decreases as the component is added.

Partial molar volumes (and partial molar quantities in general) can be measured in several ways. One method is to measure the dependence of the volume on the composition and to fit the observed volume to a function of the mole fraction x_A by using a computer curve-fitting program (that is, by finding the parameters that give a best fit of a particular function to the experimental data). Once the function has been found, its slope can be determined at any composition of interest by differentiation. For instance, if it were found that the molar volume of a mixture was described by the function

$$V_m = a + bx_A + c(x_A^2 - 1)$$

with particular values of the parameters a, b, and c, then with $n_A = x_A n$ the partial molar volume of A at any composition could be obtained from

$$V_A = \left(\frac{\partial V_m}{\partial x_A}\right)_{p,T} = b + 2cx_A$$

Example 7.1 *Using partial molar volumes*

The total volume of an ethanol solution at 25°C containing 1.000 kg of water is found to be given by the expression

$$V/\text{mL} = 1002.93 + 54.6664(m/m^{\ominus}) - 0.36394(m/m^{\ominus})^2$$
$$+ 0.028256(m/m^{\ominus})^3$$

where m is the molality and $m^{\ominus} = 1 \text{ mol kg}^{-1}$. Calculate the partial molar volumes of ethanol and water in a solution prepared by mixing 1.000 kg of water and 500.0 g of ethanol.

Method. First, calculate the molality of ethanol (its molar mass is 46.069 g mol^{-1}) and hence the total volume of the solution. Then determine the partial molar volume of ethanol by differentiation of V with respect to n_E, where n_E is the amount of C_2H_5OH in the solution. For this differentiation, note that $n_E = (m/m^{\ominus}) \times 1 \text{ mol}$ when the mass of solvent is 1.000 kg. The partial molar volume of water V_W can then be obtained by using eqn 3.

Answer. The amount of C_2H_5OH in the solution is

$$n_E = \frac{500.0 \text{ g}}{46.069 \text{ g mol}^{-1}} = 10.85 \text{ mol}$$

It follows that the molality of ethanol in the solution is 10.85 mol kg^{-1} and hence that $m/m^{\ominus} = 10.85$. The total volume of the solution is therefore

$$V/\text{mL} = 1002.93 + 54.6664 \times 10.85 - 0.36394 \times (10.85)^2$$
$$+ 0.028256 \times (10.85)^3 = 1589$$

Hence, the volume is 1589 mL. The partial molar volume of ethanol is obtained from

$$V_E = \left(\frac{\partial V}{\partial n_E}\right)_{n_W} = \frac{1}{1 \text{ mol}}\left(\frac{\partial V}{\partial(m/m^{\ominus})}\right)_{n_W}$$

$$= \frac{1 \text{ mL}}{1 \text{ mol}} \times \{54.6664 - 2 \times 0.36394(m/m^{\ominus})$$

$$+ 3 \times 0.028256(m/m^{\ominus})^2\}$$

At the molality of the solution, this expression evaluates to $56.75\ \mathrm{mL\ mol^{-1}}$. Equation 3 solves to

$$V_{\mathrm{W}} = \frac{V - n_{\mathrm{E}} V_{\mathrm{E}}}{n_{\mathrm{W}}}$$

Substitution of the values calculated previously together with $n_{\mathrm{W}} = 55.51\ \mathrm{mol}$ gives

$$V_{\mathrm{W}} = \frac{1589\ \mathrm{mL} - (10.85\ \mathrm{mol}) \times (56.75\ \mathrm{mL\ mol^{-1}})}{55.51\ \mathrm{mol}} = 17.53\ \mathrm{mL\ mol^{-1}}$$

Exercise E7.1. At 25°C, the density of a 50 per cent by mass ethanol/water solution is $0.914\ \mathrm{g\ cm^{-3}}$. Given that the partial molar volume of water in the solution is $17.4\ \mathrm{cm^3\ mol^{-1}}$, what is the partial molar volume of the ethanol? [$56.4\ \mathrm{cm^3\ mol^{-1}}$]

A word of warning is in order at this point: molar volumes (and molar entropies) are always positive, but the corresponding partial molar quantities need not be. For example, the limiting partial molar volume of $MgSO_4$ (its partial molar volume in the limit of zero concentration) is $-1.4\ \mathrm{cm^3\ mol^{-1}}$, which means that the addition of $1\ \mathrm{mol}\ MgSO_4$ to a large volume of water results in a *decrease* in volume of $1.4\ \mathrm{cm^3}$. The contraction occurs because the salt breaks up the open structure of water as the ions become hydrated, and it collapses slightly.

Partial molar Gibbs energies

The concept of a partial molar quantity can be extended to any of the extensive state functions. One already encountered, but under a different name, is the partial molar Gibbs energy, the chemical potential:

$$\mu_{\mathrm{J}} = \left(\frac{\partial G}{\partial n_{\mathrm{J}}} \right)_{p,T,n'} \tag{4}$$

By the same argument that led to eqn 3, the total Gibbs energy of a binary mixture is

$$G = n_{\mathrm{A}} \mu_{\mathrm{A}} + n_{\mathrm{B}} \mu_{\mathrm{B}} \tag{5}$$

where μ_{A} and μ_{B} are the chemical potentials at the composition of the mixture.

The Gibbs–Duhem equation

Because the total Gibbs energy of a mixture is given by eqn 5 and the chemical potentials depend on the composition, when the compositions are changed infinitesimally we might expect G to change by

$$\mathrm{d}G = \mu_{\mathrm{A}}\,\mathrm{d}n_{\mathrm{A}} + \mu_{\mathrm{B}}\,\mathrm{d}n_{\mathrm{B}} + n_{\mathrm{A}}\,\mathrm{d}\mu_{\mathrm{A}} + n_{\mathrm{B}}\,\mathrm{d}\mu_{\mathrm{B}}$$

However, we saw in Section 5.4 that at constant pressure and temperature the Gibbs energy changes by

$$\mathrm{d}G = \mu_{\mathrm{A}}\,\mathrm{d}n_{\mathrm{A}} + \mu_{\mathrm{B}}\,\mathrm{d}n_{\mathrm{B}}$$

Because G is a state function, these two equations must be equal to each other, which implies that, at constant temperature and pressure,

$$n_A \, d\mu_A + n_B \, d\mu_B = 0 \tag{6a}$$

This equation is a special case of the **Gibbs–Duhem equation**:

$$\sum_J n_J \, d\mu_J = 0 \tag{6b}$$

The significance of eqn 6 is that the chemical potentials of a mixture cannot change independently: in a binary mixture, if one increases the other must decrease. The same line of reasoning applies to all partial molar quantities (see Example 7.2).

Example 7.2 *Using the Gibbs–Duhem equation*

The experimental value of the partial molar volume of $K_2SO_4(aq)$ at 298 K is given by the expression $V_B/(\text{cm}^3 \, \text{mol}^{-1}) = 32.280 + 18.216(m/m^{\ominus})^{\frac{1}{2}}$, where m is the molality of K_2SO_4. Use the Gibbs–Duhem equation to derive an equation for the molar volume of water in the solution. The molar volume of pure water at 298 K is $18.079 \, \text{cm}^3 \, \text{mol}^{-1}$.

Method. The Gibbs–Duhem equation for the partial molar volumes of two components is found by using the procedure set out above: we know that the volume is a state function, so that when a change dV is calculated from eqn 3 it must be the same as that given by eqn 2. With the equation established, the partial molar volume of one component can be determined from the other by integrating the equation.

Answer. From eqn 3 we can write

$$dV = d(n_A V_A + n_B V_B) = n_A \, dV_A + n_B \, dV_B + V_A \, dn_A + V_B \, dn_B$$

However, we know from eqn 2 that

$$dV = V_A \, dn_A + V_B \, dn_B$$

For these two expressions to be consistent, it must be the case that

$$n_A \, dV_A + n_B \, dV_B = 0$$

This expression can be rearranged to

$$dV_A = -\left(\frac{n_B}{n_A}\right) dV_B$$

The amount of B in 1 kg of water is related to its molality by $n_B = (m/m^{\ominus}) \times 1 \, \text{mol}$, and the amount of H_2O is $n_A = (1 \, \text{kg})/M$, where M is the molar mass of H_2O. It follows from the expression for V_B that

$$dV_B = \frac{9.108 \, \text{cm}^3 \, \text{mol}^{-1}}{(m/m^{\ominus})^{\frac{1}{2}}} \times d(m/m^{\ominus})$$

the integrated form of the Gibbs–Duhem relation is therefore

$$\int_{V_A^*}^{V_A} dV_A = -\frac{9.108 \, \text{cm}^3 \times M}{1 \, \text{kg}} \int_0^{m/m^{\ominus}} \left(\frac{m}{m^{\ominus}}\right)^{\frac{1}{2}} d(m/m^{\ominus})$$

which evaluates to

$$V_A/(cm^3 \, mol^{-1}) = 18.079 - 0.1094(m/m^{\ominus})^{\frac{3}{2}}$$

Comment. This correlation of the changes in the two partial molar quantities in a binary mixture, where one increases as the other decreases $(dV_A \propto -dV_B)$ can be seen in Fig. 7.1 where increases in the partial molar volume of water are mirrored by decreases in the partial molar volume of ethanol, and vice versa.

Exercise E7.2. Show that, if the chemical potential of a component in a binary mixture increases, then that of the other component must decrease. $[n_A \, d\mu_A = -n_B \, d\mu_B]$

7.2 The thermodynamics of mixing

The dependence of the Gibbs energy of a mixture on its composition is given by eqn 5, and we know that at constant temperature and pressure systems tend towards lower Gibbs energy. This is the link we need in order to apply thermodynamics to the discussion of spontaneous changes of composition, as in the mixing of two substances. One simple example of a spontaneous mixing process is that of two gases introduced into the same container. The mixing is spontaneous, and so must correspond to a decrease in G. We shall now see how to express this idea quantitatively.

The Gibbs energy of mixing

Let the amounts of two perfect gases in the two containers be n_A and n_B; both are at a temperature T and a pressure p. At this stage, the chemical potentials of the two gases have their 'pure' values and the free energy of the total system is

$$G_i = n_A\mu_A + n_B\mu_B = n_A\left\{\mu_A^{\ominus} + RT \ln\left(\frac{p}{p^{\ominus}}\right)\right\} + n_B\left\{\mu_B^{\ominus} + RT \ln\left(\frac{p}{p^{\ominus}}\right)\right\}$$

After mixing, the partial pressures of the gases are p_A and p_B, with $p_A + p_B = p$. The total Gibbs energy changes to

$$G_f = n_A\left\{\mu_A^{\ominus} + RT \ln\left(\frac{p_A}{p^{\ominus}}\right)\right\} + n_B\left\{\mu_B^{\ominus} + RT \ln\left(\frac{p_B}{p^{\ominus}}\right)\right\}$$

The difference $G_f - G_i$, the **Gibbs energy of mixing** $\Delta_{mix}G$, is therefore

$$\Delta_{mix}G = n_A RT \ln\left(\frac{p_A}{p}\right) + n_B RT \ln\left(\frac{p_B}{p}\right)$$

We may replace n_J by $x_J n$ and use Dalton's law (Section 1.2) to write $p_J/p = x_J$ for each component, which gives

$$\Delta_{mix}G = nRT(x_A \ln x_A + x_B \ln x_B) \tag{7}°$$

Because mole fractions are never greater than 1, the logarithms in this equation are negative, and $\Delta_{mix}G < 0$ (Fig. 7.3). The conclusion that $\Delta_{mix}G$ is negative confirms that perfect gases mix spontaneously in all proportions. However, the equation extends common sense by allowing

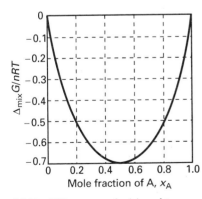

7.3 The Gibbs energy of mixing of two perfect gases and (as discussed later) of two liquids that form an ideal solution. The Gibbs energy is negative for all compositions and temperatures, so perfect gases mix spontaneously in all proportions.

us to discuss the process quantitatively. We see, for instance, that for perfect gases $\Delta_{mix}G$ is directly proportional to the temperature but is independent of the total pressure.

Example 7.3 *Calculating a Gibbs energy of mixing*

A container is divided into two compartments. One contains 3.0 mol H_2 at 1.0 atm and 25°C; the other contains 1.0 mol N_2 at 3.0 atm and 25°C. Calculate the Gibbs energy of mixing when the partition is removed. Assume perfect behaviour.

Method. We can proceed by calculating the initial Gibbs energy from the chemical potentials. Then we need to calculate the Gibbs energy for the system when both gases occupy the same volume. To do so, we need the final partial pressures of the gases. To calculate these partial pressures, we need to know the final volume, which is the sum of the two initial volumes. These initial volumes can be obtained from the perfect gas law and the initial pressures of the two separate components.

Answer. The initial Gibbs energy is

$$G_i = 3.0 \, mol \times \{\mu^{\ominus}(H_2) + RT \ln 1.0\}$$
$$+ 1.0 \, mol \times \{\mu^{\ominus}(N_2) + RT \ln 3.0\}$$

The initial volumes occupied are

$$V(H_2) = \frac{3.0 \, mol \times RT}{1.0 \, atm} \qquad V(N_2) = \frac{1.0 \, mol \times RT}{3.0 \, atm}$$

so the total volume of the container is

$$V = RT\left(3.0 + \frac{1.0}{3.0}\right) \times \frac{1 \, mol}{1 \, atm} = \frac{10.0 \, mol \times RT}{3.0 \, atm}$$

the final pressure (with $n = 4.0$ mol) is therefore

$$p = \frac{nRT}{V} = 1.2 \, atm$$

The final partial pressures, using $x(H_2) = 0.75$ and $x(N_2) = 0.25$, are $p(H_2) = 0.90$ atm and $p(N_2) = 0.30$ atm, and the final Gibbs energy is

$$G_f = 3.0 \, mol \times \{\mu^{\ominus}(H_2) + RT \ln 0.90\}$$
$$+ 1.0 \, mol \times \{\mu^{\ominus}(N_2) + RT \ln 0.30\}$$

The Gibbs energy of mixing is the difference $G_f - G_i$, or $-2.62 \, mol \times RT$, which evaluates to -6.5 kJ.

Comment. In this example, the value of $\Delta_{mix}G$ is the sum of two contributions: the mixing itself and the changes in pressure of the two gases. When 3.0 mol of H_2 mixes with 1.0 mol of N_2 at the same pressure, the change of Gibbs energy is given by eqn 7 as -5.6 kJ independent of the initial common pressure.

Exercise E7.3. Suppose that 2.0 mol H_2 at 2.0 atm and 25°C and 4.0 mol

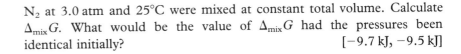

N_2 at 3.0 atm and 25°C were mixed at constant total volume. Calculate $\Delta_{mix}G$. What would be the value of $\Delta_{mix}G$ had the pressures been identical initially? $[-9.7 \text{ kJ}, -9.5 \text{ kJ}]$

Other thermodynamic mixing functions

The quantitative expression for $\Delta_{mix}G$ lets us compute the **entropy of mixing** $\Delta_{mix}S$. Because $(\partial G/\partial T)_{p,n} = -S$, it follows immediately from eqn 7 that, for a mixture of perfect gases,

$$\Delta_{mix}S = -\left(\frac{\partial \Delta_{mix}G}{\partial T}\right)_{p,n_A,n_B} = -nR(x_A \ln x_A + x_B \ln x_B) \qquad (8)°$$

Because $\ln x < 0$, it follows that $\Delta_{mix}S > 0$ for all compositions (Fig. 7.4). This increase in entropy is what we expect when one gas disperses into the other and the system becomes more chaotic. The Gibbs energy of mixing in Example 7.3 was calculated as $-2.62 \text{ mol} \times RT$, so the corresponding entropy of mixing is $+2.62 \text{ mol} \times R$, or $+22 \text{ J K}^{-1}$.

The **enthalpy of mixing** $\Delta_{mix}H$ of two perfect gases may be found from $\Delta G = \Delta H - T\Delta S$ (because $\Delta T = 0$ for the isothermal process). From eqns 7 and 8, we find

$$\Delta_{mix}H = 0 \quad (\text{constant } p, T) \qquad (9)°$$

the enthalpy of mixing is zero, as we should expect for a system in which there are no interactions between particles. It follows that the whole of the driving force for mixing comes from the increase in entropy of the system (because $\Delta_{mix}H = 0$, the entropy of the surroundings is unchanged).

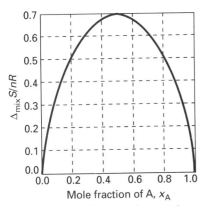

7.4 The entropy of mixing of two perfect gases and (as discussed later) of two liquids that form an ideal solution. The entropy increases for all compositions and temperatures, so perfect gases mix spontaneously in all proportions. Because there is no transfer of heat to the surroundings, when perfect gases mix the entropy of the surroundings is unchanged. Hence, the graph also shows the total entropy of the system plus the surroundings when perfect gases mix.

MOLECULAR INTERPRETATION 7.1

Before two gases A and B mix we can be certain that a molecule will be A if it is taken from one region of the container and B if it is selected from another region. After mixing, when the molecules are free to travel throughout the combined container, we cannot predict with certainty whether a particular molecule will be A or B. Hence, there is an increase in disorder on mixing.

The material presented in Section 4.2 can be used to predict the increase in entropy that accompanies mixing. When the volume occupied by molecules of species J changes from V_J to V, the change in entropy is

$$\Delta S = n_J R \ln\left(\frac{V}{V_J}\right)$$

The change in entropy when the molecules of two gases A and B each spread out in this way is

$$\Delta_{mix}S = n_A R \ln\left(\frac{V}{V_A}\right) + n_B R \ln\left(\frac{V}{V_B}\right)$$

From the perfect gas law, with $V = (n_A + n_B)RT/p$ and $V_J = n_J RT/p$, it follows that

$$\Delta_{mix}S = n_A R \ln\left(\frac{n_A + n_B}{n_A}\right) + n_B R \ln\left(\frac{n_A + n_B}{n_B}\right)$$

$$= n_A R \ln\left(\frac{1}{x_A}\right) + n_B R \ln\left(\frac{1}{x_B}\right)$$

which rearranges to eqn 8.

The lack of any enthalpy of mixing arises from the fact that, in a perfect gas, the molecules do not interact with one another. It then also follows that

$$\Delta_{mix}G = \Delta_{mix}H - T\,\Delta_{mix}S = -T\,\Delta_{mix}S$$

Substitution of the preceding expression into this gives eqn 7. We can conclude, therefore, that the Gibbs energy of mixing stems from the change in disorder of the system that arises from the mingling of the molecules of the two gases.

7.3 The chemical potentials of liquids

To discuss the equilibrium properties of liquid mixtures we need to know how the chemical potential of a liquid varies with its composition. To calculate its value, we use the fact that, at equilibrium, the chemical potential of a substance present as a vapour must be equal to its chemical potential in the liquid.

Ideal solutions

We shall denote quantities relating to pure substances by the superscript *, so the chemical potential of pure liquid A is $\mu_A^*(l)$. Because the vapour pressure of the pure liquid is p_A^*, it follows from eqn 5.17 that the chemical potential of A in the vapour is $\mu_A^{\ominus} + RT \ln(p_A^*/p^{\ominus})$. These two chemical potentials are equal at equilibrium (Fig. 7.5), so we can write

$$\mu_A^*(l) = \mu_A^{\ominus} + RT \ln\left(\frac{p_A^*}{p^{\ominus}}\right)$$

If another substance, a solute, is also present in the liquid, the chemical potential of A in the liquid is $\mu_A(l)$ and its vapour pressure is p_A. In this case

$$\mu_A(l) = \mu_A^{\ominus} + RT \ln\left(\frac{p_A}{p^{\ominus}}\right)$$

Next, we combine these two equations to eliminate the standard chemical potential of the gas, and obtain

$$\mu_A(l) = \mu_A^*(l) + RT \ln\left(\frac{p_A}{p_A^*}\right) \tag{10}°$$

The final step draws on additional experimental information about the relation between the ratio of vapour pressures and the composition of the liquid. In a series of experiments on mixtures of closely related liquids (such as benzene and methylbenzene), the French chemist François

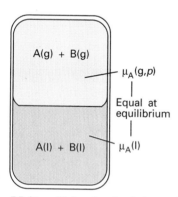

A(g) + B(g)

$\mu_A(g,p)$

Equal at
equilibrium

A(l) + B(l) $\mu_A(l)$

7.5 At equilibrium the chemical potential of the gaseous form of a substance A is equal to the chemical potential of its condensed phase. The equality is preserved if a solute is also present. Because the chemical potential of A in the vapour depends on its partial vapour pressure, it follows that the chemical potential of liquid A can be related to its partial vapour pressure.

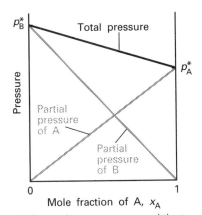

7.6 The total vapour pressure and the two partial vapour pressures of an ideal binary mixture are proportional to the mole fractions of the components.

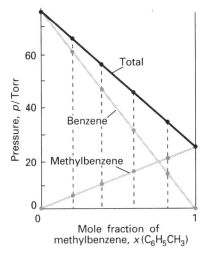

7.7 Two similar liquids, in this case benzene and toluene (methylbenzene), behave almost ideally, and the variation of their vapour pressures with composition resembles that for an ideal solution.

Raoult found that the ratio of the partial vapour pressure of each component to its vapour pressure as a pure liquid, p_A/p_A^*, is approximately equal to the mole fraction of A in the mixture. That is, he established

Raoult's law: $\quad p_A = x_A p_A^*$ $\qquad$ **(11)°**

This law is illustrated in Fig. 7.6. Some mixtures obey Raoult's law very well, especially when the components are chemically similar (Fig. 7.7). Mixtures that obey the law throughout the composition range from pure A to pure B are called **ideal solutions**. When we write equations relating to ideal solutions we shall label them with the superscript °, as in eqn 11.

For an ideal solution, it follows from eqns 10 and 11 that the chemical potential of a solvent in a solution is related to its mole fraction by

$$\mu_A(l) = \mu_A^*(l) + RT \ln x_A \qquad \textbf{(12)°}$$

This important equation can be used as the *definition* of an ideal solution (so that it implies Raoult's law rather than stemming from it). It is in fact a better definition than eqn 11 because it does not assume that the gas is perfect.

MOLECULAR INTERPRETATION 7.2

The origin of Raoult's law can be understood in molecular terms by considering the rates at which molecules leave and return to the liquid. The law reflects the fact that the presence of a second component reduces the rate at which A molecules leave the surface of the liquid but does not inhibit the rate at which they return (Fig. 7.8).

The rate at which A molecules leave the surface is proportional to the number of them at the surface, which in turn is proportional to the mole fraction of A:

$$\text{rate of vaporization} = k x_A$$

where k is a constant of proportionality. The rate at which molecules condense is proportional to their concentration in the gas phase, which in

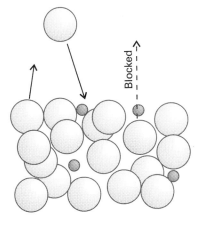

7.8 A pictorial representation of the molecular basis of Raoult's law. The large spheres represent solvent molecules at the surface of a solution (the uppermost line of spheres) and the small spheres are solute molecules. The latter hinder the escape of solvent molecules into the vapour, but do not hinder their return.

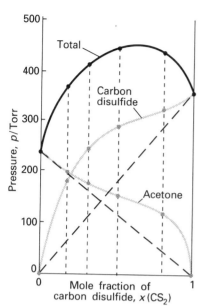

7.9 Strong deviations from ideality are shown by dissimilar liquids (in this case carbon disulfide and acetone).

turn is proportional to their partial pressure:

$$\text{rate of condensation} = k'p_A$$

At equilibrium, the rates of vaporization and condensation are equal, so

$$k'p_A = kx_A$$

It follows that

$$p_A = \frac{k}{k'} \times x_A$$

For the pure liquid, $x_A = 1$, so

$$p_A^* = \frac{k}{k'}$$

Equation 11 then follows by substitution of this relation into the previous one.

Some solutions depart significantly from Raoult's law (Fig. 7.9). Nevertheless, even in these cases the law is obeyed increasingly closely for the component in excess (the solvent) as it approaches purity. The law is therefore a good approximation for the solvent if the solution is dilute.

Ideal–dilute solutions

In ideal solutions the solute, as well as the solvent, obeys Raoult's law. However, for real solutions at low concentrations, the English chemist William Henry found experimentally that, although the vapour pressure of the solute is proportional to its mole fraction, the constant of proportionality is not the vapour pressure of the pure substance (Fig. 7.10):

Henry's law: $p_B = x_B K_B$ (13)°

In this expression x_B is the mole fraction of the solute and K_B is a constant (with the dimensions of pressure) chosen so that the plot of the vapour pressure of B against its mole fraction is tangent to the experimental curve at $x_B = 0$.

Mixtures for which (a) the solute obeys Henry's law but not Raoult's law and (b) the solvent obeys Raoult's law are called **ideal–dilute solutions**. We shall also label equations with ° when they have been derived from Henry's law.

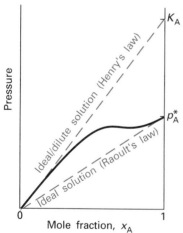

7.10 When a component (the solvent) is nearly pure, it has a vapour pressure that is proportional to mole fraction with a slope p_A^* (Raoult's law). When it is the minor component (the solute) its vapour pressure is still proportional to the mole fraction, but the constant of proportionality is now K_A (Henry's law).

MOLECULAR INTERPRETATION 7.3

The difference in behaviour of the solute and solvent at low concentrations (as expressed by Henry's and Raoult's laws, respectively) arises from the fact that, in a dilute solution, the solvent molecules are in an environment very much like the one they have in the pure liquid whereas the solute is surrounded by solvent molecules. Thus, the solvent behaves like a slightly modified pure liquid, but the solute behaves entirely differently from its pure state unless the solvent and solute molecules happen to be very similar, in which case the solute also obeys Raoult's law.

Example 7.4 *Investigating the validity of Raoult's and Henry's laws*

The vapour pressures of each component in a mixture of propanone (acetone, A) and chloroform (trichloromethane, C) were measured at 35°C with the following results:

x_C	0	0.20	0.40	0.670	0.80	1
p_C/Torr	0	35	82	142	219	293
p_A/Torr	347	270	185	102	37	0

Confirm that the mixture conforms to Raoult's law for the component in large excess and to Henry's law for the minor component. Find the Henry's law constants.

Method. Both Raoult's and Henry's laws are statements about the form of the graph of partial vapour pressure against mole fraction. Therefore, we should plot the partial vapour pressures against mole fraction. Raoult's law is tested by comparing the data with the straight line $p_J = x_J p_J^*$ for each component in the region in which it is in excess (and acting as the solvent). Henry's law is tested by finding a straight line $p = xK$ that is tangent to each partial vapour pressure at low x where the component be treated as the solute. The data are plotted in Fig. 7.11 together with the Raoult's law lines. Henry's law requires $K = 175$ Torr for acetone and $K = 165$ Torr for chloroform.

Comment. Notice how the system deviates from both Raoult's and Henry's laws even for quite small departures from $x = 1$ and $x = 0$, respectively. We deal with these deviations in Section 7.6.

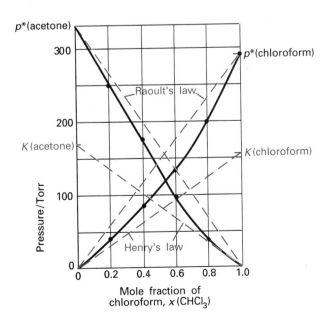

7.11 The experimental partial vapour pressures of a mixture of chloroform and acetone based on the data in Example 7.4. The values of K are obtained by extrapolating the dilute solution vapour pressures as explained in the Example.

Table 7.1* Henry's law constants for gases in water at 298 K

	K /Torr
CO_2	1.25×10^6
H_2	5.34×10^7
N_2	6.51×10^7
O_2	3.30×10^7

* More values are given in the Data section at the end of this volume.

Exercise E7.4. The vapour pressure of chloromethane at various mole fractions in a mixture at 25°C was found to be as follows:

x	0.005	0.009	0.019	0.024
p/Torr	205	363	756	946

Estimate the Henry's law constant.   [4×10^4 Torr]

Some Henry's law data are listed in Table 7.1. As well as providing a link between the mole fraction of solute and its partial pressure, the data in the table may also be used to calculate gas solubilities. The following example illustrates the procedure.

Example 7.5 *Using Henry's law*

Estimate the molar solubility (the solubility in moles per litre) of oxygen in water at 25°C and a partial pressure of 160 Torr, its partial pressure in the atmosphere at sea level.

Method. The mole fraction of solute is given by Henry's law as $x = p/K$ where p is its partial pressure. All we need do is to calculate the mole fraction that corresponds to the stated partial pressure, and then interpret that mole fraction as a molar concentration. For the latter part of the calculation, we calculate the amount of O_2 dissolved in 1.00 kg of water (which corresponds to about 1.00 L water). The solution is dilute, so the expressions for the mole fraction can be simplified.

Answer. Because the amount of O_2 dissolved is small, its mole fraction is

$$x(O_2) = \frac{n(O_2)}{n(O_2) + n(H_2O)} \approx \frac{n(O_2)}{n(H_2O)}$$

Hence,

$$n(O_2) \approx x(O_2)n(H_2O) = \frac{p}{K} \times n(H_2O)$$

$$\approx \frac{160\ \text{Torr}}{3.30 \times 10^7\ \text{Torr}} \times 55.5\ \text{mol} = 2.7 \times 10^{-4}\ \text{mol}$$

The molality of the saturated solution is therefore 2.7×10^{-4} mol kg^{-1}, corresponding to a molar concentration of approximately 2.7×10^{-4} mol L^{-1}.

Comment. Knowledge of Henry's law constants for gases in fats and lipids is important for the discussion of respiration, especially when the partial pressure of oxygen is abnormal, as in diving and mountaineering.

Exercise E7.5. Calculate the molar solubility of nitrogen in water exposed to air at 25°C; partial pressures were calculated in Example 1.6.

[5.1×10^{-4} mol L^{-1}]

THE PROPERTIES OF SOLUTIONS

In this section we shall consider the thermodynamics of mixing of liquids in terms of the material introduced earlier in the chapter. First, we consider the simple case of mixtures of liquids that mix to form an ideal solution, for this identifies the thermodynamic consequences of molecules of one species mingling randomly with molecules of the second species. This discussion provides a background for depicting the deviations from ideal behaviour that real solutions exhibit. Then we consider solutions in which the solute is involatile.

7.4 Liquid mixtures

The Gibbs energy of mixing of two liquids to form an ideal solution is calculated in much the same way as for two gases. When the liquids are separate, the total Gibbs energy is

$$G_i = n_A \mu_A^*(l) + n_B \mu_B^*(l)$$

When they are mixed, the individual chemical potentials are given by eqn 12 and the total Gibbs energy is

$$G_f = n_A\{\mu_A^*(l) + RT \ln x_A\} + n_B\{\mu_B^*(l) + RT \ln x_B\}$$

Consequently, the Gibbs energy of mixing is

$$\Delta_{mix}G = nRT(x_A \ln x_A + x_B \ln x_B) \tag{14}°$$

where $n = n_A + n_B$.

Equation 14 is the same as that for two perfect gases, and all the conclusions drawn there are valid here: the driving force for mixing is the increasing entropy of the system as the molecules mingle, and the enthalpy of mixing is zero. It should be noted, however, that solution ideality means something different from gas perfection. In a perfect gas there are *no* interactions between molecules. In ideal solutions there are interactions, but the average A–B interactions in the mixture are the same as the average A–A and B–B interactions in the pure liquids. The variation of the Gibbs energy of mixing with composition is the same as that already depicted for gases in Fig. 7.3; the same is true of the entropy of mixing, Fig. 7.4.

Real solutions are composed of particles for which A–A, A–B, and B–B interactions are all different. Not only may there be an enthalpy change when liquids mix, but there may also be an additional contribution to the entropy arising from the way in which the molecules of one type might cluster together instead of mingling freely with the others. If the enthalpy change is large and positive (and mixing is endothermic) or if the entropy change is adverse (because of a reorganization of the molecules that results in an orderly mixture), the Gibbs energy might be positive for mixing. In that case, separation is spontaneous and the liquids may be immiscible. Alternatively, the liquids might be **partially miscible**, which means that they are miscible only over a certain range of compositions.

The thermodynamic properties of real solutions may be expressed in

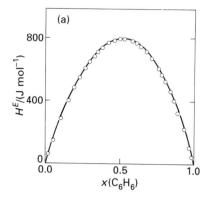

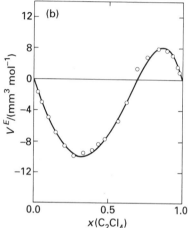

7.12 Experimental excess functions at 25°C. (a) H^E for benzene/cyclohexane; this shows that the mixing is endothermic (because $\Delta_{mix}H = 0$ for an ideal solution). (b) The excess volume V^E for tetrachloroethene/cyclopentane; this graph shows that there is a contraction at low tetrachloroethane mole fractions, but an expansion at high mole fractions (because $\Delta_{mix}V = 0$ for an ideal mixture).

terms of the **excess functions** (G^E, S^E, etc.), the difference between the observed thermodynamic function of mixing and the function for an ideal solution. The excess entropy, for example, is defined as

$$S^E = \Delta_{mix}S(\text{actual}) - \Delta_{mix}S(\text{ideal}) \tag{15}$$

where $\Delta_{mix}S$ (ideal) is obtained by differentiation of eqn 14 with respect to temperature. The excess enthalpy and volume are both equal to the observed enthalpy and volume of mixing, because the ideal values are zero in each case.

Deviations of the excess functions from zero indicate the extent to which the solutions are non-ideal. In this connection a useful model system is the **regular solution**, a solution for which $H^E \neq 0$ but $S^E = 0$. A regular solution can be thought of as one in which the two kinds of molecules are distributed randomly (as in an ideal solution) but have different energies of interaction with each other. Two examples of the composition dependence of excess functions are shown in Fig. 7.12.

7.5 Colligative properties

Now we shall see how to calculate the effect of a solute on the boiling and freezing points of mixtures. We shall also see how a solute gives rise to an osmotic pressure. All the properties we will consider depend (in dilute solutions) only on the number of solute particles present, not on their identity; for this reason they are called **colligative properties** (denoting 'depending on the collection').

The common features of colligative properties

We make two assumptions:

(1) The solute is not volatile, so it does not contribute to the vapour.

(2) The solute does not dissolve in the solid solvent.

The latter assumption is quite drastic, although it is true of many mixtures; it can be avoided at the expense of more algebra, but that introduces no new principles. Both assumptions are removed in the general but qualitative discussion of mixtures in Chapter 8.

Colligative properties have a common origin: they stem from the reduction of the chemical potential of the liquid solvent as a result of the presence of solute. The reduction is from $\mu_A^*(l)$ for the pure solvent to $\mu_A^*(l) + RT \ln x_A$ when a solute is present ($\ln x_A$ is negative because $x_A < 1$). There is no *direct* influence of the solute on the chemical potential of the solvent vapour and the solid solvent because the non-volatile, insoluble solute does not appear in either the vapour or the solid. As can be seen from Fig. 7.13, the reduction in chemical potential of the solvent implies that the liquid–vapour equilibrium occurs at a higher temperature (the boiling point is raised).

MOLECULAR INTERPRETATION 7.4

The molecular origin of the lowering of the chemical potential is not the energy of interaction of the solute and solvent particles, because the

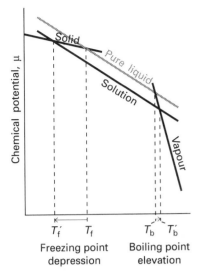

Freezing point depression   Boiling point elevation

7.13 The chemical potential of a solvent in the presence of a solute. The lowering of the liquid's chemical potential has a greater effect on the freezing point than on the boiling point because of the angles at which the lines intersect (which are determined by entropies; recall Fig. 6.10).

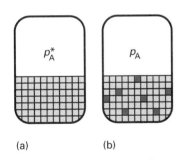

(a)   (b)

7.14 (a) The vapour pressure of a pure liquid represents a balance between the increased disorder arising from vaporization and the decreased disorder of the surroundings. Here the structure of the liquid is represented highly schematically by the grid of squares. (b) When solute (the black squares) is present, the disorder of the condensed phase is relatively higher than that of the pure liquid, and there is a decreased tendency to acquire the disorder characteristic of the vapour.

lowering occurs even in ideal solutions (which have zero enthalpy of mixing). If it is not an enthalpy effect, then it must be an entropy effect.

In the absence of a solute, the pure liquid solvent has an entropy that reflects the disorder of its molecules. Its vapour pressure reflects the tendency of the solution towards greater entropy, which can be achieved if the liquid vaporizes to form a more disordered gas (Fig. 7.14a). When a solute is present, there is an additional contribution to the entropy of the liquid, even in an ideal solution. Because the entropy of the liquid is already higher than that of pure liquid, there is a weaker tendency to form the gas (Fig. 7.14b). The effect of the solute appears as a lowered vapour pressure, and hence a higher boiling point.

Similarly, the enhanced molecular randomness of the solution opposes the tendency to freeze. Consequently, a lower temperature must be reached before equilibrium between solid and solution is achieved. Hence, the freezing point is lowered.

The strategy for the quantitative discussion of the elevation of boiling point and the depression of freezing point is to look for the temperature at which, at 1 atm, one phase (the pure solvent vapour or the pure solid solvent) has the same chemical potential as the solvent in the solution. This is the new equilibrium temperature for the phase transition at 1 atm, and hence corresponds to the new boiling point or the new freezing point of the solvent.

The elevation of boiling point

The heterogeneous equilibrium of interest when considering boiling is between the solvent vapour and the solvent in solution (Fig. 7.15), the pressure being 1 atm. We denote the solvent by A and the solute by B. The equilibrium is established at a temperature for which

$$\mu_A^*(g) = \mu_A^*(l) + RT \ln x_A$$

(The pressure of 1 atm is the same throughout, and will not be written explicitly.) We shall now show that this equation implies that the presence of a solute at a mole fraction x_B causes an increase in normal boiling point from T^* to $T^* + \Delta T$, where

$$\Delta T = \left(\frac{RT^{*2}}{\Delta_{vap}H}\right)x_B \qquad (16)°$$

JUSTIFICATION

The equation for the equality of chemical potentials of the vapour and liquid phases of the solvent

$$\mu_A^*(g) = \mu_A^*(l) + RT \ln x_A$$

rearranges into

$$\ln(1 - x_B) = \frac{\mu_A^*(g) - \mu_A^*(l)}{RT} = \frac{\Delta_{vap}G}{RT}$$

where $\Delta_{vap}G$ is the (molar) Gibbs energy of vaporization of the pure

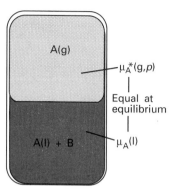

7.15 The heterogeneous equilibrium involved in the calculation of the elevation of boiling point is between A in the pure vapour and A in the mixture, A being the solvent and B an involatile solute.

solvent and x_B is the mole fraction of the solute; we have used $x_A + x_B = 1$. We now write

$$\Delta_{vap}G = \Delta_{vap}H - T\,\Delta_{vap}S$$

and ignore the small temperature dependence of $\Delta_{vap}H$ and $\Delta_{vap}S$. Then, at a general mole fraction x_B:

$$\ln(1 - x_B) = \frac{\Delta_{vap}H}{RT} - \frac{\Delta_{vap}S}{R}$$

When $x_B = 0$, the boiling point is that of the pure liquid, T^*, and

$$\ln 1 = \frac{\Delta_{vap}H}{RT^*} - \frac{\Delta_{vap}S}{R}$$

Because $\ln 1 = 0$, the difference of the two equations is

$$\ln(1 - x_B) = \frac{\Delta_{vap}H}{R}\left(\frac{1}{T} - \frac{1}{T^*}\right)$$

We now suppose that the amount of solute present is so small that $x_B \ll 1$. We can then write $\ln(1 - x_B) \approx -x_B$ and hence obtain

$$x_B = \frac{\Delta_{vap}H}{R}\left(\frac{1}{T^*} - \frac{1}{T}\right)$$

Because $T \approx T^*$, it also follows that

$$\frac{1}{T^*} - \frac{1}{T} = \frac{T - T^*}{TT^*} \approx \frac{\Delta T}{T^{*2}}, \qquad \text{where } \Delta T = T - T^*$$

and the previous equation rearranges to eqn 16.

Because eqn 16 makes no reference to the identity of the solute, only to its mole fraction, we conclude that the elevation of boiling point is a colligative property. The value of ΔT does depend on the properties of the solvent, and the biggest changes occur for solvents with high boiling points.[1] We show in Example 7.6 that, for dilute solutions, the elevation of boiling point may be written

$$\Delta T = K_b m_B \tag{17}$$

where K_b is the **ebullioscopic constant** of the solvent and m_B is the molality of the solution (the amount of solute per kilogram of solvent).

Example 7.6 *Evaluating the elevation of boiling point*

Show that, when the solution is dilute, the elevation of boiling point is given by eqn 17. Evaluate K_b for benzene as solvent.

Method. The starting point is eqn 16: we see that we need to express the mole fraction in terms of the molality. To do so, we consider a solution

---

1 By Trouton's rule (Section 4.4), $\Delta_{vap}H/T^*$ is a constant; therefore eqn 16 has the form $\Delta T \propto T^*$ and is independent of $\Delta_{vap}H$ itself.

that contains an amount n_B of B and 1 kg of solvent A. Because the solution is dilute, we can base approximations on $n_B \ll n_A$.

Answer. Because the mole fraction of B is small,

$$x_B = \frac{n_B}{n_A + n_B} \approx \frac{n_B}{n_A}$$

The amount of solvent molecules in 1 kg of solvent of molar mass $\dot{M}$ is

$$n_A = \frac{1\,\text{kg}}{M}$$

Therefore,

$$x_B = \frac{n_B}{n_A} = n_B \times \frac{M}{1\,\text{kg}} = m_B \times M$$

where m_B is the molality of B. Hence, from eqn 16,

$$\Delta T = \left(\frac{RT^{*2}M}{\Delta_{vap}H}\right)m_B$$

and we can identify the ebullioscopic constant as

$$K_b = \frac{RT^{*2}M}{\Delta_{vap}H}$$

For benzene, $T^* = 353.2\,\text{K}$, $M = 78.11\,\text{g mol}^{-1}$, and $\Delta_{vap}H = 30.8\,\text{kJ mol}^{-1}$ (Table 2.3); substitution of these values gives $K_b = 2.63\,\text{K/(mol kg}^{-1})$.

Comment. The experimental value for benzene is $2.53\,\text{K/(mol kg}^{-1})$.

Exercise E7.6. Evaluate the ebullioscopic constant for water.

$$[0.51\,\text{K/(mol kg}^{-1})]$$

In practice, ebullioscopic constants are best regarded as empirical constants to be determined experimentally, for that avoids the approximation that the solution is ideal. Once measured they can be used to determine the molar masses of solutes. The technique is called **ebullioscopy**, but it is now rarely used, partly because superheating is difficult to avoid (so the determination of ΔT is inherently inaccurate) but also because there are plenty of other methods available.

The depression of freezing point

The heterogeneous equilibrium now of interest is between pure solid solvent and the solution with solute present at a mole fraction x_B (Fig. 7.16). At the freezing point, the chemical potentials of A in the two phases are equal:

$$\mu_A^*(s) = \mu_A^*(l) + RT \ln x_A$$

The only difference between this calculation and the last is the appearance of the solid's chemical potential in place of that of the

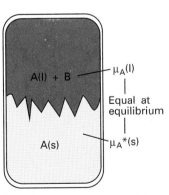

7.16 The heterogeneous equilibrium involved in the calculation of the lowering of freezing point is between A in the pure solid and A in the mixture. A being the solvent and B a solute that is insoluble in solid A.

Table 7.2* Cryoscopic and ebullioscopic constants

	$K_f/(K/(mol\ kg^{-1}))$	$K_b/(K/(mol\ kg^{-1}))$
Benzene	5.12	2.53
Camphor	40	
Phenol	7.27	3.04
Water	1.86	0.51

* More values are given in the Data section.

vapour. Therefore we can write the result directly from eqn 16:

$$\Delta T = \left(\frac{RT^{*2}}{\Delta_{\text{fus}}H}\right)x_B \tag{18}°$$

where ΔT is the freezing point depression, $T^* - T$, and $\Delta_{\text{fus}}H$ is the enthalpy of fusion of the solvent. Larger depressions are observed in solvents with low enthalpies of fusion and high melting points. When the solution is dilute, the mole fraction is proportional to the molality, and we can write (by an argument analogous to that in Example 7.6)

$$\Delta T = K_f m_B \tag{19}$$

where K_f is the **cryoscopic constant** (Table 7.2). This constant is also best regarded in practice as an empirical parameter. Once the cryoscopic constant of a solvent is known, the depression of freezing point may be used to measure the molar mass of a solute in the method known as **cryoscopy**; however, the technique is of little more than historical interest.

Solubility

Although it is not strictly a colligative property, the solubility of a solute may be estimated by the same techniques as we have been using. If a solid solute is left in contact with a solvent, it dissolves until the solution is saturated. Saturation is a state of equilibrium, with the undissolved solute in equilibrium with the dissolved solute. Therefore, in a saturated solution the chemical potential of the pure solid solute, $\mu_B^*(s)$, and the chemical potential of B in solution, μ_B, are equal (Fig. 7.17). Because the latter is

$$\mu_B = \mu_B^*(l) + RT \ln x_B$$

we can write

$$\mu_B^*(s) = \mu_B^*(l) + RT \ln x_B$$

This expression is the same as the starting equation of the last section, except that the quantities refer to the solute B, not the solvent A.

The starting point is the same but the aim is different. In the present case we want to find the mole fraction of B in solution at equilibrium when the temperature is T. Therefore, we start by rearranging the last

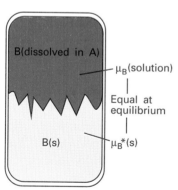

7.17 The heterogeneous equilibrium involved in the calculation of the solubility is between pure solid B and B in the mixture.

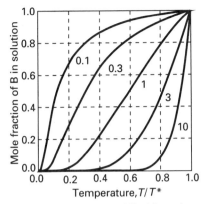

7.18 The variation of solubility (the mole fraction of solute B in a saturated solution) with temperature (T^* is the freezing temperature of the solute). Individual curves are labelled with the value of $\Delta_{fus}H/RT^*$.

equation to

$$\ln x_B = \frac{\mu_B^*(s) - \mu_B^*(l)}{RT} = -\frac{\Delta_{fus}G}{RT}$$

$$= -\frac{\Delta_{fus}H}{RT} + \frac{\Delta_{fus}S}{R}$$

At the melting point of the solute T^* we know that $\Delta_{fus}G = 0$, and thus $\Delta_{fus}G/RT^* = 0$ too; consequently, the term $\Delta_{fus}G/RT^*$ may be added to the right-hand side. If the enthalpy and entropy of melting are constant over the temperature range of interest, then the entropy terms cancel and, in much the same way as before, we obtain

$$\ln x_B = -\frac{\Delta_{fus}H}{R}\left(\frac{1}{T} - \frac{1}{T^*}\right) \tag{20}°$$

This expression (which is plotted in Fig. 7.18) shows that the solubility of B decreases exponentially as the temperature is lowered from its melting point and that solutes with high melting points and large enthalpies of melting have low solubilities at normal temperatures. However, the detailed content of eqn 20 should not be treated too seriously because it is based on highly questionable approximations, such as the ideality of the solution. One aspect of its approximate character is that it fails to predict that solutes will have different solubilities in different solvents, for no solvent properties appear in the expression.

Osmosis

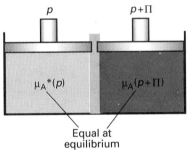

7.19 The equilibrium involved in the calculation of osmotic pressure Π is between pure solvent A at a pressure p on one side of the semipermeable membrane and A as a component of the mixture on the other side of the membrane, where the pressure is $p + \Pi$.

The phenomenon of **osmosis** (from the Greek word for 'push') is the passage of a pure solvent into a solution separated from it by a **semipermeable membrane,** a membrane permeable to the solvent but not to the solute (Fig. 7.19). The **osmotic pressure** Π is the pressure that must be applied to the solution to stop the influx of solvent. One of the most important examples of osmosis is transport of fluids through cell membranes, but it is also the basis of **osmometry,** the determination of molar mass by the measurement of osmotic pressure. Osmometry is widely used to determine the molar masses of macromolecules.

In the simple arrangement shown in Fig. 7.20, the opposing pressure arises from the head of solution that the osmosis itself produces. Equilibrium is reached when the hydrostatic pressure of the column of solution matches the osmotic pressure. The complication inherent to this arrangement is that the entry of solvent into the solution results in its dilution, so it is more difficult to treat than the arrangement in Fig. 7.19 in which there is no flow and the concentrations remain unchanged.

The thermodynamic treatment of osmosis depends on noting that, at equilibrium, the chemical potential of the solvent must be the same on each side of the membrane, as shown in Fig. 7.19. On the pure solvent side the chemical potential of the solvent, which is at a pressure p, is $\mu_A^*(p)$. On the solution side, the chemical potential is lowered by the presence of the solute at a mole fraction x_A but it is raised on account of the greater pressure $p + \Pi$ that the solution experiences. At equilibrium

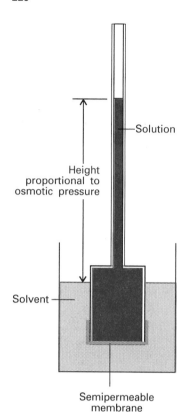

Height proportional to osmotic pressure

Solvent

Semipermeable membrane

7.20 In a simple version of the osmotic pressure experiment, A is at equilibrium on each side of the membrane when enough has passed into the solution to cause a hydrostatic pressure difference.

the two are equal:

$$\mu_A^*(p) = \mu_A(x_A, p + \Pi)$$

The presence of solute is taken into account in the normal way:

$$\mu_A(x_A, p + \Pi) = \mu_A^*(p + \Pi) + RT \ln x_A$$

We saw in Section 5.2 how to take the effect of pressure into account:

$$\mu_A^*(p + \Pi) = \mu_A^*(p) + \int_p^{p+\Pi} V_m \, dp$$

where V_m is the molar volume of the pure solvent. When the last three equations are combined we get

$$-RT \ln x_A = \int_p^{p+\Pi} V_m \, dp$$

For dilute solutions, $\ln x_A$ may be replaced by $\ln (1 - x_B) \approx -x_B$. We may also assume that the pressure range in the integration is so small that the molar volume of the solvent is a constant. That being so, V_m may be taken outside the integral, giving

$$RTx_B = \Pi V_m$$

When the solution is dilute, $x_B \approx n_B/n_A$. Moreover, because $n_A V_m = V$, the total volume of the solvent, the equation simplifies to the **van't Hoff equation**:

$$\Pi V = n_B RT \qquad (21)°$$

Because $n_B/V = [B]$, the molar concentration of the solute, a simpler form of this equation is

$$\Pi = [B]RT \qquad (22)°$$

The van't Hoff equation applies to very dilute solutions in which we can be confident of ideal behaviour.

One of the most common applications of osmometry is to the measurement of molar masses of macromolecules (proteins and synthetic polymers). As these huge molecules dissolve to produce solutions that are far from ideal, it is assumed that the van't Hoff equation is only the first term of a virial-like expansion:

$$\Pi = [B]RT\{1 + B[B] + \ldots\} \qquad (23)$$

The additional terms take the non-ideality into account. The osmotic pressure is measured at a series of concentrations, and a plot of $\Pi/[B]$ against $[B]$ is used to find the molar mass of B.

Example 7.7 *Using osmometry*

The osmotic pressures of solutions of poly(vinyl chloride), PVC, in cyclohexanone at 298 K are given below. The pressures are expressed in terms of the heights of solution (of density $\rho = 0.980 \, \text{g cm}^{-3}$) in balance with the osmotic pressure. Determine the molar mass of the polymer.

$c/(\text{g L}^{-1})$	1.00	2.00	4.00	7.00	9.00
h/cm	0.28	0.71	2.01	5.10	8.00

Comment. We use eqn 23 with $[\text{B}] = c/M$ where c is the mass concentration and M is the molar mass of the polymer; the osmotic pressure is related to the hydrostatic pressure by $\Pi = \rho g h$ (Example 1.2) with $g = 9.81 \text{ m s}^{-2}$. Because

$$\frac{h}{c} = \frac{RT}{\rho g M}\left(1 + \frac{B}{M}c + \ldots\right)$$

$$= \frac{RT}{\rho g M} + \left(\frac{RTB}{\rho g M^2}\right)c + \ldots$$

we should plot h/c against c, and expect a straight line with intercept $RT/\rho g M$ at $c = 0$.

Answer. The data give the following:

$c/(\text{g L}^{-1})$	1.00	2.00	4.00	7.00	9.00
$(h/c)/(\text{cm/g L}^{-1})$	0.28	0.36	0.503	0.729	0.889

The points are plotted in Fig. 7.21. The intercept is at 0.21. Therefore,

$$M = \frac{RT}{\rho g} \times \frac{1}{0.21 \text{ cm }(\text{g L}^{-1})^{-1}}$$

$$= \frac{(8.3145 \text{ J K}^{-1} \text{ mol}^{-1}) \times (298 \text{ K})}{(980 \text{ kg m}^{-3}) \times (9.81 \text{ m s}^{-2})} \times \frac{1}{2.1 \times 10^{-3} \text{ m}^4 \text{ kg}^{-1}}$$

$$= 1.2 \times 10^2 \text{ kg mol}^{-1}$$

Comment. Molar masses of macromolecules are often reported in daltons (Da), with $1 \text{ Da} = 1 \text{ g mol}^{-1}$; the macromolecule in this example has a molar mass of about 120 kDa. Osmometry is a very important technique for the measurement of molar masses of macromolecules, partly because the other colligative techniques give such small effects. We take up the discussion again in Chapter 23.

Exercise E7.7. Estimate the depression of freezing point of the most concentrated of these solutions, taking K_f as about $10 \text{ K}/(\text{mol kg}^{-1})$.

[0.8 mK]

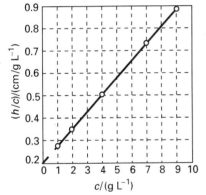

7.21 The plot involved in the determination of molar mass by osmometry (Example 7.7). The molar mass is calculated from the intercept at $c = 0$; in Chapter 23 we shall see that additional information comes from the slope.

ACTIVITIES

Now we shall see how to adjust the expressions developed earlier in the chapter to take into account deviations from ideal behaviour. Just as the fugacity was introduced in Chapter 5 to take into account the effects of gas imperfections in a manner that resulted in the least upset to the form of equations, so the expressions encountered in the treatment of ideal solutions can also be preserved almost intact by introducing the concept of activity.

7.6 The solvent activity

The general form of the chemical potential of a real or ideal solvent is given by a straightforward modification of eqn 10:

$$\mu_A(l) = \mu_A^*(l) + RT \ln \left(\frac{p_A}{p_A^*} \right) \tag{24}$$

where p_A^* is the vapour pressure of pure A and p_A is the vapour pressure of A when it is a component of a solution. In the case of an ideal solution, the solvent obeys Raoult's law at all concentrations and we write

$$\mu_A(l) = \mu_A^*(l) + RT \ln x_A$$

The standard state of the solvent or the solute is the pure liquid (at 1 bar) and is obtained when $x_A = 1$. When the solution does not obey Raoult's law, the form of the last equation can be preserved by writing

$$\mu_A(l) = \mu_A^*(l) + RT \ln a_A \tag{25}$$

The quantity a_A is the **activity** of A, a kind of 'effective' mole fraction, just as the fugacity is an effective pressure.

Because eqn 24 is true for both real and ideal solutions (the only approximation being the use of pressures rather than fugacities), we can conclude by comparing it with eqn 25 that

$$a_A = \frac{p_A}{p_A^*} \tag{26a}$$

We see there is nothing mysterious about the activity: the activity of a solvent can be determined experimentally simply by measuring its vapour pressure and using eqn 26a. For example, the vapour pressure of 0.500 M $KNO_3(aq)$ at 100°C is 749.7 Torr, so the activity of water in the solution at this temperature is

$$a(H_2O) = \frac{749.7 \, \text{Torr}}{760.0 \, \text{Torr}} = 0.9864$$

Because all solvents obey Raoult's law (that $p_A/p_A^* = x_A$) increasingly closely as the concentration of solute approaches zero, the activity of the solvent approaches the mole fraction as $x_A \to 1$:

$$a_A \to x_A \text{ as } x_A \to 1$$

As in the case of real gases, a convenient way of expressing this convergence is to introduce the **activity coefficient** γ by the definition

$$a_A = \gamma_A x_A \qquad \gamma_A \to 1 \text{ as } x_A \to 1 \tag{26b}$$

The chemical potential of the solvent is then

$$\mu_A = \mu_A^* + RT \ln x_A + RT \ln \gamma_A \tag{27}$$

an equation closely resembling that for the chemical potential of a real gas (see the unnumbered equation following eqn 5.24). The standard state of the solvent, the pure liquid solvent at 1 bar, is established when $x_A = 1$.

7.7 The solute activity

The problem with defining activity coefficients and standard states for solutes is that they approach ideal–dilute (Henry's law) behaviour as $x_B \to 0$, not as $x_B \to 1$ (corresponding to pure solute). We shall show how to set up the definitions for a solute that obeys Henry's law exactly, and then show how to allow for deviations.

Ideal–dilute solutions

A solute B that satisfies Henry's law has a vapour pressure given by $p_B = K_B x_B$, where K_B is an empirical constant. In this case, the chemical potential of B is

$$\mu_B = \mu_B^* + RT \ln \left(\frac{p_B}{p_B^*}\right) = \mu_B^* + RT \ln \left(\frac{K_B}{p_B^*}\right) + RT \ln x_B$$

Both K_B and p_B^* are constant characteristics of the solute, so the second term may be combined with the first to give a new standard chemical potential which we denote $\mu^\dagger$:

$$\mu_B^\dagger = \mu_B^* + RT \ln \left(\frac{K_B}{p_B^*}\right)$$

It then follows that

$$\mu_B = \mu_B^\dagger + RT \ln x_B \qquad (28)^\circ$$

Real solutes

We now permit deviations from ideal–dilute, Henry's law behaviour. For the *solvent* activity we introduced a_A in place of x_A into Raoult's law, and obtained eqn 25 for the chemical potential. For the *solute*, we introduce a_B in place of x_B in eqn 28 and obtain

$$\mu_B = \mu_B^\dagger + RT \ln a_B \qquad (29)$$

The standard state remains unchanged in this last stage, and all the deviations from ideality are captured in the activity a_B. The value of the activity at any concentration can be obtained in the same way as for the solvent, but in place of eqn 26a we use

$$a_B = \frac{p_B}{K_B} \qquad (30)$$

As in the case of the solvent, it is sensible to introduce an activity coefficient through

$$a_B = \gamma_B x_B \qquad (31)$$

Now all the deviations from ideality are captured in the activity coefficient γ_B. Because the solute obeys Henry's law as its concentration goes to zero, it follows that

$$a_B \to x_B \text{ and } \gamma_B \to 1 \text{ as } x_B \to 0$$

Deviations of the solute from ideality disappear at zero concentration, but for the solvent the deviations disappear as purity is aproached.

Example 7.8 *Measuring activity*

Use the information in Example 7.4 to calculate the activity and activity coefficient of chloroform in acetone at 35°C treating it as both a solvent and a solute.

Method. For the activity of chloroform as a solvent (the Raoult's law activity) form $a = p/p^*$ and $\gamma = a/x$; for its activity as a solute (the Henry's law activity) form $a = p/K$ and $\gamma = a/x$.

Answer. Because $p^* = 293$ Torr and $K = 165$ Torr, we can construct the following table:

x	0	0.20	0.40	0.60	0.80	1.00	
a	0	0.12	0.28	0.49	0.75	1.00	Raoult
γ	—	0.60	0.70	0.82	0.94	1.00	
a	0	0.21	0.50	0.86	1.33	1.78	Henry
γ	1	1.05	1.25	1.43	1.66	1.78	

Comment. Notice that $\gamma \to 1$ as $x \to 1$ in the Raoult's law case, but that $\gamma \to 1$ as $x \to 0$ in the Henry's law case.

Exercise E7.8. Calculate the activities and activity coefficients for acetone according to the two conventions.

Activities in terms of molalities

The compositions of mixtures are often expressed as molalities m in place of mole fractions. It proves convenient to introduce yet another definition of activity, but one that follows naturally from what we have done so far. First, we note that in dilute solutions the amount of solute is much less than the amount of solvent ($n_B \ll n_A$), so to a good approximation $x_B \approx n_B/n_A$. Because n_B is proportional to the molality m_B, we can write

$$x_B = \kappa \times \frac{m_B}{m^\ominus} \qquad \text{where } m^\ominus = 1 \text{ mol kg}^{-1}$$

In this expression, κ is a constant and $m^\ominus$ has been introduced so that the right-hand side is dimensionless (it cancels the units in m_B itself). For an ideal–dilute solution we can therefore write

$$\mu_B = \mu_B^\dagger + RT \ln \kappa + RT \ln \left(\frac{m_B}{m^\ominus} \right)$$

We now define a new standard chemical potential $\mu^\ominus$ by combining $RT \ln \kappa$ with $\mu^\dagger$:

$$\mu_B^\ominus = \mu_B^\dagger + RT \ln \kappa$$

This definition allows us to write

$$\mu_B = \mu_B^\ominus + RT \ln \left(\frac{m_B}{m^\ominus} \right) \qquad \qquad (32)°$$

Table 7.3 Standard states

Component	Basis	Standard state	Chemical potential	Limits
Solvent‡	Raoult	Pure solvent	$\mu = \mu^* + RT \ln a$ $a = p/p^*$ and $a = \gamma x$	$\gamma \to 1$ as $x \to 1$ (pure solvent)
Solute	Henry	(1) A hypothetical state of the pure solute	$\mu = \mu^\dagger + RT \ln a$ $a = p/K$ and $a = \gamma x$	$\gamma \to 1$ as $x \to 0$
		(2) A hypothetical state of the solute at molality $m^\ominus$	$\mu = \mu^\ominus + RT \ln a$ $a = \gamma m / m^\ominus$	$\gamma \to 1$ as $m \to 0$

‡ The implication of this entry is that the activity of a pure solid or liquid is 1.

According to this definition, the chemical potential of the solute has its standard value $\mu_B^\ominus$ when the molality of B is equal to $m^\ominus$ (that is, at $1\ \mathrm{mol\ kg^{-1}}$).

Now, as before, we incorporate deviations from ideality by introducing a dimensionless activity a_B, a dimensionless activity coefficient γ_B, and writing

$$a_B = \gamma_B \times \frac{m_B}{m^\ominus} \qquad \text{where } \gamma_B \to 1 \text{ as } m_B \to 0 \tag{33}$$

The standard state remains unchanged in this last stage and, as before, all the deviations from ideality are captured in the activity coefficient γ_B. We then arrive at the following succinct expression for the chemical potential of a real solute at any molality:

$$\mu = \mu^\ominus + RT \ln a \tag{34}$$

It is important to be aware of the different definitions of standard states and activities, and they are summarized in Table 7.3. We shall put them to work in the next few chapters, when we shall see that using them is much easier than defining them.

CHECK LIST OF KEY IDEAS

1 The definition and measurement of **partial molar properties** (Section 7.1).

2 The total volume of a mixture as the sum of **partial molar volumes** (eqn 3) and the total Gibbs energy as a sum of **chemical potentials** (eqn 5).

3 The **Gibbs–Duhem equation** relating the changes in the chemical potentials of all the substances in a mixture (eqn 6).

4 The **Gibbs energy of mixing** (eqn 7) of two perfect gases, their **entropy of mixing** (eqn 8), and their **enthalpy of mixing** (eqn 9).

5 The **chemical potential of a liquid** in terms of the partial pressure of its vapour (eqn 10) and, through **Raoult's law** (eqn 11), in terms of its mole fraction (eqn 12).

6 The definition of an **ideal solution** (Section 7.3) and of an **ideal–dilute solution** through **Henry's law** (eqn 13).

7   The **thermodynamic mixing functions** of an ideal solution and the definition and significance of **excess functions** of real solutions (Section 7.4).

8   The description of **colligative properties** in terms of the chemical potential of the solvent (Section 7.5), and the calculation of the **elevation of boiling point** (eqn 17), the **depression of freezing point** (eqn 19), and the **osmotic pressure** in terms of the **van't Hoff equation** (eqn 21).

9   The estimation of the **solubility** of a substance that forms an ideal solution in terms of its enthalpy of fusion (eqn 20).

10 The definition of the **solvent activity** and the **activity coefficient** in terms of Raoult's law (eqn 26a) and the definition of the **solute activity** in terms of Henry's law (eqn 30).

11 The determination of activities by measurement of vapour pressure (Example 7.8).

EXERCISES

7.1. The partial molar volumes of acetone and chloroform in a mixture in which the mole fraction of $CHCl_3$ is 0.4693 are 74.166 $cm^3\,mol^{-1}$ and 80.235 $cm^3\,mol^{-1}$, respectively. What is the volume of a solution of mass 1.000 kg?

7.2. At 25°C, the density of a 50 per cent by mass ethanol–water solution is 0.914 $g\,cm^{-3}$. Given that the partial molar volume of water in the solution is 17.4 $cm^3\,mol^{-1}$, calculate the partial molar volume of the ethanol.

7.3. At 300 K, the vapour pressures of dilute solutions of HCl in liquid $GeCl_4$ are as follows:

$x(HCl)$	0.005	0.012	0.019
p/kPa	32.0	76.9	121.8

Show that the solution obeys Henry's law in this range of mole fractions and calculate Henry's law constant at 300 K.

7.4. Predict the vapour pressure of HCl above its solution in liquid germanium tetrachloride of molality 0.10 mol kg^{-1}. For data, see Exercise 7.3.

7.5. Calculate the cryoscopic and ebullioscopic constants of carbon tetrachloride.

7.6. The vapour pressure of a 500-g sample of benzene was 400 Torr at 60.6°C, but it fell to 386 Torr when 19.0 g of an involatile organic compound was dissolved in it. Calculate the molar mass of the compound.

7.7. The addition of 100 g of a compound to 750 g of CCl_4 lowered the freezing point of the solvent by 10.5 K. Calculate the molar mass of the compound.

7.8. The osmotic pressure of an aqueous solution at 300 K is 120 kPa. Calculate the freezing point of the solution.

7.9. Consider a container of volume 5.0 L that is divided into two compartments of equal size. In the left compartment there is N_2 gas at 1.0 atm and 25°C; in the right compartment there is H_2 at the same temperature and pressure. Calculate the entropy and Gibbs energy of mixing when the partition is removed. Assume that the gases are perfect.

7.10. Air is a mixture with a composition given in Exercise E1.6. Calculate the entropy of mixing when it is prepared from the pure (and perfect) gases.

7.11. Calculate the Gibbs energy, entropy, and enthalpy of mixing when 1.00 mol C_6H_{14} (hexane) is mixed with 1.00 mol C_7H_{16} (heptane) at 298 K; treat the solution as ideal.

7.12. What proportions of hexane and heptane should be mixed (a) by mole fraction, (b) by mass in order to achieve the greatest entropy of mixing?

7.13. Use Henry's law and the data in Table 7.1 to calculate the solubility (as a molality) of CO_2 in water at 25°C when its partial pressure is (a) 0.10 atm, (b) 1.00 atm.

7.14. The mole fractions of N_2 and O_2 in air at sea level are approximately 0.78 and 0.21. Calculate the molalities of the solution formed in an open flask of water at 25°C.

7.15. A water-carbonating plant is available for use in the home and operates by providing carbon dioxide at 5.0 atm. Estimate the molar concentration of the soda water it produces.

7.16. Calculate the freezing point of a 250-cm^3 glass of water sweetened with 7.5 g of sucrose.

7.17. The enthalpy of fusion of anthracene is 28.8 kJ mol^{-1} and its melting point is 217°C. Calculate its ideal solubility in benzene at 25°C.

7.18. Predict the ideal solubility of lead in bismuth at

280°C given that its melting point is 327°C and its enthalpy of fusion is 5.2 kJ mol^{-1}.

7.19. The osmotic pressures of solutions of polystyrene in toluene were measured at 25°C and the pressures were expressed in terms of the height of the solvent of density 1.004 g cm^{-3}:

$c/(g\,L^{-1})$	2.042	6.613	9.521	12.602
h/cm	0.592	1.910	2.750	3.600

Calculate the molar mass of the polymer.

7.20. The molar mass of an enzyme was determined by dissolving it in water, measuring the osmotic pressure at 20°C, and extrapolating the data to zero concentration. The following data were obtained:

$c/(mg\,cm^{-3})$	3.211	4.618	5.112	6.722
h/cm	5.746	8.238	9.119	11.990

Calculate the molar mass of the enzyme.

7.21. Substances A and B are both volatile liquids with $p_A^* = 300$ Torr, $p_B^* = 250$ Torr, and $K_B = 200$ Torr (concentration expressed as mole fraction). When $x_A = 0.90$, $m_B = 2.22$ mol kg^{-1}, $p_A = 250$ Torr and $p_B = 25$ Torr. Calculate the activities and activity coefficients of A and B. Use the mole fraction, the system of calculation based on Raoult's law for A, and that based on Henry's law (both mole fractions and molalities) for B.

7.22. Given that $p^*(H_2O) = 0.02308$ atm and $p(H_2O) = 0.02239$ atm in a solution in which 0.122 kg of a non-volatile solute ($M = 241$ g mol^{-1}) is dissolved in 0.920 kg of water at 293 K, calculate the activity and activity coefficient of water in the solution.

7.23. A dilute solution of bromine in carbon tetrachloride behaves as an ideal–dilute solution. The vapour pressure of pure CCl$_4$ is 33.85 Torr at 298 K. The Henry's law constant when the concentration of Br$_2$ is expressed as a mole fraction is 122.36 Torr. Calculate the vapour pressure of each component, the total pressure, and the composition of the vapour phase when the mole fraction of Br$_2$ is 0.050, on the assumption that the conditions of the ideal–dilute solution are satisfied at this concentration.

7.24. Benzene and toluene form nearly ideal solutions. The boiling point of pure benzene is 80.1°C. Calculate the chemical potential of benzene in solution relative to that of pure benzene when $x_{benzene} = 0.30$ at its boiling point. If the activity coefficient of benzene in this solution were actually 0.93 rather than 1.00, what would be its vapour pressure?

7.25. By measuring the equilibrium between liquid and vapour phases of an acetone–methanol solution at 57.2°C at 1.00 atm, it was found that $x_{acet} = 0.400$ (the mole fraction in the liquid) when $y_{acet} = 0.516$ (the mole fraction in the vapour). Calculate the activities and activity coefficients of both components in this solution on the basis of Raoult's law. The vapour pressures of the pure components at this temperature are: $p^*(acet) = 786$ Torr and $p^*(meth) = 551$ Torr.

PROBLEMS
Numerical problems

7.1. The following table gives the mole fraction of methylbenzene (A) in liquid (x) and gaseous (y) mixtures with butanone (B) at equilibrium at 303.15 K and the total vapour pressure p. Take the vapour to be perfect and calculate the partial pressures of the two components. Plot them against their respective mole fractions in the liquid mixture and find the Henry's law constants for the two components.

x_A	y_A	p/kPa
0	0	36.066
0.0898	0.0410	34.121
0.2476	0.1154	30.900
0.3577	0.1762	28.626
0.5194	0.2772	25.239
0.6036	0.3393	23.402
0.7188	0.4450	20.693
0.8019	0.5435	18.592
0.9105	0.7284	15.496
1	1	12.295

7.2. The volume of an aqueous solution of NaCl at 25°C was measured at a series of molalities m and it was found that the volume fitted the expression

$$V/cm^3 = 1003 + 16.62(m/m^\ominus)$$
$$+ 1.77(m/m^\ominus)^{3/2} + 0.12(m/m^\ominus)^2$$

where V is the volume of a solution formed from 1.000 kg of water. Calculate the partial molar volume of the components at $m = 0.100$ mol kg^{-1}.

7.3. At 18°C the total volume of a solution formed from MgSO$_4$ and 1.000 kg of water fits the expression

$$V/cm^3 = 1001.21 + 34.69(m/m^\ominus - 0.070)^2$$

Calculate the partial molar volumes of the salt and the solvent when $m = 0.050$ mol kg^{-1}.

7.4. What amounts of ethanol and water should be mixed in order to produce 100 cm^3 of a mixture containing 50 per cent by mass of ethanol? What change in volume is brought about by adding 1.00 cm^3 of ethanol to the mixture? (Use data from Fig. 7.1.)

7.5. Potassium fluoride is very soluble in glacial acetic acid and the solutions have a number of unusual properties. In an attempt to understand them, freezing point depression data were obtained by taking a solution of known molality and then diluting it several times (J. Emsley, *J. Chem. Soc. A* 2702 (1971)). The following data were obtained:

$m/(\text{mol kg}^{-1})$	0.015	0.037	0.077	0.295	0.602
$\Delta T/\text{K}$	0.115	0.295	0.470	1.381	2.67

Calculate the apparent molar mass of the solute and suggest an interpretation. Use $\Delta_{\text{fus}}H = 11.4 \text{ kJ mol}^{-1}$ and $T_{\text{f}}^* = 290 \text{ K}$.

7.6. In a study of the properties of an aqueous solution of $\text{Th}(\text{NO}_3)_4$ (by A. Apelblat, D. Azoulay, and A. Sahar, *J. Chem. Soc. Faraday Trans.* I, 1618 (1973)), a freezing point depression of 0.0703 K was observed for a 9.6 mmol kg^{-1} aqueous solution. What is the apparent number of ions per formula unit?

7.7. The table below lists the vapour pressures of mixtures of iodoethane (I) and ethyl acetate (A) at 50°C. Find the activity coefficients of both components on the basis of (a) Raoult's law, (b) Henry's law with I as solute.

x_I	p_I/Torr	p_A/Torr
0	0	280.4
0.0579	20.0	266.1
0.1095	52.7	252.3
0.1918	87.7	231.4
0.2353	105.4	220.8
0.3718	155.4	187.9
0.5478	213.3	144.2
0.6349	239.1	122.9
0.8253	296.9	66.6
0.9093	322.5	38.2
1.0000	353.4	0

7.8. Plot the vapour pressure data for a mixture of benzene (B) and acetic acid (A) given below and plot the vapour pressure/composition curve for the mixture at 50°C. Then confirm that Raoult's and Henry's laws are obeyed in the appropriate regions. Deduce the activities and activity coefficients of the components on the basis of Raoult's law and then, taking B as the solute, its activity and activity coefficients on the basis of Henry's law.

Finally, evaluate the excess Gibbs energy of the mixture over the composition range spanned by the data.

x_A	p_A/Torr	p_B/Torr
0.0160	3.63	262.9
0.0439	7.25	257.2
0.0835	11.51	249.6
0.1138	14.2	244.8
0.1714	18.4	231.8
0.2973	24.8	211.2
0.3696	28.7	195.6
0.5834	36.3	153.2
0.6604	40.2	135.1
0.8437	50.7	75.3
0.9931	54.7	3.5

7.9. The excess molar Gibbs energy of solutions of methylcyclohexane (MCH) and tetrahydrofuran (THF) at 303.15 K were found to fit the expression

$$G^E = RTx(1-x)\{0.4857 - 0.1077(2x-1) + 0.0191(2x-1)^2\}$$

where x is the mole fraction of the methylcyclohexane. Calculate the Gibbs energy of mixing when a mixture of 1.00 mol of MCH and 3.00 mol of THF is prepared.

Theoretical problems

7.10. The excess Gibbs energy of a certain binary mixture is equal to $gRTx_A(1-x_A)$ where g is a constant. Find an expression for the chemical potential of A in the mixture and sketch its dependence on the composition.

7.11. Use the Gibbs–Duhem equation to derive the Gibbs–Duhem–Margules equation

$$\left(\frac{\partial \ln f_A}{\partial \ln x_A}\right)_{p,T} = \left(\frac{\partial \ln f_B}{\partial \ln x_B}\right)_{p,T}$$

where f is the fugacity. Use the relation to show that, when the fugacities are replaced by pressures, then, if Raoult's law applies to one component in a mixture, it must also apply to the other.

7.12. Use the Gibbs–Duhem equation to show that the partial molar volume (or any partial molar property) of a component B can be obtained if the partial molar volume (or other property) of A is known for all compositions up to the one of interest. Do this by proving that

$$V_B = V_B^* + \int_{V_A^*}^{V_A} \frac{x_A}{1-x_A} dV_A$$

7.13. Starting from the relation $\ln x_A = \Delta G/RT$, use the Gibbs–Helmholtz equation to find an expression for $d \ln x_A$ in terms of dT. Integrate $d \ln x_A$ from $x_A = 0$ to the value of interest, and integrate the right-hand side from the transition temperature for the pure liquid A to the value in the solution. Show that, if the enthalpy of transition is constant, then eqns 16 and 18 are obtained.

7.14. The osmotic coefficient, ϕ, is defined as

$$\phi = -\frac{x_A}{x_B} \ln a_A$$

By writing $r = x_B/x_A$ and using the Gibbs–Duhem equation, show that we can calculate the activity of B from the activities of A over a composition range by using the formula

$$\ln\left(\frac{a_B}{r}\right) = \phi - \phi(0) + \int_0^r \left(\frac{\phi - 1}{r}\right) dr$$

7.15. Show, that the osmotic pressure of a real solution is given by

$$\Pi V = -RT \ln a_A$$

Go on to show that, so long as the concentration of the solution is low, this takes the form

$$\Pi V = \phi RT[B]$$

and hence that the osmotic coefficient ϕ (which is defined in Problem 7.14) may be determined from osmometry.

8

Phase diagrams

Phase diagrams for pure substances were introduced in Chapter 6. Now we shall develop their use systematically and show how they are rich summaries of empirical information about a wide range of systems. To set the stage, we introduce the famous phase rule of Gibbs, which shows the extent to which various parameters can be varied yet the equilibrium between phases preserved. The phase rule limits the appearance of phase diagrams by showing, for example, that a pure substance cannot be found in a state in which four phases coexist in equilibrium. With the rule established, we shall see how it can be used to discuss the phase diagrams that we met in the two preceding chapters.

The chapter then introduces systems of gradually increasing complexity, first dealing with systems of two components and then introducing some features of systems of three components. In each case we shall see how the phase diagram for the system summarizes empirical observations on the conditions under which the various phases of the system are stable.

In this chapter we describe a systematic way of discussing the physical changes mixtures undergo when they are heated or cooled and when their compositions are changed. In particular, we see how to construct and interpret phase diagrams. These diagrams let us judge whether two or three substances are mutually miscible, whether an equilibrium can exist over a range of conditions, or whether the system must be brought to a definite pressure, temperature, and composition before equilibrium is established. Phase diagrams are of considerable commercial and industrial significance, particularly for semiconductors, ceramics, steels, and alloys. They are also the basis of separation procedures in the petroleum industry and of the formulation of foods and cosmetic preparations.

All phase diagrams can be discussed in terms of a relationship, the phase rule, derived by Gibbs. We shall derive this rule first, and then apply it to a wide variety of systems.

PHASES, COMPONENTS, AND DEGREES OF FREEDOM

The derivation and application of the phase rule requires a careful use of terms, so we shall begin by presenting a number of definitions.

8.1 Definitions

The term **phase** was introduced at the start of Chapter 6, where we saw that it signifies a state of matter that is uniform throughout, not only in chemical composition but also in physical state. (The words are Gibbs's.) Thus we speak of the solid, liquid, and gas phases of a substance, and of its various solid phases (as for black phosphorus and white phosphorus). The **number of phases** in a system is denoted P. A gas, or a gaseous mixture, is a single phase, a crystal is a single phase, and two totally miscible liquids form a single phase. Ice is a single phase ($P = 1$) even though it might be chipped into small fragments. A slurry of ice and water is a two-phase ($P = 2$) system even though it is difficult to map the boundaries between the phases. A system in which calcium carbonate undergoes thermal decomposition consists of two solid phases (one consisting of calcium carbonate and the other of calcium oxide) and one gaseous phase (consisting of carbon dioxide).

An alloy of two metals is a two-phase system ($P = 2$) if the metals are immiscible, but a single-phase system ($P = 1$) if they are miscible. This example shows that it is not always easy to decide whether a system consists of one phase or of two. A solution of solid A in solid B—a homogeneous mixture of the two substances—is uniform on a molecular scale. In a solution, atoms of A are surrounded by atoms of A and B, and any sample cut from the sample, however small, is representative of the composition of the whole.

A dispersion is uniform on a macroscopic scale but not on a microscopic scale, for it consists of grains or droplets of one substance in a matrix of the other. A small sample could come entirely from one of the minute grains of pure A and would not be representative of the whole (Fig. 8.1). Such dispersions are important because, in many advanced materials (including steels), heat treatment cycles are used to achieve the precipitation of a fine dispersion of particles of one phase (such as a carbide phase) within a matrix formed by a saturated solid solution phase. It is the ability to control this microstructure resulting from phase equilibria that makes it possible to tailor the mechanical properties of the materials to a particular application.

By a **constituent** of a system we shall mean a chemical species (an ion or a molecule) that is present. Thus, a mixture of ethanol and water has two constituents. The term constituent should be carefully distinguished from 'component', which has a more technical meaning. A **component** is a chemically independent constituent of a system. The **number of components** C in a system is the minimum number of independent species necessary to define the composition of *all* the phases present in the system.

When no reaction takes place, the number of components is equal to the number of constituents. Thus, pure water is a one-component system

(a)

(b)

8.1 The difference between (a) a single-phase solution, in which the composition is uniform on a microscopic scale, and (b) a dispersion, in which regions of one component are embedded in a matrix of a second component.

(because we need only the species H_2O to specify its composition). Similarly, a mixture of ethanol and water is a two-component system (we need the species H_2O and C_2H_5OH to specify its composition). When a reaction can occur under the conditions prevailing in the system, we need to decide the minimum number of species that, after allowing for reactions in which one species is synthesized from others, can be used to specify the composition of all the phases. Consider, for example, a system in which the equilibrium

$$CaCO_3(s) \rightleftharpoons CaO(s) + CO_2(g)$$

phase 1   phase 2   phase 3

occurs. To specify the composition of the gas we need the species CO_2 and to specify the composition of phase 2 we need the species CaO. However, we do not need an additional species to specify phase 1 because its identity ('$CaCO_3$') can be expressed in terms of the other two constituents by making use of the reaction

$$CaO + CO_2 \rightarrow CaCO_3$$

Hence, the system has two components ($C = 2$).

Example 8.1 *Counting components*

How many components are present in a system in which ammonium chloride undergoes thermal decomposition?

Method. Begin by writing down the chemical equation for the reaction and identifying the constituents of the system (all the species present) and the phases. Then decide whether, under the conditions prevailing in the system, any of the constituents can be prepared from any of the other constituents. The removal of these constituents leaves the number of *independent* constituents. Finally, identify the minimum number of these independent constituents that are needed to specify the composition of all the phases.

Answer. The chemical reaction is

$$NH_4Cl(s) \rightleftharpoons NH_3(g) + HCl(g)$$
phase 1   phase 2

There are three constituents. However, NH_3 and HCl can be prepared in the correct stoichiometric proportions by the reaction

$$NH_4Cl \rightarrow NH_3 + HCl$$

Therefore, the compositions of both phase 1 and phase 2 can be expressed in terms of the single species NH_4Cl. Therefore, there is only one component in the system ($C = 1$).

Comment. If additional HCl (or NH_3) were supplied to the system, then the decomposition of NH_4Cl would not give the correct composition of the gas phase and HCl (or NH_3) would have to be invoked as a second component. A system that consists of hydrogen, oxygen, and water at room temperature has *three* components, despite it being possible to form H_2O from H_2 and O_2: under the conditions *prevailing* in the system, hydrogen and oxygen do not react to form water, so they are independent constituents.

Exercise E8.1. Give the number of components in the following systems: (a) water, allowing for its ionization, (b) aqueous acetic acid, (c) magnesium carbonate in equilibrium with its vapour.   [(a) 1, (b) 2, (c) 2]

The **variance** F of a system is the number of intensive variables that can be changed independently without disturbing the number of phases in equilibrium. In a single-component ($C = 1$), single-phase systems, the pressure and temperature may be changed independently without changing the number of phases, so $F = 2$. We say that such a system is **bivariant** or that it has two **degrees of freedom**. On the other hand, if two phases are in equilibrium (a liquid and its vapour, for instance), then to preserve that number of phases when the temperature is changed, the pressure must be adjusted to match the new vapour pressure. The variance of the system has fallen to 1 because only the pressure (or the temperature) can be changed independently.

8.2 The phase rule

In one of the most elegant calculations in the whole of chemical thermodynamics, J. W. Gibbs[1] deduced the **phase rule,** which is a general relation between the variance F, the number of components C, and the number of phases P at equilibrium for a system of any composition:

$$F = C - P + 2 \tag{1}$$

JUSTIFICATION

We begin by counting the total number of intensive variables (properties that do not depend on the size of the system). The pressure p and temperature T count as 2. We can specify the composition of a phase by giving the mole fractions of $C - 1$ components. (We need specify only $C - 1$ and not all C mole fractions because $x_1 + x_2 + \ldots + x_C = 1$, and all mole fractions are known if all except one are specified.) Because there are P phases, the total number of composition variables is $P(C - 1)$. At this stage, the total number of intensive variables is $P(C - 1) + 2$.

At equilibrium, the chemical potential of a component J must be the same in every phase (Chapter 6):

$$\mu_{J,\alpha} = \mu_{J,\beta} = \ldots \text{ for } P \text{ phases}$$

1 Josiah Willard Gibbs spent most of his working life at Yale, and may justly be regarded as the originator of chemical thermodynamics. He reflected for years before publishing his conclusions, and then did so in precisely expressed papers in an obscure journal (*The Transactions of the Connecticut Academy of Arts and Sciences*). He needed interpreters before the power of his work was recognized and before it could be applied to industrial processes. He is regarded by many as the first great American theoretical scientist.

That is, there are $P-1$ equations to be satisfied for each component J. As there are C components, the total number of equations is $C(P-1)$. Each equation reduces our freedom to vary one of the $P(C-1)+2$ intensive variables. It follows that the total variance is

$$F = P(C-1) + 2 - C(P-1) = C - P + 2$$

which is eqn 1.

We shall now go on to see how the phase rule summarizes what we already know about one-component systems and then apply it to more complex cases.

One-component systems

For a one-component system, such as pure water,

$$F = 3 - P$$

When only one phase is present, $F = 2$ and both p and T can be varied independently without changing the number of phases. In other words, a single phase is represented by an *area* on a phase diagram. When two phases are in equilibrium, $F = 1$, which implies that pressure is not freely variable if the temperature is set; indeed, at a given temperature, a liquid has a characteristic vapour pressure. It follows that the equilibrium of two phases is represented by a *line* in the phase diagram. Instead of selecting the temperature, we could select the pressure, but having done so the two phases would be in equilibrium at a single definite temperature. Therefore, freezing (or any other phase transition) occurs at a definite temperature at a given pressure.

When three phases are in equilibrium, $F = 0$ and the system is **invariant**. This special condition can be established only at a definite temperature and pressure which is characteristic of the substance and outside our control. The equilibrium of three phases is therefore represented by a *point*, the **triple point**, on the phase diagram. Four phases cannot be in equilibrium in a one-component system because F cannot be negative. These features are summarized in Fig. 8.2.

The features summarized in the illustration can be identified in the experimentally determined phase diagram for water shown in Fig. 8.3. This diagram summarizes the changes that take place as a sample, such as that at a, is cooled at constant pressure. The sample remains entirely gaseous until the temperature reaches b, when liquid appears. Two phases are now in equilibrium and $F = 1$. Because we have decided to specify the pressure, which uses up the single degree of freedom, the temperature at which this equilibrium occurs is not under our control. Lowering the temperature takes the system to c in the one-phase, liquid region. The temperature can now be varied around the point c at will, and only when ice appears at d does the variance become 1 again.

Experimental procedures

Detecting a phase change is not always as simple as seeing a kettle boil, so special techniques have been developed. One technique is **thermal**

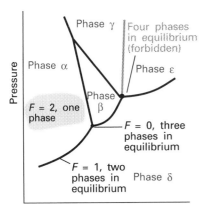

8.2 The typical regions of a one-component phase diagram. The lines represent conditions under which the two adjoining phases are in equilibrium. A point represents the unique set of conditions under which three phases coexist in equilibrium. Four phases cannot mutually coexist in equilibrium.

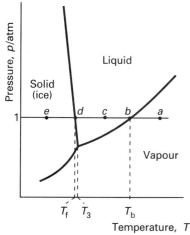

8.3 The phase diagram for water, a simplified version of Fig. 6.5. The label T_3 marks the temperature of the triple point, T_b the normal boiling point, and T_f the normal freezing point.

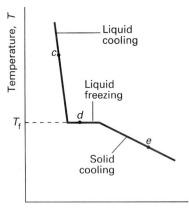

8.4 The cooling curve for the isobar *cde* in Fig. 8.3. The halt marked *d* corresponds to the pause in the fall of temperature while the first-order exothermic transition (freezing) occurs. This pause enables T_f to be located even if the transition cannot be observed visually.

8.5 Ultrahigh pressures (up to about 2 Mbar) can be achieved using a diamond anvil. The sample, together with a ruby for pressure measurement and a drop of liquid for pressure transmission, are placed between two gem-quality diamonds. The principle of its action is that of a nutcracker: the pressure is exerted by turning the screw by hand.

analysis, which takes advantage of the effect of the enthalpy change during a first-order transition (Section 6.5). In this method, a sample is allowed to cool and its temperature is monitored. At a first-order transition, heat is evolved and the cooling stops until the transition is complete. The cooling curve along the isobar *cde* in Fig. 8.3 therefore has the shape shown in Fig. 8.4. The transition temperature is obvious, and is used to mark point *d* on the phase diagram. This technique is useful for solid–solid transitions, where simple visual inspection of the sample may be inadequate.

Modern work on phase transitions often deals with systems at very high pressures and more sophisticated detection procedures must be adopted. Some of the highest pressures currently attainable are produced in a **diamond-anvil cell** like that illustrated in Fig. 8.5. The sample is placed in a minute cavity between two gem-quality diamonds, and then pressure is exerted simply by turning the screw. The advance in design this represents is quite remarkable for, with a turn of the screw, pressures of up to about 1 Mbar can be reached which a few years ago could not be reached with equipment weighing tons.

The pressure is monitored spectroscopically by observing the shift of spectral lines in small pieces of ruby added to the sample, and the properties of the sample itself are observed optically through the diamond anvils. One application of the technique is to study the transition of covalent solids to metallic solids. Iodine, for instance, becomes metallic at around 200 kbar while remaining as I_2, but makes a transition to a monatomic metallic solid at around 210 kbar. Studies such as these are relevant to the structure of material deep inside the Earth (at the centre of the Earth the pressure is around 5 Mbar) and in the interiors of the giant planets, where even hydrogen may be metallic.

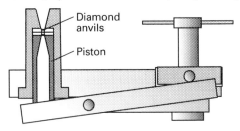

TWO-COMPONENT SYSTEMS

In this section we shall begin by showing how the general form of a phase diagram for a binary mixture of two volatile liquids can be constructed by drawing on the properties of ideal solutions introduced in Chapter 7, particularly Raoult's law (eqn 7.11). Then we shall see how deviations from ideality distort the diagrams in a way that has important implications for distillations. Finally, we shall meet phase diagrams for systems consisting of a variety of liquids and solids and see how the phase rule can be used to guide our interpretation of them. When two components are present in a system, $C = 2$ and

$$F = 4 - P$$

If the temperature is constant, the remaining variance is $F' = 3 - P$, which has a maximum value of 2. One of these two remaining degrees of freedom is the pressure and the other is the composition (as expressed by the mole fraction of one component). Hence, one form of the phase diagram is a map of pressures and compositions at which each phase is stable. Alternatively, the pressure could be held constant and the phase diagram depicted in terms of temperature and composition. We shall introduce both types of diagram.

8.3 Vapour pressure diagrams

The partial vapour pressures of the components of an ideal solution of two volatile liquids are related to the composition of the liquid mixture by Raoult's law:

$$p_A = x_A p_A^* \qquad p_B = x_B p_B^* \tag{2}°$$

where p_A^* is the vapour pressure of pure A and p_B^* that of pure B. The total vapour pressure p of the mixture is therefore

$$p = p_A + p_B = x_A p_A^* + x_B p_B^* = p_B^* + (p_A^* - p_B^*)x_A \tag{3}°$$

This expression shows that the total vapour pressure (at some fixed temperature) changes linearly with the composition from p_B^* to p_A^* (Fig. 8.6).

The composition of the vapour

The compositions of the liquid and vapour that are in equilibrium are not necessarily the same, and common sense suggests that the vapour should be richer in the more volatile component. This expectation can be confirmed as follows. The partial pressures of the components are given by eqn 2. It follows from Dalton's law that the mole fractions in the gas, y_A and y_B, are

$$y_A = \frac{p_A}{p} \qquad y_B = \frac{p_B}{p} \tag{4}°$$

The partial pressures and the total pressure may be expressed in terms of the mole fractions in the liquid by using eqn 2 for p_J and eqn 3 for the total vapour pressure p, which gives

$$y_A = \frac{x_A p_A^*}{p_B^* + (p_A^* - p_B^*)x_A} \qquad y_B = 1 - y_A \tag{5}°$$

We now have to show that, if A is the more volatile component, then its mole fraction in the vapour y_A is greater than its mole fraction in the liquid x_A. Figure 8.7 shows the composition of the vapour plotted against the composition of the liquid for various values of $p_A^*/p_B^* > 1$: we see that in all cases $y_A > x_A$, as we expect. Note that if B is non-volatile so that $p_B^* = 0$ at the temperature of interest, then it makes no contribution to the vapour ($y_B = 0$).

Equation 3 shows how the total vapour pressure of the mixture varies with the composition of the liquid. Because we can relate the composition of the liquid to the composition of the vapour through eqn 5, we can now also relate the total vapour pressure to the composition of the

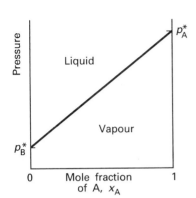

8.6 The variation of the total vapour pressure of a binary mixture with the mole fraction of A in the liquid when Raoult's law is obeyed.

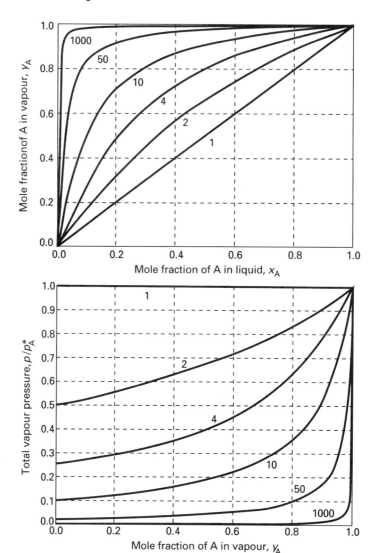

8.7 The mole fraction of A in the vapour of a binary ideal solution expressed in terms of its mole fraction in the liquid, calculated using eqn 5 for various values of p_A^*/p_B^* (the label on each curve) with A more volatile than B. In all cases the vapour is richer than the liquid in A.

8.8 The dependence of the vapour pressure of the same system as in Fig. 8.7, but expressed in terms of the mole fraction of A in the vapour by using eqn 6. Individual curves are labelled with the value of p_A^*/p_B^*.

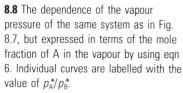

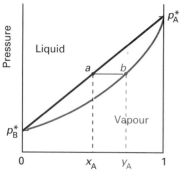

8.9 The dependence of the total vapour pressure of an ideal solution on the mole fraction of A in the entire system. A point between the two lines corresponds to both liquid and vapour being present; outside that region there is only one phase present.

vapour:

$$p = \frac{p_A^* p_B^*}{p_A^* + (p_B^* - p_A^*) y_A} \tag{6}°$$

This expression is plotted in Fig. 8.8.

The interpretation of the diagrams

If we are interested in distillation, then both the vapour and the liquid compositions are of equal interest. It is then sensible to combine the two preceding diagrams into one. The result is summarized in Fig. 8.9. The point a indicates the vapour pressure of a mixture of composition x_A and the point b indicates the composition of the vapour that is in equilibrium with the liquid at that pressure. Note that, when two phases are in equilibrium, $P = 2$ so $F' = 1$. That is, if the composition is specified (so using up the only degree of freedom), the pressure at which the two phases are in equilibrium is fixed. Hence, the compositions and pressures

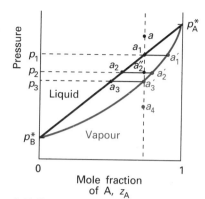

8.10 The points of the pressure–composition diagram discussed in the text. The vertical line through a is an isopleth, a line of constant composition of the entire system.

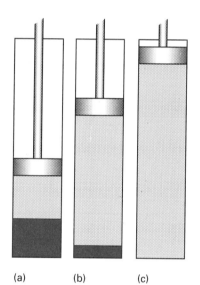

8.11 (a) A liquid in a container exists in equilibrium with its vapour. (b) When the pressure is changed by drawing out a piston, the compositions of the phases adjust. (c) When the piston is pulled so far out that all the liquid has vaporized and only the vapour is present, the pressure falls as the piston is withdrawn.

at which the two phases are in equilibrium are represented by a line on the phase diagram.

A richer interpretation of the phase diagram is obtained if we interpret the horizontal axis as showing the *overall* composition z_A of the system, allowing for the different compositions and amounts of the liquid and vapour phases. If the horizontal axis of the vapour pressure diagram is labelled with z_A, then all the points down to the solid diagonal line in Fig. 8.9 correspond to a system that is under such high pressure that it contains only a liquid phase (the applied pressure is higher than the vapour pressure), so $z_A = x_A$, the composition of the liquid. On the other hand, all points below the grey curve correspond to a system that is under such low pressure that it contains only a vapour phase (the applied pressure is lower than the vapour pressure), so $z_A = y_A$.

Points that lie *between* the two lines correspond to a system in which there are two phases present, one a liquid and the other a vapour. To see this interpretation, consider lowering the pressure on a liquid mixture of overall composition a in Fig. 8.10: the lowering of pressure can be achieved by drawing out a piston (Fig. 8.11); this degree of freedom is permitted by the phase rule because $F' = 2$ when $P = 1$ and, even if the composition is selected, one degree of freedom remains. The changes to the system do not affect the overall composition, so the state of the system moves down the vertical line that passes through a. This vertical line is called an **isopleth**, from the Greek words for 'equal abundance'. Until the point a_1 is reached (when the pressure has been reduced to p_1), the sample consists of a single liquid phase. At a_1 the liquid can exist in equilibrium with its vapour. As we have seen, the composition of the vapour phase is given by point a_1'. The horizontal line joining the two points is called a **tie line**. The composition of the liquid is the same as initially (a_1 lies on the isopleth through a), so we have to conclude that at this pressure there is virtually no vapour present; however, the tiny amount of vapour that *is* present has the composition a_1'.

Now consider the effect of lowering the pressure to p_2, so taking the system to a pressure and overall composition represented by the point a_2''. This new pressure is below the vapour pressure of the original liquid, so it vaporizes until the vapour pressure of the remaining liquid falls to p_2. Now we know that the composition of such a liquid must be a_2. Moreover, the composition of the vapour in equilibrium with that liquid must be given by the point a_2' at the other end of the tie line. Note that two phases are now in equilibrium, so $F' = 1$ for all points between the two lines; hence, for a given pressure (such as at p_2) the variance is zero, and the vapour and liquid phases have fixed compositions (Fig. 8.12). If the pressure is reduced to p_3, a similar readjustment in composition takes place, and now the compositions of the liquid and vapour are represented by the points a_3 and a_3', respectively. The latter point corresponds to a system in which the composition of the vapour is the same as the overall composition, so we have to conclude that the amount of liquid present is now virtually zero, but the tiny amount of liquid that is present has the composition a_3. A further decrease in pressure takes the system to the point a_4; at this stage, only vapour is present and its composition is the

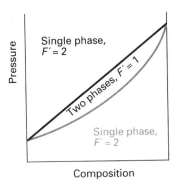

8.12 The general scheme of interpretation of a pressure–composition diagram (a vapour pressure diagram).

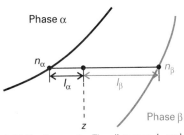

8.13 The lever rule. The distances l_α and l_β are used to find the proportions of the amounts of phases α (such as vapour) and β (for example, liquid) present at equilibrium. The lever rule is so called because a similar rule relates the masses at two ends of a lever to their distances from a pivot ($m_\alpha l_\alpha = m_\beta l_\beta$ for balance).

same as the initial overall composition of the system (the composition of the original liquid).

The lever rule

A point in the two-phase region of a phase diagram indicates not only qualitatively that both liquid and vapour are present, but represents quantitatively the relative amounts of each. To find the relative amounts of two phases α and β that are in equilibrium, we measure the distances l_α and l_β along the horizontal tie line, and then use the **lever rule** (Fig. 8.13):

$$n_\alpha l_\alpha = n_\beta l_\beta \tag{7}$$

where n_α is the amount of phase α and n_β the amount of phase β. In the case illustrated in Fig. 8.13, because $l_\beta \approx 2l_\alpha$, the amount of phase α is about twice the amount of phase β.

JUSTIFICATION

To prove the lever rule we write $n = n_\alpha + n_\beta$ and the overall amount of A as nz_A. The overall amount of A is also the sum of its amounts in the two phases:

$$nz_A = n_\alpha x_A + n_\beta y_A$$

Since also

$$nz_A = n_\alpha z_A + n_\beta z_A$$

by equating these two expressions it follows that

$$n_\alpha(z_A - x_A) = n_\beta(y_A - z_A)$$

or

$$n_\alpha l_\alpha = n_\beta l_\beta$$

as was to be proved.

To see in more detail how the lever rule is used, consider the changes in Fig. 8.10 again. At p_1, the ratio l_{vap}/l_{liq} is almost infinite for this tie line, so n_{liq}/n_{vap} is also almost infinite, and there is only a trace of vapour present. When the pressure is reduced to p_2, the value of l_{vap}/l_{liq} is about 0.7, so $n_{liq}/n_{vap} \approx 0.7$ and the amount of liquid is about 0.7 times the amount of vapour. When the pressure has been reduced to p_3, the sample is almost completely gaseous and, because $l_{vap}/l_{liq} \approx 0$, we conclude that there is only a trace of liquid present.

8.4 Temperature–composition diagrams

Reducing the pressure at constant temperature is one way of doing distillation, but it is more common to distil at constant pressure by raising the temperature. To discuss distillation in this way we need a **temperature–composition diagram**, a phase diagram in which the boundaries show the composition of the phases that are in equilibrium at various temperatures (and a given pressure, typically 1 atm). An example

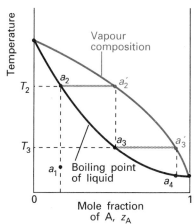

8.14 The temperature–composition diagram corresponding to an ideal mixture with the component A more volatile than component B. Successive boilings and condensations of a liquid originally of composition a_1 lead to a condensate that is pure A. The separation technique is called fractional distillation.

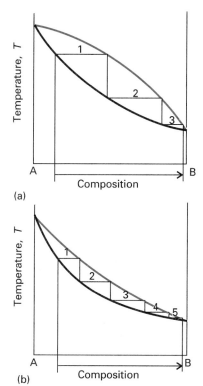

8.15 The number of theoretical plates is the number of steps needed to bring about a specified degree of separation of two components in a mixture. The two systems shown correspond to (a) three and (b) five theoretical plates.

is shown in Fig. 8.14. Note that the liquid phase region now lies in the lower part of the diagram.

The distillation of mixtures

The interpretation of a temperature–composition diagram is similar to that of the pressure–composition diagram. The region between the lines is a two-phase region where $F' = 1$ and, hence, at a given temperature the compositions of the phases in equilibrium are fixed. The regions outside the phase lines correspond to a single phase, so $F' = 2$ and the temperature and composition are both modifiable.

To understand the content of the phase diagram, consider what happens when a liquid of composition a_1 is heated. It boils when the temperature reaches T_2. Then the liquid has composition a_2 (the same as a_1) and the vapour (which is present only as a trace) has composition a_2'. The vapour is richer in the more volatile component A (the component with the lower boiling point), as common sense leads us to expect. From the location of a_2' we can state the vapour's composition at the boiling point, and from the location of the tie line joining a_2 and a_2' we can read off the boiling temperature (T_2) of the original liquid mixture.

In a simple distillation, the vapour is withdrawn and condensed. If the vapour in this example is drawn off and completely condensed, then it gives a liquid of composition a_3, which is richer in the more volatile component than the original liquid. In **fractional distillation** the boiling and condensation cycle is repeated successively. We can follow the changes that occur by seeing what happens when the condensate of composition on the vertical line through a_3 is reheated. The phase diagram shows that this mixture boils at T_3 and yields a vapour of composition a_3' which is even richer in the more volatile component. That vapour is drawn off, and it condenses to a liquid of composition a_4. The cycle can then be repeated until in due course almost pure A is obtained.

In chemical engineering applications, the efficiency of a fractionating column is expressed in terms of the number of **theoretical plates,** the number of effective vaporization and condensation steps that are required to achieve a condensate of given composition from a given distillate. Thus, to achieve the degree of separation shown in Fig. 8.15a, the fractionating column must correspond to three theoretical plates, whereas to achieve the same separation for the system shown in Fig. 8.15b the fractionating column must be designed to correspond to five theoretical plates.

Azeotropes

Although many liquids have temperature–composition phase diagrams resembling the ideal version in Fig. 8.14, in a number of important cases there are marked deviations. A maximum in the phase diagram (Fig. 8.16) may occur when the favourable interactions between A and B molecules reduce the vapour pressure of the mixture below the ideal value: in effect, the A–B interactions stabilize the liquid. In such cases the excess Gibbs energy G^E (Section 7.4), is negative (more favourable

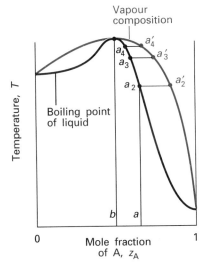

8.16 A high-boiling azeotrope. When the liquid of composition a is distilled, the composition of the remaining liquid changes towards b but no further.

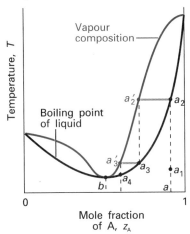

8.17 A low-boiling azeotrope. When the mixture at a is fractionally distilled, the vapour in equilibrium in the fractionating column moves towards b and then remains unchanged.

to mixing than ideal). Examples of this behaviour include chloroform/acetone and nitric acid/water mixtures. Phase diagrams showing a minimum (Fig. 8.17) indicate that the mixture is destabilized relative to the ideal solution, the A–B interactions then being unfavourable. For such mixtures G^E is positive (less favourable to mixing than ideal), and there may be contributions from both enthalpy and entropy effects. Examples include mixtures of dioxane and water and of ethanol and water.

Deviations from ideality are not always so strong as to lead to a maximum or minimum in the phase diagram, but when they do there are important consequences for distillation. Consider a liquid of composition a on the right of the maximum in Fig. 8.16. The vapour (at a_2') of the boiling mixture (at a_2) is richer in A. If that vapour is removed (and condensed elsewhere), then the remaining liquid will move to a composition that is richer in B, such as that represented by a_3, and the vapour in equilibrium with this mixture will have composition a_3'. If that vapour is removed, then the composition of the boiling liquid shifts to a point such as a_4 and the composition of the vapour shifts to a_4'. Hence, as evaporation proceeds, the composition of the remaining liquid shifts towards B as A is drawn off. The boiling point of the liquid rises, and the vapour becomes richer in B. When so much A has been evaporated that the liquid has reached the composition b, the vapour has the same composition as the liquid. Evaporation then occurs without change of composition. The mixture is said to form an **azeotrope** (which comes from the Greek words for 'boiling without changing'). When the azeotropic composition thas been reached, distillation cannot separate the two liquids because the condensate has the same composition as the azeotropic liquid. One example of azeotrope formation is hydrochloric acid/water, which is azeotropic at 80 per cent by mass of water and boils unchanged at 108.6°C.

The system shown in Fig. 8.17 is also azeotropic, but shows it in a different way. Suppose we start with a mixture of composition a_1 and follow the changes in the composition of the vapour that rises through a fractionating column (essentially a vertical glass tube packed with glass rings to give a large surface area). The mixture boils at a_2 to give a vapour of composition a_2'. This vapour condenses in the column to a liquid of the same composition (now marked a_3). That liquid reaches equilibrium with its vapour at a_3', which condenses higher up the tube to give a liquid of the same composition, which we now call a_4. The fractionation therefore shifts the vapour towards the azeotropic composition, but not beyond, and the azeotropic vapour emerges from the top of the column. An example is ethanol/water, which boils unchanged when the water content is 4 per cent and the temperature is 78°C.

Immiscible liquids

Finally we consider the distillation of two immiscible liquids, such as octane and water. As they are immiscible we can regard their 'mixture' as unscrambled with each component in a separate vessel (Fig. 8.18). If the vapour pressures of the two pure components are p_A and p_B, then the total vapour pressure is $p = p_A^* + p_B^*$ and the mixture boils (in an open container) when $p = 1$ atm. The presence of the second component means that the agitated 'mixture' boils at a lower temperature than either would alone because boiling begins when the total pressure reaches 1 atm, not when either vapour pressure reaches 1 atm. This is the basis of steam distillation, which enables some heat-sensitive, water-insoluble organic compounds to be distilled at a lower temperature than their normal boiling point. The only snag is that the composition of the condensate is in proportion to the vapour pressures of the components, so oils of low volatility distil in low abundance.

8.5 Liquid–liquid phase diagrams

Now we consider temperature–composition phase diagrams for binary systems that consist of pairs of **partially miscible liquids**, which are liquids that do not mix in all proportions at all temperatures. An example is hexane and nitrobenzene. The same principles of interpretation apply as to liquid–vapour diagrams. When $P = 2$, $F' = 1$, and the selection of a temperature implies that the compositions of the immiscible liquid phases are fixed. When $P = 1$ (corresponding to a system in which the two liquids are fully mixed), both the temperature and the composition may be adjusted.

Phase separation

Suppose a small amount of a liquid B is added to a sample of another liquid A at a temperature T'. It dissolves completely, and the binary system remains a single phase. As more B is added, a stage comes at which no more dissolves. The sample now consists of two phases in equilibrium with each other ($P = 2$), the more abundant one consisting of A saturated with B, the minor one of a trace of B saturated with A. In the temperature–composition diagram drawn in Fig. 8.19, the composition of the former is represented by the point a' and that of the latter by the point a''. The relative abundances of the two phases are given by the lever rule.

When more B is added, A dissolves in it slightly. The compositions of the two phases in equilibrium remain a' and a'' (because $P = 2$ implies that $F'' = 0$, where the double prime indicates that two intensive variables, the temperature and the pressure, are held constant, and hence that the compositions of the phases are invariant at a fixed temperature and pressure), but the amount of the second phase increases at the expense of the first. A stage is reached when so much B is present that it can dissolve all the A, and the system reverts to a single phase. The addition of more B now simply dilutes the solution, and from then on it remains a single phase.

The composition of the two phases at equilibrium varies with the temperature. For hexane and nitrobenzene, raising the temperature increases their miscibility. The two-phase system therefore becomes less extensive, because each phase in equilibrium is richer in its minor component: the A-rich phase is richer in B and the B-rich phase is richer in A. The entire phase diagram can be constructed by repeating the

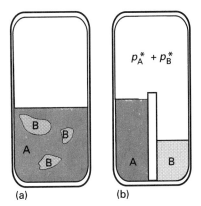

8.18 The distillation of two immiscible liquids can be regarded as the joint distillation of the separated components, and boiling occurs when the sum of the partial pressures equals the external pressure.

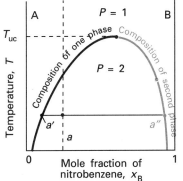

8.19 The temperature–composition diagram for hexane and nitrobenzene at 1 atm. The region below the curve corresponds to the compositions and temperatures at which the liquids are partially miscible. The upper critical temperature T_{uc} is the temperature above which the two liquids are miscible in all proportions. For this system it lies at 293 K.

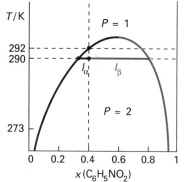

8.20 The temperature–composition diagram for hexane and nitrobenzene at 1 atm again, with the points and lengths discussed in the text.

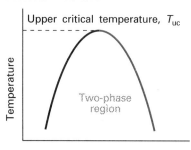

(a)

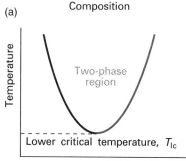

(b)

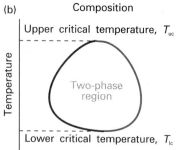

(c)

8.21 The three types of critical-solution behaviour shown by liquids. (a) A system with an upper critical temperature, (b) a system with a lower critical temperature, (c) a system with both upper and lower critical temperatures.

observations at different temperatures and drawing the envelope of the two-phase region. This is usually done at fixed compositions by lowering the temperature to determine the envelope.

Example 8.2 *Interpreting a liquid–liquid phase diagram*

A mixture of 50 g (0.59 mol) of hexane and 50 g (0.41 mol) of nitrobenzene was prepared at 290 K. What are the compositions of the phases, and in what proportions do they occur? To what temperature must the sample be heated in order to obtain a single phase?

Method. The compositions of phases in equilibrium are given by the points where the tie line through the point representing the temperature and overall composition of the system intersect the phase boudary. Their proportions are given by the lever rule (eqn 7). The temperature at which the components are completely miscible is found by following the isopleth upwards and noting the temperature at which it enters the one-phase region of the phase diagram.

Answer. We denote hexane by H and nitrobenzene by N; refer to Fig. 8.20, which is a slightly simplified version of Fig. 8.19. The point $x_N = 0.41$, $T = 290$ K occurs in the two-phase region of the phase diagram. The horizontal tie line cuts the phase boundary at $x_N = 0.35$ and $x_N = 0.83$, so those are the compositions of the two phases. The ratio of amounts of each phase is equal to the ratio of distances l_α and l_β:

$$\frac{l_\beta}{l_\alpha} = \frac{0.83 - 0.41}{0.41 - 0.35} = \frac{0.42}{0.06} = 7$$

Heating the sample to 292 K takes it into the single-phase region.

Comment. Because the phase diagram has been constructed experimentally, these conclusions are exact. They would be modified if the system were subjected to a different pressure.

Exercise E8.2. Repeat the problem for 50 g hexane and 100 g nitrobenzene at 273 K. $[x_N = 0.09$ and 0.95 in ratio 1:8; 290 K]

Critical temperatures

The **upper critical temperature** T_{uc} is the highest temperature at which phase separation occurs (Fig. 8.21a). Above the upper critical temperature the two components are fully miscible. This temperature exists because the greater thermal motion overcomes any potential energy advantage for molecules of one type being close together. One example is the nitrobenzene–hexane system shown in Fig. 8.19; another is the palladium–hydrogen system, which shows two phases, one a solid solution of H_2 in palladium and the other a palladium hydride, up to 300°C, but forms a single phase at higher temperatures (Fig. 8.22).

Some systems show a **lower critical temperature** T_{lc}, below which they mix in all proportions and above which they form two phases (Fig. 8.21b). An example is water and triethylamine (Fig. 8.23). In this case, at low temperatures, the two components are more miscible because they

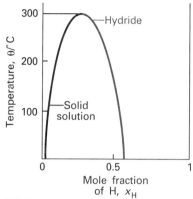

8.22 The phase diagram for palladium and palladium hydride, which has an upper critical temperature at 300°C.

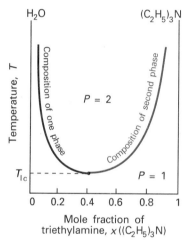

8.23 The temperature–composition diagram for water and triethylamine. This system shows a lower critical temperature at 292 K. The labels indicate the interpretation of the boundaries.

form a weak complex; at higher temperatures, the complexes break up and the two components are less miscible.

Some systems have both upper and lower critical temperatures (Fig. 8.21c). They occur because, after the weak complexes have been disrupted leading to partial miscibility, the thermal motion at higher temperatures homogenizes the mixture again, just as in the case of ordinary partially miscible liquids. The most famous example is that of nicotine and water, which are partially miscible between 61°C and 210°C (Fig. 8.24).

The distillation of partially miscible liquids

We now consider what happens when the conditions are such that a vapour may be present too. We shall consider a pair of liquids that are partially miscible and form a low-boiling azeotrope. This combination of properties is quite common because both properties reflect the tendency of the two kinds of molecule to avoid each other. There are two possibilities: one in which the liquids become fully miscible before they boil; the other in which boiling occurs before mixing is complete.

Figure 8.25 shows the phase diagram for two components that become fully miscible before they boil. Distillation of a mixture of composition a_1 leads to a vapour of composition b_1 which condenses to the completely miscible single-phase solution at b_2. Phase separation occurs only when this distillate is cooled to a point in the two-phase liquid region, such as b_3. This description applies only to the first drop of distillate. If distillation continues, the composition of the remaining liquid changes. In the end, when the whole sample has evaporated and condensed, the composition is back to a_1.

Figure 8.26 shows the second possibility, in which there is no upper critical temperature. The distillate obtained from a liquid initially of composition a_1 has composition b_3 and is a two-phase mixture. One phase has composition b_3' and the other has composition b_3''.

The behaviour of a system of composition represented by the isopleth e is interesting. A system at e_1 forms two phases, which persist (but with changing proportions) up to the boiling point at e_2. The vapour of this mixture has the same composition as the liquid (the liquid is an azeotrope). Similarly, condensing a vapour of composition e_3 gives a liquid of the same composition. At a fixed temperature, the mixture vaporizes and condenses like a single substance.

Example 8.3 *Interpreting a phase diagram*

State the changes that occur when a mixture of composition $x_B = 0.95$ (a_1 in Fig. 8.27) is boiled and the vapour condensed.

Method. The area occupied by the point gives the number of phases; the compositions of the phases are given by the points at the intersections of the horizontal tie line with the phase boundaries; the relative abundances are given by the lever rule (eqn 7).

Answer. The initial point is in the one-phase region. When heated it boils

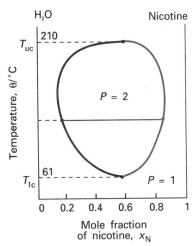

8.24 The temperature–composition diagram for water and nicotine, which has both upper and lower critical temperatures. Note the high temperatures for the liquid (especially the water): the diagram corresponds to a sample under pressure.

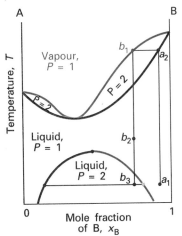

8.25 The temperature–composition diagram for a binary system in which the upper critical temperature is less than the boiling point at all compositions. The mixture forms a low-boiling azeotrope.

at 370 K (a_2 in Fig. 8.27) giving a vapour of composition $x_B = 0.66$ (b_1 in Fig. 8.27). The liquid gets richer in B and the last drop (of pure B) evaporates at 392 K. The boiling range of the liquid is therefore 370 to 392 K. If the initial vapour is drawn off, it has a composition $x_B = 0.66$. This composition would be maintained if the sample were very large, but for a finite sample it shifts to higher values and ultimately to $x_B = 0.95$. Cooling the distillate corresponds to moving down the $x_B = 0.66$ isopleth. At 350 K, for instance, the liquid phase has composition $x_B = 0.87$, the vapour $x_B = 0.49$, in relative proportions $1:1.3$. At 340 K the sample consists of three phases, the vapour, and two liquids, one of composition $x_B = 0.30$, the other of composition $x_B = 0.80$ in the ratio $0.62:1$. Further cooling moves the system into the two-phase liquid region, and at 298 K the compositions are 0.20 and 0.90 in the ratio $0.82:1$. As further distillate boils over, the overall composition of the distillate becomes richer in B. When the last drop has been condensed the phase composition is the same as at the beginning.

Exercise E8.3. Repeat the discussion, beginning at the point $x_B = 0.4$, $T = 298$ K.

8.6 Liquid–solid phase diagrams

Solid and liquid phases may both be present in a system at temperatures below the boiling point. An example is a pair of metals that are almost completely immiscible right up to their melting points (such as antimony and bismuth). The phase diagram is shown in Fig. 8.28; note how closely it resembles Fig. 8.26, but instead of liquid and vapour phases the system has solid and liquid phases.

Consider the two-component liquid of composition a_1. The changes that occur may be expressed as follows.

(1) $a_1 \rightarrow a_2$. The system enters the two-phase region labelled 'Liquid + B'. Almost pure solid B begins to come out of solution and the remaining liquid becomes richer in A.

(2) $a_2 \rightarrow a_3$. More of the solid forms, and the relative amounts of the solid and liquid (which are in equilibrium) are given by the lever rule: at a_3 there are roughly equal amounts of each. The liquid phase is richer in A than before (its composition is given by b_3) because some B has been deposited.

(3) $a_3 \rightarrow a_4$. At the end of this step, at a_4, there is less liquid than at a_3, and its composition is given by e. This liquid now freezes to give a two-phase system of almost pure B and almost pure A. At a_5, for example, the compositions of the two phases are a_5' and a_5''.

Eutectics

The isopleth at e in Fig. 8.28 corresponds to the **eutectic** composition, the name coming from the Greek words for 'easily melted'. A liquid with the eutectic composition freezes at a single temperature, without previously depositing solid A or B. A solid with the eutectic composition melts, without change of composition, at the lowest temperature of any

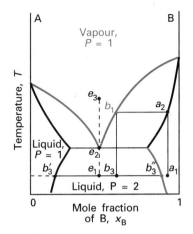

8.26 The temperature–composition diagram for a binary system in which boiling occurs before the two liquids are fully miscible.

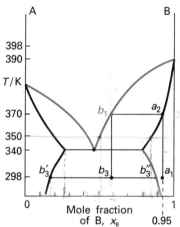

8.27 The points of the phase diagram in Fig. 8.26 that are discussed in Example 8.3.

mixture. Solutions of composition to the right of e deposit B as they cool, and solutions to the left deposit A: only the eutectic mixture (apart from pure A or pure B) solidifies at a single definite temperature ($F' = 0$ when $C = 2$ and $P = 3$) without gradually unloading one or other of the components from the liquid.

One technologically important eutectic is solder, which has mass composition 67 per cent tin and 33 per cent lead and melts at 183°C. The eutectic formed by 23 per cent NaCl and 77 per cent H_2O melts at -21.1°C. When salt is added to ice under isothermal conditions (for example, when spread on an icy road) the mixture melts if the temperature is above -21.1°C (and the eutectic composition has been achieved). When salt is added to ice under adiabatic conditions (for example, when added to ice in a vacuum flask) the ice melts, but in doing so it absorbs heat from the rest of the mixture. The temperature of the system falls and, if enough salt is added, cooling continues down to the eutectic temperature. Eutectic formation occurs in the great majority of binary alloy systems, and is of great importance for the microstructure of solid materials. Although a eutectic solid is a two-phase system, it crystallizes out in a nearly homogeneous mixture of microcrystals. The two microcrystalline phases can be distinguished by microscopy and structural techniques such as X-ray diffraction.

Thermal analysis is a very useful practical way of detecting eutectics. We can see how it is used by considering the rate of cooling down the isopleth through a_1 in Fig. 8.28. The liquid cools steadily (Fig. 8.29) until it reaches a_2, when B begins to be deposited. Cooling is now slower because the solidification of B is exothermic and retards the cooling. When the remaining liquid reaches the eutectic composition, the temperature remains constant ($F' = 0$) until the whole sample has solidified: this region of constant temperature is the **eutectic halt**. If the liquid has the eutectic composition e initially, then the liquid cools steadily down to the freezing temperature of the eutectic, when there is a long eutectic halt as the entire sample solidifies (like the freezing of a pure liquid).

Monitoring the cooling curves at different overall compositions gives a clear indication of the structure of the phase diagram. The solid–liquid boundary is given by the points at which the rate of cooling changes. The longest eutectic halt gives the location of the eutectic composition and its melting temperature.

Reacting systems

Many binary mixtures react to produce compounds, and technologically important examples of this behaviour include the III/V semiconductors, such as gallium arsenide, which forms the compound GaAs. Although three constituents are present, there are only two components because GaAs is formed from the reaction Ga + As → GaAs. We shall illustrate some of the principles involved with a system that forms a compound C that also forms eutectic mixtures with the species A and B.

A system prepared by mixing an excess of B with A consists of C and unreacted B. This is a binary C, B system, which we suppose form

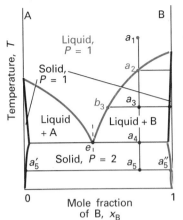

8.28 The temperature–composition phase diagram for two almost immiscible solids and their completely miscible liquids. Note the similarity to Fig. 8.26. The isopleth through e corresponds to the eutectic composition, the mixture with lowest melting point.

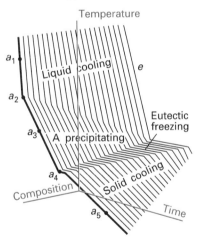

8.29 The cooling curves for the system shown in Fig. 8.28. For isopleth a, the rate of cooling slows at a_2 because solid B deposits from solution. There is a complete halt at a_4 while the eutectic solidifies. This halt is longest for the eutectic isopleth e. The eutectic halt shortens again for compositions beyond e (richer in A). Cooling curves are used to construct the phase diagram.

eutectic. The principal change from the eutectic phase diagram in Fig. 8.28 is that the whole of the diagram is squeezed into the range of compositions lying between equal amounts of A and B ($x_B = 0.5$, marked C in Fig. 8.30) and pure B. The interpretation of the information in the diagram is obtained in the same way as for Fig. 8.28: the solid deposited on cooling along the isopleth at a_1 in Fig. 8.30 is the compound C slightly contaminated with B, and the two-phase solid that exists when the temperature is below e consists of C and B (each one slightly contaminated by the other).

Incongruent melting

In some cases the compound C is not stable as a liquid. An example is the alloy Na_2K, which survives only as a solid (Fig. 8.31).

Consider what happens as a liquid at a_1 is cooled:

(1) $a_1 \rightarrow a_2$. Some solid Na (slightly contaminated with K) is deposited, and the remaining liquid is richer in K.

(2) $a_2 \rightarrow$ just below a_3. The sample is now entirely solid, and consists of solid Na and solid Na_2K (each slightly contaminated by the other).

Now consider the isopleth at b_1:

(1) $b_1 \rightarrow b_2$. No obvious change occurs until the phase boundary is reached at b_2 when solid Na begins to deposit.

(2) $b_2 \rightarrow b_3$. Solid Na deposits, but at b_3 a reaction occurs to form Na_2K: this compound is formed by the K atoms diffusing into the solid Na.

At this stage the liquid Na/K mixture is in equilibrium with a little solid Na_2K, but there is still no liquid compound.

(3) $b_3 \rightarrow b_4$. As cooling continues, the amount of solid compound increases until at b_4 the liquid reaches its eutectic composition. It then solidifies to give a two-phase solid consisting of solid K and solid Na_2K.

If the solid is reheated, the sequence of events is reversed. No liquid Na_2K forms at any stage because it is too unstable to exist as a liquid. This behaviour is an example of **incongruent melting**, in which a compound melts into its components and does not itself form a liquid phase.

8.7 Ultrapurity and controlled impurity

Advances in technology have called for materials of extreme purity. For example, semiconductor devices consist of almost perfectly pure silicon or germanium doped to a precisely controlled extent. For these materials to operate successfully, the impurity level must be kept down to less than 1 in 10^9 (which corresponds to about one grain of salt in 5 tons of sugar).

Consider a liquid of composition on the isopleth through a in Fig. 8.32: it is mainly A with some B impurity. On cooling to a_1, a solid of composition b_1 appears. Removing that solid gives a slightly purer material than the original, but not much of it (by the lever rule). That

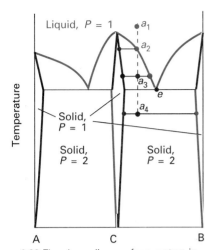

8.30 The phase diagram for a system in which A and B react to form a compound C = AB. This diagram resembles two versions of Fig. 8.26 in each half of the diagram. The constituent C is a true compound, not just an equimolar mixture.

solid could be used as the starting substance for a second stage of this **fractional crystallization** process. In each stage, the composition is shifted towards pure A, in the manner of fractional distillation, but the procedure is slow and wasteful.

We should recognize, however, that Fig. 8.32 applies when the freezing is so slow that the composition of the solid is uniform and has its equilibrium composition. In a real system equilibrium is not achieved because B does not have time to disperse throughout the whole solid sample. The technique of **zone refining** makes use of the non-equilibrium properties of the system. It relies on the impurities being more soluble in the molten sample than in the solid, and sweeps them up by passing a molten zone repeatedly from one end to the other along a sample.

Consider a liquid (this represents the molten zone) on the isopleth through a, and let it cool without the entire sample coming to overall equilibrium. If the temperature falls to a_2 a solid of composition b_2 is deposited and the remaining liquid (the zone where the heater has moved on) is at a_2'. Cooling that liquid down an isopleth passing through a_2' deposits solid of composition b_3 and leaves liquid at a_3'. The process continues until the last drop of liquid to solidify is heavily contaminated with B. There is plenty of everyday evidence that impure liquids freeze in this way. For example, an ice cube is clear near the surface but misty in the core: the water used to make ice normally contains dissolved air; freezing proceeds from the outside, and air is accumulated in the retreating liquid phase. It cannot escape from the interior of the cube, so when that freezes it occludes the air in a mist of tiny bubbles.

In the technique of zone refining the sample is in the form of a narrow cylinder. This cylinder is heated in a thin disk-like zone that is swept from one end of the sample to the other. The advancing liquid zone accumulates the impurities as it passes. In practice a train of hot and cold zones are swept repeatedly from one end to the other (Fig. 8.33). The zone at the end of the sample is the impurity dump: when the heater has gone by, it cools to a dirty solid which can be discarded.

8.31 (a) The phase diagram for an actual system (sodium and potassium) like that shown in Fig. 8.30, but with two differences. One is that the compound is Na$_2$K, corresponding to A$_2$B and not AB as in that illustration. The second is that the compound exists only as a solid, not as a liquid. The transformation of the compound at its melting point is an example of incongruent melting. (b) The species present in each region of the phase diagram.

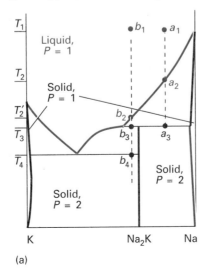

(a)

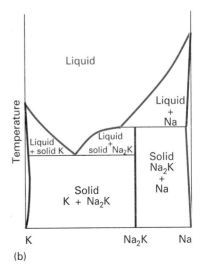

(b)

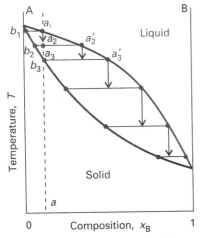

8.32 A binary temperature–composition diagram can be used to discuss zone refining, as explained in the text.

A modification of zone refining is **zone levelling**. It is used to introduce controlled amounts of impurity (for example, of indium into germanium). A sample rich in the required dopant is put at the head of the main sample, and made molten. The zone is then dragged repeatedly in alternate directions through the sample, where it deposits a uniform distribution of the impurity.

THREE-COMPONENT SYSTEMS

For a three-component system, $F = 5 - P$, so the variance may reach 4. Holding the temperature and pressure constant leaves two degrees of freedom (the mole fractions of two of the components). One of the best ways of showing how phase equilibria vary with the composition of the system is to use a triangular phase diagram. This section explains how these diagrams are constructed and interpreted and gives two simple examples.

8.8 Triangular phase diagrams

The mole fractions of the three components of a ternary system ($C = 3$) satisfy

$$x_A + x_B + x_C = 1$$

A phase diagram drawn as an equilateral triangle ensures that this property is satisfied automatically because the sum of the distances to a point inside an equilateral triangle measured parallel to the edges is equal to the length of the side of the triangle (Fig. 8.34), and that side may be taken to have unit length.

Figure 8.34 shows how this approach works in practice. The edge AB corresponds to $x_C = 0$, and likewise for the other two edges. Hence, each of the three edges corresponds to one of the three binary systems (A, B), (B, C), and (C, A). An interior point corresponds to a system in which all thr_ substances are present. The point P, for instance, represents $x_A = 0.50$, $x_B = 0.10$, and $x_C = 0.40$.

Any point on a straight line joining an apex to a point on the opposite edge (the broken line in Fig. 8.34) represents a composition that is progressively richer in A the closer the point is to the A apex but which

8.33 The procedure for zone refining. (a) Initially, impurities are distributed uniformly along the sample. (b) After a molten zone is passed along the rod, the impurities are more concentrated at the right. In practice, a series of molten zones are passed along the rod from left to right.

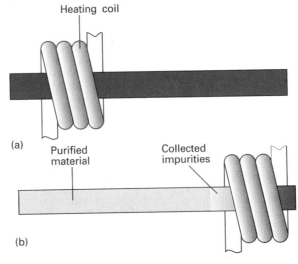

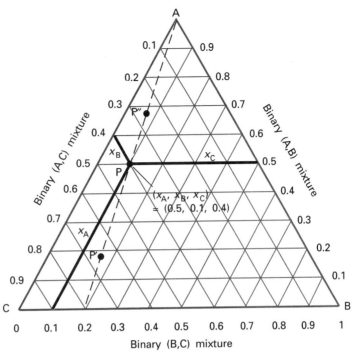

8.34 The triangular coordinates used for the discussion of three-component systems. The edges correspond to binary systems. All points along the broken line correspond to mole fractions of C and B in the same ratio.

has the same proportions of B and C. Therefore, if we wish to represent the changing composition of a system as A is added, we draw a line from the A apex to the point on BC representing the initial binary system. Any ternary system formed by adding A then lies at some point on this line.

Example 8.4 *Marking points on a ternary phase diagram*

Mark the following points on a triangular composition diagram:

(a) $x_A = 0.20$, $x_B = 0.80$, $x_C = 0$
(b) $x_A = 0.42$, $x_B = 0.26$, $x_C = 0.32$
(c) $x_A = 0.80$, $x_B = 0.10$, $x_C = 0.10$
(d) $x_A = 0.10$, $x_B = 0.20$, $x_C = 0.70$
(e) $x_A = 0.20$, $x_B = 0.40$, $x_C = 0.40$
(f) $x_A = 0.30$, $x_B = 0.60$, $x_C = 0.10$

Method. The mole fraction x_A is measured along either edge leading to apex A; likewise for x_B and B; x_C takes care of itself (but it is sensible to check).

Answer. The points are plotted in Fig. 8.35.

Comment. Note that the points (d), (e), and (f) have $x_A/x_B = 0.50$, and fall on a straight line, as stated in the text.

Exercise E8.4. Plot the following points:

(a') $x_A = 0.25$, $x_B = 0.25$, $x_C = 0.50$
(b') $x_A = 0.50$, $x_B = 0.25$, $x_C = 0.25$
(c') $x_A = 0.80$, $x_B = 0$, $x_C = 0.20$
(d') $x_A = 0.60$, $x_B = 0.25$, $x_C = 0.15$
(e') $x_A = 0.20$, $x_B = 0.75$, $x_C = 0.05$

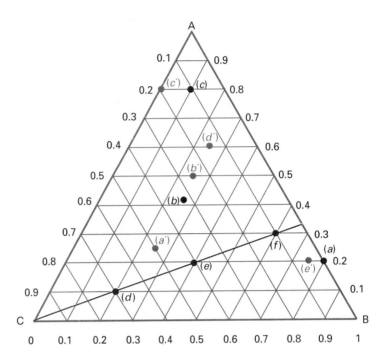

8.35 The points referred to in Example 8.4 (black) and Exercise E8.4 (blue).

8.9 Partially miscible liquids

Water and acetic acid are fully miscible, as are chloroform and acetic acid. Water and chloroform are only partially miscible. What happens when all three are present together?

The phase diagram for this ternary system at room temperature and pressure is shown in Fig. 8.36. It shows that the two fully miscible pairs

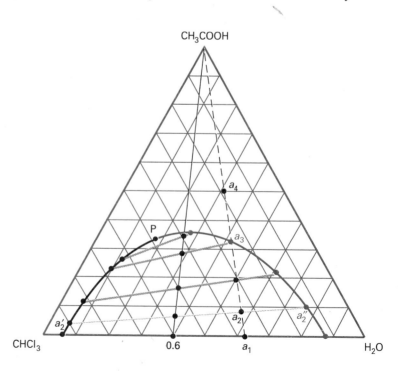

8.36 The phase diagram, at fixed temperature and pressure, of the three-component system acetic acid, chloroform, and water. Only some of the tie lines have been drawn in the two-phase region. All points along the line *a* correspond to chloroform and water present in the same ratio.

form single-phase regions and that the water/chloroform system (along the base of the triangle) has a two-phase region. The base of the triangle corresponds to one of the horizontal lines in a two-component phase diagram. The tie lines in the two-phase regions are constructed experimentally by determining the compositions of the two phases that are in equilibrium, marking them on the diagram, and then joining them with a straight line.

A single-phase system is formed when enough acetic acid is added to the binary water-chloroform mixture. This effect is shown by following the line a_1, a_4 in Fig. 8.36.

(1) At a_1. The system consists of two phases and the relative amounts of the two phases can be read off in the usual way (by using the lever rule).

(2) $a_1 \rightarrow a_2$. The addition of acetic acid takes the system along the line joining a_1 to the acetic acid apex. At a_2 the solution still has two phases, but there is more water in the chloroform phase (a_2') and more chloroform in the water (a_2'') because the acid helps both to dissolve. The phase diagram shows that there is more acetic acid in the water-rich phase than in the other (a_2'' is closer than a_2' to the acetic acid apex).

(3) $a_2 \rightarrow a_3$. At a_3 two phases are present, but the chloroform-rich layer is present only as a trace.

(4) $a_3 \rightarrow a_4$. Further addition of acid takes the system towards a_4, and only a single phase is present.

Example 8.5 *Interpreting a ternary phase diagram* (1)

A mixture is prepared consisting of chloroform ($x_C = 0.60$) and water ($x_W = 0.40$). Describe the changes that occur when acetic acid is added to the mixture.

Method. We base the answer on Fig. 8.36. The relative proportions of chloroform and water remain constant, and so the addition of acetic acid (A) corresponds to motion along the line from the point $x_C = 0.60$ on the base line opposite the A apex to the apex itself. The tie lines give the compositions of the phases at their intersections with the boundaries; the lever rule gives their proportions. We shall denote compositions in the order (x_C, x_W, x_A).

Answer. The initial composition is (0.60, 0.40, 0). This point lies in the two-phase region, the phase compositions being (0.95, 0.05, 0) and (0.12, 0.88, 0) with relative proportions 1.3:1. Addition of acetic acid takes the system along the straight line to A. When sufficient acid has been added to raise its mole fraction to 0.18 the overall composition is (0.49, 0.33, 0.18) and the system consists of two phases of compositions (0.82, 0.06, 0.12) and (0.17, 0.60, 0.23) in almost equal abundance. When enough acid has been added to raise its mole fraction to 0.37 the system consists of a trace of a phase of composition (0.64, 0.11, 0.25) and a dominating phase of composition (0.35, 0.28, 0.37). Further

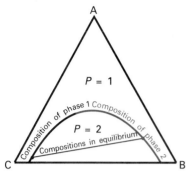

8.37 The interpretation of a triangular phase diagram. The region inside the curved line consists of two phases, and the compositions of the two phases in equilibrium are given by the points at the ends of the tie lines (the tie lines are determined experimentally).

addition of acid takes the system into the single-phase region, where it remains right up to the point corresponding to pure acid.

Exercise E8.5. Repeat the question, starting with a mixture of composition $(0.70, 0, 0.30)$ to which water is added.

The point marked P in Fig. 8.36 is called the **plait point**. It is yet another example of a critical point. At the plait point, the compositions of the two phases in equilibrium become identical. For convenience, the general interpretation of a triangular phase diagram is summarized in Fig. 8.37.

8.10 The role of added salts

The presence of one solute may affect the solubility of another. The **salting-out effect** is the reduction of the solubility of a gas (or other non-electrolyte) in water when a salt is added. A **salting-in effect** may also occur, in which the ternary system is more concentrated (in the sense of having less water) than in the binary system. A salt may also affect the solubility of another electrolyte, as we can see by examining the ternary system consisting of ammonium chloride, ammonium sulfate, and water (Fig. 8.38).

The phase diagram has the following interpretation.

(1) b. This point indicates the solubility of the chloride in water, and a mixture of composition b_1 consists of the undissolved chloride and a saturated solution of composition b.

(2) c. This point similarly indicates the solubility of the sulfate.

(3) a_1. This point corresponds to a single phase; as water is evaporated, the composition moves along the line a_1 to a_2.

(4) $a_1 \rightarrow a_2$. At a_2 the system enters the two-phase region, and some solid chloride crystallizes (all the tie lines ending on that boundary also end at the pure chloride apex). The liquid becomes richer in sulfate, and its composition moves towards d.

(5) $a_2 \rightarrow a_3$. When enough water has been removed to bring the overall composition to a_3 the liquid composition is d. At this point (which is joined to the chloride apex and to the sulfate apex), the system consists of saturated solution in equilibrium with the two solids.

Note that this point, which corresponds to the joint solubility of the two solids, corresponds to a smaller mole fraction of water than in either of the binary systems b and c. This means that the two salts form a more concentrated solution overall than either does alone.

(6) $a_3 \rightarrow a_4$. If more water is removed after the system has arrived at d, the amount of solution decreases, but its composition remains constant (at d, the saturated solution).

Both solids are precipitated, and the system has three phases: each point in the three-phase region is tied to d and the two solid apexes.

(7) a_4. All the water has been evaporated; the system is now binary and consists of a mixture of the two solids.

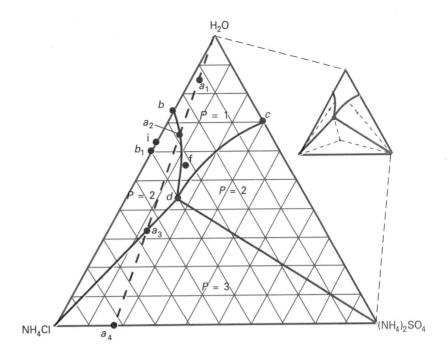

8.38 The phase diagram, at constant temperature and pressure, for the ternary system NH$_4$Cl/(NH$_4$)$_2$SO$_4$/H$_2$O. The points i and f are the ones mentioned in Example 8.6. All tie lines in the two-phase regions terminate at an apex, and all tie lines in the three-phase region terminate at the three corners of the triangular area, as shown in the insert.

Example 8.6 *Interpreting a ternary phase diagram* (2)

A solution of 50 g of ammonium chloride in 30 g of water is prepared at room temperature, and then 45 g of ammonium sulfate is added. Describe the initial and final states.

Method. We use Fig. 8.38 after converting compositions to mole fractions. We shall write compositions in the order (x_W, x_C, x_S) for water (W), chloride (C), and sulfate (S) and use the lever rule for proportions of each phase.

Answer. Molar masses are as follows: H$_2$O, 18.02; NH$_4$Cl, 53.49; (NH$_4$)$_2$SO$_4$, 132.1 g mol^{-1}. The initial amounts are $n_W = 1.66$ mol and $n_C = 0.93$ mol, so the initial composition (point i in Fig. 8.38) is (0.64, 0.36, 0). In the final state the amounts of W and S are the same but $n_S = 0.34$ mol. Therefore the final composition (point f in Fig. 8.38) is (0.57, 0.31, 0.12). From Fig. 8.38 we see that (0.64, 0.36, 0) corresponds to a two-phase system consisting of solid C with saturated solution of composition (0.74, 0.26, 0) in relative proportions 0.16:1. After addition of S there is only one phase.

Exercise E8.6. Describe the initial and final states of a solution of compositions (0.80, 0, 0.20) and (0.40, 0.50, 0.10).

Although phase diagrams might look complicated, they convey simple, experimentally established information. To interpret them it is helpful to think operationally. That is, definite processes should be imagined, and the diagram should be considered bearing in mind how it was constructed originally: phases come and go, systems boil and freeze, and relative amounts of different phases change. It is also wise to concentrate on the lines rather than the areas. The points at the ends of the tie lines give the compositions of the phases in equilibrium with each other. The distances of the points from the isopleth give (through the lever rule) their relative abundances.

CHECK LIST OF KEY IDEAS

1 The definition of the terms **phase, component,** and **variance** of a system (Section 8.1).

2 The statement of the **phase rule** (eqn 1) and its application to the **phase diagrams** of one-component systems.

3 The use of **thermal analysis** to determine phase boundaries (Section 8.2).

4 The derivation of the phase rule from the equality of chemical potentials of a substance in every phase (Section 8.2).

5 The description of mixtures of volatile liquids in terms of **vapour pressure diagrams** (Section 8.3) and their interpretation in terms of the **lever rule** (eqn 7).

6 The construction and interpretation of **temperature–composition diagrams** and their use in the description of **distillation** and **fractional distillation** (Section 8.4).

7 The formation of **azeotropes** and their consequences for distillation (Section 8.4).

8 The description of **partially miscible liquids** in terms of phase diagrams (Section 8.5) and the use of the lever rule in their interpretation.

9 The significance of the **upper critical temperature** and the **lower critical temperature** of a binary liquid mixture (Section 8.5).

10 The description of the **distillation of partially miscible liquids** in terms of their phase diagrams (Section 8.5).

11 The phase diagrams of binary solid systems and the formation of **eutectics** (Section 8.6).

12 The phase diagrams of mixtures that form compounds and the process of **incongruent melting** (Section 8.6).

13 The use of **fractional crystallization** and **zone refining** to purify solids and of **zone levelling** to dope them selectively (Section 8.7).

14 The use of **triangular phase diagrams** for the description of ternary systems (Section 8.8).

EXERCISES

8.1. At 90°C, the vapour pressure of toluene is 400 Torr and that of o-xylene is 150 Torr. What is the composition of the liquid mixture that boils at 90°C when the pressure is 0.50 atm? What is the composition of the vapour produced?

8.2. The vapour pressure of pure liquid A at 300 K is 575 Torr and that of pure liquid B is 390 Torr. These two compounds from ideal liquid and gaseous mixtures. Consider the equilibrium composition of a mixture in which the mole fraction of A in the vapour is 0.350. Calculate the total pressure of the vapour and the composition of the liquid mixture.

8.3. It is found that the boiling point of a binary solution of A and B with mole fraction A, $x_A = 0.6589$, is 88°C. At this temperature the vapour pressures of pure A and B are 957 and 379.5 Torr, respectively. (a) Is this solution ideal? (b) What is the initial composition of the vapour above the solution?

8.4. Dibromoethene (de, $p_{de}^* = 172$ Torr at 358 K) and dibromopropene (dp, $p_{dp}^* = 128$ Torr at 358 K) form a nearly ideal solution. If $z_{de} = 0.60$, what is (a) p_{total} when the system is all liquid, (b) the composition of the vapour when the system is still almost all liquid?

8.5. Benzene and toluene form nearly ideal solutions. At 20°C the vapour pressures of pure benzene and toluene are 74 Torr and 22 Torr, respectively. A solution consisting of 1.00 mol of each component is boiled by reducing the external pressure below the vapour pressure. Calculate (a) the pressure when boiling begins, (b) the composition of each component in the vapour, and (c) the vapour pressure when only a few drops of liquid remain. Assume that the rate of vaporization is low enough for the temperature to remain constant at 20°C.

8.6. The following temperature/composition data were obtained for a mixture of octane (O) and toluene (T) at 760 Torr, where x is the mole fraction in the liquid and y the mole fraction in the vapour at equilibrium.

$\theta/°C$	x_T	y_T
110.9	0.908	0.923
112.0	0.795	0.836
114.0	0.615	0.698
115.8	0.527	0.624
117.3	0.408	0.527
119.0	0.300	0.410
121.1	0.203	0.297
123.0	0.097	0.164

The boiling points are 110.6°C for T and 125.6°C for O. Plot the temperature/composition diagram for the mixture. What is the composition of the vapour in equilibrium with the liquid of composition (a) $x_T = 0.250$ and (b) $x_O = 0.250$?

8.7. State the number of components in the following systems. (a) NaH_2PO_4 in water at equilibrium with water vapour but disregarding the fact that the salt is ionized. (b) The same, but taking into account the ionization of the salt. (c) $AlCl_3$ in water, noting that hydrolysis and precipitation of $Al(OH)_3$ occur.

8.8. Blue $CuSO_4 \cdot 5H_2O$ crystals release their water of hydration when heated. How many phases and components are present in an otherwise empty heated container?

8.9. Ammonium carbonate decomposes when it is heated. (a) How many components and phases are present when the salt is heated in an otherwise empty container? (b) Now suppose that additional ammonia is also present. How many components and phases are present?

8.10. A saturated solution of Na_2SO_4, with excess of the solid, is present at equilibrium with its vapour in a closed vessel. (a) How many phases and components are present? (b) What is the variance of the system? Identify the independent variables.

8.11. Now suppose that the solution referred to in Exercise 8.10 is not saturated. (a) How many phases and components are present? (b) What is the variance of the system? Identify the independent variables.

8.12. A research paper reports that substances A and B form two different coexisting liquid phases. The overall composition is changed at constant temperature and pressure, and the composition of each of the coexisting liquid phases is also reported to change. (a) Was the system at equilibrium? (b) What would have been observed if the system had been at equilibrium? Explain your answers.

8.13. Sketch phase diagrams for the following types of systems. Label the regions and intersections of the diagrams, stating what materials (possibly compounds or azeotropes) are present and whether they are solid, liquid, or gas. (a) One-component, pressure–temperature diagram, liquid density greater than that of solid; (b) two-component, temperature–composition, solid–liquid diagram, one compound AB formed that melts congruently, negligible solid–solid solubility; (c) two-component, temperature–composition, solid–liquid diagram, one compound of formula AB_2 that melts incongruently, negligible solid–solid solubility; (d) two-component, temperature–composition, liquid–vapour diagram, formation of an azeotrope at $x_B = 0.333$, complete miscibility.

8.14. Label the regions of the phase diagram in Fig. 8.39. State what substances (if compounds, give their formulae) exist in each region. Label each substance in each region as solid, liquid, or gas.

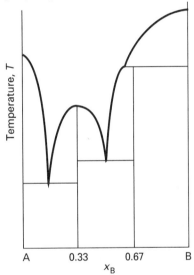

8.39 The phase diagram for Exercise 8.14.

8.15. Methylethyl ether (A) and diborane, B_2H_6, (B) form a compound AB that melts congruently at 133 K. The system exhibits two eutectics, one at 25 mole per cent B and 123 K and a second at 90 mole per cent B and 104 K. The melting points of pure A and B are 131 K and 110 K, respectively. Sketch the phase diagram for this system. Assume negligible solid–solid solubility.

8.16. Sketch the phase diagram of the system NH_3/N_2H_4, given that the two substances do not form a compound with each other, that NH_3 freezes at $-78°C$ and N_2H_4 freezes at $+2°C$, that a eutectic is formed when the mole fraction of N_2H_4 is 0.07, and that the eutectic melts at $-80°C$.

8.17. Figure 8.19 shows the phase diagram for two partially miscible liquids, which can be taken to be that for water (A) and 2-methyl-1-propanol (B). Describe what will be observed when a mixture of composition $x_B = 0.8$ is heated, at each stage giving the number, composition, and relative amounts of the phase present.

8.18. Figure 8.40 is the phase diagram for silver and tin. Label the regions, and describe what will be observed when liquids of composition a and b are cooled to 200°C.

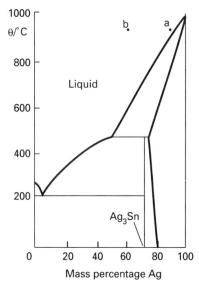

8.40 The phase diagram for Exercises 8.18–21

8.19. Indicate on the phase diagram in Fig. 8.40 the feature that denotes incongruent melting. What is the composition of the eutectic mixture and at what temperature does it melt?

8.20. Sketch the cooling curves for the isopleths a and b in Fig. 8.40.

8.21. Use the phase diagram in Fig. 8.40 to state (a) the solubility of Ag in Sn at 800°C, (b) the solubility of Ag_3Sn in Ag at 460°C, (c) the solubility of Ag_3Sn in Ag at 300°C.

8.22. Figure 8.41 shows the experimentally determined phase diagrams for the nearly ideal solution of hexane and heptane. (a) Label the regions of the diagrams to show which phases are present. (b) For a solution containing 1 mol each of hexane and heptane, estimate the vapour pressure at 70°C when vaporization on reduction of the external pressure just begins. (c) What is the vapour pressure of the solution at 70°C when just one drop of liquid remains? (d) Estimate from the figures the mole fraction of hexane in the liquid and vapour phases for the conditions of part b. (e) What are the mole fractions for the conditions of part c? (f) At 85°C and 760 Torr, what are the amounts of substance in the liquid and vapour phases when $z_{Heptane} = 0.40$?

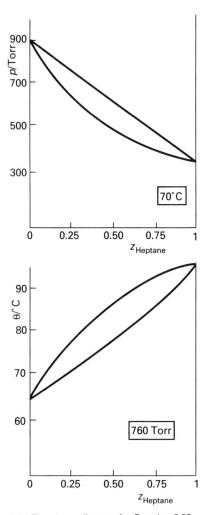

8.41 The phase diagram for Exercise 8.22

8.23. Uranium tetrafluoride and zirconium tetrafluoride melt at 1035°C and 912°C, respectively. They form a continuous series of solid solutions with a minimum melting temperature of 765°C and composition $x(ZrF_4) = 0.77$. At 900°C, the liquid solution of composition $x(ZrF_4) = 0.28$ is in equilibrium with a solid solution of composition $x(ZrF_4) = 0.14$. At 850°C The two compositions are 0.870 and 0.980, respectively. Sketch the phase diagram for this system and state what is observed when a liquid of composition $x(ZrF_4) = 0.40$ is cooled slowly from 900°C to 500°C.

8.24. Methane (melting point 91 K) and tetrafluoromethane (melting point 89 K) do not form solid solutions with each other and as liquids they are only partially miscible. The upper critical temperature of the liquid mixture is 94 K at $x(CF_4) = 0.43$ and the eutectic temperature is 84 K at $x(CF_4) = 0.88$. At 86 K the phase in equilibrium with the tetrafluoromethane-rich solution changes from solid methane to a methane-rich liquid. At that temperature, the two liquid solutions that are in mutual equilibrium have the compositions $x(CF_4) = 0.10$ and $x(CF_4) = 0.80$. Sketch the phase diagram.

8.25. Describe the phase changes that take place when a liquid mixture of 4.0 mol of B_2H_6 (melting point 131 K) and 1.0 mol of CH_3OCH_3 (melting point 135 K) is cooled from 140 K to 90 K. These substances form a compound $(CH_3)_2OB_2H_6$ which melts congruently at 133 K. The system exhibits one eutectic at $x(B_2H_6) = 0.25$ and 123 K and another at $x(B_2H_6) = 0.90$ and 104 K.

8.26. Refer to the information in Exercise 8.25 and sketch the cooling curves for liquid mixtures in which $x(B_2H_6)$ is (a) 0.10, (b) 0.30, (c) 0.50, (d) 0.80, (e) 0.95.

8.27. Hexane and perfluorohexane show partial miscibility below 22.70°C. The critical concentration at the upper critical temperature is $x = 0.355$, where x is the mole fraction of C_6F_{14}. At 22.0°C the two solutions in equilibrium have $x = 0.24$ and $x = 0.48$, respectively, and at 21.5°C the mole fractions are 0.22 and 0.51, respectively. Sketch the phase diagram. Describe the phase changes that occur when perfluorohexane is added to a fixed amount of hexane at (a) 23°C, (b) 22°C.

8.28. Mark the following features on triangular coordinates: (a) the point (0.2, 0.2, 0.6), (b) the point (0, 0.2, 0.8), (c) the point at which all three mole fractions are the same.

8.29. Mark the following points on a ternary phase diagram for the system $NaCl/Na_2SO_4 \cdot 10H_2O/H_2O$: (a) 25 per cent by mass NaCl, 25 per cent $Na_2SO_4 \cdot 10H_2O$,

and the rest H_2O, (b) the line denoting the same relative composition of the two salts but with changing amounts of water.

8.30. Refer to the ternary phase diagram in Fig. 8.36. How many phases are present, and what are their compositions and relative abundances, in a mixture that contains 2.3 g of water, 9.2 g of chloroform, and 3.1 g of acetic acid? Describe what happens when (a) water, (b) acetic acid is added to the mixture.

8.31. Figure 8.38 shows the phase diagram for the ternary system $NH_4Cl/(NH_4)_2SO_4/H_2O$ at 25°C. Identify the number of phases present for mixtures of compositions (a) (0.2, 0.4, 0.4), (b) (0.4, 0.4, 0.2), (c) (0.2, 0.1, 0.7), (d) (0.4, 0.16, 0.44). The numbers are mole fractions of the three components in the order (NH_4Cl, $(NH_4)_2SO_4$, H_2O).

8.32. Referring to Fig. 8.38, deduce the molar solubility of (a) NH_4Cl, (b) $(NH_4)_2SO_4$ in water at 25°C.

8.33. Describe what happens when (a) $(NH_4)_2SO_4$ is added to a saturated solution of NH_4Cl in water in the presence of excess NH_4Cl, (b) water is added to a mixture of 25 g of NH_4Cl and 75 g of $(NH_4)_2SO_4$.

8.34. At a certain temperature, the solubility of I_2 in liquid CO_2 is $x(I_2) = 0.03$. At the same temperature its solubility in nitrobenzene is 0.04. Liquid carbon dioxide and nitrobenzene are miscible in all proportions, and the solubility of I_2 in the mixture varies linearly with the proportion of nitrobenzene. Sketch a phase diagram for the ternary system.

PROBLEMS
Numerical problems
8.1. The compound p-azoxyanisole forms a liquid crystal. 5.0 g of the solid was placed in a tube, which was then evacuated and sealed. Use the phase rule to prove that the solid will melt at a definite temperature and that the liquid crystal phase will make a transition to a normal liquid phase at a definite temperature.

8.2. Magnesium oxide and nickel oxide withstand high temperatures. However, they do melt when the temperature is high enough, and the behaviour of mixtures of the two is of considerable interest to the ceramics industry. Draw the temperature–composition diagram for the system using the data below, where x is the mole fraction of MgO in the solid and y its mole fraction in the liquid.

$\theta/°C$	1960	2200	2400	2600	2800
x	0	0.35	0.60	0.83	1.00
y	0	0.18	0.38	0.65	1.00

State (a) the temperature at which a mixture with $x = 0.30$ begins to melt, (b) the composition and proportion of the phases present when a solid of composition $x = 0.30$ is heated to 2200°C, (c) the temperature at which a liquid of composition $y = 0.70$ will begin to solidify.

8.3. The bismuth/cadmium phase diagram is of interest in metallurgy, and its general form can be estimated from expressions for the depression of freezing point. Construct the diagram using the following data: $T_f(Bi) = 544.5$ K, $T_f(Cd) = 594$ K, $\Delta_{fus}H(Bi) = 10.88$ kJ mol^{-1}, $\Delta_{fus}H(Cd) = 6.07$ kJ mol^{-1}. The metals are mutually insoluble as solids. Use the phase diagram to state what would be observed when a liquid of composition $x(Bi) = 0.70$ is cooled slowly from 550 K. What are the relative abundances of the liquid and solid at (a) 460 K and (b) 350 K? Sketch the cooling curve for the mixture.

8.4. Phosphorus and sulfur form a series of compounds. The best characterized are P_4S_3, P_4S_7, and P_4S_{10}, all of which melt congruently at 174, 308, and 288°C. Assuming that only these three binary compounds of the two elements exist, (a) draw schematically only the P/S phase diagram. Label each region of the diagram with the substance that exists in that region and indicate its phase. Label the horizontal axis as x_S and give the numerical values of x_S that correspond to the compounds. The melting point of pure phosphorus is 44°C and that of pure sulfur is 119°C. (b) Draw, schematically, the cooling curve for a mixture of composition $x_S = 0.28$. Assume that a eutectic occurs at $x_S = 0.2$ and that there is negligible solid–solid solubility.

8.5. The table below gives the break and halt temperatures found in the cooling curves of two metals A and B. Construct a phase diagram consistent with the data of these curves. Label the regions of the diagram, stating what phases and substances are present. Give the probable formulae of any compounds that form.

$100x_B$	Break, $\theta/°C$	First, halt, $\theta/°C$	Second halt, $\theta/°C$
0.0		1100	
10.0	1060	700	
20.0	1000	700	
30.0	940	700	400
40.0	850	700	400
50.0	750	700	400
60.0	670	400	
70.0	550	400	
80.0		400	
90.0	450	400	
100.0		500	

8.6. Consider the phase diagram in Fig. 8.42, which

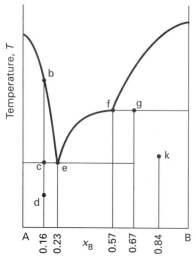

8.42 The phase diagram for Problem 8.6

represents a solid–liquid equilibrium. Label all regions of the diagram by the chemical species that exist in that region and their phases. Indicate the number of species and phases present at the points labelled b, d, e, f, g, and k. Sketch cooling curves for compositions $x_B = 0.16$, 0.23, 0.57, 0.67, and 0.84.

8.7. Solutions of *m*-toluidine (3-methylphenylamine) were made up in glycerol and then warmed from room temperature. The mixture became turbid at θ_1 and then cleared at θ_2. Plot the phase diagram using the data below, and find the upper and lower critical temperatures.

Mass%	18	20	40	60	80	85
$\theta_1/°C$	48	18	8	10	19	25
$\theta_2/°C$	53	90	120	118	83	53

Mass% denotes the mass percentage composition of *m*-toluidine. State what happens as *m*-toluidine is added dropwise to glycerol at 60°C. State the number of phases present at each composition and their relative amounts.

8.8. At 25°C, aqueous solutions containing Li_2SiF_6 and $(NH_4)_2SiF_6$ can be in equlibrium with the two virtually pure solids or with solid $Li(NH_4)SiF_6$. It is also found that solid Li_2SiF_6 can be in equilibrium with solutions ranging from mass percentage 29.5 per cent Li_2SiF_6 and 70.5 per cent H_2O to 29.3 per cent Li_2SiF_6 and 67.6 per cent H_2O. Also $(NH_4)_2SiF_6$ can be in equilibrium with solutions ranging from 18.7 per cent $(NH_4)_2SiF_6$ and 81.3 per cent H_2O to 16.5 per cent $(NH_4)_2SiF_6$ and 70.4 per cent H_2O. Sketch the phase diagram for this ternary system and include some tie lines. Describe the changes that occur when a solution that is initially 3.0 per cent Li_2SiF_6 and 83.0 per cent H_2O is slowly dehydrated.

8.9. Sketch the phase diagram for the Mg/Cu system using the following information: $\theta_f(Mg) = 648°C$, $\theta_f(Cu) = 108.5°C$; two intermetallic compounds are formed with $\theta_f(MgCu_2) = 800°C$ and $\theta_f(Mg_2Cu) = 580°C$; eutectics of mass percentage Mg composition and melting points 10 per cent (690°C), 33 per cent (560°C), and 65 per cent (380°C). A sample of Mg/Cu alloy containing 25 per cent Mg by mass was prepared in a crucible heated to 800°C in an inert atmosphere. Describe what will be observed if the melt is cooled slowly to room temperature. Specify the composition and relative abundances of the phases and sketch the cooling curve.

8.10. Methanol (M), diethyl ether (E), and water (W) form a partially miscible ternary system. The phase diagram at 20°C was determined by adding methanol to various binary ether/water mixtures of mole fraction x in ether and noting the mole fractions of methanol y at which complete miscibility occurred. Plot the phase diagram using the following data.

x	0.10	0.20	0.30	0.40	0.50	0.60	0.70	0.80	0.90
y	0.20	0.27	0.30	0.28	0.26	0.22	0.17	0.12	0.07

How many phases will be present in a mixture that consists of 5.0 g of methanol, 30.0 g of ether, and 50.0 g of water? What mass of water would have to be added or removed to change the number of phases?

8.11. Iron(II) chloride (melting point 677°C) and potassium chloride (melting point 776°C) form the compounds $KFeCl_3$ and K_2FeCl_4 at elevated temperatures. $KFeCl_3$ melts congruently at 380°C and K_2FeCl_4 melts incongruently at 399°C. Eutectics are formed with compositions $x = 0.38$ (melting point 351°C) and $x = 0.54$ (melting point 393°C), where x is the mole fraction of $FeCl_2$. The KCl solubility curve intersects the K_2FeCl_4 curve at $x = 0.34$. Sketch the phase diagram. State the phases that are in equilibrium when a mixture of composition $x = 0.36$ is cooled from 400°C to 300°C.

8.12. The binary system nitroethane/decahydronaphthalene (DEC) shows partial miscibility, with the two-phase region lying between $x = 0.08$ and $x = 0.84$, where x is the mole fraction of nitroethane. The binary system liquid carbon dioxide/DEC is also partially miscible, with its two-phase region lying between $y = 0.36$ and $y = 0.80$, where y is the mole fraction of DEC. Nitroethane and liquid carbon dioxide are miscible in all proportions. The addition of liquid carbon dioxide to mixtures of nitroethane and DEC increases the range of miscibility, and the plait point is reached when z, the mole fraction of CO_2, is 0.18 and $x = 0.53$. The addition of nitroethane to mixtures of carbon dioxide and DEC also results in another plait point at $x = 0.08$ and $y = 0.52$. (a) Sketch the phase diagram for the ternary system. (b) For some binary mixtures of nitroethane and liquid carbon dioxide, the addition of arbitrary amounts of DEC will not cause phase separation. Find the range of concentration for such binary mixtures.

Theoretical problems

8.13. Show that two phases are in thermal equilibrium only if their temperatures are the same and that they are in mechanical equilibrium only if their pressures are equal.

8.14. Prove that a straight line from the apex A of a ternary phase diagram to the opposite edge BC represents mixtures of constant ratio of B and C, however much A is present.

9

Chemical equilibrium

This chapter develops the concept of chemical potential and shows how it can be used to account for the equilibrium composition of chemical reactions. First, we consider a simple gas phase reaction and see that its Gibbs energy can be expressed in terms of a standard reaction Gibbs energy and a function of the composition. The latter represents the contribution of the mixing of the reactants and products. Then the expression is developed to obtain a general expression for the reaction Gibbs energy at an arbitrary composition. The equilibrium composition corresponds to the minimum in the Gibbs energy plotted against the extent of reaction, and by locating this minimum we establish the relation between the equilibrium constant and the standard reaction Gibbs energy. At that point we shall be able to calculate equilibrium constants from tables of thermodynamic data.

Next, we explore how equilibria respond to changes in the conditions. A useful rule of thumb is Le Chatelier's principle. However, the thermodynamic formulation of equilibrium enables us to establish the quantitative effects of changes in pressure and temperature.

The final section of the chapter applies the information established earlier to the discussion of three important types of equilibria. One concerns the feasibility of using carbon as a reducing agent in an industrial process. A second is the role of proton transfer equilibria in the description of solutions of acids, bases, and salts. The third is the assessment of the driving power of biochemically significant reactions.

Chemical reactions move towards a dynamic equilibrium in which both reactants and products are present but have no further tendency to undergo net change. In some cases the concentration of products is so much greater than the concentration of unchanged reactants in the equilibrium mixture that for all practical purposes the reaction is 'complete'. However, in many important cases the equilibrium mixture has significant concentrations of both reactants and products. In this chapter we shall see how to use thermodynamics to predict the equilibrium composition under any reaction conditions.

In industry it is obviously worse than useless to build a sophisticated plant if the overall reaction has a tendency to run in the wrong direction. If a plant is to be run economically we must know how to maximize yields. Does that mean raising or lowering the temperature or the pressure? Thermodynamics gives a very simple recipe for deciding what to do. We might also be interested in the way that food is used in the complicated series of biochemical reactions involved in warming the body, powering muscular contraction, and energizing the nervous system. Some reactions (such as the oxidation of carbohydrates) are spontaneous and may be coupled to others to drive them in non-spontaneous but necessary directions (as in the biosynthesis of proteins). With thermodynamics we can sort out the reactions that need to be driven and calculate the spare driving force of reactions that occur spontaneously. Another field in which equilibria are of great importance is that of the reactions of acids and bases, and we shall deal with it towards the end of the chapter.

As always in thermodynamics, though, we must bear in mind the possibility that kinetic factors might upset our predictions. Thermodynamics can tell us the direction of spontaneous change and the composition at equilibrium. It cannot tell us whether a kinetically viable pathway exists for that change to occur or for the equilibrium to be reached.

SPONTANEOUS CHEMICAL REACTIONS

We have seen that the direction of spontaneous change at constant temperature and pressure is towards lower values of the Gibbs energy G. The idea is entirely general, and in this chapter we apply it to the discussion of reactions. For instance, we shall see that it is possible to decide whether a reaction is spontaneous by calculating the Gibbs energy of the mixture at various compositions (Fig. 9.1) and identifying the location of the minimum of the graph. Because the minimum in G occurs at the composition corresponding to equilibrium, we can go on to predict whether the reaction proceeds effectively to completion (as in Fig. 9.1a), to an intermediate composition (Fig. 9.1b), or virtually not at all (Fig. 9.1c).

9.1 The Gibbs energy minimum

We begin with the simplest possible chemical equilibrium: $A \rightleftharpoons B$. Even though this looks trivial, there are many examples of it, such as the isomerization of pentane to 2-methylbutane and the conversion of L-alanine to D-alanine.

The extent of reaction

Suppose an infinitesimal amount $d\xi$ of A turns into B, then we can write

Change in amount of A present: $dn_A = -d\xi$
Change in amount of B present: $dn_B = +d\xi$

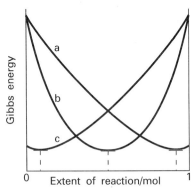

9.1 The direction of spontaneous chemical change is towards the minimum Gibbs energy. This illustration shows three possibilities: (a) equilibrium lies close to pure B (the reaction 'goes to completion'), (b) equilibrium corresponds to A and B present in similar proportions, and (c) equilibrium lies close to pure A (the reaction 'does not go').

The **extent of reaction** ξ is a measure of the progress of the reaction. We shall choose it so that (for the reaction $A \rightarrow B$) pure A corresponds to $\xi = 0$ and $\xi = 1$ mol signifies that 1 mol of A has been destroyed and 1 mol of B has been formed.

In general, when the reaction advances by $d\xi$, the change in composition of the reaction mixture results in a change in the Gibbs energy of the system. The change dG can be expressed in terms of the chemical potentials (the partial molar Gibbs energies) of the species in the mixture. At constant temperature and pressure

$$dG = \mu_A \, dn_A + \mu_B \, dn_B = -\mu_A \, d\xi + \mu_B \, d\xi$$

This equation can be reorganized into

$$\left(\frac{\partial G}{\partial \xi}\right)_{p,T} = \mu_B - \mu_A$$

This equation gives the slope of a graph of the Gibbs energy plotted against the extent of reaction.

Because the chemical potentials vary with composition, the slope of the graph of G against ξ changes as the reaction proceeds. Moreover, because the reaction proceeds in the direction of decreasing G,

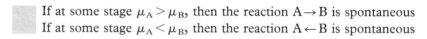

If at some stage $\mu_A > \mu_B$, then the reaction $A \rightarrow B$ is spontaneous
If at some stage $\mu_A < \mu_B$, then the reaction $A \leftarrow B$ is spontaneous

This pattern is illustrated in Fig. 9.2. When $\mu_A = \mu_B$, the slope of the graph is zero. This condition occurs at the minimum of the curve in Fig. 9.2, and corresponds to the position of chemical equilibrium:

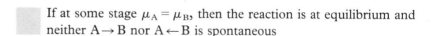

If at some stage $\mu_A = \mu_B$, then the reaction is at equilibrium and neither $A \rightarrow B$ nor $A \leftarrow B$ is spontaneous

It follows that if we can find the composition of the reaction mixture that ensures $\mu_A = \mu_B$, then we can identify the composition of the reaction mixture at equilibrium.

The reaction Gibbs energy

The **reaction Gibbs energy** $\Delta_r G$ for the reaction $A \rightarrow B$ is the change in G when 1 mol A forms 1 mol B at a *fixed composition* of the reaction mixture. Thus, it can be imagined as the change in G when the reaction takes place in such a huge amount of material that the consumption of 1 mol A and production of 1 mol B leaves the composition virtually unchanged. More formally, the reaction Gibbs energy is defined as the slope of the graph of the Gibbs energy plotted against the extent of reaction:

$$\Delta_r G = \left(\frac{\partial G}{\partial \xi}\right)_{p,T} \tag{1}$$

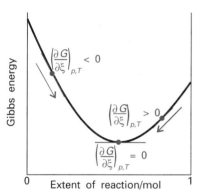

9.2 As the reaction advances (represented by motion from left to right along the horizontal axis) the slope of the Gibbs energy changes. Equilibrium corresponds to zero slope, at the foot of the valley.

Although Δ normally signifies a difference in values, in this formal (IUPAC) definition Δ_r signifies a *derivative*, the slope of G with respect to ξ. The relation between this formal definition and the interpretation given above can be seen by writing eqn 1 as $dG = \Delta_r G \times d\xi$. Then, if the volume of reaction material is very large so that its composition is constant when ξ increases by 1 mol, this equation integrates to $\Delta G = \Delta_r G \times 1\,\text{mol}$.

We have seen that for $A \rightarrow B$ the derivative in eqn 1 is equal to the difference $\mu_B - \mu_A$ at a specified composition. Therefore

$$\Delta_r G = \mu_B - \mu_A$$

It follows that we can write the condition for equilibrium at constant temperature and pressure as

$$\Delta_r G = 0 \tag{2}$$

This equation states that the slope of the graph of the Gibbs energy plotted against the extent of reaction is zero when products are formed from reactants at the equilibrium composition of the reaction mixture.

We must be careful to distinguish the reaction Gibbs energy, the slope of G at a specified composition, from the standard molar reaction Gibbs energy $\Delta_r G^\ominus$:

$$\Delta_r G^\ominus = \mu_B^\ominus - \mu_A^\ominus$$

In this case, Δ_r has its normal meaning as the difference of two quantities for it denotes the difference between the molar Gibbs energy of pure B in its standard state and the molar Gibbs energy of pure A in its standard state (Section 4.9). If A and B are gases, liquids, or solids their standard states correspond to the pure substances at 1 bar; if they are solutions, then their standard states are those described in Sections 7.6 and 7.7. As we saw in Section 4.9, the difference in standard molar Gibbs energies of the products and reactants is equal to the difference in their standard Gibbs energies of formation, and in practice we calculate $\Delta_r G^\ominus$ from

$$\Delta_r G^\ominus = \Delta_f G^\ominus(B) - \Delta_f G^\ominus(A)$$

Exergonic and endergonic reactions

The reaction Gibbs energy, $\Delta_r G$, can be used to rewrite the conditions for the spontaneity of a reaction:

> If at some stage $\Delta_r G < 0$, then the reaction $A \rightarrow B$ is spontaneous
> If at some stage $\Delta_r G > 0$, then the reaction $A \leftarrow B$ is spontaneous

Reactions for which $\Delta_r G < 0$ are called **exergonic** (from the Greek words for work-producing). The name signifies that, because they are spontaneous, they can be used to drive other processes, such as other reactions, or used to do non-expansion work. Reactions for which $\Delta_r G > 0$ are called **endergonic** (signifying work-consuming). They are spontaneous in the reverse direction. Reactions at equilibrium are spontaneous in neither direction: they are neither exergonic nor endergonic.

9.2 The composition of reactions at equilibrium

We have seen that the condition for equilibrium can be expressed in terms of the chemical potentials of the reactants and products. We saw in Chapter 5 how to express chemical potentials in terms of concentrations and partial pressures. In this section we show how to combine these relations and predict the composition of the reaction mixture at equilibrium.

Perfect gas equilibria

Consider first the reaction $A \rightarrow B$ and suppose that A and B are perfect gases. Then, with eqn 5.17 used to express the chemical potentials of A and B in terms of their partial pressures, we obtain

$$\Delta_r G = \left(\mu_B^{\ominus} + RT \ln \left(\frac{p_B}{p^{\ominus}} \right) \right) - \left(\mu_A^{\ominus} + RT \ln \left(\frac{p_A}{p^{\ominus}} \right) \right)$$

$$= \Delta_r G^{\ominus} + RT \ln \left(\frac{p_B}{p_A} \right)$$

If the ratio of partial pressures is denoted Q, we obtain

$$\Delta_r G = \Delta_r G^{\ominus} + RT \ln Q, \qquad \text{where } Q = \frac{p_B}{p_A} \qquad \text{(3)}^{\circ}$$

MOLECULAR INTERPRETATION 9.1

Some insight into the origin of eqn 3 can be obtained by considering the contributions to it. Consider a hypothetical reaction in which the A molecules changed to B molecules without mingling together. Then the Gibbs energy of the system would change from $G^{\ominus}(A)$ to $G^{\ominus}(B)$ in proportion to the amount of B that had been formed, and the slope of the graph of G against the extent of reaction would be constant and equal to $\Delta_r G^{\ominus}$ at all stages of the reaction (Fig. 9.3). However, in fact, as the reaction proceeds the newly produced B molecules mix with the surviving A molecules. We have seen that the contribution of a mixing process to the change in Gibbs energy (eqn 7.7) is

$$\Delta_{mix} G = nRT(x_A \ln x_A + x_B \ln x_B)$$

This expression makes a U-shaped contribution to the total change in Gibbs energy, and the sum of this term and the linear change arising from the change in identity of A to B gives rise to a minimum in the graph (Fig. 9.3). As can be seen from the illustration, there is now a minimum in the Gibbs energy, and its position corresponds to the equilibrium composition of the reaction mixture. It is quite easy to verify that the slope of the graph at any point is given by eqn 3, with the second term stemming from the Gibbs energy of mixing.

The important message of this discussion is that, in molecular terms, the minimum in the Gibbs energy curve stems from the Gibbs energy of mixing of the two perfect gases. Hence, an important contribution to the position of chemical equilibrium is the mixing of the products with the reactants as the products are formed.

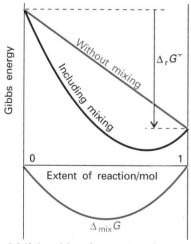

9.3 If the mixing of reactants and products is ignored, the Gibbs energy changes linearly from its initial value (pure reactants) to its final value (pure products). However, as products are formed, there is a further contribution to the Gibbs energy arising from their mixing (lowest curve). The sum of the two contributions has a minimum. That minimum corresponds to the equilibrium of the system.

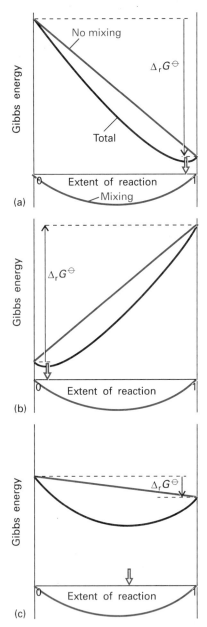

9.4 The position of the minimum (as summarized in Fig. 9.1) depends on the size of the standard Gibbs energy of reaction. In each case shown here, the observed variation in Gibbs energy is the sum of $\Delta_r G^\ominus$ and the same mixing contribution. The position of the minimum varies with $\Delta_r G^\ominus$ and is indicated by the open arrow: (a) large negative $\Delta_r G^\ominus$, (b) large positive $\Delta_r G^\ominus$, (c) small negative $\Delta_r G^\ominus$.

At equilibrium $\Delta_r G = 0$. The ratio of partial pressures at equilibrium is denoted K, and eqn 3 becomes

$$0 = \Delta_r G^\ominus + RT \ln K$$

which rearranges to

$$RT \ln K = -\Delta_r G^\ominus \tag{4}$$

This relation is a special case of one of the most important equations in chemical thermodynamics: it is the link between tables of thermodynamic data, such as those in the Data section at the end of this volume, and the chemically important equilibrium constant K.

We see from eqn 4 that, when $\Delta_r G^\ominus > 0$, $K < 1$, so at equilibrium the partial pressure of A exceeds that of B, which means that the reactant A is favoured in the equilibrium. When $\Delta_r G^\ominus < 0$, $K > 1$ so that at equilibrium the partial pressure of B exceeds that of A. Now the product B is favoured in the equilibrium.

MOLECULAR INTERPRETATION 9.2

We saw in *Molecular interpretation 9.1* that the minimum in the Gibbs energy plot arises from the Gibbs energy of mixing. When $\Delta_r G^\ominus$ is large and negative (a steeply sloping downwards slope on the graph), the addition of the contribution from mixing gives rise to a minimum that lies close to pure products (Fig. 9.4a); consequently, the equilibrium composition corresponds to almost pure products. When $\Delta_r G^\ominus$ is large and positive, the mixing term results in a minimum that lies very close to the reactants (Fig. 9.4b), so that now the equilibrium composition corresponds to the formation of only a tiny amount of product. If $\Delta_r G^\ominus$ is close to zero (Fig. 9.4c), the molecular mixing term is the dominant contribution and, as it goes through a minimum when equal amounts of A and B are present, in such cases the equilibrium composition corresponds to A and B present in similar amounts.

The general case of a reaction

The derivation of the relation between G and the equilibrium constant of a general reaction runs in a similar way. We shall develop the special and general cases in parallel and consider the reaction

$$2A + 3B \rightarrow C + 2D$$

This chemical equation is a special case of the general equation (Section 2.5)

$$0 = \sum_J \nu_J J$$

in which $\nu_A = -2$, $\nu_B = -3$, $\nu_C = +1$, and $\nu_D = +2$.

When the reaction advances by $d\xi$, the amounts of reactants and products change as follows:

$$dn_A = -2\,d\xi \qquad dn_B = -3\,d\xi \qquad dn_C = +d\xi \qquad dn_D = +2\,d\xi$$

In general: $dn_J = \nu_J\,d\xi$

The resulting change in the Gibbs energy at constant temperature and pressure is

$$dG = \mu_A \, dn_A + \mu_B \, dn_B + \mu_C \, dn_C + \mu_D \, dn_D$$
$$= (-2\mu_A - 3\mu_B + \mu_C + 2\mu_D) \, d\xi$$

In general: $\quad dG = \left(\sum_J \nu_J \mu_J \right) d\xi$

Therefore, the reaction Gibbs energy, the slope of G as ξ changes, is

$$\Delta_r G = \left(\frac{\partial G}{\partial \xi} \right)_{p,T} = -2\mu_A - 3\mu_B + \mu_C + 2\mu_D$$

In general: $\quad \Delta_r G = \sum_J \nu_J \mu_J$

The chemical potentials can be expressed in terms of activities by using

$$\mu_J = \mu_J^{\ominus} + RT \ln a_J$$

where a_J is interpreted as a fugacity if J is a gas (specifically, $a_J = f_J / p^{\ominus}$), and it is easy to deduce that

$$\Delta_r G = \Delta_r G^{\ominus} + RT \ln Q \tag{5}$$

where the standard molar reaction Gibbs energy is

$$\Delta_r G^{\ominus} = -2\mu_A^{\ominus} - 3\mu_B^{\ominus} + \mu_C^{\ominus} + 2\mu_D^{\ominus}$$

In general: $\quad \Delta_r G^{\ominus} = \sum_J \nu_J \mu_J^{\ominus}$

and the **reaction quotient** Q is

$$Q = \frac{a_C a_D^2}{a_A^2 a_B^3}$$

To write the general expression for Q it is convenient to introduce the symbol $\prod$ to denote the product of what follows it (just as $\sum$ denotes the sum), and to write

In general: $\quad Q = \prod_J a_J^{\nu_J} \tag{6}$

This expression signifies that each activity (or fugacity if the species is a gas) is raised to the power equal to its stoichiometric number, and then all such terms are multiplied together. Because reactants have negative stoichiometric numbers, they automatically appear as the denominator when the product is written out explicitly. Recall from Table 7.3 that, for pure solids and liquids, the activity is 1, so such substances make no contribution to Q even though they may appear in the chemical equation.

Example 9.1 *Writing a general reaction quotient*

Use the general formula for Q to write the reaction quotient for

$$4NH_3(g) + 5O_2(g) \rightarrow 4NO(g) + 6H_2O(g)$$

Method. The simplest approach is to write the activities of all the product species in the numerator and those of all the reactant species in the denominator and then to raise each activity to a power given by its stoichiometric number. However, to show the content of eqn 6 more fully, we shall adopt a lengthier procedure. First, we write the chemical equation in the form $0 = $ *sum over species* and identify the stoichiometric numbers. Then we insert these values into eqn 6. Finally, we replace the activities of gas phase species by fugacities, by using $a_J = f_J/p^{\ominus}$.

Answer. We write the equation in the form

$$0 = 4NO(g) + 6H_2O(g) - 4NH_3(g) - 5O_2(g)$$

which lets us identify the stoichiometric numbers as

$$v(NO) = +4 \qquad v(H_2O) = +6 \qquad v(NH_3) = -4 \qquad v(O_2) = -5$$

and hence to write

$$Q = \prod_J a_J^{v_J} = a(NO)^4 a(H_2O)^6 a(NH_3)^{-4} a(O_2)^{-5}$$

$$= \frac{a(NO)^4 a(H_2O)^6}{a(NH_3)^4 a(O_2)^5}$$

As all the substances are gases, we replace the activities by fugacities and obtain

$$Q = \frac{\left(\dfrac{f_{NO}}{p^{\ominus}}\right)^4 \left(\dfrac{f_{H_2O}}{p^{\ominus}}\right)^6}{\left(\dfrac{f_{NH_3}}{p^{\ominus}}\right)^4 \left(\dfrac{f_{O_2}}{p^{\ominus}}\right)^5} = \frac{f_{NO}^4 f_{H_2O}^6}{f_{NH_3}^4 f_{O_2}^5 p^{\ominus}}$$

Comment. At low pressure, it may be acceptable to replace the fugacities by the partial pressures of the gases. Note that, in common with all reaction quotients, the Q in this example is dimensionless (fugacities have the dimensions of pressure).

Exercise E9.1. Repeat the question for the reaction

$$2H_2S(g) + 3O_2(g) \rightarrow 2SO_2(g) + 2H_2O(g) \qquad \left[Q = \frac{f_{SO_2}^2 f_{H_2O}^2 p^{\ominus}}{f_{H_2S}^2 f_{O_2}^3}\right]$$

Now we conclude the argument based on eqn 5. At equilibrium the slope of G is zero: $\Delta_r G = 0$. The activities then have their equilibrium values, and we can write

$$\left(\frac{a_C a_D^2}{a_A^2 a_B^3}\right)_{equilibrium} = K$$

In general: $$\left(\prod_J a_J^{v_J}\right)_{equilibrium} = K$$

From now on, we shall not write the 'equilibrium' subscript explicitly and will rely on the context to make it clear that:

For K we use equilibrium values.

For Q we use the values at the specified stage of the reaction.

Setting $\Delta_r G = 0$ and replacing Q by K in eqn 5 leads at once to

$$RT \ln K = -\Delta_r G^{\ominus} \qquad (7\mathbf{a})$$

with

$$K = \prod_J a_J^{v_J} \qquad (7\mathbf{b})$$

An equilibrium constant K expressed, as here, in terms of activities is called a **thermodynamic equilibrium constant**, and eqn 7a is then an exact thermodynamic relation. Note that, because activities are dimensionless numbers, the thermodynamic equilibrium constant is also dimensionless. In elementary applications, the activities that occur in eqn 7b are often replaced by molalities or molar concentrations, and fugacities are replaced by partial pressures. In either case, the resulting expressions are only approximations. The approximation is particularly severe for electrolyte solutions, for in them activity coefficients differ from 1 even in very dilute solutions.

To use eqn 7, the standard molar reaction Gibbs energy is evaluated in the usual way from tables of standard Gibbs energies of formation at the temperature of the reaction:

$$\Delta_r G^{\ominus} = \sum_J v_J \Delta_f G^{\ominus}(J) \qquad (8)$$

Example 9.2 *Calculating an equilibrium constant*

Calculate the equilibrium constant for the reaction

$$N_2(g) + 3H_2(g) \rightarrow 2NH_3(g)$$

at 298 K and show how K is related to the partial pressures of the species at equilibrium when the overall pressure is low enough for the gases to be treated as perfect.

Method. The systematic approach involves identifying the stoichiometric numbers (by writing the chemical equation in the form $0 = sum$ *over species*) and then calculating the standard molar reaction Gibbs energy from eqn 8 and converting it to the value of the equilibrium constant by using eqn 7a. The expression for the equilibrium constant is obtained from eqn 7b and, because the gases are taken to be perfect, we replace each fugacity by a partial pressure.

Answer. The reaction is

$$0 = 2NH_3(g) - N_2(g) - 3H_2(g)$$

so $v(NH_3) = +2$, $v(N_2) = -1$, and $v(H_2) = -3$. Therefore

$$\Delta_r G^{\ominus} = 2\Delta_f G^{\ominus}(NH_3) - \Delta_f G^{\ominus}(N_2) - 3\Delta_f G^{\ominus}(H_2)$$
$$= 2\Delta_f G^{\ominus}(NH_3) = 2 \times (-16.5 \text{ kJ mol}^{-1})$$

Then, because $RT = 2.48 \text{ kJ mol}^{-1}$,

$$\ln K = -\frac{2 \times (-16.5 \text{ kJ mol}^{-1})}{2.48 \text{ kJ mol}^{-1}} = 13.3$$

Hence, $K = 6.0 \times 10^5$. This result is exact. The thermodynamic equilibrium constant for the reaction is

$$K = \frac{f_{NH_3}^2 \, p^{\ominus 2}}{f_{N_2} f_{H_2}^3}$$

and this ratio has exactly the value we have just calculated. However, at low overall pressures when fugacities can be replaced by partial pressures, the approximate form of the equilibrium constant is

$$K = \frac{p_{NH_3}^2 \, p^{\ominus 2}}{p_{N_2} p_{H_2}^3}$$

Exercise E9.2. Evaluate the equilibrium constant for $N_2O_4(g) \rightleftharpoons 2NO_2(g)$ at 298 K.   [0.15]

Example 9.3 *Estimating the degree of dissociation at equilibrium*

The standard Gibbs energy of reaction for the decomposition $H_2O(g) \rightarrow H_2(g) + \frac{1}{2}O_2(g)$ is $+118.08 \text{ kJ mol}^{-1}$ at 2300 K. What is the degree of dissociation of H_2O at 2257 K and 1.00 bar?

Method. We know that the equilibrium constant can be obtained from the standard reaction Gibbs energy by using eqn 7, so the task is to relate the degree of dissociation α to K and then to find its numerical value. Proceed by expressing the equilibrium compositions in terms of α, and solve for α in terms of K. Because the standard reaction Gibbs energy is large and positive, we can anticipate that K will be small, and hence that $\alpha \ll 1$, which opens the way to making approximations to obtain its numerical value.

Answer. The equilibrium constant is obtained from eqn 7 in the form

$$\ln K = -\frac{\Delta_r G^{\ominus}}{RT} = -\frac{118.08 \times 10^3 \, \text{J mol}^{-1}}{(8.3145 \, \text{J K}^{-1} \, \text{mol}^{-1}) \times (2300 \, \text{K})} = -6.175$$

It follows that $K = 2.08 \times 10^{-3}$. The equilibrium composition can be expressed in terms of α by drawing up the following table:

	H_2O	H_2	O_2
Amount	$(1-\alpha)n$	αn	$\frac{1}{2}\alpha n$
Mole fraction	$\dfrac{1-\alpha}{1+\frac{1}{2}\alpha}$	$\dfrac{\alpha}{1+\frac{1}{2}\alpha}$	$\dfrac{\frac{1}{2}\alpha}{1+\frac{1}{2}\alpha}$
Partial pressure	$\dfrac{(1-\alpha)p}{1+\frac{1}{2}\alpha}$	$\dfrac{\alpha p}{1+\frac{1}{2}\alpha}$	$\dfrac{\frac{1}{2}\alpha p}{1+\frac{1}{2}\alpha}$

The equilibrium constant is therefore

$$K = \frac{\left(\dfrac{p_{H_2}}{p^{\ominus}}\right)\left(\dfrac{p_{O_2}}{p^{\ominus}}\right)^{\frac{1}{2}}}{\left(\dfrac{p_{H_2O}}{p^{\ominus}}\right)} = \frac{\alpha^{\frac{3}{2}} p^{\frac{1}{2}}}{\sqrt{2}(1-\alpha)(1+\frac{1}{2}\alpha)^{\frac{1}{2}}(p^{\ominus})^{\frac{1}{2}}}$$

Now make the approximation that α can be neglected compared with 1, and hence obtain

$$K = \frac{\alpha^{\frac{3}{2}}}{\sqrt{2}} \left(\frac{p}{p^{\ominus}} \right)^{\frac{1}{2}}$$

Under the stated condition, $p/p^{\ominus} = 1.00$, so

$$\alpha \approx (\sqrt{2} \times K)^{\frac{2}{3}} = (\sqrt{2} \times 2.08 \times 10^{-3})^{\frac{2}{3}} = 0.0205$$

Comment. Always check that the approximation is consistent with the final answer. In this case $\alpha \ll 1$, in accord with the original assumption.

Exercise E9.3. Given that the standard Gibbs energy of reaction at 2000 K is $+135.2 \, \text{kJ mol}^{-1}$ for the same reaction, suppose that steam at 200 kPa is passed through a furnace tube at that temperature. Calculate the mole fraction of O_2 present in the output gas stream.   [0.00222]

The relation between thermodynamic and practical equilibrium constants

The only remaining problem is to express the thermodynamic equilibrium constant in terms of the mole fractions x_J or molalities m_J of the species. To do so, we need to know the activity coefficients, and then to use $a_J = \gamma_J x_J$ or $\gamma_J m_J / m^{\ominus}$ (recalling that the activity coefficients depend on the choice). For example, in the latter case, for an equilibrium of the form $A + B \rightleftharpoons C + D$, where all four species are solutes, we write

$$K = \frac{a_C a_D}{a_A a_B} = \frac{\gamma_C \gamma_D}{\gamma_A \gamma_B} \times \frac{m_C m_D}{m_A m_B} = K_{\gamma} \times K_m \tag{9}$$

The activity coefficients must be evaluated at the equilibrium composition of the mixture, which may involve a complicated calculation, because the latter is known only if the equilibrium composition is already known. In elementary applications, and to begin the iterative calculation of the concentrations in a real example, the assumption is often made that the activity coefficients are all so close to unity that $K_{\gamma} = 1$. Then we obtain the result widely used in elementary chemistry that $K \approx K_m$, and equilibria are discussed in terms of molalities (or molar concentrations) themselves. In Chapter 10 we shall see a way of making better estimates of activity coefficients for equilibria involving ions.

MOLECULAR INTERPRETATION 9.3

A deeper insight into the origin and significance of the equilibrium constant can be obtained by considering the Boltzmann distribution of molecules over the available states of a system composed of reactants and products. When atoms can exchange partners, as in a reaction, the available states of the system include arrangements in which the atoms are present in the form of reactants and in the form of products: these arrangements have their characteristic sets of energy levels, but the Boltzmann distribution does not distinguish between their identities, only their energies. The atoms distribute themselves over both sets of energy levels in accord with the Boltzmann distribution (Fig. 9.5). At a given

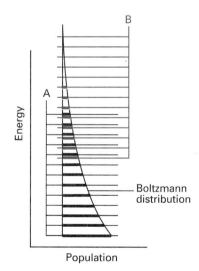

9.5 The Boltzmann distribution of populations over the energy levels of two species A and B with similar densities of energy levels; the reaction $A \rightarrow B$ is endothermic in this example. The bulk of the population is associated with the species A, so that species is dominated at equilibrium.

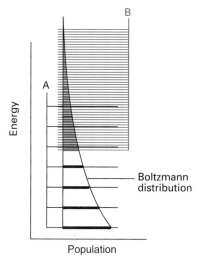

9.6 Even though the reaction $A \rightarrow B$ is still endothermic, the density of energy levels in B is so much greater than that in A that the population associated with B is greater than that associated with A; thus B is dominant at equilibrium.

temperature, there will be a specific distribution of populations and hence a specific composition of the reaction mixture.

It can be appreciated from Fig. 9.5 that, if the reactants and products both have similar arrays of molecular energy levels, then the dominant species in a reaction mixture at equilibrium will be the species with the lower set of energy levels. However, the fact that the Gibbs energy occurs in the expression is a signal that entropy plays a role as well as energy. Its role can be appreciated by referring to Fig. 9.6. There we see that, although the B energy levels lie higher than the A energy levels, in this instance they are much more closely spaced. As a result, their total population may be considerable and B could even dominate in the reaction mixture at equilibrium. Closely spaced energy levels correlate with a high entropy, so in this case we see that entropy effects dominate adverse energy effects. This competition is mirrored in eqn 7a, as can be seen most clearly by using $\Delta_r G^\ominus = \Delta_r H^\ominus - T\,\Delta_r S^\ominus$ and writing it in the form

$$K = e^{-(\Delta_r H^\ominus / RT) + (\Delta_r S^\ominus / R)}$$

Note that a positive reaction enthalpy results in a lowering of the equilibrium constant (that is, an endothermic reaction can be expected to have an equilibrium composition that favours the reactants). However, if there is a positive reaction entropy, then the equilibrium composition may favour the products despite the endothermic character of the reaction.

THE RESPONSE OF EQUILIBRIA TO THE CONDITIONS

The equilibrium constant for a reaction is unaffected by the presence of a catalyst or an enzyme (a biological catalyst): catalysts increase the rate at which equilibrium is attained but do not affect its position. However, in industry reactions rarely reach equilibrium, partly on account of the rates at which reactants mix, and under these non-equilibrium conditions catalysts can have some unexpected effects and may change the composition of the reaction mixture. For example, in the commercially important Fischer–Tropsch synthesis of hydrocarbon fuels from carbon monoxide and hydrogen, the choice of catalyst influences the distribution of chain lengths and molar masses in the product.

9.3 How equilibria respond to pressure

The equilibrium constant depends on the value of $\Delta_r G^\ominus$, which is defined at a single, standard pressure. The value of $\Delta_r G^\ominus$, and therefore of K, is a constant, independent of the pressure at which the equilibrium is established.[1] Formally we may express this independence as

$$\left(\frac{\partial K}{\partial p}\right)_T = 0 \tag{10}$$

1 This is not quite true for reactions in solution, for which

$$\left(\frac{\partial \ln K}{\partial p}\right)_T = -\frac{\Delta_r V^\ominus}{RT}$$

where $\Delta_r V^\ominus$ is the standard volume of reaction, the change in volume between the standard states of the reactants and products, which is usually small. For our purposes we shall suppose that $\Delta_r V^\ominus = 0$.

The conclusion that K is independent of pressure does not necessarily mean that the equilibrium composition is independent of the pressure. However, before considering the consequences of pressure, we need to distinguish between the two ways in which pressure may be applied. Pressure can be applied by injecting an inert gas into the reaction container. However, so long as the gases are perfect, this addition of gas leaves all the partial pressures of the reacting gases unchanged: partial pressures of perfect gases are the pressures that each one would exert if it were alone in the container, so the presence of another gas has no effect. Put another way, the addition of an inert gas leaves the molar concentration of the original gases unchanged, as they continue to occupy the same volume. It follows that pressurization by the addition of an inert gas has no effect on the equilibrium composition of the system (so long as the gases are perfect). Alternatively, the pressure of the system may be increased by physical compression, by confining the gases to a smaller volume. Now the partial pressures are changed. Put another way, their molar concentrations are modified because the volume the gases occupy is reduced.

We need to consider the role of compression and to understand how changes in partial pressures can be consistent with the general result expressed in eqn 10 that the equilibrium constant itself is independent of the pressure. We shall see that compression can adjust the individual partial pressures of the reactants and products in such a way that, although each one changes, their ratio (as it appears in the equilibrium constant) remains the same. Consider, for instance, the perfect-gas equilibrium $A \rightleftharpoons 2B$, for which the equilibrium constant is

$$K = \frac{p_B^2}{p_A p^{\ominus}}$$

The right-hand side of this expression remains constant only if an increase in p_A cancels an increase in the *square* of p_B. This relatively steep increase of p_A compared to p_B will occur if the equilibrium composition shifts in favour of A at the expense of B. Then the number of A molecules will increase as the volume of the container is decreased and its partial pressure will rise more rapidly than can be ascribed to a simple change in volume alone (Fig. 9.7).

The increase in the number of A molecules and the corresponding decrease in the number of B molecules brought about by compression is a special case of a principle proposed by the French chemist (and inventor of oxyacetylene welding) Henri Le Chatelier:

Le Chatelier's principle. A system at equilibrium, when subjected to a disturbance, responds in a way that tends to minimize the effect of the disturbance.

The principle implies that, if a system at equilibrium is compressed (by reduction of volume), the reaction will adjust so as to minimize the increase in pressure. This it can do by reducing the number of particles in the gas phase, which implies a shift $A \leftarrow 2B$. The quantitative treatment of the effect of compression leads to the conclusion that the

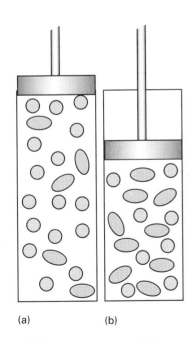

(a)   (b)

9.7 When a reaction at equilibrium is compressed (from a to b), the reaction responds by reducing the number of molecules in the gas phase (in this case by producing the dimers represented by the ellipses).

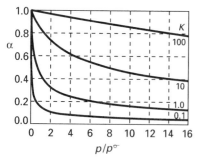

9.8 The pressure dependence of the degree of dissociation α at equilibrium for an $A(g) \rightleftharpoons 2B(g)$ reaction for different values of the equilibrium constant K. The value $\alpha = 0$ corresponds to pure A; $\alpha = 1$ corresponds to pure B.

extent of dissociation α of A into 2B is

$$\alpha = \left(\frac{1}{1 + \dfrac{4p}{Kp^{\ominus}}} \right)^{\frac{1}{2}} \tag{11}°$$

This formula shows that, even though K is independent of pressure, the amounts of A and B do depend on pressure (Fig. 9.8). It also shows that, as p is increased, α decreases in accord with Le Chatelier's principle.

JUSTIFICATION

Suppose that there is an amount n of A present initially (and no B). At equilibrium the amount of A is $(1 - \alpha)n$ and the amount of B is $2\alpha n$. It follows that the mole fractions present at equilibrium are

$$x_A = \frac{(1 - \alpha)n}{(1 - \alpha)n + 2\alpha n} = \frac{1 - \alpha}{1 + \alpha} \qquad x_B = \frac{2\alpha}{1 + \alpha}$$

The equilibrium constant for the reaction is

$$K = \frac{\left(\dfrac{p_B}{p^{\ominus}} \right)^2}{\left(\dfrac{p_A}{p^{\ominus}} \right)} = \frac{\left(\dfrac{x_B p}{p^{\ominus}} \right)^2}{\left(\dfrac{x_A p}{p^{\ominus}} \right)} = \frac{x_B^2}{x_A} \times \frac{p}{p^{\ominus}} = \frac{4\alpha^2}{(1 + \alpha)(1 - \alpha)} \times \frac{p}{p^{\ominus}}$$

$$= \frac{4\alpha^2}{1 - \alpha^2} \times \frac{p}{p^{\ominus}}$$

This expression rearranges into eqn 11.

Example 9.4 *Predicting the effect of pressure on an equilibrium*

Predict the effect of an increase in pressure on the composition of the ammonia synthesis at equilibrium. Assume perfect-gas behaviour.

Method. We use Le Chatelier's principle to predict the qualitative effect, and identify the direction of the shift in equilibrium as the one that corresponds to a decrease in the number of gas-phase molecules. The quantitative effect can be evaluated by expressing the equilibrium constant in terms of the mole fractions of the three species, as illustrated in the preceding *Justification*.

Answer. In the forward direction of the reaction

$$N_2(g) + 3H_2(g) \rightarrow 2NH_3(g)$$

the number of gas molecules decreases (from 4 to 2). Le Chatelier's principle then predicts that an increase in pressure will favour the product. For the quantitative analysis of the effect, we carry out the following calculation:

$$K = \frac{\left(\dfrac{p_{NH_3}}{p^\ominus}\right)^2}{\left(\dfrac{p_{N_2}}{p^\ominus}\right)\left(\dfrac{p_{H_2}}{p^\ominus}\right)^3} = \frac{\left(\dfrac{x_{NH_3}p}{p^\ominus}\right)^2}{\left(\dfrac{x_{N_2}p}{p^\ominus}\right)\left(\dfrac{x_{H_2}p}{p^\ominus}\right)^3} = \frac{x_{NH_3}^2}{x_{N_2}x_{H_2}^3} \times \left(\frac{p^\ominus}{p}\right)^2$$

That is,

$$K = K_x \times \left(\frac{p^\ominus}{p}\right)^2 \quad \text{where } K_x = \frac{x_{NH_3}^2}{x_{N_2}x_{H_2}^3}$$

Because K is independent of pressure, K_x increases 100-fold when the pressure is increased 10-fold.

Comment. The Haber synthesis of ammonia is run at high pressure in order to make use of this result. In precise work it is necessary to base the argument on the pressure independence of the thermodynamic equilibrium constant, and therefore to take note of the pressure dependence of the fugacity coefficients when discussing K_x.

Exercise E9.4. Predict the effect of a 10-fold pressure increase on the equilibrium composition of the reaction $3N_2(g) + H_2(g) \rightarrow 2HN_3(g)$.

[100-fold increase in K_x]

9.4 The response of equilibria to temperature

Le Chatelier's principle predicts that a system at equilibrium will tend to shift in the endothermic direction if the temperature is raised, for then energy is absorbed as heat. Likewise, an equilibrium can be expected to shift in the exothermic direction if the temperature is lowered, for then the reduction in temperature is opposed. These conclusions can be summarized as follows:

> **Exothermic reactions:** increased temperature favours the reactants.
> **Endothermic reactions:** increased temperature favours the products.

We shall now justify these remarks and see how to express the changes quantitatively.

The van't Hoff equation

The **van't Hoff equation** is an expression for the slope of a graph of the equilibrium constant (specifically, $\ln K$) plotted against the temperature. It may be expressed in either of two ways:

$$\frac{d \ln K}{dT} = \frac{\Delta_r H^\ominus}{RT^2} \tag{12a}$$

$$\frac{d \ln K}{d(1/T)} = -\frac{\Delta_r H^\ominus}{R} \tag{12b}$$

JUSTIFICATION

From eqn 7a, we know that

$$\ln K = -\frac{\Delta_r G^{\ominus}}{RT}$$

Differentiation of $\ln K$ with respect to temperature then gives

$$\frac{d \ln K}{dT} = -\frac{1}{R}\frac{d}{dT}\left(\frac{\Delta_r G^{\ominus}}{T}\right)$$

The differentials are complete because K and $\Delta_r G^{\ominus}$ depend only on temperature, not on pressure. To develop this equation we use the Gibbs–Helmholtz equation (eqn 5.11) in the form

$$\frac{d}{dT}\left(\frac{\Delta_r G^{\ominus}}{T}\right) = -\frac{\Delta_r H^{\ominus}}{T^2}$$

where $\Delta_r H^{\ominus}$ is the standard reaction enthalpy at the temperature T. Combining the two equations gives the van't Hoff equation:

$$\frac{d \ln K}{dT} = \frac{\Delta_r H^{\ominus}}{RT^2}$$

The second form of the equation is obtained by noting that

$$\frac{d}{dT}\left(\frac{1}{T}\right) = -\frac{1}{T^2}, \quad \text{or } dT = -T^2\, d\left(\frac{1}{T}\right)$$

It then follows that

$$\frac{d \ln K}{d(1/T)} = -\frac{\Delta_r H^{\ominus}}{R}$$

The first form of the van't Hoff equation shows that $d \ln K/dT < 0$ (and therefore $dK/dT < 0$) for a reaction that is exothermic under standard conditions ($\Delta_r H^{\ominus} < 0$). A negative slope means that $\ln K$, and therefore K itself, decreases as the temperature rises. Therefore, as asserted above, in the case of an exothermic reaction the equilibrium shifts away from products. The opposite occurs in the case of endothermic reactions.

Some insight into the thermodynamic basis of this behaviour can be found in the expression $\Delta_r G = \Delta_r H - T\,\Delta_r S$ written in the form $-\Delta_r G/T = -\Delta_r H/T + \Delta_r S$. When the reaction is exothermic, $-\Delta_r H/T$ corresponds to a positive change of entropy of the surroundings of the reaction system and is a driving force for the formation of products. When the temperature is raised, $-\Delta_r H/T$ decreases, and the increasing entropy of the surroundings is a less potent driving force; as a result, the equilibrium lies less to the right. When the reaction is endothermic, the principal driving force is the increasing entropy of the reaction system. The importance of the unfavourable change of entropy of the surroundings is reduced if the temperature is raised (because then $\Delta_r H/T$ is smaller), so the reaction is able to shift towards products.

MOLECULAR INTERPRETATION 9.4

The general arrangement of energy levels for an endothermic reaction is shown in Fig. 9.9a. When the temperature is increased, the Boltzmann distribution adjusts, and the populations change as shown and correspond to an increased population of the higher energy states at the expense of the population of the lower energy states. We see that the states that arise from the B molecules become more populated at the expense of the A molecules. Therefore, the total population of B states increases, and B becomes more abundant in the equilibrium mixture. On the other hand, if the reaction is exothermic (Fig. 9.9b), then an increase in temperature increases the population of the A states at the expense of the B states, so the reactants become more abundant.

9.9 The effect of temperature on a chemical equilibrium can be interpreted in terms of the change in the Boltzmann distribution with temperature and the effect of that change in populations of the species. (a) In an endothermic reaction, the population of B increases at the expense of A as the temperature is raised. (b) In an exothermic reaction, the opposite happens.

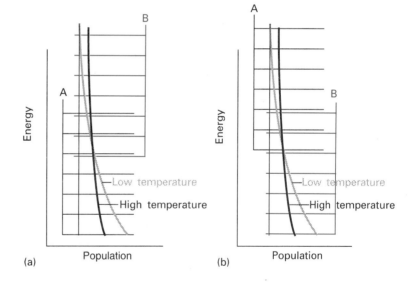

Example 9.5 *Measuring a reaction enthalpy*

The data below show the temperature variation of the equilibrium constant of the reaction

$$Ag_2CO_3(s) \rightleftharpoons Ag_2O(s) + CO_2(g)$$

Calculate the standard reaction enthalpy of the decomposition.

T/K	350	400	450	500
K	3.98×10^{-4}	1.41×10^{-2}	1.86×10^{-1}	1.48

Method. It follows from eqn 12b that, so long as the reaction enthalpy is largely independent of temperature, a plot of $-\ln K$ against $1/T$ should be a straight line of slope $\Delta_r H^{\ominus}/R$.

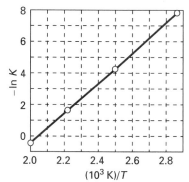

9.10 When $-\ln K$ is plotted against $1/T$, a straight line is expected with slope equal to $\Delta_r H^{\ominus}/R$. This is a non-calorimetric method for the measurement of reaction enthalpies. The data are those for Example 9.5.

Answer. We draw up the following table:

T/K	350	400	450	500
$(10^3\,\mathrm{K})/T$	2.86	2.50	2.22	2.00
$-\ln K$	7.83	4.26	1.68	-0.39

These points are plotted in Fig. 9.10. The slope of the graph is $+9.6 \times 10^3$, so

$$\Delta_r H^{\ominus} = (+9.6 \times 10^3\,\mathrm{K}) \times R = +80\,\mathrm{kJ\,mol^{-1}}$$

Comment. This is a non-calorimetric method of determining $\Delta_r H^{\ominus}$. A drawback is that the reaction enthalpy is temperature-dependent, so the plot is not expected to be perfectly linear. However, the temperature dependence is weak in many cases, so the plot is reasonably straight. In practice, the method is not very accurate, but it is often the only method available.

Exercise E9.5. The equilibrium constant of the reaction $2SO_2(g) + O_2(g) \rightleftharpoons 2SO_3(g)$ is 4.0×10^{24} at 300 K, 2.5×10^{10} at 500 K, and 3.0×10^4 at 700 K. Estimate the standard reaction enthalpy at 500 K. $[-200\,\mathrm{kJ\,mol^{-1}}]$

The value of K at different temperatures

If we wish to find the value of the equilibrium constant at a temperature T_2 in terms of its value K_1 at another temperature T_1, we would integrate eqn 12b:

$$\int_{\ln K_1}^{\ln K_2} \mathrm{d}\ln K = -\frac{1}{R}\int_{1/T_1}^{1/T_2} \Delta_r H^{\ominus}\, \mathrm{d}\!\left(\frac{1}{T}\right)$$

The integral on the left evaluates easily to $\ln K_2 - \ln K_1$. If we suppose that $\Delta_r H^{\ominus}$ varies only slightly with temperature over the temperature range of interest, then we may take it outside the integral. It follows that

$$\ln K_2 = \ln K_1 - \frac{\Delta_r H^{\ominus}}{R}\left(\frac{1}{T_2} - \frac{1}{T_1}\right) \tag{13}$$

Example 9.6 *Calculating the equilibrium constant at another temperature*

The equilibrium constant for the synthesis of ammonia at 298 K was calculated in Example 9.2. Estimate its value at 500 K.

Method. This example is a straightforward application of eqn 13. To use that formula, we need the standard molar enthalpy of reaction, which can be obtained from Table 2.12 in the Data section. We assume that the value is constant over the range of temperatures.

Answer. The standard molar reaction enthalpy for

$$N_2(g) + 3H_2(g) \rightarrow 2NH_3(g)$$

is obtained from Table 2.12:

$$\Delta_r H^{\ominus} = 2\Delta_f H^{\ominus}(NH_3) = -92.2\,\mathrm{kJ\,mol^{-1}}$$

Because at 298 K, $K = 6.0 \times 10^5$ (which we take as K_1 in eqn 13), it follows from eqn 13 that

$$\ln K_2 = \ln (6.0 \times 10^5) - \frac{(-92.2 \,\text{kJ}\,\text{mol}^{-1})}{8.3145 \,\text{J}\,\text{K}^{-1}\,\text{mol}^{-1}} \times \left(\frac{1}{500 \,\text{K}} - \frac{1}{298 \,\text{K}} \right)$$

$$= -1.73$$

Therefore, $K_2 = 0.18$.

Comment. Note the considerable decrease in the value of the equilibrium constant for this exothermic reaction. This is in accord with Le Chatelier's principle, but now expressed quantitatively.

Exercise E9.6. The equilibrium constant for $N_2O_4(g) \rightleftharpoons 2NO_2(g)$ at 298 K was calculated in Exercise E9.2. Estimate its value at 100°C. [15]

APPLICATIONS TO SELECTED SYSTEMS

In this section we look at some of the conclusions that can be drawn from the existence of equilibrium constants and from the equation $\Delta_r G^\ominus = -RT \ln K$. Because thermodynamic data are often listed for 298 K, equilibrium constants are often calculated at that temperature. It is helpful to note that $RT = 2.48 \,\text{kJ}\,\text{mol}^{-1}$ at 298 K. Then, if we write $\Delta_r G^\ominus = g \,\text{kJ}\,\text{mol}^{-1}$, it follows that $\ln K = -g/2.48$. Furthermore, because $\log K = (\ln K)/2.303$:

$$\text{At } 25°\text{C:} \quad K = 10^{-g/5.71} \tag{14}$$

This formula gives a quick way of calculating K near room temperature. Note that when $\Delta_r G^\ominus$ (and therefore g) is negative, $K > 1$ and products dominate reactants.

9.5 The extraction of metals from their oxides

Metals can be obtained from their oxides by reduction with carbon if either of the equilibria

$$MO(s) + \tfrac{1}{2}C(s) \rightleftharpoons M(s) + \tfrac{1}{2}CO_2(g)$$

$$MO(s) + C(s) \rightleftharpoons M(s) + CO(g)$$

lies to the right. These equilibria can be discussed in terms of the thermodynamic functions for the reactions

(i) $M(s) + \tfrac{1}{2}O_2(g) \rightarrow MO(s)$
(ii) $\tfrac{1}{2}C(s) + \tfrac{1}{2}O_2(g) \rightarrow \tfrac{1}{2}CO_2(g)$
(iii) $C(s) + \tfrac{1}{2}O_2(g) \rightarrow CO(g)$
(iv) $CO(g) + \tfrac{1}{2}O_2(g) \rightarrow CO_2(g)$

The temperature dependences of the standard Gibbs energies of these reactions depend on the reaction entropy through $d\Delta_r G^\ominus/dT = -\Delta_r S^\ominus$. Because in reaction (iii) there is a net *increase* in the amount of gas, the standard reaction entropy is large and positive; therefore, $\Delta_r G^\ominus$ decreases sharply with increasing temperature. In reaction (iv) there is a similar net decrease in the amount of gas, so that $\Delta_r G^\ominus$ increases sharply with increasing temperature. In reaction (ii) the amount of gas is

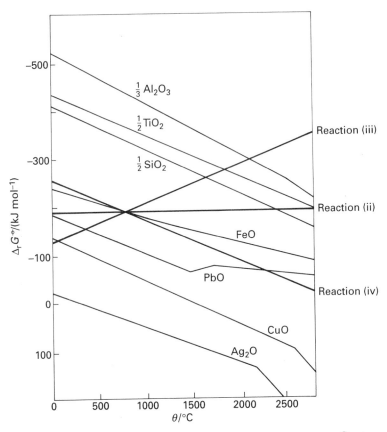

9.11 An Ellingham diagram for the discussion of metal ore reduction. Note that $\Delta_r G^{\ominus}$ is most negative at the top of the diagram.

constant, and so the entropy change is small and $\Delta_r G^{\ominus}$ changes only slightly with temperature. These remarks are summarized in Fig. 9.11, which is called an **Ellingham diagram** (note that $\Delta_r G^{\ominus}$ decreases upwards!).

The standard Gibbs energy of reaction (i) indicates the metal's affinity for oxygen. At room temperature the contribution of the reaction entropy to $\Delta_r G^{\ominus}$ is dominated by the reaction enthalpy, so the order of increasing $\Delta_r G^{\ominus}$ is the same as the order of increasing $\Delta_r H^{\ominus}$. This gives the order of values on the left of the diagram (Al_2O_3 is most exothermic; Ag_2O is least). The standard reaction entropy is similar for all metals because in each case gaseous oxygen is eliminated and a compact, solid oxide is formed. As a result, the temperature dependence of the standard Gibbs energy of oxidation should be similar for all metals, as is shown by the similar slopes of the lines in the diagram. The kinks at high temperatures correspond to the evaporation of the metals; less pronounced kinks occur at the melting temperatures of the metals and the oxides.

Successful reduction of the oxide depends on the outcome of the competition of the carbon for the oxygen bound to the metal. The standard Gibbs energies for the reductions can be expressed in terms of the standard Gibbs energies for reactions (i) to (iv) above:

$$MO(s) + \tfrac{1}{2}C(s) \rightarrow M(s) + \tfrac{1}{2}CO_2(g) \qquad \Delta_r G^{\ominus} = \Delta_r G^{\ominus}(ii) - \Delta_r G^{\ominus}(i)$$
$$MO(s) + C(s) \rightarrow M(s) + CO(g) \qquad \Delta_r G^{\ominus} = \Delta_r G^{\ominus}(iii) - \Delta_r G^{\ominus}(i)$$
$$MO(s) + CO(g) \rightarrow M(s) + CO_2(g) \qquad \Delta_r G^{\ominus} = \Delta_r G^{\ominus}(iv) - \Delta_r G^{\ominus}(i)$$

The equilibrium lies to the right if $\Delta_r G^{\ominus} < 0$. This is the case when the line for reaction (i) lies below (is more positive than) the line for one of the carbon reactions (ii) to (iv).

The spontaneity of a reduction at any temperature can be predicted simply by looking at the diagram: a metal oxide is reduced by any carbon reaction lying above it because the overall reaction then has $\Delta_r G^{\ominus} < 0$. For example, CuO can be reduced to Cu at any temperature above room temperature. Even in the absence of carbon, Ag_2O decomposes when heated above 200°C because then the standard Gibbs energy for reaction (i) becomes positive (so that the reverse reaction is then spontaneous). On the other hand, Al_2O_3 is not reduced by carbon until the temperature has been raised to above 2000°C.

9.6 Acids and bases

One of the most important examples of chemical equilibrium is the one that exists when acids and bases are present in solution. According to the **Brønsted–Lowry classification**:

> An **acid** HA is a proton donor, $HA \rightarrow H^+ + A^-$
>
> A **base** B is a proton acceptor, $B + H^+ \rightarrow BH^+$

Hydrogen chloride, HCl, is an acid because it can donate a proton (that is, an H^+ ion) to another molecule. Ammonia, NH_3, is a base, because it can accept a proton from another molecule and become NH_4^+. These definitions make no mention of the solvent (and apply even if no solvent is present); however, by far the most important medium is aqueous solution, and we confine our attention to that.

Acid–base equilibria in water

An acid HA takes part in the following equilibrium in water:

$$HA(aq) + H_2O(l) \rightleftharpoons H_3O^+(aq) + A^-(aq)$$

The H_3O^+ ion, the hydrated proton, is called a **hydronium ion** and the proton acceptor A^- is called the **conjugate base** of the acid HA. An example of an acid ionization equilibrium is the one established when HF is present in water:

$$HF(aq) + H_2O(l) \rightleftharpoons H_3O^+(aq) + F^-(aq)$$

The equilibrium constant for the proton transfer reaction is

$$K = \frac{a(H_3O^+)a(A^-)}{a(HA)a(H_2O)}$$

However, if we confine attention to dilute solutions, the activity of water is close to 1 (the value for pure water), and the equilibrium can be expressed in terms of the **acidity constant**

$$K_a = \frac{a(H_3O^+)a(A^-)}{a(HA)} \tag{15}$$

For a base B in water, the characteristic proton transfer equilibrium is

$$B(aq) + H_2O(l) \rightleftharpoons HB^+(aq) + OH^-(aq)$$

The proton donor formed when the base accepts a proton and becomes HB^+ is called the **conjugate acid** of the base B. An example of a base in water is NH_3:

$$NH_3(aq) + H_2O(l) \rightleftharpoons NH_4^+(aq) + OH^-(aq)$$

The conjugate acid of NH_3 is the ammonium ion, NH_4^+. Because HB^+ is a Brønsted acid in its own right, it also participates in a proton-transfer equilibrium

$$HB^+(aq) + H_2O(l) \rightleftharpoons H_3O^+(aq) + B(aq)$$

An example is the equilibrium when NH_4^+ ions are present in water:

$$NH_4^+(aq) + H_2O(l) \rightleftharpoons H_3O^+(aq) + NH_3(aq)$$

The importance of the concept of conjugate species is that we can treat bases in the same way as we treat acids, so long as we focus on the properties of the conjugate acid, HB^+ (e.g. on NH_4^+ when dealing with aqueous ammonia).

The general form of the **Brønsted equilibrium** for proton transfer reactions is

$$Acid(aq) + H_2O(l) \rightleftharpoons H_3O^+(aq) + Base(aq)$$

where Acid and Base are a conjugate pair (such as HF and F^- or NH_4^+ and NH_3). The general form of the acidity constant is

$$K_a = \frac{a(H_3O^+)a(Base)}{a(Acid)}$$

where 'Acid' may be a substance that we conventionally regard as an acid (such as HF) or the conjugate acid of a base (such as NH_4^+). In the Brønsted–Lowry theory, there is no fundamental distinction between acids and conjugate acids or between bases and conjugate bases: an acid is an acid and a base is a base. Values of acidity constants fall over a wide range (e.g. $K_a = 5.6 \times 10^{-10}$ for NH_4^+ and 0.16 for HIO_3 at 298 K). It is therefore convenient to list them as their logarithms, and we introduce

$$pK_a = -\log K_a \tag{16}$$

In this notation, $pK_a = 9.25$ for NH_4^+ and 0.80 for HIO_3. It is important to remember that the higher the pK_a of an acid, the smaller its K_a and hence the weaker its proton-donating power to water. A list of values is given in Table 9.1.

We shall see that the use of pK_a in place of K_a simplifies the appearance of a number of equations. This simplification stems from the fact that the K_a of the acid ionization equilibrium is related to the Gibbs energy of the proton donation reaction by

$$\Delta_r G^\ominus = -RT \ln K_a = 2.303RT \times pK_a$$

Hence, manipulations of pK_a values are in fact manipulations of $\Delta_r G^\ominus$ values in disguise.

Table 9.1* Acidity constants in water at 298 K

	pK_{a1}	pK_{a2}	pK_{a3}
Acetic acid, CH_3COOH	4.75		
Ammonium ion, NH_4^+	9.25		
Carbonic acid, H_2CO_3	6.37	10.25	
Phosphoric acid, H_3PO_4	2.12	7.21	12.67

* More values are given in the Data section at the end of this volume.

Autoprotolysis and pH

Water is **amphiprotic**, which means that it can act as both an acid and a base:

$$HF(aq) + H_2O(l) \rightarrow H_3O^+(aq) + F^-(aq)$$
Acid   Base

$$H_2O(l) + NH_3(aq) \rightarrow NH_4^+(aq) + OH^-(aq)$$
Acid   Base

A special case of this amphiprotic character occurs with water itself, for one H_2O molecule may act as an acid by donating a proton to another H_2O molecule acting as a base. This is an example of an **autoprotolysis equilibrium**, a proton transfer equilibrium involving a single substance:

$$H_2O(l) + H_2O(l) \rightleftharpoons H_3O^+(aq) + OH^-(aq)$$
Acid   Base

The equilibrium constant of this reaction, with the approximation that the solution is so dilute that the activity of water may be set equal to 1, is called the **autoprotolysis constant of water**:

$$K_w = a(H_3O^+)a(OH^-) \qquad pK_w = -\log K_w \qquad (17)$$

At 25°C, $K_w = 1.008 \times 10^{-14}$ ($pK_w = 14.00$), showing that only a few of the water molecules are ionized. Because the molar concentrations of H_3O^+ and OH^- are equal in pure water, it is reasonable to suppose that their activities are also equal.[2] It then follows that

$$a(H_3O^+) = K_w^{\frac{1}{2}} = 1.004 \times 10^{-7} \text{ at } 298 \text{ K}$$

In very dilute solutions, the activity of a species is approximately equal to the numerical value of its molality (that is, $a = m/m^{\ominus}$) and to the numerical value of its molar concentration. Because the activity is so low, so long as no other ions are present in the solution, the molar concentrations of H_3O^+ and OH^- ions in pure water are each about $1.0 \times 10^{-7} \text{ mol L}^{-1}$.

The hydronium ion activity plays a central role in many processes, and its magnitude can vary over a wide range. For instance, in 1 M HCl(aq), $a(H_3O^+) = 0.81$, in pure water it is about 10^{-7}, and in 1 M NaOH(aq) it is about 10^{-14}. The wide span of values is compressed by the **pH scale**,

---

2 The activities of individual ions and the sharing of the departures from ideality between them will be clarified when we consider mean ionic activity coefficients in Section 10.2.

where

$$pH = -\log a(H_3O^+) \tag{18a}$$

It should be noted that the higher the pH of a solution, the lower the hydronium ion activity. A pH may be negative: that corresponds to an activity of greater than 1. For example, in 2.00 M HCl(aq), in which the hydronium ion activity is 2.02, pH $= -0.31$. We shall see that it is also convenient to use the **pOH scale**, which is defined as

$$pOH = -\log a(OH^-) \tag{18b}$$

By taking logarithms of the expression for K_w and changing signs throughout, we find that the pH of a solution and its pOH are related by

$$pK_w = pH + pOH \tag{19}$$

Hence, if the pH of a solution increases, the pOH must decrease to keep their sum equal to pK_w (which is 14.00 at 25°C). In pure water pH = pOH (because the activities of the two ions are equal), so at 298 K,

$$pH = \tfrac{1}{2}pK_w = 7.00$$

Hence, pH = 7.00 corresponds to neutrality at 298 K. (At blood temperature, 37°C, when $pK_w = 13.68$, neutrality corresponds to pH = 6.84.) In an acidic aqueous solution, $a(H_3O^+)$ is greater than in pure water, so that pH < 7 at 25°C; in basic solutions, pH > 7 at 25°C.

pH calculations

In very dilute aqueous solutions,

$$a_J \approx m_J/m^\ominus \approx [J]/(\text{mol L}^{-1})$$

where [J] is the molar concentration of species J. When it simplifies the discussion, we shall make the approximation of replacing the activities in acidity constants by the numerical values of the molar concentrations and writing

$$K_a \approx \frac{[H_3O^+][\text{Base}]}{[\text{Acid}]}$$

where [J] should be understood as $[J]/(\text{mol L}^{-1})$. This approximation is legitimate only if all the ions are present at low concentration, not merely the ions of interest, because (as we shall see quantitatively in Section 10.2) all ions contribute to the departures from ideality. Concentrations must be very low for this approximation to be permissible; for a 10^{-3} M solution of a 1:1 electrolyte in water at 25°C, activity coefficients are about 0.96, and their neglect introduces an error of approaching 10 per cent into the interpretation of equilibrium constants. When concentrations are not low enough for concentrations to be used, activity coefficients can be found from tables (or estimated from the equations we give in Section 10.2).

Example 9.7 *Calculating the pH of an acid*

Calculate the pH of a 0.20 M HCN(aq) solution. Make the approximation that the activity coefficient of the HCN is close to 1, but do not make such an assumption about the ions present in the solution.

Method. It is always a good idea to begin by writing down the chemical equation for the equilibrium and the expression for the acidity constant. The stoichiometry of the reaction, and the assumption that the activity coefficients of the cations and anions are equal, lets us equate the activities of H_3O^+ and CN^-. Then, by taking logarithms, the expression can be written in terms of pK_a and pH.

Answer. The equilibrium to consider is

$$HCN(aq) + H_2O(l) \rightleftharpoons H_3O^+(aq) + CN^-(aq) \quad K_a = \frac{a(H_3O^+)a(CN^-)}{a(HCN)}$$

The activity of the HCN molecules is approximately equal to the numerical value of their molar concentration, so we write $a(HCN) \approx [HCN]/(mol\,L^{-1})$. The activities of the two types of ions are equal, so we set $a(CN^-) = a(H_3O^+)$ and rearrange the expression for K_a to

$$a(H_3O^+) = \left(K_a \times \frac{[HCN]}{mol\,L^{-1}}\right)^{\frac{1}{2}}$$

On taking logarithms and changing the sign throughout, this expression turns into

$$pH = \tfrac{1}{2}pK_a - \tfrac{1}{2}\log([HCN]/mol\,L^{-1})$$
$$= \tfrac{1}{2} \times 9.31 - \tfrac{1}{2}\log 0.20 = 5.0$$

Comment. Several approximations have been made, so the numerical conclusions are not reliable to more than one decimal place, and even that may be overoptimistic. The approximations include the neglect of the activity coefficient of HCN, the assumption that the activities of H_3O^+ and CN^- are equal, and the neglect of the small contribution of hydronium ions from the autoprotolysis of water.

Exercise E9.7. Calculate the pH of $0.10\,M$ $NH_3(aq)$. (*Hint.* Treat the solution as involving the ionization of the acid NH_4^+.) [11.1]

Strong and weak acids

A **strong acid** is a strong proton donor that is virtually completely ionized in solution:

$$HCl(aq) + H_2O(l) \rightarrow H_3O^+(aq) + Cl^-(aq) \text{ virtually completely}$$

$$H_2SO_4(aq) + H_2O(l) \rightarrow H_3O^+(aq) + HSO_4^-(aq) \text{ virtually completely}$$

The acidity constant of a strong acid is greater than 1, so its pK_a is negative. However, the pK_a of a strong acid is only rarely measurable, and it is best to treat the acid as fully ionized in typical solutions. A **strong base** is a strong proton acceptor, and is virtually fully protonated in solution. An example is the O^{2-} ion, which does not exist as such in water because it is fully protonated:

$$O^{2-}(aq) + H_2O(l) \rightarrow 2OH^-(aq) \quad \text{virtually completely}$$

Saying that a base is strong is equivalent to saying that its conjugate acid is a very weak proton donor; hence its acidity constant is very small indeed. Thus, OH^- is an extremely weak proton donor (to H_2O).

A **weak acid** is a Brønsted acid that is incompletely ionized in solution and in particular has $K_a < 1$ and hence a positive pK_a. An example is acetic acid in water:

$$CH_3COOH(aq) + H_2O(l) \rightleftharpoons H_3O^+(aq) + CH_3CO_2^-(aq) \quad pK_a = 4.75$$

Another example is the HSO_4^- ion in water:

$$HSO_4^-(aq) + H_2O(l) \rightleftharpoons H_3O^+(aq) + SO_4^{2-}(aq) \quad pK_a = 1.92$$

Thus H_2SO_4 is a strong acid in water but its conjugate base HSO_4^- is only a weak acid. Because the pK_a of HSO_4^- is smaller than that of CH_3COOH, it follows that HSO_4^- is a stronger weak acid than CH_3COOH in water (in the sense that it is more fully ionized in a solution of the same molar concentration).

A **weak base** is a Brønsted base that is only partly protonated. An example is NH_3 in water:

$$NH_3(aq) + H_2O(l) \rightleftharpoons NH_4^+(aq) + OH^-(aq)$$

The fact that both NH_3 and NH_4^+ coexist in the solution implies that NH_4^+ (and, in general, the conjugate acid of any weak base) is a weak acid. For NH_4^+ in water, for example, $pK_a = 9.25$.

Example 9.8 *Calculating the pH of a solution*

Calculate the pH of $0.15\,M\ NH_4Cl(aq)$. The activity coefficient of a univalent ion in a solution of a $1:1$ electrolyte of this concentration is about 0.6.

Method. The cation NH_4^+ is a weak acid, so we can expect its presence to lower the pH of the solution. The Cl^- ion is far too weak a base to have any effect on the pH. To proceed, write down the chemical equation for the ionization of the weak acid (NH_4^+) and reorganize it into an expression for $\log a(H_3O^+)$ by making use of the reaction stoichiometry. Because NH_4^+ is a weak acid, it is permissible to suppose that its molar concentration in solution is virtually unchanged from the concentration of the salt.

Answer. The equilibrium to consider is

$$NH_4^+(aq) + H_2O(l) \rightleftharpoons H_3O^+(aq) + NH_3(aq) \quad K_a = \frac{a(H_3O^+)a(NH_3)}{a(NH_4^+)}$$

If we suppose that the activity coefficients of the ions are the same, then they cancel in the expression for K_a; we can also suppose that the activity coefficient of the neutral NH_3 molecule is close to 1. Then,

$$K_a = \frac{m(H_3O^+)m(NH_3)}{m(NH_4^+)m^\ominus}$$

An NH_3 molecule is produced for each H_3O^+ ion formed, so $m(NH_3) = m(H_3O^+)$. Therefore

$$m(H_3O^+) = \{K_a \times m(NH_4^+)m^\ominus\}^{\frac{1}{2}}$$

The NH_4^+ ion is such a weak acid that the molality of NH_4^+ is insignificantly different from the molality of the salt and the molality is numerically almost the same as the molar concentration. Therefore, with $K_a = 5.6 \times 10^{-10}$,

$$m(H_3O^+) = \{(5.6 \times 10^{-10}) \times (0.15\ \text{mol kg}^{-1}) \times (1\ \text{mol kg}^{-1})\}^{\frac{1}{2}}$$
$$= 9.2 \times 10^{-6}\ \text{mol kg}^{-1}$$

The activity coefficient for the ions in a 0.15 M aqueous solution of a $1:1$ electrolyte is given as 0.6, so the activity of hydronium ions in the solution is

$$a(H_3O^+) \approx 0.6 \times (9.2 \times 10^{-6}) = 6 \times 10^{-6}$$

It then follows that the pH of the solution is 5.2.

Comment. We have ignored any hydronium ions that might come from the water itself; as we shall see below, this contribution is usually negligible except in pure water.

Exercise E9.8. Calculate the pH of 0.15 M $NaCH_3CO_2(aq)$; the activity coefficient is about 0.6. (*Hint.* Express the equilibrium constant for the reaction $CH_3CO_2^-(aq) + H_2O(l) \rightleftharpoons CH_3COOH(aq) + OH^-(aq)$ in terms of K_a by using $K = K_w/K_a$, and calculate pOH initially.) [8.9]

The distinction between weak and strong acids and bases is an illustration of the different types of behaviour shown in Fig. 9.1: strong acids and bases are those for which the minimum Gibbs energy of the solution lies close to the (ionized) products; weak acids and bases are those for which the minimum lies close to the (non-ionized) reactants.

Acid–base titrations

One application in which acidity constants play an important role is in acid–base titrations, for they are used to decide the value of the pH that signals the **stoichiometric point** (the 'equivalence point'), the stage at which a stoichiometric amount of an acid has been added to the solution of a base.

In a titration of a strong acid with a strong base (or vice versa), the ions present at the stoichiometric point (the cations from the strong base, such as Na^+ from NaOH, and the anions from the strong acid, such as Cl^- from HCl) barely affect the pH. The solution consists of these ions, water, and H_3O^+ and OH^- ions from the autoprotolysis of water.

Because the autoprotolysis stoichiometry guarantees that pH = pOH, it follows that pH = 7 at the stoichiometric point.

At the stoichiometric point of a titration of a weak acid (such as CH_3COOH) and strong base (NaOH), the analyte (the solution being titrated) has become an aqueous solution of the weak acid–strong base salt (sodium acetate). At this point, the solution contains $CH_3CO_2^-$ and Na^+ ions together with any ions stemming from autoprotolysis. The presence of the Brønsted base $CH_3CO_2^-$ means that we can expect a pH of greater than 7. At the stoichiometric point of a titration of a weak base (such as NH_3) and a strong acid (HCl), the analyte is a solution of a strong acid–weak base salt (ammonium chloride) and contains NH_4^+ and Cl^- ions. Because Cl^- is a negligibly weak Brønsted base and NH_4^+ is a weak Brønsted acid (Example 9.8), the solution is acidic and its pH will be less than 7.

The next few paragraphs develop this argument and show how to predict the pH at any stage of an acid–base titration. We shall do this in two stages. In the first, we show how to plot the complete pH curve during a titration. This procedure gives unwieldy equations. Therefore, we shall also show how to make a series of approximations that are useful in practice.

The complete pH curve

We shall suppose that we are titrating a volume V_A of a solution of a weak acid of nominal molar concentration A_0 (the analyte) with a solution of a strong base MOH of molar concentration B (the titrant). When a volume V_B of the titrant has been added to the analyte, the total volume of the analyte is $V = V_A + V_B$. At such a stage in the titration, the molar concentration H of hydronium ions is given by the solution of the equation

$$\frac{V_B}{V_A} = \frac{(K_a + H)(K_w - H^2) + K_w A_0 H}{(K_a + H)(BH + H^2 - K_w)} \tag{20}$$

Some of the curves obtained using this formula are shown in Fig. 9.12.

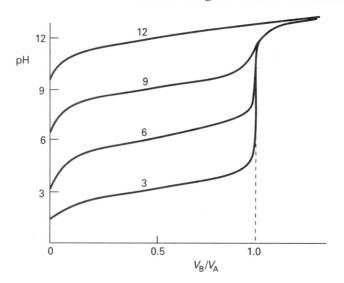

9.12 The variation of pH during the titration of a weak acid with a strong base for different values of pK_a.

JUSTIFICATION

Because the solution is electrically neutral at all stages, we know that at all times

$$[M^+] + [H_3O^+] = [A^-] + [OH^-]$$

We also know that the total number of A groups (as HA and A^-) is constant and equal to $A_0 \times V_A$. However, their concentration is changing because the volume of the solution is changing. At any stage, therefore,

$$[HA] + [A^-] = A \qquad \text{with } A = \frac{A_0 V_A}{V_A + V_B}$$

The molar concentration of M^+ cations also changes because the amount added at any stage is $B \times V_B$ and the total volume of the solution is $V_A + V_B$. Therefore, at any stage

$$[M^+] = S \qquad \text{with } S = \frac{BV_B}{V_A + V_B}$$

The quantity S denotes the current concentration of the salt that is the product of the reaction. At all stages of the titration we know that [HA] and $[A^-]$ are related by

$$K_a = \frac{a(H_3O^+)a(A^-)}{a(HA)} \approx \frac{[H_3O^+][A^-]}{[HA]}$$

Moreover, the molar concentrations of $[H_3O^+]$ and $[OH^-]$ are related by the water autoprotolysis constant:

$$K_w = a(H_3O^+)a(OH^-) \approx [H_3O^+][OH^-]$$

Now we bring all these features together. The conservation of A groups can be expressed in terms of $[A^-]$ alone by making use of the acidity constant:

$$[A^-] = \frac{AK_a}{K_a + [H_3O^+]} = \frac{AK_a}{K_a + H}$$

Next, we express the electrical neutrality condition in terms of $[H_3O^+]$ alone, for we now have expressions for $[M^+]$ in terms of the volume of added base, for $[A^-]$ in terms of $H = [H_3O^+]$, and for $[OH^-]$ in terms of K_w. The condition becomes

$$H + \frac{BV_B}{V_A + V_B} = \frac{A_0 V_A K_a}{(V_A + V_B)(K_a + H)} + \frac{K_w}{H}$$

This is an equation for H in terms of the volume of base added. Unfortunately, it is a cubic equation in H which is very awkward to solve. However, it can be rearranged into eqn 20 for V_B as a function of H, which enables us to find the volume of base needed to achieve any pH.

Although eqn 20 is very general, and can be used to find the relation between pH and the volume of base added at all stages of a titration, it is far from transparent and, without a computer, not very easy to use. We shall therefore seek to identify the main features of the curve by making a series of approximations.

The pH in the course of a weak acid–strong base titration

The approximations we shall make are based on the fact that the acid is weak, and therefore that HA is more abundant than any A^- ions in the solution. Furthermore, when HA is present, it provides hydronium ions that greatly outnumber any that stem from the autoprotolysis of water. Finally, when excess base is present, the OH^- ions it provides dominate any that come from the water autoprotolysis.

Example 9.9 *Estimating the pH at the start of a titration*

Estimate the pH of a solution of a weak acid of concentration A_0. What is the pH of 0.010 M HClO(aq)?

Method. We can proceed as in Example 9.7. The approximations we employ are that the acid is mostly present as HA molecules (because it is weak), the activity coefficient of HA is 1, and any H_3O^+ ions that are produced by the acid are much more numerous than those produced by the autoprotolysis of water.

Answer. The first two approximations let us write $a(HA) \approx [HA] \approx A_0$. We can also write $a(A^-) \approx a(H_3O^+)$ because HA is the only important source of both ions and their activity coefficients are equal. Hence,

$$K_a = \frac{a(H_3O^+)a(A^-)}{a(HA)} \approx \frac{a(H_3O^+)^2}{A_0}$$

This expression can be rearranged to

$$a(H_3O^+) = (K_a A_0)^{\frac{1}{2}}$$

and hence to

$$pH = \tfrac{1}{2}pK_a - \tfrac{1}{2}\log A_0 \tag{21}$$

From Table 9.1, for HClO $pK_a = 7.43$, so

$$pH = \tfrac{1}{2} \times 7.43 - \tfrac{1}{2} \log 0.010 = 4.7$$

Exercise E9.9. Find an expression for the pH of a weak base of concentration B_0 and calculate the pH of 0.010 M NH_3(aq). (*Hint.* See the remark in Exercise E9.8.) [$pH = \tfrac{1}{2}pK_a + \tfrac{1}{2}pK_w + \tfrac{1}{2}\log B_0$, pH = 10.6]

After the addition of some base (but before the stoichiometric point is reached), the concentration of A^- ions stems almost entirely from the salt that is present, for the weak acid present provides only a few A^- ions. Therefore $[A^-] \approx S$. The number of HA molecules that remain is the original number $A_0 V_A$ less the number of HA molecules that have been converted to salt by the addition of base, so the molar concentration of HA is $A' = A - S$. This calculation ignores the small additional loss of HA as a result of its ionization in solution. Hence

$$K_a = \frac{a(H_3O^+)a(A^-)}{a(HA)} \approx \frac{a(H_3O^+)S}{A'}$$

The derivation has made the doubtful approximation that the activity coefficient of the A^- ions is close to 1. It follows that

$$pH = pK_a - \log\left(\frac{A'}{S}\right) \qquad (22a)$$

This expression is called the **Henderson–Hasselbalch equation**. The general form of this equation, once we recognize that A' is the concentration of acid in the solution and S is the concentration of base, is

$$pH = pK_a - \log\left(\frac{[\text{Acid}]}{[\text{Base}]}\right) \qquad (22b)$$

When the molar concentrations of acid and salt are equal,

$$pH = pK_a \quad \text{when} \quad S = A' \qquad (23)$$

Hence the pK_a of the acid can be measured directly from the pH of the mixture. In practice this is done by recording the pH during a titration and then examining the record for the pH half-way to the stoichiometric point.

At the stoichiometric point the H_3O^+ ions in the solution stem from the influence of the OH^- ions on the autoprotolysis equilibrium, and the OH^- ions are produced by the Brønsted equilibrium

$$A^-(aq) + H_2O(l) \rightleftharpoons HA(aq) + OH^-(aq)$$
$$K = \frac{a(HA)a(OH^-)}{a(A^-)} \approx \frac{[HA][OH^-]}{[A^-]}$$

Example 9.10 *Calculating the pH at the stoichiometric point*

Find an approximate expression for the pH at the stoichiometric point of the titration of a weak acid with a strong base. Calculate its value for the titration of 25.00 mL of 0.100 M HClO(aq) with 0.100 M NaOH(aq).

Method. We need to identify the appropriate approximations when the solution is that of the salt alone and only M^+ and A^- ions are nominally present. Because only a small amount of HA is formed in this way, the concentration of A^- ions is almost exactly that of the salt, and we can write $[A^-] \approx S$. The number of OH^- ions that arises from the Brønsted equilibrium written above greatly outnumber those produced by the water autoprotolysis, so $[HA] \approx [OH^-]$. The equilibrium constant for the base protolysis can be expressed in terms of the acidity constant of the conjugate acid HA by using the autoprotolysis constant of water.

Answer. It follows from the expression for the equilibrium constant that

$$K \approx \frac{[OH^-]^2}{S}, \quad \text{so } [OH^-] \approx (SK)^{\frac{1}{2}}$$

Next, note that

$$K = \frac{a(HA)a(OH^-)}{a(A^-)} = \frac{a(HA)a(OH^-)a(H_3O^+)}{a(A^-)a(H_3O^+)} = \frac{K_w}{K_a}$$

Therefore,

$$[OH^-] \approx \left(\frac{SK_w}{K_a}\right)^{\frac{1}{2}}$$

So, after taking negative logarithms of both sides,

$$pOH = \tfrac{1}{2}pK_w - \tfrac{1}{2}pK_a - \tfrac{1}{2}\log S$$

Finally, because $pH = pK_w - pOH$, it follows that

$$pH = \tfrac{1}{2}pK_a + \tfrac{1}{2}pK_w + \tfrac{1}{2}\log S \tag{24}$$

At the stoichiometric point of the titration described, the concentration of NaClO is $0.050 \, mol \, L^{-1}$ (because the volume of the solution has increased from 25.00 mL to 50.00 mL), so

$$pH = \tfrac{1}{2} \times 7.43 + \tfrac{1}{2} \times 14.00 + \tfrac{1}{2}\log 0.050 = 10.1$$

Exercise E9.10. Show that the pH at the stoichiometric point of a titration of a strong acid with a weak base is given by

$$pH = \tfrac{1}{2}pK_a - \tfrac{1}{2}\log S \tag{25}$$

where pK_a is the acidity constant of the conjugate acid of the weak base. Calculate the pH at the stoichiometric point of a titration of 25.00 mL of $0.200 \, M \, NH_3(aq)$ with $0.300 \, M \, HCl(aq)$. [5.1]

When so much strong base has been added that the titration has been carried well past the stoichiometric point, the pH is controlled by the excess base present. Then,

$$[H_3O^+] \approx \frac{K_w}{[OH^-]}$$

If we write the molar concentration of excess base as B' this expression can be written

$$pH = pK_w + \log B' \tag{26}$$

In this expression, as in all the preceding ones, the molar concentrations must take into account the change of volume that occurs as the titrant is added to the analyte.

The general form of the pH curve throughout a titration is illustrated in Fig. 9.13. The pH rises slowly from the value given by the 'weak acid alone' formula (eqn 21) following the values given by the Henderson–Hasselbalch equation (eqn 22) until the stoichiometric point is ap-

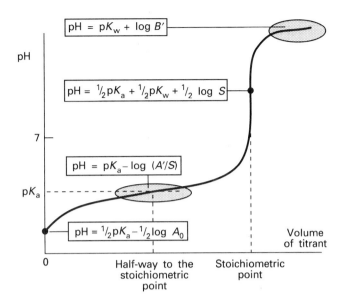

9.13 A summary of the regions of the pH curve for the titration of a weak acid with a strong base and the equations used in different regions.

proached. It then changes rapidly to and through the value given by the 'salt alone' formula (eqn 24). It then climbs less rapidly towards the value given by the 'base in excess' formula (eqn 26). The stoichiometric point can be detected easily by observing where the pH changes rapidly through the value given by the 'salt alone formula' (eqn 24).

Buffers and indicators

The slow variation of the pH in the vicinity of $S = A'$, when the molar concentrations of the salt and acid are equal, is the basis of **buffer action**, the ability of a solution to oppose changes in pH when small amounts of strong acids and bases are added to the solution. The mathematical basis of buffer action is the logarithmic dependence given by the Henderson–Hasselbalch equation (eqn 22), which is quite flat near $pH = pK_a$. The physical basis of buffer action is that the existence of an abundant supply of A^- ions (because a salt is present) can remove any H_3O^+ ions brought by additional strong acid; moreover, the numerous HA molecules can supply H_3O^+ ions to react with any strong base that is added. This stabilization of a dynamic equilibrium against an outside perturbation can be regarded as another example of Le Chatelier's principle.

Example 9.11. *Estimating the pH of a buffer solution*

Estimate the pH of an aqueous buffer solution that contains $0.200 \, mol \, L^{-1} \, KH_2PO_4$ and $0.100 \, mol \, L^{-1} \, K_2HPO_4$.

Method. The pH of a solution of a weak acid and its salt can be estimated from the Henderson–Hasselbalch equation. To do so, we must identify the acid HA and its conjugate base A^-.

Answer. In this example, the acid is the anion $H_2PO_4^-$ and its conjugate base is the anion HPO_4^{2-}:

$$H_2PO_4^-(aq) + H_2O(l) \rightleftharpoons H_3O^+(aq) + HPO_4^{2-}(aq)$$

The acidity constant we require is therefore pK_{a2} for H_3PO_4, which from Table 9.1 is 7.21. Then, with $A' = 0.200 \, mol \, L^{-1}$ and $S = 0.100 \, mol \, L^{-1}$, eqn 22 gives the pH of the solution as

$$pH \approx 7.21 - \log\left(\frac{0.200}{0.100}\right) = 6.91$$

Hence, the solution should buffer close to $pH = 7$.

Exercise E9.11. Calculate the pH of an aqueous buffer solution which contains $0.100 \, mol \, L^{-1} \, NH_3$ and $0.200 \, mol \, L^{-1} \, NH_4Cl$.

[8.95; more realistically, 9]

The rapid change of pH near the stoichiometric point in a titration is the basis of indicator detection. An **acid–base indicator** is normally some large, water-soluble, weakly acidic organic molecule which can exist as acid (HIn) or conjugate base (In^-) forms that differ in colour. The two forms are in equilibrium in solution:

$$HIn(aq) + H_2O(l) \rightleftharpoons H_3O^+(aq) + In^-(aq)$$

and, if we make the usual assumption that the solution is so dilute that the activity of water is 1, the equilibrium is described by the constant

$$K_{In} = \frac{a(H_3O^+)a(In^-)}{a(HIn)}$$

The ratio of acid and base forms at a given pH is found by rearranging this expression to

$$\log\left(\frac{[HIn]}{[In^-]}\right) \approx pK_{In} - pH \tag{27}$$

Therefore, when the pH is less than pK_{In}, the indicator is predominantly in its acidic form and has the corresponding colour; when the pH is greater than pK_{In}, the indicator is mainly in its basic form. The **end-point** is the pH of the solution when both forms are present in equal abundance, which occurs when $pH = pK_{In}$.

At the stoichiometric point of an acid–base titration, the pH changes sharply through several units and, if the pH passes through pK_{In}, then there is a pronounced colour change. With a well-chosen indicator, the end-point coincides with the stoichiometric point of the titration.

Care must be taken to use an indicator that changes colour at the pH appropriate to the type of titration. Thus, in a weak acid–strong base titration, the stoichiometric point lies at the pH given by eqn 24, so that an indicator that changes at that pH must be selected. Broadly speaking, an indicator with $pK_{In} > 7$ is required because the stoichiometric point lies at $pH > 7$. Similarly, in a strong acid–weak base titration, an indicator changing near the pH given by eqn 25 should be used. The stoichiometric points of such titrations lie at pH < 7, so that an indicator with $pK_{In} < 7$ is required.

9.7 Biological activity: the thermodynamics of ATP

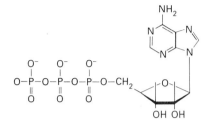

1 ATP

An important biochemical is adenosine triphosphate, ATP (**1**). Its function is to store the energy made available when food is metabolized and then to supply it on demand to a wide variety of processes, including muscular contraction, reproduction, and vision. The essence of ATP's action is its ability to lose its terminal phosphate group by hydrolysis and to form adenosine diphosphate (ADP):

$$ATP(aq) + H_2O(l) \rightarrow ADP(aq) + P_i^-(aq) + H_3O^+(aq)$$

(P_i^- denotes an inorganic phosphate group, such as $H_2PO_4^-$.) This reaction is exergonic and can drive an endergonic reaction if suitable enzymes are available.

Biological standard states

The conventional standard state of hydrogen ions (unit activity, pH $= 0$) is not appropriate to normal biological conditions. Therefore, in biochemistry it is common to adopt the **biological standard state**, in which pH $= 7$ (an activity of 10^{-7}, neutral solution). We shall adopt this convention in this section, and label the corresponding standard thermodynamic functions as $G^{\oplus}$, $H^{\oplus}$, and $S^{\oplus}$ (some texts use $X^{\ominus\prime}$). The relation between the thermodynamic and biological standard Gibbs energies for a reaction of the form

$$A + \nu H^+(aq) \rightarrow P$$

is

$$\Delta_r G^{\oplus} = \Delta_r G^{\ominus} + 16.12 \times \nu RT \tag{28}$$

Note that there is no difference between the two standard values if protons are not involved in the reaction ($\nu = 0$).

JUSTIFICATION

The reaction Gibbs energy is

$$\Delta_r G = \mu(P) - \mu(A) - \nu\mu(H^+)$$

If all the species other than H^+ are in their standard states, this expression becomes

$$\Delta_r G = \mu^{\ominus}(P) - \mu^{\ominus}(A) - \nu\mu(H^+)$$

Then, because

$$\mu(H^+) = \mu^{\ominus}(H^+) + RT \ln a(H^+) = \mu^{\ominus}(H^+) - 2.303 RT \times pH$$

the expression becomes

$$\Delta_r G = \mu^{\ominus}(P) - \mu^{\ominus}(A) - \nu\mu^{\ominus}(H^+) + 2.303\nu RT \times pH$$
$$= \Delta_r G^{\ominus} + 2.303\nu RT \times pH$$

It follows that at pH = 7.00

$$\Delta_r G^{\oplus} = \Delta_r G^{\ominus} + 16.12 \times vRT$$

As an illustration of eqn 28, consider the reaction

$$NADH(aq) + H^+(aq) \rightarrow NAD^+(aq) + H_2(g)$$

at 37°C, for which $\Delta_r G^{\ominus} = -21.8 \, kJ \, mol^{-1}$. NADH is the reduced form of nicotinamide adenine dinucleotide and NAD^+ is its oxidized form; the molecules play an important role in the later stages of the respiratory process. It follows that

$$\Delta_r G^{\oplus} = -21.8 \, kJ \, mol^{-1} + 16.1 \times (8.314 \, J \, K^{-1} \, mol^{-1}) \times (310 \, K)$$
$$= +19.7 \, kJ \, mol^{-1}$$

Note that the biological standard value is opposite in sign (in this example) to the thermodynamic standard value.

The standard values for the ATP hydrolysis at 37°C (310 K, blood temperature) are $\Delta_r G^{\oplus} = -30 \, kJ \, mol^{-1}$, $\Delta_r H^{\oplus} = -20 \, kJ \, mol^{-1}$, and $\Delta_r S^{\oplus} = +34 \, J \, K^{-1} \, mol^{-1}$. The hydrolysis is therefore exergonic ($\Delta_r G^{\oplus} < 0$) under these conditions, and $30 \, kJ \, mol^{-1}$ is available for driving other reactions. Moreover, because the reaction entropy is large, the reaction Gibbs energy is sensitive to temperature. On account of its exergonicity the ADP–phosphate bond has been called a **high-energy phosphate bond**. The name is intended to signify a high tendency to undergo reaction, and should not be confused with a 'strong' bond. In fact, even in the biological sense it is not of very 'high energy'. The action of ATP depends on it being intermediate in activity. Thus it acts as a phosphate donor to a number of acceptors (e.g. glucose), but is recharged by more powerful phosphate donors in the respiration cycle.

Anaerobic and aerobic metabolism

The efficiency of some biological processes can be gauged in terms of the value of $\Delta_r G^{\oplus}$ given above, as we shall see by considering aerobic and anaerobic metabolism. Aerobic metabolism is a series of reactions in which inhaled oxygen plays a role; anaerobic metabolism is a form of metabolism in which inhaled oxygen plays no role. The energy source of anaerobic cells is glycolysis, the partial oxidation of glucose to lactic acid, and at blood temperature $\Delta_r G^{\oplus} = -218 \, kJ \, mol^{-1}$. The standard reaction enthalpy is $-120 \, kJ \, mol^{-1}$, the exergonicity exceeding the exothermicity on account of the large increase of entropy accompanying the fracture of the glucose molecule. The glycolysis is coupled to a reaction in which two ADP molecules are converted into two ATP molecules:

$$Glucose + 2P_i^- + 2ADP \rightarrow 2Lactate^- + 2ATP + 2H_2O$$

The standard reaction Gibbs energy is $(-218 \, kJ \, mol^{-1}) - 2(-30 \, kJ \, mol^{-1}) = -158 \, kJ \, mol^{-1}$. The reaction is exergonic, and therefore spontaneous: the metabolism of the food has been used to 'recharge' the ATP.

Metabolism by aerobic respiration is much more efficient. The standard Gibbs energy of combustion of glucose is $-2880 \, kJ \, mol^{-1}$, so terminating its oxidation at lactic acid is a poor use of resources. In aerobic respiration the oxidation is carried out to completion, and an extremely complex set of reactions preserves as much of the energy released as possible. In the overall reaction, 38 ATP molecules are generated for each glucose molecule consumed. Each mole of ATP extracts 30 kJ from the 2880 kJ supplied by 1 mol of $C_6H_{12}O_6$ (180 g of glucose), so 1140 kJ has been stored for later use.

Each ATP molecule can be used to drive an endergonic reaction for which $\Delta_r G^{\ominus}$ does not exceed $+30$ kJ mol^{-1}. For example, the biosynthesis of sucrose from glucose and fructose can be driven (if a suitable enzyme system is available) because the reaction is endergonic to the extent $\Delta_r G^{\ominus} = +23$ kJ mol^{-1}. The biosynthesis of proteins is strongly endergonic, not only on account of the enthalpy change but also on account of the large decrease in entropy that occurs when many amino acids are assembled into a precisely determined sequence. For instance, the formation of a peptide link is endergonic, with $\Delta_r G^{\ominus} = +17$ kJ mol^{-1}, but the biosynthesis can only occur indirectly and requires the consumption of three ATP molecules for each link. In a moderately small protein like myoglobin, with about 150 peptide links, the construction alone requires 450 ATP molecules, and therefore about 12 mol of glucose molecules for 1 mol of protein molecules.

CHECK LIST OF KEY IDEAS

1 The definition of the **extent of reaction** and of the **reaction Gibbs energy** (eqn 1).

2 The meaning of the terms **exergonic** and **endergonic** (Section 9.1).

3 The reaction Gibbs energy in terms of the **reaction quotient** and the standard reaction Gibbs energy (eqn 5), and the **equilibrium constant** in terms of the standard reaction Gibbs energy (eqn 7).

4 The lack of variation of the equilibrium constant with pressure, **Le Chatelier's principle**, and the variation of the equilibrium composition with pressure (Section 9.3).

5 The **van't Hoff equation** for the variation of the equilibrium constant with temperature (eqn 12) and its use for the determination of reaction enthalpy (Example 9.5).

6 Thermodynamic considerations in the extraction of metals from their ores and the interpretation of an **Ellingham diagram** (Section 9.5).

7 The discussion of acids and bases in terms of **proton-transfer equilibria** (Section 9.6) and the definition of **acidity constants** (eqn 15).

8 The distinction between **weak acids** and **strong acids** and the role of the **autoprotolysis equilibrium** of water (Section 9.6).

9 The use of equilibrium constants to calculate the pH in the course of an acid/base titration and the derivation of the **Henderson–Hasselbalch equation** (eqn 22).

10 The equilibria responsible for **buffer action** and indicator detection of the stoichiometric point (Section 9.6).

11 The definition of the **biological standard state** (Section 9.7) and its relation to the conventional standard state (eqn 28).

12 Thermodynamic aspects of metabolism and respiration involving the reactions of ATP (Section 9.7).

EXERCISES

9.1. The equilibrium constant for the isomerization of *cis*-2-butene to *trans*-2-butene is $K = 2.07$ at 400 K. Calculate the standard reaction Gibbs energy.

9.2. The standard reaction Gibbs energy of the isomerization of *cis*-2-pentene to *trans*-2-pentene at 400 K is -3.67 kJ mol^{-1}. Calculate the equilibrium constant of the isomerization.

9.3. At 2257 K and 1 atm total pressure, water is 1.77 per cent dissociated at equilibrium by way of the reaction $2H_2O(g) \rightleftharpoons 2H_2(g) + O_2(g)$. Calculate (a) K, (b) $\Delta_r G^{\ominus}$, and (c) $\Delta_r G$ at this temperature.

9.4. Nitrogen tetroxide, which is present in the equilibrium $N_2O_4(g) \rightleftharpoons 2NO_2(g)$, is 18.46 per cent dissociated at 25°C and 1.00 bar. Calculate (a) K, (b) $\Delta_r G^{\ominus}$, (c) $\Delta_r G$ for the production of $NO_2(g)$ at 1.00 bar and 25°C from $N_2O_4(g)$ at 10.0 bar and 25°C, (d) K at 100°C given that $\Delta_r H^{\ominus} = +57.2$ kJ mol^{-1} over the temperature range.

9.5. For the equilibrium, $N_2O_4(g) \rightleftharpoons 2NO_2(g)$, the degree of dissociation α_e at 298 K is 0.201 at 1.00 bar total pressure. Calculate (a) the temperature at which $\alpha = 0.50$, (b) $\Delta_r G$, (c) $\Delta_r G^{\ominus}$ at that temperature.

9.6. From information in the Data section at the end of this volume, calculate the standard Gibbs energy and the equilibrium constant at (a) 298 K and (b) 400 K for the reaction $PbO(s, red) + CO(g) \rightleftharpoons Pb(s) + CO_2(g)$. Assume that the reaction enthalpy is independent of temperature.

9.7. From information in the Data section, calculate the standard Gibbs energy and the equilibrium constant at (a) 25°C and (b) 50°C for the reaction $CH_4(g) + 3Cl_2(g) \rightleftharpoons CHCl_3(l) + 3HCl(g)$. Assume that the reaction enthalpy is independent of temperature. For $CHCl_3$, $\Delta_f G^{\ominus} = -134.47$ kJ mol^{-1} and $\Delta_f H^{\ominus} = -73.66$ kJ mol^{-1}.

9.8. In the gas-phase reaction $2A + B \rightleftharpoons 3C + 2D$ it was found that when 1.00 mol A, 2.00 mol B, and 1.00 mol D were mixed and allowed to come to equilibrium at 25°C, the resulting mixture contained 0.90 mol C at a total pressure of 1.00 bar. Calculate (a) the mole fractions of each species at equilibrium, (b) K_x, (c) K, and (d) $\Delta_r G^{\ominus}$.

9.9. The standard reaction enthalpy of $Zn(s) + H_2O(g) \rightarrow ZnO(s) + H_2(g)$ is approximately constant at $+224$ kJ mol^{-1} from 920 K up to 1600 K. The standard reaction Gibbs energy is $+33$ kJ mol^{-1} at 1280 K. Assuming that $\Delta_r H^{\ominus}$ and $\Delta_r S^{\ominus}$ remain constant, estimate the temperature at which the equilibrium constant becomes greater than 1.

9.10. The equilibrium constant of the reaction $2C_3H_6(g) \rightleftharpoons C_2H_4(g) + C_4H_8(g)$ is found to fit the expression

$$\ln K = -1.04 - \frac{1088 \text{ K}}{T} + \frac{15.1 \times 10^5 \text{ K}^2}{T^2}$$

between 300 K and 600 K. Calculate the standard reaction enthalpy and standard reaction entropy at 400 K.

9.11. The standard reaction Gibbs energy of the isomerization of borneol ($C_{10}H_{17}OH$) to isoborneol in the gas phase at 503 K is $+9.4$ kJ mol^{-1}. Calculate the reaction Gibbs energy in a mixture consisting of 0.15 mol of borneol and 0.30 mol of isoborneol when the total pressure is 600 Torr.

9.12. The equilibrium pressure of H_2 over solid uranium and uranium hydride, UH_3, at 500 K is 1.04 Torr. Calculate the standard Gibbs energy of formation of $UH_3(s)$ at 500 K.

9.13. Calculate the percentage change in the equilibrium constant K_x of the following reactions when the total pressure is increased from 1.0 bar to 2.0 bar at constant temperature: (a) $H_2CO(g) \rightleftharpoons CO(g) + H_2(g)$, (b) $CH_3OH(g) + NOCl(g) \rightleftharpoons HCl(g) + CH_3NO_2(g)$.

9.14. The equilibrium constant for the gas-phase isomerization of borneol ($C_{10}H_{17}OH$) to isoborneol at 503 K is 0.106. A mixture consisting of 7.50 g of borneol and 14.0 g of isoborneol in a 5.0-L container is heated to 503 K and allowed to come to equilibrium. Calculate the mole fractions of the two substances at equilibrium.

9.15. Use the data in Table 2.12 of the Data section to decide which of the following reactions have $K > 1$ at 298 K.

(a) $HCl(g) + NH_3(g) \rightleftharpoons NH_4Cl(s)$
(b) $2Al_2O_3(s) + 3Si(s) \rightleftharpoons 3SiO_2(s) + 4Al(s)$
(c) $Fe(s) + H_2S(g) \rightleftharpoons FeS(s) + H_2(g)$
(d) $FeS_2(s) + 2H_2(g) \rightleftharpoons Fe(s) + 2H_2S(g)$
(c) $2H_2O_2(l) + H_2S(g) \rightleftharpoons H_2SO_4(l) + 2H_2(g)$

9.16. Which of the equilibria in Exercise 9.15 are favoured (in the sense of K increasing) by a rise in temperature at constant pressure?

9.17. What is the standard enthalpy of a reaction for which the equilibrium constant is (a) doubled, (b) halved when the temperature is increased by 10 K at 298 K?

9.18. The standard Gibbs energy of formation of $NH_3(g)$ is -16.5 kJ mol^{-1} at 298 K. What is the reaction Gibbs energy when the partial pressure of N_2, H_2, and NH_3 (treated as perfect gases) are 3.0 bar, 1.0 bar, and 4.0 bar, respectively? What is the spontaneous direction of the reaction in this case?

9.19. The dissociation vapour pressure of NH_4Cl at 427°C is 608 kPa but at 459°C it has risen to 1115 kPa. Calculate (a) the equilibrium constant, (b) the standard reaction Gibbs energy, (c) the standard enthalpy, (d) the standard entropy of dissociation, all at 427 °C. Assume that the vapour behaves as a perfect gas and that $\Delta H^{\ominus}$ and $\Delta S^{\ominus}$ are independent of temperature in the range given.

9.20. At 20°C, $pK_w = 14.17$, at 25°C it is 14.00, and at 30°C it is 13.84. Calculate the standard enthalpy of the autoprotolysis reaction at 25°C.

9.21. Estimate the temperature at which (a) $CaCO_3$ decomposes and (b) $CuSO_4 \cdot 5H_2O$ undergoes dehydration.

9.22. At the half-way point in the titration of a weak acid with a strong base the pH was measured as 5.40. What is the acidity constant and the pK_a of the acid? What is the pH of the solution that is 0.015 M in the acid?

9.23. Calculate the pH of (a) 0.10 M $NH_4Cl(aq)$, (b) 0.10 M $NaCH_3CO_2$, (c) 0.100 M $CH_3COOH(aq)$.

9.24. Calculate the pH at the stoichiometric point of the titration of 25.00 mL of 0.100 M lactic acid with 0.150 M NaOH(aq).

9.25. Sketch the pH curve of a solution containing 0.10 M $NaCH_3CO_2(aq)$ and a variable amount of acetic acid.

9.26. From the information in Table 9.1 select suitable buffers for (a) pH = 2.2 and (b) pH = 7.0.

PROBLEMS
Numerical problems

9.1. The equilibrium constant for the reaction $I_2(s) + Br_2(g) \rightleftharpoons 2IBr(g)$ is 0.164 at 25°C. (a) Calculate $\Delta_r G^{\ominus}$ for this reaction. (b) Bromine gas is introduced into a container with excess solid iodine. The pressure and temperature are held at 0.164 atm and 25°C. Find the partial pressure of IBr(g) at equilibrium. Assume that the vapour pressure of iodine is negligible. (c) In fact, solid iodine has a measurable vapour pressure at 25°C. In this case how would the calculation have to be modified?

9.2. Consider the decomposition of methane, $CH_4(g)$, into the elements $H_2(g)$ and C(s, graphite). (a) Given that $\Delta_f H^{\ominus}(CH_4, g) = -74.85 \text{ kJ mol}^{-1}$ and that $\Delta_r S^{\ominus}(CH_4, g) = -80.67 \text{ J K}^{-1} \text{mol}^{-1}$ at 298 K, calculate the value of the equilibrium constant at 298 K. (b) Assuming that $\Delta_r H^{\ominus}$ is independent of temperature, calculate K at 50°C. (c) Calculate the degree of dissociation α_e of methane at 25°C and a total pressure of 0.010 bar. (d) Without doing any numerical calculations, explain how the degree of dissociation for this reaction will change as the pressure and temperature are varied.

9.3. The equilibrium pressure of H_2 over U(s) and $UH_3(s)$ between 450 K and 715 K fits the expression

$$\ln (p/\text{Pa}) = 69.32 - \frac{14.64 \times 10^3 \text{ K}}{T} - 5.65 \ln (T/\text{K})$$

Find an expression for the standard enthalpy of formation of $UH_3(s)$ and from it calculate $\Delta_r C_p$.

9.4. The degree of dissociation α_e of $CO_2(g)$ into CO(g) and $O_2(g)$ at high temperatures and 1 bar total pressure was found to vary with temperature as follows:

T/K	1395	1443	1498
$\alpha_e/10^{-4}$	1.44	2.50	4.71

Assuming $\Delta_r H^{\ominus}$ to be constant over this temperature range, calculate K, $\Delta_r G^{\ominus}$, $\Delta_r H^{\ominus}$, and $\Delta_r S^{\ominus}$. Make any justifiable approximations.

9.5. The standard reaction enthalpy of the decomposition of $CaCl_2 \cdot NH_3(s)$ into $CaCl_2(s)$ and $NH_3(g)$ is nearly constant at $+78 \text{ kJ mol}^{-1}$ between 350 K and 470 K. The equilibrium pressure of NH_3 in the presence of $CaCl_2 \cdot NH_3$ is 12.8 Torr at 400 K. Find an expression for the temperature dependence of $\Delta_r G^{\ominus}$ in the same range.

9.6. Calculate the equilibrium constant of the reaction $CO(g) + H_2(g) \rightleftharpoons H_2CO(g)$ given that, for the production of liquid formaldehyde, $\Delta_r G^{\ominus} = +28.95 \text{ kJ mol}^{-1}$ at 298 K and that the vapour pressure of formaldehyde is 1500 Torr at that temperature.

9.7. Acetic acid was evaporated in a container of volume 21.45 cm³ at 437 K and at an external pressure of 764.3 Torr, and the container was then sealed. The mass of acid present in the sealed container was 0.0519 g. The experiment was repeated with the same container but at 471 K, and it was found that 0.0380 g of acetic acid was present. Calculate the equilibrium constant for the dimerization of the acid in the vapour and the enthalpy of vaporization.

9.8. Hydrogen and carbon monoxide have been investigated for use in fuel cells, so their solubilities in molten salts are of interest. Their solubilities in a molten $NaNO_3/KNO_3$ mixture was examined (E. Desimoni and P. G. Zambonin, *J. Chem. Soc. Faraday Trans.* I, 2014 (1973)) with the following results:

$$\log s(H_2) = -5.39 - \frac{768 \text{ K}}{T}$$

$$\log s(CO) = -5.98 - \frac{980 \text{ K}}{T}$$

where s is the solubility in mol cm⁻³ bar⁻¹. Calculate the standard molar enthalpies of solution of the two gases at 570 K.

9.9. A sealed container was filled with 0.300 mol $H_2(g)$, 0.400 mol $I_2(g)$, and 0.200 mol HI(g) at 870 K and total pressure 1.00 bar. Calculate the amounts of the components in the mixture at equilibrium given that $K = 870$ for the reaction $H_2(g) + I_2(g) \rightleftharpoons 2HI(g)$.

9.10. Triethylamine (TEA) and 2,4-dinitrophenol (DNP) form a complex in chlorobenzene, and the equilibrium constant for its formation has been measured over a range of temperatures (K. J. Ivin, J. J. McGarvey, E. L. Simmons, and R. Small, *J. Chem. Soc. Faraday Trans.* I, 1016 (1973)):

$\theta/°C$	17.5	25.2	30.0	35.5	39.5	45.0
K	29670 ± 1230	14450 ± 560	9270 ± 70	5870 ± 120	3580 ± 30	2670 ± 70

Calculate the standard enthalpy and entropy of formation of the complex from TEA and DNP at 20°C.

9.11. The dissociation of I_2 can be monitored by measuring the total pressure, and three sets of results are as follows:

T/K	973	1073	1173
$100p/atm$	6.244	7.500	9.181
$10^4 n_I$	2.4709	2.4555	2.4366

where n_I is the amount of I atoms in the mixture, which occupied 342.68 cm³. Calculate the equilibrium constants of the dissociation and the standard enthalpy of dissociation at the mean temperature.

9.12. Boron trifluoride acts as a catalyst for the equilibrium between acetaldehyde (CH_3CHO) and paraldehyde, a trimer of acetaldehyde. The partial pressures of the components in the trimerization reaction are too low for the accurate determination of the equilibrium constant by direct measurement, but this problem can be overcome by ensuring that liquid forms of the two substances are always present (W. K. Busfield, R. M. Lee, and D. Merigold, *J. Chem. Soc. Faraday Trans.* I, 936 (1973)). Assume that the gases are perfect, and show that the equilibrium constant for the trimerization can be written

$$K = \frac{p_P(p_A - p)(p_A - p_P)^2 p^{\ominus 2}}{p_A^3(p - p_P)^3}$$

where p_A is the vapour pressure of acetaldehyde, p_P that of paraldehyde, and p is the total pressure. Use the enthalpies of vaporization of acetaldehyde and paraldehyde, which are 25.6 kJ mol⁻¹ and 41.5 kJ mol⁻¹, respectively, and the following data to calculate the standard enthalpy and entropy of trimerization of acetaldehyde in the gas phase.

$\theta/°C$	20.0	22.0	26.0	28.0	30.0	32.0	34.0	36.0	38.0	40.0
p/kPa	23.9	27.3	36.5	42.6	49.9	56.9	65.1	74.3	85.0	96.2

You also need to know that the vapour pressures of the two components are given by

$$\ln(p/kPa) = a - \frac{\Delta_{vap}H^{\ominus}}{RT}$$

with $a = 15.5$ for acetaldehyde and $a = 17.2$ for paraldehyde. Given that the boiling points of acetaldehyde and paraldehyde are 294 K and 398 K, respectively, calculate the enthalpy and entropy of the trimerization in the liquid phase.

Theoretical problems

9.13. Show that if K_p (the equilibrium constant in terms of partial pressures) increases with temperature, then K_ϕ (the matching combination of fugacity coefficients) must decrease, where $K = K_p K_\phi$.

9.14. Express the equilibrium constant of a gas-phase reaction $A + 3B \rightleftharpoons 2C$ in terms of the equilibrium value of the extent of reaction ξ, given that initially A and B were present in stoichiometric proportions. Find an expression for ξ as a function of the total pressure p of the reaction mixture and sketch a graph of the expression obtained.

9.15. When light passes through a cell of length l containing an absorbing gas at a pressure p, the absorption is proportional to pl. Consider the equilibrium $2NO_2 \rightleftharpoons N_2O_4$, with NO_2 the absorbing species. Show that when two cells of lengths l_1 and l_2 are used, and the pressures needed to obtain equal absorptions are p_1 and p_2, respectively, then the equilibrium constant is given by

$$K = \frac{(p_1 \rho^2 - p_2)^2}{\rho(\rho - 1)(p_2 - p_1 \rho)p^{\ominus}}$$

with $\rho = l_1/l_2$. The following data were obtained (R. J. Nordstrum and W. H. Chan, *J. Phys. Chem.* **80**, 847 (1976))

Absorbance	p_1/Torr	p_2/Torr
0.05	1.00	5.47
0.10	2.10	12.00
0.15	3.15	18.65

with $l_1 = 395$ mm and $l_2 = 75$ mm. Determine the equilibrium constant of the reaction.

9.16. Find an expression for the standard reaction Gibbs energy at a temperature T' in terms of its value at another temperature T and the coefficients a, b, and c in the expression for the molar heat capacity listed in Table 2.2. Evaluate the standard Gibbs energy of formation of $H_2O(l)$ at 372 K from its value at 298 K.

10

Equilibrium electrochemistry

The principles of thermodynamics established in the preceding chapters, particularly Chapter 9, can be applied to solutions of electrolytes. Thus, with a suitable definition, it is possible to measure and employ the standard Gibbs energies of ions in solution in much the same way as they are used for other species. One important extension of the previous material however, is the need to take into account activity coefficients, for they differ significantly from 1 on account of the strong ionic interactions in electrolyte solutions. These coefficients are best treated as empirical quantities, but we shall see that it is possible to estimate them in very dilute solutions.

The bulk of the chapter is concerned with the description of the thermodynamic properties of reactions that take place in electro-chemical cells, in which, as the reaction proceeds, it drives electrons through an external circuit. We shall see that thermodynamic arguments can be used to derive an expression for the electric potential of such cells and that the potential can be related to their composition. There are two major topics developed in this connection. One is the definition and tabulation of the contribution of individual electrodes to the overall potential of a cell, which leads to the formulation of standard potentials of species, the principal experimental quantities employed in the discussion of electrochemical systems at equilibrium. The second feature is the use of these standard potentials to predict the equilibrium constants of chemical reactions.

The other matters dealt with in this chapter complete the discussion of earlier topics. We shall see, for instance, how electrochemical techniques are used to measure the pH of solutions, the pK_a of weak acids, the solubilities of sparingly soluble salts, and the standard Gibbs energy, enthalpy, and entropy of reactions.

In the final chapter of this part of the text we turn to a consideration of the special properties of electrolyte solutions. Although the thermodynamic properties of electrolyte solutions can be discussed in terms of chemical potentials and activities in much the same way as solutions of non-electrolytes, there are a number of distinctive features. One is the presence of strong interactions between ions in solution, which means that deviations from ideality are marked even in quite dilute systems. Therefore, we must equip ourselves with a means of dealing with activity coefficients that differ significantly from 1. A second feature is of much greater utility. Because many reactions of ions involve the transfer of electrons, they can be studied (and utilized) by allowing them to take place in an electrochemical cell, when the reactions produce an electric current in an external circuit or can be caused to take place by passing an electric current through the cell. Measurements like the ones we describe in this chapter lead to a collection of data that are very useful for discussing the characteristics of electrolyte solutions and of ionic equilibria in solution, as we shall see.

In common with the preceding chapters, we concentrate here on the thermodynamics of electrochemical processes. Their kinetic aspects are described in Chapter 29.

THE THERMODYNAMIC PROPERTIES OF IONS IN SOLUTION

Many of the concepts described in previous chapters carry over without change into the discussion of electrolyte solutions. However, departures from ideality are important, even at very low concentrations (they cannot be ignored even at $10^{-3} \, \text{mol L}^{-1}$), and we must know how to take them into account.

10.1 Thermodynamic functions of formation

The standard enthalpy and Gibbs energy of a reaction involving ions in solution are expressed in terms of standard enthalpies and Gibbs energies of formation, which are listed in Table 2.12. These properties are used in exactly the same way as those for neutral compounds, as the following example illustrates.

Example 10.1 *Using standard Gibbs energies of formation*

Calculate the solubility (s, in moles of solute per kilogram of solvent) of silver chloride in water at 25°C.

Method. The solubility of a substance is the molality of the substance when the solution is saturated; that is, when undissolved solute and dissolved solute are in dynamic equilibrium. Therefore, solubility can be discussed in terms of equilibrium constants. Thus we proceed by writing down the chemical equation for the solubility equilibrium and the corresponding equilibrium constant. Then we express the activities that occur in the equilibrium constant in terms of s. Because silver chloride is only very sparingly soluble, it is (guardedly) permissible to ignore activity coefficients. Finally, we express the equilibrium constant in terms of the standard Gibbs energy of the (dissolution) reaction and evaluate the latter from the standard Gibbs energies of formation of the species.

Answer. The dissolution reaction is

$$AgCl(s) \rightarrow Ag^+(aq) + Cl^-(aq)$$

and, in the saturated solution (at equilibrium),

$$K = a(Ag^+)a(Cl^-)$$

Because one Ag^+ ion is formed for each Cl^- ion, their molalities are equal; moreover, because silver chloride is only very sparingly soluble, activities may be replaced by molalities, giving

$$K = \frac{m(Ag^+)}{m^\ominus} \times \frac{m(Cl^-)}{m^\ominus}$$

Each AgCl formula unit that dissolves gives rise to one Ag^+ ion and one Cl^- ion, so the molalities of both species can be set equal to s. The expression for K therefore becomes

$$K = \frac{s^2}{m^{\ominus 2}}$$

which rearranges to

$$s = K^{\frac{1}{2}} \times m^\ominus$$

We can calculate the equilibrium constant from

$$RT \ln K = -\Delta_r G^\ominus$$

and obtain $\Delta_r G^\ominus$ from

$$\begin{aligned}
\Delta_r G^\ominus &= \Delta_f G^\ominus(Ag^+, aq) + \Delta_f G^\ominus(Cl^-, aq) - \Delta_f G^\ominus(AgCl, s) \\
&= 77.11 \text{ kJ mol}^{-1} + (-131.23 \text{ kJ mol}^{-1}) - (-109.79 \text{ kJ mol}^{-1}) \\
&= +55.67 \text{ kJ mol}^{-1}
\end{aligned}$$

Therefore, because $RT = 2.4790 \text{ kJ mol}^{-1}$,

$$\ln K = -\frac{55.67 \text{ kJ mol}^{-1}}{2.4790 \text{ kJ mol}^{-1}} = -22.46$$

so $K = 1.77 \times 10^{-10}$ and $s = 1.33 \times 10^{-5} \text{ mol kg}^{-1}$.

Exercise E10.1. Use the data in Table 2.12 to calculate the solubility of Hg_2Cl_2 in water at 25°C. (*Hint.* The solubility equilibrium is $Hg_2Cl_2(s) \rightleftharpoons Hg_2^{2+}(aq) + 2Cl^-(aq)$.) $[3.07 \times 10^{-7} \text{ mol kg}^{-1}]$

The values of $\Delta_f H^\ominus$ and $\Delta_f G^\ominus$ refer to the formation of solutions of ions from the reference states of the parent elements (this point was made in Section 4.9). However, a special problem with ions arises from the fact that solutions of cations cannot be prepared without their accompanying anions. Thus, although the standard enthalpy of an overall reaction such as

$$Ag(s) + \tfrac{1}{2}Cl_2(g) \rightarrow Ag^+(aq) + Cl^-(aq)$$

$$\Delta_r H^\ominus = \Delta_f H^\ominus(Ag^+, aq) + \Delta_f H^\ominus(Cl^-, aq)$$

is meaningful and measurable (and found to be $-61.58\,\text{kJ}\,\text{mol}^{-1}$), the enthalpies of the individual formation reactions

$$\text{Ag(s)} - e^- \rightarrow \text{Ag}^+(\text{aq}) \qquad \text{and } \tfrac{1}{2}\text{Cl}_2(\text{g}) + e^- \rightarrow \text{Cl}^-(\text{aq})$$

are not measurable.

The enthalpies of formation of ions

The problem is solved by *defining* one ion, conventionally the hydrogen ion, to have zero standard enthalpy of formation:

$$\Delta_f H^\ominus(\text{H}^+, \text{aq}) = 0 \text{ at all temperatures}$$

In essence, this definition adjusts the actual values of the enthalpies of formation of ions all by a fixed amount which is chosen so that the standard enthalpy of formation of one of them, $\text{H}^+(\text{aq})$, has the value zero. Then in the reaction

$$\tfrac{1}{2}\text{H}_2(\text{g}) + \tfrac{1}{2}\text{Cl}_2(\text{g}) \rightarrow \text{H}^+(\text{aq}) + \text{Cl}^-(\text{aq}) \qquad \Delta_r H^\ominus = -167.16\,\text{kJ}\,\text{mol}^{-1}$$

we can write

$$\Delta_r H^\ominus = \Delta_f H^\ominus(\text{H}^+, \text{aq}) + \Delta_f H^\ominus(\text{Cl}^-, \text{aq}) = \Delta_f H^\ominus(\text{Cl}^-, \text{aq})$$

and hence identify $\Delta_f H^\ominus(\text{Cl}^-, \text{aq})$ as $-167.16\,\text{kJ}\,\text{mol}^{-1}$. Then, with $\Delta_f H^\ominus(\text{Cl}^-, \text{aq})$ established, we can find the value of $\Delta_f H^\ominus(\text{Ag}^+, \text{aq})$ from

$$\text{Ag(s)} + \tfrac{1}{2}\text{Cl}_2(\text{g}) \rightarrow \text{Ag}^+(\text{aq}) + \text{Cl}^-(\text{aq}) \qquad \Delta_r H^\ominus = -61.58\,\text{kJ}\,\text{mol}^{-1}$$

The procedure may then be extended to other ions, and the resulting values are given in Tables 10.1 and 2.12 of the Data section at the end of this volume.

Table 10.1* Standard thermodynamic functions of ions in solution at 298 K

Ion	$\Delta_f H^\ominus/(\text{kJ mol}^{-1})$	$S^\ominus/(\text{J K}^{-1}\,\text{mol}^{-1})$	$\Delta_f G^\ominus/(\text{kJ mol}^{-1})$
Cl^-	-167.2	$+56.5$	-131.3
Cu^{2+}	$+64.8$	-99.6	$+65.5$
H^+	0	0	0
K^+	-252.4	$+102.5$	-283.3
Na^+	-240.1	$+59.0$	-261.9
PO_4^{3-}	-1277.0	-221.8	-1019.0

* More values are given in the Data section at the end of this volume; see Table 2.12.

Example 10.2 *Calculating a standard enthalpy of formation of an ion*

Calculate the standard enthalpy of formation of $\text{Ag}^+(\text{aq})$ from the information given above.

Method. To do the calculation we need to express the standard enthalpy of reaction in terms of the individual standard enthalpies of formation of the ions, then insert the data, and finally solve the resulting equation for the one unknown.

Answer. The thermochemical equation is

$$Ag(s) + \tfrac{1}{2}Cl_2(g) \rightarrow Ag^+(aq) + Cl^-(aq) \qquad \Delta_r H^\ominus = -61.58 \text{ kJ mol}^{-1}$$

Because

$$\Delta_r H^\ominus = \Delta_f H^\ominus(Ag^+, aq) + \Delta_f H^\ominus(Cl^-, aq) = -61.58 \text{ kJ mol}^{-1}$$

it follows that

$$\begin{aligned}
\Delta_f H^\ominus(Ag^+, aq) &= -61.58 \text{ kJ mol}^{-1} - \Delta_f H^\ominus(Cl^-, aq) \\
&= -61.58 - (-167.16) \text{ kJ mol}^{-1} \\
&= +105.58 \text{ kJ mol}^{-1}
\end{aligned}$$

Exercise E10.2. The standard enthalpy of formation of $AgNO_3(aq)$ is $-99.4 \text{ kJ mol}^{-1}$ at 298 K. Calculate the standard enthalpy of formation of the nitrate ion in water. $\qquad [-205.0 \text{ kJ mol}^{-1}]$

The Gibbs energies of formation of ions

We can use the same procedure to define the standard Gibbs energies of formation and the standard entropies of ions in solution. The standard Gibbs energy of formation of the H^+ ion in water is defined as zero:

$$\Delta_f G^\ominus(H^+, aq) = 0 \text{ at all temperatures}$$

Then in the reaction

$$\tfrac{1}{2}H_2(g) + \tfrac{1}{2}Cl_2(g) \rightarrow H^+(aq) + Cl^-(aq) \qquad \Delta_r G^\ominus = -131.23 \text{ kJ mol}^{-1}$$

we can write

$$\Delta_r G^\ominus = \Delta_f G^\ominus(H^+, aq) + \Delta_f G^\ominus(Cl^-, aq) = \Delta_f G^\ominus(Cl^-, aq)$$

and hence identify $\Delta_f G^\ominus(Cl^-, aq)$ as $-131.23 \text{ kJ mol}^{-1}$ just as we did for its enthalpy of formation. Then, with $\Delta_f G^\ominus(Cl^-, aq)$ established, we can find the value of $\Delta_f G^\ominus(Ag^+, aq)$ from

$$Ag(s) + \tfrac{1}{2}Cl_2(g) \rightarrow Ag^+(aq) + Cl^-(aq) \qquad \Delta_r G^\ominus = -54.12 \text{ kJ mol}^{-1}$$

which leads to $\Delta_f G^\ominus(Ag^+, aq) = +77.11 \text{ kJ mol}^{-1}$, the value used in Example 10.1. All the values in Tables 10.1 and 2.12 are calculated in the same way. Later in the chapter we shall see how to measure the standard reaction Gibbs energies on which they are based.

Contributions to the Gibbs energy of formation

The factors responsible for the magnitude of the Gibbs energy of formation of an ion in solution can be identified by analysing it in terms of a thermodynamic cycle. As an illustration, we consider the reasons for the difference between the standard Gibbs energies of formation of Cl^- and Br^- in water, which are -131 and -104 kJ mol^{-1}, respectively. We do so by treating their formation in the reaction

$$\tfrac{1}{2}H_2(g) + \tfrac{1}{2}X_2(g \text{ or } l) \rightarrow H^+(aq) + X^-(aq) \qquad (g \text{ for } Cl_2, \ l \text{ for } Br_2)$$

as the outcome of the following sequence (with values taken from Table 2.12):[1]

		$\Delta_r G^{\ominus}/(\text{kJ mol}^{-1})$	
		X = Cl	X = Br
Dissociation of H_2	$\frac{1}{2}H_2(g) \rightarrow H(g)$	+203	+203
Ionization of H	$H(g) \rightarrow H^+(g) + e^-(g)$	+1318	+1318
Hydration of H^+	$H^+(g) \rightarrow H^+(aq)$	x	x
Formation of X	$\frac{1}{2}X_2(g \text{ or } l) \rightarrow X(g)$	+106	+82
Electron gain by X	$X(g) + e^-(g) \rightarrow X^-(g)$	−349	−325
Hydration of X^-	$X^-(g) \rightarrow X^-(aq)$	y	y'
Overall:	$\frac{1}{2}H_2(g) + \frac{1}{2}X_2(g \text{ or } l) \rightarrow H^+(aq) + X^-(aq)$	$\Delta_f G^{\ominus}(Cl^-)$	$\Delta_f G^{\ominus}(Br^-)$

According to the convention we have adopted, the overall standard Gibbs energy is equal to $\Delta_f G^{\ominus}(X^-, aq)$. Its value is the sum of the standard Gibbs energies of the individual steps:

$$\Delta_f G^{\ominus}(Cl^-, aq) = x + y + 1278 \text{ kJ mol}^{-1}$$
$$\Delta_f G^{\ominus}(Br^-, aq) = x + y' + 1278 \text{ kJ mol}^{-1}$$

(It is coincidental that the number 1278 appears for both ions for, by chance, the difference in electron affinities is cancelled by the difference in dissociation enthalpies.) An important point to note is that the value of $\Delta_f G^{\ominus}(X^-, aq)$ is not determined by the properties of X alone but includes contributions from the dissociation, ionization, and hydration of hydrogen.

The difference between the two values is due (in this example) to the difference in the hydration of the ions:

$$\Delta_f^{\ominus} G(Cl^-, aq) - \Delta_f^{\ominus} G(Br^-, aq) = y - y'$$

The two unknown quantities y and y' are the **Gibbs energies of hydration** $\Delta_{hyd} G^{\ominus}$ of the ions, and in general (for nonaqueous solvents) their **Gibbs energies of solvation** $\Delta_{solv} G^{\ominus}$. The latter is the standard Gibbs energy for

$$M^+(g) \rightarrow M^+(\text{solution}) \qquad \Delta_{solv} G^{\ominus}$$

Gibbs energies of solvation may be estimated from an equation derived by Max Born, who identified $\Delta_{solv} G^{\ominus}$ with the electrical work of transferring an ion from a vacuum into the solvent treated as a continuous dielectric of relative permittivity ε_r (Table 10.2). The resulting **Born equation** is

$$\Delta_{solv} G^{\ominus} = -\frac{z_i^2 e^2 N_A}{8\pi\varepsilon_0 r_i}\left(1 - \frac{1}{\varepsilon_r}\right) \tag{1}$$

where z_i is the charge number of the ion (the number of charges it has) and r_i its radius (N_A is Avogadro's constant). Note that $\Delta_{solv} G^{\ominus} < 0$, and

Table 10.2* Relative permittivities (dielectric constants) at 298 K

	ε_r
Ammonia	16.9
	22.4 (at −33°C)
Benzene	2.274
Ethanol	24.30
Water	78.54

* More values are given in the Data section.

1 The standard Gibbs energies of formation of the gas-phase ions are unknown. We have therefore used their enthalpies of formation, and have assumed that the entropy of ionization of H is largely cancelled by the entropy of electron gain of Cl. Partly for this reason, we have reduced the numbers of significant figures in the calculation from those available in Table 2.12.

that it is strongly negative for small, highly charged ions in media of high relative permittivity.

JUSTIFICATION

The strategy of the calculation is to identify the Gibbs energy of solvation with the work of transferring an ion from a vacuum into the solvent. That work is calculated by taking the difference of the work of charging an ion when it is in the solution and the work of charging the same ion when it is in a vacuum. The derivation considers an ion to be a sphere of radius r_i immersed in a medium of permittivity ε. When the charge of the sphere is q, the electric potential at its surface is

$$\phi = \frac{q}{4\pi\varepsilon r_i}$$

(The electrical concepts required in this chapter are reviewed in *Further information 3*.) The work of bringing up a charge dq to the sphere is $\phi \, dq$. Therefore, the total work of charging the sphere from 0 to $z_i e$ is

$$w = \int_0^{z_i e} \phi \, dq = \frac{1}{4\pi\varepsilon r_i} \int_0^{z_i e} q \, dq = \frac{1}{4\pi\varepsilon r_i} \times \frac{(z_i e)^2}{2}$$

This electrical work of charging, when multiplied by the Avogadro constant, can be identified with the standard molar Gibbs energy for charging the ions.

The work of charging an ion in a vacuum is obtained by setting $\varepsilon = \varepsilon_0$, the vacuum permittivity. The corresponding value for charging the ion in a medium is obtained by setting $\varepsilon = \varepsilon_r \varepsilon_0$, where ε_r is the relative permittivity of the medium. It follows that the change in standard molar Gibbs energy that accompanies the transfer of ions from a vacuum to a solvent is the difference of these two quantities, and is

$$\Delta_{solv}G^{\ominus} = \frac{z_i^2 e^2 N_A}{8\pi\varepsilon_r \varepsilon_0 r_i} - \frac{z_i^2 e^2 N_A}{8\pi\varepsilon_0 r_i}$$

which can easily be rearranged into eqn 1.

Example 10.3 *Accounting for the difference in standard Gibbs energies of formation*

Account for the difference in the values of $\Delta_f G^{\ominus}$ for Cl^- and Br^- in water at 25°C given that their radii are 181 pm and 196 pm, respectively.

Method. The simplest approach is to consider the implication of the Born equation by noting that the standard Gibbs energy of solvation is proportional to z^2/r. For a quantitative assessment, use the Born equation to estimate the difference in solvation energies of the two ions, and identify that difference with the difference in their standard Gibbs energies of formation.

Answer. We know that Cl^- has the more negative value (by $27\,kJ\,mol^{-1}$), which suggests that the standard Gibbs energy of hydration is greater than for Br^-, which is consistent with the smaller radius of Cl^-. For water at 25°C, when $\varepsilon_r = 78.54$, the Born equation is

$$\Delta_{hyd}G^{\ominus} = -\frac{z_i^2}{r_i/pm} \times (6.86 \times 10^4\,kJ\,mol^{-1})$$

Hence, because $z_i^2 = 1$ in both cases, we find $\Delta_{hyd}G^{\ominus}(Cl^-) = -379\,kJ\,mol^{-1}$ and $\Delta_{hyd}G^{\ominus}(Br^-) = -350\,kJ\,mol^{-1}$. Therefore

$$\Delta_f G^{\ominus}(Cl^-, aq) - \Delta_f G^{\ominus}(Br^-, aq) = -29\,kJ\,mol^{-1}$$

in good agreement with the experimental difference ($-27\,kJ\,mol^{-1}$).

Comment. The standard Gibbs energy of formation of an ion is a subtle balance of many contributions, as the thermodynamic cycle shows, and it is possible to predict trends only by taking all of them into account.

Exercise E10.3 Estimate the value of $\Delta_f G^{\ominus}(I^-, aq)$ from the value for Cl^- given the radius of I^- as 220 pm and other data as in Table 2.12.

$$[-67\,kJ\,mol^{-1}\ (-52\,kJ\,mol^{-1}\ actual)]$$

The entropies of ions in solution

Although the partial molar entropies of the solute in an electrolyte can be measured, there is no experimental way of ascribing a part of that entropy to the cations and a part to the anions. Therefore, yet again, we are forced to *define* the partial molar entropy of one species and set up a table of values for other ions on that basis. The entropies of ions in solution are reported on a scale in which the standard entropy of H^+ ions in water is taken as zero:

$$S^{\ominus}(H^+, aq) = 0 \text{ at all temperatures}$$

Some values based on this choice are listed in Tables 10.1 and 2.12 in the Data section at the end of this volume.

Because the entropies of ions in water are values relative to that of the hydrogen ion in water, they may be either positive or negative. A positive entropy means that an ion has a higher partial molar entropy than H^+ in water and a negative entropy means that the ion has a lower partial molar entropy than H^+ in water. For instance, the entropy of $Cl^-(aq)$ is $+57\,J\,K^{-1}\,mol^{-1}$ and that of $Mg^{2+}(aq)$ is $-128\,J\,K^{-1}\,mol^{-1}$. Partial molar ionic entropies vary as expected on the basis that they are related by the degree to which the ions order the water molecules around them in the solution. Small, highly charged ions induce local structure in the surrounding water, and the disorder of the solution is decreased more than in the case of large, singly charged ions. The absolute, Third-Law partial molar entropy of the proton in water can be estimated by proposing a model of the structure it induces, and there is some agreement on the value $-21\,J\,K^{-1}\,mol^{-1}$. The negative value indicates that the proton induces order in the solvent.

10.2 Ion activities

For dilute solutions of non-electrolytes (Chapter 7, as summarized in Table 7.3) it is generally safe to make the approximation that solute activities can be replaced by their molalities (in the sense $a \approx m/m^{\ominus}$, with $m^{\ominus} = 1 \, \text{mol kg}^{-1}$). However, in ionic solutions, the interactions between ions are so strong that this approximation is valid only in very dilute solutions (less than $10^{-3} \, \text{mol kg}^{-1}$ total ion concentration) and, in precise work, activities themselves must be used. For example, we have seen that the thermodynamic equilibrium constant K is related to the equilibrium constant in terms of molalities K_m (or the corresponding constants in terms of concentrations or mole fractions) by

$$K = K_{\gamma} K_m \tag{2}$$

(this is eqn 9.9). Therefore, to interpret K correctly we must know how the activity coefficients depend on the molality of the solution.

The definition of activity

We saw in Section 7.7 that the chemical potential of a solute in a real solution is related to its activity a by

$$\mu = \mu^{\ominus} + RT \ln a \tag{3}$$

where the standard state is a hypothetical solution with molality $m^{\ominus}$ in which the ions are behaving ideally. The activity is related to the molality m by

$$a = \gamma \times \frac{m}{m^{\ominus}} \tag{4}$$

where the activity coefficient γ depends on the composition, molality, and temperature of the solution. As the solution approaches ideality (in the sense of obeying Henry's law) at low molalities, the activity coefficient tends towards 1:

$$\gamma \to 1 \text{ and } a \to m/m^{\ominus} \qquad \text{as } m \to 0$$

Because all deviations from ideality are carried in the activity coefficient, the chemical potential can be written

$$\mu = \mu^{\ominus} + RT \ln \frac{m}{m^{\ominus}} + RT \ln \gamma$$

$$= \mu^{\text{id}} + RT \ln \gamma$$

where μ^{id} is the chemical potential of the ideal–dilute solution of the same molality.

Mean activity coefficients

If the chemical potential of a univalent cation M^+ is denoted μ_+ and that of a univalent anion X^- is denoted μ_-, then the total Gibbs energy of the ions in the electrically neutral solution is the sum of these partial molar quantities. The Gibbs energy of an ideal solution is

$$G^{\text{id}} = \mu_+^{\text{id}} + \mu_-^{\text{id}}$$

However, for a real solution of M^+ and X^- of the same molality,

$$G = \mu_+ + \mu_- = \mu_+^{id} + \mu_-^{id} + RT \ln \gamma_+ + RT \ln \gamma_-$$
$$= G^{id} + RT \ln \gamma_+ \gamma_-$$

All the deviations from ideality are contained in the last term.

There is no *experimental* way of separating the product $\gamma_+ \gamma_-$ into contributions from the cations and the anions. The best we can do experimentally is to assign responsibility for the non-ideality equally to both kinds of ion. Therefore, for a 1,1-electrolyte, we introduce the **mean activity coefficient**

$$\gamma_\pm = (\gamma_+ \gamma_-)^{\frac{1}{2}}$$

and express the individual chemical potentials of the ions as

$$\mu_+ = \mu_+^{id} + RT \ln \gamma_\pm \qquad \mu_- = \mu_-^{id} + RT \ln \gamma_\pm$$

The sum of these two chemical potentials is the same as before, but now the non-ideality is shared equally.

This approach can be generalized to the case of a compound $M_p X_q$ that dissolves to give a solution of p cations and q anions. The total Gibbs energy of the ions is the sum of the partial molar Gibbs energies:

$$G = p\mu_+ + q\mu_- = G^{id} + pRT \ln \gamma_+ + qRT \ln \gamma_-$$

If we introduce the mean activity coefficient

$$\gamma_\pm = (\gamma_+^p \gamma_-^q)^{1/s} \qquad s = p + q \tag{5a}$$

and write the chemical potential of each ion as

$$\mu_i = \mu_i^{id} + RT \ln \gamma_\pm \tag{5b}$$

we get the same expression as above for G when we write

$$G = p\mu_+ + q\mu_- \tag{5c}$$

However, now both types of ion share equal responsibility for the non-ideality.

The Debye–Hückel limiting law

The long range and strength of the Coulombic interaction between ions means that it is likely to be primarily responsible for the departures from ideality in ionic solutions and to dominate all the other contributions to non-ideality. This domination is the basis of the **Debye–Hückel theory** of ionic solutions, which was devised by Peter Debye and Erich Hückel in 1923. We give here a qualitative account of the theory and its principal conclusions. The calculations are outlined in *Further information 4* at the end of this volume.

Oppositely charged ions attract each other. As a result, cations and anions are not uniformly distributed in solutions: anions are more likely to be found near cations, and vice versa (Fig. 10.1). Overall the solution is electrically neutral, but near any given ion there is an excess of **counter-ions**, the ions of opposite charge. Averaged over time, counter-ions are more likely to be found by any given ion. This time-averaged,

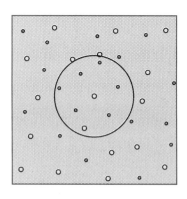

10.1 The picture underlying the Debye–Hückel theory is of a tendency for anions to be found around cations, and of cations to be found around anions (one such local clustering region is shown by the circle). The ions are in ceaseless motion, and the diagram represents the time average of their motion.

spherical haze, in which counter-ions outnumber ions of the same charge as the central ion, has a net charge equal in magnitude but opposite in sign to that on the central ion and is called its **ionic atmosphere**. The energy, and therefore the chemical potential, of any given central ion is lowered as a result of its electrostatic interaction with its ionic atmosphere. This lowering of energy appears as the difference between the Gibbs energy G and the ideal value G^{id} of the solution, and hence can be identified with $RT \ln \gamma_{\pm}$.

As shown in *Further information 4*, the model leads to the result that, at very low concentrations, the activity coefficient can be calculated from the **Debye–Hückel limiting law**

$$\log \gamma_{\pm} = -|z_+ z_-| A(I/m^{\ominus})^{\frac{1}{2}} \tag{6}$$

where $A = 0.509$ for an aqueous solution at 25°C (in general, A depends on the relative permittivity and the temperature) and I is the **ionic strength** of the solution

$$I = \frac{1}{2} \sum_i z_i^2 m_i \tag{7a}$$

Table 10.3 Ionic strength and molality, $I = k \times m$

	X^-	X^{2-}	X^{3-}	X^{4-}
M^+	1	3	6	10
M^{2+}	3	4	15	12
M^{3+}	6	15	9	42
M^{4+}	10	12	42	16

For example, the ionic strength of an M_2X_3 solution of molality m, which is understood to give M^{3+} and X^{2-} ions in solution, is $15m$.

In this expression and eqn 6, z_i is the charge number of an ion i (positive for cations and negative for anions) and m_i is its molality. The ionic strength occurs widely wherever ionic solutions are discussed, as we shall see. The sum extends over all the ions present in the solution. For solutions consisting of two types of ion at molalities m_+ and m_-

$$I = \frac{1}{2}(m_+ z_+^2 + m_- z_-^2) \tag{7b}$$

The ionic strength emphasizes the charges of the ions because the charge numbers occur as their squares. Table 10.3 summarizes the relation of ionic strength and molality in an easily usable form.

Example 10.4 *Estimating the mean ionic activity coefficient*

Estimate the mean activity coefficient of $0.0050 \, \text{mol kg}^{-1}$ KCl(aq) at 25°C.

Method. First, the ionic strength of the solution should be evaluated, and then the mean activity coefficient should be estimated from eqn 6.

Answer. For a completely dissociated (1,1)-electrolyte, $z_+ = 1$ and $z_- = -1$; it follows that the ionic strength is

$$I = \frac{1}{2}(m_+ + m_-) = m$$

where m is the molality of the solution (and $m_+ = m_- = m$). Then, from eqn 6,

$$\log \gamma_{\pm} = -0.509 \times (0.00500)^{\frac{1}{2}} = -0.0360$$

Hence, $\gamma_{\pm} = 0.920$.

Comment. The experimental value is 0.927.

Exercise E10.4. Calculate the ionic strength and the mean activity coefficient of $0.00100 \, \text{mol kg}^{-1}$ $CaCl_2(aq)$ at 25°C.

$$[0.00300 \, \text{mol kg}^{-1}, 0.880]$$

The name 'limiting law' is applied to eqn 6 because ionic solutions of moderate molalities may have activity coefficients that differ from the values given by this expression, yet all solutions are expected to conform in the limit of arbitrarily low molalities. Some experimental values of activity coefficients for salts of various valence types are listed in Table 10.4. Figure 10.2 shows some of these values plotted against $I^{\frac{1}{2}}$ and compares them with the theoretical straight lines calculated from eqn 6. The agreement at very low molalities (less than 0.01 to 0.001 mol kg^{-1}, depending on charge type) is impressive, and convincing evidence in support of the model. Nevertheless, the departures from the theoretical curves above these molalities are large, and show that the approximations are valid only at very low concentrations.

Table 10.4* Mean activity coefficients in water at 298 K

$m/(\text{mol kg}^{-1})$	KCl	CaCl$_2$
0.001	0.966	0.888
0.01	0.902	0.732
0.1	0.770	0.524
1.0	0.607	0.725

* More values are given in the Data section.

Example 10.5 *Using the limiting law to analyse an equilibrium*

The solubility of silver chloride in water at 25°C is $1.274 \times 10^{-5} \, \text{mol kg}^{-1}$. Calculate (a) the standard reaction Gibbs energy for $AgCl(s) \rightarrow Ag^+(aq) + Cl^-(aq)$ and (b) the solubility of silver chloride in $0.020 \, \text{mol kg}^{-1}$ $K_2SO_4(aq)$.

Method. The thermodynamic equilibrium constant is expressed in terms of activities; these may be obtained from the molalities of the ions (each of which is equal to s multiplied by the mean activity coefficient, which in turn is obtained from eqn 6). Then the standard reaction Gibbs energy is obtained by using eqn 9.7, such that $\Delta_r G^{\ominus} = -RT \ln K$. For part (b) calculate the ionic strength of the solution (the contribution of the AgCl can be ignored as its solubility is so low), and interpret the equilibrium constant obtained in part (a) in terms of a new solubility by using the mean activity coefficient calculated for this solution. Because the K^+ and SO_4^{2-} ions produce a favourable ionic atmosphere around the Ag^+ and Cl^- ions, and thus lower their Gibbs energy, we can expect the solubility of AgCl to be greater in the second solution than in pure water.

Answer. (a) The equilibrium constant for the dissolution of AgCl is

$$K = a(Ag^+)a(Cl^-) = \gamma_{\pm}^2 (s/m^{\ominus})^2$$

The ionic strength of the solution (a 1:1 electrolyte) is

$$I = m = 1.274 \times 10^{-5} \, \text{mol kg}^{-1}$$

It follows from eqn 6 that

$$\log \gamma_{\pm} = -0.509 \times (1.274 \times 10^{-5})^{\frac{1}{2}} = -1.82 \times 10^{-3}$$

which implies that $\gamma_{\pm} = 0.996$. Therefore,

$$K = 0.996^2 \times (1.274 \times 10^{-5})^2 = 1.61 \times 10^{-10}$$

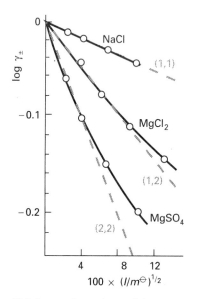

10.2 An experimental test of the Debye–Hückel limiting law. Although there are marked deviations for moderate ionic strengths, the limiting slopes as $I \rightarrow 0$ are in good agreement with the theory, so it can be used for extrapolating data to very low molalities.

The standard reaction Gibbs energy is therefore

$$\Delta_r G^\ominus = -RT \ln K = -(2.48 \text{ kJ mol}^{-1}) \times \ln (1.61 \times 10^{-10})$$
$$= +55.9 \text{ kJ mol}^{-1}$$

(b) The ionic strength of the potassium sulfate solution is

$$I = \tfrac{1}{2}(1^2 \times 2 \times 0.020 + 2^2 \times 0.020) \text{ mol kg}^{-1} = 0.060 \text{ mol kg}^{-1}$$

The mean activity coefficient of the Ag^+ and Cl^- ions in such a solution is obtained from

$$\log \gamma_\pm = -0.509 \times (0.060)^{\frac{1}{2}} = -0.12, \quad \text{so} \quad \gamma_\pm = 0.76$$

The solubility of AgCl in this solution is therefore

$$\frac{s}{m^\ominus} = \frac{(K)^{\frac{1}{2}}}{\gamma_\pm} = \frac{(1.61 \times 10^{-10})^{\frac{1}{2}}}{0.76} = 1.7 \times 10^{-5}$$

The new solubility is therefore $1.7 \times 10^{-5} \text{ mol kg}^{-1}$.

Comment. The contribution of silver chloride to the ionic strength is only just negligible. For a more accurate answer, the ionic strength of the solution should be recalculated using this first estimate, including the contribution from the AgCl in solution, and a new value of s calculated. The cycle is then continued until the results are self-consistent.

Exercise E10.5. The acidity constant of acetic acid is 1.75×10^{-5} at 25°C. Use the limiting law to estimate the percentage deprotonation of the acid when its molality is $0.100 \text{ mol kg}^{-1}$. [1.4 per cent]

When the ionic strength of the solution is too high for the limiting law to be valid, it is found that the activity coefficient may be estimated from the **extended Debye–Hückel law**:

$$\log \gamma_\pm = -\frac{A |z_+ z_-| (I/m^\ominus)^{\frac{1}{2}}}{1 + B(I/m^\ominus)^{\frac{1}{2}}} \tag{8}$$

where B is another dimensionless constant. Although B can be interpreted as a measure of the closest approach of the ions, it is best regarded as an adjustable empirical parameter. A curve drawn in this way is shown in Fig 10.3. It is clear that eqn 8 accounts for some activity coefficients over a moderate range of dilute solutions (up to about 0.1 mol kg^{-1}); nevertheless it remains very poor near 1 mol kg^{-1}. Current theories of activity coefficients take an indirect route. They set up a theory for the change in activity of the *solvent*, and then use the Gibbs–Duhem equation (eqn 7.6) which relates the chemical potentials of the components of a solution:

$$n_A \, d\mu_A + n_B \, d\mu_B + \ldots = 0$$

to arrive at an estimate of the activity coefficient of the solute. The results are reasonably reliable for solutions with molalities greater than about 0.1 mol kg^{-1}.

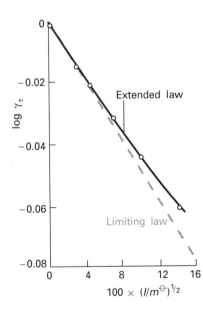

10.3 The extended Debye–Hückel law gives agreement with experiment over a wider range of molalities (as shown here for a 1,1-electrolyte), but fails at higher molalities.

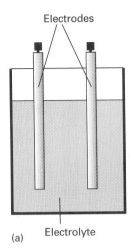

Electrodes

Electrolyte

(a)

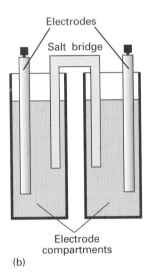

Electrodes

Salt bridge

Electrode
compartments

(b)

10.4 Two basic types of cell: (a) a cell
with a single electrode compartment and
a shared electrolyte, (b) two separate
electrode compartments joined by a salt
bridge.

ELECTROCHEMICAL CELLS

Now we turn to the investigation of reactions in solution in terms of
electrical measurements. The basic apparatus is an **electrochemical cell**
(Fig. 10.4). A cell consists of two **electrodes**, or metallic conductors, in
contact with an **electrolyte**, an ionic conductor (which may be a
solution, a liquid, or a solid). An electrode and its electrolyte comprise
an **electrode compartment**. The two electrodes may share the same
compartment. If the electrolytes are different, the two compartments may
be joined by a **salt bridge**, which is a concentrated electrolyte solution in
agar jelly that completes the electrical circuit and enables the cell to
function.

There are two principal types of electrochemical cell. A **galvanic cell**
is an electrochemical cell that produces electricity as a result of the
spontaneous reaction occurring inside it. An **electrolytic cell** is an
electrochemical cell in which a non-spontaneous reaction is driven by an
external source of current. (Electrolytic cells have a common electrolyte
and no salt bridge.) We consider the thermodynamic aspects of electrochemical cells in this section. The kinetic problems, which govern the
current output of galvanic cells and play an important role in the
functioning of electrolytic cells, are considered in Chapter 29.

10.3 Half-reactions and electrodes

A **redox reaction** is a reaction in which there is a transfer of electrons
from one species to another. The **reducing agent** (or 'reductant') is the
electron donor and the **oxidizing agent** (or 'oxidant') is the electron
acceptor. The electron transfer may be accompanied by other events,
such as atom or ion transfer, but the key effect is electron transfer and
hence a change in oxidation number of an element. Examples of redox
reactions include the reaction

$$2Mg(s) + O_2(g) \rightarrow 2MgO(s)$$

in which magnesium is the reducing agent and oxygen the oxidizing
agent, the reaction

$$CuO(s) + H_2(g) \rightarrow Cu(s) + H_2O(g)$$

in which hydrogen is the reducing agent and copper(II) oxide is the
oxidizing agent, and the reaction

$$Cu^{2+}(aq) + Zn(s) \rightarrow Cu(s) + Zn^{2+}(aq)$$

in which a Cu^{2+} ion is the oxidizing agent and zinc metal is the reducing
agent.

Half-reactions

Any redox reaction may be expressed in terms of two **half-reactions**,
which are conceptual reactions showing the loss and gain of electrons.
For example, the reduction of Cu^{2+} ions by zinc can be expressed as the

sum of the following two half-reactions:

Reduction of Cu^{2+}: $Cu^{2+}(aq) + 2e^- \rightarrow Cu(s)$

Oxidation of Zn: $Zn(s) \rightarrow Zn^{2+}(aq) + 2e^-$

Overall (sum): $Cu^{2+}(aq) + Zn(s) \rightarrow Cu(s) + Zn^{2+}(aq)$

It is common practice, however, to write all half-reactions as reductions; then the overall reaction is the *difference* of the two:

Reduction of Cu^{2+}: $Cu^{2+}(aq) + 2e^- \rightarrow Cu(s)$

Reduction of Zn^{2+}: $Zn^{2+}(aq) + 2e^- \rightarrow Zn(s)$

Overall (difference): $Cu^{2+}(aq) + Zn(s) \rightarrow Cu(s) + Zn^{2+}(aq)$

The reduced and oxidized substances in a half-reaction form a **redox couple**, denoted Ox/Red. Thus, the redox couples mentioned so far are Cu^{2+}/Cu and Zn^{2+}/Zn. In general we shall write a couple as Ox/Red and the corresponding reduction half-reaction as

$$Ox + \nu e^- \rightarrow Red \tag{9}$$

We shall often find it useful to express the composition of an electrode compartment in terms of the reaction quotient Q for the half-reaction. This quotient is defined like the reaction quotient for the overall reaction, but the electrons are ignored. Thus, for the two half-reactions given above, we would write

$$Cu^{2+}(aq) + 2e^- \rightarrow Cu(s) \qquad Q = \frac{1}{a(Cu^{2+})}$$

$$Zn^{2+}(aq) + 2e^- \rightarrow Zn(s) \qquad Q = \frac{1}{a(Zn^{2+})}$$

In each case we have used the fact that the pure metal (the standard state of the element) has unit activity (recall Table 7.3).

The overall reaction need not be a redox reaction for it to be expressed in terms of half-reactions. For instance, the expansion of a gas

$$H_2(g, p_i) \rightarrow H_2(g, p_f)$$

is not a redox reaction but it can be expressed as the difference of two reductions:

$$2H^+(aq) + 2e^- \rightarrow H_2(g, p_f)$$
$$2H^+(aq) + 2e^- \rightarrow H_2(g, p_i)$$

The two couples in this case are both H^+/H_2. Another important example of an overall reaction that can be expressed as the difference of half-reactions, but which is not itself a redox reaction, is the process of dissolving.

Example 10.6 *Expressing a reaction in terms of half-reactions*

Express the dissolution of silver chloride in water as the difference of two reduction half-reactions.

Method. First, write the overall chemical equation. Then select one of the reactants, and write a half-reaction in which it is reduced to one of the products. Next, subtract that half reaction from the overall reaction to identify the second half-reaction. Finally, write the second half-reaction as a reduction.

Answer. The chemical equation of the overall reaction is

$$AgCl(s) \rightarrow Ag^+(aq) + Cl^-(aq)$$

We select as one half-reaction the reduction of AgCl (more precisely, the reduction of the Ag(I) in AgCl to Ag(0)):

$$AgCl(s) + e^- \rightarrow Ag(s) + Cl^-(aq)$$

Subtraction of this equation from the overall reaction leaves

$$-e^- \rightarrow Ag^+(aq) - Ag(s)$$

which rearranges to

$$Ag^+(aq) + e^- \rightarrow Ag(s)$$

Comment. In the dissolution of AgCl there is no net change of oxidation number, so it is not a redox reaction.

Exercise E10.6. Express the formation of H_2O from H_2 and O_2 in acidic solution (a true redox reaction) as the difference of two reduction half-reactions.

$$[4H^+(aq) + 4e^- \rightarrow 2H_2(g), \ O_2(g) + 4H^+(aq) + 4e^- \rightarrow 2H_2O(l)]$$

Reactions at electrodes

In an electrochemical cell the reduction and oxidation processes responsible for the overall reaction are separated in space: one half-reaction takes place in one electrode compartment and the other takes place in the other compartment. As the reaction proceeds, the electrons released in the half-reaction

$$Red_1 \rightarrow Ox_1 + \nu e^-$$

in one compartment travel through the external circuit and re-enter the cell through the other electrode. There they are used to reduce the oxidized member of the couple in that compartment:

$$Ox_2 + \nu e^- \rightarrow Red_2$$

The electrode at which oxidation occurs is called the **anode**; the electrode at which reduction occurs is called the **cathode**:

Anode reaction (oxidation): $Red_1 \rightarrow Ox_1 + \nu e^-$

Cathode reaction (reduction): $Ox_2 + \nu e^- \rightarrow Red_2$

In a galvanic cell, the cathode has a higher potential than the anode: the species undergoing reduction, Ox_2, withdraws electrons from its electrode (the cathode, Fig. 10.5), so leaving a relative positive charge on

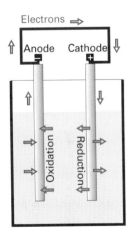

10.5 When a spontaneous reaction takes place in a galvanic cell, electrons are deposited in one electrode (the site of oxidation, the anode) and collected from another (the site of reduction, the cathode), so there is a net flow of current that can be used to do work.

it (corresponding to a high potential). At the anode, oxidation results in the transfer of electrons to the electrode, so giving it a relative negative charge (corresponding to a low potential). In an electrolytic cell, the anode is also the location of oxidation (by definition), but now electrons must be withdrawn from the species in that compartment since that process does not occur spontaneously, and at the cathode there must be a supply of electrons to drive the reduction. Therefore, in an electrolytic cell the anode must be made relatively positive to the cathode.

Varieties of electrodes

A **metal/metal-ion electrode** consists of a metal in contact with a solution of one of its salts, such as copper in contact with an aqueous solution of Cu^{2+} ions (as copper(II) sulfate). This type of electrode is denoted $M\,|\,M^+(aq)$, where M is the metal and the vertical bar denotes an interface between two phases. An example is $Cu\,|\,Cu^{2+}(aq)$. Note that the electrode description runs in the order Red | Ox, which is opposite to the order in which the couple is written. In a **gas electrode** (Fig. 10.6), a gas is in equilibrium with a solution of its ions in the presence of an inert metal. The inert metal (which is often platinum) acts as a source or sink of electrons, but takes no other part in the reaction other than acting as a catalyst for it. One example is the **hydrogen electrode**, in which hydrogen is bubbled through a solution of hydrogen ions and the redox couple is H^+/H_2. This electrode is denoted $Pt\,|\,H_2(g)\,|\,H^+(aq)$.

The hydrogen electrode (like any electrode) may be either a cathode or an anode, depending on the other electrode in the cell and the spontaneous direction of the overall reaction. The reaction at the electrode when it is acting as a cathode is

$$2H^+(aq) + 2e^- \rightarrow H_2(g) \qquad Q = \frac{f_{H_2}/p^{\ominus}}{a(H^+)^2}$$

where f denotes the fugacity (Section 5.6). In elementary work, the fugacity is replaced by the pressure p.

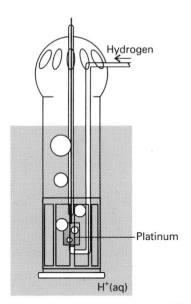

10.6 In a gas electrode the gas is bubbled over the inert (but catalytic) metal surface, and the equilibrium is between it and its ions (e.g. between H_2 and H^+ or between Cl_2 and Cl^-).

Example 10.7 *Writing the half-reaction and reaction quotient for a gas electrode*

Write the half-reaction and the reaction quotient for the reduction of oxygen to water in dilute acidic solution.

Method. The first step is a simple balancing exercise: use H^+ ions to balance the H atoms and electrons to balance the charge. For the reaction quotient, include activities of products in the numerator and reactants (other than electrons) in the denominator.

Answer. The reduction of O_2 in acidic solution produces H_2O according to the half-reaction

$$O_2(g) + 4H^+(aq) + 4e^- \rightarrow 2H_2O(l)$$

The reaction quotient for the half-reaction is therefore

$$Q = \frac{a(H_2O)^2}{a(H^+)^4 \times \dfrac{f_{O_2}}{p^{\ominus}}} \approx \frac{1}{a(H^+)^4(p_{O_2}/p^{\ominus})}$$

The approximations used in the second step are that the activity of water is 1 (because the solution is dilute and the water almost pure) and the oxygen behaves like a perfect gas.

Exercise E10.7. Write the half-reaction and the reaction quotient for a chlorine gas electrode. $[Cl_2(g) + 2e^- \rightarrow 2Cl^-(aq), \ Q = a(Cl^-)^2 p^{\ominus}/f_{Cl_2}]$

A **metal/insoluble-salt electrode** consists of a metal M covered by a porous layer of insoluble salt MX with the whole immersed in a solution containing X^- ions. The electrode is denoted $M \mid MX \mid X^-$; an example is the **silver/silver chloride electrode** $Ag \mid AgCl \mid Cl^-$. The reduction half-reaction for the electrode is typically

$$MX(s) + e^- \rightarrow M(s) + X^-(aq) \qquad Q = a(X^-)$$

The half-reaction for the silver/silver chloride electrode is

$$AgCl(s) + e^- \rightarrow Ag(s) + Cl^-(aq) \qquad Q = a(Cl^-)$$

Example 10.8 *Writing the half-reaction for a metal/insoluble-salt electrode*

Write the half-reaction and the reaction quotient for the lead/lead sulfate electrode of the lead–acid battery.

Method. First, identify the species that is undergoing reduction and its product and then write the half-reaction: use H_2O to balance the O atoms (if necessary), H^+ to balance the H atoms, and, finally, electrons to balance the charge. Write the half-reaction as a reduction. The reaction quotient is then constructed in the same way as in Example 10.7.

Answer. The electrode is $Pb \mid PbSO_4(s) \mid HSO_4^-(aq), \ H^+(aq)$, in which Pb(II) is reduced to metallic lead. The reduction half-reaction is therefore

$$PbSO_4(s) + H^+(aq) + 2e^- \rightarrow Pb(s) + HSO_4^-(aq)$$

and the reaction quotient, noting that the two pure solids have unit activity, is $Q = a(HSO_4^-)/a(H^+)$.

Exercise E10.8. Write the half-reaction and the reaction quotient for the calomel electrode $Hg(l) \mid Hg_2Cl_2(s) \mid Cl^-(aq)$.

$$[Hg_2Cl_2(s) + 2e^- \rightarrow 2Hg(l) + 2Cl^-(aq), \ Q = a(Cl^-)^2]$$

All electrodes depend on oxidation and reduction, but the term **oxidation–reduction electrode**, or **redox electrode**, is normally reserved for the case in which a species exists in solution in two oxidation states. The equilibrium is

$$Ox + ve^- \rightarrow Red \qquad Q = \frac{a(Red)}{a(Ox)}$$

A redox electrode is denoted $M \mid Red, Ox$, where M is an inert metallic conductor that makes electrical contact with the solution. An example is

$$Pt \mid Fe^{2+}(aq), Fe^{3+}(aq) \qquad Fe^{3+}(aq) + e^- \rightarrow Fe^{2+}(aq)$$

$$Q = \frac{a(Fe^{2+})}{a(Fe^{3+})}$$

10.4 Varieties of cells

The simplest type of cell has a single electrolyte common to both electrodes (Fig. 10.7). In some cases it is necessary to immerse the electrodes in different electrolytes, as in the **Daniell cell** (Fig. 10.8) in which the redox couple at one electrode is Cu^{2+}/Cu and that at the other is Zn^{2+}/Zn. In an **electrolyte concentration cell** (Fig. 10.9), the electrode compartments are identical except for the concentrations of the electrolytes. In an **electrode concentration cell** the electrodes themselves have different concentrations, either because they are gas electrodes operating at different pressures or because they are amalgams (solutions in mercury) with different concentrations.

Liquid junction potentials

In a cell with two different electrolyte solutions in contact, as in the Daniell cell, there is an additional source of potential difference, the **liquid junction potential** E_{lj} across the interface of the two electrolytes. Another example of a junction potential is that between different concentrations of hydrochloric acid. At the junction, the mobile H^+ ions diffuse into the more dilute solution. The bulkier Cl^- ions follow, but initially do so more slowly, which results in a potential difference at the junction. The potential then settles down to a value such that, after that brief initial period, the ions diffuse at the same rates. Electrolyte concentration cells always have a liquid junction; electrode concentration cells do not.

The contribution of the liquid junction to the potential can be reduced (to about 1 to 2 mV) by joining the electrolyte compartments through a salt bridge (Fig. 10.10). The reason for the success of the salt bridge is that the liquid junction potentials at either end are largely independent of the concentrations of the two dilute solutions, so they nearly cancel.

Notation

In the notation for cells, phase boundaries are denoted by a vertical bar. For example, the cell in Fig. 10.7 is denoted

$$Pt \mid H_2(g) \mid HCl(aq) \mid AgCl(s) \mid Ag$$

A liquid junction is denoted by ⦂, so the cell in Fig. 10.8 is written

$$Zn(s) \mid ZnSO_4(aq) \vdots CuSO_4(aq) \mid Cu(s)$$

A double vertical line $\parallel$ denotes an interface for which it is assumed that

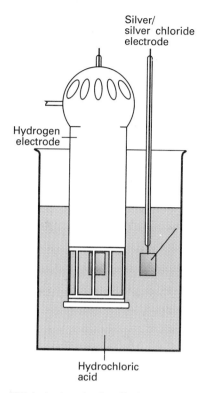

Silver/silver chloride electrode

Hydrogen electrode

Hydrochloric acid

10.7 A simple galvanic cell without a liquid junction. Hydrogen is bubbled over the platinum electrode, which shares a common electrolyte (hydrochloric acid) with the other electrode (a silver/silver chloride electrode).

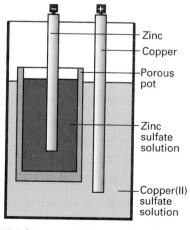

Zinc
Copper
Porous pot
Zinc sulfate solution
Copper(II) sulfate solution

10.8 One version of the Daniell cell.

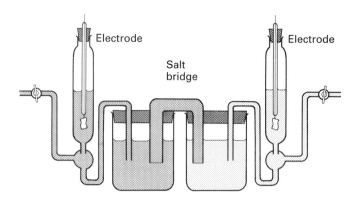

10.9 In an electrolyte concentration cell, the two electrolytes (e.g. hydrochloric acid) are at different concentrations, but the electrodes are otherwise the same. The bridge (a concentrated salt solution) completes the electrical circuit by allowing ions to migrate between the two solutions.

the junction potential has been eliminated. Thus the cell in Fig. 10.10 is denoted

$$Zn(s) \mid ZnSO_4(aq) \parallel CuSO_4(aq) \mid Cu(s)$$

An electrolyte concentration cell in which the liquid junction potential is assumed to be eliminated is denoted

$$Pt \mid H_2(g) \mid HCl(aq, m_1) \parallel HCl(aq, m_2) \mid H_2(g) \mid Pt$$

The cell reaction

The current produced by a galvanic cell arises from the spontaneous chemical reaction taking place inside it. The **cell reaction** is the reaction in the cell written on the assumption that the right-hand electrode is the cathode, and hence that the spontaneous reaction is one in which reduction is taking place in the right-hand compartment. Later we see how to predict if the right-hand electrode is in fact the cathode; if it is, then the cell reaction is spontaneous as written. If the left-hand electrode turns out to be the cathode, the reverse of the cell reaction is spontaneous.

To write the cell reaction corresponding to the cell diagram, we first write the right-hand half-reaction as a reduction (because we have assumed that to be spontaneous). Then we subtract from it the left-hand reduction half-reaction (for, by implication, that electrode is the site of oxidation). Thus, in the cell

$$Zn(s) \mid ZnSO_4(aq) \parallel CuSO_4(aq) \mid Cu(s)$$

the two electrodes and their reduction half-reactions are

Right: $Cu^{2+}(aq) + 2e^- \rightarrow Cu(s)$

Left: $Zn^{2+}(aq) + 2e^- \rightarrow Zn(s)$

Hence, the overall cell reaction is the difference

Overall (R − L): $Cu^{2+}(aq) + Zn(s) \rightarrow Cu(s) + Zn^{2+}(aq)$

The cell potential

A cell in which the overall cell reaction has not reached chemical equilibrium can do electrical work as the reaction drives electrons through an external circuit. The work that a given transfer of electrons

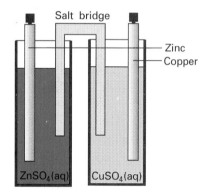

10.10 The salt bridge, essentially an inverted U-tube full of concentrated salt solution in a jelly, has two opposing liquid junction potentials that almost cancel.

can accomplish depends on the potential difference between the two electrodes. This potential difference is called the **cell potential** and is measured in volts (V). When the cell potential is large, a given number of electrons travelling between the electrodes can do a large amount of electrical work; when the cell potential is small, the same number of electrons can do only a small amount of work. A cell in which the overall reaction is at equilibrium can do no work, and then the cell potential is zero.

According to the discussion in Section 4.8, we know that the maximum electrical work that a system (the cell) can do is given by the value of ΔG and, in particular, that, for a spontaneous process (in which both ΔG and w are negative),

$$w_{e,max} = \Delta G \text{ at constant temperature and pressure} \tag{10}$$

Therefore, to make thermodynamic measurements on the cell by measuring the work it can do, we must ensure that it is operating reversibly. Only then is it producing maximum work and only then can eqn 10 be used to relate that work to ΔG. Therefore, to measure ΔG we must ensure that the cell is operating reversibly at a specific, constant composition. Both these conditions are achieved by measuring the cell potential when it is balanced by an exactly opposing source of potential so that the cell reaction occurs reversibly and the composition is constant (in effect, the cell reaction is poised for change, but not actually changing). The resulting potential difference is called the **zero-current cell potential** E (formerly, the 'electromotive force', or emf, of the cell).

The relation between E and $\Delta_r G$

The relation between the reaction Gibbs energy and the zero-current cell potential is

$$-\nu FE = \Delta_r G \tag{11}$$

The quantity F is the product of e and N_A and is called **Faraday's constant**:

$$F = 96.485 \text{ kC mol}^{-1}$$

F is the magnitude of the charge per mole of electrons. Equation 11 is the key connection between electrical measurements on the one hand and thermodynamic properties on the other, and will be the basis of all that follows.

JUSTIFICATION

We consider the change in G when the cell reaction advances by an infinitesimal amount $d\xi$ at some composition. We saw in Section 9.2 that, at constant temperature and pressure, G changes by

$$dG = \sum_J \mu_J \, dn_J = \sum_J \nu_J \mu_J \, d\xi$$

The reaction Gibbs energy $\Delta_r G$ at the specified composition is

$$\Delta_r G = \left(\frac{\partial G}{\partial \xi}\right)_{p,T} = \sum_J \nu_J \mu_J$$

so we can write

$$dG = \Delta_r G \times d\xi$$

The maximum work that the reaction can do as it advances by $d\xi$ at constant temperature and pressure is therefore

$$dw_e = \Delta_r G \times d\xi$$

This work is infinitesimal, and the composition of the system is virtually constant when it occurs.

Suppose that the reaction advances by an amount $d\xi$; then an amount $\nu\, d\xi$ of electrons must travel from the anode to the cathode. The total charge transported between the electrodes when this change occurs is $-\nu e N_A\, d\xi$ (because $\nu\, d\xi$ is the amount of electrons and the charge per mole of electrons is $-e N_A$). Hence, the total charge transported is $-\nu F\, d\xi$ because $e N_A = F$.

The work done when an infinitesimal charge $-\nu F\, d\xi$ travels from the anode to the cathode is equal to the product of the charge and the potential difference E (see Table 2.1 and *Further information 3*):

$$dw_e = -\nu F\, d\xi \times E$$

When this relation is equated to the expression for dG, the advancement $d\xi$ cancels and eqn 11 is obtained.

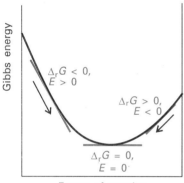

10.11 As explained in Chapter 9, a spontaneous reaction occurs in the direction of decreasing Gibbs energy. When expressed in terms of a cell potential, the spontaneous direction of change corresponds to a positive value of E. When the cell reaction is at equilibrium, the cell potential is zero.

It follows from eqn 11 that, by knowing the reaction Gibbs energy at a specified composition, we can state the zero-current potential at that composition. Note that a *negative* reaction Gibbs energy, corresponding to a spontaneous cell reaction, corresponds to a *positive* zero-current cell potential. Another way of looking at the content of eqn 11 is that it shows that the driving power of a cell (that is, the cell potential) is proportional to the *slope* of the Gibbs energy with respect to the extent of reaction. It is plausible that a reaction that is far from equilibrium (when the slope is steep) has a strong tendency to drive electrons through an external circuit (Fig. 10.11). When the slope is close to zero (when the cell reaction is close to equilibrium), the cell potential is small.

It is now quite easy to estimate the potential that can be expected for a typical cell by combining eqn 11 with a plausible estimate of the reaction Gibbs energy. The latter varies with composition but, in a typical reaction far from equilibrium, it is of the order of the standard reaction Gibbs energies, which are of the order of $-100\ \mathrm{kJ\,mol^{-1}}$. Therefore, taking ν as 1, we can expect a typical cell potential to be of the order of

$$E \approx -\frac{(-100 \times 10^3\ \mathrm{J\,mol^{-1}})}{1 \times (96 \times 10^3\ \mathrm{C\,mol^{-1}})} \approx 1\ \mathrm{V}$$

We have used $1\ \mathrm{J} = 1\ \mathrm{C\,V}$.

The Nernst equation

We can go on to relate the zero-current cell potential to the activities of the participants in the cell reaction. We know from eqn 9.5 that the reaction Gibbs energy is related to the composition of the reaction mixture by

$$\Delta_r G = \Delta_r G^\ominus + RT \ln Q \qquad \text{with } Q = \prod_J a_J^{v_J}$$

It follows, on division of both sides by $-vF$, that

$$E = -\frac{\Delta_r G^\ominus}{vF} - \frac{RT}{vF} \ln Q$$

The first term on the right of this equation is called the **standard cell potential** and is denoted $E^\ominus$:

$$-vFE^\ominus = \Delta_r G^\ominus \tag{12}$$

That is, *the standard cell potential is the standard Gibbs energy of the reaction expressed as a potential (in volts)*. It follows that

$$E = E^\ominus - \frac{RT}{vF} \ln Q \tag{13a}$$

We see from this expression that the standard cell potential (which will shortly move to centre stage of the exposition) can be interpreted as the zero-current cell potential when all the reactants and products are in their standard states, for then all activities are 1 and so $Q = 1$ and $\ln Q = 0$. However, the fact that it is merely a disguised form of the standard reaction Gibbs energy should always be kept in mind and underlies all its applications.

Equation 13a is the **Nernst equation** for the zero-current cell potential at any cell composition. Because $RT/F = 25.7\,\text{mV}$ at 25°C, a practical form of this equation is

$$E = E^\ominus - \frac{25.7\,\text{mV}}{v} \ln Q \tag{13b}$$

For a reaction in which $v = 1$, if Q is increased by a factor of 10, then the cell potential decreases by 59.2 mV (Fig. 10.12).

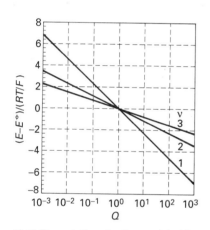

10.12 The variation of cell potential with the value of the reaction quotient for the cell reaction for different values of v (the number of electrons transferred). At 25°C, $RT/F = 25.69$ mV, so the vertical scale refers to multiples of this value.

Concentration cells

The Nernst equation can be used to derive an expression for the potential of an electrolyte concentration cell. Consider the cell

$$M \,|\, M^+(aq, L) \,\|\, M^+(aq, R) \,|\, M$$

where the solutions L and R have different molalities. The cell reaction is

$$M^+(aq, R) \rightarrow M^+(aq, L) \qquad Q = \frac{a_L}{a_R} \qquad v = 1$$

The standard cell potential is zero, because a cell cannot drive a current through a circuit when the two electrode compartments are identical (and, specifically, $\Delta_r G^\ominus = 0$ for the cell reaction). Therefore, the cell potential when the compartments have different concentrations is

$$E = -\frac{RT}{F}\ln\frac{a_L}{a_R} \approx -\frac{RT}{F}\ln\frac{m_L}{m_R} \tag{14}$$

If R is the more concentrated solution, then $E > 0$. Physically, the positive potential arises because positive ions tend to be reduced, thus withdrawing electrons from the electrode, and this process is dominant in the more concentrated right-hand electrode compartment.

One important example of a membrane system that resembles this description is the biological cell wall, which is more permeable to K^+ ions than to either Na^+ or Cl^- ions. The concentration of K^+ inside the cell is about 20 to 30 times that on the outside, and is maintained at that level by a specific pumping operation fuelled by ATP and governed by enzymes. It follows from eqn 14 that the potential difference between the two sides is predicted to be

$$E = -(25.7\ \text{mV}) \times \ln\tfrac{1}{20} = +77\ \text{mV}$$

This estimate accords quite well with the measured value.

The transmembrane potential difference plays a particularly interesting role in the transmission of nerve impulses. Potassium and sodium ion pumps occur throughout the nervous system and, when the nerve is inactive, there is a high K^+ concentration inside the cells and a high Na^+ concentration outside. The potential difference across the cell wall is about 70 mV. When the cell wall is subjected to a pulse of about 20 mV, the structure of the membrane adjusts and it becomes permeable to Na^+. This causes a decrease in membrane potential as the Na^+ ions flood into the interior of the cell. The change in potential difference triggers the adjacent part of the cell wall, and the pulse of collapsing potential passes along the nerve. Behind the pulse the sodium and potassium pumps restore the concentration difference ready for the next pulse.

Cells at equilibrium

A special case of the Nernst equation has great importance in electrochemistry. Suppose the reaction has reached equilibrium; then $Q = K$, where K is the equilibrium constant of the cell reaction. However, a chemical reaction at equilibrium cannot do work and hence it generates zero potential difference between the electrodes of a galvanic cell. Therefore, setting $E = 0$ and $Q = K$ in the Nernst equation gives

$$\ln K = \frac{\nu F E^\ominus}{RT} \tag{15}$$

This very important equation lets us predict equilibrium constants from measured standard cell potentials. For example, because the standard potential of the Daniell cell is

$$Zn(s)\,|\,ZnSO_4(aq)\,\|\,CuSO_4(aq)\,|\,Cu(s) \qquad E^\ominus = +1.10\ \text{V}$$

the equilibrium constant for the cell reaction (for which $\nu = 2$) is

$$Cu^{2+}(aq) + Zn(s) \rightarrow Cu(s) + Zn^{2+}(aq) \qquad K = 1.5 \times 10^{37}$$

We conclude that the displacement of copper by zinc goes virtually to completion.

10.5 Standard potentials

A galvanic cell is a combination of two electrodes, each one of which can be considered as making a characteristic contribution to the overall cell potential. Although it is not possible to measure the contribution of a single electrode, we can define the potential of one of the electrodes as having a zero potential and then assign values to others on that basis. The specially selected electrode is the **standard hydrogen electrode** (SHE):

$$\text{Pt} \mid \text{H}_2(\text{g}) \mid \text{H}^+(\text{aq}) \qquad E^{\ominus} = 0 \text{ at all temperatures}$$

The **standard potential** $E^{\ominus}$ of another couple is then assigned by constructing a cell in which it is the right-hand electrode and the standard hydrogen electrode is the left-hand electrode. For example, the standard potential of the Ag^+/Ag couple is the standard potential of the following cell:

$$\text{Pt} \mid \text{H}_2(\text{g}) \mid \text{H}^+(\text{aq}) \parallel \text{Ag}^+(\text{aq}) \mid \text{Ag(s)}$$
$$E^{\ominus}(\text{Ag}^+/\text{Ag}) = E^{\ominus} = +0.80 \text{ V at } 25°\text{C}$$

Likewise, the standard potential of the AgCl/Ag, Cl^- couple is the standard potential of the following cell:

$$\text{Pt} \mid \text{H}_2(\text{g}) \mid \text{H}^+(\text{aq}) \parallel \text{Cl}^-(\text{aq}) \mid \text{AgCl(s)} \mid \text{Ag(s)}$$
$$E^{\ominus}(\text{AgCl/Ag, Cl}^-) = E^{\ominus} = +0.22 \text{ V at } 25°\text{C}$$

(From now on, all values will refer to 298 K.) Although the standard potential is often written as though it refers to a half-reaction such as

$$\text{AgCl(s)} + \text{e}^- \rightarrow \text{Ag(s)} + \text{Cl}^-(\text{aq})$$
$$E^{\ominus}(\text{AgCl/Ag, Cl}^-) = +0.22 \text{ V}$$

it should be understood that these equations are only shorthand for writing

$$\text{AgCl(s)} + \tfrac{1}{2}\text{H}_2(\text{g}) \rightarrow \text{Ag(s)} + \text{H}^+(\text{aq}) + \text{Cl}^-(\text{aq}) \qquad E^{\ominus} = +0.22 \text{ V}$$

and that the standard potential is determined by properties of the hydrogen electrode as well as the species to which the potential refers.

An important feature of standard cell potentials and standard potentials is that they are *intensive* properties and are unchanged if the chemical equation for the cell reaction or a half-reaction is multiplied by a numerical factor. A numerical factor increases the value of the standard Gibbs energy of the reaction, but it also increases the number of electrons transferred by the same factor, and, by eqn 12, the value of $E^{\ominus}$ remains unchanged. It follows that we can write

$$2\text{AgCl(s)} + 2\text{e}^- \rightarrow 2\text{Ag(s)} + 2\text{Cl}^-(\text{aq})$$
$$E^{\ominus}(\text{AgCl/Ag, Cl}^-) = +0.22 \text{ V}$$

as well as the expression given above.

The standard potential of a cell in terms of reduction potentials

The standard potential of a cell formed from any two electrodes can be calculated by taking the difference of their standard potentials. This rule follows from the fact that a cell such as

$$Ag(s) \,|\, Ag^+(aq) \,\|\, Cl^-(aq) \,|\, AgCl(s) \,|\, Ag(s)$$

is equivalent to two cells joined back-to-back:

$$Ag(s) \,|\, Ag^+(aq) \,\|\, H^+(aq) \,|\, H_2(g) \,|\, Pt \urcorner$$
$$\llcorner Pt \,|\, H_2(g) \,|\, H^+(aq) \,\|\, Cl^-(aq) \,|\, AgCl(s) \,|\, Ag(s)$$

The overall potential of this composite cell, and therefore of the cell of interest, is

$$E^\ominus = E^\ominus(AgCl/Ag, Cl^-) - E^\ominus(Ag^+/Ag) = -0.58 \text{ V}$$

The standard reduction potentials in Table 10.5 can all be used in the same way, and the standard cell potential is the difference right − left of the corresponding standard potentials. Because $\Delta G^\ominus = -\nu F E^\ominus$, it then follows that, if the result gives $E^\ominus > 0$, then the corresponding cell reaction is spontaneous in the direction written (in the sense that $K > 1$).

Table 10.5* Standard potentials at 298 K

Couple	$E^\ominus$/V
$Ce^{4+}(aq) + e^- \rightarrow Ce^{3+}(aq)$	+1.61
$Cu^{2+}(aq) + 2e^- \rightarrow Cu(s)$	+0.34
$AgCl(s) + e^- \rightarrow Ag(s) + Cl^-(aq)$	+0.22
$2H^+(aq) + 2e^- \rightarrow H_2(g)$	0
$Zn^{2+}(aq) + 2e^- \rightarrow Zn(s)$	−0.76
$Na^+(aq) + e^- \rightarrow Na(s)$	−2.71

* More values are given in the Data section.

Example 10.9 *Identifying the spontaneous direction of a reaction*

One of the reactions important in corrosion in an acidic environment is

$$Fe(s) + 2H^+(aq) + \tfrac{1}{2}O_2(g) \rightarrow Fe^{2+}(aq) + H_2O(l)$$

Does the equilibrium constant favour the formation of $Fe^{2+}(aq)$?

Method. We need to decide whether the standard potential for the reaction as written is positive, for a positive value would imply that $\Delta G^\ominus$ is negative and hence that $K > 1$. The sign of the cell potential is found by identifying the half-reactions that make up the overall reaction, and then taking their standard potentials from Table 10.5 in the Data section at the end of this volume.

Answer. The two reduction half-reactions are

(a) $Fe^{2+}(aq) + 2e^- \rightarrow Fe(s)$ $\qquad E^\ominus = -0.44 \text{ V}$

(b) $2H^+(aq) + \tfrac{1}{2}O_2(g) + 2e^- \rightarrow H_2O(l)$ $\qquad E^\ominus = +1.23 \text{ V}$

The difference (b) − (a) is

$$Fe(s) + 2H^+(aq) + \tfrac{1}{2}O_2(g) \rightarrow Fe^{2+}(aq) + H_2O(l) \qquad E^\ominus = +1.67 \text{ V}$$

Therefore, since $E^\ominus > 0$, the reaction has $K > 1$, favouring products.

Comment. Recall the point made earlier that the chemical equation for a half-reaction can be multiplied by any common factor without affecting its standard potential. Therefore, if either half-reaction needs to be multiplied by a factor before forming the difference, the standard potentials are *not* affected.

Exercise E10.9. Can zinc displace copper from solution (that is, reduce Cu^{2+} ions to metallic copper) when the ions are at unit activity?   [Yes]

Example 10.10 *Calculating an equilibrium constant*

Calculate the equilibrium constants for the disproportionation $2Cu^+(aq) \rightarrow Cu(s) + Cu^{2+}(aq)$ at 298 K.

Method. The strategy is to calculate the standard potential for the cell in which the reaction of interest is the cell reaction, and then to use eqn 15. To proceed, express the overall reaction as the difference of two reduction half-reactions and then find the corresponding standard potentials by referring to Table 10.5 in the Data section.

Answer. The half-reactions and standard potentials we require are

$$R: Cu(s) \,|\, Cu^+(aq) \qquad Cu^+(aq) + e^- \rightarrow Cu(s) \qquad E^\ominus = +0.52\,V$$
$$L: Pt \,|\, Cu^{2+}(aq), Cu^+(aq) \quad Cu^{2+}(aq) + e^- \rightarrow Cu^+(aq) \quad E^\ominus = +0.15\,V$$

The standard cell potential is therefore

$$E^\ominus = +0.52\,V - 0.15\,V = +0.37\,V$$

Then, because $v = 1$,

$$\ln K = \frac{0.37\,V}{0.02569\,V} = 14.4$$

Hence, $K = 1.8 \times 10^6$.

Comment. The equilibrium lies strongly towards the right of the reaction as written, so Cu^+ disproportionates almost totally in solution.

Exercise E10.10. Calculate the equilibrium constant for the reaction $Sn^{2+}(aq) + Pb(s) \rightarrow Sn(s) + Pb^{2+}(aq)$ at 298 K.   [2.0]

The composition dependence of individual potentials

We know that the overall cell potential depends on composition in accord with the Nernst equation, eqn 13. Similar equations may be written for the individual reduction potentials. For example, the potential of a Ag^+/Ag electrode at an arbitrary Ag^+ ion activity is given by the expression

$$E(Ag^+/Ag) = E^\ominus(Ag^+/Ag) - \frac{RT}{F} \ln Q$$

where Q is the reaction quotient for the half-reaction

$$Ag^+(aq) + e^- \rightarrow Ag(s) \qquad Q = \frac{1}{a(Ag^+)} \qquad v = 1$$

JUSTIFICATION

Consider a cell in which the left-hand electrode is a standard hydrogen electrode. For example, the reaction in the cell

$$Pt \mid H_2(g) \mid H^+(aq) \parallel Ag^+(aq) \mid Ag(s)$$

is

$$H_2(g) + 2Ag^+(aq) \rightarrow 2H^+(aq) + 2Ag(s) \qquad Q = \frac{a(H^+)^2}{(f_{H_2}/p^\ominus)a(Ag^+)^2} \qquad \nu = 2$$

The Nernst equation is

$$E = E^\ominus - \frac{RT}{2F}\ln Q = E^\ominus(Ag^+/Ag) - \frac{RT}{2F}\ln Q$$

When the hydrogen electrode has its standard composition, $a(H^+) = 1$ and $f_{H_2} = p^\ominus$, but the Ag^+/Ag electrode has an arbitrary composition. Under these conditions

$$Q = \frac{1}{a(Ag^+)^2}$$

and the cell potential can be regarded as arising solely from the Ag^+/Ag electrode. We can then write

$$E(Ag^+/Ag) = E^\ominus(Ag^+/Ag) - \frac{RT}{2F}\ln\left(\frac{1}{a(Ag^+)^2}\right)$$

$$= E^\ominus(Ag^+/Ag) - \frac{RT}{F}\ln\left(\frac{1}{a(Ag^+)}\right)$$

The same expression is obtained by writing the half-reaction at the right-hand electrode as the reduction

$$Ag^+(aq) + e^- \rightarrow Ag(s) \qquad Q = \frac{1}{a(Ag^+)} \qquad \nu = 1$$

and using the Nernst equation directly

$$E(Ag^+/Ag) = E^\ominus(Ag^+/Ag) - \frac{RT}{F}\ln Q$$

as in the text.

Example 10.11 *Calculating the potential difference at an electrode*

Calculate the change that takes place in the potential difference of a silver/silver chloride electrode when an excess of $0.010 \, mol \, kg^{-1}$ KCl(aq) is added at 298 K. The activity of Cl^- ions is 1.3×10^{-5} in a saturated silver chloride solution at this temperature.

Method. The calculation hinges on the use of the Nernst equation for a silver/silver chloride electrode, and on expressing the change in potential when the activity of the Cl^- ions is changed. Their activity in saturated silver chloride solution is given; the activity of Cl^- ions in aqueous potassium chloride solution can be found by reference to Table 10.4. Begin by writing the half-reaction for the electrode and the corresponding Nernst equation; then set up the equation for the difference in potentials requested.

Answer. The half-reaction is

$$AgCl(s) + e^- \rightarrow Ag(s) + Cl^-(aq) \qquad Q = a(Cl^-) \qquad v = 1$$

and the Nernst equation is therefore

$$E(AgCl/Ag, Cl^-) = E^{\ominus}(AgCl/Ag, Cl^-) - \frac{RT}{F} \ln a(Cl^-)$$

The change in potential when the activity of Cl^- ions changes from a_1 to a_2 is therefore

$$\Delta E = -\frac{RT}{F} \ln\left(\frac{a_2}{a_1}\right)$$

The mean activity coefficient is 0.906 for $0.010 \, mol \, kg^{-1}$ KCl(aq), so $a = 0.00906$. The change in electrode potential is therefore

$$\Delta E = -(25.69 \, mV) \times \ln\left(\frac{0.00906}{1.3 \times 10^{-5}}\right) = -0.17 \, V$$

Comment. The activity of Cl^- ions in saturated AgCl(aq) is inferred from the solubility constant: $a(Cl^-) = (K_s)^{\frac{1}{2}}$.

Exercise E10.11. Calculate ΔE when an $0.05 \, mol \, kg^{-1}$ KCl(aq) solution is added to a calomel electrode compartment at 298 K (for which, initially, $a = 8.7 \times 10^{-7}$). $\qquad\qquad$ [$-0.28 \, V$]

Example 10.12 *Calculating the pressure dependence of a potential difference*

Chlorine gas is bubbled over a platinum electrode dipping into aqueous sodium chloride at 298 K. Calculate the change in the potential of the electrode when the chlorine pressure is increased from 1.0 atm to 2.0 atm.

Method. The procedure is similar to that in the preceding example. As before, write down the half-reaction and the reaction quotient and then set up the Nernst equation for the electrode potential. Express the difference in potential arising from a change in pressure in terms of the Nernst equation, and then substitute the data. For simplicity, assume that the gas behaves perfectly at the pressures specified, and so replace fugacity by pressure.

Answer. The electrode half-reaction is

$$Cl_2(g) + 2e^- \rightarrow 2Cl^-(aq) \qquad Q = \frac{a(Cl^-)^2}{f_{Cl_2}/p^{\ominus}} \qquad v = 2$$

Therefore

$$E(Cl_2/Cl^-) = E^{\ominus}(Cl_2/Cl^-) - \frac{RT}{2F} \ln Q$$

$$= E^{\ominus}(Cl_2/Cl^-) - \frac{RT}{2F} \ln\left(\frac{a(Cl^-)^2}{f_{Cl_2}/p^{\ominus}}\right)$$

When the pressure (fugacity) of chlorine changes from p_1 to p_2, the change in potential is

$$\Delta E = \frac{RT}{2F} \ln\left(\frac{p_2}{p_1}\right)$$

Therefore, the change in potential at 298 K (when $RT/F = 25.69$ mV) is

$$\Delta E = \tfrac{1}{2} \times (25.69\text{ mV}) \times \ln 2.0 = +8.9\text{ mV}$$

Comment. The potential increases when the gas pressure is increased because the equilibrium shifts away from $Cl_2(g)$ towards $Cl^-(aq)$, and the resulting withdrawal of electrons from the electrode makes it more positive.

Exercise E10.12. Derive an expression for the potential of an electrode at which the half-reaction is $G_2(g) + 4e^- \rightarrow 2G^{2-}(aq)$.

$$[E = E^{\ominus} - (RT/2F) \ln a(p^{\ominus}/f)^{\frac{1}{2}}]$$

Finally, for a redox electrode, the Nernst equation gives

$$E(Fe^{3+}, Fe^{2+}) = E^{\ominus}(Fe^{3+}, Fe^{2+}) - \frac{RT}{F} \ln\left(\frac{a(Fe^{2+})}{a(Fe^{3+})}\right)$$

The equation implies that an increase in Fe^{3+} tends to increase the electrode potential and an increase in Fe^{2+} tends to lower it. However, the wider importance of this equation is that, instead of treating the potential difference as arising from the equilibrium, we can imagine *controlling* the equilibrium by modifying the electrode potential. If we can ensure that the potential of the cell

$$\text{SHE} \parallel Fe^{3+}(aq), Fe^{2+}(aq) \mid Pt$$

is $E(Fe^{3+}, Fe^{2+})$, or the appropriate value if the electrode on the left is something other than a SHE, then the concentrations of reduced and oxidized species will adjust so that $a(Fe^{2+})/a(Fe^{3+})$ has a value that satisfies the equation for E. This gives us electrical control over the composition of solutions.

The measurement of standard reduction potentials

The procedure for measuring a standard potential can be illustrated by considering a specific case, the silver chloride electrode. The measurement is made on the **Harned cell**:

$$Pt \mid H_2(g) \mid HCl(aq) \mid AgCl(s) \mid Ag(s)$$
$$\tfrac{1}{2}H_2(g) + AgCl(s) \rightarrow HCl(aq) + Ag(s)$$

for which

$$E = E^{\ominus}(\text{AgCl/Ag, Cl}^-) - \frac{RT}{F}\ln\left(\frac{a(\text{H}^+)a(\text{Cl}^-)}{(f_{\text{H}_2}/p^{\ominus})^{\frac{1}{2}}}\right) \tag{16}$$

We shall set $f_{\text{H}_2} = p^{\ominus}$ from now on. The activities can be expressed in terms of the molality m and the mean activity coefficient $\gamma_{\pm}$ through eqn 6:

$$E = E^{\ominus}(\text{AgCl/Ag, Cl}^-) - \frac{RT}{F}\ln\left(\frac{m}{m^{\ominus}}\right)^2 - \frac{RT}{F}\ln\gamma_{\pm}^2$$

This expression rearranges to

$$E + \frac{2RT}{F}\ln\left(\frac{m}{m^{\ominus}}\right) = E^{\ominus}(\text{AgCl/Ag, Cl}^-) - \frac{2RT}{F}\ln\gamma_{\pm} \tag{17a}$$

From the Debye–Hückel limiting law for a 1,1-electrolyte,

$$\ln\gamma_{\pm} \propto (m/m^{\ominus})^{\frac{1}{2}}$$

(the natural logarithm used here is proportional to the common logarithm that appears in eqn 6). Therefore, with the constant of proportionality in this relation written A',

$$E + \frac{2RT}{F}\ln\left(\frac{m}{m^{\ominus}}\right) = E^{\ominus}(\text{AgCl/Ag, Cl}^-) + \frac{2A'RT}{F} \times \left(\frac{m}{m^{\ominus}}\right)^{\frac{1}{2}} \tag{17b}$$

(In precise work, the $m^{\frac{1}{2}}$ term is brought to the left, and a higher-order correction term from the extended Debye–Hückel law is used on the right.) The expression on the left is evaluated at a range of molalities, plotted against $m^{\frac{1}{2}}$, and extrapolated to $m = 0$. The intercept at $m^{\frac{1}{2}} = 0$ is the value of $E^{\ominus}(\text{AgCl/Ag, Cl}^-)$.

Example 10.13 *Determining the standard potential of a cell*

The potential of the cell $\text{Zn}\,|\,\text{ZnCl}_2(\text{aq}, m)\,|\,\text{AgCl(s)}\,|\,\text{Ag}$ at 25°C has the following values:

$m/(10^{-3}m^{\ominus})$	0.772	1.253	1.453	3.112	6.022
E/V	1.2475	1.2289	1.2235	1.1953	1.1742

Determine the standard potential of the cell.

Method. Proceed as described above. Start by writing the Nernst equation for the cell, and then express the activities that occur in Q in terms of the mean activity coefficient. The latter can be written as proportional to $(m/m^{\ominus})^{\frac{1}{2}}$ by using the Debye–Hückel limiting law. However, there is no need to write all the constants because the standard cell potential is obtained by extrapolation, as explained in the text.

Answer. The cell reaction is

$$\text{Zn(s)} + 2\text{AgCl(s)} \rightarrow 2\text{Ag(s)} + \text{ZnCl}_2(\text{aq}) \qquad \nu = 2$$

The Nernst equation is therefore

$$E = E^{\ominus} - \frac{RT}{2F} \ln a(Zn^{2+})a(Cl^-)^2$$

The activities are related to the molality m of $ZnCl_2$ by

$$a(Zn^{2+})a(Cl^-)^2 = \gamma_{\pm}^3 m(Zn^{2+})m(Cl^-)^2/m^{\ominus 3} = 4\gamma_{\pm}^3 m^3/m^{\ominus 3}$$

Only a little work is then needed to convert the Nernst equation to

$$E + \frac{3RT}{2F} \ln\left(\frac{m}{m^{\ominus}}\right) + \frac{RT}{2F} \ln 4 = E^{\ominus} + C(m/m^{\ominus})^{\frac{1}{2}}$$

where C is a collection of constants that come from the limiting law. We now draw up the following table, using $RT/2F = 0.01285$ V (from inside front cover):

$m/(10^{-3}m^{\ominus})$	0.772	1.253	1.453	3.112	6.022
$\{m/(10^{-3}m^{\ominus})\}^{\frac{1}{2}}$	0.879	1.119	1.205	1.764	2.454
$E/V + 0.03854 \times \ln(m/m^{\ominus})$ $+ 0.01285 \ln 4$	0.9891	0.9892	0.9895	0.9906	0.9950

The data are plotted in Fig. 10.13; as can be seen, they extrapolate to $E^{\ominus} = +0.9886$ V.

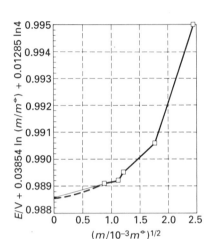

10.13 The plot and the extrapolation used for the experimental measurement of a standard cell potential (see Example 10.13 for the data). The intercept at $m^{\frac{1}{2}} = 0$ is $E^{\ominus}$.

Exercise E10.13. The data below are for the cell $Pt \mid H_2(g, p^{\ominus}) \mid HBr(aq, m) \mid AgBr(s) \mid Ag$ at 25°C. Determine the standard cell potential.

$m/(10^{-4}m^{\ominus})$	4.042	8.444	37.19	
E/V	0.47381	0.43636	0.36173	[+0.071 V]

The measurement of activity coefficients

Once the standard potential of an electrode in a cell is known, the activities of the ions with respect to which it is reversible can be determined simply by measuring the cell potential with the ions at the concentration of interest. For example, the mean activity coefficient of the ions in hydrochloric acid of molality m is obtained from eqn 17a in the form

$$\ln \gamma_{\pm} = \frac{E^{\ominus}(AgCl/Ag, Cl^-) - E}{2RT/F} - \ln\left(\frac{m}{m^{\ominus}}\right) \tag{17c}$$

once E has been measured.

APPLICATIONS OF STANDARD POTENTIALS

The measurement of zero-current cell potential is a convenient source of data on the Gibbs energies, enthalpies, and entropies of reactions. In practice the standard values of these quantities are the ones normally determined.

10.6 The electrochemical series

We have seen that for two redox couples, Ox_1/Red_1 and Ox_2/Red_2, and the cell

$$Red_1, Ox_1 \ \| \ Red_2, Ox_2 \qquad E^{\ominus} = E_2^{\ominus} - E_1^{\ominus}$$

the cell reaction

$$Red_1 + Ox_2 \rightarrow Ox_1 + Red_2$$

is spontaneous as written if $E^{\ominus} > 0$ and therefore if $E_2^{\ominus} > E_1^{\ominus}$. Because, in the cell reaction, Red_1 is reducing Ox_2, we can conclude that

> Red_1 has a thermodynamic tendency to reduce Ox_2 if $E_1^{\ominus}$ is lower than $E_2^{\ominus}$.

More briefly: *low reduces high*. For example,

$$E^{\ominus}(Zn^{2+}, Zn) = -0.76\,V < E^{\ominus}(Cu^{2+}, Cu) = +0.34\,V$$

and Zn has a thermodynamic tendency to reduce Cu^{2+}. Hence the reaction

$$Zn(s) + CuSO_4(aq) \rightarrow ZnSO_4(aq) + Cu(s)$$

can be expected to have $K > 1$ (in fact, as we have seen, $K = 1.5 \times 10^{37}$ at 298 K).

Table 10.6 shows a part of the **electrochemical series** of the elements, their redox couples arranged in the order of their reducing power. The reduced member of a couple with a lower standard potential can reduce the oxidized member of couples with higher standard potentials. This is a qualitative conclusion. The quantitative value of K is then obtained by doing the calculations we have introduced previously. For example, to determine whether zinc can displace magnesium from aqueous solutions at 298 K, we note that the displacement of magnesium is the reduction of its ions. However, from Table 10.6 we see that the Zn^{2+}/Zn couple lies above Mg^{2+}/Mg. Therefore, zinc cannot reduce magnesium ions. Even for reactions that are thermodynamically favourable there may be kinetic factors that result in very slow rates of reaction.

Table 10.6* The electrochemical series of the metals

Least strongly reducing

Gold
Platinum
Silver
Mercury
Copper
(Hydrogen)
Lead
Tin
Nickel
Iron
Zinc
Chromium
Aluminium
Magnesium
Sodium
Calcium
Potassium

Most strongly reducing

---

* The complete series can be inferred from Table 10.5 in the Data section.

10.7 Solubility constants

We can discuss the solubility s (the molality of the saturated solution) of a sparingly soluble salt MX in terms of the equilibrium

$$MX(s) \rightleftharpoons M^+(aq) + X^-(aq) \qquad K_s = a(M^+)a(X^-)$$

where the activities are those at equilibrium (that is, in the saturated solution) and we have used $a = 1$ for a pure solid. The equilibrium constant K_s is called the **solubility constant** (formerly, and still commonly, the **solubility product**) of the salt. When the solubility is so low that $\gamma_{\pm} \approx 1$ even in the saturated solution we can write $a = m/m^{\ominus}$, moreover, because both molalities are equal to s in the saturated solution, we can conclude that

$$K_s \approx \left(\frac{s}{m^\ominus}\right)^2$$

and hence that

$$s \approx (K_s)^{\frac{1}{2}} \times m^\ominus \tag{18}°$$

It follows that we can estimate s from the standard potential of a cell with a reaction corresponding to the solubility equilibrium.

Example 10.14 *Evaluating a solubility from electrochemical data*

Evaluate the solubility of AgCl(s) from cell potential data at 298 K.

Method. We need to find an electrode combination that reproduces the solubility equilibrium, and then identify the solubility constant with the equilibrium constant of the cell reaction. The solubility itself is obtained from an equation like eqn 18.

Answer. The solubility equilibrium is

$$AgCl(s) \rightleftharpoons Ag^+(aq) + Cl^-(aq) \qquad K_s = a(Ag^+)a(Cl^-)$$

and we saw in Example 10.6 that it can be expressed as the difference of the following two half-reactions:

$$AgCl(s) + e^- \rightarrow Ag(s) + Cl^-(aq) \qquad E^\ominus = +0.22\ V$$
$$Ag^+(aq) + e^- \rightarrow Ag(s) \qquad E^\ominus = +0.80\ V$$

The cell potential is therefore $-0.58\ V$. Then, because $v = 1$,

$$\ln K_s = \frac{vE^\ominus}{RT/F} = \frac{1 \times (-0.58\ V)}{2.5693 \times 10^{-2}\ V} = -23$$

Therefore, $K_s = 1.8 \times 10^{-10}$ and $s = 1.3 \times 10^{-5}\ mol\ kg^{-1}$.

Exercise E10.14. Calculate the solubility constant and the solubility of mercury(I) chloride at 298 K. (*Hint.* The mercury(I) ion is the diatomic species Hg_2^{2+}.) $\qquad$ [1.3×10^{-18}, $6.9 \times 10^{-7}\ mol\ kg^{-1}$]

10.8 The measurement of pH and pK

The potential of a hydrogen electrode

$$2H^+(aq) + 2e^- \rightarrow H_2(g) \qquad Q = \frac{f_{H_2}/p^\ominus}{a(H^+)^2} \qquad v = 2 \tag{19a}$$

is

$$E(H^+/H_2) = E^\ominus(H^+/H_2) - \frac{RT}{2F}\ln Q = \frac{RT}{F}\ln \frac{a(H^+)}{(f_{H_2}/p^\ominus)^{\frac{1}{2}}} \tag{19b}$$

This expression makes sense physically. Increasing the activity of the hydrogen ions (by decreasing the pH, because $pH = -\log a(H^+)$) increases the tendency of the positive ions to discharge at the electrode, so we should expect its potential to become more positive. Because the ion activity occurs in the numerator of the logarithmic term, the equation does predict that E increases as $a(H^+)$ is increased.

The potential of the hydrogen electrode is directly proportional to the pH of the solution. Thus, if $f = p^{\ominus}$ (so that the hydrogen fugacity is 1 bar), we can use

$$\ln a(H^+) = \ln 10 \times \log a(H^+) = -2.303\,\text{pH}$$

to obtain

$$E(H^+/H_2) = -\frac{2.303RT}{F} \times \text{pH} \tag{20a}$$

At 25°C, when $RT/F = 25.69$ mV, this relation becomes

$$E(H^+/H_2) = -59.16\,\text{mV} \times \text{pH} \tag{20b}$$

Hence, each unit decrease in pH increases the electrode potential by 59 mV.

The determination of pH

The measurement of the pH of a solution is simple in principle, for it is based on the measurement of the potential of a hydrogen electrode immersed in the solution. The left-hand electrode of the cell is typically a saturated calomel reference electrode with potential $E(\text{cal})$; the right-hand electrode is the hydrogen electrode with potential given by eqn 20b. The potential of the cell is therefore

$$E = (-59.16\,\text{mV}) \times \text{pH} - E(\text{cal}) \quad \text{(at 25°C)}$$

and therefore[2]

$$\text{pH} = \frac{E + E(\text{cal})}{(-59.16\,\text{mV})} \tag{21}$$

In practice, indirect methods are much more convenient, and the hydrogen electrode is replaced by the **glass electrode**. This electrode (Fig. 10.14) is sensitive to hydrogen ion activity, and has a potential proportional to pH. It is filled with a phosphate buffer containing Cl^- ions, and conveniently has $E = 0$ when the external medium is at pH = 7. The glass electrode is much more convenient to handle than the gas electrode itself, and can be calibrated using solutions of known pH.

The responsiveness of a glass electrode to the hydronium ion activity is a result of complex processes at the interface between the glass

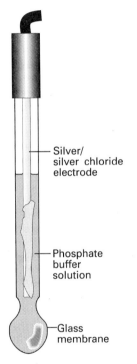

10.14 The glass electrode. It is usually used in conjunction with a calomel electrode that makes contact with the test solution through a salt bridge.

— Silver/ silver chloride electrode

— Phosphate buffer solution

— Glass membrane

---

2 The practical definition of the pH of a solution X is

$$\text{pH(X)} = \text{pH(S)} + \frac{FE}{2.3026RT}$$

where E is the potential of the cell

$$\text{Pt} \mid H_2(g) \mid X(aq) \parallel 3.5\,\text{M KCl(aq)} \parallel S(aq) \mid H_2(g) \mid \text{Pt}$$

and S is a solution of standard pH. The currently recommended primary standards include a saturated aqueous solution of potassium hydrogentartrate, which has pH = 3.557 at 25°C, and 0.0100 mol kg^{-1} disodium tetraborate, which has pH = 9.180 at that temperature.

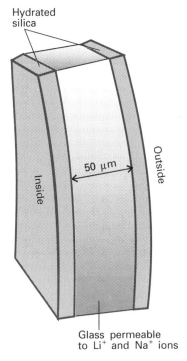

Hydrated silica

Inside

50 μm

Outside

Glass permeable to Li⁺ and Na⁺ ions

10.15 A section through the wall of a glass electrode.

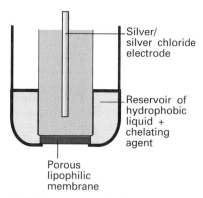

Silver/ silver chloride electrode

Reservoir of hydrophobic liquid + chelating agent

Porous lipophilic membrane

10.16 The structure of an ion-selective electrode. Chelated ions are able to migrate through the lipophilic membrane.

membrane and the solutions on either side of it. The membrane itself is permeable to Na^+ and Li^+ ions but not to H^+ ions. Therefore, the potential difference across the glass membrane must arise by a mechanism that is different from that responsible for biological transmembrane potentials. A clue to the mechanism comes from a detailed inspection of the glass membrane, for each face is coated with a thin layer of hydrated silica (Fig. 10.15). The hydrogen ions in test solution modify this layer to an extent that depends on their activity in the solution, and the charge modification of the outside layer is transmitted to the inner layer by the Na^+ and Li^+ ions in the glass. The hydronium ion activity gives rise to a membrane potential by this indirect mechanism.

The electrochemical determination of pH opens up a route to the electrochemical determination of pK_a, for we saw in Section 9.6 that the pK_a of an acid is equal to the pH of a solution containing equal amounts of the acid and its conjugate base.

Species-selective electrodes

A suitably adapted glass electrode can be used to detect the presence of certain gases. A simple form of a **gas-sensing electrode** consists of a glass electrode contained in an outer sleeve filled with an aqueous solution and separated from the test solution by a membrane that is permeable to gas. When a gas such as sulfur dioxide or ammonia diffuses into the aqueous solution, it modifies its pH, which in turn affects the potential of the gas electrode.

Somewhat more sophisticated devices are used as **ion-selective electrodes** that give potentials according to the presence of specific ions present in a test solution. In one arrangement, a porous lipophilic (hydrocarbon-attracting) membrane is attached to a small reservoir of a hydrophobic (water-repelling) liquid, such as dioctylphenylphosphonate, that saturates it (Fig. 10.16). The liquid contains a chelating agent, such as $(RO)_2PO_2^-$ with R a C_8 to C_{18} chain, that acts as a kind of solubilizing agent for the ions with which it can form a complex. The chelated ions are able to migrate through the lipophilic membrane, and hence give rise to a transmembrane potential, which is detected by a silver/silver-chloride electrode in the interior of the assembly. Electrodes of this construction can be designed to be sensitive to a variety of ionic species, including calcium, zinc, iron, lead, and copper ions.

10.9 Potentiometric titrations

In a **redox titration** the reduced form of an ion (e.g. Fe^{2+}) is oxidized by the addition of an oxidant (e.g. Ce^{4+}). In a **potentiometric titration** the stoichiometric point of a redox titration is detected by monitoring the potential of the cell formed by a platinum electrode in the mixture and another electrode in electrical contact with the mixture through a salt bridge. As we shall see, there is a sharp change in cell potential at the stoichiometric point when exactly enough oxidant has been added to oxidize all the reduced ion.

We shall illustrate the technique with the redox reaction

$$Fe^{2+}(aq) + Ce^{4+}(aq) \rightarrow Fe^{3+}(aq) + Ce^{3+}(aq)$$

for which the two half-reactions are

$$Ce^{4+}(aq) + e^- \rightarrow Ce^{3+}(aq) \qquad E^{\ominus}(Ce^{4+}, Ce^{3+}) = +1.61 \text{ V}$$

$$Fe^{3+}(aq) + e^- \rightarrow Fe^{2+}(aq) \qquad E^{\ominus}(Fe^{3+}, Fe^{2+}) = +0.77 \text{ V}$$

The equilibrium constant for the reaction (which has $v = 1$) is therefore 1.52×10^{14}. It follows that Ce^{4+} is a good choice because it ensures that the equilibrium lies strongly towards products (Fe^{3+}). During the course of the titration, the presence of the iron couple gives rise to an electrode potential $E(Fe^{3+}, Fe^{2+})$ and the cerium couple gives rise to the potential $E(Ce^{4+}, Ce^{3+})$. However, these couples share the same electrode, and therefore the potentials must be equal (a single electrode cannot be at two different potentials simultaneously). Consequently, at any stage of the titration the electrode potential may be expressed in either of the following ways:

$$E = E^{\ominus}(Ce^{4+}, Ce^{3+}) - \frac{RT}{F} \ln\left(\frac{a(Ce^{3+})}{a(Ce^{4+})}\right)$$

$$E = E^{\ominus}(Fe^{3+}, Fe^{2+}) - \frac{RT}{F} \ln\left(\frac{a(Fe^{2+})}{a(Fe^{3+})}\right)$$

and whichever expression is more convenient may be used. We shall simplify the following discussion by supposing that activities may be replaced by molalities, and we shall also ignore the dilution of the sample as the cerium solution is added: these are minor technical complications that do not affect the main conclusions.

Suppose that initially the amount of Fe(II) present in the sample is f. As Ce(IV) is added this amount is reduced to $(1 - x)f$; simultaneously, the amount of Fe(III) rises to xf, where x depends on the amount of Ce(IV) added (if K were infinite, xf would be equal to the amount of oxidant added). It follows that at an intermediate stage of the titration, the electrode's potential is

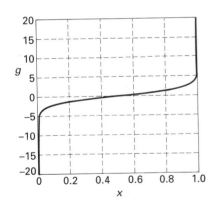

$$E = E^{\ominus}(Fe^{3+}, Fe^{2+}) + \frac{RT}{F} \times g(x) \qquad \text{where } g(x) = \ln\left(\frac{x}{1-x}\right) \qquad (22)$$

10.17 The function g in eqn 22. This quick–slow–quick variation occurs widely in electrochemistry.

The function g occurs widely in electrochemistry, and is plotted in Fig. 10.17. It has a large negative value when $x \ll 1$, is zero when $x = \frac{1}{2}$, and rises to a large positive value as $x \rightarrow 1$. An initial rapid rise, a slow variation around $x = \frac{1}{2}$, and a further rapid rise are characteristic of many electrochemical measurements and can usually be traced back to the properties of g. We saw another example in the discussion of acid–base titrations in Section 9.6.

In the context of a potentiometric titration, the shape of g indicates that the electrode potential is initially strongly negative, rises rapidly towards $E^{\ominus}(Fe^{3+}, Fe^{2+})$ as Ce(IV) is added, and is exactly equal to that value when enough has been added to equalize the molalities (actually the activities) of Fe(II) and Fe(III). As more Ce(IV) is added, the electrode potential rises slowly at first, but then the rapid second rise of g takes over. The line in Fig. 10.17 barely leaves the zero axis (on the scale used) until x has reached about 0.98, so the stoichiometric point of the

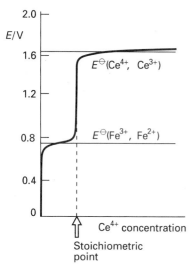

10.18 The variation of an electrode potential during the titration of Fe(II) and Ce(IV). The stoichiometric point is recognized by the sudden change from the standard potential of the Fe^{3+}/Fe^{2+} couple to that of the Ce^{4+}/Ce^{3+} couple.

titration is signalled by the abrupt rise in potential away from $E^{\ominus}(Fe^{3+}, Fe^{2+})$.

When enough Ce(IV) has been added to eliminate all but a trace of the Fe(II), it is more convenient to discuss the potential in terms of the Ce(IV) molality. If an amount c of Ce(IV) has been added, the amount present in solution after the stoichiometric point is about $c - f$ because virtually all the Fe(II) has been oxidized (since the equilibrium constant is so large). The amount of Ce(III) present is about f however much Ce(IV) is added after the stoichiometric point. Therefore, the potential of the electrode is

$$E = E^{\ominus}(Ce^{4+}, Ce^{3+}) + \frac{RT}{F} \times h(f) \qquad \text{where } h(f) = \ln\left(\frac{c - f}{f}\right) \qquad (23)$$

The potential E rises to $E^{\ominus}(Ce^{4+}, Ce^{3+})$ at $c = 2f$ (about twice as much cerium as was needed to reach the stoichiometric point), and then climbs only very slowly as more cerium is added because, when $c \gg f$, the function $h \approx \ln(c/f)$, which increases only very slowly as c increases.

The complete curve is sketched in Fig. 10.18. It is clear that the stoichiometric point can be detected by noting where there is a rapid change of potential from near $E^{\ominus}(Fe^{3+}, Fe^{2+})$ to near $E^{\ominus}(Ce^{4+}, Ce^{3+})$ or, in general, for other redox systems, from the standard potential of the couple in the analyte to the standard potential of the couple in the titrant.

10.10 Thermodynamic functions from cell potential measurements

The standard cell potential is related to the standard reaction Gibbs energy by eqn 12:

$$\Delta_r G^{\ominus} = -\nu F E^{\ominus}$$

Therefore, by measuring $E^{\ominus}$ we can obtain this important thermodynamic quantity. Its value can then be used to calculate the Gibbs energy of formation of ions using the convention explained in Section 10.1. For example, the cell reaction of

$$H_2 \mid H^+(aq) \parallel Ag^+(aq) \mid Ag \qquad E^{\ominus} = +0.7996 \text{ V}$$

is

$$Ag^+(aq) + \tfrac{1}{2}H_2(g) \rightarrow H^+(aq) + Ag(s) \qquad \Delta_r G^{\ominus} = -\Delta_f G^{\ominus}(Ag^+, aq)$$

Therefore, with $\nu = 1$, we find

$$\Delta_f G^{\ominus}(Ag^+, aq) = -(-FE^{\ominus}) = +77.10 \text{ kJ mol}^{-1}$$

as in Table 10.1.

Example 10.15 *Evaluating a reduction potential from two others*

Given that the standard potentials of the Cu^{2+}/Cu and Cu^+/Cu couples are $E^{\ominus}(Cu^{2+}/Cu) = +0.340 \text{ V}$ and $E^{\ominus}(Cu^+/Cu) = +0.522 \text{ V}$, evaluate $E^{\ominus}(Cu^{2+}, Cu^+)$.

Method. First, we note that reaction Gibbs energies may be added (as in

Hess's law analyses of reaction enthalpies). Therefore, we should convert the $E^{\ominus}$ values to $\Delta_r G^{\ominus}$ values by using eqn 12, add them appropriately, and then convert the overall $\Delta_r G^{\ominus}$ to the required $E^{\ominus}$ by using eqn 12 again. This roundabout procedure is necessary because, as we shall see, although the factor F cancels the factor ν in general does not.

Answer. The electrode reactions are as follows:

(a) $Cu^{2+}(aq) + 2e^- \rightarrow Cu(s)$

$E^{\ominus} = +0.340\,V$, so $\Delta_r G^{\ominus} = -2F \times 0.340\,V = -0.680\,V \times F$

(b) $Cu^+(aq) + e^- \rightarrow Cu(s)$

$E^{\ominus} = +0.522\,V$, so $\Delta_r G^{\ominus} = -0.522\,V \times F$

The required reaction is

(c) $Cu^{2+}(aq) + e^- \rightarrow Cu^+(aq)$ $E^{\ominus}$

Because (c) = (a) − (b), the standard Gibbs energy of reaction (c) is

$$\Delta_r G^{\ominus} = \Delta_r G^{\ominus}(a) - \Delta_r G^{\ominus}(b) = -0.158\,V \times F$$

Therefore, since now $\nu = 1$,

$$E^{\ominus} = -\frac{\Delta_r G^{\ominus}}{F} = +0.158\,V$$

Comment. Note that we cannot combine the $E^{\ominus}$ values directly, as they are not extensive properties; we must always work via $\Delta_r G^{\ominus}$.

Exercise E10.15. Calculate the standard potential of the Fe^{3+}/Fe couple from the values for the Fe^{3+}/Fe^{2+} and Fe^{2+}/Fe couples. $[-0.016\,V]$

The temperature coefficient of the cell potential gives the entropy of the cell reaction. This conclusion follows from the thermodynamic relation $(\partial G/\partial T)_p = -S$ and eqn 12, which combine to give

$$\frac{dE^{\ominus}}{dT} = \frac{\Delta_r S^{\ominus}}{\nu F} \tag{24}$$

(The derivative is complete because $E^{\ominus}$, like $\Delta_r G^{\ominus}$, is independent of the pressure.) Hence we have an electrochemical technique for obtaining standard reaction entropies and through them the entropies of ions in solution.

Finally, we can combine the results obtained so far and use them to obtain the standard reaction enthalpy:

$$\Delta_r H^{\ominus} = \Delta_r G^{\ominus} + T\,\Delta_r S^{\ominus} = -\nu F\left(E^{\ominus} - T\frac{dE^{\ominus}}{dT}\right) \tag{25}$$

This expression provides a non-calorimetric method for measuring $\Delta_r H^{\ominus}$ and, through the convention $\Delta_f H^{\ominus}(H^+, aq) = 0$, the standard enthalpies of formation of ions in solution. Thus, electrical measurements can be used to calculate all the thermodynamic properties with which this chapter began.

Example 10.16 *Using the temperature coefficient of the cell potential*

The standard cell potential of

$$Pt \,|\, H_2(g) \,|\, HBr(aq) \,|\, AgBr(s) \,|\, Ag(s)$$

was measured over a range of temperatures, and the data were fitted to the following polynomial:

$$E^{\ominus}/V = 0.07131 - 4.99 \times 10^{-4}(T/K - 298)$$
$$- 3.45 \times 10^{-6}(T/K - 298)^2$$

Evaluate the standard reaction Gibbs energy, enthalpy, and entropy at 298 K.

Method. The standard Gibbs energy of reaction is obtained by using eqn 12 after evaluating $E^{\ominus}$ at 298 K. The standard entropy of reaction is obtained by using eqn 24, which involves differentiating the polynomial with respect to T and then setting $T = 298$ K. The standard reaction enthalpy is obtained by combining the values of the standard Gibbs energy and entropy.

Answer. At $T = 298$ K, $E^{\ominus} = 0.07131$ V, so

$$\Delta_r G^{\ominus} = -vFE^{\ominus} = -(1) \times (96.485 \text{ kC mol}^{-1}) \times (+0.07131 \text{ V})$$
$$= -6.880 \text{ kJ mol}^{-1}$$

The temperature coefficient of the cell potential is

$$\frac{dE^{\ominus}}{dT} = -4.99 \times 10^{-4} \text{ V K}^{-1}$$
$$- 2 \times (3.45 \times 10^{-6}) \times (T - 298 \text{ K}) \text{ V K}^{-2}$$

At $T = 298$ K this expression evaluates to

$$\frac{dE^{\ominus}}{dT} = -4.99 \times 10^{-4} \text{ V K}^{-1}$$

So, from eqn 24, the standard reaction entropy is

$$\Delta_r S^{\ominus} = 1 \times (9.6485 \times 10^4 \text{ C mol}^{-1}) \times (-4.99 \times 10^{-4} \text{ V K}^{-1})$$
$$= -48.1 \text{ J K}^{-1} \text{ mol}^{-1}$$

It then follows that

$$\Delta_r H^{\ominus} = \Delta_r G^{\ominus} + T \Delta_r S^{\ominus}$$
$$= -6.880 \text{ kJ mol}^{-1} + (298 \text{ K}) \times (-0.0481 \text{ kJ K}^{-1} \text{ mol}^{-1})$$
$$= -21.2 \text{ kJ mol}^{-1}$$

Comment. One difficulty with this procedure lies in the accurate measurement of small temperature coefficients of cell potential. Nevertheless, it is another example of the striking ability of thermodynamics to relate the apparently unrelated, in this case to relate electrical measurements to thermal properties.

Exercise E10.16. Predict the standard potential of the Harned cell at 303 K from tables of thermodynamic data. [0.2191 V]

CHECK LIST OF KEY IDEAS

1 The use of and definition of the standard thermodynamic functions of formation of ions in solution (Section 10.1).

2 The contributions to the **standard Gibbs energy of formation** of ions in solution and the use of the **Born equation** (eqn 1) to judge trends.

3 The **activities** of ions in solution and the significance of the **mean activity coefficient** (Section 10.2 and eqn 5).

4 The physical basis of the **Debye–Hückel limiting law** (eqn 6) for calculating mean activity coefficients and the definition and significance of **ionic strength** (eqn 7).

5 The use of **half-reactions** to discuss **redox reactions** (Section 10.3) and oxidation and reduction half-reactions at electrodes (Section 10.3).

6 The half-reactions at different types of electrode, including the **metal/metal-ion electrode**, the **gas electrode**, the **metal/insoluble-salt electrode**, and the **redox electrode** (Section 10.3).

7 The classification of **electrochemical cells** and the role of the **salt bridge** (Section 10.4).

8 The **cell reaction** and the **cell potential**, and the link between the **zero-current cell potential** and the **reaction Gibbs energy** (eqn 11).

9 The derivation of the **Nernst equation** (eqn 13) for the dependence of the zero-current cell potential on the composition of the cell.

10 The definition of the **standard cell potential** and its relation to the equilibrium constant of the cell reaction (eqn 15).

11 The definition of **standard potential** and the dependence of electrode potentials on the composition (Section 10.5).

12 The **measurement** of standard potentials and activity coefficients (Section 10.5).

13 The formulation and significance of the **electrochemical series** (Section 10.6).

14 The calculation of **solubility constants** and of **solubilities** from standard potentials (Section 10.7).

15 The use of electrochemical measurements to determine pH and pK_a (Section 10.8).

16 The principles of **potentiometric titrations** (Section 10.9) and the variation of potential as an oxidizing agent is added to a solution (eqns 22 and 23).

17 The electrochemical measurement of the standard Gibbs energy, enthalpy, and entropy of reaction from the **temperature dependence** of the standard cell potential (Section 10.10).

EXERCISES

10.1. Calculate $\Delta_r H^\ominus$ for the reaction $Zn(s) + CuSO_4(aq) \rightarrow ZnSO_4(aq) + Cu(s)$ from the information in Table 2.12 in the Data section.

10.2. Calculate the molar solubility of mercury(II) chloride at 25°C from standard Gibbs energies of formation.

10.3. Estimate the standard Gibbs energy of formation of $F^-(aq)$ from the value for $Cl^-(aq)$, taking the radius of F^- as 131 pm.

10.4. Confirm that the ionic strengths of KCl, $MgCl_2$, $FeCl_3$, $Al_2(SO_4)_3$, and $CuSO_4$ solutions are related to the molalities m of the solution by $I(KCl) = m$, $I(MgCl_2) = 3m$, $I(FeCl_3) = 6m$, $I(Al_2(SO_4)_3) = 15m$, and $I(CuSO_4) = 4m$.

10.5. Calculate the ionic strength of a solution that is $0.10\,\mathrm{mol\,kg^{-1}}$ in KCl(aq) and $0.20\,\mathrm{mol\,kg^{-1}}$ $CuSO_4$(aq).

10.6. Calculate the ionic strength of a solution that is $0.040\,\mathrm{mol\,kg^{-1}}$ in $K_3[Fe(CN)_6]$(aq), $0.030\,\mathrm{mol\,kg^{-1}}$ in KCl(aq), and $0.050\,\mathrm{mol\,kg^{-1}}$ in NaBr(aq).

10.7. Calculate the masses of (a) $Ca(NO_3)_2$ and, separately, (b) NaCl to add to a $0.150\,\mathrm{mol\,kg^{-1}}$ solution of KNO_3(aq) containing $500\,\mathrm{g}$ of solvent to raise its ionic strength to $0.250\,\mathrm{mol\,kg^{-1}}$.

10.8. What molality of $CuSO_4$ has the same ionic strength as $1.00\,\mathrm{mol\,kg^{-1}}$ KCl(aq)?

10.9. Express the mean activity coefficient of the ions in a solution of $CaCl_2$ in terms of the activity coefficients of the individual ions.

10.10. Estimate the mean ionic activity coefficient of $CaCl_2$ in a solution that is $0.010\,\mathrm{mol\,kg^{-1}}$ $CaCl_2$(aq) and $0.030\,\mathrm{mol\,kg^{-1}}$ NaF(aq).

10.11. The mean activity coefficient in an $0.500\,\mathrm{mol\,kg^{-1}}$ $LaCl_3$(aq) solution is 0.303 at $25\,°C$. What is the percentage error in the value predicted by the Debye–Hückel limiting law?

10.12. The mean activity coefficient of HBr in three dilute aqueous solutions at $25\,°C$ are 0.930 (at $5.0\,\mathrm{mmol\,kg^{-1}}$), 0.907 (at $10.0\,\mathrm{mmol\,kg^{-1}}$), and 0.879 (at $20.0\,\mathrm{mmol\,kg^{-1}}$). Estimate the value of B in the extended Debye–Hückel law.

10.13. For CaF_2, $K_s = 3.9 \times 10^{-11}$ at $25\,°C$ and the standard Gibbs energy of formation of CaF_2(s) is $-1167\,\mathrm{kJ\,mol^{-1}}$. Calculate the standard Gibbs energy of formation of CaF_2(aq).

10.14. Consider a hydrogen electrode in aqueous HBr solution at $25\,°C$ operating at $1.15\,\mathrm{atm}$. Calculate the change in the electrode potential when the molality of the acid is changed from $5.0\,\mathrm{mmol\,kg^{-1}}$ to $20.0\,\mathrm{mmol\,kg^{-1}}$. Activity coefficients are given in Exercise 10.12.

10.15. Devise a cell in which the cell reaction is $Mn(s) + Cl_2(g) \rightarrow MnCl_2$(aq). Give the half-reactions for the electrodes and from the standard cell potential of $2.54\,\mathrm{V}$ deduce the standard potential of the Mn^{2+}/Mn couple.

10.16. Write the cell reactions and electrode half-reactions for the following cells:

(a) $Zn \mid ZnSO_4(aq) \parallel AgNO_3(aq) \mid Ag$
(b) $Cd \mid CdCl_2(aq) \parallel HNO_3(aq) \mid H_2(g) \mid Pt$
(c) $Pt \mid K_3[Fe(CN)_6](aq), K_4[Fe(CN)_6](aq) \parallel$
$CrCl_3(aq) \mid Cr$
(d) $Pt \mid Cl_2(g) \mid HCl(aq) \parallel K_2CrO_4(aq)$
$\mid Ag_2CrO_4(s) \mid Ag$
(e) $Pt \mid Fe^{3+}(aq), Fe^{2+}(aq) \parallel Sn^{4+}(aq), Sn^{2+}(aq) \mid Pt$
(f) $Cu \mid Cu^{2+}(aq) \parallel Mn^{2+}(aq), H^+(aq) \mid MnO_2(s) \mid Pt$

10.17. Devise cells in which the following are the reactions:

(a) $Zn(s) + CuSO_4(aq) \rightarrow ZnSO_4(aq) + Cu(s)$
(b) $2AgCl(s) + H_2(g) \rightarrow 2HCl(aq) + 2Ag(s)$
(c) $2H_2(g) + O_2(g) \rightarrow 2H_2O(l)$
(d) $2Na(s) + 2H_2O(l) \rightarrow 2NaOH(aq) + H_2(g)$
(e) $H_2(g) + I_2(s) \rightarrow 2HI(aq)$

10.18. Use standard potentials to calculate the standard potential of the cells in Exercises 10.16 and 10.17.

10.19. (a) Calculate the standard cell potential of $Hg \mid HgCl_2(aq) \parallel TlNO_3(aq) \mid Tl$ at $25\,°C$. (b) Calculate the cell potential when the activity of the Hg^{2+} ion is 0.150 and that of the Tl^+ ion is 0.93.

10.20. Calculate the standard Gibbs energies at $25\,°C$ of the following reactions from the standard potential data in Table 10.5:

(a) $2Na(s) + 2H_2O(l) \rightarrow 2NaOH(aq) + H_2(g)$
(b) $2K(s) + 2H_2O(l) \rightarrow 2KOH(aq) + H_2(g)$
(c) $K_2S_2O_8(aq) + 2KI(aq) \rightarrow I_2(s) + 2K_2SO_4(aq)$
(d) $Pb(s) + Zn(NO_3)_2(aq) \rightarrow Pb(NO_3)_2(aq) + Zn(s)$

10.21. The standard reaction Gibbs energy for $K_2CrO_4(aq) + 2Ag(s) + 2FeCl_3(aq) \rightarrow Ag_2CrO_4(s) + 2FeCl_2(aq) + 2KCl(aq)$ is $-62.5\,\mathrm{kJ\,mol^{-1}}$ at $298\,\mathrm{K}$. (a) Calculate the standard potential of the corresponding galvanic cell and (b) the standard potential of the Ag_2CrO_4/Ag, CrO_4^{2-} couple.

10.22. Two half-cell reactions may be combined in such a way as to form (a) a new half-cell reaction or (b) a complete cell reaction. Illustrate both (a) and (b) by using the half-cell reactions listed below and calculate $E^\ominus$ for both the new half-cell and complete cell reaction.

(i) $2H_2O(l) + 2e^- \rightarrow H_2(g) + 2OH^-(aq)$
$ E_1^\ominus = -0.828\,\mathrm{V}$
(ii) $Ag^+(aq) + e^- \rightarrow Ag(s)\qquad E_2^\ominus = +0.799\,\mathrm{V}$

10.23. Calculate the standard potential of the couple $Ag_2S, H_2O/Ag, S^{2-}, O_2, H^+$ from the following data:

$Ag_2S(s) + 2e^- \rightarrow 2Ag(s) + S^{2-}(aq)\qquad E^\ominus = -0.69\,\mathrm{V}$
$O_2(g) + 4H^+(aq) + 4e^- \rightarrow 2H_2O(l)\qquad E^\ominus = +1.23\,\mathrm{V}$

10.24. Consider the cell, $Pt \mid H_2(g, p^\ominus) \mid HCl(aq) \mid AgCl \mid Ag$, for which the cell reaction is $2AgCl(s) + H_2(g) \rightarrow 2Ag(s) + 2HCl(aq)$. At $25\,°C$ and a molality of HCl of $0.010\,\mathrm{mol\,kg^{-1}}$, $E = +0.4658\,\mathrm{V}$. (a) Write the Nernst equation for the cell reaction. (b) Calculate $\Delta_r G$ for the cell reaction. (c) Assuming that the Debye–Hückel limiting law holds at this concentration, calculate $E^\ominus(AgCl, Ag)$.

10.25. Use the Debye–Hückel limiting law and the Nernst equation to estimate the potential of the cell $Ag(s) \mid AgBr(s) \mid KBr(aq, 0.050\,\mathrm{mol\,kg^{-1}}) \parallel Cd(NO_3)_2(aq, 0.010\,\mathrm{mol\,kg^{-1}}) \mid Cd$ at $25\,°C$.

10.26. Use the information in Table 10.5 to calculate the standard potential of the cell $Ag \mid AgNO_3(aq) \parallel Fe(NO_3)_2(aq) \mid Fe$ and the standard Gibbs energy and enthalpy of the cell reactions at 25°C. Estimate the value of $\Delta_r G^\ominus$ at 35°C.

10.27. The solubility constant of $Cu_3(PO_4)_2$ is 1.3×10^{-37}. Calculate (a) the solubility of $Cu_3(PO_4)_2$, (b) the potential of the cell $Pt \mid H_2(g) \mid HCl(aq, pH = 0) \parallel Cu_3(PO_4)(aq, satd.) \mid Cu$ at 25°C.

10.28. Calculate the equilibrium constants of the following reactions at 25°C from standard potential data:

   (a) $Sn(s) + Sn^{4+}(aq) \rightleftharpoons 2Sn^{2+}(aq)$
   (b) $Sn(s) + 2AgCl(s) \rightleftharpoons SnCl_2(aq) + 2Ag(s)$
   (c) $2Ag(s) + Cu(NO_3)_2(aq) \rightleftharpoons Cu(s) + 2AgNO_3(aq)$
   (d) $Sn(s) + CuSO_4(aq) \rightleftharpoons Cu(s) + SnSO_4(aq)$
   (e) $Cu^{2+}(aq) + Cu(s) \rightleftharpoons 2Cu^+(aq)$

10.29. Use the standard potentials of the couples Au^+/Au $(+1.69\ V)$, Au^{3+}/Au $(+1.40\ V)$, and Fe^{3+}/Fe^{2+} $(+0.77\ V)$ to calculate $E^\ominus$ and the equilibrium constant for the reaction $2Fe^{2+}(aq) + Au^{3+}(aq) \rightleftharpoons 2Fe^{3+}(aq) + Au^+(aq)$.

10.30. Determine the standard potential of a cell in which the reaction is $Co^{3+}(aq) + 3Cl^-(aq) + 3Ag(s) \rightarrow 3AgCl(s) + Co(s)$ from the standard potentials of the couples $Ag/AgCl, Cl^-$ $(+0.22\ V)$, Co^{3+}/Co^{2+} $(+1.81\ V)$, and Co^{2+}/Co $(-0.28\ V)$.

10.31. The solubilities of AgCl and $BaSO_4$ in water are $1.34 \times 10^{-5}\ mol\ kg^{-1}$ and $9.51 \times 10^{-4}\ mol\ kg^{-1}$, respectively, at 25°C. Calculate their solubility constants. Is there any significant effect when activity coefficients are ignored?

10.32. Derive an expression for the potential of an electrode for which the half-reaction is the reduction of $Cr_2O_7^{2-}$ ions to Cr^{3+} ions in acidic solution.

10.33. The zero-current potential of the cell $Pt \mid H_2(g) \mid HCl(aq) \mid AgCl(s) \mid Ag$ was 0.322 V at 25°C. What is the pH of the electrolyte solution?

10.34. The solubility of AgBr is $2.6\ \mu mol\ kg^{-1}$ at 25°C. What is the standard potential of the cell $Ag \mid AgBr(aq) \mid AgBr(s) \mid Ag$ at that temperature?

10.35. The standard potential of the cell $Ag \mid AgI(s) \mid AgI(aq) \mid Ag$ is 0.9509 V at 25°C. Calculate (a) the solubility of AgI and (b) its solubility constant.

PROBLEMS
Numerical problems

10.1. Devise a cell in which the overall reaction is $Pb(s) + Hg_2SO_4(s) \rightarrow PbSO_4(s) + 2Hg(l)$. What is its potential when the electrolyte is saturated with both salts at 25°C?

10.2. Given that $\Delta_r G^\ominus = -212.7\ kJ\ mol^{-1}$ for the cell reaction in the Daniell cell at 25°C and that $m(CuSO_4) = 1.0 \times 10^{-3}\ mol\ kg^{-1}$ and $m(ZnSO_4) = 3.0 \times 10^{-3}\ mol\ kg^{-1}$, calculate (a) the ionic strengths of the solutions, (b) the mean ionic activity coefficients in the compartments, (c) the reaction quotient, (d) the standard cell potential, and (e) the cell potential. (Take $\gamma_+ = \gamma_- = \gamma_\pm$ in the respective compartments.)

10.3. Although the hydrogen electrode may be conceptually the simplest electrode and is the basis for the reference state of electrical potential in electrochemical systems, it is cumbersome to use. Therefore, several substitutes for it have been devised. One of these alternatives is the quinhydrone electrode (quinhydrone, $Q \cdot QH_2$, is a complex of quinone, $C_6H_4O_2 = Q$, and hydroquinone, $C_6H_4O_2H_2 = QH_2$). The electrode half-reaction is $Q(aq) + 2H^+(aq) + 2e^- \rightarrow QH_2(aq)$, $E^\ominus = +0.6994\ V$. If the cell $Hg \mid Hg_2Cl_2(s) \mid HCl(aq) \mid Q \cdot QH_2 \mid Au$ is prepared and the measured cell potential is $+0.190\ V$, what is the pH of the HCl solution? Assume that the Debye–Hückel limiting law is applicable.

10.4. A fuel cell develops an electric potential from the chemical reaction between reagents supplied from an outside source. What is the zero-current potential of a cell fuelled by (a) hydrogen and oxygen, (b) the combustion of butane at 1.0 atm and 298 K?

10.5. The fugacity of a gas can be determined electrochemically from the pressure dependence of the potential of a gas electrode. The zero-current potential of the cell $Pt \mid H_2(g, p^\ominus) \mid HCl(aq, 0.010\ mol\ kg^{-1}) \mid Cl_2(g, p) \mid Pt$ was as follows at 298 K:

p/bar	1.000	50.00	100.0
E/V	1.5962	1.6419	1.6451

Calculate the fugacities of chlorine at the three pressures. (Use activity coefficients from Table 10.4.)

10.6. Consider the cell, $Zn(s) \mid ZnCl_2(0.0050\ mol\ kg^{-1}) \mid Hg_2Cl_2(s) \mid Hg(l)$, for which the cell reaction is $Hg_2Cl_2(s) + Zn(s) \rightarrow 2Hg(l) + 2Cl^-(aq) + Zn^{2+}(aq)$. Given that $E^\ominus(Zn^{2+}, Zn) = -0.7628\ V$, $E^\ominus(Hg_2Cl_2, Hg) = +0.2676\ V$, and that the measured value of the cell potential is $+1.2272\ V$, (a) write down the Nernst equation for the cell. Also determine (b) the standard cell potential, (c) $\Delta_r G$, $\Delta_r G^\ominus$, and K for the cell reaction, (d) the mean ionic activity and activity coefficient of $ZnCl_2$ from the measured cell potentials, and (e) the mean ionic activity coefficient of $ZnCl_2$ from the Debye–Hückel limiting law. (f) Given that $dE/dT = -4.52 \times 10^{-4} \, V \, K^{-1}$, calculate $\Delta_r S^\ominus$ and $\Delta_r H^\ominus$.

10.7. The zero-current potential of the cell $Pt \mid H_2(g, p^\ominus) \mid HCl(aq, m) \mid Hg_2Cl_2(s) \mid Hg(l)$ has been measured with high precision (G. J. Hills and D. J. G. Ives, *J. Chem. Soc.* 311 (1951)) with the following results at 25°C:

$m/(\text{mmol kg}^{-1})$	1.6077	3.0769	5.0403	7.6938	10.9474
E/V	0.60080	0.56825	0.54366	0.52267	0.50532

Determine the standard potential of the cell and the mean activity coefficient of HCl at these molalities. (Make a least-squares fit of the data to the best straight line.)

10.8. Careful measurements of the potential of the cell $Pt \mid H_2(g, p^\ominus) \mid NaOH(aq, \, 0.0100 \, mol \, kg^{-1})$, $NaCl(aq, 0.01125 \, mol \, kg^{-1}) \mid AgCl(s) \mid Ag$ have been reported (C. P. Bezboruah, M. F. G. F. C. Camoes, A. K. Covington, and J. V. Dobson, *J. Chem. Soc. Faraday Trans.* I, **69**, 949 (1973)). Among the data is the following information:

$\theta/°C$	20.0	25.0	30.0
E/V	1.04774	1.04864	1.04942

Calculate pK_w at these temperatures and the standard enthalpy and entropy of the autoprotolysis of water at 25.0°C.

10.9. Measurements of the potentials of cells of the type $Ag \mid AgX(s) \mid MX(m_1) \mid M_xHg \mid MX(m_2) \mid AgX(s) \mid Ag$, where M_xHg denotes an amalgam and the electrolyte is an alkali metal halide dissolved in ethylene glycol, have been reported (U. Sen, *J. Chem. Soc. Faraday Trans.* I, **69**, 2006 (1973)) and a selection of values for LiCl are given below. Estimate the activity coefficient at the concentration marked * and then use this value to calculate activity coefficients from the measured cell potential at the other concentrations. Base your answer on the following version of the extended Debye–Hückel law:

$$\log \gamma_\pm = -\frac{A(m/m^\ominus)^{\frac{1}{2}}}{1 + B(m/m^\ominus)^{\frac{1}{2}}} + km/m^\ominus$$

with $A = 1.461$, $B = 1.70$, and $k = 0.20$. For $m_2 = 0.09141 \, mol \, kg^{-1}$:

$m_1/(\text{mol kg}^{-1})$	0.0555	0.09141*	0.1652	0.2171	1.040	1.350
E/V		-0.0220	0.0000	0.0263	0.0379	0.1156	0.1336

10.10. Suppose the extended Debye–Hückel law for a 1,1-electrolyte is written in the simplified form

$$\log \gamma_\pm = -0.509(m/m^\ominus)^{\frac{1}{2}} + k(m/m^\ominus)$$

where k is a constant. Show that a plot of y against $m/m^\ominus$, where

$$y = E + 0.1183 \log(m/m^\ominus) - 0.0602(m/m^\ominus)^{\frac{1}{2}}$$

should give a straight line with intercept $E^\ominus$ and slope

$-0.1183k$. Apply the technique to the following data (at 25°C) on the cell $Pt \mid H_2(g, p^\ominus) \mid HCl(aq, m) \mid AgCl(s) \mid Ag$:

$m/(\text{mmol kg}^{-1})$	123.8	25.63	9.138	5.619	3.215
E/mV	341.99	418.24	468.60	492.57	520.53

(a) Find the standard cell potential and the standard potential of the AgCl/Ag, Cl^- couple. (b) The zero-current cell potential was measured as 352.4 mV when $m = 100.0 \, mmol \, kg^{-1}$. What are the pH and the mean ionic activity coefficient?

10.11. The mean activity coefficients for aqueous solutions of NaCl at 25°C are given below. Confirm that they support the Debye–Hückel limiting law and that an improved fit is obtained with the extended law.

$m/(\text{mmol kg}^{-1})$	1.0	2.0	5.0	10.0	20.0
$\gamma_\pm$	0.9649	0.9519	0.9275	0.9024	0.8712

10.12. The standard potential of the AgCl/Ag, Cl^- couple has been measured very carefully over a range of temperatures (R. G. Bates and V. E. Bowers, *J. Res. Nat. Bur. Stand.* **53**, 283 (1954)) and the results were found to fit the expression

$$E^\ominus/V = 0.23659 - 4.8564 \times 10^{-4}(\theta/°C)$$
$$- 3.4205 \times 10^{-6}(\theta/°C)^2 + 5.869 \times 10^{-9}(\theta/°C)^3$$

Calculate the standard Gibbs energy and enthalpy of formation of Cl^-(aq) and its entropy at 298 K.

10.13. Use the data below to confirm that the Debye–Hückel limiting law correctly predicts the limiting values of the mean activity coefficient of acetic acid by demonstrating that pK_a', where $K_a = K_a' K_\gamma$, plotted against $(\alpha m)^{\frac{1}{2}}$, where m is the molality of the acid and α its degree of ionization, should be a straight line.

$m/(\text{mmol kg}^{-1})$	0.0280	0.1114	0.2184	1.0283	2.414	5.9115
α	0.5393	0.3277	0.2477	0.1238	0.0829	0.0540

10.14. The $Sb \mid Sb_2O_3(s) \mid OH^-$(aq) electrode is reversible with respect to OH^- ions. Derive an expression for its potential in terms of (a) the pOH and (b) the pH of the solution. By how much does the potential change when the molality of NaOH(aq) in the electrode compartment is increased from $0.010 \, mol \, kg^{-1}$ to $0.050 \, mol \, kg^{-1}$ at 25°C? Use the Debye–Hückel limiting law to estimate any activity coefficients required.

10.15. Superheavy elements are now of considerable interest. Shortly before it was (falsely) believed that the first had been discovered, an attempt was made to predict the chemical properties of ununpentium (element 115, O. L. Keller, C. W. Nestor, and B. Fricke, *J. Phys. Chem.* **78**, 1945 (1974)). In one part of the paper the standard enthalpy and entropy of the reaction

$$Uup^+(aq) + \tfrac{1}{2}H_2(g) \rightarrow Uup(s) + H^+(aq)$$

were estimated from the following data: $\Delta_{sub}H^\ominus(Uup) =$

1.5 eV, $E_i(\text{Uup}) = 5.52$ eV, $\Delta_{\text{hyd}}H^{\ominus}(\text{Uup}^+, \text{aq}) = -3.22$ eV, $S^{\ominus}(\text{Uup}^+, \text{aq}) = 1.34$ meV K^{-1}, $S^{\ominus}(\text{Uup}, \text{s}) = 0.69$ meV K^{-1}. Estimate the expected standard potential of the Uup$^+$/Uup couple.

Theoretical problems

10.16. Show that the solubility s of a sparingly soluble 1:1 salt is related to its solubility product by

$$s = K_s^{\frac{1}{2}} e^{1.172\sqrt{s}}$$

10.17. Suppose that a sparingly soluble salt MX has solubility constant K_s and solubility s. Show that in an ideal solution that is of concentration C of a freely soluble salt NX the solubility of MX is changed to

$$s' = \tfrac{1}{2}(C^2 + 4K_s)^{\frac{1}{2}} - \tfrac{1}{2}C$$

and that $s' = K_s/C$ when K_s is small (in a sense to be specified).

10.18. Show that, if the ionic strength of a solution of the sparingly soluble salt MX and the freely soluble salt NX is dominated by the concentration C of the latter and if it is valid to use the Debye–Hückel limiting law, then the solubility s' in the mixed solution is given by

$$s' = \frac{K_s}{C} \times e^{4.606A\sqrt{C}}$$

when K_s is small (in a sense to be specified).

10.19. Show that the freezing-point depression of a real solution in which the solvent of molar mass M has activity a_A obeys

$$\frac{\text{d}\ln a_A}{\text{d}(\Delta T)} = -\frac{M}{K_f}$$

and use the Gibbs–Duhem equation to show that

$$\frac{\text{d}\ln a_B}{\text{d}(\Delta T)} = -\frac{1}{m_B K_f}$$

where a_B is the solute activity and m_B is its molality. Use the Debye–Hückel limiting law to show that the osmotic coefficient (ϕ, Problem 7.14) is given by

$$\phi = 1 - \tfrac{1}{3}A'(m/m^{\ominus})^{\frac{1}{2}}$$

with $A' = 2.303A$.

PART TWO

Structure

In Part 1 we examined the properties of bulk matter from the viewpoint of thermodynamics. In Part 2 we turn to the study of individual atoms and molecules from the viewpoint of quantum mechanics. The two viewpoints merge in Chapter 19.

11

Quantum theory: introduction and principles

This chapter introduces some of the basic principles of quantum mechanics. First, it reviews the principal experimental results that overthrew the concepts of classical physics. These experiments led to the conclusions that particles may not have an arbitrary energy and that the concepts of 'particle' and 'wave' blend together. The overthrow of classical mechanics inspired the development of a new set of concepts and the formulation of quantum mechanics. In quantum mechanics, all the properties of a system are expressed in terms of a wavefunction that is obtained by solving the Schrödinger equation. We shall see how to interpret wavefunctions. Finally, we shall introduce some of the techniques of quantum mechanics in terms of operators, and see that they lead to the uncertainty principle, one of the most profound departures from classical mechanics.

In the *Molecular interpretations* of Part 1 we saw that bulk properties could be related to the behaviour of individual atoms and molecules. Now these individual species move to the centre of the stage and we consider the laws that govern their behaviour. To understand the structures of individual atoms and molecules and the behaviour of their component electrons and nuclei we need to know how such subatomic particles move in response to the forces they experience. It was once thought that the motion of atoms and subatomic particles could be expressed using the laws of **classical mechanics** introduced in the seventeenth century by Isaac Newton, for these laws were very successful at explaining the motion of everyday objects and planets. However, towards the end of the nineteenth century, experimental evidence accumulated showing that classical mechanics failed when it was applied to very small particles, and it took until about 1926 to discover the appropriate concepts and equations for describing them. We describe the concepts of this new mechanics, which is called **quantum mechanics**, in this chapter, and apply them throughout the remainder of the text.

THE ORIGINS OF QUANTUM MECHANICS

The basic principles of classical mechanics are reviewed in *Further information 5*. In brief, they show that classical physics: (1) predicts a precise trajectory for particles; and (2) allows the translational, rotational, and vibrational modes of motion to be excited to any energy simply by controlling the forces that are applied. These conclusions agree with everyday experience. Everyday experience, however, does not extend to individual atoms, and careful experiments of the type described in Section 11.1 have shown that the laws of classical mechanics fail when applied to the transfers of very small quantities of energy. Classical mechanics is in fact only an approximate description of the motion of particles and fails when small masses and small transfers of energy are involved.

11.1 The failures of classical physics

In this section we review some of the experimental evidence that showed that several concepts of classical mechanics are untenable. In particular, we shall see that observations on black-body radiation, heat capacities, and atomic and molecular spectra indicate that systems can take up energy only in discrete amounts.

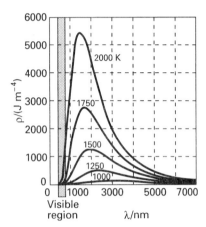

11.1 The energy density per unit wavelength range in a black-body cavity at several temperatures. Note how the energy density increases in the visible region as the temperature is raised, and how the peak shifts to shorter wavelengths. The total energy density (the area under the curve) increases as the temperature is increased (as T^4).

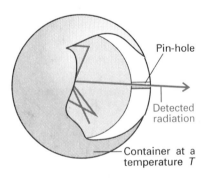

Black-body radiation

A hot object emits electromagnetic radiation. At high temperatures an appreciable proportion of the radiation is in the visible region of the spectrum, and a higher proportion of short-wavelength blue light is generated as the temperature is raised. This behaviour is seen when a heated iron bar glowing red hot becomes white hot when heated further. The precise dependence is illustrated in Fig. 11.1, which shows how the energy output varies with wavelength at several temperatures. The curves are those of an ideal emitter called a **black body,** which is an object capable of emitting and absorbing all frequencies of radiation uniformly. A good approximation to a black body is a pin-hole in a container, because any radiation leaking out of the hole has been absorbed and re-emitted inside so many times that it has come to thermal equilibrium with the walls (Fig. 11.2).

Figure 11.1 shows that the peak in the energy output shifts to shorter wavelengths as the temperature is raised. As a result, the short-wavelength tail of the energy distribution strengthens in the visible region and the perceived colour shifts towards the blue, as already mentioned. An analysis of the data led Wilhelm Wien (in 1893) to summarize this behaviour as follows:

Wien's displacement law: $T\lambda_{max} = \frac{1}{5}c_2$, where $c_2 = 1.44\,\text{cm K}$ (1)

The constant c_2 is called the **second radiation constant.** Using its value, we can predict that $\lambda_{max} = 2900\,\text{nm}$ at $1000\,\text{K}$.

A second feature of black-body radiation had been noticed in 1879 by Josef Stefan, who considered the **total energy density** $\mathscr{E}$, the total energy per unit volume in the electromagnetic field. The electromagnetic field inside the container in Fig. 11.2 has a definite total energy, which increases as the temperature is increased; the total energy density is this total energy divided by the volume of the interior of the container. Stefan concluded that the total energy density is proportional to the fourth power of the temperature

Stefan–Boltzmann law: $\mathscr{E} = aT^4$ (2a)

Boltzmann's name is attached to this law because he explained it theoretically. An alternative form of the law is in terms of the **excitance** M, the power (the energy per unit time, in watts) emitted per unit area (broadly speaking, the brightness of the emission). Because the excitance is proportional to the energy density in the container, M is also proportional to T^4 and we can write

$$M = \sigma T^4 \qquad \sigma = 5.67 \times 10^{-8}\,\mathrm{W\,m^{-2}\,K^{-4}} \tag{2b}$$

The constant σ is called the **Stefan–Boltzmann constant**. The Stefan–Boltzmann law implies that $1\,\mathrm{cm^2}$ of the surface of the black body at 1000 K radiates about 6 W when all wavelengths of the emitted radiation are taken into account.

The physicist Lord Rayleigh studied black-body radiation from a classical viewpoint. He thought of the electromagnetic field as a collection of oscillators, and regarded the presence of radiation of a frequency ν (and therefore of wavelength $\lambda = c/\nu$) as the result of exciting the electromagnetic oscillator of that frequency. Rayleigh used the equipartition principle (see the *Introduction*, p. 15) to calculate the average energy of the oscillators. With minor help from James Jeans, he arrived at the **Rayleigh–Jeans law**:

$$\mathrm{d}\mathscr{E} = \rho\,\mathrm{d}\lambda, \qquad \text{where } \rho = \frac{8\pi kT}{\lambda^4} \tag{3}$$

In this expression k is Boltzmann's constant, $k = 1.381 \times 10^{-23}\,\mathrm{J\,K^{-1}}$. Unfortunately (for Rayleigh, Jeans, and classical physics), although the Rayleigh–Jeans formula is quite successful at long wavelengths and low frequencies, it fails badly at high frequencies. Thus, as λ decreases, ρ increases without going through a maximum (Fig. 11.3). The equation therefore predicts that oscillators of very short wavelength (high frequency, corresponding to ultraviolet radiation, X-rays, and even γ-rays) are strongly excited even at room temperature. This absurd result, which implies that a large amount of energy is radiated in the high-frequency region of the electromagnetic spectrum, is called the **ultraviolet catastrophe**. According to classical physics, even relatively cool objects should radiate in the visible and ultraviolet regions; that is, objects should glow in the dark: there should in fact be no darkness.

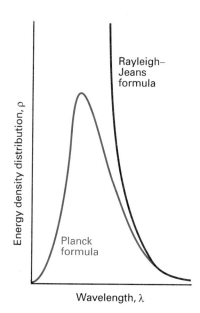

11.3 Theoretical attempts to account for black-body radiation. The Rayleigh–Jeans law (eqn 3) leads to an infinite energy density at short wavelengths called the ultraviolet catastrophe. The Planck distribution (eqn 5) is in good agreement with experiment.

The Planck distribution

The German physicist Max Planck studied black-body radiation from the viewpoint of thermodynamics. In 1900 he found that he could account for the experimental observations by proposing that *the energy of each electromagnetic oscillator is limited to discrete values and cannot be varied arbitrarily*. The limitation of energy to discrete values is called the **quantization of energy** (from the Latin word *quantum*, amount). In particular, he proposed that the permitted energies of an oscillator of frequency v are integer multiples of hv:

$$E = nhv, \qquad \text{with } n = 0, 1, 2, \ldots \tag{4}$$

where h is a fundamental constant now known as **Planck's constant**:

$$h = 6.62608 \times 10^{-34}\,\text{J s}$$

It is quite easy to see why Rayleigh's approach was unsuccessful and Planck's hypothesis was successful. The thermal motion of the atoms in the walls of the black body excites the oscillators of the electromagnetic field. According to classical mechanics, all the oscillators of the field share equally in the energy supplied by the walls so that even the highest frequencies are excited. The excitation of high-frequency oscillators leads to the ultraviolet catastrophe. According to Planck's hypothesis, however, oscillators are excited only if they can acquire an energy of at least hv. This energy is too large for the walls to supply in the case of the high-frequency oscillators, so the latter remain unexcited. *The effect of quantization is to eliminate the contribution from the high-frequency oscillators, for they cannot be excited with the energy available.*

Detailed calculation (of the kind illustrated in Chapter 19) shows that the energy density in the range λ to $\lambda + d\lambda$ is given by the **Planck distribution**:

$$d\mathscr{E} = \rho\,d\lambda, \qquad \text{where } \rho = \frac{8\pi hc}{\lambda^5}\left(\frac{1}{e^{hc/\lambda kT} - 1}\right) \tag{5}$$

This expression fits the experimental curve very well at all wavelengths, and the value of h, which at the time of its introduction was an undetermined parameter in the theory. The value of h may be obtained by varying its value until a best fit is obtained.

The Planck distribution resembles the Rayleigh–Jeans law (eqn 3) apart from the all-important exponential factor. For short wavelengths, $hc/\lambda kT$ is large and $e^{hc/\lambda kT} \to \infty$; therefore

$$\rho \to 0 \text{ as } \lambda \to 0 \text{ and } v \to \infty$$

Hence the energy density approaches zero at high frequencies, in agreement with observation. For long wavelengths, $hc/\lambda kT \ll 1$, and the denominator in the Planck distribution can be replaced by

$$e^{hc/\lambda kT} - 1 = \left(1 + \frac{hc}{\lambda kT} + \ldots\right) - 1 \approx \frac{hc}{\lambda kT}$$

It is then easy to show (by substitution of this relation into eqn 5) that the Planck formula reduces to the Rayleigh–Jeans law for long-wavelength radiation.

The Planck distribution also accounts for the Stefan–Boltzmann and Wien laws. The former is obtained by integrating the energy density over all wavelengths from $\lambda = 0$ to $\lambda = \infty$, which gives

$$\mathscr{E} = \int_0^\infty \rho\,d\lambda = aT^4, \qquad \text{with } a = \frac{4\sigma}{c}, \quad \sigma = \frac{2\pi^5 k^4}{15c^2 h^3}$$

Substitution of the values of the fundamental constants then gives $\sigma = 5.67 \times 10^{-8}\,\mathrm{W\,m^{-2}\,K^{-4}}$, in accord with the experimental value. The Wien law is obtained by looking for the wavelength at which $\mathrm{d}\mathscr{E}/\mathrm{d}\lambda = 0$, the condition for the maximum in the distribution. At high temperatures this derivative is zero at λ_{max}, where

$$T\lambda_{\mathrm{max}} = \frac{hc}{5k}$$

This result lets us identify the second radiation constant as $c_2 = hc/k$. The value of c_2 works out as $1.439\,\mathrm{cm\,K}$, in good agreement with experiment.

At this stage, we have to conclude that the energy of the oscillators that make up the electromagnetic field are quantized.

Heat capacities

Now we consider experimental data on the heat capacities of solids, which also led to the view that energy is quantized. If classical physics were valid, the equipartition theorem[1] would imply that the mean vibrational energy of each atom in a solid is $3kT$. For a solid composed of N atoms, the total vibrational energy of all the atoms is expected to be $3NkT$. The contribution of the vibrational energy to the molar internal energy is therefore

$$U_{\mathrm{m}} = 3N_{\mathrm{A}}kT = 3RT$$

because $N_{\mathrm{A}}k = R$, the gas constant. The molar constant volume heat capacity (eqn 2.16) is then predicted to be

$$C_{V,\mathrm{m}} = \left(\frac{\partial U_{\mathrm{m}}}{\partial T}\right)_V = 3R \tag{6}$$

This result is known as **Dulong and Petit's law**, for it had been proposed by them on the basis of some somewhat slender experimental evidence.

Significant deviations from Dulong and Petit's law were observed when technological advances made it possible to measure heat capacities at low temperatures. It was then found that the molar heat capacities of all metals are lower than $3R$ at low temperatures and that the values approach zero as $T \to 0$. To account for these observations Einstein assumed (in 1905) that each atom oscillated about its equilibrium position with a single frequency v. He then invoked Planck's hypothesis to assert that the energy of any oscillation is nhv, where n is an integer. First he calculated the molar vibrational energy of the metal (by a method described in Section 19.3) and obtained

$$U_{\mathrm{m}} = \frac{3N_{\mathrm{A}}hv}{e^{hv/kT} - 1}$$

1 See the *Introduction*. The energy of a one-dimensional oscillator is determined by *two* quadratic terms, the kinetic energy $p^2/2m$ and the potential energy, which is proportional to x^2. According to the theorem, the average contribution of each term is $\frac{1}{2}kT$, so that jointly they contribute kT. For an atom free to vibrate in three dimensions there are three such contributions giving a total average energy of $3kT$.

in place of the classical expression $3RT$. Then he found the heat capacity by differentiating U_m with respect to T. The resulting expression is now known as the **Einstein formula**:

$$C_{V,m} = 3Rf^2, \qquad \text{where } f = \frac{h\nu}{kT}\left(\frac{e^{h\nu/2kT}}{e^{h\nu/kT} - 1}\right) \qquad (7)$$

At high temperatures (when $h\nu/kT \ll 1$) the exponentials in f can be expanded using $e^x = 1 + x + \ldots$ and higher terms ignored. The result is

$$f = \frac{h\nu}{kT} \times \frac{1 + \dfrac{h\nu}{2kT} + \ldots}{\left(1 + \dfrac{h\nu}{kT} + \ldots\right) - 1} = 1 + \frac{h\nu}{2kT} + \ldots \approx 1$$

Consequently, the classical result ($C_{V,m} = 3R$) is obtained at high temperatures. At low temperatures, when $h\nu/kT \gg 1$,

$$f \approx \frac{h\nu}{kT} \times \frac{e^{h\nu/2kT}}{e^{h\nu/kT}} = \frac{h\nu}{kT} \times e^{-h\nu/2kT}$$

and the strongly decaying exponential function goes to zero more rapidly than the term that multiplies it goes to infinity, so that $f \to 0$ as $T \to 0$ and the heat capacity therefore also approaches zero. Einstein's formula therefore accounts for the decrease of heat capacity at low temperatures. The physical reason for this success is that, as in the Planck calculation, at low temperatures only a few oscillators possess enough energy to begin oscillating. At higher temperatures, there is enough energy available for all the oscillators to become active: all $3N$ oscillators contribute, and the heat capacity approaches its classical value.

The overall temperature dependence predicted by the Einstein formula is plotted in Fig. 11.4: the general shape of the curve is satisfactory, but the numerical agreement is in fact quite poor. The poor fit arises from Einstein's assumption that all the atoms oscillate with the same frequency, whereas in fact they oscillate with a range of frequencies. This complication is taken into account by averaging over all the frequencies present, the final result being the **Debye formula**. The details of this modification which, as Fig. 11.4 shows, gives improved agreement with experiment, need not distract us at this stage from the main conclusion, which is that *quantization must be introduced in order to explain the thermal properties of solids*.

At this point, we have to conclude that not only are the electromagnetic oscillators quantized, but so too are the oscillations of atoms in matter.

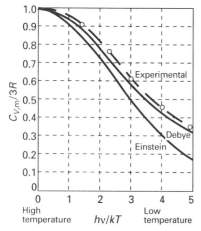

11.4 Experimental low-temperature heat capacities and theoretical predictions. The Dulong and Petit law predicts no falling off at low temperatures (to the right of the graph): the Einstein equation (eqn 7) predicts the general temperature dependence fairly well, but is everywhere too low. Debye's modification of Einstein's calculation gives very good agreement with experiment. For copper, $h\nu/kT = 2$ corresponds to about 170 K, so the detection of deviations from Dulong and Petit's law had to await advances in low-temperature physics.

Atomic and molecular spectra

The most directly compelling evidence for the quantization of energy comes from the observation of the frequencies of radiation absorbed and emitted by atoms and molecules.

A typical atomic spectrum is shown in Fig. 11.5, and a typical molecular spectrum is shown in Fig. 11.6. The obvious feature of both is that radiation is emitted or absorbed at a series of discrete frequencies.

11.5 The spectrum of light emitted by excited mercury atoms consists of radiation at a series of discrete frequencies (which increases to the right). Photographic recording of spectra has been replaced by electronic detection and graphical portrayal, but this photograph and the next show the origin of the term 'spectral line' for an absorption or emission spectrum.

11.6 When a molecule changes its state, it does so by absorbing light at definite frequencies. This suggests that it can possess only discrete energies, not an arbitrary energy. This photograph is a part of the ultraviolet absorption spectrum of ScF (provided by Dr R. F. Barrow).

This can be understood if the energy of the atoms or molecules is also confined to discrete values, for then energy can be discarded or absorbed only in discrete amounts (Fig. 11.7). Then, if the energy of an atom decreases by ΔE, the energy is carried away as radiation of frequency $v = \Delta E/h$, and a line appears in the spectrum.

11.2 Wave–particle duality

In this section we shall see the experimental evidence that led to the formulation of a fundamental revision of two basic concepts about the nature of the world. Through the photoelectric effect we see that electromagnetic radiation—which classical physics treats as wave-like—actually also displays the characteristics of particles. Then, through experiments on electron diffraction, we shall see that electrons—which classical physics treats as particles—also display the characteristics of waves.

The particle character of electromagnetic radiation

The information we have met so far strongly suggests that electromagnetic radiation has a particle-like character. Thus, because an oscillator of frequency v can possess only the energies $0, hv, 2hv, \ldots$, Planck's work suggests that radiation of that frequency can be thought of as consisting of $0, 1, 2, \ldots$ particles, each particle having an energy hv. These particles of electromagnetic radiation are now called **photons**. The observation of discrete spectra from atoms and molecules can be pictured as the atom or molecule generating a photon of energy hv when it discards an energy of magnitude ΔE, with $\Delta E = hv$.

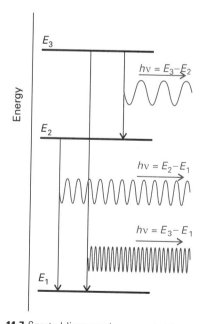

11.7 Spectral lines can be accounted for if we assume that a molecule emits a photon as it changes between discrete energy levels. Note that high-frequency radiation is emitted when the energy change is large.

Example 11.1 *Calculating the number of photons*

Calculate the number of photons emitted by a 100-W yellow lamp in 1.0 s. Take the wavelength of yellow light as 560 nm and assume 100 per cent efficiency.

Method. Each photon has an energy $h\nu$, so the total number of photons needed to produce an energy E is $E/h\nu$. To use this equation, we need to know the frequency of the radiation (from $\nu = c/\lambda$) and the total energy emitted by the lamp. The latter is given by the product of the power (in watts) and the time for which the source acts ($E = Pt$).

Answer. The number of photons is

$$N = \frac{E}{h\nu} = \frac{Pt}{h \times \dfrac{c}{\lambda}} = \frac{\lambda Pt}{hc}$$

Substitution of the data gives

$$N = \frac{(5.60 \times 10^{-7}\,\text{m}) \times (100\,\text{W}) \times (1.0\,\text{s})}{(6.626 \times 10^{-34}\,\text{J s}) \times (2.998 \times 10^8\,\text{m s}^{-1})} = 2.7 \times 10^{20}$$

Exercise E11.1. How many 1000-nm photons does a 1-mW monochromatic infrared rangefinder emit in 0.1 s? $[5 \times 10^{14}]$

Further evidence for the particle-like character of radiation (and historically the motivation for Einstein's proposal of the existence of particles of electromagnetic radiation) comes from the measurement of the energies of electrons produced by the **photoelectric effect**. This effect is the ejection of electrons from metals when they are exposed to ultraviolet radiation. The experimental characteristics of the photoelectric effect are as follows:

(1) No electrons are ejected, regardless of the intensity of the radiation, unless its frequency exceeds a threshold value characteristic of the metal.

(2) The kinetic energy of the ejected electrons varies linearly with the frequency of the incident radiation but is independent of its intensity.

(3) Even at low light intensities, electrons are ejected immediately if the frequency is above threshold.

These observations strongly suggest that the photoelectric effect depends on the ejection of an electron when it is involved in a collision with a particle-like projectile that carries enough energy to expel it from the metal. If we suppose that the projectile is a photon of energy $h\nu$, where ν is the frequency of the radiation, then the conservation of energy requires that the kinetic energy of the ejected electron should obey

$$\tfrac{1}{2}m_e v^2 = h\nu - \Phi \tag{8}$$

In this expression Φ is a characteristic of the metal called its **work function**, the energy required to remove an electron (the analogue of ionization energy for atoms, Fig. 11.8). If $h\nu < \Phi$, then photoejection cannot occur because the photon brings insufficient energy: this accounts for observation (1). Equation 8 predicts that the kinetic energy of an ejected electron should vary linearly with the frequency, in agreement

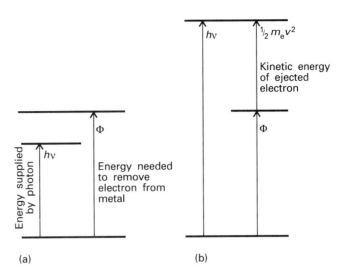

11.8 The photoelectric effect can be explained if it is supposed that the incident radiation is composed of photons that have energy proportional to the frequency of the radiation. (a) The energy of the photon is insufficient to drive an electron out of the metal. (b) The energy of the photon is more than enough to eject an electron, and the excess energy is carried away as the kinetic energy of the photoelectron (the ejected electron).

with observation (2). When a photon collides with an electron, it gives up all its energy, so we should expect electrons to appear as soon as the collisions begin, provided they have sufficient energy: this agrees with observation (3).

At this stage, we have to conclude that electromagnetic radiation, while possessing the properties of waves, also possesses properties characteristic of particles.

The wave character of particles

Although contrary to the long-established wave theory of light, the view that light consists of particles had been held before, but discarded. No significant scientist, however, had taken the view that matter is wave-like. Nevertheless, experiments carried out in 1925 forced people to even that conclusion. The crucial experiment was performed by the American physicists Clinton Davisson and Lester Germer, who observed the diffraction of electrons by a crystal (Fig. 11.9). Diffraction is a characteristic property of waves because it occurs when there is interference between their peaks and troughs. Depending on whether the interference is constructive or destructive, it leads to regions of enhanced and diminished intensity. Davisson and Germer's success was a lucky accident, because a chance rise of temperature caused their polycrystalline sample to anneal, and the ordered planes of atoms then acted as a diffraction grating. At almost the same time G. P. Thomson, working in Aberdeen, showed that a beam of electrons was diffracted when passed through a thin gold foil (Fig. 11.10).

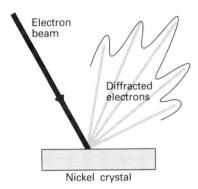

11.9 The Davisson–Germer experiment. The scattering of an electron beam from a nickel crystal shows a variation of intensity characteristic of a diffraction experiment in which waves interfere constructively and destructively in different directions.

The Davisson–Germer experiment, which has since been repeated with other particles (including molecular hydrogen), shows clearly that particles have wave-like properties. We have also seen that waves of electromagnetic radiation have particle properties. Thus we are brought to the heart of modern physics. *When examined on an atomic scale, the concepts of particle and wave melt together, particles taking on the characteristics of waves, and waves the characteristics of particles.*

Some progress towards coordinating these properties had already been

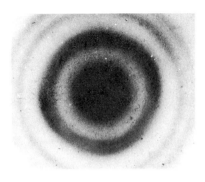

11.10 The Thomson experiment. This is the actual photograph obtained by G. P. Thomson when he directed a beam of electrons through a thin gold foil and detected rings of alternating intensity on a photographic plate that responded to the arrival of fast electrons.

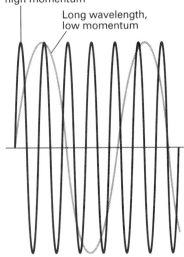

Short wavelength, high momentum

Long wavelength, low momentum

11.11 An illustration of the de Broglie relation between momentum and wavelength. A particle with high momentum has a wavefunction with a short wavelength, and vice versa.

made by Louis de Broglie when, in 1924, he had suggested on theoretical grounds that any particles, not only photons, travelling with a momentum p should have (in some sense) a wavelength given by the **de Broglie relation**:

$$\lambda = \frac{h}{p} \tag{9}$$

That is, a particle with a high linear momentum has a short wavelength (Fig. 11.11). Conversely, long-wavelength electromagnetic radiation has a low momentum. Macroscopic bodies have such high momenta (even when they are moving slowly) that their wavelengths are undetectably small, and the wave-like properties cannot be observed. Electromagnetic radiation in the visible and ultraviolet regions has such long wavelengths that the momenta of the individual photons are very small and undetectable except in special arrangements. Thus, the particle-like properties of electromagnetic radiation went unnoticed for a long time.

Example 11.2 *Estimating the de Broglie wavelength*

Estimate the wavelength of electrons that have been accelerated from rest through a potential difference of 1.00 kV.

Method. To use the de Broglie relation, we need to know the linear momentum of the electrons. To calculate it, we note that the energy acquired by an electron that falls through a potential difference $\Delta\phi$ is $e \times \Delta\phi$, where e is the magnitude of its charge. At the end of the period of acceleration all the acquired energy is in the form of kinetic energy, $E_K = \frac{1}{2}m_e v^2$. Therefore, by expressing the kinetic energy in terms of the linear momentum $p = m_e v$, which gives $E_K = p^2/2m_e$, and equating the kinetic energy to the energy acquired by acceleration, we can find p and hence, through eqn 9, the wavelength.

Answer. After the period of acceleration,

$$\frac{p^2}{2m_e} = e\,\Delta\phi$$

which implies first that

$$p = (2m_e e\,\Delta\phi)^{\frac{1}{2}}$$

and then that the de Broglie wavelength is

$$\lambda = \frac{h}{(2m_e e\,\Delta\phi)^{\frac{1}{2}}}$$

Substitution of the data and the fundamental constants (from inside the front cover) gives

$$\lambda = \frac{6.626 \times 10^{-34}\,\text{J s}}{\{2 \times (9.109 \times 10^{-31}\,\text{kg}) \times (1.602 \times 10^{-19}\,\text{C}) \times (1.00 \times 10^{3}\,\text{V})\}^{\frac{1}{2}}}$$

$$= 3.88 \times 10^{-11}\,\text{m}$$

Comment. We have used $1\,C\,V = 1\,J$ and $1\,J = 1\,kg\,m^2\,s^{-2}$. The wavelength of 38.8 pm is comparable to typical bond lengths in molecules (about 100 pm). Electrons accelerated in this way are used in the technique of electron diffraction (Section 21.10) for the determination of molecular structure.

Exercise E11.2. Calculate the wavelength of an electron in a 10-MeV particle accelerator ($1\,MeV = 10^6\,eV$). [0.39 pm]

We now have to conclude that, not only has electromagnetic radiation the character of particles, but electrons (and other particles) have the characteristics of waves. This joint particle and wave character of matter and radiation is called **wave–particle duality**. Duality strikes at the heart of classical physics, where particles and waves are treated as entirely separate entities. We have also seen that the energies of electromagnetic radiation and of matter cannot be varied continuously and that, for small objects, the discreteness of energy is highly significant. In classical mechanics, in contrast, energies could be varied continuously. Such total failure of classical physics for small objects implies that its basic concepts are false. A new mechanics had to be devised to take its place.

THE DYNAMICS OF MICROSCOPIC SYSTEMS

From now on, we shall combine the characteristics of particles and waves and suppose that the position of a particle is distributed through space like the amplitude of a wave. This remark, and the blending of particle and wave properties that it entails, will probably seem mysterious at this stage: it will be interpreted more fully shortly. The wave that in quantum mechanics replaces the classical concept of trajectory is called a **wavefunction** and denoted ψ.

The following sections will build up an understanding of the significance of the wavefunctions of particles. First, we shall see how to calculate the form of the wavefunction, and then we deal with its interpretation. The features that we shall develop, which it would be helpful to bear in mind throughout the following discussion, are as follows:

(1) A wavefunction is just a mathematical function (such as $\sin x$ or e^{-x}), which may be large in one region, small in others, and zero elsewhere.

(2) A wavefunction contains all the information it is possible to know about the location and motion of the particle it describes.

(3) If a wavefunction is large at a particular point, then the particle has a high probability of being at that point; if the wavefunction is zero at a point, then the particle will not be found there.

(4) The more rapidly a wavefunction changes from place to place, the higher the kinetic energy of the particle it describes.

11.3 The Schrödinger equation

In 1926, the Austrian physicist Erwin Schrödinger proposed an equation for finding the wavefunction of any system. The **Schrödinger equation** for a particle of mass m moving in one dimension with energy E is

$$-\frac{\hbar^2}{2m}\frac{d^2\psi}{dx^2} + V(x)\psi = E\psi \tag{10}$$

The term $V(x)$, which in general depends on the position x, is the potential energy of the particle; $\hbar$ (which is read h-cross or h-bar) is a convenient modification of Planck's constant:

$$\hbar = \frac{h}{2\pi} = 1.05457 \times 10^{-34}\,\text{J s}$$

Various ways of expressing this equation, of incorporating the time-dependence of the wavefunction, and of extending it to more dimensions are collected in Table 11.1.

Table 11.1 The Schrödinger equation

For one-dimensional systems:

$$-\frac{\hbar^2}{2m}\frac{d^2\psi}{dx^2} + V(x)\psi = E\psi$$

where $V(x)$ is the potential energy of the particle and E is its energy. For three-dimensional systems:

$$-\frac{\hbar^2}{2m}\nabla^2\psi + V\psi = E\psi$$

where V may depend on position and ∇^2 ('del squared') is

$$\nabla^2 = \frac{\partial^2}{\partial x^2} + \frac{\partial^2}{\partial y^2} + \frac{\partial^2}{\partial z^2}$$

In systems with spherical symmetry:

$$\nabla^2 = \frac{\partial^2}{\partial r^2} + \frac{2}{r}\frac{\partial}{\partial r} + \frac{1}{r^2}\Lambda^2$$

where

$$\Lambda^2 = \frac{1}{\sin^2\theta}\frac{\partial^2}{\partial\phi^2} + \frac{1}{\sin\theta}\frac{\partial}{\partial\theta}\sin\theta\frac{\partial}{\partial\theta}$$

In the general case the Schrödinger equation is written

$$H\psi = E\psi$$

where H is the Hamiltonian operator for the system:

$$H = -\frac{\hbar^2}{2m}\nabla^2 + V$$

For the evolution of a system with time, it is necessary to solve the time-dependent Schrödinger equation:

$$H\psi = i\hbar\frac{\partial\psi}{\partial t}$$

JUSTIFICATION

Although the Schrödinger equation should be regarded as a postulate, like Newton's equations of motion, it can be seen to be plausible by noting that it implies the de Broglie relation for a freely moving particle. First, eqn 10 can be rearranged into

$$\frac{d^2\psi}{dx^2} = -\frac{2m}{\hbar^2}\{E - V(x)\}\psi$$

If the potential is constant at V, a solution of this equation is

$$\psi = e^{ikx} = \cos kx + i \sin kx, \qquad \text{where } k = \left(\frac{2m(E - V)}{\hbar^2}\right)^{\frac{1}{2}}$$

Cos kx (or $\sin kx$) is a wave of wavelength $\lambda = 2\pi/k$, as can be seen by comparing $\cos kx$ with the standard form of a harmonic wave, $\cos(2\pi x/\lambda)$. The quantity $E - V$ is equal to the kinetic energy of the particle E_K, so $k = (2mE_K/\hbar^2)^{\frac{1}{2}}$, which implies that $E_K = k^2\hbar^2/2m$. Because $E_K = p^2/2m$, it follows that

$$p = k\hbar$$

Therefore, the linear momentum is related to the wavelength of the wavefunction by

$$p = \frac{2\pi}{\lambda} \times \frac{h}{2\pi} = \frac{h}{\lambda}$$

which is de Broglie's relation.

We shall now start to establish the interpretation of a wavefunction. To do so, we shall use the de Broglie relation (or, equivalently, the solution established in the *Justification*) that

$$\lambda = \frac{h}{\{2m(E - V)\}^{\frac{1}{2}}} \tag{11}$$

This equation shows that *the greater the difference between the total energy and the potential energy, the shorter the wavelength of the wavefunction*. In other words, *the greater the kinetic energy, the shorter the wavelength*. A stationary particle, one with zero kinetic energy (and hence $E = V$), has infinite wavelength, which means that its wavefunction has the same value everywhere. That is, for a particle at rest, $\psi = $ constant.

The only wavefunctions we have seen so far are true waves (as in Fig. 11.11). However, in due course we shall encounter wavefunctions that do not spread harmonically through space, and might resemble those shown in Fig. 11.12. Such waves do not have a 'wavelength' and, to interpret them in terms of the kinetic energy they represent, we need a more general feature of their shape. This general feature is the *curvature* of the wavefunction, which for our purposes we shall interpret as the second derivative, $d^2\psi/dx^2$. The curvature of a wavefunction in general

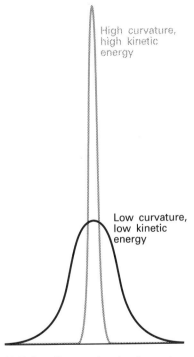

High curvature, high kinetic energy

Low curvature, low kinetic energy

11.12 Even if a wavefunction does not have the form of a periodic wave, it is still possible to infer from it the average kinetic energy of a particle by noting its average curvature. This illustration shows two wavefunctions: the sharply curved function corresponds to a higher kinetic energy than the less sharply curved function.

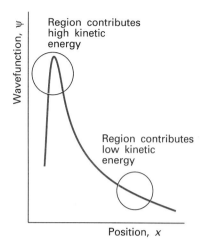

11.13 The observed kinetic energy of a particle is an average of contributions from the entire space covered by the wavefunction. Sharply curved regions contribute a high kinetic energy to the average; slightly curved regions contribute only a small kinetic energy.

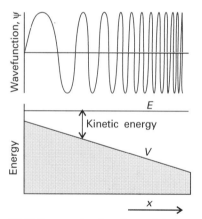

11.14 The wavefunction of a particle in a potential decreasing towards the right and hence subjected to a constant force to the right. Only the real part of the wavefunction is shown: the imaginary part is similar, but displaced to the right.

varies from place to place, so the kinetic energy of the particle varies similarly. Wherever a wavefunction is sharply curved, its contribution to the total kinetic energy is large (Fig. 11.13). Wherever the wavefunction is not sharply curved, its contribution to the overall kinetic energy is low. The *observed* kinetic energy of the particle is the integral of all the contributions of the kinetic energy from each region. Hence, we can expect the particle to have a high kinetic energy if the average curvature of the wavefunction is high. Thus, a wavefunction with a short wavelength (and hence, on average, a high curvature) has a high kinetic energy, and a wavefunction with short wavelength (and hence a low average curvature) has a low kinetic energy.

The association of sharp curvature with high kinetic energy will turn out to be a valuable guide to the interpretation of wavefunctions and the prediction of their shapes. For example, suppose we need to know the wavefunction of a particle with a given total energy and a potential energy that decreases with increasing x, as in the lower half of Fig. 11.14. Because the difference $E - V = E_K$ increases from left to right, the wavefunction must become more sharply curved as x increases: its wavelength decreases as the local contribution to its kinetic energy increases. We can therefore guess that the wavefunction will look like the function drawn in the upper half of Fig. 11.14, and more detailed calculation confirms this to be so.

11.4 The Born interpretation of the wavefunction

The interpretation of the wavefunction in terms of the location of the particle it describes is based on a suggestion made by Max Born. He made use of an analogy with the wave theory of light, in which the square of the amplitude of an electromagnetic wave in a region is interpreted as its intensity and therefore (in quantum terms) as a measure of the probability of finding a photon present in the region. The **Born interpretation** of the wavefunction is that the square of the wavefunction (or $\psi^*\psi$ if ψ is complex) at a point is proportional to the probability of finding the particle at that point. Specifically, for a one-dimensional system (Fig. 11.15):

> If the amplitude of the wavefunction of a particle is ψ at some point x, then the probability of finding the particle between x and $x + dx$ is proportional to $\psi^*\psi\, dx$.

Thus $\psi^*\psi$ is a **probability density** (because it must be multiplied by the length of the infinitesimal region dx to obtain the probability). The wavefunction ψ itself is called a **probability amplitude**. For a particle free to move in three dimensions (e.g. an electron near a nucleus in an atom), the wavefunction depends on the point $\boldsymbol{r}$ with coordinates x, y, and z, and the interpretation of $\psi(\boldsymbol{r})$ is then (Fig. 11.16):

> If the amplitude of the wavefunction of a particle is ψ at some point $\boldsymbol{r}$, then the probability of finding the particle in an infinitesimal volume $d\tau = dx\, dy\, dz$ at the point $\boldsymbol{r}$ is proportional to $\psi^*\psi\, d\tau$.

The Born interpretation does away with any worry about the significance of a *negative* (and, in general, complex) value of ψ: the value of

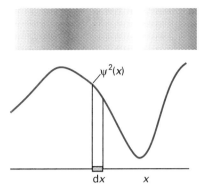

11.15 The interpretation of the wavefunction. The probability of finding the particle at points on the x-axis is represented by the density of shading in the upper half of the diagram. This density is determined by the *square* of the wavefunction at each point.

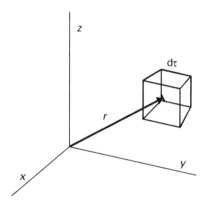

11.16 The Born interpretation of the wavefunction in three-dimensional space implies that the probability of finding the particle in the volume element $d\tau = dx\, dy\, dz$ at some location r is proportional to the product of $d\tau$ and the value of $\psi^*\psi$ at that location.

$\psi^*\psi$, which is often written $|\psi|^2$ and called the *square modulus* of ψ, is real and never negative. There is no *direct* significance in the negative value of a wavefunction. For example, the sign of the wavefunction has nothing whatsoever to do with the electric charge of the particle. Nor does a negative value imply that a particle is in any sense missing from a region: only the square, a positive quantity, is directly physically significant. A wavefunction is therefore quite different from a wave in water, where a negative displacement corresponds to a low water level and a positive displacement to a high water level: both negative and positive regions of a wavefunction may correspond to a high probability of finding a particle in a region (Fig. 11.17) because only the square modulus of ψ is physically significant. However, later we shall see that the presence of positive and negative regions of a wavefunction is of great *indirect* significance, because it gives rise to the possibility of constructive and destructive interference between wavefunctions belonging to different atoms.

Example 11.3 *Interpreting a wavefunction*

The wavefunction of an electron in the lowest energy state of a hydrogen atom is $\psi = Ne^{-r/a_0}$, with N a constant, $a_0 = 52.9$ pm, and r the distance from the nucleus. (Notice that this wavefunction depends only on this distance, not the angular position.) Calculate the relative probabilities of finding the electron inside a small volume of magnitude 1.0 pm^3 located at (a) the nucleus, (b) a distance a_0 from the nucleus. The total volume of a hydrogen atom is of the order of 10^7 pm^3.

Method. The probability is equal to $\psi^2\, d\tau$ evaluated at the location in question. The volume 1.0 pm^3 is so small (even on the scale of the atom) that we can ignore the variation of ψ within it and write the probability as equal to the probability density (ψ^2) evaluated at the point of interest multiplied by the volume of interest V. That is, we make the approximation that

$$\text{Probability} = \int_{\text{Volume}} \psi^2\, d\tau \approx \psi^2 \int_{\text{Volume}} d\tau = \psi^2 V$$

Answer. In each case $V = 1.0$ pm^3. (a) At the nucleus, $r = 0$, so there $\psi^2 = 1.0 \times N^2$ and

$$\text{Probability} = (1.0 \times N^2) \times 1.0\,\text{pm}^3$$

(b) At a distance $r = a_0$ in an arbitrary direction,

$$\psi^2 = N^2 e^{-2} = 0.14 \times N^2$$

and

$$\text{Probability} = (0.14 \times N^2) \times 1.0\,\text{pm}^3$$

Therefore, the ratio of probabilities is $1.0/0.14 = 7.4$.

Comment. Note that it is more probable (by a factor of 7.4) that the electron will be found at the nucleus than in the same volume element located at a distance a_0 from the nucleus.

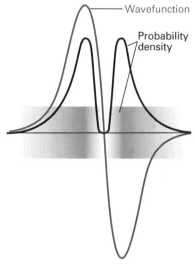

Wavefunction

Probability density

11.17 The sign of a wavefunction has no direct physical significance: the positive and negative regions of this wavefunction both correspond to the same probability distribution (as given by the square of ψ and depicted by the density of shading).

Exercise E11.3. The wavefunction for the lowest energy orbital in the ion He^+ is $\psi = Ne^{-2r/a_0}$. Repeat the calculation for this ion. Any comment?

[55; more compact wavefunction]

Normalization

It is a mathematical feature of the Schrödinger equation that, if ψ is a solution, then so is $N\psi$, where N is any constant. This feature is confirmed by noting that ψ occurs in every term in eqn 10, so that any constant factor can be cancelled. This freedom to vary the wavefunction by a constant factor means that it is always possible to find a **normalization constant,** such that the proportionality of the Born interpretation becomes an equality.

We find the normalization constant by noting that, for a normalized wavefunction $N\psi$, the probability that a particle is in the region dx is equal to $(N\psi^*)(N\psi)\,dx$. Furthermore, the sum over all space of these individual probabilities must be 1 (the probability of the particle being somewhere in the system is 1). Expressed mathematically the latter requirement is

$$N^2 \int \psi^*\psi \, dx = 1 \tag{12}$$

where the integral is over all the space accessible to the particle. It follows that the constant that normalizes an arbitrary wavefunction ψ is

$$N = \frac{1}{\left(\int \psi^*\psi \, dx \right)^{\frac{1}{2}}} \tag{13}$$

Therefore, by evaluating the integral, we can find the value of N. From now on, unless we state otherwise, we always use wavefunctions that have been normalized to 1; that is, from now on we assume that ψ already includes a factor that ensures that

$$\int \psi^*\psi \, dx = 1$$

In three dimensions, the wavefunction is normalized if

$$\int \psi^*\psi \, dx \, dy \, dz = 1$$

or, more succinctly, if

$$\int \psi^*\psi \, d\tau = 1 \tag{14}$$

where $d\tau = dx \, dy \, dz$. In all such integrals, the integration is over all space. For systems with spherical symmetry, it is best to work in spherical polar coordinates (Fig. 11.18). In these coordinates,

$$x = r \sin \theta \cos \phi \qquad y = r \sin \theta \sin \phi \qquad z = r \cos \theta \tag{15}$$

$$d\tau = r^2 \sin \theta \, dr \, d\theta \, d\phi \tag{16}$$

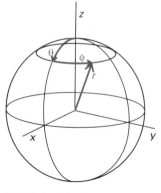

11.18 Spherical polar coordinates. The radius r ranges from 0 to ∞; the co-latitude θ ranges from 0 (north pole) to π (south pole), and the azimuth ϕ ranges from 0 to 2π.

To cover all space, the radius r ranges from 0 to ∞, the co-latitude θ ranges from 0 to π, and the azimuth ϕ ranges from 0 to 2π.

Example 11.4 *Normalizing a wavefunction*

Normalize the wavefunction used for the hydrogen atom in Example 11.3.

Method. We need to find the factor N that guarantees that the integral in eqn 14 is equal to 1. Because the wavefunction is spherically symmetrical, it is sensible to work in spherical polar coordinates.

Answer. The integration we require is

$$\int \psi^* \psi \, d\tau = N^2 \left(\int_0^\infty r^2 e^{-2r/a_0} \, dr \right) \left(\int_0^\pi \sin\theta \, d\theta \right) \left(\int_0^{2\pi} d\phi \right)$$

$$= N^2 \times \frac{a_0^3}{4} \times 2 \times 2\pi = \pi a_0^3 N^2$$

Therefore, in order for this integral to equal 1,

$$N = \left(\frac{1}{\pi a_0^3} \right)^{\frac{1}{2}}$$

and the normalized wavefunction is

$$\psi = \left(\frac{1}{\pi a_0^3} \right)^{\frac{1}{2}} e^{-r/a_0}$$

Comment. If Example 11.3 is now repeated, we can obtain the actual probabilities of finding the electron in the volume element at each location, not just their relative values. The results are (a) 2.2×10^{-6}, corresponding to 1 chance in about 500 000 inspections of finding the electron in the test volume, and (b) 3.1×10^{-7}, corresponding to 1 chance in 3 million.

Exercise E11.4. Normalize the wavefunction given in Exercise E11.3.

$$[N = (8/\pi a_0^3)^{\frac{1}{2}}]$$

Finally, it should be noted that the dimensions of a normalized wavefunction are $1/(\text{length})^{n/2}$, where n is the number of physical dimensions. Thus, in one dimension, $n = 1$ and a normalized wavefunction has the dimensions of $1/(\text{length})^{\frac{1}{2}}$. For a three-dimensional system, it has the dimension of $1/(\text{length})^{\frac{3}{2}}$.

Quantization

The Born interpretation puts severe restrictions on the acceptability of wavefunctions. The principal constraint is that ψ must not be infinite anywhere.[2] If it were, then the integral in eqn 14 would be infinite and

---

2 Infinitely sharp spikes are acceptable so long as they have zero width. The true constraint is that the wavefunction must not be infinite over any finite region. In elementary quantum mechanics the simpler restriction, to finite ψ, is sufficient.

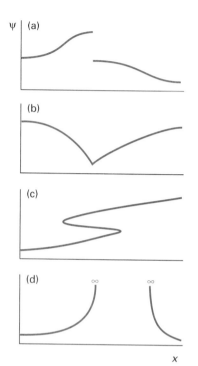

11.19 The wavefunction must satisfy stringent conditions for it to be acceptable. (a) Unacceptable because it is not continuous; (b) unacceptable because its slope is discontinuous; (c) unacceptable because it is not single-valued; (d) unacceptable because it is infinite over a finite region.

the normalization constant would be zero. The normalized function would be zero everywhere, except where it is infinite, which would be unacceptable. The requirement that ψ is finite everywhere rules out many possible solutions of the Schrödinger equation, because many mathematically acceptable solutions rise to infinity. We shall see examples shortly.

The requirement that ψ is finite everywhere is not the only restriction implied by the Born interpretation. We could imagine (and will shortly meet) a function that gives rise to more than one value of $\psi^*\psi$ at a single point. The Born interpretation implies that such functions are unacceptable, because it would be absurd to have more than one probability that a particle is at some point. This restriction is expressed by saying that the wavefunction must be **single-valued**, that is, have only one value at each point of space.

The Schrödinger equation itself also implies some mathematical restrictions on the type of functions that will occur. Because it is a second-order differential equation, the second derivative of ψ must be well-defined if the equation is to be applicable everywhere. We can take the second derivative of a function only if it is continuous (so there are no sharp steps in it, Fig. 11.19) and if its first derivative, its slope, is continuous (so there are no kinks).[3] Therefore, wavefunctions must (a) be continuous, (b) have continuous first derivatives.

At this stage we see that ψ must be continuous, have a continuous slope, be single-valued, and be finite everywhere. An acceptable wavefunction cannot be zero everywhere, because the particle it describes must be somewhere. These are such severe restrictions that acceptable solutions of the Schrödinger equations do not in general exist for arbitrary values of the energy E. In other words, *a particle may possess only certain energies*, for otherwise its wavefunction would be physically unacceptable. That is, *the energy of a particle is quantized*. We can find the acceptable energies by solving the Schrödinger equation for motion of various kinds, and selecting the solutions that conform to the restrictions listed above. That is the work we do in the next chapter.

Summary

It may be helpful to have the following summary of what we have established so far. First, classical mechanics, in which the central concept is the trajectory of a particle, fails badly when it is applied to bodies as small as atoms and subatomic particles. It has to be replaced by quantum mechanics, in which the central concept is the wavefunction. The wavefunction contains all the information about the system, and allows us to predict the probability of finding a particle at any location. Whereas Newton's equations of motion are the basis of classical mechanics and are used to calculate the trajectory of a particle, quantum mechanics is

---

3 There are cases, and we shall meet them, where acceptable wavefunctions have kinks. These cases arise when the potential energy has peculiar properties, such as rising abruptly to infinity. When the potential energy is smoothly well-behaved and finite, the slope of the wavefunction must be continuous; if the potential energy becomes infinite, then the slope of the wavefunction need not be continuous.

based on Schrödinger's equation, which is used to calculate the wavefunction. Although there are an infinite number of mathematically acceptable solutions of the Schrödinger equation, the Born interpretation of the wavefunction implies that physically acceptable solutions of the equation exist only for certain values of the energy. That is, the restrictions on the wavefunction lead automatically to the quantization of energy.

QUANTUM MECHANICAL PRINCIPLES

Some of the principal concepts of quantum mechanics can now be introduced by considering a free particle in one dimension. For a particle of mass m free to move parallel to the x-axis with zero potential energy ($V = 0$ everywhere, so the energy of the particle is independent of its position) the Schrödinger equation is

$$-\frac{\hbar^2}{2m}\frac{d^2\psi}{dx^2} = E\psi$$

The solutions of this equation have the form

$$\psi = Ae^{ikx} + Be^{-ikx} \qquad E = \frac{k^2\hbar^2}{2m} \tag{17}$$

where A and B are constants.[4] Because the total energy of the particle is its kinetic energy, $p^2/2m$, we can infer that the constant k also tells us the magnitude of the linear momentum through the relation

$$p = k\hbar \tag{18}$$

One of the questions we must now address is the significance of the coefficients A and B. We shall make progress by setting up quantum mechanics in a more general and powerful way than we have presented it so far.

11.5 Operators and observables

The Schrödinger equation (eqn 10) may be rewritten in the succinct form

$$H\psi = E\psi \tag{19a}$$

with

$$H = -\frac{\hbar^2}{2m}\frac{d^2}{dx^2} + V(x) \tag{19b}$$

The quantity H is an **operator**, something that carries out an operation on the function ψ. In this case, the operation is to take the second derivative of ψ and (after multiplication by $-\hbar^2/2m$), to add the result to

---

4 A useful mathematical relation is that

$$e^{ikx} = \cos kx + i \sin kx$$

It follows that an alternative way of writing the solutions in eqn 17 is

$$\psi = C \cos kx + D \sin kx$$

where C and D are constants.

the outcome of multiplying ψ by V. The operator H plays a special role in quantum mechanics, and is called the **Hamiltonian operator** after the nineteenth-century mathematician William Hamilton. Hamilton developed a form of classical mechanics that, it subsequently turned out, could readily be converted into quantum mechanics and which shows very clearly the relation between the two theories.

Eigenvalues and eigenfunctions

When the Schrödinger equation is written as in eqn 19a, it is seen to be in an **eigenvalue equation**, an equation of the form

$$(\text{Operator})(\text{function}) = (\text{constant factor}) \times (\text{same function})$$

In symbols, denoting the function by f, the operator $\hat{\Omega}$, and the constant factor by ω

$$\hat{\Omega}f = \omega f \tag{20}$$

The factor ω is called the **eigenvalue** of the operator $\hat{\Omega}$. In eqn 19a, the eigenvalue is the energy. The function f (which must be the same on each side in an eigenvalue equation) is called an **eigenfunction** and is different for each eigenvalue. In eqn 19a, the eigenfunction is the wavefunction corresponding to the energy E.

Example 11.5 *Identifying an eigenfunction*

Show that e^{ax} is an eigenfunction of the operator d/dx, and find the corresponding eigenvalue. Show that e^{ax^2} is not an eigenfunction of d/dx.

Method. We need to operate on the function with the operator and check whether the result is a constant factor times the original function.

Answer. For $\hat{\Omega} = d/dx$ and $f = e^{ax}$:

$$\hat{\Omega}f = \frac{de^{ax}}{dx} = ae^{ax} = af$$

Therefore e^{ax} is an eigenfunction of d/dx, and its eigenvalue is a. For $f = e^{ax^2}$,

$$\hat{\Omega}f = \frac{d}{dx}e^{ax^2} = 2axe^{ax^2} = 2ax \times f$$

which is not an eigenvalue equation even though the same function f occurs on the right, because f is multiplied by a variable factor ($2ax$), not by a constant factor.

Comment. Much of quantum mechanics involves looking for functions that are eigenfunctions of a given operator, especially of the Hamiltonian operator for the energy.

Exercise E11.5. Is the function $\cos ax$ an eigenfunction of (a) d/dx, (b) d^2/dx^2?   [(a) No, (b) yes]

The importance of eigenvalue equations is that the pattern

$$(\text{Energy operator})(\text{wavefunction}) = (\text{energy}) \times (\text{wavefunction})$$

exemplified by the Schrödinger equation is repeated for other **observables**, or measurable properties of a system. Thus, it is often the case that we can write

(Operator corresponding to an observable)(wavefunction)

= (value of observable) × (wavefunction)

The symbol $\hat{\Omega}$ in eqn 20 is then interpreted as an operator (e.g. the Hamiltonian H) corresponding to an observable (e.g. the energy), and the eigenvalue ω is the value of that observable (e.g. the value of the energy E). Therefore, if we know both the wavefunction ψ and the operator $\hat{\Omega}$ corresponding to the observable Ω of interest, and the wavefunction is an eigenfunction of the operator $\hat{\Omega}$, then we can predict the outcome of an observation of the property Ω (e.g. an atom's energy) by picking out the factor ω in the eigenvalue equation

$$\hat{\Omega}\psi = \omega\psi$$

Shortly we shall see to what extent the value of an observable can be predicted when the wavefunction of the system is not an eigenfunction of the operator $\hat{\Omega}$.

Operators

The first step, however, is to make the procedure concrete by setting up and using the operator corresponding to a given observable. According to the basic postulates of quantum mechanics (*Further information 6*), the form of the operator for linear momentum parallel to the x-axis is

$$\hat{p} = \frac{\hbar}{i}\frac{d}{dx} \qquad (21)$$

That is, to find the linear momentum of a particle parallel to the x-axis from the eigenvalue equation

$$\hat{p}\psi = p\psi$$

we differentiate the wavefunction with respect to x, and then pick out the momentum p from the eigenvalue equation

$$\frac{\hbar}{i}\frac{d\psi}{dx} = p\psi$$

Similarly, the operator for position along the x-axis is multiplication by the coordinate x:

$$\hat{x} = x \times \qquad (22)$$

This expression is also justified in *Further information 6*.

Now suppose the wavefunction of a free particle, eqn 17, has $B = 0$, then

$$\psi = Ae^{ikx}$$

It follows that the equation for finding the linear momentum of a particle with this wavefunction is

$$\frac{\hbar}{i}\frac{d\psi}{dx} = \frac{\hbar}{i}A\frac{de^{ikx}}{dx} = \frac{\hbar}{i}A \times ike^{ikx} = \hbar kAe^{ikx} = k\hbar\psi$$

Hence $p = k\hbar$, as we already knew. Now, however, suppose instead that the wavefunction had $A = 0$ so that

$$\psi = Be^{-ikx}$$

Then, by the same reasoning, $p = -k\hbar$. It follows that a particle described by the second wavefunction has the same magnitude of momentum (and the same kinetic energy) as before, but directed towards $-x$.

At this stage, we can summarize the features we have established as follows:

(1) To find the value of an observable from a wavefunction, act on the wavefunction with the operator for the property of interest. If the result is an eigenvalue equation, then the required value is the eigenvalue of the operator.

(2) The wavefunction for a particle travelling towards $+x$ is proportional to e^{ikx} and the wavefunction for a particle travelling towards $-x$ is proportional to e^{-ikx}. In each case the magnitude of the linear momentum is $k\hbar$.

11.6 Superposition and expectation values

Suppose now that the wavefunction in eqn 17 has $A = B$. What is the linear momentum of the particle it describes? We quickly run into trouble if we use the operator technique. The wavefunction is

$$\psi = A(e^{ikx} + e^{-ikx}) = 2A \cos kx$$

(see footnote 4) which is a perfectly respectable wave-like wavefunction. However, when we operate with p, we find

$$\frac{\hbar}{i} \frac{d\psi}{dx} = \frac{2A\hbar}{i} \frac{d(\cos kx)}{dx} = -\frac{2kA\hbar}{i} \sin kx$$

which is not an eigenvalue equation because the function on the right is different from the original function. We need to add to the interpretation given above a remark about how to interpret a wavefunction that is not an eigenfunction of the operator of interest.

Linear superposition of wavefunctions

When the wavefunction of a particle is not an eigenfunction of an operator, the property to which the operator corresponds does not have a definite value. However, in the current example the momentum is not completely indefinite because the cosine wavefunction is a **linear superposition**, or sum, of e^{ikx} and e^{-ikx}, and these two functions, as we have seen, individually correspond to definite momentum states. Symbolically we can write the linear superposition as

$$\psi = \underset{\substack{\text{Particle with} \\ \text{linear momentum} \\ +k\hbar}}{\psi_{\rightarrow}} + \underset{\substack{\text{Particle with} \\ \text{linear momentum} \\ -k\hbar}}{\psi_{\leftarrow}}$$

The interpretation of this composite wavefunction is that, if the momentum of the particle is repeatedly measured in a long series of observations, then its *magnitude* will be found to be $k\hbar$ in all the measurements (because that is the value for each component of the wavefunction). However, because the two component wavefunctions occur equally in the superposition, *half* of the measurements will show that the particle is moving to the right, and *half* of the measurements will show that it is moving to the left. According to quantum mechanics, we cannot predict in which direction the particle will in fact be found to be travelling; all we can say is that, in a long series of observations, there are equal probabilities of finding the particle to be travelling to the right and to the left.

The same interpretation applies to any wavefunction written as a linear superposition of eigenfunctions of an operator. Thus, suppose the wavefunction is known to be a superposition of many different linear momentum eigenfunctions and written in the form

$$\psi = c_1\psi_1 + c_2\psi_2 + \ldots = \sum_k c_k\psi_k \tag{23}$$

where the c_k are numerical coefficients and the ψ_k correspond to different momentum states. Then according to quantum mechanics,

(1) When the momentum is measured, in a single observation one of the values corresponding to the ψ_k that contribute to the superposition will be found.

(2) Which of these possible values will be found is unpredictable. However, the probability of measuring a particular value in a series of observations is proportional to the square modulus of its coefficient in the superposition, $|c_k|^2$.

(3) The average value of a large number of observations is given by the expectation value $\langle\Omega\rangle$ of the operator $\hat{\Omega}$.

The **expectation value** of an operator $\hat{\Omega}$ is defined as

$$\langle\Omega\rangle = \int \psi^*\hat{\Omega}\psi \, d\tau \tag{24}$$

This formula is valid only for normalized wavefunctions.

JUSTIFICATION

If ψ is an eigenfunction of $\hat{\Omega}$ with eigenvalue ω, the expectation value is

$$\langle\Omega\rangle = \int \psi^*\hat{\Omega}\psi \, d\tau = \int \psi^*\omega\psi \, d\tau = \omega \int \psi^*\psi \, d\tau = \omega$$

because ω is a constant and may be taken outside the integral, and the resulting integral is equal to 1 for a normalized wavefunction. The interpretation of this expression is that, because every observation of the property Ω results in the value ω (because the wavefunction is an eigenfunction of the observation), the mean value of all the observations is also ω.

If ψ is not an eigenfunction of the operator of interest, we can still write it as a linear superposition of eigenfunctions. For simplicity, suppose it is the sum of two eigenfunctions (the general case, eqn 23, can easily be developed). Then

$$\langle \Omega \rangle = \int (c_1\psi_1 + c_2\psi_2)^* \hat{\Omega}(c_1\psi_1 + c_2\psi_2)\, d\tau$$

$$= \int (c_1\psi_1 + c_2\psi_2)^* (c_1\omega_1\psi_1 + c_2\omega_2\psi_2)\, d\tau$$

$$= c_1^* c_1 \omega_1 \int \psi_1^*\psi_1\, d\tau + c_2^* c_2 \omega_2 \int \psi_2^*\psi_2\, d\tau$$

$$\qquad + c_1^* c_2 \omega_2 \int \psi_1^*\psi_2\, d\tau + c_2^* c_1 \omega_1 \int \psi_2^*\psi_1\, d\tau$$

$$= |c_1|^2\, \omega_1 + |c_2|^2\, \omega_2$$

In the last line we have made use of the fact that the individual wavefunctions are normalized (so the integrals over their squares are each equal to 1) and the fact that the integrals of products of wavefunctions corresponding to different eigenvalues are zero. The latter general result is established in *Further information 6*. The expression we have derived, namely,

$$\langle \Omega \rangle = |c_1|^2\, \omega_1 + |c_2|^2\, \omega_2$$

shows that the expectation value is the sum of the two eigenvalues weighted by the probabilities that each one will be found in a series of measurements. Hence, the expectation value is the mean value of a series of observations.

Example 11.6 *Calculating an expectation value*

Calculate the average value of the distance of an electron from the nucleus in the hydrogen atom in its state of lowest energy.

Method. The average radius is the expectation value of the operator corresponding to the distance from the nucleus, which is multiplication by r. To evaluate $\langle r \rangle$, we need to know the normalized wavefunction (from Example 11.4) and then evaluate the integral in eqn 24. A useful integral for calculations on atomic wavefunctions is

$$\int_0^\infty x^n \mathrm{e}^{-ax}\, dx = \frac{n!}{a^{n+1}}$$

where $n!$ denotes factorial n: $n! = n(n-1)(n-2)\ldots 1$.

Answer. The average value is given by the expectation value

$$\langle r \rangle = \int \psi^* r \psi\, d\tau$$

which we evaluate using spherical polar coordinates. Using the normalized function in Example 11.4 gives

$$\langle r \rangle = \frac{1}{\pi a_0^3} \left(\int_0^\infty r^3 e^{-2r/a_0} \, dr \right) \left(\int_0^\pi \sin\theta \, d\theta \right) \left(\int_0^{2\pi} d\phi \right)$$

$$= \frac{1}{\pi a_0^3} \times \frac{3! \, a_0^4}{2^4} \times 2 \times 2\pi = \tfrac{3}{2} a_0$$

Because $a_0 = 52.9$ pm (Example 11.3), $\langle r \rangle = 79$ pm.

Comment. The result that $\langle r \rangle = 79$ pm means that, if a very large number of measurements of the distance of the electron from the nucleus are made, then the mean value will be 79 pm. However, each different observation will give a different individual result, because the wavefunction is not an eigenfunction of the operator corresponding to r.

Exercise E11.6. Evaluate the root mean square distance, $\langle r^2 \rangle^{\frac{1}{2}}$, of the electron from the nucleus in the hydrogen atom. $[\sqrt{3}a_0]$

The uncertainty principle

We have seen that, if the wavefunction is Ae^{ikx}, then the particle it describes has a definite state of linear momentum, namely travelling to the right (positive x) with momentum $k\hbar$. But we might also ask for the position of the particle when it is in this state. The Born interpretation instructs us to answer this question by forming the probability density $\psi^*\psi$. In this case:

$$\psi^*\psi = (Ae^{ikx})^*(Ae^{ikx}) = A^2(e^{-ikx})(e^{ikx}) = A^2$$

This probability density is a constant A^2 independent of x. Therefore, the particle has an equal probability of being found anywhere. In other words, *if the momentum is specified precisely, it is impossible to predict the location of the particle.* This statement is one-half of **Heisenberg's uncertainty principle**, one of the most celebrated results of quantum mechanics:

> It is impossible to specify simultaneously, with arbitrary precision, both the momentum and the position of a particle.

Before discussing the principle further, we must establish its other half: that if the position of a particle is specified exactly, then we can say nothing about its momentum. The argument draws on the idea of expressing a wavefunction as a superposition of eigenfunctions, and runs as follows. If we know that the particle is at a definite location, then its wavefunction must be large there and zero everywhere else (Fig. 11.20). Such a wavefunction can be created by adding together a large number of harmonic (sine and cosine) functions or, what is equivalent, a number of e^{ikx} functions. In other words, we can create a sharply localized wavefunction by forming a linear superposition of wavefunctions that correspond to many different linear momenta. The superposition of a few harmonic functions gives a broad, ill-defined wavefunction (Fig. 11.21a). However, as the number of wavefunctions in the superposition increases,

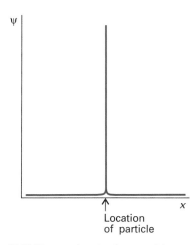

11.20 The wavefunction for a particle at a well-defined location is a sharply spiked function that has zero amplitude everywhere except at the particle's position.

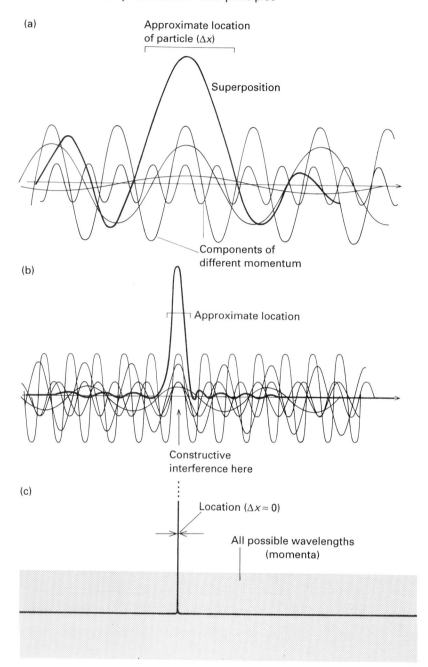

11.21 (a) The wavefunction for a particle with an ill-defined location can be regarded as the sum (superposition) of several wavefunctions of definite wavelength that interfere constructively in one place but destructively elsewhere. (b) As more waves are used in the superposition, the location becomes more precise at the expense of uncertainty in the particle's momentum. (c) An infinite number of waves are needed to construct the wavefunction of a perfectly localized particle.

the wavefunction becomes sharper because of the more complete interference between the positive and negative regions of the individual waves (Fig. 11.21b). When an infinite number of components is used, the wavefunction is a sharp, infinitely narrow spike (Fig. 11.21c), which corresponds to perfect localization of the particle. Now the particle is perfectly localized, but at the expense of discarding all information about its momentum. This is because, as we saw above, a measurement of the momentum will give a result corresponding to any one of the infinite number of waves in the superposition, and which one it will give is unpredictable. Hence, if we know the location of the particle precisely, its momentum is completely unpredictable.

A quantitative version of this result may be obtained by considering the expectation values of position and momentum:

$$\Delta p \, \Delta q \geq \tfrac{1}{2} \hbar \qquad (25)$$

In this expression Δp is the 'uncertainty' in the linear momentum (strictly, it is the root mean square (r.m.s.) deviation of the momentum from its mean value) parallel to the axis q, and Δq is the uncertainty in position along that axis (the r.m.s. deviation of the position from the mean position, essentially the half-width of the superposition in Fig. 11.21). If there is complete certainty about the position of the particle ($\Delta q = 0$), then the only way that eqn 25 can be satisfied is for $\Delta p = \infty$, which implies complete uncertainty about the momentum. Conversely, if the momentum is known exactly ($\Delta p = 0$), then the position must be completely uncertain ($\Delta q = \infty$).

The p and q that appear in eqn 25 refer to the same direction in space. Therefore, whereas position on the x-axis and momentum parallel to the x-axis are restricted by the uncertainty relation, simultaneous location of position on x and motion parallel to y or z are not restricted.

Example 11.7 *Using the uncertainty principle*

The speed of a 1.0-g projectile is known to within $1 \times 10^{-6} \, \mathrm{m \, s^{-1}}$. Calculate the minimum uncertainty in its position.

Method. Estimate Δp from $m\Delta v$, where Δv is the uncertainty in the speed; then use eqn 25 to estimate the minimum uncertainty in position, Δq.

Answer. The minimum uncertainty in position is

$$\Delta q = \frac{\hbar}{2m\Delta v}$$

$$= \frac{1.055 \times 10^{-34} \, \mathrm{J \, s}}{2 \times (1.0 \times 10^{-3} \, \mathrm{kg}) \times (1 \times 10^{-6} \, \mathrm{m \, s^{-1}})}$$

$$= 5 \times 10^{-26} \, \mathrm{m}$$

Comment. The uncertainty is completely negligible for all practical purposes concerning macroscopic objects. However, if the mass is that of an electron, then the same uncertainty in speed implies an uncertainty in position far larger than the diameter of an atom, so the concept of a trajectory, the simultaneous possession of a precise position and momentum, is untenable.

Exercise E11.7. Estimate the minimum uncertainty in the speed of an electron in a hydrogen atom (taking its diameter as $2a_0$).   [500 km s^{-1}]

The Heisenberg uncertainty principle applies to a number of pairs of observables called **complementary observables**, which are defined in terms of the properties of their operators.[5] Other than position and momentum along the same axis, complementary observables include properties related to angular momentum (which we meet in the next chapter). With the discovery that some pairs of observables are complementary we are at the heart of the difference between classical and quantum mechanics. Classical mechanics supposed, falsely as we now know, that the position and momentum of a particle could be specified simultaneously with arbitrary precision. However, quantum mechanics shows that position and momentum are complementary, and that we have to make a choice: we can specify position at the expense of momentum, or momentum at the expense of position.

The realization that some observables are complementary allows us to make considerable progress with the calculation of atomic and molecular properties; but it does away with some of classical physics' most cherished concepts.

---

5 Specifically, two observables Ω_1 and Ω_2 are complementary if their operators satisfy the relation

$$\hat{\Omega}_1\hat{\Omega}_2 \neq \hat{\Omega}_2\hat{\Omega}_1$$

Thus, position x and linear momentum along x are complementary because (noting that operators always operate on a wavefunction)

$$\hat{x}\hat{p}\psi = x \times \frac{\hbar}{i}\frac{d\psi}{dx}$$

whereas

$$\hat{p}\hat{x}\psi = \frac{\hbar}{i}\frac{d(x\psi)}{dx} = \frac{\hbar}{i}\psi + x \times \frac{\hbar}{i}\frac{d\psi}{dx}$$

which is not the same.

CHECK LIST OF KEY IDEAS

1 The characteristics of **black-body radiation**, the failure of the classical **Rayleigh–Jeans law** (eqn 3) to account for it, and the success of the **Planck distribution** (eqn 5) based on the **quantization** of energy.

2 The classical interpretation of **Dulong and Petit's law** for heat capacity (eqn 6), its failure at low temperatures, and the success of the **Einstein formula** (eqn 7) based on the quantization of material oscillators.

3 The interpretation of the **photoelectric effect** in terms of **photons** of radiation (eqn 8).

4 The **de Broglie relation** (eqn 9) between the wavelength of a particle and its linear momentum and its confirmation by the diffraction of electrons.

5 The **wavefunction** of a system and the **Schrödinger equation** for calculating the wavefunction of a system (eqn 10).

6 The interpretation of the curvature of the wavefunction in terms of the **kinetic energy** of the particle (Section 11.3).

7 The **Born interpretation** of the wavefunction as a **probability amplitude** for the location of a particle (Section 11.4).

8 The **constraints** on physically acceptable wavefunctions and their implication that the energy of a system is quantized (Section 11.4).

9 The role of **operators** in quantum mechanics and the significance of **eigenfunctions** and **eigenvalues** (Section 11.5).

10 The construction and interpretation of **linear superpositions** (eqn 23) and the significance of an **expectation value** (eqn 24) as a quantum mechanical average value.

11 The **uncertainty principle** (eqn 25), its sig-nificance, and its interpretation in terms of superpositions.

12 The concept of **complementary observables** in quantum mechanics (Section 11.6).

EXERCISES

11.1. Calculate the power radiated by a $2.0 \, m \times 3.0 \, m$ section of the surface of a hot body at 1500 K.

11.2. The power delivered to a photodetector that col-lects 8.0×10^7 photons in 3.8 ms from monochromatic light is $0.72 \, \mu W$. What is the frequency of the radiation?

11.3. Determine the wavelength of the radiation of the most intense electromagnetic radiation emitted from the surface of the star Sirius, which has a surface temperature of 11 000 K.

11.4. Calculate the speed of an electron of wavelength 3.0 cm.

11.5. The fine-structure constant α plays a special role in the structure of matter; its approximate value is 1/137. What is the wavelength of an electron travelling at a speed αc, where c is the speed of light? (Note that the circum-ference of the first Bohr orbit in the hydrogen atom is 332 pm.)

11.6. A certain diffraction experiment requires the use of electrons of wavelength 0.45 nm. Calculate the speed of the electrons.

11.7. Calculate the linear momentum of photons of wavelength 750 nm. What speed does an electron need to have the same linear momentum?

11.8. The energy required for the ionization of a certain atom is $3.44 \times 10^{-18} \, J$. The absorption of a photon of unknown wavelength ionizes the atom and ejects an electron with velocity $1.03 \times 10^6 \, m \, s^{-1}$. Calculate the wavelength of the incident radiation.

11.9. The speed of a certain proton is $4.5 \times 10^5 \, m \, s^{-1}$. If the uncertainty in its momentum is to be reduced to 0.0100 per cent, what uncertainty in its location must be tolerated?

11.10. Calculate the energy per photon and the energy per mole of photons for radiation of wavelength (a) 600 nm (red), (b) 550 nm (yellow), (c) 400 nm (blue), (d) 200 nm (ultraviolet), (e) 150 pm (X-ray), (f) 1 cm (microwave).

11.11. Calculate the speed to which a stationary H atom would be accelerated if it absorbed each of the photons used in Exercise 11.10.

11.12. A glow-worm of mass 5.0 g emits red light (650 nm) with a power of 0.10 W entirely in the backward direction. To what speed will it have accelerated after 10 years if released into free space and assumed to live?

11.13. A sodium lamp emits yellow light (550 nm). How many photons does it emit each second if its power is (a) 1.0 W, (b) 100 W?

11.14. The peak of the sun's emission occurs at about 480 nm; estimate the temperature of its surface.

11.15. The work function for metallic caesium is 2.14 eV. Calculate the kinetic energy and the speed of the electrons ejected by light of wavelength (a) 700 nm, (b) 300 nm.

11.16. Calculate the size of the quantum involved in the excitation of (a) an electronic motion of period 10^{-15} s, (b) a molecular vibration of period 10^{-14} s, (c) a pendu-lum of period 1 s. Express the results in J and in kJ mol^{-1}.

11.17. Calculate the de Broglie wavelength of (a) a mass of 1.0 g travelling at $1.0 \, cm \, s^{-1}$, (b) the same, travelling at $100 \, km \, s^{-1}$, (c) an He atom travelling at $1000 \, m \, s^{-1}$ (a typical speed at room temperature).

11.18. Calculate the de Broglie wavelength of an electron accelerated from rest through a potential difference of (a) 100 V, (b) 100 kV.

11.19. Calculate the minimum uncertainty in the speed of a ball of mass 500 g that is known to be within $1.0 \, \mu m$ of a certain point on a bat. What is the minimum uncertainty in the position of a bullet of mass 5.0 g that is known to have a speed somewhere between 350.00001 and 350.00000 $m \, s^{-1}$?

11.20. An electron is confined to a linear region with a length of the same order as the diameter of an atom (ca. 100 pm). Calculate the minimum uncertainties in its position and speed.

11.21. In an X-ray photoelectron experiment, a photon of wavelength 150 pm ejects an electron from the inner shell of an atom and it emerges with a speed of $2.14 \times 10^7 \, m \, s^{-1}$. Calculate the binding energy of the electron.

PROBLEMS

Numerical problems

11.1. The Planck distribution gives the energy in the wavelength range $d\lambda$ at the wavelength λ. Calculate the energy density in the range 650–655 nm inside a cavity of volume $100\,cm^3$ when its temperature is (a) 25°C, (b) 3000°C.

11.2. The wavelength of the emission maximum from a small pin-hole in an electrically heated container was determined at a series of temperatures and the results are given below. Deduce a value for Planck's constant.

$\theta/°C$	1000	1500	2000	2500	3000	3500
λ_{max}/nm	2181	1600	1240	1035	878	763

11.3. Write a computer program to evaluate the Planck distribution at any temperature and wavelength or frequency, and add to it a routine for evaluating integrals for the energy density of the radiation between any two wavelengths. Use it to calculate the total energy density in the visible region (600 nm to 350 nm) for a black body at (a) 100°C, (b) 500°C, (c) 700 K. What are the classical values at these temperatures?

11.4. The Einstein frequency is often expressed in terms of an equivalent temperature θ_E, where $\theta_E = h\nu/k$. Confirm that θ_E has the dimensions of temperature, and express the criterion for the validity of the high-temperature form of the Einstein equation in terms of it. Evaluate θ_E for (a) diamond, for which $\nu = 4.65 \times 10^{13}$ Hz and (b) for copper, for which $\nu = 7.15 \times 10^{12}$ Hz. What fraction of the Dulong and Petit value of the heat capacity does each substance reach at 25°C?

11.5. The ground-state wavefunction for a particle confined to a one-dimensional box of length L is

$$\psi = \left(\frac{2}{L}\right)^{\frac{1}{2}} \sin\left(\frac{\pi x}{L}\right)$$

Suppose the box is 10.0 nm long. Calculate the probability that the particle is (a) between $x = 4.95$ nm and 5.05 nm, (b) between $x = 1.95$ nm and 2.05 nm, (c) between $x = 9.90$ and 10.00 nm, (d) in the right half of the box, (e) in the central third of the box.

11.6. The ground-state wavefunction of a hydrogen atom is

$$\psi = \left(\frac{1}{\pi a_0^3}\right)^{\frac{1}{2}} e^{-r/a_0}$$

where $a_0 = 53$ pm (the Bohr radius). Calculate the probability that the electron will be found somewhere within a small sphere of radius 1.0 pm centred on the nucleus. Now suppose that the same sphere is relocated at $r = a_0$. What is the probability that the electron is inside it?

Theoretical problems

11.7. Derive Wien's law, that $\lambda_{max}T$ is a constant, from the Planck distribution, and deduce an expression for the constant.

11.8. Normalize the following wavefunctions: (a) $\sin(n\pi x/L)$ in the range $0 \leq x \leq L$, (b) a constant in the range $-L \leq x \leq L$, (c) $e^{-r/a}$ in three-dimensional space, (d) $xe^{-r/2a}$ in three-dimensional space. *Hint*: The volume element in three dimensions is $d\tau = r^2\,dr\,\sin\theta\,d\theta\,d\phi$, with $0 \leq r < \infty$, $0 \leq \theta \leq \pi$, $0 \leq \phi \leq 2\pi$. A useful integral is given in Example 11.6.

11.9. Two (unnormalized) excited state wavefunctions of the H atom are

(a) $\psi = \left(2 - \dfrac{r}{a_0}\right)e^{-r/a_0}$

(b) $\psi = r\sin\theta\cos\theta\,e^{-r/2a_0}$

Normalize both functions to 1.

11.10. Identify which of the following functions are eigenfunctions of the operator d/dx: (a) e^{ikx}, (b) $\cos kx$, (c) k, (d) kx, (e) $e^{-\alpha x^2}$. Give the corresponding eigenvalue where appropriate.

11.11. Determine which of the following functions are eigenfunctions of the inversion operator $\hat{\imath}$ (which has the effect of making the replacement $x \to -x$): (a) $x^3 - kx$, (b) $\cos kx$, (c) $x^2 + 3x - 1$. State the eigenvalue of $\hat{\imath}$ when relevant.

11.12. Which of the functions in Problem 11.10 are (a) also eigenfunctions of d^2/dx^2 and (b) only eigenfunctions of d^2/dx^2? Give the eigenvalues where appropriate.

11.13. A particle is in a state described by the wavefunction

$$\psi = (\cos\chi)e^{ikx} + (\sin\chi)e^{-ikx}$$

where χ is a parameter. What is the probability that the particle will be found with a linear momentum (a) $+k\hbar$, (b) $-k\hbar$? What form would the wavefunction have if it were 90 per cent certain that the particle had linear momentum $+k\hbar$?

11.14. Evaluate the kinetic energy of the particle with wavefunction given in Problem 11.13.

11.15. Calculate the average linear momentum of a particle described by the following wavefunctions: (a) e^{ikx}, (b) $\cos kx$, (c) $e^{-\alpha x^2}$, where in each one x ranges from $-\infty$ to $+\infty$.

11.16. Evaluate the expectation values of r and r^2 for a hydrogen atom with wavefunctions given in Problem 11.9.

11.17. Calculate (a) the mean potential energy and (b) the mean kinetic energy of an electron in the ground state of a hydrogenic atom.

11.18. Write a computer program for constructing superpositions of cosine functions like those drawn in Fig. 11.21 and explore how the wavefunction described becomes more localized as more components are included. Include routines that determine the probability that a given momentum will be observed. If you plot the superposition (which you should), set $x = 0$ at the centre of the screen and build the superposition there. Include a routine that includes the evaluation of the root mean square location of the packet, $\langle x^2 \rangle^{\frac{1}{2}}$.

11.19. Determine the value of $\hat{\Omega}_1 \hat{\Omega}_2 - \hat{\Omega}_2 \hat{\Omega}_1$ for the operators (a) d/dx and x, (b) d/dx and x^2, (c) a and $a^\dagger$, where $a = (\hat{x} + i\hat{p})/\sqrt{2}$ and $a^\dagger = (\hat{x} - i\hat{p})/\sqrt{2}$.

Quantum theory: techniques and applications

To find the properties of particles according to quantum mechanics we need to solve the appropriate Schrödinger equation for the system. This chapter presents the essentials of the solutions for three basic types of motion: translation, vibration, and rotation. We shall see that, when certain conditions are imposed on the solutions of the Schrödinger equation, only certain wavefunctions and their corresponding energies are acceptable. Hence, quantization emerges as a natural consequence of the equation. Moreover, we shall also see that the forms of the acceptable wavefunctions can often be anticipated in terms of the de Broglie relation and the shapes of the waves that fit into the system. The solutions will bring to light a number of highly non-classical, and therefore surprising, features of particles, particularly their ability to tunnel into and through regions where classical physics would forbid them to be found, and the limitation of a rotating body to certain spatial orientations. We shall also encounter a property of the electron, its spin, that has no classical counterpart. This property, though, will later prove to be the key to understanding the periodic table and the formation of chemical bonds.

In the course of the chapter we shall encounter a number of technical features of quantum mechanics (such as the orthogonality of wavefunctions and the relation of degeneracy to symmetry) that will be employed in the discussion of atomic and molecular structure later in the text.

The three basic modes of motion—translation, vibration, and rotation—all play an important role in chemistry because they are ways in which molecules can store energy. Gas molecules, for instance, undergo translational motion, and their kinetic energy is a contribution to the total internal energy of a sample. Molecules can also store energy as rotational kinetic energy, and transitions between their rotational energy levels are responsible for their rotational spectra. Energy is also stored as molecular vibration, and transitions between vibrational energy levels give rise to vibrational spectra.

In this chapter we see how the concepts of quantum mechanics introduced in Chapter 11 can be developed into a powerful set of techniques for dealing with these types of motion. In later chapters we shall see how these calculations are used to account for atomic structure, molecular structure, and spectroscopy.

The Schrödinger equation is a differential equation, so to understand its content and implications we need to understand a little about the solutions of such equations. The necessary material is summarized in *Further information 7*. In most cases the solutions of the equation will be quoted; it is then sensible to verify that the solutions are correct by substituting them into the equation.

TRANSLATIONAL MOTION

The quantum mechanical description of free motion was introduced in Section 11.3. We saw there that the Schrödinger equation is

$$-\frac{\hbar^2}{2m}\frac{d^2\psi}{dx^2} = E\psi \tag{1a}$$

and that the general solutions are

$$\psi = Ae^{ikx} + Be^{-ikx} \qquad E = \frac{k^2\hbar^2}{2m} \tag{1b}$$

That these functions are solutions can be verified by substituting them into the left-hand side of the differential equation and showing that the result is equal to $E\psi$. In this case, all values of k, and therefore all values of the energy E, are permitted. It follows that the translational energy of a free particle is not quantized.

We saw in Section 11.5 that a wavefunction of the form e^{ikx} describes a particle with linear momentum $p = k\hbar$ travelling towards positive x (to the right) and that a wavefunction of the form e^{-ikx} describes a particle with the same momentum travelling towards negative x (to the left). In either state, $\psi^*\psi$ is independent of x, which implies that the position of the particle is completely unpredictable. This conclusion is consistent with the uncertainty principle, because, if the momentum is certain, then the position cannot be specified.

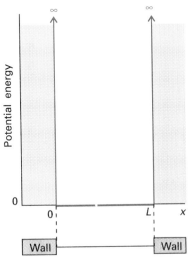

Potential energy

∞ ∞

0

0 L x

Wall   Wall

12.1 A particle in a one-dimensional region with impenetrable walls. Its potential energy is zero between $x = 0$ and $x = L$, and rises abruptly to infinity as soon as it touches the walls (represented by the grey regions).

12.1 The particle in a box

In this section we consider the problem of a **particle in a box**, in which a particle of mass m is confined between two walls at $x = 0$ and $x = L$. In the **infinite square well**, the potential energy of the particle is zero inside the box but rises abruptly to infinity at the walls (Fig. 12.1). This potential energy is an idealization of the potential energy of a gas molecule that is free to move in a one-dimensional container. Later, we shall generalize the calculation to a particle in a three-dimensional container.

The Schrödinger equation

The Schrödinger equation for the region between the walls (where $V = 0$) is

$$-\frac{\hbar^2}{2m}\frac{d^2\psi}{dx^2} = E\psi \tag{2a}$$

This equation is the same as that for the free particle, so the general solutions given in eqn 1b are also the same. It is convenient[1] to write them as

$$\psi = C \sin kx + D \cos kx \qquad E = \frac{k^2 \hbar^2}{2m} \tag{2b}$$

For a free particle, all such solutions are acceptable. However, the presence of the walls restricts the acceptable solutions to wavefunctions with wavelengths that fit between them. In other words, acceptable wavefunctions must satisfy certain **boundary conditions**, which in this case are that $\psi = 0$ at each wall. It is physically impossible for the particle to be found with an infinite potential energy, so the wavefunction must be zero outside the box; then the continuity of the wavefunction requires it to vanish at the walls.

The acceptable solutions

At this stage we know that the wavefunction has the general form given in eqn 2b but with the additional requirement that it must satisfy the two boundary conditions

$$\psi = 0 \text{ at } x = 0 \text{ and } x = L$$

Consider the wall at $x = 0$. According to eqn 2b, $\psi(0) = D$ (because $\sin 0 = 0$ and $\cos 0 = 1$). But the boundary condition there is that $\psi(0) = 0$, which requires $D = 0$. It follows that the wavefunction must be of the form

$$\psi = C \sin kx$$

The amplitude at the other wall (at $x = L$) is

$$\psi = C \sin kL$$

which must also be zero. Taking $C = 0$ would give $\psi = 0$ for all x, which would conflict with the Born interpretation (the particle must be somewhere). Therefore, kL must be chosen so that $\sin kL = 0$, which is satisfied by

$$kL = n\pi \qquad n = 1, 2, \ldots$$

($n = 0$ is ruled out because it implies $k = 0$ and $\psi = 0$ everywhere, which is unacceptable, and negative values of n merely change the sign of $\sin kL$.) Because k and E are related by eqn 2b, it follows that the energy of the particle is limited to the values

$$E_n = \frac{n^2 \hbar^2 \pi^2}{2mL^2} = \frac{n^2 h^2}{8mL^2} \qquad n = 1, 2, \ldots \tag{3}$$

We see that the energy of the particle is quantized, and that the quantization arises from the boundary conditions that ψ must satisfy if it is to be an acceptable wavefunction. This is a general conclusion: *the need to satisfy boundary conditions implies that only certain wavefunctions are*

---

1 We use $e^{i\theta} = \cos\theta + i\sin\theta$ and $e^{-i\theta} = \cos\theta - i\sin\theta$, and absorb all numerical factors into the coefficients C and D.

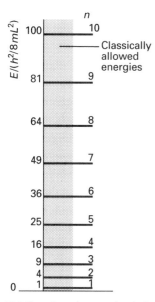

12.2 The allowed energy levels for a particle in a box. Note that the energy levels increase as n^2, and that their separation increases as the quantum number increases.

acceptable and hence restricts observables to discrete values. So far only energy has been quantized; shortly we shall see other physical observables may also be quantized.

Normalization

Before discussing the solution in more detail, we shall complete the derivation of the wavefunctions (which are real) by finding the normalization constant (here written C). To do so, we look for the value of C that ensures that the integral of ψ^2 over all x is equal to 1:

$$C^2 \int_0^L \sin^2\left(\frac{n\pi x}{L}\right) \mathrm{d}x = C^2 \times \frac{L}{2} = 1, \qquad \text{so } C = \left(\frac{2}{L}\right)^{\frac{1}{2}} \tag{4}$$

Therefore, the complete solution to the problem is

$$\text{Energies: } E_n = \frac{n^2 h^2}{8mL^2} \tag{5a}$$

$$\text{Wavefunctions: } \psi_n = \left(\frac{2}{L}\right)^{\frac{1}{2}} \sin\left(\frac{n\pi x}{L}\right) \qquad \text{for } 0 \le x \le L \tag{5b}$$

with $n = 1, 2, \ldots$.

The energies and wavefunctions are labelled with the quantum number n. A **quantum number** is an integer (in some cases, as we shall see, a half-integer) that labels the state of the system. For a particle in a box there is an infinite number of acceptable solutions, and the quantum number n specifies the one of interest (Fig. 12.2). As well as acting as a label, a quantum number is used to calculate the energy corresponding to that state (through eqn 5a) and to write down the wavefunction explicitly (using eqn 5b).

The properties of the solutions

Figure 12.3 shows the shapes of some of the wavefunctions of a particle in a box. With these images in mind it is easy to see the origin of the

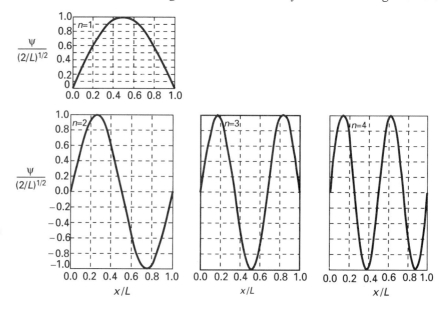

12.3 The first four normalized wavefunctions of a particle in a box. Each wavefunction is a standing wave, and successive functions possess one more half-wave and a correspondingly shorter wavelength.

quantization: each wavefunction is a standing wave and, to fit into the cavity, successive functions must possess one more half-wavelength. Shortening the wavelength so that more half-wavelengths can fit into the box results in a sharper average curvature of the wavefunction and therefore an increase in the kinetic energy of the particle it describes.

Example 12.1 *Deriving the energies of a particle in a box*

Derive the energy levels of a particle in a box from the de Broglie relation.

Method. We see from Fig. 12.3 that successive wavefunctions possess one more half-wavelength. Therefore, the first thing to do is to find an expression for the permitted wavelengths. To convert those permitted wavelengths to energies, we use the de Broglie relation to express wavelength as a linear momentum. Finally, we use the expression for the kinetic energy in terms of the momentum to find the permitted energies.

Answer. The permitted wavelengths satisfy

$$L = n \times \tfrac{1}{2}\lambda \qquad n = 1, 2, \ldots$$

and therefore

$$\lambda = \frac{2L}{n}, \qquad \text{with } n = 1, 2, \ldots .$$

According to the de Broglie relation, these wavelengths correspond to the momenta

$$p = \frac{h}{\lambda} = \frac{nh}{2L}$$

The particle has only kinetic energy inside the box (where $V = 0$), so the permitted energies are

$$E_n = \frac{p^2}{2m} = \frac{n^2 h^2}{8mL^2}$$

as obtained more formally earlier.

Exercise E12.1. What is the average value of the linear momentum of a particle in a box with quantum number n? $[\langle p \rangle = 0]$

The linear momentum of a particle in a box is not well defined because the wavefunction $\sin kx$ is a standing wave and not an eigenfunction of the linear momentum operator (Section 11.6). However, each wavefunction is a superposition of momentum eigenfunctions:

$$\psi_n = \left(\frac{2}{L}\right)^{\frac{1}{2}} \sin\left(\frac{n\pi x}{L}\right) = \left(\frac{2}{L}\right)^{\frac{1}{2}} \times \frac{1}{2i}(e^{ikx} - e^{-ikx}), \qquad \text{with } k = \frac{n\pi}{L} \qquad (6)$$

It follows that measurement of the linear momentum will give the value $k\hbar$ for half the measurements of momentum and $-k\hbar$ for the other half. This detection of opposite directions of travel with equal probability is the quantum mechanical version of the classical picture that a particle in a box rattles from wall to wall, and in any given period spends half its time travelling to the left and half travelling to the right.

Because n cannot be zero, the lowest energy that the particle may possess is not zero (as would be allowed by classical mechanics) but

$$E_1 = \frac{h^2}{8mL^2} \tag{7}$$

This lowest, irremovable energy is called the **zero-point energy**. The physical origin of the zero-point energy can be explained in two ways. First, the uncertainty principle requires a particle to possess kinetic energy if it is confined to a finite region: the particle's location is not completely indefinite, so its momentum cannot be precisely zero. Hence it has non-zero kinetic energy. Alternatively, if the wavefunction is to be zero at the walls, but smooth, continuous, and not zero everywhere, then it must be curved, and curvature in a wavefunction implies the possession of kinetic energy.

The separation between adjacent energy levels is

$$E_{n+1} - E_n = \frac{(n+1)^2 h^2}{8mL^2} - \frac{n^2 h^2}{8mL^2} = (2n+1)\frac{h^2}{8mL^2} \tag{8}$$

This separation decreases as the length of the container increases and is very small when the container has macroscopic dimensions. The separation of adjacent levels becomes zero when the walls are infinitely far apart. Atoms and molecules free to move in laboratory-sized vessels may therefore be treated as though their translational energy is not quantized. The translational energy of truly free particles is not quantized.

Example 12.2 *Using the particle in a box solutions (1)*

An electron is confined to a molecule of length 1.0 nm (about five atoms long). What is (a) its minimum energy and (b) the minimum excitation energy from that state?

Method. The minimum energy is the zero-point energy of a particle of mass m_e, the energy of the state with $n = 1$. The minimum excitation energy is the energy separation of the states with $n = 1$ and $n = 2$.

Answer. For a container of length $L = 1.0$ nm, $h^2/8m_e L^2 = 6.02 \times 10^{-20}$ J. Therefore, $E_1 = 6.0 \times 10^{-20}$ J (corresponding to 0.37 eV). The minimum excitation energy is

$$E_2 - E_1 = \frac{4h^2}{8m_e L^2} - \frac{h^2}{8m_e L^2} = \frac{3h^2}{8m_e L^2} = 1.8 \times 10^{-19} \text{ J}$$

which corresponds to 1.1 eV.

Comment. Typical energy-level separations in molecules are of the order of a few electronvolts. The electron-in-a-box is a crude model of molecular structure; it can be used for estimating rough values of transition energies and hence used to predict the colours of dye and indicator molecules.

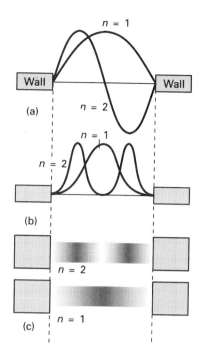

$n = 1$

Wall   Wall

(a) $n = 2$

$n = 1$

$n = 2$

(b)

$n = 2$

$n = 1$

(c)

12.4 (a) The first two wavefunctions, (b) the corresponding probability distributions, and (c) a representation of the probability distribution in terms of the darkness of shading.

Exercise E12.2. Estimate a typical nuclear excitation energy by calculating the first excitation energy of a proton confined to a region roughly equal to the diameter of a nucleus (10^{-14} m). [6 MeV]

The probability density of a particle in a box is

$$\psi^2 = \frac{2}{L}\sin^2\left(\frac{n\pi x}{L}\right) \tag{9}$$

and varies with position within the box. The non-uniformity is pronounced when n is small (Fig. 12.4) but ψ^2 becomes more uniform as n increases. The distribution at high quantum numbers reflects the classical result that a particle bouncing between the walls spends, on the average, equal times at all points. That the quantum result corresponds to the classical prediction at high quantum numbers is an aspect of the **correspondence principle**, which states that classical mechanics emerges from quantum mechanics as high quantum numbers are reached.

Example 12.3 *Using the particle in a box solutions (2)*

What is the probability P of locating the electron between $x = 0$ and $x = 0.2$ nm in its lowest energy state in the box described in Example 12.2?

Method. The value of $\psi^2\,dx$ is the probability of finding the particle in the small region dx located at x; therefore, the total probability of finding the electron in the specified region is the integral of $\psi^2\,dx$ over that region. The wavefunction of the electron is given by eqn 5b with $n = 1$.

Answer. The probability of finding the particle in a region between $x = 0$ and $x = l$ is

$$P = \frac{2}{L}\int_0^l \sin^2\left(\frac{n\pi x}{L}\right) dx = \frac{l}{L} - \frac{1}{2n\pi}\sin\left(\frac{2n\pi l}{L}\right)$$

We then set $n = 1$ and $l = 0.2$ nm, which gives $P = 0.05$, or a chance of 1 in 20 of finding the electron in that region. As n becomes infinite, the second term makes no contribution to P and the classical result, $P = l/L$, is obtained.

Exercise E12.3. For the proton in Exercise E12.2, calculate the probability that in its ground state it will be found between $x = 0.25L$ and $x = 0.75L$. [0.82]

A further property of wavefunctions can now be illustrated. Two wavefunctions are **orthogonal** if the integral of their product vanishes. Specifically, the functions ψ_n and $\psi_{n'}$ are orthogonal if

$$\int \psi_n^* \psi_{n'}\,d\tau = 0 \tag{10}$$

where the integration is over all space. It is a general feature of quantum

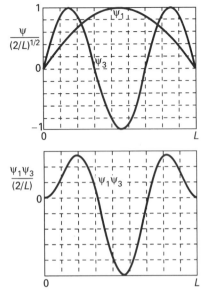

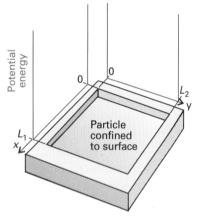

12.5 Two functions are orthogonal if the integral of their product is zero. Here the calculation of the integral is illustrated graphically for two wavefunctions of a particle in a square well. The integral is equal to the total area beneath the graph of the product, and is zero.

12.6 A two-dimensional square well. The particle is confined to the plane bounded by impenetrable walls. As soon as it touches the walls, its potential energy rises to infinity.

mechanics that *wavefunctions that correspond to different energies are orthogonal*. We can verify it in the case of wavefunctions of a particle in a box by considering, for example, the wavefunctions with $n = 1$ and $n' = 3$ (Fig. 12.5):

$$\int_0^L \psi_1^* \psi_3 \, dx = \frac{2}{L} \int_0^L \sin\left(\frac{\pi x}{L}\right) \sin\left(\frac{3n\pi}{L}\right) dx = 0$$

The property of orthogonality is of great importance in quantum mechanics because it enables us to eliminate a large number of integrals from calculations. More pragmatically, orthogonality plays a central role in the theory of chemical bonding (Chapter 14) and spectroscopy (Chapters 16 and 17).

12.2 Motion in two dimensions

Next, we consider a two-dimensional version of the particle in a box. Now the particle is confined to a rectangular surface of length L_1 in the x-direction and L_2 in the y-direction and the potential energy is zero everywhere except at the walls, where it is infinite (Fig. 12.6). The Schrödinger equation is

$$-\frac{\hbar^2}{2m}\left(\frac{\partial^2 \psi}{\partial x^2} + \frac{\partial^2 \psi}{\partial y^2}\right) = E\psi \tag{11}$$

The wavefunction ψ is now a function of both x and y; as a result, the derivatives are partial and the equation is an example of a *partial differential equation*.

Separation of variables

Some partial differential equations can be simplified by the **separation of variables technique** that divides the equation into two or more ordinary differential equations, one for each variable. The method works in this case, as we can see by testing whether a solution of eqn 11 can be found by writing the wavefunction as a product of functions, one depending only on x and the other only on y:

$$\psi = X(x)Y(y)$$

The notation $X(x)Y(y)$ reminds us that the two functions into which the wavefunction is factored depend only on x and only on y for X and Y, respectively. With this substitution, we find that the Schrödinger equation separates into two ordinary differential equations, one for each coordinate:

$$-\frac{\hbar^2}{2m}\frac{d^2X}{dx^2} = E_X X \qquad -\frac{\hbar^2}{2m}\frac{d^2Y}{dy^2} = E_Y Y \qquad E = E_X + E_Y$$

The quantity E_X is the energy associated with the motion of the particle parallel to the x-axis, and likewise for E_Y and motion parallel to the y-axis.

JUSTIFICATION

The first step in the justification of the separability of the wavefunction into the product of two functions X and Y is to note that, because X is

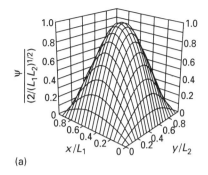

(a)

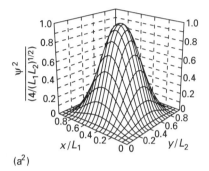

(a²)

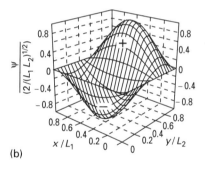

(b)

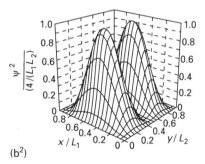

(b²)

12.7 The wavefunctions and probability densities for a particle confined to a rectangular surface. (a) $n_1 = 1$, $n_2 = 1$, the state of lowest energy, and (a²), the corresponding probability distribution. (b) $n_1 = 1$, $n_2 = 2$ and (b²) the corresponding probability distribution. (c, next page) $n_1 = 2$, $n_2 = 2$ and (c²) the corresponding probability distribution.

independent of y and Y is independent of x, we can write

$$\frac{\partial^2 \psi}{\partial x^2} = Y \frac{d^2 X}{dx^2} \qquad \frac{\partial^2 \psi}{\partial y^2} = X \frac{d^2 Y}{dy^2}$$

Then the Schrödinger equation becomes

$$-\frac{\hbar^2}{2m}\left(Y\frac{d^2 X}{dx^2} + X\frac{d^2 Y}{dy^2}\right) = EXY$$

When both sides are divided by XY the resulting equation can be rearranged into

$$\frac{1}{X}\frac{d^2 X}{dx^2} + \frac{1}{Y}\frac{d^2 Y}{dy^2} = -\frac{2mE}{\hbar^2}$$

The first term on the left is independent of y, so if y is varied only the *second* term can change. But the sum of these two terms is a constant given by the right-hand side of the equation; therefore, even the second term cannot change when y is changed. In other words, the second term is a constant, which we write $-2mE_Y/\hbar^2$. By a similar argument, the first term is a constant when x changes, and we write it $-2mE_X/\hbar^2$, and $E = E_X + E_Y$. Therefore, we can write

$$\frac{1}{X}\frac{d^2 X}{dx^2} = -\frac{2mE_X}{\hbar^2} \qquad \frac{1}{Y}\frac{d^2 Y}{dy^2} = -\frac{2mE_Y}{\hbar^2}$$

which rearrange into the two ordinary differential equations quoted in the text.

Each of the two ordinary differential equations is the same as the one-dimensional square-well Schrödinger equation; hence we can adapt the results in eqn 5 without further calculation:

$$X_{n_1} = \left(\frac{2}{L_1}\right)^{\frac{1}{2}}\sin\left(\frac{n_1\pi x}{L_1}\right) \qquad Y_{n_2} = \left(\frac{2}{L_2}\right)^{\frac{1}{2}}\sin\left(\frac{n_2\pi y}{L_2}\right)$$

Then, because $\psi = XY$ and $E = E_X + E_Y$, we obtain

$$\psi_{n_1,n_2} = \frac{2}{(L_1 L_2)^{\frac{1}{2}}}\sin\left(\frac{n_1\pi x}{L_1}\right)\sin\left(\frac{n_2\pi y}{L_2}\right) \quad \text{for } 0 \le x \le L_1, 0 \le y \le L_2$$

$$\tag{12a}$$

$$E_{n_1,n_2} = \left(\frac{n_1^2}{L_1^2} + \frac{n_2^2}{L_2^2}\right)\frac{h^2}{8m} \tag{12b}$$

with the quantum numbers taking the values $n_1 = 1, 2, \ldots$ and $n_2 = 1, 2, \ldots$ independently. Some of these functions are plotted in Fig. 12.7; they are the two-dimensional versions of the wavefunctions shown in Fig. 12.3.

A particle in a three-dimensional box can be treated in the same way. The wavefunctions have another factor (for the z-dependence), and the energy has an additional term in n_3^2/L_3^2.

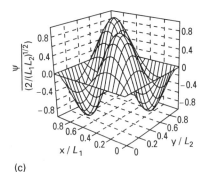

(c)

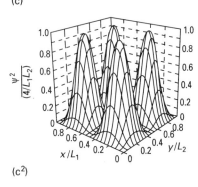

(c²)

12.7 (continued)

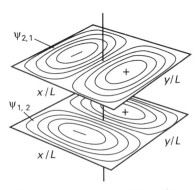

12.8 It is often easier to represent the functions in terms of contour diagrams. Here we show contour diagrams for $\psi_{2,1}$ and $\psi_{1,2}$ in a square square-well. Note that one can be converted into the other by a 90° rotation: we say that they are related by a 'symmetry transformation'. These two functions are also degenerate (i.e. have the same energy).

Degeneracy

An interesting feature of the solutions is obtained when the plane surface is square, i.e. when $L_1 = L$ and $L_2 = L$. Then eqn 12 becomes

$$\psi_{n_1,n_2} = \frac{2}{L}\sin\left(\frac{n_1\pi x}{L}\right)\sin\left(\frac{n_2\pi y}{L}\right) \tag{13a}$$

$$E_{n_1,n_2} = (n_1^2 + n_2^2)\frac{h^2}{8mL^2} \tag{13b}$$

Consider the cases $n_1 = 1$, $n_2 = 2$ and $n_1 = 2$, $n_2 = 1$:

$$\psi_{1,2} = \frac{2}{L}\sin\left(\frac{\pi x}{L}\right)\sin\left(\frac{2\pi y}{L}\right) \qquad E_{1,2} = \frac{5h^2}{8mL^2}$$

$$\psi_{2,1} = \frac{2}{L}\sin\left(\frac{2\pi x}{L}\right)\sin\left(\frac{\pi y}{L}\right) \qquad E_{2,1} = \frac{5h^2}{8mL^2}$$

We see that different wavefunctions correspond to the same energy, the condition called **degeneracy**. In this case, in which there are two degenerate wavefunctions, we say that the level with energy $5(h^2/8mL^2)$ is **doubly degenerate**.

The occurrence of degeneracy is related to the symmetry of the system. Contour diagrams of the two degenerate functions $\psi_{1,2}$ and $\psi_{2,1}$ are shown in Fig. 12.8: because the box is square, we see that we can convert one into the other simply by rotating the plane by 90°. Interconversion by rotation through 90° is not possible when the plane is not square, and $\psi_{1,2}$ and $\psi_{2,1}$ are then not degenerate. We shall see many examples of degeneracy in the pages that follow (e.g. in the hydrogen atom), and all of them can be traced to the symmetry properties of the system (see Section 15.4). In general, *if one wavefunction can be transformed into another by a symmetry transformation of the system, then the two wavefunctions are degenerate.*

12.3 Tunnelling

If the potential energy of a particle does not rise to infinity when it is in the walls of the container, and $E < V$, the wavefunction does not decay abruptly to zero. If the walls are thin (so the potential energy falls to zero again after a finite distance), then the exponential decay of the wavefunction stops, and it begins to oscillate again like the wavefunctions inside the box (Fig. 12.9). Hence the particle might be found on the outside of a container even though according to classical mechanics it has insufficient energy to escape. Such leakage through classically forbidden zones is called **tunnelling**.

We can use the Schrödinger equation to calculate the probability of tunnelling of a particle of mass m incident on the barrier from the left. Inside a barrier of constant height (a region where $V > 0$ and constant), the Schrödinger equation for a particle free to move in one dimension is

$$\frac{d^2\psi}{dx^2} = \frac{2m}{\hbar^2}(V - E)\psi \tag{14a}$$

We shall suppose that the potential energy inside the barrier is so large

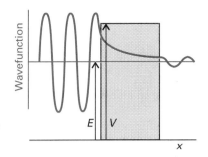

12.9 A particle incident on a barrier from the left has an oscillating wavefunction, but inside the barrier its amplitude varies exponentially (for $E < V$). If the barrier is not too thick, the wavefunction is non-zero at its opposite face, and so oscillates there, which corresponds to the particle penetrating the barrier. (Only the real component is shown.)

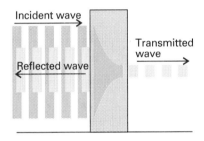

12.10 When a particle is incident on a barrier from the left, the wavefunctions consist of a wave representing linear momentum to the right, a reflected component representing momentum to the left, a varying but not oscillating component inside the barrier, and a (weak) wave representing motion to the right on the far side of the barrier.

that V is larger than E, so $V - E$ is positive. The general solutions of this equation are

$$\psi = Ae^{\kappa x} + Be^{-\kappa x} \qquad \kappa = \left(\frac{2m(V - E)}{\hbar^2}\right)^{\frac{1}{2}} \tag{14b}$$

as can readily be verified by differentiating ψ twice with respect to x. The important feature to note is that the two exponentials are now *real* functions (as distinct from the complex, oscillating functions for the interior of the box). On the far edge of the barrier where $V = 0$, the wavefunction becomes oscillatory again.

The complete wavefunction for a particle incident from the left consists of an incident wave, a wave reflected from the barrier, the exponentially changing amplitudes inside the barrier, and an oscillating wave representing the propagation of the particle to the right after successfully tunnelling through the barrier (Fig. 12.10). The probability P that the particle penetrates through the barrier of length L can be found from the amplitude of the wave on the right of the barrier, and the result is that

$$P = 16\varepsilon(1 - \varepsilon)e^{-2L/D} \qquad \text{where } \varepsilon = \frac{E}{V}, \qquad D = \frac{\hbar}{\{2m(V - E)\}^{\frac{1}{2}}} \tag{14c}$$

This result is valid for high, wide barriers (in the sense $L \gg D$). The probability decreases exponentially with the thickness of the barrier and with $m^{\frac{1}{2}}$. It follows that particles of low mass are more able to tunnel through barriers than heavy ones (Fig. 12.11). Tunnelling is very important for electrons, and moderately important for protons; for heavier particles it is less important. A number of effects in chemistry (e.g. the isotope dependence of some reaction rates) depend on the ability of the proton to tunnel more readily than the deuteron. The very rapid equilibration of proton transfer reactions (which were discussed in Chapter 9) is also a manifestation of the ability of protons to tunnel through barriers and transfer quickly from an acid to a base.

Example 12.4 *Estimating a tunnelling probability*

Estimate the relative probabilities that a proton and a deuteron can tunnel through the same barrier of height 1.0 eV and length 100 pm when their energy is 0.9 eV.

Method. We should first calculate D to determine whether the condition $L \gg D$ for the validity of eqn 14c is satisfied. Then calculate the ratio of the values of P for the two particles.

Answer. The value of D for a particle of mass m is

$$D = \frac{1.055 \times 10^{-34}\,\text{J s}}{\{2 \times (m/u) \times (1.673 \times 10^{-27}\,\text{kg}) \times (1.602 \times 10^{-20}\,\text{J})\}^{\frac{1}{2}}}$$

$$= \frac{14.4\,\text{pm}}{(m/u)^{\frac{1}{2}}}$$

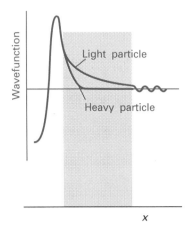

12.11 The wavefunction of a heavy particle decays more rapidly inside a barrier than that of a light particle. Consequently, a light particle has a greater probability of tunnelling through the barrier.

The values of D for a proton ($m = 1.0\,\mathrm{u}$) and a deuteron ($m = 2.0\,\mathrm{u}$) are $14\,\mathrm{pm}$ and $10\,\mathrm{pm}$, respectively, which are both much smaller than the width of the barrier, so eqn 14c can be used. The ratio of probabilities is then

$$\frac{P_\mathrm{H}}{P_\mathrm{D}} = \mathrm{e}^{-2L(1/D_\mathrm{H} - 1/D_\mathrm{D})} = \mathrm{e}^{-200(1/14 - 1/10)} = 3.0 \times 10^2$$

Comment. The result shows that the tunnelling probability of a proton (in the system specified) is much greater than that of a deuteron. The probability itself is 9×10^{-7} for a proton.

Exercise E12.4. Calculate the relative tunnelling probabilities when the barrier is twice as long, the other conditions being unchanged.   [9×10^4]

VIBRATIONAL MOTION

A particle undergoes **harmonic motion** if it experiences a restoring force that is proportional to its displacement:

$$F = -kx \tag{15a}$$

where k is the force constant. As explained in *Further information 5*, because force is related to potential energy by $F = -\mathrm{d}V/\mathrm{d}x$, a force of this form corresponds to a potential energy

$$V = \tfrac{1}{2}kx^2 \tag{15b}$$

This expression, which is the equation of a parabola (Fig. 12.12), is the origin of the term **parabolic potential energy** for the potential energy characteristic of a harmonic oscillator. The Schrödinger equation for the particle is therefore

$$-\frac{\hbar^2}{2m}\frac{\mathrm{d}^2\psi}{\mathrm{d}x^2} + \tfrac{1}{2}kx^2\psi = E\psi \tag{16}$$

12.4 The energy levels

The solution of eqn 16 is outlined in *Further information 8*. The most important feature of the result is that the permitted energy levels of a harmonic oscillator are

$$E_v = (v + \tfrac{1}{2})\hbar\omega \qquad \text{with } v = 0, 1, 2, \ldots \text{ and } \omega = \left(\frac{k}{m}\right)^{\frac{1}{2}} \tag{17}$$

It follows that the separation between adjacent levels is

$$E_{v+1} - E_v = \hbar\omega$$

which is the same for all v. Therefore, the energy levels form a uniform ladder of spacing $\hbar\omega$ (Fig. 12.13).

The energy separation $\hbar\omega$ is negligibly small for macroscopic objects but is of great importance for objects with mass similar to that of atoms. For instance, the force constant of a typical chemical bond is around $500\,\mathrm{N\,m^{-1}}$ and, because the mass of a proton is about $1.7 \times 10^{-27}\,\mathrm{kg}$, the frequency is $\omega \approx 5 \times 10^{14}\,\mathrm{s^{-1}}$ and the separation of adjacent levels is

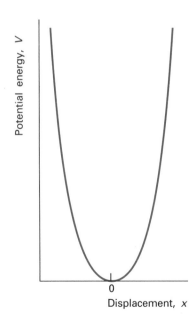

12.12 The parabolic potential energy $V = \tfrac{1}{2}kx^2$ of a harmonic oscillator, where x is the displacement from equilibrium. The narrowness of the curve depends on the force constant k.

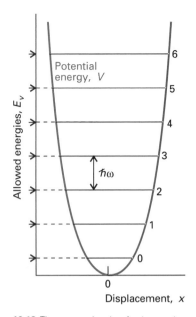

12.13 The energy levels of a harmonic oscillator are evenly spaced. Even in its lowest state an oscillator has an energy greater than zero.

$\hbar\omega \approx 6 \times 10^{-20}$ J (about 0.4 eV). This energy separation corresponds to 30 kJ mol^{-1}, which is chemically significant.

The excitation of a harmonic oscillator from one level to the one immediately above requires an energy 6×10^{-20} J (0.4 eV) and hence, if it is caused by a photon, requires radiation of frequency

$$v = \frac{\Delta E}{h} = \frac{6 \times 10^{-20} \, \text{J}}{6.626 \times 10^{-34} \, \text{J s}} \approx 9 \times 10^{13} \, \text{Hz}$$

and therefore of wavelength

$$\lambda = \frac{c}{v} = \frac{9 \times 10^{13} \, \text{s}^{-1}}{2.998 \times 10^8 \, \text{m s}^{-1}} \approx 3 \, \mu\text{m}$$

It follows that transitions between the vibrational energy levels of molecules require infrared radiation, as we shall describe in Chapter 16.

Because the smallest permitted value of v is 0, an oscillator has a zero-point energy

$$E_0 = \tfrac{1}{2}\hbar\omega \tag{18}$$

For the typical molecular oscillator specified above, the zero-point energy is about 3×10^{-20} J, which corresponds to 0.2 eV, or 15 kJ mol^{-1}. The mathematical reason for the zero-point energy is that v cannot take negative values, for if it did the wavefunction would be ill-behaved. The physical reason is the same as that for the particle in a square well: the particle is confined, its position is not completely uncertain, and therefore its momentum, and hence its kinetic energy, cannot be exactly zero. We can picture this zero-point state as one in which the particle fluctuates incessantly around its equilibrium position; classical mechanics would allow the particle to be perfectly still.

12.5 The wavefunctions

It is helpful at the outset to identify the similarities between the harmonic oscillator and the particle in a box, for then we shall be able to anticipate the form of the oscillator wavefunctions without detailed calculation. As for the particle in a box, a particle undergoing harmonic motion is trapped in a symmetrical well in which the potential energy rises to large values (and ultimately to infinity) for sufficiently large displacements (compare Figs 12.1 and 12.12). However, there are two important differences. First, the wavefunction approaches zero more slowly at large displacements, because the potential energy climbs towards infinity only as x^2 and not abruptly. Second, as the kinetic energy of the particle depends on the displacement in a more complex way (on account of the variation of the potential energy), the curvature of the wavefunction also varies in a more complex way.

The form of the wavefunctions

As we show in *Further information 8*, the wavefunction for a harmonic oscillator in a state with quantum number v has the form

$$\psi_v = N \times (\text{polynomial in } x) \times (\text{bell-shaped Gaussian function})$$

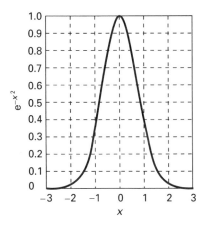

12.14 The graph of the Gaussian function, $y = e^{-x^2}$.

where N is a normalization constant. A **Gaussian function** is a function of the form e^{-x^2} (Fig. 12.14). The precise form of the wavefunctions is

$$\psi_v = N_v H_v e^{-y^2/2}, \qquad \text{where } y = \left(\frac{mk}{\hbar^2}\right)^{\frac{1}{4}} \times x \qquad (19)$$

The factor H_v is a **Hermite polynomial**, some of which are listed in Table 12.1. For instance, because $H_0 = 1$, the wavefunction for the ground state (the lowest energy state) of the harmonic oscillator is

$$\psi_0 = N_0 e^{-y^2/2} \qquad (20a)$$

It follows that the probability density is the bell-shaped Gaussian function

$$\psi_0^2 = N_0^2 e^{-y^2} \qquad (20b)$$

The wavefunction and the probability distribution are shown in Fig. 12.15. Both curves have their largest values at zero displacement (at $x = 0$, which corresponds to $y = 0$), and so capture the classical picture of the zero-point motion as arising from the fluctuation of the particle about its equilibrium position (Fig. 12.16). The wavefunction for the first excited state of the oscillator, the state with $v = 1$, is obtained by noting that $H_1 = 2y$ (note that some of the Hermite polynomials are *very* simple functions!):

$$\psi_1 = N_1 \times 2y e^{-y^2/2}$$

This function is zero at zero displacement, and the probability density has a maximum on either side of zero displacement (Fig. 12.17).

The shapes of several wavefunctions are shown in Fig. 12.18; the shading in Fig. 12.16 that represents the probability density is based on the squares of these functions. At high quantum numbers, harmonic oscillator wavefunctions have their largest amplitudes near the turning points of the classical motion (where $V = E$, so the kinetic energy is zero). We see classical properties emerging in the correspondence limit of high quantum numbers, for a classical particle is most likely to be found at the turning points (where it travels most slowly) and is least likely to be found at zero displacement (where it travels with maximum velocity).

Table 12.1 The Hermite polynomials $H_v(y)$

v	H_v
0	1
1	$2y$
2	$4y^2 - 2$
3	$8y^3 - 12y$
4	$16y^4 - 48y^2 + 12$
5	$32y^5 - 160y^3 + 120y$
6	$64y^6 - 480y^4 + 720y^2 - 120$

The Hermite polynomials (which continue up to infinite v) satisfy the equation

$$H_v'' - 2yH_v' + 2vH_v = 0$$

and the recursion relation

$$H_{v+1} = 2yH_v - 2vH_{v-1}$$

An important integral is

$$\int_{-\infty}^{\infty} H_{v'} \cdot H_v e^{-y^2}\, dy = \begin{cases} 0 & \text{if } v' \neq v \\ \pi^{\frac{1}{2}} 2^v v! & \text{if } v' = v \end{cases}$$

Example 12.5 *Normalizing a harmonic oscillator wavefunction*

Find the normalization constant for the harmonic oscillator wavefunctions. The integrals required are given in Table 12.1.

Method. Normalization is always carried out by evaluating the integral of $|\psi|^2$ over all space and then finding the normalization factor from

$$N = \left(\frac{1}{\int \psi^\star \psi\, d\tau}\right)^{\frac{1}{2}}$$

The normalized wavefunction is then equal to $N\psi$. In this one-dimensional problem, the volume element is dx and the integration is

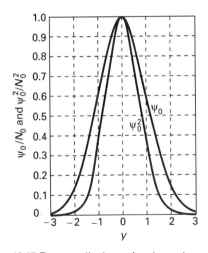

12.15 The normalized wavefunction and probability distribution for the lowest energy state of a harmonic oscillator.

from $-\infty$ to $+\infty$. The wavefunctions are expressed in terms of the dimensionless variable $y = x/\alpha$ with $\alpha = (\hbar^2/mk)^{\frac{1}{4}}$, so we begin by expressing the integral in terms of y, by using $dx = \alpha\, dy$.

Answer. The unnormalized wavefunction is

$$\psi_v = H_v e^{-y^2/2}$$

It follows from the integrals given in Table 12.1 that

$$\int_{-\infty}^{\infty} \psi_v^* \psi_v \, dx = \alpha \int_{-\infty}^{\infty} \psi_v^* \psi_v \, dy = \alpha \int_{-\infty}^{\infty} H_v^2 e^{-y^2} \, dy = \alpha \pi^{\frac{1}{2}} 2^v v!$$

Therefore,

$$N = \frac{1}{(\alpha \pi^{\frac{1}{2}} 2^v v!)^{\frac{1}{2}}}$$

and is different for each value of v.

Comment. The Hermite polynomials are one of a class of functions called *orthogonal polynomials*. These polynomials have a wide range of important properties that allow a number of quantum mechanical calculations to be done with relative ease. See *Further reading* for a reference to their properties.

Exercise E12.5. Confirm, by explicit evaluation of the integral, that ψ_0 and ψ_1 are orthogonal.

$$\left[\int_{-\infty}^{\infty} \psi_0 \psi_1 \, dx = 0 \right]$$

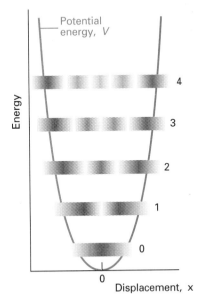

12.16 The probability distributions for the first five states of a harmonic oscillator represented by the density of shading. Note how the regions of highest probability (the regions of densest shading) move towards the turning points of the classical motion as v increases.

The properties of the oscillator

Once the wavefunctions are available, we can start calculating the properties of the harmonic oscillator. For instance, we can calculate the expectation values of an observable Ω by evaluating integrals of the type

$$\langle \Omega \rangle = \int_{-\infty}^{\infty} \psi_v^* \hat{\Omega} \psi_v \, dx$$

When the explicit wavefunctions are substituted the integrals look fearsome, but the Hermite polynomials have many simplifying features. For instance, we show in the following example that the mean displacement $\langle x \rangle$ and the mean square displacement $\langle x^2 \rangle$ of the oscillator when it is in the state with quantum number v are

$$\langle x \rangle = 0 \qquad \langle x^2 \rangle = (v + \tfrac{1}{2}) \frac{\hbar}{(mk)^{\frac{1}{2}}} \tag{21}$$

The result for $\langle x \rangle$ shows that the oscillator is as likely to be found on either side of $x = 0$ (like a classical oscillator). The result for $\langle x^2 \rangle$ shows that the mean square displacement increases with v. This increase is apparent from the probability densities in Fig. 12.16, and corresponds to the classical amplitude of swing increasing as the oscillator becomes more highly excited.

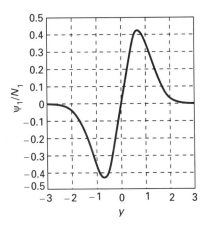

12.17 The normalized wavefunction for the first excited state of a harmonic oscillator. The probability distribution is the square of this function. Notice that the wavefunction is orthogonal to the wavefunction for $v = 0$ (Fig. 12.15).

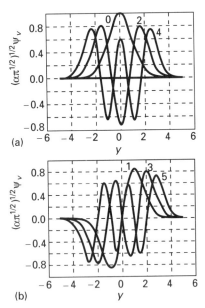

12.18 The normalized wavefunctions for states of a harmonic oscillator. (a) ψ for even v; (b) ψ for odd v. All wavefunctions with v even are symmetrical around zero displacement; all wavefunctions with v odd are antisymmetric and have a node at $x = 0$. The number of nodes is equal to the value of v.

Example 12.6 *Calculating properties of a harmonic oscillator*

Calculate the mean displacement of the oscillator $\langle x \rangle$ from equilibrium when it is in a quantum state v.

Method. Normalized wavefunctions must be used to calculate the expectation value. The operator for position along x is multiplication by the value of x (Section 11.5). The resulting integral can be evaluated either by inspection (the integrand is the product of an odd and an even function), or by explicit evaluation using the formulae in Table 12.1. We illustrate the latter procedure.

Answer. The integral we require is

$$\langle x \rangle = \int_{-\infty}^{\infty} \psi_v^* x \psi_v \, dx = N_v^2 \int_{-\infty}^{\infty} (H_v e^{-y^2/2}) x (H_v e^{-y^2/2}) \, dx$$

$$= \alpha^2 N_v^2 \int_{-\infty}^{\infty} (H_v e^{-y^2/2}) y (H_v e^{-y^2/2}) \, dy$$

$$= \alpha^2 N_v^2 \int_{-\infty}^{\infty} H_v y H_v e^{-y^2} \, dy$$

Now use the recursion relation in Table 12.1 to form

$$yH_v = vH_{v-1} + \tfrac{1}{2}H_{v+1}$$

which turns the integral into

$$\int_{-\infty}^{\infty} H_v y H_v e^{-y^2} \, dy = v \int_{-\infty}^{\infty} H_v H_{v-1} e^{-y^2} \, dy + \tfrac{1}{2} \int_{-\infty}^{\infty} H_v H_{v+1} e^{-y^2} \, dy$$

Both integrals are zero (Table 12.1), so $\langle x \rangle = 0$.

Comment. The result is in fact obvious from the probabilities depicted in Fig. 12.16: they are all symmetrical about $x = 0$, and so displacements to the right are as probable as displacements to the left. The reason for going through the calculation in detail even though the result is obvious is that the same technique is applicable to other observables for which the result is not obvious.

Exercise E12.6. Calculate the mean square displacement $\langle x^2 \rangle$ of the particle from its equilibrium position. (Use the recursion relation twice.)

[eqn 21]

We can now calculate the mean potential energy of an oscillator very simply:

$$\langle V \rangle = \langle \tfrac{1}{2}kx^2 \rangle = \tfrac{1}{2}k\langle x^2 \rangle = \tfrac{1}{2}(v + \tfrac{1}{2})\hbar \left(\frac{k}{m}\right)^{\frac{1}{2}}$$

$$= \tfrac{1}{2}(v + \tfrac{1}{2})\hbar\omega$$

Because the total energy in the state with quantum number v is

$(v + \frac{1}{2})\hbar\omega$, it follows that

$$\langle V \rangle = \tfrac{1}{2}E_v \tag{22a}$$

The total energy is the sum of the potential and kinetic energies, so it follows at once that the mean kinetic energy of the oscillator is

$$\langle E_\mathrm{K} \rangle = \tfrac{1}{2}E_v \tag{22b}$$

The result that the mean potential and kinetic energies are equal (and therefore that both are equal to half the total energy) is a special case of the **virial theorem**:

> If the potential energy of a particle has the form $V = ax^b$, then its mean potential and kinetic energies are related by
>
> $$2\langle E_\mathrm{K} \rangle = b\langle V \rangle \tag{23}$$

For a harmonic oscillator $b = 2$, so $\langle E_\mathrm{K} \rangle = \langle V \rangle$, as we have found. The virial theorem is a short cut to the establishment of a number of useful results, and we shall use it again.

An oscillator may be found in classically forbidden regions, where the potential energy is greater than the total energy ($V > E$). For example, it follows from the shape of the wavefunction (see the *Justification* that follows) that in its lowest energy state there is about 8 per cent chance of finding it stretched beyond its classical limit and 8 per cent chance of finding it with a classically forbidden compression. The probability of being found in classically forbidden regions decreases quickly with increasing v, and vanishes entirely as v approaches infinity, as we would expect from the correspondence principle. Macroscopic oscillators (such as pendulums) are in states with very high quantum numbers, so that the probability that they will be found in a classically forbidden region is wholly negligible. Molecules, however, are normally in their vibrational ground states, and for them the probability is very significant.

JUSTIFICATION

According to classical mechanics, the turning point x_tp of an oscillating particle occurs when its kinetic energy is zero, which is when its potential energy $\tfrac{1}{2}kx^2$ is equal to its total energy E. This equality occurs when

$$x_\mathrm{tp}^2 = \frac{2E}{k} \qquad \text{or } x_\mathrm{tp} = \pm\left(\frac{2E}{k}\right)^{\frac{1}{2}}$$

The probability of finding the oscillator stretched beyond a displacement x_tp is the sum of the probabilities $\psi^2\,\mathrm{d}x$ of finding it in any of the intervals $\mathrm{d}x$ lying between x_tp and infinity:

$$P = \int_{x_\mathrm{tp}}^{\infty} \psi^2\,\mathrm{d}x$$

The variable of integration is best expressed in terms of $x = \alpha y$, and then the turning point lies at

$$y_\mathrm{tp} = x_\mathrm{tp}/\alpha = (2v + 1)^{\frac{1}{2}}$$

Table 12.2* The error function

z	erf z
0	0
0.01	0.0113
0.05	0.0564
0.10	0.1125
0.50	0.5205
1.00	0.8427
1.50	0.9661
2.0	0.9953

* More values are given in the Data section at the end of this volume.

For the state of lowest energy ($v = 0$), $y_{tp} = 1$ and the probability is

$$P = \int_{x_{tp}}^{\infty} \psi_0^2 \, dx = \alpha N_0^2 \int_1^{\infty} e^{-y^2} \, dy = \frac{1}{\pi^{\frac{1}{2}}} \int_1^{\infty} e^{-y^2} \, dy$$

The integral is a special case of the **error function**, erf z, which is defined as follows:

$$\text{erf } z = 1 - \frac{2}{\pi^{\frac{1}{2}}} \int_z^{\infty} e^{-y^2} \, dy$$

The values of this function are tabulated (just like sine and cosine functions), and a small selection of values is given in Table 12.2. In the present case

$$P = \tfrac{1}{2}(1 - \text{erf } 1) = \tfrac{1}{2}(1 - 0.843) = 0.079$$

It follows that, in 7.9 per cent of a large number of observations, an oscillator in the state $v = 0$ (whatever its mass and the value of the force constant) will be found stretched into a classically forbidden region. There is the same probability of finding the oscillator with a classically forbidden compression. The total probability of finding the oscillator in a classically forbidden region (stretched or compressed) is about 16 per cent.

ROTATIONAL MOTION

The treatment of rotational motion can be broken down into two parts. The first deals with motion in two dimensions and the second with rotation in three dimensions. It may be helpful to review the classical description of rotational motion given in *Further information 5*, particularly the concepts of moment of inertia and angular momentum.

12.6 Rotation in two dimensions

We consider a particle of mass m constrained to move in a circular path of radius r in the xy-plane (Fig. 12.19). The total energy is equal to the kinetic energy, because the potential energy is zero everywhere. We can therefore write $E = p^2/2m$. According to classical mechanics, the angular momentum around the z-axis (which lies perpendicular to the xy-plane) is $J_z = pr$, so the energy can be expressed as $J_z^2/2mr^2$. Because mr^2 is the moment of inertia I of the mass on its path, it follows that

$$E = \frac{J_z^2}{2I} \tag{24}$$

We shall now see that not all values of the angular momentum are permitted in quantum mechanics and therefore that both angular momentum and rotational energy are quantized.

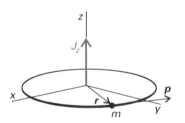

12.19 The angular momentum of a particle of mass m on a circular path of radius r in the xy-plane is represented by a vector of magnitude pr perpendicular to the plane.

The qualitative origin of quantized rotation

Because $J_z = pr$, and, from the de Broglie relation, $p = h/\lambda$, the angular momentum about the z-axis is

$$J_z = \frac{hr}{\lambda}$$

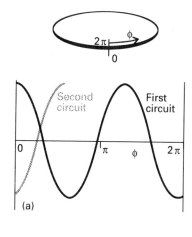

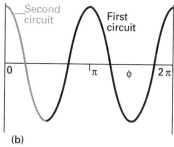

12.20 Two solutions of the Schrödinger equation for a particle on a ring. The circumference has been opened out into a straight line; the points at $\phi = 0$ and 2π are identical. The solution in (a) is unacceptable because it is not single-valued. Moreover, on successive circuits it interferes destructively with itself, and does not survive. The solution in (b) is acceptable: it is single-valued and on successive circuits it reproduces itself.

This equation shows that the shorter the wavelength of the particle on a circular path of a given radius, the greater the angular momentum of the particle. It follows that, if we can see why the wavelength is restricted to discrete values, then we shall understand why the angular momentum is quantized.

Suppose for the moment that λ can take an arbitrary value. In that case, the wavefunction depends on the azimuthal angle ϕ as shown in Fig. 12.20a. When ϕ increases beyond 2π, the wavefunction continues to change but, for an arbitrary wavelength, it gives rise to a different value at each point, which is unacceptable (Section 11.4). An acceptable solution is obtained if the wavefunction reproduces itself on successive circuits, as in Fig. 12.20b. Because only some wavefunctions have this property, it follows that only some angular momenta are acceptable, and therefore that only certain energies exist. Hence, the energy of the particle is quantized. In particular, the wavelength must be a whole-number fraction of the circumference if its ends are to match after each circuit. That is,

$$\lambda = \frac{2\pi r}{m_l}$$

with m_l (the conventional notation for this quantum number) a number with integer values including 0. The angular momentum is therefore limited to the values

$$J_z = \frac{h}{\lambda} \times r = \frac{m_l h}{2\pi r} \times r = m_l \times \frac{h}{2\pi}$$

That is,

$$J_z = m_l \hbar \qquad m_l = 0, \pm 1, \pm 2, \ldots \tag{25}$$

Positive values of m_l correspond to rotation in a clockwise sense around the z-axis (as viewed in the direction of z, Fig. 12.21) and negative values of m_l correspond to counterclockwise rotation around z. Furthermore, it then also follows that the energy is limited to the values

$$E = \frac{J_z^2}{2I} = m_l^2 \times \frac{\hbar^2}{2I} \tag{26a}$$

We shall shortly see that the corresponding wavefunctions are

$$\psi_{m_l} = \left(\frac{1}{2\pi}\right)^{\frac{1}{2}} e^{im_l\phi} \tag{26b}$$

The wavefunction with $m_l = 0$ is $\psi_0 = (1/2\pi)^{\frac{1}{2}}$, and has the same value at all points on the circle.

We have arrived at a number of conclusions about rotational motion by cobbling together some classical notions and the de Broglie relation. Such a procedure can be very useful for establishing the general form (and, as in this case, the exact energies) for a quantum mechanical system. However, to be sure that the correct solutions have been obtained, and to obtain practice for more complex problems where this less formal approach is inadequate, we need to solve the Schrödinger

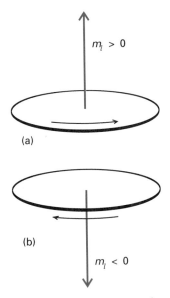

12.21 The angular momentum of a particle confined to a plane can be represented by a vector of length $|m_l|$ units along the z-axis and with an orientation that indicates the direction of motion of the particle.

equation explicitly. The formal solution is obtained in the following *Justification*.

JUSTIFICATION

The two-dimensional Schrödinger equation for a particle in a plane (with $V = 0$) is the same as in eqn 11:

$$-\frac{\hbar^2}{2m}\left(\frac{\partial^2 \psi}{\partial x^2} + \frac{\partial^2 \psi}{\partial y^2}\right) = E\psi$$

Instead of trying to solve this equation as it stands, it is better to transform it to polar coordinates and to write

$$x = r\cos\phi \qquad y = r\sin\phi$$

because then the condition $r = $ constant is much easier to impose. (In general, it is always a good idea to use coordinates that reflect the full symmetry of the system.) Since r is constrained to be a constant, by standard manipulations we can write

$$\frac{\partial^2 \psi}{\partial x^2} + \frac{\partial^2 \psi}{\partial y^2} = \frac{1}{r^2}\frac{d^2 \psi}{d\phi^2}$$

and hence transform the Schrödinger equation into

$$-\frac{\hbar^2}{2mr^2}\frac{d^2 \psi}{d\phi^2} = E\psi$$

The moment of inertia $I = mr^2$ has appeared automatically and the equation may be written

$$\frac{d^2 \psi}{d\phi^2} = -\frac{2IE}{\hbar^2}\psi$$

The normalized general solutions of the equation are

$$\psi_{m_l} = \left(\frac{1}{2\pi}\right)^{\frac{1}{2}} e^{im_l\phi} \qquad m_l = \pm\frac{(2IE)^{\frac{1}{2}}}{\hbar}$$

as in eqn 26b. The quantity m_l is just a dimensionless number at this stage.

We now select the acceptable solutions from among these general solutions by imposing the condition that the wavefunction should be single-valued. That is, the wavefunction ψ must satisfy the **cyclic boundary condition** and match at points separated by a complete revolution.

$$\psi(\phi + 2\pi) = \psi(\phi)$$

On substituting the general wavefunction into this condition, we find

$$\psi(\phi + 2\pi) = \left(\frac{1}{2\pi}\right)^{\frac{1}{2}} e^{im_l(\phi + 2\pi)} = \left(\frac{1}{2\pi}\right)^{\frac{1}{2}} e^{im_l\phi}e^{2\pi im_l} = \psi(\phi)e^{2\pi im_l}$$

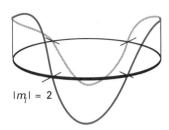

$|m_l| = 2$

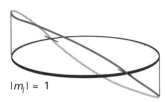

$|m_l| = 1$

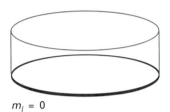

$m_l = 0$

12.22 The real parts of the wavefunctions of a particle on a ring. As shorter wavelengths are achieved, the angular momentum grows in steps of $\hbar$.

As $e^{i\pi} = -1$, this relation is equivalent to

$$\psi(\phi + 2\pi) = (-1)^{2m_l}\psi(\phi)$$

Hence, $2m_l$ must be a positive or negative even integer and therefore m_l must be an integer:

$$m_l = 0, \pm 1, \pm 2, \ldots$$

Quantization of rotation

We can summarize the conclusions so far as follows. The energy is quantized and restricted to the values given in eqn 26a ($E = m_l^2\hbar^2/2I$). The occurrence of m_l as its square means that the energy of rotation is independent of the sense of rotation (the sign of m_l), as we expect physically. In other words, states with a given value of $|m_l|$ are doubly degenerate, except for $m_l = 0$, which is non-degenerate. Although the result has been derived for the rotation of a single mass point, it also applies to any body of moment of inertia I constrained to rotate about one axis.

We have also seen that the angular momentum is quantized and confined to the values given in eqn 25 ($J_z = m_l\hbar$). The increasing angular momentum is associated with the increasing number of nodes in the wavefunction: the wavelength decreases stepwise as $|m_l|$ increases, so the momentum with which the particle travels round the ring increases (Fig. 12.22). As shown in the following *Justification*, the same conclusion can be obtained formally by using the relation between eigenvalues and the values of observables that was established in Section 11.5.

JUSTIFICATION

In the discussion of translational motion in one dimension, we saw that the opposite signs in the wavefunctions e^{ikx} and e^{-ikx} correspond to opposite directions of travel, and that the linear momentum is given by the eigenvalue of the linear momentum operator. The same conclusions can be drawn here, but now we need the eigenvalues of the angular momentum operator. In classical mechanics the orbital angular momentum l_z about the z-axis is defined as

$$l_z = xp_y - yp_x$$

where p_x is the component of linear motion parallel to the x-axis and p_y is the component parallel to the y-axis. The operators for the two linear momentum components are proportional to differentiation with respect to x and y, so in quantum mechanics the operator for angular momentum about the z-axis, which we denote $\hat{l}_z$, is

$$\hat{l}_z = \frac{\hbar}{i}\left(x\frac{\partial}{\partial y} - y\frac{\partial}{\partial x}\right)$$

When expressed in terms of polar coordinates, this equation becomes

$$\hat{l}_z = \frac{\hbar}{i} \frac{\partial}{\partial \phi}$$

With the angular momentum operator available, we can test the wavefunction in eqn 26b. Disregarding the normalization constant, we find

$$\hat{l}_z \psi_{m_l} = \frac{\hbar}{i} \frac{d}{d\phi} e^{im_l\phi} = im_l \times \frac{\hbar}{i} e^{im_l\phi} = m_l\hbar \psi_{m_l}$$

That is, ψ_{m_l} is an eigenfunction of the orbital angular momentum operator and corresponds to an angular momentum $m_l\hbar$. When m_l is positive, the angular momentum is positive (clockwise when seen from below); when m_l is negative, the angular momentum is negative (counterclockwise when seen from below). These features are the origin of the vector representation of the angular momentum, in which the magnitude is represented by the length of a vector and the direction of motion by its orientation (Fig. 12.21).

Finally, we can explore the question of the position of the particle when it is in a state of definite angular momentum. As usual, we form the probability density:

$$\psi_{m_l}^* \psi_{m_l} = \left\{ \left(\frac{1}{2\pi} \right)^{\frac{1}{2}} e^{-im_l\phi} \right\} \left\{ \left(\frac{1}{2\pi} \right)^{\frac{1}{2}} e^{im_l\phi} \right\} = \frac{1}{2\pi}$$

Because this ψ^*c is independent of ϕ, the probability of locating the particle at any point on the ring is also independent of ϕ. Hence the location of the particle is completely indefinite, and knowing the angular momentum precisely eliminates the possibility of specifying the particle's location. *Angular momentum and angle are a pair of complementary observables* (in the sense defined in Section 11.6), and the inability to specify them simultaneously to arbitrary precision is another example of the uncertainty principle.

12.7 Rotation in three dimensions

We now consider a particle of mass m that is constrained to move on the surface of a sphere of radius r. We shall use the results of this calculation when we come to describe the states of electrons in atoms (Chapter 13) and of rotating molecules (Chapter 16). The latter application arises from the fact that the rotation of a solid body of moment of inertia I can be represented by a single point of mass M rotating at a radius R, the **radius of gyration** of the body, which is defined so that $I = MR^2$. The requirement that the wavefunction should match as a path is traced over the poles as well as round the equator of the sphere surrounding the central point (Fig. 12.23) introduces a second cyclic boundary condition and therefore a second quantum number.

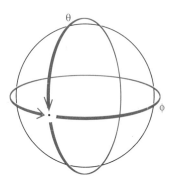

12.23 A particle on the surface of a sphere must satisfy two cyclic boundary conditions, and this leads to two quantum numbers for its state of angular momentum.

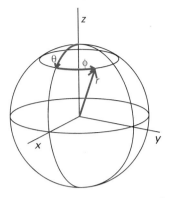

12.24 Spherical polar coordinates. For a particle confined to the surface of a sphere, only the co-latitude θ and the azimuth ϕ can change.

The Schrödinger equation

The Schrödinger equation in three dimensions (Table 11.1) is

$$-\frac{\hbar^2}{2m}\nabla^2\psi + V\psi = E\psi \tag{27}$$

We show in *Further information 9* that this partial differential equation can be simplified by the separation of variables procedure by expressing the wavefunction (for constant r) as the product

$$\psi = \Theta\Phi \tag{28}$$

where Θ is a function solely of the angle θ (the co-latitude, Fig. 12.24) and Φ is a function only of the angle ϕ (the azimuth).

Solution of the Schrödinger equation shows that the acceptable wavefunctions are specified by two quantum numbers l and m_l, which are restricted to the values

$$l = 0, 1, 2, \ldots \tag{29a}$$

$$m_l = 0, \pm 1, \pm 2, \ldots, \pm l \tag{29b}$$

Equivalently:

$$m_l = l, l-1, l-2, \ldots, -l \tag{29c}$$

Note that the quantum number l is positive and that, for a given value of l, there are $2l + 1$ permitted values of m_l. The normalized wavefunctions are usually denoted Y_{l,m_l} and are called the **spherical harmonics**. Some of the spherical harmonics are listed in Table 12.3, and their amplitudes at different points on the spherical surface are illustrated in Fig. 12.25.

The energy E of the particle is restricted to the values

$$E = l(l+1)\frac{\hbar^2}{2I} \qquad l = 0, 1, 2, \ldots \tag{30}$$

We see that the energy is quantized and that it is independent of m_l. Because there are $2l + 1$ different wavefunctions (one for each value of m_l) that correspond to the same energy, a level with quantum number l is $(2l + 1)$-fold degenerate.

Angular momentum

The energy of a rotating particle is related classically to its angular momentum by $E = J^2/2I$ (see *Further information 5*). Therefore, by comparing this equation with eqn 30, we can deduce that the magnitude of the angular momentum is quantized and confined to the values

Magnitude of angular momentum $= \{l(l+1)\}^{\frac{1}{2}}\hbar \qquad l = 0, 1, 2, \ldots$

(31a)

We have already seen (in the context of two-dimensional rotation) that the angular momentum about the z-axis is quantized, and that it has the values

z-component of angular momentum $= m_l\hbar$

$$m_l = 0, \pm 1, \pm 2, \ldots, \pm l \tag{31b}$$

A feature of the wavefunction is that the higher the value of l the larger

Table 12.3 The spherical harmonics $Y_{l,m_l}(\theta, \phi)$

l	m_l	Y_{l,m_l}
0	0	$\left(\dfrac{1}{4\pi}\right)^{\frac{1}{2}}$
1	0	$\left(\dfrac{3}{4\pi}\right)^{\frac{1}{2}}\cos\theta$
	± 1	$\mp\left(\dfrac{3}{8\pi}\right)^{\frac{1}{2}}\sin\theta\, e^{\pm i\phi}$
2	0	$\left(\dfrac{5}{16\pi}\right)^{\frac{1}{2}}(3\cos^2\theta - 1)$
	± 1	$\mp\left(\dfrac{15}{8\pi}\right)^{\frac{1}{2}}\cos\theta\sin\theta\, e^{\pm i\phi}$
	± 2	$\left(\dfrac{15}{32\pi}\right)^{\frac{1}{2}}\sin^2\theta\, e^{\pm 2i\phi}$
3	0	$\left(\dfrac{7}{16\pi}\right)^{\frac{1}{2}}(5\cos^3\theta - 3\cos\theta)$
	± 1	$\mp\left(\dfrac{21}{64\pi}\right)^{\frac{1}{2}}(5\cos^2\theta - 1)\sin\theta\, e^{\pm i\phi}$
	± 2	$\left(\dfrac{105}{32\pi}\right)^{\frac{1}{2}}\sin^2\theta\cos\theta\, e^{\pm 2i\phi}$
	± 3	$\mp\left(\dfrac{35}{64\pi}\right)^{\frac{1}{2}}\sin^3\theta\, e^{\pm 3i\phi}$

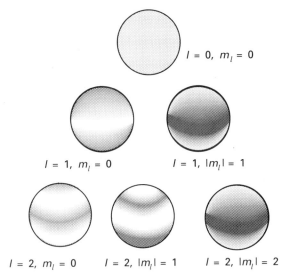

$l = 0$, $m_l = 0$

$l = 1$, $m_l = 0$ $l = 1$, $|m_l| = 1$

$l = 2$, $m_l = 0$ $l = 2$, $|m_l| = 1$ $l = 2$, $|m_l| = 2$

12.25 The probability densities for a particle on the surface of a sphere. Note that the number of nodes increases as the value of l increases, and that their location is determined by the value of m_l. There are no equator-cutting nodes in the *probability densities* of states with definite values of m_l, and the densities of $+|m_l|$ and $-|m_l|$ are the same. The wavefunctions of these states differ, not the probabilities.

the number of nodal lines (the positions at which $\psi = 0$) in the wavefunction. This feature reflects the fact that higher angular momentum implies higher kinetic energy, and therefore a more sharply buckled wavefunction. We can also see that the states corresponding to high angular momentum around the z-axis are those in which most nodes cut the equator (recall Fig. 12.22): this indicates a high kinetic energy arising from motion parallel to the equator because the curvature is greatest in that direction.

Example 12.7 *Calculating the energy levels of a rotating molecule*

Calculate the energies of the first five rotational levels of H_2 (for which $I = 4.603 \times 10^{-48}$ kg m^2), the corresponding magnitudes of the angular momentum, and the number of different values of m_l in each case.

Method. For the energies, we use eqn 30 with $l = 0, 1, 2, 3$, and 4. For the magnitude of the angular momentum, use eqn 31 and express the result as multiples of $\hbar$. For a state of given l there are $2l + 1$ values of m_l in integral steps from $-l$ to $+l$. In the case of rotating molecules the angular momentum quantum number is normally denoted J and we shall use that notation in this example.

Answer. We need

$$\frac{\hbar^2}{2I} = \frac{(1.05457 \times 10^{-34}\,\text{J s})^2}{2 \times (4.603 \times 10^{-48}\,\text{kg m}^2)} = 1.208 \times 10^{-21}\,\text{J}$$

This energy corresponds to 0.727 kJ mol^{-1}. Draw up the following table; $|\boldsymbol{J}|$ denotes the magnitude of the angular momentum.

J	0	1	2	3	4
$J(J+1)$	0	2	6	12	20
$E/(\text{kJ mol}^{-1})$	0	1.45	4.36	8.72	14.54
$|\boldsymbol{J}|/\hbar$	0	1.41	2.45	3.46	4.47
Number of components	1	3	5	7	9

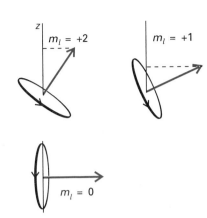

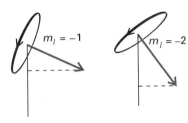

12.26 The permitted orientations of angular momentum when $l = 2$. We shall see soon that this representation is too specific because the azimuthal orientation of the vector (its angle around z) is indeterminate.

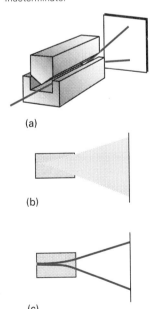

(a)

(b)

(c)

12.27 (a) The experimental arrangement for the Stern–Gerlach experiment: the magnet provides an inhomogeneous field. (b) The classically expected result. (c) The observed outcome using silver atoms.

Comment. Note that the smaller the moment of inertia of the molecule, the greater the separation of the rotational energy levels. Since all components of a given J correspond to the same energy, the rotational level with quantum number J is $(2J + 1)$-fold degenerate.

Exercise E12.7. Repeat the calculation for a deuterium molecule (same bond length, approximately twice the mass).

[Energies smaller by a factor of two; same angular momenta and numbers of components]

Space quantization

The result that m_l is confined to the discrete values $l, l - 1, \ldots, -l$ for a given value of l means that the component of angular momentum about the z-axis may take only $2l + 1$ values. If the angular momentum is represented by a vector of length proportional to its magnitude (i.e. of length $\{l(l + 1)\}^{\frac{1}{2}}$ units), then to represent correctly the value of the component of angular momentum, the vector must be oriented so that its projection on the z-axis is of length m_l units. In classical terms, this means that the plane of rotation of the particle can take only a discrete range of orientations (Fig. 12.26). The remarkable implication is that the *orientation* of a rotating body is quantized.

The quantum mechanical result that a rotating body may not take up an arbitrary orientation with respect to some specified axis (for example, an axis defined by the direction of an externally applied electric or magnetic field) is called **space quantization**. It was confirmed by an experiment first performed by Otto Stern and Walther Gerlach in 1921, who shot a beam of silver atoms through an inhomogeneous magnetic field (Fig. 12.27a). The idea behind the experiment was that a rotating, charged body behaves like a magnet and interacts with the applied field. According to classical mechanics, because the orientation of the angular momentum can take any value, the associated magnet can take any orientation. Because the direction in which the magnetic is driven by the inhomogeneous field depends on the orientation, it follows that a broad band of atoms is expected to emerge from the region where the magnetic field acts (Fig. 12.27b). According to quantum mechanics, since the angular momentum is quantized, the associated magnet lies in a number of discrete orientations, so that several sharp bands of atoms are expected (Fig. 12.27c).

In their first experiment. Stern and Gerlach appeared to confirm the classical prediction. However, the experiment is difficult because collisions between the atoms in the beam blur the bands. When the experiment was repeated with a beam of very low intensity (so that collisions were less frequent) they observed discrete bands, and so confirmed the quantum prediction.

The vector model

Throughout the preceding discussion, we have referred to the z-component of angular momentum (the component about an arbitrary axis, which is conventionally denoted z), and have made no reference to

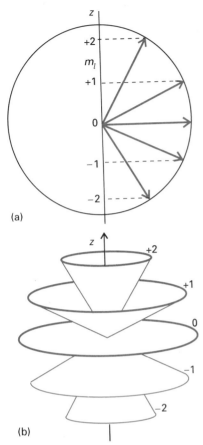

(a)

(b)

12.28 (a) A summary of Fig. 12.26. However, since the azimuthal angle around z of the vector is indeterminate, a better representation is as in (b), where each vector lies at an unspecified azimuthal angle on its cone.

the x- and y-components (the components about the two axes perpendicular to z). The reason for this omission is that the uncertainty principle forbids the simultaneous, exact specification of more than one component. Therefore, if l_z is known, it is impossible to ascribe values to the other two components. It follows that the illustration in Fig. 12.26, which is summarized in Fig. 12.28a, gives a false impression of the state of the system, because it suggests definite values for the x- and y-components. A better picture must reflect the impossibility of specifying l_x and l_y if l_z is known.

The **vector model** of angular momentum uses pictures like that in Fig. 12.28b. The cones are drawn with side $\{l(l+1)\}^{\frac{1}{2}}$ units, and represent the magnitude of the angular momentum. Each cone has a definite projection (of m_l units) on the z-axis, representing the system's precise value of l_z. The l_x and l_y projections, however, are indefinite. The vector representing the state of angular momentum can be thought of as lying with its tip on any point on the mouth of the cone. At this stage it should not be thought of as sweeping round the cone; that aspect of the model will be added later when we allow the picture to convey more information.

The vector model of angular momentum, although only a pictorial representation of aspects of the quantum mechanical properties, turns out to be surprisingly useful when we turn to the structure and spectra of atoms.

12.8 Spin

Stern and Gerlach observed *two* bands of Ag atoms in their experiment. This observation seems to conflict with one of the predictions of quantum mechanics, because an angular momentum l gives rise to $2l+1$ orientations, which is equal to 2 only if $l = \frac{1}{2}$, contrary to the conclusion that l must be an integer. The conflict was resolved by the suggestion that the angular momentum they were observing was not due to orbital angular momentum (the motion of an electron around the atomic nucleus) but arose instead from the motion of the electron about its own axis. The internal angular momentum of the electron is called its **spin**.

The wavefunction of an electron spinning at a single point in space does not have to satisfy the same boundary conditions as those for a particle circulating around a central point, so the quantum number for spin angular momentum is subject to different restrictions. To distinguish this spin angular momentum from orbital angular momentum we use the quantum number s (in place of l) and m_s for the projection on the z-axis. The magnitude of the spin angular momentum is $\{s(s+1)\}^{\frac{1}{2}}\hbar$ and the component $m_s\hbar$ is restricted to the $2s+1$ values

$$m_s = s, s-1, s-2, \ldots, -s \tag{32}$$

The detailed analysis of the spin of a particle is quite sophisticated (it is rooted in special relativity), and shows that the property should not be taken to be an actual spinning motion. However, that picture can be very useful when used with care. For an electron it turns out that only one value of s is allowed, $s = \frac{1}{2}$, corresponding to an angular momentum of magnitude $\frac{1}{2} \times \sqrt{3}\hbar = 0.866\hbar$. The spin angular momentum is an intrinsic

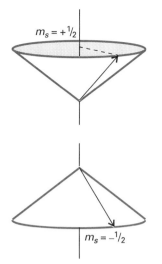

12.29 An electron spin ($s = \frac{1}{2}$) can take only two orientations with respect to a specified axis. An α electron is an electron with $m_s = +\frac{1}{2}$; a β electron is an electron with $m_s = -\frac{1}{2}$. The vector representing the magnitude of the spin angular momentum lies at an angle of 55° to the z-axis (more precisely, at $\arccos(1/\sqrt{3}\,)$).

property of the electron, like its rest mass and its charge, and every electron has exactly the same value. The spin may lie in any of $2s + 1 = 2$ different orientations (Fig. 12.29). One orientation corresponds to $m_s = +\frac{1}{2}$ (this state is often denoted α or $\uparrow$); the other orientation corresponds to $m_s = -\frac{1}{2}$ (this state is denoted β or $\downarrow$).

The outcome of the Stern–Gerlach experiment can now be explained if we suppose that each Ag atom possesses an angular momentum due to the spin of a single electron, because the two bands of atoms then correspond to the two spin orientations. Why the atoms behave like this will be explained in Chapter 13.

Like the electron, other elementary particles have characteristic spins. For example, protons and neutrons are **spin-$\frac{1}{2}$ particles** (i.e. $s = \frac{1}{2}$) and so invariably spin with angular momentum $\frac{1}{2} \times \sqrt{3}\hbar = 0.866\hbar$. Because the masses of a proton and a neutron are so much greater than the mass of an electron, yet they all have the same spin angular momentum, the classical picture would be of particles spinning much more slowly than an electron. Some elementary particles have $s = 1$, and so have an intrinsic angular momentum of magnitude $\sqrt{2}\hbar = 1.414\hbar$. Some mesons are **spin-1 particles** (as are some atomic nuclei), but for our purposes the most important spin-1 particle is the photon. We shall see the importance of photon spin in the next chapter.

Particles with half-integral spin are called **fermions** and those with integral spin (including 0) are called **bosons**. Thus, electrons and protons are fermions and photons are bosons. It is a very deep feature of nature that all the elementary particles that constitute matter are fermions, whereas the fundamental particles that are responsible for the forces that bind fermions together are all bosons. (Photons, for example, transmit the electromagnetic force that binds together electrically charged particles.) Matter, therefore, is an assembly of fermions held together by forces conveyed by bosons.

The properties of angular momentum that we have developed are set out in Table 12.4. As mentioned there, when we use the quantum numbers l and m_l we shall mean orbital angular momentum; when we use s and m_s we shall mean spin angular momentum; and when we use j and m_j we shall mean either (or, in some contexts to be described in Chapter 13, a combination of orbital and spin momenta).

Table 12.4 Angular momentum

The quantum numbers:
    Orbital angular momentum quantum number: $l = 0, 1, 2$
    Orbital magnetic quantum number: $m_l = 0, \pm 1, \ldots, \pm l$
    Spin angular momentum quantum number: $s = \frac{1}{2}$
    Spin magnetic quantum number: $m_s = \pm \frac{1}{2}$

In general:
    Angular momentum quantum number j
    Magnetic quantum number m_j

The magnitude of the angular momentum is equal to $\{j(j + 1)\}^{\frac{1}{2}}\hbar$ and the z-component of angular momentum is equal to $m_j \hbar$ with the $2j + 1$ values $j, j - 1, \ldots, -j$.

For the total angular momentum of a composite system see Section 13.8.

CHECK LIST OF KEY IDEAS

1 The **boundary conditions** that must be satisfied by the wavefunctions of a **particle in a box** and the resulting quantized **energy levels** (eqn 5).

2 The properties of the wavefunctions of a particle in a box (Section 12.1) and the **zero-point energy**.

3 The concept of **quantum number** for labelling a state and specifying an observable (Section 12.1).

4 The **correspondence principle** and the emergence of classical behaviour at high quantum numbers (Section 12.1).

5 The technique of **separation of variables** for separating the Schrödinger equation in several variables into separate equations (Section 12.2) for each variable.

6 The wavefunctions and energies of a **particle in a two-dimensional square well** (eqn 12) and the relation of **degeneracy** of the levels to the symmetry of the system.

7 The **tunnelling** of a particle into classically forbidden regions (Section 12.3).

8 The quantum mechanical description of **vibrational motion** (Section 12.4) and the wavefunctions and energy levels of a **harmonic oscillator** (eqns 19 and 17, respectively).

9 The mean potential energy and the mean kinetic energy of a harmonic oscillator as a special case of the **virial theorem** (eqn 23).

10 The calculation of the extent to which a harmonic oscillator may be found in classically forbidden regions of extension and contraction (Section 12.5).

11 The quantum mechanical description of **rotational motion in two dimensions** and the role of **cyclic boundary conditions** (Section 12.6).

12 The **quantized energy levels** of a particle moving on a circle (eqn 26) and the **quantization of angular momentum** about an axis (eqn 25).

13 The **vector representation** of angular momentum (Fig. 12.21).

14 The quantum mechanical description of **rotational motion in three dimensions** (Section 12.7).

15 The **quantized energy levels** of a particle free to rotate in three dimensions (eqn 30) and the **spherical harmonic** wavefunctions (Section 12.7).

16 The **quantization of angular momentum** (eqn 31), **space quantization** (Section 12.7), and the representation of angular momentum in terms of the **vector model** (Fig. 12.28).

17 The **Stern–Gerlach experiment** and the property of electron **spin** (Section 12.8).

EXERCISES

12.1. Calculate the energy separations in J, $kJ\,mol^{-1}$, eV, and cm^{-1} between the levels (a) $n = 2$ and $n = 1$, (b) $n = 6$ and $n = 5$ of an electron in a box of length 1.0 nm.

12.2. Calculate the probability that a particle will be found between $0.49L$ and $0.51L$ in a box of length L when it has (a) $n = 1$, (b) $n = 2$. Take the wavefunction to be a constant in this range.

12.3. (a) Write the Schrödinger equation for a particle in a square well of length L. (b) Calculate the expectation values of p and p^2 for a particle in the state $n = 1$.

12.4. What are the most likely locations of a particle in a box of length L in the state $n = 3$?

12.5. Consider a particle in a cubic box. What is the degeneracy of the level that has an energy three times that of the lowest level?

12.6. Calculate the percentage change in a given energy level of a particle in a cubic box when the edge of the cube is decreased by 10 per cent in each direction.

12.7. A nitrogen molecule is confined in a cubic box of volume $1.00\,m^3$. Assuming that the molecule has an energy equal to $\frac{3}{2}kT$ at $T = 300\,K$, what is the value of $n = (n_x^2 + n_y^2 + n_z^2)^{\frac{1}{2}}$, for this particle? What is the energy separation between the levels n and $n + 1$? What is its de Broglie wavelength? Would it be appropriate to describe this particle as a classical particle?

12.8. Calculate the zero-point energy of a harmonic oscillator consisting of a particle of mass 2.33×10^{-26} kg and force constant 155 N m^{-1}.

12.9. For a harmonic oscillator consisting of a particle of mass 1.33×10^{-25} kg, the difference in adjacent energy levels is 4.82×10^{-21} J. Calculate the force constant of the oscillator.

12.10. Calculate the wavelength of a photon needed to excite a transition between neighbouring energy levels of a harmonic oscillator of mass equal to that of a proton and force constant 855 N m^{-1}.

12.11. Refer to the preceding exercise, and calculate the change in wavelength that would result from doubling the mass of the particle.

12.12. Calculate the minimum excitation energies of (a) a pendulum of length 1.0 m on the surface of the Earth, (b) the balance-wheel of a clockwork watch ($v = 5$ Hz), (c) the 33-Hz quartz crystal of a watch, and (d) the bond between two O atoms in O_2, for which $k = 1177$ N m^{-1}.

12.13. Confirm that the wavefunction for the ground state of a one-dimensional linear harmonic oscillator given in Table 12.1 is a solution of the Schrödinger equation for the oscillator and that its energy is $\frac{1}{2}\hbar\omega$. Assuming that the vibrations of a $^{35}Cl_2$ molecule are equivalent to those of a harmonic oscillator with a force constant $k = 329$ N m^{-1}, what is the zero-point energy of vibration of this molecule?

12.14. The wavefunction, $\Phi(\phi)$, for the motion of a particle in a ring is of the form $\Phi = Ne^{im\phi}$. Determine the normalization constant N.

12.15. An Ar atom rotates in a circle about a fixed centre with orbital angular momentum quantum number $m_l = 2$. If its energy of rotation is 2.47×10^{-23} J, calculate the distance of the mass from the centre of rotation.

12.16. A point mass rotates in a sphere with $l = 1$. Calculate the magnitude of its angular momentum and the possible projections of the angular momentum on an arbitrary axis.

12.17. Draw scale vector diagrams to represent the states (a) $s = \frac{1}{2}$, $m_s = +\frac{1}{2}$, (b) $l = 1$, $m_l = +1$, (c) $l = 2$, $m_l = 0$.

12.18. Draw the vector diagram for all the permitted states of a particle with $l = 6$.

PROBLEMS
Numerical problems

12.1. Calculate the separation between the two lowest levels for an O_2 molecule in a one-dimensional container of length 5.0 cm. At what value of n does the energy of the molecule reach $\frac{1}{2}kT$ at 300 K, and what is the separation of this level from the one immediately below?

12.2. To a crude first approximation, a π electron in a linear polyene may be considered to be a particle in a one-dimensional box. The polyene β-carotene contains 22 conjugated C atoms, and the average internuclear distance is 140 pm. Each state up to $n = 11$ is occupied by two electrons. Calculate (a) the separation in energy between the ground state and the first excited state in which one electron occupies the state with $n = 12$, (b) the frequency of the radiation required to produce a transition between these two states, and (c) the *total* probability of finding an electron between C atoms 11 and 12 in the ground state of the 22-electron molecule.

12.3. The mass to use in the expression for the vibrational frequency of a diatomic molecule is the effective mass $\mu = m_A m_B/(m_A + m_B)$, where m_A and m_B are the masses of the individual atoms. The following data on the infrared absorption wavefunctions (in cm^{-1}) of molecules is taken from G. Herzberg, *Spectra of diatomic molecules*, van Nostrand (1950):

$H^{35}Cl$	$H^{81}Br$	HI	CO	NO
2990	2650	2310	2170	1904

Calculate the force constants of the bonds and arrange them in order of increasing stiffness.

12.4. The rotation of an HI molecule can be pictured as the orbital motion of an H atom at a distance 160 pm from a stationary I atom. (This is quite a good picture; to be precise, both atoms rotate around their common centre of mass, which is very close to the I nucleus.) Suppose that the molecule rotates only in a plane. Calculate the energy needed to excite the molecule into rotation. What, apart from 0, is the minimum angular momentum of the molecule?

12.5. Calculate the energies of the first four rotational levels of HI free to rotate in three dimensions, using for its moment of inertia $I = \mu R^2$, with $\mu = m_H m_I/(m_H + m_I)$ and $R = 160$ pm.

Theoretical problems

12.6. Set up the Schrödinger equation for a particle of mass m in a three-dimensional square well with sides L_1, L_2, and L_3. Show that the wavefunction is defined by three quantum numbers and that the Schrödinger equation is separable. Find the energy levels, and specialize the result to a cubic box of side L.

12.7. The wavefunction inside a long barrier of height V is $\psi = N e^{-\kappa x}$. Calculate (a) the probability that the particle is inside the barrier and (b) the average penetration depth of the particle into the barrier.

12.8. Confirm that a function of the form e^{-gx^2} is a solution of the Schrödinger equation for the ground state of a harmonic oscillator and find an expression for g in terms of the mass and force constant of the oscillator.

12.9. Calculate the mean kinetic energy of a harmonic oscillator by using the relations in Table 12.1.

12.10. Calculate the values of $\langle x^3 \rangle$ and $\langle x^4 \rangle$ for a harmonic oscillator by using the relations in Table 12.1.

12.11. Determine the values of $\Delta x = (\langle x^2 \rangle - \langle x \rangle^2)^{\frac{1}{2}}$ and $\Delta p = (\langle p^2 \rangle - \langle p \rangle^2)^{\frac{1}{2}}$ for (a) a particle in a box of length L and (b) a harmonic oscillator. Discuss these quantities with reference to the uncertainty principle.

12.12. We shall see in Chapter 16 that the intensity of spectroscopic transitions between the vibrational states of a molecule are proportional to the square of the integral $\int \psi_{v'} x \psi_v \, dx$ over all space. Use the relations between Hermite polynomials given in Table 12.1 to show that the only permitted transitions are those for which $v' = v \pm 1$ and evaluate the integral in these cases.

12.13. Use the virial theorem to obtain an expression for

the relation between the mean kinetic and potential energies of an electron in a hydrogen atom.

12.14. Evaluate the z-component of the angular momentum and the kinetic energy of a particle on a ring that is described by the (unnormalized) wavefunctions (a) $e^{i\phi}$, (b) $e^{-2i\phi}$, (c) $\cos \phi$, and (d) $(\cos \chi) e^{i\phi} + (\sin \chi) e^{-i\phi}$.

12.15. Confirm that the spherical harmonics (a) $Y_{0,0}$, (b) $Y_{2,-1}$, and (c) $Y_{3,3}$ satisfy the Schrödinger equation for a particle free to rotate in three dimensions, and find its energy and angular momentum in each case.

12.16. Confirm that $Y_{3,3}$ is normalized to 1. (The integration required is over the surface of a sphere.)

12.17. Derive an expression in terms of l and m_l for the half-angle of the apex of the cone used to represent an angular momentum according to the vector model. Evaluate the expression for an α spin. Show that the minimum possible angle approaches 0 as $l \rightarrow \infty$.

12.18. Show that the function $f = \cos ax \cos by \cos cz$ is an eigenfunction of ∇^2 and determine its eigenvalue.

12.19. Derive (in Cartesian coordinates) the quantum mechanical operators for the three components of angular momentum starting from the classical definition of angular momentum, $\boldsymbol{l} = \boldsymbol{r} \times \boldsymbol{p}$. Show that any two of the components do not commute and find their commutator, the value of $\hat{\Omega}_1 \hat{\Omega}_2 - \hat{\Omega}_2 \hat{\Omega}_1$.

12.20. Starting from the operator $\hat{l}_z = \hat{x}\hat{p}_y - \hat{y}\hat{p}_x$, prove that in spherical polar coordinates $\hat{l}_z = -i\hbar \, \partial/\partial\phi$.

13 Atomic structure and atomic spectra

The principles of quantum mechanics introduced in the preceding two chapters are now used to describe the internal structures of atoms. First, we see what experimental information is available from a study of the spectrum of atomic hydrogen. Then we set up and solve the Schrödinger equation for an electron near a nucleus. The symmetry of the problem lets us separate the full equation into an angular part and a radial part. The solutions of the angular part are the same as the wavefunctions of angular momentum derived in Chapter 12. The radial solutions introduce another quantum number that plays a principal role in the discussion of the structures of atoms. The wavefunctions obtained are the 'atomic orbitals' of hydrogenic atoms, and we shall see their shapes and significance.

Next, we use the hydrogenic atomic orbitals to construct a description of the structures of many-electron atoms and see that, in conjunction with the Pauli exclusion principle, it is reasonably straightforward to account for the periodic table and the periodicity of atomic properties. The spectra of many-electron atoms are more complicated than those of hydrogen, but the same principles apply. We shall see in the closing sections of the chapter how such spectra are described by using term symbols, the origin of the finer details of their appearance, and the effects on them of an applied magnetic field.

In this chapter we see how to use quantum mechanics to describe the **electronic structure** of an atom, the arrangement of electrons around its nucleus. The concepts we shall meet are of central importance for understanding the structures and reactions of atoms and molecules, and hence have extensive chemical applications. Right from the outset, we need to distinguish between hydrogenic atoms and many-electron atoms. A **hydrogenic atom** is a one-electron atom or ion of general atomic number Z; examples of hydrogenic atoms are H, He^+, Li^{2+}, and U^{91+}. A **many-electron atom** is an atom or ion with more than one electron;

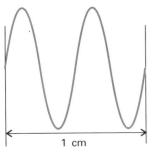

(a) $\tilde{\nu}$ = 2 cm^{-1}

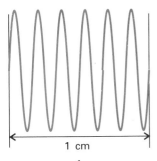

(b) $\tilde{\nu}$ = 6 cm^{-1}

13.1 The wavenumber of electromagnetic radiation is the number of complete waves per unit length (typically, per centimetre). This illustration shows two examples with wavenumbers (a) 2 cm^{-1}, (b) 6 cm^{-1}. Visible light has wavenumbers of the order of 10^4 cm^{-1}.

examples include all neutral atoms other than H, so that even He, with only two electrons, is a many-electron atom. Hydrogenic atoms are important because their structures can be discussed exactly. They also provide a set of concepts that are used to describe the structures of many-electron atoms and, as we shall see in the next chapter, the structures of molecules too.

One of the principal experimental techniques for determining the structures of atoms is **spectroscopy**, the detection and analysis of the electromagnetic radiation absorbed or emitted by a species. In discussions of spectroscopy we shall often refer to the **wavenumber $\tilde{\nu}$** of the radiation emitted or absorbed by an atom. The wavenumber is related to the wavelength λ and frequency ν by

$$\tilde{\nu} = \frac{1}{\lambda} = \frac{\nu}{c}$$

where c is the speed of light. It is common in spectroscopy to express wavenumbers in reciprocal centimetres, cm^{-1}; a wavenumber of 1000 cm^{-1} then signifies that there are 1000 complete wavelengths per centimetre (Fig. 13.1). The record of spectral intensity as a function of wavenumber, frequency, or wavelength of the radiation emitted or absorbed by an atom or a molecule is called its **spectrum** (from the Greek word for appearance).

THE STRUCTURE AND SPECTRA OF HYDROGENIC ATOMS

When an electric discharge is passed through gaseous hydrogen, the H_2 molecules are dissociated and the energetically excited H atoms that are produced emit light of discrete frequencies (Fig. 13.2). The first important contribution to the interpretation of this spectrum was made by the Swiss schoolteacher Johann Balmer, who pointed out in 1885 that (in modern terms) the wavenumbers of the lines in the visible region fit the expression

$$\tilde{\nu} \propto \frac{1}{2^2} - \frac{1}{n^2} \qquad n = 3, 4, \ldots$$

The lines this formula describes are now called the **Balmer series**. When further lines were discovered in the ultraviolet, giving the **Lyman series**,

13.2 The spectrum of atomic hydrogen. The observed spectrum and its resolution into overlapping series are shown. Note that the Balmer series lies in the visible region.

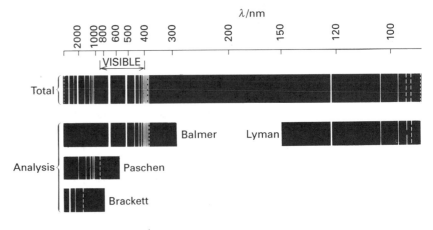

and in the infrared, the **Paschen series**, the Swedish spectroscopist Johannes Rydberg noted (in 1890) that all of them could be fitted to the expression

$$\tilde{v} = \mathfrak{R}_H \left(\frac{1}{n_1^2} - \frac{1}{n_2^2} \right) \qquad \mathfrak{R}_H = 109\,677\;\text{cm}^{-1} \qquad (1)$$

with $n_1 = 1$ (the Lyman series), 2 (the Balmer series), and 3 (the Paschen series), and that in each case $n_2 = n_1 + 1, n_1 + 2, \ldots$. The constant $\mathfrak{R}_H$ is now called the **Rydberg constant** for the hydrogen atom.

Example 13.1 *Calculating the wavelengths of spectral lines*

Calculate the longest wavelength transition in the Lyman series of atomic hydrogen.

Method. The Lyman series is given by eqn 1 by setting $n_1 = 1$. The transition with the longest wavelength is the one with the smallest wavenumber, which (for this series) is obtained by setting $n_2 = 2$.

Answer. The wavenumber of the transition with $n_1 = 1$ and $n_2 = 2$ is

$$\tilde{v} = \mathfrak{R}_H \times \left(\frac{1}{1^2} - \frac{1}{2^2} \right) = \tfrac{3}{4} \times 109\,677\;\text{cm}^{-1} = 82\,258\;\text{cm}^{-1}$$

The wavelength is the reciprocal of the wavenumber, so

$$\lambda = \frac{1}{8.2258 \times 10^6\;\text{m}^{-1}} = 1.2157 \times 10^{-7}\;\text{m}$$

Comment. The transition occurs at 121.57 nm, in the vacuum ultraviolet region of the spectrum.

Exercise E13.1. Calculate the shortest wavelength transition in the Paschen series.
                                                                    [821 nm]

The form of eqn 1 strongly suggests that the wavenumber of each spectral line can be written as the difference of two **terms**, each of the form

$$T = \frac{\mathfrak{R}_H}{n^2} \qquad (2)$$

The concept of a spectroscopic term is the content of the **Ritz combination principle**:

The wavenumber of any spectral line is the difference of two terms.

We say that two terms T_1 and T_2 **combine** to produce a spectral line of wavenumber

$$\tilde{v} = T_1 - T_2 \qquad (3)$$

The Ritz combination principle applies to all types of atoms and molecules, but only for hydrogenic atoms do the terms have the simple form $(\text{constant})/n^2$.

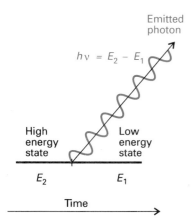

$h\nu = E_2 - E_1$

Emitted photon

High energy state

Low energy state

E_2 E_1

Time

13.3 Energy is conserved when a photon is emitted, so the difference in energy of the atom before and after the emission event must be equal to the energy of the photon emitted.

The Ritz combination principle is readily explained in terms of photons and the conservation of energy. Thus, a spectroscopic line arises from the transition of an atom from one energy level (a term) to another (another term) with the emission of the difference in energy as a photon (Fig. 13.3). This interpretation leads to the **Bohr frequency condition**:

> When an atom changes its energy by ΔE, the difference is carried away as a photon of frequency ν, where
>
> $$\Delta E = h\nu \qquad (4)$$

Thus, if each spectroscopic term represents an energy hcT, then the difference in energy when the atom undergoes a transition between two terms is

$$\Delta E = hcT_1 - hcT_2$$

and the frequency of the light emitted is given by

$$\nu = cT_1 - cT_2$$

This expression rearranges into the Ritz formula when expressed in terms of wavenumbers (on division by c).

Because radiation is absorbed and emitted by atoms only at certain wavenumbers, it follows that only certain energy states of atoms are permitted. Our tasks in this chapter are to determine the origin of this energy quantization, to find the permitted energy levels, and to account for the value of $\mathfrak{R}_H$.

13.1 The structure of hydrogenic atoms

We can find a description of the structure of a hydrogenic atom by adopting Rutherford's nuclear model, which consists of an electron and a central, massive nucleus of charge Ze, and then solving the Schrödinger equation for that electron's wavefunction. The (Coulomb) potential energy of the electron is

$$V = -\frac{Ze^2}{4\pi\varepsilon_0 r} \qquad (5)$$

where r is the distance of the electron from the nucleus and ε_0 is the vacuum permittivity (see *Further information 3*). Hence, the Schrödinger equation for the joint system of an electron and a nucleus of mass m_N is

$$-\frac{\hbar^2}{2m_e}\nabla_e^2\psi - \frac{\hbar^2}{2m_N}\nabla_N^2\psi - \frac{Ze^2}{4\pi\varepsilon_0 r}\psi = E\psi$$

(The subscript on ∇^2 indicates differentiation with respect to the electron or nuclear coordinates.)

At first sight the calculation would seem to be of considerable difficulty, because it involves two particles, the electron and the nucleus, and hence six coordinates. However, physical intuition suggests that the full Schrödinger equation ought to separate into two equations, one for the motion of the atom as a whole and the other for the motion of the electron relative to the nucleus. We have already solved the first of these equations, because it corresponds to the free motion of a particle of mass

m (the mass of the atom) in space (Section 11.3). The equation for the motion of the electron relative to the nucleus still involves three coordinates of the electron relative to the nucleus. However, because the Coulomb potential energy is independent of angle, we can suspect that even that equation is separable.

The strategy of the calculation is to separate the relative motion of the electron and the nucleus from the motion of the atom as a whole. Then we separate the relative motion of the electron and the nucleus into angular and radial parts. We shall see that *three* quantum numbers are required for the latter three-dimensional system. Two quantum numbers, l and m_l, have already been encountered in the discussion of the angular momentum of a particle around a central point; the third quantum number n arises from the boundary conditions associated with the radial motion of the electron.

The separation of internal motion

The motion of the electron relative to the nucleus can be separated from the motion of the atom as a whole by a series of straightforward steps that are described in *Further information 10*. The resulting Schrödinger equation for the internal motion of the electron relative to the nucleus is

$$-\frac{\hbar^2}{2\mu}\nabla^2\psi + V\psi = E\psi \qquad (6)$$

In this equation, V is the Coulomb potential energy given in eqn 5 and μ is the **reduced mass**:

$$\frac{1}{\mu} = \frac{1}{m_e} + \frac{1}{m_N} \qquad (7)$$

The reduced mass is very similar to the electron mass because m_N, the mass of the nucleus, is much larger than the mass of an electron, and the second term on the right of eqn 7 is very small. In all except the most precise work, the reduced mass can be replaced by m_e.

The next step is to separate the angular variation of the wavefunction from its radial dependence. We do this by writing

$$\psi(r, \theta, \phi) = R(r)Y(\theta, \phi)$$

and testing whether the Schrödinger equation can then be separated into two equations, one for R and the other for Y. The details of the separation are given in *Further information 11*. The equation for the angular part of the wavefunction is the same as the Schrödinger equation for a particle free to move round a central point, and we considered it in Section 12.7. The solutions are the spherical harmonics (Table 12.3), and are specified by the quantum numbers l and m_l; we shall consider them in more detail shortly.

The radial wave equation

The novel part of the calculation is the **radial wave equation**, which is

$$-\frac{\hbar^2}{2\mu}\frac{d^2\Pi}{dr^2} + V_{\text{eff}}\Pi = E\Pi \qquad (8a)$$

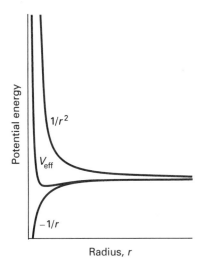

13.4 The effective potential energy of an electron in the hydrogen atom. When it has zero angular momentum the effective potential is the Coulombic potential energy. When the electron has angular momentum, the centrifugal effect gives rise to a positive contribution that is very large close to the nucleus. We can expect the $l = 0$ and $l \neq 0$ wavefunctions to be very different near the nucleus.

To simplify the appearance of this equation we have written $\Pi = rR$ and have introduced the **effective potential energy** of the particle:

$$V_{\text{eff}} = -\frac{Ze^2}{4\pi\varepsilon_0 r} + \frac{l(l+1)\hbar^2}{2\mu r^2} \tag{8b}$$

The radial wave equation is the description of the motion of a particle of mass μ in a one-dimensional region where the potential energy is V_{eff}.

We can anticipate some features of the shapes of the radial wavefunctions by analysing the shape of the effective potential energy. The first term in eqn 8b is the Coulomb potential energy of the electron in the field of the nucleus. The second term stems from the centrifugal force that arises from the angular momentum of the electron around the nucleus. When $l = 0$, the electron has no angular momentum, and the effective potential energy is purely Coulombic and attractive at all radii (Fig. 13.4). When $l \neq 0$, the centrifugal term gives a positive contribution to the effective potential energy. When the electron is close to the nucleus (r small), this repulsive term dominates the attractive Coulombic component and the net effect is an effective repulsion of the electron from the nucleus. The two effective potential energies, the one for $l = 0$ and the one for $l \neq 0$, are qualitatively very different close to the nucleus; however, they are similar at large distances because the centrifugal contribution tends to zero more rapidly than the Coulombic contribution. Therefore, we can expect the solutions with $l = 0$ and $l \neq 0$ to be quite different near the nucleus but similar far away from it.

We shall not go through the technical steps of solving the radial wave equation. It is sufficient to know that acceptable solutions can be found only for integral values of a quantum number n. The radial wave equation depends on l, and the radial wavefunctions, which depend on the values of both n and l, all have the form

$$R = (\text{polynomial in } r) \times (\text{decaying exponential in } r)$$

These functions are most simply written in terms of the dimensionless variable ρ:

$$\rho = \frac{2Z}{n} \times \frac{r}{a_0}, \qquad \text{where } a_0 = \frac{4\pi\varepsilon_0\hbar^2}{m_e e^2} \tag{9}$$

The quantity a_0 is called the **Bohr radius** and has the value 52.9 pm; the same quantity appeared in Bohr's early model of the hydrogen atom as the radius of the orbit of lowest energy. Specifically, the radial wavefunctions for an electron with quantum numbers n and l are

$$R_{n,l} = N_{n,l}\rho^l L_{n,l}(\rho)\mathrm{e}^{-\rho/2} \tag{10}$$

where $L(\rho)$ is a polynomial in ρ and $N_{n,l}$ is a normalization constant. An example of one of these polynomials[1] is the one with $n = 3$ and $l = 0$, which is

$$L_{3,0} = 6 - 6\rho + \rho^2$$

The first few polynomials are listed in Table 13.1 and the appearance of the radial wavefunction R is illustrated in Fig. 13.5. Note that, because R

1 The technical name for the polynomials is *associated Laguerre polynomials*.

Table 13.1 Hydrogenic radial wavefunctions*

Orbital	n	l	$R_{n,l}$
$1s$	1	0	$2\left(\dfrac{Z}{a_0}\right)^{\frac{3}{2}} e^{-\frac{1}{2}\rho}$
$2s$	2	0	$\dfrac{1}{2(2)^{\frac{1}{2}}}\left(\dfrac{Z}{a_0}\right)^{\frac{3}{2}} (2 - \rho)e^{-\frac{1}{2}\rho}$
$2p$	2	1	$\dfrac{1}{2(6)^{\frac{1}{2}}}\left(\dfrac{Z}{a_0}\right)^{\frac{3}{2}} \rho e^{-\frac{1}{2}\rho}$
$3s$	3	0	$\dfrac{1}{9(3)^{\frac{1}{2}}}\left(\dfrac{Z}{a_0}\right)^{\frac{3}{2}} (6 - 6\rho + \rho^2)e^{-\frac{1}{2}\rho}$
$3p$	3	1	$\dfrac{1}{9(6)^{\frac{1}{2}}}\left(\dfrac{Z}{a_0}\right)^{\frac{3}{2}} (4 - \rho)\rho e^{-\frac{1}{2}\rho}$
$3d$	3	2	$\dfrac{1}{9(30)^{\frac{1}{2}}}\left(\dfrac{Z}{a_0}\right)^{\frac{3}{2}} \rho^2 e^{-\frac{1}{2}\rho}$

* The full wavefunction is $\psi = RY$, where Y is given in Table 12.3. In the table, $\rho = 2Zr/na_0$.

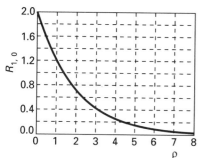

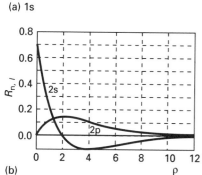

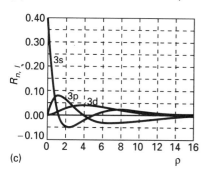

13.5 (a–c) The radial wavefunctions of the first few states of the hydrogen atom and (a²–c², next page) the probability densities. Note that the s orbitals have a non-zero and finite value at the nucleus.

is proportional to ρ^l, all radial wavefunctions are zero at the nucleus unless $l = 0$.

13.2 Atomic orbitals and their energies

The wavefunctions of hydrogenic atoms are examples of **atomic orbitals**, the name expressing something less definite than the 'orbits' of classical mechanics. In general, an **orbital** is a one-electron wavefunction, and an atomic orbital is a one-electron wavefunction that describes the distribution of an electron in an atom. All hydrogenic atomic orbitals have the form

$$\psi_{n,l,m_l} = R_{n,l}Y_{l,m_l} \tag{11}$$

where R is one of the functions in Table 13.1 and Y is one of the spherical harmonics listed in Table 12.3.

The quantum numbers

Each hydrogenic atomic orbital is defined by three quantum numbers (Table 13.2), n, l, and m_l. When an electron is described by one of these wavefunctions, we say that it **occupies** that orbital. Thus, an electron described by the wavefunction $\psi_{1,0,0}$ is said to occupy the orbital with $n = 1$, $l = 0$, and $m_l = 0$.

Two of the quantum numbers, l and m_l, come from the angular solutions, and specify the angular momentum of the electron around the nucleus:

An electron in an orbital with quantum number l has an angular momentum of magnitude $\{l(l + 1)\}^{\frac{1}{2}}\hbar$, with

$$l = 0, 1, 2, \ldots , n - 1$$

An electron in an orbital with quantum number m_l has a z-

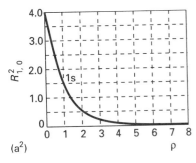

(a²)

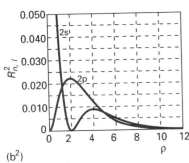

(b²)

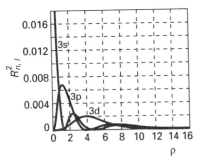

13.5 (Continued)

Table 13.2 Hydrogenic atoms

The wavefunctions of hydrogenic atoms depend on three quantum numbers:

Principal quantum number: $n = 1, 2, 3, \ldots$

Angular momentum quantum number: $l = 0, 1, 2, \ldots, n - 1$

Magnetic quantum number: $m_l = l, l - 1, l - 2, \ldots, -l$

the energy is related to n by

$$E_n = -\frac{hc\Re}{n^2} \qquad hc\Re = \frac{Z^2 \mu e^4}{32\pi^2 \varepsilon_0^2 \hbar^2}$$

The magnitude of the orbital angular momentum of the electron is $\{l(l + 1)\}^{\frac{1}{2}}\hbar$ and its component on an arbitrary axis is $m_l\hbar$. Each energy level is n^2-fold degenerate.

The wavefunctions are products of radial and angular components:

$$\psi = R(r)Y(\theta, \phi)$$

The angular wavefunctions Y are the spherical harmonics (Table 12.3) and the radial wavefunctions R are the normalized associated Laguerre polynomials (Table 13.1).

The selection rules for spectroscopic transitions are

Δn unrestricted $\Delta l = \pm 1$

component of angular momentum $m_l\hbar$, with

$$m_l = 0, \pm 1, \pm 2, \ldots, \pm l$$

The third quantum number n is called the **principal quantum number**. It can take the values

$$n = 1, 2, 3, \ldots$$

and determines the energy through

$$E_n = -\frac{Z^2 \mu e^4}{32\pi^2 \varepsilon_0^2 \hbar^2} \times \frac{1}{n^2} \tag{12}$$

This result is consistent with the experimental information described earlier, specifically eqn 2, with the Rydberg constant for an atom of general Z given by

$$hc\Re = \frac{Z^2 \mu e^4}{32\pi^2 \varepsilon_0^2 \hbar^2} \tag{13}$$

(The negative sign in eqn 12 will be explained shortly.) The energy is independent of l and m_l and depends only on n. Therefore, *all the orbitals of a given n have the same energy* whatever their values of l and m_l.

To define fully the state of an electron in a hydrogen atom we need to specify not only the orbital it occupies but also its spin state. We saw in Section 12.8 that an electron possesses an intrinsic angular momentum that is described by the two quantum numbers s and m_s (the analogues of l and m_l). The value of s is fixed at $\frac{1}{2}$ for an electron, so we do not need to consider it further at this stage. However, m_s may be either $+\frac{1}{2}$ or $-\frac{1}{2}$, and to specify the electron's state in a hydrogen atom we need to specify which of these values describes it. It follows that to specify the state of an

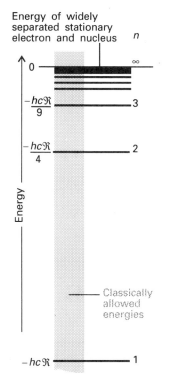

Energy of widely separated stationary electron and nucleus

13.6 The energy levels of the hydrogen atom. The values are relative to an infinitely separated, stationary electron and a proton.

electron in a hydrogen atom, we need to give the values of *four* quantum numbers:

n, which specifies the energy of the electron.

l, which specifies the magnitude of the electron's orbital angular momentum.

m_l, which specifies the orientation of the electron's orbital angular momentum.

m_s, which specifies the orientation of the electron's spin angular momentum.

The energy levels

The energy levels predicted by eqn 12 are depicted in Fig. 13.6. The separation of neighbouring levels is proportional to Z^2, so the levels are four times as wide apart (and the ground state four times deeper in energy) in He^+ than in H. All the energies given by eqn 12 are negative. They refer to the **bound states** of the atom, in which the energy of the atom is lower than that of the infinitely separated, relatively stationary electron and nucleus (which corresponds to the zero of energy). There are also solutions of the Schrödinger equation with positive energies. These solutions correspond to **unbound states** of the electron, the states to which an electron is raised when it is ejected from the atom by a high-energy collision or photon. The energies of the unbound electron are not quantized and form the **continuum states** of the atom.

For hydrogen itself, $Z = 1$ and the bound-state energies are given by

$$E_n = -\frac{hc\Re_H}{n^2} \qquad \text{with } hc\Re_H = \frac{\mu e^4}{32\pi^2 \varepsilon_0^2 \hbar^2} \tag{14}$$

The emission spectrum of atomic hydrogen can now be explained by supposing that an electron undergoes a transition from a state with principal quantum number n_2 (and energy $-hc\Re_H/n_2^2$) to one with principal quantum number n_1 (and energy $-hc\Re_H/n_1^2$). As it does so, it discards the energy difference as a photon of energy $h\nu$ and frequency ν, with

$$h\nu = \frac{hc\Re_H}{n_1^2} - \frac{hc\Re_H}{n_2^2}$$

When this expression is divided by hc it gives exactly the experimentally determined expression, eqn 1. Moreover, insertion of the values of the fundamental constants into the expression for $\Re_H$ gives almost exact agreement with the experimental values of the Rydberg constant. The only discrepancies arise from the neglect of relativistic corrections, which the non-relativistic Schrödinger equation ignores.

Ionization energies

We now return to the generally larger-scale features of atomic energies. The **ionization energy** I of an element is the minimum energy required to remove an electron from the **ground state**, the state of lowest energy, of one of its atoms. The ground state of hydrogen is the state with $n = 1$, which has energy

$$E_1 = -hc\mathfrak{R}_H$$

The atom is ionized when the electron has been excited to the level corresponding to $n = \infty$ (see Fig. 13.6). Therefore, the energy that must be supplied is

$$I = hc\mathfrak{R}_H = 2.179 \times 10^{-18}\,\text{J}$$

which corresponds to 13.60 eV.

The spectroscopic determination of ionization energies depends on the determination of the **series limit**, the wavenumber at which the series terminates and becomes a continuum. If the upper state lies at an energy E_{upper} then, when the atom makes a transition to E_{lower}, a photon is emitted of wavenumber $\tilde{\nu}$, where $hc\tilde{\nu} = E_{upper} - E_{lower}$. Because the upper energy levels are given by an expression of the form $E_{upper} = -hc\mathfrak{R}/n^2$, the emission lines will occur at

$$\tilde{\nu} = -\frac{\mathfrak{R}}{n^2} - \frac{E_{lower}}{hc}$$

The separation between successive lines decreases as n increases so the lines converge. Moreover, a plot of the wavenumbers against $1/n^2$ should give a straight line of slope $-\mathfrak{R}$ and intercept $-E_{lower}/hc$. The value of $-E_{lower}$ is the ionization energy of the atom in that state (and is the ground-state ionization energy when the lower state is the ground state).

Example 13.2 *Measuring an ionization energy spectroscopically*

The spectrum of atomic hydrogen shows lines at the wavenumbers 82 258, 97 492, 102 824, 105 292, 106 632, 107 440 cm^{-1}. Determine (a) the ionization energy of the lower state, (b) the value of the Rydberg constant.

Method. We need to plot the wavenumber against $1/n^2$ with $n = 2, 3, \ldots$ (this choice for n assumes that the lower state is $n = 1$; if a straight line is not obtained, try $n = 2, 3, \ldots$, etc.). According to the equation above, the intercept at $1/n^2 = 0$ is E_{lower}/hc, or $-I/hc$, and the slope is $-\mathfrak{R}_H$. Use a computer (or a calculator) to make a least-squares fit of the data to get a result that reflects the precision of the data.

Answer. The wavenumbers are plotted against $1/n^2$ in Fig. 13.7. The (least-squares) intercept lies at $-109\,677$ cm^{-1}, and so the ionization energy is $-hc$ times this quantity, or 2.1788×10^{-18} J (1312.1 kJ mol^{-1}). The slope is, in this instance, numerically the same, so $\mathfrak{R}_H = 109\,677$ cm^{-1}.

Comment. A similar extrapolation procedure can be used for non-hydrogenic atoms because, even though their energies are not in general given by an equation like eqn 14, the energy of an electron in an excited state does have that form, particularly when n is large. When n is large, the excited electron is so far from the other electrons present in the atom

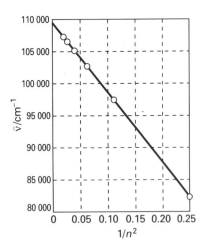

13.7 The plot of the data in Example 13.2 used to determine the ionization energy of an atom (in this case, of H).

that the latter behave like a point-like collection of negative charge at the nucleus and reduce the apparent atomic number from Z to 1.

Exercise E13.2. The spectrum of atomic deuterium shows lines at the following wavenumbers: 15 238, 20 571, 23 039, 24 380 cm^{-1}. Determine (a) the ionization energy of the lower state, (b) the ionization energy of the ground state, (c) the mass of the deuteron (by expressing the Rydberg constant in terms of the reduced mass of the electron and the deuteron, and solving for the mass of the deuteron).

[(a) 328.1 kJ mol^{-1}, (b) 1312.4 kJ mol^{-1}, (c) 3.4×10^{-27} kg]

Shells and subshells

All the orbitals of a given value of n form a single **shell** of the atom. In a hydrogenic atom, all orbitals of given n, and therefore belonging to the same shell, have the same energy. It is common to refer to successive shells by letters:

$$n = 1\ 2\ 3\ 4\ \ldots$$
$$K\ L\ M\ N\ \ldots$$

Thus, all the orbitals of the shell with $n = 2$ form the L shell of the atom.

The orbitals with the same value of n but different values of l form the **subshells** of a given shell. These subshells are generally referred to by the letters $s, p, \ldots$ by using the correspondence

$$l = 0\ 1\ 2\ 3\ 4\ \ldots$$
$$s\ p\ d\ f\ g\ \ldots$$

(The letters then run alphabetically with the omission of j.) Thus, the subshell with $l = 1$ of the shell with $n = 2$ is called the **2p subshell**, and its three orbitals (corresponding to $m_l = +1$, 0, and -1, respectively) are called the **2p orbitals**. Figure 13.8 is a version of Fig. 13.6 which shows the subshells explicitly. An electron that occupies one of the 2p orbitals is called a **2p electron**. Because l can range from 0 to $n - 1$, giving n values in all, it follows that there are n subshells of a shell with principal quantum number n. Thus, when $n = 1$, there is only one subshell, the one with $l = 0$. When $n = 2$, there are two subshells, the 2s subshell (with $l = 0$) and the 2p subshell (with $l = 1$).

When $n = 1$ there is only one subshell, that with $l = 0$, and that subshell contains only one orbital, with $m_l = 0$ (the only value of m_l permitted). When $n = 2$, there are four orbitals, one in the s subshell with $l = 0$ and $m_l = 0$, and three in the $l = 1$ subshell with $m_l = +1$, 0, and -1. When $n = 3$ there are nine orbitals (one with $l = 0$, three with $l = 1$, and five with $l = 2$). The organization of orbitals in the shells is summarized in the marginal table. In general, the number of orbitals in a shell of principal quantum number n is n^2, so in a hydrogenic atom each shell is n^2-fold degenerate.

s orbitals

The **ground state** of an atom is the state of lowest energy, in which the electron is most tightly bound to the nucleus. The orbital occupied in the

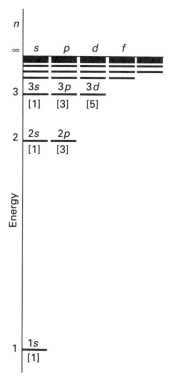

13.8 The energy levels of the hydrogen atom showing the subshells and (in square brackets) the numbers of orbitals in each subshell. All orbitals of a given shell have the same energy in hydrogenic atoms.

	Number of orbitals				
n l: 0	1	2	3	4	Total
s	p	d	f	g	
1 1					1
2 1	3				4
3 1	3	5			9
4 1	3	5	7		16
5 1	3	5	7	9	25

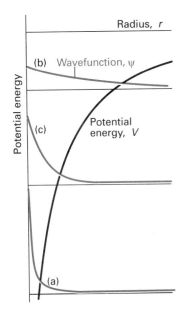

13.9 The balance of kinetic and potential energies that accounts for the structure of the ground state of hydrogen (and similar atoms). (a) The sharply curved but localized orbital has high mean kinetic energy, but low mean potential energy; (b) the mean kinetic energy is low, but the potential energy is not very favourable; (c) the compromise of moderate kinetic energy and moderately favourable potential energy.

ground state is the one with $n = 1$ (and therefore necessarily with $l = 0$ and $m_l = 0$, the only possible values of these quantum numbers when $n = 1$). From Tables 13.1 and 12.3 we can write (for $Z = 1$):

$$\psi = \left(\frac{1}{\pi a_0^3}\right)^{\frac{1}{2}} e^{-r/a_0} \tag{15}$$

As this wavefunction is independent of angle it has the same value at all points of constant radius; that is, the $1s$ orbital is **spherically symmetrical**. The wavefunction decays exponentially from a maximum value of $(1/\pi a_0^3)^{\frac{1}{2}}$ at the nucleus (at $r = 0$). It follows that the most probable point at which the electron will be found is at the nucleus itself.

We can understand the general form of the ground-state wavefunction by considering the contributions of the potential and kinetic energies to the total energy of the atom. The closer the electron is to the nucleus on average, the lower is its average potential energy. This suggests that the lowest potential energy should be obtained with a sharply peaked wavefunction that has a large amplitude at the nucleus and is zero everywhere else (Fig. 13.9a). However, this shape implies a high kinetic energy, because such a wavefunction has a very high average curvature. The electron would have a very low kinetic energy if its wavefunction had only a very low average curvature (Fig. 13.9b). However, such a wavefunction spreads to great distances from the nucleus and the average potential energy of the electron will be high. The actual ground-state wavefunction is a compromise (Fig. 13.9c) between these two extremes: it spreads some way from the nucleus (so the potential energy is not as low as in the first example, but nor is it very high) and has a reasonably low average curvature (so the kinetic energy is not very low, but nor is it as high as in the first example).

One way of depicting the probability density of the electron is to represent ψ^2 by the density of shading (Fig. 13.10). A simpler procedure is to show only the **boundary surface**, the shape that captures about 90 per cent of the electron probability. For the $1s$ orbital, the boundary surface is a sphere (Fig. 13.11).

All s orbitals are spherically symmetric, but differ in the number of **radial nodes**, points at which the radial wavefunction passes through zero. For instance, the $2s$ orbital has radial nodes where the polynomial $L_{2,0}$ is equal to zero:

$$L_{2,0} = 2 - \rho = 0 \text{ at } \rho = 2, \text{ which implies } r = 2a_0$$

Hence, the $2s$ orbital has a radial node at $2a_0$ (see Fig. 13.5b). Similarly, the $3s$ orbitals has two nodes which are found by solving

$$L_{3,0} = 6 - 6\rho + \rho^2 = 0 \qquad \text{with } r = \tfrac{3}{2}\rho a_0$$

One radial node is at $1.90a_0$ and the other is at $7.10a_0$ (see Fig. 13.5c).

The energies of the s orbitals increase (the electron becomes less tightly bound) as n increases because the average distance of the electron from the nucleus increases.

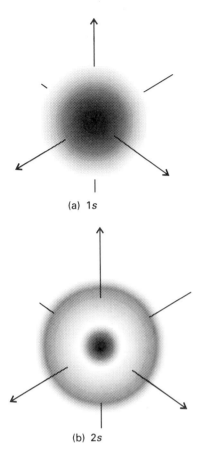

(a) 1s

(b) 2s

13.10 Representations of the (a) 1s and (b) 2s hydrogenic atomic orbitals in terms of their electron densities (as represented by the density of shading).

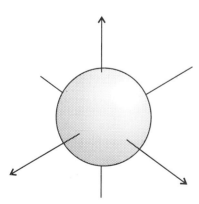

13.11 The boundary surface of an s orbital, within which there is a 90 per cent probability of finding the electron.

Example 13.3 *Calculating the mean radius of an orbital*

Use the hydrogenic orbitals to calculate the mean radius of a 1s orbital.

Method. The mean value is the expectation value

$$\langle r \rangle = \int r\psi^2 \, d\tau$$

We therefore need to evaluate the integral using the wavefunctions given in Table 13.1 and $d\tau = r^2 \, dr \sin\theta \, d\theta \, d\phi$. The integral required was given in Example 11.6.

Answer. The integration over the angular part gives 1 in each case (because the spherical harmonics are normalized). It follows that

$$\langle r \rangle = \int_0^\infty rR^2 \times r^2 \, dr$$

For a 1s orbital,

$$R = \frac{2}{a_0^{\frac{3}{2}}} e^{-r/a_0}$$

Hence

$$\langle r \rangle = \frac{4}{a_0^3} \int_0^\infty r^3 e^{-2r/a_0} \, dr = \tfrac{3}{2} a_0$$

Comment. The general expression for the mean radius of an orbital with quantum numbers l and n is

$$\langle r \rangle_{n,l} = n^2 \left\{ 1 + \frac{1}{2} \left(1 - \frac{l(l+1)}{n^2} \right) \right\} \times \frac{a_0}{Z}$$

The variation with n and l is shown in Fig. 13.12. Note that, for a given principal quantum number, the mean radius *decreases* as l increases.

Exercise E13.3. Evaluate the mean radius of a 3s orbital by integration, and of a 3p orbital by using the general formula. $[27a_0/2, 25a_0/2]$

Radial distribution functions

The wavefunction tells us, through the value of ψ^2, the probability of finding an electron in any region. We can imagine a probe with a volume $d\tau$ and sensitive to electrons, which we can move around near the nucleus of a hydrogen atom. Because the probability density in the ground state of the atom is

$$\psi^2 \propto e^{-2r/a_0}$$

the reading from the detector decreases exponentially as the probe is moved out along any radius (Fig. 13.13) but is constant if the probe is removed on a circle of constant radius.

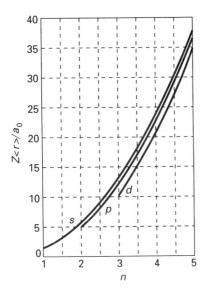

13.12 The variation of the mean radius of a hydrogenic atom with the principal and orbital angular momentum quantum numbers. Although the dependence is shown as continuous curves, only integer values of n are physically significant. Note that the mean radius lies in the order $d < p < s$ for a given value of n (on a vertical line).

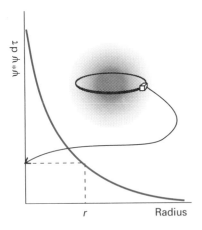

13.13 A constant-volume electron-sensitive detector would show its greatest reading at the nucleus, and a smaller reading elsewhere. The same reading would be obtained anywhere on a circle of given radius: the s orbital is spherically symmetrical.

Now consider the probability of finding the electron anywhere on a spherical shell of thickness dr at a radius r. The sensitive volume of the probe is now the volume of the shell (Fig. 13.14), which is $4\pi r^2 \, dr$. The probability that the electron will be found between the inner and outer surfaces of this shell is the probability density at the radius r multiplied by the volume of the probe:

$$\psi^2 \times 4\pi r^2 \, dr = P \, dr, \qquad \text{with } P = 4\pi r^2 \psi^2$$

The **radial distribution function** P is a probability density in the sense that, when it is multiplied by dr, it gives the probability of finding the electron anywhere in a shell of thickness dr at the radius r.

For a $1s$ orbital,

$$P \propto r^2 e^{-2r/a_0}$$

Because r^2 increases with radius from zero at the nucleus, and ψ^2 decreases towards zero at infinity, P is zero at the nucleus and at infinity and passes through a maximum at an intermediate radius (see Fig. 13.14). The maximum of P marks the **most probable radius** (as distinct from point) at which the electron will be found and, for a $1s$ orbital, occurs at $r = a_0$, the Bohr radius. When we carry through the same calculation for the radial distribution function of the $2s$ orbital (by forming $4\pi r^2 \psi^2$), we find that the most probable radius at which the electron will be found is $5.2a_0 = 275$ pm. This larger value reflects the expansion of the atom as its energy increases.

Example 13.4 *Calculating the most probable radius*

Calculate the most probable radius at which an electron will be found when it occupies a $1s$ orbital of a hydrogenic atom of atomic number Z, and tabulate the values for the one-electron species from H to Ne^{9+}.

Method. We can find the radius at which the radial distribution function of the hydrogenic $1s$ orbital has a maximum value by solving $dP/dr = 0$.

Answer. The radial distribution function is

$$P = 4\pi r^2 \psi^2 = 4\pi r^2 \times \frac{Z^3}{\pi a_0^3} e^{-2Zr/a_0} = \frac{4Z^3}{a_0^3} \times r^2 e^{-2Zr/a_0}$$

It then follows that

$$\frac{dP}{dr} = \frac{4Z^3}{a_0^3} \left(2r - \frac{2Zr^2}{a_0}\right) e^{-2Zr/a_0} = 0 \text{ at } r = r*$$

Therefore,

$$r* = \frac{a_0}{Z}$$

Then, with $a_0 = 52.9$ pm,

	H	He	Li	Be	B	C	N	O	F	Ne
$r*$/pm	52.9	26.5	17.6	13.2	10.6	8.82	7.56	6.61	5.88	5.29

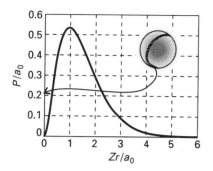

13.14 The radial distribution function P gives the probability that the electron will be found anywhere in a shell of radius r. For a $1s$ electron in hydrogen it is a maximum when r is equal to the Bohr radius a_0.

Comment. Notice how the $1s$ orbital is drawn towards the nucleus as the nuclear charge increases. At uranium the most probable radius is only 0.58 pm, almost 100 times closer than for hydrogen. (On a scale where $r^* = 10$ cm for H, $r^* = 1$ mm for U.) The electron then experiences strong accelerations, and relativisitic effects are important.

Exercise E13.4. Find the most probable distance of a $2s$ electron from the nucleus in a hydrogenic atom. $\qquad [(3 + \sqrt{5})a_0/Z]$

p orbitals

A p electron has non-zero angular momentum (its actual magnitude is $2^{\frac{1}{2}}\hbar$). This momentum has a profound effect on the shape of the wavefunction close to the nucleus, for p orbitals have zero amplitude at $r = 0$. This difference from s orbitals can be understood classically in terms of the centrifugal effect of the angular momentum, which tends to fling the electrons away from the nucleus. It is also what we expect from the form of the effective potential energy shown in Fig. 13.4, which rises to infinity as $r \to 0$ and excludes the wavefunction from the nucleus. The same centrifugal effect appears in all orbitals with $l > 0$ (such as the d orbitals and the f orbitals). As remarked previously, all orbitals with $l > 0$ have zero amplitude at the nucleus, and consequently zero probability of finding the electron there.

The three $2p$ orbitals are distinguished by the three different values that m_l can take when $l = 1$. Because the quantum number m_l tells us the angular momentum around an axis, these different values of m_l denote orbitals in which the electron has different angular momenta around an arbitrary z-axis but the same magnitude of momentum (because l is the same for all three). The orbital with $m_l = 0$, for instance has zero angular momentum around the z-axis. Its angular variation is proportional to $\cos \theta$ so the probability density, which is proportional to $\cos^2 \theta$, has its maximum value on either side of the nucleus along the z-axis (at $\theta = 0$ and 180°). Because in spherical polar coordinates $z = r \cos \theta$, this orbital may also be written

$$p_z \propto \cos \theta = \frac{z}{r}$$

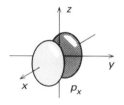

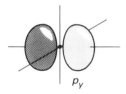

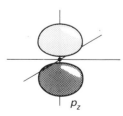

13.15 The boundary surfaces of p orbitals. The nodal plane passes through the nucleus and separates the two lobes of each orbital. The light and dark tinted areas denote regions of opposite sign of the wavefunction.

This way of writing the orbital is the origin of the name **p_z orbital** for this orbital: its boundary surface is shown in Fig. 13.15. The wavefunction is zero at $\theta = 90°$ (everywhere in the xy-plane, where $z = 0$), so the xy-plane is a **nodal plane** of the orbital.

The orbitals with $m_l = \pm 1$ (which are proportional to $\sin \theta e^{\pm i\phi}$) do have angular momentum about the z-axis. As we have seen (in Section 12.6), wavefunctions with this ϕ dependence correspond to a particle with angular momentum either clockwise or counterclockwise around the z-axis: $e^{+i\phi}$ corresponds to counterclockwise rotation when viewed from above, and $e^{-i\phi}$ corresponds to clockwise rotation. They have zero amplitude where $\theta = 0$ and 180° (along the z-axis) and maximum amplitude at 90°, which is in the xy-plane. To draw the functions it is

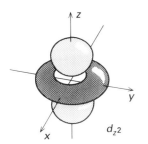

d_{z^2}

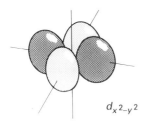

$d_{x^2-y^2}$

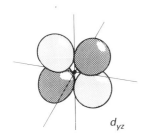

d_{yz}

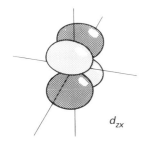

d_{zx}

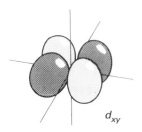

d_{xy}

13.16 The boundary surfaces of d orbitals. Two nodal planes in each orbital intersect at the nucleus and separate the four lobes of each orbital. The light and dark tinted areas denote regions of opposite sign of the wavefunction.

usual to take the real linear combinations

$$e^{i\phi} + e^{-i\phi} \propto \cos\phi \qquad e^{i\phi} - e^{-i\phi} \propto \sin\phi$$

and then the angular parts of the two functions have the form

$$p_x \propto \sin\theta\cos\phi \propto \frac{x}{r} \qquad p_y \propto \sin\theta\sin\phi \propto \frac{y}{r}$$

These linear combinations of the original orbitals are called the **p_x and p_y orbitals**, respectively. They are standing waves with no net angular momentum around the z-axis, as they are composed of equal and opposite values of m_l. The p_x orbital has the same shape as a p_z orbital, but it is directed along the x-axis (see Fig. 13.15); the p_y orbital is similarly directed along the y-axis.

JUSTIFICATION

In this remark, we justify the step of taking linear combinations of orbitals when we want to indicate a particular point. The freedom to do so rests on the fact that, whenever two or more wavefunctions have the same energy, then any linear combination of them also has the same energy and is an equally valid solution of the Schrödinger equation.

Suppose ψ_1 and ψ_2 are both solutions of the Schrödinger equation with energy E; then we know that

$$H\psi_1 = E\psi_1 \quad \text{and} \quad H\psi_2 = E\psi_2$$

Now consider the linear combination

$$\psi = c_1\psi_1 + c_2\psi_2$$

where c_1 and c_2 are arbitrary coefficients. Then it follows that

$$H\psi = H(c_1\psi_1 + c_2\psi_2) = c_1 H\psi_1 + c_2 H\psi_2$$
$$= c_1 E\psi_1 + c_2 E\psi_2 = E\psi$$

Hence, the linear combination is also a solution corresponding to the same energy E.

d orbitals

When $n = 3$, l can be 0, 1, or 2. As a result, this shell consists of one $3s$ orbital, three $3p$ orbitals, and five $3d$ orbitals. The five d orbitals have $m_l = 2$, 1, 0, -1, -2 and correspond to five different angular momenta around the z-axis (but the same magnitude of angular momentum, because $l = 2$ in each case). As for the p orbitals, d orbitals with opposite values of m_l (and hence opposite senses of motion around the z-axis) may be combined in pairs to give standing waves, and the boundary surfaces of the resulting shapes are shown in Fig. 13.16.

13.3 Spectroscopic transitions and selection rules

The energies of the hydrogenic atoms are given by eqns 12 and 13. When the electron undergoes a **transition**, a change of state, from an orbital with quantum numbers n_2, l_2, m_{l2} to another (lower-energy) orbital with quantum numbers n_1, l_1, m_{l1}, it undergoes a change of energy ΔE and discards the excess energy as a photon of electromagnetic radiation with a frequency v given by the Bohr frequency condition (eqn 4).

It is tempting to think that all possible transitions are permissible and that the spectrum of the atom arises from the transition of an electron from any initial orbital to any other orbital. However, this is not so, because a photon has an intrinsic spin angular momentum corresponding to $s = 1$. If a photon is generated by an electron undergoing a transition, then the angular momentum of the electron must change to compensate for the angular momentum carried away by the photon as its spin. Thus an electron in a d orbital with $l = 2$ cannot make a transition into an s orbital with $l = 0$ because the photon cannot carry away enough angular momentum. Similarly, an s electron cannot make a transition to another s orbital, because there would then be no change in the electron's angular momentum to make up for the angular momentum carried away by the photon. It follows that some spectroscopic transitions are **allowed**, meaning that they can occur, while others are **forbidden**, meaning that they cannot occur.

A **selection rule** is a statement about which transitions are allowed. They are derived (for atoms) by identifying the transitions that conserve angular momentum when a photon is emitted or absorbed. The selection rules for hydrogenic atoms are

$$\Delta l = \pm 1 \qquad \Delta m_l = 0, \pm 1 \tag{16}$$

The principal quantum number n can change by any amount consistent with the Δl for the transition because it does not relate directly to the angular momentum.

Example 13.5 *Using the selection rules*

To what orbitals may a $4d$ electron make radiative transitions?

Method. We first identify the value of l and then apply the selection rule for this quantum number.

Answer. Because $l = 2$, the final orbital must have $l = 1$ or 3. Thus, an electron may make a transition from a $4d$ orbital to any np orbital (subject to $\Delta m_l = 0$ or ± 1) and to any nf orbital (subject to the same rule). However, it cannot undergo a transition to any other orbital, so a transition to any ns orbital or to another nd orbital is forbidden.

Exercise E13.5. To what orbitals may a $4s$ electron make radiative transitions?

[To np orbitals only]

The selection rules account for the structure of a **Grotrian diagram** (Fig. 13.17), which summarizes the energies of the states and the transitions between them. The thicknesses of the transition lines in the diagram denote their relative intensities in the spectrum. The intensities may also be calculated from the wavefunctions of the two states, but we shall not deal with this aspect here.

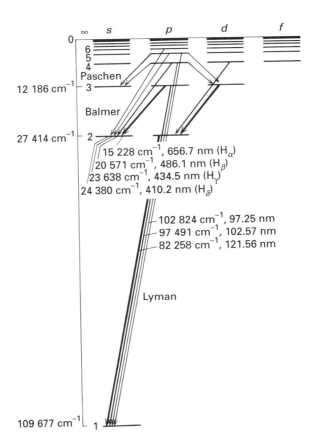

13.17 A Grotrian diagram that summarizes the appearance and analysis of the spectrum of atomic hydrogen. The thicker the line, the more intense the transition.

THE STRUCTURE OF MANY-ELECTRON ATOMS

The Schrödinger equations for many-electron atoms are extremely complicated because all the electrons interact with each other. Even in the case of a helium atom, with its two electrons, no analytical expression for the orbitals and energies can be given, and we are forced to make approximations. We shall adopt a simple approach based on what we already know about the structure of hydrogenic atoms. Later we shall see the kind of numerical computations that are currently used to obtain accurate wavefunctions and energies.

13.4 The orbital approximation

The actual wavefunction of a many-electron atom is a very complicated function of the coordinates of all the electrons, and we should write it $\Psi(r_1, r_2, \ldots)$. However, in the **orbital approximation** we suppose that a reasonable first approximation to this exact wavefunction is obtained by thinking of each electron as occupying its 'own' ψ, and writing

$$\Psi(r_1, r_2, \ldots) = \psi(r_1)\psi(r_2) \ldots \tag{17}$$

We can think of the individual orbitals as resembling the hydrogenic orbitals but with nuclear charges that are modified by the presence of all the other electrons in the atom. This description is only approximate, but it is a useful model for discussing the chemical properties of atoms and is the starting point for more sophisticated descriptions of atomic structure.

JUSTIFICATION

The orbital approximation would be exact if there were no interactions between electrons. To demonstrate the validity of this remark, we need to consider a system in which the Hamiltonian operator for the energy is the sum of two contributions, one for electron 1 and the other for electron 2:

$$H = H_1 + H_2$$

In an actual atom (such as helium atom), there is an additional term corresponding to the interaction of the two electrons, but we are ignoring that term. We shall now show that, if $\psi(\mathbf{r}_1)$ *is an eigenfunction of* H_1 with energy E_1 and $\psi(\mathbf{r}_2)$ *is an eigenfunction of* H_2 with energy E_2, then the product $\Psi(\mathbf{r}_1, \mathbf{r}_2) = \psi(\mathbf{r}_1)\psi(\mathbf{r}_2)$ is an eigenfunction of the combined Hamiltonian H. To do so we write

$$H\Psi(\mathbf{r}_1, \mathbf{r}_2) = (H_1 + H_2)\psi(\mathbf{r}_1)\psi(\mathbf{r}_2) = \{H_1\psi(\mathbf{r}_1)\}\psi(\mathbf{r}_2) + \psi(\mathbf{r}_1)\{H_2\psi(\mathbf{r}_2)\}$$
$$= \{E_1\psi(\mathbf{r}_1)\}\psi(\mathbf{r}_2) + \psi(\mathbf{r}_1)\{E_2\psi(\mathbf{r}_2)\} = (E_1 + E_2)\psi(\mathbf{r}_1)\psi(\mathbf{r}_2)$$
$$= E\Psi(\mathbf{r}_1, \mathbf{r}_2)$$

where $E = E_1 + E_2$. This is the result we need to prove. However, if the electrons interact (as they do in fact), then the proof fails. Therefore, the orbital approximation is indeed only an approximation.

The helium atom

The orbital approximation allows us to express the electronic structure of an atom by reporting its **configuration**, the list of occupied orbitals (usually, but not necessarily, in its ground state). Thus, as the ground state of a hydrogenic atom consists of the single electron in a $1s$ orbital, we report its configuration as $1s^1$.

The He atom has two electrons. We can imagine forming the atom by adding the electrons in succession to the orbitals of the bare nucleus (of charge $2e$). The first electron occupies a $1s$ hydrogenic orbital; but, because $Z = 2$, that orbital is more compact than in H itself. The second electron joins the first in the $1s$ orbital, so the electron configuration of the ground state of He is $1s^2$.

The Pauli principle

Lithium, with $Z = 3$, has three electrons. The first two occupy a $1s$ orbital drawn in even more closely than in He around the more highly charged nucleus. The third electron, however, does not join the first two in the $1s$ orbital because that configuration is forbidden by the **Pauli exclusion principle**:

No more than two electrons may occupy any given orbital, and if two do occupy one orbital, then their spins must be paired.

Electrons with paired spins, which we denote ↑↓, have zero net spin

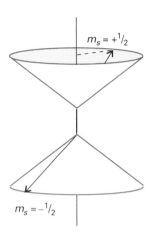

angular momentum because the spin of one electron is cancelled by the spin of the other. Specifically, one electron has $m_s = +\frac{1}{2}$, the other has $m_s = -\frac{1}{2}$, and they are orientated on their respective precessional cones so that the resultant spin is zero (Fig. 13.18). The exclusion principle is the key to the structure of complex atoms, to chemical periodicity, and to molecular structure. It was proposed by Wolfgang Pauli in 1924 when he was trying to account for the absence of some lines in the spectrum of helium. Later he was able to derive a very general form of the principle from theoretical considerations. The Pauli exclusion principle in fact applies to any pair of identical fermions (particles with half-integral spin). Thus it applies to protons, neutrons, and ^{13}C nuclei (all of which have spin $\frac{1}{2}$) and to ^{35}Cl nuclei (which have spin $\frac{3}{2}$). It does not apply to identical bosons (particles with integral spin), and which include photons (spin 1) and ^{12}C nuclei (spin 0), and any number of identical bosons may occupy the same orbital. A deeper form of the principle is given in *Further information 12*.

In Li ($Z = 3$), the third electron cannot enter the $1s$ orbital because that orbital is already full: we say the K shell is **complete** and that the two electrons form a **closed shell**. Because a similar closed shell is characteristic of the He atom, we denote it [He]. The third electron is excluded from the K shell and must occupy the next available orbital, which is one with $n = 2$ and hence belonging to the L shell. However, we now have to decide whether the next available orbital is the $2s$ orbital or a $2p$ orbital, and therefore whether the lowest energy configuration of the atom is $[He]2s^1$ or $[He]2p^1$.

13.18 Electrons with paired spins have zero resultant spin angular momentum. They can be represented by two vectors that lie at an indeterminate position on the cones shown here, but, wherever one lies on its cone, the other points in the opposite direction so that their resultant is zero.

Penetration and shielding

Unlike in hydrogenic atoms, the $2s$ and $2p$ orbitals (and, in general, all subshells of a given shell) are not degenerate in many-electron atoms. For reasons we shall now explain, s electrons generally lie lower in energy than p electrons of a given shell, and p electrons lie lower than d electrons.

An electron in a many-electron atom experiences a Coulombic repulsion from all the other electrons present. If it is at a distance r from the nucleus, it experiences a repulsion that can be represented by a point negative charge located at the nucleus and equal in magnitude to the total charge of the electrons within a sphere of radius r (Fig. 13.19). The effect of this point negative charge, when averaged over all the locations of the electron, is to reduce the full charge of the nucleus from Ze to $Z_{eff}e$, where Z_{eff} is the **effective atomic number**. We say that the electron experiences a **shielded nuclear charge**, and that the true atomic number is reduced from Z to Z_{eff} by an amount called the **shielding constant**, σ:

$$Z_{eff} = Z - \sigma \tag{18}$$

The electrons do not actually 'block' the full Coulombic attraction of the nucleus: the effective charge is simply a way of expressing the net outcome of the nuclear attraction and the electronic repulsions in terms of a single equivalent charge at the centre of the atom.

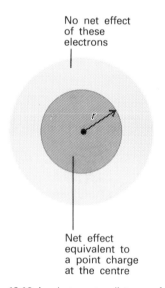

13.19 An electron at a distance r from the nucleus experiences a Coulombic repulsion from all the electrons within a sphere of radius r which is equivalent to the effect of a point negative charge located on the nucleus. The negative charge reduces the effective nuclear charge of the nucleus from Ze to $Z_{eff}e$.

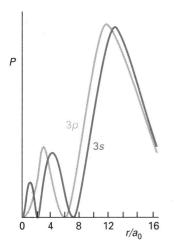

13.20 An electron in an s orbital (here a 3s orbital) is more likely to be found close to the nucleus than an electron in a p orbital of the same shell (note the closeness of the innermost peak of the 3s orbital to the nucleus at $r = 0$). Hence it experiences less shielding and is more tightly bound.

Table 13.3* Effective atomic numbers Z_{eff}

He 1s	1.69
C 1s	5.67
2s	3.22
2p	3.14

* More values are given in the Data section at the end of this volume.

The effective atomic number is different for s and p electrons because they have different radial wavefunctions (Fig. 13.20). An s electron has a greater penetration through inner shells than a p electron in the sense that it is more likely to be found close to the nucleus than a p electron of the same shell (the p orbital, remember, has a node at the nucleus). Because only electrons inside the sphere defined by the location of the electron (in effect, the core electrons) contribute to shielding, an s electron experiences less shielding than a p electron and therefore experiences a larger Z_{eff}. Consequently, by the combined effects of penetration and shielding, an s electron is more tightly bound than a p electron of the same shell. Similarly, a d electron penetrates less than a p electron of the same shell, and therefore experiences more shielding and an even smaller Z_{eff}.

Effective atomic numbers for different types of electrons in atoms have been calculated from their wavefunctions (which are obtained by numerical solution of the Schrödinger equation for the atom, see Section 13.5). Some values are given in Table 13.3. We see that, in general, valence-shell s electrons do experience higher effective atomic numbers than p electrons, although there are some discrepancies. We shall return to this point shortly.

The consequence of penetration and shielding is that the energies of subshells in a many-electron atom in general lie in the order

$$s < p < d$$

The individual orbitals of a given subshell (such as the three p orbitals of the p subshell) remain degenerate because they all have the same radial characteristics and so experience the same effective nuclear charge.

We can now complete the Li story. Because the shell with $n = 2$ consists of two non-degenerate subshells, with the 2s orbital lower in energy than the three 2p orbitals, the third electron occupies the 2s orbital. This results in the ground-state configuration $1s^2 2s^1$, with the central nucleus surrounded by a complete helium-like shell of two 1s electrons, and around that a more diffuse 2s electron. The electrons in the outermost shell of an atom in its ground state are called the **valence electrons** because they are largely responsible for the chemical bonds that the atom forms. Thus, the valence electron in Li is a 2s electron and its other two electrons belong to its **core**.

The building-up principle

The extension of the procedure used for H, He, and Li to other atoms is called the **building-up principle**, or the *Aufbau* principle, from the German word for building up. The building-up principle proposes an order of occupation of the hydrogenic orbitals that accounts for the experimentally determined ground-state configuration of neutral atoms.

We imagine the bare nucleus of atomic number Z, and then feed into the orbitals Z electrons in succession. The order of occupation is

$$1s\ 2s\ 2p\ 3s\ 3p\ 4s\ 3d\ 4p\ 5s\ 4d\ 5p\ 6s$$

and each orbital may accommodate up to two electrons. This order of occupation is approximately the order of energies of the individual orbitals since, in general, the lower the energy of the orbital, the lower the total energy of the atom as a whole when that orbital is occupied. However, there are complicating effects arising from electron–electron repulsions that are important when the orbitals have very similar energies (such as the $4s$ and $3d$ orbitals near Ca and Sc), and we must take special care then.

We feed the Z electrons in succession into the orbitals subject to the demand of the exclusion principle that no more than two electrons can occupy any one orbital. Because an s subshell consists of only one orbital, up to two electrons may occupy it. A p subshell consists of three orbitals, so it can accommodate up to six electrons; a d subshell consists of five orbitals and can accommodate up to 10 electrons.

As an example, consider the carbon atom, for which $Z = 6$, and there are six electrons to accommodate. Two electrons enter and fill the $1s$ orbital and two enter and fill the $2s$ orbital, leaving two electrons to occupy the orbitals of the $2p$ subshell. Hence the ground configuration of C is $1s^2 2s^2 2p^2$, or more succinctly $[He]2s^2 2p^2$, with [He] the helium-like $1s^2$ core. However, we can be more precise. On electrostatic grounds, we can expect the last two electrons to occupy different $2p$ orbitals because they will then be further apart on average and repel each other less than if they were in the same orbital. Thus one electron can be thought of as occupying the $2p_x$ orbital and the other the $2p_y$ orbital (the x, y, z designation is arbitrary), and the lowest energy configuration of the atom is $[He]2s^2 2p_x^1 2p_y^1$. The same rule applies whenever degenerate orbitals of a subshell are available for occupation. Thus, another rule of the building-up principle is:

Electrons occupy different orbitals of a given subshell before doubly occupying any one of them.

Thus nitrogen ($Z = 7$) has the configuration $[He]2s^2 2p_x^1 2p_y^1 2p_z^1$, and only when we get to oxygen ($Z = 8$) is a $2p$ orbital doubly occupied, giving $[He]2s^2 2p_x^2 2p_y^1 2p_z^1$.

An additional point arises when electrons occupy orbitals singly, for there is then no requirement that their spins should be paired. We need to know whether the lowest energy is achieved when the electron spins are the same (both α, for instance, denoted ↑↑) or when they are paired (↑↓). This question is resolved by an empirical observation known as **Hund's rule**:

An atom in its ground state adopts a configuration with the greatest number of unpaired electrons.

The explanation of Hund's rule is complicated, but it reflects the quantum mechanical property of **spin correlation**, that electrons with parallel spins behave as if they have a tendency to stay well apart and hence repel each other less.[2]

---

2 The effect of spin correlation is to allow the atom to shrink slightly, so the electron–nucleus interaction is improved when the spins are parallel.

JUSTIFICATION

The origin of spin correlation makes use of the material on the Pauli principle described in *Further information 12*. Suppose electron 1 is described by a wavefunction $\psi_a(r_1)$ and electron 2 is described by a wavefunction $\psi_b(r_2)$; then, in the orbital approximation the joint wavefunction of the electrons is the product $\Psi = \psi_a(r_1)\psi_b(r_2)$. However, this wavefunction is not acceptable, because it suggests that we know which electron is in which orbital, whereas we cannot keep track of electrons. According to quantum mechanics, the correct description is either of the two following wavefunctions:

$$\Psi_+ = \psi_a(r_1)\psi_b(r_2) + \psi_b(r_1)\psi_a(r_2) \quad \Psi_- = \psi_a(r_1)\psi_b(r_2) - \psi_b(r_1)\psi_a(r_2)$$

According to the Pauli principle (as described in *Further information 12*), the combination Ψ_+ is the appropriate one for a spin-paired state whereas Ψ_- should be used for electrons with parallel spins.

Now consider the values of the two combinations when one electron approaches another, so that $r_1 = r_2$. We see that Ψ_- vanishes, which means that there is zero probability of finding the two electrons at the same point in space when they have parallel spins. The other combination does not vanish when the two electrons are at the same point in space. Because the two electrons have different *relative* spatial distributions depending on whether their spins are parallel or not, it follows that their Coulombic interaction is different, and hence that the two states have different energies.

We can now conclude that, in the ground state of the carbon atom, the two $2p$ electrons have the same spin, that all three $2p$ electrons in the N atoms have the same spin, and that the two $2p$ electrons in different orbitals in the O atom have the same spin (the two in the $2p_x$ orbital are necessarily paired).

Neon, with $Z = 10$, has the configuration $[He]2s^2 2p^6$, which completes the L shell. This closed-shell configuration is denoted $[Ne]$ and acts as a core for subsequent elements. The next electron must enter the $3s$ orbital and begin a new shell, so an Na atom, with $Z = 11$, has the configuration $[Ne]3s^1$. Like lithium with the configuration $[He]2s^1$, sodium has a single s electron outside a complete core.

This analysis has brought us to the origin of chemical periodicity. The L shell is completed by eight electrons, so the element with $Z = 3$ (Li) should have similar properties to the element with $Z = 11$ (Na). Likewise, Be $(Z = 4)$ should be similar to Mg $(Z = 12)$, and so on up to the noble gases He $(Z = 2)$, Ne $(Z = 10)$, and Ar $(Z = 18)$.

Argon has complete $3s$ and $3p$ subshells and, as the $3d$ orbitals are high in energy, it counts as having a closed-shell configuration. Indeed, the $3d$ orbitals are so high in energy that the next electron (for K) occupies the $4s$ orbital and the K atom resembles an Na atom. The same is true of a Ca atom, which has the configuration $[Ar]4s^2$. However, at this point, the $3d$ orbitals become comparable in energy to the $4s$ orbitals (Fig. 13.21), and they commence to be filled.

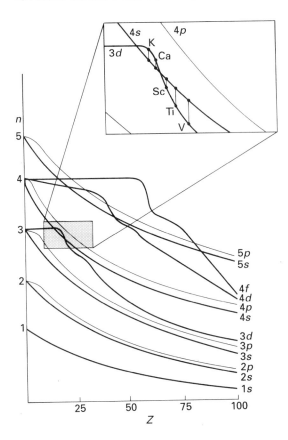

13.21 The orbital energies of the elements. Note the relative energies of the 3d and 4s orbitals close to potassium (see inset).

Ten electrons can be accommodated in the five 3d orbitals, which accounts for the electron configurations of scandium to zinc. However, the building-up principle has less clear-cut predictions about ground-state configurations of these elements because electron–electron repulsions are comparable to the energy difference between the 4s and 3d orbitals, and a simple analysis no longer works. At gallium, the energy of the 3d orbitals has fallen so far below those of the 4s and 4p orbitals that they (the 3d orbitals) can be largely ignored, and the building-up principle can be used in the same way as in preceding periods. Now the 4s and 4p subshells constitute the valence shell, and the period terminates with krypton. Because 18 electrons have intervened since argon, this period is the first **long period** of the periodic table. The existence of the d-block elements (the 'transition metals') reflects the stepwise occupation of the 3d orbitals, and the subtle shades of energy differences along this series give rise to the rich complexity of inorganic d-metal chemistry. A similar intrusion of the f orbitals in periods 6 and 7 accounts for the existence of the f block of the periodic table (the lanthanides and actinides).

We derive the configurations of cations of elements in the s, p, and d blocks of the periodic table by removing electrons from the ground-state configuration of the neutral atom in a specific order. First, we remove p electrons (if any are present), then s electrons, and then as many d electrons as are necessary to achieve the stated charge. For instance, since the configuration of Fe is $[Ar]3d^6 4s^2$, the Fe^{3+} cation has the configuration $[Ar]3d^5$. We derive the configurations of anions simply by continuing the building-up procedure and adding electrons to the neutral atom until the configuration of the next noble gas has been reached. Thus, the configuration of the O^{2-} ion (which exists in solid ionic oxides, but is unstable in the gas phase) is achieved by adding two electrons to $[He]2s^2 2p^4$, giving $[He]2s^2 2p^6$, the same as the configuration of Ne.

The periodicity of ionization energies

The minimum energy necessary to remove an electron from a many-electron atom is the **first ionization energy** I_1, of the element. The **second ionization energy** I_2 is the minimum energy needed to remove a second electron (from the singly charged cation). The variation of the first ionization energy through the periodic table is shown in Fig. 13.22 and some numerical values have already been given in Table 2.6 (see the Data section at the end of this volume).

Lithium has a low first ionization energy: its outermost electron is well-shielded from the nucleus by the core ($Z_{eff} = 1.3$, compared with $Z = 3$) and it is easily removed. Beryllium has a higher nuclear charge than lithium, and its outermost electron (one of the two 2s electrons) is more difficult to remove: its ionization energy is higher. The ionization energy decreases between beryllium and boron because in the latter the outermost electron occupies a 2p orbital and is less strongly bound than if it had been a 2s electron. The ionization energy increases between boron and carbon because the latter's outermost electron is also 2p and the nuclear charge has increased. Nitrogen has a still higher ionization energy because of the further increase in nuclear charge.

There is now a kink in the curve that reduces the ionization energy of oxygen below what would be expected by simple extrapolation. The explanation is that at oxygen a 2p orbital must become doubly occupied, and the electron–electron repulsions are increased above what would be expected by simple extrapolation along the row. (The kink is less pronounced in the next row, between phosphorus and sulfur, because their orbitals are more diffuse.) The values for oxygen, fluorine, and neon fall roughly on the same line, the increase of their ionization energies

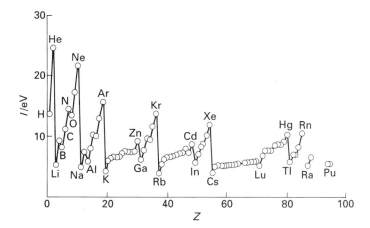

13.22 The first ionization energies of the elements plotted against atomic number.

reflecting the increasing attraction of the more highly charged nuclei for the outermost electrons.

The outermost electron in sodium is $3s$. It is far from the nucleus, and the latter's charge is shielded by the compact, complete neon-like core. As a result, the ionization energy of sodium is substantially lower than that of neon. The periodic cycle starts again along this row, and the variation of the ionization energy can be traced to similar reasons.

13.5 Self-consistent field orbitals

The central difficulty of the Schrödinger equation is the presence of the electron–electron interaction terms. The potential energy of the electrons is

$$V = \sum_i \left(-\frac{Ze^2}{4\pi\varepsilon_0 r_i} \right) + \sum_{i,j \, \text{pairs}} \left(\frac{e^2}{4\pi\varepsilon_0 r_{ij}} \right) \tag{19}$$

The first term on the right is the total attractive interaction between the electrons and the nucleus. The second term is the total repulsive interaction; r_{ij} is the distance between electrons i and j, and the sum is over all pairs of electrons. It is hopeless to expect to find analytical solutions of the Schrödinger equation with such a complicated potential energy term, but computational techniques are available that give very detailed and reliable numerical solutions for the wavefunctions and energies. The techniques were originally introduced by Douglas Hartree (before computers were available) and then modified by Vladimir Fock to take into account the Pauli principle correctly. In broad outline, the **Hartree–Fock procedure** is as follows.

Imagine that we have a rough idea of the structure of the atom. In the Na atom, for instance, the orbital approximation suggests the configuration $1s^2 2s^2 2p^6 3s^1$ with the orbitals approximated by hydrogenic atomic orbitals. Now consider the $3s$ electron. A Schrödinger equation can be written for this electron by ascribing to it a potential energy that arises from the nuclear attraction and the average electronic repulsion from the other electrons in their approximate orbitals. This equation has the form

$$-\frac{\hbar^2}{2m_e}\nabla^2\psi_{3s} - \frac{Ze^2}{4\pi\varepsilon_0 r}\psi_{3s} + V_{ee}\psi_{3s} = E\psi_{3s} \tag{20}$$

where V_{ee}, which depends on the wavefunctions of all the other electrons, is the average repulsion term. The equation may be solved for ψ_{3s} (by numerical integration), and the solution obtained will be different from the solution guessed initially.

The procedure is then repeated for another orbital, such as $2p$. The Schrödinger equation is written in a form like eqn 20 but with the improved $3s$ orbital used in setting up the electron–electron repulsion term. The equation is then solved, giving an improved version of $2p$. This procedure is repeated for the $2s$ and $1s$ orbitals, each time using the improved orbitals found at the earlier stage. Then the whole procedure is repeated using the improved orbitals, and a second improved set of orbitals is obtained. The recycling continues until the orbitals and energies obtained are insignificantly different from those used at the start

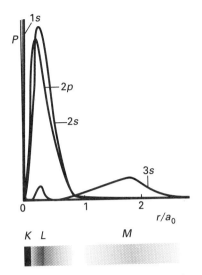

13.23 The radial distribution functions for the orbitals of Na based on SCF calculations. Note the shell-like structure, with the 3s orbital outside the inner K and L shells.

of the latest cycle. The solutions are then **self-consistent** and accepted as solutions of the problem.

Some of the self-consistent field (SCF) Hartree–Fock (HF) atomic orbitals (AO) for sodium are shown in Fig. 13.23. They show the grouping of electron density into shells, as was anticipated by the early chemists, and the differences of penetration as discussed above. These SCF calculations therefore support the qualitative discussions that are used to explain chemical periodicity. They also considerably extend that discussion by providing detailed wavefunctions and precise energies.

THE SPECTRA OF COMPLEX ATOMS

The spectra of atoms rapidly become very complicated as the number of electrons increases, but there are some important and moderately simple features. The general idea is straightforward: lines in the spectrum (in either emission or absorption) occur when the atom undergoes a change of state with a change of energy ΔE, and emits or absorbs a photon of frequency $\nu = \Delta E/h$ and wavenumber $\tilde{\nu} = \Delta E/hc$. Hence, we can expect the spectrum to give information about the energies of electrons in atoms. However, the actual energy levels are not given solely by the energies of the orbitals, because the electrons interact with each other in various ways, and there are contributions to the energy in addition to those we have already considered.

13.6 Singlet and triplet states

Suppose we are interested in the energy levels of a He atom with its two electrons. We know that the ground configuration is $1s^2$ and can anticipate that an excited configuration will be one in which one of the electrons has been promoted into a $2s$ orbital, giving the configuration $1s^1 2s^1$. The two electrons need not be paired because they occupy different orbitals. According to Hund's rule, the state of the atom with the spins parallel ($\uparrow\uparrow$) lies lower in energy than the state in which they are paired ($\uparrow\downarrow$). Both states are permissible, and can contribute to the spectrum of the atom. The significance of the expression 'paired spins' was represented in Fig. 13.18. The corresponding representation of parallel spins is illustrated in Fig. 13.24.

Parallel and antiparallel (paired) spins differ in their overall spin angular momentum. In the paired case, the two spin momenta cancel each other, and there is zero net spin. The paired-spin arrangement is called a **singlet**. The angular momenta of two parallel spins add together to give a non-zero total spin, and the resulting state is called a **triplet**. As illustrated in Fig. 13.24, there are three ways of achieving a non-zero total spin but only one way to achieve zero spin.

The fact that the parallel arrangement of spins in the $1s^1 2s^1$ configuration of the He atom lies lower in energy than the antiparallel arrangement can now be expressed by saying that the triplet state of the $1s^1 2s^1$ configuration of He lies lower in energy than the singlet state. This is a general conclusion that applies to other atoms (and molecules) and, for states arising from the same configuration, *the triplet state generally lies lower than the singlet state*. The origin of the energy difference lies in the

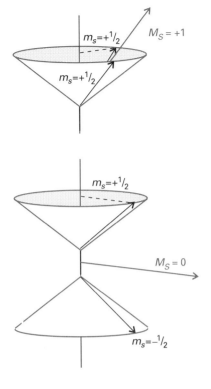

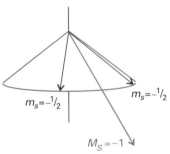

13.24 When two electrons have parallel spins, they have a non-zero total spin angular momentum. There are three ways of achieving this resultant, which are shown by these vector representations. Note that, although we cannot know the orientation of the spin vectors on the cones, the angle between the vectors is the same in all three cases, for all three cases have the same total spin angular momentum (that is, the resultant of the two vectors has the same length in each case, but points in different directions). Compare this diagram with Fig. 13.18, which shows the antiparallel case. Note that, whereas two paired spins are precisely antiparallel, two 'parallel' spins are not strictly parallel.

effect of spin correlation on the Coulombic interactions between electrons, as we saw in the case of Hund's rule for ground-state configurations. Because the Coulombic interaction between electrons in an atom is strong, the difference in energies between singlet and triplet states of the same configuration can be very large. The two states of $1s^1 2s^1$ He, for instance, differ by 6421 cm^{-1} (corresponding to 77 kJ mol^{-1} or 0.80 eV).

The spectrum of atomic helium is more complicated than that of atomic hydrogen, but there are two simplifying features. One is that the only excited configurations it is necessary to consider are of the form $1s^1 nl^1$: that is, only *one* electron is excited. Excitation of two electrons requires an energy that exceeds the ionization energy of the atom, so the He$^+$ ion is formed instead of the doubly excited atom. Second, no transitions take place between singlet and triplet states because the relative orientation of the two electron spins cannot change during a transition. Thus, there is a spectrum arising from transitions between singlet states (including the ground state) and between triplet states, but not between the two. Spectroscopically, helium behaves like two distinct species, and the early spectroscopists actually thought of helium as consisting of 'parahelium' and 'orthohelium'. The Grotrian diagram for helium is shown in Fig. 13.25, which shows the two sets of transitions. Note also that the triplet states lie lower than the corresponding singlet state of the same configuration.

13.7 Spin–orbit coupling

Electron spin has a further implication for the energies of atoms. Because an electron has spin angular momentum and because moving charges generate magnetic fields, an electron has a magnetic moment that arises from its spin (Fig. 13.26). Similarly, an electron with orbital angular momentum (that is, an electron in an orbital with $l \neq 0$) is in effect a circulating current, and possesses a magnetic moment that arises from its orbital momentum. The interaction of the spin and orbital magnetic moments is called **spin–orbit coupling**. The strength of the coupling, and its effect on the energy levels of the atom, depend on the relative orientations of the spin and orbital magnetic moments, and therefore on the relative orientations of the two angular momenta (Fig. 13.27).

The total angular momentum

One way of expressing the dependence of the spin–orbit interaction on the relative orientation of the spin and orbital momenta is to say that it depends on the **total angular momentum** of the electron, the vector sum of its spin and orbital momenta. Thus, when the spin and orbital angular momenta are parallel, the total angular momentum is high; when the two angular momenta are opposed, the total angular momentum is low.

The total angular momentum of a spinning, orbiting electron is quantized. It is described by the quantum numbers j and m_j with $j = l + \frac{1}{2}$ (when the two angular momenta are in the same direction) or $j = l - \frac{1}{2}$ (when they are opposed, Fig. 13.28). The different values of j that can arise for a given value of l constitute a **level**. For $l = 0$, the only permitted

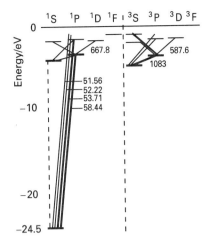

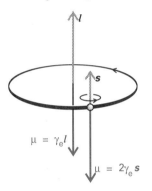

13.25 The Grotrian diagram for a helium atom. Note that there are no transitions between the singlet and triplet levels. Wavelengths are given in picometres.

13.26 Angular momentum gives rise to a magnetic moment (μ). In the case of an electron, the magnetic moment is antiparallel to the orbital angular momentum, but proportional to it. In the case of spin angular momentum, there is a factor 2, which increases the magnetic moment to twice its expected value (as explained in Section 13.9).

value is $j = \tfrac{1}{2}$ (the total angular momentum is the same as the spin angular momentum since there is no other angular momentum in the atom). When $l = 1$, j may be either $\tfrac{3}{2}$ (the spin and orbital angular momenta are in the same sense) or $\tfrac{1}{2}$ (the spin and orbital momenta are in opposite senses).

Example 13.6 *Identifying the levels of a configuration*

Identify the levels that may arise from the configurations (a) d^1, (b) s^1.

Method. In each case, we must identify the value of l and then the possible values of j, which are the sum and difference of the orbital and spin momenta.

Answer. For a d electron, $l = 2$ and there are two levels in the configuration, one with $j = 2 + \tfrac{1}{2} = \tfrac{5}{2}$ and the other with $j = 2 - \tfrac{1}{2} = \tfrac{3}{2}$. (b) For an s electron $l = 0$, so only one level is possible, and $j = \tfrac{1}{2}$.

Exercise E13.6. Identify the levels of the configurations (a) p^1 and (b) f^1.

$[\text{(a) } \tfrac{3}{2}, \tfrac{1}{2}; \text{ (b) } \tfrac{7}{2}, \tfrac{5}{2}]$

The dependence of the spin–orbit interaction on the value of j is expressed in terms of the **spin–orbit coupling constant** A (which is typically expressed as a wavenumber). A quantum mechanical calculation leads to the result that the energies of the levels with quantum numbers s, l, and j are given by

$$E_{l,s,j} = \tfrac{1}{2}hcA\{j(j+1) - l(l+1) - s(s+1)\} \tag{21}$$

JUSTIFICATION

The energy of a magnetic moment μ in a magnetic field B is equal to their scalar product $-\mu \cdot B$. If the magnetic field arises from the orbital angular momentum of the electron, it is proportional to l; if the magnetic moment μ is that of the electron spin, then it is proportional to s. In each case the field and momenta have directions that are determined by the directions of the two momenta, which is why we have written the vectors l and s rather than just their magnitudes. It then follows that the energy of interaction is proportional to the scalar product $s \cdot l$:

$$\text{Energy of interaction} = -\mu \cdot B \propto s \cdot l$$

Next, we note that the total angular momentum is the vector sum of the orbital and spin momenta:

$$j = l + s$$

The magnitude of the vector j is calculated by evaluating

$$j \cdot j = (l + s) \cdot (l + s) = l \cdot l + s \cdot s + 2s \cdot l$$

However, we know that the square of the magnitude of an angular momentum vector is proportional to $j(j+1)$, $l(l+1)$, and $s(s+1)$, where

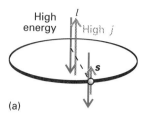

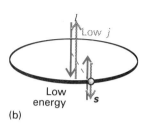

13.27 Spin–orbit coupling is a magnetic interaction between spin and orbital magnetic moments. When the angular momenta are parallel as in (a), the magnetic moments are aligned unfavourably; when they are opposed as in (b), the interaction is favourable. This magnetic coupling is the cause of the splitting of a configuration into levels.

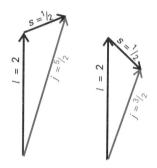

13.28 The coupling of the spin and orbital angular momenta of a *d* electron ($l = 2$) gives two possible values of *j* depending on the relative orientation of the spin and orbital angular momenta of the electron.

j, *l*, and *s* are the corresponding quantum numbers. Therefore, we can rearrange the last expression to

$$s \cdot l \propto j(j + 1) - l(l + 1) - s(s + 1)$$

It then follows that the interaction energy is

$$E \propto j(j + 1) - l(l + 1) - s(s + 1)$$

Then, taking the constant of proportionality as $\frac{1}{2}hcA$ gives eqn 21.

We can now assess the effect of spin–orbit coupling on the states of any atom with a single electron outside a closed core (as in an alkali metal atom). In the ground state of an alkali metal atom the electron has $l = 0$, so $j = \frac{1}{2}$. Because the orbital angular momentum is zero in this state, the spin–orbit coupling energy is zero (as is confirmed by setting $j = s$ and $l = 0$ in eqn 21). If the electron is excited to an orbital with $l = 1$, it has orbital angular momentum and can give rise to a magnetic field that interacts with its spin. In this configuration the electron can have $j = \frac{3}{2}$ or $j = \frac{1}{2}$, and the energies are

$$E_{\frac{3}{2}} = \frac{1}{2}hcA\{\frac{3}{2} \times \frac{5}{2} - 1 \times 2 - \frac{1}{2} \times \frac{3}{2}\} = \frac{1}{2}hcA$$
$$E_{\frac{1}{2}} = \frac{1}{2}hcA\{\frac{1}{2} \times \frac{3}{2} - 1 \times 2 - \frac{1}{2} \times \frac{3}{2}\} = -hcA$$

The corresponding energies are shown in Fig. 13.29.

The strength of the spin–orbit coupling depends on the nuclear charge. We can understand why this is so by imagining that we are riding on the orbiting electron. We then see a charged nucleus apparently orbiting around us, like the Sun rising and setting, and as a result we find ourselves at the centre of a ring of current. The greater the nuclear charge the greater this current, and therefore the stronger the magnetic field we detect. Since the spin magnetic moment of the electron interacts with this orbital magnetic field, the greater the nuclear charge, the stronger the spin–orbit interaction. The coupling increases sharply with atomic number (as Z^4) and, whereas it is only small in H (giving rise to shifts of energy levels of no more than about $0.4\ \text{cm}^{-1}$), in heavy atoms like Pb it is very large (giving shifts of the order of thousands of cm^{-1}).

Fine structure

When an electronically excited alkali metal atom undergoes a transition and the *p* electron falls into a lower *s* orbital, two spectral lines are observed, depending on which of the two levels of the $[X]p^1$ configuration was occupied initially. The two lines are an example of **fine structure** in a spectrum. Fine structure can be clearly seen in the emission spectrum from sodium vapour excited by an electric discharge (for example, in one kind of street lighting). The yellow line at 589 nm ($17\,000\ \text{cm}^{-1}$) is actually a doublet composed of one line at 589.76 nm ($16\,956\ \text{cm}^{-1}$) and another at 589.16 nm ($16\,973\ \text{cm}^{-1}$). The transitions (Fig. 13.30) are from the $j = \frac{3}{2}$ and $\frac{1}{2}$ levels of the $[\text{Ne}]3p^1$ configuration to the ground configuration $[\text{Ne}]3s^1$. Therefore, in Na, the spin–orbit coupling affects the energies by about $17\ \text{cm}^{-1}$.

Energy

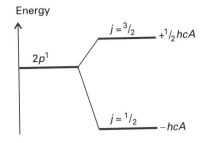

13.29 The levels of a ^2P term arising from spin–orbit coupling. Note that the low-j level lies below the high-j level.

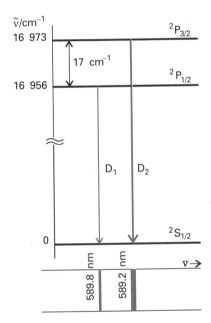

13.30 The energy-level diagram for the formation of the sodium D lines. The splitting of the spectral lines (by 17 cm^{-1}) reflects the splitting of the levels of the ^2P term.

Example 13.7 *Analysing a spectrum for the spin–orbit coupling constant*

The origin of the D lines in the spectrum of atomic sodium is shown in Fig. 13.30. They lie at 16 956.2 cm^{-1} and 16 973.4 cm^{-1}. Calculate the spin–orbit coupling constant for the upper configuration of the Na atom.

Method. We see from Fig. 13.30 that the splitting of the lines is equal to the energy separation of the $j = \frac{3}{2}$ and $\frac{1}{2}$ levels of the excited configuration, which can be expressed in terms of A by using eqn 21. Therefore, we set the observed splitting equal to the energy separation calculated from eqn 21 and solve for A.

Answer. The two levels are split by

$$\Delta \tilde{\nu} = \tfrac{1}{2}A\{\tfrac{3}{2}(\tfrac{3}{2} + 1) - \tfrac{1}{2}(\tfrac{1}{2} + 1)\} = \tfrac{3}{2}A$$

The experimental value is 17.2 cm^{-1}; therefore

$$A = \tfrac{2}{3} \times 17.2 \text{ cm}^{-1} = 11.5 \text{ cm}^{-1}$$

Comment. The same calculation repeated for the other alkali metal atoms gives Li: 0.23 cm^{-1}; K: 38.5 cm^{-1}; Rb: 158 cm^{-1}; Cs: 370 cm^{-1}.

Exercise E13.7. The configuration $\ldots 4p^65d^1$ of rubidium has two levels at 25 700.56 cm^{-1} and 25 703.52 cm^{-1} above the ground configuration. What is the spin–orbit coupling constant in this excited state?

[1.18 cm^{-1}]

13.8 Term symbols and selection rules

We have used expressions such as 'the $j = \frac{3}{2}$ level of a configuration'. A **term symbol**, which is a symbol looking like $^2P_{\frac{3}{2}}$ or $^3D_2$, conveys this information much more succinctly. The convention of using lower-case letters to label orbitals and upper-case letters to label overall states applies throughout spectroscopy, not just to atoms.

A term symbol gives three pieces of information:

(1) The letter (e.g. P or D in the examples) indicates the **total orbital angular momentum quantum number** L.

(2) The left superscript in the term symbol (e.g. the 2 in $^2P_{\frac{3}{2}}$) gives the **multiplicity** of the term.

(3) The right subscript on the term symbol (e.g. the $\frac{3}{2}$ in $^2P_{\frac{3}{2}}$) is the value of the **total angular momentum quantum number** J.

We shall now say what each of these statements means; the contributions to the energies which we are about to discuss are summarized in Fig. 13.31.

The total orbital angular momentum

When several electrons are present, it is necessary to judge how their individual orbital angular momenta add together or oppose each other.

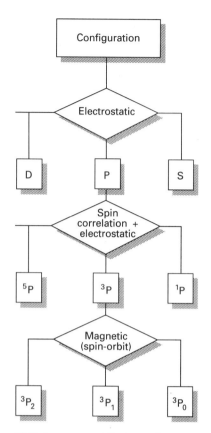

13.31 A summary of the types of interaction that are responsible for the various kinds of splitting of energy levels in atoms. For light atoms, magnetic interactions are small, but in heavy atoms they may dominate the electrostatic (charge–charge) interactions.

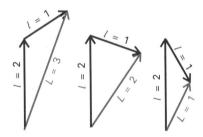

13.32 The total orbital angular momenta of a p electron and a d electron correspond to $l = 3$, 2 and 1 and reflect the different relative orientations of the two momenta.

The total orbital angular momentum quantum number[3] L (a non-negative integer) is obtained by coupling the individual orbital angular momenta using the **Clebsch–Gordan series**:

$$L = l_1 + l_2, l_1 + l_2 - 1, \ldots, |l_1 - l_2| \qquad (22)$$

The maximum value $L = l_1 + l_2$ is obtained when the two orbital angular momenta are in the same direction; the lowest value $|l_1 - l_2|$ is obtained when they are in opposite directions. The intermediate values represent possible intermediate relative orientations of the two momenta (Fig. 13.32). For two p electrons (for which $l_1 = l_2 = 1$), $L = 2$, 1, 0. The code for converting the value of L into a letter is the same as for the $s, p, d, f, \ldots$ designation of orbitals, but uses upper-case letters:

$$\begin{aligned} L{:}\ &0\ \ 1\ \ 2\ \ 3\ \ 4\ \ \ldots \\ &S\ \ P\ \ D\ \ F\ \ G\ \ldots \end{aligned} \qquad (23)$$

Thus, a p^2 configuration can give rise to D, P, and S terms. The terms differ in energy on account of the different spatial distribution of the electrons and the consequent differences in repulsion between them.

A closed shell has zero orbital angular momentum because all the individual orbital angular momenta sum to zero. Therefore, when working out term symbols, we need consider only the electrons of the unfilled shell. In the case of a single electron outside a closed shell, the value of L is the same as the value of l, so the configuration $[\text{Ne}]3s^1$ has only an S term.

Example 13.8 *Deriving the total orbital angular momentum of a configuration*

Find the terms that can arise from the configurations (a) d^2, (b) p^3.

Method. We use the Clebsch–Gordan series and begin by finding the minimum value of L (so that we know where the series terminates). When there are more than two electrons to couple together we use two series in succession: first we couple two electrons, and then we couple the third to each combined state, and so on.

Answer. (a) Minimum value: $|l_1 - l_2| = |2 - 2| = 0$. Therefore,

$$L = 2 + 2, 2 + 2 - 1, \ldots, 0 = 4, 3, 2, 1, 0$$

corresponding to G, F, D, P, S terms, respectively.
(b) First coupling: Minimum value: $|1 - 1| = 0$. Therefore,

$$L' = 1 + 1, 1 + 1 - 1, \ldots, 0 = 2, 1, 0$$

Now couple l_3 with $L' = 2$, to give $L = 3$, 2, 1; with $L' = 1$, to give $L = 2$, 1, 0; and with $L' = 0$, to give $L = 1$. The overall result is

$$L = 3, 2, 2, 1, 1, 1, 0$$

giving one F, two D, three P, and one S terms.

---

3 The total orbital angular momentum is quantized and, like the other momenta we have encountered, its magnitude is given by the value of L, and is $\{L(L+1)\}^{\frac{1}{2}}\hbar$. It also has $2L + 1$ orientations distinguished by the quantum number M_L, which can take the values $L, L - 1, \ldots, -L$. Similar remarks apply to the total spin S (its orientations are denoted M_S), and the total angular momentum J (with its orientations M_J).

Exercise E13.8. Repeat the question for the configurations (a) f^1d^1 and (b) d^3. [(a) H, G, F, D, P; (b) I, 2H, 3G, 4F, 5D, 3P, S]

___

The multiplicity

When there are several electrons to be taken into account, we must assess their **total spin angular momentum quantum number** S (a non-negative integer or half-integer). Once again, we use the Clebsch–Gordan series to decide on the value of S, noting that each electron has $s = \frac{1}{2}$, which gives (Fig. 13.33)

$$S = 1, 0$$

If there are three electrons, the total spin angular momentum is obtained by coupling the third spin to each of the values of S for the first two spins:

$$S = \tfrac{3}{2}, \tfrac{1}{2} \quad \text{and} \quad S = \tfrac{1}{2}$$

The **multiplicity** of a term is the value of $2S + 1$. When $S = 0$ (as for a closed shell) the electrons are all paired and there is no net spin; this gives a singlet term, such as ^1S. (Take care not to confuse the italic S of the spin and the roman S of the term symbol.) A single electron has $S = s = \frac{1}{2}$, so a configuration such as $[\text{Ne}]3s^1$ can give rise to a doublet term ^2S. The configuration $[\text{Ne}]3p^1$ likewise is a doublet ^2P. When there are two unpaired electrons, $S = 1$ so $2S + 1 = 3$, giving a triplet term, such as ^3D. We discussed the relative energies of singlets and triplets in Section 13.4 and saw that their energies differ on account of the different effects of spin correlation.

The total angular momentum

As we have seen, the quantum number j tells us the relative orientation of the spin and orbital angular momenta of a single electron. The total angular momentum quantum number J (a non-negative integer or half-integer) does the same for several electrons. If there is a single electron outside a closed shell, then $J = j$, with j either $l + \frac{1}{2}$ or $|l - \frac{1}{2}|$. The $[\text{Ne}]3s^1$ configuration has $j = \frac{1}{2}$ (because $l = 0$ and $s = \frac{1}{2}$), so the ^2S term has a single level, $^2S_{\frac{1}{2}}$. The $[\text{Ne}]3p^1$ configuration has $l = 1$, and therefore $j = \frac{3}{2}$ and $\frac{1}{2}$; the ^2P term therefore has two levels, $^2P_{\frac{3}{2}}$ and $^2P_{\frac{1}{2}}$, and these lie at different energies on account of the magnetic spin–orbit interaction.

If there are several electrons outside a closed shell we have to consider the coupling of all the spins and all the orbital angular momenta. This complicated problem can be simplified when the spin–orbit coupling is weak (for atoms of low atomic number) for then we can use the **Russell–Saunders coupling scheme**. This scheme is based on the view that, if spin–orbit coupling is weak, then it is effective only when all the orbital momenta are operating cooperatively. We therefore imagine that all the orbital angular momenta of the electrons couple to give a total L, and that all the spins are similarly coupled to give a total S. Only at this stage do we imagine the two kinds of momenta coupling through the spin–orbit interaction to give a total J. The permitted values of J are

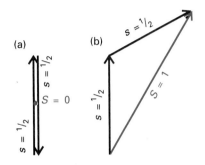

13.33 For two electrons (which have $s = \frac{1}{2}$), only two total spin states are permitted ($S = 0, 1$). The state with $S = 0$ can have only one value of M_S ($M_S = 0$) and is a singlet; the state with $S = 1$ can have any of three values of M_S (+1, 0, −1) and is a triplet. The vector representations of the singlet and triplet states are shown in Figs. 13.18 and 13.24, respectively.

given by the Clebsch–Gordan series

$$J = L + S, L + S - 1, \ldots, |L - S| \qquad (24)$$

For example, in the case of the ^3D term of the configuration $[He]2p^13p^1$, the permitted values of J are 3, 2, 1 (because ^3D has $L = 2$ and $S = 1$), so the term has three levels, $^3D_3$, $^3D_2$, and $^3D_1$.

When $L \geqslant S$, the multiplicity is equal to the number of levels. For example, a ^2P term has the two levels $^2P_{\frac{3}{2}}$ and $^2P_{\frac{1}{2}}$, and ^3D has the three levels $^3D_3$, $^3D_2$, and $^3D_1$. However, this is not the case when $L < S$: the term ^2S, for example, has only the one level $^2S_{\frac{1}{2}}$.

Example 13.9 *Deriving term symbols*

Write the term symbols for the ground configuration of Na and F, and the excited configuration $1s^22s^22p^13p^1$ of C.

Method. Begin by writing the configurations, but ignore inner closed shells. Then couple the orbital momenta to find L and the spins to find S. Next, couple L and S to find J. Finally, express the term as $^{2S+1}\{L\}_J$, where $\{L\}$ is the appropriate letter. For F, for which the valence configuration is $2p^5$, treat the single gap in the closed-shell $2p^6$ configuration as a single particle.

Answer. For Na the configuration is $[Ne]3s^1$, and we consider the single $3s$ electron. Since $L = l = 0$ and $S = s = \frac{1}{2}$, it is possible for $J = j + s = \frac{1}{2}$ only. Hence the term symbol is $^2S_{\frac{1}{2}}$. For F the configuration is $[He]2s^22p^5$, which we can treat as $[Ne]2p^{-1}$ (where the notation $2p^{-1}$ signifies the absence of a $2p$ electron). Hence $L = 1$, and $S = s = \frac{1}{2}$. Two values of $J = j$ are allowed: $J = \frac{3}{2}, \frac{1}{2}$. Hence the term symbols for the two levels are $^2P_{\frac{3}{2}}$, $^2P_{\frac{1}{2}}$. For C the configuration is effectively $2p^13p^1$. This is a two-electron problem, and $l_1 = l_2 = 1$, $s_1 = s_2 = \frac{1}{2}$. It follows that $L = 2, 1, 0$ and $S = 1, 0$. The terms are therefore ^3D and ^1D, ^3P and ^1P, and ^3S and ^1S. For ^3D, $L = 2$ and $S = 1$; hence $J = 3, 2, 1$ and the levels are $^3D_3$, $^3D_2$, and $^3D_1$. For ^1D, $L = 2$ and $S = 0$, so that the single level is $^1D_2$. The triplet of levels of ^3P is $^3P_2$, $^3P_1$, and $^3P_0$, and the singlet is $^1P_1$. For the ^3S term there is only a single level, $^3S_1$ (because $J = 1$ only), and the singlet term is $^1S_0$.

Comment. The reason why we have treated an excited configuration of carbon is that in the ground configuration, $2p^2$, the Pauli principle forbids some terms, and deciding which survive (^1D, ^3P, ^1S) is quite complicated. That is, there is a distinction between 'equivalent electrons', which are electrons that occupy the same orbitals, and 'inequivalent electrons', which are electrons that occupy different orbitals.

Exercise E13.9. Write down the terms arising from the configurations (a) $2s^12p^1$, (b) $2p^13d^1$.

[(a) $^3P_2$, $^3P_1$, $^3P_0$, $^1P_1$; (b) $^3F_4$, $^3F_3$, $^3F_2$, $^1F_3$, $^3D_3$, $^3D_2$, $^3D_1$, $^1D_2$, $^3P_2$, $^3P_1$, $^3P_0$, $^1P_1$]

Russell–Saunders coupling fails when the spin–orbit coupling is large (in heavy atoms). In that case, the individual spin and orbital momenta of the electrons are occupied into individual j values; then these momenta are combined into a grand total J. This scheme is called **jj-coupling**. For example, in a p^2 configuration, the individual values of j are $\frac{3}{2}$ and $\frac{1}{2}$ for each electron. If the spin and the orbital angular momenta of each electron are coupled together strongly, it is best to consider each electron as a particle with angular momentum $j = \frac{3}{2}$ or $\frac{1}{2}$. These individual total momenta then couple as follows:

$$j_1 = \tfrac{3}{2} \text{ and } j_2 = \tfrac{3}{2} \quad \text{give } J = 3, 2, 1, 0$$
$$j_1 = \tfrac{3}{2} \text{ and } j_2 = \tfrac{1}{2} \quad \text{give } J = 2, 1$$
$$j_1 = \tfrac{1}{2} \text{ and } j_2 = \tfrac{3}{2} \quad \text{give } J = 2, 1$$
$$j_1 = \tfrac{1}{2} \text{ and } j_2 = \tfrac{1}{2} \quad \text{give } J = 1, 0$$

For heavy atoms, in which jj-coupling is appropriate, it is best to discuss their energies using these quantum numbers.

Although jj-coupling should be used for assessing the energies of heavy atoms, the term symbols derived from Russell–Saunders coupling can still be used as labels. To see why this procedure is valid, we need to examine how the energies of the atomic states change as the spin–orbit coupling increases in strength. Such a **correlation diagram** is shown in Fig. 13.34. It shows that there is a correspondence between the low spin–orbit coupling (Russell–Saunders coupling) and high spin–orbit coupling (jj-coupling) schemes, so the labels derived by using the Russell–Saunders scheme can be used to label the states of the jj-coupling scheme.

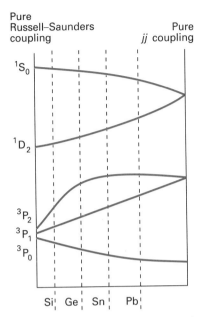

13.34 The correlation diagram for some of the states of a two-electron system. All atoms lie between the two extremes, but the heavier the atom, the closer it lies to the pure jj-coupling case.

Selection rules

Any state of the atom, and any spectral transition, can be specified using term symbols. For example, the transitions giving rise to the yellow sodium doublet (which were shown in Fig. 13.30) are

$$3p^1\,{}^2P_{\frac{3}{2}} \rightarrow 3s^1\,{}^2S_{\frac{1}{2}} \quad \text{and} \quad 3p^1\,{}^2P_{\frac{1}{2}} \rightarrow 3s^1\,{}^2S_{\frac{1}{2}}$$

Note that, by convention, the upper term precedes the lower. The corresponding absorptions would therefore be denoted

$${}^3P_{\frac{3}{2}} \leftarrow {}^2S_{\frac{1}{2}} \quad \text{and} \quad {}^2P_{\frac{1}{2}} \leftarrow {}^2S_{\frac{1}{2}}$$

We have seen that selection rules arise from the conservation of angular momentum during a transition and from the fact that a photon has a spin of 1. They can therefore be expressed in terms of the term symbols, because the latter carry information about angular momentum. A detailed analysis leads to the following rules:

$$\Delta S = 0 \qquad \Delta L = 0, \pm 1 \qquad \text{with } \Delta l = \pm 1$$
$$\Delta J = 0, \pm 1 \qquad \text{but } J = 0 \not\leftrightarrow J = 0 \tag{25}$$

The rule about ΔS (no change of overall spin) stems from the fact that the light does not affect the spin directly. The rules about ΔL and Δl express the fact that the orbital angular momentum of an individual electron must change (so $\Delta l = \pm 1$), but whether or not this results in an overall change of orbital momentum depends on the coupling.

The selection rules given above apply when Russell–Saunders coupling is valid (in light atoms). If we insist on labelling the terms of heavy atoms with symbols like ^3D, then we shall find that the selection rules progressively fail as the atomic number increases because the quantum numbers S and L become ill defined as jj-coupling becomes more appropriate. As explained at the end of the last section, Russell–Saunders term symbols are only a convenient way of labelling the terms of heavy atoms: they do not bear any direct relation to the actual angular momenta of the electrons in a heavy atom. For this reason, transitions between singlet and triplet states (for which $\Delta S = \pm 1$), while forbidden in light atoms, are allowed in heavy atoms.

13.9 The effect of magnetic fields

Orbital and spin angular momenta give rise to magnetic moments (recall the evidence provided for electron spin by the Stern–Gerlach experiment, Section 12.8). It can be expected that the application of a magnetic field should modify an atom's spectrum. We shall first establish how the energies of an atom depend on the strength of an external field and then see how the spectrum is affected.

The magnetic moment of an electron

The orbital angular momentum of an electron around the z-axis (which we shall now take as the direction of the applied field) is $m_l\hbar$. Because the component of magnetic moment on the z-axis, μ_z, is proportional to the angular momentum around that axis, we can write

$$\mu_z = \gamma_e m_l \hbar \tag{26a}$$

where γ_e is a constant called the **magnetogyric ratio** of the electron. If the magnetic moment is treated as arising from the circulation of an electron of charge $-e$, then standard electromagnetic theory gives

$$\gamma_e = -\frac{e}{2m_e} \tag{26b}$$

The negative sign (arising from the sign of the electron's charge) shows that the orbital magnetic moment of the electron is antiparallel to its orbital angular momentum (as was depicted in Fig. 13.26). It follows that the possible values of μ_z are

$$\mu_z = -\frac{e}{2m_e} \times m_l \hbar = -\mu_B m_l \tag{27a}$$

where the **Bohr magneton** μ_B is

$$\mu_B = \frac{e\hbar}{2m_e} \tag{27b}$$

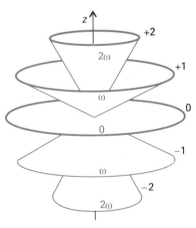

13.35 The different energies of the m_l states in a magnetic field are represented by different rates of precession of the vectors representing the angular momentum.

Its numerical value is $9.274 \times 10^{-24}\,\mathrm{J\,T^{-1}}$. The Bohr magneton is often regarded as the fundamental quantum of magnetic moment.

The energy of a magnetic moment in a magnetic field B is[4]

$$E = -\mu_z B \tag{28a}$$

Therefore, in the presence of a magnetic field, an electron in a state with quantum number m_l has an additional contribution to its energy given by

$$E = \mu_B m_l B \tag{28b}$$

The same expression, but with m_l replaced by M_L, applies when the orbital magnetic moment arises from several electrons.

A p electron has $l = 1$ and $m_l = 0, \pm 1$. In the absence of a magnetic field, these three states are degenerate. When a field is present, the degeneracy is removed: the state with $m_l = +1$ moves up in energy by an amount $\mu_B B$, the state with $m_l = 0$ is unchanged, and the state with $m_l = -1$ moves down by an amount $\mu_B B$:

$$E_{+1} = \mu_B B \qquad E_0 = 0 \qquad E_{-1} = -\mu_B B$$

The different energies arising from an interaction with an external field are sometimes represented on the vector model by picturing the vectors as **precessing**, or sweeping round their cones (Fig. 13.35) with the rate of precession proportional to the energy of the state.

The spin magnetic moment of an electron is also proportional to its angular momentum. However, it is not given by $\gamma_e m_s \hbar$ but by about twice this value:

$$\mu_z = g_e \gamma_e m_s \hbar \qquad g_e = 2.0023 \tag{29}$$

The extra factor g_e is called the **g-factor of the electron**. The factor 2 (as distinct from 2.0023) is derived from the Dirac equation; the additional 0.0023 arises from interactions of the electron with the electromagnetic fluctuations in the vacuum that surrounds the electron. The energy of an electron in a state m_s in a magnetic field B is

$$E = -g_e \gamma_e m_s \hbar B = g_e \mu_B m_s B \tag{30}$$

The same expression, but with m_s replaced by M_S, applies to the magnetic moment arising from the spin of several electrons.

The Zeeman effect

The **Zeeman effect** is the modification of an atomic spectrum by the application of a strong magnetic field. In particular, the **normal Zeeman effect** is the observation of three lines in the spectrum where, in the absence of the field, there is only one (Fig. 13.36). The splitting is in fact extremely small: a field of 2 T (20 kG) is needed to produce a splitting of about 1 cm^{-1}, which should be compared with typical optical transition wavenumbers of $20\,000 \text{ cm}^{-1}$ and more.

Much more common than the normal Zeeman effect is the **anomalous Zeeman effect**, in which the original line splits into more than three components. The origin of this complexity is the anomalous magnetic moment of electron spin, which results in a more complicated splitting pattern.

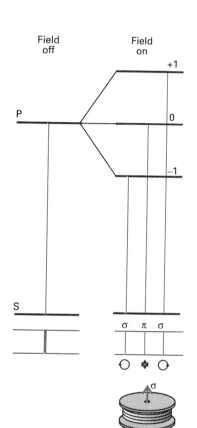

13.36 The normal Zeeman effect. On the left, when the field is off, a single spectral line is observed. When the field is on, the line splits into three, with different polarizations. The circularly polarized lines are called the 'σ-lines'; the plane-polarized lines are called 'π-lines'. Which line is observed depends on the orientation of the observer.

---

4 This is a result from standard magnetic theory. B is actually the magnetic induction, and is measured in tesla T; the unit gauss, G, is also still widely used: $1 \text{ T} = 10^4 \text{ G}$.

CHECK LIST OF KEY IDEAS

1 The **spectrum of atomic hydrogen**, and the classification of the **spectral lines** into **series** (Section 13.1 and eqn 1).

2 The **Ritz combination principle** (eqn 3) for expressing transitions using **terms** and the **Bohr frequency condition** (eqn 4).

3 The Schrödinger equation for **hydrogenic atoms**, the separation of the internal motions, and the separation of the **angular wavefunction** from the **radial wavefunction**.

4 The **energy levels** of hydrogenic atoms (eqn 12) in terms of the **principal quantum number**.

5 The measurement of the **ionization energy** of an atom from the **series limit** (Example 13.2).

6 The **shells** and **subshells** of atoms and the notion of **atomic orbitals** (Section 13.2).

7 The representation of atomic orbitals by their **boundary surfaces** and the **radial distribution function** (Section 13.2).

8 The form of *s* **orbitals**, *p* **orbitals**, and *d* **orbitals** (Section 13.2).

9 The **selection rules** of spectroscopic transitions (Section 13.3) and the representation of transitions on a **Grotrian diagram** (Fig. 13.17).

10 The **orbital approximation** for many-electron atoms and the **Pauli exclusion principle** (Section 13.4).

11 The concept of a **closed shell** and of **paired spins** (Section 13.4).

12 The effect of **penetration** and **shielding** on the **effective atomic number** of an atom, and the relative order of orbitals in a many-electron atom (Section 13.4).

13 The **building-up principle** for predicting the ground-state configuration of many-electron atoms (Section 13.4).

14 **Hund's rule** and the role of **spin correlation** (Section 13.4).

15 The periodicity of atomic properties, particularly the **ionization energy** (Section 13.4).

16 The concept of **self-consistent field** orbitals of many-electron atoms (Section 13.5) and the **Hartree–Fock procedure**.

17 **Singlet and triplet states** of atoms (Section 13.6) and the origin and effects of **spin–orbit coupling** and the **fine structure** of spectra (Section 13.7).

18 The use of the **Clebsch–Gordan series** to deduce the total angular momentum of a combined system (Section 13.8), and the use of **term symbols** to specify the state of an atom.

19 The **Russell–Saunders** and *jj*-**coupling** schemes for light and heavy atoms, and the **selection rules** for transitions when Russell–Saunders coupling is appropriate (Section 13.8).

20 The **magnetogyric ratio** of an electron (eqn 26), the **Bohr magneton** (eqn 27), and the *g*-**factor** of an electron (eqn 29).

21 The energy of an electron in a magnetic field (eqn 30) and the normal and anomalous **Zeeman effects** (Section 13.9).

EXERCISES

13.1. When 58.4-nm ultraviolet radiation from a helium lamp is directed on to a sample of krypton, electrons are ejected at $1.59 \times 10^6 \, \text{m s}^{-1}$. Calculate the ionization energy of krypton.

13.2. By differentiation of the 2*s* radial wavefunction, show that it has a minimum in its amplitude and locate it.

13.3. Locate the radial nodes in the 3*s* orbital of an H atom.

13.4. The radial wavefunction for the ground state of a hydrogen atom is Ne^{-r/a_0}. Determine the normalization constant for this wavefunction.

13.5. Calculate the average kinetic and potential energies of an electron in the ground state of a hydrogen atom.

13.6. Write down the expression for the radial distribution function of a 2*s* electron in a hydrogenic atom and determine the radius at which the electron is most likely to be found.

13.7. What is the orbital angular momentum of an electron in the orbitals (a) $1s$, (b) $3s$, (c) $3d$, (d) $2p$, (e) $3p$? Give the numbers of angular and radial nodes in each case.

13.8. Calculate the permitted values of j for (a) a d electron, (b) an f electron.

13.9. An electron in two different states of an atom is known to have $j = \frac{3}{2}$ and $\frac{1}{2}$. What is its orbital angular momentum quantum number in each case?

13.10. What are the allowed total angular momentum quantum numbers of a composite system in which $j_1 = 5$ and $j_2 = 3$?

13.11. State the orbital degeneracy of the levels in the hydrogen atom that have energy (a) $-\mathfrak{R}_H$, (b) $-\frac{1}{9}\mathfrak{R}_H$, and (c) $-\mathfrak{R}_H/25$.

13.12. What information does the term symbol $^1D_2$ provide about the angular momentum of an atom?

13.13. At what radius does the probability of finding an electron at a point in the H atom fall to 50 per cent of its maximum value?

13.14. At what radius in the H atom does the radial distribution function of the ground state have (a) 50 per cent, (b) 75 per cent of its maximum value?

13.15. Which of the following transitions are allowed in the normal electronic emission spectrum of an atom: (a) $2s \rightarrow 1s$, (b) $2p \rightarrow 1s$, (c) $3d \rightarrow 2p$, (d) $5d \rightarrow 2s$, (e) $5p \rightarrow 3s$?

13.16. How many electrons can occupy the following subshells: (a) $1s$, (b) $3p$, (c) $3d$, and (d) $6g$?

13.17. (a) Write the ground-state electronic configuration of the Ni^{2+} ion. (b) What are the possible values of the total spin quantum numbers S and M_S for this ion?

13.18. Suppose that an atom has (a) two, (b) three, (c) four electrons in different orbitals. What are the possible values of the total spin quantum number S? What is the multiplicity in each case?

13.19. What atomic terms are possible for the electron configuration ns^1nd^1? Which set of terms is likely to lie lowest in energy?

13.20. What values of J may occur in the terms (a) ^1S, (b) ^2P, (c) ^3P, (d) ^3D, (e) ^4D? How many states (distinguished by the quantum number M_J) belong to each level?

13.21. Give the possible term symbols for (a) Li $[He]2s^1$, (b) Na $[Ne]3p^1$, (c) Sc $[Ar]3d^14s^2$, and (d) Br $[Ar]3d^{10}4s^24p^5$.

13.22. The energy of an electron increases by 2.23×10^{-22} J when a magnetic field of 12.0 T is applied. What is the value of m_l for this electron?

13.23. Calculate the magnetic induction B required to produce a splitting of $1.0\ cm^{-1}$ between the states of a P term.

PROBLEMS
Numerical problems

13.1. The 'Humphreys series' is another group of lines in the spectrum of atomic hydrogen. It begins at 12 368 nm and has been traced to 3281.4 nm. What are the transitions involved? What are the wavelengths of the intermediate transitions?

13.2. A series of lines in the spectrum of atomic hydrogen lies at 656.46 nm, 486.27 nm, 434.17 nm, and 410.29 nm. What is the wavelength of the next line in the series? What is the ionization energy of the atom when it is in the lower state of the transitions?

13.3. The Li^{2+} ion is hydrogenic and has a Lyman series at 740 747 cm^{-1}, 877 924 cm^{-1}, 925 933 cm^{-1}, and beyond. Show that the energy levels are of the form $-\mathfrak{R}/n^2$ and find the value of $\mathfrak{R}$ for this ion. Go on to predict the wavenumbers of the two longest-wavelength transitions of the Balmer series of the ion and find the ionization energy of the ion.

13.4. A series of lines in the spectrum of neutral Li atoms arise from combinations of $1s^22p^1\ ^2P$ with $1s^2nd^1\ ^2D$ and occur at 610.36 nm, 460.29 nm, and 413.23 nm. The d orbitals are hydrogenic. It is known that the ^2P term lies at 670.78 nm above the ground state, which is $1s^22s^1\ ^2S$. Calculate the ionization energy of the ground-state atom.

13.5. The characteristic emission from K atoms when heated is purple and lies at 770 nm. On close inspection, the line is found to have two closely spaced components, one at 766.70 nm and the other at 770.11 nm. Account for this observation, and deduce what information you can.

13.6. Calculate the mass of the deuteron given that the first line in the Lyman series of H lies at 82 259.098 cm^{-1}, whereas that of D lies at 82 281.476 cm^{-1}. Calculate the ratio of the ionization energies of D and H.

13.7. Positronium consists of an electron and a positron (same mass, opposite charge) orbiting round their common centre-of-mass. The broad features of the spectrum are therefore expected to be hydrogen-like, the differences arising largely from the mass differences. Predict the wavenumbers of the first three lines of the Balmer series of positronium. What is the binding energy of the ground state of positronium?

13.8. In 1976 it was mistakenly believed that the first of the 'superheavy' elements had been discovered in a sample of mica. Its atomic number was believed to be 126. What is the most probable distance of the innermost electrons from the nucleus of an atom of this element? (In such elements, relativistic effects are very important, but ignore them here.)

Theoretical problems

13.9. Is an electron further from the nucleus on average when it is in a $2s$ orbital or a $2p$ orbital of a hydrogenic atom?

13.10. What is the most probable point (not radius) that a $2p$ electron will be found in the hydrogen atom?

13.11. Show by explicit integration that (a) hydrogenic $1s$ and $2s$ orbitals, (b) $2p_x$ and $2p_y$ orbitals are mutually orthogonal.

13.12. Determine whether the p_x and p_y orbitals are eigenfunctions of l_z. If not, does a linear combination exist that is an eigenfunction of l_z?

13.13. Show that l_z and l^2 both commute with the Hamiltonian for the H atom. What is the significance of this result?

13.14. The 'size' of an atom is sometimes considered to be measured by the radius of a sphere that contains 90 per cent of the charge density of the electrons in the outermost occupied orbital. Calculate the 'size' of a hydrogen atom in its ground state according to this definition.

13.15. One of the most famous of the obsolete theories of the hydrogen atom was proposed by Bohr. It has been replaced by quantum mechanics, but, by a remarkable coincidence (not the first one where the Coulomb potential is concerned), the energies it predicts agree exactly with those obtained from the Schrödinger equation. In the Bohr atom, an electron travels in a circle around the nucleus. The Coulombic force of attraction ($Ze^2/4\pi\varepsilon_0 r^2$) is balanced by the centrifugal effect of the orbital motion. Bohr proposed that the angular momentum is limited to integral values of $\hbar$. When the two forces are balanced, the atom remains in a 'stationary state' until it makes a spectral transition. Calculate the energies of a hydrogenic atom using the Bohr model.

13.16. The Bohr model of the atom is specified in Problem 13.15. What features of it are untenable according to quantum mechanics? How does the Bohr ground state differ from the actual ground state? Is there an experimental distinction between the Bohr and quantum mechanical models of the ground state?

13.17. Atomic units of length and energy may be based on the properties of a particular atom. The usual choice is that of a hydrogen atom, with the unit of length being the Bohr radius a_0 and the unit of energy being the energy of the $1s$ orbital. If the positronium atom (e^+, e^-) were used instead, with analogous definitions of units of length and energy, what would be the relation between these two sets of atomic units?

14

Molecular structure

The concepts developed in Chapter 13, particularly those of orbitals, can be extended to a description of the electronic structures of molecules. There are two principal quantum mechanical theories of molecular electronic structure. In valence-bond theory the starting point is the concept of the shared electron pair. We see how to write the wavefunction for such a pair, and how it may be extended to account for the structures of a wide variety of molecules. The theory introduces the concepts of σ and π bonds, promotion, and hybridization that are used widely in chemistry. In molecular orbital theory (with which the bulk of the chapter is concerned) the concept of atomic orbital is extended to that of molecular orbital, which is a wavefunction that spreads over all the atoms in a molecule. We see how to construct such molecular orbitals, how to interpret them, and how to use a version of the building-up principle to predict the electron configurations of molecules.

Molecular orbital theory can be used to discuss the properties of a wide variety of substances in a uniform manner. We see how to apply it to polyatomic molecules, where the theory exhibits some of the features that account for their shapes, and to conjugated molecules. The latter can be described quantitatively by invoking a simple approximation scheme. The extreme versions of such electronically delocalized systems are solids, in which electron delocalization is virtually infinite and results in the properties of electrical conduction and semiconduction.

In this chapter we consider the origin of the strengths, numbers, and three-dimensional arrangement of chemical bonds between atoms. The quantum mechanical description of chemical bonding has become highly developed through the use of computers and it is now possible to consider the structures of molecules of almost any complexity. We shall concentrate on the quantum mechanical description of the **covalent**

bond, which was identified by G. N. Lewis (in 1916, before quantum mechanics was established fully) as an electron pair shared between two neighbouring atoms. We shall see, however, that the other principal type of bond, an **ionic bond**, in which the cohesion arises from the Coulombic attraction between ions of opposite charge, is also captured as a limiting case of a covalent bond between dissimilar atoms.

In Chapter 13 we took the hydrogen atom as the primitive species for discussing atomic structure, and based our discussion of complex atoms on what we learned from it. In this chapter we use the simplest molecules of all, the hydrogen molecule-ion H_2^+ and the hydrogen molecule H_2, to introduce the essential features of bonding, and then use them as guides to the structures of more complex systems.

All theories of molecular structure make the same simplification at the outset. Whereas the Schrödinger equation for a hydrogen atom can be solved exactly, an exact solution is not possible for any molecule, even H_2^+, because the simplest molecule consists of three particles (two nuclei and one electron). The **Born–Oppenheimer approximation** is therefore adopted, in which it is supposed that the nuclei, being so much heavier than an electron, move relatively slowly and may be treated as stationary while the electrons move relative to them. We can therefore think of the nuclei as being fixed at an arbitrary separation R, and then solve the Schrödinger equation for the wavefunction of the electrons alone. The approximation is quite good for ground-state molecules, for calculations suggest that the nuclei in H_2 move through only about 1 pm while the electron speeds through 1000 pm, so that the error of assuming that the nuclei are stationary is small. Exceptions to its validity include certain excited states of polyatomic molecules and the ground states of cations; both types of species are important when considering photoelectron spectroscopy (Section 17.8) and mass spectrometry.

The Born–Oppenheimer approximation allows us to select an internuclear separation, and (in principle) to solve the Schrödinger equation for the electrons for that nuclear separation. Then we can choose a different separation and repeat the calculation, and so on. In this way we can explore how the energy of the molecule varies with bond length (and, in more complex molecules, with angles too), and obtain a **molecular potential energy curve**. A typical example of such a curve is illustrated in Fig. 14.1. It is called a *potential* energy curve because the kinetic energy of the nuclei is zero (as they are stationary). Once the curve has been calculated or determined experimentally (by using the spectroscopic techniques described in Chapter 16), we can identify the **equilibrium bond length** (the internuclear separation at the minimum of the curve) and the **bond dissociation energy** D_0, which is closely related to the depth of the minimum below the energy of the infinitely widely separated atoms.[1]

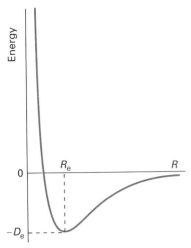

14.1 A molecular potential energy curve. The equilibrium bond length corresponds to the energy minimum.

---

1 The dissociation energy differs from the depth of the well by an energy equal to the zero-point vibrational energy of the bonded atoms. If the depth of the well is denoted D_e and the dissociation energy D_0, then $D_0 = D_e - \frac{1}{2}\hbar\omega$ where ω is the vibrational frequency of the bond.

VALENCE-BOND THEORY

The **valence-bond theory** of bonding describes each electron pair in a molecule by a wavefunction that allows each electron to be found on both atoms joined by the bond. The language we shall introduce in this section, which includes concepts such as spin pairing, σ and π bonds, and hybridization, is widely used throughout chemistry, particularly in the description of the properties and reactions of organic compounds.

14.1 The hydrogen molecule

The simplest molecule with an electron pair bond is H_2 and we shall use it to introduce the basic concepts of the theory.

The spatial wavefunction

The wavefunction for an electron on each of two widely separated H atoms is

$$\psi = \psi_{H1sA}(1)\psi_{H1sB}(2)$$

if electron 1 is on atom A and electron 2 is on atom B. This wavefunction is exact for atoms so widely separated that the electrons do not interact with one another (Section 13.4). When the two atoms are within bonding distance of each other, the electrons interact with one another and the wavefunction written above is only an approximation to the true wavefunction, but it is a reasonable starting point for the description of the bond.

However, there is one major difference that must be taken into account when the atoms are close: it is not possible to know whether it is electron 1 that is on A or electron 2. An equally valid description is therefore

$$\psi = \psi_{H1sA}(2)\psi_{H1sB}(1)$$

in which electron 2 is on A and electron 1 is on B. When two outcomes are equally probable, quantum mechanics instructs us to describe the true state of the system as a linear superposition of the wavefunctions for each possibility (Section 11.6), so a better description of the molecule than either wavefunction alone is

$$\psi = \psi_{H1sA}(1)\psi_{H1sB}(2) \pm \psi_{H1sA}(2)\psi_{H1sB}(1)$$

At this stage, the + and − combinations are equally valid descriptions. However, it turns out that (as shown in the following *Justification*) the combination with lower energy is the one with a + sign, so the **valence-bond wavefunction** of the H_2 molecule is

$$\psi = \psi_{H1sA}(1)\psi_{H1sB}(2) + \psi_{H1sA}(2)\psi_{H1sB}(1) \tag{1}$$

In general, for orbitals that we can symbolize as A and B on atoms A and B, respectively, the valence-bond wavefunction for an A—B bond is

$$\psi = A(1)B(2) + A(2)B(1) \tag{2}$$

with the linear combination taken with the positive sign.

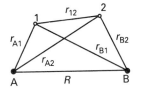

1

JUSTIFICATION

The valence-bond wavefunction for H_2 is an approximate solution of the Schrödinger equation in which the potential energy of the two electrons (**1**) is

$$V = -\frac{e^2}{4\pi\varepsilon_0}\left(\frac{1}{r_{A1}} + \frac{1}{r_{A2}} + \frac{1}{r_{B1}} + \frac{1}{r_{B2}}\right) + \frac{e^2}{4\pi\varepsilon_0 r_{12}}$$

The first four terms (in parentheses) are the attractive contributions from the interaction between the electrons and the nuclei. The remaining term is the repulsive interaction between the two electrons. The energy of the molecule is calculated by evaluating the expectation value

$$E = \frac{\int \psi^* H\psi \, d\tau}{\int \psi^* \psi \, d\tau}$$

using for the Hamiltonian the expression

$$H = -\frac{\hbar^2}{2m_e}\nabla_1^2 - \frac{\hbar^2}{2m_e}\nabla_2^2 + V$$

where V is the potential energy term given above. When the valence-bond expression is used in this expression for the energy, the outcome is

$$E_{\pm} = 2E_H + \frac{J \pm K}{1 \pm S^2} + \frac{e^2}{4\pi\varepsilon_0 R}$$

where the $+$ and $-$ signs match the $+$ or $-$ sign in the linear superposition. In this expression, E_H is the energy of a hydrogen atom, and J and K are collections of integrals over the wavefunctions that represent the interaction of the electrons with the nuclei and the repulsion of each electron by the other. The quantity S is the overlap integral, $\int \psi_A^* \psi_B \, d\tau$, which is discussed in more detail shortly. The final term is the nucleus–nucleus repulsion energy. The integrals J and K are both *negative*, so it follows that the lower energy is achieved with the $+$ sign, and hence with the valence-bond combination given in eqn 1.

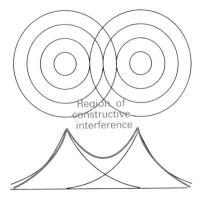

14.2 The orbital overlap responsible for bonding in the hydrogen molecule and the constructive interference in the internuclear region.

The formation of the bond in H_2 can be pictured as the consequence of an electron that is initially on one atom being able to escape to the other atom. As a result there is a high probability that the two electrons that participate in this atom-swapping will be found between the two nuclei and hence will bind them together. More formally, the wave pattern represented by the term $\psi_{H1sA}(1)\psi_{H1sB}(2)$ interferes *constructively* with the wave pattern represented by the contribution $\psi_{H1sA}(2)\psi_{H1sB}(1)$, and there is an enhancement in the value of the wavefunction in the internuclear region (Fig. 14.2). Hence, there is an enhanced probability of finding the electrons in the internuclear region. This merging of two contributing wavefunctions to give a region of enhanced amplitude is called the **overlap** of wavefunctions; it will play a central role throughout the chapter.

The electron distribution we have described is called a **σ bond**. A σ bond has cylindrical symmetry around the internuclear axis and is so called because, when viewed along the internuclear axis, it resembles a pair of electrons in an s orbital (and σ is the Greek equivalent of s). More precisely, the electrons in a σ bond have zero orbital angular momentum about the internuclear axis. (It should be recalled from Section 12.6 that the orbital angular momentum of an electron is related to the number of angular nodes in its wavefunction; there are, however, no angular nodes in the wavefunction of a σ bond, so it has zero orbital angular momentum around the internuclear axis.)

The molecular potential energy curve for H_2 is calculated by changing the internuclear separation R and evaluating the expectation value of the energy at each selected separation. The resulting graph is shown in Fig. 14.3. The energy falls below that of two separated H atoms as the two atoms are brought within bonding distance, and each electron is free to migrate to the other atom. However, the energy reduction that follows from this process is counteracted by an increase in energy from the Coulombic repulsion between the two positively charged nuclei. This positive contribution to the energy becomes large as R becomes small; consequently the total potential energy curve passes through a minimum and then climbs to a strongly positive value at small internuclear separations.

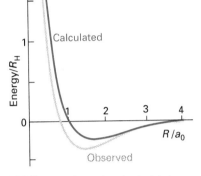

14.3 The experimental and calculated molecular potential energy curves for H_2. The latter is obtained from the valence-bond description of the molecule.

The role of electron spin

So far, the electron spin has not played a role in the argument, yet a chemist's picture of a covalent bond is one in which the spins of two electrons are paired. The origin of the role of spin is that *the wavefunction given in eqn 2 can be formed only by a pair of electrons with opposed spins.* Thus, spin-pairing is not an end in itself, but it is a means of achieving a wavefunction (and the probability distribution it implies) that corresponds to a low energy.

JUSTIFICATION

We saw in *Further information 12* that the Pauli principle requires the wavefunction of two electrons to change sign when the labels of the electrons are interchanged. The total valence-bond wavefunction for two electrons is

$$\Psi(1, 2) = \{A(1)B(2) + A(2)B(1)\}\sigma(1, 2)$$

where σ represents the spin component of the wavefunction. When the labels 1 and 2 are interchanged, this wavefunction becomes

$$\Psi(2, 1) = \{A(2)B(1) + A(1)B(2)\}\sigma(2, 1)$$
$$= \{A(1)B(2) + A(2)B(1)\}\sigma(2, 1)$$

However, the Pauli principle requires that $\Psi(2, 1) = -\Psi(1, 2)$, which is satisfied only if $\sigma(2, 1) = -\sigma(1, 2)$. The combination of two spins that has this property is

$$\sigma(1, 2) = \alpha(1)\beta(2) - \alpha(2)\beta(1)$$

which corresponds to *paired* electron spins (see *Further information 12*). Therefore, we conclude that the state of lower energy (and hence the formation of a chemical bond) is achieved if the electron spins are paired.

14.2 Homonuclear diatomic molecules

The essential features of valence-bond theory are the pairing of the electrons and the formation of a wavefunction that allows both electrons to be found on either atom. The same description can be applied to more complex molecules, such as **homonuclear diatomic molecules**, which are diatomic molecules in which both atoms belong to the same element. Nitrogen, N_2, is an example. To construct the valence-bond description of N_2 we consider the *valence* electron configuration of each atom:

$$N \qquad 2s^2 2p_x^1 2p_y^1 2p_z^1$$

It is conventional to take the z-axis to be the internuclear axis, so we can imagine each atom as having a $2p_z$ orbital pointing towards a $2p_z$ orbital on the other atom (Fig. 14.4), with the $2p_x$ and $2p_y$ orbitals perpendicular to the axis. A σ bond is then formed by spin pairing between the two electrons in the opposing $2p_z$ orbitals. Its spatial wavefunction has the form

$$\psi = \psi_{N2p_zA}(1)\psi_{N2p_zB}(2) + \psi_{N2p_zA}(2)\psi_{N2p_zB}(1)$$

and there is constructive interference between the two components in the internuclear region.

The remaining p orbitals cannot merge to give σ bonds as they do not have cylindrical symmetry around the internuclear axis. Instead, the electrons in them merge to form two π **bonds**. A π bond arises from the spin pairing of electrons in two p orbitals that approach side-by-side, and is so called because, viewed along the internuclear axis, a π bond resembles a pair of electrons in a p orbital (and π is the Greek equivalent of p). More precisely, an electron in a π bond has one unit of orbital angular momentum about the internuclear axis, for the wavefunction has one angular node.

In N_2 there are two π bonds: one is formed by spin pairing in two neighbouring $2p_x$ orbitals, and the other is formed by spin pairing in two neighbouring $2p_y$ orbitals. The overall bonding pattern in N_2 is therefore a σ bond plus two π bonds (Fig. 14.5), which is consistent with $:N\equiv N:$ for its Lewis structure.

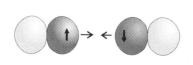

14.4 Spin pairing of two electrons in p orbitals also results in the formation of a σ bond as a result of the overlap of orbitals with cylindrical symmetry around the internuclear axis.

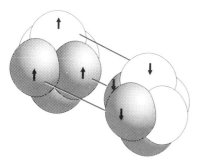

14.5 The valence-bond description of the bonding in N_2. The electrons in two $N2p_z$ orbitals pair to form a σ bond, and the electrons in the $2p_x$ and $2p_y$ orbitals on each N atom (perpendicular to the bond) pair to form two π bonds.

Example 14.1 *Describing the valence-bond structure of a diatomic molecule*

Describe the valence-bond ground state of the Cl_2 molecule.

Method. Begin by writing the valence electron configuration of the atoms and then decide which electrons can be paired when the atoms are within bonding distance of one another.

Answer. The ground-state electron configuration of a Cl atom is $[\text{Ne}]3s^2 3p_x^2 3p_y^2 3p_z^1$. This configuration suggests that a σ bond can be formed between two atoms by spin pairing of the electrons in the $3p_z$ orbitals. The remaining six valence electrons on each atom are not involved in the bonding. This description is consistent with the Lewis structure $:\ddot{\text{Cl}}\!-\!\ddot{\text{Cl}}:$. The valence-bond wavefunction for the bonding pair is

$$\psi = \psi_{\text{Cl}3p_z\text{A}}(1)\psi_{\text{Cl}3p_z\text{B}}(2) + \psi_{\text{Cl}3p_z\text{A}}(2)\psi_{\text{Cl}3p_z\text{B}}(1)$$

Exercise E14.1. Describe the ground state of HCl in valence-bond terms.

$$[\psi = \psi_{\text{H}1s}(1)\psi_{\text{Cl}3p_z}(2) + \psi_{\text{H}1s}(2)\psi_{\text{Cl}3p_z}(1)]$$

14.3 Polyatomic molecules

The concept of bond formation by electron pairing and the overlapping of orbitals to form σ and π bonds can be extended readily to polyatomic species. Each σ bond in a polyatomic molecule is formed by the spin pairing of electrons in any atomic orbitals with cylindrical symmetry about the relevant internuclear axis. Likewise, π bonds are formed by pairing electrons that occupy atomic orbitals of the appropriate symmetry. A consideration of the electronic structure of H_2O will make this clear.

The valence electron configuration of an O atom is $2s^2 2p_x^2 2p_y^1 2p_z^1$. The two unpaired electrons in the O$2p$ orbitals can each pair with an electron in an H$1s$ orbital, and each combination results in the formation of a σ bond (each bond has cylindrical symmetry about the respective O—H internuclear direction). Because the $2p_y$ and $2p_z$ orbitals lie at 90° to each other, the two σ bonds also lie at 90° to each other (Fig. 14.6). We can predict, therefore, that H_2O should be an angular molecule, which it is. However, the model predicts a bond angle of 90°, whereas the actual bond angle is 104°.

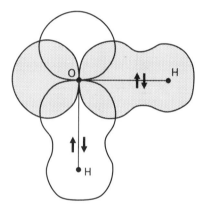

14.6 The simplest valence-bond description of H_2O in which the two σ bonds are formed by spin pairing between electrons in H$1s$ and O$2p$ orbitals. Although the model predicts an angular molecule, the bond angle is incorrect.

Example 14.2 *Predicting the shape of a molecule using valence-bond theory*

Give a valence-bond description of NH_3 and predict the bond angle of the molecule on the basis of this description.

Method. Write down the ground-state electron configuration of an N atom and decide which electrons and orbitals can be used to form bonds. Then, from the spatial arrangement of those orbitals in the atom, infer the shape of the resulting molecule.

Answer. The valence electron configuration of an N atom is

N $2s^2 2p_x^1 2p_y^1 2p_z^1$

This configuration suggests that three H atoms can form bonds by spin pairing with the electrons in the three half-filled $2p$ orbitals. The latter are perpendicular to each other, so we predict a trigonal pyramidal molecule with a bond angle of 90°.

Comment. The molecule is trigonal pyramidal, but the experimental bond angle is 107°. The origin of this discrepancy is discussed below.

Exercise E14.2. Use valence-bond arguments to suggest a shape for the hydrogen peroxide molecule H_2O_2.   [Each H—O—O bond 90°]

Promotion

An apparent deficiency of valence-bond theory is its inability to account for carbon's tetravalence (its ability to form four bonds). The ground-state configuration of C is $2s^2 2p_x^1 2p_y^1$, which suggests that a carbon atom should be capable of forming only *two* bonds, not four. This deficiency is overcome by allowing for **promotion**, the excitation of an electron to an orbital of higher energy. Although electron promotion requires an invest-ment of energy, it is worthwhile if that energy can be more than recovered in the greater strength or number of bonds that it allows to be formed.

In carbon, for example, the promotion of a $2s$ electron to a $2p$ orbital leads to the configuration $2s^1 2p_x^1 2p_y^1 2p_z^1$, with *four* unpaired electrons in separate orbitals. These electrons may pair with four electrons in orbitals provided by four other atoms (such as four H$1s$ orbitals if the molecule is CH_4), and hence form four electron-pair σ bonds. Although energy was required to promote the electron, it is more than recovered by the atom's ability to form four bonds in place of the two bonds of the unpromoted atom. Promotion, and the formation of four bonds, is a characteristic feature of carbon because the promotion energy is quite small: the promoted electron leaves a doubly occupied $2s$ orbital and enters a vacant $2p$ orbital, hence significantly relieving the electron–electron repulsion it experiences in the former.

Hybridization

The description of the bonding in CH_4 (and its homologues) is still incomplete because it appears to imply the presence of three σ bonds of one type (formed from H$1s$ and C$2p$ orbitals) and a fourth σ bond of a distinctly different character (formed from H$1s$ and C$2s$). This problem is overcome by realizing that the electron density distribution in the promoted atom is equivalent to the electron density in which each electron occupies a **hybrid orbital** formed by the interference between the C$2s$ and C$2p$ orbitals. The origin of the hybridization can be appreciated by thinking of the four atomic orbitals, which are waves centred on a nucleus, as being like ripples spreading from a single point on the surface of a lake: the waves interfere destructively and construc-tively in different regions, and give rise to four new shapes.

The specific linear combinations that give rise to four equivalent hybrid orbitals are

$$h_1 = s + p_x + p_y + p_z \qquad h_2 = s - p_x - p_y + p_z$$
$$h_3 = s - p_x + p_y - p_z \qquad h_4 = s + p_x - p_y - p_z$$

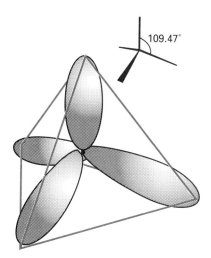

14.7 The four tetrahedrally oriented equivalent sp^3 hybrid atomic orbitals. The angle between the axes of the bonds is 109.47°.

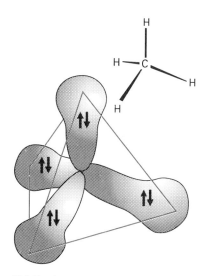

14.8 The four equivalent σ bonds that are formed when an electron in one of the hybrid orbitals shown in Fig. 14.7 pairs with an electron in an H1s orbital. The resulting shape agrees with the known shape of a CH_4 molecule.

As a result of the constructive and destructive interference between the component orbitals, each hybrid orbital consists of a large lobe pointing in the direction of one corner of a regular tetrahedron (Fig. 14.7); the angle between the axes of the hybrid orbitals is the tetrahedral angle 109.47°. Because each hybrid is built from one s orbital and three p orbitals, it is called an **sp^3 hybrid orbital**.

It is now easy to see how the valence-bond description of the methane molecule leads to a tetrahedral molecule containing four *equivalent* C—H bonds. Each hybrid orbital of the promoted C atom contains a single unpaired electron; an H1s electron can pair with each one, giving rise to a σ bond pointing in a tetrahedral direction. For example, the wavefunction for the bond formed by the hybrid orbital h_1 and the 1s_A orbital (with wavefunction that we shall denote A) is

$$\psi = h_1(1)A(2) + h_1(2)A(1)$$

Because each sp^3 hybrid orbital has the same composition, all four σ bonds are identical apart from their orientation in space (Fig. 14.8).

A further feature of hybridization is that a hybrid orbital has pronounced directional character in the sense that it has an enhanced amplitude in the internuclear region where overlap is important. This directional character arises from the constructive interference between the s orbital and the positive lobes of the p orbitals (Fig. 14.9). As a result of this directional character, the bond strength is greater than for an s or p orbital alone. This increased bond strength is another factor that helps to repay the promotion energy.

Hybridization can also be used to describe the structure of the ethene molecule, $H_2C{=}CH_2$, and the torsional rigidity of double bonds. The ethene molecule is planar, with HCH and HCC bond angles close to 120°. To reproduce the σ bonding structure that the arrangement of electron pairs requires we promote each C atom to a $2s^1 2p^3$ configuration. However, instead of using all four orbitals to form hybrids, we form **sp^2 hybrid orbitals** by the superposition of an s orbital and *two* p orbitals. As shown in Fig. 14.10, the three hybrid orbitals

$$h_1 = s + \sqrt{2}p_y \qquad h_2 = s + \sqrt{\tfrac{3}{2}}p_x - \sqrt{\tfrac{1}{2}}p_y \qquad h_3 = s - \sqrt{\tfrac{3}{2}}p_x - \sqrt{\tfrac{1}{2}}p_y$$

lie in a plane and point towards the corners of an equilateral triangle. (The detailed form of these hybrids is justified below.) The third $2p$ orbital ($2p_z$) is not included in the hybridization, and its axis is perpendicular to the plane in which the hybrids lie.

The structure of the ethene molecule can now be described as follows. The sp^2-hybridized C atoms each form three σ bonds by spin pairing with either the h_1 hybrid of the other C atom or with H1s orbitals. The σ framework therefore consists of C—H and C—C σ bonds at 120° to each other. When the two CH_2 groups lie in the same plane, the two electrons in the unhybridized p orbitals can pair and form a π bond (Fig. 14.11). The formation of this π bond locks the framework into the planar arrangement, for any rotation of one CH_2 group relative to the other leads to a weakening of the π bond (and consequently to an increase in energy of the molecule).

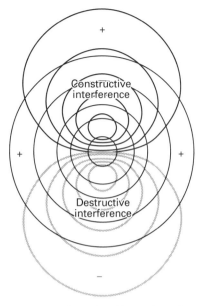

14.9 The regions of positive and negative wavefunction that give rise to constructive and destructive interference and hence the shapes of hybrid orbitals.

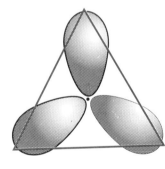

(a)

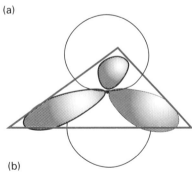

(b)

14.10 (a) The three equivalent sp^2 orbitals formed from mixing an s orbital and two p orbitals. The angle between neighbouring hybrids is 120°. (b) The unhybridized p orbital is perpendicular to the plane in which the three hybrids lie.

A similar description applies to the linear acetylene molecule, H—C≡C—H, but now the C atoms are **sp hybridized** and the σ bonds are formed by using hybrid atomic orbitals of the form

$$h_1 = s + p_z \qquad h_2 = s - p_z$$

These two orbitals lie along the internuclear axis. The electrons in them pair either with an electron in the corresponding hybrid orbital on the other C atom or with an electron in one of the H1s orbitals. Electrons in the two remaining p orbitals on each atom, which are perpendicular to the molecular axis, pair to form two perpendicular π bonds (as in Fig. 14.12).

Other hybridization schemes, particularly those involving d orbitals, are often invoked to account for (or at least be consistent with) other molecular geometries. For example, sp^3d^2 hybridization results in six equivalent hybrid orbitals pointing towards the corners of a regular octahedron. (The hybridization of N atomic orbitals always results in the formation of N hybrid orbitals.) This octahedral hybridization scheme is sometimes invoked to account for the structure of octahedral molecules, such as SF_6. These schemes are summarized in Table 14.1.

The 'pure' schemes in the table are not the only possibilities: it is possible to form hybrid orbitals with intermediate proportions of orbitals. For example, as more p orbital character is included in an sp hybridization scheme, the hybridization changes towards sp^2 and the angle between the hybrids changes continuously from 180° for pure sp hybridization to 120° for pure sp^2 hybridization. If the proportion of p character continues to be increased (by reducing the proportion of s orbital), then the hybrids eventually become pure p orbitals making 90° to each other.

The possibility of varying the composition of hybrids opens the way to accounting for the shape of H_2O with its bond angle of 104°. Each O—H σ bond is formed from an O atom hybrid orbital with a composition that lies between pure p (which would lead to a bond angle of 90°) and pure sp^3 (which would lead to a bond angle of 109.5°). Why one particular bond angle and hybridization are adopted is answered by calculating the energy of the molecules as the bond angle is varied, for it corresponds to the minimum energy of the molecule.

The construction of hybrid orbitals

The question we address in this section is how to combine s and p orbitals on a single atom into identical hybrids with a specific angle between them. Our aim is to construct hybrids that are (a) equivalent but (b) spatially distinct. These requirements translate into forming hybrids of the s, p_x, and p_y orbitals (where the xy-plane is the plane in which the hybrid orbitals are to lie) that (a) have the same composition (in the sense that they have the same proportions of s and p character) and (b) are **orthogonal** in the sense that one hybrid orbital has zero overlap with another

$$S = \int \psi_A^* \psi_B \, d\tau = 0 \tag{3}$$

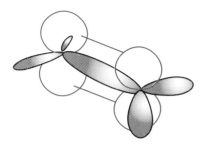

14.11 The pattern of bonding in a carbon–carbon double bond. An sp^2 hybrid on each atom overlaps with its neighbour to form a σ bond, and the remaining sp^2 hybrids form σ bonds with neighbouring atoms. The formation of the carbon–carbon σ bond brings the two unhybridized $2p$ orbitals into a position where they can overlap to form a π bond.

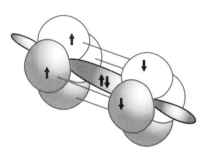

14.12 The pattern of bonding in a carbon–carbon triple bond. An sp hybrid on each atom overlaps with its neighbour to form a σ bond, and the remaining sp hybrids form σ bonds with neighbouring atoms. The two perpendicular $2p$ orbitals on each atom are brought into a position where they can overlap to form two π bonds.

Table 14.1* Some hybridization schemes

Coordination number	Shape	Hybridization
2	Linear	sp
3	Trigonal planar	sp^2
4	Tetrahedral	sp^3
5	Bipyramidal	sp^3d
6	Octahedral	sp^3d^2

* More schemes are given in the Data section at the end of this volume.

where S is the **overlap integral**. The value of S can be regarded as a measure of the similarity of the wavefunctions ψ_A and ψ_B. If the two wavefunctions are identical (and normalized), then $S = 1$. If one orbital is an s orbital and the other is a p orbital on the same atom, then $S = 0$ and the two wavefunctions are entirely different (that is, orthogonal, like two perpendicular lines).

Example 14.3 *Judging the orthogonality of orbitals*

Confirm that a $2p_x$ orbital is orthogonal to a $2p_y$ orbital of the same atom.

Method. We need to evaluate the integral

$$S = \int \psi_{2p_x} \psi_{2p_y} \, d\tau$$

and can do so by noting that

$$\psi_{2p_x} = xf(r) \qquad \psi_{2p_y} = yf(r)$$

and expressing the orbitals in spherical polar coordinates.

Answer. In terms of polar coordinates,

$$\psi_{2p_x} = rf(r) \sin \theta \cos \phi \qquad \psi_{2p_y} = rf(r) \sin \theta \sin \phi$$

All we need to show is that the integral over one of the angles vanishes. The two orbitals differ in their dependence on ϕ, so we begin with that integration:

$$S \propto \int_0^{2\pi} \cos \phi \sin \phi \, d\phi = 0$$

Comment. All the spherical harmonics are mutually orthogonal, so all hydrogenic atomic orbitals are mutually orthogonal as long as they belong to the same atom.

Exercise E14.3. Show that $3d_{xy}$ is orthogonal to $2p_x$.

Two $O2p$ orbitals in H_2O are mutually orthogonal and point at 90° to each other. When $O2s$ character is mixed into them in the process of hybridization, they are no longer orthogonal (they are similar to the extent of having a common $O2s$ character). Only if the hybrids bend further away from each other, thus reducing their spatial resemblance, can they become distinct again. Therefore, we can anticipate that the greater the angle between the hybrids, the greater their s orbital content.

The relation between the proportion of s character and the angle between the hybrids can be expressed quantitatively. If the two hybrid orbitals are written

$$h = as + bp \qquad h' = as + bp'$$

then the coefficient a is related to the angle Φ between the hybrids by

$$a^2 = \frac{\cos \Phi}{\cos \Phi - 1} \tag{4}$$

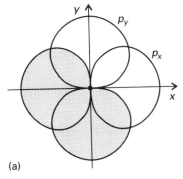

(a)

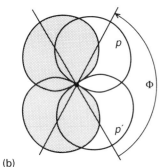

(b)

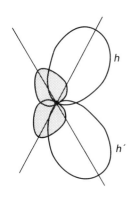

(c)

14.13 (a) Two orthogonal p orbitals. (b) The orbitals p and p' can be expressed as linear combinations of p_x and p_y, but the combinations are not orthogonal. (c) The orthogonal hybrids, obtained by mixing s-character into p and p'.

In common with the conventional interpretation of wavefunctions, a^2 is the probability that, if inspected, the electron in either of the hybrid orbitals would be found to have s-orbital character. (The probability that the electron has p character is $b^2 = 1 - a^2$.)

JUSTIFICATION

Suppose we want to construct a p orbital that points along a line making an angle $\frac{1}{2}\Phi$ to the x-axis (Fig. 14.13). Then the combination of p orbitals required is[2]

$$p = p_x \cos \tfrac{1}{2}\Phi + p_y \sin \tfrac{1}{2}\Phi$$

The equivalent orbital pointing along $-\frac{1}{2}\Phi$ is

$$p' = p_x \cos \tfrac{1}{2}\Phi - p_y \sin \tfrac{1}{2}\Phi$$

Although the p_x and p_y orbitals are mutually orthogonal and normalized, these two combinations are not orthogonal:

$$\int pp' \, d\tau = \int (p_x \cos \tfrac{1}{2}\Phi + p_y \sin \tfrac{1}{2}\Phi)(p_x \cos \tfrac{1}{2}\Phi - p_y \sin \tfrac{1}{2}\Phi) \, d\tau$$

$$= \cos^2 \tfrac{1}{2}\Phi \int p_x^2 \, d\tau - \sin^2 \tfrac{1}{2}\Phi \int p_y^2 \, d\tau$$

$$= \cos^2 \tfrac{1}{2}\Phi - \sin^2 \tfrac{1}{2}\Phi = \cos \Phi$$

The integral vanishes only if $\Phi = 90°$.

Now we add s character. The same proportion must be added to both (so that the hybrids remain equivalent). We therefore write

$$h = as + bp \qquad h' = as + bp'$$

We can find the coefficients a and b by requiring the hybrids to satisfy two conditions. One is the requirement that the hybrids are normalized:

$$\int h^2 \, d\tau = a^2 \int s^2 \, d\tau + b^2 \int p^2 \, d\tau + 2ab \int sp \, d\tau = a^2 + b^2 = 1$$

(The s and p orbitals are individually normalized and mutually orthogonal.) The second requirement is for the hybrids to be distinct, that is, orthogonal:

$$\int hh' \, d\tau = \int (as + bp)(as + bp') \, d\tau$$

$$= a^2 \int s^2 \, d\tau + b^2 \int pp' \, d\tau + ab \int (sp + sp') \, d\tau$$

$$= a^2 + b^2 \int pp' \, d\tau = a^2 + b^2 \cos \Phi = 0$$

2 A p_x orbital is proportional to $x = r \cos \phi$ and a p_y orbital is proportional to $y = r \sin \phi$. A p orbital directed along the line at $\frac{1}{2}\Phi$ is proportional to $\cos (\phi - \frac{1}{2}\Phi)$ because it is like a p_x orbital but rotated through an angle $\frac{1}{2}\Phi$. Now note that

$$p(\Phi) \propto \cos (\phi - \tfrac{1}{2}\Phi) \propto \cos \phi \cos \tfrac{1}{2}\Phi + \sin \phi \sin \tfrac{1}{2}\Phi \propto p_x \cos \tfrac{1}{2}\Phi + p_y \sin \tfrac{1}{2}\Phi$$

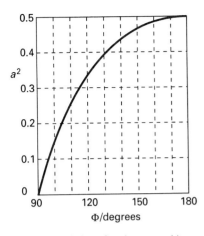

14.14 The variation of s-character with the angle between two equivalent hybrids calculated using eqn 4. The p character is $1 - a^2$.

On combining these two results we get

$$a^2 = \frac{\cos \Phi}{\cos \Phi - 1}$$

Equation 4 is plotted in Fig. 14.14. It shows that, as the angle between the hybrids Φ increases from 90° to 180°, their s character increases from $a^2 = 0$ (pure p) to $a^2 = 0.5$ (a 50:50 mixture of s and p).

Example 14.4 *Calculate the composition of hybridized orbitals*

Calculate the hybridization of the O–H bonds in H_2O, which has a bond angle of 104°.

Method. We can calculate the s character of a hybrid from eqn 4 and its p character from $1 - a^2$. Hybrids are often denoted $s^{a^2}p^{b^2}$. When $\Phi = 104°$, $\cos \Phi = -0.24$; hence $a^2 = 0.19$. The p character of the hybrid is therefore 0.81, and it can be denoted $s^{0.19}p^{0.81}$.

Comment. Note that hybrids need not be of integral composition (such as sp^3). The coefficients are the square roots of 0.19 and 0.81, and so the hybrid wavefunctions are of the form $0.44s + 0.90p$, where p is a p orbital directed from O to H.

Exercise E14.4. What is the hybridization for a 120° bond angle?

$$[s^{\frac{1}{3}}p^{\frac{2}{3}}, \text{ or } sp^2 \text{ in terms of ratios}]$$

The linear combination of three atomic orbitals leads to three orthogonal hybrids (Fig. 14.15). The third hybrid is $a's + b'p''$, where p'' is an orbital directed along $-x$ (so $p'' = -p_x$). Normalization requires that

$$a'^2 + b'^2 = 1$$

and the orthogonality of this hybrid to the other two requires

$$aa' + bb' \cos \tfrac{1}{2}\Phi = 0$$

After a little rearrangement, these conditions lead to

$$a'^2 = \frac{1 + \cos \Phi}{1 - \cos \Phi} \tag{5}$$

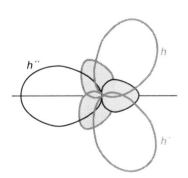

14.15 Three orthogonal atomic orbitals hybridize to give three orthogonal hybrid orbitals. The orbital on the x-axis is the third; its composition is given by eqn 5.

for the s character of this non-bonding, lone-pair orbital. For H_2O, $a'^2 = 0.61$, implying 61 per cent s character and a hybrid orbital of the form $0.78s + 0.62p$.

The sp^2 hybrids discussed earlier are a special case of the set of three hybrid orbitals we have been discussing. All three sp^2 hybrid orbitals are equivalent, and the angles between them are 120°. It follows that the composition of each hybrid has the form $s + \sqrt{2}p$ and directed towards the corners of an equilateral triangle (as illustrated earlier in Fig. 14.10a). A similar approach can be used to show that the basis ($2s$, $2p_x$, $2p_y$, $2p_z$) leads to four equivalent hybrids of composition $s + \sqrt{3}p$ making an angle of 109.47° (that is, $2 \arccos{(1/\sqrt{3})}$). These combinations are the sp^3 hybrids introduced in connection with the discussion of CH_4, and which point towards the corners of a regular tetrahedron (as in Fig. 14.7).

MOLECULAR ORBITAL THEORY

Valence-bond theory focuses its attention on individual bonds in molecules. In an alternative approach, that of **molecular orbital theory**, it is accepted that electrons should not be regarded as belonging to particular bonds but should be treated as spreading throughout the entire molecule. This theory has been more fully developed than valence-bond theory and provides the language that is widely used in modern discussions of bonding in small inorganic molecules, d-metal complexes, and solids. To introduce it, we shall follow the same strategy as in Chapter 13 where the one-electron H atom was taken as the fundamental species for discussing atomic structure and then developed into a description of many-electron atoms. In this chapter we use the simplest molecule of all, the one-electron hydrogen molecule-ion H_2^+ to introduce the essential features of bonding, and then use it as a guide to the structures of more complex systems. These applications include diatomic molecules, polyatomic molecules, and finally solids consisting of effectively infinite numbers of atoms.

14.4 The hydrogen molecule-ion

The Schrödinger equation for the electron in a hydrogen molecule-ion is

$$-\frac{\hbar^2}{2m_e}\nabla^2\psi + V\psi = E\psi \qquad \text{with } V = -\frac{e^2}{4\pi\varepsilon_0}\left(\frac{1}{r_{A1}} + \frac{1}{r_{B1}}\right) \tag{6}$$

where r_{A1} and r_{B1} are the distances of the electron from the two nuclei. The one-electron wavefunctions obtained by solving this equation are called **molecular orbitals**. A molecular orbital ψ gives, through the value of ψ^2, the distribution of the electron in the molecule. A molecular orbital is like an atomic orbital, but spreads throughout the molecule.

Exact, analytical molecular orbitals may be obtained for H_2^+ (within the Born–Oppenheimer approximation), but they are very complicated functions and do not give much insight into the form of the orbitals and the contributions to the energy. Therefore, we shall adopt a simpler procedure that, while more approximate, gives more insight.

Linear combinations of atomic orbitals

The approximation we shall adopt is based on the fact that, when the electron is very close to nucleus A, the term $1/r_{A1}$ in V is very much bigger than $1/r_{B1}$. That being so, the potential energy in eqn 6 reduces to

$$V = -\frac{e^2}{4\pi\varepsilon_0 r_{A1}}$$

The Schrödinger equation for the electron in the molecule is then the same as that for an isolated H atom, and its lowest energy solution is a $1s$

orbital on A, which we write $\psi_{1s}(A)$. Thus, close to A, the molecular orbital resembles an atomic 1s orbital. Likewise, close to B the molecular orbital resembles a 1s orbital on B, $\psi_{1s}(B)$. This discussion suggests that we can approximate the overall wavefunction ψ as a sum of the two atomic orbitals:

$$\psi = N\{\psi_{1s}(A) + \psi_{1s}(B)\} \tag{7}$$

where N is a normalization factor. In accord with the preceding discussion, when the electron is close to A its distance from B is large, $\psi_{1s}(B)$ is small, and therefore the wavefunction is almost pure $\psi_{1s}(A)$. Similarly, ψ is almost pure $\psi_{1s}(B)$ close to B.

The technical term for a sum of the kind in eqn 7 is a **linear combination of atomic orbitals** (LCAO). An approximate molecular orbital formed from a linear combination of atomic orbitals is called an LCAO–MO. A molecular orbital that has cylindrical symmetry around the internuclear axis, such as the one we are discussing, is called a **σ orbital** (because it resembles an s orbital when viewed along the axis and has zero orbital angular momentum around the internuclear axis).

Example 14.5 *Normalizing a molecular orbital*

Normalize the molecular orbital in eqn 7.

Method. We need to find the factor N in the expression

$$\psi = N\{\psi_{1s}(A) + \psi_{1s}(B)\}$$

that ensures that

$$\int \psi^2 \, d\tau = 1$$

To proceed, we substitute the LCAO into this integral, and make use of the fact that the atomic orbitals are individually normalized. The additional integral required is the overlap integral, eqn 3.

Answer. When we substitute the wavefunction, we find

$$N^2\left\{ \int \psi_{1s}(A)^2 \, d\tau + \int \psi_{1s}(B)^2 \, d\tau + 2\int \psi_{1s}(A)\psi_{1s}(B) \, d\tau \right\} = N^2\{1 + 1 + 2S\}$$

$$= 1$$

Therefore, the normalization factor is

$$N = \left(\frac{1}{2(1 + S)} \right)^{\frac{1}{2}}$$

Comment. In H_2^+, S is about 0.59, so $N = 0.56$.

Exercise E14.5. Normalize an orbital of the form $\psi_{1s}(A) - \psi_{1s}(B)$.

$$[N = 1/2^{\frac{1}{2}}(1 - S)^{\frac{1}{2}} = 1.10 \text{ in } H_2^+]$$

We should keep in mind the approximations we have made so far. The Born–Oppenheimer approximation separates the electronic and nuclear motions and allows us to talk in terms of the molecular orbitals of the electron in the field of the stationary nuclei. The LCAO approximation goes one step further and approximates a molecular orbital as a sum of atomic orbitals. It allows us to use atomic orbitals to discuss the distribution of electrons in molecules.

σ orbitals

According to the Born interpretation, the probability density of the electron in H_2^+ is proportional to the square of its wavefunction. The probability density of the LCAO–MO $1s\sigma$ orbital in eqn 7 is

$$\psi^2 = N^2\{\psi_{1s}(A) + \psi_{1s}(B)\}^2$$
$$= N^2\{\psi_{1s}(A)^2 + \psi_{1s}(B)^2 + 2\psi_{1s}(A)\psi_{1s}(B)\} \qquad (8)$$

From Example 14.5 we know that $N^2 = 0.31$. Because $\psi_{1s}(A)$ is the function

$$\psi_{1s}(A) = \left(\frac{1}{\pi a_0^3}\right)^{\frac{1}{2}} e^{-r_A/a_0}$$

with r_A the distance of the electron from A, and similarly for $\psi_{1s}(B)$, it is easy to evaluate ψ and hence give the probability density at any point. The result is shown in Fig. 14.16.

An important feature of eqn 8 becomes apparent when we examine the probability density in the internuclear region, where both atomic orbitals have similar amplitudes. According to eqn 8, the total probability density is proportional to the sum of

(1) $\psi_{1s}(A)^2$, the probability density if the electron were confined to the orbital on A.

(2) $\psi_{1s}(B)^2$, the probability density if the electron were confined to the orbital on B.

(3) $2\psi_{1s}(A)\psi_{1s}(B)$, an extra contribution to the density.

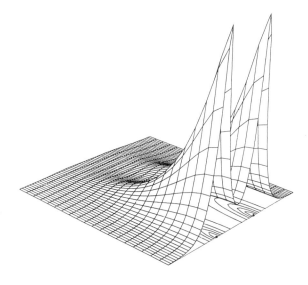

14.16 A representation of the amplitude of the bonding orbital formed from the overlap of two H1s orbitals and a contour diagram (beneath the surface). The probability density (the square of the amplitude) has a similar appearance.

This last contribution, the **overlap density**, is crucial, because it represents an enhancement of the probability of finding the electron in the internuclear region above what it would be if it were confined to one of the two atoms. That is, because the electron is free to move from one nucleus to the other, the electron density in the internuclear region is increased, much as in valence-bond theory. The enhancement can be traced to the constructive interference of the atomic orbitals: each has a positive amplitude in the internuclear region, so the total amplitude is greater there than if the electron were confined to a single atomic orbital.

We shall constantly use the result that *electrons accumulate in regions where atomic orbitals overlap and interfere constructively*. The accumulation of electron density between the nuclei puts the electron in a position where it interacts strongly with both nuclei. Hence the energy of the molecule is lower than that of the separate atoms, where each electron can interact strongly with only one nucleus.[3]

Bonding orbitals

The σ orbital we have described is an example of a **bonding orbital**, an orbital which, if occupied, helps to bind two atoms together. An electron that occupies a σ orbital is called a **σ electron** and, if that is the only electron present in the molecule (as in the ground state of H_2^+), then we report the configuration of the molecule as $1\sigma^1$.

The energy of the 1σ orbital decreases as R decreases from large values because electron density accumulates in the internuclear region as the two atomic orbitals increasingly overlap. However, at small separations there is too little space between the nuclei for significant accumulation of electron density there. In addition, the nucleus–nucleus repulsion $V_{nuc,nuc}$ (which is proportional to $1/R$) becomes large. As a result, the energy of the molecule rises at short distances, and there is a minimum in the potential energy curve. As in the valence-bond theory, the internuclear separation at the minimum of the curve is the equilibrium bond length R_e and the depth of the minimum D_e is closely related to the bond dissociation energy D_0. Calculations on H_2^+ give $R_e = 130$ pm and $D_e = 1.77$ eV (171 kJ mol^{-1}); the experimental values are 106 pm and 2.6 eV, so this simple LCAO–MO description of the molecule, while inaccurate, is not absurdly wrong.

3 Unfortunately, this neat explanation is probably incorrect in the case of H_2^+ (at least). This is because shifting an electron away from a nucleus into the internuclear region *raises* its potential energy. The modern explanation is more subtle and does not emerge from the simple LCAO treatment given here. It seems that, at the same time as the electron shifts into the internuclear region, the atomic orbitals shrink. This orbital shrinkage improves the electron–nucleus attraction more than it is damaged by the migration to the internuclear region, so there is a net lowering of potential energy. The kinetic energy of the electron is also modified, but it is dominated by the potential energy.

Even though the details are obscure for more complex molecules, it is generally found that bonding occurs when electrons accumulate between nuclei even though the actual cause of the bonding may be an accompanying shrinkage of the orbitals. Although orbital shrinkage reduces orbital overlap, it increases the electron–nucleus attraction and results in a net lowering of the energy of the molecule. Therefore, throughout the following discussion we ascribe the strength of chemical bonds to the accumulation of electron density in the internuclear region and leave open the question whether in molecules more complicated than H_2^+ the true source of energy lowering is that accumulation itself or some indirect but related effect.

Antibonding orbitals

The argument that led to the expression of ψ as the sum of two atomic orbitals is equally well satisfied by writing a molecular orbital as the difference

$$\psi' = N\{\psi_{1s}(A) - \psi_{1s}(B)\} \tag{9}$$

because this wavefunction also resembles one or other of the atomic orbitals close to the two nuclei. However, this linear combination corresponds to a higher energy than the orbital in eqn 7 and is in fact a good approximation to the next-higher exact solutions of the Schrödinger equation for H_2^+. Because ψ' is cylindrically symmetrical around the internuclear axis it is also a σ orbital; it will be labelled 2σ (or $2\sigma^*$). The normalization factor is given in the exercise of Example 14.5 and for H_2^+ is 1.10.

We can see from eqn 9 that the 2σ orbital has a nodal plane where $\psi_{1s}(A)$ and $\psi_{1s}(B)$ cancel. Consequently, there is zero probability of finding the electron half-way between the nuclei if it occupies this orbital. We express this reduction in amplitude by saying that, because the atomic orbitals are superimposed with opposite signs, they interfere destructively where they overlap (Fig. 14.17). This reduction is obvious when we write the probability density:

$$\begin{aligned}
\psi'^2 &= N^2\{\psi_{1s}(A) - \psi_{1s}(B)\}^2 \\
&= N^2\{\psi_{1s}(A)^2 + \psi_{1s}(B)^2 - 2\psi_{1s}(A)\psi_{1s}(B)\}
\end{aligned} \tag{10}$$

The third term reduces the probability of finding the electron between the nuclei relative to its value if the electron were confined to one of the atomic orbitals.

The 2σ orbital is an example of an **antibonding orbital**, an orbital that, if occupied, contributes to a reduction in the cohesion between two atoms and helps to raise the energy of the molecule relative to the separated atoms. Antibonding orbitals are often labelled with an asterisk (*), so this particular orbital could also be denoted $2\sigma^*$.

The destabilizing effect of an antibonding electron stems partly because it is excluded from the internuclear region and hence is

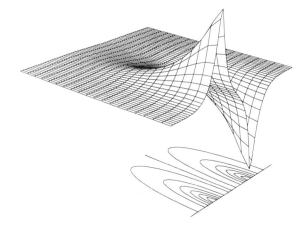

14.17 A representation of the amplitude of the antibonding orbital formed from the overlap of two H1s orbitals and a contour diagram of the amplitude (beneath the surface). Note the presence of an internuclear node.

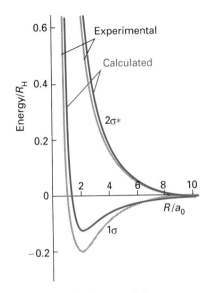

14.18 The molecular potential energy curve for the hydrogen molecule-ion showing the variation of the energy of the molecule as the bond length is changed and the electron is in either the bonding or the antibonding orbital. The experimental curves are in grey.

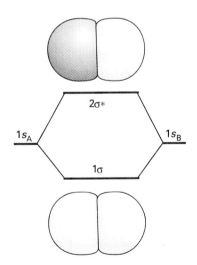

14.19 A molecular orbital energy level diagram for orbitals constructed from (1s, 1s)-overlap, and the corresponding orbitals (dark and light shading represent different signs of the wavefunction). The separation of the levels corresponds to the equilibrium bond length.

distributed largely outside the bonding region. In effect, whereas a bonding electron pulls two nuclei together, an antibonding electron pulls the nuclei apart. The combined effect of the electron distribution and the internuclear repulsion results in an antibonding orbital being more strongly antibonding than the corresponding bonding orbital is bonding.

The experimentally determined variation of the bonding and antibonding orbital energies with internuclear separation is shown in Fig. 14.18. Superimposed on the illustration are the curves calculated from simple LCAO–MO theory.

14.5 The structures of diatomic molecules

In Chapter 13 we used the hydrogenic atomic orbitals and the building-up principle to deduce the ground electronic configurations of many-electron atoms. We can do the same for many-electron diatomic molecules (such as H_2 with two electrons and Br_2 with 70), but using the H_2^+ molecular orbitals instead. As we shall illustrate in the following sections, first with H_2 and then with heavier molecules, the general procedure is to construct molecular orbitals by combining the atomic orbitals supplied by the atoms. The electrons supplied by the atoms are then accommodated in the orbitals so as to achieve the lowest overall energy subject to the constraint of the Pauli exclusion principle that no more than two electrons may occupy a single orbital (and then must be paired). As in the case of atoms, if several degenerate molecular orbitals are available, we add the electrons to each individual orbital before doubly occupying any one orbital (because that minimizes electron–electron repulsions). We also take note of Hund's rule (Section 13.4), that if electrons do occupy different degenerate orbitals, then they do so with parallel spins.

The hydrogen and helium molecules

We shall illustrate the general procedure by considering H_2, the simplest many-electron diatomic molecule. First, we need to build the molecular orbitals. Because each H atom of H_2 contributes a $1s$ orbital (as in H_2^+), we can form the 1σ and $2\sigma^*$ orbitals from them, as we have seen already. At the experimental internuclear separation these orbitals will have the energies shown in Fig. 14.19, which is called a **molecular orbital energy-level diagram**. Note that from two atomic orbitals we can build two molecular orbitals. In general, *from N atomic orbitals we can build N molecular orbitals*.

There are two electrons to accommodate, and both can enter 1σ by pairing their spins. The ground-state configuration is therefore $1\sigma^2$ (Fig. 14.20) and the atoms are joined by a bond consisting of an electron pair in a bonding σ orbital. This approach shows that an electron pair, which was the focus of Lewis's account of chemical bonding, represents the maximum number of electrons that can enter a bonding molecular orbital.

The same argument shows why He does not form diatomic molecules. Each He atom contributes a $1s$ orbital, so the same two molecular orbitals can be constructed. (They differ in detail from those in H_2

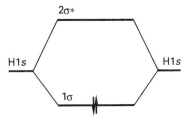

14.20 The ground electronic configuration of H_2 is obtained by accommodating the two electrons in the lowest available orbital (the bonding orbital).

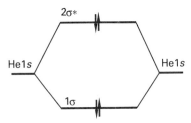

14.21 The ground electronic configuration of the hypothetical four-electron molecule He_2 has two bonding electrons and two antibonding electrons. It has a higher energy than the separated atoms, so He_2 is unstable relative to the separated atoms.

because the He1s orbitals are more compact, but the general shape is the same, and we can use the same energy-level diagram in the discussion.)

There are four electrons to accommodate. Two can enter the 1σ orbital, but then it is full, and the next two must enter the $2\sigma^*$ orbital (Fig. 14.21). The ground electronic configuration of He_2 is therefore

$$He_2 \qquad 1\sigma^2 2\sigma^{*2}$$

We see that there is one bond and one antibond. Because an antibond is slightly more antibonding than a bond is bonding, the He_2 molecule has a higher energy than the separated atoms, so it is unstable relative to the individual atoms.

Example 14.6 *Judging the stability of diatomic molecules*

Are Li_2 and Be_2 likely to exist if only the valence s orbitals contribute to molecular orbitals?

Method. We need to assess the electron configuration of the molecules. To do so, we note that each molecular orbital is built from $2s$ atomic orbitals, which give one bonding and one antibonding combination. Then consider the effect of adding the electron in accord with the Pauli exclusion principle.

Answer. Each Li atom supplies one valence electron, which fills the bonding σ orbital, to give a bonding configuration. Each Be atom provides two valence electrons, which fill the bonding and antibonding combinations, resulting in no net bond.

Comment. In fact, as we shall see, Be_2 does exist, because the $2p$ orbitals contribute to the orbitals and provide another bonding orbital.

Exercise E14.6. Is LiH likely to exist if the Li atom uses only its $2s$ orbital for bonding?   [Yes, (Li2s, H1s)σ^2]

Bond order

A measure of the net bonding in a diatomic molecule is its **bond order** b defined as

$$b = \tfrac{1}{2}(n - n^*) \tag{11}$$

where n is the number of electrons in bonding orbitals and n^* is the number in antibonding orbitals. Thus each electron pair in a bonding orbital increases the bond order by 1 and each pair in an antibonding orbital decreases it by 1. For H_2, $b = 1$, corresponding to a single bond, H—H, between the two atoms. In He_2, $b = 0$ and there is no bond.

As we shall see, the bond order is a useful parameter for discussing the characteristics of bonds, because it correlates with bond length: *the greater the bond order between atoms of a given pair of elements, the shorter the bond.* It also correlates with bond strength: *the greater the bond order, the greater the strength.*

Period 2 diatomic molecules

We shall now see how the concepts we have introduced apply to **homonuclear diatomic molecules** in general, which are diatomic molecules formed from identical atoms, such as N_2 and Cl_2. In line with the building-up procedure, we first consider the molecular orbitals that may be formed and do not (at this stage) trouble about how many electrons are available. The atomic orbitals available are:

1. The **core orbitals**, the orbitals that form the inner, closed shells.
2. The **valence orbitals**, the orbitals of the valence shell.
3. The **virtual orbitals**, the orbitals of the atom that are unoccupied in its ground state.

In elementary treatments (but not in modern sophisticated treatments) the core orbitals are ignored as being too compact to have significant overlap with orbitals on other atoms. The virtual orbitals are ignored on the grounds that they are too high in energy to participate in bonding. Therefore, we form molecular orbitals using only the valence orbitals.

In Period 2, the valence orbitals are $2s$ and $2p$. A general principle of molecular orbital theory is that *all orbitals of the appropriate symmetry contribute to a molecular orbital*. Thus, to build σ orbitals, we form linear combinations of *all* atomic orbitals that have cylindrical symmetry about the internuclear axis. These orbitals include the $2s$ orbitals on each atom and the $2p_z$ orbitals on the two atoms (Fig. 14.22). Thus, the general form of the σ orbitals that may be formed is

$$\psi = c_{2s(A)}\psi_{2s(A)} + c_{2s(B)}\psi_{2s(B)} + c_{2p_z(A)}\psi_{2p_z(A)} + c_{2p_z(B)}\psi_{2p_z(B)}$$

From these four atomic orbitals we can form *four* molecular orbitals of σ symmetry by an appropriate choice of the coefficients c. The procedure for calculating the coefficients will be described in Section 14.7. At this stage we shall adopt a simpler route and suppose that, because the $2s$ and $2p_z$ orbitals have distinctly different energies, they may be treated separately. That is, the four σ orbitals fall approximately into two sets, one consisting of two molecular orbitals of the form

$$\psi = c_{2s(A)}\psi_{2s(A)} + c_{2s(B)}\psi_{2s(B)}$$

and another consisting of two orbitals of the form

$$\psi = c_{2p_z(A)}\psi_{2p_z(A)} + c_{2p_z(B)}\psi_{2p_z(B)}$$

Because atoms A and B are identical, the energies of their $2s$ orbitals are the same, so the coefficients are equal (apart from a possible difference in sign); the same is true of the $2p$ orbitals. Therefore, the two sets of orbitals have the form

$$\psi = N\{\psi_{2s(A)} \pm \psi_{2s(B)}\}$$
$$\psi = N\{\psi_{2p_z(A)} \pm \psi_{2p_z(B)}\}$$

where the Ns are normalization constants (and are different for each orbital).

The $2s$ orbitals on the two atoms overlap to give a bonding and an

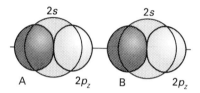

14.22 In a Period 2 diatomic molecule there are four atomic orbitals with σ symmetry. The four are shown here. Four σ molecular orbitals can be formed from these four orbitals.

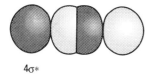

4σ*

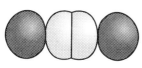

3σ

14.23 The interference leading to the formation of a $2p\sigma$ bonding orbital (labelled 3σ) and the corresponding antibonding orbital ($4\sigma^*$).

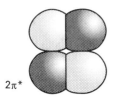

2π*

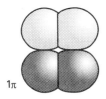

1π

14.24 The interference leading to the formation of a $2p\pi$ bonding orbital (labelled 1π) and the corresponding antibonding orbital ($2\pi^*$).

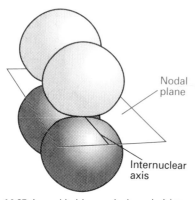

Nodal plane

Internuclear axis

14.25 A π orbital has a single node lying in the plane passing through the two nuclei.

antibonding σ orbital (1σ and $2\sigma^*$, respectively) in exactly the same way as we have already seen for $1s$ orbitals. The two $2p_z$ orbitals directed along the internuclear axis overlap strongly, and may do so either constructively or destructively, to give a bonding or antibonding σ orbital, respectively (Fig. 14.23). These two σ orbitals are labelled 3σ and $4\sigma^*$, respectively. In general, note how the numbering follows the order of increasing energy.

π orbitals

Now consider the $2p_x$ and $2p_y$ orbitals of each atom, which are perpendicular to the internuclear axis and may overlap broadside-on. This overlap may be constructive or destructive, and results in a bonding or an antibonding π orbital (Fig. 14.24). The notation π is the analogue of p in atoms for, when viewed along the axis of the molecule, a π orbital looks like a p orbital (Fig. 14.25) and has one unit of orbital angular momentum around the internuclear axis. The two $2p_x$ orbitals overlap to give bonding and antibonding π_x orbitals, and the two $2p_y$ orbitals overlap to give two π_y orbitals. The π_x and π_y bonding orbitals are degenerate (have the same energy); so too are their antibonding partners.

In some cases, π orbitals are less strongly bonding than σ orbitals because their maximum overlap occurs off-axis, away from the optimum bonding region. This relative weakness suggests that the molecular orbital energy level diagram ought to be as shown in Fig. 14.26. However, we must remember that we have constructed the diagram on the assumption that the $2s$ and $2p_z$ orbitals contribute to different sets of molecular orbitals, whereas in fact all four atomic orbitals contribute jointly to the four σ orbitals. Hence, there is no guarantee that this order of energies should prevail, and it is found experimentally (by spectroscopy) and by detailed calculation that the order varies along Period 2 (Fig. 14.27). The order shown in Fig. 14.28 is therefore appropriate as far as N_2, and Fig. 14.26 applies for O_2 and F_2. The relative order is controlled by the separation of the $2s$ and $2p$ orbitals in the atoms, which increases across the group. The consequent switch in order occurs at about N_2.

There is no need to consider the overlap of orbitals with different symmetry properties relative to the internuclear axis. That is, $(2s, 2p_x)$ overlap makes no contribution to bonding because the effect of the constructive overlap in one region is exactly cancelled by the effect of the destructive overlap in another, so there is no net overlap and no net contribution to bonding (Fig. 14.29).

Example 14.7 *Assessing the contribution of d orbitals*

Can d orbitals contribute to σ and π orbitals in diatomic molecules?

Method. We need to assess the symmetry of d orbitals with respect to the internuclear z-axis. A d_{z^2} orbital has cylindrical symmetry around z and so can contribute to σ orbitals. The d_{zx} and d_{yz} orbitals have π symmetry with respect to the axis (Fig. 14.30) and so can contribute to π orbitals.

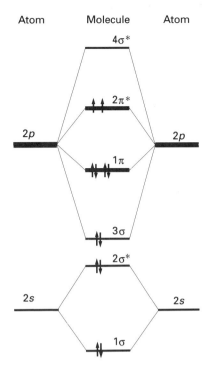

14.26 The molecular orbital energy-level diagram for homonuclear diatomic molecules. As remarked in the text, this diagram should be used for O_2 (for which the electron configuration is shown) and F_2.

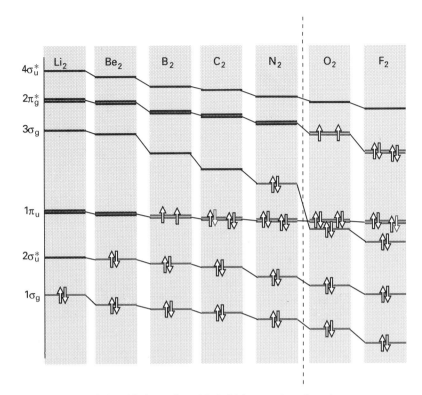

14.27 The variation of the orbital energies of Period-2 homonuclear diatomics.

Exercise E14.7. Sketch the 'δ orbitals' that may be formed by the remaining two d orbitals (and which contribute to bonding in some d-metal cluster compounds).   [See Fig. 14.30]

The overlap integral

The extent to which two atomic orbitals overlap is measured by the overlap integral, S:

$$S = \int \psi(A)^* \psi(B)\, d\tau \tag{12}$$

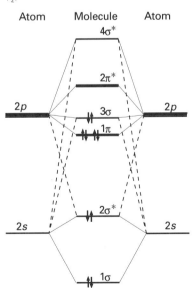

14.28 An alternative molecular orbital energy-level diagram for homonuclear diatomic molecules. As remarked in the text, this diagram should be used for diatomics as far as N_2. The electron configuration shown is that for N_2.

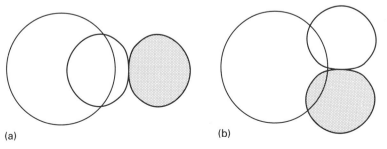

14.29 Overlapping s- and p-orbitals. (a) End-on overlap leads to non-zero overlap and to the formation of a σ orbital. (b) Broadside overlap leads to no net accumulation of electron density.

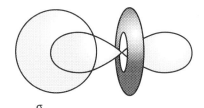

σ

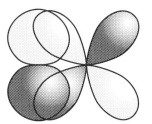

π (plus its partner)

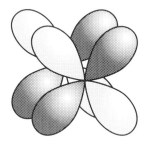

δ

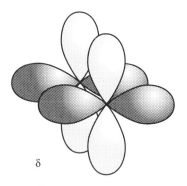

δ

14.30 The types of molecular orbital to which d orbitals can contribute. The σ and π combinations can be formed with s, p, and d orbitals of the appropriate symmetry but the δ orbitals arise from the overlap of d orbitals on different atoms.

If the atomic orbital $\psi(A)$ on A is small wherever the orbital $\psi(B)$ on B is large, or vice versa, then the product of their amplitudes is everywhere small and the integral—the sum of these products—is small. If $\psi(A)$ and $\psi(B)$ are simultaneously large in some region of space, then S may be large. If the two normalized atomic orbitals are identical (e.g. $1s$ orbitals on the same nucleus), then $S = 1$. This is illustrated in Fig. 14.31a and b. In some cases, simple formulas can be given for overlap integrals and their variation with bond length plotted (Fig. 14.32). It follows that $S = 0.59$ for two $1s$ orbitals at the equilibrium bond length in H_2^+, which is an unusually large value. Typical values for orbitals with $n = 2$ are in the range 0.2 to 0.3.

Now consider the arrangement in Fig. 14.31c in which an s orbital is superimposed on a p_x orbital of a different atom. At some point $\mathbf{r}$ the product $\psi(A)^*\psi(B)$ may be large. However, there is a point $\mathbf{r'}$ where $\psi(A)^*\psi(B)$ has exactly the same magnitude but an opposite sign. When the integral is evaluated, these two contributions are added together and cancel. For every point in the upper half of the diagram, there is a point in the lower that cancels, so $S = 0$. Therefore, there is no net overlap between the s and p orbitals in this arrangement: in the language introduced in Section 14.3, orbitals for which $S = 0$ are orthogonal.

The structures of homonuclear diatomic molecules

We show the general layout of the valence-shell atomic orbitals of Period 2 atoms on the left and right of the molecular orbital energy-level diagrams in Figs 14.26 and 14.28. The lines in the middle are an indication of the energies of the molecular orbitals that can be formed by overlap of atomic orbitals: from the eight valence shell orbitals (four from each atom), we can form eight molecular orbitals. The core ($1s$) orbitals overlap so little with one another (because they are so compact) that their contribution to bonding can be ignored. If they were shown on the molecular orbital energy level diagram, then they would form a bonding and an antibonding combination that would be represented by two narrowly spaced horizontal lines that differed insignificantly in energy from the $1s$ orbitals themselves.

With the orbitals established, we can deduce the ground configurations of the molecules by adding the appropriate number of electrons to the orbitals and following the building-up rules. Anionic species (such as the peroxide ion, O_2^{2-}) need more electrons than the parent neutral molecules and cationic species (such as O_2^+) need fewer.

We shall illustrate the procedure with N_2, which has 10 valence electrons. Two electrons pair, enter, and fill the 1σ orbital; the next two enter and fill the $2\sigma^*$ orbital. Six electrons remain. There are two 1π orbitals, so four electrons can be accommodated in them. The last two enter the 3σ orbital. The ground-state configuration of N_2 is therefore

$$N_2 \qquad 1\sigma^2 2\sigma^{*2} 1\pi^4 3\sigma^2$$

and the bond order is

$$b = \tfrac{1}{2}(8 - 2) = 3$$

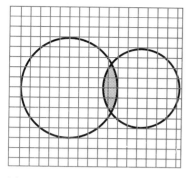

(a)

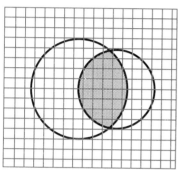

(b)

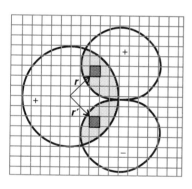

(c)

14.31 A schematic representation of the contributions to the overlap integral. (a) $S \approx 0$ because the orbitals are far apart and their product is always small. (b) S is large (but less than 1) because $\psi(A)\psi(B)$ is large over a substantial region. (c) $S = 0$ because the positive region of overlap is exactly cancelled by the negative region.

This bond order accords with the Lewis structure of the molecule $(:N{\equiv}N:)$ and is consistent with its high dissociation energy $(942 \text{ kJ mol}^{-1})$.

Example 14.8 *Writing the electron configuration of a diatomic molecule*

Write the ground-state electron configuration of O_2.

Method. First, we decide which of the two molecular orbital energy-level diagrams to adopt, and the total number of valence electrons to accommodate. Then, those electrons are fed into the available orbitals in order of increasing energy and in accord with the building-up principle.

Answer. The O_2 molecule $(2Z = 16)$ has two more electrons than N_2, and 12 of the 16 are in its valence shell. The first 10 recreate the N_2 configuration (with a reversal of the order of the 3σ and 1π orbitals); the last two must enter the $2\pi^*$ orbitals. Its configuration and bond order are therefore

$$O_2 \qquad 1\sigma^2 2\sigma^{*2} 3\sigma^2 1\pi^4 2\pi^{*2} \qquad b = 2$$

Comment. The bond order accords with the classical view that oxygen has a double bond and may be denoted $\ddot{O}{=}\ddot{O}$. We see below that the two outermost electrons are unpaired; to emphasize this feature the structure is sometimes written $\ddot{O}{\cdot}{\cdot}\ddot{O}$.

Exercise E14.8. Write the electron configuration of F_2 and deduce its bond order.

$$[1\sigma^2 2\sigma^{*2} 3\sigma^2 1\pi^4 2\pi^{*4}, \ 1]$$

According to Hund's rule, the two $2\pi^*$ electrons in O_2 will occupy different orbitals: one will enter $2\pi_x^*$ and the other will enter $2\pi_y^*$. Because the electrons are in different orbitals, they will have parallel spins. Therefore, we can predict that an O_2 molecule will have a net spin angular momentum $(S = 1)$ and, in the language introduced in Section 13.6, be in a triplet state. Because electron spin is the source of a magnetic moment, we can go on to predict that oxygen should be paramagnetic,[4] which is in fact the case.

An F_2 molecule has two more electrons than an O_2 molecule and the configuration and bond order

$$F_2 \qquad 1\sigma^2 2\sigma^{*2} 3\sigma^2 1\pi^4 2\pi^{*4} \qquad b = 1$$

We conclude that F_2 is a singly-bonded molecule, in agreement with its Lewis structure $:\ddot{F}{-}\ddot{F}:$. The low bond order is consistent with its low dissociation energy $(154 \text{ kJ mol}^{-1})$. The hypothetical molecule dineon, Ne_2, has two further electrons:

$$Ne_2 \qquad 1\sigma^2 2\sigma^{*2} 3\sigma^2 1\pi^4 2\pi^{*4} 4\sigma^{*2} \qquad b = 0$$

The zero bond order is consistent with the monatomic nature of Ne.

---

4 A paramagnetic substance tends to move into a magnetic field; a diamagnetic substance tends to move out of one. Paramagnetism, the rarer property, arises when the molecules have unpaired electron spins. Both properties are discussed in more detail in Section 22.6.

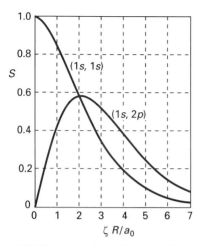

14.32 The overlap integral between $1s$ orbitals and between $1s$ and $2p_z$ orbitals as a function of separation. The $1s$ orbital is H1s and the $2p$ orbital is proportional to $e^{-\zeta R/a_0}$, where ζ is a variable parameter.

Example 14.9 *Judging the relative bond strengths of molecules and ions*

Judge whether N_2^+ is likely to have a larger or smaller dissociation energy than N_2.

Method. Because the molecule with the larger bond order is likely to have the larger dissociation energy, we should compare their electronic configurations and assess their bond orders.

Answer. From Fig. 14.28:

$$N_2 \qquad 1\sigma^2 2\sigma^{*2} 1\pi^4 3\sigma^2 \qquad b = 3$$
$$N_2^+ \qquad 1\sigma^2 2\sigma^{*2} 1\pi^4 3\sigma^1 \qquad b = 2.5$$

Because the cation has the smaller bond order, we expect it to have the smaller dissociation energy.

Comment. The experimental dissociation energies are 942 kJ mol^{-1} for N_2 and 842 kJ mol^{-1} for N_2^+.

Exercise E14.9. Which can be expected to have the higher dissociation energy, F_2 or F_2^+? [F_2^+]

14.6 More about notation

So far, we have seen how to label molecular orbitals by taking note of their symmetries with respect to rotation around the internuclear axis. Certain other features of their symmetry can also be used. As we shall see in later chapters, these symmetry designations are used, among other things, to formulate selection rules in molecular spectroscopy.

Parity

The molecular orbitals of homonuclear diatomic molecules are labelled with a subscript g or u that specifies their **parity**, their behaviour under the process known as **inversion**. To decide on the parity, we consider any point in a homonuclear diatomic, and note the sign of the orbital. Then we travel through the centre of the molecule (the 'centre of inversion') and go to the corresponding point on the other side; this process is the 'operation of inversion' (Fig. 14.33). If the orbital has the same sign, it has even parity and is denoted g (from *gerade*, the German for even). If the orbital has opposite sign, then it has odd parity and is denoted u (from *ungerade*, uneven). The parity designation applies only to homonuclear diatomic molecules, because heteronuclear diatomic molecules (such as HCl) do not have a centre of inversion.

We see from Fig. 14.33 that a bonding σ orbital has even parity, so we write it σ_g; a σ^* orbital has odd parity and is written σ_u. A bonding π orbital has odd parity and is denoted π_u and a π^* orbital has even parity, denoted π_g.

Term symbols

The term symbols of molecules (the analogues of the symbols ^2P, etc. for atoms) are constructed in a similar way to those for atoms, but now we

Centre of inversion

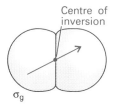

σ_g

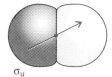

σ_u

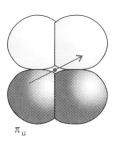

π_u

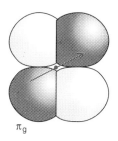

π_g

14.33 The parity of an orbital is even (g) if its amplitude is unchanged under inversion in the centre of symmetry of the molecule, but odd (u) if the amplitude changes sign. Heteronuclear diatomic molecules do not have a centre of inversion, so the g, u classification is irrelevant.

must pay attention to the component of **total orbital angular momentum** about the internuclear axis. This angular momentum is denoted by the symbols $\Sigma, \Pi, \Delta, \ldots$ corresponding to the S, P, D, ... of atoms:

> Component of total
> orbital angular momentum   0   ±1   ±2 ...
> about internuclear axis/$\hbar$
> Symbol: Σ Π Δ ...

A single electron in a σ orbital has zero orbital angular momentum: the orbital is cylindrically symmetrical and has no angular nodes when viewed along the internuclear axis. The term symbol for H_2^+ is therefore Σ. As in atoms, we use a superscript with the value of $2S + 1$ to denote the multiplicity of the term. In this case, since there is only one electron, $S = \frac{1}{2}$ and the term symbol is $^2\Sigma$, a doublet term. The overall parity of the term is added as a right subscript and (if there are several electrons) is calculated using

$$g \times g = g \qquad u \times u = g \qquad u \times g = u$$

(The rules can be generated by interpreting g as $+1$ and u as -1.) For H_2^+ the parity of the only occupied orbital is g, so the term itself is also g, and in full dress is $^2\Sigma_g$. The term symbol for any closed-shell homonuclear diatomic molecule is $^1\Sigma_g$ because the spin is zero (all electrons paired), there is no orbital angular momentum from a closed shell, and the overall parity is g.

A π electron has one unit of orbital angular momentum about the internuclear axis and, if it is the only electron outside a closed shell, gives rise to a Π term. If there are two π electrons (as in O_2) then the term symbol may be either Σ (if the electrons are orbiting in opposite directions, which is the case if they occupy different π orbitals) or Δ (if they are orbiting in the same direction, which is the case if they occupy the same π orbital). For O_2 the two π electrons occupy different orbitals with parallel spins, so the ground term is $^3\Sigma$. The overall parity of the molecule is

$$(\text{closed shell}) \times g \times g = g$$

The term symbol is therefore $^3\Sigma_g$.

Finally, a superscript $+$ or $-$ on a Σ term symbol denotes the behaviour of the molecular wavefunction under reflection in a plane containing the nuclei (Fig. 14.34). For O_2, one electron is $2\pi_x$, which changes sign under reflection in the yz-plane, and the other is $2\pi_y$, which does not. The overall reflection symmetry is therefore

$$(\text{closed shell}) \times (+) \times (-) = (-)$$

and the full term symbol is $^3\Sigma_g^-$. The need for all this dressing of a basic symbol will become apparent when we deal with the spectroscopic selection rules in Chapter 17.

14.7 Heteronuclear diatomic molecules

A **heteronuclear diatomic molecule** is a diatomic molecule formed from atoms of two different elements, such as CO and HCl. The electron

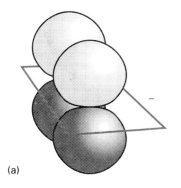

(a)

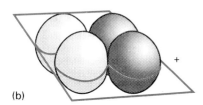

(b)

14.34 The $\pm$ symmetry refers to the symmetry of an orbital when it is reflected in a plane containing the two nuclei.

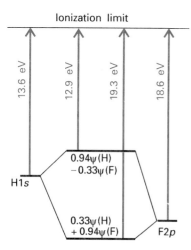

14.35 The atomic orbital energy levels of H and F atoms and the molecular orbitals they form. The bonding orbital has predominantly (88 per cent) F atom character and the antibonding orbital has predominantly (88 per cent) H atom character.

distribution in the covalent bond between the atoms is not evenly shared because it is energetically favourable for the electron pair to be found closer to one atom than the other. This imbalance results in a **polar bond**, which is a covalent bond in which the electron pair is shared unequally by the two atoms. The bond in HF, for instance, is polar, with the electron pair closer to the more electronegative F atom. The accumulation of the electron pair near the F atom results in that atom having a net negative charge, which is called a **partial negative charge** and denoted $\delta-$. There is a compensating **partial positive charge** $\delta+$ on the H atom.

Polar bonds

A polar bond consists of two electrons in an orbital of the form

$$\psi = c_A\psi(A) + c_B\psi(B) \tag{13}$$

with unequal coefficients. The proportion of the atomic orbital $\psi(A)$ in the bond is c_A^2 and that of $\psi(B)$ is c_B^2. A non-polar bond has $c_A^2 = c_B^2$, and a pure ionic bond has one coefficient zero (so that A^+B^- would have $c_A = 0$ and $c_B = 1$). The atomic orbital with the lower energy makes the larger contribution to the bonding molecular orbital. The opposite is true of the antibonding orbital, for which the dominant component comes from the atomic orbital with higher energy.

These points can be illustrated by considering HF and judging the energies of the atomic orbitals from the ionization energies of the atoms. The general form of the molecular orbitals is

$$\psi = c_H\psi(H) + c_F\psi(F)$$

where $\psi(H)$ is an H1s orbital and $\psi(F)$ is an F2p orbital. The H1s orbital lies at 13.6 eV below the zero of energy (the separated proton and electron) and the F2p orbital lies at 18.6 eV below the zero of energy (Fig. 14.35). Hence, the bonding σ orbital in HF is mainly F2p and the antibonding σ orbital is mainly H1s orbital, so there is a partial negative charge on the F atom and a partial positive charge on the H atom.

The variation principle

A systematic way of finding the coefficients in the linear combinations used to build molecular orbitals is provided by the **variation principle**:

> If an arbitrary wavefunction is used to calculate the energy, then the value calculated is never less than the true energy.

The arbitrary wavefunction is called the **trial wavefunction**. The principle implies that, if we vary the coefficients in the trial wavefunction until it achieves its lowest energy, then those coefficients will be the best. We might get a lower energy if we use a more complicated wavefunction (for example, by taking a linear combination of several atomic orbitals on each atom), but we shall have the optimum molecular orbital that can be built from the given set of atomic orbitals.

The method can be illustrated by the trial wavefunction

$$\psi = c_A\psi(A) + c_B\psi(B)$$

We show in the following *Justification* that the coefficients are given by the solutions of the two **secular equations**[5]

$$(\alpha_A - E)c_A + (\beta - ES)c_B = 0$$
$$(\beta - ES)c_A + (\alpha_B - E)c_B = 0 \tag{14}$$

where α is called a **Coulomb integral**. It can be interpreted as the energy of the electron when it occupies $\psi(A)$ (for α_A) or $\psi(B)$ (for α_B), and is negative. In a homonuclear diatomic molecule, $\alpha_A = \alpha_B$. The integral β is called a **resonance integral** (for classical reasons). It vanishes when the orbitals do not overlap, and at equilibrium bond lengths it is normally negative.

JUSTIFICATION

The trial wavefunction

$$\psi = c_A\psi(A) + c_B\psi(B)$$

is real but not normalized because at this stage the coefficient can take arbitrary values, so in the following we write $\psi^* = \psi$ but do not assume that $\int \psi^2 \, d\tau = 1$. The energy of the orbital is the expectation value of the energy operator (the Hamiltonian H, Section 11.5):

$$E = \frac{\int \psi^* H\psi \, d\tau}{\int \psi^* \psi \, d\tau}$$

We must search for values of the coefficients in the trial function that minimize the value of E. This is a standard problem in calculus, and is solved by finding the coefficients for which

$$\frac{\partial E}{\partial c_A} = 0 \quad \text{and} \quad \frac{\partial E}{\partial c_B} = 0$$

The first step is to express the two integrals in terms of the coefficients. The denominator is

$$\int \psi^2 \, d\tau = \int \{c_A\psi(A) + c_B\psi(B)\}^2 \, d\tau$$
$$= c_A^2 \int \psi(A)^2 \, d\tau + c_B^2 \int \psi(B)^2 \, d\tau + 2c_Ac_B \int \psi(A)\psi(B) \, d\tau$$
$$= c_A^2 + c_B^2 + 2c_Ac_BS$$

because the individual atomic orbitals are normalized and the third

---

5 'Secular' is derived from the Latin word for age or generation. The term comes via astronomy, where the same equations appear in connection with slowly accumulating modifications of planetary orbits.

integral is the overlap integral S (eqn 12). The numerator is

$$\int \psi H \psi \, d\tau = \int \{c_A \psi(A) + c_B \psi(B)\} H \{c_A \psi(A) + c_B \psi(B)\} \, d\tau$$

$$= c_A^2 \int \psi(A) H \psi(A) \, d\tau + c_B^2 \int \psi(B) H \psi(B) \, d\tau$$

$$+ 2 c_A c_B \int \psi(A) H \psi(B) \, d\tau$$

There are some complicated integrals in this expression, but we can denote them by the constants

$$\alpha_A = \int \psi(A) H \psi(A) \, d\tau \qquad \alpha_B = \int \psi(B) H \psi(B) \, d\tau$$

$$\beta = \int \psi(A) H \psi(B) \, d\tau$$

Then

$$\int \psi H \psi \, d\tau = c_A^2 \alpha_A + c_B^2 \alpha_B + 2 c_A c_B \beta$$

The complete expression for E is

$$E = \frac{c_A^2 \alpha_A + c_B^2 \alpha_B + 2 c_A c_B \beta}{c_A^2 + c_B^2 + 2 c_A c_B S}$$

Its minimum is found by differentiation with respect to the two coefficients. This involves elementary but slightly tedious work, and the end result is eqn 14.

To solve the secular equations for the coefficients we need to know the energy E of the orbital. As for any set of simultaneous equations the secular equations have a solution if the **secular determinant**, the determinant of the coefficients, is zero; that is, if

$$\begin{vmatrix} \alpha_A - E & \beta - ES \\ \beta - ES & \alpha_B - E \end{vmatrix} = 0 \tag{15}$$

This determinant expands to a quadratic equation in E (see Example 14.10). Its two roots give the energies of the bonding and antibonding molecular orbitals formed from the atomic orbitals and, according to the variation principle, these roots are the best energies for the given basis set.

Example 14.10 *Finding the roots of a secular determinant*

Find the energies E of the bonding and antibonding orbitals of a homonuclear diatomic molecule by solving eqn 15.

Method. We need to know that a 2×2 determinant expands as follows:

$$\begin{vmatrix} A & B \\ C & D \end{vmatrix} = AD - BC$$

Answer. When we apply the determinant expansion rule to eqn 15 with $\alpha_A = \alpha_B = \alpha$ we get

$$\begin{vmatrix} \alpha - E & \beta - ES \\ \beta - ES & \alpha - E \end{vmatrix} = (\alpha - E)^2 - (\beta - ES)^2 = 0$$

The solutions of this equation are

$$E_+ = \frac{\alpha + \beta}{1 + S} \qquad E_- = \frac{\alpha - \beta}{1 - S}$$

Exercise E14.10. Find the coefficients corresponding to these two energies. [See below, eqn 16]

The values of the coefficients in the linear combination are obtained by solving the secular equations using the two energies obtained from the secular determinant: the lower energy gives the coefficients for the bonding molecular orbitals, the upper energy gives the coefficients for the antibonding molecular orbital. The secular equations give expressions for the ratio of the coefficients in each case, and so we need a further equation in order to find their individual values. This equation is obtained by demanding that the best wavefunction should also be normalized, which means that, at this final stage, we must also ensure (from eqn 12) that

$$\int \psi^2 \, d\tau = c_A^2 + c_B^2 + 2 c_A c_B S = 1$$

Two simple cases

The complete solutions of the secular equations are very cumbersome even for 2×2 determinants, but there are two cases where the roots can be written down very simply.

We saw in Example 14.10 that, when the two atoms are the same and we can write $\alpha_A = \alpha_B = \alpha$, the solutions are

$$E_+ = \frac{\alpha + \beta}{1 + S} \qquad c_A = \left\{ \frac{1}{2(1 + S)} \right\}^{\frac{1}{2}} \qquad c_B = c_A$$

$$E_- = \frac{\alpha - \beta}{1 - S} \qquad c_A = \left\{ \frac{1}{2(1 - S)} \right\}^{\frac{1}{2}} \qquad c_B = -c_A$$

In this case, the best bonding function has the form

$$\psi_+ = \left\{ \frac{1}{2(1 + S)} \right\}^{\frac{1}{2}} \{\psi(A) + \psi(B)\} \tag{16a}$$

and the corresponding antibonding function is

$$\psi_- = \left\{ \frac{1}{2(1 - S)} \right\}^{\frac{1}{2}} \{\psi(A) - \psi(B)\} \tag{16b}$$

in agreement with the discussion of homonuclear diatomics we have already given.

When it is justifiable to neglect overlap, the secular determination is

$$\begin{vmatrix} \alpha_A - E & \beta \\ \beta & \alpha_B - E \end{vmatrix} = 0$$

and its solutions can be expressed in terms of the parameter ζ, with

$$\tan 2\zeta = \frac{2\beta}{\alpha_A - \alpha_B} \tag{17a}$$

The solutions are

$$E = \alpha_A - \beta \cot \zeta \qquad \psi = -(\sin \zeta)\psi(A) + (\cos \zeta)\psi(B)$$
$$E = \alpha_B + \beta \cot \zeta \qquad \psi = (\cos \zeta)\psi(A) + (\sin \zeta)\psi(B) \tag{17b}$$

An important feature revealed by these solutions is that, as the difference in energy $\alpha_A - \alpha_B$ between the two atomic orbitals increases, the value of ζ decreases.[6] When the energy difference is large the energies of the molecular orbitals differ only slightly from those of the atomic orbitals, which implies in turn that the bonding and antibonding effects are small. That is, *the strongest bonding and antibonding effects are obtained when the two contributing orbitals have closely similar energies.*

Now we apply these general points to hydrogen fluoride. In HF, the ionization energies of the valence orbitals are as follows:

$$H1s: \ 13.6 \text{ eV} \qquad F2s: \ 40.2 \text{ eV} \qquad F2p: \ 18.6 \text{ eV}$$

Because the F2p and H1s orbitals are much closer in energy than the F2s and H1s orbitals, to a first approximation we can neglect the contribution of the F2s orbital. More generally, the difference in energy is the justification for neglecting the contribution to bonding of core orbitals, for these differ in energy markedly from the valence orbitals. The core orbitals of one atom may have a similar energy to the core orbitals of the other atom; but core–core interaction is largely negligible because, as we have seen, the overlap between them (and hence the value of β) is so small.

Example 14.11 *Calculating the molecular orbitals of HF*

Calculate the wavefunctions and energies of the σ orbitals in the HF molecule, taking $\beta = -2.0$ eV.

Method. To use eqn 17, we need to know the values of the Coulomb integrals α_H and α_F. Since they represent the energies of the H1s and F2p electrons, respectively, we can estimate that they are approximately equal to (the negative of) the ionization energies of the atoms. Calculate ζ and then write the wavefunction using eqn 17.

---

6 Since $\tan x \approx x$ and $\cot x \approx 1/x$ when $x \ll 1$, when $\alpha_A - \alpha_B$ is large $\tan \zeta = \beta/(\alpha_A - \alpha_B)$. Then the energies of the two molecular orbitals are

$$E = \alpha_A - \beta \cot \zeta \approx \alpha_B \text{ and } E \approx \alpha_B + \beta \cot \zeta \approx \alpha_A$$

since $\sin x \approx x$ and $\cos x \approx 1$ when $x \ll 1$, the orbitals are respectively almost pure $\psi(B)$ and almost pure $\psi(A)$.

Answer. It follows that we should use $\alpha_H = -13.6$ eV and $\alpha_F = -18.6$ eV, which gives $\tan 2\zeta = -0.800$, so $\zeta = -19.33°$. Then

$$E = -19.3 \text{ eV} \qquad \psi = 0.33\psi(H) + 0.94\psi(F)$$
$$E = -12.9 \text{ eV} \qquad \psi = 0.94\psi(H) - 0.33\psi(F)$$

Comment. Notice how the lower energy orbital (the one with energy -19.3 eV) has a composition that is more F$2p$ orbital than H$1s$, and that the opposite is true of the higher-energy, antibonding orbital (Fig. 14.35).

Exercise E14.11. The ionization energy of Cl is 13.1 eV; find the form and energies of the σ orbitals in the HCl molecule using $\beta = -2.0$ eV.

$$[E = -11.3 \text{ eV}, \ \psi = -0.66\psi(H) + 0.75\psi(Cl);$$
$$E = -15.4 \text{ eV}, \ \psi = 0.75\psi(H) + 0.66\psi(Cl)]$$

Semi-empirical and ab initio methods

We can now see how the LCAO coefficients for diatomic molecules are found: the secular equations are solved for the energies, and those energies are used to obtain the optimum coefficients. There is still the problem of knowing the values of the Coulomb and resonance integrals. One approach has been to estimate them from spectroscopic information (such as ionization energies, as in Example 14.11). This combination of empirical data and quantum mechanical calculation has given rise to the **semi-empirical methods** of molecular structure calculation. The modern tendency, however, particularly for small molecules but increasingly for bigger ones too, is to calculate the integrals from first principles. These **ab initio** methods demand extensive numerical computation, and for this reason theoretical chemists are among the heaviest users of computers.

MOLECULAR ORBITALS FOR POLYATOMIC SYSTEMS

The bonds in polyatomic molecules are built in the same way as in diatomic molecules, the only difference being that we use more atomic orbitals to construct the molecular orbitals, and these molecular orbitals spread over the entire molecule. In general, a molecular orbital has the form

$$\psi = \sum_i c_i \psi_i \tag{18}$$

where the ψ_i are the atomic orbitals of a given symmetry type (for instance, of cylindrical σ symmetry in a linear molecule). In H_2O, for instance, the atomic orbitals are the two H$1s$ orbitals, the O$2s$ orbital, and the three O$2p$ orbitals (if we consider only the valence shell). From these six orbitals we can construct six molecular orbitals that spread over the three atoms. The molecular orbitals differ in energy: the lowest energy, most strongly bonding orbitals have the least number of nodes between adjacent atoms and the highest energy, most strongly antibonding orbitals have the greatest numbers of nodes between neighbouring atoms.

The principal difference between diatomic and polyatomic molecules lies in the greater range of shapes that are possible: a diatomic molecule is necessarily linear, but a triatomic molecule, for instance, may be linear or angular with a certain bond angle. In principle, the shape of a polyatomic molecule—the specification of all its symmetry, its bond lengths, and its bond angles—can be predicted by calculating the total energy of the molecule for a variety of conformations, and then identifying the one that leads to the lowest energy. However, more insight into the features that control molecular geometry can be obtained by analysing the orbitals and their energies in a more pictorial fashion. We shall illustrate what is involved by considering H_2O, which has an experimental bond angle of $104°$.

14.8 Walsh diagrams

The molecular orbitals of H_2O (and of H_2X molecules in general) have the form

$$\psi = c_1\psi(H_A) + c_2\psi(H_B) + c_3\psi_{2s}(O) + c_4\psi_{2p_x}(O)$$
$$+ c_5\psi_{2p_y}(O) + c_6\psi_{2p_z}(O)$$

There are six such orbitals (because they are built from six atomic orbitals) and eight valence electrons to accommodate in them. The simplest way of proceeding is to consider two conformations of the molecule, the linear $180°$ molecule and the angular $90°$ molecule, and then to identify how the molecular orbitals of one shape turn into the molecular orbitals of the other as the bond angle changes from $180°$ to $90°$. The procedure results in the construction of a **Walsh diagram**, a diagram showing the variation of orbital energy with molecular geometry.

The Walsh diagram for H₂X molecules

We can identify the general form of the molecular orbitals we need to consider in the angular H_2O molecule by deciding which linear combinations have the correct symmetry for orbital formation. At this stage we shall proceed intuitively; in Chapter 15 we shall see how to select the combinations by calculation.

In the hypothetical linear HOH molecule, the molecular orbitals (Fig. 14.36) are classified as either σ or π and have the form

$$\sigma_g: \quad \psi = c_1\psi_{2s}(O) + c_2\{\psi(H_A) + \psi(H_B)\} \quad \text{(two orbitals)}$$
$$\pi_u: \quad \psi = \psi_{2p_x}(O) \text{ and } \psi_{2p_z}(O) \quad \text{(one orbital each)}$$
$$\sigma_u: \quad \psi = c_1\psi_{2p_y}(O) + c_2\{\psi(H_A) - \psi(H_B)\} \quad \text{(two orbitals)}$$

We have added the parity labels, but they no longer tell us which is bonding and antibonding. Thus, there are two σ_g orbitals, one bonding (with the two coefficients the same sign) and the other antibonding (with the coefficients of opposite sign). There are no orbitals of π symmetry on the H atoms, so the $O2p_x$ and $O2p_z$ orbitals do not form bonding and antibonding molecular orbitals. They are examples of **non-bonding orbitals**, orbitals that do not contribute directly to the bonding between atoms. The coefficients in the molecular orbitals may be found in the normal way, by setting up and solving the secular determinants using

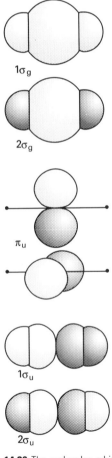

$1\sigma_g$

$2\sigma_g$

π_u

$1\sigma_u$

$2\sigma_u$

14.36 The molecular orbitals that can be constructed from the H1s, O2s, and O2p atomic orbitals in a hypothetical linear H_2O molecule.

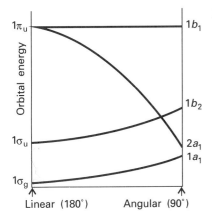

14.37 The Walsh diagram for H_2O. The energies of the linear molecule are shown on the left (see Fig. 14.36 for their compositions) and those of the 90° molecule are shown on the right (see Fig. 14.38). The actual molecule has a bond angle of 104°.

estimates of the Coulomb and resonance integrals, and the energies of the orbitals are shown on the left of the diagram in Fig. 14.37.

In the hypothetical 90° angular molecule, the molecular orbitals are formed from the following groupings of atomic orbitals (Fig. 14.38):

$$a_1: \quad \psi = c_1\psi_{2s}(O) + c_2\psi_{2p_z}(O) + c_3\{\psi(H_A) + \psi(H_B)\}$$
$$b_1: \quad \psi = \psi_{2p_x}(O)$$
$$b_2: \quad \psi = c_1\psi_{2p_y}(O) + c_2\{\psi(H_A) - \psi(H_B)\}$$

We can no longer classify the orbitals as σ and π because those labels apply only when there is an axis of symmetry; the labels used here will be explained in Chapter 15 (as will be the choice of the orbitals from which each molecular orbital is built). There are three a_1 orbitals (because we are combining three atomic orbitals), two b_2 orbitals, and one (non-bonding) b_1 orbital. The central feature of molecular orbital theory is the formation of molecular orbitals from all the atomic orbitals available that have the same symmetry, and the linear combinations listed above can be regarded as a grouping of the atomic orbitals into different symmetry classes. This grouping is the subject of Chapter 15.

The lowest-energy orbital in 90° H_2O is the one labelled $1a_1$, which is built from the overlap of the $O2p_z$ orbital with the $\psi(A) + \psi(B)$ combination of $H1s$ orbitals. As the bond angle changes to 180°, the two $H1s$ orbitals overlap less but the contribution to the molecular orbital of the $O2s$ orbital increases. In the 180° molecule, $O2s$ is the only contribution from the O atom to the $1a_1$ orbital. The replacement of $O2p_z$ by $O2s$ lowers the energy of the orbital. The energy of the $1b_2$ orbital is also lowered because the adverse H–H overlap decreases and the $H1s$ orbitals move into a better position for overlap with the $O2p_y$ orbital. The biggest change occurs for the $2a_1$ orbital. It is a pure $O2s$ orbital in the 90° molecule, but correlates with a pure $O2p_z$ orbital in the 180° molecule. Hence, it shows a steep rise in energy as the bond angle increases. The $1b_1$ orbital is a non-bonding $O2p$ orbital perpendicular to the molecular plane in the 90° molecule and remains non-bonding in the linear molecule. Hence, its energy barely changes with angle.

The shape of the H₂O molecule

The principal feature that determines whether or not the H_2O molecule is bent is whether the $2a_1$ orbital is occupied. This is the orbital that has considerable $O2s$ character in the bent molecule but not in the linear molecule. Hence, a lower total energy is achieved if, when it is occupied, the molecule is bent. The shape adopted by an H_2O molecule therefore depends on the number of electrons that occupy the orbitals.

Example 14.12 *Using a Walsh diagram to predict a shape*

Predict the shape of the H_2O molecule from the Walsh diagram.

Method. We choose an intermediate bond angle along the horizontal axis of the H_2O diagram in Fig. 14.37 and accommodate eight electrons. Then we consider whether the energy can be reduced by a modification

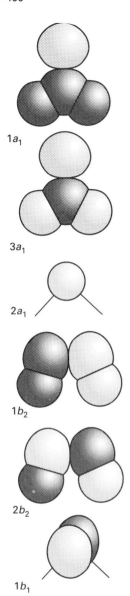

$1a_1$

$3a_1$

$2a_1$

$1b_2$

$2b_2$

$1b_1$

14.38 The molecular orbitals that can be constructed from the H1s, O2s, and O2p atomic orbitals in a hypothetical 90° H_2O molecule.

of the bond angle. To do so, we look at the effect upon the energies of the *occupied* orbitals of a change in bond angle.

Answer. The resulting configuration is $1a_1^2 2a_1^2 1b_2^2 1b_1^2$. The $2a_1$ orbital is occupied, so we expect the non-linear molecule to have a lower energy than the linear molecule.

Exercise E14.12. Predict the shape of the BeH_2 molecule.   [Linear]

14.9 The Hückel approximation

Molecular orbital theory takes large molecules and extended aggregates of atoms, such as solid materials, in its stride. We shall consider two examples: **conjugated molecules**, in which there is an alternation of single and double bonds along a chain of carbon atoms, and solids.

The π molecular orbital energy-level diagrams of conjugated molecules can be constructed using a set of approximations suggested by Erich Hückel in 1931. In his approach, the π orbitals are treated separately from the σ orbitals, and the latter form a rigid framework that determines the shape of the molecule. All the C atoms are treated identically, so all the Coulomb integrals α are set equal. For example, in ethene, we take the σ bonds as fixed, and concentrate on finding the energies of the single π bond and its companion antibond. In butadiene (CH_2=CH—CH=CH_2), the σ framework H_2C—CH—CH—CH_2 is taken as fixed, and we concentrate on finding the π orbitals spreading across the four C atoms.

The secular determinant

We express the π orbitals as LCAOs of the C2p orbitals. In ethene we would write

$$\psi = c_A \psi(A) + c_B \psi(B)$$

and in butadiene

$$\psi = c_A \psi(A) + c_B \psi(B) + c_C \psi(C) + c_D \psi(D)$$

where the $\psi(A)$ is a C2p orbital on atom A, and so on. Next, the optimum coefficients and energies are found by the variation principle as explained in Section 14.7. That is, we have to solve the secular determinant, which in the case of ethene is eqn 15 with $\alpha_A = \alpha_B = \alpha$. The determinant for butadiene is similar, but more atoms contribute and, being at various distances from each other, they have different overlap and resonance integrals.

Ethene:

$$\begin{vmatrix} \alpha - E & \beta - ES \\ \beta - ES & \alpha - E \end{vmatrix} = 0$$

Butadiene:

$$\begin{vmatrix} \alpha - E & \beta_{AB} - ES_{AB} & \beta_{AC} - ES_{AC} & \beta_{AD} - ES_{AD} \\ \beta_{BA} - ES_{BA} & \alpha - E & \beta_{BC} - ES_{BC} & \beta_{BD} - ES_{BD} \\ \beta_{CA} - ES_{CA} & \beta_{CB} - ES_{CB} & \alpha - E & \beta_{CD} - ES_{CD} \\ \beta_{DA} - ES_{DA} & \beta_{DB} - ES_{DB} & \beta_{DC} - ES_{DC} & \alpha - E \end{vmatrix} = 0$$

The roots of the ethene determinant can be found very easily (they are the same as those in Example 14.10). However, for elementary calculations, the roots of the butadiene determinant are obviously going to prove difficult to find. In a modern computation all the resonance integrals and overlap integrals would be computed, but a rough idea of the molecular orbital energy level diagram can be obtained very readily if we make the following additional **Hückel approximations**:

> All overlap integrals are set equal to zero.
> All resonance integrals between non-neighbours are set equal to zero.
> All remaining resonance integrals are set equal (to β).

These approximations are obviously very severe, but they let us calculate at least a general picture of the molecular orbital energy levels with very little work. The assumptions result in the following structure of the secular determinant:

> All diagonal elements: $\alpha - E$
> Off-diagonal elements between neighbouring atoms: β
> All other elements: 0

Ethene and frontier orbitals

For ethene, the Hückel approximation leads to

$$\begin{vmatrix} \alpha - E & \beta \\ \beta & \alpha - E \end{vmatrix} = 0$$

The determinant expands to

$$(\alpha - E)^2 - \beta^2 = 0$$

and its roots are

$$E_{\pm} = \alpha \pm \beta$$

The + sign corresponds to the bonding combination (β is negative) and the − sign corresponds to the antibonding combination (Fig. 14.39). The building-up principle then leads to the configuration $1\pi^2$, because each carbon atom supplies one electron to the π system, in agreement with the previous qualitative discussion. However, it goes beyond, because we have an estimate of the bond energy (β), and can see that an excited state of the molecule, when an electron is excited into the π^* orbital, lies about 2β above the ground state. The constant β is often left as an adjustable parameter; an approximate value for $(C2p, C2p)$-overlap π bonds is about -75 kJ mol^{-1}, corresponding to -0.8 eV.

The **highest occupied molecular orbital** in ethene, its HOMO, is the 1π orbital; the **lowest unoccupied molecular orbital**, its LUMO, is the $2\pi^*$ orbital. These two orbitals jointly form the **frontier orbitals** of the molecules. The frontier orbitals are important because they are largely responsible for the chemical and spectroscopic properties of the molecule.

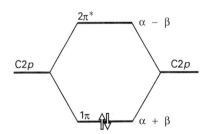

14.39 The Hückel molecular orbital energy levels of ethene (ethylene). Two electrons occupy the lower π orbital.

Butadiene and π-electron binding energy

For butadiene, the approximations result in the determinant

$$\begin{vmatrix} \alpha - E & \beta & 0 & 0 \\ \beta & \alpha - E & \beta & 0 \\ 0 & \beta & \alpha - E & \beta \\ 0 & 0 & \beta & \alpha - E \end{vmatrix} = 0$$

Example 14.13 *Finding the roots of a determinant*

Find the roots of the butadiene secular determinant.

Method. A 4×4 determinant is expanded in a series of moves like the 2×2 determinant treated in Example 14.10. After expansion, the terms are grouped to give a polynomial in E, which is set equal to 0 and then solved for E. A 4×4 determinant expands into a quartic equation, but we shall see that it may be expressed as a quadratic equation that can be solved by elementary methods.

Answer.

$$\begin{vmatrix} \alpha - E & \beta & 0 & 0 \\ \beta & \alpha - E & \beta & 0 \\ 0 & \beta & \alpha - E & \beta \\ 0 & 0 & \beta & \alpha - E \end{vmatrix}$$

$$= (\alpha - E) \begin{vmatrix} \alpha - E & \beta & 0 \\ \beta & \alpha - E & \beta \\ 0 & \beta & \alpha - E \end{vmatrix}$$

$$- \beta \begin{vmatrix} \beta & \beta & 0 \\ 0 & \alpha - E & \beta \\ 0 & \beta & \alpha - E \end{vmatrix}$$

$$= (\alpha - E)^2 \begin{vmatrix} \alpha - E & \beta \\ \beta & \alpha - E \end{vmatrix} - \beta (\alpha - E) \begin{vmatrix} \beta & \beta \\ 0 & \alpha - E \end{vmatrix}$$

$$- \beta^2 \begin{vmatrix} \alpha - E & \beta \\ \beta & \alpha - E \end{vmatrix} + \beta^2 \begin{vmatrix} 0 & \beta \\ 0 & \alpha - E \end{vmatrix}$$

$$= (\alpha - E)^4 - (\alpha - E)^2 \beta^2 - (\alpha - E)^2 \beta^2 - (\alpha - E)^2 \beta^2 + \beta^4$$

$$= (\alpha - E)^4 - 3(\alpha - E)^2 \beta^2 + \beta^4 = 0$$

The expanded determinant has the form of a quadratic equation

$$x^2 - 3x + 1 = 0, \quad \text{with } x = \left(\frac{\alpha - E}{\beta} \right)^2$$

The roots are $x = 2.62$ and 0.38. Therefore, the energies of the four LCAO–MOs are

$$E = \alpha \pm 1.62\beta, \ \alpha \pm 0.62\beta$$

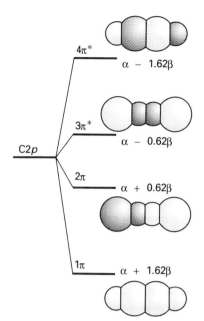

14.40 The Hückel molecular orbital energy levels of butadiene and the view from above of the corresponding π orbitals. The four p electrons (one supplied by each C atom) occupy the two lower π orbitals. Note that the orbitals are delocalized.

Exercise E14.13. Write down and expand the secular determinant for cyclobutadiene. [See Example 14.14]

We show in Example 14.13 that the energies of the four LCAO–MOs are

$$E = \alpha \pm 1.62\beta, \ \alpha \pm 0.62\beta$$

These orbitals and their energies are drawn in Fig. 14.40. Note that the greater the number of internuclear nodes, the higher the energy of the orbital. There are four electrons to accommodate, so the ground-state configuration is $1\pi^2 2\pi^2$. The frontier orbitals of butadiene are the 2π orbital (the HOMO, which is largely bonding) and the $3\pi^*$ orbital (the LUMO, which is largely antibonding).

There is an important point that emerges when we calculate the total **π-electron binding energy** E_π, the sum of the energies of each π electron, and compare it with what we find in ethene. In ethene the total energy is

$$E_\pi = 2(\alpha + \beta) = 2\alpha + 2\beta$$

In butadiene it is

$$E = 2(\alpha + 1.62\beta) + 2(\alpha + 0.62\beta) = 4\alpha + 4.48\beta$$

Therefore, the energy of the molecule lies lower by 0.48β (about -36 kJ mol^{-1}) than the sum of two individual π bonds. This extra stabilization of a conjugated system is called the **delocalization energy**.

Example 14.14 *Estimating the delocalization energy*

Use the Hückel approximation to find the energies of the π orbitals of cyclobutadiene, and estimate the delocalization energy.

Method. We need to set up the secular determinant using the same basis as for butadiene, but noting that A and D are also now neighbours. Then solve for the roots of the secular equation and assess the total π-bond energy. For the delocalization energy, subtract from the total π-bond energy the energy of two π bonds.

Answer. The secular determinant is

$$\begin{vmatrix} \alpha - E & \beta & 0 & \beta \\ \beta & \alpha - E & \beta & 0 \\ 0 & \beta & \alpha - E & \beta \\ \beta & 0 & \beta & \alpha - E \end{vmatrix} = 0$$

This determinant expands to

$$x(x-4) = 0, \quad \text{with } x = \left(\frac{\alpha - E}{\beta}\right)^2$$

The solutions are $x = 0$ and $x = 4$, so the energies of the orbitals are

$$E = \alpha + 2\beta, \ \alpha, \ \alpha, \ \alpha - 2\beta$$

Four electrons must be accommodated. Two occupy the lowest orbital (of energy $\alpha + 2\beta$), and two occupy the doubly degenerate orbitals (of energy α). The total energy, disregarding electron–electron repulsions, is therefore $4\alpha + 4\beta$. Two isolated π bonds would have an energy $4\alpha + 4\beta$; therefore, in this case, the delocalization energy is zero.

Exercise E14.14. Repeat the calculation for benzene.   [Next subsection]

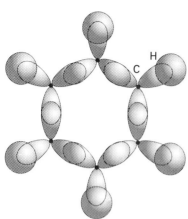

14.41 The σ framework of benzene is formed by the overlap of s and p orbitals that lie in the plane of the ring and which are depicted here as Csp^2 hybrids. These orbitals fit without strain into a hexagonal arrangement.

Benzene and aromatic stability

The most notable example of delocalization conferring extra stability is benzene and the aromatic molecules based on its structure. Benzene is often expressed in a mixture of valence-bond and molecular orbital terms, with typically valence-bond language used for its σ framework and molecular orbital language used to describe its π electrons.

First, the valence-bond component. The six C atoms are regarded as sp^2 hybridized with a single perpendicular $2p$ orbital. One H atom is bonded by $(Csp^2, H1s)$ overlap to each C atom, and the remaining hybrids overlap to give a regular hexagon of atoms (Fig. 14.41). The internal angle of a regular hexagon is $120°$ so the sp^2 hybridization is ideally suited for forming σ bonds. We see that benzene's hexagonal shape permits strain-free σ bonding.

Now consider the molecular orbital component of the description. The six $C2p$ orbitals overlap to give six π orbitals that spread all round the ring. Their energies are calculated within the Hückel approximation by solving the secular determinant

$$\begin{vmatrix} \alpha - E & \beta & 0 & 0 & 0 & \beta \\ \beta & \alpha - E & \beta & 0 & 0 & 0 \\ 0 & \beta & \alpha - E & \beta & 0 & 0 \\ 0 & 0 & \beta & \alpha - E & \beta & 0 \\ 0 & 0 & 0 & \beta & \alpha - E & \beta \\ \beta & 0 & 0 & 0 & \beta & \alpha - E \end{vmatrix} = 0$$

When this determinant is expanded in the same way as in Examples 14.10 and 14.13, the roots are found to be simply

$$E = \alpha \pm 2\beta, \alpha \pm \beta, \alpha \pm \beta$$

as shown in Fig. 14.42. The orbitals there have been given special labels that we explain in Chapter 15. Note that the lowest-energy orbital is bonding between all neighbouring atoms, the highest-energy orbital is antibonding between each pair of neighbours, and the intermediate orbitals are a mixture of bonding, non-bonding, and antibonding between adjacent pairs.

We now apply the building-up principle to the π system. There are six electrons to accommodate (one from each C atom), so the three lowest orbitals (a_{2u} and the doubly degenerate pair e_{1g}) are fully occupied, giving the ground-state configuration $a_{2u}^2 e_{1g}^4$. A significant point is that the only molecular orbitals occupied are those with net bonding character.

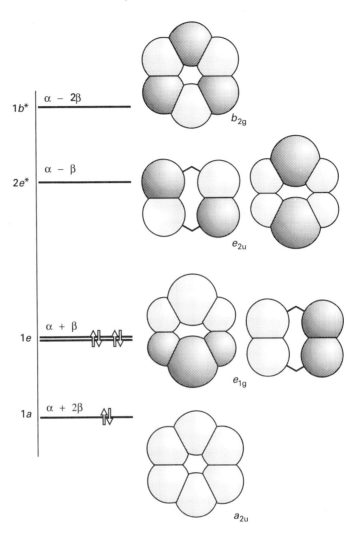

$1b^*$ $\alpha - 2\beta$ b_{2g}

$2e^*$ $\alpha - \beta$ e_{2u}

$1e$ $\alpha + \beta$ e_{1g}

$1a$ $\alpha + 2\beta$ a_{2u}

14.42 The Hückel π orbitals of benzene and the corresponding energy levels. The labels are explained in Chapter 15. The bonding and antibonding character of the delocalized orbitals reflects the numbers of nodes between the atoms. In the ground state, only the net bonding orbitals are occupied.

The π-electron energy of benzene is

$$E_\pi = 2(\alpha + 2\beta) + 4(\alpha + \beta) = 6\alpha + 8\beta$$

If we ignored delocalization and thought of the molecule as having three isolated π bonds, then it would be ascribed a π-electron energy of only $3(2\alpha + 2\beta) = 6\alpha + 6\beta$. The delocalization energy is therefore $2\beta \approx -150$ kJ mol^{-1}, which is considerably more than for butadiene.

This discussion suggests that aromatic stability can be traced to two main contributions. First, the shape of the regular hexagon is ideal for the formation of strong σ bonds: the σ framework is relaxed and without strain. Second, the π orbitals are such as to be able to accommodate all the electrons in bonding orbitals, and the delocalization energy is large.

14.10 The band theory of solids

The extreme case of delocalization is a solid, in which atom after atom lies in a three-dimensional array and takes part in bonding spreading throughout the sample. We shall distinguish two types of solid according to the variation of their electrical conductivity with temperature:

> A **metallic conductor** is a substance with a conductivity that decreases as the temperature is raised.
>
> A **semiconductor** is a substance with a conductivity that increases as the temperature is raised.

A semiconductor generally has a lower conductivity than that typical of metals, but the magnitude of the conductivity is not the criterion of the distinction. It is conventional to classify semiconductors with very low electrical conductivities as **insulators**. We shall use the latter term, but it should be appreciated that it is one of convenience rather than one of fundamental significance.

We shall consider a one-dimensional solid initially, which consists of a single, infinitely long line of atoms, each one having one s orbital available for forming molecular orbitals. We can construct the LCAO–MOs of the solid by adding atoms to a line and then find the electronic structure using the building-up principle.

The formation of bands

One atom contributes one s orbital at a certain energy (Fig. 14.43a). When a second atom is brought up it overlaps the first, and forms a bonding and antibonding orbital (Fig. 14.43b). The third atom overlaps its nearest neighbour (and only slightly the next-nearest) and, from these

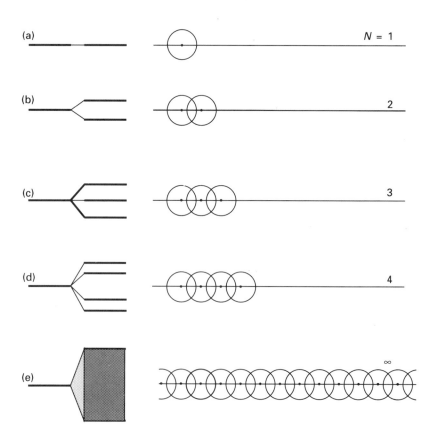

14.43 The formation of a band of N molecular orbitals by successive addition of N atoms to a line. Note that the band remains of finite width and, although it looks continuous when N is very large, it consists of N different orbitals.

three atomic orbitals, three molecular orbitals are formed (Fig. 14.43c): one is fully bonding, one fully antibonding, and the intermediate orbital is non-bonding between neighbours. The fourth atom leads to the formation of a fourth molecular orbital (Fig. 14.43d). At this stage we can begin to see that the general effect of bringing up successive atoms is to spread the range of energies covered by the molecular orbitals, and also to fill in the range of energies with more and more orbitals (one more for each atom). When N atoms have been added to the line, there are N molecular orbitals covering a band of finite width, and the Hückel secular determinant is

$$
\begin{vmatrix}
\alpha - E & \beta & 0 & 0 & 0 & \dots & 0 \\
\beta & \alpha - E & \beta & 0 & 0 & \dots & 0 \\
0 & \beta & \alpha - E & \beta & 0 & \dots & 0 \\
0 & 0 & \beta & \alpha - E & \beta & \dots & 0 \\
0 & 0 & 0 & \beta & \alpha - E & \dots & 0 \\
\vdots & \vdots & \vdots & \vdots & \vdots & \dots & \vdots \\
0 & 0 & 0 & 0 & 0 & \dots & \alpha - E
\end{vmatrix} = 0
$$

where β is now the (s, s) resonance integral. The theory of determinants applied to such a symmetrical example as this (technically a 'tridiagonal determinant') leads to the following expression for the roots:

$$
E_k = \alpha + 2\beta \cos\left(\frac{k\pi}{N+1}\right) \qquad k = 1, 2, \dots, N \tag{19}
$$

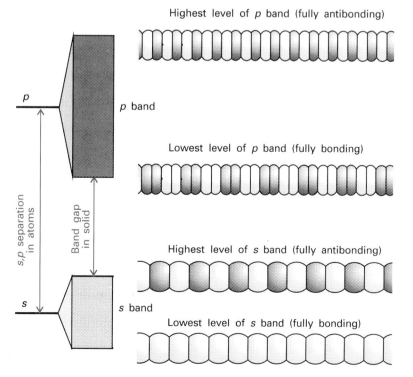

14.44 The overlap of s orbitals gives rise to an s band and the overlap of p orbitals gives rise to a p band. In this case the s and p orbitals of the atoms are so widely spaced that there is a band gap. In many cases the separation is less, and the bands overlap.

Highest level of p band (fully antibonding)

Lowest level of p band (fully bonding)

Highest level of s band (fully antibonding)

Lowest level of s band (fully bonding)

p

p band

s

s band

s,p separation in atoms

Band gap in solid

When N is infinitely large, the difference between neighbouring energy levels (the energies corresponding to k and $k + 1$) is infinitely small, but the band still has finite width:

$$E_N - E_1 \rightarrow 4\beta \qquad \text{as } N \rightarrow \infty$$

We can think of this band as consisting of N different molecular orbitals, the lowest-energy orbital ($k = 1$) being fully bonding, and the highest-energy orbital ($k = N$) being fully antibonding between adjacent atoms (Fig. 14.44).

The band formed from overlap of s orbitals is called the **s band**. If the atoms have p orbitals available, then the same procedure leads to a **p band** (as shown in the upper half of Fig. 14.44). If the atomic p orbitals lie higher in energy than the s orbitals, then the p band lies higher than the s band, and there may be a **band gap**, a range of energies to which no orbital corresponds.

The occupation of orbitals at $T = 0$

Now consider the electronic structure of a solid formed from atoms each able to contribute one electron (e.g. the alkali metals). There are N atomic orbitals and therefore N molecular orbitals squashed into an apparently continuous band. There are N electrons to accommodate. Because the orbitals in the bands are so close together, electrons close to the top of the filled orbitals can be excited out of them by thermal motion of the atoms. This is a complication that we can avoid by considering the solid at $T = 0$ when there is no such motion, and all the electrons occupy the lowest available orbitals.

At $T = 0$, only the lowest $\frac{1}{2}N$ molecular orbitals are occupied (Fig. 14.45), and the HOMO is called the **Fermi level**. However, unlike in the discrete molecules we have considered so far, there are empty orbitals very close in energy to the Fermi level, so it requires hardly any energy to excite the uppermost electrons. Some of the electrons are therefore very mobile, and give rise to electrical conductivity.

The occupation of orbitals at $T > 0$

At temperatures above absolute zero, there is no sharp distinction between occupied and unoccupied orbitals in a band because electrons can be excited by the thermal motion of the atoms. The population P of the orbitals is given by the **Fermi–Dirac distribution**, a version of the Boltzmann distribution that takes into account the effect of the Pauli principle:

$$P = \frac{1}{e^{(E - E_F)/kT} + 1} \tag{20}$$

The quantity E_F is the **Fermi energy**, the energy of the level for which $P = \frac{1}{2}$ (note that the Fermi energy changes as the temperature changes). The shape of the Fermi–Dirac distribution is shown in Fig. 14.46. For energies well above the Fermi energy, the 1 in the denominator can be neglected, and

$$P \approx e^{-(E - E_F)/kT} \tag{21}$$

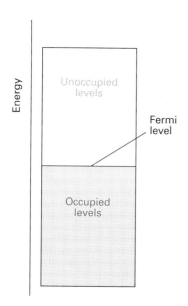

14.45 When N electrons occupy a band of N orbitals, it is only half full and the electrons near the Fermi level (the top of the filled levels) are mobile.

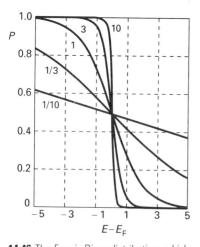

14.46 The Fermi–Dirac distribution, which gives the population of the levels at a temperature T. The high-energy tail decays exponentially towards zero. The numbers labelling the curves are the values of $(E - E_F)/kT$.

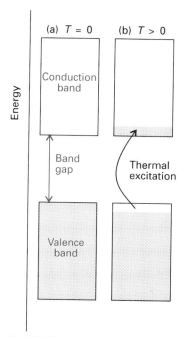

14.47 (a) When 2N electrons are present, the band is full and the material is an insulator at $T = 0$. (b) At temperatures above $T = 0$, electrons populate the levels of the upper 'conduction' band at the expense of the filled 'valence' band and the solid is a semiconductor.

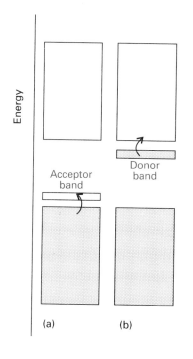

The population now resembles a Boltzmann distribution, decaying exponentially with increasing energy. The higher the temperature, the longer the exponential tail of the Fermi–Dirac distribution.

The electrical conductivity of a metallic solid decreases with increasing temperature even though more electrons are excited into empty orbitals. This apparent paradox is resolved by noting that the increase in temperature causes more vigorous thermal motion of the atoms, so collisions between the moving electrons and an atom are more likely. That is, the electrons are scattered out of their paths through the solid and are less efficient at transporting charge.

Insulators and semiconductors

When each atom provides two electrons, the $2N$ electrons fill the N orbitals of the s band. The Fermi level now lies at the top of the band (at $T = 0$), and there is a gap before the next band begins (Fig. 14.47a). As the temperature is increased, the tail of the Fermi–Dirac distribution extends across the gap, and electrons populate the empty orbitals of the upper band (Fig. 14.47b). They are now mobile, and the solid is an electric conductor. In fact, it is a semiconductor, because the electrical conductivity depends on the number of electrons that are promoted across the gap, and that number increases as the temperature is raised. If the gap is large, though, very few electrons will be promoted at ordinary temperatures, and the conductivity will remain close to zero, giving an insulator. Thus, the conventional distinction between an insulator and a semiconductor is related to the size of the band gap and is not absolute like the distinction between a metal (incomplete bands at $T = 0$) and a semiconductor (full bands at $T = 0$).

Another method of increasing the number of charge carriers and enhancing the semiconductivity of a solid is to implant foreign atoms into an otherwise pure material. If these **dopants** can trap electrons, they withdraw electrons from the filled band, leaving holes which allow the remaining electrons to move (Fig. 14.48a). This procedure gives rise to **p-type semiconductivity**, the p indicating that the holes are relatively positive to the electrons in the band. Alternatively, a dopant might carry excess electrons (e.g. P atoms introduced into germanium), and these additional electrons occupy otherwise empty bands, giving **n-type semiconductivity** (Fig. 14.48b), where n denotes the negative charge of the carriers. The preparation of doped but otherwise ultrapure materials was described in Section 8.7.

14.48 (a) A dopant with fewer electrons than its host can form a narrow band that accepts electrons from the valence band. The holes in the band are mobile, and the substance is a p-type semiconductor. (b) A dopant with more electrons than its host forms a narrow band that can supply electrons to the conduction band. The electrons it supplies are mobile, and the substance is an n-type conductor.

CHECK LIST OF KEY IDEAS

1 The **Born–Oppenheimer approximation** for the separation of electronic and nuclear motion.

2 The description of the chemical bond in terms of **valence-bond theory** (Section 14.1) and the formation of σ and π bonds.

3 The valence-bond explanation of the role of electron spin in the formation of a covalent bond (Section 14.1).

4 The significance of electron **promotion** in the formation of bonds, and particularly in increasing the number of bonds that carbon may form (Section 14.3).

5 The formation of **hybrid orbitals** and their use in the description of the shapes of molecules (Section 14.3).

6 The description of the **carbon–carbon double bond** (Section 14.3) and its **torsional rigidity**.

7 The definition and significance of the **overlap integral** of orbitals on the same atom (eqn 3) and the significance of **orthogonality** (Section 14.3).

8 The approximation of molecular orbitals as **linear combinations of atomic orbitals** (Section 14.4).

9 The formation of **bonding orbitals** and **antibonding orbitals** in the hydrogen molecule ion (eqns 7 and 9) and the description of the corresponding electron distributions.

10 The role of **constructive interference** and **destructive interference** between atomic orbitals in bond formation (Section 14.4).

11 The use of a **molecular orbital energy-level diagram** to derive the electron configurations of molecules (Section 14.5).

12 The definition of **bond order** (eqn 11) and a description of how bond lengths and strengths correlate with it.

13 The formation of σ **orbitals** and π **orbitals** in homonuclear diatomic molecules (Section 14.5)

and the assessment of overlap in terms of the **overlap integral** (eqn 12).

14 The ground-state electron configurations of **homonuclear diatomic molecules** and their bond orders (Section 14.5).

15 The **notation** used to specify atomic orbitals, including the use of g and u to denote orbital **parity** (Section 14.6).

16 The construction of **term symbols** used to classify overall molecular states (Section 14.6).

17 The ground-state electron configurations of **heteronuclear diatomic molecules** and the **polarity** of covalent bonds (Section 14.7).

18 The **variation principle** for calculating the optimum orbitals and energies (Section 14.7) and the form and solution of the **secular equations** (eqn 14) using a **secular determinant** (eqn 15).

19 The use of ionization energies to judge the composition and energies of molecular orbitals (Example 14.11).

20 The use of a **Walsh diagram** to show the dependence of orbital energy on molecular shape, and to judge the shapes of polyatomic molecules (Section 14.8).

21 The **Hückel approximation** for the description of delocalized bonding, the identification of **frontier orbitals**, and the calculation of the π-**electron binding energy** (Section 14.9).

22 The molecular orbital description of **aromatic stability** (Section 14.9).

23 The classification of substances according to their electrical conductivity and the **band theory** of solids (Section 14.10).

24 The **band gap** in solids and the **Fermi–Dirac distribution** (eqn 20) for the population of bands.

25 The classification of **semiconductors** as **p-type** and **n-type** and the role of **dopants** (Section 14.10).

EXERCISES

14.1. Write down the spatial part of the valence-bond wavefunction for the bond between an H atom and an O atom (formed by using H1s and O2p_z orbitals).

14.2. Write the spatial part of the wavefunction for an OH bond treated (a) as an ionic entity and (b) as a superposition of covalent and ionic contributions.

14.3. Write down the spatial part of the valence-bond wavefunction for the structure of an H_2O molecule (a) without assuming hybridization on the O atom, (b) with hybridization assumed.

14.4. Show that the sp^3 hybrids h_1 and h_2 given on p. 468 are orthogonal to one another.

14.5. Show that the sp^2 hybrid orbital $(s + \sqrt{2}p)/\sqrt{3}$ is normalized to 1 if the s and p orbitals are normalized to 1.

14.6. Write down the spatial part of the valence-bond wavefunction of an N_2 molecule. How can the pure covalent description of the bond be improved?

14.7. What is the composition of the hybrid orbitals in H_2S in which the bond angle is 92°?

14.8. Give the ground-state molecular orbital electron configurations and bond orders of (a) Li_2, (b) Be_2, and (c) C_2.

14.9. Give the ground-state electron configurations of (a) H_2^-, (b) N_2, and (c) O_2.

14.10. Give the ground-state electron configurations of (a) CO, (b) NO, and (c) CN^-.

14.11. From the ground-state electron configurations of B_2 and C_2 predict which molecule should have the greater bond dissociation energy.

14.12. Which of the molecules N_2, NO, O_2, C_2, F_2, and CN would you expect to be stabilized by (a) the addition of an electron to form AB^-, (b) the removal of an electron to form AB^+?

14.13. Sketch the molecular orbital energy level diagrams for (a) CO and (b) XeF and deduce their ground-state electron configurations. Is XeF likely to have a shorter bond length than XeF^+?

14.14. Where it is appropriate, give the parity of (a) π^* in F_2, (b) σ^* in NO, (c) δ in Tl_2, (d) δ^* in Fe_2.

14.15. Confirm the parities of the six π molecular orbitals of benzene illustrated in Fig. 14.42.

14.16. The bonding molecular orbital in H_2^+ is $\psi \propto \psi_{1s}(A) + \psi_{1s}(B)$. Determine the relative probability of finding the electron at one of the protons compared to the

midpoint between the two protons. Use $\psi_{1s} = (1/\pi a_0^3)^{\frac{1}{2}} \exp(-r/a_0)$ and an internuclear distance of 74 pm.

14.17. Consider a bonding electron in a diatomic molecule from the molecular orbital point of view. If the probabilities of finding the electron in atomic orbitals ψ_A and ψ_B are 0.25 and 0.75, respectively, what is the LCAO wavefunction for the electron? (Neglect overlap.)

14.18. The term symbol for the ground state of the N_2^+ ion is $^2\Sigma_g$. What is the total spin and total orbital angular momentum of the molecule? Show that the term symbol agrees with the electron configuration that would be predicted by using the building-up principle.

14.19. One of the excited states of the C_2 molecule has the valence electron configuration $1\sigma_g^2 2\sigma_u^2 1\pi_u^3 2\pi_g^1$. Give the multiplicity and parity of the term.

14.20. To what extent can the electron configurations of NO and N_2 be used to predict which is likely to have the shorter bond length?

14.21. One of the excited states of H_2 is $^3\Pi_u$, and can be considered to be formed from one H atom in its ground state and the other in an excited state. Give the electron configuration of the molecule.

14.22. Normalize the molecular orbital $\psi_s(A) + \lambda\psi_s(B)$ in terms of the parameter λ and the overlap integral S.

14.23. Confirm that the bonding and antibonding combinations $\psi_s(A) \pm \psi_s(B)$ are mutually orthogonal.

14.24. Account for the following observations:

    (a) The bond length in Li_2 is 267 pm, whereas that in Na_2 is 308 pm.
    (b) The dissociation energy of N_2 is 7.38 eV, whereas for N_2^+ it is 6.35 eV.
    (c) The dissociation energy of O_2 is 5.08 eV, whereas for O_2^+ it is 6.48 eV.
    (d) The dissociation energies of N_2^+ and O_2^+ are about the same.

14.25. Which of the following triatomic molecules and ions are expected to be linear: (a) CO_2, (b) NO_2, (c) NO_2^+, (d) NO_2^-, (e) SO_2, (f) H_2O? Give reasons in each case.

14.26. Construct the molecular orbital energy level diagrams of (a) ethene (ethylene) and (b) ethyne (acetylene) on the basis that the molecules are formed from overlap of the appropriately hybridized CH_2 or CH fragments.

14.27. Write down the secular determinants for (a) linear H_3, (b) cyclic H_3 within the Hückel approximation.

14.28. Predict the electronic configurations of (a) the benzene anion, (b) the benzene cation. Estimate the π-bond energy in each case.

14.29. Based on Hückel theory, which member of the following sets is expected to have the lowest energy? (a) The cation, neutral species, or anion of butadiene. (b) The cation, neutral species, or anion of cyclobutadiene. (c) Butadiene or cyclobutadiene.

14.30. On the basis of Hückel theory, would you predict the existence of the dianion of benzene? What factors beyond those contained in simple Hückel theory would need to be taken into consideration to answer this question?

14.31. Determine the probability that an electron in a metal will be found in a state with energy (a) $E = E_F - kT$, (b) $E = E_F$, (c) $E = E_F + kT$.

PROBLEMS
Numerical problems

14.1. Show that if a wave $\cos kx$ centred on A (so that x is measured from A) interferes with a similar wave $\cos k'x$ centred on B (with x measured from B) a distance R away, then constructive interference occurs in the intermediate region when $k = k' = \pi/2R$ and destructive interference if $kR = \frac{1}{2}\pi$ and $k'R = \frac{3}{2}\pi$.

14.2. Calculate the total amplitude of the normalized bonding and antibonding LCAO–MOs that may be formed from two H1s orbitals at a separation of 106 pm. Plot the two amplitudes for positions along the molecular axis both inside and outside the internuclear region. The overlap integral required is given by the expression

$$S = \left\{1 + \frac{R}{a_0} + \frac{1}{3}\left(\frac{R}{a_0}\right)^2\right\} e^{-R/a_0}$$

14.3. Repeat the calculation in Problem 14.2 but plot the probability densities of the two orbitals. Then form the 'difference density', the difference between ψ^2 and $\frac{1}{2}\{\psi_s(A)^2 + \psi_s(B)^2\}$.

14.4. Imagine a small electron-sensitive probe of volume 1.00 pm^3 inserted into an H$_2^+$ molecule-ion in its ground state. Calculate the probability that it will register the presence of an electron at the following positions: (a) at nucleus A, (b) at nucleus B, (c) half-way between A and B, (d) at a point 20 pm along the bond from A to 10 pm perpendicularly. Do the same for the molecule-ion the instant after the electron has been excited into the antibonding LCAO–MO.

14.5. The energy of H$_2^+$ with internuclear separation R is given by the expression

$$E = E_H - \frac{V_1 + V_2}{1 + S} + \frac{e^2}{4\pi\varepsilon_0 R}$$

where E_H is the energy of an isolated H atom, V_1 is the attractive potential energy between the electron centred on one nucleus and the charge of the other nucleus, V_2 is the attraction between the overlap density and one of the nuclei, and S is the overlap integral. The values are given below. Plot the molecular potential energy curve and find the bond dissociation energy (in eV) and the equilibrium bond length.

R/a_0	0	1	2	3	4
V_1/R_H	1.000	0.729	0.473	0.330	0.250
V_2/R_H	1.000	0.736	0.406	0.199	0.092
S	1.000	0.858	0.587	0.349	0.189

where $R_H = 27.3$ eV, $a_0 = 52.9$ pm, and $E_H = -\frac{1}{2}R_H$.

14.6. The same data as in Problem 14.5 may be used to calculate the molecular potential energy curve for the antibonding orbital, which is given by

$$E = E_H - \frac{V_1 - V_2}{1 - S} + \frac{e^2}{4\pi\varepsilon_0 R}$$

Plot the curve.

14.7. In the 'free electron molecular orbital' (FEMO) theory, the electrons in a conjugated molecule are treated as independent particles in a box of length L. Sketch the form of the two occupied orbitals in butadiene predicted by this model and predict the minimum excitation energy of the molecule. The tetraene CH$_2$=CHCH=CHCH=CHCH=CH$_2$ can be treated as a box of length $8R$, where $R = 140$ pm (as in this case, an extra half bond length is often added at each end of the box). Calculate the minimum excitation energy of the molecule and sketch the HOMO and LUMO.

14.8. One theory of the bonding in noble gas compounds is a molecular orbital approach featuring delocalized three-centre σ bonds. For example, in the linear FXeF molecule orbitals are formed by overlap of Xe5p and F2p orbitals. A typical three-centre σ orbital would then be $\psi = c_1 p_1 + c_{Xe} p_{Xe} + c_2 p_2$. Carry out a simple Hückel treatment of this system, obtaining energy levels and coefficients for the molecular orbitals. Discuss their bonding or antibonding character and deduce the ground-state electron configuration of the molecule. Is the molecule predicted to be stable? Initially you might make approximations such as $\alpha_{Xe} = \alpha_F$, but go on to use the facts that the ionization energies of xenon and fluorine are 12.13 eV and 17.42 eV, respectively.

Theoretical problems

14.9. An sp^2 hybrid orbital that lies in the xy-plane and makes an angle of $120°$ to the x-axis has the form

$$h = (\tfrac{1}{3})^{\frac{1}{2}}\{s - (\tfrac{1}{2})^{\frac{1}{2}}p_x + (\tfrac{3}{2})^{\frac{1}{2}}p_y\}$$

Use hydrogenic atomic orbitals to write the explicit form of the hybrid orbital. Show that it has its maximum amplitude in the direction specified.

14.10. Show that the strongest bonding interactions between atomic orbitals are likely when their energies are similar. *Hint*: Develop an expression for the energies of the bonding and antibonding orbitals by looking for the roots of the secular determinant.

14.11. Use the expressions in Problems 14.5 and 14.6 to show that the antibonding orbital is more antibonding than the bonding orbital is bonding at all internuclear separations.

14.12. Derive the expressions used in Problems 14.5 and 14.6 using the normalized LCAO–MOs for the H_2^+ molecule-ion. Proceed by evaluating the expectation value of the Hamiltonian for the ion. Make use of the fact that $\psi_s(A)$ and $\psi_s(B)$ each individually satisfy the Schrödinger equation for an isolated H atom.

14.13. Construct the Walsh diagram for an AH_3 molecule, and use it to predict the shapes of (a) NH_3, (b) CH_3^+.

14.14. Consider the allyl radical (**1**) and the cyclopropenyl radical (**2**). Set up and solve the Hückel secular determinant for the energies, draw a π-electron energy level diagram for each radical, and give their electronic

(1)   **(2)**

configurations. Estimates of β from experimental data give $\beta = 22\,000\ \text{cm}^{-1}$. From this estimate calculate the approximate wavelengths of π-electron transitions in the radicals. Also, determine the delocalization energies, and charge densities in these two radicals. Which radical is expected to be the more stable?

14.15. Write down expressions for the electron probability distributions in H_2 on the basis of valence-bond and molecular orbital (LCAO–MO) theories. Interpret the differences between the two wavefunctions and show that the valence-bond description can be brought closer to the molecular orbital wavefunction by the incorporation of ionic contributions. Go on to show that the molecular orbital description can be brought closer to the valence-bond description by configuration interaction; that is, by the inclusion of a contribution to the wavefunction of a wavefunction corresponding to another electron configuration. Specifically, add to the initial MO wavefunction a wavefunction in which both electrons occupy the antibonding orbital (that is, form the wavefunction $\sigma(1)\sigma(2) + \lambda\sigma^*(1)\sigma^*(2)$). Show that the valence-bond description with ionic–covalent resonance can be made to coincide with the molecular orbital description with configuration interaction.

15

Molecular symmetry

In this chapter we sharpen the concept of 'shape' into a precise definition of 'symmetry', and show that it may be discussed systematically. We shall see how to classify any molecule according to its symmetry, and to use this classification to discuss molecular properties without the need for detailed calculation. We shall see how to judge from its symmetry classification alone whether a molecule is polar or chiral.

After describing the symmetry properties of molecules themselves, we turn to a consideration of the effects of symmetry transformations on the orbitals that are associated with the atoms in a molecule, and see that their transformation properties can be used to set up a labelling scheme. These symmetry labels are very powerful, because they can be used to identify which integrals necessarily vanish. One important integral is the overlap integral between two orbitals. By knowing which atomic orbitals have non-zero overlap we can decide which ones can contribute to the formation of molecular orbitals and how to select linear combinations of atomic orbitals that match the symmetry of the nuclear framework. As we remarked in Chapter 14, this grouping of orbitals is one of the central features of molecular orbital theory. Finally, by considering the symmetry properties of integrals, we shall see that it is possible to derive the selection rules that govern the presence or absence of spectroscopic transitions.

The systematic discussion of symmetry is called **group theory**. Much of group theory is a summary of common sense about the symmetries of objects. However, because group theory is systematic, its rules can be applied in a straightforward, mechanical way, and in some cases it gives unexpected results. In most cases the theory gives a simple, direct method for arriving at useful conclusions with the minimum of calculation, and this is the aspect we stress here.

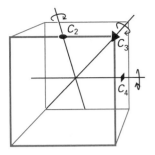

15.1 Some of the symmetry elements of a cube. The twofold, threefold, and fourfold axes are labelled with the conventional symbols.

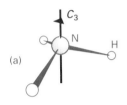

(a)

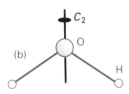

(b)

15.2 (a) An NH_3 molecule has a threefold (C_3) axis, and (b) an H_2O molecule has a twofold (C_2) axis. Both have other symmetry elements too.

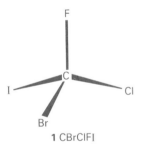

1 CBrClFI

THE SYMMETRY ELEMENTS OF OBJECTS

Some objects are 'more symmetrical' than others. A sphere is more symmetrical than a cube because it looks the same after it has been rotated through any angle about any diameter. A cube looks the same if it is rotated through 90°, 180°, or 270° about an axis passing through the centres of any of its opposite faces (Fig. 15.1), or by 120° or 240° about an axis passing through any of its opposite corners. Similarly, an NH_3 molecule is 'more symmetrical' than an H_2O molecule because NH_3 looks the same after rotations of 120° or 240° about the axis shown in Fig. 15.2, whereas H_2O looks the same only after a rotation of 180°.

An action that leaves an object looking the same after it has been carried out is called a **symmetry operation**. Symmetry operations include rotations, reflections, and inversions. There is a corresponding **symmetry element** for each symmetry operation, which is the point, line, or plane with respect to which the symmetry operation is performed. For instance, a rotation (a symmetry operation) is carried out around an axis (the corresponding symmetry element). We shall see that we can classify molecules by identifying all their symmetry elements, and grouping together molecules that possess the same set of symmetry elements. This procedure puts the trigonal pyramidal species NH_3 and SO_3^{2-} into one group and the angular species H_2O and SO_2 into another group.

15.1 Operations and elements

There are five kinds of symmetry operation (and five kinds of symmetry element) that leave at least a single point unchanged. The classification of objects according to symmetry operations that leave a single common point unchanged gives rise to the **point groups**. When we consider crystals (Chapter 21), we shall meet symmetries arising from translation through space. These more extensive groups are called **space groups**.

The **identity** E consists of doing nothing; the corresponding symmetry element is the entire object. Because every molecule is indistinguishable from itself if nothing is done to it, every object possesses at least the identity element. One reason for including it is that some molecules (e.g. CBrClFI, **1**) have only this symmetry element; another reason is technical and connected with the logical completeness of group theory (see *Further information 13*).

An **n-fold rotation** (the operation) about an **n-fold axis of symmetry** C_n (the corresponding element) is a rotation through $360°/n$. The operation C_1 is a rotation through 360°, and is equivalent to the identity operation E. An H_2O molecule has one twofold axis C_2. An NH_3 molecule has one threefold axis C_3, with which are associated two symmetry operations, one being 120° rotation in a clockwise sense and the other 120° rotation in a counterclockwise sense. (There is only one twofold rotation associated with a C_2 axis because clockwise and counterclockwise 180° rotations are identical.) A pentagon has a C_5 axis, with two (clockwise and counterclockwise) rotations through 72° associated with it. Also associated with it are the operations C_5^2 cor-

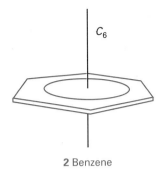

2 Benzene

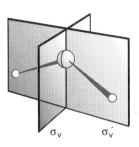

15.3 The H_2O molecule has two mirror planes. They are both vertical (i.e. contain the principal axis) and so are denoted σ_v and σ_v'.

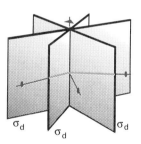

15.4 Dihedral mirror planes (σ_d) bisect the C_2 axes perpendicular to the principal axis.

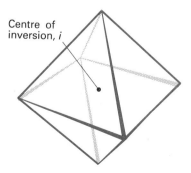

15.5 A regular octahedron has a centre of inversion, i.

responding to two successive C_5 rotations; there are two such operations, one through 144° in a clockwise sense and the other through 144° in a counterclockwise sense. A cube has three C_4 axes, four C_3 axes, and six C_2 axes. However, even this high symmetry is exceeded by a sphere, which possesses an infinite number of symmetry axes (along any diameter) of all possible integral values of n. If a molecule possesses several rotation axes, then the one (or more) with the greatest value of n is called the **principal axis**. The principal axis of a benzene molecule is the sixfold axis perpendicular to the hexagonal ring (**2**).

A **reflection** (the operation) in a **plane of symmetry** or a **mirror plane** σ (the element) may be either parallel or perpendicular to the principal axis of a molecule. If the plane is parallel to the principal axis, it is called **vertical** and denoted σ_v. An H_2O molecule has two vertical planes of symmetry (Fig. 15.3) and an NH_3 molecule has three. A vertical mirror plane that bisects the angle between two C_2 axes (Fig. 15.4) is called a **dihedral plane** and is denoted σ_d. When the plane of symmetry is perpendicular to the principal axis it is called **horizontal** and denoted σ_h. A C_6H_6 molecule has a C_6 principal axis and a horizontal mirror plane (as well as several other symmetry elements).

In an **inversion** (the operation) through a **centre of inversion** i (the element) we imagine taking each point in a molecule, moving it to its centre, and then moving it out the same distance on the other side; that is, the point (x, y, z) is taken into the point $(-x, -y, -z)$. Neither an H_2O molecule nor an NH_3 molecule has a centre of inversion, but a sphere and a cube do have one. A C_6H_6 molecule does have a centre of inversion, as does a regular octahedron (Fig. 15.5); a regular tetrahedron and a CH_4 molecule do not.

An **n-fold improper rotation** or an **n-fold rotary-reflection** (the operation) about an **n-fold axis of improper rotation** or an **n-fold rotary-reflection axis** S_n (the symmetry element) is composed of two successive transformations. The first component is a rotation through $360°/n$, and the second is a reflection through a plane perpendicular to the axis of that rotation; neither operation alone is a symmetry operation—only the overall outcome of the two operations. A CH_4 molecule (Fig. 15.6a) has three S_4 axes.

15.2 The symmetry classification of molecules

To classify molecules according to their symmetries, we list their symmetry elements and collect together those with the same list of elements. This procedure puts CH_4 and CCl_4, which both possess the same symmetry elements as a regular tetrahedron, into the same group and H_2O into another group.

The name of the group to which a molecule belongs is determined by the symmetry elements it possesses. There are two systems of notation (Table 15.1). The **Schoenflies system** is more common for the discussion of individual molecules, and the **Hermann–Mauguin system**, or **International system**, is used almost exclusively in the discussion of crystal symmetry.

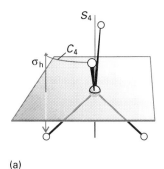

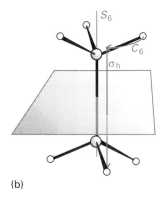

(a)

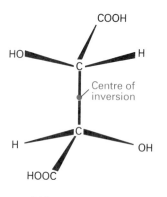

(b)

15.6 (a) A CH_4 molecule has a fourfold improper rotation axis, S_4: the molecule is indistinguishable after a 90° rotation followed by a reflection across the horizontal plane, but neither operation alone is a symmetry operation. (b) The staggered form of ethane has an S_6 axis composed of a 60° rotation followed by a reflection.

3 Meso-tartaric acid

4 Quinoline

Table 15.1 The notation for point groups*

C_i	$\bar{1}$								
C_s	m								
C_1	1	C_2	2	C_3	3	C_4	4	C_6	6
		C_{2v}	$2mm$	C_{3v}	$3m$	C_{4v}	$4mm$	C_{6v}	$6mm$
		C_{2h}	$2/m$	C_{3h}	$\bar{6}$	C_{4h}	$4/m$	C_{6h}	$6/m$
		D_2	222	D_3	32	D_4	422	D_6	622
		D_{2h}	mmm	D_{3h}	$\bar{6}2m$	D_{4h}	$4/mmm$	D_{6h}	$6/mmm$
		D_{2d}	$\bar{4}2m$	D_{3d}	$\bar{3}m$	S_4	$\bar{4}$	S_6	$\bar{3}$
T	23	T_d	$\bar{4}3m$	T_h	$m3$				
O	432	O_h	$m3m$						

* In the International system (or Hermann–Mauguin system) for point groups, a number n denotes the presence of an n-fold axis and m denotes a mirror plane. A diagonal line / indicates that the mirror plane is perpendicular to the symmetry axis. It is important to distinguish symmetry elements of the same type but of different classes, as in $4/mmm$, in which there are three classes of mirror plane (σ_v, σ_h, and σ_d). A bar over a number indicates that the element is combined with an inversion. The only groups listed in this table are the so-called crystallographic point groups (Section 21.1).

The groups C_1, C_i, C_s

A molecule belongs to the group C_1 if it has no element other than the identity (e.g. CBrClFI, **1**). It belongs to C_i if it has the identity and the inversion (e.g. *meso*-tartaric acid, **3**). It belongs to C_s if it has the identity and a plane of reflection (e.g. the quinoline molecule, **4**).

The groups C_n, C_{nv}, and C_{nh}

A molecule belongs to the group C_n if it possesses an n-fold axis. (Note that the symbol C_n is now playing a triple role: as the label of a symmetry element, a symmetry operation, and a group; the notation is summarized at the end of the chapter.) An H_2O_2 molecule in the configuration shown as (**5**) has the elements E and C_2, so it belongs to the group C_2.

If, in addition to the identity and a C_n axis, a molecule has n vertical mirror planes σ_v, then it belongs to the group C_{nv}. An H_2O molecule, for example, has the symmetry elements E, C_2, and $2\sigma_v$, so it belongs to the group C_{2v}. An NH_3 molecule has the elements E, C_3, and $3\sigma_v$, so it belongs to the group C_{3v}. A heteronuclear diatomic molecule such as HCl belongs to the group $C_{\infty v}$ because all rotations around the axis and reflections across it are symmetry operations. Other members of the group $C_{\infty v}$ include the linear OCS molecule and a cone.

Objects that, in addition to the identity and an n-fold principal axis, also have a horizontal mirror plane σ_h belong to the groups C_{nh}. An example is *trans*-CHCl=CHCl (**6**), which has the elements E, C_2, and σ_h and so belongs to the group C_{2h}; the molecule $B(OH)_3$ in the conformation shown in (**7**) belongs to the group C_{3h}. The presence of certain symmetry elements may be implied by the presence of others: thus, in C_{2h} the operations C_2 and σ_h jointly imply the presence of a centre of inversion (Fig. 15.7).

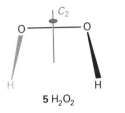

5 H_2O_2

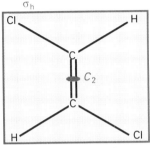

6 *trans*-CHCl=CHCl

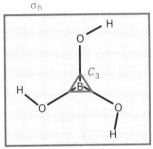

7 $B(OH)_3$

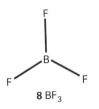

8 BF_3

The groups D_n, D_{nh}, and D_{nd}

A molecule that has an *n*-fold principal axis and *n* twofold axes perpendicular to C_n (Fig. 15.8) belongs to the group D_n. A molecule belongs to D_{nh} if it also possesses a horizontal mirror plane (Fig. 15.9). The trigonal planar BF_3 molecule (**8**) has the elements E, C_3, $3C_2$, and σ_h (with one C_2 axis along each B—F bond), and so belongs to D_{3h}. The C_6H_6 molecule has the elements E, C_6, $3C_2$, $3C_2'$, and σ_h together with some others that these imply, so it belongs to D_{6h}. All homonuclear diatomic molecules, such as N_2, belong to the group $D_{\infty h}$ because all rotations around the axis are symmetry operations, as are end-to-end rotation and end-to-end reflection; $D_{\infty h}$ is also the group of the linear OCO and HCCH molecules and of a uniform cylinder. Other examples of D_{nh} molecules are C_2H_4 (D_{2h}, **9**), PCl_5 (D_{3h}, **10**), and $[AuCl_4]^-$ (D_{4h}, **11**).

A molecule belongs to the group D_{nd} if, in addition to the elements of D_n, it possesses *n* dihedral mirror planes σ_d. The twisted, 90° allene (**12**) belongs to D_{2d}, and the staggered conformation of ethane (**13**) belongs to D_{3d}.

The groups S_n

Molecules that have not been classified into one of the groups we have mentioned so far but which possess one S_n axis belong to the group S_n. An example is tetraphenylmethane (**14**). Molecules belonging to S_n with $n > 4$ are rare. Note that the group S_2 is the same as C_i, so such a molecule will already have been classified as C_i.

The cubic groups

A number of very important molecules (e.g. CH_4 and SF_6) possess more than one principal axis. They all belong to the **cubic groups** and, in particular, to the **tetrahedral groups** T, T_d, and T_h or to the **octahedral groups** O, O_h (Fig. 15.10). A few icosahedral (20-faced) molecules belonging to the **icosahedral group** I (Fig. 15.11) are also known: they include some of the boranes and buckminsterfullerene, C_{60} (**15**). The groups T_d and O_h are the groups of the regular tetrahedron (e.g. CH_4) and the regular octahedron (e.g. SF_6), respectively. If the object possesses the rotational symmetry of the tetrahedron or the octahedron,

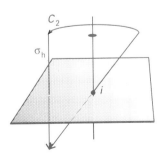

15.7 The presence of a twofold axis and a horizontal mirror plane jointly imply the presence of a centre of inversion in the molecule.

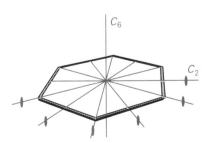

15.8 A molecule with *n* twofold rotation axes perpendicular to an *n*-fold rotation axis belongs to the group D_n.

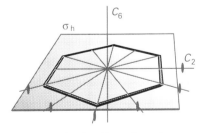

15.9 A molecule with a mirror plane perpendicular to a C_n axis, and with *n* twofold axes in the plane, belongs to the group D_{nh}.

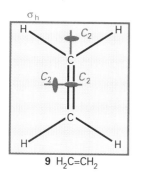

9 $H_2C=CH_2$

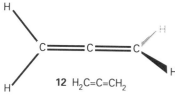

12 $H_2C=C=CH_2$

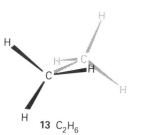

13 C_2H_6

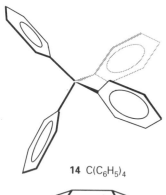

14 $C(C_6H_5)_4$

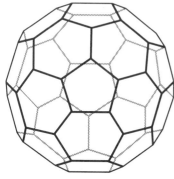

15 C_{60}, buckminsterfullerene

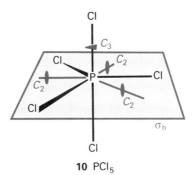

10 PCl_5

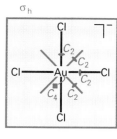

11 $[AuCl_4]^-$

but none of their planes of reflection, then it belongs to the simpler groups T or O (Fig. 15.12). The group T_h is based on T but also contains a centre of inversion (Fig. 15.13).

The full rotation group

The **full rotation group** R_3 (the 3 refers to rotation in three dimensions) consists of an infinite number of rotation axes with all possible values of n. A sphere and an atom belong to R_3, but no molecule does. Exploring the consequences of R_3 is a very important way of applying symmetry arguments to atoms and is an alternative approach to the theory of orbital angular momentum.

Example 15.1 *Identifying a point group of a molecule*

Identify the point group to which the sandwich molecule ruthenocene (two eclipsed cyclopentadienyl rings, **16**) belongs.

Method. The identification of a molecule's point group is simplified by referring to the flow diagram in Fig. 15.14 and the shapes shown in Fig. 15.15.

Answer. The path we trace through the flow diagram in Fig. 15.14 is shown by a blue line; it ends at D_{nh}. Because the molecule has a fivefold axis, it belongs to the group D_{5h}.

Comment. If the rings were staggered, as they are in an excited state of ferrocene that lies at $4\,\text{kJ mol}^{-1}$ above the ground state (**17**), then the horizontal reflection plane would be absent, but the dihedral planes would be present.

Exercise E15.1. Classify the pentagonal antiprismatic excited state of ferrocene.
$[D_{5d}]$

15.3 Some immediate consequences of symmetry

We can make some statements about the properties of a molecule as soon as we have identified its point group.

Polarity

A **polar molecule** is one with a permanent electric dipole moment (HCl, O_3, and NH_3 are examples). If the molecule belongs to the group C_n with

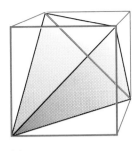

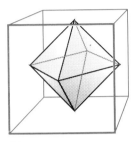

15.10 (a) Tetrahedral and (b) octahedral molecules are drawn in a way that shows their relation to a cube: they belong to the cubic groups T_d and O_h, respectively.

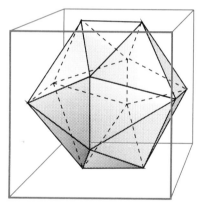

15.11 The relation of an icosahedron to a cube. The buckminsterfullerene molecule (**15**) is related to this object by cutting off each apex to form a regular pentagon.

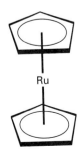

16 Ruthenocene, $Ru(C_5H_5)_2$

$n > 1$, then it cannot possess a charge distribution with a dipole moment perpendicular to the symmetry axis (Fig. 15.16a) because the symmetry of the molecule implies that any dipole that exists in one direction perpendicular to the axis is cancelled by an opposing dipole. For example, the perpendicular component of the dipole associated with one OH bond in H_2O is cancelled by an equal but opposite component of the dipole of the second OH bond, so any dipole that the molecule has must be parallel to the twofold symmetry axis. However, as the group makes no reference to operations relating the two ends of the molecules, a charge distribution may exist that results in a dipole along the axis (Fig. 15.16b); H_2O does indeed have a dipole moment parallel to its twofold symmetry axis. The same remarks apply to the group C_{nv}, so molecules belonging to any of the C_{nv} groups may be polar. In all the other groups, such as C_{3h}, D, etc., there are symmetry operations that take one end of the molecule into the other. Therefore, as well as having no dipole perpendicular to the axis, such molecules can have none along the axis, for otherwise these additional operations would not be symmetry operations.

We can conclude that only molecules belonging to the groups C_n, C_{nv}, and C_s may have a permanent electric dipole moment, and, in the case of C_n and C_{nv}, that dipole moment must lie along the rotation axis. Thus ozone O_3, which is angular and belongs to the group C_{2v}, may be polar, but carbon dioxide CO_2, which is linear and belongs to the group $D_{\infty h}$, is not.

Chirality

A **chiral molecule** (from the Greek word for 'hand') is a molecule that cannot be superimposed on its mirror image. Chiral molecules are said to be **optically active** because they rotate the plane of polarized light (a property discussed in more detail in Section 22.2). A chiral molecule and its mirror-image partner constitute an **enantiomeric pair** and rotate the plane of polarization in equal but opposite directions.

It follows from the theory of optical activity that a molecule may be chiral only if it does not possess an axis of improper rotation S_n. However, we need to be aware that such an axis may be present under a different name and be implied by other symmetry elements that are present. For example, molecules belonging to the groups C_{nh} possess an S_n axis implicitly because they possess both C_n and σ_h, which are the two components of an improper rotation axis. Any molecule containing a centre of inversion i also possesses an S_2 axis because i is equivalent to C_2 in conjunction with σ_h, and that combination of elements is S_2 (Fig. 15.17). It follows that all molecules with centres of inversion are achiral and hence optically inactive. Similarly, because $S_1 = \sigma$, it follows that any molecule with a mirror plane is achiral.

A molecule may be chiral if it does not have a centre of inversion or a mirror plane, which is the case with the amino acid alanine $NH_2CH(CH_3)COOH$ (**18**) but not with glycine NH_2CH_2COOH (**19**). However, a molecule may be achiral even though it does not have a centre of inversion. For example, the S_4 species (**20**) is achiral and optically inactive for, though it lacks i, it does have an S_4 axis.

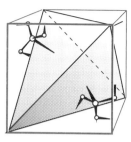

(a) *T*

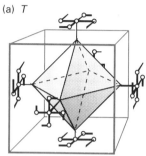

(b) *O*

15.12 The shapes corresponding to the point groups (a) *T* and (b) *O*. The presence of the windmill-like structures reduce the symmetry of the object from T_d and O_h, respectively.

C_2	(i.e. rotation by 180°)
σ +1	(i.e. no change of sign)
π −1	(i.e. change of sign)

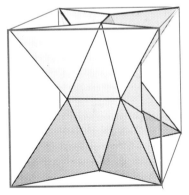

15.13 The shape of an object belonging to the group T_h.

CHARACTER TABLES

We shall now turn our attention away from the symmetries of molecules themselves and direct it at the symmetry characteristics of orbitals that belong to the various atoms in a molecule. This material will enable us to discuss the formulation and labelling of molecular orbitals and selection rules in spectroscopy.

15.4 Character tables and symmetry labels

We saw in Chapter 14 that molecular orbitals of diatomic and linear polyatomic molecules are labelled σ, π, etc. These labels refer to the symmetries of the orbitals with respect to rotations around the principal axis of the molecule. Thus, a σ orbital does not change sign under a rotation through any angle, a π orbital changes sign when rotated by 180°, and so on (Fig. 15.18). The symmetry classification σ and π can also be assigned to individual *atomic* orbitals in a linear molecule. For example, we can speak of an individual p_z orbital as having σ symmetry if the z-axis lies along the bond, because p_z is cylindrically symmetrical about the bond. This labelling of orbitals according to their behaviour under rotations can be generalized and extended to non-linear polyatomic molecules, where there may be reflections and inversions to take into account as well as rotations.

Labels analogous to σ and π are also used to denote the symmetries of orbitals in polyatomic molecules. These labels look like a, a_1, e, e_g, and we first encountered them in Section 14.8 in connection with H_2O and in Fig. 14.42 in connection with the molecular orbitals of benzene. As we shall see, these labels indicate the behaviour of the orbitals under the symmetry operations of the relevant group of the molecule.

The structure of character tables

A label is assigned to an orbital by referring to the **character table** of the group, which is a table that characterizes the different symmetry types possible in the point group. Thus, to assign the labels σ and π, we use the table shown in the margin. This table is a fragment of the full character table for a linear molecule. The entry +1 shows that the orbital remains the same and the entry −1 shows that the orbital changes sign under the operation C_2 at the head of the column (as illustrated in Fig. 15.18). Thus, to assign the label σ or π to a particular orbital, all we do is to compare the orbital's behaviour with the information in the character table.

The entries in a complete character table are derived using the formal techniques of group theory and are called **characters**, χ. These numbers characterize the essential features of each symmetry type in a way that we shall illustrate using the C_{3v} character table (Table 15.2). Character tables for other groups are given at the end of the Data section at the end of this volume and are used in exactly the same way.

The columns in a character table are labelled with the symmetry operations of the group, which in C_{3v} are E, C_3, and σ_v. The numbers multiplying each symmetry element are the numbers of operations of the same kind; more formally, they are the number of operations of each

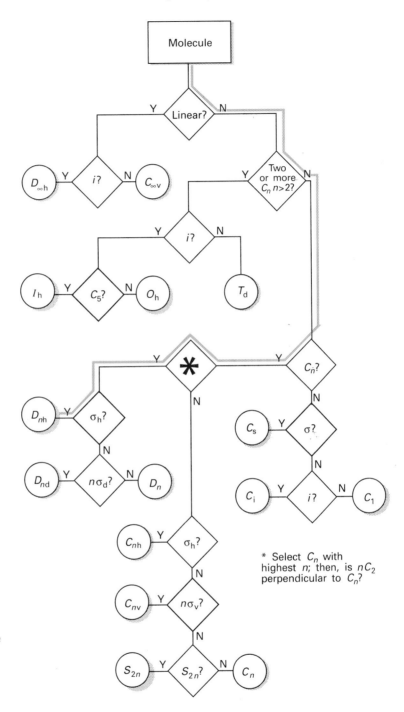

15.14 A flow diagram for determining the point group of a molecule. Start at the top and answer the question posed in each diamond (Y = yes, N = no).

class. Thus, we see from the top row of the table that there is only one identity operation, but there are two threefold rotations (clockwise and counterclockwise rotations by 120°) and three reflections (one through each of the three vertical mirror planes). The total number of operations in a group is called the **order** of the group and is denoted h.

The subsequent rows in the table summarize the symmetry properties of the orbitals. They are labelled with the **symmetry species** (the

15.15 A summary of the shapes corresponding to different point groups. The group to which a molecule belongs can often be identified from this diagram without going through the formal procedure in Fig. 15.14.

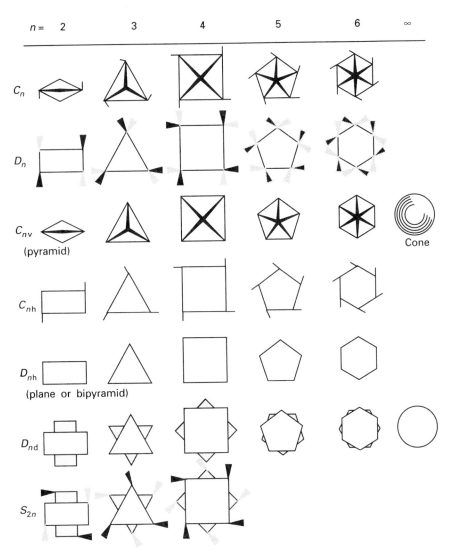

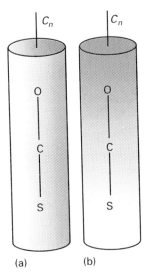

15.16 (a) A molecule with a C_n axis cannot have a dipole perpendicular to the axis, but (b) it may have one parallel to the axis. The shading represents regions of different electric charge.

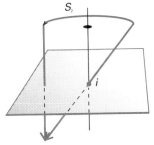

15.17 Some elements are implied by the other elements in a group. Any molecule containing an inversion also possesses at least an S_2 element because i and S_2 are equivalent.

analogues of the labels σ and π). More formally, the symmetry species label the **irreducible representations** of a group, which (as explained in the lengthy *Justification* that follows) are the basic types of behaviour that orbitals may show when subjected to the symmetry operations of the group. They are the analogues of the -1 and $+1$ that occur in the fragment of the character table given earlier and which showed, respectively, whether an orbital changed sign or did not change sign when the molecule was subjected to a rotation of $180°$ about its internuclear axis. By convention, irreducible representations are labelled with upper-case roman letters (such as A_1 and E) but the orbitals to which they apply are labelled with the lower-case italic equivalents (so an orbital of symmetry type A_1 is called an a_1 orbital). Note that care must be taken to distinguish the identity element E (italic, a column heading) from the symmetry label E (roman, a row label). Examples of each type of orbital are shown in Fig. 15.19.

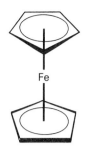

17 Excited ferrocene, $Fe(C_5H_5)_2{}^*$

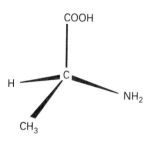

18 L-Alanine, $NH_2CH(CH_3)COOH$

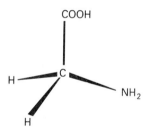

19 Glycine, NH_2CH_2COOH

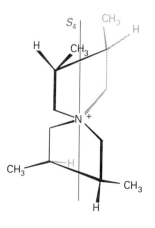

20 $N(CH_2CH(CH_3)CH(CH_3)CH_2)_2^+$

JUSTIFICATION

The origin of character tables is the representation of the effects of symmetry operations by matrices. As an illustration, consider the C_{2v} molecule SO_2 and the valence p_x orbitals on each atom (Fig. 15.20) which we shall denote p_S, p_A, and p_B. Under σ_v the change

$$(p_S, p_B, p_A) \leftarrow (p_S, p_A, p_B)$$

takes place. We can express this transformation using matrix multiplication:

$$(p_S, p_B, p_A) = (p_S, p_A, p_B)\begin{pmatrix} 1 & 0 & 0 \\ 0 & 0 & 1 \\ 0 & 1 & 0 \end{pmatrix}$$

This relation can be expressed more succinctly as

$$(p_S, p_B, p_A) = (p_S, p_A, p_B)\boldsymbol{D}(\sigma_v) \qquad \text{where } \boldsymbol{D}(\sigma_v) = \begin{pmatrix} 1 & 0 & 0 \\ 0 & 0 & 1 \\ 0 & 1 & 0 \end{pmatrix}$$

The matrix $\boldsymbol{D}(\sigma_v)$ is called a **representative** of the operation σ_v. Representatives take different forms according to the basis (the set of orbitals) that has been adopted.

We can use the same technique to find matrices that reproduce the other symmetry operations. For instance, C_2 has the effect

$$(-p_S, -p_B, -p_A) \leftarrow (p_S, p_A, p_B)$$

and its representative is

$$\boldsymbol{D}(C_2) = \begin{pmatrix} -1 & 0 & 0 \\ 0 & 0 & -1 \\ 0 & -1 & 0 \end{pmatrix}$$

The effect of σ_v' is

$$(-p_S, -p_A, -p_B) \leftarrow (p_S, p_A, p_B)$$

and its representative is

$$\boldsymbol{D}(\sigma_v') = \begin{pmatrix} -1 & 0 & 0 \\ 0 & -1 & 0 \\ 0 & 0 & -1 \end{pmatrix}$$

The identity operation has no effect on the basis, so its representative is the unit matrix:

$$\boldsymbol{D}(E) = \begin{pmatrix} 1 & 0 & 0 \\ 0 & 1 & 0 \\ 0 & 0 & 1 \end{pmatrix}$$

The set of matrices that represents all the operations of the group is called a **matrix representation** of the group for the particular basis we

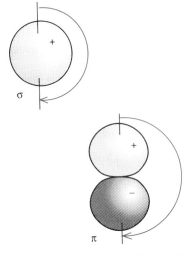

15.18 A rotation through 180° leaves the sign of a σ orbital unchanged but the sign of a π orbital is changed. In the language introduced in this chapter, the characters of the C_2 rotation are $+1$ and -1 for the σ and π orbitals, respectively.

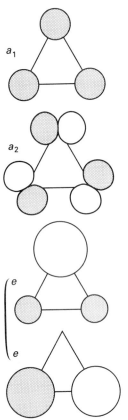

15.19 Typical symmetry-adapted linear combinations of orbitals in a C_{3v} molecule.

Table 15.2* The C_{3v} character table

$C_{3v}, 3m$	E	$2C_3$	$3\sigma_v$	$h = 6$	
A_1	1	1	1	z	$z^2, x^2 + y^2$
A_2	1	1	-1		
E	2	-1	0	(x, y)	$(xy, x^2 - y^2), (xz, yz)$

* More character tables are given at the end of the Data section at the end of this volume.

have chosen. We denote this three-dimensional representation by the symbol $\Gamma^{(3)}$. The discovery of a matrix representation of the group means that we have found a link between the symbolic manipulations of the operations and algebraic manipulations involving numbers. It may readily be verified that the matrices, when multiplied together, reproduce the group multiplication table (*Further information 13*).

The **character** χ of an operation in a particular matrix representation is the sum of the diagonal elements of the representative of each operation. Thus, in the basis we are illustrating, the characters of the representatives are

$$\begin{array}{cccc} \boldsymbol{D}(E) & \boldsymbol{D}(C_2) & \boldsymbol{D}(\sigma_v) & \boldsymbol{D}(\sigma_v') \\ 3 & -1 & 1 & -3 \end{array}$$

The character of an operation depends on the basis.

The representatives in the basis we have chosen are three-dimensional (i.e. they are 3×3 matrices), but inspection shows that they are all of the form

$$\begin{pmatrix} \blacksquare & 0 & 0 \\ 0 & \blacksquare & \\ 0 & & \end{pmatrix}$$

and that the symmetry operations never mix p_S with the other two functions. This suggests that the basis can be cut into two parts, one consisting of p_S alone and the other of (p_A, p_B). It is readily verified that the p_S orbital itself is a basis for the one-dimensional representation

$$\boldsymbol{D}(E) = 1 \qquad \boldsymbol{D}(C_2) = -1 \qquad \boldsymbol{D}(\sigma_v) = 1 \qquad \boldsymbol{D}(\sigma_v') = -1$$

which we shall call $\Gamma^{(1)}$. The remaining two basis functions are a basis for the two-dimensional representation $\Gamma^{(2)}$:

$$\boldsymbol{D}(E) = \begin{pmatrix} 1 & 0 \\ 0 & 1 \end{pmatrix} \qquad \boldsymbol{D}(C_2) = \begin{pmatrix} 0 & -1 \\ -1 & 0 \end{pmatrix}$$

$$\boldsymbol{D}(\sigma_v) = \begin{pmatrix} 0 & 1 \\ 1 & 0 \end{pmatrix} \qquad \boldsymbol{D}(\sigma_v') = \begin{pmatrix} -1 & 0 \\ 0 & -1 \end{pmatrix}$$

These matrices are the same as those of the original three-dimensional representation, except for the loss of the first row and column. We say that the original three-dimensional representation has been **reduced** to the **direct sum** of a one-dimensional representation **spanned** by p_S and a

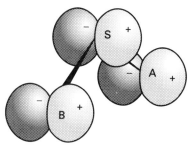

15.20 The three p_x orbitals that are used to illustrate the construction of a matrix representation in a C_{2v} molecule (SO_2).

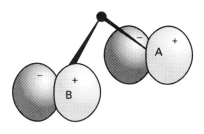

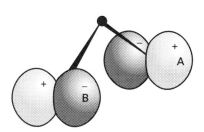

15.21 Two symmetry-adapted linear combinations of the basis orbitals shown in Fig. 15.20. The two combinations each span a one-dimensional irreducible representation, and their symmetry species are different.

two-dimensional representation spanned by (p_A, p_B). This reduction is consistent with the common-sense view that the central orbital plays a role different from the other two. The reduction is denoted symbolically by

$$\Gamma^{(3)} = \Gamma^{(1)} + \Gamma^{(2)}$$

The one-dimensional representation cannot be reduced any further, and is called an **irreducible representation** of the group. We can demonstrate that the two-dimensional representation is reducible (for the basis) by switching attention to the linear combinations $p_1 = p_A + p_B$ and $p_2 = p_A - p_B$. These combinations are sketched in Fig. 15.21.

The representatives in the new basis can be constructed from the old. For example, because under σ_v

$$(p_B, p_A) \leftarrow (p_A, p_B)$$

it follows (by applying these transformations to the linear combinations) that

$$(p_1, -p_2) \leftarrow (p_1, p_2)$$

The transformation is achieved by writing

$$(p_1, -p_2) = (p_1, p_2)\boldsymbol{D}(\sigma_v) \qquad \text{with } \boldsymbol{D}(\sigma_v) = \begin{pmatrix} 1 & 0 \\ 0 & -1 \end{pmatrix}$$

which gives us the representative $\boldsymbol{D}(\sigma_v)$ in the new basis. The remaining three representatives may be found similarly, and the complete representation is

$$\boldsymbol{D}(E) = \begin{pmatrix} 1 & 0 \\ 0 & 1 \end{pmatrix} \qquad \boldsymbol{D}(C_2) = \begin{pmatrix} -1 & 0 \\ 0 & 1 \end{pmatrix}$$

$$\boldsymbol{D}(\sigma_v) = \begin{pmatrix} 1 & 0 \\ 0 & -1 \end{pmatrix} \qquad \boldsymbol{D}(\sigma_v') = \begin{pmatrix} -1 & 0 \\ 0 & -1 \end{pmatrix}$$

The new representatives are all in block diagonal form

$$\begin{pmatrix} \blacksquare & 0 \\ 0 & \blacksquare \end{pmatrix}$$

and the two combinations are not mixed with one another by any operation of the group. We have therefore achieved the reduction of $\Gamma^{(2)}$ to the direct sum of two one-dimensional representations. Thus, p_1 spans

$$\boldsymbol{D}(E) = 1 \qquad \boldsymbol{D}(C_2) = -1 \qquad \boldsymbol{D}(\sigma_v) = 1 \qquad \boldsymbol{D}(\sigma_v') = -1$$

which is the same one-dimensional representation as that spanned by p_S, and p_2 spans

$$\boldsymbol{D}(E) = 1 \qquad \boldsymbol{D}(C_2) = 1 \qquad \boldsymbol{D}(\sigma_v) = -1 \qquad \boldsymbol{D}(\sigma_v') = -1$$

which is a different one-dimensional representation; we shall denote it $\Gamma^{(1)'}$. It is easy to check that either set of 1×1 matrices is a representation by multiplying pairs together and seeing that they reproduce the original group multiplication table.

Now we can make the final link to the material in the text. The **character table** of a group is the list of the characters of all its

Table 15.3* The C_{2v} character table

$C_{2v}, 2mm$	E	C_2	σ_v	σ_v'	$h = 4$	
A_1	1	1	1	1	z	z^2, y^2, x^2
A_2	1	1	−1	−1		xy
B_1	1	−1	1	−1	x	xz
B_2	1	−1	−1	1	y	yz

* More character tables are given at the end of the Data section.

	E	C_2	σ_v	σ_v'
$\Gamma^{(1)}$	1	−1	1	−1
$\Gamma^{(1)'}$	1	1	−1	−1

irreducible representations. At this point we have found two irreducible representations of the group C_{2v}. Their characters are shown in the margin. The two irreducible representations are normally labelled B_1 and A_2, respectively. An A or a B is used to denote a one-dimensional representation: A is used if the character under the principal rotation is $+1$ and B is used if the character is -1. (The letter E denotes a two-dimensional irreducible representation and a T a three-dimensional irreducible representation: all the irreducible representations of C_{2v} are one-dimensional.) There are in fact only two more species of irreducible representations of this group, for it is a surprising theorem of group theory that

Number of symmetry species = number of classes

In C_{2v} there are four classes (four columns in the character table), so there are only four species of irreducible representation. The full character table of the group is shown in Table 15.3.

Character tables and orbital degeneracy

The entry in the column headed by the identity operation E gives the degeneracy of the orbitals. Thus, in a C_{3v} molecule, any orbital with a symmetry label a_1 or a_2 must be non-degenerate and have a character 1 in the column headed E (the + sign is normally omitted in character tables). Conversely, if we know that we are dealing with a non-degenerate orbital in a C_{3v} molecule, then its symmetry type must be either A_1 or A_2 and the orbital will be labelled either a_1 or a_2. Similarly *any* doubly degenerate pair of orbitals in C_{3v} must be labelled e and have a character 2 in the column labelled E.

Because there are no characters with the value 3 in the column headed E, we know at a glance that there can be no triply degenerate orbitals in a C_{3v} molecule. This last point is a powerful result of group theory for it means that, with a glance at the character table of a molecule (which we can identify from the Data section once we know the point group of the molecule), we can state the maximum possible degeneracy of its orbitals.

JUSTIFICATION

The representative matrix for the identity operation is always the unit matrix. For a one-dimensional irreducible representation it is the 1×1

unit matrix, for a two-dimensional irreducible representation it is the 2×2 unit matrix, and so on. Since all the diagonal elements of a unit matrix are equal to 1, the character, the sum of the diagonal elements is equal to the dimension of the matrix.

Example 15.2 *Using a character table to judge degeneracy*

Can a trigonal planar molecule such as BF_3 have triply degenerate orbitals?

Method. First, we identify the point group, and then refer to the corresponding character table. The maximum number in the column headed by the identity E is the maximum orbital degeneracy possible in a molecule of that symmetry group.

Answer. Trigonal planar molecules belong to the point group D_{3h}. Reference to the character table for this group (Data section) shows that the maximum degeneracy is 2, as no character exceeds 2 in the column headed E. Therefore, the orbitals cannot be triply degenerate.

Exercise E15.2. The buckminsterfullerene molecule C_{60} belongs to the icosahedral point group. What is the maximum possible degree of degeneracy of its orbitals? [5]

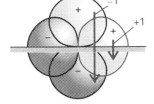

15.22 The two orbitals shown here have different properties under reflection through the mirror plane: one changes sign (character -1); the other does not (character $+1$).

Character	Significance
$+1$	The orbital is unchanged
-1	The orbital changes sign

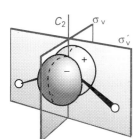

15.23 A p_x orbital on the central atom of a C_{2v} molecule and the symmetry elements of the group.

Characters and operations

The characters in the rows labelled A and B and in the columns headed by symmetry operations other than the identity E indicate the behaviour of an orbital under the corresponding operations. The specific interpretation is as shown in the margin. It follows that we can identify the symmetry label of the orbital by comparing the changes that occur to an orbital under each operation and then comparing the resulting $+1$ or -1 with the entries in a row of the character table for the point group concerned.

For the rows labelled E or T (which refer to the behaviour of sets of doubly and triply degenerate orbitals respectively), the characters in a row of the table are the *sums* of the characters summarizing the behaviour of the individual orbitals. Thus, if one member of a doubly degenerate pair remains unchanged under a symmetry operation but the other changes sign (Fig. 15.22), then the entry is reported as $\chi = 1 - 1 = 0$. Care has to be exercised with these characters because the transformations of orbital can be quite complicated; nevertheless, the sums of the characters are usually integers.

As an example, consider a $2p_x$ orbital on the O atom of H_2O. Because H_2O belongs to the point group C_{2v}, we know by referring to the C_{2v} character table (see Table 15.3) that the labels available for the orbitals are A_1, A_2, B_1, and B_2. We can decide the appropriate label for $2p_x$ by noting that under a 180° rotation (C_2) the orbital changes sign (Fig. 15.23), so it must be either B_1 or B_2 as only these two symmetry types have character -1 under C_2. The $2p_x$ orbital also changes sign under the reflection σ'_v, which identifies it as B_1. As we shall see, any molecular orbital built from this atomic orbital will also be a b_1 orbital. Similarly, $2p_y$ changes sign under C_2 but not under σ'_v, and so, it contributes to b_2 orbitals.

The behaviour of s, p, and d orbitals on a central atom under symmetry operations of the molecule is so important that the symmetry species of these orbitals are generally included in a character table. Thus, the right-hand column of Table 15.2 shows that p_z (which is proportional to $zf(r)$), has symmetry species A_1 in C_{3v}, whereas p_x and p_y (which are proportional to $xf(r)$ and $yf(r)$, respectively) are jointly of E symmetry. In technical terms, we say that p_x and p_y jointly **span** an irreducible representation of symmetry species E. An s orbital on the central atom always spans the fully symmetrical irreducible representation (usually A_1) of a group as it is unchanged under all symmetry operations.

The five d orbitals of a shell are represented by xy for d_{xy}, etc., and are listed on the right of the character table opposite the symmetry species to which they belong. Thus, we can see at a glance that, in C_{3v}, d_{xy} and $d_{x^2-y^2}$ on a central atom jointly belong to E.

The classification of linear combinations of orbitals

So far, we have dealt with the symmetry classification of individual orbitals. The same technique may be applied to linear combinations of orbitals on atoms that are related by symmetry transformations of the molecule, such as the combination $\psi_1 = \psi(A) + \psi(B) + \psi(C)$ of the three H1s orbitals in the C_{3v} molecule NH_3 (Fig. 15.24). Because ψ_1 is non-degenerate, we know that it is either A_1 or A_2. It remains unchanged under a C_3 rotation and under any of the three vertical reflections of the group, so its characters are

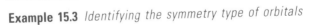

E	$2C_3$	$3\sigma_v$
1	1	1

Comparison with the C_{3v} character table shows that ψ_1 is of symmetry type A_1 and therefore that it contributes to a_1 molecular orbitals in NH_3.

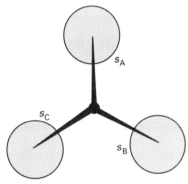

15.24 The three H1s orbitals used to construct symmetry-adapted linear combinations in a C_{3v} molecule such as NH_3.

Example 15.3 *Identifying the symmetry type of orbitals*

Identify the symmetry type of the orbital

$$\psi = \psi_A - \psi_B$$

in a C_{2v} NO_2 molecule, where ψ_A is a $2p_x$ orbital on one O atom and ψ_B that on the other.

Method. The negative sign in ψ indicates that the sign of ψ_B is opposite to that of ψ_A. We need to consider how the combination changes under each operation of the group and then write a character of $+1$, -1, or 0 as specified above. Then we compare the resulting characters with each row in the character table for the point group, and hence identify the symmetry species.

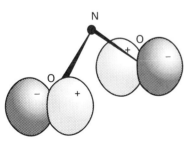

15.25 One symmetry-adapted linear combination of O2p_x orbitals in the C_{2v} NO_2 molecule.

Answer. The combination is shown in Fig. 15.25. Under a C_2 rotation ψ

changes into itself, implying a character of $+1$. Under the reflection σ_v' both orbitals change sign, so $\psi \to -\psi$, implying a character of -1. Under σ_v ψ also changes sign, so the character for this operation is also -1. The characters are therefore

E	C_2	σ_v	σ_v'
1	1	-1	-1

These values match the characters of the A_2 symmetry label, so ψ can contribute to an a_2 orbital.

Exercise E15.3. Identify the symmetry type of the combination $\psi(A) - \psi(B) + \psi(C) - \psi(D)$ in a square planar array of H atoms (of point group D_{4h}). $\hfill [B_{2g}]$

15.5 Vanishing integrals and orbital overlap

Suppose we had to evaluate the integral

$$I = \int f_1 f_2 \, d\tau \qquad (1)$$

where f_1 and f_2 are functions. For example, f_1 might be an atomic orbital ψ_A on one atom and f_2 an atomic orbital ψ_B on another atom, in which case the integral would be their overlap integral S. If we knew that the integral was zero, then we could say at once that a molecular orbital does not result from (ψ_1, ψ_2) overlap in that molecule. We shall now see that character tables provide a quick way of judging whether an integral is necessarily zero.

The key point in dealing with the integral I is that the value of any integral, and of an overlap integral in particular, is independent of the orientation of the molecule (Fig. 15.26). In group theoretical language we express this by saying that I is unchanged by any symmetry operation of the molecule, and that each operation brings about the trivial transformation

$$I \to I$$

Because the volume element $d\tau$ is unchanged by a symmetry operation, it follows that the integral is non-zero only if the integrand itself, the product $f_1 f_2$, is unchanged by any symmetry operation of the molecular point group. If the integral changed sign under a symmetry operation, then the integral would be the sum of equal and opposite contributions, and hence would be zero. It follows that the only contribution to a non-zero integral comes from functions for which, under any symmetry operation of the molecular point group,

$$f_1 f_2 \to f_1 f_2$$

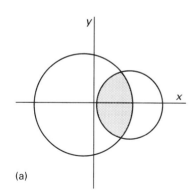

(a)

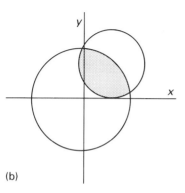

(b)

15.26 The value of an integral I (e.g. an area) is independent of the orientation of the object. That is, I is a basis of a representation of symmetry species A_1.

and hence for which the characters of the operations are all equal to $+1$. Therefore, for I not to be zero, the integrand $f_1 f_2$ must have symmetry species A_1 in the molecular point group.

We use the following procedure to deduce the symmetry species spanned by the product $f_1 f_2$ and hence to see whether it does indeed span A_1 (or whatever symbolizes the fully symmetric irreducible representation).

> (1) Decide on the symmetry species of the individual functions f_1 and f_2 by reference to the character table, and write their characters in two rows in the same order as in the table.

For example, if f_1 is the s_N orbital in NH_3 and f_2 is the linear combination $s_3 = s_B - s_C$ (Fig. 15.27), then, since s_N spans A_1 and s_3 is a member of the basis spanning E, we write

$$f_1: \qquad 1 \qquad 1 \qquad 1$$
$$f_2: \qquad 2 \qquad -1 \qquad 0$$

> (2) Multiply the numbers in each column, writing the results in the same order. For the NH_3 calculation,

$$f_1 f_2: \qquad 2 \qquad -1 \qquad 0$$

> (3) Inspect the row so produced, and see if it can be expressed as a sum of characters from each column of the group. If this sum does not contain A_1 then the integral must be zero.

In C_{3v}, for instance, any set of characters for the three symmetry operations can always be expressed as the sum

$$\chi = c_1 \chi(A_1) + c_2 \chi(A_2) + c_3 \chi(E)$$

and the integral must be zero if $c_1 = 0$. In the present example, the characters $2, -1, 0$ are those of E alone, so the integrand does not span A_1. It follows that the integral must be zero. Inspection of the form of the functions (see Fig. 15.27) shows why this is so: s_3 has a node running through s_N. Had we taken $f_1 = s_N$ and $f_2 = s_1$ instead, where $s_1 = s_A + s_B + s_C$, then, since each spans A_1 with characters $1, 1, 1$,

$$f_1: \qquad 1 \qquad 1 \qquad 1$$
$$f_2: \qquad 1 \qquad 1 \qquad 1$$
$$f_1 f_2: \qquad 1 \qquad 1 \qquad 1$$

The characters of the product are those of A_1 itself. Therefore, s_1 and s_N may have non-zero overlap.

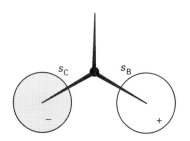

15.27 A symmetry-adapted linear combination that belongs to the symmetry species E in a C_{3v} molecule such as NH_3. This combination can form a molecular orbital by overlapping with the p_x orbital on the central atom (the orbital with its axis parallel to the width of the page; see Fig. 15.28c).

JUSTIFICATION

The procedure described above is based on the **little orthogonality theorem** of group theory:

$$\sum_C g(C) \chi^{(\Gamma)}(C) \chi^{(\Gamma')}(C) = 0$$

The sum is over the classes of operation (the columns in the character table), g is the number of operations in each class (such as the 2 in $2C_3$), and Γ and Γ' are two *different* irreducible representations. If Γ and Γ' are the same irreducible representations, then

$$\sum_C g(C)\chi^{(\Gamma)}(C)\chi^{(\Gamma)}(C) = h$$

where h is the order of the group.

To find out whether a reducible representation contains a given irreducible representation, we use an expression derived from the little orthogonality theorem. Thus, for a given operation the character of a reducible representation is a linear combination of the characters of the irreducible representations of the group:

$$\chi(C) = \sum_\Gamma c_\Gamma \chi^{(\Gamma)}(C)$$

To find the coefficients for a given irreducible representation Γ', we multiply both sides by $g(C)\chi^{(\Gamma')}$ and sum over all the classes of operation C:

$$\sum_C g(C)\chi^{(\Gamma')}(C)\chi(C) = \sum_C \sum_\Gamma g(C)\chi^{(\Gamma')}(C)c_\Gamma \chi^{(\Gamma)}(C)$$

When the right-hand side is summed over C, the little orthogonality theorem gives 0 for all terms for which Γ is not equal to Γ'. However, because we are summing over all Γ, a term with $\Gamma = \Gamma'$ is guaranteed to be present. Only that term contributes, and the right-hand side is then equal to $hc_{\Gamma'}$. It then follows that

$$c_{\Gamma'} = \frac{1}{h}\sum_C g(C)\chi^{(\Gamma')}(C)\chi(C)$$

This formula is exceptionally important for finding the decomposition of a reducible representation, because Γ' can be set equal to each irreducible representation in turn, and the coefficients c determined. It takes an even easier form if we only want to know if the totally symmetric irreducible representation is present, because all the characters of the representation are 1, so that

$$c_{A_1} = \frac{1}{h}\sum_C g(C)\chi(C)$$

It is important to note that group theory is specific about when an integral must be zero, but that integrals that it allows to be non-zero may be zero for reasons unrelated to symmetry. For example, the N—H distance may be so great that the s_1, s_N overlap integral is zero simply because the orbitals are so far apart.

Example 15.4 *Deciding if an integral must be zero* (1)

May the integral of the function $f = xy$ be non-zero when evaluated over a region the shape of an equilateral triangle centered on the origin?

Method. First, we note that an integral over a single function f is included in the previous discussion if we take $f_1 = f$ and $f_2 = 1$. Therefore, we need

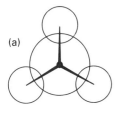

(a)

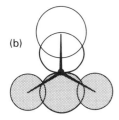

(b)

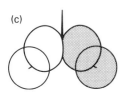

(c)

15.28 Orbitals of the same symmetry species may have non-vanishing overlap. This diagram illustrates the three bonding orbitals that may be constructed from (N2s, H1s) and (N2p, H1s) overlap in a C_{3v} molecule. (a) a_1; (b) and (c) the two components of the doubly degenerate e orbitals. (There are also antibonding orbitals of the same species.)

to judge whether f alone belongs to the symmetry species A_1 in the point group of the system. To decide that, we identify the point group and then examine the character table to see whether f belongs to A_1.

Answer. An equilateral triangle has the point-group symmetry D_{3h}. If we refer to the character table of the group, we see that xy is a member of a basis that spans the irreducible representation E'. Therefore, its integral must be zero, because the integrand has no component that spans A_1'.

Exercise E15.4. Can the function $x^2 + y^2$ have a non-zero integral when integrated over a regular pentagon centred on the origin?   [Yes]

Orbitals with non-zero overlap

The rules just given let us decide which atomic orbitals may have non-zero overlap in a molecule. We have seen that s_N may have non-zero overlap with s_1 (the combination $1s_A + 1s_B + 1s_C$), so that (s_N, s_1)-overlap bonding and antibonding molecular orbitals can form (Fig. 15.28). The general rule is that *only orbitals of the same symmetry species may have non-zero overlap*, so that *only orbitals of the same symmetry species form bonding and antibonding combinations*. It should be recalled from Chapter 14 that the selection of atomic orbitals that have mutual non-zero overlap is the central and initial step in the construction of molecular orbitals by the LCAO procedure. We are therefore at the point of contact between group theory and the material introduced in that chapter. The molecular orbitals formed from a particular set of atomic orbitals with non-zero overlap are labelled with the lower-case letter corresponding to the symmetry species. Thus, the (s_N, s_1)-overlap orbitals are called a_1 orbitals (and a_1^* if we wish to emphasize that they are antibonding).

The s_2 and s_3 linear combinations have symmetry species E. Does the N atom have orbitals that have non-zero overlap with them (and give rise to an e orbital)? Intuition (as supported by Fig. 15.28b and c) suggest that N$2p_x$ and N$2p_y$ should be suitable. We can confirm this conclusion by noting that the character table shows that, in C_{3v}, x and y jointly belong to the symmetry species E. Therefore, N$2p_x$ and N$2p_y$ also belong to E, and so may have non-zero overlap with s_2 and s_3. (Verify this conclusion by multiplying the characters as $E \times E$ and finding that the product of characters can be expressed as $A_1 + A_2 + E$ by using the formula given in the previous *Justification*.) The two e orbitals that result are shown in Fig. 15.28 (there are also two antibonding e orbitals).

The power of the method can be illustrated by exploring whether any d orbitals on the central atom can take part in bonding. The d orbitals have the forms

$$d_{z^2} = (3z^2 - r^2)f \qquad d_{x^2-y^2} = (x^2 - y^2)f$$
$$d_{xy} = xyf \qquad d_{yz} = yzf \qquad d_{zx} = zxf$$

where f is a function of the radius alone and therefore symmetrical under all symmetry operations of a point group. Their symmetries can be taken from the character tables by noting how xy, yz, etc. transform. Reference to the C_{3v} table shows that d_{z^2} has A_1 symmetry and that the pairs $(d_{x^2-y^2}, d_{xy})$ and (d_{yz}, d_{zx}) each transform as E. It follows that molecular orbitals may be formed by (s_1, d_{z^2}) overlap and by overlap of the s_2, s_3 combinations with the E d orbitals. Whether or not the d orbitals are in fact important is a question that group theory cannot answer because the extent of their involvement depends on energy considerations, not symmetry.

Although we have illustrated the technique with the group C_{3v}, it is entirely general, and the importance of knowing which s, p, and d orbitals overlap is one of the reasons why the transformation properties of x, xz, etc. are listed in the character tables.

Example 15.5 *Determining which orbitals can contribute to bonding*

The four H1s orbitals of methane span $A_1 + T_2$. With which of the C atom orbitals can they overlap? What if the C atom had d orbitals available?

Method. We need to refer to the T_d character table (in the Data section at the end of this volume) and look for s, p, and d orbitals spanning A_1 or T_2.

Answer. An s orbital spans A_1, so it may have non-zero overlap with the A_1 combination of H1s orbitals. The C2p orbitals span T_2, so they may have non-zero overlap with the T_2 combination. The d_{xy}, d_{yz}, and d_{zx} orbitals span T_2 so they may overlap the same combination. Neither of the other two d orbitals span A_1 (they span E), so they remain non-bonding orbitals.

Comment. It follows that, in methane, there are (C2s, H1s)-overlap a_1 orbitals and (C2p, H1s)-overlap t_2 orbitals. The C3d orbitals might contribute to the latter. The lowest energy configuration is probably $a_1^2 t_2^6$, with all bonding orbitals occupied.

Exercise E15.5. Consider the octahedral SF_6 molecule, with the bonding arising from overlap of S orbitals and a 2p orbital on each F directed towards the central S. The latter span $A_{1g} + E_g + T_{1u}$. What S orbitals have non-zero overlap? Suggest what the ground configuration is likely to be

$$[3s(A_{1g}), 3p(T_{1u}), 3d(E_g); a_{1g}^2 t_{1u}^6 e_g^4]$$

Symmetry-adapted linear combinations

So far we have only asserted the forms of the linear combinations (such as s_1, etc.) that have a particular symmetry. Group theory also provides machinery that takes an arbitrary **basis**, or set of atomic orbitals (s_A, etc.), as input and generates combinations of the specified symmetry. Because these combinations are adapted to the symmetry of the molecule, they are called **symmetry-adapted linear combinations** (SALC). Symmetry-adapted linear combinations are the building blocks of LCAO molecular orbitals, for they include combinations such as the $\psi_{1s}(A) \pm \psi_{1s}(B)$ used to construct molecular orbitals in H_2O (Section

14.8) and some of the more complex examples that we have seen since then. The selection of symmetry-adapted linear combinations of atomic orbitals is the first step in any molecular orbital treatment of molecules, and is central, for instance, to the construction and analysis of Walsh diagrams and to the description of d-metal complexes.

The technique for building symmetry-adapted linear combinations is derived using the full power of group theory. We shall not show the derivation, which is very lengthy, but present the main conclusions as a set of rules:

(1) Construct a table showing the effect of each operation on each orbital of the original basis.

For example, from the (s_N, s_A, s_B, s_C) basis in NH_3 we form the table shown in the margin.

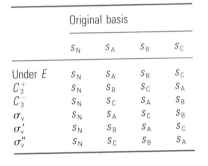

	Original basis			
	s_N	s_A	s_B	s_C
Under E	s_N	s_A	s_B	s_C
C_3^+	s_N	s_B	s_C	s_A
C_3^-	s_N	s_C	s_A	s_B
σ_v	s_N	s_A	s_C	s_B
σ_v'	s_N	s_B	s_A	s_C
σ_v''	s_N	s_C	s_B	s_A

(2) To generate the combination of a specified symmetry species, take each column in turn and:
   (i) Multiply each member of the column by the character of the corresponding operation.
   (ii) Add together all the orbitals in each column with the factors as determined in (i).
   (iii) Divide the sum by the order of the group.

In our example, in order to generate the A_1 combination we take the characters for A_1 $(1,1,1,1,1,1)$, so that rules (i) and (ii) lead to

$$\psi \propto s_N + s_N + \ldots = 6s_N$$

The order of the group (the number of elements) is 6, so the combination of A_1 symmetry that can be generated from s_N is s_N itself. Applying the same technique to the column under s_A gives

$$\psi = \tfrac{1}{6}(s_A + s_B + s_C + s_A + s_B + s_C) = \tfrac{1}{3}(s_A + s_B + s_C)$$

The same combination is built from the other two columns, so they give no further information. The combination we have just formed is the s_1 combination we used before (apart from the numerical factor).

JUSTIFICATION

It is possible to express the rules given here in a succinct formula derived from group theory. In this case, to form an orbital of symmetry species Γ we form $P\psi$, where

$$P\psi = \frac{1}{h}\sum_R \chi^{(\Gamma)}(R)R\psi$$

where R is an operation of the group. Note that the actual operations occur in the formula, not the classes as in the earlier expressions. The quantity P is called a **projection operator**. As an example of its form, to project out a B_1 symmetry-adapted linear combination in the group C_{2v}

we would use

$$P = \tfrac{1}{4}\{\chi^{(B_1)}(E)E + \chi^{(B_1)}(C_2)C_2 + \chi^{(B_1)}(\sigma_v)\sigma_v + \chi^{(B_1)}(\sigma_v')\sigma_v'\}$$
$$= \tfrac{1}{4}\{E - C_2 + \sigma_v - \sigma_v'\}$$

We now form the overall molecular orbital by forming a linear combination of all the symmetry-adapted linear combinations of the specified symmetry species. In this case, therefore, the a_1 molecular orbital is

$$\psi = c_N s_N + c_1 s_1$$

This is as far as group theory can take us. The coefficients must be found by solving the Schrödinger equation because they do not come directly from the symmetry of the system.

Suppose we try to generate a symmetry-adapted linear combination of species A_2 despite the fact that the previous work has shown that there is no such combination. The characters for A_2 are 1, 1, 1, -1, -1, -1. The column under s_N generates zero, and so do the other three. Therefore, we find that we generate no combination of A_2 symmetry.

When we try to generate the E symmetry-adapted combinations we run into a problem because, for representations of dimension 2 or more, where the characters are the sums of the numbers that represent the effect of the symmetry operations on the individual orbitals of the basis, the rules generate sums of the symmetry-adapted combinations. This problem can be illustrated as follows. The E characters are 2, -1, -1, 0, 0, 0, so the column under s_N gives

$$\psi = \tfrac{1}{6}\{2s_N - s_N - s_N + 0 + 0 + 0\} = 0$$

The other columns give

$$\tfrac{1}{6}(2s_A - s_B - s_C) \qquad \tfrac{1}{6}(2s_B - s_A - s_C) \qquad \tfrac{1}{6}(2s_C - s_B - s_A)$$

However, any one of these combinations can be expressed as a sum of the other two (they are not linearly independent). The difference of the second and third gives $\tfrac{1}{2}(s_B - s_C)$, and this and the first, $\tfrac{1}{6}(2s_A - s_B - s_C)$, are the two (now linearly independent) symmetry-adapted combinations we have used in the discussion of e orbitals.

15.6 Vanishing integrals and selection rules

Integrals of the form

$$I = \int f_1 f_2 f_3 \, d\tau \tag{2}$$

are also common in quantum mechanics, and it is important to know when they are necessarily zero. For the integral to be non-zero it is necessary that the product $f_1 f_2 f_3$ spans A_1. To test whether this is so, the characters of all three functions are multiplied together in the same way as in the rules set out above.

Example 15.6 *Deciding if an integral must be zero* (2)

Does the integral $\int (3d_{z^2}) x (3d_{xy}) \, d\tau$ vanish in a C_{2v} molecule?

Method. We must refer to the C_{2v} character table (Table 15.3) and the characters of the irreducible representations spanned by $3z^2 - r^2$ (the form of the d_{z^2} orbital), x, and xy; then we can use the procedure set out above (with one more row of multiplication). Note that $3z^2 - r^2 = 2z^2 - x^2 - y^2$.

Answer. We draw up the following table:

	E	C_2	σ_v	σ_v'	
$f_3 = d_{xy}$	1	1	−1	−1	A_2
$f_2 = x$	1	−1	1	−1	B_1
$f_1 = d_{z^2}$	1	1	1	1	A_1
$f_1 f_2 f_3$	1	−1	−1	1	

The characters are those of B_2. Therefore, the integral is necessarily zero.

Comment. In more complicated cases it may be necessary to use the expression given in the *Justification* on p. 528 to decide whether the characters generated in the last line of the procedure can be expressed as a sum that includes the characters of the totally symmetric irreducible representation A_1.

Exercise 15.6. Does the integral $\int (2p_x)(2p_y)(2p_z) \, d\tau$ necessarily vanish in an octahedral environment? [No; it spans A_1]

In Chapters 16 and 17 we shall see that the intensity of a spectral line arising from a molecular transition between some initial state with wavefunction ψ_i and a final state with wavefunction ψ_f depends on the (electric) transition dipole moment $\boldsymbol{\mu}$. The z-component of this vector is defined through

$$\mu_z = -e \int \psi_f^* z \psi_i \, d\tau \qquad (3)$$

where $-e$ is the charge of the electron. Stating the conditions for this quantity (and the x- and y-components) to be zero amounts to specifying the **selection rules** for the transition, the rules that specify allowed transitions. The transition moment has the form of the integral in eqn 2, so that, once we know the symmetry species of the states, we can use group theory to decide which transitions have zero transition dipole moment and are therefore forbidden.

As an example, we investigate whether an electron in an a_1 orbital in H_2O (which belongs to C_{2v}) can make an electric dipole transition to a b_1 orbital (Fig. 15.29). We must examine all three components of the transition dipole, and take f_2 in eqn 2 as x, y, and z in turn. Reference to

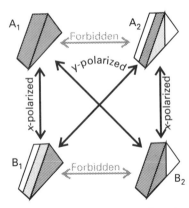

15.29 The polarizations of the allowed transitions in a C_{2v} molecule. The shading indicates the structure of the orbitals of the specified symmetry species.

the C_{2v} character table shows that these components transform as B_1, B_2, and A_1, respectively. The three calculations run as follows.

	x-component				y-component				z-component				
	E	C_2	σ_v	σ_v'	E	C_2	σ_v	σ_v'	E	C_2	σ_v	σ_v'	
f_1	1	1	1	1	1	1	1	1	1	1	1	1	A_1
f_2	1	−1	1	−1	1	−1	−1	1	1	1	1	1	
f_3	1	−1	1	−1	1	−1	1	−1	1	−1	1	−1	B_1
$f_1 f_2 f_3$	1	1	1	1	1	1	−1	−1	1	−1	1	−1	
	A_1				A_2				B_1				

Only the first product (with $f_2 = x$) spans A_1, so only the x-component of the transition dipole may be non-zero. Therefore, we conclude that the electric dipole transitions $a_1 \leftrightarrow b_1$ are allowed. We can go on to state that the radiation emitted (or absorbed) is x-polarized and has its electric field vector in the x-direction, because that form of radiation couples with the x-component of a transition dipole.

Example 15.7 *Deducing a selection rule*

Is $p_x \leftrightarrow p_y$ an allowed transition in a tetrahedral molecule?

Method. We must decide whether the product $p_y q p_x$, with $q = x$, y, or z, spans A_1 using the T_d character table.

Answer. The procedure works out as follows:

	E	$8C_3$	$3C_2$	$6\sigma_d$	$6S_4$	
$f_1(p_x)$	3	0	−1	−1	1	T_2
$f_2(q)$	3	0	−1	−1	1	T_2
$f_3(p_y)$	3	0	−1	−1	1	T_2
$f_1 f_2 f_3$	27	0	−1	−1	1	

A_1 occurs (once) in this set of characters, so $p_x \leftrightarrow p_y$ is allowed.

Comment. A more detailed analysis (by using the matrix representatives rather than the characters) shows that only $q = z$ gives a non-zero integral, so the transition is z-polarized. That is, the electromagnetic radiation involved in the transition has its electric vector aligned in the z-direction.

Exercise E15.7. What are the allowed transitions, and their polarizations, of a b_1 electron in a C_{4v} molecule? $[b_1 \rightarrow b_1(z); \ b_1 \rightarrow e(x, y)]$

The following chapters will show many more examples of how the systematic use of symmetry using the techniques of group theory can greatly simplify the analysis of molecular structure and spectra.

CHECK LIST OF KEY IDEAS

1 The significance of **symmetry operations** and of the corresponding **symmetry elements** (Section 15.1).

2 The **classification** of molecules into **point groups** according to the symmetry elements they possess (Section 15.2).

3 The identification of **polar molecules** and **chiral molecules** from their point groups (Section 15.3).

4 The **character** of symmetry operations and the information contained in **character tables** (Section 15.4).

Table 15.4a Key definitions in group theory

Term	Meaning
Symmetry operation	An action that leaves an object looking the same
Symmetry element	A point, line, or plane with respect to which a symmetry operation is performed
Point group	A group of symmetry operations that leave a point unchanged
Order	The number of symmetry operations in a group
Basis	A set of functions or objects to which a symmetry operation is applied.
Matrix representative	A matrix that, when applied to a basis, reproduces the effect of a symmetry operation on that basis
Matrix representation	A set of matrices that multiply in the same way as all the elements of the group
Character	The sum of the diagonal elements of a matrix representative
Reducible representation	A matrix representation that can be cast into block-diagonal form by forming a linear combination of the basis functions
Irreducible representation	A matrix representation that cannot be cast into block-diagonal form by forming a linear combination of the basis functions
Symmetry species	The label for an irreducible representation with a specific set of characters
Character table	A table displaying the characters corresponding to each symmetry operation and each symmetry species of the irreducible representations of a group
Projection operator	An operator that, when applied to a basis, generates a linear combination of functions that act as a basis for an irreducible representation of a given symmetry species

Table 15.4b Notation

Symbol	Meaning
A, B	Labels for one-dimensional irreducible representations
C_n	n-fold rotation, n-fold axis of symmetry, name of group
$\mathbf{D}$	Matrix representative
d	Dihedral
E	Identity operation
E	Label for two-dimensional irreducible representations
h	Horizontal
i	Inversion operation, centre of inversion
S_n	n-fold improper rotation (or rotary-reflection), n-fold axis of improper rotation (or rotary-reflection axis)
Γ	General symmetry species label
v	Vertical
χ	Character of an operation
σ	Reflection, mirror plane

5 The use of a character table to identify the **symmetry species** of an orbital and a linear combination of orbitals (Section 15.5).

6 The use of character tables to decide if an integral is necessarily zero (Section 15.5).

7 The use of character tables to decide if orbitals have zero overlap (Section 15.5).

8 The construction of **symmetry-adapted linear combinations** of atomic orbitals for the construction of molecular orbitals (Section 15.5).

9 The deduction of **selection rules** from character tables (Section 15.6).

10 The key terms introduced in this chapter are summarized in Table 15.4.

EXERCISES

15.1. The CH_3Cl molecule belongs to the point group C_{3v}. List the symmetry elements of the group and locate them in the molecule.

15.2. Which of the following molecules may be polar: (a) pyridine (C_{2v}), (b) nitroethane (C_s), (c) gas-phase $HgBr_2$ ($D_{\infty h}$), (d) $B_3N_3H_6$ (D_{3h}), (e) CH_3Cl (C_{3v}), (f) $HW_2(CO)_{10}$ (D_{4h}), (g) $SnCl_4$ (T_d)?

15.3. Use symmetry properties to determine whether or not the integral $\int p_x z p_z \, d\tau$ is necessarily zero in a molecule with symmetry C_{4v}.

15.4. Show that the transition $A_1 \rightarrow A_2$ is forbidden for electric dipole transitions in a C_{3v} molecule.

15.5. Show that the function xy has symmetry species B_2 in the group C_{4v}.

15.6. Molecules belonging to the point groups D_{2h}, C_{3h}, T_h, and T_d cannot be chiral. Which elements of these groups rule out chirality?

15.7. The group D_2 consists of the elements E, C_2, C_2', and C_2'', where the three twofold rotations are around mutually perpendicular axes. Construct the group multiplication table. *Hint:* See *Further information 13*.

15.8. Identify the point groups to which the following objects belong: (a) a sphere, (b) an isosceles triangle, (c) an equilateral triangle, (d) an unsharpened cylindrical pencil, (e) a sharpened cylindrical pencil, (f) a three-balanced propellor, (g) a four-legged table, (h) yourself (approximately).

15.9. List the symmetry elements of the following molecules and name the point groups to which they belong: (a) NO_2, (b) N_2O, (c) $CHCl_3$, (d) $CH_2=CH_2$, (e) *cis*-CHCl=CHCl, (f) *trans*-CHCl=CHCl.

15.10. List the symmetry elements of the following molecules and name the point groups to which they belong: (a) naphthalene, (b) anthracene, (c) the three dichlorobenzenes.

15.11. Assign (a) dichloromethane and (b) sulfur tetrafluoride to point groups.

15.12. Assign the following molecules to point groups: (a) HF, (b) IF_7 (pentagonal bipyramid), (c) XeO_2F_2 (seesaw), (d) $Fe_2(CO)_9$ (**1**) (e) cubane, C_8H_8, (f) tetrafluorocubane, $C_8H_4F_4$ (**2**).

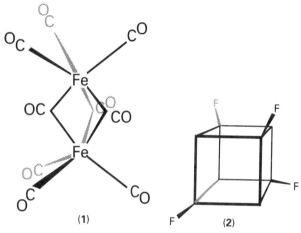

(1)   (2)

15.13. Which of the molecules in Exercises 15.11 and 15.12 can be (a) polar, (b) chiral?

15.14. Consider the C_{2v} molecule NO_2. The combination $p_x(A) - p_x(B)$ of the two O atoms (with x perpendicular to the plane) spans A_2. Is there any orbital of the central N atom that can have a non-zero overlap with that combination of O orbitals? What would be the case in SO_2 where $3d$ orbitals might be available?

15.15. The ground state of NO_2 is A_1 in the group C_{2v}. To what excited states may it be excited by electric dipole transitions, and what polarization of light is it necessary to use?

15.16. The ClO_2 molecule (which belongs to the group C_{2v}) was trapped in a solid. Its ground state is known to be B_1. Light polarized parallel to the y-axis (parallel to the OO separation) excited the molecule to an upper state. What is the symmetry of that state?

15.17. What states of (a) benzene, (b) naphthalene may be reached by electric dipole transitions from their (totally symmetrical) ground states?

15.18. Write $f_1 = \sin \theta$ and $f_2 = \cos \theta$, and show by symmetry arguments using the group C_s that the integral of their product over a symmetrical range around $\theta = 0$ is zero.

PROBLEMS

15.1. List the symmetry elements of the following molecules and name the point groups to which they belong: (a) staggered CH_3CH_3, (b) chair and boat cyclohexane, (c) B_2H_6, (d) $[Co(en)_3]^{3+}$ where en is ethylenediamine (ignore its detailed structure), (e) crown-shaped S_8. Which of these molecules can be (i) polar, (ii) chiral?

15.2. The group C_{2h} consists of the elements E, C_2, σ_h, i. Construct the group multiplication table and find an example of a molecule that belongs to the group.

15.3. The group D_{2h} has a C_2 axis perpendicular to the principal axis and a horizontal mirror plane. Show that the group must therefore have a centre of inversion.

15.4. Consider the H_2O molecule, which belongs to the group C_{2v}. Take as a basis the two $H1s$ orbitals and the four valence orbitals of the O atom and set up the 6×6 matrices that represent the group in this basis. Confirm by explicit matrix multiplication the group multiplications (a) $C_2\sigma_v = \sigma_v'$ and (b) $\sigma_v\sigma_v' = C_2$. Confirm by calculating the traces of the matrices: (a) that symmetry elements in the same class have the same character, (b) that the representation is reducible, and (c) that the basis spans $3A_1 + B_1 + 2B_2$.

15.5. Confirm that the z-component of orbital angular momentum is a basis for an irreducible representation of A_2 symmetry in C_{3v}

15.6. The (one-dimensional) matrices $\boldsymbol{D}(C_3) = 1$ and $\boldsymbol{D}(C_2) = 1$, and $\boldsymbol{D}(C_3) = 1$, and $\boldsymbol{D}(C_2) = -1$ both represent the group multiplication $C_3C_2 = C_6$ in the group C_{6v} with $\boldsymbol{D}(C_6) = +1$ and -1, respectively. Use the character tale to confirm these remarks. What are the representatives of σ_v and σ_d in each case?

15.7. Construct the multiplication table of the Pauli spin matrices and the 2×2 unit matrix:

$$\sigma_x = \begin{pmatrix} 0 & 1 \\ 1 & 0 \end{pmatrix} \qquad \sigma_y = \begin{pmatrix} 0 & -i \\ i & 0 \end{pmatrix}$$

$$\sigma_z = \begin{pmatrix} 1 & 0 \\ 0 & -1 \end{pmatrix} \qquad \boldsymbol{1} = \begin{pmatrix} 1 & 0 \\ 0 & 1 \end{pmatrix}$$

Do the four matrices form a group under multiplication?

15.8. What irreducible representations do the five $F2p_z$ orbitals of PF_5 span (with z lying along each P—F bond)? Are there s and p orbitals of the central P atom that may form molecular orbitals with them? Could d orbitals, even if they were present on the P atom, play a role in orbital formation in PF_5?

15.9. Suppose that a methane molecule became distorted to (a) C_{3v} symmetry by the lengthening of one bond, (b) C_{2v} symmetry, by a kind of scissors action in which one bond angle opened and another closed slightly. Would more d orbitals become available for bonding?

15.10. The algebraic forms of the f orbitals are a radial function multiplied by one of the factors

$$\text{(a) } z(5z^2 - 3r^2), \qquad \text{(b) } y(5y^2 - 3r^2),$$

$$\text{(c) } x(5x^2 - 3r^2), \qquad \text{(d) } z(x^2 - y^2),$$

$$\text{(e) } y(x^2 - z^2), \qquad \text{(f) } x(z^2 - y^2), \qquad \text{(g) } xyz$$

Identify the irreducible representations spanned by these orbitals in (a) C_{2v}, (b) C_{3v}, (c) T_d, (d) O_h. Consider a lanthanide ion at the centre of (a) a tetrahedral complex, (b) an octahedral complex. What sets of orbitals do the seven f orbitals split into?

15.11. Does the product xyz necessarily vanish when integrated over (a) a cube, (b) a tetrahedron, (c) a hexagonal prism, each centred on the origin?

15.12. Treat the naphthalene molecule as belonging to the group C_{2v} with the C_2 axis perpendicular to the plane. Classify the irreducible representations spanned by the carbon $2p_z$ orbitals and find their symmetry-adapted linear combinations.

15.13. The NO_2 molecule belongs to the group C_{2v}, with the C_2 axis bisecting the ONO angle. Taking as a basis the $N2s$, $N2p$, and $O2p$ orbitals, identify the irreducible representations they span, and construct the symmetry-adapted linear combinations.

15.14. Construct the symmetry-adapted linear combinations of $C2p_z$ orbitals for benzene, and use them to calculate the Hückel secular determinant. This procedure leads to equations that are much easier to solve than those obtained by using the original orbitals, and show that the Hückel orbitals are those specified in Section 14.9.

16

Spectroscopy 1: rotational and vibrational spectra

Spectroscopy provides a wealth of detailed information about the identities, structures, and energy levels of species. Certain features are common to all spectroscopic measurements, and the opening section of the chapter explores the general arrangement of a spectrometer and the types of measurement that are made. An important feature of spectra is the intensity of the transitions, and we encounter the molecular features, such as populations and transition moments, that govern intensities.

The general strategy we adopt in the chapter is to set up expressions for the energy levels of molecules, and then apply selection rules and considerations of populations to infer the form of the spectrum. Rotational energy levels are considered first, and we see that simple expressions for their values can be obtained by importing quantum mechanical characteristics into classical expressions for the energies of rotating bodies. These expressions are then used to interpret spectra in terms of molecular dimensions and rigidities. One important aspect of rotating molecules is that not all molecules can occupy all rotational states: we see the experimental evidence for this restriction and its explanation in terms of nuclear spin.

Next, we consider the vibrational energy levels of diatomic molecules and see that we can use the properties of harmonic oscillators developed in Chapter 12. Although the harmonic approximation is quite good, precise measurements, and measurements on highly excited molecules, require us to take anharmonicities into account. We see how this may be done and how vibrational spectra may be used to obtain dissociation energies of molecules. When we turn to polyatomic molecules, we find that their vibrations may be discussed as though they consisted of a set of independent harmonic oscillators; thus the same approach as employed for diatomic molecules may be used. However, the symmetry properties of the complex modes of vibration of polyatomic molecules are helpful for deciding which modes of vibration can be studied spectroscopically, and we draw on some of the material treated in Chapter 15.

Exercises

Problems

Throughout the chapter we describe absorption spectroscopy, in which photons are absorbed, and Raman spectroscopy, in which photons collide with molecules and are scattered with a different energy. The energy levels explored by the two techniques are the same, but different transitions are observed and hence different information may be obtained.

The origin of spectral lines in molecular spectroscopy is the emission or absorption of a photon when the energy of a molecule changes. The difference from atomic spectroscopy is that the energy of a molecule can change not only as a result of electronic transitions but also because the molecule can undergo changes of rotational and vibrational state. Molecular spectra are therefore more complex than atomic spectra. However, they also contain information relating to more properties, and their analysis leads to values of bond strengths, lengths, and angles. They also provide a way of determining a variety of molecular properties, particularly molecular dimensions, shapes, and dipole moments.

Pure rotational spectra (in which only the rotational state of a molecule changes) can be observed, but vibrational spectra of gaseous samples show features that arise from rotational transitions that accompany the vibrational transitions. Similarly, electronic spectra (Chapter 17) show features arising from simultaneous vibrational and rotational transitions. The simplest way of dealing with these complexities is to tackle each type of transition in turn, and then to see how simultaneous changes affect the appearance of the spectrum.

GENERAL FEATURES OF SPECTROSCOPY

All types of spectra have some features in common, and we examine these first.

16.1 Experimental techniques

In **emission spectroscopy** a molecule undergoes a transition from a state of high energy E_1 to a state of lower energy E_2 and emits the excess energy as a photon. In **absorption spectroscopy** the net absorption[1] of nearly monochromatic incident radiation is monitored as it is swept over a range of frequencies. The energy $h\nu$ of the photon emitted or absorbed and, therefore, the frequency ν of the radiation emitted or absorbed is given by the Bohr frequency condition

$$h\nu = E_1 - E_2 \tag{1}$$

This relation is often expressed in terms of the **vacuum wavelength** λ, where

$$\lambda = \frac{c}{\nu} \tag{2a}$$

---

1 We say *net* absorption, because it will become clear that, when a sample is irradiated, both absorption and emission at a given frequency are stimulated, and the detector measures the difference, the net absorption.

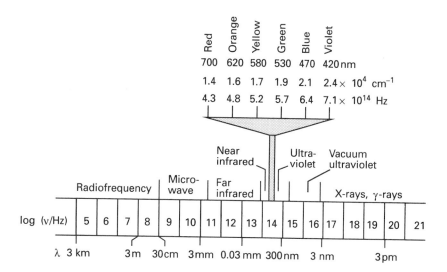

16.1 The electromagnetic spectrum and the classification of the spectral regions.

or the **vacuum wavenumber** $\tilde{\nu}$:

$$\tilde{\nu} = \frac{\nu}{c} \tag{2b}$$

The units of the latter are almost always chosen as reciprocal centimetres (cm^{-1}). Figure 16.1 summarizes the frequencies, wavelengths, and wavenumbers of the various regions of the electromagnetic spectrum.

Emission and absorption spectroscopy give the same information about energy level separations, but practical considerations generally determine which technique is employed. In practice, emission spectroscopy, if it is used at all, is used only for visible and ultraviolet spectroscopy; absorption spectroscopy is much more widely employed, and we shall concentrate on it. Absorption spectra are also often easier to interpret than emission spectra.

All absorption spectrometers consist of a source of radiation, a sample cell, and a detector (Fig. 16.2). The characteristics of each component depend on the region of the electromagnetic spectrum being considered. Most spectrometers also include a monochromator, a device for achieving monochromatic (single-frequency) radiation.

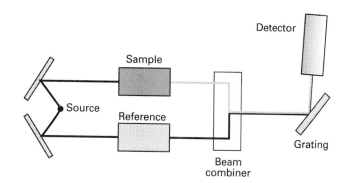

16.2 The layout of a typical absorption spectrometer. The beams pass alternately through the sample and reference cells, and the detector is synchronized with them so that the relative absorption can be determined.

Sources of radiation

The source generally produces radiation spanning a range of frequencies. For the far infrared, the source is a mercury arc inside a quartz envelope, most of the radiation being generated by the hot quartz. A **Nernst filament** is used to generate radiation in the near infrared. This device consists of a heated ceramic filament containing rare-earth oxides, which emits radiation closely resembling that of a true black body. For the visible region of the spectrum, a tungsten/iodine lamp is used, which gives out intense white light. A discharge through deuterium gas or xenon in quartz is still widely used for the near ultraviolet. In a few cases the source generates monochromatic radiation that can be swept over a range of values. One such generator is the **klystron**, an electronic device used to generate microwaves. Lasers, which are discussed in more detail in Chapter 17, generate monochromatic electromagnetic radiation that can often be tuned over a range of frequencies; different types of laser are used to cover different regions of the electromagnetic spectrum.

For certain applications **synchrotron radiation** from a synchrotron storage ring is appropriate. A synchrotron storage ring consists of an electron beam (actually a series of closely spaced packets of electrons) travelling in a circular path of several metres in diameter. Accelerated charges emit electromagnetic radiation and, as electrons travelling in a circle are constantly accelerated by the forces that constrain them to their path, they generate radiation (Fig. 16.3). Synchrotron radiation spans a wide range of frequencies, up to and including the far ultraviolet. In all except the microwave region, it is much more intense than can be obtained by most conventional sources. The disadvantage of the source is that it so large and costly that it is essentially a national facility, not a laboratory commonplace.

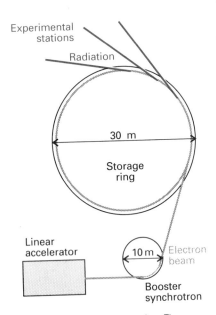

16.3 A synchrotron storage ring. The electrons injected into the ring from the linear accelerator and booster synchrotron are accelerated to high speed in the main ring. An electron in a curved path is subject to acceleration, and an accelerated charge radiates electromagnetic energy.

The dispersing element

In all but specialized techniques using monochromatic microwave radiation and lasers, absorption spectrometers include a component for separating the frequencies of the radiation so that the variation of the absorption with frequency can be monitored. In conventional spectrometers, this component is a **dispersing element** that separates different frequencies into different spatial directions.

The simplest dispersing element is a glass or quartz prism that utilizes the variation of refractive index with the frequency of the incident radiation (Fig. 16.4). Materials generally have a higher refractive index for high-frequency than for low-frequency radiation, and therefore high-frequency radiation undergoes a greater deflection when passing through a prism. Problems of absorption by the prism can be avoided by replacing it by a **diffraction grating**. A diffraction grating consists of a glass or ceramic plate into which fine grooves have been cut about 1000 nm apart (which is comparable to the wavelength of visible light) and covered with a reflective aluminium coating. The grating causes interference between waves reflected from its surface, and constructive

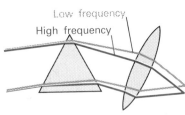

16.4 One simple dispersing element is a prism, which separates frequencies spatially by making use of the higher refractive index of matter for high-frequency radiation. The shortest wavelength for which a glass prism can be used is about 400 nm, but quartz can be used down to 180 nm.

interference occurs at specific angles that depend on the frequency of the radiation being used. By shaping the grooves appropriately, a process called **blazing** (Fig. 16.5), the intensity of the interference pattern can be enhanced.

Fourier transform techniques

Modern spectrometers, particularly those operating in the infrared, now almost always use **Fourier transform techniques** of spectral detection and analysis. The heart of a Fourier transform spectrometer is a **Michelson interferometer**, which is a device for analysing the frequencies present in a composite signal. The total signal from a sample is the analogue of a chord played on a piano, and the Fourier transformation of the signal is equivalent to the separation of the chord into its individual notes, its spectrum.

A Michelson interferometer works by splitting the beam from the sample into two (Fig. 16.6) and introducing a varying path difference p into one of them. When the two components recombine, there is a phase difference between them, and they interfere either constructively or destructively depending on the extra path that one has taken. The detected signal oscillates as the two components alternately come into and out of phase as the path difference is changed (Fig. 16.7). If the radiation has wavenumber $\tilde{v}$, the detected signal varies with p as

$$I(p) = I(\tilde{v})(1 + \cos 2\pi\tilde{v}p)$$

hence, the interferometer converts the presence of a particular component in the signal into a variation in intensity of the radiation reaching the detector. An actual signal consists of radiation spanning a large number of wavenumbers, and the total intensity at the detector is the sum of all their oscillating intensities (Fig. 16.8):

$$I(p) = \int_0^\infty I(\tilde{v})(1 + \cos 2\pi\tilde{v}p)\, d\tilde{v} \tag{3a}$$

The problem is to find $I(\tilde{v})$, the variation of intensity with wavenumber, which is the spectrum we require, from the record of values of $I(p)$. This step is a standard technique of mathematics, and is the 'Fourier transformation' step from which this form of spectroscopy takes its name. Specifically:

$$I(\tilde{v}) = 4\int_0^\infty \{I(p) - \tfrac{1}{2}I(0)\}\cos(2\pi\tilde{v}p)\, dp \tag{3b}$$

This integration is carried out in a computer that is interfaced to the spectrometer, and the output $I(\tilde{v})$ is the absorption spectrum of the sample (Fig. 16.9).

A major advantage of this procedure is that all the radiation emitted by the source is monitored continuously. This is in contrast to a spectrometer in which a monochromator discards most of the generated radiation. As a result, Fourier spectrometers have a higher sensitivity than conventional spectrometers. The resolution they can achieve is

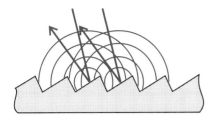

16.5 A diffraction grating is 'blazed' as shown here in order to enhance the intensity of the diffracted radiation in each direction. A diffraction grating works on the principle of interference between waves reflected from the grating, as indicated schematically by the spherical wave fronts. Waves from different facets interfere with one another, and the direction of the resulting wave front depends on the wavelength of the radiation.

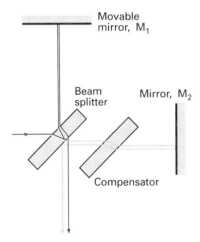

16.6 A Michelson interferometer. The beam-splitting element divides the incident beam into two beams with a path difference that depends on the location of the mirror M_1. The compensator ensures that both beams pass through the same thickness of material.

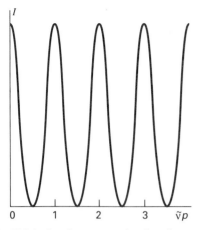

16.7 An interferogram produced as the path length p is changed in the interferometer shown in Fig. 16.6. Only a single frequency component is present in the radiation.

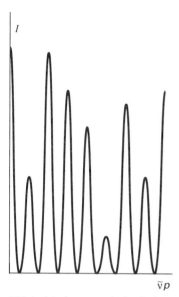

16.8 An interferogram obtained when several frequencies are present in the radiation.

determined by the maximum path length difference of the interferometer:

$$\Delta \tilde{\nu} = \frac{1}{2p_{max}} \qquad (4)$$

Thus, to achieve a resolution of 0.1 cm^{-1} requires a maximum path length difference of 5 cm.

Detectors

The third component of a spectrometer is the **detector**, the device that converts incident radiation into an electric current for the appropriate signal processing or plotting. Radiation-sensitive semiconductor devices, such as a **charge-coupled device** (CCD), are increasingly dominating this role in the spectrometer. In the optical and ultraviolet region a **photomultiplier** is widely used. In this device, each incident photon ejects an electron from a photosensitive surface; the electron is accelerated by a potential difference and ejects a shower of electrons where it strikes a screen. These electrons are accelerated, and each one releases a further shower on impact with another screen. Thus the impact of the initial photon is converted into a cascade of electrons, which is converted into a current in an external circuit.

Although semiconductor detectors are increasingly being used in the infrared, thermocouples are still widely used. A thermocouple detector consists typically of a blackened gold foil to which are attached thermoelectric alloys. A **thermistor bolometer** is essentially a resistance thermometer, and is typically formed from a mixture of oxides deposited on quartz. In each case the radiation is chopped by a shutter that rotates in the beam so that an alternating signal is obtained from the detector (which is easier to amplify than a steady signal). A microwave detector is typically a **crystal diode** consisting of a tungsten tip in contact with a semiconductor, such as germanium, silicon, or gallium arsenide.

The sample

The highest resolution is obtained when the sample is gaseous and of such low pressure that collisions between the molecules are infrequent. Gaseous samples are essential for rotational (microwave) spectroscopy, for only then can molecules rotate freely. To achieve sufficient absorption, the path lengths through gaseous samples must be very long, of the order of metres; long path lengths are achieved by multiple passage of the beam between two parallel mirrors at each end of the sample cavity (Fig. 16.10).

The most common range for infrared spectroscopy is 4000 to 625 cm^{-1}. Ordinary glass and quartz absorb over most of this range, so other material must be used as windows. Thus, the sample is typically a liquid held between windows of sodium chloride (which is transparent down to 625 cm^{-1}) or potassium bromide (which is transparent down to 400 cm^{-1}). Other ways of preparing the sample include grinding it into a paste with 'Nujol', a hydrocarbon oil, or pressing it into a solid disc, perhaps with powdered potassium bromide.

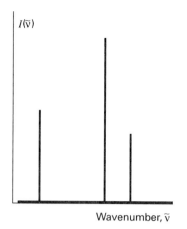

$I(\tilde{v})$

Wavenumber, $\tilde{v}$

16.9 The three frequency components and their intensities that account for the appearance of the interferogram in Fig. 16.8. This spectrum is the Fourier transform of the interferogram, and is a depiction of the contributing frequencies.

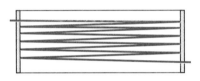

16.10 When a molecule absorbs only very weakly, a detectable absorption may be found only if the path length through the sample is very long. A long path length can be achieved by reflecting the incident beam back and forth through the sample.

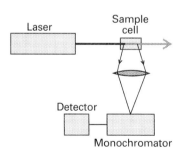

16.11 The arrangement adopted in laser Raman spectroscopy. The scattered radiation is monitored at right angles to the incident radiation.

Raman spectroscopy

In **Raman spectroscopy** the energy levels of molecules are explored by examining the frequencies present in the radiation scattered by molecules. In a typical experiment, a monochromatic incident beam, typically in the visible region of the spectrum, is passed through the sample and the radiation scattered perpendicular to the beam is monitored (Fig. 16.11). About 1 in 10^7 of the incident photons collide with the molecules, give up some of their energy, and emerge with a lower energy. These scattered photons constitute the lower-frequency **Stokes radiation** from the sample. Other incident photons may collect energy from the molecules (if they are already excited), and emerge as higher-frequency **anti-Stokes radiation**. The shifts in frequency of the scattered radiation from the incident radiation are quite small, and the latter must be very monochromatic if the shifts are to be observed. Moreover, the intensity of scattered radiation is low, so very intense incident beams are needed. Lasers are ideal in both respects, and have entirely displaced the mercury arcs used originally. Detection is usually with a photomultiplier. An advantage of Raman spectroscopy over infrared spectroscopy is that the radiation can be entirely in the visible region, so the complications arising from needing to select a range of infrared-transparent sample cells are avoided.

16.2 The intensities of spectral lines

It is found that the intensity of absorption by a sample varies with the length l of the sample in accord with the **Beer–Lambert law**:

$$\log \frac{I}{I_0} = -\varepsilon [\text{J}] l \tag{5}$$

where I_0 is the incident intensity (at a particular wavenumber), I is the intensity after passage through a sample of length l, and $[\text{J}]$ is the molar concentration of absorbing species J. The quantity ε is called the **molar absorption coefficient** (formerly, and still widely, the 'extinction coefficient'). The molar absorption coefficient depends on the frequency of the incident radiation and is greatest where the absorption is most intense. Its dimensions are 1/(concentration × length), and it is normally convenient[2] to express it in $\text{L mol}^{-1} \text{cm}^{-1}$. The dimensionless product $A = \varepsilon [\text{J}] l$ is called the **absorbance** (formerly the 'optical density') of the sample, and the ratio I/I_0 is the **transmittance** T. These two quantities are related as follows:

$$\log T = -A \tag{6}$$

Hence, the absorbance can be measured experimentally by determining the ratio of the incident to emergent intensities and taking the logarithm.

2 Alternative units are $\text{cm}^2 \text{mol}^{-1}$ (or $\text{cm}^2 \text{mmol}^{-1}$, with $1 \text{L mol}^{-1} \text{cm}^{-1} = 1 \text{cm}^2 \text{mmol}^{-1}$). This change of units emphasizes the point that ε is a molar cross-section for absorption, and, the greater the cross-section of the molecule for absorption, the greater the attenuation of the intensity of the beam.

JUSTIFICATION

The Beer–Lambert law is an empirical result. However, it is simple to account for its form. The reduction in intensity dI that occurs when light passes through a layer of thickness dl containing an absorbing species J at a molar concentration $[J]$ is proportional to the thickness of the layer, the concentration, and the incident intensity I (because the rate of absorption is proportional to the intensity, see below). We can therefore write

$$dI = -\kappa [J] I \, dl$$

where κ is the proportionality coefficient or, equivalently,

$$\frac{dI}{I} = -\kappa [J] \, dl$$

These expressions apply to each successive layer into which the sample can be regarded as being divided. Therefore, to obtain the intensity that emerges from a sample of thickness l when the incident intensity is I_0 we sum all the successive changes:

$$\int_{I_0}^{I} \frac{dI}{I} = -\kappa \int_{0}^{l} [J] \, dl$$

If the concentration is uniform, $[J]$ is independent of location, and the expression integrates to

$$\ln \frac{I}{I_0} = -\kappa [J] l$$

The expression gives the Beer–Lambert law when the logarithm is converted to base 10 by using $\ln x = (\ln 10) \times \log x$ and replacing κ by $\varepsilon \ln 10$.

The Beer–Lambert law implies that the intensity of electromagnetic radiation transmitted through a sample at a given wavenumber decreases exponentially with the sample thickness and the concentration. If the transmittance is 0.1 for a path length of 1 cm (corresponding to a 90 per cent reduction in intensity), then it would be $(0.1)^2 = 0.01$ for a path of double the length (corresponding to a 99 per cent reduction in intensity overall).

The maximum value of the molar absorption coefficient ε_{max} is an indication of the intensity of a transition. However, as absorption bands generally spread over a range of wavenumbers, quoting the absorption coefficient at a single wavenumber might not give a true indication of the intensity of a transition. The **integrated absorption coefficient** $\mathcal{A}$ is the sum of the absorption coefficients over the entire band (Fig. 16.12) and corresponds to the area under the absorption band:

$$\mathcal{A} = \int \varepsilon(\tilde{\nu}) \, d\tilde{\nu} \tag{7}$$

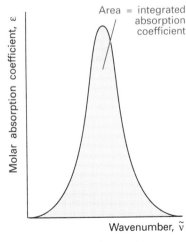

16.12 The intensity of a transition is the area under a plot of the molar absorption coefficient against the wavenumber of the incident radiation.

For lines of similar widths, the integrated absorption coefficients are proportional to the height of the line.

Absorption intensities

A glance at the spectra (whether emission, absorption, or Raman) illustrated in this chapter and the next shows that their lines occur with a variety of intensities. We shall also see that some lines that might be expected to occur do not appear at all. To account for these features, we must see how the intensities of spectral lines depend on the population of molecular states and the strength of the interaction of molecules with the electromagnetic field. Einstein considered the question of the rates of transitions between two states in the presence of an electromagnetic field and identified one process for absorption and two processes for emission.

The process of **stimulated absorption** is the transition from a low-energy state to one of higher energy that is driven by oscillations of the electromagnetic field at the transition frequency. The more intense the electromagnetic field (the more intense the incident radiation), then the greater the rate at which transitions are induced (Fig. 16.13) and hence the stronger the absorption by the sample. (This effect was invoked in the justification of the Beer–Lambert law.) Einstein wrote the transition rate[3] w from the lower to the upper state as

$$w = B\rho \tag{8}$$

The constant B is the **Einstein coefficient of stimulated absorption**

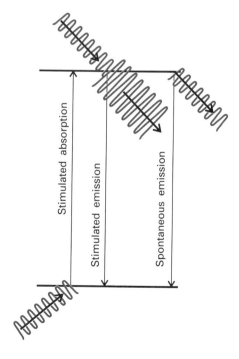

16.13 The processes that account for absorption and emission of radiation and the attainment of thermal equilibrium. The excited state can return to the lower state spontaneously as well as by a process stimulated by radiation already present at the transition frequency.

Stimulated absorption

Stimulated emission

Spontaneous emission

---

3 Specifically, w is the rate of change of probability of the molecule being found in the upper state: $w = \mathrm{d}P/\mathrm{d}t$.

and ρ is the energy density of radiation at the frequency of the transition. If the molecule is exposed to black-body radiation from a source of temperature T, then ρ would be given by the Planck distribution (Section 11.1):

$$\rho = \frac{8\pi h v^3}{c^3} \times \frac{1}{e^{hv/kT} - 1} \tag{9}$$

For the time being we can treat B as an empirical parameter that characterizes the transition: if B is large, then a given intensity of incident radiation will induce transitions strongly and the sample will be strongly absorbing. The **total rate of absorption** W is the transition rate of a single molecule multiplied by the number of molecules N in the lower state:

$$W = Nw \tag{10}$$

Einstein considered that the radiation was also able to induce a molecule in the upper state to undergo a transition to the lower state, and hence to generate a photon of frequency v. Thus, he wrote the rate of this **stimulated emission** as

$$w' = B'\rho \tag{11}$$

where B' is the **Einstein coefficient of stimulated emission**. Note that *only radiation of the same frequency as the transition can stimulate an excited state to fall to a lower state*. However, Einstein realized that stimulated emission was not the only means by which the excited state could generate radiation and return to the lower state, and concluded that an excited state could undergo **spontaneous emission** at a rate that was independent of the intensity of the radiation (of any frequency) that is already present. He therefore wrote the total rate of transition to a lower state as

$$w' = A + B'\rho \tag{12a}$$

The constant A is the **Einstein coefficient of spontaneous emission**. The overall rate of emission is

$$W' = N'(A + B'\rho) \tag{12b}$$

where N' is the population of the upper state.

Einstein was able to show that the two coefficients of stimulated absorption and emission are equal, and that the coefficient of spontaneous emission is related to them by

$$A = \left(\frac{8\pi h v^3}{c^3}\right) \times B \tag{13}$$

JUSTIFICATION

At thermal equilibrium, the rates of emission and absorption are equal, so that

$$NB\rho = N'(A + B'\rho)$$

This expression rearranges into

$$\rho = \frac{N'A}{NB - N'B'} = \left(\frac{A}{B}\right) \times \frac{1}{\left(\dfrac{N}{N'}\right) - \left(\dfrac{B'}{B}\right)} = \left(\frac{A}{B}\right) \times \frac{1}{e^{h\nu/kT} - \left(\dfrac{B'}{B}\right)}$$

We have used the Boltzmann expression (see the *Introduction*) for the ratio of populations of states of energies E and E' in the last step:

$$\frac{N'}{N} = e^{-h\nu/kT}, \qquad \text{where } h\nu = E' - E$$

This result has the same form as the Planck distribution (eqn 9), which describes the radiation density at thermal equilibrium. Indeed, when we compare the two expressions for ρ, we can conclude that

$$B' = B$$
$$A = \left(\frac{8\pi h\nu^3}{c^3}\right) \times B$$

as in eqn 13.

The growth of the importance of spontaneous emission with increasing frequency is a very important conclusion, as we shall see when we consider the operation of lasers in the next chapter. The equality of the coefficients of stimulated emission and absorption implies that, if two states happen to have equal populations, then the rate of stimulated emission is exactly equal to the rate of stimulated absorption, and there is then no net absorption.

Spontaneous emission can be largely ignored at the relatively low frequencies of rotational and vibrational transitions, and the intensities of these transitions can be discussed in terms of stimulated emission and absorption. Then the net rate of absorption is given by

$$W_{\text{net}} = NB\rho - N'B'\rho = (N - N')B\rho \tag{14}$$

and is proportional to the population difference of the two states involved in the transition.

Example 16.1 *Estimating relative transition intensities*

Estimate the relative intensities at 25°C of absorptions originating in the ground state and the first excited state when the energy levels involved are separated by (a) 10 000 cm^{-1}, (b) 1000 cm^{-1}, and (c) 1.0 cm^{-1}.

Method. The intensities are proportional to the population difference between the two states, which is given by the Boltzmann formula quoted in the text.

Answer. At 25°C, $kT/hc = 207$ cm^{-1}, so the ratios of populations are

(a) $e^{-10000/207} = e^{-48} = 14.4 \times 10^{-21}$
(b) $e^{-1000/207} = e^{-4.8} = 0.0082$
(c) $e^{-1.0/207} = e^{-0.0048} = 0.99$

Because the population of the upper state is negligible in (a) and (b), the only significant absorption is from the lower state. Moreover, in these two cases, the stimulated emission from the upper state is also negligible, and we need consider only stimulated absorption from the ground state when assessing the net intensity of absorption. However, for (c) we can draw neither conclusion. Because adjacent states are almost equally populated, transitions can originate with significant intensity from many states, and stimulated emission from upper states makes a significant contribution to the net absorption intensity.

Exercise E16.1. Repeat the analysis for a temperature of 1500 K.

[(a) As before; upper-state populations are significant for (b) and (c)]

It follows from eqn 14, so long as the intrinsic intensities of the transitions are the same, that the relative intensities of two lines corresponding to transitions originating from two different states should be proportional to the relative populations of the two initial states. Because the first electronically excited state of a molecule is usually of the order of $10^4 \, cm^{-1}$ above the ground state, it is not populated at room temperature (see Example 16.1). Therefore, an electronic absorption spectrum is normally due entirely to transitions originating from the ground electronic state. Vibrational energy levels are separated by about 500 to 4000 cm^{-1}, so the principal transitions are also normally those from the ground vibrational state, and stimulated emission makes a negligible contribution to the net absorption. In contrast, rotational energy levels are separated by only 1 to $10^2 \, cm^{-1}$, and many states are occupied even at room temperature; consequently, rotational transitions occur from a wide range of initial states, not only the lowest, and stimulated emission from the occupied higher states is important.

Molecules are often prepared in short-lived excited states as a result of chemical reaction, electric discharge, or photolysis. In these cases the populations may be quite different from those at thermal equilibrium. The resulting spectra, if they can be taken quickly enough, then arise from transitions from all the populated levels.

Selection rules and transition moments

We met the concept of a 'selection rule' in Section 13.3 as a rule that determines whether a transition is forbidden or allowed. Selection rules also apply to molecular spectra, and the form they take depends on the type of transition. The underlying classical idea is that, for the molecule to be able to interact with the electromagnetic field and absorb or create a photon of frequency v, it must possess, at least transiently, a dipole oscillating at that frequency. For emission and absorption spectra this transient dipole is expressed quantum mechanically in terms of the **transition dipole moment**, and for a transition between states with wavefunctions ψ_i and ψ_f is defined as

$$\boldsymbol{\mu}_{fi} = \int \psi_f^* \boldsymbol{\mu} \psi_i \, d\tau \tag{15}$$

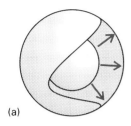

(a)

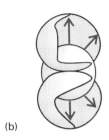

(b)

16.14 (a) When a $1s$ electron becomes a $2s$ electron, there is a spherical migration of charge. There is no dipole moment associated with this migration of charge: this transition is electric-dipole forbidden. (b) In contrast, when a $1s$ electron becomes a $2p$ electron, there is a dipole associated with the charge migration; this transition is allowed. (There are subtle effects arising from the sign of the wavefunction that give the charge migration a dipolar character, which this diagram does not attempt to convey.)

where $\boldsymbol{\mu}$ is the electric dipole moment operator. The size of the transition dipole can be regarded as a measure of the charge redistribution that accompanies a transition: a transition will be active (and generate or absorb photons strongly) only if the accompanying charge redistribution is dipolar (Fig. 16.14).

The coefficient of stimulated absorption (and emission), and therefore the intensity of the transition, is proportional to the square of the transition dipole moment, and a detailed analysis gives

$$B = \frac{|\boldsymbol{\mu}_{\mathrm{fi}}|^2}{6\varepsilon_0\hbar^2} \tag{16}$$

so that only if the transition moment is non-zero does the transition contribute to the spectrum. We see that, to identify the selection rules, we must establish the conditions for which $\boldsymbol{\mu}_{\mathrm{fi}} \neq 0$.

A **gross selection rule** specifies the general features that a molecule must have if it is to have a spectrum of a given kind. For instance, we shall see that a rotational transition dipole moment (the transition moment accompanying the change in rotational state of a molecule) is zero unless the molecule has a permanent electric dipole. This rule, and others like it for other types of transition, will be explained in the relevant sections of the chapter.

A detailed study of the transition moment leads to the **specific selection rules** that express the allowed transitions in terms of the changes in quantum numbers. We have already encountered examples of specific selection rules when discussing atomic spectra (Section 13.3), such as the rule $\Delta l = \pm 1$ for the angular momentum quantum number. Specific selection rules can often be interpreted in terms of the change of angular momentum when a photon (with its intrinsic spin angular momentum $s = 1$) enters or leaves a molecule, and we shall discuss them once we have set up the quantum numbers needed to describe rotation and vibration.

16.3 Linewidths

Spectral lines are not infinitely narrow, and a number of effects contribute to their observed widths.

Doppler broadening

One important broadening process in gaseous samples is the **Doppler effect**, in which radiation is shifted in frequency when the source is moving towards or away from the observer. When a source emitting electromagnetic radiation of frequency ν recedes with a speed v, the observer detects radiation of frequency

$$\nu' = \frac{\nu}{1 + \dfrac{v}{c}}$$

where c is the speed of light. A source approaching the observer appears

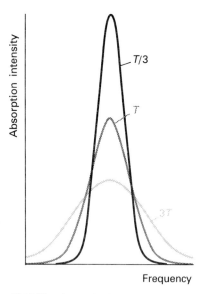

16.15 The shape of a Doppler-broadened spectral line reflects the Maxwell distribution of speeds in the sample at the temperature of the experiment. Notice that the line broadens as the temperature is increased.

to be emitting radiation of frequency

$$\nu' = \frac{\nu}{1 - \dfrac{v}{c}}$$

Molecules reach high speeds in all directions in a gas, and a stationary observer detects the corresponding Doppler-shifted range of frequencies. Some molecules approach the observer, some move away; some move quickly, others slowly. The detected spectral 'line' is the absorption or emission profile arising from all the resulting Doppler shifts. The profile reflects the Maxwell distribution of molecular speeds parallel to the line of sight (see Section 1.3), which is a bell-shaped Gaussian curve (of the form e^{-x^2}). The Doppler line shape is therefore also a Gaussian curve (Fig. 16.15), and calculation shows that, when the temperature is T and the mass of the molecule is m, the width of the line at half-height is

$$\delta\nu = \frac{2\nu}{c}\left(\frac{2kT\ln 2}{m}\right)^{\frac{1}{2}} \tag{17a}$$

In terms of the wavelength,

$$\delta\lambda = \frac{2\lambda}{c}\left(\frac{2kT\ln 2}{m}\right)^{\frac{1}{2}} \tag{17b}$$

For a molecule like N_2 at room temperature ($T \approx 300$ K), we find

$$\frac{\delta\nu}{\nu} \approx 2.3 \times 10^{-6}$$

For a typical transition wavenumber of $1\,\mathrm{cm}^{-1}$ (corresponding to a frequency of 30 GHz), the linewidth is of the order of 70 kHz.

Doppler broadening increases with temperature because the molecules acquire a wider range of speeds. Therefore, to obtain spectra of maximum sharpness it is best to work with cold gaseous samples.

Lamb-dip spectroscopy

A novel approach to the elimination of Doppler broadening has become available with the advent of lasers and their extremely high monochromaticity and of radiofrequency techniques with precise frequency control. The precise location of absorption frequencies in this way is called **Lamb-dip spectroscopy**, which is named after its discoverer, W. Lamb.

When an intense, monochromatic beam with a frequency slightly higher than that of the absorption maximum passes through a gaseous sample, only the molecules that happen to be moving away from the source at some precise speed absorb radiation. If the beam is then reflected back through the sample, more radiation is absorbed, but this time by the molecules that happen to be moving at the same precise speed but away from the mirror. The detector therefore observes a double dose of absorption. However, when the incident radiation is at the absorption peak, only those molecules moving perpendicular to the line

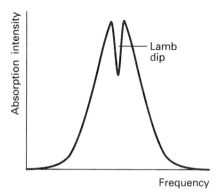

16.16 A Lamb dip. The origin is explained in the text. Lamb-dip spectroscopy enables the positions of the centres of absorption lines of gas-phase samples to be pinpointed very precisely even if there is Doppler broadening.

of the beam (and therefore having no Doppler shift) absorb on the first passage, and the same ones absorb on the reflected path. Because some of those molecules were excited on the first passage, fewer are available to absorb the light on its second passage, so a less intense absorption is observed (Fig. 16.16). This reduction in intensity appears as a dip, the **Lamb dip**, in the absorption curve, and its position gives a very precise location of the transition frequency.

Lifetime broadening

It is found that spectroscopic lines from gas-phase samples are still not infinitely sharp even when Doppler broadening has been largely eliminated, either by working at low temperatures or by Lamb-dip spectroscopy. The same is true of the spectra of samples in condensed phases and solutions. This residual broadening can be traced to certain intrinsic properties of quantum mechanics. Specifically, when the Schrödinger equation is solved for a system that is changing with time, it is found that it is impossible to specify the energy levels exactly. If, on average, a system survives in a state for a time τ, the lifetime of the state, then its energy levels are blurred to an extent of order δE, where

$$\delta E \approx \frac{\hbar}{\tau} \tag{18a}$$

Equation 18a is reminiscent of the Heisenberg uncertainty principle (eqn 11.25) and, although the connection is tenuous, this **lifetime broadening** is often called 'uncertainty broadening'. Expressing wavenumbers through $\delta E = hc\delta\tilde{\nu}$ and using the values of the fundamental constants gives the practical form of the relation as

$$\delta\tilde{\nu} \approx \frac{5.31\ \mathrm{cm}^{-1}}{(\tau/\mathrm{ps})} \tag{18b}$$

No excited state has an infinite lifetime; therefore, all states are subject to some lifetime broadening and, the shorter the lifetimes of the states involved in a transition, the broader the spectral lines.

Two processes are responsible for the finite lifetimes of excited states. The dominant one for low-frequency transitions is **collisional deactivation**, which arises from collisions between molecules or with the walls of the container. If the **collisional lifetime**, the mean time between collisions, is τ_{col}, then the resulting collisional linewidth is $\delta E_{\mathrm{col}} \approx \hbar/\tau_{\mathrm{col}}$. The collisional lifetime can be lengthened in gaseous samples, and the broadening minimized, by working at low pressures.

The rate of spontaneous emission cannot be changed. Hence it is a natural limit to the lifetime of an excited state, and the resulting lifetime broadening is the **natural linewidth** of the transition. The natural linewidth is an intrinsic property of the transition and cannot be changed by modifying the conditions. Natural linewidths depend strongly on the transition frequency (they increase with the coefficient of spontaneous emission A and therefore as ν^3), so low-frequency transitions (such as the microwave transitions of rotational spectroscopy) have very small

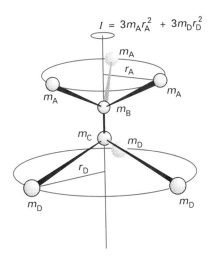

$$I = 3m_A r_A^2 + 3m_D r_D^2$$

16.17 The definition of moment of inertia. In this molecule there are three identical atoms attached to the B atom and three different but mutually identical atoms attached to the C atom. In this example, the centre of mass lies on the C_3 axis, and the perpendicular distances are measured from the axis passing through the B and C atoms.

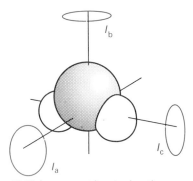

16.18 An asymmetric rotor has three different moments of inertia; all three rotation axes coincide at the centre of mass of the molecule.

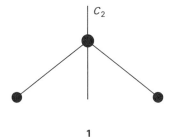

1

natural linewidths, and collisional and Doppler line-broadening processes are dominant. The natural lifetimes of electronic transitions are very much shorter than for vibrational and rotational transitions, so the natural linewidths of electronic transitions are much greater than those of vibrational and rotational transitions. For example, a typical electronic excited state natural lifetime is about 10^{-8} s (10^4 ps), corresponding to a natural width of about 5×10^{-4} cm^{-1} (15 MHz). A typical rotational natural lifetime is about 10^3 s, corresponding to a natural linewidth of only 5×10^{-15} cm^{-1} (of the order of 10^{-4} Hz).

PURE ROTATION SPECTRA

The general strategy that we shall adopt for discussing molecular spectra and the information they contain is to find expressions for the energy levels of molecules and then to calculate the transition frequencies by applying the selection rules. We then predict the appearance of the spectrum by taking into account the populations of the states. In this section we illustrate the strategy by considering the rotational states of molecules.

The key molecular parameter we shall need is the **moment of inertia I** of the molecule (this property was first encountered in Section 12.6 and *Further information 5*, where we saw that the kinetic energy of a body rotating at an angular velocity ω is $\frac{1}{2}I\omega^2$). The moment of inertia of a molecule is defined as the mass of each atom multiplied by the square of its perpendicular distance from the rotational axis, which passes through the centre of mass of the molecule (Fig. 16.17):

$$I = \sum_i m_i r_i^2 \tag{19}$$

where r_i is the perpendicular distance of the atom i from the axis of rotation. The moment of inertia depends on the masses of the atoms present and the molecular geometry, so we can suspect (and later shall see explicitly) that rotational spectroscopy will give information about bond lengths and bond angles. In general, the rotational properties of any molecule can be expressed in terms of the moments of inertia about three perpendicular axes set in the molecule (Fig. 16.18). The convention is to label the moments of inertia I_a, I_b, and I_c, with $I_c \geq I_b \geq I_a$. For linear molecules, the moment of inertia around the internuclear axis is zero. The explicit expressions for the moments of inertia of some symmetrical molecules are given in Table 16.1.

Example 16.2 *Calculating the moment of inertia of a molecule*

Calculate the moment of inertia of an H_2O molecule around its twofold axis (the bisector of the HOH angle, **1**).

Method. According to eqn 19, the moment of inertia is the sum of the masses multiplied by the squares of their distances from the axis of rotation. The latter can be expressed using trigonometry and the bond angle and bond length.

Table 16.1 Moments of inertia†

1. Diatomics

$$I = \frac{m_A m_B}{m} R^2 = \mu R^2$$

2. Linear rotors

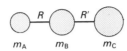

$$I = m_A R^2 + m_C R'^2$$
$$- \frac{(m_A R - m_C R')^2}{m}$$

$$I = 2 m_A R^2$$

3. Symmetric rotors

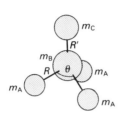

$$I_\parallel = 2 m_A R^2 (1 - \cos\theta)$$
$$I_\perp = m_A R^2 (1 - \cos\theta)$$
$$+ \frac{m_A}{m}(m_B + m_C)R^2(1 + 2\cos\theta)$$
$$+ \frac{m_C R'}{m}\{(3 m_A + m_B)R'$$
$$+ 6 m_A R[\tfrac{1}{3}(1 + 2\cos\theta)]^{\frac{1}{2}}\}$$

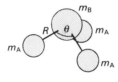

$$I_\parallel = 2 m_A R^2 (1 - \cos\theta)$$
$$I_\perp = m_A R^2 (1 - \cos\theta)$$
$$+ \frac{m_A m_B}{m} R^2(1 + 2\cos\theta)$$

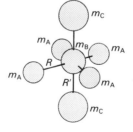

$$I_\parallel = 4 m_A R^2$$
$$I_\perp = 2 m_A R^2 + 2 m_C R'^2$$

4. Spherical rotors

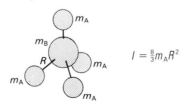

$$I = \tfrac{8}{3} m_A R^2$$

$$I = 4 m_A R^2$$

† In each case m is the total mass of the molecule.

Answer. From eqn 19,

$$I = \sum_i m_i r_i^2 = m_H r_H^2 + 0 + m_H r_H^2 = 2m_H r_H^2$$

If the bond angle of the molecule is denoted 2ϕ and the bond length is R, trigonometry gives

$$r_H = R \sin \phi$$

It follows that

$$I = 2m_H R^2 \sin^2 \phi$$

For H_2O, with bond angle $104.5°$ and bond length $95.7 \, \text{pm}$, we obtain

$$I = 2 \times (1.67 \times 10^{-27} \, \text{kg}) \times (9.57 \times 10^{-11} \, \text{m})^2 \times \sin^2 52.3°$$
$$= 1.91 \times 10^{-47} \, \text{kg m}^2$$

Comment. The mass of the O atom makes no contribution to the moment of inertia for this mode of rotation as it is immobile while the H atoms rotate around it.

Exercise E16.2. Calculate the moment of inertia of a $CH^{35}Cl_3$ molecule around its threefold axis. The C—Cl bond length is $177 \, \text{pm}$ and the HCCl angle is $107°$; $m(^{35}Cl) = 34.97 \, \text{u}$. $[4.99 \times 10^{-45} \, \text{kg m}^2]$

We shall suppose initially that molecules are **rigid rotors** that do not distort under the stress of rotation. Rigid rotors can be classified into four types:

Spherical rotors have three equal moments of inertia (CH_4, SiH_4, and SF_6 are examples).

Symmetric rotors have two equal moments of inertia (such as NH_3, CH_3Cl, and CH_3CN).

Linear rotors have one moment of inertia (the one about the axis) equal to zero (such as CO_2, HCl, OCS, and $HC{\equiv}CH$).

Asymmetric rotors have three different moments of inertia (H_2O, H_2CO, and CH_3OH are examples).

In group theoretical language, a spherical rotor is a molecule that belongs to a cubic or icosahedral point group; a symmetric rotor is a molecule with at least a threefold axis of symmetry. All diatomic molecules are linear rotors. The energy levels of asymmetric rotors are complicated and we shall not consider them.

16.4 The rotational energy levels

The rotational energy levels of a rigid rotor may be obtained by solving the appropriate Schrödinger equation. Fortunately, there is a much less onerous short cut that depends on noting the classical expression for the energy of a rotating body, expressing it in terms of the angular momentum, and then importing the quantum mechanical properties of angular momentum into the equations.

The energy of a body rotating about an axis a is

$$E = \tfrac{1}{2}I_a\omega_a^2$$

where ω_a is the angular velocity (in radians per second, rad s^{-1}) about that axis and I_a is the corresponding moment of inertia. A body free to rotate about three axes has an energy

$$E = \tfrac{1}{2}I_a\omega_a^2 + \tfrac{1}{2}I_b\omega_b^2 + \tfrac{1}{2}I_c\omega_c^2$$

Because the classical angular momentum about the axis a is $J_a = I_a\omega_a$, with similar expressions for the other directions, it follows that

$$E = \frac{J_a^2}{2I_a} + \frac{J_b^2}{2I_b} + \frac{J_c^2}{2I_c} \tag{20}$$

This is the key equation. We described the quantum mechanical properties of angular momentum in Section 12.7, and we can now make use of them in conjunction with this equation to obtain the rotational energy levels.

Spherical rotors

When all three momenta of inertia are equal to some value I, as in CH_4 and SF_6, the classical expression for the energy is

$$E = \frac{J_a^2 + J_b^2 + J_c^2}{2I} = \frac{J^2}{2I}$$

where J is the magnitude of the angular momentum. We can immediately find the quantum expression by making the replacement

$$J^2 \rightarrow J(J+1)\hbar^2 \qquad \text{with } J = 0, 1, 2, \ldots$$

Therefore, the energy of a spherical rotor is confined to the values

$$E = J(J+1)\frac{\hbar^2}{2I}, \qquad \text{with } J = 0, 1, 2, \ldots \tag{21}$$

The resulting ladder of energy levels is illustrated in Fig. 16.19. The energy is normally expressed in terms of the **rotational constant** B of the molecule, where

$$hcB = \frac{\hbar^2}{2I}, \qquad \text{so } B = \frac{\hbar}{4\pi cI} \tag{22a}$$

The expression for the energy is then

$$E = hcBJ(J+1) \qquad J = 0, 1, 2, \ldots \tag{22b}$$

The rotational constant as defined by eqn 22a has the dimensions of a wavenumber[4] and is normally expressed in reciprocal centimeters

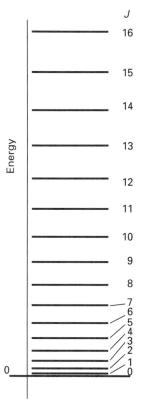

16.19 The rotational energy levels of a linear or spherical rotor. Note that the energy separation between neighbouring levels increases as J increases.

4 The definition of B as a wavenumber is convenient when we come to vibration–rotation spectra. However, for pure rotational spectroscopy it is more common to define B as a frequency and to report it in MHz or GHz. The appropriate definition is then $B = \hbar/4\pi I$ and the energy is $E = hBJ(J+1)$.

(cm^{-1}). The energy of a rotational state is normally reported as the **rotational term** $F(J)$, a wavenumber, by division by hc:

$$F(J) = BJ(J+1) \tag{22c}$$

The separation of adjacent levels is

$$F(J) - F(J-1) = 2BJ \tag{23}$$

Because the rotational constant decreases as I increases, we see that *large molecules have closely spaced rotational energy levels.* We can estimate the magnitude of the separation by considering CCl_4: from the bond lengths and masses of the atoms we find $I = 4.85 \times 10^{-45}$ kg m^2, and hence $B = 0.0577$ cm^{-1}.

Symmetric rotors

In symmetric rotors, two moments of inertia are equal but different from the third (as in CH_3Cl, NH_3, and C_6H_6); the unique axis of the molecule is its **figure axis**. We shall write the unique moment of inertia (that about the figure axis) as $I_\parallel$ and the other two as $I_\perp$. If $I_\parallel > I_\perp$ then the rotor is **oblate** (like a pancake, and C_6H_6); if $I_\parallel < I_\perp$ it is classified as **prolate** (like a cigar, and CH_3Cl). The classical expression for the energy becomes

$$E = \frac{J_b^2 + J_c^2}{2I_\perp} + \frac{J_a^2}{2I_\parallel}$$

This expression can be written in terms of $J^2 = J_a^2 + J_b^2 + J_c^2$:

$$E = \frac{J^2 - J_a^2}{2I_\perp} + \frac{J_a^2}{2I_\parallel} = \frac{J^2}{2I_\perp} + \left(\frac{1}{2I_\parallel} - \frac{1}{2I_\perp}\right)J_a^2$$

Now we generate the quantum expression by replacing J^2 by $J(J+1)\hbar^2$, where J is the angular momentum quantum number. We also know from the quantum theory of angular momentum (Section 12.7) that the component of angular momentum about any axis is restricted to the values $K\hbar$, with $K = 0, \pm 1, \ldots, \pm J$. ($K$ is the quantum number used to signify a component on the figure axis; M_J is reserved for a component on a laboratory axis.) Consequently, we also replace J_a^2 by $K^2\hbar^2$. The rotational terms are therefore

$$F(J, K) = BJ(J+1) + (A-B)K^2$$

$$J = 0, 1, 2, \ldots; \qquad K = 0, \pm 1, \ldots, \pm J \tag{24a}$$

$$A = \frac{\hbar}{4\pi c I_\parallel} \qquad B = \frac{\hbar}{4\pi c I_\perp} \tag{24b}$$

Equation 24a matches what we should expect for the dependence of the energy levels on the two moments of inertia of the molecule. When $K = 0$, there is no component of angular momentum about the figure axis (Fig. 16.20) and the energy levels depend only on $I_\perp$. When $K = \pm J$, almost all the angular momentum arises from rotation around the figure axis, and the energy levels are determined largely by $I_\parallel$. Moreover, the

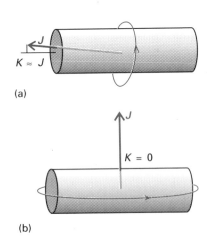

16.20 The significance of the quantum number K. (a) When $|K|$ is close to its maximum value J, most of the molecular rotation is around the figure axis. (b) When $K = 0$ the molecule has no angular momentum about its figure axis: it is undergoing end-over-end rotation.

sign of K does not affect the energy because opposite values of K correspond to opposite senses of rotation, and the energy does not depend on the sense of rotation.

Example 16.3 *Calculating the rotational energy levels of a molecule*

An $^{14}NH_3$ molecule is a symmetric rotor with bond length 101.2 pm and HNH bond angle 106.7°. Calculate its rotational terms.

Method. We should begin by calculating the rotational constants A and B by using the expressions for moments of inertia given in Table 16.1. Then we use eqn 24a to find the rotational terms.

Answer. Substitution of $m_A = 1.0078$ u, $m_B = 14.0031$ u, $R = 101.2$ pm, and $\theta = 106.7°$ into the second of the symmetric rotor expressions in Table 16.1 gives

$$I_\parallel = 4.4128 \times 10^{-47}\ \text{kg m}^2 \qquad I_\perp = 2.8059 \times 10^{-47}\ \text{kg m}^2$$

Hence, $A = 6.344$ cm^{-1} and $B = 9.977$ cm^{-1}. It follows from eqn 24a that

$$F(J, K)/\text{cm}^{-1} = 9.977J(J + 1) - 3.633K^2$$

Comment. For $J = 1$, the energy needed for the molecule to rotate mainly about its figure axis ($K = J$) is equivalent to 16.32 cm^{-1}, but end-over-end rotation ($K = 0$) corresponds to 19.95 cm^{-1}.

Exercise E16.3. The CH$_3{}^{35}$Cl molecule has a C—Cl bond length of 178 pm, a C—H bond length of 111 pm, and an HCH angle of 110.5°. Calculate its rotational energy levels.
$$[F(J, K)/\text{cm}^{-1} = 0.444J(J + 1) + 4.58K^2]$$

Linear rotors

For a linear rotor (such as CO_2, HCl, and C_2H_2) in which the nuclei are regarded as mass points, the rotation occurs only about an axis perpendicular to the line of atoms and there is zero angular momentum around the line. Therefore the component of angular momentum around the figure axis of a linear rotor is identically zero, so we set $K \equiv 0$ in eqn 24a. The rotational terms of a linear molecule are therefore

$$F(J) = BJ(J + 1) \qquad J = 0, 1, 2, \ldots \tag{25}$$

Degeneracies and the Stark effect

A symmetric rotor has an energy that depends on J and K, and each level except those with $K = 0$ is doubly degenerate, for states with K and $-K$ have the same energy. However, we must not forget that the angular momentum of the molecule has a component on a laboratory axis. This component is quantized, and its permitted values are $M_J \hbar$ with $M_J = 0, \pm 1, \ldots, \pm J$, giving $2J + 1$ values in all (Fig. 16.21). The quantum number M_J does not appear in the expression for the energy, but it is still necessary for a complete specification of the state of the

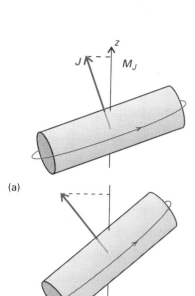

(a)

(b)

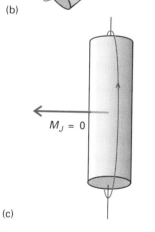

(c)

16.21 The significance of the quantum number M_J. (a) When M_J is close to its maximum value J, most of the molecular rotation is around the laboratory-axis. (b) An intermediate value of M_J. (c) When $M_J = 0$ the molecule has no angular momentum about the z-axis. All three diagrams correspond to a state with $K = 0$; there are corresponding diagrams for different values of K, in which the angular momentum makes a different angle to the molecule's figure axis.

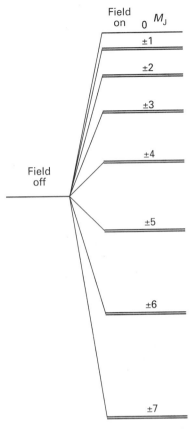

16.22 The effect of an electric field on the energy levels of a polar linear rotor. All levels are doubly degenerate except that with $M_J = 0$.

rotor. Consequently, all $2J + 1$ orientations of the rotating molecule have the same energy. It follows that a symmetric rotor level is $2(2J + 1)$-fold degenerate for $K \neq 0$ and $(2J + 1)$-fold degenerate if $K = 0$. A linear rotor has K fixed at 0, but the angular momentum may still have $2J + 1$ components on the laboratory axis, so its degeneracy is $2J + 1$.

A spherical rotor can be regarded as a version of a symmetric rotor in which A is equal to B: The quantum number K may still take any one of $2J + 1$ values, but the energy is independent of which value it takes. Therefore, as well as having a $(2J + 1)$-fold degeneracy arising from the orientation in space, the rotor also has a $(2J + 1)$-fold degeneracy arising from the orientation with respect to an axis selected in the molecule. The overall degeneracy of a spherical rotor with quantum number J is therefore $(2J + 1)^2$. This degeneracy increases very rapidly: when $J = 10$, for instance, there are 441 states of the same energy.

The degeneracy associated with the quantum number M_J (the orientation of the rotation in space) is partly removed when an electric field is applied to a polar molecule (e.g. HCl or NH_3), as illustrated in Fig. 16.22. The splitting of states by an electric field is called the **Stark effect**. For a linear rotor in an electric field $\mathscr{E}$, the energy is given by

$$E = hcBJ(J + 1) + \frac{\mu^2 \mathscr{E}^2 \{J(J + 1) - 3M_J^2\}}{2hcBJ(J + 1)(2J - 1)(2J + 3)} \tag{26}$$

The details of this complicated expression are unimportant here, but it should be noted that the energy depends on the square of the permanent electric dipole moment μ. The observation of the Stark effect can therefore be used to measure this property, but the technique is limited to molecules that are sufficiently volatile to be studied by microwave spectroscopy. However, as spectra can be taken for samples at pressures of only about 10 mTorr, even some quite non-volatile substances may be studied. Sodium chloride, for example, can be studied as diatomic NaCl molecules at high temperatures.

Centrifugal distortion

We have treated molecules as rigid rotors. However, the atoms of rotating molecules are subject to centrifugal forces that tend to distort the molecular geometry and change the moments of inertia (Fig. 16.23). The effect of centrifugal distortion on a diatomic molecule is to stretch the bond and hence to increase the moment of inertia. As a result, centrifugal distortion reduces the rotational constant and, consequently, the energy levels are slightly closer than the rigid-rotor expressions predict. The effect is usually taken into account largely empirically by subtracting a term from the energy and writing

$$F(J) = BJ(J + 1) - D_J J^2 (J + 1)^2 \tag{27}$$

The parameter D_J is the **centrifugal distortion constant**. It is large when the bond is easily stretched. The centrifugal distortion constant of a diatomic molecule is related to the vibrational wavenumber of the bond $\tilde{v}$ (which, as we shall see later, is a measure of its stiffness) through the

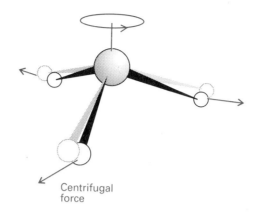

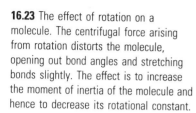

16.23 The effect of rotation on a molecule. The centrifugal force arising from rotation distorts the molecule, opening out bond angles and stretching bonds slightly. The effect is to increase the moment of inertia of the molecule and hence to decrease its rotational constant.

approximate relation:

$$D_J = \frac{4B^3}{\tilde{v}^2} \tag{28}$$

Here the observation of the convergence of the rotational levels as J increases can be interpreted in terms of the rigidity (specifically, the force constant) of the bond.

16.5 Rotational transitions

Typical values of B for small molecules are in the region of 0.1 to 10 cm^{-1} (e.g. 0.356 cm^{-1} for NF_3 and 10.59 cm^{-1} for HCl), so rotational transitions lie in the microwave region of the spectrum. The transitions are detected by monitoring the net absorption of microwave radiation generated either by a klystron or, in modern instruments, by a backward wave oscillator or semiconductor Gunn diode, which are tunable over a wide range of frequencies. For technical reasons related to the detection system, it is desirable to **modulate** the energy levels (that is, vary them in an oscillatory manner) so that the absorption intensity, and therefore the detected signal, oscillates: it is easier to amplify an alternating signal than a steady one. The oscillation is achieved by **Stark modulation**, in which an alternating electric field (of strength of order 10^5 V m^{-1} and frequency between 10 and 100 kHz) is applied to the sample to modulate the energies of the rotational states.

Rotational selection rules

We have already remarked (Section 16.2) that the gross selection rule for the observation of a pure rotation spectrum is that a molecule must have a permanent electric dipole moment. That is, *for a molecule to give a pure rotational spectrum, it must be polar*. The classical basis of this rule is that a polar molecule appears to possess a fluctuating dipole when rotating (Fig. 16.24) but a non-polar molecule does not. The permanent dipole can be regarded as a handle with which the molecule stirs the electromagnetic field into oscillation (and vice versa for absorption). Consequently, homonuclear diatomic molecules and symmetrical ($D_{\infty h}$) linear molecules such as CO_2 are rotationally inactive. Spherical rotors cannot have

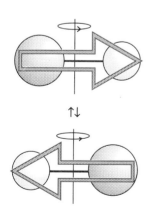

16.24 To a stationary observer, a rotating polar molecule looks like an oscillating dipole that can stir the electromagnetic field into oscillation. This picture is the classical origin of the gross selection rule for rotational transitions.

electric dipole moments unless they become distorted by rotation, so they are also inactive except in special cases. An example of a spherical rotor that does become sufficiently distorted for it to acquire a dipole moment is SiH_4, which has a dipole moment of about $8.3\,\mu D$ by virtue of its rotation when $J \approx 10$ (for comparison, HCl has a permanent dipole moment of $1.1\,D$; molecular dipole moments and their units are discussed in Section 22.1). The pure rotational spectrum of SiH_4 has been detected by using long path lengths (10 m) through high pressure (4 atm) samples.

Example 16.4 *Using the gross selection rule*

State which of the following molecules have rotational absorption spectra: N_2, CO_2, OCS, H_2O, $CH_2{=}CH_2$, C_6H_6.

Method. We need to decide which of the molecules are polar.

Answer. Only OCS and H_2O are polar, so only these two give rise to a rotational absorption spectrum.

Exercise E16.4. Repeat the question for H_2, NO, N_2O, CH_4.

[(a) NO, N_2O]

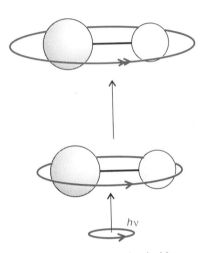

16.25 When a photon is absorbed by a molecule, the angular momentum of the combined system is conserved. If the molecule is rotating in the same sense as the spin of the incoming photon, then J increases by 1.

The specific selection rules are found by evaluating the transition dipole moment between the states. For a linear molecule, the transition moment vanishes unless the following conditions are fulfilled:

$$\Delta J = \pm 1 \qquad \Delta M_J = 0,\ \pm 1$$

The transition $\Delta J = +1$ corresponds to absorption and the transition $\Delta J = -1$ corresponds to emission. The allowed change in J in each case arises from the conservation of angular momentum when a photon, a spin-1 particle, is emitted or absorbed (Fig. 16.25). The change in M_J is also a consequence of the conservation of angular momentum, and takes into account the direction in which the photon leaves or enters the molecule.

When the transition moment is evaluated for all possible orientations of the molecule relative to the line of flight of the photon, it is found that the total $J+1 \leftrightarrow J$ transition intensity is proportional to

$$|\mu_{J+1,J}|^2 = \mu^2 \times \frac{J+1}{2J+1} \to \tfrac{1}{2}\mu^2 \qquad \text{for } J \gg 1 \tag{29}$$

where μ is the permanent electric dipole moment of the molecule. Although the intensity of the line varies with J, the dependence is weak and the dominant effect on intensities is the population of the states. It should be noted that the intensity is proportional to the square of the permanent electric dipole moment, so strongly polar molecules give rise to much more intense rotational lines than less polar molecules.

A selection rule for K is needed for symmetric rotors. If a symmetric rotor has an electric dipole, then it must lie parallel to the figure axis, as in NF_3 (recall Fig. 15.16). Such a molecule cannot be accelerated into different states of rotation around the figure axis by the absorption of radiation, so $\Delta K = 0$ for a symmetric rotor.

The appearance of rotational spectra

When these selection rules are applied to the expressions for the energy levels of a rigid rotor, it follows that the wavenumbers of the allowed $J + 1 \leftarrow J$ absorptions are

$$\tilde{\nu} = 2B(J + 1) \qquad J = 0, 1, 2, \ldots \tag{30a}$$

When centrifugal distortion is taken into account, the corresponding expression is

$$\tilde{\nu} = 2B(J + 1) - 4D_J(J + 1)^3 \tag{30b}$$

However, because the second term is typically very small compared with the first, the appearance of the spectrum closely resembles that predicted from eqn 30a.

Example 16.5 *Predicting the appearance of a rotational spectrum*

Predict the form of the rotational spectrum of NH_3.

Method. We calculated the energy levels in Example 16.3. The NH_3 molecule is a polar symmetric rotor, so the selection rules $\Delta J = \pm 1$ and $\Delta K = 0$ apply. For absorption, $\Delta J = +1$ and we can use eqn 30a.

Answer. Because $B = 9.977 \text{ cm}^{-1}$, we can draw up the following table for the $J + 1 \leftarrow J$ transitions.

$J =$	0	1	2	3 ...
$\tilde{\nu}/\text{cm}^{-1}$	19.95	39.91	59.86	79.82

The line spacing is 19.95 cm^{-1}.

Exercise E16.5. Repeat the problem for $C^{35}ClH_3$ (see Exercise E16.3 for details).
[Lines of separation 0.888 cm^{-1}]

The form of the spectrum predicted by eqn 30a is shown in Fig. 16.26. The most significant feature is that it consists of a series of lines with wavenumbers $2B, 4B, 6B, \ldots$, and separation $2B$. The intensities increase with increasing J and pass through a maximum before tailing off as J becomes large. It should be recalled from Section 16.2 that the observed absorption is the net outcome of the stimulated absorption less the stimulated emission, and that the intensity of each transition depends on the value of J. Hence, the value of J corresponding to the most intense line is not quite the same as the value of J for the most highly populated level. The value of J for the most highly populated rotational energy level in a linear molecule is

$$J_{\text{max}} \approx \left(\frac{kT}{2hcB}\right)^{\frac{1}{2}} - \frac{1}{2} \tag{31}$$

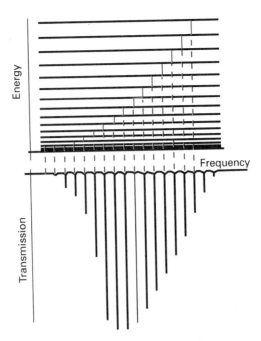

16.26 The rotational energy levels of a linear rotor, the transitions allowed by the selection rule $\Delta J = \pm 1$, and a typical pure rotational absorption spectrum. The intensities reflect the populations of the initial level in each case and the strengths of the transition dipole moments.

For a typical molecule (for example, OCS, with $B = 0.2$ cm^{-1}) at room temperature, $kT \approx 1000hcB$, so $J_{max} \approx 20$.

JUSTIFICATION

There is a maximum in population because the Boltzmann distribution decays exponentially with increasing J, but the degeneracy of the levels, the number of states with a given energy, increases. Specifically, the population of a rotational energy level J is given by the Boltzmann expression

$$\frac{N_J}{N} \propto \text{Number of states of the level } J \times e^{-E_J/kT}$$

$$\propto (2J+1)e^{-hcBJ(J+1)/kT}$$

The value of J corresponding to a maximum of this expression may be found by treating J as a continuous variable, differentiating with respect to J, and then setting the result equal to zero. The result is eqn 31.

The measurement of the line spacing gives B, and hence the moment of inertia perpendicular to the figure axis of the molecule. Because the masses of the atoms are known, it is a simple matter to deduce the bond length of a diatomic molecule. However, in the case of a polyatomic molecule such as OCS or NH$_3$, the analysis gives only a single quantity $I_\perp$ and it is not possible to infer both bond lengths (in OCS) or the bond length and bond angle (in NH$_3$). This difficulty can be overcome by using isotopically substituted molecules, such as ABC and A'BC; then, by assuming that $R(A\!-\!B) = R(A'\!-\!B)$, both A—B and B—C bond lengths can be extracted from the two moments of inertia. A famous example of this procedure is the study of OCS; the actual calculation is worked through in Problem 16.11. The assumption that bond lengths are unchanged by isotopic substitution is only an approximation, but it is a good approximation in most cases.

16.6 Rotational Raman spectra

The gross selection rule for rotational Raman transitions is that *the molecule must be anisotropically polarizable*. We shall begin by explaining what this means.

The distortion of a molecule in an electric field is determined by its **polarizability** α (we deal with polarizabilities in detail in Section 22.1). More precisely, if the strength of the field is $\mathscr{E}$, then the molecule acquires an induced dipole moment

$$\mu = \alpha \mathscr{E} \tag{32}$$

in addition to any permanent dipole moment it may have. We see that the greater the polarizability, the greater the dipole induced by a given field. A Xe atom, for example, has a greater polarizability than a He atom because its outer electrons are less tightly under the control of the more distant central nucleus and are more easily displaced by an externally applied field.

An atom is isotropically polarizable. That is, the same distortion is induced whatever the direction of the applied field. The polarizability of a spherical rotor is also isotropic. However, non-spherical rotors have polarizabilities that do depend on the direction of the field and hence are anisotropically polarizable (Fig. 16.27). The electron distribution in H_2, for example, is more distorted when the field is applied parallel to the bond than when it is applied perpendicular to it, and we write $\alpha_{\parallel} > \alpha_{\perp}$.

All linear molecules and diatomics (whether homonuclear or heteronuclear) have anisotropic polarizabilities and so are rotationally Raman active. This activity is one reason for the importance of rotational Raman spectroscopy, for the technique can be used to study many of the molecules that are inaccessible to pure rotational microwave spectroscopy. Spherical rotors such as CH_4 and SF_6, however, are rotationally Raman inactive as well as rotationally microwave inactive.[5]

The specific rotational Raman selection rules are

Linear rotors: $\Delta J = 0, \pm 2$

Symmetric rotors: $\Delta J = 0, \pm 1, \pm 2; \Delta K = 0$

The $\Delta J = 0$ transitions do not lead to a shift of the scattered photon's frequency in pure rotational Raman spectroscopy, and contribute to the unshifted **Rayleigh scattered** light.

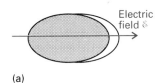

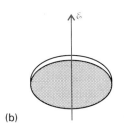

16.27 An electric field applied to a molecule results in its distortion, and the distorted molecule acquires a contribution to its dipole moment (even if it is non-polar initially). The polarizability may be different when the field is applied (a) parallel or (b) perpendicular to the molecular axis (or, in general, in different directions relative to the molecule); if that is so, then the molecule has an anisotropic polarizability.

5 Rotational inactivity does not mean that the molecules are never found in rotationally excited states. Molecular collisions do not have to obey such restrictive selection rules, and hence collisions between molecules can lead to the population of any rotational state.

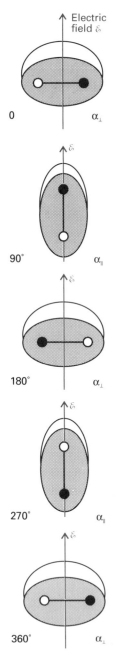

16.28 The distortion induced in a molecule by an applied electric field returns to its initial value after a rotation of only 180° (i.e. twice a revolution). This is the origin of the $\Delta J = \pm 2$ selection rule in rotational Raman spectroscopy.

JUSTIFICATION

The classical origin of the 2 in the selection rule is as follows. In an electric field $\mathscr{E}$, a molecule acquires a dipole moment of magnitude $\alpha\mathscr{E}$, where α is the polarizability. If the electric field is that of a light wave of frequency ω_i, then the induced dipole moment is time-dependent and has the form

$$\mu = \alpha\mathscr{E} = \alpha\mathscr{E}_i \cos \omega_i t$$

If the molecule is rotating, then to an external observer its polarizability is also time-dependent (if it is anisotropic), and we can write

$$\alpha = \alpha_0 + \Delta\alpha \cos 2\omega_R t$$

The 2 appears because the polarizability returns to its initial value twice each revolution (Fig. 16.28). Substituting this expression into the expression for the induced dipole moment gives

$$\begin{aligned}\mu &= (\alpha_0 + \Delta\alpha \cos 2\omega_R t) \times (\mathscr{E}_i \cos \omega_i t)\\ &= \alpha_0\mathscr{E}_i \cos \omega_i t + \mathscr{E}_i\Delta\alpha \cos 2\omega_R t \cos \omega_i t\\ &= \alpha_0\mathscr{E}_i \cos \omega_i t + \tfrac{1}{2}\mathscr{E}_i\Delta\alpha\{\cos(\omega_i + 2\omega_R)t + \cos(\omega_i - 2\omega_R)t\}\end{aligned}$$

This calculation shows that the induced dipole has a component oscillating at the incident light frequency (so that it radiates Rayleigh radiation), and that it also has two components at $\omega_i \pm 2\omega_R$, which give rise to the shifted Raman lines. Note that these lines appear only if $\Delta\alpha \neq 0$; hence the polarizability must be anisotropic for there to be Raman lines.

We can predict the form of the Raman spectrum of a linear rotor (Fig. 16.29) by applying the selection rule $\Delta J = \pm 2$ to the rotational energy levels. When the molecule makes a transition with $\Delta J = +2$, the scattered radiation leaves it in a higher rotational state, so the wavenumber of the incident radiation, initially $\tilde{\nu}_i$, is decreased (Fig. 16.30). These transitions account for the **Stokes lines** in the spectrum:

$$\tilde{\nu}(J+2\leftarrow J) = \tilde{\nu}_i - \{F(J+2) - F(J)\} = \tilde{\nu}_i - 2B(2J+3) \qquad \textbf{(33a)}$$

The Stokes lines appear to low frequency of the incident light and at displacements $6B, 10B, 14B, \ldots$, from $\tilde{\nu}_i$ for $J = 0, 1, 2, \ldots$ When the molecule makes a transition with $\Delta J = -2$, the scattered photon emerges with increased energy. These transitions account for the **anti-Stokes lines** of the spectrum:

$$\tilde{\nu}(J\rightarrow J-2) = \tilde{\nu}_i + \{F(J) - F(J-2)\} = \tilde{\nu}_i + 2B(2J-1) \qquad \textbf{(33b)}$$

The anti-Stokes lines occur at displacements of $6B, 10B, 14B, \ldots$, (for $J = 2, 3, \ldots$; $J = 2$ is the lowest state that can contribute under the selection rule $\Delta J = -2$) to high frequency of the incident radiation. The separation of adjacent lines in both the Stokes and the anti-Stokes regions

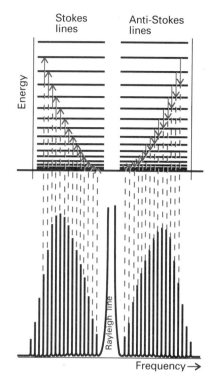

Stokes lines | Anti-Stokes lines

Energy

Rayleigh line

Frequency →

16.29 The rotational energy levels of a linear rotor and the transitions allowed by the $\Delta J = 2$ Raman selection rules. The form of a typical rotational Raman spectrum is also shown.

is $4B$, so from its measurement $I_\perp$ can be determined and then used to find the bond lengths exactly as in the case of microwave spectroscopy.

Example 16.6 *Predicting the form of a Raman spectrum*

Predict the form of the rotational Raman spectrum of $^{14}N_2$, for which $B = 1.99$ cm^{-1} when it is exposed to monochromatic 336.732-nm laser radiation.

Method. The molecule is rotationally Raman active because end-over-end rotation modulates its polarizability as viewed by a stationary observer. The Stokes and anti-Stokes lines are given by the expressions above.

Answer. Because $\lambda_i = 336.732$ nm corresponds to $\tilde{\nu}_i = 29\,697.2$ cm^{-1}, eqns 33a and 33b give the following line positions:

	J			
	0	1	2	3
Stokes lines				
$\tilde{\nu}$/cm^{-1}	29 685.3	29 677.3	29 669.3	29 661.4
λ/nm	336.868	336.958	337.048	337.139
Anti-Stokes lines				
$\tilde{\nu}$/cm^{-1}			29 709.1	29 717.1
λ/nm			336.597	336.507

Comment. There will be a strong central line at 336.732 nm accompanied on either side by lines of increasing and then decreasing intensity (as a result of transition moment and population effects). The spread of the entire spectrum is very small (about 300 cm^{-1} at room temperature), so the incident light must be highly monochromatic.

Exercise E16.6. Repeat the calculation for the rotational Raman spectrum of NH_3 ($B = 9.977$ cm^{-1}).

16.7 Nuclear statistics and rotational states

If eqn 33 were used in conjunction with the rotational Raman spectrum of CO_2, the rotational constant would be inconsistent with other measurements of C—O bond lengths. The results are consistent only if it is supposed that the molecule can exist in states with even values of J, so the Stokes lines are $2 \leftarrow 0$, $4 \leftarrow 2$, etc. and not $2 \leftarrow 0$, $3 \leftarrow 1$, etc.

The explanation of the missing lines is the Pauli principle and the fact that O nuclei are spin-0 bosons: just as the Pauli principle excludes certain electronic states, so too does it exclude certain molecular rotational states. The form of the Pauli principle given in *Further information 12* states that, when two identical bosons are exchanged, the overall wavefunction must remain unchanged in every respect, including sign. In particular, when a CO_2 molecule rotates through 180°, two

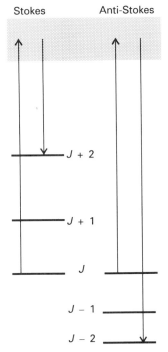

16.30 In a Stokes transition, a molecule is raised to an unspecified excited electronic state by the incident photon, and immediately emits a photon as it falls to a rotational state of higher energy than the initial state. The net effect is the rotational excitation of the molecule at the expense of the scattered photon. In an anti-Stokes transition, the molecule falls to a rotational state lower than its initial state, so the scattered photon emerges with more energy than it had initially. Because the lifetime of the excited electronic state is zero, there is no need to worry about its infinitely ill-defined energy: the transition to it is a *virtual* transition, not a real transition.

identical O nuclei are interchanged, so the overall wavefunction of the molecule must remain unchanged. However, inspection of the form of the rotational wavefunctions (which have the same form as the s, p, etc. orbitals of atoms) shows that they change sign by $(-1)^J$ under such a rotation (Fig. 16.31). Therefore, *only even values of J are permissible for CO_2*, and hence the Raman spectrum shows only alternate lines.

The selective occupation of rotational states that stems from the Pauli principle is termed **nuclear statistics**. Nuclear statistics must be taken into account whenever a rotation interchanges equivalent nuclei. However, the consequences are not always as simple as for CO_2 because there are complicating features when the nuclei have non-zero spin. For molecular hydrogen and fluorine, for instance, with their two identical spin-$\frac{1}{2}$ nuclei, the populations of odd J and even J rotation states are in the ratio of 3:1, and there is an alternation in intensity in their rotational Raman spectra (Fig. 16.32). In general, for a homonuclear diatomic molecule with nuclei of spin I, the ratio of populations is given by

$$\frac{\text{Population of states with odd } J}{\text{Population of states with even } J} = \begin{cases} \dfrac{I+1}{I} & \text{for half-integral spin nuclei} \\[2ex] \dfrac{I}{I+1} & \text{for integral spin nuclei} \end{cases}$$

(34)

For hydrogen, $I = \frac{1}{2}$, and the ratio is 3:1. For N_2, with $I = 1$, the ratio is 1:2.

JUSTIFICATION

Hydrogen nuclei are fermions, so the Pauli principle requires the overall wavefunction to change sign under particle interchange. However, the rotation of an H_2 molecule through 180° has a more complicated effect than merely relabelling the nuclei, because it interchanges their spin states too if the spins are paired ($\uparrow\downarrow$) but not if they are parallel ($\uparrow\uparrow$).

If the spins are parallel, then for the overall wavefunction of the molecule to change sign, the rotational wavefunction must change sign. Hence, *only odd values of J are allowed if the nuclear spins are parallel*. In contrast, if the nuclear spins are paired, then their wavefunction is $\alpha(A)\beta(B) - \alpha(B)\beta(A)$ (see *Further information 12*), which changes sign when A and B are interchanged. Therefore, for the overall wavefunction to change sign requires the rotational wavefunction not to change sign. Hence, *only even values of J are allowed if the nuclear spins are paired*.

As there are three nuclear spin states with parallel spins (just like the triplet state of two parallel electrons), but only one state with paired spins (the analogue of the singlet state of two electrons, see Fig. 13.24), it follows that the populations of the odd J and even J states should be in the ratio of 3:1, and hence the intensities of transitions originating in these levels will be in the same ratio.

Nuclear spin states change into each other only very slowly, so an H_2

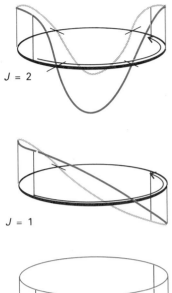

$J = 2$

$J = 1$

$J = 0$

16.31 The symmetries of rotational wavefunctions (shown here, for simplicity as a two-dimensional rotor) under a rotation through 180°. Wavefunctions with J even do not change sign: those with J odd do change sign.

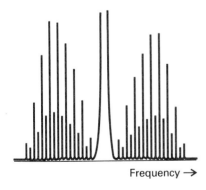

Frequency →

16.32 The rotational Raman spectrum of a diatomic molecule with two identical spin-$\frac{1}{2}$ nuclei shows an alternation in intensity as a result of nuclear statistics.

molecule with parallel nuclear spins remains distinct from one with paired nuclear spins for long periods. The two forms of hydrogen can be separated by physical techniques, and stored. The form with parallel nuclear spins is called **ortho*-hydrogen** and the form with paired nuclear spins is called **para*-hydrogen**. Because *ortho*-hydrogen cannot exist in a state with $J = 0$, it continues to rotate at very low temperatures and has an effective rotational zero-point energy (Fig. 16.33). This is of some concern to manufacturers of liquid hydrogen, for the slow conversion of *ortho*-hydrogen into *para*-hydrogen (which can exist with $J = 0$) as nuclear spins slowly realign releases rotational energy which vaporizes the liquid. Techniques are used to accelerate the conversion of *ortho*-hydrogen to *para*-hydrogen to avoid this problem. One such technique is to pass hydrogen over a metal surface: the molecules adsorb on to the surface as atoms, which then recombine in the lower-energy *para*-hydrogen form.

THE VIBRATIONS OF DIATOMIC MOLECULES

In this section, we adopt the same strategy of finding expressions for the energy levels, establishing the selection rules, and then discussing the form of the spectrum. We shall also see how the simultaneous excitation of rotation modifies the appearance of a vibrational spectrum.

16.8 Molecular vibrations

We shall base our discussion on Fig. 16.34, which shows a typical potential energy curve (Fig. 14.1) of a diatomic molecule.

In regions close to R_e (at the minimum of the curve) the potential energy can be approximated by a parabola, so we can write

$$V = \tfrac{1}{2}k(R - R_e)^2 \tag{35}$$

where k is the **force constant** of the bond. The steeper the walls of the potential (the stiffer the bond), the greater the force constant. The Schrödinger equation for the motion of the two atoms of masses m_1 and m_2 with this potential energy is

$$-\frac{\hbar^2}{2\mu}\frac{d^2\psi}{dx^2} + V\psi = E\psi \tag{36a}$$

where μ is the **effective mass**:

$$\frac{1}{\mu} = \frac{1}{m_1} + \frac{1}{m_2} \tag{36b}$$

These equations are derived in the same way as in *Further information 10*, where the separation of variables procedure was used to separate the relative motion of the atoms from the motion of the molecule as a whole. (In that context, the effective mass is called the *reduced mass*, and the name is widely used in this context too.)

The Schrödinger equation in eqn 36 is the same as the one discussed in Section 12.4 in connection with a particle of mass μ undergoing harmonic motion. Therefore, we can use the results of that section

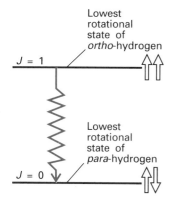

16.33 When hydrogen is cooled, the molecules with parallel nuclear spins accumulate in their lowest available rotational state, the one with $J = 1$. They can enter the lowest rotational state only if the spins change their relative orientation and become antiparallel. This is a slow process under normal circumstances, so energy is slowly released.

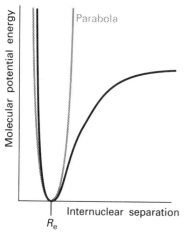

16.34 A molecular potential energy curve can be approximated by a parabola near the bottom of the well. The parabolic potential leads to harmonic oscillations. At high excitation energies the parabolic approximation is poor (the true potential is less confining), and is totally wrong near the dissociation limit.

directly, and immediately write down the permitted vibrational energy levels:

$$E_v = (v + \tfrac{1}{2})\hbar\omega \qquad \text{with } \omega = \left(\frac{k}{\mu}\right)^{\frac{1}{2}} \qquad \text{and } v = 0, 1, 2, \ldots \qquad (37)$$

The **vibrational terms** of a molecule, the energies of its vibrational states expressed in wavenumbers, are denoted G, so

$$G(v) = (v + \tfrac{1}{2})\tilde{v}, \qquad \text{with } \tilde{v} = \frac{\omega}{2\pi c} \qquad (38)$$

The vibrational wavefunctions are the same as those discussed in Section 12.5.

It is important to note that the vibrational terms depend on the effective mass of the molecule, not directly on its total mass. This dependence is physically reasonable, for if atom 1 were as heavy as a brick wall, then we would find $\mu \approx m_2$, the mass of the lighter atom. The vibration would then be that of a light atom relative to that of a stationary wall (this is approximately the case in HI, for example, where the I atom barely moves and $\mu \approx m_H$). For a homonuclear diatomic molecule $m_1 = m_2$, and the effective mass is half the total mass: $\mu = \tfrac{1}{2}m$.

An HCl molecule has a force constant of $516\,\text{N m}^{-1}$, a reasonably typical value. The effective mass of $^1H^{35}Cl$ is $1.63 \times 10^{-27}\,\text{kg}$ (note that this mass is very close to the mass of the hydrogen atom, $1.67 \times 10^{-27}\,\text{kg}$, so the Cl atom is like a brick wall). These values imply

$$\omega = 5.63 \times 10^{14}\,\text{s}^{-1} \quad v = 8.95 \times 10^{13}\,\text{Hz} \quad \tilde{v} = 2990\,\text{cm}^{-1} \quad \lambda = 3.35\,\mu\text{m}$$

These characteristics correspond to electromagnetic radiation in the infrared region, so vibrational spectroscopy is an infrared technique.

16.9 Selection rules

The gross selection rule for a molecular vibration is that *the electric dipole moment of the molecule must change when the atoms are displaced*. The classical basis of this rule is that the molecule can shake the electromagnetic field into oscillation if its dipole changes as it vibrates (Fig. 16.35), and vice versa. Note that the molecule need not have a permanent dipole: the rule requires only a *change* in dipole moment, possibly from zero. Some vibrations do not affect the molecule's dipole moment (e.g. the stretching motion of a homonuclear diatomic molecule), so they neither absorb nor generate radiation: such vibrations are said to be **infrared inactive**. Homonuclear diatomic molecules are infrared inactive because their dipole moments remain zero however long the bond; heteronuclear diatomic molecules are infrared active.

Example 16.7 *Using the gross selection rules*

State which of the following molecules have vibrational absorption spectra: N_2, CO_2, OCS, H_2O, $CH_2{=}CH_2$, C_6H_6.

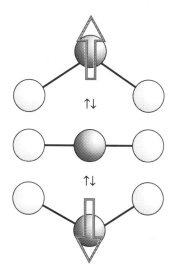

16.35 The oscillation of a molecule, even if it is non-polar, may result in an oscillating dipole that can interact with the electromagnetic field.

Method. Molecules that give rise to vibrational spectra have dipole moments that change during the course of a vibration. Therefore, judge whether a distortion of the molecule can change its dipole moment (including changing it from zero). Take into account both bending and stretching motions of polyatomic molecules.

Answer. All the molecules except N_2 possess at least one vibrational mode that results in a change of dipole moment, so all except N_2 can show a vibrational absorption spectrum.

Comment. Not all the modes of complex molecules are vibrationally active. For example, the symmetric stretch of CO_2, in which the O—C—O bonds stretch and contract symmetrically is inactive because it leaves the dipole moment unchanged (at zero).

Exercise E16.7. Repeat the question for H_2, NO, N_2O, CH_4.

[NO, N_2O, CH_4]

The specific vibrational selection rule, which is obtained from an analysis of the expression for the transition moment and the properties of integrals over harmonic oscillator wavefunctions, is

$$\Delta v = \pm 1$$

Transitions for which $\Delta v = +1$ correspond to absorption and those with $\Delta v = -1$ correspond to emission. It follows from these selection rules that the wavenumbers of allowed vibrational transitions, which are denoted $\Delta G_{v+\frac{1}{2}}$ for the transition $v + 1 \leftarrow v$, are

$$\Delta G_{v+\frac{1}{2}} = G(v + 1) - G(v) = \tilde{v} \tag{39}$$

As we have seen, $\tilde{v}$ lies in the infrared region of the electromagnetic spectrum, so vibrational transitions absorb and generate infrared radiation.

At room temperature $kT/hc \approx 200 \text{ cm}^{-1}$, and most vibrational wavenumbers are significantly greater than 200 cm^{-1}. It follows from the Boltzmann distribution that almost all the molecules will be in their vibrational ground states initially. Hence, the dominant spectral transition will be the **fundamental transition**, $1 \leftarrow 0$. As a result, the spectrum is expected to consist of a single absorption line. If the molecules are formed in a vibrationally excited state, such as when vibrationally excited HF is formed in the reaction $H_2 + F_2 \rightarrow 2HF^*$, then the transitions $5 \rightarrow 4$, $4 \rightarrow 3$, etc. may also appear (in emission). In the harmonic approximation, all these lines lie at the same frequency, and the spectrum is also a single line. However, as we shall now show, the breakdown of the harmonic approximation causes the transitions to lie at slightly different frequencies, so several lines are observed.

16.10 Anharmonicity

The vibrational terms in eqn 38 are only approximate because they are based on the parabolic approximation to the actual potential energy curve. A parabola cannot be correct at all extensions because it does not allow a bond to dissociate; moreover, it suggests that one atom can move

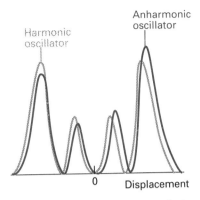

16.36 The probability distribution (ψ^2) of a slightly anharmonic oscillator compared with the probability distribution of a harmonic oscillator (in each case for $v = 3$). The anharmonic oscillator (for a typical diatomic molecule) is more likely to be found at large extensions and less likely to found significantly compressed than is a harmonic oscillator.

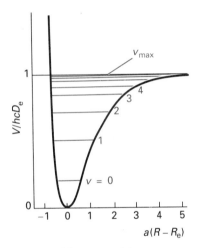

16.37 The Morse potential energy curve reproduces the general shape of a molecular potential energy curve. The corresponding Schrödinger equation can be solved, and the values of the energies obtained. The number of bound levels is finite.

through another. At high vibrational excitations the swing of the atoms (more precisely, the spread of the vibrational wavefunction, Fig. 16.36) allows the molecule to explore regions of the curve where the parabolic approximation is poor. The motion then becomes **anharmonic** in the sense that the restoring force is no longer proportional to the displacement. Because the actual curve is less confining than a parabola, we can anticipate that the energy levels become less widely spaced at high excitations.

The convergence of energy levels

One approach to the calculation of the energy levels in the presence of anharmonicity is to use a function that resembles the true potential energy more closely. The **Morse potential energy** is

$$V = hcD_e\{1 - e^{-a(R-R_e)}\}^2 \tag{40a}$$

where D_e is the depth of the potential minimum and

$$a = \left(\frac{\mu}{2hcD_e}\right)^{\frac{1}{2}} \times \omega \tag{40b}$$

Equation 40a is plotted in Fig. 16.37. Near the well minimum it resembles a parabola (as can be checked by expanding the exponential as far as the first term), but unlike a parabola it allows for dissociation at high energies. The Schrödinger equation can be solved for the Morse potential and the permitted energy levels are

$$G(v) = (v + \tfrac{1}{2})\tilde{v} - (v + \tfrac{1}{2})^2 x_e\tilde{v}, \quad \text{with } x_e = \frac{a^2\hbar}{2\mu\omega} = \frac{\tilde{v}}{4D_e} \tag{41}$$

The parameter x_e is called the **anharmonicity constant**. The number of vibrational levels of a Morse oscillator is finite, and $v = 0, 1, 2, \ldots, v_{max}$, as shown in Fig. 16.37. The second term in the expression for G in eqn 41 substracts from the first with increasing effect as v increases, and gives rise to the convergence of the levels at high quantum numbers.

Although the Morse oscillator is quite useful theoretically, in practice the more general expression

$$G(v) = (v + \tfrac{1}{2})\tilde{v} - (v + \tfrac{1}{2})^2 x_e\tilde{v} + (v + \tfrac{1}{2})^3 y_e\tilde{v} + \cdots \tag{42}$$

where $x_e, y_e, \ldots$, are empirical constants characteristic of the molecule, is used to fit the experimental data and to find the dissociation energy of the molecule. When anharmonicities are present, the wavenumbers of transitions with $\Delta v = +1$ are

$$\Delta G_{v+\frac{1}{2}} = \tilde{v} - 2(v + 1)x_e\tilde{v} + \cdots \tag{43}$$

This equation shows that the transitions move to lower wavenumbers as v increases. (In the harmonic approximation all lines lie at the same wavenumber $\tilde{v}$.)

Anharmonicity also accounts for the appearance of additional weak absorption lines corresponding to the transitions $2 \leftarrow 0$, $3 \leftarrow 0$, etc., even though these first, second, ..., **overtones** are forbidden by the selection rule $\Delta v = \pm 1$. The first overtone (which is also called the second harmonic), for example, gives rise to an absorption at

$$G(v + 2) - G(v) = 2\tilde{v} - 2(2v + 3)x_e\tilde{v} + \cdots \qquad (44)$$

The reason for the appearance of overtones is that the selection rule is derived using harmonic oscillator wavefunctions, and these are only approximately valid when anharmonicity is present. Therefore, the selection rule is also only an approximation. For an anharmonic oscillator, all values of Δv are allowed, but $\Delta v > 1$ is allowed only weakly if the anharmonicity is slight.

The Birge–Sponer extrapolation

When several vibrational transitions are detectable, a graphical technique called **Birge–Sponer extrapolation** may be used to determine the dissociation energy D_0 of the bond (Fig. 16.38). The depth of the potential well D_e differs from D_0 by the zero-point energy:

$$D_e = D_0 + \tfrac{1}{2}(1 - \tfrac{1}{2}x_e)\tilde{v} \approx D_0 + \tfrac{1}{2}\tilde{v} \qquad (45a)$$

The basis of the Birge–Sponer extrapolation is that the sum of successive energy separations $\Delta G_{v+\frac{1}{2}}$ from the zero-point level to the dissociation limit is the dissociation energy:

$$D_0 = \Delta G_{\frac{1}{2}} + \Delta G_{1+\frac{1}{2}} + \cdots = \sum_v \Delta G_{v+\frac{1}{2}} \qquad (45b)$$

just as the height of a ladder is the sum of the separation of its rungs. The construction in Fig. 16.39 shows that the area under the plot of $\Delta G_{v+\frac{1}{2}}$ against v is equal to the sum, and therefore to D_0. The successive terms decrease linearly when only the x_e anharmonicity constant is taken into account and the inaccessible part of the spectrum can be estimated by linear extrapolation. Most actual plots differ from the linear plot as shown in the illustration, so the value of D_0 obtained in this way is usually an overestimate of the true value.

Example 16.8 *Using a Birge–Sponer extrapolation*

The observed vibrational energy level separations of H_2^+ lie at the following values for $1 \leftarrow 0$, $2 \leftarrow 1$, ..., respectively (in cm^{-1}): 2191, 2064, 1941, 1821, 1705, 1591, 1479, 1368, 1257, 1145, 1033, 918, 800, 677, 548, 411. Determine the dissociation energy of the molecule.

Method. We need to plot the separations against v, extrapolate linearly to the point cutting the v-axis, and then measure the area under the curve.

Answer. The points are plotted in Fig. 16.40; the extended line is the Birge–Sponer linear extrapolation. The area under the curve (use the formula for the area of a triangle or count the squares) is 214. Each square corresponds to $100\ cm^{-1}$ (refer to the scale of the vertical axis); hence the dissociation energy is $21\,400\ cm^{-1}$ (corresponding to $256\ kJ\ mol^{-1}$).

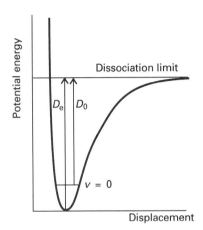

16.38 The relation between the dissociation energy D_0 and the minimum energy D_e of a molecular potential energy curve.

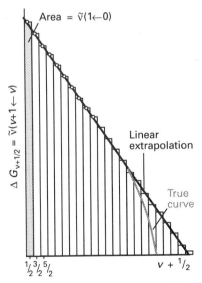

16.39 The area under a plot of transition wavenumber against vibrational quantum number is equal to the dissociation energy of the molecule. The assumption that the differences approach zero linearly is the basis of the Birge–Sponer extrapolation.

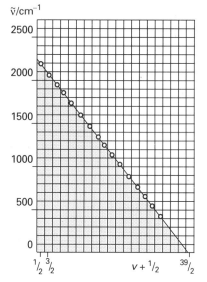

16.40 The Birge–Sponer plot used in Example 16.8. The area is obtained simply by counting the squares beneath the line or using the formula for the area of a right-angled triangle.

16.41 A high-resolution vibration–rotation spectrum of HCl. The lines appear in pairs because $H^{35}Cl$ and $H^{37}Cl$ both contribute (their abundance ratio is 3:1). There is no Q branch, because $\Delta J = 0$ is forbidden for this molecule.

Exercise E16.8. The vibrational levels of HgH converge rapidly, and successive separations are 1203.7, 965.6, 632.4, and 172 cm^{-1}. Estimate the dissociation energy. [40 kJ mol^{-1}]

16.11 Vibration–rotation spectra

At high resolution, each line of the vibrational spectrum of a gas-phase heteronuclear diatomic molecule is found to consist of a large number of closely spaced components (Fig. 16.41). For this reason, molecular spectra are often called **band spectra**. The separation between the components is of the order of 10 cm^{-1}, which suggests that the structure is due to rotational transitions accompanying the vibrational transition. A rotational change should be expected because classically we can think of the transition as leading to a sudden increase or decrease in the instantaneous bond length. Just as ice-skaters rotate more rapidly when they bring their arms in, and more slowly when they throw them out, so the molecular rotation is either accelerated or retarded by a vibrational transition.

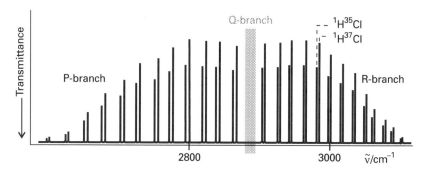

Spectral branches

A detailed analysis of the quantum mechanics of the process shows that the rotational quantum number J changes by ± 1 during the vibrational transition of a diatomic molecule. If the molecule also possesses angular momentum about its axis, as in the case of the electronic orbital angular momentum of the $^2\Pi$ molecule NO, then the selection rules also allow $\Delta J = 0$.

The appearance of the vibration–rotation spectrum of a diatomic molecule can be discussed in terms of the combined vibration–rotation terms S:

$$S(v, J) = G(v) + F(J)$$

If we ignore anharmonicity and centrifugal distortion,

$$S(v, J) = (v + \tfrac{1}{2})\tilde{v} + BJ(J + 1)$$

In a more detailed treatment B is allowed to depend on the vibrational state because, as v increases, the molecule swells slightly and the moment of inertia changes. Initially we shall continue with the simple expression.

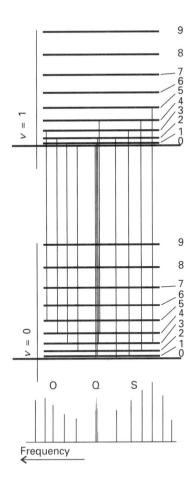

16.42 The formation of P, Q, and R branches in a vibration–rotation spectrum. The intensities reflect the populations of the initial rotational levels.

When the vibrational transition $v + 1 \leftarrow v$ occurs, J changes by ± 1 and in some cases by 0 (when $\Delta J = 0$ is allowed). The absorptions then fall into groups called **branches** of the spectrum. The **P branch** consists of all transitions with $\Delta J = -1$:

$$\tilde{\nu}_P(J) = S(v + 1, J - 1) - S(v, J) = \tilde{\nu} - 2BJ \qquad \textbf{(46a)}$$

This branch (Fig. 16.42) consists of lines at $\tilde{\nu} - 2B$, $\tilde{\nu} - 4B, \ldots$, with an intensity distribution reflecting both the populations of the rotational levels and the $J - 1 \leftarrow J$ transition moment.

The **Q branch** consists of all lines with $\Delta J = 0$, and its wavenumbers are all

$$\tilde{\nu}_Q(J) = S(v + 1, J) - S(v, J) = \tilde{\nu} \qquad \textbf{(46b)}$$

for all values of J. This branch, when it is allowed (as in the case of NO), forms a single line at the vibrational transition wavenumber. In practice, since the rotational constants of the two vibrational levels are slightly different, the Q branch appears as a cluster of closely spaced lines. In Fig. 16.41 there is a gap at the expected location of the Q branch because it is forbidden in HCl.

The **R branch** consists of lines with $\Delta J = +1$:

$$\tilde{\nu}_R(J) = S(v + 1, J + 1) - S(v, J) = \tilde{\nu} + 2B(J + 1) \qquad \textbf{(46c)}$$

This branch consists of lines displaced from $\tilde{\nu}$ to high wavenumber by $2B$, $4B, \ldots$, (Fig. 16.42).

The separation between the lines in the P and R branches of a vibrational transition gives the value of B, so the bond length can be deduced without needing to take pure rotational microwave spectrum (although the latter is more precise).

Combination differences

The rotational constant of the vibrationally excited state B_1 (in general, B_v) is in fact slightly smaller than that of the ground vibrational state B_0 because the anharmonicity of the vibration results in a slightly extended bond in the upper state. As a result, the Q branch (if it exists) consists of a series of closely spaced lines, the lines of the R branch converge slightly as J increases, and those of the P branch diverge (see Fig. 16.41):

$$\tilde{\nu}_P(J) = \tilde{\nu} - (B_1 + B_0)J + (B_1 - B_0)J^2$$
$$\tilde{\nu}_Q(J) = \tilde{\nu} + (B_1 - B_0)J(J + 1) \qquad \textbf{(47)}$$
$$\tilde{\nu}_R(J) = \tilde{\nu} + (B_1 + B_0)(J + 1) + (B_1 - B_0)(J + 1)^2$$

To determine the two rotational constants individually we use the method of **combination differences**. This procedure is used widely in spectroscopy to extract information about a particular state. It involves setting up expressions for the difference in the wavenumbers of transitions to a common state; the resulting expression then depends solely on properties of the other state.

As can be seen from Fig. 16.43, the transitions $\tilde{\nu}_R(J - 1)$ and

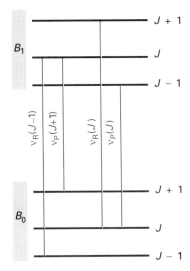

16.43 The method of combination differences makes use of the fact that some transitions share a common level.

$\tilde{\nu}_P(J+1)$ have a common upper state, and hence can be anticipated to have a common dependence on B_1. Indeed, it is easy to show from eqn 47 that

$$\tilde{\nu}_R(J-1) - \tilde{\nu}_P(J+1) = 4B_0(J+\tfrac{1}{2}) \tag{48a}$$

Therefore, a plot of the combination difference against $J+\tfrac{1}{2}$ should be a straight line of slope $4B_0$, so the rotational constant of the molecule in the state $v = 0$ can be determined. (Any deviation from a straight line is a consequence of centrifugal distortion, so that effect can be investigated too.) Similarly, the transitions $\tilde{\nu}_R(J)$ and $\tilde{\nu}_P(J)$ have a common lower state, and hence their combination difference gives information about the upper state:

$$\tilde{\nu}_R(J) - \tilde{\nu}_P(J) = 4B_1(J+\tfrac{1}{2}) \tag{48b}$$

16.12 Vibrational Raman spectra of diatomic molecules

The gross selection rule for vibrational Raman transitions is that *the polarizability should change as the molecule vibrates*. As homonuclear and heteronuclear diatomic molecules swell and contract during a vibration, the control of the nuclei over the electrons varies, and hence the molecular polarizability changes. Both types of diatomic molecule are therefore vibrationally Raman active.

The specific selection rule for vibrational Raman transitions is $\Delta v = \pm 1$. The lines to high frequency of the incident light, the anti-Stokes lines, are those for which $\Delta v = -1$. They are usually weak because very few molecules are in an excited vibrational state initially. The lines to low frequency, the Stokes lines, correspond to $\Delta v = +1$. Superimposed on these lines, in gas-phase spectra, is a branch structure arising from the simultaneous rotational transitions that accompany the vibrational excitation (Fig. 16.44). The selection rules are $\Delta J = 0, \pm 2$ (as in pure rotational Raman spectroscopy), and give rise to the **O branch** ($\Delta J = -2$), the Q branch ($\Delta J = 0$), and the **S branch** ($\Delta J = +2$):

$$\tilde{\nu}_O(J) = \tilde{\nu} + 2B - 4BJ \qquad \tilde{\nu}_Q(J) = \tilde{\nu} \qquad \tilde{\nu}_S(J) = \tilde{\nu} + 6B + 4BJ \tag{49}$$

Note that, unlike in infrared spectroscopy, a Q branch is obtained for all linear molecules. The spectrum of CO, for instance, is shown in Fig. 16.45: the structure of the Q branch arises from the differences in rotational constants of the upper and lower vibrational states.

The information available from vibrational Raman spectra adds to that from infrared spectroscopy because homonuclear diatomics can also be studied. The spectra can be interpreted in terms of the force constants, dissociation energies, and bond lengths, and some of the information obtained is included in Table 16.2.

Table 16.2* Properties of diatomic molecules

	$\tilde{\nu}/cm^{-1}$	B/cm^{-1}	$k/(N\,m^{-1})$
$^1H_2$	4400	60.86	575
$^1H^{35}Cl$	2991	10.59	516
$^1H^{127}I$	2309	6.61	313
$^{35}Cl_2$	560	0.244	323

* More values are given in the Data section at the end of this volume.

THE VIBRATIONS OF POLYATOMIC MOLECULES

There is only one mode of vibration for a diatomic molecule, the bond stretch. In polyatomic molecule there are several modes because bonds may stretch and angles may bend.

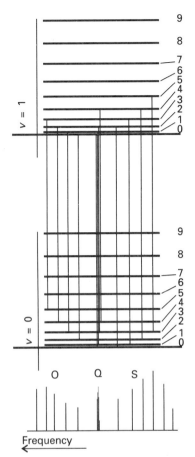

16.44 The formation of O, Q, and S branches in a vibration–rotation Raman spectrum of a linear rotor. Note that the frequency scale runs in the opposite direction to that in Fig. 16.42, because the higher-energy transitions (on the right) extract more energy from the incident beam and leave it at lower frequency.

16.13 Normal modes

We begin by calculating the total number of vibrational modes of a polyatomic molecule consisting of N atoms. We then see that we can choose combinations of these atomic displacements that give the simplest description of the vibrations of the molecule.

The number of vibrational modes

For a non-linear molecule that consists of N atoms, there are $3N - 6$ independent modes of vibration. If the molecule is linear, then there are $3N - 5$ independent vibrational modes.

JUSTIFICATION

The total number of coordinates needed to specify the locations of N atoms is $3N$. Each atom may change its location by varying one of its three coordinates (x, y, and z), so the total number of displacements available is $3N$. These displacements can be grouped together in a physically sensible way. For example, three coordinates are needed to specify the location of the centre of mass of the molecule, so three of these displacements correspond to the translational motion of the molecule as a whole. The remaining $3N - 3$ are non-translational 'internal' modes of the molecule.

Two angles are needed to specify the orientation of a linear molecule in space: in effect, we need to give only the latitude and longitude of the direction in which the molecular axis is pointing (Fig. 16.46a). However, three angles are needed for a non-linear molecule because we also need to specify the orientation of the molecule around the direction defined by the latitude and longitude (Fig. 16.46b). Therefore two (linear) or three (non-linear) of the $3N - 3$ internal displacements are rotational. This leaves $3N - 5$ (linear) or $3N - 6$ (non-linear) displacements of the atoms relative to each other: these are the vibrational modes. It follows that the number of modes of vibration N_{vib} is $3N - 5$ for linear molecules and $3N - 6$ for non-linear molecules.

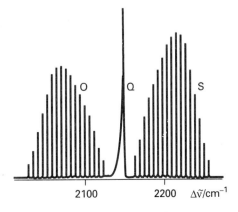

16.45 The structure of a vibrational line in the vibrational Raman spectrum of carbon monoxide, showing the O, Q, and S branches.

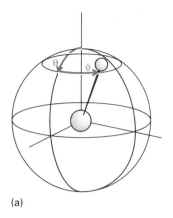

(a)

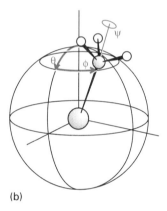

(b)

16.46 The specification of the centre of mass of a molecule uses up three degrees of freedom. (a) The orientation of a linear molecule requires the specification of two angles. (b) The orientation of a non-linear molecule requires the specification of three angles.

As an illustration, H_2O is a non-linear triatomic molecule, and has three modes of vibration (and three modes of rotation); CO_2 is a linear triatomic molecule, and has four modes of vibration (and only two modes of rotation). Even a middle-sized molecule such as naphthalene ($C_{10}H_8$) has 48 distinct modes of vibration.

Combinations of displacements

The next step is to find the best description of the modes. One choice for the four modes of CO_2, for example, might be the ones in Fig. 16.47a. This illustration shows the stretching of one bond (the mode ν_L), the stretching of the other (ν_R), and (in Fig. 16.47c) the two perpendicular bending modes (ν_2). The description, while permissible, has a disadvantage: when one CO bond vibration in Fig. 16.47a is excited, the motion of the C atom sets the other CO bond in motion, so energy flows backwards and forwards between ν_L and ν_R. Moreover, the position of the centre of mass of the molecule varies in the course of either vibration.

The description of the vibrational motion is much simpler if linear combinations of ν_L and ν_R are taken. For example, one combination is ν_1 in Fig. 16.47b: this is the **symmetric stretch**: in it, the C atom is buffeted simultaneously from each side and the motion continues indefinitely. Another mode is ν_3, the **antisymmetric stretch**, in which the two O atoms always move in the same direction and against the motion of the C atom. Both modes are independent in the sense that, if one is excited, then it does not excite the other. They are two of the normal modes of the molecule, its independent, collective vibrational displacements. The two other normal modes are the bending modes ν_2. In general, a **normal mode** is an independent, synchronous motion of atoms or groups of atoms that may be excited without leading to the excitation of any other normal mode.

The four normal modes of CO_2, and the N_{vib} normal modes of polyatomics in general, are the key to the description of molecular vibrations. Each normal mode behaves like an independent harmonic oscillator (if anharmonicities are neglected), so each has a series of terms

$$G_Q(v) = (v + \tfrac{1}{2})\tilde{\nu}_Q \tag{50a}$$

where $\tilde{\nu}_Q$ is the wavenumber of mode Q and depends on the force constant k_Q for the mode and on the effective mass μ_Q of the mode, with

$$\tilde{\nu}_Q = \frac{\omega_Q}{2\pi c} \quad \text{and} \quad \omega_Q = \left(\frac{k_Q}{\mu_Q}\right)^{\frac{1}{2}} \tag{50b}$$

The effective mass of the mode is a measure of the mass that is swung about by the vibration and in general is a complicated function of the masses of the atoms. For example, in the symmetric stretch of CO_2 the C atom is stationary, and the effective mass depends on the masses of only the O atoms. In the antisymmetric stretch and in the bends, all three atoms move, and so all three atoms contribute to the effective mass. The three normal modes of H_2O are shown in Fig. 16.48: note that the predominantly bending mode (ν_2) has a lower frequency than the others,

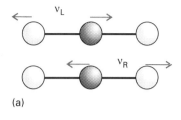

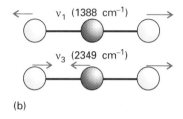

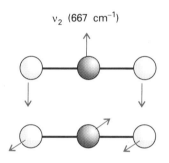

16.47 Alternative descriptions of the vibrations of CO_2. (a) The stretching modes are not independent, and if one CO group is excited the other begins to vibrate. (b) The symmetric and antisymmetric stretches are independent, and one can be excited without affecting the other: they are normal modes. (c) The two perpendicular bending motions are also normal modes.

which are predominantly stretching modes. It is generally the case that the frequencies of bending motions are lower than those of stretching modes. One point that must be appreciated is that only in special cases (such as the CO_2 molecule) are the normal modes purely stretches or purely bends. In general, a normal mode is a composite motion of simultaneous stretching and bending of bonds. Another point in this connection is that heavy atoms generally move less than light atoms in normal modes.

The symmetry species of normal modes

One of the most powerful ways of dealing with normal modes, especially of complex molecules, is to classify them according to their symmetries. Each normal mode must belong to one of the symmetry species discussed in Chapter 15.

Example 16.9 *Identifying the symmetry species of a normal mode*

Establish the symmetry species of the normal mode vibrations of CH_4.

Method. The procedure begins by deciding on the symmetry species of the irreducible representations spanned by all the $3N$ displacements of the atoms, using the characters of the molecular point group. We find these characters (as explained in Example 15.3) by counting 1 if the displacement is unchanged under a symmetry operation, -1 if it changes sign, and 0 if it is changed into some other displacement. Next, we subtract the symmetry species of the translations. Translational displacements span the same symmetry species as x, y, and z, so they can be obtained from the right-most columns of the character table. Finally, we subtract the symmetry species of the rotations, which are also given in the character table.

Answer. There are $3 \times 5 = 15$ possible displacements, of which $3 \times 5 - 6 = 9$ are vibrations. Refer to Fig. 16.49. Under E no displacement coordinates are changed, so the character is 15. Under C_3 no diplacements are left unchanged, so the character is 0. Under a C_2 rotation the z-displacement of the central atom is left unchanged, while its x- and y-components both change sign. Therefore $\chi(C_2) = 1 - 1 - 1 + 0 + 0 + \cdots = -1$. Under S_4 the z-displacement of the central atom is reversed, so $\chi(S_4) = -1$. Under σ_d the x- and z-displacements of C, H_3, and H_4 are left unchanged and the y-displacements are reversed; hence $\chi(\sigma_d) = 3 + 3 - 3 = 3$. The characters are therefore 15 0 −1 −1 3, corresponding to $A_1 + E + T_1 + 3T_2$. The translations span T_2; the rotations span T_1. Hence the vibrations span $A_1 + E + 2T_2$.

Comment. The modes themselves are shown in Fig. 16.50. We shall soon see that symmetry analysis gives a quick way of deciding which modes are active.

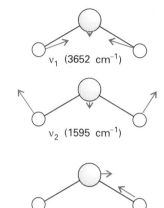

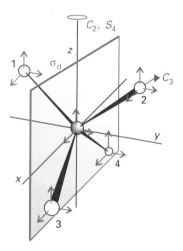

16.48 The three normal modes of H_2O. The mode v_2 is predominantly bending, and occurs at lower wavenumber than the other two.

v_1 (3652 cm^{-1})

v_2 (1595 cm^{-1})

v_3 (3756 cm^{-1})

16.49 The atomic displacements of CH_4 and the symmetry elements used to calculate the characters.

Exercise E16.9. Establish the symmetry species of the normal modes of H_2O. [$2A_1 + B_2$]

16.14 The vibrational spectra of polyatomic molecules

The gross selection rule for infrared activity is that *the motion corresponding to a normal mode should be accompanied by a change of dipole moment.* Deciding whether this is so can sometimes be done by inspection. For example, the symmetric stretch of CO_2 leaves the dipole moment unchanged (at zero, see Fig. 16.47), so this mode is infrared inactive. The antisymmetric stretch, however, changes the dipole moment because the molecule becomes unsymmetrical as it vibrates, so this mode is infrared active. Because the dipole moment change is parallel to the figure axis, the transitions arising from this mode are classified as **parallel bands** in the spectrum. Both bending modes are infrared active: they are accompanied by a changing dipole perpendicular to the figure axis, so transitions involving them lead to a **perpendicular band** in the spectrum. The latter bands eliminate the linearity of the molecule, and as a result a Q branch is observed; the parallel bands do not show a Q branch.

Symmetry and normal mode activity

It is best to use group theory to judge the activities of more complex modes of vibration. This is easily done by checking the character table of the molecular point group of the symmetry species of the irreducible representations spanned by x, y, and z, for these are also the symmetry species of the components of the electric dipole moment. Then the rule to apply is as follows:

> If the symmetry species of a normal mode is the same as any of the symmetry species of x, y, or z, then the mode is infrared active.

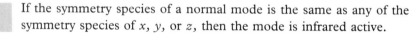

JUSTIFICATION

The rule hinges on the form of the transition dipole between the ground-state vibrational wavefunction ψ_0, and the first excited state ψ_1:

$$\mu_{10} = -e \int \psi_1 x \psi_0 \, d\tau$$

for the x-component, with similar expressions for the two other components of the transition moment. The ground-state vibrational wavefunction is a Gaussian function of the form e^{-x^2}, so it is symmetrical in x. The wavefunction for the first excited state will give a non-vanishing integral only if it is proportional to x, for then the integrand is proportional to x^2 rather than to xy or xz. Consequently, the excited state wavefunction must have the same symmetry as the displacement x.

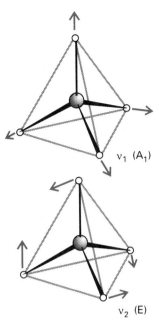

ν_1 (A$_1$)

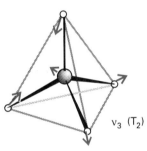

ν_2 (E)

ν_3 (T$_2$)

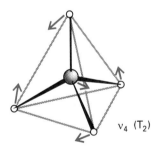

ν_4 (T$_2$)

16.50 Typical normal modes of vibration of a tetrahedral molecule. There are in fact two modes of symmetry species E and three modes of each T$_2$ symmetry species.

Example 16.10 *Identifying infrared active modes*

Which modes of CH_4 are infrared active?

Method. Refer to the T_d character table to establish the symmetry species of x, y, and z, and then use the rule given above.

Answer. The symmetry species of x, y, and z is T$_2$. We found in Example 16.9 that the symmetry species of the normal modes are A$_1$ + E + 2T$_2$. Therefore, only the T$_2$ modes are infrared active.

Comment. The distortions accompanying these modes lead to a changing dipole moment. The A$_1$ mode, which is inactive, is the symmetrical 'breathing' mode of the molecule.

Exercise E16.10. Which of the normal modes of H_2O are infrared active?

[All three]

The appearance of the spectrum

The active modes are subject to the specific selection rule $\Delta v_Q = \pm 1$, so the wavenumber of the **fundamental transition** (the first harmonic) of each active mode is $\tilde{v}_Q$. From the analysis of the spectrum a picture may be constructed of the stiffness of various parts of the molecule: that is, we can establish its **force field**, the set of force constants corresponding to all the displacements of the atoms.

Superimposed on this simple scheme are the complications arising from anharmonicities and the effects of molecular rotation. Very often the sample is a liquid or a solid, and the molecules are unable to rotate freely. In a liquid, for example, a molecule may be able to rotate through only a few degrees before it is struck by another, so it changes its rotational state frequently. This random changing of orientation is called **tumbling**.

The lifetimes of rotational states in liquids are very short, so the rotational energies are ill-defined. Collisions occur at a rate of about $10^{13}\,\text{s}^{-1}$, and, even allowing for only a 10 per cent success rate in knocking the molecule into another rotational state, a lifetime broadening (eqn 18) of more than $1\,\text{cm}^{-1}$ can easily result. The rotational structure of the vibrational spectrum is blurred by this effect, so the infrared spectrum of molecules in condensed phases usually consist of broad lines spanning the entire range of the resolved gas-phase spectrum, and show no branch structure.

One very important application of infrared spectroscopy to condensed phase samples, for which the blurring of the rotational structure by random collisions is a welcome simplification, is to chemical analysis. The vibrational spectra of different groups in a molecule give rise to absorptions at characteristic frequencies. Their intensities are also approximately transferable between molecules. Consequently, the molecules in a sample can often be identified by examining its infrared spectrum and accounting for all the bands by referring to a table of characteristic frequencies and intensities (Table 16.3 and Fig. 16.51).

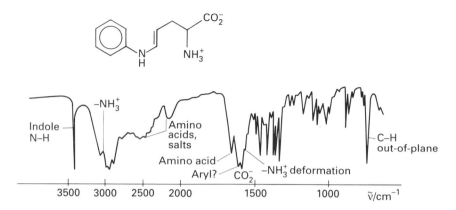

16.51 The infrared absorption spectrum of an amino acid, and a partial assignment.

Table 16.3* Typical vibrational wavenumbers, $\tilde{v}/cm^{-1}$

C—H stretch	2850–2960
C—H bend	1340–1465
C—C stretch	700–1250
C≡C stretch	1620–1680

* More values are given in the Data section.

16.15 Vibrational Raman spectra of polyatomic molecules

The study of the Raman spectra of polyatomic species has undergone a resurgence in recent years as a result of the use of lasers. Inorganic chemists often use Raman spectra to identify the species present in a reaction mixture, and the intensities can be used to determine concentrations from which equilibrium constants and rate constants can be calculated. Geochemists make use of the ability of Raman spectroscopy to study species at high temperatures and pressures, and electrochemists have found it possible to make detailed studies of electrolyte solutions, such as the extent of ion-pairing, and the structure of solution–electrode interfaces. The use of optical-fibre Raman probes has enabled the technique to be applied to well defined areas in places that are difficult to approach.

The normal modes of vibration of molecules are Raman active if they are accompanied by a changing polarizability. It is sometimes quite difficult to judge by inspection when this is so. The symmetric stretch of CO_2, for example, alternately swells and contracts the molecule: this motion changes its polarizability, so the mode is Raman active. The other modes of CO_2 leave the polarizability unchanged, so they are Raman inactive.

Symmetry aspects of Raman transitions

Group theory provides an explicit recipe for judging the Raman activity of a normal mode. In this case, the symmetry species of the quadratic forms (x^2, xy, etc.) listed in the character table are noted (they transform in the same way as the polarizability), and then we use the following rule:

> If the symmetry species of a normal mode is the same as the symmetry species of a quadratic form, then the mode is Raman active.

Example 16.11 *Identifying Raman-active normal nodes*

Which of the vibrations of CH_4 are Raman active?

Method. We adopt the rule specified above, and refer to the T_d character table. It was established in Example 16.9 that the symmetry species of the normal modes are $A_1 + E + 2T_2$.

Answer. Because the quadratic forms span $A_1 + E + T_2$, all the normal modes are Raman active.

Comment. All totally symmetric vibrations, whatever the point group of the molecule, are Raman active (and polarized; see below).

Exercise E16.11. Which of the vibrational modes of H_2O are Raman active?
[All three]

The **exclusion rule** also helps us to decide which modes are active:

> If the molecule has a centre of inversion, then no modes can be both infrared and Raman active.

(A mode may be inactive in both.) Because it is often possible to judge intuitively if a mode changes the molecular dipole moment, we can use this rule to identify modes that are not Raman active. The rule applies to CO_2 but to neither H_2O nor CH_4 because they have no centre of inversion.

Depolarization

The assignment of Raman lines to particular vibrational modes is aided by noting the state of polarization of the scattered light. The **depolarization ratio** ρ of a line is the ratio of the intensities of the scattered light with a polarization perpendicular and parallel to the plane of polarization of the incident radiation (Fig. 16.52):

$$\rho = \frac{I_\perp}{I_\parallel} \tag{51}$$

If the emergent light is not polarized, then both intensities are the same and ρ is close to 1; if the light retains its initial polarization, then $I_\perp = 0$ so $\rho = 0$. We classify a line as **depolarized** if it has ρ close to 0.75 and as **polarized** if it has ρ less than 0.75. A general rule is that only totally symmetrical vibrations give rise to polarized lines in which the incident polarization is largely preserved. Vibrations that are not totally symmetrical give rise to depolarized lines because the incident radiation can give rise to radiation in the perpendicular direction too. This means that, if we observe the Raman spectrum with a polarizing filter (a 'half-wave plate') first parallel and then perpendicular to the polarization of the incident beam, the intensity of the polarized lines will appear significantly reduced and hence these lines can be ascribed to symmetrical vibrations.

Applications

One application of vibrational Raman spectroscopy is the determination of the shapes of symmetrical molecules such as XeF_4 (D_{4h}) and SF_6 (O_h). Another application makes use of the fact that the intensity characteris-

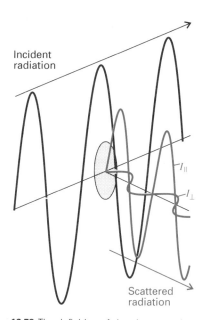

Incident radiation

$I_\parallel$

$I_\perp$

Scattered radiation

16.52 The definition of the planes used for the specification of the depolarization ratio ρ in Raman scattering.

16.53 The vibrational Raman spectrum of lysozyme in water and the superposition of the Raman spectra of the constituent amino acids. (From *Raman Spectroscopy*, D. A. Long. Copyright 1977, McGraw-Hill Inc. Used with the permission of the McGraw-Hill Book Company.)

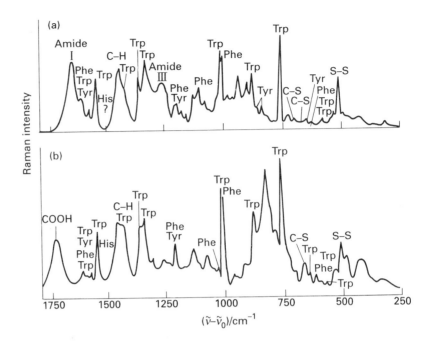

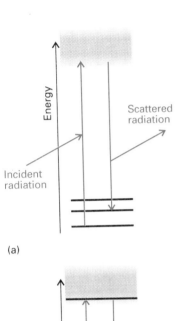

tics of Raman transitions, which depend on molecular polarizabilities, are more readily transferred from molecule to molecule than the intensities of infrared spectra, which depend on dipole moments and are more sensitive to the other groups present in a molecule and to the solvent. An example of the technique is shown in Fig. 16.53, which shows the vibrational Raman spectrum of an aqueous solution of lysozyme and, for comparison, a superposition of the Raman spectra of the constituent amino acids. The differences are indications of the effects of conformation, environment, and specific interactions (such as S–S linking) in the enzyme molecule.

Resonance Raman spectra

A modification of the basic Raman effect involves using incident radiation that nearly coincides with the frequency of an electronic transition of the sample (Fig. 16.54). The technique is then called **resonance Raman spectroscopy**. It is characterized by a much greater intensity in the scattered radiation and, because only a few vibrational modes contribute to the scattering, to a greatly simplified Raman spectrum. The resonance Raman spectrum shown in Fig. 16.55, for example, is of solid potassium chromate. The nine peaks that are identified are the Stokes lines that correspond to the excitation of the symmetric breathing mode of the tetrahedral CrO_4^{2-} ion and the transfer of up to nine vibrational quanta during the photon–ion collision. The high intensity of the resonance Raman transitions is employed to examine the metal ions in biological macromolecules (such as the iron in haemoglobin and cytochromes or the cobalt in vitamin B), which are present in such low abundances that conventional Raman spectroscopy cannot detect them.

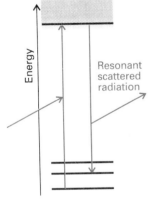

16.54 (a) In conventional Raman spectroscopy, the incident radiation does not match an absorption frequency of the molecule, (b) However, in the resonance Raman effect, the incident radiation has a frequency that coincides with a molecular transition.

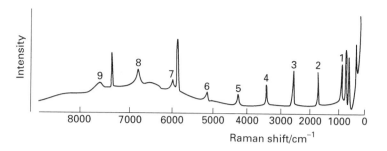

16.55 The resonance Raman spectrum of solid K_2CrO_4. The peaks are due to the totally symmetric stretching mode of the CrO_4^{2-} anion. (W. Kiefer and H. J. Bernstein, *Molecular Physics*, **23**, 835 (1972).)

Coherent anti-Stokes Raman spectroscopy

The intensity of Raman transitions may be enhanced by **coherent anti-Stokes Raman spectroscopy** (CARS, Fig. 16.56). The technique relies on the fact that, if two laser beams of frequencies v_1 and v_2 pass through a sample, then they mix together and give rise to radiation of several different frequencies, one of which is

$$v' = 2v_1 - v_2$$

Suppose that v_2 is varied until it matches any Stokes line from the sample, such as the one with frequency $v_1 - \Delta v$; then the coherent emission will have frequency

$$v' = 2v_1 - (v_1 - \Delta v) = v_1 + \Delta v$$

which is the frequency of the corresponding anti-Stokes line. This coherent radiation forms a narrow beam of high intensity.

An advantage of CARS is that it can be used to study Raman transitions in the presence of competing incoherent background radiation, and so can be used to observe the Raman spectra of species in flames. The intensities of the transitions can then be interpreted in terms of the temperatures of different regions of the flame.

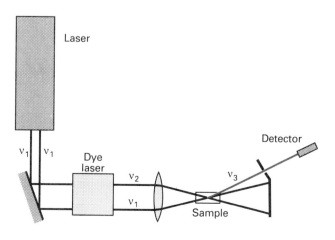

16.56 The arrangement for the CARS experiment.

CHECK LIST OF KEY IDEAS

1 The classification of spectra as **emission, absorption,** and **Raman** and the experimental techniques used for their study (Section 16.1).

2 The general principles of **Fourier transform spectroscopy** (Section 16.1).

3 The **Beer–Lambert law** (eqn 5) for the reduction in intensity when radiation passes through an absorbing medium.

4 The **Einstein transition probabilities,** and the coefficients of **stimulated absorption and emission** and of **spontaneous emission** (Section 16.2).

5 The **gross selection rules** and **specific selection rules** of transitions (Section 16.2).

6 The contributions to spectral **linewidths** (Section 16.3), particularly **Doppler broadening** (eqn 17) and **lifetime broadening** (eqn 18).

7 The principles of **Lamb-dip spectroscopy** (Section 16.3).

8 The **rotational energy levels of rigid rotors** (Section 16.4) in terms of the **moments of inertia** of molecules and their **rotational constants** (spherical rotors: eqn 22; symmetric rotors: eqn 24; linear rotors: eqn 25).

9 The **Stark effect** on the rotational energies of polar molecules in electric fields (eqn 26).

10 The effect of **centrifugal distortion** on the rotational energy levels of molecules (eqn 27).

11 The **selection rules** and transition moments for pure rotational transitions and the contribution of state populations to the intensities of spectral lines (Section 16.5).

12 The **electric polarizabilities** of molecules and their contribution to the detection of rotational Raman transitions (Section 16.6).

13 The **Stokes** and **anti-Stokes lines** in a Raman spectrum (Section 16.6).

14 The role of **nuclear statistics** in limiting the rotational states that molecules may occupy (Section 16.7), and the existence of **ortho-** and **para-hydrogen**.

15 The **harmonic approximation** for the description of the vibrations of molecules (Section 16.8) and the **vibrational terms** of a diatomic molecule (eqn 38).

16 The **selection rules** for vibrational transitions and the appearance of vibrational spectra (Section 16.9).

17 The **anharmonicity** of molecular vibrations, its description in terms of the **Morse potential energy** (eqn 40), and its effect on the vibrational spectrum (Section 16.10).

18 The use of the **Birge–Sponer extrapolation** to determine dissociation energies (Section 16.10 and Example 16.8).

19 The form of **P, Q,** and **R branches** in **vibration–rotation spectra** (Section 16.11).

20 The **vibrational Raman spectra** of diatomic molecules (Section 16.12).

21 The number of vibrations of polyatomic molecules and their description in terms of **normal modes** (Section 16.13).

22 The **symmetry analysis** of normal modes and their **infrared and Raman activities** (Section 16.13 and Example 16.9).

23 The features characteristic of the vibrational spectra of polyatomic molecules (Section 16.14).

24 The **exclusion rule** for centrosymmetric molecules (Section 16.15) and information from the **depolarization ratios** (eqn 51) of Raman transitions about the symmetries of normal modes (Section 16.15).

25 The enhancement of Raman intensities in **resonance Raman spectroscopy** and **coherent anti-Stokes Raman spectroscopy** (Section 16.15).

EXERCISES

16.1. Calculate the ratio of the quotients of the Einstein coefficients of spontaneous and stimulated emission, A/B, relative to their value for 70.8 pm X-rays for transitions with the following characteristics: (a) 500 nm visible light, (b) 3000 cm^{-1} infrared radiation, (c) 3 cm microwave radiation, (d) 500 MHz radiofrequency radiation.

16.2. Calculate the frequency of the $J = 4 \leftarrow 3$ transition in the pure rotational spectrum of $^{14}N^{16}O$. The equilibrium bond length is 115 pm.

16.3. If the wavenumber of the $J = 1$ rotational state of $^1H^{35}Cl$ considered as a rigid rotator is 20.68 cm^{-1}, what is (a) the moment of inertia of the molecule, (b) the bond length?

16.4. Given that the spacing of lines in the microwave spectrum of AlH is constant at 12.604 cm^{-1}, calculate the moment of inertia and bond length of the molecule.

16.5. An object of mass 1.0 kg suspended from the end of a rubber band has a vibrational frequency of 2.0 Hz. Calculate the force constant of the rubber band.

16.6. The rotational constant of $^{127}I^{35}Cl$ is 0.1142 cm^{-1}. Calculate the bond length of the molecule.

16.7. Determine the HC and CN bond lengths in HCN from the rotational constants $B(^1H^{12}C^{14}N) = 44.316$ GHz; $B(^2H^{12}C^{14}N) = 36.208$ GHz. Assume that replacement of ^1H by ^2H leaves the bond lengths unchanged.

16.8. Determine the CO and CS bond lengths in OCS from the rotational constants $B(^{16}O^{12}C^{32}S) = 6081.5$ MHz; $B(^{16}O^{12}C^{34}S) = 5932.8$ MHz.

16.9. The wavenumber of the incident radiation in a Raman spectrometer is 20 487 cm^{-1}. What is the wavenumber of the scattered Stokes radiation for the $J = 2 \leftarrow 0$ transition of $^{14}N_2$?

16.10. Infrared absorption by $^1H^{81}Br$ gives rise to an R branch from $v = 0$. What is the wavenumber of the line originating from the rotational state with $J = 2$? Use the information in Table 16.2.

16.11. Calculate the percentage difference between the fundamental vibrational wavenumbers of $^{23}Na^{35}Cl$ and $^{23}Na^{37}Cl$ on the assumption that their force constants are the same.

16.12. The wavenumber of the fundamental vibrational transition of $^{35}Cl_2$ is 564.9 cm^{-1}. Calculate the force constant of the bond.

16.13. For $^{127}I^{35}Cl$, $\tilde{v} = 384.3$ cm^{-1} and $x_e\tilde{v} = 1.5$ cm^{-1}. Calculate the wavenumber of the pure vibrational transition with the highest wavenumber and that of the next highest. Assume $y_e = 0$.

16.14. The bond dissociation energy of $^{127}I^{35}Cl$ is 2.153 eV. Use the information in Exercise 16.13 to calculate the depth of the molecular potential energy curve of this molecule.

16.15. The molecule CH_2Cl_2 belongs to the point group C_{2v}. The displacements of the atoms span $5A_1 + 2A_2 + 4B_1 + 4B_2$. What are the symmetries of the normal modes of vibration?

16.16. Which of the following molecules may show a pure rotational microwave absorption spectrum: (a) H_2, (b) HCl, (c) CH_4, (d) CH_3Cl, (e) CH_2Cl_2, (f) H_2O, (g) H_2O_2, (h) NH_3?

16.17. Which of the following molecules may show infra-red absorption spectra: (a) H_2, (b) HCl, (c) CO_2, (d) H_2O, (e) CH_3CH_3, (f) CH_4, (g) CH_3Cl, (h) N_2?

16.18. Which of the following molecules may show a pure rotational Raman spectrum: (a) H_2, (b) HCl, (c) CH_4, (d) CH_3Cl, (e) CH_2Cl_2, (f) CH_3CH_3, (g) SF_6?

16.19. What is the Doppler-shifted wavelength of a red (660 nm) traffic light approached at 80 km h^{-1}? At what speed would it appear green (520 nm)?

16.20. A spectral line of $^{48}Ti^{8+}$ in a distant star was found to be shifted from 654.2 to 706 nm and to be broadened to 61.8 pm. What is the speed of recession and the surface temperature of the star?

16.21. Estimate the lifetime of a state that gives rise to a line of width (a) 0.1 cm^{-1}, (b) 1 cm^{-1}, (c) 100 MHz.

16.22. A molecule in a liquid undergoes about 1×10^{13} collisions in each second. Suppose that (a) every collision is effective in deactivating the molecule vibrationally and (b) that one collision in 100 is effective. Calculate the width (in cm^{-1}) of vibrational transitions in the molecule.

16.23. Calculate the relative numbers of Cl_2 molecules ($\tilde{v} = 559.7$ cm^{-1}) in the ground and first excited vibrational states at (a) 298 K, (b) 500 K.

16.24. The pure rotational spectrum of $^1H^{131}I$ consists of a series of lines separated by 13.10 cm^{-1}. Calculate the bond length of the molecule.

16.25. The hydrogen halides have the following fundamental vibrational wavenumbers:

	HF	H^{35}Cl	H^{81}Br	H^{127}I
$\tilde{v}$/cm^{-1}	4141.3	2988.9	2649.7	2309.5

Calculate the force constants of the hydrogen–halogen bonds.

16.26. From the data in Exercise 16.25, predict the

fundamental vibrational wavenumbers of the deuterium halides.

16.27. For $^{16}O^{16}O$, ΔG values for the transitions $v = 1 \leftarrow 0$, $2 \leftarrow 0$, and $3 \leftarrow 0$ are, respectively, 1556.22, 3088.28, and 4596.21 cm^{-1}. Calculate $\tilde{\nu}$ and x_e. Assume y_e to be zero.

16.28. The first five vibrational energy levels of HCl are at 1481.86, 4367.50, 7149.04, 9826.48, and 12 399.8 cm^{-1}. Calculate the dissociation energy of the molecule in cm^{-1} and eV.

16.29. The rotational Raman spectrum of $^{35}Cl_2$ shows a series of Stokes lines separated by 0.9752 cm^{-1} and a similar series of anti-Stokes lines. Calculate the bond length of the molecule.

16.30. How many normal modes of vibration are there

for the following molecules: (a) H_2O, (b) H_2O_2, (c) C_2H_4, (d) C_6H_6?

16.31. Which of the three vibrations of an AB_2 molecule are infrared or Raman active when it is (a) bent, (b) linear?

16.32. Consider the vibrational mode that corresponds to the uniform expansion of the benzene ring. Is the mode (a) Raman, (b) infrared active?

16.33. Predict the shape of the nitronium ion, NO_2^+, from its Lewis structure and the VSEPR model. It has one Raman but not IR active vibrational mode at 1400 cm^{-1} and two strong infrared active modes at 2360 and 540 cm^{-1}. Are these data consistent with the predicted shape of the molecule? Assign the vibrational wavenumbers to the modes from which they arise.

PROBLEMS
Numerical problems

16.1. Calculate the Doppler width (as a fraction of the transition wavelength) for any kind of transition in (a) HCl, (b) ICl at 25°C. What would be the widths of the rotational and vibrational transitions in these molecules (in MHz and cm^{-1}, respectively), given $B(ICl) = 0.1142$ cm^{-1} and $\tilde{\nu}(ICl) = 384$ cm^{-1} and additional information from Table 16.2.

16.2. The number of collisions that a molecule undergoes per unit time in a gas of pressure p is

$$z = 4\sigma \left(\frac{kT}{\pi m}\right)^{\frac{1}{2}} \times \frac{p}{kT}$$

where σ is the collision cross-section. Find an expression for the collision-limited lifetime of an excited state assuming that every collision is effective. Estimate the width of rotational transition in HCl ($\sigma = 0.30$ nm^2) at 25°C and 1.0 atm. To what value must the pressure of the gas be reduced in order to ensure that collision broadening is less important than Doppler broadening?

16.3. The rotational constant of NH_3 is equivalent to 298 GHz. Compute the separation of the pure rotational spectrum lines in GHz, cm^{-1}, and mm, and show that the value of B is consistent with an N—H bond length of 101.4 pm and a bond angle of 106.78°.

16.4. The rotational constant for $^{12}C^{16}O$ is 1.9314 cm^{-1} and 1.6116 cm^{-1} in the ground and first excited vibrational states, respectively. By how much does the internuclear distance change as a result of this transition?

16.5. Pure rotational Raman spectra of gaseous C_6H_6 and

C_6D_6 yield the following rotational constants: $B(C_6H_6) = 0.18960$ cm^{-1}, $B(C_6D_6) = 0.15681$ cm^{-1}. The moments of inertia of the molecules about any axis perpendicular to the C_6 axis were calculated from these data as $I(C_6H_6) = 147.59 \times 10^{-47}$ kg m^2, $I(C_6D_6) = 178.45 \times 10^{-47}$ kg m^2. Calculate the CC, CH, and CD bond lengths. Assume $R(CH) = R(CD)$, and that $R(CC)$ is the same in both molecules.

16.6. The vibrational energy levels of NaI lie at the wavenumbers 142.81, 427.31, 710.31, and 991.81 cm^{-1}. Show that they fit the expression $(v + \frac{1}{2})\tilde{\nu} - (v + \frac{1}{2})^2 x_e \tilde{\nu}$ and deduce the force constant, zero-point energy, and dissociation energy of the molecule.

16.7. At low resolution, the strongest absorption band in the infrared absorption spectrum of $^{12}C^{16}O$ is centred at 2150 cm^{-1}. Upon closer examination at higher resolution, this band is observed to be split into two sets of closely spaced peaks, one on each side of the centre of the spectrum at 2143.26 cm^{-1}. The separation between the peaks immediately to the right and left of the centre is 7.655 cm^{-1}. Make the harmonic oscillator and rigid rotor approximations and calculate from these data: (a) the vibrational wavenumber of a CO molecule, (b) its molar zero-point vibrational energy, (c) the force constant of the CO bond, (d) the rotational constant B, and (e) the bond length of CO.

16.8. Rotational absorption lines from $^1H^{35}Cl$ gas were found at the following wavenumbers (R. L. Hausler and R. A. Oetjen, *J. Chem. Phys.* **21**, 1340 (1953)): 83.32, 104.13, 124.73, 145.37, 165.89, 186.23, 206.60, 226.86 cm^{-1}. Calculate the moment of inertia and the bond length of the molecule. Predict the positions of the corresponding lines in $^2H^{35}Cl$.

16.9. Is the bond length in HCl the same as that in DCl? The wavenumbers of the $J = 1 \leftarrow 0$ rotational transitions for $H^{35}Cl$ and $^2H^{35}Cl$ are 20.8784 and 10.7840 cm^{-1}, respectively. Accurate atomic masses are 1.007825 u and 2.0140 u for ^1H and ^2H, respectively. The mass of ^{35}Cl is 34.96885 u. Based on this information alone, can you conclude that the bond lengths are the same or different in the two molecules?

16.10. Thermodynamic considerations suggest that the copper monohalides CuX should exist mainly as polymers in the gas phase, and indeed it proved difficult to obtain the monomers in sufficient abundance to detect spectroscopically. This difficulty was overcome by flowing the halogen gas over copper heated to 1100 K (E. L. Manson, F. C. de Lucia, and W. Gordy, *J. Chem. Phys.* **64**, 2724 (1975)). For CuBr the $J = 14 \leftarrow 13$, $15 \leftarrow 14$, and $16 \leftarrow 15$ transitions occurred at 84 421.34, 90 449.25, and 96 476.72 MHz, respectively. Calculate the rotational constant and bond length of CuBr.

16.11. The microwave spectrum of $^{16}O^{12}CS$ (C. H. Townes, A. N. Holden, and F. R. Merritt, *Phys. Rev.* **74**, 1113 (1948)) gave absorption lines (in GHz) as follows:

J	1	2	3	4
^{32}S	24.325 92	36.488 82	48.651 64	60.814 08
^{34}S	23.732 33		47.462.40	

Using the expressions for moments of inertia in Table 16.1 and assuming that the bond lengths are unchanged by substitution, calculate the CO and CS bond lengths in OCS.

16.12. The HCl molecule is quite well described by the Morse potential with $D_e = 5.33$ eV, $\tilde{v} = 2989.7$ cm^{-1}, and $x_e\tilde{v} = 52.05$ cm^{-1}. Assuming that the potential is unchanged on deuteration, predict the dissociation energies (D_0) of (a) HCl, (b) DCl.

16.13. The Morse potential (eqn 40) is very useful as a simple representation of the actual molecular potential energy. When RbH was studied it was found that $\tilde{v} = 936.8$ cm^{-1} and $x_e\tilde{v} = 14.15$ cm^{-1}. Plot the potential energy curve from 50 pm to 800 pm around $R_e = 236.7$ pm. Then go on to explore how the rotation of a molecule may weaken its bond by allowing for the kinetic energy of rotation of a molecule and plotting

$$V^* = V + hcBJ(J+1) \qquad B = \frac{\hbar}{4\pi c\mu R^2}$$

Plot these curves on the same diagram for $J = 40$, 80, and 100, and observe how the dissociation energy is affected by the rotation. (Taking $B = 3.020$ cm^{-1} at the equilibrium bond length will greatly simplify the calculation.)

Theoretical problems

16.14. Show that the moment of inertia of a diatomic molecule composed of atoms of masses m_A and m_B and bond length R is equal to μR^2, where μ is the effective mass of the molecule.

16.15. Derive an expression for the value of J corresponding to the most highly populated rotational energy level of a diatomic rotor at a temperature T remembering that the degeneracy of each level is $2J + 1$. Evaluate the expression for ICl (for which $B = 0.1142$ cm^{-1}) at 25°C. Repeat the problem for the most highly populated level of a spherical rotor, taking note of the fact that each level is $(2J + 1)^2$-fold degenerate. Evaluate the expression for CH$_4$ (for which $B = 5.24$ cm^{-1}) at 25°C.

16.16. The moments of inertia of the linear mercury(II) halides are very large, so the O and S branches of their vibrational Raman spectra show little rotational structure. Nevertheless, the peaks of both branches can be identified and have been used to measure the rotational constants of the molecules (R. J. H. Clark and D. M. Rippon, *J. Chem. Soc. Faraday Soc.* II, **69**, 1496 (1973)). Show, from a knowledge of the value of J corresponding to the intensity maximum, that the separation of the peaks of the O and S branches is given by the 'Placzek–Teller relation':

$$\delta\tilde{v} = \left(\frac{32BkT}{hc}\right)^{\frac{1}{2}}$$

The following widths were obtained at the temperature stated:

	HgCl$_2$	HgBr$_2$	HgI$_2$
$\theta/°C$	282	292	292
$\delta\tilde{v}/cm^{-1}$	23.8	15.2	11.4

Calculate the bond lengths in the three molecules.

16.17. Show that the internuclear separation of a rotating diatomic molecule as a result of centrifugal distortion can be expressed approximately as

$$R_J = R_0 + \frac{\hbar^2 J(J+1)}{k\mu R_0^3}$$

Assume that the force constant k does not change as a result of rotation.

17

Spectroscopy 2: electronic transitions

In this second chapter on spectroscopy we examine the transitions of highest energy that normally concern chemists, in which electrons are shifted from one region of a molecule to another. In contrast to rotational and vibrational transitions, simple analytical expressions for electronic energy levels cannot be given, so we concentrate in this chapter on the qualitative features of electronic transitions. Foremost among these features is the vibrational structure of electronic transitions, which we shall see can be rationalized in terms of the slowness with which nuclei can change their positions. A common theme throughout the chapter is that electronic transitions occur within a stationary nuclear framework.

A very important feature of electronic transitions is the manner by which excitation energy is lost. We shall pay particular attention in this chapter to the radiative decay processes, which include fluorescence and phosphorescence. Both processes are examples of spontaneous radiative decay. A specially important example of stimulated radiative decay is that responsible for the action of lasers, and we shall see how this stimulated emission may be achieved and employed.

An extreme case of photon absorption is photoionization, in which an electron is expelled completely from a molecule. Photoelectron spectroscopy uses measurements of the energies with which electrons are ejected from molecules to build up detailed pictures of orbital energies and the role of particular electrons in bonding. Hence it is an experimental technique for exploring the concepts encountered in Chapter 14.

The energies needed to change the electron distributions of molecules are of the order of several electronvolts (1 eV is equivalent to about $8000 \, cm^{-1}$ or $100 \, kJ \, mol^{-1}$). Consequently, the photons emitted or absorbed when such changes occur lie in the visible and ultraviolet regions of the spectrum, which spreads from about $14\,000 \, cm^{-1}$ for red

Table 17.1* Colour, frequency, and energy of light

Colour	λ/nm	$\nu/(10^{14}\ Hz)$	$E/(kJ\ mol^{-1})$
Infrared	>1000	<3.0	<120
Red	700	4.3	170
Yellow	580	5.2	210
Blue	470	6.4	250
Ultraviolet	<300	>10	>400

* More values are given in the Data section at the end of this volume.

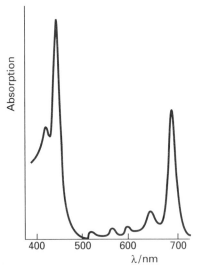

17.1 The absorption spectrum of chlorophyll in the visible region. Note that it absorbs in the red and blue regions, and that green light is not absorbed.

light to 25 000 cm^{-1} for violet light, and on to 50 000 cm^{-1} for ultraviolet radiation (Table 17.1). In some cases the relocation of electrons may be so extensive that it results in the breaking of a bond and the dissociation of the molecule.

THE CHARACTERISTICS OF ELECTRONIC TRANSITIONS

The nuclei in a molecule are subjected to different forces after an electronic transition has occurred and the molecule may respond by bursting into vibration. The resulting **vibrational structure** of electronic transitions can be resolved for gaseous samples, but in a liquid or solid the lines usually merge together and result in a broad, almost featureless band (Fig. 17.1). Superimposed on the vibrational transitions that accompany the electronic transition of a molecule in the gas phase is an additional branch structure that arises from rotational transitions. The electronic spectra of gaseous samples are therefore very complicated, but rich in information.

17.1 The vibrational structure

The width of electronic absorption bands in liquid samples can be traced to their vibrational structure, which is usually unresolved in solution. This structure, which can be resolved in gases and weakly interacting solvents, arises from the vibrational transitions that accompany electronic excitation.

The Franck–Condon principle

The distribution of relative intensities of the vibrational structure of an electronic transition is explained by the Franck–Condon principle:

> Because the nuclei are so much more massive than the electrons, an electronic transition takes place very much faster than the nuclei can respond.

As a result of the transition, electron density is rapidly built up in new regions of the molecule and removed from others, and the initially stationary nuclei suddenly experience a new force field. They respond to the new force by beginning to vibrate, and (in classical terms) swing backwards and forwards from their original separation, which had been maintained during the rapid electronic excitation. The stationary equilibrium separation of the nuclei in the initial electronic state therefore becomes a stationary turning point (the points of a vibration when the nuclei are at the end-points of their swing) in the final electronic state (Fig. 17.2). Because the nuclear framework remains constant during this excitation, we may imagine the transition as being up the vertical line in Fig. 17.2. The vertical line is the origin of the expression **vertical transition**, which is used to denote an electronic transition that occurs without change of nuclear geometry.

The quantum mechanical version of the Franck–Condon principle refines this picture to the point of letting us calculate the intensities of the transitions to different vibrational levels of the electronically excited

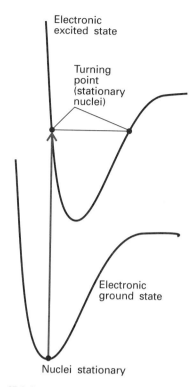

Electronic excited state

Turning point (stationary nuclei)

Electronic ground state

Nuclei stationary

17.2 According to the Franck–Condon principle, the most intense transition is from the ground vibrational state to the vibrational state lying vertically above it. Transitions to other vibrational levels also occur, but with lower intensity.

molecule. Before the absorption, the molecule is in the lowest vibrational state of its lowest electronic state (Fig. 17.3). The form of the vibrational wavefunction shows that the most probable location of the nuclei is at their equilibrium separation R_e. Consequently, the electronic transition is most likely to take place when the nuclei have this separation. When the transition occurs, the molecule is excited to the state represented by the upper curve.

The vertical transition cuts through several vibrational levels of the upper electronic state. The level marked * is the one in which the nuclei are most probably at the same initial separation R_e (because the vibrational wavefunction has maximum amplitude there), so this vibrational state is the most probable state for the termination of the transition. However, it is not the only final vibrational state because several nearby states have an appreciable probability of the nuclei being at the separation R_e. Therefore, transitions occur to all the vibrational states in this region, but most intensely to the state with a vibrational wavefunction that peaks most strongly near R_e.

The vibrational structure of the spectrum depends on the relative displacement of the two potential energy curves, and a long **progression** of vibrations (a lot of vibrational structure) is stimulated if the two potential energy curves are appreciably displaced from one another. The upper curve is usually displaced to greater equilibrium bond lengths because electronically excited states usually have more antibonding character than electronic ground states.

The separation of the vibrational lines of an electronic absorption spectrum depends on the vibrational energies of the upper electronic state. Hence electronic absorption spectra may be used to assess the force fields and dissociation energies of electronically excited molecules (e.g. via a Birge–Sponer plot, Section 16.10).

Franck–Condon factors

The quantitative form of the Franck–Condon principle is derived from the expression for the transition dipole moment, which, as we saw in Section 16.2, is an integral of the form

$$\boldsymbol{\mu}_{\text{fi}} = -e \int \psi_{\text{f}}^* \boldsymbol{r} \psi_{\text{i}} \, d\tau \tag{1}$$

where ψ_{i} and ψ_{f} are the initial and final wavefunctions, respectively, of the system. The intensity of the transition is proportional to the square modulus, $|\boldsymbol{\mu}_{\text{fi}}|^2$, of the transition dipole moment. When considering the intensity of a joint electronic and vibrational transition we use the fact that the wavefunctions of the initial and final states are the products of the respective electronic and vibrational wavefunctions. That is, the total wavefunction for an electronic state ψ_ε and vibrational state ψ_v is $\psi_\varepsilon(\boldsymbol{r})\psi_v(\boldsymbol{R})$, where $\boldsymbol{r}$ stands for the electronic coordinates and $\boldsymbol{R}$ for the nuclear coordinates; ε labels the electronic state and v labels the vibrational state. The transition dipole moment for the excitation

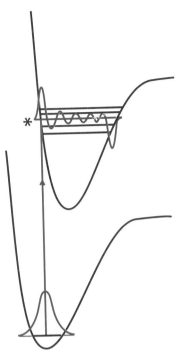

17.3 In the quantum mechanical version of the Franck–Condon principle, the molecule undergoes a transition to the upper vibrational state that most closely resembles the vibrational wavefunction of the vibrational ground state of the lower electronic state. The two wavefunctions shown here have the greatest overlap integral of all the vibrational states of the upper electronic state and hence are most closely similar.

ε', $v' \leftarrow \varepsilon$, v is therefore approximately

$$\boldsymbol{\mu}_{\mathrm{fi}} = -e \int \psi_{\varepsilon'}^*(\boldsymbol{r}) \psi_{v'}^*(\boldsymbol{R}) \boldsymbol{r} \psi_\varepsilon(\boldsymbol{r}) \psi_v(\boldsymbol{R})\, \mathrm{d}\tau_{\mathrm{elec}}\, \mathrm{d}\tau_{\mathrm{nuc}}$$

$$= -e \int \psi_{\varepsilon'}^*(\boldsymbol{r}) \boldsymbol{r} \psi_\varepsilon(\boldsymbol{r})\, \mathrm{d}\tau_{\mathrm{elec}} \int \psi_{v'}^*(\boldsymbol{R}) \psi_v(\boldsymbol{R})\, \mathrm{d}\tau_{\mathrm{nuc}}$$

The first factor is a measure of the extent of electron redistribution that occurs during the transition: it is a measure of the dipolar 'kick' that the transition delivers to the electromagnetic field. The second factor, which we write

$$S_{v',v} = \int \psi_{v'}^*(\boldsymbol{R}) \psi_v(\boldsymbol{R})\, \mathrm{d}\tau_{\mathrm{nuc}} \tag{2}$$

is a measure of the match between the arrangement of nuclei in the upper and lower electronic states. The expression is an overlap integral between the initial and final vibrational state wavefunctions, and we saw in Section 14.3 that overlap integrals are a measure of the degree of similarity of wavefunctions.

Because the transition from a vibrational level v of the ground electronic state to a vibrational level v' of the excited electronic state is proportional to the square of the transition dipole moment, that intensity is proportional to $S_{v',v}^2$. The quantity $S_{v',v}^2$ is known as the **Franck–Condon factor** for the ε', $v' \leftarrow \varepsilon$, v transition. It follows that the greater the overlap of the vibrational state wavefunction in the upper electronic state with the vibrational wavefunction in the lower electronic state, the greater the absorption intensity of that particular simultaneous electronic and vibrational transition.

Example 17.1 *Calculating a Franck–Condon factor*

Consider the transition from one electronic state to another, their bond lengths being R_e and R_e' and their force constants equal. Calculate the Franck–Condon factor for the 0–0 transition and show that the transition is most intense when the bond lengths are equal.

Method. We need to calculate $S_{0,0}$, the overlap integral of the two ground-state vibrational wavefunctions. The difference between harmonic and anharmonic vibrational wavefunctions is negligible for $v = 0$, so harmonic oscillator wavefunctions (eqn 12.9) can be used.

Answer. We use the wavefunctions

$$\psi_0 = \left(\frac{1}{\alpha \pi^{\frac{1}{2}}}\right)^{\frac{1}{2}} \mathrm{e}^{-y^2/2} \qquad \psi_0' = \left(\frac{1}{\alpha \pi^{\frac{1}{2}}}\right)^{\frac{1}{2}} \mathrm{e}^{-y'^2/2}$$

where $y = (R - R_e)/\alpha$ and $y' = (R - R_e')/\alpha$, with $\alpha^2 = \hbar/(mk)^{\frac{1}{2}}$ (Section 12.5). The overlap integral is

$$S_{0,0} = \int_{-\infty}^{\infty} \psi_0'(R) \psi_0(R)\, \mathrm{d}R = \frac{1}{\alpha \pi^{\frac{1}{2}}} \int_{-\infty}^{\infty} \mathrm{e}^{-\frac{1}{2}(y^2 + y'^2)}\, \mathrm{d}R$$

It is not difficult to manipulate this expression into

$$S_{0,0} = \frac{1}{\pi^{\frac{1}{2}}} e^{-(R_e - R'_e)/4\alpha^2} \int_{-\infty}^{\infty} e^{-z^2} dz, \qquad \text{where } \alpha z = R - \tfrac{1}{2}(R_e + R'_e)$$

The value of the integral is $\pi^{\frac{1}{2}}$. Therefore, the overlap integral is

$$S_{0,0} = e^{-(R_e - R'_e)^2/4\alpha^2}$$

and the intensity is proportional to $S_{0,0}^2$. This quantity is equal to 1 when $R_e = R'_e$ and decreases as the equilibrium bond lengths diverge.

Comment. For Br_2, $R_e = 228$ pm and there is an upper state with $R'_e = 266$ pm. Taking the vibrational wavenumber as 250 cm^{-1}, gives $S_{0,0}^2 = 5 \times 10^{-10}$, so the intensity of the 0–0 transition is only 5×10^{-10} of what it would have been if the potential curves had been directly above each other, and thus it is unlikely to be observed.

Exercise E17.1. Suppose the vibrational wavefunctions can be approximated by rectangular functions of width W and W', centred on the equilibrium bond lengths. Find the corresponding Franck–Condon factor when the centres are coincident and $W' < W$. $\quad [W'^2/W^2]$

17.2 Specific types of transitions

The absorption of a photon can often be traced to the excitation of specific electrons or electrons that belong to a small group of atoms. For example, when a carbonyl group ($>C{=}O$) is present, an absorption at about 290 nm is normally observed, although its precise location depends on the nature of the rest of the molecule. Groups with characteristic optical absorptions are called **chromophores** (from the Greek for 'colour bringer'), and their presence often accounts for the colours of substances.

d–d transitions

All five d orbitals of a given shell are degenerate in a free atom. In a d-metal complex, where the immediate environment of the atom is no longer spherical, the d orbitals are not all degenerate, and electrons can absorb energy by making transitions between them. In an octahedral complex, such as $[\text{Ti}(OH_2)_6]^{3+}$, the five d orbitals of the central atom are split into two sets (**1**), a triply degenerate set labelled t_{2g} and a doubly degenerate set labelled e_g. The three t_{2g} orbitals lie below the two e_g orbitals; their separation in energy is denoted Δ_O and called the **ligand-field splitting parameter** (for octahedral symmetry). The d orbitals also divide into two sets in a tetrahedral complex, but in this case the e orbitals lie below the t_2 orbitals and their separation is written Δ_T. Neither separation is large, so transitions between the two sets of orbitals typically occur in the visible region of the spectrum even though they are electronic. The transitions are responsible for many of the colours that are so characteristic of d-metal complexes. As an example, the spectrum of $[\text{Ti}(OH_2)_6]^{3+}$ near $20\,000 \text{ cm}^{-1}$ (500 nm) is shown in Fig. 17.4, and can be ascribed to the promotion of its single d electron from a t_{2g} orbital

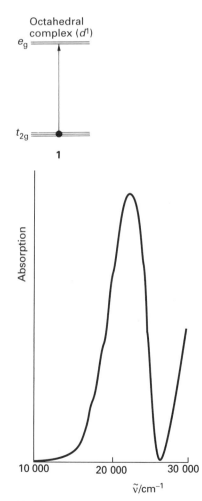

Octahedral complex (d^1)

e_g

t_{2g}

1

Absorption

10 000   20 000   30 000

$\tilde{\nu}/\text{cm}^{-1}$

17.4 The electronic absorption spectrum of $[\text{Ti}(OH_2)_6]^{3+}$ in aqueous solution.

to an e_g orbital. The wavenumber of the absorption maximum suggests that $\Delta_O \approx 20\,000\ cm^{-1}$ for this complex, which corresponds to about 2.5 eV.

Vibronic transitions

A major problem with the interpretation of visible spectra of octahedral complexes is that $d-d$ transitions are forbidden in them. The **Laporte selection rule** for centrosymmetric complexes (those with a centre of inversion) and atoms states that:

> The only allowed transitions are transitions that are accompanied by a change of parity.

That is, $u \leftrightarrow g$ and $g \leftrightarrow u$ transitions are allowed, but $g \leftrightarrow g$ and $u \leftrightarrow u$ transitions are forbidden.

JUSTIFICATION

The group theoretical basis of the Laporte selection rule is as follows. The transition dipole moment in eqn 1 vanishes unless the integral is symmetric. Hence, in a centrosymmetric complex it must have even (g) parity (in O_h it needs to be A_{1g}, but only the g symmetry is important for this argument). The three components of the dipole moment operator transform like x, y, and z and are all u. Therefore, for a $d-d$ transition (a $g \leftrightarrow g$ transition), the overall parity of the transition dipole is $g \times u \times g = u$, so it must be zero. Likewise, for a $u \leftrightarrow u$ transition, the overall parity is $u \times u \times u = u$, so it must also vanish. Hence, transitions without a change of parity are forbidden.

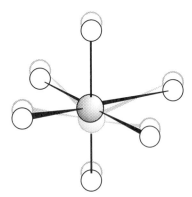

17.5 A $d-d$ transition is parity-forbidden because it corresponds to a g–g transition. However, a vibration of the molecule can destroy the inversion symmetry of the molecule so that the g, u classification no longer applies. The removal of the centre of symmetry gives rise to a vibronically allowed transition.

A forbidden $g \leftrightarrow g$ transition can become allowed if the centre of symmetry is eliminated by an asymmetrical vibration, such as the one shown in Fig. 17.5. In a non-centrosymmetric molecule, $d-d$ transitions are not parity-forbidden, so the $e_g \leftarrow t_{2g}$ transition becomes weakly allowed if the complex vibrates asymmetrically. A transition that derives its intensity from a vibration of a molecule is called a **vibronic transition**.

Charge-transfer transitions

A complex may absorb light as a result of the transfer of an electron from the ligands into the d orbitals of the central atom, or vice versa. In such **charge-transfer transitions** the electron moves through a considerable distance, which means that the transition dipole moment may be large and, because the transitions are not parity-forbidden, the absorption is correspondingly intense. This mode of chromophore activity is shown by the manganate(VII) ion, MnO_4^-, and accounts for its intense violet colour (which arises from strong absorption within the range 420–700 nm). In this oxoanion, the electron migrates from an orbital that is largely confined to the O atom ligands to an orbital that is largely confined to the Mn atom.

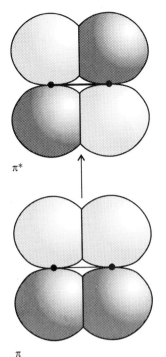

π^*

π

17.6 A C=C bond acts as a chromophore. One of its important transitions is the $\pi^* \leftarrow \pi$ transition illustrated here, in which an electron is promoted from a π orbital to the corresponding antibonding orbital.

2

3

$\pi^* \leftarrow \pi$ and $\pi^* \leftarrow n$ transitions

Absorption by a C=C double bond excites a π electron into an antibonding π^* orbital (Fig. 17.6). The chromophore activity is therefore due to a **$\pi^* \leftarrow \pi$ transition** (which is normally read 'π to π-star transition'). Its energy is around 7 eV for an unconjugated double bond, which corresponds to an absorption at 180 nm (in the ultraviolet). When the double bond is part of a conjugated chain, the energies of the molecular orbitals lie closer together and the $\pi^* \leftarrow \pi$ transition moves to longer wavelength and may even lie in the visible region if the conjugated system is long enough. We shall see an important example of such a transition in a moment.

The transition responsible for absorption in carbonyl compounds can be traced to the lone pairs of electrons on the O atom. One of these electrons may be excited into an empty π^* orbital of the carbonyl group (Fig. 17.7), which gives rise to a **$\pi^* \leftarrow n$ transition** (an 'n to π-star transition'). Typical absorption energies are about 4 eV (290 nm). Because $\pi^* \leftarrow n$ transitions in carbonyls are symmetry forbidden, the absorptions are weak (Table 17.2).

An important example of $\pi^* \leftarrow \pi$ and $\pi^* \leftarrow n$ transitions is provided by the photochemical mechanism of vision. The retina of the eye contains 'visual purple', which is a protein in combination with 11-*cis*-retinal (**2**). The 11-*cis*-retinal acts as a chromophore, and is the primary receptor for photons entering the eye. A solution of 11-*cis*-retinal absorbs at about 380 nm, but, in combination with the protein (a link which might involve the elimination of the terminal carbonyl), the absorption maximum shifts to about 500 nm and tails into the blue. The conjugated double bonds are responsible for the ability of the molecule to absorb over the entire visible region, but they also play another important role. In its electronically excited state the conjugated chain can isomerize, one-half of the chain being able to twist about an excited C=C bond and forming all-*trans*-retinal (**3**). On account of its different shape, the new isomer cannot fit into the protein. The primary step in vision therefore appears to be photon absorption followed by isomerization: the uncoiling of the molecule then triggers a nerve impulse to the brain.

THE FATES OF ELECTRONICALLY EXCITED STATES

The energy of an electronically excited state may be lost in a variety of ways. A **radiative decay process** is a process in which a molecule discards its excitation energy as a photon. A more common fate is **non-radiative decay**, in which the excess energy is transferred into the vibration, rotation, and translation of the surrounding molecules. This **thermal degradation** converts the excitation energy into thermal motion of the environment (i.e. to 'heat'). An excited molecule may also take part in a chemical reaction, as we shall discuss in Part 3.

17.3 Fluorescence and phosphorescence

In this section we consider radiative decay by spontaneous emission, which, as explained in Section 16.2, involves transitions that take place

Table 17.2* Absorption characteristics of some groups and molecules

Group	$\tilde{\nu}_{max}/cm^{-1}$	λ_{max}/nm	$\varepsilon_{max}/(L\ mol^{-1}\ cm^{-1})$
$C{=}C(\pi^* \leftarrow \pi)$	61 000	163	15 000
	57 300	174	5 500
$C{=}O(\pi^* \leftarrow n)$	37–35 000	270–290	10–20
$H_2O(\pi^* \leftarrow n)$	60 000	167	7 000

* More values are given in the Data section.

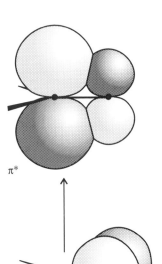

π^*

n

17.7. A $>C{=}O$ group acts as a chromophore primarily on account of the excitation of a non-bonding O lone-pair electron to an antibonding CO π^* orbital.

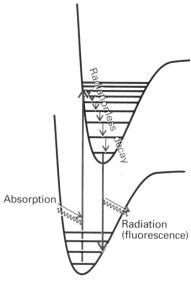

17.8 The sequence of steps leading to fluorescence. After the initial absorption the upper vibrational states undergo radiationless decay by giving up energy to the surroundings. A radiative transition then occurs from the ground state of the upper electronic state.

without the need for photons of the same frequency to be present already. In **fluorescence**, the spontaneously emitted radiation ceases immediately after the exciting radiation is extinguished. In **phosphorescence**, the spontaneous emission may persist for long periods (even hours, but characteristically seconds or fractions of seconds). The difference suggests that fluorescence is an immediate conversion of absorbed light into re-emitted energy and that phosphorescence involves the storage of energy in a reservoir from which it slowly leaks.

Fluorescence

Figure 17.8 shows the sequence of steps involved in fluorescence. The initial absorption takes the molecule to an excited electronic state and, if the absorption spectrum were monitored, it would look like the one shown in Fig. 17.9a. The excited molecule is subjected to collisions with the surrounding molecules, and as it gives up energy it steps down the ladder of vibrational levels, the process of **radiationless decay**, to the lowest vibrational level of the excited electronic state. The surrounding molecules, however, might be unable to accept the larger energy difference needed to lower the molecule to the ground electronic state. It might therefore survive long enough to undergo spontaneous emission, and emit the remaining excess energy as radiation. The downward electronic transition is vertical (in accord with the Franck–Condon principle) and the **fluorescence spectrum** (Fig. 17.9b) has a vibrational structure characteristic of the *lower* electronic state.

The 0–0 absorption and downward fluorescence transitions are not always exactly coincident because the solvent may interact differently with the solute in the ground and excited states. Because the solvent molecules do not have time to rearrange during the transition, the absorption occurs in an environment characteristic of the solvated ground state; however, the fluorescence occurs in an environment characteristic of the solvated excited state.

Fluorescence occurs at a lower frequency than the incident radiation because the emissive transition occurs after some vibrational energy has been discarded into the surroundings. The vivid oranges and greens of fluorescent dyes are an everyday manifestation of this effect: they absorb in the ultraviolet and blue, and fluoresce in the visible. The mechanism also suggests that the intensity of the fluorescence ought to depend on the ability of the solvent molecules to accept the electronic and vibrational

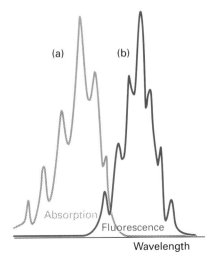

17.9 An absorption spectrum (a) shows a vibrational structure characteristic of the upper state. A fluorescence spectrum (b) shows a structure characteristic of the lower state; it is also displaced to lower frequencies (but the 0–0 transitions are coincident) and resembles a mirror image of the absorption.

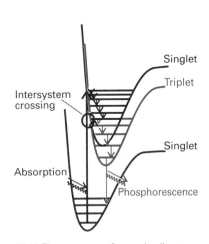

17.10 The sequence of steps leading to phosphorescence. The important step is the intersystem crossing, the switch from singlet to triplet state brought about by spin–orbit coupling. The triplet state acts as a slowly radiating reservoir because the return to the ground state is spin-forbidden.

quanta. It is indeed found that a solvent composed of molecules with widely spaced vibrational levels (such as water) can in some cases accept the large quantum of electronic energy and so extinguish, or 'quench', the fluorescence.

Phosphorescence

Figure 17.10 shows the sequence of events leading to phosphorescence. The first steps are the same as in fluorescence, but the presence of a triplet excited state plays a decisive role. (We first encountered triplet states in Section 13.6: they are states in which two electrons have parallel spins.)

The singlet and triplet excited states share a common geometry at the point where their potential energy curves intersect. Hence, if there is a mechanism for unpairing two electron spins (and achieving the conversion of $\uparrow\downarrow$ to $\uparrow\uparrow$), the molecule may undergo **intersystem crossing** and become a triplet state. We saw in the discussion of atomic spectra (Section 13.8) that singlet–triplet transitions may occur in the presence of spin–orbit coupling, and the same is true in molecules. We can expect intersystem crossing to be important when a molecule contains a moderately heavy atom (such as S), because then the spin–orbit coupling is large.

If an excited molecule crosses into a triplet state, it continues to deposit energy into the surroundings and to step down the vibrational ladder. However, it is now stepping down the triplet's ladder, and at the lowest vibrational energy level it is trapped because the triplet state is at a lower energy than the corresponding singlet (recall Hund's rule, Section 13.4). The solvent cannot absorb the final, large quantum of electronic excitation energy, and the molecule cannot radiate its energy because return to the ground state is spin-forbidden. The radiative transition, however, is not totally forbidden because the spin–orbit coupling that was responsible for the intersystem crossing also breaks the selection rule. The molecules are therefore able to emit weakly, and the emission may continue long after the original excited state was formed.

The mechanism accounts for the observation that the excitation energy seems to get trapped in a slowly leaking reservoir. It also suggests (as is confirmed experimentally) that phosphorescence should be most intense from solid samples: energy transfer is then less efficient and the intersystem crossing has time to occur as the singlet excited state steps slowly past the intersection point. The mechanism also suggests that the phosphorescence efficiency should depend on the presence of a moderately heavy atom (with strong spin–orbit coupling), which is in fact the case. The confirmation of the mechanism is the experimental observation (using the sensitive resonance techniques described in Chapter 18) that the sample is paramagnetic while the reservoir state, with its unpaired electron spins, is populated.

The various types of non-radiative and radiative transitions that can occur in molecules are often represented on a **Jablonski diagram** of the type shown in Fig. 17.11.

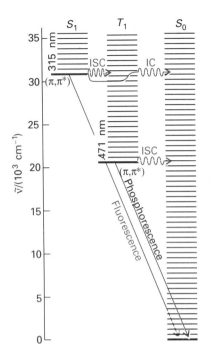

17.11 A Jablonski diagram (here, for naphthalene) is a simplified portrayal of the relative positions of the electronic energy levels of a molecule. Vibrational levels of states of a given electronic state lie above each other, but the relative horizontal locations of the columns bear no relation to the nuclear separations in the states. The ground vibrational states of each electronic state are correctly located vertically but the other vibrational states are shown only schematically. (IC, internal conversion; ISC, intersystem crossing.)

17.4 Dissociation and predissociation

Another fate for an electronically excited molecule is dissociation (Fig. 17.12). The onset of dissociation can be detected in an absorption spectrum by seeing that the vibrational structure of a band terminates at a certain energy. Absorption occurs in a continuous band above this dissociation limit because the final state is an unquantized translational motion of the fragments. Locating the dissociation limit is a valuable way of determining the bond dissociation energy.

In some cases, the vibrational structure disappears but resumes at higher photon energies. This **predissociation** can be interpreted in terms of the molecular potential energy curves shown in Fig. 17.13. When a molecule is excited to a vibrational level, its electrons may undergo a reorganization that results in it undergoing an **internal conversion**, a conversion to another state of the same multiplicity. An internal conversion occurs most readily at the point of intersection of the two molecular potential energy curves, because there the nuclear geometries of the two states are the same. The state into which the molecule converts may be dissociative, so the states near the intersection have a finite lifetime and hence their energies are imprecisely defined. As a result, the absorption spectrum is blurred in the vicinity of the intersection. When the incoming photon brings enough energy to excite the molecule to a vibrational level high above the intersection, the internal conversion does not occur (the nuclei are unlikely to have the same geometry). Consequently, the levels resume their well-defined, vibrational character with correspondingly well-defined energies and the line structure resumes on the high-frequency side of the blurred region.

LASERS

Lasers have transformed chemistry as much as they have the everyday world. In this section, we see some of the principles of their operation, and then explore their applications in chemistry. Lasers lie very much on the frontier of physics and chemistry, for their operation depends on details of optics and, in some cases, of solid-state processes. We shall concentrate on the more chemical aspects of their operation, particularly the materials from which they are made and the events taking place within them, and largely ignore the engineering aspects of their design. Similarly, when we discuss their applications, we shall limit ourselves to purely chemical applications, such as spectroscopy and photochemistry.

17.5 General principles of laser action

The word laser is an acronym formed from **l**ight **a**mplification by **s**timulated **e**mission of **r**adiation. As this name suggests, it is a process that depends on stimulated emission as distinct from the spontaneous emission processes characteristic of fluorescence and phosphorescence. In stimulated emission (Section 16.2) an excited state is stimulated to emit a photon by the presence of radiation of the same frequency, and

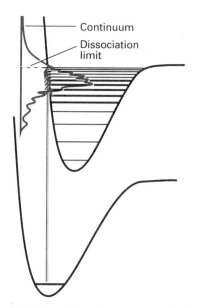

17.12 When absorption occurs to unbound states of the upper electronic state, the molecule dissociates and the absorption is a continuum. Below the dissociation limit the electronic spectrum shows a normal vibrational structure.

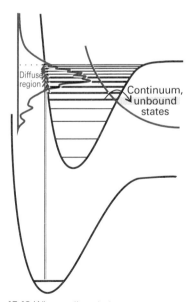

17.13 When a dissociative state crosses a bound state, as in the upper part of the illustration, molecules excited to levels near the crossing may dissociate. This process is called predissociation, and is detected in the spectrum as a loss of vibrational structure that resumes at higher frequencies.

the more photons that are present, the greater the probability of the emission. The essential feature of laser action is positive feedback and the strong gain, or growth of intensity, that results: the more photons present of the appropriate frequency, the more photons of that frequency will be stimulated to form.

The theory of stimulated emission and stimulated absorption described in Section 16.2 shows that, for a given intensity of illumination, the probability of any individual molecule undergoing a transition between two states 1 and 2 is exactly the same in emission as it is in absorption. Therefore, if there are more molecules in the lower-energy state, then there will be a net absorption of the incident radiation. However, if there are more molecules in the upper state, then illumination of the sample will result in a net emission of radiation, and the intensity of the incident light will be enhanced (Fig. 17.14). This enhancement is the light amplification by stimulated emission of radiation that a laser achieves.

Population inversion

One requirement of laser action is the existence of a **metastable excited state**, an excited state with a long enough lifetime for it to participate in stimulated emission. Another requirement is the existence of a greater population in the metastable state than in the lower state where the transition terminates. Because at thermal equilibrium the opposite is true (by the Boltzmann distribution), we see that it is necessary to achieve a **population inversion** in which there are more molecules in the upper state than in the lower.

One way of achieving population inversion is illustrated in Fig. 17.15. The inversion is achieved indirectly through an intermediate state I. Thus, the molecule is excited to I, which then gives up some of its energy non-radiatively and changes into a lower state A; the laser transition is the return of A to the ground state X. Because three energy levels are involved overall, this arrangement leads to a **three-level laser**. In practice, I consists of many states, all of which can convert to the upper of the two laser states A. The $I \leftarrow X$ transition is stimulated with an intense flash of light in the process called **pumping**. In some cases the pumping flash is achieved with an electric discharge through xenon or with the light of another laser. The conversion of I to A should be rapid, and the laser transitions from A to X should be relatively slow.

The disadvantage of the three-level arrangement described here is that it is difficult to achieve the population inversion, because so many ground-state molecules must be converted to the excited state by the pumping action. The four-level laser simplifies this task by having the laser transition terminate in a state A' other than the ground state (Fig. 17.16). Because A' is unpopulated initially, any population in A corresponds to a population inversion, and we can expect laser action if A is sufficiently metastable. Moreover, this population inversion can be maintained if the $A' \rightarrow X$ transitions are rapid, for these will deplete any population in A' that stems from the laser transition, and keep the state relatively empty.

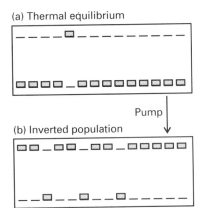

(a) Thermal equilibrium

(b) Inverted population

Pump

(c) Laser action

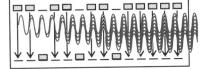

17.14 A schematic illustration of the steps leading to laser action. (a) The Boltzmann population of states, with more atoms in the ground state. (b) When the initial state absorbs, the populations are inverted (the atoms are pumped to the excited state). (c) A cascade of radiation then occurs, as one emitted photon stimulates another atom to emit, and so on. The radiation is coherent (phases in step).

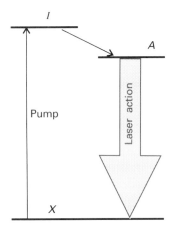

17.15 The transitions involved in one kind of three-level laser. The pumping pulse populates the intermediate state *I*, which in turn populates the laser state *A*. The laser transition is the stimulated emission $A \rightarrow X$.

Cavity and mode characteristics

The laser medium is confined to a cavity that ensures that only certain types of photon are generated abundantly. (By type of photon is meant its frequency, direction of travel, and state of polarization.) The cavity is essentially a region between two mirrors, which reflect the light back and forth. The only wavelengths that can be sustained by the cavity satisfy

$$N \times \tfrac{1}{2}\lambda = L$$

where N is an integer and L is the length of the cavity. That is, only an integral number of half-wavelengths fit into the cavity, all other waves undergoing destructive interference with themselves. The frequency difference between neighbouring values of N is

$$\Delta \nu = \frac{c}{2L} \tag{3}$$

(We are supposing that the refractive index of the medium is 1.) In addition, not all wavelengths that can be sustained by the cavity are amplified by the laser medium (many fall outside the range of frequencies of the laser transitions), so only a few contribute to the laser radiation. These wavelengths are the **resonant modes** of the laser.

Photons with the correct wavelength for the resonant modes of the cavity and the correct frequency to stimulate the laser transition are highly amplified. One photon might be generated spontaneously, and travel through the medium. It stimulates the emission of another photon, which in turn stimulates more (as was depicted in Fig. 17.14). The cascade of energy builds up rapidly, and soon the cavity is an intense reservoir of radiation at all the resonant modes that it can sustain. Some of this light can be withdrawn if one of the mirrors is partially transmitting.

The resonant modes of the cavity have various natural characteristics and to some extent may be selected. In the first place, only photons that are travelling strictly parallel to the axis of the cavity undergo more than a couple of reflections, so that only they are amplified; all others simply vanish into the surroundings. Hence, laser light is generally highly **collimated** and forms a beam with very low divergence. It may also be **polarized**, with its electric vector in a particular plane (or in some other state of polarization), by including a polarizing filter into the cavity. Polarization is also achieved in certain lasers by the shape of the cell containing the laser medium. Thus, to cut down reflections that result where a refractive index changes abruptly (as at the walls of a glass container), the walls are arranged to make a special angle, known as the 'Brewster angle', to the direction of propagation of the light (Fig. 17.17). The Brewster windows are more transparent to light of one polarization than the other. Consequently, the gain is greater for that polarization, and the laser emits polarized light.

Laser light is also **coherent** in the sense that the electromagnetic waves are all in step. It is important to distinguish between two types of coherence. In **spatial coherence** we are concerned with the waves being

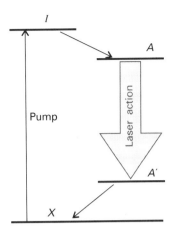

17.16 The transitions involved in a four-level laser. Since the laser transition terminates in an excited state (A'), the population inversion between A and A' is much easier to achieve.

in step across the cross-section of the beam emerging from the cavity. In **temporal coherence** we are concerned with the extent to which the waves remain in step along the beam. The latter is normally expressed in terms of a **coherence length** l_C, and is related to the range of wavelengths $\Delta\lambda$ present in the beam:

$$l_C = \frac{\lambda^2}{2\Delta\lambda} \tag{4}$$

If the beam were perfectly monochromatic, with strictly one wavelength present, then $\Delta\lambda$ would be zero, and the waves would remain in step for infinite distances. If many wavelengths are present, then the waves get out of step in a short distance and the coherence length is small. A typical light bulb gives out light with a coherence length of only about 400 nm; a He–Ne laser with $\Delta\lambda \approx 2$ pm has a coherence length of 10 cm or so.

Q-switching

A laser can generate light for as long as the population inversion is maintained. When heat is easily dissipated, the laser may act continuously, for the population of the upper level can be replenished by pumping. In some cases, practical considerations govern whether or not continuous pumping is feasible, as we shall see when we consider some particular lasers. In other cases, especially when overheating is a problem, the laser can be operated only in pulses, perhaps of microsecond or millisecond duration, so that the medium has a chance to cool or the lower state discard its population.

It is sometimes desirable to have pulses of radiation rather than a continuous output, with a lot of energy concentrated into a brief pulse. One way of achieving pulses is by **Q-switching**, the modification of the resonance characteristics of the laser cavity. (The name comes from the 'Q-factor' used as a measure of the quality of a resonance cavity in microwave engineering.)

17.17 The Brewster window at the end of a laser leads to polarized emission. If light strikes a transparent material of refractive index n, no light polarized parallel to the plane of incidence is reflected if the light strikes at an angle θ_B given by $\tan\theta_B = n$. For glass with $n = 1.5$, the Brewster angle is $\theta_B \approx 57°$. Some light with perpendicular polarization is reflected, so the gain is greater for the light of parallel polarization, which therefore dominates the output.

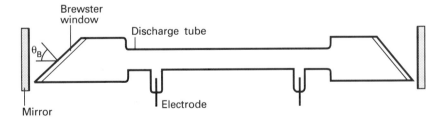

Example 17.2 *Relating the power and energy of a laser*

A laser rated at 0.10 J can generate radiation in 3.0-ns pulses. What is the average power output per pulse?

Method. The power output P is the energy per unit time, and is expressed in watts (1 W = 1 J s^{-1}). So, to calculate the power, we divide the energy output by the time over which the pulse is generated.

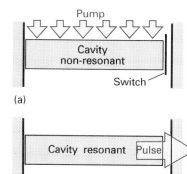

(a)

(b)

17.18 The principle of Q-switching. The excited state is populated while the cavity is non-resonant. Then the resonance characteristics are suddenly restored, and the stimulated emission emerges in a giant pulse.

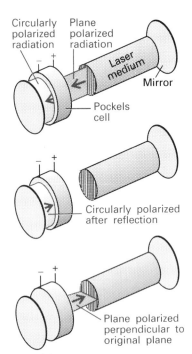

17.19 When light passes through a Pockels cell that is 'on', its plane of polarization is rotated and so the laser cavity is non-resonant (its Q-factor, a measure of its resonant quality, is reduced). When the cell is turned off, no change of polarization occurs, and the cavity is resonant.

Answer. From the data,

$$P = \frac{0.10\,\text{J}}{3.0 \times 10^{-9}\,\text{s}} = 3.3 \times 10^{7}\,\text{J s}^{-1}$$

That is, the pulses deliver 33 MW of power.

Comment. The answer gives the average power; the peak power will be larger. If the power consumption of the laser is 100 W, then the 0.10 J required for each pulse can be supplied in 0.1 ms assuming 100 per cent conversion efficiency.

Exercise E17.2. Calculate the average power output of a laser in which a 2.0-J pulse can be delivered in 1.0 ns.   [2.0 GW]

The aim of Q-switching is to achieve a healthy population inversion in the absence of the resonant cavity, then to plunge the population-inverted medium into a cavity, and hence to obtain a sudden pulse of radiation. The switching may be achieved by damaging the resonance characteristics of the cavity in some way while the pumping pulse is active, and then suddenly to restore them (Fig. 17.18). In practice, Q-switching can give pulses of about 10-ns duration.

There are several ways of Q-switching a laser. The earliest technique was to rotate one of the cavity mirrors so that it periodically came back into correct alignment with its partner. A faster method is to use a **Pockels cell**, which is an electro-optical device based on the ability of crystals of ammonium dihydrogenphosphate to convert plane-polarized light to circularly polarized light when a potential difference is applied. If a Pockels cell is made part of a laser cavity, then its action and the change of polarization that occurs when light is reflected from a mirror convert light polarized in one plane into reflected light polarized in the perpendicular plane (Fig. 17.19). As a result, the reflected light does not stimulate more emission. However, if the cell is suddenly turned off, the polarization effect is extinguished and all the energy stored in the cavity can emerge as an intense pulse of stimulated radiation. An alternative technique is to use a **saturable dye** that loses its power to absorb when many of its molecules have been excited by intense radiation. It then suddenly becomes transparent, and the cavity becomes resonant.

Mode locking

The technique of mode locking can produce pulses of picosecond duration and less. We have seen that a laser radiates at a number of different frequencies, depending on the precise details of the resonance characteristics of the cavity and in particular on the number of half-wavelengths of radiation that can be trapped between the mirrors (the cavity modes). The resonant modes differ in frequency by multiples of $c/2L$ (eqn 3). Normally, these modes have random phases relative to each other. However, it is possible to lock their phases together so that they interfere with each other. The constructive interference occurs at a

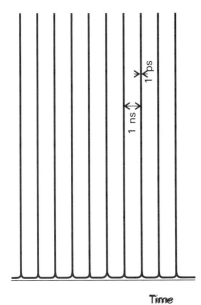

17.20 The output of a mode-locked laser consists of a stream of very narrow pulses separated by an interval equal to the time it takes for light to make a round trip inside the cavity.

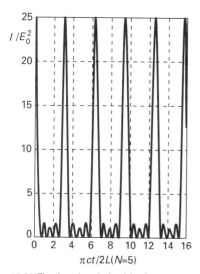

17.21 The function derived in the *Justification* showing in more detail the structure of the pulses generated by a mode-locked laser.

series of sharp peaks separated by regions of destructive interference, and the power of the laser is obtained in picosecond bursts (Fig. 17.20). The sharpness of the peaks depends on the range of modes superimposed and, the wider the range, the narrower the pulses. In a laser with a cavity of length 30 cm, the peaks will be separated by 2 ns. If 10^3 modes contribute, the width of the pulses will be 4 ps.

JUSTIFICATION

Here we shall show that, if N modes with frequencies differing by $c/2L$ are superimposed, then they give rise to a series of peaks separated by $2L/c$.

The general expression for a wave of amplitude E_0 and frequency ω is $E_0 e^{i\omega t}$. Therefore, each wave has the form

$$\psi_n = E_0 e^{2\pi i(v + nc/2L)t}$$

where v is the lowest frequency and $n = 0, 1, \ldots, N-1$. The total wave has the form

$$\psi = \sum_n \psi_n = E_0 e^{2\pi i v t} \sum_0^{N-1} e^{i\pi nct/L}$$

The sum we require is a geometrical progression:

$$\sum_0^{N-1} e^{i\pi nct/L} = 1 + e^{i\pi ct/L} + e^{2i\pi ct/L} + \cdots + e^{(N-1)i\pi ct/L}$$

$$= \frac{\sin(N\pi ct/2L)}{\sin(\pi ct/2L)} \times e^{(N-1)i\pi ct/2L}$$

The intensity of the radiation is equal to the square modulus of the total amplitude, so

$$I = \psi^*\psi = E_0^2 \times \frac{\sin^2(N\pi ct/2L)}{\sin^2(\pi ct/2L)}$$

The function I is shown in Fig. 17.21. We see that it is a series of peaks with maxima separated by $t = 2L/c$, the round-trip transit time of the light in the cavity.

Mode locking is achieved by varying the loss of the cavity periodically at the frequency $c/2L$. The modulation can be pictured as the opening of a shutter in synchrony with the round-trip travel time of the photons in the cavity, so that only photons making the journey in that time are amplified. The modulation can be achieved by linking a prism in the cavity to a transducer driven by a radiofrequency source at a frequency $c/2L$. The transducer sets up standing-wave vibrations in the prism and modulates the loss it introduces into the cavity. Mode locking may also be accomplished passively by including a saturable dye. This procedure makes use of the fact that the gain is very sensitive to amplification and,

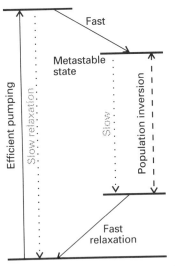

17.22 A summary of the features needed for efficient laser action.

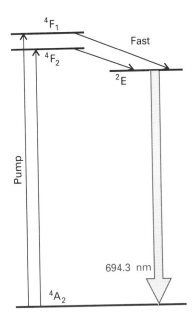

17.23 The transitions involved in the ruby laser. The laser medium, ruby, consists of Al_2O_3 doped with Cr^{3+} ions.

once a particular frequency begins to grow, it can quickly dominate. If a saturable dye is included in the cavity, a spontaneous fluctuation in intensity—a bunching of photons—may result in its becoming transparent, and the bunch can pass through and travel to the far end of the cavity, amplifying as it goes. The dye immediately shuts down again (if it is well chosen), but opens when the intense pulse returns from the mirror at the far end and saturates it. In this way, that particular bunch of photons may grow to considerable intensity since it alone is stimulating emission in the cavity.

17.6 Practical lasers

Figure 17.22 summarizes the requirements for an efficient laser. In practice, the requirements can be satisfied by using a variety of different systems, and in this section we review some that are commonly available.

Solid-state lasers

A **solid-state laser** is one in which the active medium is in the form of a single crystal or a glass. The first successful laser, the **ruby laser** built by Theodore Maiman in 1960, is an example (Fig. 17.23). Ruby is Al_2O_3 containing a small proportion of Cr^{3+} ions. (The normal green of Cr^{3+} is modified to red by the distortion of the local crystal field stemming from the replacement of an Al^{3+} ion by a slightly larger Cr^{3+} ion.) Ruby is a three-level laser, and the ground state, which is also the lower level of the laser transition, is $^4A_2$ with three unpaired spins on each Cr^{3+} ion. The population inversion results from pumping a majority of the Cr^{3+} ions into an excited state using an intense flash from another source, followed by a radiationless transition to another excited state. The pumping flash need not be monochromatic because the upper level actually consists of several states spanning a band of frequencies. The transition from the lower of the two excited states to the ground state ($^2E \rightarrow {}^4A_2$) is the laser transition, and gives rise to red 694-nm radiation. The population inversion is very difficult to sustain continuously and, in practice, the ruby laser is pulsed. Typical pulses from a Q-switched ruby laser might consist of 2-J pulses persisting for 10 ns, corresponding to an average power of 0.2 GW.

The **neodymium laser** is an example of a four-level laser (Fig. 17.24). In one form it consists of Nd^{3+} ions at low concentration in yttrium aluminium garnet (YAG, specifically $Y_3Al_5O_{12}$), and is then known as a Nd–YAG laser. A cheaper medium is glass, but glass is a poorer thermal conductor than YAG and, if glass is used, the laser must be pulsed. A neodymium laser operates at a number of wavelengths in the infrared, the band at 1064 nm being most common. The transition at 1064 nm is very efficient and the laser is capable of substantial power output. The power is great enough for frequency doubling to be used efficiently. **Frequency doubling** is a technique in which the laser beam is converted to radiation with twice (and in general a multiple) of its initial frequency as it passes through a transparent material. A frequency-doubled Nd–YAG laser has a wavelength 532 nm, which corresponds to green light.

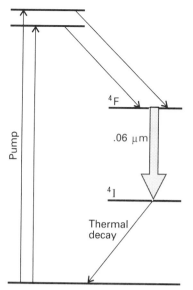

17.24 The transitions involved in the neodymium laser. The laser action takes place between two excited states, and the population inversion is easier to achieve.

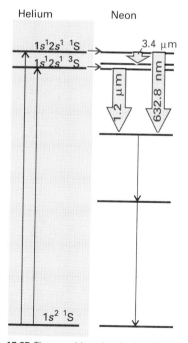

17.25 The transitions involved in the helium–neon laser. The pumping (of the neon) depends on a coincidental matching of the helium and neon energy separations, so that excited He atoms can transfer their excess energy to Ne atoms during a collision.

Example 17.3 *Accounting for multiphoton phenomena*

Show that if a substance responds non-linearly to incident radiation of frequency ω, then it may act as the source of radiation of twice the incident frequency.

Method. The key idea is that radiation of a particular frequency arises from oscillations of an electric dipole at that frequency. Therefore, we express the induced electric dipole moment of the system in terms of powers of the applied electric field, and then express powers of harmonic (cosine) terms as sums and differences of cosine terms. We then inspect the sum to see if $\cos 2\omega t$ is present.

Answer. The incident electric field E induces an electric dipole μ, and allowing for non-linear response, we can write

$$\mu = \alpha E + \beta E^2 + \cdots$$

The non-linear terms can be expanded as follows if we suppose that the incident electric field is $E_0 \cos \omega t$:

$$\beta E^2 = \beta E_0^2 \cos^2 \omega t = \tfrac{1}{2}\beta E_0^2(1 + \cos 2\omega t)$$

Hence, the non-linear term contributes an induced electric dipole that oscillates at the frequency 2ω and which can act as a source of radiation of that frequency.

Exercise E17.3. Show that, if a substance responds non-linearly to two sources of radiation, one of frequency ω_1 and the other of frequency ω_2 then it may give rise to radiation of the sum and difference of the two frequencies. $[\beta E^2 \propto \cos(\omega_1 + \omega_2)t + \cos(\omega_1 - \omega_2)t]$

Gas lasers

Gas lasers are widely used and, since they can be cooled by a rapid flow of the gas through the cavity, they can be used to generate high powers. The pumping is normally achieved using a gas that is different from the gas responsible for the laser emission itself.

In the **helium–neon laser** (Fig. 17.25) the active medium is a mixture of helium and neon in a mole ratio of about 5:1. The initial step is the excitation of an He atom to the metastable $1s^12s^1$ configuration using an electric discharge (the collisions of electrons and ions cause transitions that are not restricted by electric-dipole selection rules). The excitation energy of this transition happens to match an excitation energy of neon and, during an He–Ne collision, efficient transfer of energy may occur, leading to the production of highly excited, metastable Ne atoms with unpopulated intermediate states. Laser action generating 633-nm radiation (among about 100 other lines) then occurs.

The **argon-ion laser** (Fig. 17.26), one of a number of 'ion lasers', consists of argon at about 1 Torr, through which is passed an electric discharge. The discharge results in the formation of Ar^+ and Ar^{2+} ions in excited states, which undergo a laser transition to a lower state. These

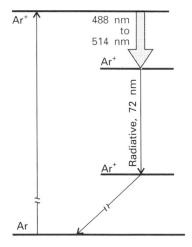

17.26 The transitions involved in an argon-ion laser.

ions then revert to their ground states by emitting hard ultraviolet radiation (at 72 nm), and are then neutralized by a series of electrodes in the laser cavity (Fig. 17.27). One of the design problems is to find materials that can withstand this damaging residual radiation. There are many lines in the laser transition because the excited ions may make transitions to many lower states, but two strong emissions from Ar^+ are at 488 nm (blue) and 514 nm (green); other transitions occur elsewhere in the visible region, in the infrared, and in the ultraviolet. The **krypton-ion laser** works similarly. It is less efficient, but gives a wider range of wavelengths, the most intense being at 647 nm (red), but it can also generate a yellow line. Both lasers are widely used in laser light shows (for this application argon and krypton are often used simultaneously in the same cavity) as well as laboratory sources of high-power radiation.

The **carbon dioxide laser** (Fig. 17.28) works on a slightly different principle, for its radiation (between 9.2 and 10.8 μm, with the strongest emission at 10.6 μm, in the infrared) arises from vibrational transitions. Most of the working gas is nitrogen, which becomes vibrationally excited by electronic and ionic collisions in an electric discharge. The vibrational levels happen to coincide with the ladder of antisymmetric stretching (v_3, Fig. 16.47) vibrational energy levels of CO_2, which pick up the energy during a collision. Laser action then occurs from the lowest excited level of v_3 to the lowest excited level of the symmetric stretch (v_1), which has remained unpopulated during the collisions. The transition is allowed by anharmonicites in the molecular potential energy. Some helium is included in the gas to help remove energy from this state and maintain the population inversion.

In the **nitrogen laser**, the efficiency of the stimulated transition (at 337 nm, in the ultraviolet, the transition $C^3\Pi_u \rightarrow B^3\Pi_g$) is so great that a single passage of a pulse of radiation is enough to generate laser radiation and mirrors are unnecessary: such lasers are said to be **superradiant**.

Chemical and exciplex lasers

Chemical reactions may also be used to generate molecules with non-equilibrium, inverted populations. For example, the photolysis of Cl_2 leads to the formation of Cl atoms that attack H_2 molecules in the

17.27 The construction of an argon-ion laser. An electric discharge results in the formation of Ar^+ and Ar^{2+} ions in excited states which are neutralized by the electrodes inside the cavity. High currents are passed through the laser plasma, and in some designs, as shown here, it is held away from the walls by magnetic fields from electromagnets that surround the cavity.

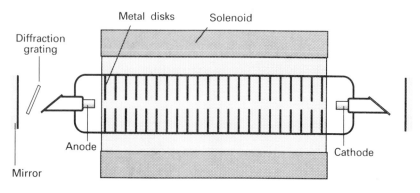

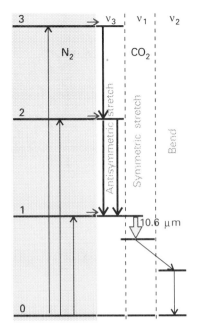

17.28 The transitions involved in the carbon dioxide laser. The laser transition is from $\nu_3 = 1$ to $\nu_1 = 1$.

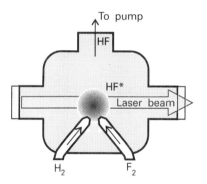

17.29 The arrangement used to achieve laser action in a chemical hydrogen/fluorine laser. The emission is from vibrationally excited HF molecules produced in the reaction.

mixture and produce HCl and H. The latter then attacks Cl_2 to produce vibrationally excited ('hot') HCl molecules. Because the newly formed HCl molecules have non-equilibrium vibrational populations, laser action can result as they return to lower states. Such processes are remarkable examples of the direct conversion of chemical energy into coherent electromagnetic radiation.

Chemical lasers have been under consideration as sources of very intense radiation (the Strategic Defense Initiative had a major interest in the technology). The reaction that fuels the hydrogen fluoride laser is between H_2 and F atoms, which produces hot HF that can produce infrared radiation in the range 2.6 to 3.0 μm; the production of DF from D_2 and F_2 generates radiation of slightly longer wavelength, 3.6 to 4.0 μm. For deployment in space, the F atoms could be present as F_2, and the reaction is between H_2 and F_2 directly, the laser emission taking place just downstream of the mixing chamber (Fig. 17.29). An alternative source of F atoms is an electric discharge through SF_6. Although the population inversion in HF does not survive for very long, the ground-state molecules are swept rapidly out of the laser zone and replaced by excited molecules.

The population inversion needed for laser action is achieved in a more underhand way in **exciplex lasers**,[1] for in these (as we shall see) the lower state does not effectively exist. This odd situation is achieved by forming an **exciplex**, a combination of two atoms that survives only in an excited state and which dissociates as soon as the excitation energy has been discarded. An example of an exciplex laser is a mixture of xenon, chlorine, and neon (which acts as a buffer gas). An electric discharge through the mixture produces excited Cl atoms, which attach to the Xe atoms to give the exciplex XeCl*. The exciplex survives for about 10 ns, which is enough time for it to participate in laser action at 308 nm (in the ultraviolet). As soon as XeCl* has discarded a photon, the atoms separate because the molecular potential energy curve of the ground state is dissociative, and the ground state of the exciplex cannot become populated (Fig. 17.30). The KrF* exciplex laser is another example: it produces radiation at 249 nm.

Dye lasers

A solid-state laser and a gas laser operate at discrete frequencies and, although the frequency required may be selected by suitable optics, the laser cannot be tuned continuously. The tuning problem is overcome by using a **dye laser**, which has such broad spectral characteristics (because the solvent broadens the vibrational structure of the transitions into bands, as we have seen). Hence, it is possible to scan the wavelength continuously (by adjusting the orientation of the diffraction grating in the cavity) and achieve laser action at any chosen wavelength. A commonly used dye is Rhodamine 6G in methanol (Fig. 17.31). As the gain is very

---

1 The term 'excimer laser' is also widely encountered and used loosely when 'exciplex laser' is more appropriate. An exciplex has the form AB* whereas an excimer, an excited dimer, is AA*.

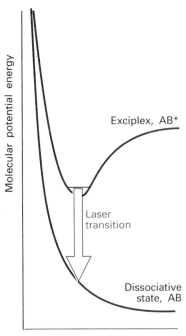

17.30 The molecular potential energy curves for an exciplex. The species can survive only as an excited state, because on discarding its energy it enters the lower, dissociative state. Since only the upper state can exist, there is never any population in the lower state.

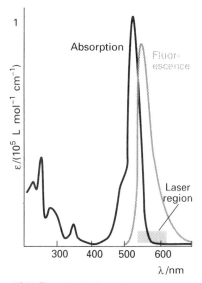

17.31 The optical absorption spectrum of the dye Rhodamine G and the region used for laser action.

high, only a short length of the optical path need be through the dye. The excited states of the active medium, the dye, are sustained by another laser or a flash lamp, and the dye solution is made to flow through the laser cavity (Fig. 17.32).

Light-emitting diodes and semiconductor lasers

We have seen (in Section 14.10) that a semiconductor is classified as 'n-type' if its conduction band is partly populated and as 'p-type' if its valence band has a small number of holes. In this section we need to consider the properties of a **p–n junction**, the interface of the two types of semiconductor.

The band structure at the junction is shown in Fig. 17.33. When a 'forward bias' is applied to the junction, in the sense that electrons are supplied through an external circuit to the n side of the junction, the electrons in the conduction band of the n-type semiconductor fall into the holes in the valence band of the p-type semiconductor. As they fall, they emit energy. In silicon semiconductors this energy is largely in the form of heat because the transition can occur only if the electron transfers linear momentum to the lattice, and the device becomes warm. However, in some materials, most notably gallium arsenide, GaAs, the transition can occur without the lattice needing to be involved (because the electron has the same linear momentum in the ground state as in the upper state), and the energy is emitted as light. Practical **light-emitting diodes** of this kind are widely used in electronic displays. Gallium arsenide itself emits infrared light, but the band gap is widened by incorporating phosphorus, and a material of composition approximately $GaAs_{0.6}P_{0.4}$ emits light in the red region of the spectrum.

A light-emitting diode is not a laser, because no resonance cavity and stimulated emission are involved. However, it is easy (in principle) to employ the light emission of electron–hole recombination as the basis of laser action. The population inversion can be sustained by sweeping away the electrons that fall into the holes of the p-type semiconductor, and a resonant cavity can be formed by using the high refractive index of the semiconducting material and cleaving single crystals so that the light is trapped by the sudden variation of refractive index. One widely used material is $Ga_{1-x}Al_xAs$, which produces infrared laser radiation and is widely used in compact-disc (CD) players.

17.7 Applications of lasers in chemistry

Laser radiation has five striking characteristics (Table 17.3). Each of them (sometimes in combination with the others) opens up interesting opportunities in spectroscopy, giving rise to **laser spectroscopy** and, in photochemistry, giving rise to **laser photochemistry**.

Spectroscopy at high photon fluxes

The high spectral power density of a laser—the high intensity of the radiation it produces at well-defined frequencies—is an aid to conventional spectroscopy in that it reduces the problem of detector noise and

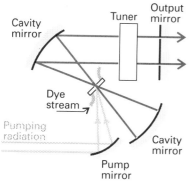

17.32 The configuration used for dye laser action. The dye is flowed through the cell inside the laser cavity. The flow helps to keep it cool.

Table 17.3 Characteristics of laser radiation and their chemical applications

Characteristic	Advantages	Applications
High power density	Multiphoton processes	Non-linear spectroscopy Saturation spectroscopy
	Low detector noise High scattering intensity	Improved sensitivity Raman spectroscopy
Monochromatic	High resolution State selection	Spectroscopy Isotope separation Photochemically precise State-to-state reaction dynamics
Collimated beam	Long path lengths Forward-scattering observable	Sensitivity Non-linear Raman spectroscopy
Coherent	Interference between separate beams	CARS
Pulsed	Precise timing of excitation	Fast reactions Relaxation Energy transfer

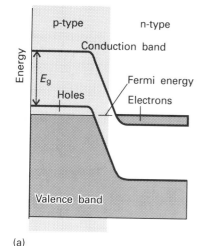

(a)

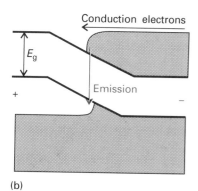

(b)

17.33 The structure of a diode junction (a) without bias and (b) with bias.

the interfering effects of background radiation. The high intensity is particularly advantageous in Raman spectroscopy, which until the introduction of lasers was plagued by the low intensity of the scattered radiation (which could be overcome only by using long exposures) and by interference from background scattering (which obscured the signal).

The high power in a very narrow linewidth of an incident laser beam also enhances the intensity of the radiation caused by fluorescence, because higher populations of the excited state may be achieved. The achievement of higher populations enhances the intensity of fluorescence and simplifies the observation of fluorescence spectra. It therefore makes more substances open to study by that technique.

The large number of photons in an incident beam generated by a laser also gives rise to a qualitatively different branch of spectroscopy, for the photon density is so high that more than one photon may be absorbed by a single molecule giving rise to **multiphoton processes**. One application of multiphoton processes is that states inaccessible by conventional one-photon spectroscopy become observable because the overall transition occurs with no change of parity. For example, in one-photon spectroscopy, only g↔u transitions are observable; in two-photon spectroscopy, however, the overall outcome of absorbing two photons is a g↔g or a u↔u transition.

High powers and monochromatic beams make possible the technique of **saturation spectroscopy**, which permits the very precise location of absorption maxima. As illustrated in Fig. 17.34, the output of a tunable laser is divided into an intense saturating beam and a less intense probe beam that pass through the sample cavity in nearly opposite directions. The chopped saturating beam periodically excites molecules that are Doppler-shifted to its frequency. The probe beam gives a modulated signal at the detector, but only if it is interacting with the same Doppler-shifted molecules despite the fact that it is coming from an

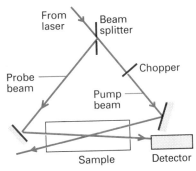

17.34 The configuration of laser radiation used for saturation spectroscopy.

opposite direction. Since those molecules must be ones that are not moving parallel to the beams, the technique selects molecules that have essentially zero Doppler shift and hence gives very high resolution. Lamb-dip spectroscopy (Section 16.3) is a version of saturation spectroscopy.

Collimated beams

The collimated beams generated by most kinds of lasers are also useful in spectroscopy for a number of reasons, one of which is that they permit the use of very long path lengths through samples, because the beam has only a small cross-section even after many reflections. A well-defined beam also implies that the detector can be designed to collect only the radiation that has passed through a sample, and can be screened much more effectively against stray scattered light. Moreover, with a collimated beam, the interaction zone in Raman spectroscopy is much more well-defined than in conventional spectroscopy, so the optics of the spectrometer can be optimized.

As with the high power output of lasers, the availability of a non-divergent beam makes possible a qualitatively different kind of spectroscopy. The beam is so well-defined that it is possible to observe Raman transitions very close to the direction of propagation of the incident beam (rather than perpendicular to it). This configuration is employed in the technique called **stimulated Raman spectroscopy**. In this form of Raman spectroscopy, the Stokes and anti-Stokes radiation in the forward direction are powerful enough to undergo more scattering and hence give up or acquire more quanta of energy from the molecules in the sample. This multiple scattering results in lines of frequency $\nu_i \pm 2\nu_M$, $\nu_i \pm 3\nu_M$, and so on, where ν_i is the frequency of the incident radiation and ν_M the frequency of a molecular excitation.

Raman spectroscopy was revitalized by the introduction of lasers. We have already commented on the enhancements of the technique that stem from the high powers and collimation of the incident beam. Its monochromaticity is also a great advantage, for it is now possible to observe scattered light that differs by only fractions of cm^{-1} from the incident radiation. Such high resolution is particularly useful for observing the rotational structure of Raman lines because rotational transitions are of the order of a few reciprocal centimetres. Monochromaticity also allows observations to be made very close to absorption frequencies, giving rise to the technique of resonance Raman spectroscopy (Section 16.15).

As before, the laser opens up qualitatively different varieties of the technique, not merely enhancements of the established technique. For instance, the intensity of Raman transitions may be enhanced by coherent anti-Stokes Raman spectroscopy (CARS, Section 16.15).

Precision-specified transitions

The monochromatic character of laser radiation is a very powerful characteristic because it allows us to excite specific states with very high precision. One consequence of state specificity for photochemistry is that the illumination of a sample may be photochemically precise and hence efficient in stimulating a reaction, since its frequency can be tuned exactly to an absorption. This is in contrast to radiation from a conventional broad-band source, much of which is not absorbed and is therefore wasted.

The specific excitation of a particular excited state of a molecule may greatly enhance the rate of a reaction even at low temperatures. The rate of a reaction is generally increased by raising the temperature because the energy of the various modes of motion of the molecule is enhanced. However, this enhancement increases the energy of all the modes, even those that do not contribute appreciably to the reaction rate. With a laser we can excite the kinetically significant mode, so that rate enhancement is achieved most efficiently. An example is the reaction

$$BCl_3 + C_6H_6 \rightarrow C_6H_5\!-\!BCl_2 + HCl$$

which normally proceeds only above 600°C in the presence of a catalyst; exposure to 10.6-μm CO_2 laser radiation results in the formation of products at room temperature without a catalyst. The commercial potential of this procedure is considerable (provided that laser photons can be produced sufficiently cheaply), since heat-sensitive compounds, such as pharmaceuticals, may be made at lower temperatures than in conventional reactions.

A related application is the study of **state-to-state reaction dynamics**, in which a specific state of a reactant molecule is excited and we monitor not only the rate at which it forms products but also the states in which they are produced. Studies such as these give highly detailed information about the deployment of energy in chemical reactions (Chapter 27).

Isotope separation

The precision state-selectivity of lasers is also of considerable potential for **laser isotope separation**. Isotope separation is possible because two **isotopomers**, or species that differ only in their isotopic composition, have slightly different energy levels and hence slightly different absorption frequencies.

One approach is to use **photoionization**, the ejection of an electron by the absorption of electromagnetic radiation. Direct photoionization by the absorption of a single photon does not distinguish between isotopomers because the upper level belongs to a continuum; to distinguish isotopomers it is necessary to deal with discrete states. As a result, at least two absorption processes are required. In the first step, a photon excites an atom to a higher state and then in the second step a photon achieves photoionization from that state (Fig. 17.35). The separation of the two states involved in the first step depends on the nuclear mass and, if the laser radiation is tuned to that frequency, only one of the isotopomers will undergo excitation and hence be available for photoionization in the second step. An example of this procedure is the

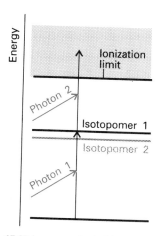

17.35 In one method of isotope separation, one photon excites an isotopomer to an excited state, and then a second photon achieves photoionization. The success of the first step depends on the nuclear mass.

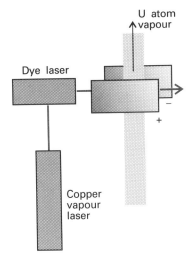

17.36 The experimental arrangement for isotope separation. The dye laser, which is pumped by a copper-vapour laser, photoionizes the U atoms selectively according to their mass, and the ions are deflected by the electric field applied between the plates.

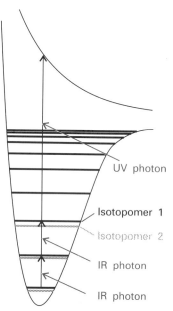

17.37 Isotopomers may be separated by making use of their selective absorption of infrared (IR) photons followed by photodissociation with an ultraviolet (UV) photon.

photoionization of uranium vapour, in which the incident laser is tuned to excite ^{235}U but not ^{238}U. The ^{235}U atoms in the atomic beam are ionized in the two-step process; they are then attracted to a negative electrode, and may be collected (Fig. 17.36). This procedure is being used in the latest generation of uranium separation plants.

Molecular isotopomers are used in techniques based on **photodissociation**, the fragmentation of a molecule following absorption of electromagnetic radiation. The key problem is to achieve both mass selectivity (which requires excitation to take place between discrete states) and dissociation (which requires excitation to continuum states). In one approach, two lasers are used: an infrared photon excites one isotopomer selectively to a higher vibrational level, and then an ultraviolet photon completes the process of photodissociation (Fig 17.37). An alternative procedure is to make use of multiphoton absorption within the ground electronic state (Fig. 17.38): the first few photons are mass-sensitive, and they open the door to a subsequent influx of enough photons to complete the dissociation process. The isotopomers $^{32}SF_6$ and $^{34}SF_6$ have been separated in this way.

In a third approach, a selectively vibrationally excited species may react with another species and give rise to products that can be separated chemically. This procedure has been employed successfully to separate isotopes of B, N, O, and, most efficiently, H. A variation on this procedure is to achieve selective **photoisomerization**, the conversion of a species to one of its isomers (particularly a geometrical isomer) on absorption of electromagnetic radiation. Once again, the initial absorption, which is isotope-selective, opens the way to subsequent further absorption and the formation of a geometrical isomer that can be separated chemically. The approach has been used with the photoisomerization of CH_3NC to CH_3CN.

A different, more physical approach, that of **photodeflection**, is based on the recoil that occurs when a photon is absorbed by an atom and the linear momentum of the photon (which is equal to h/λ) is transferred to the atom. The atom is deflected from its original path only if the absorption actually occurs, and the incident radiation can be tuned to a particular isotope. The deflection is very small, so an atom must absorb dozens of photons before its path is changed sufficiently to allow collection. For instance, if a Ba atom absorbs about 50 photons of 550-nm light, it will be deflected by only about 1 mm after a flight of 1 m.

Pulsed techniques

The ability of lasers to produce pulses of very short duration is particularly useful in chemistry when we want to monitor processes in time. Q-switched lasers produce nanosecond pulses, which are generally fast enough to study reactions with rates controlled by the speed with which reactants can move through a fluid medium. However, when we want to study the rates at which energy is converted from one mode to another within a molecule, we need the shorter time scale of picosecond

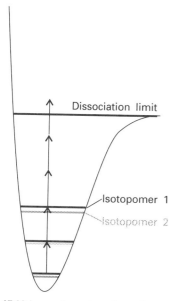

17.38 In an alternative scheme for separating isotopomers, multiphoton absorption of infrared photons is used to reach the dissociation limit of a ground electronic state.

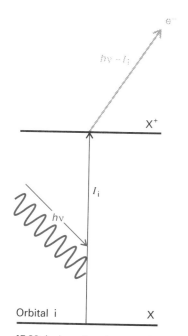

17.39 An incoming photon carries an energy hv; an energy I_i is needed to remove an electron from an orbital i, and the difference appears as the kinetic energy of the electron.

pulses. These time scales are available from mode-locked lasers, and modern techniques have reduced time scales of pulses to the femtosecond region (1 fs $= 10^{-15}$ s), the shortest pulse currently reported being about 6 fs, corresponding to a packet of electromagnetic radiation of only a few wavelengths long. We shall see some of the information obtained from this femtosecond spectroscopy in Section 27.5. Pulse techniques are used to study ultrafast dynamical processes such as energy transfer and conversion from one mode of motion to another. They are used to study relaxation of a disturbed set of level populations to thermal equilibrium, and, of particular importance in chemistry, to study the rates of fast reactions.

PHOTOELECTRON SPECTROSCOPY

The technique of photoelectron spectroscopy (PES) measures the ionization energies of molecules when electrons are ejected from different orbitals, and uses the information to infer the orbital energies. The technique is also used to study solids, and in Chapter 28 we shall see the important information that it gives about species at or on surfaces.

17.8 The technique

Because energy is conserved when a photon ionizes a sample, the energy of the incident photon hv must be equal to the sum of the ionization energy I of the sample and the kinetic energy of the **photoelectron**, the ejected electron (Fig. 17.39):

$$hv = \tfrac{1}{2}m_e v^2 + I$$

This equation (which is like the one used for the photoelectric effect, Section 11.2) can be refined in two ways. First, photoelectrons may originate from one of a number of different orbitals, and each one requires a different ionization energy. Hence, a series of different kinetic energies of the photoelectrons will be obtained, each one satisfying

$$hv = \tfrac{1}{2}m_e v^2 + I_i \tag{5}$$

where I_i is the ionization energy for ejection of an electron from an orbital i. Therefore, by measuring the kinetic energies of the photoelectrons, and knowning v, these ionization energies can be determined. Photoelectron spectra are interpreted in terms of an approximation called **Koopmans' theorem** which states that the ionization energy I_i is equal to the orbital energy of the ejected electron (formally: $I_i = -\varepsilon_i$). That is, we can identify the ionization energy with the energy of the orbital from which it is ejected. However, the theorem is only an approximation because it ignores the fact that the remaining electrons rearrange their distributions when ionization occurs.

The second refinement is that the ejection of an electron may leave an ion in a vibrationally excited state. Then not all the excess energy of the photon appears as kinetic energy of the photoelectron, and we should write

$$hv = \tfrac{1}{2}m_e v^2 + I_i + E_{vib}^+ \tag{6}$$

where E_{vib}^+ is the energy used to excite the ion into vibration. Each vibrational quantum that is excited leads to a different kinetic energy of the photoelectron, and gives rise to the **vibrational structure** in the photoelectron spectrum.

The ionization energies of molecules are several electronvolts even for valence electrons, so it is essential to work in at least the ultraviolet region of the spectrum and with wavelengths of less than about 200 nm. A great deal of work has been done with radiation generated by a discharge through helium: the He(I) line ($1s^1 2p^1 \rightarrow 1s^2$) lies at 58.43 nm, corresponding to a photon energy of 21.22 eV. Its use gives rise to the technique of **ultraviolet photoelectron spectroscopy** (UPS). If the core electrons are being studied, then photons of even higher energy are needed to expel them: X-rays are used, and the technique is denoted XPS. A modern version of PES makes use of synchrotron radiation (Section 16.1) which may be continuously tuned between UV and X-ray energies. The additional information that stems from the variation of the photoejection probability with wavelength is a valuable guide to the identity of the element and the orbital from which photoionization occurs.

Example 17.4 *Interpreting a photoelectron spectrum*

Photoelectrons ejected from N_2 with He(I) radiation had kinetic energies of 5.63 eV. What is their ionization energy?

Method. We use eqn 5 with 1 eV = 8065.5 cm^{-1}.

Answer. Helium(I) radiation of wavelength 58.43 nm has wavenumber 1.711×10^5 cm^{-1} and therefore corresponds to an energy of 21.22 eV. Then, from eqn 5,

$$21.22 \text{ eV} = 5.63 \text{ eV} + I_i$$

so $I_i = 15.59$ eV.

Comment. This ionization energy (corresponding to 1504 kJ mol^{-1}) is the energy needed to remove an electron from the HOMO of the N_2 molecule, the $3\sigma_g$ bonding orbitals (see Fig. 14.27).

Exercise E17.4. Under the same circumstances, 4.53-eV electrons are also detected. To what ionization energy does that correspond? Suggest an origin.
[16.7 eV, $1\pi_u$]

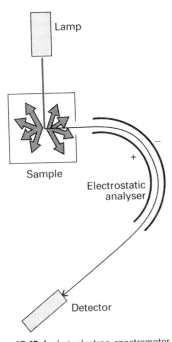

17.40 A photoelectron spectrometer consists of a source of ionizing radiation (such as a helium discharge lamp for UPS and an X-ray source for XPS), an electrostatic analyser, and an electron detector. The deflection of the electron path caused by the analyser depends on the speed of the electrons.

The kinetic energies of the photoelectrons are measured using an electrostatic deflector (Fig. 17.40) which produces different deflections in the paths of the photoelectrons as they pass between charged plates. As the field strength is increased, electrons of different speeds, and therefore kinetic energies, reach the detector. Thus the electron flux can be recorded and plotted against kinetic energy to obtain the photoelectron spectrum. Present techniques give energy resolutions of about 5 meV, corresponding to about 40 cm^{-1}.

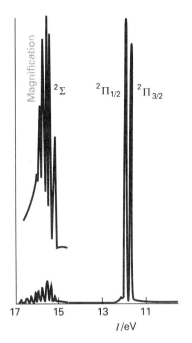

17.41 The photoelectron spectrum of HBr. The lowest ionization energy bands (Π) correspond to the ionization of a Br lone-pair electron. The higher ionization energy band (Σ) corresponds to the ionization of a bonding electron. The structure on the latter is due to the vibrational excitation of HBr⁺ that results from the ionization.

17.9 Ultraviolet photoelectron spectroscopy

A typical photoelectron spectrum (of HBr) is shown in Fig. 17.41. If we disregard the fine structure, we see that the HBr lines fall into two main groups. The least tightly bound electrons (with the lowest ionization energies and hence highest kinetic energies when ejected) are those in the non-bonding lone pairs of Br (with $I = 11.88$ eV). The next ionization energy lies at 15.2 eV, and corresponds to the removal of an electron from the H—Br σ bond.

The HBr spectrum shows that ejection of a σ electron is accompanied by a long vibrational progression. The Franck–Condon principle would account for this progression if ejection were accompanied by an appreciable change of equilibrium bond length between HBr and HBr⁺ because the ion is formed in a bond-compressed state, which is consistent with the important bonding effect of the σ electrons. The lack of much vibrational structure in the two bands labelled $^2\Pi$ is consistent with the non-bonding role of the Br$2p\pi$ lone-pair electrons, for the equilibrium bond length is little changed when one is removed.

Example 17.5 *Interpreting a UV photoelectron spectrum*

The highest kinetic-energy electrons in the spectrum of H_2O using 21.22-eV He radiation are at about 9 eV and show a large vibrational spacing of 0.41 eV. The symmetric stretching mode of the neutral H_2O molecule lies at 3652 cm⁻¹. What conclusions can be drawn from the nature of the orbital from which the electron is ejected?

Method. We need to interpret the vibrational fine structure, which indicates the vibrational characteristics of the ion, in relation to the information about the vibrational characteristics of the neutral molecule.

Answer. Because 0.41 eV corresponds to 3310 cm⁻¹, which is similar to the 3652 cm⁻¹ of the non-ionized molecule, we can suspect that the electron is ejected from an orbital that has little influence on the bonding in the molecule. That is, photoejection is from a largely non-bonding orbital.

Exercise E17.5. In the same spectrum of H_2O, the band near 7.0 eV shows a long vibrational series with spacing 0.125 eV. The bending mode of H_2O lies at 1596 cm⁻¹. What conclusions can you draw about the characteristics of the orbital occupied by the photoelectron?

[The electron contributed to non-neighbour H—H bonding]

17.10 X-ray photoelectron spectroscopy

In XPS, the energy of the incident photon is so great that electrons are ejected from inner cores of atoms. As a first approximation, we would not expect core ionization energies to be sensitive to the bonds between atoms because they are too tightly bound to be greatly affected by the changes that accompany bond formation. This turns out to be largely true, and the core ionization energies are characteristic of the individual atom rather than the overall molecule. Consequently, XPS gives lines characteristic of the elements present in a compound or alloy. For instance, the K-shell ionization energies of the second row elements are

Li	Be	B	C	N	O	F
50	110	190	280	400	530	690 eV

Detection of one of these values (and values corresponding to ejection from other inner shells) indicates the presence of the corresponding element. Figure 17.42, for example, shows the result of an X-ray photoelectron analysis of lunar rock brought back on an Apollo Mission. This chemical analysis application is responsible for the alternative name, **electron spectroscopy for chemical analysis** or ESCA. The technique is mainly limited to the study of surface layers (as we shall explore in Chapter 28) because, even though X-rays may penetrate into the bulk sample, the ejected electrons cannot escape except from within a few nanometers of the surface. Despite (or because of) this limitation, the technique is very useful for studying the surface state of heterogeneous catalysts, the differences between surface and bulk structures, and the processes that can cause damage to high-temperature superconductors and semiconductor wafers.

Whereas it is largely true that core ionization energies are unaffected by bond formation, it is not entirely true, and small shifts can be detected and interpreted in terms of the environments of the atoms. For example, the azide ion N_3^- gives the spectrum shown in Fig. 17.43. Although the spectrum lies in the region of 400 eV (and hence is typical of N$1s$ electrons), it has a doublet structure with splitting 6 eV. This splitting can be understood by noting that the structure of the ion is $\ddot{N}=N=\ddot{N}$,

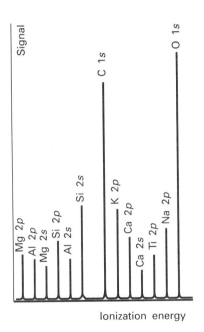

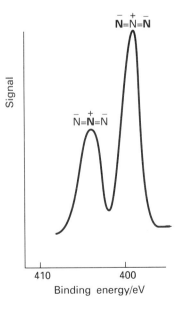

17.42 (left) The X-ray photoelectron spectrum of a sample of moon dust, with the assignment. (Adapted from *Physical methods and molecular structure*, The Open University Press (1977).)

17.43 (right) The photoelectron spectrum of solid NaN₃ excited by Al K-radiation showing the region of N core ionization and the assignment. (K. Siegbahn *et al.*, *Science*, **176**, 245 (1972).)

17.42   **17.43**

with charge distribution $(-, +, -)$. Hence the presence of the negative charges on the terminal atoms lowers the core ionization energies, while the positive charge on the central atom raises it. This inequivalence of the atoms results in two lines in the spectrum with intensities in the ratio 2:1. Observations like this can be used to obtain valuable information about the presence of chemically inequivalent atoms of the same element.

CHECK LIST OF KEY IDEAS

1 The use of the **Franck–Condon principle** to account for the vibrational structure of electronic transitions and the concept of a **vertical transition** (Section 17.1).

2 The **Laporte selection rule** and the **vibronic character** of *d–d* transitions in complexes (Section 17.2).

3 **Charge-transfer transitions** and $\pi^* \leftarrow \pi$ and $\pi^* \leftarrow n$ transitions (Section 17.2).

4 The mechanisms of **fluorescence** and **phosphorescence** and the characteristics of a **fluorescence spectrum** (Section 17.3).

5 The mechanism of **intersystem crossing** (Section 17.3) and **internal conversion** leading to **predissociation** (Section 17.4).

6 The principles of **laser action**, including **population inversion**, **pumping**, and the difference between **three-level** and **four-level** lasers (Section 17.5).

7 The characteristics of laser radiation and the formation of pulses by *Q*-**switching** and **mode locking** (Section 17.5).

8 Examples of practical lasers, including **solid-state lasers**, **gas lasers**, **ion lasers**, **chemical lasers**, **exciplex lasers**, **dye lasers**, and **semiconductor lasers** (Section 17.6).

9 The applications of lasers in chemistry, particularly to **multiphoton spectroscopy**, **laser-Raman spectroscopy**, **precision state-selection**, and **fast reactions** (Section 17.7).

10 The techniques of **ultraviolet photoelectron spectroscopy** and **X-ray photoelectron spectroscopy** (Section 17.8).

EXERCISES

17.1. The molar absorption coefficient of a substance dissolved in hexane is known to be $855 \, \text{L mol}^{-1} \text{cm}^{-1}$ at 270 nm. Calculate the percentage reduction in intensity when light of that wavelength passes through 2.5 mm of a solution of concentration $3.25 \times 10^{-3} \, \text{mol L}^{-1}$.

17.2. A solution of an unknown component of a biological sample when placed in an absorption cell of 1.00 cm path length transmits 20.1 per cent of light of 340 nm incident upon it. If the concentration of the component is $1.11 \times 10^{-4} \, \text{mol L}^{-1}$, what is the molar absorption coefficient?

17.3. When light of wavelength 400 nm passes through 3.5 mm of a solution of an absorbing substance at a concentration of $6.67 \times 10^{-4} \, \text{mol L}^{-1}$, the transmission is 65.5 per cent. Calculate the molar absorption coefficient

of the solute at this wavelength and express the answer in $\text{cm}^2 \text{mol}^{-1}$.

17.4. The molar absorption coefficient of a solute at 540 nm is $286 \, \text{L mol}^{-1} \text{cm}^{-1}$. When light of that wavelength passes through a 6.5-mm cell containing a solution of the solute, 46.5 per cent of the light was absorbed. What is the concentration of the solution?

17.5. The absorption associated with a particular transition begins at $43\,480 \, \text{cm}^{-1}$, peaks sharply at $38\,460 \, \text{cm}^{-1}$, and ends at $34\,480 \, \text{cm}^{-1}$. The maximum value of the molar absorption coefficient is $1.21 \times 10^4 \, \text{L mol}^{-1} \text{cm}^{-1}$. Estimate the integrated absorption coefficient of the transition assuming a triangular lineshape (use eqn 16.7).

17.6. An excited state of the Mn^{2+} ion has the configuration $t_2^2 e^3$. Is the transition from the ground state spin-allowed?

17.7. The two compounds, 2,3-dimethyl-2-butene and 2,5-dimethyl-2,4-hexadiene, are to be distinguished by their ultraviolet absorption spectra. The maximum absorption in one compound occurs at 192 nm and in the other at 243 nm. Match the maxima to the compounds and justify the assignment.

17.8. The compound $CH_3CH{=}CHCHO$ has a strong absorption in the ultraviolet at 46 950 cm^{-1} and a weak absorption at 30 000 cm^{-1}. Justify these features in terms of the structure of the compound.

17.9. The photoionization of H_2 by 21-eV photons produces H_2^+. Explain why the intensity of the $v = 2 \leftarrow 0$ transition is stronger than that of the $0 \leftarrow 0$ transition.

17.10. The following data were obtained for the absorption by Br_2 in carbon tetrachloride using a 2.0-mm cell. Calculate the molar absorption coefficient of bromine at the wavelength employed:

$[Br_2]/(mol\,L^{-1})$	0.0010	0.0050	0.0100	0.0500
T/per cent	81.4	35.6	12.7	3.0×10^{-3}

17.11. A 2.0-mm cell was filled with a solution of benzene in a non-absorbing solvent. The concentration of the benzene was 0.010 mol L^{-1} and the wavelength of the radiation was 256 nm (where there is a maximum in the absorption). Calculate the molar absorption coefficient of benzene at this wavelength given that the transmission was 48 per cent. What will the transmittance be in a 4.0-mm cell at the same wavelength?

17.12. A swimmer enters a gloomier world (in one sense) on diving to greater depths. Given that the mean molar absorption coefficient of sea water in the visible region is $6.2 \times 10^{-5}\,L\,mol^{-1}\,cm^{-1}$, calculate the depth at which a diver will experience (a) half the surface intensity of light, (b) one-tenth the surface intensity.

17.13. The electronic absorption bands of many molecules in solution have half-widths at half-height of about 5000 cm^{-1}. Estimate the integrated absorption coefficients of bands for which (a) $\varepsilon_{max} \approx 1 \times 10^4\,L\,mol^{-1}\,cm^{-1}$, (b) $\varepsilon_{max} \approx 5 \times 10^2\,L\,mol^{-1}\,cm^{-1}$.

PROBLEMS
Numerical problems

17.1. The vibrational wavenumber of the oxygen molecule in its electronic ground state is 1580 cm^{-1}, whereas that in the first excited state ($B^3\Sigma_u^-$), to which there is an allowed electronic transition, is 700 cm^{-1}. If the separation in energy between the minima in their respective potential energy curves of these two electronic states is 6.175 eV, what is the wavenumber of the lowest energy transition in the band of transitions originating from the $v = 0$ vibrational state of the electronic ground state ($X^3\Sigma_g^-$) to this excited state? Ignore any rotational structure or anharmonicity.

17.2. A Birge–Sponer extrapolation yields 7760 cm^{-1} as the area under the curve for the B state of the oxygen molecule described in Problem 17.1. Given that the B state dissociates to ground-state atoms (at zero energy, ^{3}P) and atoms at 15 870 cm^{-1} (^{1}D) and that the lowest vibrational state of the B state is 49 363 cm^{-1} above the lowest vibrational state of the ground electronic state, calculate the dissociation energy of the molecular ground state to the ground-state atoms (^{3}P).

17.3. The electronic spectrum of the IBr molecule shows two low-lying, well-defined convergence limits at 14 660 and 18 345 cm^{-1}. Energy levels for the iodine and bromine atoms occur at 0, 7598 cm^{-1} and 0, 3685 cm^{-1}, respectively. Other atomic levels are at much higher energies. What possibilities exist for the numerical value of the dissociation energy of IBr? Decide which is the correct possibility by calculating this quantity from $\Delta_f H^{\ominus}$(IBr, g) $= +40.79$ kJ mol^{-1} and the dissociation energies of I_2(g) and Br_2(g), which are 146 and 190 kJ mol^{-1}, respectively.

17.4. In many cases it is possible to assume that an absorption band has a Gaussian line shape (one proportional to e^{-x^2}) centred on the band maximum. Assume such a line shape, and show that

$$\mathscr{A} = 1.0645 \varepsilon_{max}\Delta\tilde{\nu}_{\frac{1}{2}}$$

where $\Delta\tilde{\nu}_{\frac{1}{2}}$ is the width at half-height. The absorption spectrum of azoethane ($CH_3CH_2N_2$) between 24 000 and 34 000 cm^{-1} is shown in Fig. 17.44. First, estimate A for the band by assuming that it is Gaussian. Then integrate the absorpion band graphically. (The latter can be done either by ruling and counting squares, or by tracing the line shape on to paper and weighing.)

17.5. A lot of information about the energy levels and wavefunctions of small inorganic molecules can be obtained from their ultraviolet spectra. An example of a spectrum with considerable vibrational structure, that of gaseous SO_2 at 25°C, is shown in Fig. 17.45. Estimate the integrated absorption coefficient for the transition. What

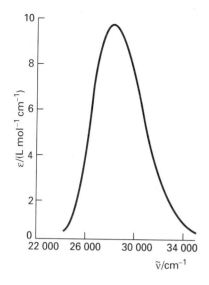

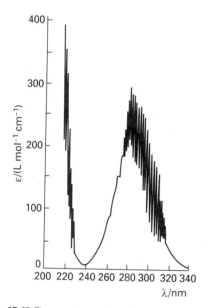

17.45 The gas-phase absorption spectrum of SO₂.

electronic states are accessible from the A_1 ground state of this C_{2v} molecule by electric dipole transitions?

17.6. A transition of particular importance in O_2 gives rise to the 'Schumann–Runge band' in the ultraviolet region. The wavenumbers (in cm⁻¹) of transitions from the ground state to the vibrational levels of the first excited state ($^3\Sigma_u^-$) are 49 357.6, 50 045.6, 50 710.7, 51 352.2, 51 969.8, 52 561.6, 53 122.8, 53 656.8, 54 158.9, 54 624.4, 55 053.3, 55 441.5, 55 784.6, 56 085.5, 56 340.5. What is the dissociation energy of the upper electronic state? (Use a Birge–Sponer plot.) The same excited state is known to dissociate into one ground-state O atom and one excited-state atom with an

energy 190 kJ mol⁻¹ above the ground state. (This excited atom is responsible for a great deal of photochemical mischief in the atmosphere.) Ground state O_2 dissociates into two ground-state atoms. Use this information to calculate the dissociation energy of ground-state O_2 from the Schumann–Runge data.

17.7. A certain molecule fluoresces at a wavelength of 400 nm with a half-life of 1.0×10^{-9} s. It phosphoresces at 500 nm. If the ratio of the transition probabilities for stimulated emission for the $S^* \rightarrow S$ to the $T \rightarrow S$ transitions is 10^5, what is the half-life of the phosphorescent state?

17.8. The photoelectron spectra of N_2 and CO are shown in Fig. 17.46. Ascribe the lines to the ionization processes involved and classify the orbitals from which the electrons are ejected as bonding, non-bonding, or antibonding in light of the extent of vibrational structure in the band. Analyse the bands near 4 eV in terms of the vibrational energy levels of the ions.

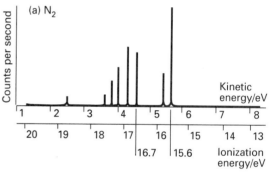

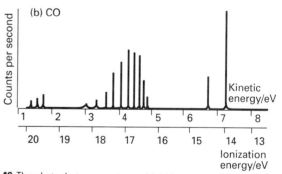

17.46 The photoelectron spectrum of (a) N_2 and (b) CO.

17.9. The photoelectron spectrum of NO can be described as follows (D. W. Turner in *Physical methods in advanced inorganic chemistry* (ed. H. A. O. Hill and P. Day), Wiley, Chichester, (1968)). Using He 584 pm (21.21 eV) radiation there is a single strong peak at kinetic energy 4.69 eV and a long series of 24 lines starting at 5.56 eV and ending at 2.2 eV. A shorter series of six lines begins at 12.0 eV and ends at 10.7 eV. Account for this spectrum.

Theoretical problems

17.10. Assume that the electronic states of the π electrons of a conjugated molecule can be approximated by the wave functions of a particle in a one-dimensional box, and that the dipole moment can be related to the displacement along this length by $\mu = -ex$. Show that the transition probability connecting states 1 and 2 is non-zero, whereas that connecting states 1 and 3 is zero.

17.11. Use a group theoretical argument to decide which of the following transitions are electric-dipole allowed: (a) the $\pi^* \leftarrow \pi$ transition in ethene, (b) the $\pi^* \leftarrow n$ transition in a carbonyl group in a C_{2v} environment.

17.12. The line marked A in Fig. 17.47 is the fluorescence spectrum of benzophenone in solid solution in ethanol at low temperatures observed when the sample is illuminated with 360-nm light. What can be said about the vibrational energy levels of carbonyl group in (a) its ground electronic state and (b) its excited electronic state? When naphthalene is illuminated with 360-nm light it does not absorb, but the line marked B in the illustration is the phosphorescence spectrum of a solid solution of a mixture of naphthalene and benzophenone in ethanol. Now a component of fluorescence from naphthalene can be detected. Account for this observation.

17.13. The fluorescence spectrum of anthracene vapour shows a series of peaks of increasing intensity with individual maxima at 440, 410, 390, and 370 nm followed by a sharp cut-off at shorter wavelengths. The absorption spectrum rises sharply from zero to a maximum at 360 nm with a trail of peaks of lessening intensity at 345, 330, and 305 nm. Account for these observations.

17.14. One measure of the intensity of a transition of frequency v is the oscillator strength f, which is defined as

$$f = \frac{8\pi^2 m_e v}{3he^2} |\boldsymbol{\mu}_{\mathrm{fi}}|^2$$

Consider an electron in an atom to be oscillating harmonically in one dimension (the three-dimensional version of this model was used in early attempts to describe atomic structure). The wavefunctions for such an electron are those in Table 12.1. Show that the oscillator strength for the transition of this electron from its ground state is exactly $\frac{1}{3}$.

17.15. Estimate the oscillator strength (see Problem 17.14) of a charge-transfer transition modelled as the migration of an electron from a hydrogen $1s$ orbital on one atom to another hydrogen $1s$ orbital on an atom a distance R away. Approximate the transition moment by $-eRS$ where S is the overlap integral of the two orbitals. Sketch the oscillator strength as a function of R using the curve for S given in Fig. 14.32. Why does the intensity fall to zero as R approaches 0 and infinity?

17.16. Suppose that you are a colour chemist and had been asked to intensify the colour of a dye without changing the type of compound, and that the dye in question was a polyene. Would you choose to lengthen or to shorten the chain? Would the modification to the length shift the apparent colour of the dye towards the red or the blue?

17.17. Spin angular momentum is conserved when a molecule dissociates into atoms. What atom multiplicities are permitted when (a) an O_2 molecule, (b) an N_2 molecule dissociates into atoms?

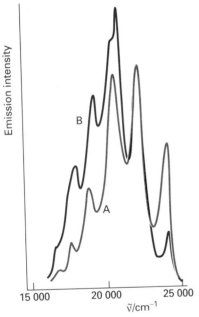

17.47 The phosphorescence spectra of naphthalene and benzophenone.

18

Spectroscopy 3: magnetic resonance

One of the most widely used spectroscopic procedures in chemistry makes use of the classical concept of resonance. In this technique, an array of energy levels is adjusted by the application of a magnetic field until the energy separations match the frequency of an electromagnetic field, and a strong absorption is observed.

The chapter begins with an account of conventional nuclear magnetic resonance (NMR) which shows how the resonance frequency of a magnetic nucleus is affected by its electronic environment and the presence of magnetic nuclei in its vicinity. Then we turn to the modern versions of NMR, which are based on the use of pulses of electromagnetic radiation and the processing of the resulting signal by Fourier transform techniques. Apart from their much greater sensitivity, Fourier transform NMR techniques give rise to procedures for simplifying spectra and obtaining dynamical information from them by an analysis of spin relaxation times. They are also applicable to solid-state samples, where the greatest problem is the linewidth.

Much less widely used than NMR is the analogous technique of electron spin resonance. Here the experimental techniques still resemble those used in the early days of NMR. The information obtained from ESR, however, is very useful for the determination of the properties of radicals and d-metal complexes.

When two pendulums share a slightly flexible support and one is set in motion, the other is forced into oscillation by the motion of the common axle. As a result, energy flows between the two pendulums. The energy transfer occurs most efficiently when the frequencies of the two pendulums are identical. The condition of strong effective coupling when the frequencies of two oscillators are identical is called **resonance**, and the excitation energy is said to **resonate** between the coupled oscillators.

Resonance is the basis of a number of everyday phenomena, including the response of radios to the weak oscillations of the electromagnetic field generated by a distant transmitter. In this chapter we explore some spectroscopic applications that, as originally developed (and in some cases still), depend on matching a set of energy levels to a source of monochromatic radiation and observing the strong absorption that occurs at resonance.

NUCLEAR MAGNETIC RESONANCE

Many nuclei possess spin angular momentum. A nucleus with spin quantum number I (which may be an integer or a half-integer and is never negative) has the following properties:

1. An angular momentum of magnitude $\sqrt{\{I(I+1)\}}\hbar$
2. A component of angular momentum $m_I\hbar$ around an arbitrary axis, where

$$m_I = I, I-1, \ldots, -I$$

3. If $I > 0$, a magnetic moment with a constant magnitude and an orientation that is determined by the value of m_I.

To say that a nucleus has a magnetic moment means that, to some extent, it behaves like a small bar magnet.

According to the second property, the spin, and hence the magnetic moment, of the nucleus may lie in $2I+1$ different orientations relative to the axis. A proton has $I = \frac{1}{2}$ and its spin may adopt either of two orientations; a ^{14}N nucleus has $I = 1$ and its spin may adopt any of three orientations. For much of this chapter we shall consider **spin-$\frac{1}{2}$ nuclei**, which are nuclei with $I = \frac{1}{2}$. As well as protons, spin-$\frac{1}{2}$ nuclei include ^{13}C, ^{19}F, and ^{31}P nuclei. The state with $m_I = +\frac{1}{2}(\uparrow)$ is denoted α and the state with $m_I = -\frac{1}{2}(\downarrow)$ is denoted β. It is worth bearing in mind that two very common nuclei, ^{12}C and ^{16}O, have zero spin, and hence zero magnetic moment, and so are invisible in magnetic resonance.

18.1 The energies of nuclei in magnetic fields

The nuclear magnetic moment of a nucleus is denoted μ (the same symbol is also used for the electric dipole moment of a molecule, but there is little chance of confusion). The component of the nuclear magnetic moment on the z-axis, μ_z, is proportional to the component of spin angular momentum on that axis $m_I\hbar$, and we write

$$\mu_z = \gamma\hbar m_I \tag{1}$$

The coefficient of proportionality γ is called the **magnetogyric ratio** of the nucleus. Theories of nuclear structure are not yet sufficiently advanced for γ to be calculated reliably, and it is treated as an empirical quantity (Table 18.1). The magnetic moment is often expressed in terms of the **nuclear g-factor** g_I and the **nuclear magneton** μ_N by using

$$\gamma\hbar = g_I\mu_N \qquad \mu_N = \frac{e\hbar}{2m_p} = 5.051 \times 10^{-27}\,\text{J T}^{-1} \tag{2}$$

where m_p is the mass of the proton. Nuclear g-factors are numbers of the order of 1 (Table 18.1): positive values of g denote a magnetic moment that is parallel to the spin, negative values indicate that the magnetic

Table 18.1* Nuclear spin properties

Nuclide	Natural abundance/%	Spin I	g-value g_I
^{1}n		$\frac{1}{2}$	−3.826
^{1}H	99.98	$\frac{1}{2}$	5.586
^{2}H	0.02	1	0.857
^{13}C	1.11	$\frac{1}{2}$	1.405
^{14}N	99.64	1	0.404

* More values are given in the Data section at the end of this volume.

moment and spin are antiparallel. The nuclear magneton is about 2000 times smaller than the Bohr magneton, so nuclear magnetic moments are about 2000 times weaker than the electron spin magnetic moment.

The basic resonance experiment

Each value of m_I corresponds to a different orientation of the nuclear spin and therefore of the nuclear magnetic moment too. In a magnetic field B, the $2I + 1$ orientations of the nucleus have different energies, which are given by

$$E_{m_I} = -\mu_z B = -\gamma \hbar B m_I \tag{3a}$$

These energies are often expressed in terms of the **Larmor frequency** ν_L:

$$E_{m_I} = -m_I h \nu_L, \qquad \text{where } \nu_L = \frac{\gamma B}{2\pi} \tag{3b}$$

The stronger the magnetic field, the higher the Larmor frequency. A field of 12 T corresponds to a Larmor frequency of about 500 MHz for protons.[1]

The energy separation of the two states of spin-$\frac{1}{2}$ nuclei (Fig. 18.1) is

$$\Delta E = E_\beta - E_\alpha = \tfrac{1}{2}\gamma \hbar B - (-\tfrac{1}{2}\gamma \hbar B) = \gamma \hbar B = h \nu_L \tag{3c}$$

For most nuclei γ is positive (and $g_I > 0$). In such cases, the β state lies above the α state, and there are slightly more α spins than β spins. If the sample is bathed in radiation of frequency ν, the energy separations come into resonance with the radiation when the frequency satisfies the **resonance condition**:

$$h\nu = \gamma \hbar B = h \nu_L$$

That is, there is resonance when $\nu = \nu_L$ and the radiation has the Larmor frequency. At resonance there is strong coupling between the nuclear spins and the radiation, and strong absorption occurs as the spins make the transition $\beta \leftarrow \alpha$. At 12 T, protons come into resonance at about 500 MHz (the Larmor frequency at that magnetic field).

The technique

In its simplest form, **nuclear magnetic resonance** (NMR) is the study of the properties of molecules containing magnetic nuclei by applying a magnetic field and observing the frequency at which they come into resonance with an oscillating electromagnetic field. Larmor frequencies of nuclei at the fields normally employed typically lie in the radiofrequency region of the electromagnetic spectrum, so NMR is a radiofrequency technique. When applied to proton spins, the technique is called **proton magnetic resonance** (^{1}H-NMR). In the early days of the technique, only protons and ^{19}F nuclei (which have relatively large

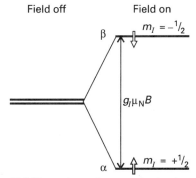

18.1 The nuclear spin energy levels of a spin-$\frac{1}{2}$ nucleus (e.g. ^{1}H or ^{13}C) in a magnetic field. Resonance occurs when the energy separation of the levels matches the energy of the photons in the electromagnetic field.

---

1 Throughout this chapter, the strength of the magnetic field, more formally the magnetic induction, will be expressed in tesla, T, where $1\,\text{T} = 1\,\text{kg s}^{-2}\,\text{A}^{-1}$. In older systems of units, the magnetic induction is expressed in gauss, G, and $1\,\text{T} = 10^4\,\text{G}$.

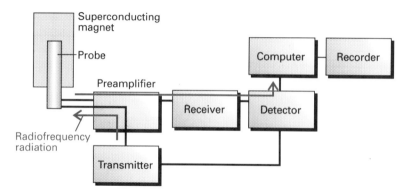

18.2 The layout of a typical NMR spectrometer. The link from the transmitter to the detector indicates that the high frequency of the transmitter is subtracted from the high-frequency received signal to give a low-frequency signal for processing in the computer.

magnetic moments) could be studied, but now a wide variety of nuclei (especially ^{13}C and ^{31}P) are investigated routinely.

An NMR spectrometer consists of a magnet that can produce a uniform, intense field and the appropriate sources of radiofrequency electromagnetic radiation. In simple instruments, the magnetic field is provided by an electromagnet or a permanent magnet. For serious work, a superconducting magnet capable of producing fields of the order of 10 T and more is used (Fig. 18.2). The sample is placed in the cylindrically wound magnet, and is rotated at about 15 Hz. The spinning helps to remove magnetic inhomogeneities and to ensure that all the magnetic nuclei experience the same average field. Although a superconducting magnet operates at the temperature of liquid helium (4 K), the sample itself is normally at room temperature or somewhat below.

The use of high magnetic fields has several advantages. The most important advantage is that it simplifies the appearance of spectra and so allows them to be interpreted more readily. We shall see how this simplification is achieved in Section 18.3. A further advantage is that the rate of energy uptake by the sample is greater in a high field. There are two contributions to this increase. One comes from the greater population difference between the α and β spin states at high fields, for the population difference is approximately proportional to B. The second contribution stems from the greater energy of each absorbed photon, which is also proportional to B. It follows that overall the signal is proportional to B^2.

JUSTIFICATION

According to the Boltzmann distribution, the ratio of populations is

$$\frac{N_\beta}{N_\alpha} = e^{-\Delta E/kT} \approx 1 - \frac{\Delta E}{kT}$$

It follows that

$$\frac{N_\alpha - N_\beta}{N_\alpha + N_\beta} \approx \frac{\Delta E}{2kT} = \frac{\gamma \hbar B}{2kT}$$

and the population difference is proportional to B. The energy of the photon absorbed when a nucleus makes a transition from its lower state to its higher state is $h\nu$; at resonance, ν is equal to ν_L and ν_L is proportional to B. Hence, at resonance each photon has an energy that is proportional to B.

18.2 The chemical shift

Nuclear magnetic moments interact with the *local* magnetic field. The local field may differ from the applied field because the latter induces electronic orbital angular momentum (that is, the circulation of electronic currents) which gives rise to a small additional magnetic field δB at the nuclei. This additional field is proportional to the applied field, and it is conventional to express it as

$$\delta B = -\sigma B \tag{4}$$

where the dimensionless quantity σ is called the **shielding constant** for the nucleus of interest (σ is usually positive but may be negative). The ability of the applied field to induce an electronic current in the molecule, and the strength of the resulting local magnetic field experienced by the nucleus, depend on the details of the electronic structure near the magnetic nucleus of interest, and nuclei in different chemical groups have different shielding constants. The calculation of reliable values of the shielding constant is very difficult, but trends in it are quite well understood and we shall concentrate on this aspect.

The δ scale of chemical shifts

Because the total local field is

$$B_{\mathrm{loc}} = B + \delta B = (1 - \sigma)B$$

The Larmor frequency is

$$\nu_L = \frac{\gamma B_{\mathrm{loc}}}{2\pi} = (1 - \sigma)\frac{\gamma B}{2\pi}$$

This frequency is different for nuclei in different environments. Hence, different nuclei, even of the same element, come into resonance at different frequencies. Although these different resonance frequencies could be reported in terms of the shielding constants, it is more convenient to express them in terms of an empirical quantity called the chemical shift.

The **chemical shift** of a nucleus is the difference between the resonance frequency of the nucleus in question and that of a reference standard. The standard for protons is the proton resonance in tetramethylsilane ($Si(CH_3)_4$, commonly referred to as TMS), which bristles with protons and dissolves without reaction in many liquids. Other references are used for other nuclei. For ^{13}C, the reference frequency is the ^{13}C resonance in TMS; for ^{31}P it is the ^{31}P resonance in 85 per cent $H_3PO_4(aq)$. The separation of the resonance of a particular nucleus from the standard increases with the strength of the applied magnetic field because the additional local field is proportional to the applied field.

Chemical shifts are reported on the δ **scale**, which is defined as

$$\delta = \frac{\nu - \nu^\circ}{\nu^\circ} \times 10^6 \qquad (5)$$

In this expression, ν° is the resonance frequency of the standard. The advantage of the δ-scale is that shifts reported on it are independent of the applied field (because both numerator and denominator are proportional to the applied field). The resonance frequencies themselves, however, do depend on the applied field through

$$\nu - \nu^\circ = \nu^\circ \delta \times 10^{-6}$$

For example, a nucleus with $\delta = 1.00$ in a spectrometer operating at 500 MHz will resonate when

$$\nu - \nu^\circ = (500 \text{ MHz}) \times (1.00) \times 10^{-6} = 500 \text{ Hz}$$

In a spectrometer operating at 100 MHz, the shift relative to the reference signal would be only 100 Hz.

A positive δ indicates that the resonance frequency of the group of nuclei in question is higher than that of the standard. Hence, $\delta > 0$ indicates that the local magnetic field is stronger than that experienced by the nuclei in the standard under the same conditions. Some typical chemical shifts are given in Fig. 18.3. As can be seen from the illustration, the nuclei of different elements have very different ranges of

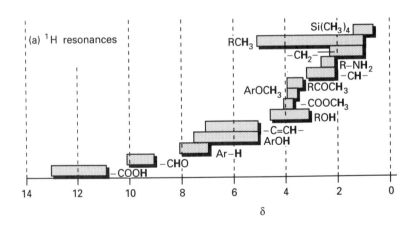

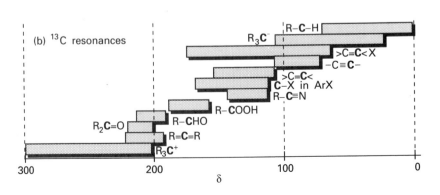

18.3 The range of typical chemical shifts for (a) ^{1}H resonances and (b) ^{13}C resonances.

chemical shifts. The ranges exhibit the variety of electronic environments of the nuclei in molecules. The shielding constant σ is an indication of the shielding or deshielding of a proton in an absolute sense; the value of the chemical shift δ is an indication of the shielding of a nucleus relative to a standard (TMS for protons).

Resonance of different groups of nuclei

The existence of a chemical shift explains the general features of the spectrum of ethanol shown in Fig. 18.4. The CH_3 protons form one group of nuclei with $\delta = 1$. The two CH_2 protons are in a different part of the molecule, experience a different local magnetic field, and hence resonate at $\delta \approx 3$. Finally, the OH proton is in another environment, and has a chemical shift of $\delta = 4$.

The relative intensities of the signal (the areas under the absorption lines) can be used to help distinguish which group of lines corresponds to which chemical group. The determination of the area under an absorption line is referred to as the **integration** of the signal (just as any area under a curve may be determined by mathematical integration). Spectrometers can integrate the absorption automatically (as indicated in Fig. 18.4). In ethanol the group intensities are in the ratio 3:2:1 because there are three CH_3 protons, two CH_2 protons, and one OH proton in each molecule. Counting the number of magnetic nuclei as well as noting their chemical shifts is valuable analytically because it helps us identify a compound present in a sample.

The origin of shielding constants

The calculation of shielding constants is a very difficult problem, even for small molecules, for it requires detailed information about the distribution of electron density in the ground and excited states of the molecule

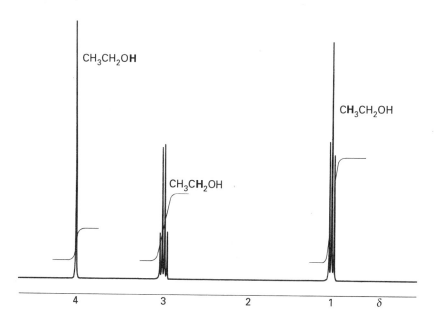

18.4 The ^{1}H-NMR spectrum of ethanol. The bold letters denote the protons giving rise to the resonance peak, and the step-like curve is the integrated signal.

and its excitation energies. Some success has been achieved with the calculation for diatomic molecules and small polyatomic molecules such as H_2O and CH_4, but large molecules are much more difficult. Nevertheless, a considerable body of useful empirical information about a variety of contributions to chemical shifts in large molecules has been compiled, and has been used to understand and interpret observations reasonably systematically.

The empirical approach supposes that the observed shielding constant is the sum of three contributions:

$$\sigma = \sigma(\text{loc}) + \sigma(\text{molecule}) + \sigma(\text{solvent})$$

The **local contribution**, $\sigma(\text{loc})$, is essentially the contribution of the electrons of the atom that contains the nucleus in question. The **molecular contribution**, $\sigma(\text{molecule})$, is the contribution from the groups of atoms that form the rest of the molecule. The **solvent contribution**, $\sigma(\text{solvent})$, is the contribution from the solvent molecules.

The local contribution

It is convenient to regard the local contribution to the shielding constant as the sum of a positive **diamagnetic contribution**, σ_d, and a negative **paramagnetic contribution**, σ_p:

$$\sigma(\text{loc}) = \sigma_d(\text{loc}) + \sigma_p(\text{loc})$$

The total local contribution is positive if the diamagnetic contribution dominates and is negative if the paramagnetic contribution dominates.

The diamagnetic contribution arises from the ability of the applied field to generate a circulation of charge in the ground state electron distribution of the atom. The circulation generates a magnetic field that opposes the applied field and hence shields the nucleus. The magnitude of σ_d depends on the electron density close to the nucleus and can be calculated from the **Lamb formula**:

$$\sigma_d = \frac{e^2 \mu_0}{3 m_e} \int_0^\infty r\rho(r)\, dr \qquad (6)$$

In this expression ρ is the electron probability density ($|\psi|^2$) at a distance r from the nucleus and μ_0 is the vacuum permeability (a fundamental constant, see inside the front cover).

Example 18.1 *Using the Lamb formula*

Calculate the shielding constant for the proton in a free H atom.

Method. To use the Lamb formula, we need to express the electron probability density in terms of the ground state $1s$ wavefunction of the atom. Wavefunctions are given in Table 13.1, and a useful integral is given in Example 11.6.

Answer. The electron density at a distance r from the nucleus of this one-electron atom is equal to ψ^2, where ψ is the $1s$ orbital of the H atom

(Section 13.2):

$$\psi = \left(\frac{1}{\pi a_0^3}\right)^{\frac{1}{2}} e^{-r/a_0}$$

Then, from the Lamb formula,

$$\sigma_d = \frac{e^2 \mu_0}{3\pi m_e a_0^3} \int_0^\infty r e^{-2r/a_0} \, dr = \frac{e^2 \mu_0}{3\pi m_e a_0^3} \times \frac{a_0^2}{4}$$

$$= \frac{e^2 \mu_0}{12\pi m_e a_0}$$

With the values of the fundamental constants inside the front cover, this expression evaluates to 1.78×10^{-5}.

Comment. The shielding constant is inversely proportional to the Bohr radius. This distance dependence can be understood as arising from the classical result that the magnetic moment of a current loop is proportional to its area (which for a hydrogen atom is of the order of a_0^2) and the magnetic field that it generates at the nucleus is inversely proportional to the cube of the latter's distance (a_0^3). Hence, the local field is proportional to $a_0^2 \times 1/a_0^3 = 1/a_0$.

Exercise E18.1. Calculate σ_d for a hydrogenic atom with atomic number Z.

$$[Z\sigma_d(H)]$$

The diamagnetic contribution is the *only* contribution in closed-shell free atoms. It is also the only contribution to the local shielding for distributions of charge that have spherical or cylindrical symmetry. Thus, it is the only contribution to the local shielding from inner cores of atoms, for these remain spherical even though the atom may be a component of a molecule and its valence electron distribution highly distorted.

The diamagnetic contribution is broadly proportional to the electron density of the atom containing the nucleus of interest. It follows that the shielding is decreased if the electron density on the atom is reduced as a result of the influence of an electronegative atom nearby.

The local paramagnetic contribution, σ_p, arises from the ability of the applied field to force the electrons to circulate through the molecule by making use of orbitals that are unoccupied in the ground state. It is zero in free atoms and around the axes of linear molecules (such as $HC{\equiv}CH$) where the electrons can circulate freely and a field applied along the internuclear axis is unable to force them into other orbitals.

The magnitude of the paramagnetic contribution depends on the ease with which the applied field can promote electrons into unoccupied orbitals. Hence, it is inversely proportional to the energy separation of the highest filled and lowest empty orbitals of the molecule, a quantity denoted Δ. The strength of the magnetic field generated by the magnetic moment of the resulting circulation of charge is inversely proportional to the cube of the distance of the nucleus from the circulating current, so the field at the nucleus is proportional to $\langle r^{-3} \rangle$. Overall, therefore,

$$\sigma_p(\text{loc}) \propto -\frac{\langle r^{-3} \rangle}{\Delta}$$

We can therefore expect large paramagnetic contributions from small atoms in molecules with low lying excited states. In fact, it is found that the paramagnetic contribution is the dominant local contribution for atoms other than hydrogen. The shielding constants of the nuclei of d-metal ions in complexes correlate quite well with spectroscopic data if Δ is identified with the ligand field splitting parameter Δ_O.

Neighbouring group contributions

The **neighbouring group contribution** arises from the currents induced in nearby groups of atoms. The effect of either kind of current (diamagnetic or paramagnetic) is to shield or deshield the nucleus depending on the relative location of the nucleus to the neighbouring group.

The applied field generates currents in the electron distribution of the neighbouring group and gives rise to a magnetic moment that is proportional to the applied field; the constant of proportionality is the **magnetic susceptibility** χ of the group. This induced magnetic moment gives rise to a magnetic field at the nucleus with a strength that is inversely proportional to the cube of the distance of the nucleus from the group of atoms. The field varies with the orientation of the molecule, but it does not average to zero because the magnetic susceptibility also changes as the molecule presents different orientations to the applied field. The result is that the shielding constant depends on three quantities: the difference in the magnetic susceptibilities parallel and perpendicular to the group (we are assuming that the group has cylindrical symmetry), the angle θ that the vector to the magnetic nucleus makes to the axis of symmetry of the group, and the distance r of the nucleus from the group (**1**):

$$\sigma(\text{neighbour}) \propto (\chi_\parallel - \chi_\perp)\left(\frac{1 - 3\cos^2\theta}{r^3}\right) \tag{7}$$

This expression shows that the neighbouring group contribution may be positive or negative according to the relative magnitudes of the two magnetic susceptibilities and the position of the nucleus. The latter effect is easy to anticipate: if $54° < \theta < 144°$, then $1 - 3\cos^2\theta$ is positive, but it is negative otherwise.

A $-\text{C}{\equiv}\text{C}-$ group is linear, and an applied field cannot induce a paramagnetic current when it is parallel to the group's axis. The pattern of shielding and deshielding resulting from the diamagnetic current is shown in Fig. 18.5. Protons lying on the axis of the group (as in acetylene itself) are shielded, but a proton that lies perpendicular to the bond (as part of a larger molecule) is deshielded (Fig. 18.5a). The opposite is true for protons near a C=C double bond (Fig. 18.5b) because in this non-linear group the applied field can induce a paramagnetic current when it lies parallel to the axis.

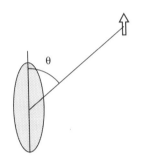

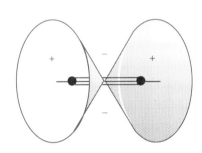

(a)

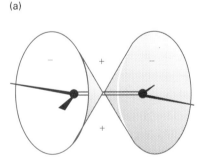

(b)

18.5 The neighbouring group effect in NMR. (a) The protons in HC≡CH are shielded by the currents induced in the triple bond, but a proton perpendicular to the bond is deshielded. (b) The opposite is true for protons near a C=C double bond because the applied field can induce a paramagnetic current parallel to the axis of a double bond.

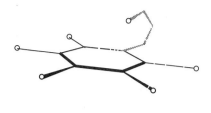

18.6 The shielding and deshielding effects of the ring current induced in the benzene ring by the applied field. Protons attached to the ring are deshielded but a proton attached to a substituent that projects above the ring is shielded.

A special case of a neighbouring group effect is found in aromatic compounds. The strong anisotropy of the magnetic susceptibility of the benzene ring is ascribed to the ability of the field to induce a **ring current**, a circulation of electrons around the ring, when it is applied perpendicular to the molecular plane. Protons in the plane are deshielded (Fig. 18.6), but any that happen to lie above or below the plane (as members of substituents of the ring) are shielded.

Neighbouring groups can also have an indirect *electrical* effect on the chemical shift, for a polar group in a molecule can give rise to an electric field that distorts the electron distribution near a nucleus. The contribution is written

$$\sigma(\text{neighbour}) = -A\mathscr{E}_{\parallel} - B_{\parallel}\mathscr{E}_{\parallel}^2 - B_{\perp}\mathscr{E}_{\perp}^2 \tag{8}$$

where A, $B_{\parallel}$, and $B_{\perp}$ are empirical constants. The terms $\mathscr{E}$ represent the electric field arising from a neighbouring polar group; $\mathscr{E}_{\parallel}$ is the component of the electric field parallel to the bond and $\mathscr{E}_{\perp}$ is the component perpendicular to the bond. The linear term represents the movement of electron density and the consequent reduction in shielding of the nucleus. The quadratic term represents the distortion of the spherical electron distribution of the atom and the consequent appearance of a paramagnetic contribution to the chemical shift.

The solvent contribution

A solvent can influence the local magnetic field experienced by a nucleus in a variety of ways, including effects that arise from specific interactions between the solute and the solvent (such as hydrogen-bond formation and other forms of Lewis acid–base complex formation).

If the solvent is polar, then an effect like the electric field neighbouring group effect can modify the electron distribution near the magnetic nucleus. Even if the solvent molecules are non-polar, a polar solute can induce an electric dipole in them, and the polarized molecules can distort the electron distribution of the solute molecule.

The magnetic susceptibility of the solvent molecules, especially if they are aromatic, can also be the source of a local magnetic field. Moreover, if there are steric interactions that result in a loose but specific interaction between a solute molecule and a solvent molecule, then protons in the solute molecule may experience shielding or deshielding effects according to their location relative to the solvent molecule (Fig. 18.7). We shall see that the NMR spectra of species that contain protons with widely different chemical shifts are easier to interpret than those in which the shifts are similar, so the appropriate choice of solvent may help to simplify the appearance and interpretation of a spectrum.

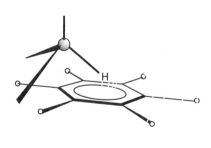

18.7 An aromatic solvent (benzene here) can give rise to local currents that shield or deshield a proton in a solute molecule. In this relative orientation of the solvent and solute, the proton on the solute molecule is shielded.

18.3 The fine structure

The splitting of resonances into individual lines in Fig. 18.4 is called the **fine structure** of the spectrum. It arises because each magnetic nucleus may contribute to the local field experienced by the other nuclei and so modify their resonance frequencies. The strength of the interaction is

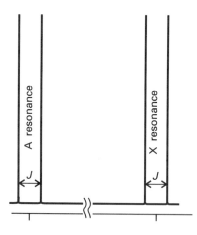

18.8 The effect of spin–spin coupling on an AX spectrum. Each resonance is split into two lines separated by J. The pairs of resonances are centred on the chemical shifts of the protons in the absence of spin–spin coupling.

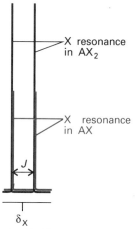

18.9 The X resonance of an AX_2 species is also a doublet, because the two equivalent X nuclei behave like a single nucleus; however, the overall absorption is twice as intense as that of an AX species.

n	Intensity distribution
0	1
1	1 1
2	1 2 1
3	1 3 3 1
4	1 4 6 4 1
.	.
.	.
.	.
n	Expand $(1 + x)^n$ and select the coefficients

expressed in terms of the **scalar coupling constant** J and reported in hertz (Hz).[2] Coupling constants are independent of the strength of the applied field because they do not depend on the latter's ability to generate local fields. If the resonance line of a particular nucleus is split by a certain amount by a second nucleus, then the resonance line of the second nucleus is split by the first to the same extent.

Patterns of coupling

In NMR, letters far apart in the alphabet (typically A and X) are used to indicate nuclei with very different chemical shifts; letters close together (such as A and B) are used for nuclei with similar chemical shifts. We shall consider first an AX system, a molecule that contains two spin-$\frac{1}{2}$ nuclei A and X with very different chemical shifts.

Suppose the spin of X is α; then the spin of A will have a Larmor frequency as a result of the combined effect of the external field, the shielding constant, and the spin–spin interaction of the nucleus A with X. The spin–spin coupling will result in one line in the spectrum of A being shifted by $\frac{1}{2}J$ from the frequency it would have in the absence of coupling. If the spin of X is β, then the spin of A will have a Larmor frequency shifted by $-\frac{1}{2}J$. Therefore, instead of a single line from A, we get a doublet of lines separated by J (Fig. 18.8) and centred on the chemical shift characteristic of A (δ_A in the illustration). The same splitting occurs in the X resonance: instead of a single line it is a doublet with splitting J (the same value as for the splitting of A) centred on the chemical shift characteristic of X (δ_X in the illustration).

A subtle point is that the X resonance in an AX_n species (such as an AX_2 or AX_3 species) is also a doublet with splitting J. As we shall explain below, a *group of equivalent nuclei resonates like a single nucleus*. The only difference for the X resonance of an AX_n species is that the intensity is n times as great as that of an AX species (Fig. 18.9). The A resonance in an AX_n species, though, is quite different from the A resonance in an AX species. For example, consider an AX_2 species with two equivalent X nuclei. The resonance of A is split into a doublet of separation J by one X, and each line of that doublet is split again by the same amount by the second X (Fig. 18.10). This splitting results in three lines in the intensity ratio 1:2:1 (because the central frequency can be obtained in two ways). The A resonance of an A_nX_2 species would also be a 1:2:1 triplet of splitting J, the only difference being that the intensity of the A resonance would be n times as great as that of AX_2.

Three equivalent X nuclei (an AX_3 species) split the resonance of A into four lines of intensity ratio 1:3:3:1 and separation J (Fig. 18.11). The X resonance, though, is still a doublet of separation J. In general, N equivalent spin-$\frac{1}{2}$ nuclei split the resonance of a nearby spin or group of equivalent spins into $n + 1$ lines with an intensity distribution given by **Pascal's triangle** (shown in the margin). Subsequent rows of the triangle

2 The scalar coupling constant is so called because the interaction it describes is proportional to the scalar product of the two interacting spins: $E \propto \mathbf{I}_1 \cdot \mathbf{I}_2$. The constant of proportionality in this expression is J. (More precisely, it is $hJ/\hbar^2$, because each angular momentum is proportional to $\hbar$.)

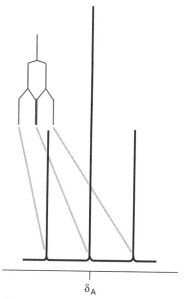

18.10 The origin of the 1:2:1 triplet in the A resonance of an AX$_2$ species. The resonance of A is split into two by coupling with one X nucleus (as shown in the inset), and then each of those two lines is split into two by coupling to the second X nucleus. Because each X nucleus causes the same splitting, the two central transitions are coincident and give rise to an absorption line of double the intensity of the outer lines.

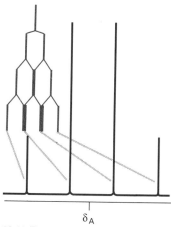

18.11 The origin of the 1:3:3:1 quartet in the A resonance of an AX$_3$ species. The third X nucleus splits each of the lines shown in Fig. 18.10 for an AX$_2$ species into a doublet, and the intensity distribution reflects the number of transitions that have the same energy.

are formed by adding together the two adjacent numbers in the line above. The easiest way of constructing the pattern of fine structure is to draw a tree-like diagram, in which each successive row shows the splitting of a subsequent proton. The procedure is illustrated in Fig. 18.12 and was used in Figs 18.10 and 18.11. It is easily extended to molecules containing nuclei with $I > \frac{1}{2}$ (Fig. 18.13).

Example 18.2 *Accounting for the fine structure in a spectrum*

Account for the fine structure in the ^{1}H-NMR spectrum of the C–H protons of ethanol.

Method. We need to consider how each group of equivalent protons (e.g. three methyl protons) split the resonance of the other groups of protons. There is no splitting *within* groups of equivalent protons. Each splitting pattern can be decided by referring to Pascal's triangle.

Answer. The three protons of the CH$_3$ group split the resonance of the CH$_2$ protons into a 1:3:3:1 quartet with a splitting J. Likewise, the two protons of the CH$_2$ group split the resonance of the CH$_3$ protons into a 1:2:1 triplet with the same splitting J. Each of these lines is split into a doublet by the OH proton but the doublet splitting cannot be detected because the OH protons migrate rapidly from molecule to molecule and their effect averages to zero.

Exercise E18.2. What fine-structure can be expected for the protons in NH$_4^+$? The spin quantum number of nitrogen is 1. [1:1:1 triplet from N]

The energy levels of coupled systems

It will be useful for later discussions to consider an NMR spectrum in terms of the energy levels of the nuclei and the transitions between them. The energy level diagram for a single spin-$\frac{1}{2}$ nucleus and its single transition were shown in Fig. 18.1, and nothing more needs to be said. For a spin-$\frac{1}{2}$ AX system there are four spin states:

$$\alpha_A \alpha_X \qquad \alpha_A \beta_X \qquad \beta_A \alpha_X \qquad \beta_A \beta_X$$

The energy depends on the orientation of the spins in the external magnetic field, and if spin–spin coupling is neglected we can write

$$E = -\gamma(1 - \sigma_A)\hbar B m_A - \gamma(1 - \sigma_X)\hbar B m_X = -h\nu_A m_A - h\nu_X m_X$$

where ν_A and ν_X are the Larmor frequencies of A and X and m_A and m_X are their quantum numbers. This expression gives the four lines on the left of Fig. 18.14. The spin–spin coupling depends on the relative orientation of the two nuclear spins and so is proportional to the product $m_A m_X$; the constant of proportionality is hJ. Therefore, the energy including spin–spin coupling is

$$E = -h\nu_A m_A - h\nu_X m_X + hJ m_A m_X$$

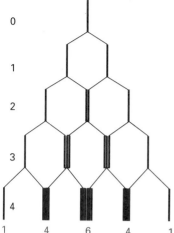

1 4 6 4 1

18.12 The intensity distribution of the A resonance of an AX_n resonance can be constructed by considering the splitting caused by 1, 2, . . . n protons, as in Figs 18.10 and 18.11. The resulting intensity distribution has a binomial distribution and is given by the integers in the corresponding row of Pascal's triangle. Note that, although the lines have been drawn side-by-side for clarity, the members of each group are coincident. Four protons, in AX_4, split the A resonance into a 1:4:6:4:1 quintet.

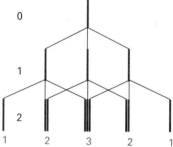

1 2 3 2 1

18.13 The intensity distribution arising from spin–spin interaction with nuclei with $I = 1$ can be constructed similarly, but each successive nucleus splits the lines into three equal intensity components. Two equivalent spin-1 nuclei give rise to a 1:2:3:2:1 quintet.

18.14 The energy levels of an AX system. The four levels on the left are those of the two spins in the absence of spin–spin coupling. The four levels on the right show how a positive spin–spin coupling constant affects the energies. The transitions shown are for $\beta \leftarrow \alpha$ of A or X, the other nucleus (X or A, respectively) remaining unchanged.

If $J > 0$, then a lower energy is obtained when $m_A m_X < 0$, which is the case if one spin is α and the other is β. A higher energy is obtained if both spins are α or both spins are β. The opposite is true if $J < 0$. The resulting energy level diagram (for $J > 0$) is shown on the right of Fig. 18.14. We see that the $\alpha\alpha$ and $\beta\beta$ states are both raised by $\frac{1}{4}hJ$ and that the $\alpha\beta$ and $\beta\alpha$ states are both lowered by $\frac{1}{4}hJ$.

When a transition of nucleus A occurs, nucleus X remains unchanged. Therefore, the A resonant absorption is a transition for which

$$\Delta m_A = +1 \qquad \Delta m_X = 0$$

There are two such transitions, one in which $\beta_A \leftarrow \alpha_A$ occurs when the X nucleus is α_X, and the other in which $\beta_A \leftarrow \alpha_A$ occurs when the X nucleus is β_X. They are shown in Fig. 18.14 and in a slightly different form in Fig. 18.15. The energies of the transitions are

$$\Delta E = h\nu_A + \tfrac{1}{2}hJ \qquad \text{and} \qquad \Delta E = h\nu_A - \tfrac{1}{2}hJ$$

Therefore, the A resonance consists of a doublet of separation J centred on the chemical shift of A (as in Fig. 18.8).

Similar remarks apply to the X resonance, which consists of two transitions according to whether the A nucleus is α or β (Fig. 18.15). The transition energies are

$$\Delta E = h\nu_X + \tfrac{1}{2}hJ \qquad \text{and} \qquad \Delta E = h\nu_X - \tfrac{1}{2}hJ$$

It follows that the X resonance also consists of two lines of separation J, but they are centred on the chemical shift of X (as shown in Fig. 18.8).

The magnitudes of coupling constants

The scalar coupling constant of two nuclei separated by N bonds is denoted ^NJ, with subscripts for the types of nuclei involved. Thus, $^1J_{CH}$ is the coupling constant for a proton joined directly to a ^{13}C atom, and $^2J_{CH}$ is the coupling constant when the same two nuclei are separated by two bonds (as in ^{13}C—C—H). A typical value of $^1J_{CH}$ is in the range 120 to 250 Hz; $^2J_{CH}$ is between 0 and 10 Hz. Both ^3J and ^4J give detectable effects in a spectrum, but couplings over larger numbers of bonds can

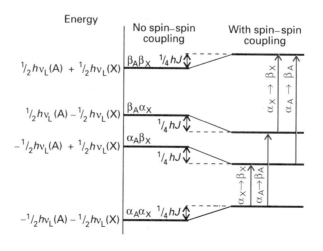

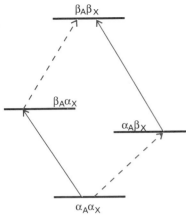

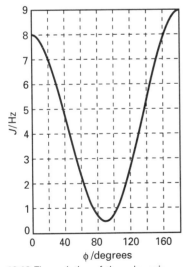

18.15 An alternative depiction of the energy levels and transitions shown in Fig. 18.14.

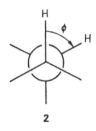

2

18.16 The variation of the spin–spin coupling constant with angle predicted by the Karplus equation, eqn 9, with $A = +4.0\,\text{Hz}$, $B = -0.5\,\text{Hz}$, and $C = +4.5\,\text{Hz}$. Note that the sign as well as the magnitude of the constant changes as the angle changes.

generally be ignored. One of the longest couplings that has been detected is $^9J_{HH} = 0.4\,\text{Hz}$ for CH_3 and CH_2 protons in the molecule $CH_3C{\equiv}CC{\equiv}CC{\equiv}CCH_2OH$.

The sign of J_{XY} indicates whether the energy of two spins is lower when they are parallel ($J < 0$) or when they are antiparallel ($J > 0$). It is found that $^1J_{CH}$ is often positive, $^2J_{HH}$ is often negative, $^3J_{HH}$ is often positive, and so on. An additional point is that the value of J varies with the angle between the bonds (Fig. 18.16). Thus, a $^3J_{HH}$ coupling constant is often found to depend on the angle ϕ (2) according to the **Karplus equation**:

$$J = A\cos 2\phi + B\cos\phi + C \tag{9}$$

with A, B, and C empirical constants with values close to $+4.0$, -0.5, and $+4.5\,\text{Hz}$, respectively. It follows that the measurement of $^3J_{HH}$ in a series of related compounds can be used to determine their conformations. The coupling constant $^1J_{CH}$ also depends on the hybridization of the C atom, as the following values indicate:

	sp	sp^2	sp^3
$^1J_{CH}/\text{Hz}$:	250	160	125

The origin of spin–spin coupling

Spin–spin coupling is a very subtle phenomenon, and it is better to treat J as an empirical parameter than to use calculated values. However, it is possible to get some insight into its origins, if not its precise magnitude or always reliably its sign, by considering the magnetic interactions within molecules.

A nucleus with spin projection m_I gives rise to a magnetic field with z-component B_{nuc} at a distance R, where

$$B_{nuc} = -\frac{g_I\mu_N\mu_0}{4\pi R^3}(1 - 3\cos^2\theta)m_I \tag{10}$$

The angle θ, the angle between the z-axis and the vector that joins the two spins, is defined in Fig. 18.17. The magnitude of this field is about $0.1\,\text{mT}$ when $R = 0.3\,\text{nm}$, corresponding to a splitting of resonance signal of about $10^4\,\text{Hz}$, and is of the order of magnitude of the splitting observed in solid samples (see Section 18.8).

In a liquid, the angle θ sweeps over all values as the molecule tumbles, and $1 - 3\cos^2\theta$ averages to zero.[3] Hence the direct dipolar interaction between spins cannot account for the fine structure of the spectra of rapidly tumbling molecules. The direct interaction does make an important contribution to the spectra of solid samples and to the spectra of molecules, such as biological and synthetic macromolecules, that tumble only slowly in solution.

Spin–spin coupling in molecules in solution can be explained in terms of the **polarization mechanism**, in which the interaction is transmitted

---

3 The volume element in polar coordinates is proportional to $\sin\theta\,d\theta$ and θ ranges from 0 to π. Therefore the average value of B_{nuc} for a tumbling molecule is proportional to

$$\int_0^\pi (1 - 3\cos^2\theta)\sin\theta\,d\theta = 0$$

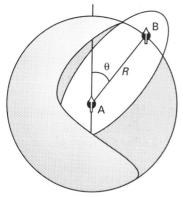

18.17 The interaction between two parallel magnetic dipoles, A and B, depends on the angle θ, the angle between the z-axis and the internuclear vector. When the molecule rotates, θ sweeps through all values. As a result, $1 - 3\cos^2\theta$ averages to zero, and the dipole–dipole interaction averages to zero.

through the electrons of the bonds. The simplest case to consider is that of $^1J_{XY}$ where X and Y are spin-$\frac{1}{2}$ nuclei joined by an electron-pair bond (Fig. 18.18). The coupling mechanism depends on the fact that in some atoms it is favourable for the nucleus and a nearby electron spin to be parallel (both α or both β), but in others it is favourable for them to be antiparallel (one α and the other β). The electron–nucleus coupling is magnetic in origin, and may be either a dipolar interaction between the magnetic moments of the electron and nucleus spins or a **Fermi contact interaction**. The latter depends on the very close approach of an electron to the nucleus and hence can occur only if the electron occupies an s orbital. We shall suppose that it is energetically favourable for an electron spin and a nuclear spin to be antiparallel (as is the case for a proton and an electron in a hydrogen atom).

JUSTIFICATION

A pictorial description of the Fermi contact interaction is as follows. First, we regard the magnetic moment of the nucleus as arising from the circulation of a current in a tiny loop with a radius similar to that of the nucleus (Fig. 18.19). Far from the nucleus the field generated by this loop is indistinguishable from the field generated by a point magnetic dipole. Close to the loop, however, the field differs from that of a point dipole. The magnetic interaction between this non-dipolar field and the electron's magnetic moment is the contact interaction.

The lines of force depicted in Fig. 18.19 correspond to those for a proton with α spin. The lower energy state of an *electron* spin in such a field is the β state.

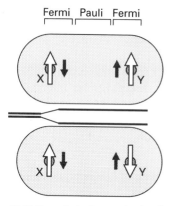

18.18 The polarization mechanism for spin–spin coupling ($^1J_{HH}$). The two arrangements have slightly different energies. In this case J is positive, corresponding to a lower energy when the nuclear spins are antiparallel.

If the X nucleus is α, a β electron of the bonding pair will tend to be found nearby (since that is energetically favourable for it). The second electron in the bond, which by the Pauli principle must have α spin if the other is β, will be found mainly at the far end of the bond (because electrons tend to stay apart to reduce their mutual repulsion). Therefore, a β electron will tend to be found near the Y nucleus. Because it is energetically favourable for the Y nucleus to be antiparallel to an electron spin, the β spin of the Y nucleus has a lower energy, and hence a lower Larmor frequency, than a Y nucleus with α spin. The opposite is true when the X nucleus is β, for now the α spin of Y has the lower energy. In other words, the antiparallel arrangement of nuclear spins lies lower in energy than the parallel arrangement as a result of their magnetic coupling with the bond electrons. Similarly, the Y nucleus has a lower Larmor frequency when it is antiparallel to X than when it is parallel. That is, $^1J_{HH}$ is positive.

To account for the value of $^2J_{XY}$, as in H—C—H, we need a mechanism that can transmit the spin alignments through the central C atom (which may be ^{12}C with no nuclear spin of its own). In this case (Fig. 18.20), an X nucleus with β spin polarizes the electrons in its bond, and the β electron is likely to be found closer to the C nucleus. The more favourable arrangement of two electrons on the same atom is with their

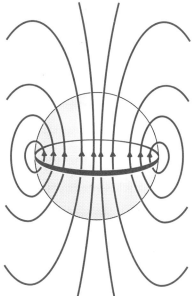

18.19 The origin of the Fermi contact interaction. From far away, the magnetic field pattern arising from a ring of current is that of a point dipole. However, if an electron can sample the field close to the region indicated by the sphere (the nucleus), then the field distribution differs significantly from that of a point dipole. For example, if the electron can penetrate the sphere, then the spherical average of the field it experiences is not zero.

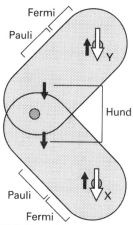

18.20 The polarization mechanism for $^2J_{HH}$ spin–spin coupling. The spin information is transmitted from one bond to the next by a version of the mechanism that accounts for the lower energy of electrons with parallel spins in different atomic orbitals (Hund's rule of maximum multiplicity). In this case J is negative, corresponding to a lower energy when the nuclear spins are parallel.

spins parallel (Hund's rule, Section 13.4), so the more favourable arrangement is for the β electron of the neighbouring bond to be close to the C nucleus. Consequently, the α electron of that bond is more likely to be found close to the Y nucleus, and therefore that nucleus will have a lower energy if it is β. Hence, according to this mechanism, the lower Larmor frequency of Y will be obtained if its spin is parallel to X. That is, $^2J_{HH}$ is negative.

The coupling of nuclear spin to electron spin by the Fermi contact interaction is most important for proton spins, but it is not necessarily the most important mechanism for other nuclei. These nuclei may also interact by a dipolar mechanism with the electron magnetic moments and with their orbital motion, and there is no simple way of specifying whether J will be positive or negative.

Equivalent nuclei

A group of nuclei are **chemically equivalent** if they are (a) related by a symmetry operation of the molecule and (b) have the same chemical shifts. They are **magnetically equivalent** if, as well as being chemically equivalent, they also have identical spin–spin interactions with any other magnetic nuclei in the molecule.

The difference between chemical and magnetic equivalence is illustrated by CH_2F_2 and $H_2C{=}CF_2$, in both of which the protons are chemically equivalent. However, although the protons in CH_2F_2 are magnetically equivalent, those in $CH_2{=}CF_2$ are not. One proton in the latter has *cis* and *trans* coupling interactions with *cis* and *trans* F nuclei ($I = \frac{1}{2}$), which might be α and β, respectively. However, the second proton in the same molecule will be *cis* to the β F nucleus and *trans* to the α F nucleus. In CH_2F_2 both protons are equally distant from the two F nuclei, so there is no distinction between them. Strictly speaking, the CH_3 protons in ethanol (and other compounds) are magnetically inequivalent on account of their different interactions with the CH_2 protons in the next group. However, they are in practice made magnetically equivalent by the rapid rotation of the CH_3 group, which averages out any differences.

An important feature of chemically equivalent magnetic nuclei is that, although they do couple together, the coupling has no effect on the appearance of the spectrum. The reason for the invisibility of the coupling is that all allowed nuclear spin transitions are *collective* reorientations of groups of equivalent nuclear spins that do not change the relative orientations of the spins within the group (Fig. 18.21). Then, because the relative orientations of nuclear spins are not changed in any transition, the magnitude of the coupling between them is undetectable. Hence, an isolated CH_3 group gives a single, unsplit line because all the allowed transitions of the group of three protons occur without change of their relative orientations.

JUSTIFICATION

Consider an A_2 system of two equivalent spin-$\frac{1}{2}$ nuclei; our aim is to see that although the nuclei may be coupled together by a spin–spin

18.21 (a) A group of two equivalent nuclei realigns as a group, without change of angle between the spins, when a resonant absorption occurs. Hence it behaves like a single nucleus and the spin–spin coupling between the individual spins of the group is undetectable. (b) Three equivalent nuclei also realign as a group without change of relative orientations.

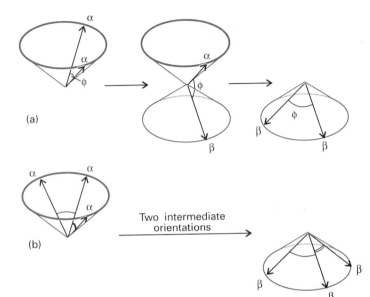

(a)

(b)

Two intermediate orientations

No spin–spin coupling   With spin–spin coupling

$\beta\beta$ $I = 1,\ M_I = -1$ $+\frac{1}{4}J$

$\alpha\beta + \beta\alpha$
$\alpha\beta - \beta\alpha$

$I = 1,\ M_I = 0$ $+\frac{1}{4}J$

$I = 0,\ M_I = 0$ $-\frac{3}{4}J$

$I = 1,\ M_I = +1$ $+\frac{1}{4}J$

$\alpha\alpha$

18.22 The energy levels of an A_2 system in the absence of spin–spin coupling are shown on the left. When spin–spin coupling is taken into account, the energy levels on the right are obtained. Note that the three states with total nuclear spin $I = 1$ correspond to parallel spins and give rise to the same increase in energy (J is positive); the one state with $I = 0$ (antiparallel nuclear spins) has a lower energy in the presence of spin–spin coupling. The only allowed transitions are those that preserve the angle between the spins, and so take place between the three states with $I = 1$. They occur at the same resonance frequency as they would have in the absence of spin–spin coupling.

interaction, that interaction does not affect the appearance of the spectrum.

First, consider the energy levels in the absence of spin–spin coupling. There are four spin states which (just as for two electrons) can be classified according to their total spin I (the analogue of S for two electrons) and their total projection M_I on the z-axis:

Spins parallel, $I = 1$: $M_I = +1$ $\alpha\alpha$
                        $M_I = 0$ $\alpha\beta + \beta\alpha$
                        $M_I = -1$ $\beta\beta$
Spins paired, $I = 0$: $M_I = 0$ $\alpha\beta - \beta\alpha$

The effect of a magnetic field on these four states is shown on the left in Fig. 18.22: the energies of the two states with $M_I = 0$ are unchanged by the field because they are composed of equal proportions of α and β spins.

The spin–spin coupling energy is proportional to the scalar product of the vectors representing the spins, and we write

$$E = (hJ/\hbar^2)\mathbf{I}_1 \cdot \mathbf{I}_2$$

The scalar product can be expressed in terms of the total nuclear spin by noting that

$$I^2 = (\mathbf{I}_1 + \mathbf{I}_2)^2 = I_1^2 + I_2^2 + 2\mathbf{I}_1 \cdot \mathbf{I}_2$$

and replacing the magnitudes by their quantum mechanical values:

$$\mathbf{I}_1 \cdot \mathbf{I}_2 = \tfrac{1}{2}\{I(I+1) - I_1(I_1+1) - I_2(I_2+1)\}\hbar^2$$

Then, because $I_1 = I_2 = \tfrac{1}{2}$, it follows that

$$E = \tfrac{1}{2}hJ\{I(I+1) - \tfrac{3}{2}\}$$

For parallel spins, $I = 1$ and $E = +\tfrac{1}{4}hJ$; for antiparallel spins $I = 0$ and

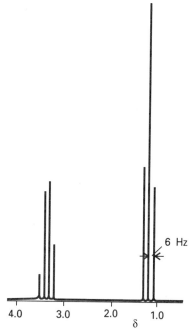

18.23 The NMR spectrum referred to in Example 18.3. The field increases to the right.

$E = -\frac{3}{4}hJ$, as in the illustration. We see that three of the states move in energy in one direction and the fourth (the one with antiparallel spins) moves three times as much in the opposite direction. The resulting energy levels are shown on the right in Fig. 18.22.

The NMR spectrum of the A_2 species arises from transitions between the levels. However, the radiofrequency field affects the two equivalent protons equally, so it cannot change the orientation of one proton relative to the other; therefore, the transitions take place *within* the set of states that correspond to parallel spins (those labelled $I = 1$), and no spin-parallel state can change to a spin-antiparallel state (the state with $I = 0$). Put another way, the allowed transitions are subject to the selection rule $\Delta I = 0$. This selection rule is in addition to the rule $\Delta M_I = \pm 1$ that arises from the conservation of angular momentum and the unit spin of the photon. The allowed transitions are shown in Fig. 18.22: we see that there are only two transitions, and that they occur at the same resonance frequency that the nuclei would have in the absence of spin–spin coupling. Hence, the spin–spin coupling interaction does not affect the appearance of the spectrum.

Suggest an interpretation of the ^{1}H-NMR spectrum in Fig. 18.23.

Method. We need to look for groups with characteristic chemical shifts (by using Fig. 18.3) and to account for the fine structure as was done for ethanol.

Answer. The resonance at $\delta = 3.4$ corresponds to CH_2 in an ether; that at $\delta = 1.2$ corresponds to CH_3 in CH_3CH_2. The fine structure of the CH_2 group (a 1:3:3:1 quartet) is characteristic of splitting caused by CH_3; the fine structure of the CH_3 resonance is characteristic of splitting caused by CH_2. The scalar coupling constant is $J = \pm 6$ Hz (the same for each group). The compound is probably $(CH_3CH_2)_2O$.

Exercise E18.3. What changes in the spectrum would be observed on recording it at five times the magnetic field?

> [Groups of lines five times further apart in frequency (but the same δ values); no change in spin–spin splitting]

Second-order spectra

NMR spectra are usually much more complex than the foregoing simple analysis suggests. We have described the extreme case in which the differences in chemical shifts are much greater than the spin–spin coupling constants. In such cases it is simple to identify groups of chemically equivalent nuclei and to think of the groups of nuclear spins as reorientating relative to each other. The spectra that result are called **first-order spectra**.

Nuclei cannot be allocated to definite groups when the differences in their chemical shifts are comparable to their spin–spin coupling interactions. The complicated spectra that are then obtained are called

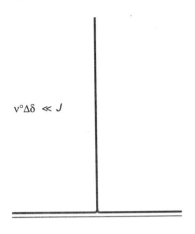

$v°\Delta\delta \ll J$

$v°\Delta\delta \approx J$

$v°\Delta\delta \approx 5J$

$v°\Delta\delta \gg J$

18.24 NMR spectra are complicated when spin–spin coupling constants and chemical shifts are comparable. This diagram shows how the resonance of a two-spin system changes as J increases relative to the difference of chemical shifts, $v° \Delta \delta$. At the top $v° \Delta \delta \ll J$, and the spectrum is that of an A_2 species. At the bottom $v° \Delta \delta \gg J$, and the (first-order) spectrum is that of an AX species. The intermediate illustrations are for $v° \Delta \delta$ not very different from J, and are the second-order spectra typical of an AB species.

second-order spectra, and are much more difficult to analyse (Fig. 18.24). Because the difference in resonance frequencies increases with field, but spin–spin coupling constants are independent of it, a second-order spectrum may become simpler (and first-order) at high fields because individual groups of nuclei become identifiable again.

A clue to the type of analysis that is appropriate is given by the notation for the types of spins involved. Thus, an AX spin system (which consists of two nuclei with a large chemical shift difference) has a first-order spectrum. An AB system, on the other hand (with two nuclei of similar chemical shifts), gives a second-order spectrum. An AX system may have widely different chemical shifts because A and X are nuclei of different elements (such as ^{13}C and ^1H), in which case they form a **heteronuclear spin system**. AX may also denote a **homonuclear spin system** in which the nuclei are of the same element but in markedly different environments.

Dilute and abundant spins: spin decoupling

Carbon-13 is a **dilute spin species** in the sense that it is unlikely that more than one ^{13}C nucleus will be found in any given small molecule (provided the sample has not been enriched with that isotope; the natural abundance of ^{13}C is only 1.1 per cent). Even in large molecules, although more than one ^{13}C nucleus may be present, it is unlikely that they will be close enough to give an observable splitting. Hence, it is not normally necessary to take into account ^{13}C–^{13}C spin–spin coupling within a molecule.

Protons are **abundant spin species** in the sense that a molecule is likely to contain many of them. If we were observing a ^{13}C-NMR spectrum, we would obtain a very complex spectrum on account of the coupling of the one ^{13}C nucleus with all the protons that are present. To avoid this difficulty, ^{13}C-NMR spectra are normally observed using the technique of **proton decoupling**. Thus, if the CH_3 protons of ethanol are irradiated with a second, strong, resonant radiofrequency source, then they undergo rapid spin reorientations and the ^{13}C nucleus senses an average orientation. As a result, its resonance is a single line and not a 1:3:3:1 quartet. Proton decoupling has the additional advantage of enhancing sensitivity, because the intensity is concentrated into a single transition frequency instead of being spread over several transition frequencies. If care is taken to ensure that the other parameters on which the strength of the signal depends are kept constant, then the intensity of proton-decoupled spectra are proportional to the number of ^{13}C nuclei present. The technique is widely used to characterize synthetic polymers.

PULSE TECHNIQUES IN NMR

Modern methods of detecting the energy separation ΔE between nuclear spin states are more sophisticated than simply looking for the frequency at which resonance occurs. One of the best analogies that has been suggested to illustrate the difference between the old and new ways of observing an NMR spectrum is that of detecting the spectrum of

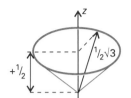

18.25 The vector model of angular momentum for a single spin-$\frac{1}{2}$ nucleus. The angle around the z-axis is indeterminate.

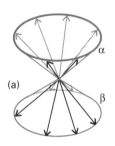

(a)

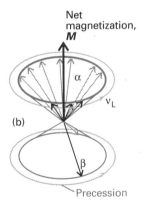

Net magnetization, **M**

(b)

18.26 The magnetization of a sample of spin-$\frac{1}{2}$ nuclei is the resultant of all their magnetic moments. (a) In the absence of an externally applied field, there are equal numbers of α and β spins at random angles around the z-axis (the field direction) and the magnetization is zero. (b) In the presence of a field, the spins precess around their cones (that is, there is an energy difference between the α and β states) and there are slightly more α spins than β spins. As a result, there is a net magnetization along the z-axis.

vibrations of a bell. We could stimulate the bell with a gentle vibration at a gradually increasing frequency, and note the frequencies at which it resonated with the stimulation. A lot of time would be spent getting zero response when the stimulating frequency was between the bell's vibrational modes. However, if we were simply to hit the bell with a hammer, we would immediately obtain a clang composed of all the frequencies that the bell can produce. The equivalent in NMR is to monitor the radiation nuclear spins emit as they return to the ground state after the appropriate stimulation.

We need to understand how the equivalent of the hammer blow is delivered and how the signal is monitored and interpreted. These features are generally expressed in terms of the vector model of angular momentum introduced in Section 12.7.

18.4 The magnetization vector

We consider a sample composed of many identical nuclei with spin quantum number $I = \frac{1}{2}$. As we saw in Section 12.7, an angular momentum can be represented by a vector of length $\{I(I + 1)\}^{\frac{1}{2}}$ units with a component of length m_I units along the z-axis. As the uncertainty principle does not allow us to specify the x- and y-components of the angular momentum, all we know is that the vector lies somewhere on a cone around the z-axis. For $I = \frac{1}{2}$, the length of the vector is $\frac{1}{2}\sqrt{3}$ and it makes an angle of 55° to the z-axis (Fig. 18.25).

In the absence of a magnetic field, the sample consists of equal numbers of α and β nuclear spins with their vectors lying at random angles on the cones (Fig. 18.26a). These angles are unpredictable, and at this stage we picture the spin vectors as stationary. The **magnetization**, **M**, of the sample, its net nuclear magnetic moment, is zero.

The effect of the static field

Two changes occur in the magnetization when a magnetic field is present. First, the energies of the two orientations change, the α spins moving to low energy and the β spins to high energy (provided $g_I > 0$). At 10 T, the Larmor frequency for protons is 427 MHz, and in the vector model the individual vectors are pictured as **precessing**, or sweeping round their cones, at this rate. This motion is a *pictorial representation* of the change in energy of the spin states (it is not an actual representation of reality). As the field is increased, the Larmor frequency increases and the precessional motion becomes faster. Second, the populations of the two spin states (the numbers of α and β spins) change, and there will be more α spins than β spins. Because the β state lies at an energy $h\nu_L$ above the α state, at a temperature T the populations are related by

$$\frac{N_\beta}{N_\alpha} = e^{-h\nu_L/kT} = 1 - \frac{h\nu_L}{kT} + \cdots$$

Because $h\nu_L/kT \approx 7 \times 10^{-5}$ for protons at 300 K and 10 T, there is only a tiny imbalance of populations, and it is even smaller for other nuclei with their smaller magnetogyric ratios. However, despite its smallness,

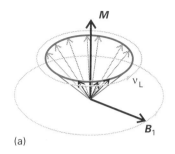

(a)

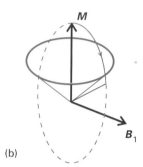

(b)

18.27 (a) In a resonance experiment, a circularly polarized radiofrequency magnetic field B_1 is applied in the xy-plane (the magnetization vector $\boldsymbol{M}$ lies along the z-axis). (b) If we step into a frame rotating at the Larmor frequency, the radiofrequency field appears to be stationary if its frequency is the same as the Larmor frequency. When the two frequencies coincide, the magnetization vector of the sample, $\boldsymbol{M}$, begins to rotate around the direction of the B_1 field.

the imbalance means that there is a net magnetization that we can represent by a vector $\boldsymbol{M}$ pointing in the z-direction and with a length proportional to the population difference (Fig. 18.26b).

The effect of the radiofrequency field

We now consider the effect of a circularly polarized radiofrequency field in the xy-plane. We have considered the electric component of this field in the other forms of spectroscopy that we have treated. However, in this chapter we consider only the magnetic component, for it is this component that interacts with the nuclear magnetic moment. The strength of the oscillating magnetic field is B_1.

Suppose we choose the frequency of the oscillating field to be equal to the Larmor frequency of the spins. This is equivalent to choosing the resonance condition in the conventional experiment. The nuclei now experience a steady B_1 field because the rotating magnetic field is in step with the precessing spins (Fig. 18.27a). Under the influence of this effectively steady field, the magnetization vector begins to precess around its direction at a rate that is proportional to B_1 (Fig. 18.27b). If we apply the B_1 field in a pulse of a certain duration, then the magnetization precesses into the xy-plane (Fig. 18.28a) and we say that we have applied a **90° pulse** (or a '$\pi/2$ pulse'). The duration of the pulse depends on the strength of the B_1 field, but is typically of the order of microseconds. To a stationary external observer (a radiofrequency coil), the magnetization vector is now rotating in the xy-plane at the Larmor frequency (at about 430 MHz). The rotating magnetization induces a 430-MHz signal in the coil, which can be amplified and processed (Fig. 18.28b). In practice, the processing takes place after subtraction of a constant high frequency so that all the signal manipulation takes place at frequencies of a few kilohertz.

As time passes, the individual spins move out of step (partly because they are precessing at slightly different rates, as we shall explain later) so that the magnetization vector shrinks exponentially with a time constant T_2 and induces an ever weaker signal in the detector coil. The form of the signal that we can expect is therefore the oscillating-decaying **free-induction decay** (FID) shown in Fig. 18.29, and the y-component of the magnetization varies as

$$M_y(t) = M_0 \cos(2\pi\nu_L t)\mathrm{e}^{-t/T_2} \tag{11}$$

We have considered the effect of a pulse of precise frequency. However, virtually the same effects are obtained if we expose the sample to a 90° pulse of radiation that spans a range of frequencies.[4] Although all the different B_1 fields are not equally efficient at rotating the magnetization vector into the xy-plane, they do not differ very much and we can be

---

4 In fact, any pulse must cover a range of frequencies: we can have a monochromatic wave only if it continues for ever. For exactly the same reason as there is an inverse relation between the width of a spectral line and the lifetime of a state (Section 16.3), the length of the pulse τ_P and its frequency spread $\Delta\nu$ are related by $\Delta\nu \approx 1/2\pi\tau_P$. A pulse lasting 1 μs has a frequency spread of nearly 200 kHz.

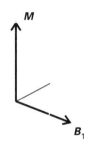

(a)

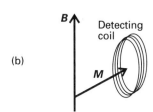

(b)

18.28 (a) If the radiofrequency field is applied for a certain time, the magnetization vector, **M**, is rotated into the *xy*-plane. (b) To an external stationary observer (the coil), the magnetization vector is rotating at the Larmor frequency, and can induce a signal in the coil.

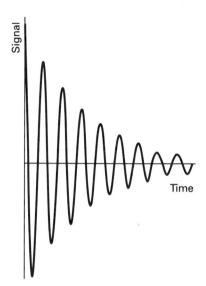

18.29 A simple free induction decay of a sample of spins with a single resonance frequency.

confident that the magnetization will end up largely in the plane. Once there, it rotates at the Larmor frequency, and generates an FID signal. Note that we do not need to know this frequency initially: the broad-band pulse is the analogue of the hammer blow on the bell, and the detected signal is a sign that a particular frequency (in our case, a nuclear spin energy level separation) is present.

Time- and frequency-domain spectra

We can think of the magnetization vector of a homonuclear AX spin system with $J = 0$ as consisting of two parts, one formed by the A spins and the other by the X spins. When the 90° pulse is applied, both magnetization vectors are rotated into the *xy*-plane. However, because the A and X nuclei precess at different frequencies, they induce two signals in the detector coils, and the overall FID curve may resemble that in Fig. 18.30a. The composite FID curve is the analogue of the struck bell emitting a rich tone composed of all the frequencies at which it can vibrate.

The problem we must address is how to recover the resonance frequencies present in a free-induction decay. We encountered a similar problem when discussing Fourier-transform infrared spectra in Section 16.1, where all the vibrational frequencies were detected at once. The same technique is used here. We know that the FID curve is a sum of oscillating functions, so the problem is to analyse it into its harmonic components.

The analysis of the FID curve is achieved by the standard mathematical technique of Fourier transformation. We start by noting that the signal $S(t)$ in the time domain, the total FID curve, is the sum (more precisely, the integral) over all the contributing frequencies

$$S(t) = \int_{-\infty}^{\infty} I(v)e^{-2\pi i v t}\, dv \tag{12a}$$

Signal oscillating with frequency v

Intensity of the contribution of the frequency v

Sum over all possible frequencies

We need $I(v)$, the spectrum in the frequency domain; it is obtained by evaluating the integral

$$I(v) = 2\,\mathrm{Re} \int_{0}^{\infty} S(t)e^{2\pi i v t}\, dt \tag{12b}$$

Where Re means take the real part of the following expression. The integration is carried out at a series of frequencies v on a computer that is built into the spectrometer. When the signal in Fig. 18.30a is transformed in this way, we get the frequency-domain spectrum shown in Fig. 18.30b. One line represents the Larmor frequency of the A nuclei and the other that of the X nuclei.

The FID curve in Fig. 18.31 was obtained from a sample of ethanol. The frequency-domain spectrum obtained from it by Fourier transformation is the one that we have already discussed (Fig. 18.4). We can now

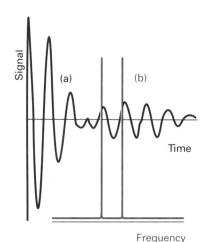

18.30 (a) A free-induction decay signal of a sample of AX species and (b) its analysis into its frequency components.

see why the FID curve in Fig. 18.31 is so complex: it arises from the precession of a magnetization vector that is composed of eight components, each with a characteristic frequency.

18.5 Linewidths and rate processes

The linewidths of NMR spectra, in common with other spectroscopic techniques, provide information about the rates of processes relating to the molecules in the sample. We have seen that the FID-signal decays with time, which implies that the component of the magnetization vector in the xy-plane must be shrinking. In this section we see some of the processes involved.

Spin relaxation

There are two reasons why the component of the magnetization vector in the xy-plane shrinks. Both reflect the fact that the nuclear spins are not in thermal equilibrium with their surroundings (for then **M** lies parallel to z) and the return to equilibrium is the process called **spin relaxation**.

At thermal equilibrium the spins have a Boltzmann distribution, with more α spins than β spins; however, a magnetization vector in the xy-plane immediately after a 90° pulse has equal numbers of α and β spins. The populations revert to their thermal equilibrium values exponentially, and as they do so the z-component of magnetization reverts to its equilibrium value M_0 with a time constant called the **longitudinal relaxation time** T_1 (Fig. 18.32):

$$M_z(t) - M_0 \propto e^{-t/T_1} \tag{13}$$

Because this relaxation process involves giving up energy to the surroundings (the 'lattice') as β spins revert to α spins, the time constant T_1 is also called the **spin–lattice relaxation time**. Spin–lattice relaxation is caused by fluctuating local magnetic fields arising from the motion of the molecules. These fluctuations can stimulate the spins to change from β to α, and vice versa, and hence to relax towards the thermal equilibrium population.

A second aspect of spin relaxation is the fanning-out of the spins in the xy-plane if they precess at different rates (Fig. 18.33). The magnetization

18.31 A free induction decay signal of a sample of ethanol. Its Fourier transform is the frequency-domain spectrum shown in Fig. 18.4.

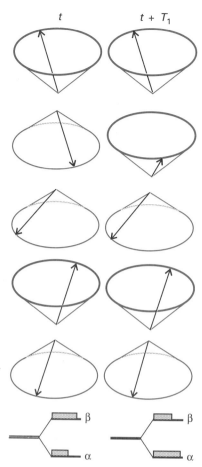

18.32 In longitudinal relaxation the spins relax back towards their thermal equilibrium populations. On the left we see the precessional cones representing spin-$\frac{1}{2}$ angular momenta, and they do not have their thermal equilibrium populations (there are more β-spins than α-spins). On the right, which represents the sample after a time T_1, the populations are those characteristic of a Boltzmann distribution.

vector is large when all the spins are bunched together immediately after a 90° pulse. However, this orderly bunching of spins is not at equilibrium, and, even if there were no spin–lattice relaxation, then we would expect the individual spins to spread out until they were uniformly distributed with all possible angles around the z-axis. At that stage, the component of magnetization vector in the plane would be zero. The randomization of the spin directions occurs exponentially with a time constant called the **transverse relaxation time**, T_2:

$$M_x(t) \propto e^{-t/T_2} \tag{14}$$

Because the relaxation involves the relative orientation of the spins, T_2 is also known as the **spin–spin relaxation time**.

If the y-component of magnetization decays with a time constant T_2, then the spectral line is broadened (Fig. 18.34) and its width at half-height becomes

$$\Delta \nu_{\frac{1}{2}} = \frac{1}{\pi T_2} \tag{15}$$

Typical values of T_2 in proton NMR are of the order of seconds, so linewidths of around 0.1 Hz can be anticipated, in broad agreement with observation. In mobile liquids, $T_2 \approx T_1$.

So far, we have assumed that the equipment, and in particular the magnet, are perfect, and that the differences in Larmor frequencies arise solely from interactions within the sample. In practice, the magnet is not perfect, and the field is different at different locations in the sample despite the sample spinning. The inhomogeneity broadens the resonance, and in most cases this **inhomogeneous broadening** of the lines dominates the broadening that we have discussed so far. It is common to express the extent of inhomogeneous broadening in terms of an **effective transverse relaxation time**, T_2^*, by using a relation like eqn 15 but writing

$$T_2^* = \frac{1}{\pi \Delta \nu_{\frac{1}{2}}} \tag{16}$$

where $\Delta \nu_{\frac{1}{2}}$ is the observed width at half-height. For example, if a line in a ^{1}H-NMR spectrum has a width of 10 Hz, then the effective transverse relaxation time is

$$T_2^* = \frac{1}{\pi \times (10 \text{ s}^{-1})} = 32 \text{ ms}$$

Exchange processes

The appearance of an NMR spectrum is changed if magnetic nuclei can jump between different environments rapidly. Consider a fluxional molecule that can jump between conformations, such as the inversion of a substituted cyclohexane (Fig. 18.35). In one conformation the substituent is axial; in another it is equatorial. The chemical shifts of the axial and equatorial ring protons are different, so that they are in effect

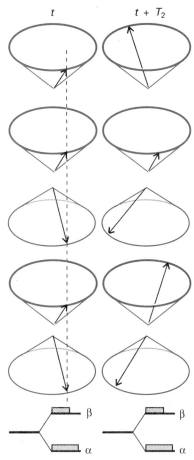

t $t + T_2$

— β   — β
α   α

18.33 The transverse relaxation time T_2 is the time it takes for the phases of the spins to become randomized (another condition for equilibrium) and to change from the orderly arrangement shown on the left to the disorderly arrangement on the right. Note that the populations of the states remain the same; only the relative phases of the spins relax.

jumping between two magnetic environments. When the inversion rate is slow, the spectrum shows two sets of lines, one from molecules with an axial substituent, and one from molecules with an equatorial substituent. When the inversion is fast, the spectrum shows a single line at the mean of the two chemical shifts.

At intermediate inversion rates, the line is very broad. This maximum broadening occurs when the lifetime τ of a conformation gives rise to a linewidth that is comparable to the difference of resonance frequencies, δv, and both broadened lines blend together into a very broad line. That is, according to eqn 16.18, the greatest broadening occurs when

$$\tau \approx \frac{1}{2\pi\delta v} \tag{17}$$

For example, if the chemical shifts differ by 100 Hz, then the spectrum collapses into a single line when the conformation lifetime is less than about 2 ms.

Example 18.4 *Interpreting lifetime broadening*

The NO group in N,N-dimethylnitrosamine, $(CH_3)_2N–NO$, rotates and, as a result, the magnetic environments of the two CH_3 groups are interchanged. In a 60-MHz spectrometer the two CH_3 resonances are separated by 39 Hz. At what rate of interconversion will the resonance collapse to a single line?

Method. We use eqn 17 for the average lifetimes of the tautomers (short-lived isomers). The rate of interconversion is the inverse of their lifetime.

Answer. With $\delta v = 39$ Hz,

$$\tau \approx \frac{1}{2\pi \times (39 \text{ s}^{-1})} = 4.1 \text{ ms}$$

It follows that the signal will collapse to a single line when the interconversion rate exceeds 250 s^{-1}.

Comment. The dependence of the rate of collapse on the temperature is used to determine the energy barrier to the interconversion.

Exercise E18.4. What would you deduce from the observation of a single line from the same molecule in a 300-MHz spectrometer?

[Rotational lifetime less than 0.8 ms]

A similar explanation accounts for the loss of structure in solvents able to exchange protons with the sample. For example, hydroxyl protons are able to exchange with water protons. When this **chemical exchange** occurs, a molecule ROH with an α-spin proton (we write this ROH_α) rapidly converts to ROH_β and then perhaps to ROH_α again because the protons provided by the solvent molecules in successive exchanges have random spin orientations. Therefore, instead of seeing a spectrum composed of contributions from both ROH_α and ROH_β molecules (that

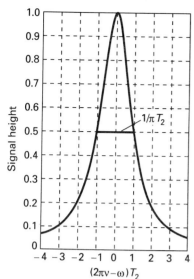

18.34 A Lorentzian absorption line. The width at half-height is inversely proportional to the parameter T_2: the longer the transverse relaxation time, the narrower the line.

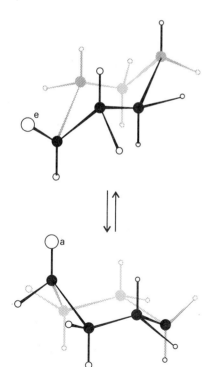

18.35 When a substituted cyclohexane molecule undergoes inversion, the axial (a) and equatorial (e) protons are interchanged and jump between magnetically distinct environments.

is, a spectrum showing the doublet structure due to the OH proton) we see a single, unsplit line at the mean position (as in Fig. 18.4). The effect is observed when the lifetime of a molecule due to this chemical exchange is so short that the lifetime broadening is greater than the doublet splitting. Because this splitting is often very small (about 1 Hz), a proton must remain attached to the same molecule for longer than about 0.1 s for the splitting to be observable. In water the exchange rate is much faster than that, so alcohols show no splitting from the OH protons. In very dry alcohol the exchange rate may be slow enough for the splitting to be detected.

The measurement of T_1

The longitudinal relaxation time can be measured by the **inversion recovery technique**. The first step is to apply a 180° pulse to the sample. A 180° pulse is achieved by applying the B_1 field for twice as long as for a 90° pulse, so that the magnetization vector precesses through 180° and points in the $-z$ direction (Fig. 18.36). No signal can be seen at this stage because there is no component of magnetization in the xy-plane where the detection coils are sensitive. The β spins begin to relax back into α spins, and the magnetization vector shrinks exponentially back towards its thermal equilibrium value M_z. After an interval τ, a 90° pulse is applied that rotates the magnetization into the xy-plane, where it starts to generate a free-induction decay signal. The frequency-domain spectrum is then obtained by Fourier transformation.

The intensity of the spectrum obtained in this way depends on the length of the magnetization vector that is rotated into the xy-plane. The length of the vector returns exponentially to its equilibrium value as the interval between the two pulses is increased (Fig. 18.36), so the intensity of the spectrum also changes similarly with increasing τ. We can therefore measure T_1 by fitting an exponential curve to the series of spectra obtained after different values of τ.

Spin echoes

The measurement of T_2 (as distinct from T_2^*) depends on being able to eliminate the effects of inhomogeneous broadening. The cunning required is at the root of some of the most important advances that have been made in NMR since its introduction.

A **spin echo** is a magnetic analogue of an audible echo: a pulse of magnetization is formed, allowed to spread, is reflected, and is then detected as another pulse a short time later. The sequence of events is illustrated in Fig. 18.37. First, a 90° pulse is applied to the sample in the x-direction, and the magnetization rotates into the xy-plane. The spins now begin to fan out because they have different Larmor frequencies, with some going faster than the mean and others going more slowly, and the signal decays with a time constant T_2^*. We shall consider the overall magnetization as being made up of a number of different magnetizations, each one of which arises from a **spin packet** of nuclei with closely similar precession frequencies.

18.36 The result of applying a 180° pulse to the magnetization in the rotating frame and the effect of a subsequent 90° pulse. The amplitude of the frequency-domain spectrum varies with the interval between the two pulses because spin–lattice relaxation has time to occur.

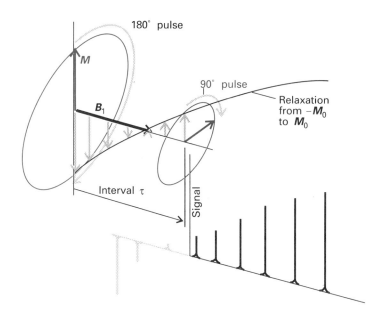

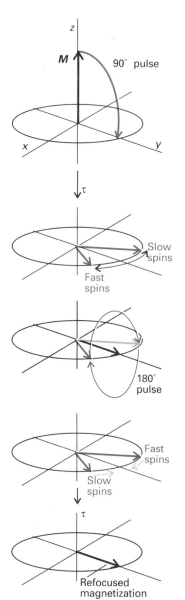

18.37 The sequence of pulses leading to the observation of a spin echo (see text).

A 180° pulse is applied in the y-direction after an interval τ. The pulse rotates the magnetization vectors of the spin packets, moving the magnetization vectors of the fast packets into the angles occupied by the slow packets, and vice versa. The vectors continue to precess, but after an interval τ they will all be found to lie in the y-direction again. The regrouping is called **refocusing**, and when the spins are aligned along y a signal will be observed, which then decays again with a time constant T_2^*.

The important feature of the technique is that the size of the echo is independent of any local fields that remain constant during the two τ intervals. If a spin packet is 'fast' because it happens to be composed of spins in a region of the sample that experiences higher than average fields, then it remains fast throughout both intervals, and what it loses on the first interval it makes up on the second interval. Hence, the size of the echo is independent of inhomogeneities in the magnetic field, for these remain constant. The true transverse relaxation arises from fields that fluctuate on a molecular time scale, and there is no guarantee that an individual 'fast' spin will remain 'fast' in the refocusing phase: the spins within the packets therefore spread with a time constant T_2. Hence, the effects of the true relaxation are not refocused, and the size of the echo decays with the time constant T_2.

In an actual experiment, the **Carr–Purcell–Meiboom–Gill sequence** of pulses (the CPMG sequence) is used. This sequence consists of a series of 180° pulses after the initial 90° pulse:

$$90_x^\circ - [\tau - 180_y^\circ - \tau - \text{Echo}] - [\tau - 180_y^\circ - \tau - \text{Echo}]$$
$$- [\tau - 180_y^\circ - \tau - \text{Echo}] - \cdots$$

The notation 180_y° signifies a 180° pulse of radiofrequency power that is applied after the phase of the radiation has been advanced by 90° from the phase used to apply the original 90° pulse (the pulse denoted 90_x°). The amplitudes of echoes die away with a time constant T_2 as spin–spin

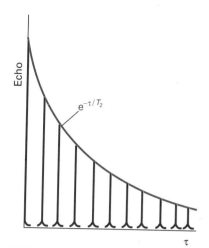

18.38 The exponential decay of the spin echoes in a CPMG pulse sequence gives the transverse relaxation time.

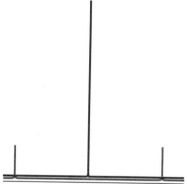

18.39 The nuclear Overhauser effect can simplify and enhance a spectrum as indicated here for an AX system, where the blue spectrum is that of ^{13}C split by coupling to a proton. When the proton absorption is saturated, the ^{13}C resonance collapses to a single line with an intensity characteristic of the Boltzmann population difference of the proton spin.

relaxation occurs, and the value of T_2 is obtained by fitting an exponential curve to their decay (Fig. 18.38).

18.6 The nuclear Overhauser effect

Spin relaxation can be used constructively to enhance the intensities of resonance lines. For example, it can be used to increase the intensity of ^{13}C resonances in proteins as an aid to the determination of their structures. The enhancement is brought about by the **nuclear Overhauser effect** (NOE) which we shall explain by considering a simple AX system in which A is a ^{13}C nucleus and X is a proton.

We have seen already that one advantage of protons in NMR is their high magnetogyric ratio, one effect of which is the existence of relatively large Boltzmann population differences and hence appreciable resonance intensities. In the NOE, relaxation processes are used to transfer this population advantage to another nucleus (^{13}C in the case we are considering), so that its resonances are enhanced. A detailed calculation (which we do not reproduce here) shows that the signal enhancement of a nucleus A that is coupled to a nucleus X that is saturated by strong irradiation at its resonance frequency is

$$\frac{S_A}{S_A^\circ} = 1 + \frac{\gamma_X}{2\gamma_A} \tag{18}$$

For ^{13}C coupled to a saturated proton, the ratio evaluates to 2.99, which shows that an enhancement of about a factor of 3 can be achieved (Fig. 18.39).

Nuclear Overhauser enhancement is used routinely in ^{13}C-NMR. To use the technique, a period of broad-band proton irradiation precedes the radiofrequency excitation pulse and achieves the Overhauser enhancement of the ^{13}C signal. The NOE is also used to determine interproton distances. This application makes use of the fact that, when the dipole–dipole mechanism is not the only relaxation mechanism, the NOE is given by

$$\frac{S_A}{S_A^\circ} = 1 + \frac{\gamma_X}{2\gamma_A} \times \frac{T_1}{T_{1,\text{dip–dip}}} \tag{19}$$

where T_1 is the total relaxation time and $T_{1,\text{dip–dip}}$ is the relaxation time that can be ascribed to a dipolar mechanism. Because the latter depends on the inverse sixth power of the distance between the two coupled nuclei, an analysis of the NOE can be used to infer internuclear distances and to distinguish conformations.

18.7 Two-dimensional NMR

An NMR spectrum contains a great deal of information and, if many protons are present, is very complex. Even a first-order spectrum is complex, for the fine structure of different groups of lines can overlap. The complexity would be reduced if we could use two axes to display the data, with resonances belonging to different groups lying at different locations on the second axis. This separation is essentially what is achieved in **two-dimensional NMR**.

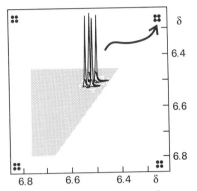

18.40 A typical two-dimensional ^{13}C-NMR spectrum obtained by correlation spectroscopy.

We have seen that a spin-echo experiment refocuses spins that are in a constant environment. Hence, if two spins are in environments with different chemical shifts, then they will be refocused and a single line will be obtained. That is, we can eliminate chemical shifts from a spectrum. Since we saw earlier that we can also remove the effects of spin–spin couplings by decoupling techniques, we can separate the two contributions to the spectrum. In practice, a clever choice of pulses and Fourier transformation techniques makes it possible to display spin coupling in one dimension and the chemical shifts in another, and so greatly simplify the appearance of a spectrum.

Much modern NMR work makes use of **correlation spectroscopy** (COSY) in which the basic pulse sequence is

$$90^\circ_x - t_1 - 90^\circ_x - \text{acquire}(t_2)$$

A series of acquisitions is taken with a variable delay t_1, much as in a spin-echo experiment. The double Fourier transform is then performed on the real time-domain variable t_2 and then on the interferograms arising from the time delay t_1. A typical outcome for an AX system is shown in Fig. 18.40: the diagram shows contours of signal intensity.

The detailed analysis of the appearance of the contour plot is quite difficult, and a simple vector diagram of the processes involved cannot be given. However, the general rules of interpretation are quite straightforward (in simple cases, at least). The peaks across the diagonal constitute the normal four peaks of a one-dimensional NMR spectrum of an AX system, so they add nothing new. The interesting information is in the off-diagonal peaks, for they indicate that the protons to which they correlate by vertical and horizontal lines are spin–spin coupled. Although this information is trivial in this AX system, it can be of enormous help in the interpretation of more complex spectra. A complex spectrum that would be impossible to interpret in one-dimensional NMR can be interpreted reasonably rapidly by two-dimensional NMR. The techniques themselves are described in the books listed in the *Further reading* section.

18.8 Solid-state NMR

The principal difficulty with the application of NMR to solids is the low resolution that is characteristic of solid samples. Nevertheless, there are good reasons for seeking to overcome these difficulties. They include the possibility that a compound of interest is unstable in solution or that it is insoluble, so that conventional solution NMR cannot be employed. Moreover, many species are intrinsically interesting as solids and it is important to determine their structure and dynamics. Synthetic polymers are particularly interesting in this regard, and information can be obtained about the arrangement of molecules, their conformations, and the motion of different parts of the chain. This kind of information is crucial to an interpretation of the bulk properties of the polymer in terms of its molecular characteristics. Similarly, inorganic substances, such as the zeolites that are used as molecular sieves and shape-selective catalysts, can be studied using solid-state NMR, and structural problems can be resolved that cannot be tackled by X-ray diffraction.

Problems of resolution and linewidth are not the only features that plague NMR studies of solids. Because molecular rotation has almost ceased (except in special cases, including 'plastic crystals' in which the molecules continue to tumble), spin–lattice relaxation times are very long but spin–spin relaxation times are very short. Hence, in a pulse experiment, there need to be lengthy delays—of several seconds—between successive pulses so that the spin system has time to revert to equilibrium. Even gathering the murky information may therefore be a lengthy process. Moreover, because lines are so broad, very high powers of radiofrequency radiation may be required to achieve saturation. Whereas solution pulse NMR uses transmitters of a few watts, solid-state NMR may require transmitters rated at kilowatts.

The origins of linewidths in solids

There are two principal contributions to the linewidths of solids. One is the direct magnetic dipolar interaction between nuclear spins. As we saw in the discussion of spin–spin coupling, a nuclear magnetic moment will give rise to a local magnetic field

$$B_{\text{loc}} = -\frac{g_I \mu_N m_I \mu_0}{4\pi R^3}(1 - 3\cos^2\theta) \tag{20}$$

Unlike in solution, this field is not motionally averaged to zero. Many nuclei may contribute to the total local field experienced by a nucleus of interest, and different nuclei in a sample may experience a wide range of fields. Typical dipole–dipole fields are of the order of 10^{-3} T, which corresponds to splittings and linewidths of the order of 10^3 Hz.

A second source of linewidth is the anisotropy of the chemical shift. We have seen that chemical shifts arise from the ability of the applied field to generate electron currents in molecules. In general, this ability depends on the orientation of the molecule relative to the applied field. In solution, when the molecule is tumbling rapidly, only the average value of the chemical shift is relevant. However, the anisotropy is not averaged to zero for stationary molecules in a solid, and molecules in different orientations have resonances at different frequencies. The chemical shift anisotropy also varies with the angle between the applied field and the principal axis of the molecule as $1 - 3\cos^2\theta$.

The reduction of linewidths

Fortunately, there are techniques available for reducing the linewidths of solid samples. One technique, **magic-angle spinning** (MAS), takes note of the $1 - 3\cos^2\theta$ dependence of both the dipole–dipole interaction and the chemical shift anisotropy. The 'magic angle' is the angle at which $1 - 3\cos^2\theta = 0$, and corresponds to 54.74°. In the technique, the sample is spun at high speed at the magic angle to the applied field (Fig. 18.41). All the dipolar anisotropies average to the value they would have at the magic angle, but at that angle they are zero. The difficulty with MAS is

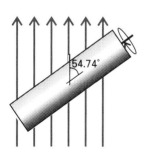

18.41 In magic-angle spinning, the sample spins at 54.74° (that is, arccos $1/\sqrt{3}$) to the applied magnetic field. Rapid motion at this angle averages dipole–dipole interactions and chemical shift anisotropies to zero.

Magnetic field

that the spinning frequency must not be less than the width of the spectrum, which is of the order of kilohertz. However, gas-driven sample spinners that can be rotated at 4 to 5 kHz are now routinely available, and a considerable body of work has been done.

The saturation and pulse techniques that we have described earlier in the chapter may also be used to reduce linewidths. The dipolar field of protons, for instance, may be averaged to zero by a decoupling procedure. However, because the range of coupling strengths is so large, radiofrequency power of the order of 1 kW is required. Elaborate pulse sequences have also been devised that eliminate linewidths to a considerable extent (by an averaging procedure that makes use of twisting the magnetization vector through an elaborate series of contortions). An example is the engagingly coined WAHUHA sequence devised by Waugh, Huber, and Haberlen:

$$\{\tau - 90_x^\circ - \tau - 90_{-y}^\circ - 2\tau - 90_y^\circ - \tau - 90_{-x}^\circ - \tau\}_n$$

with FID acquisition during one of the 2τ episodes.

ELECTRON SPIN RESONANCE

The energy levels of an electron spin in a magnetic field B (Fig. 18.42) are

$$E_{m_s} = g_e \mu_B m_s B, \qquad m_s = \pm \tfrac{1}{2} \tag{21}$$

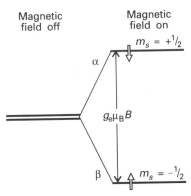

where μ_B is the Bohr magneton (Section 13.9) This equation shows that the energy of an α electron ($m_s = +\tfrac{1}{2}$) increases and the energy of a β electron ($m_s = -\tfrac{1}{2}$) decreases as the field is increased, and that the separation of the levels is

$$\Delta E = E_\alpha - E_\beta = g_e \mu_B B$$

When the sample is exposed to electromagnetic radiation of frequency ν, resonant absorption occurs when the resonance condition

$$h\nu = g_e \mu_B B \tag{22}$$

18.42 Electron spin levels in a magnetic field. Note that the β state is lower in energy than the α state (because the magnetogyric ratio of an electron is negative). Resonance is achieved when the frequency of the incident radiation matches the frequency corresponding to the energy separation.

is fulfilled. **Electron spin resonance** (ESR; or electron paramagnetic resonance, EPR), is the study of molecules containing unpaired electrons by observing the magnetic fields at which they come into resonance with monochromatic radiation. Magnetic fields of about 0.3 T (the value used in most commercial ESR spectrometers) correspond to resonance with an electromagnetic field of frequency 10 GHz (10^{10} Hz) and wavelength 3 cm. Because 3-cm radiation falls in the X-band of the microwave region of the electromagnetic spectrum, ESR is a microwave technique.

The layout of an ESR spectrometer is shown in Fig. 18.43. It consists of a microwave source (a klystron), a cavity in which the sample is inserted in a glass or quartz container, a microwave detector, and an electromagnet with a field that can be varied in the region of 0.3 T. The ESR spectrum is obtained by monitoring the microwave absorption as the field is changed, and a typical spectrum (of the benzene radical anion, $C_6H_6^-$) is shown in Fig. 18.44. The peculiar appearance of the spectrum,

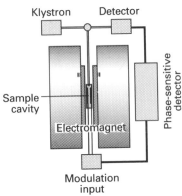

18.43 The layout of an ESR spectrometer. A typical magnetic field is 0.3 T, which requires 9-GHz (3 cm) microwaves for resonance.

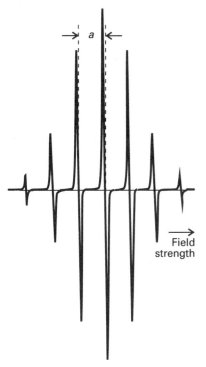

18.44 The ESR spectrum of the benzene radical anion, $C_6H_6^-$, in fluid solution. *a* is the hyperfine splitting of the spectrum; the centre of the spectrum is determined by the *g*-value of the radical.

which is in fact the first derivative of the absorption, arises from the detection technique, which is sensitive to the slope of the absorption curve (Fig. 18.45).

The sample molecules must possess unpaired electron spins, so ESR is less widely applicable than NMR. It is used to study radicals formed during chemical reactions or by radiation, many *d*-metal complexes, and molecules in triplet states (such as those involved in phosphorescence, Section 17.3). It is insensitive to normal, spin-paired molecules. The sample may be a gas, a liquid, or a solid, but the free rotation of molecules in the gas phase gives rise to complications.

18.9 The *g*-value

As in NMR, the spin magnetic moment interacts with the local magnetic field, so the resonance condition should be written

$$h\nu = g_e \mu_B B_{loc} = g_e \mu_B (1 - \sigma)B$$

However, in ESR it is conventional to write $g = (1 - \sigma)g_e$, where g is the **g-value** of the radical or complex. Then the resonance condition is

$$h\nu = g\mu_B B \tag{23}$$

Example 18.5 *Measuring a g-value*

The centre of the ESR spectrum of the methyl radical occurred at 329.4 mT in a spectrometer operating at 9.233 GHz. What is the *g*-value?

Method. We use eqn 23 in the form

$$g = \frac{h\nu}{\mu_B B}$$

A convenient factor is

$$\frac{h}{\mu_B} = 71.4448 \, mT \, GHz^{-1}$$

Answer. From the data,

$$g = 71.4448 \, mT \, GHz^{-1} \times \frac{9.233 \, GHz}{329.4 \, mT} = 2.0026$$

Comment. Many organic radicals have *g*-values close to 2.0026; inorganic radicals have *g*-values typically in the range 1.9 to 2.1; *d*-metal complexes have *g*-values in a wider range (e.g. 0 to 4).

Exercise E18.5. At what magnetic field would the methyl radical come into resonance in a spectrometer operating at 9.468 GHz?   [337.8 mT]

The deviation of g from $g_e = 2.0023$ depends on the ability of the applied field to induce local electron currents in the radical, and

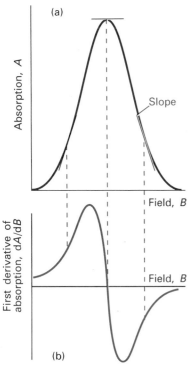

18.45 When phase-sensitive detection is used, the signal is the first derivative of the absorption intensity. (a) The absorption; (b) the signal, the slope of the absorption signal at each point. Note that the peak of the absorption corresponds to the point where the derivative passes through zero.

therefore its value gives some information about electronic structure. However, because g-values differ very little from g_e in many radicals (e.g. 2.003 for H, 1.999 for NO_2, 2.01 for ClO_2), its main use in chemical applications is to aid the identification of the species present in a sample.

18.10 Hyperfine structure

The most important feature of ESR spectra is their **hyperfine structure**, the splitting of individual resonance lines into components. In general in spectroscopy, the term 'hyperfine structure' means the structure of a spectrum that can be traced to interactions of the electrons with nuclei other than as a result of the latter's point electric charge. The source of the hyperfine structure in ESR is the magnetic interaction between the electron spin and the magnetic dipole moments of the nuclei present in the radical.

The effects of nuclear spin

Consider the effect on the ESR spectrum of a single H nucleus located somewhere in a radical. The proton spin is a source of magnetic field and, depending on the orientation of the nuclear spin, the field it generates adds to or subtracts from the applied field. The total local field is therefore

$$B_{loc} = B + am_I \qquad m_I = \pm\tfrac{1}{2}$$

where a is the **hyperfine coupling constant**. Half the radicals in a sample have $m_I = +\tfrac{1}{2}$, so half resonate when the applied field satisfies the condition

$$h\nu = g\mu_B(B + \tfrac{1}{2}a), \qquad \text{or } B = \frac{h\nu}{g\mu_B} - \tfrac{1}{2}a$$

The other half (which have $m_I = -\tfrac{1}{2}$) resonate when

$$h\nu = g\mu_B(B - \tfrac{1}{2}a), \qquad \text{or } B = \frac{h\nu}{g\mu_B} + \tfrac{1}{2}a$$

Therefore, instead of a single line, the spectrum shows two lines of half the original intensity separated by a and centred on the field determined by g (Fig. 18.46).

If the radical contains an N atom ($I = 1$), its ESR spectrum is split into three lines of equal intensity because the ^{14}N nucleus has three possible spin orientations, and each spin orientation is possessed by one third of all the radicals in the sample. In general, a spin-I nucleus splits the spectrum into $2I + 1$ hyperfine lines of equal intensity.

When there are several magnetic nuclei present in the radical, each one contributes to the hyperfine structure. In the case of equivalent protons (for example the two CH_2 protons in the radical $CH_3CH_2\cdot$) some of the hyperfine lines are coincident. It is not hard to show, if the radical contains N equivalent protons, that there are $N + 1$ hyperfine lines with a binomial intensity distribution (i.e. the intensity distribution given by Pascal's triangle). The spectrum of the benzene radical anion in Fig.

18.44, which has seven lines with intensity ratio 1:6:15:20:15:6:1, is consistent with a radical containing six equivalent protons.

Example 18.6 *Predicting the hyperfine structure of an ESR spectrum*

A radical contains one ^{14}N nucleus ($I = 1$) with hyperfine constant 1.61 mT and two equivalent protons ($I = \frac{1}{2}$) with hyperfine constant 0.35 mT. Predict the form of the ESR spectrum.

Method. We should consider the hyperfine structure that arises from each type of nucleus or group of equivalent nuclei in succession. So, split a line with one nucleus, then each of those lines is split by a second nucleus (or group of nuclei), and so on. It is best to start with the nucleus with the largest hyperfine splitting; however, any choice could be made, and the order in which nuclei are considered does not affect the conclusion.

Answer. The ^{14}N nucleus gives three hyperfine lines of equal intensity separated by 1.61 mT. Each line is split into doublets of spacing 0.35 mT by the first proton, and each line of these doublets is split into doublets with the same 0.35-mT splitting (Fig. 18.47). The central lines of each split doublet coincide, and so the proton splitting gives 1:2:1 triplets of internal splitting 0.35 mT. Therefore, the spectrum consists of three equivalent 1:2:1 triplets.

Comment. Often it is quicker to realize that a group of equivalent protons gives characteristic hyperfine patterns (two giving a 1:2:1 triplet, in this case), and to superimpose the patterns directly.

Exercise E18.6. Predict the form of the ESR spectrum of a radical containing three equivalent ^{14}N nuclei.   [Fig. 18.48]

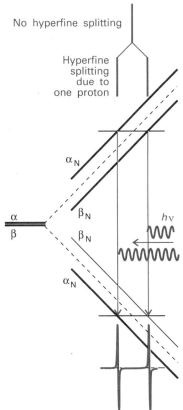

18.46 The hyperfine interaction between an electron and a spin-$\frac{1}{2}$ nucleus results in four energy levels in place of the original two. As a result, the spectrum consists of two lines (of equal intensity) instead of one. The intensity distribution can be summarized by a simple stick diagram.

The hyperfine structure of an ESR spectrum is a kind of fingerprint that helps to identify the radicals present in a sample. Moreover, because the magnitude of the splitting depends on the distribution of the unpaired electron near the magnetic nuclei present, the spectrum can be used to map the molecular orbital occupied by the unpaired electron. For example, because the hyperfine splitting in $C_6H_6^-$ is 0.375 mT, and one proton is close to a C atom with one-sixth the unpaired electron spin density (because the electron is spread uniformly around the ring), the hyperfine splitting caused by a proton in the electron spin entirely confined to a single adjacent C atom should be 6×0.375 mT = 2.25 mT. If in another aromatic radical we find a hyperfine splitting constant a, then the spin density there can be calculated from the **McConnell equation**:

$$a = Q\rho, \qquad \text{with } Q = 2.25 \text{ mT} \tag{24}$$

In this equation, ρ is the unpaired electron spin density on a C atom and a is the hyperfine splitting observed for the H atom to which it is attached.

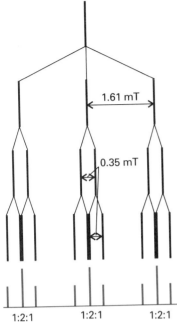

1.61 mT

0.35 mT

1:2:1   1:2:1   1:2:1

18.47 The analysis of the hyperfine structure of radicals containing one ^{14}N nucleus ($I = 1$) and two equivalent protons.

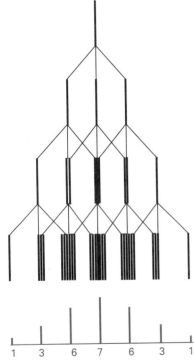

1 3 6 7 6 3 1

18.48 The analysis of the hyperfine structure of radicals containing three equivalent ^{14}N nuclei.

Example 18.7 *Mapping the unpaired spin density*

The hyperfine structure of the ESR spectrum of (naphthalene)$^-$ can be interpreted as arising from two groups of four equivalent protons. Those at the α positions in the ring have $a = 0.490$ mT and for those in the β positions $a = 0.183$ mT. Map the unpaired spin density round the ring.

Method. The spin densities are obtained by using the McConnell equation and calculating ρ at each C atom of the ring (provided it is attached to an H atom).

Answer. The data convert to $\rho = 0.22$ for the α positions and $\rho = 0.08$ for the β positions (**3**).

```
           0.22   0.22
     0.08 ⬡⬡ 0.08
     0.08 ⬡⬡ 0.08
           0.22   0.22
```

3

Comment. Notice how the unpaired electron density accumulates in the α positions. The same value of Q may be used for the approximate mapping of spin density in heterocyclic radicals.

Exercise E18.7. The spin density in (anthracene)$^-$ is shown in (**4**). Predict the form of its ESR spectrum.

```
     0.193  0.097
    ⬡⬡⬡ 0.048
```

4

[A 1:2:1 triplet of splitting 0.43 mT split into a 1:4:6:4:1 quintet of splitting 0.22 mT, split into a 1:4:6:4:1 quintet of splitting 0.11 mT, $3 \times 5 \times 5 = 75$ lines in all]

The origin of the hyperfine interaction

The hyperfine interaction is an interaction between the magnetic moments of the unpaired electron and the nuclei. There are two contributions to the interaction.

An electron in a p orbital does not approach the nucleus very closely, so it experiences a field that appears to arise from a point magnetic dipole. The resulting interaction is called the **dipole–dipole interaction**. The contribution of a magnetic nucleus to the local field experienced by the unpaired electron is given by an expression like that in eqn 10. A characteristic of this type of interaction is that it is anisotropic: that is, its magnitude (and sign) depends on the orientation of the radical with respect to the applied field. Furthermore, just as in the case of NMR, the dipole–dipole interaction averages to zero when the radical is free to tumble. Therefore, hyperfine structure due to the dipole–dipole interaction is observed only for radicals trapped in solids.

Table 18.2* Hyperfine coupling constants for atoms, a/mT

Nuclide	Isotropic coupling	Anisotropic coupling
^1H	50.8 (1s)	
^2H	7.8 (1s)	
^{14}N	55.2 (2s)	3.4 (2p)
^{19}F	1720 (2s)	108.4 (2p)

* More values are given in the Data section.

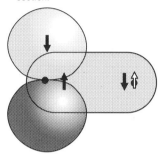

(b) High energy

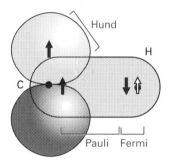

(a) Low energy

18.49 The polarization mechanism for the hyperfine interaction in π-electron radicals. The arrangement in (a) is lower in energy than that in (b), and so there is an effective coupling between the unpaired electron and the proton.

An s electron is spherically distributed around a nucleus and so has zero average dipole–dipole interaction with it even in a solid sample. However, because an s electron has a non-zero probability of being at the nucleus, it is incorrect to treat the interaction as one between two point dipoles. An s electron has a Fermi contact interaction with the nucleus, which as we saw earlier is a magnetic interaction that occurs when the point dipole approximation fails. The contact interaction is isotropic (that is, independent of the radical's orientation), and consequently is shown even by rapidly tumbling molecules in fluids (so long as the spin density has some s-character).

The dipole–dipole interactions of p electrons and the Fermi contact interaction of s electrons can be quite large. For example, a $2p$ electron in a nitrogen atom experiences an average field of about 3.4 mT from the ^{14}N nucleus. A $1s$ electron in a hydrogen atom experiences a field of about 50 mT as a result of its Fermi contact interaction with the central proton. More values are listed in Table 18.2. The magnitudes of the contact interactions in radicals can be interpreted in terms of the s orbital character of the molecular orbital occupied by the unpaired electron, and the dipole–dipole interaction can be interpreted in terms of the p character. The analysis of hyperfine structure therefore gives information about the composition of the orbital, and especially the hybridization of the atomic orbitals (see Problem 18.6).

We still have the source of the hyperfine structure of the $C_6H_6^-$ anion and other aromatic radical anions to explain. The sample is fluid and, as the radicals are tumbling, the hyperfine structure cannot be due to the dipole–dipole interaction. Moreover, the protons lie in the nodal plane of the π orbital occupied by the unpaired electron, so the structure cannot be due to a Fermi contact interaction. The explanation lies in a **polarization mechanism** similar to the one responsible for spin–spin coupling in NMR. There is a magnetic interaction between a proton and the σ electrons that results in one of the electrons tending to be found with a greater probability nearby (Fig. 18.49). The other electron is therefore more likely to be close to the C atom at the other end of the bond. The unpaired electron on the C atom has a lower energy if it is parallel to that electron (Hund's rule favours parallel electrons on atoms), so the unpaired electron can detect the spin of the proton indirectly. Calculation using this model leads to a hyperfine interaction in agreement with the observed value of 2.8 mT.

CHECK LIST OF KEY IDEAS

1 The **magnetic moments** of nuclei, their energies in magnetic fields (eqn 3), and the **resonance condition** (Section 18.1).

2 The technique of **nuclear magnetic resonance** (Section 18.1).

3 The **shielding constant** and the **chemical shift** (Section 18.2), and the δ-scale of chemical shifts (eqn 5).

4 The origin of chemical shifts, including the **neighbouring group effect** and **ring currents** (Section 18.2).

5 The **fine structure** of NMR spectra, the **scalar coupling constant**, and the patterns expected from particular groups of nuclei (Section 18.3).

6 The origin of the spin–spin coupling, including the **polarization mechanism** and the **Fermi contact interaction** (Section 18.3).

7 The concepts of **chemically equivalent nuclei** and **magnetically equivalent nuclei** (Section 18.3).

8 **First-order** and **second-order** spectra, **dilute spins** and **abundant spins**, and **proton decoupling** (Section 18.3).

9 The **magnetization vector**, its **precession**, and the effect of a **radiofrequency field** (Section 18.4).

10 The effect of a **90° pulse** and the **free-induction decay** signal (Section 18.4).

11 **Fourier transforms** and the relation between **time-domain spectra** and **frequency-domain spectra** (eqn 12).

12 **Spin relaxation** and the significance of the longitudinal **relaxation time** and the **transverse relaxation time** (Section 18.5).

13 The effect of **inhomogeneities** and of **exchange processes** on the appearance of spectra (Section 18.5).

14 The measurement of relaxation times by **inversion recovery** and **spin echoes** (Section 18.5).

15 The **nuclear Overhauser effect** and its use for the enhancement of resonance intensities (Section 18.6).

16 The basis of **two-dimensional NMR** and its use for the interpretation of complex spectra (Section 18.7).

17 The application of NMR to the study of solid samples and the reduction in linewidth by **magic-angle spinning** (Section 18.8).

18 The **electron spin resonance** technique and the electron *g***-value** (Section 18.9).

19 The **hyperfine structure** of ESR spectra, its interpretation, and its origin (Section 18.10).

EXERCISES

18.1. From their atomic numbers and mass numbers, decide whether the following nuclei are likely to have zero, half-integral, or integral spin: (a) ^{18}O, (b) ^{10}B, (c) ^{14}N, (d) ^{19}F, (e) ^{35}Cl.

18.2. What is the resonance frequency of a proton in a magnetic field of 14.1 T?

18.3. ^{32}S has a nuclear spin of $\frac{3}{2}$ and a nuclear *g*-factor of 0.4289. Calculate the energies of the nuclear spin states in a magnetic field of 7.500 T.

18.4. In which of the following systems is the energy level separation the largest: (a) a proton in a 600-MHz NMR spectrometer, (b) a deuteron in the same spectrometer, (c) a ^{14}N nucleus in the same spectrometer, (d) an electron in a radical in a field of 0.300 T?

18.5. Calculate the energy difference between the lowest and highest nuclear spin states of a ^{14}N nucleus ($I = 1$, $g = 0.4036$) in a 15.00-T magnetic field.

18.6. Calculate the magnetic field needed to satisfy the resonance condition for unshielded protons in a 150.0-MHz radiofrequency field.

18.7. Use Table 18.1 to predict the magnetic fields at which ^{1}H, ^{2}H, ^{13}C, ^{14}N, ^{19}F, and ^{31}P come into resonance at (a) 250 MHz, (b) 500 MHz.

18.8. Calculate the relative population differences ($\delta N/N$) for protons in fields of (a) 0.3 T, (b) 1.5 T, and (c) 10 T at 25°C.

18.9. The first generally available NMR spectrometers operated at a frequency of 60 MHz; today it is not uncommon to use a spectrometer that operates at 600 MHz. What are the relative population differences of ^{13}C spin states in these two spectrometers at 25°C? What are the relative values of the chemical shifts observed for nuclei in these spectrometers in terms of (a) δ values, (b) frequencies?

18.10. The chemical shift of the CH_3 protons in acetaldehyde (ethanal) is $\delta = 2.20$ and that of the CHO proton is 9.80. What is the difference in local magnetic field between the two regions of the molecule when the applied field is (a) 1.5 T, (b) 15 T?

18.11. The absolute values of the proton shielding constants for HF and HI are 2.871×10^{-5} and 4.447×10^{-5}, respectively. What is the frequency separation of the resonances in a 500-MHz NMR spectrometer?

18.12. Use the information in Exercise 18.10 to state the splitting (in Hz) between the methyl and aldehydic proton resonances in a spectrometer operating at (a) 250 MHz, (b) 500 MHz.

18.13. Sketch the appearance of the ^{1}H-NMR spectrum of acetaldehyde using $J = 2.90$ Hz and the data in Exercise 18.10 in a spectrometer operating at (a) 250 MHz, (b) 500 MHz.

18.14. Two groups of protons are made equivalent by the isomerization of a fluxional molecule. At low temperatures, where the interconversion is slow, one group has $\delta = 4.0$ and the other has $\delta = 5.2$. At what rate of interconversion will the two signals merge in a spectrometer operating at 250 Hz?

18.15. Sketch the form of the ^{19}F-NMR spectra of a natural sample of tetrafluoroborate ions, BF_4^-, allowing for the relative abundances of $^{10}BF_4^-$ and $^{11}BF_4^-$.

18.16. From the data in Table 18.1, predict the frequency needed for ^{31}P-NMR in an NMR spectrometer designed to observe proton resonance at 500 MHz. Sketch the proton and ^{31}P resonances in the NMR spectrum of PH_4^+.

18.17. Sketch the form of an $A_3M_2X_4$ spectrum, where A, M, and X are protons with distinctly different chemical shifts and $J_{AM} > J_{AX} > J_{MX}$.

18.18. Which of the following molecules have sets of nuclei that are chemically but not magnetically equivalent: (a) CH_3CH_3, (b) $CH_2{=}CH_2$, (c) $CH_2{=}C{=}CF_2$, (d) *cis*- and *trans*-$[Mo(CO)_4(PH_3)_2]$?

18.19. The duration of a 90° or 180° pulse depends on the strength of the B_1 field. If a 90° pulse in proton NMR requires 10 μs, what is the strength of the B_1 field? How long would the corresponding 180° pulse require?

18.20. What magnetic field would be required in order to use an ESR X-band spectrometer (9 GHz) to observe ^{1}H-NMR and a 300-MHz spectrometer to observe ESR?

18.21. Some commercial ESR spectrometers use 8-mm microwave radiation (the Q band). What magnetic field is needed to satisfy the resonance condition?

18.22. It is possible to produce very high magnetic fields over small volumes by special techniques. What would be the resonance frequency of an electron spin in an organic radical in a field of 1.0 kT? How does this frequency compare to typical molecular rotational, vibrational, and electronic energy-level separations?

18.23. The centre of the ESR spectrum of atomic hydrogen lies at 329.12 T in a spectrometer operating at 9.2231 GHz. What is the g-value of the atom?

18.24. A radical containing two equivalent protons shows a three-line spectrum with an intensity distribution 1:2:1. The lines occur at 330.2, 332.5, and 334.8 mT. What is the hyperfine coupling constant for each proton? What is the g-value of the radical given that the spectrometer is operating at 9.319 GHz?

18.25. A radical containing two inequivalent protons with hyperfine constants 2.0 and 2.6 mT gives a spectrum centred on 332.5 mT. At what fields do the hyperfine lines occur and what are their relative intensities?

18.26. Predict the intensity distribution in the hyperfine lines of the ESR spectra of (a) CH_3, (b) CD_3.

18.27. The benzene radical anion has $g = 2.0025$. At what field should you search for resonance in a spectrometer operating at (a) 9.302 GHz, (b) 33.67 GHz?

18.28. The fluxional lifetime of a molecule is 200 ms. At 100 MHz, the difference between the two forms is reflected in a separation of 90.0 Hz between the two resonances in the NMR spectrum. Determine whether or not the two lines will be merged or resolved in the observed spectrum.

18.29. The ESR spectrum of a radical with a single magnetic nucleus is split into four lines of equal intensity. What is the nuclear spin of the nucleus?

18.30. Sketch the form of the hyperfine structures of radicals XH_2 and XD_2, where the nucleus X has $I = \frac{5}{2}$.

PROBLEMS
Numerical problems

18.1. A scientist investigates the possibility of neutron spin resonance, and has available a commercial NMR spectrometer operating at 300 MHz. What field is required for resonance? What is the relative population difference at room temperature? Which is the lower energy spin state of the neutron?

18.2. Two groups of protons have $\delta = 4.0$ and $\delta = 5.2$, respectively, and are interconverted by a conformational change of a fluxional molecule. In a 60-MHz spectrometer the spectrum collapsed into a single line at 280 K but at 300 MHz the collapse did not occur until the temperature had been raised to 300 K. What is the activation energy of the interconversion?

18.3. The angular NO_2 molecule has a single unpaired electron and can be trapped in a solid matrix or prepared inside a nitrite crystal by radiation damage of NO_2^- ions. When the applied field is parallel to the OO direction the centre of the spectrum lies at 333.64 mT in a spectrometer operating at 9.302 GHz. When the field lies along the bisector of the ONO angle, the resonance lies at 331.94 mT. What are the g-values in the two orientations?

18.4. The hyperfine coupling constant in CH_3 is 2.3 mT. Use the information in Table 18.2 to predict the splitting between the hyperfine lines of the spectrum of CD_3. What are the overall widths of the hyperfine spectra in each case?

18.5. The p-dinitrobenzene radical anion can be prepared by reduction of p-dinitrobenzene. The radical anion has two equivalent N nuclei ($I = 1$) and four equivalent protons. Predict the form of the ESR spectrum using $a(N) = 0.148$ mT and $a(H) = 0.112$ mT.

18.6. When an electron occupies a $2s$ orbital on an N atom it has a hyperfine interaction of 55.2 mT with the nucleus. The spectrum of NO_2 shows an isotropic hyperfine interaction of 5.7 mT. For what proportion of its time is the unpaired electron of NO_2 occupying a $2s$ orbital? The hyperfine coupling constant for an electron in a $2p$ orbital of an N atom is 3.4 mT. In NO_2 the anisotropic part of the hyperfine coupling is 1.3 mT. What proportion of its time does the unpaired electron spend in the $2p$ orbital of the N atom in NO_2? What is the total probability that the electron will be found on (a) the N atom, (b) the O atoms? What is the hybridization ratio of the N atom? Does the hybridization support the view that NO_2 is angular? Use the discussion of the hybridization ratio in Section 14.3 to interpret the hybridization as a bond angle. (The experimental bond angle is 134°.)

18.7. The hyperfine coupling constants observed in the radical anions (**1′**, **2′**, and **3′**) are shown (in mT). Use the value for the benzene radical anion to map the probability of finding the unpaired electron in the π orbital on each C atom.

Theoretical problems

18.8. The z-component of the magnetic field at a distance R from a magnetic moment parallel to the z-axis is given by eqn 20. In a solid, a proton at a distance R from another proton can experience such a field and the measurement of the splitting it causes in the spectrum can be used to calculate R. In gypsum, for instance, the splitting in the H_2O resonance can be interpreted in terms of a magnetic field of 0.715 mT generated by one proton and experienced by the other. What is the separation of the protons in the H_2O molecule?

18.9. In a liquid the dipolar magnetic field averages to zero: show this by evaluating the average of the field given in eqn 20. (*Hint.* The volume element is proportional to $\sin\theta \, d\theta \, d\phi$ in polar coordinates.)

18.10. In a liquid crystal a molecule might not rotate freely in all directions and the dipolar interaction might not average to zero. Suppose a molecule is trapped so that, although the vector separating two protons may rotate freely around the z-axis, the colatitude may vary only between 0 and θ'. Average the dipolar field over this restricted range of orientations and confirm that the average vanishes when θ' is equal to π (corresponding to free rotation over a sphere). What is the average value of the local dipolar field for the H_2O molecule in Problem 18.8 if it is dissolved in a liquid crystal that enables it to rotate up to $\theta' = 30°$?

18.11. The shape of a spectral line $I(\omega)$ is related to the free induction decay signal $G(t)$ by

$$I(\omega) = A \operatorname{Re} \int_0^\infty G(t) e^{i\omega t} \, dt$$

where A is a constant and Re means take the real part of what follows. Calculate the line shape corresponding to an oscillating, decaying function

$$G(t) = \cos \omega_0 t \, e^{-t/\tau}$$

18.12. In the language of Problem 18.11, show that if

$$G(t) = (a \cos \omega_1 t + b \cos \omega_2 t) e^{-t/\tau}$$

then the spectrum consists of two lines with intensities proportional to a and b and located at $\omega = \omega_1$ and $\omega = \omega_2$, respectively.

18.13. Write a computer program to construct $G(t)$ for spectra with absorption lines at arbitrary frequencies, and explore the different free induction decay signals that are associated with different numbers of protons and patterns of coupling.

19

Statistical thermodynamics: the concepts

Statistical thermodynamics provides the link between the microscopic properties of matter and its bulk properties. In the course of the next two chapters we establish relations which, by the end of Chapter 20, will allow us to calculate equilibrium constants from spectroscopic data on molecules.

Two key ideas are introduced in this chapter. The first is the Boltzmann distribution. This enormously important result was introduced in the 'Introduction', where we saw that it can be used to predict the populations of states. In this chapter we see its derivation in terms of some general concepts about the distribution of particles over available states. The derivation leads naturally to the introduction of the partition function, which is the central mathematical concept of these two chapters. The partition function is like a thermodynamic wavefunction, in the sense that it contains all the thermodynamic information about the system, just as the quantum mechanical wavefunction contains all the dynamic information. We see how to interpret the partition function and how to calculate it in a number of simple cases.

The next part of the chapter shows how to extract thermodynamic information from the partition function. In this chapter we introduce two fundamental calculations relating to the First and Second Laws of thermodynamics: the calculation of the internal energy and the entropy of a system. The chapter shows how the temperature dependence of these two fundamental thermodynamic properties can be calculated for a few simple cases.

In the final part of the chapter, we generalize the discussion to include systems that are composed of assemblies of interacting particles. Very similar equations are developed to those in the first part of the chapter, but they are much more widely applicable. One application is to the calculation of the entropy of a gas of indistinguishable molecules, which concludes the chapter and provides a basis for the chemical applications in the next chapter.

The preceding chapters of this part of the text have shown how the energy levels of molecules can be calculated, measured spectroscopically, and related to their sizes and shapes. The next major step is to see how a knowledge of these energy levels can be used to account for the properties of matter in bulk. To do so, we now introduce the concepts of **statistical thermodynamics**, the link between molecular properties and bulk thermodynamic properties.

The crucial step in going from the quantum mechanics of individual molecules to the thermodynamics of bulk samples is to recognize that the latter deals with *average* behaviour. For example, the pressure of a gas is the average force per unit area exerted by its molecules, and there is no need to specify which molecules happen to be striking the wall at any instant. Nor is it necessary to consider the fluctuations in the pressure as different numbers of molecules collide with the wall at different moments. The fluctuations in pressure are very small compared with the steady pressure: it is highly improbable that there will be a sudden lull in the number of collisions, or a sudden storm.

This chapter introduces statistical thermodynamics in two stages. The first, the derivation of the Boltzmann distribution for individual particles, is of restricted applicability, but it has the advantage of taking us directly to a result of central importance in a straightforward and elementary way. We can *use* statistical thermodynamics once we have deduced the Boltzmann distribution. Then (in Section 19.5) we elaborate the arguments a little so as to be able to cope with systems composed of interacting particles.

THE DISTRIBUTION OF MOLECULAR STATES

We shall consider a system composed of N molecules. Although the total energy is constant at E, it is not possible to be definite about how that energy is shared between the molecules. Collisions take place, and they result in the ceaseless redistribution of energy not only between the molecules but also among their different modes of motion. The closest we can come to a description of the distribution of energy is the statement of the **population** of a state, the average number of molecules that occupy it, and to say that, on average, there are n_i molecules in a state of energy ε_i. The populations of the states remain almost constant, but the precise identities of the molecules in each state may change at every collision.

The problem we address in this section is the calculation of the populations of states for any type of molecule in any mode of motion at any temperature. The only restriction is that the molecules should be independent, in the sense that the total energy of the system is a sum of their individual energies. We are discounting (at this stage) the possibility that in a real system a contribution to the total energy may arise from interactions between molecules. We shall also adopt the **principle of equal *a priori* probabilities**,[1] the assumption that all possibilities for the

---

1 *A priori* means in this context loosely 'as far as one knows'. We have no reason to presume otherwise than that all states are equally likely to be occupied whatever their nature.

distribution of energy are equally probable. That is, we assume that vibrational states of a certain energy, for instance, are as likely to be populated as rotational states of the same energy.

19.1 Configurations and weights

Any individual molecule may exist in states with energies $\varepsilon_0, \varepsilon_1, \ldots$. We shall always take ε_0, the lowest state, as the zero of energy ($\varepsilon_0 = 0$), and measure all other energies relative to that state. As a result, we may have to add a constant to the calculated energy of the system in order to obtain the internal energy, U. For example, if we are considering the vibrational contribution to the internal energy, then we must add the total zero-point energy of the oscillators in the sample because the energy of the lowest state is not in fact zero.

Instantaneous configurations

At any instant there will be n_0 molecules in the state with energy ε_0, n_1 with ε_1, and so on. The specification of the set of populations $n_0, n_1, \ldots$ in the form $\{n_0, n_1, \ldots\}$ is a statement of the instantaneous **configuration** of the system. The instantaneous configuration fluctuates with time because the populations change. We can picture a large number of different instantaneous configurations. One, for example, might be $\{N, 0, 0, \ldots\}$, corresponding to every molecule being in its ground state. Another might be $\{N-2, 2, 0, 0, \ldots\}$, in which two molecules are in the first excited state. The latter configuration is intrinsically more likely to be found than the former because it can be achieved in more ways: $\{N, 0, 0, \ldots\}$ can be achieved in only one way, but $\{N-2, 2, 0, \ldots\}$ can be achieved in $\frac{1}{2}N(N-1)$ different ways (Fig. 19.1).[2]

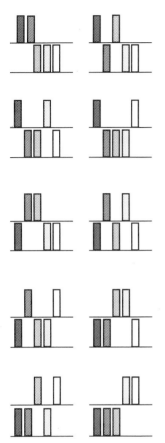

19.1 Whereas a configuration $\{5, 0, 0, \ldots\}$ can be achieved in only one way, a configuration $\{3, 2, 0, \ldots\}$ can be achieved in the 10 different ways shown here, where the tinted blocks represent different molecules.

JUSTIFICATION

One candidate for promotion to an upper state can be selected in N ways. There are $N-1$ candidates for the second choice, so the total number of choices is $N(N-1)$. However, we should not distinguish the choice (Jack, Jill) from the choice (Jill, Jack) because they lead to the same arrangements. Therefore, only half of the choices lead to distinguishable arrangements, and the total number of distinguishable choices is $\frac{1}{2}N(N-1)$.

If, as a result of collisions, the system were to fluctuate between the configurations $\{N, 0, 0, \ldots\}$ and $\{N-2, 2, 0, \ldots\}$, then it would almost always be found in the second, more likely state (especially if N were large). In other words, a system free to switch between the two configurations would show properties characteristic almost exclusively of the second configuration.

2 At this stage in the argument, we are ignoring the requirement that the total energy of the system should be constant (the second configuration has a higher energy than the first). The constraint of total energy is imposed later in this section.

In general, the configuration $\{n_0, n_1, \ldots\}$ can be achieved in W different ways, where W is called the **weight** of the configuration. The weight of the configuration $\{n_0, n_1, \ldots\}$ is given by the expression

$$W = \frac{N!}{n_0! \, n_1! \, n_2! \, \ldots} \tag{1}$$

where $x!$, x factorial, denotes $x(x-1)(x-2)\ldots 1$, and by definition $0! = 1$. This expression is a generalization of the formula $W = \frac{1}{2}N(N-1)$, and reduces to it for the configuration $\{N-2, 2, 0, \ldots\}$.

JUSTIFICATION

Consider the number of ways of distributing N balls into bins. The first ball can be selected in N different ways, the next ball in $N-1$ different ways for the balls remaining, and so on. Therefore, there are $N(N-1)\ldots 1 = N!$ ways of selecting the balls for distribution over the bins. However, if there are n_0 balls in the bin labelled ε_0, there would be $n_0!$ different ways in which the same balls could have been chosen (Fig. 19.2). Similarly, there are $n_1!$ ways in which the n_1 balls in the bin labelled ε_1 can be chosen, and so on. Therefore, the total number of distinguishable ways of distributing the balls so that there are n_0 in bin ε_0, n_1 in bin ε_1, etc. regardless of the order in which the balls were chosen is $N!/n_0! \, n_1! \, \ldots$, which is the content of eqn 1.

19.2 The 18 molecules shown here can be distributed into four receptacles (distinguished by the three vertical lines) in 18! different ways. However, 3! of the selections that put three molecules in the first receptacle are equivalent, 6! that put six molecules into the second receptacle are equivalent, and so on. Hence the number of distinguishable arrangements is 18!/3! 6! 5! 4!.

$N = 18$

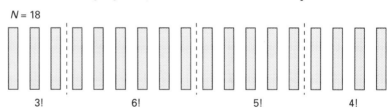

3!   6!   5!   4!

Example 19.1 *Calculating the weight of a configuration*

Calculate the number of ways of distributing 20 identical objects into six boxes, with the arrangement 1, 0, 3, 5, 10, 1.

Method. The calculation is a straightforward application of eqn 1. We begin by writing the configuration of the system, and then use eqn 1. Note that $0! = 1$.

Answer. The configuration is $\{1, 0, 3, 5, 10, 1\}$ with $N = 20$; therefore the weight is

$$W = \frac{20!}{1! \times 0! \times 3! \times 5! \times 10! \times 1!} = 9.31 \times 10^8$$

Exercise E19.1. Calculate the weight of the configuration in which 20 objects are distributed in the arrangement 0, 1, 5, 0, 8, 0, 3, 2, 0, 1.

$$[4.19 \times 10^{10}]$$

It will turn out to be more convenient to deal with the natural logarithm of the weight, ln W, rather than with the weight itself. We shall therefore need the expression

$$\ln W = \ln\left(\frac{N!}{n_0!\, n_1!\, n_2!\, \ldots}\right) = \ln N! - \ln(n_0!\, n_1!\, n_2!\, \ldots)$$

$$\left[\text{because } \ln\left(\frac{x}{y}\right) = \ln x - \ln y\right]$$

$$= \ln N! - (\ln n_0! + \ln n_1! + \ln n_2!\, \ldots)$$

$$[\text{because } \ln(xy) = \ln x + \ln y]$$

$$= \ln N! - \sum_i \ln n_i!$$

One reason for introducing ln W is that it is easier to make approximations. In particular, we can simplify the factorials by using **Stirling's approximation** in the form[3]

$$\ln x! \approx x \ln x - x \tag{2}$$

Then the approximate expression for the weight is

$$\ln W = (N \ln N - N) - \sum_i (n_i \ln n_i - n_i)$$

$$= N \ln N - \sum_i n_i \ln n_i \tag{3}$$

The second line is derived by noting that the sum of n_i is equal to N, so that the second and fourth terms on the right in the first line cancel.

The dominating configuration

We have seen that the configuration $\{N-2, 2, 0, \ldots\}$ dominates $\{N, 0, 0, \ldots\}$, and it should be easy to believe that there may be other configurations that greatly dominate both. We shall see, in fact, that there is a configuration with so great a weight that it overwhelms all the rest in importance to such an extent that the system will almost always be found in it. The properties of the system will therefore be characteristic of that particular dominating configuration. This dominating configuration can be found by looking for the values of n_i that lead to a maximum value of W. Because W is a function of all the n_i, we can do this search by varying the n_i and looking for the values that correspond to $dW = 0$ (just as in the search for the maximum of any function). However, there are two difficulties with this procedure.

The first difficulty is that the only permitted configurations are those corresponding to the specified, constant, total energy of the system. This requirement rules out many configurations; $\{N, 0, 0, \ldots\}$ and

---

3 The precise form of the approximation is

$$x! = (2\pi)^{1/2} x^{x+1/2} e^{-x}$$

and is reliable when x is greater than about 10. We deal with far larger values of x, and the simplified version in eqn 2 is normally adequate.

$\{N-2, 2, 0, \ldots\}$, for instance, have different energies, so both cannot occur in the same isolated system. It follows that in looking for the configuration with the greatest weight, we must ensure that the configuration also satisfies the condition

Constant total energy: $\sum_i n_i \varepsilon_i = E$

where E is the total energy of the system.

The second constraint is that, because the total number of molecules present is also fixed (at N), we cannot arbitrarily vary all the populations simultaneously. Thus, increasing the population of one state by 1 demands that the population of another state must be reduced by 1. Therefore, the search for the maximum value of W is also subject to the condition

Constant total number of molecules: $\sum_i n_i = N$

The Boltzmann distribution

We are looking for the set of numbers $n_0, n_1, \ldots$ for which W has its maximum value. We show in the following *Justification* that the populations in the configuration of greatest weight depend on the energy of the state according to the expression

$$\frac{n_i}{N} = e^{\alpha - \beta \varepsilon_i} \tag{4}$$

where α and β are constants. Already the exponential dependence on the energy that is typical of the Boltzmann distribution is apparent.

JUSTIFICATION

Although we are looking for the maximum value of W, it is equivalent, and it turns out to be simpler, to find the condition for $\ln W$ being a maximum. Because $\ln W$ depends on all the n_i, when a configuration changes and the n_i change to $n_i + dn_i$, the function $\ln W$ changes to $\ln W + d \ln W$, where

$$d \ln W = \sum_i \left(\frac{\partial \ln W}{\partial n_i} \right) dn_i$$

At a maximum, $d \ln W = 0$. However, when the n_i change, they do so subject to the two constraints

$$\sum_i \varepsilon_i \, dn_i = 0 \quad \text{and} \quad \sum_i dn_i = 0$$

The first states that the total energy must not change, and the second states that the total number of molecules must not change. These two constraints prevent us from solving $d \ln W = 0$ simply by setting all $(\partial \ln W / \partial n_i) = 0$ because the dn_i are not all independent.

The way to take constraints into account was devised by the French mathematician Lagrange, and is called the **method of undetermined multipliers**. The technique is described in *Further information 14*. All we need here is the rule that *a constraint should be multiplied by a constant and then added to the main variation equation*. The variables are then treated as though they were all independent, and the constants are evaluated at the end of the calculation.

We employ the technique as follows. The two constraints are multiplied by the constants $-\beta$ and α, respectively (the minus sign in $-\beta$ has been included for future convenience), and then added to the expression for d ln W:

$$\mathrm{d} \ln W = \sum_i \left(\frac{\partial \ln W}{\partial n_i}\right) \mathrm{d}n_i + \alpha \sum_i \mathrm{d}n_i - \beta \sum_i \varepsilon_i \, \mathrm{d}n_i$$

$$= \sum_i \left\{\left(\frac{\partial \ln W}{\partial n_i}\right) + \alpha - \beta \varepsilon_i\right\} \mathrm{d}n_i$$

All the dn_i are now treated as independent. Hence the only way of satisfying d ln $W = 0$ is to require that, for each i,

$$\left(\frac{\partial \ln W}{\partial n_i}\right) + \alpha - \beta \varepsilon_i = 0 \qquad \qquad \textbf{(A)}$$

when the n_i have their most probable values.

The expression for ln W is given in eqn 3. Differentiation of it with respect to n_i gives[4]

$$\frac{\partial \ln W}{\partial n_i} = -\ln\left(\frac{n_i}{N}\right)$$

---

4 The argument runs as follows. First, from eqn 3 we can write the required derivative as

$$\left(\frac{\partial \ln W}{\partial n_i}\right) = \frac{\partial (N \ln N)}{\partial n_i} - \sum_j \left(\frac{\partial (n_j \ln n_j)}{\partial n_i}\right)$$

and deal with the two derivatives separately. The derivative of the $N \ln N$ term is obtained as follows:

$$\frac{\partial (N \ln N)}{\partial n_i} = \left(\frac{\partial N}{\partial n_i}\right) \ln N + \frac{\partial N}{\partial n_i} = \ln N + 1$$

because $N = n_1 + n_2 + \ldots$, and so its derivative with respect to any of the n_i is 1. The derivative of the remaining term is

$$\sum_j \left(\frac{\partial (n_j \ln n_j)}{\partial n_i}\right) = \sum_j \left\{\left(\frac{\partial n_j}{\partial n_i}\right) \ln n_j + n_j \left(\frac{\partial \ln n_j}{\partial n_i}\right)\right\} = \sum_j \left(\frac{\partial n_j}{\partial n_i}\right)(\ln n_j + 1) = \ln n_i + 1$$

because

$$\frac{\partial \ln n_j}{\partial n_i} = \frac{1}{n_j} \frac{\partial n_j}{\partial n_i}$$

Moreover, if $i \neq j$, n_j is independent of n_i,

$$\frac{\partial n_j}{\partial n_i} = 0$$

However, if $i = j$, then

$$\frac{\partial n_j}{\partial n_i} = \frac{\partial n_j}{\partial n_j} = 1$$

It therefore follows that

$$\frac{\partial \ln W}{\partial n_i} = -(\ln n_i + 1) + (\ln N + 1) = -\ln\left(\frac{n_i}{N}\right)$$

It follows from eqn A that

$$-\ln\left(\frac{n_i}{N}\right) + \alpha - \beta\varepsilon_i = 0$$

The most probable population of the state of energy ε_i is therefore

$$\frac{n_i}{N} = e^{\alpha - \beta\varepsilon_i}$$

as in eqn 4.

The final step is to evaluate the constants α and β in eqn 4. First, we note that

$$N = \sum_j n_j = N e^{\alpha} \sum_j e^{-\beta\varepsilon_j}$$

(We are free to label the states with j instead of i.) Because the N cancels on each side of this equality, it follows that

$$e^{\alpha} = \frac{1}{\sum_j e^{-\beta\varepsilon_j}}$$

Therefore, we can write

$$\frac{n_i}{N} = \frac{e^{-\beta\varepsilon_i}}{\sum_j e^{-\beta\varepsilon_j}} \tag{5}$$

This expression is called the **Boltzmann distribution** (recall the 'Introduction', where it was first encountered, without proof). We shall confirm shortly that $\beta = 1/kT$, where T is the thermodynamic temperature.

19.2 The molecular partition function

From now on we shall write the Boltzmann distribution as

$$p_i = \frac{e^{-\beta\varepsilon_i}}{q} \tag{6}$$

where p_i is the fraction of molecules in the state i, $p_i = n_i/N$, and q is the **molecular partition function**:

$$q = \sum_j e^{-\beta\varepsilon_j}, \quad \text{where } \beta = \frac{1}{kT} \tag{7}$$

The sum in q is sometimes expressed slightly differently. It may happen that several states have the same energy, and so give the same contribution to the sum. If, for example, g_j states have the same energy ε_j (so the energy level is g_j-fold degenerate), then we could write

$$q = \sum_{\text{levels}} g_j e^{-\beta\varepsilon_j} \tag{8}$$

where the sum is now over energy *levels* (sets of states with the same energy), not individual *states*.

Example 19.2 *Writing a partition function*

Write an expression for the partition function of a linear molecule (such as HCl) treated as a rigid rotor.

Method. To use eqn 8 we need to know (a) the energies of the levels, (b) the degeneracies, the number of states that belong to each level. Whenever calculating a partition function, the energies of the levels are expressed relative to 0 for the state of lowest energy. The energy levels of a rigid linear rotor were derived in Section 16.4.

Answer. From Section 16.4, the energy levels of a linear rotor are $hcBJ(J+1)$, with $J = 0, 1, 2, \ldots$ The state of lowest energy has zero energy, so no adjustment need be made to the energies given by this expression. Each J level consists of $2J + 1$ degenerate states. Therefore,

$$q = \sum_{J=0}^{\infty} (2J + 1)e^{-\beta hcBJ(J+1)}$$

Comment. The sum can be evaluated numerically by supplying the value of B (from spectroscopy or calculation) and the temperature. For reasons explained in Section 20.2, this expression applies only to unsymmetrical linear rotors (e.g. HCl, not CO_2; in general, to $C_{\infty v}$ and not $D_{\infty h}$ species).

Exercise E19.2. Write down the partition function for a two-level system, the lower state (at energy 0) being non-degenerate, and the upper state (at an energy ε) doubly degenerate. $\qquad$ $[q = 1 + 2e^{-\beta\varepsilon}]$

An interpretation of the partition function

Some insight into the significance of a partition function can be obtained by considering how it depends on the temperature. When T is close to zero, the parameter $\beta = 1/kT$ is close to infinity. Then every term except one in the sum defining q is zero because each one has the form e^{-x} with $x \to \infty$. The exception is the term with $\varepsilon_0 \equiv 0$ (or the g_0 terms at zero energy if the ground state is g_0-fold degenerate), because then $\varepsilon_0/kT \equiv 0$ whatever the temperature, including zero. As there is only one surviving term when $T = 0$, and its value is g_0, it follows that

As $T \to 0$, $\qquad q \to g_0$

That is, at $T = 0$, the partition function is equal to the degeneracy of the ground state.

At the other extreme, consider the case when T is so high that for each term in the sum $\varepsilon_j/kT \approx 0$. Because $e^{-x} = 1$ when $x = 0$, each term in the sum now contributes 1. It follows that the sum is equal to the number of molecular states, which in general is infinite:

As $T \to \infty$, $\qquad q \to \infty$

In some idealized cases, the molecule may have only a finite number of states; then the upper limit of q is equal to the number of states. For example, if we were considering only the spin energy levels of a radical in

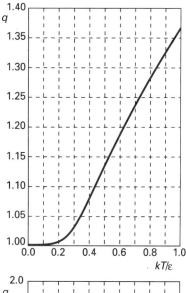

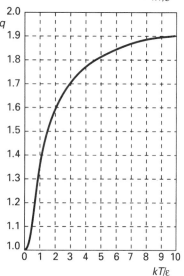

19.3 The partition function for a two-level system as a function of temperature. The two graphs differ in the scale of the temperature axis to show the approach to 1 as $T \to 0$ (top) and the slow approach to 2 as $T \to \infty$ (bottom).

a magnetic field, then there would be only two states ($m_s = \pm\frac{1}{2}$). The partition function for such a system can therefore be expected to rise towards 2 as T is increased towards infinity. Specifically, it follows from eqn 7 that the partition function for a system with a state at zero energy and another state at the energy ε is

$$q = 1 + e^{-\beta\varepsilon} \tag{9}$$

This function is plotted in Fig. 19.3 as a function of T (specifically, of kT/ε which is equal to $1/\beta\varepsilon$): notice how it does indeed rise from 1 to 2 as the temperature is increased.

We see that *the molecular partition function gives an indication of the average number of states that are thermally accessible to a molecule at the temperature of the system.* At $T = 0$, only the ground level is accessible and $q = g_0$. At very high temperatures, virtually all states are accessible, and q is correspondingly large. The two-level calculation, as illustrated in Fig. 19.3, shows that 'high temperature' means $kT \gg \varepsilon$, and q has a high value; a 'low temperature' means $kT \ll \varepsilon$, and then q is close to its minimum value (g_0).

The partition function for a uniform ladder of energy levels

As an explicit example, consider a molecule with an infinite number of equally spaced non-degenerate energy levels (Fig. 19.4). These levels can be thought of as the vibrational energy levels of a diatomic molecule in the harmonic approximation. We can expect the partition function to increase from 1 at $T = 0$ and approach infinity as $T \to \infty$. This expectation can be confirmed by noting that, if the separation of neighbouring levels is ε, then the partition function is

$$q = 1 + e^{-\beta\varepsilon} + e^{-2\beta\varepsilon} + \ldots = 1 + e^{-\beta\varepsilon} + (e^{-\beta\varepsilon})^2 + \ldots$$

This expression has the form of a geometrical progression,[5] and hence

$$q = \frac{1}{1 - e^{-\beta\varepsilon}} \tag{10}$$

Equation 10 is plotted as a function of T (specifically, of kT/ε, or $1/\beta\varepsilon$) in Fig. 19.5: notice that it rises from 1 to infinity as the temperature is raised, just as we anticipated. It follows from eqn 6 that the fraction of molecules in the state with energy ε_i is

$$p_i = (1 - e^{-\beta\varepsilon})e^{-\beta\varepsilon_i} \tag{11}$$

The variation of p_i with temperature is illustrated in Fig. 19.6. We see

5 The sum of the infinite series

$$S = 1 + x + x^2 + \ldots$$

is obtained by multiplying both sides by x, which gives

$$xS = x + x^2 + x^3 + \ldots = S - 1$$

This relation reorganizes to

$$S = \frac{1}{1 - x}$$

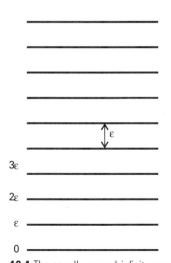

3ε

2ε

ε

0

19.4 The equally spaced infinite array of energy levels used in the calculation of the partition function. A harmonic oscillator has the same spectrum of levels.

that at very low temperatures q is close to 1, and only the lowest state is significantly populated. As the temperature is raised, the population breaks out of the lowest state, and the upper states become progressively more highly populated. At the same time the partition function rises from 1, and its value gives an indication of the range of states populated. The name 'partition function' reflects the sense in which q measures how the total number of molecules is distributed—partitioned—over the available states.

The corresponding expressions for a two-level system are

$$p_0 = \frac{1}{1 + e^{-\beta\varepsilon}} \qquad p_1 = \frac{e^{-\beta\varepsilon}}{1 + e^{-\beta\varepsilon}} \tag{12}$$

These functions are plotted as a function of T (specifically, of kT/ε, or $1/\beta\varepsilon$) in Fig. 19.7. Notice how the populations tend towards equality ($p_0 = 0.5$, $p_1 = 0.5$) as $T \to \infty$. A common error is to suppose that all the molecules in the system will be found in the upper energy state when $T = \infty$; however, we see from eqn 12 that, as $T \to \infty$, the populations of states become equal. The same conclusion is true of multilevel systems too: *as $T \to \infty$, all states become equally populated.*

Example 19.3 *Using the partition function to calculate a population*

Calculate the proportion of I_2 molecules in their ground, first excited, and second excited vibrational states at 25°C. The vibrational wavenumber is 214.6 cm^{-1}.

Method. Vibrational energy levels have a constant separation (in the harmonic approximation, Section 16.8), so the partition function is given by eqn 10 and the fraction of populations by eqn 11. To use the latter equation, we identify the index i with the quantum number v, and calculate p_v for $v = 0$, 1, and 2. The following quantity is useful in this and similar calculations:

At 298.15 K, $\qquad \dfrac{kT}{hc} = 207.223 \text{ cm}^{-1}$

Answer. First, we note that

$$\beta\varepsilon = \frac{hc\tilde{v}}{kT} = \frac{214.6 \text{ cm}^{-1}}{207.223 \text{ cm}^{-1}} = 1.036$$

Then it follows from eqn 11 that the populations are

$$p_v = (1 - e^{-\beta\varepsilon})e^{-v\beta\varepsilon} = 0.645\, e^{-1.036v}$$

Therefore, $p_0 = 0.645$, $p_1 = 0.229$, $p_2 = 0.081$.

Comment. The I—I bond is not stiff and the atoms are heavy: as a result, the vibrational energy separations are small and at room temperature several vibrational levels are significantly populated. The value of the partition function, $q = 1.55$, reflects this small but significant spread of populations.

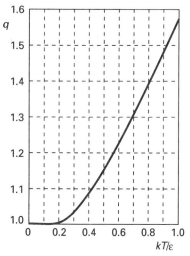

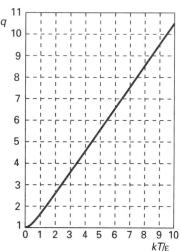

19.5 The partition function for the system shown in Fig. 19.4 (a harmonic oscillator) as a function of temperature. The two graphs differ in the scale of the temperature axis to show the approach to 1 as $T \to 0$ (top) and the straight-line approach to infinity as $T \to \infty$ (bottom).

Exercise E19.3. At what temperature would the $v = 1$ level have (a) half the population of the ground state, (b) the same population as the ground state?   [(a) 445 K, (b) infinite]

The translational partition function

A second example of the calculation of q is that for a molecule of mass m free to move in a container of volume V. We shall take the calculation in two stages. First, we evaluate the partition function for a particle that is free to move only in the x-direction. Then we extend the result to a particle that is free to travel in three dimensions.

It is confirmed in the *Justification* below that the translational partition function for a particle that is free to move in a one-dimensional container of length X is

$$q_x = \left(\frac{2\pi m}{h^2 \beta}\right)^{\frac{1}{2}} X \tag{13}$$

JUSTIFICATION

The energy levels of a molecule of mass m in a container of length X are given by eqn 12.3 with $L = X$:

$$E_n = \frac{n^2 h^2}{8mX^2}, \qquad \text{with } n = 1, 2, \dots$$

The lowest level ($n = 1$) has energy $h^2/8mX^2$, so the energies relative to that level are

$$\varepsilon_n = (n^2 - 1)\varepsilon, \qquad \text{where } \varepsilon = \frac{h^2}{8mX^2}$$

The sum to evaluate is therefore

$$q_x = \sum_{n=1}^{\infty} e^{-(n^2-1)\beta\varepsilon}$$

This expression is exact. However, the sum cannot be evaluated explicitly (other than by doing it numerically for a specific value of $\beta\varepsilon$), so we need to make an approximation. The translational energy levels are very close together in a container the size of a typical laboratory vessel; therefore, the sum can be approximated by an integral:

$$q_x = \int_1^{\infty} e^{-(n^2-1)\beta\varepsilon} \, dn$$

The extension of the lower limit to $n = 0$ and the replacement of $n^2 - 1$ by n^2 turns the integral into standard form. We make the substitution $x^2 = n^2\beta\varepsilon$, implying $dn = dx/(\beta\varepsilon)^{\frac{1}{2}}$, and therefore that

$$q_x = \left(\frac{1}{\beta\varepsilon}\right)^{\frac{1}{2}} \int_0^{\infty} e^{-x^2} \, dx = \left(\frac{1}{\beta\varepsilon}\right)^{\frac{1}{2}} \times \frac{\pi^{\frac{1}{2}}}{2} = \left(\frac{2\pi m}{h^2 \beta}\right)^{\frac{1}{2}} X$$

Low temperature		High temperature	

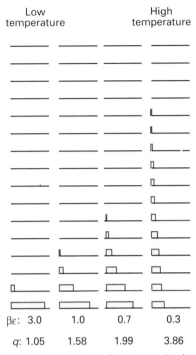

$\beta\varepsilon$:	3.0	1.0	0.7	0.3
q:	1.05	1.58	1.99	3.86

19.6 The populations of the energy levels of the system shown in Fig. 19.4 at different temperatures, and the corresponding values of the partition function calculated using eqn 10. Note that $\beta = 1/kT$.

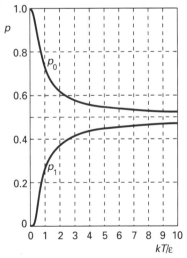

19.7 The fraction of populations of the two states of a two-level system as a function of temperature. Note that as the temperature approaches infinity, the populations of the two states become equal (and the fractions both approach 0.5).

Now consider the extension of the calculation to three dimensions. We take the length of the container in the y direction to be Y and that in the z direction to be Z, with $XYZ = V$, the volume of the container. The total energy of a molecule ε_i is the sum of its translational energies in all three directions:

$$\varepsilon_i = \varepsilon_{n_1}^{(X)} + \varepsilon_{n_2}^{(Y)} + \varepsilon_{n_3}^{(Z)}$$

where n_1, n_2, and n_3 are the quantum numbers for motion in the x-, y-, and z-directions, respectively. Therefore, because $e^{a+b+c} = e^a e^b e^c$, the partition function factorizes as follows:

$$q = \sum_{\text{all } n} e^{-\beta\varepsilon_{n_1}^{(X)} - \beta\varepsilon_{n_2}^{(Y)} - \beta\varepsilon_{n_3}^{(Z)}} = \sum_{\text{all } n} e^{-\beta\varepsilon_{n_1}^{(X)}} e^{-\beta\varepsilon_{n_2}^{(Y)}} e^{-\beta\varepsilon_{n_3}^{(Z)}}$$

$$= \left(\sum_{n_1} e^{-\beta\varepsilon_{n_1}^{(X)}}\right)\left(\sum_{n_2} e^{-\beta\varepsilon_{n_2}^{(Y)}}\right)\left(\sum_{n_3} e^{-\beta\varepsilon_{n_3}^{(Z)}}\right) = q_x q_y q_z$$

In general, if the energy is a *sum* of independent contributions, then the partition function is a *product* of partition functions for each mode of motion.

Equation 13 gives the partition function for translational motion in the x-direction. The only change for the other two directions is to replace the length X by the lengths Y or Z. Hence the partition function for motion in three dimensions is

$$q = \left(\frac{2\pi m}{h^2 \beta}\right)^{\frac{3}{2}} XYZ$$

The product of lengths XYZ is the volume V of the container, so we can write

$$q = \frac{V}{\Lambda^3}, \qquad \text{where } \Lambda = h\left(\frac{\beta}{2\pi m}\right)^{\frac{1}{2}} \tag{14}$$

The quantity Λ has the dimensions of length and is called the **thermal wavelength** of the molecule.

Example 19.4 *Calculating a translational partition function*

Calculate the translational partition function of an H_2 molecule confined to a 100-cm^3 vessel at 25°C.

Method. Because the partition function is a measure of the number of thermally accessible states, and large numbers of translational states are accessible even at low temperatures, we expect q to be very large. To find its value, substitute the fundamental constants and the data into eqn 14 in the form obtained by writing $\beta = 1/kT$:

$$\Lambda = \frac{h}{(2\pi mkT)^{\frac{1}{2}}}$$

Answer. We use $kT = 4.116 \times 10^{-21}$ J and $m = 2.016$ u; then the thermal wavelength of H_2 is

$$\Lambda = \frac{6.626 \times 10^{-34}\, \mathrm{J\, s}}{\{2\pi \times (2.016 \times 1.6605 \times 10^{-27}\, \mathrm{kg}) \times (4.116 \times 10^{-21}\, \mathrm{J})\}^{\frac{1}{2}}}$$

$$= 7.12 \times 10^{-11}\, \mathrm{m}$$

Therefore,

$$q = \frac{1.00 \times 10^{-4}\, \mathrm{m}^3}{(7.12 \times 10^{-11}\, \mathrm{m})^3} = 2.77 \times 10^{26}$$

Comment. We see that about 10^{26} quantum states are thermally accessible, even at room temperature and for this light molecule. Many states are occupied if the thermal wavelength (which in this case is 71.2 pm) is small compared with the linear dimensions of the container.

Exercise E19.4. Calculate the translational partition function for the D_2 molecule under the same conditions.   [$q = 7.9 \times 10^{26}$, $2^{\frac{3}{2}}$ times larger]

THE INTERNAL ENERGY AND THE ENTROPY

The importance of the molecular partition function is that it contains all the information needed to calculate the thermodynamic properties of a system of independent molecules at thermal equilibrium.

19.3 The internal energy

We shall begin to unfold the importance of q by showing that we can derive the internal energy of the system from it.

The relation between U and q

The total energy of the system is

$$E = \sum_i n_i \varepsilon_i \tag{15}$$

Because the most probable configuration is so strongly dominating, we can use the Boltzmann expression for the populations and write

$$E = \frac{N}{q} \sum_i \varepsilon_i e^{-\beta \varepsilon_i}$$

This expression can be manipulated into a form involving only q. To do so, we note that

$$\frac{\mathrm{d}}{\mathrm{d}\beta} e^{-\beta \varepsilon_i} = -\varepsilon_i e^{-\beta \varepsilon_i} \tag{16}$$

It follows that we can write

$$E = -\frac{N}{q} \sum_i \frac{\mathrm{d}}{\mathrm{d}\beta} e^{-\beta \varepsilon_i} = -\frac{N}{q} \frac{\mathrm{d}}{\mathrm{d}\beta} \sum_i e^{-\beta \varepsilon_i} = -\frac{N}{q} \frac{\mathrm{d}q}{\mathrm{d}\beta} \tag{17}$$

As an illustration of the application of this equation, we can use the two-level partition function in eqn 9, $q = 1 + e^{-\beta \varepsilon}$, to deduce the total

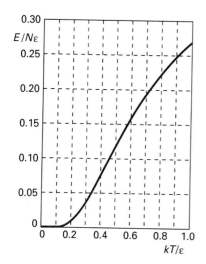

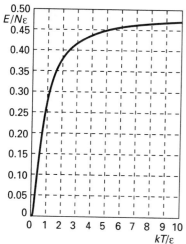

19.8 The total energy of a two-level system (expressed as a multiple of $N\varepsilon$) as a function of temperature, on two temperature scales. The upper graph shows the slow rise away from zero energy at low temperatures; the slope of the graph at $T = 0$ is 0 (that is, the heat capacity is zero at $T = 0$). The lower graph shows the slow rise to 0.5 as $T \to \infty$, as both states become equally populated (see Fig. 19.7).

energy of N two-level systems:

$$E = -\left(\frac{N}{1 + e^{-\beta\varepsilon}}\right)\frac{d}{d\beta}(1 + e^{-\beta\varepsilon}) = \frac{N\varepsilon e^{-\beta\varepsilon}}{1 + e^{-\beta\varepsilon}}$$

This expression can be written more neatly as

$$E = \frac{N\varepsilon}{1 + e^{\beta\varepsilon}} \tag{18}$$

It is plotted in Fig. 19.8. Notice how the energy is zero at $T = 0$, when only the lower state (at the zero of energy) is occupied, and rises to $\frac{1}{2}N\varepsilon$ as T approaches infinity, when the two levels are equally populated.

There are several points in relation to eqn 17 that need to be made. Because $\varepsilon_0 = 0$ (remember that we measure all energies from the lowest available level), E should be interpreted as the value of the internal energy relative to its value at $T = 0$. Therefore, to obtain the conventional internal energy U, we must add the internal energy at $T = 0$:

$$U = U(0) + E \tag{19}$$

Second, because the partition function may depend on variables other than the temperature (e.g. the volume), the derivative with respect to β in eqn 17 is actually a *partial* derivative with these other variables held constant. The complete expression relating the molecular partition function to the thermodynamic internal energy of a system of independent molecules is therefore

$$U = U(0) - \frac{N}{q}\left(\frac{\partial q}{\partial \beta}\right)_V \tag{20a}$$

An equivalent form is obtained by noting that $dx/x = d \ln x$:

$$U = U(0) - N\left(\frac{\partial \ln q}{\partial \beta}\right)_V \tag{20b}$$

These two equations confirm that we need know only the partition function (as a function of temperature) to calculate the internal energy. In due course we shall see that *all* thermodynamic functions can be calculated once we know q as a function of the temperature and volume of the system. In this respect, q plays a role in statistical thermodynamics very similar to that played by the wavefunction in quantum mechanics: q is a kind of thermal wavefunction.

The relation $\beta = 1/kT$

We shall now confirm that the parameter β, which we have anticipated is equal to $1/kT$, does indeed have that value. To do so, we shall compare the equipartition expression for the internal energy of a perfect gas, which from the Introduction (p. 15) we know to be

$$U = U(0) + \tfrac{3}{2}nRT \tag{21a}$$

with the value calculated from the translational partition function, which is

$$U = U(0) + \frac{3N}{2\beta} \qquad \qquad (21b)$$

JUSTIFICATION

To use eqn 20, we introduce the translational partition function from eqn 14:

$$\left(\frac{\partial q}{\partial \beta}\right)_V = \frac{d}{d\beta} \frac{V}{\Lambda^3} = V \frac{d}{d\beta} \frac{1}{\Lambda^3} = -3 \frac{V}{\Lambda^4} \frac{d\Lambda}{d\beta}$$

Then we note from the formula for Λ in eqn 14 that

$$\frac{d\Lambda}{d\beta} = \frac{d}{d\beta}\left(\frac{h\beta^{\frac{1}{2}}}{(2\pi m)^{\frac{1}{2}}}\right) = \frac{1}{2\beta^{\frac{1}{2}}} \times \frac{h}{(2\pi m)^{\frac{1}{2}}} = \frac{\Lambda}{2\beta}$$

and so obtain

$$\left(\frac{\partial q}{\partial \beta}\right)_V = -\frac{3V}{2\beta\Lambda^3}$$

Then, by eqn 20a,

$$U = U(0) - N \times \frac{\Lambda^3}{V} \times \left(-\frac{3V}{2\beta\Lambda^3}\right) = U(0) + \frac{3N}{2\beta}$$

It now follows, by equating eqns 21a and 21b, that

$$\beta = \frac{N}{nRT} = \frac{nN_A}{nN_A kT} = \frac{1}{kT}$$

as was to be proved. (We have used $N = nN_A$, where n is the amount of gas molecules, N_A is Avogadro's constant, and $R = N_A k$.)

Example 19.5 *Calculating the heat capacity from the partition function*

Calculate the constant-volume heat capacity of a monatomic gas.

Method. The answer depends on being able to relate the property required (the heat capacity, C_V) to the partition function. To do so, we use the definition of the heat capacity in terms of the internal energy, and then express the latter in terms of the partition function by using eqn 21.

Answer. The definition of constant-volume heat capacity is

$$C_V = \left(\frac{\partial U}{\partial T}\right)_V$$

Therefore, because

$$U = U(0) + \frac{3N}{2\beta} = U(0) + \tfrac{3}{2}nRT$$

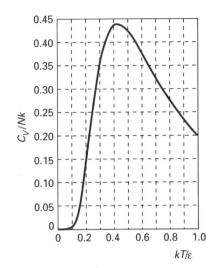

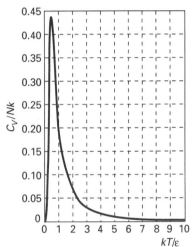

19.9. The heat capacity of a two-level system expressed as a multiple of Nk. The upper graph shows the approach to zero as $T \to 0$. The heat capacity is zero at $T = 0$. The lower graph shows that the heat capacity passes through a maximum and then falls to zero as $T \to \infty$.

it follows that

$$C_V = \tfrac{3}{2}nR$$

The molar heat capacity is therefore equal to $\tfrac{3}{2}R$, which has the value $12.5 \, \mathrm{J\,K^{-1}\,mol^{-1}}$.

Comment. This value agrees almost exactly with experimental data on monatomic gases at normal pressures. In more complex molecules, other modes of motion contribute (Section 20.4). In similar calculations, such as in the following exercise, differentiation with respect to T can often more conveniently be treated in terms of differentiation with respect to β by using

$$\frac{\mathrm{d}}{\mathrm{d}T} = \frac{\mathrm{d}\beta}{\mathrm{d}T} \times \frac{\mathrm{d}}{\mathrm{d}\beta} = -\frac{1}{kT^2}\frac{\mathrm{d}}{\mathrm{d}\beta}$$

Exercise E19.5. Calculate and plot the heat capacity of N two-level systems. $[C = Nk(\beta\varepsilon)^2 e^{\beta\varepsilon}/(1 + e^{\beta\varepsilon})^2$, see Fig. 19.9]

19.4 The statistical entropy

If it is true that the partition function contains all thermodynamic information, then it must be possible to use it to calculate the entropy as well as the internal energy. Since we know (Section 4.2) that entropy is related to the dispersal of energy and that the partition function is a measure of the number of states that are thermally accessible, we can be confident that the two are indeed related.

We shall develop a relation between the entropy and the partition function in two stages. In the first stage we justify one of the most celebrated equations in statistical thermodynamics, the **Boltzmann formula** for the entropy:

$$S = k \ln W \tag{22}$$

In this expression, W is the weight of the most probable configuration of the system. Then, in the following stage, we express W in terms of the partition function.

JUSTIFICATION

The justification of eqn 22 begins by noting that a change in the internal energy

$$U = U(0) + \sum_i n_i \varepsilon_i$$

may arise from either a modification of the energy levels of a system (when ε_i changes to $\varepsilon_i + \mathrm{d}\varepsilon_i$) or from a modification of the populations (when n_i changes to $n_i + \mathrm{d}n_i$). The most general change is therefore

$$\mathrm{d}U = \mathrm{d}U(0) + \sum_i n_i \, \mathrm{d}\varepsilon_i + \sum_i \varepsilon_i \, \mathrm{d}n_i$$

Because the energy levels do not change when a system is heated at

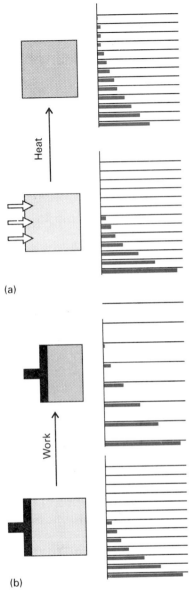

constant volume (Fig. 19.10), in the absence of all changes other than heating

$$dU = \sum_i \varepsilon_i \, dn_i$$

We know from thermodynamics (and specifically from eqn 5.1) that under the same conditions

$$dU = dq_{rev} = T \, dS$$

Therefore,

$$dS = \frac{dU}{T} = k\beta \sum_i \varepsilon_i \, dn_i$$

We also know that for changes in the most probable configuration (the only one we need consider)

$$\left(\frac{\partial \ln W}{\partial n_i}\right) + \alpha - \beta\varepsilon_i = 0$$

(This expression is eqn A in the *Justification* following eqn 4.) After rearranging this expression to

$$\beta\varepsilon_i = \left(\frac{\partial \ln W}{\partial n_i}\right) + \alpha$$

we find that

$$dS = k \sum_i \left(\frac{\partial \ln W}{\partial n_i}\right) dn_i + k\alpha \sum_i dn_i$$

But the sum over the dn_i is zero, because the number of molecules is constant. Hence

$$dS = k \sum_i \left(\frac{\partial \ln W}{\partial n_i}\right) dn_i = k(d \ln W)$$

This relation strongly suggests the definition $S = k \ln W$, as in eqn 22.

19.10 (a) When a system is heated, the energy levels are unchanged but their populations are changed. (b) When work is done on a system, the energy levels themselves are changed. The levels in this case are the one-dimensional particle-in-a-box energy levels of Chapter 12: they depend on the size of the container and move apart as its length is decreased.

The statistical entropy behaves in exactly the same way as the thermodynamic entropy. Thus, as the temperature is lowered, the value of W, and hence of S, decreases because fewer configurations are compatible with the total energy. In the limit $T \to 0$, $W = 1$, so $\ln W = 0$, because only one configuration (every molecule in the lowest level) is compatible with $E = 0$. It follows that $S \to 0$ as $T \to 0$, which is compatible with the Third Law of thermodynamics, that the entropies of all perfect crystals approach the same value as $T \to 0$ (Section 4.5).

In the final step, we relate the Boltzmann formula for the entropy to the partition function. To do so, we use the expression for $\ln W$ in eqn 3, which gives

$$S = \frac{U - U(0)}{T} + Nk \ln q \qquad (23)$$

JUSTIFICATION

The first stage is to write

$$S = k \sum_i (n_i \ln N - n_i \ln n_i) = -k \sum_i n_i \ln\left(\frac{n_i}{N}\right) = -Nk \sum_i p_i \ln p_i \qquad \textbf{(B)}$$

where $p_i = n_i/N$, the fraction of molecules in state i. It follows from eqn 6 that

$$\ln p_i = -\beta \varepsilon_i - \ln q$$

and therefore that

$$S = -Nk\left(-\beta \sum_i p_i \varepsilon_i - \sum_i p_i \ln q\right) = k\beta\{U - U(0)\} + Nk \ln q$$

We have used the fact that the sum over the p_i is equal to 1 and the sum over $Np_i\varepsilon_i$ is equal to $U - U(0)$. We have already established that $\beta = 1/kT$, so eqn 23 immediately follows.

Example 19.6 *Calculating the entropy of a collection of oscillators*

Calculate the entropy of a collection of N independent harmonic oscillators, and evaluate it using vibrational data for I_2 at 25°C (Example 19.3).

Method. To use eqn 23 we need the partition function for a molecule with evenly spaced vibrational energy levels. This partition function is given in eqn 10. With the partition function available, the internal energy can be found by differentiation (as in eqn 20), and the two expressions then combined to give S.

Answer. The molecular partition function as given in eqn 10 is

$$q = \frac{1}{1 - e^{-\beta\varepsilon}}$$

The internal energy is obtained using eqn 20a:

$$U - U(0) = -\frac{N}{q}\left(\frac{\partial q}{\partial \beta}\right)_V = \frac{N\varepsilon e^{-\beta\varepsilon}}{1 - e^{-\beta\varepsilon}} = \frac{N\varepsilon}{e^{\beta\varepsilon} - 1}$$

The entropy is therefore

$$S = Nk\left\{\frac{\beta\varepsilon}{e^{\beta\varepsilon} - 1} - \ln(1 - e^{-\beta\varepsilon})\right\}$$

This function is plotted in Fig. 19.11. For I_2 at 25°C, $\beta\varepsilon = 1.036$ (Example 19.3), so $S_m = 8.38\,\mathrm{J\,K^{-1}\,mol^{-1}}$.

Comment. We shall see in Section 20.2 that the total entropy of a molecule may be ascribed to its different modes of motion, and the value we have just calculated will turn out to be the contribution arising from the vibrations of the I_2 molecule.

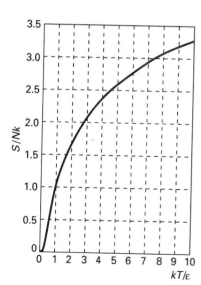

19.11 The temperature variation of the entropy of the system shown in Fig. 19.4 (expressed here as a multiple of Nk). The entropy approaches zero as $T \to 0$, and increases without limit as $T \to \infty$.

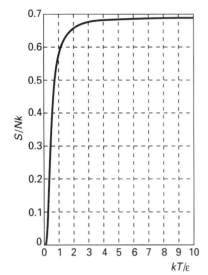

19.12 The temperature variation of the entropy of a two-level system (expressed as a multiple of Nk). As $T \rightarrow \infty$, the two states become equally populated and S approaches $Nk \ln 2$.

Exercise E19.6. Evaluate the molar entropy of N two-level systems and plot the resulting expression. What is the entropy when the two states are equally thermally accessible?

$$[S/Nk = \beta\varepsilon/(1 + e^{\beta\varepsilon}) + \ln(1 + e^{-\beta\varepsilon}); \text{ see Fig. 19.12}; S = Nk \ln 2]$$

THE CANONICAL PARTITION FUNCTION

In this section we see how to generalize our conclusions to include systems composed of interacting molecules and molecules free to exchange positions with each other.

19.5 The canonical ensemble

The crucial new concept we need when treating systems of interacting particles is the **ensemble**. The term has basically its normal meaning of 'collection', but, like so many scientific terms, it has been sharpened and refined into a precise significance.

The concept of ensemble

To set up an ensemble, we take a closed system of specified volume, composition, and temperature, and think of it as replicated $\mathbb{N}$ times (Fig. 19.13). All the identical closed systems are regarded as being in thermal contact with each other, and hence they can exchange energy with each other. The total energy of all the systems is $\mathbb{E}$ and, because they are in thermal equilibrium with each other, they all have the same temperature T. This imaginary collection of replications of the actual system is called the **canonical ensemble**.[6]

The important point about an ensemble is that it is a collection of *imaginary* replications of the system, so that we are free to let the number of members be as large as we like; when appropriate, we can let $\mathbb{N}$ become infinite.[7] The number of members of the ensemble in a state with energy E_i is denoted $\mathbb{n}_i$ and we can speak of the configuration of the ensemble (by analogy with the configuration of the system used in Section 19.1) and its weight, $\mathbb{W}$.

Dominating configurations

Just as in Section 19.1, some of the configurations of the ensemble will be very much more probable than others. For instance, it is very unlikely

---

6 The word 'canon' means 'according to a rule'. There are two other important ensembles. In the *microcanonical ensemble* the condition of constant temperature is replaced by the requirement that all the systems should have exactly the same energy: each system is individually isolated. In the *grand canonical ensemble* the volume and temperature of each system is the same, but they are open. This means that matter can be imagined as able to pass between the systems; the composition of each one may fluctuate, but now the chemical potential is the same in each system. We can summarize these ensembles as follows:

Microcanonical: N, V, E common

Canonical: N, V, T common

Grand canonical: μ, V, T common

7 Note that $\mathbb{N}$ is unrelated to N, the number of molecules in the actual system; $\mathbb{N}$ is the number of *imaginary* replications of that system.

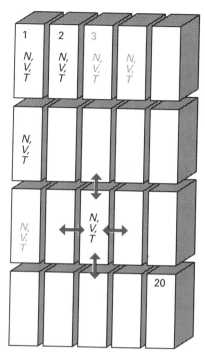

19.13 A representation of the canonical ensemble, in this case for $\mathbb{N} = 20$. The individual replications of the actual system all have the same composition and volume. They are all in mutual thermal contact, and so all have the same temperature. Energy may be transferred between them as heat, and so they do not all have the same energy. The total energy ($\mathbb{E}$) of all 20 replications is a constant because the ensemble is isolated overall.

that the whole of the energy $\mathbb{E}$ will accumulate in one system. By analogy with the earlier discussion, we can anticipate that there will be a dominating configuration, and that we can evaluate the thermodynamic properties by taking the average over the ensemble using that single, most probable, configuration. In the thermodynamic limit of $\mathbb{N} \to \infty$, this dominating configuration is overwhelmingly the most probable, and it dominates the properties of the system virtually completely.

The quantiative discussion follows the argument in Section 19.1 with the modification that N and n_i are replaced by $\mathbb{N}$ and $\mathbb{n}_i$. The weight of a configuration $\{\mathbb{n}_0, \mathbb{n}_1, \ldots\}$ is

$$\mathbb{W} = \frac{\mathbb{N}!}{\mathbb{n}_0! \, \mathbb{n}_1! \ldots}$$

The configuration of greatest weight, subject to the constraints that the total energy of the ensemble is constant and that the total number of members is fixed at $\mathbb{N}$, is given by the **canonical distribution**

$$\frac{\mathbb{n}_i}{\mathbb{N}} = \frac{e^{-\beta E_i}}{Q}, \qquad \text{where } Q = \sum_i e^{-\beta E_i} \qquad (24)$$

The quantity Q, which is a function of the temperature, is called the **canonical partition function**.

Fluctuations from the most probable distribution

The shape of the canonical distribution is only *apparently* an exponentially decreasing function of the energy of the system. We must appreciate that eqn 24 gives the probability of occurrence of members in a single state i *of the entire system* of energy E_i. There may in fact be numerous states with almost identical energies. For example, in a gas the identities of the molecules moving slowly or quickly can change without necessarily affecting the total energy. The **density of states**, the number of states per unit energy range, is a very sharply increasing function of energy. It follows that the probability of a member of an ensemble having a specified energy (as distinct from being in a specified state) is given by eqn 24, a sharply decreasing function, multiplied by a sharply increasing function (Fig. 19.14). Therefore, the overall distribution is a sharply peaked function near some energy. We conclude that most members of the ensemble have an energy very close to the mean value.

19.6 The thermodynamic information in the partition function

Like the molecular partition function, the canonical partition function carries all the thermodynamic information about a system. However, Q is more general than q because it is not based on the assumption that the molecules are independent. We can therefore use Q to discuss the properties of condensed phases and real gases where molecular interactions are important.

The internal energy

If the total energy of the ensemble is $\mathbb{E}$, and there are $\mathbb{N}$ members, then the average energy of any one member is the arithmetic mean over the

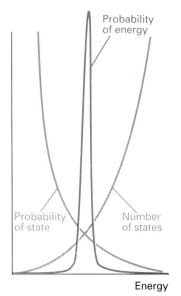

19.14 To construct the form of the distribution of members of the canonical ensemble in terms of their energies, we multiply the probability that any one is in a state of given energy, eqn 27, by the number of states corresponding to that energy (a steeply rising function). The product is a sharply peaked function at the mean energy, which shows that almost all the members of the ensemble have that energy.

entire ensemble

$$E = \frac{\mathbb{E}}{\mathbb{N}} \tag{25}$$

We use this quantity to calculate the internal energy of the system in the limit of $\mathbb{N}$ (and $\mathbb{E}$) approaching infinity:

$$U = U(0) + E = U(0) + \frac{\mathbb{E}}{\mathbb{N}} \qquad \text{as } \mathbb{N} \to \infty \tag{26}$$

The fraction $\mathbb{p}_i$ of members of the ensemble in a state i with energy E_i is given by the analogue of eqn 6 as

$$\mathbb{p}_i = \frac{e^{-\beta E_i}}{Q} \tag{27}$$

It follows that the internal energy is given by

$$U = U(0) + \sum_i \mathbb{p}_i E_i = U(0) + \frac{1}{Q} \sum_i E_i e^{-\beta E_i}$$

By the same argument that led to eqn 20,

$$U = U(0) - \frac{1}{Q} \left(\frac{\partial Q}{\partial \beta} \right)_V = U(0) - \left(\frac{\partial \ln Q}{\partial \beta} \right)_V \tag{28}$$

The entropy

The total weight $\mathbb{W}$ of a configuration of the ensemble is the product of the average weight W of each member of the ensemble,

$$\mathbb{W} = W^{\mathbb{N}}$$

Hence, we can calculate S from

$$S = k \ln W = k \ln \mathbb{W}^{1/\mathbb{N}} = \frac{k}{\mathbb{N}} \ln \mathbb{W} \tag{29}$$

It follows, by the same argument used in Section 19.4, that

$$S = \frac{U - U(0)}{T} + k \ln Q \tag{30}$$

19.7 Independent molecules

We shall now see how to recover the molecular partition function from the more general canonical partition function when the molecules are independent. When the molecules are distinguishable (in the sense to be described), the relation between Q and q is $Q = q^N$.

JUSTIFICATION

The total energy of a collection of N independent molecules is the sum of the energies of the molecules. Therefore, we can write the total energy of a state i of the system as

$$E_i = \varepsilon_i(1) + \varepsilon_i(2) + \ldots + \varepsilon_i(N)$$

In this expression, $\varepsilon_i(1)$ is the energy of molecule 1 when the system is in the state i, $\varepsilon_i(2)$ the energy of molecule 2 when the system is in the same state i, and so on. The canonical partition function is then

$$Q = \sum_i e^{-\beta\varepsilon_i(1)-\beta\varepsilon_i(2)-\ldots-\beta\varepsilon_i(N)}$$

The sum over the states of the system can be reproduced by letting each molecule enter all its own individual states (although we meet an important proviso shortly). Therefore, instead of summing over the states i of the system, we can sum over all the individual states i of molecule 1, all the states i of molecule 2, and so on. This rewriting of the original expression leads to

$$Q = \left(\sum_i e^{-\beta\varepsilon_i}\right)\left(\sum_i e^{-\beta\varepsilon_i}\right)\ldots\left(\sum_i e^{-\beta\varepsilon_i}\right) = \left(\sum_i e^{-\beta\varepsilon_i}\right)^N$$

which has the form

$$Q = q^N$$

Distinguishable and indistinguishable molecules

If all the molecules are identical and free to move through space, we cannot distinguish them and the relation $Q = q^N$ is not valid. Suppose that molecule 1 is in some state a, molecule 2 is in b, and molecule 3 is in c, then one member of the ensemble has an energy $E = \varepsilon_a + \varepsilon_b + \varepsilon_c$. This member, however, is indistinguishable from one formed by putting molecule 1 in state b, molecule 2 in state c, and molecule 3 in state a, or some other permutation. There are six such permutations in all, and $N!$ in general. In the case of indistinguishable molecules, it follows that we have counted too many states in going from the sum over system states to the sum over molecular states, so that writing $Q = q^N$ overestimates the value of Q. The detailed argument is quite involved, but at all except very low temperatures it turns out that the correction factor is $1/N!$. Therefore:

For distinguishable molecules: $Q = q^N$ (31a)

For indistinguishable molecules: $Q = \dfrac{q^N}{N!}$ (31b)

For molecules to be indistinguishable, they must be of the same kind: an Ar atom is never indistinguishable from a Ne atom. Their identity, however, is not the only criterion. If identical molecules are in a crystal lattice, then each one can be 'named' with a set of coordinates. Identical molecules in a lattice are therefore distinguishable because their sites are distinguishable, and we use eqn 31a. On the other hand, identical molecules in a gas are free to move to different locations, and there is no way of keeping track of the identity of a given molecule; we therefore use eqn 31b.

The entropy of a monatomic gas

An important application of the previous material is the derivation, as shown in the *Justification* below, of the **Sackur–Tetrode equation** for the entropy of a monatomic gas:

$$S = nR \ln\left(\frac{e^{\frac{5}{2}}V}{nN_A\Lambda^3}\right), \quad \text{where } \Lambda = \frac{h}{(2\pi mkT)^{\frac{1}{2}}} \quad \text{(32a)}$$

Because the gas is perfect, we can use the relation $V = nRT/p$ to express the entropy in terms of the pressure as

$$S = nR \ln\left(\frac{e^{\frac{5}{2}}kT}{p\Lambda^3}\right) \quad \text{(32b)}$$

JUSTIFICATION

For a gas of independent molecules, Q may be replaced by $q^N/N!$, with the result that eqn 30 becomes

$$S = \frac{U - U(0)}{T} + kN \ln q - k \ln N!$$

Because the number of molecules ($N = nN_A$) in a typical sample is large, we can use Stirling's approximation (and $R = kN_A$):

$$S = \frac{U - U(0)}{T} + nR \ln q - k\,(N \ln N - N)$$

$$= \frac{U - U(0)}{T} + nR(\ln q - \ln N + 1)$$

The only mode of motion for a gas of atoms is translation, and the partition function is $q = V/\Lambda^3$, eqn 14. The internal energy is given by eqn 21, so the entropy is

$$S = \tfrac{3}{2}nR + nR\left(\ln \frac{V}{\Lambda^3} - \ln nN_A + 1\right)$$

$$= nR\left(\ln e^{\frac{3}{2}} + \ln \frac{V}{\Lambda^3} - \ln nN_A + \ln e\right)$$

which rearranges into eqn 32.

Example 19.7 *Using the Sackur–Tetrode equation*

Calculate the standard molar entropy of gaseous argon at 25°C.

Method. To calculate the *molar* entropy from eqn 32b we divide both sides by n. To calculate the *standard* molar entropy, we then set $p = p^{\ominus}$:

$$S_m^{\ominus} = R \ln\left(\frac{e^{\frac{5}{2}}kT}{p^{\ominus}\Lambda^3}\right)$$

Answer. The mass of an Ar atom is $m = 39.95\,u$. At 25°C, its thermal wavelength is 16.0 pm (by the same kind of calculation as in Example

19.4). Therefore,

$$S_m^\ominus = R \ln\left\{\frac{e^{\frac{5}{2}} \times (4.12 \times 10^{-21} J)}{(10^5 \, N \, m^{-2}) \times (1.60 \times 10^{-11} \, m)^3}\right\}$$

$$= 18.6R = 155 \, J \, K^{-1} \, mol^{-1}$$

Comment. We can anticipate, on the basis of the number of accessible states for a lighter molecule, that the standard molar entropy of Ne is likely to be smaller than for Ar, and its actual value is $17.60R$.

Exercise E19.7. Calculate the translational contribution to the standard molar entropy of H_2 at 25°C.   [14.2R]

We can now make another connection with the material in Part 1. The Sackur–Tetrode equation implies that, when a perfect gas expands isothermally from V_i to V_f, its entropy changes by

$$\Delta S = nR \ln(aV_f) - nR \ln(aV_i) = nR \ln\left(\frac{V_f}{V_i}\right)$$

where aV is the collection of quantities inside the logarithm of eqn 32a. This is exactly the expression we obtained by using classical thermodynamics (eqn 4.12).

CHECK LIST OF KEY IDEAS

1 The concepts of a **configuration** of a system and of its **weight** (Section 19.1 and eqn 1).

2 The use of **Stirling's approximation** to find the most probable configuration and hence to deduce the **Boltzmann distribution** (eqn 6) for the most probable distribution of populations.

3 The definition of the **molecular partition function** (eqn 7) and its interpretation as a measure of the number of thermally accessible states (Section 19.2).

4 The calculation of the molecular partition function for a two-level system (eqn 9), a uniform ladder of states (eqn 10), and for translational motion (eqn 14).

5 The relation between the **internal energy** of non-interacting molecules and the molecular partition function (eqn 20), and the internal energy of a perfect gas (eqn 21).

6 The calculation of **heat capacity** from the molecular partition function (Example 19.5).

7 The justification of the **Boltzmann formula** for the entropy (eqn 22) and the relation between the **statistical entropy** and the molecular partition function (eqn 23).

8 The definition and significance of the **canonical ensemble** of systems composed of interacting molecules (Section 19.5).

9 The **canonical distribution** and the definition of the **canonical partition function** (eqn 24).

10 The **internal energy** (eqn 28) and the **entropy** (eqn 30) in terms of the canonical partition function.

11 The relation between the canonical and molecular partition functions for non-interacting molecules (eqn 31).

12 The derivation of the **Sackur–Tetrode equation** for the entropy of a monatomic gas (eqn 32).

13 The key equations of this chapter are summarized in Table 19.1.

Table 19.1 Key equations, with $\beta = 1/kT$

Definition of molecular partition function:

$$q = \sum_i e^{-\beta \varepsilon_i} \quad \text{or} \quad q = \sum_{\text{levels}} g_i e^{-\beta \varepsilon_i}$$

Two-level system, energies 0, ε:

$$q = 1 + e^{-\beta \varepsilon}$$

Evenly spaced, infinite level system, energies 0, ε, 2ε, ... :

$$q = (1 - e^{-\beta \varepsilon})^{-1}$$

Translational motion of particle of mass m in volume V:

$$q = \frac{V}{\Lambda^3} \quad \Lambda = \left(\frac{h^2 \beta}{2\pi m}\right)^{\frac{1}{2}} = \frac{h}{(2\pi m kT)^{\frac{1}{2}}}$$

Boltzmann distribution:

$$p_i = \frac{e^{-\beta \varepsilon_i}}{q}, \quad p_i = \frac{n_i}{N}$$

Canonical partition function:

$$Q = \sum_i e^{-\beta E_i}$$

Boltzmann formula:

$$S = k \ln W$$

Internal energy:

$$U = U(0) - \frac{N}{q}\left(\frac{\partial q}{\partial \beta}\right)_V = U(0) - N\left(\frac{\partial \ln q}{\partial \beta}\right)_V \quad \text{(independent particles)}$$

$$U = U(0) - \frac{1}{Q}\left(\frac{\partial Q}{\partial \beta}\right)_V = U(0) - \left(\frac{\partial \ln Q}{\partial \beta}\right)_V \quad \text{(in general)}$$

Entropy:

$$S = \frac{U - U(0)}{T} + Nk \ln q \quad \text{(independent particles)}$$

$$S = \frac{U - U(0)}{T} + k \ln Q \quad \text{(in general)}$$

Relation between q and Q:

For distinguishable molecules: $Q = q^N$

For indistinguishable molecules: $Q = \dfrac{q^N}{N!}$

EXERCISES

19.1. What are the relative populations of the states of a two-level system when the temperature is infinite?

19.2. Calculate the translational partition function at (a) 300 K and (b) 600 K of a molecule of molar mass 120 g mol^{-1} in a container of volume 2.00 cm^3.

19.3. Calculate (a) the thermal wavelength, (b) the translational partition function of an Ar atom in a cubic box of side 1.00 cm at (i) 300 K and (ii) 3000 K.

19.4. Calculate the ratio of the translational partition functions of D_2 and H_2 at the same temperature and volume.

19.5. A certain atom has a threefold degenerate ground level, a non-degenerate electronically excited level at 3500 cm^{-1}, and a threefold degenerate level at 4700 cm^{-1}. Calculate the partition function of these electronic states at 1900 K.

19.6. Calculate the electronic contribution to the molar internal energy at 1900 K for a sample composed of the atoms specified in Exercise 19.5.

19.7. A certain molecule has a non-degenerate excited state lying at 540 cm^{-1} above the non-degenerate ground state. At what temperature will 10 per cent of the molecules be in the upper state?

19.8. Test Stirling's approximation, that $\ln x! \approx x \ln x - x$, for $x = 5$, 10, and 15 by comparing its predictions with the exact values. Repeat the comparison for the more exact form of the approximation

$$\ln x! = (x + \tfrac{1}{2}) \ln x - x + \tfrac{1}{2} \ln 2\pi$$

19.9 An electron spin can adopt either of two orientations in a magnetic field, and its energies are $\pm \mu_B B$ where μ_B is the Bohr magneton. Deduce an expression for the partition function and mean energy of the electron and sketch the variation of the functions with B. Calculate the relative populations of the spin states at (a) 40 K, (b) 298 K when $B = 1.0$ T.

19.10. Derive an expression for (a) the partition function, (b) the mean energy of a ^{14}N nucleus ($I = 1$) in a magnetic field.

19.11. Consider a system of distinguishable particles having only two non-degenerate energy levels separated by an energy which is equal to the value of kT at 10 K. Calculate (a) the ratio of populations in the two states at (1) 1.0 K, (2) 10 K, and (3) 100 K, (b) the molecular partition function at 10 K, (c) the molar energy at 10 K, (d) the molar heat capacity at 10 K, (e) the molar entropy at 10 K.

19.12. At what temperature would the population of the

first excited vibrational state of HCl be $1/e$ times its population of the ground state?

19.13. Calculate the standard molar entropy of neon gas at (a) 200 K, (b) 298.15 K.

19.14. Calculate the vibrational contribution to the entropy of Cl_2 at 500 K given that the wavenumber of the vibration is 560 cm^{-1}.

19.15. Identify the systems for which it is essential to include a factor of $1/N!$ on going from Q to q: (a) a sample of helium gas, (b) a sample of carbon monoxide gas, (c) a solid sample of carbon monoxide, (d) water vapour, (e) ice.

19.16. Confirm that the Sackur–Tetrode equation accounts for (a) the pressure dependence and (b) the temperature dependence of the entropy of a monatomic perfect gas as derived from thermodynamics.

19.17. The Sackur–Tetrode equation for the molar entropy of a perfect monatomic gas may be written $S_m = R \ln(aT^{5/2}/p)$ where a is a combination of constants. Evaluate the change in molar Gibbs energy that accompanies heating a sample from 273 to 373 K at constant pressure.

PROBLEMS
Numerical problems

19.1. A certain atom has a doubly degenerate ground level pair and an upper level of four degenerate states at 450 cm^{-1} above the ground level. In an atomic beam study of the atoms it was observed that 30 per cent of the atoms were in the upper level and that the translational temperature of the beam was 300 K. Are the electronic states of the atoms in thermal equilibrium?

19.2. Explore the conditions under which the 'integral' approximation for the translational partition function is not valid by considering the translational partition function of an Ar atom in a cubic box of side 1.00 cm. Estimate the temperature at which, according to the integral approximation, $q = 10$ and evaluate the exact partition function at that temperature. Are the electronic states of the atoms in thermal equilibrium?

19.3. (a) Calculate the electronic partition function of a Te atom at (i) 298 K, (ii) 5000 K by direct summation using the following data:

Term	Degeneracy	Wavenumber/cm^{-1}
Ground	5	0
1	1	4 707
2	3	4 751
3	5	10 559

(b) What proportion of the Te atoms are in the ground term and in the term labelled 2 at the two temperatures? (c) Calculate the electronic contribution to the standard molar entropy of gaseous Te atoms.

19.4. The four lowest electronic levels of a Ti atom are: $^3F_2$, $^3F_3$, $^3F_4$, and $^4F_1$, at 0, 170, 387, and 6557 cm^{-1}, respectively. There are many other electronic states at higher energies. The boiling point of titanium is 3287°C. What are the relative populations of these levels at the boiling point? (*Hint.* The degeneracies of the levels are $2J + 1$.)

19.5. The NO molecule has a doubly degenerate excited level 121.1 cm^{-1} above the doubly degenerate ground term. Calculate and plot the electronic partition function of NO from $T = 0$ to 1000 K. Evaluate (a) the term populations and (b) the electronic contribution to the molar internal energy at 300 K. Calculate the electronic contribution to the molar entropy of the NO molecule at 300 K and 500 K.

19.6. Calculate, by explicit summation, the vibrational partition function and the vibrational contribution to the molar internal energy of I_2 molecules at (a) 100 K, (b) 298 K given that its vibrational energy levels lie at the following wavenumbers above the zero-point energy level: 0, 213.30, 425.39, 636.27, 845.93 cm^{-1}. What proportion of I_2 molecules are in the ground and first two excited levels at the two temperatures? Calculate the vibrational contribution to the molar entropy of I_2 at the two temperatures.

Theoretical problems

19.7. A sample consisting of five molecules has a total energy 5ε. Each molecule is able to occupy states of energy $j\varepsilon$ with $j = 0, 1, 2, \ldots$ (a) Calculate the weight of the configuration in which the molecules share the energy equally. (b) Draw up a table with columns headed by the energy of the states and write beneath them all configurations that are consistent with the total energy. Calculate the weights of each configuration and identify the most probable configurations.

19.8. A sample of nine molecules is numerically tractable but on the verge of being thermodynamically significant.

Draw up a table of configurations for $N = 9$, total energy 9ε in a system with energy levels $j\varepsilon$ (as in Problem 19.7). Before evaluating the weights of the configurations, guess (by looking for the most 'exponential' distribution of populations) which of the configurations will turn out to be the most probable. Go on to calculate the weights and identify the most probable configuration.

19.9. The most probable configuration is characterized by a parameter we know as the 'temperature'. The temperatures of the system specified in Problems 19.7 and 19.8 must be such as to give a mean value of ε for the energy of each molecule and a total energy $N\varepsilon$ for the system. (a) Show that the temperature can be obtained by plotting p_j against j, where p_j is the (most probable) fraction of molecules in the state with energy $j\varepsilon$. Apply the procedure to the system in Problem 19.8. What is the temperature of the system when ε corresponds to $50 \, cm^{-1}$? (b) Choose configurations other than the most probable, and show that the same procedure gives a worse straight line, indicating that a temperature is not well-defined for them.

19.10. A certain molecule can exist in either a non-degenerate singlet state or a triplet state (with degeneracy 3). The energy of the triplet exceeds that of the singlet by ε. Assuming that the molecules are distinguishable (localized) and independent, (a) obtain the expression for the molecular partition function. (b) Find expressions in terms of ε for the molar energy, molar heat capacity, and molar entropy of such molecules and calculate their values at $T = \varepsilon/k$.

19.11. Consider a system with energy levels $\varepsilon_j = j\varepsilon$ and N molecules. (a) Show that if the mean energy per molecule is $a\varepsilon$, then the temperature is given by

$$\beta = \frac{1}{\varepsilon} \ln\left(1 + \frac{1}{a}\right)$$

Evaluate the temperature for a system in which the mean energy is ε, taking ε equivalent to $50 \, cm^{-1}$. (b) Calculate the molecular partition function q for the system when its mean energy is $a\varepsilon$. (c) Show that the entropy of the system is

$$S/k = (1 + a) \ln (1 + a) - a \ln a$$

and evaluate this expression for a mean energy ε.

19.12. Suppose that by some means we contrive to invert the population of a two-level system in the sense that the upper and lower levels have the populations of the lower and upper levels respectively in the system in thermal equilibrium at a temperature T. Show that the relative populations are still given by a Boltzmann-like expression but with a temperature $-T$. Under what circumstances is it possible to speak of negative temperatures of an evenly spaced three-level system?

19.13. Consider Stirling's approximation for $\ln N!$ in the derivation of the Boltzmann distribution. What difference would it make if (a) a cruder approximation, $N! \approx N^N$, (b) the better approximation in Footnote 3 were used instead?

20

Statistical thermodynamics: the machinery

In this chapter we apply the concepts of statistical thermodynamics to the calculation of chemically significant quantities. In most cases we deal with independent molecules, so we can use the molecular partition function.

First, we establish the relations between thermodynamic functions and partition functions. The principal relations, those between the internal energy and entropy and the partition function, were established in Chapter 19, and we use classical thermodynamics to establish the relations we require from these basic relations. Next, we show that the molecular partition function can be factorized into contributions from each mode of motion, and establish the formulae for the partition functions for translational, rotational, vibrational, and electronic modes of motion. These contributions can be calculated from spectroscopic data. Finally, we turn to specific calculations. These applications include the mean energies of modes of motion (which justifies the equipartition theorem used earlier in the text), the heat capacities of substances, and residual entropies. Those three applications are all quite simple. However, we need to draw on all the material of these two chapters to derive the relation between equilibrium constants and spectroscopic data, which is done at the end of the chapter. That final calculation also elucidates some of the molecular features that determine the magnitudes of equilibrium constants and their temperature dependence.

A partition function is the bridge between thermodynamics, spectroscopy, and quantum mechanics. Once it is known, we can calculate thermodynamic functions, heat capacities, entropies, and equilibrium constants and, in the process, discover new ways of thinking about their significance. The principal aim of this chapter is to show how to calculate as chemical a quantity as an equilibrium constant from spectroscopic data on bond lengths and angles, force constants, and dissociation energies.

Table 20.1 Statistical thermodynamic relations

In terms of the canonical partition function Q

$$U - U(0) = -\left(\frac{\partial \ln Q}{\partial \beta}\right)_V$$

$$S = \frac{U - U(0)}{T} + k \ln Q$$

$$A - A(0) = -kT \ln Q$$

$$p = kT\left(\frac{\partial \ln Q}{\partial V}\right)_T$$

$$H - H(0) =$$
$$-\left(\frac{\partial \ln Q}{\partial \beta}\right)_V + kTV\left(\frac{\partial \ln Q}{\partial V}\right)_T$$

$$G - G(0) = -kT \ln Q + kTV\left(\frac{\partial \ln Q}{\partial V}\right)_T$$

For indistinguishable, independent particles $Q = q^N/N!$

$$U - U(0) = -N\left(\frac{\partial \ln q}{\partial \beta}\right)_V$$

$$S = \frac{U - U(0)}{T} + nR(\ln q - \ln N + 1)$$

$$G - G(0) = -nRT \ln\left(\frac{q_m}{N_A}\right)$$

where q_m is the molar partition function. For distinguishable, independent particles $Q = q^N$

$$U - U(0) = -N\left(\frac{\partial \ln q}{\partial \beta}\right)_V$$

$$S = \frac{U - U(0)}{T} + nR \ln q$$

$$G - G(0) = -nRT \ln q$$

In general, for indistinguishable independent particles,

$$Q = \frac{(q_{external}q_{internal})^N}{N!}$$
$$= \frac{(q_{external})^N}{N!} \times (q_{internal})^N$$

The thermodynamic functions are then the sums of internal and external (translational) contributions.

FUNDAMENTAL RELATIONS

We do two things in this section. First, we show how all the thermodynamic functions can be obtained once we know the partition function. Then we show how to calculate the molecular partition function, and through that the thermodynamic functions, from spectroscopic data. In the following expressions, we often use the parameter $\beta = 1/kT$ which was introduced in Chapter 19.

20.1 The thermodynamic functions

We have already derived (in Section 19.6) the two formulae for calculating the internal energy and the entropy of a system from its partition function Q:

$$U - U(0) = -\left(\frac{\partial \ln Q}{\partial \beta}\right)_V \tag{1a}$$

$$S = \frac{U - U(0)}{T} + k \ln Q \tag{1b}$$

If the molecules are independent, we can go on to make the substitutions $Q = q^N$ (if the molecules are distinguishable, as in a solid) or $Q = q^N/N!$ (if they are indistinguishable, as in a perfect gas). All the thermodynamic functions introduced in Part 1 are related to U and S, so we have a route to their calculation from Q. We shall now show how the relations can be constructed by drawing on some of the results established in Part 1. For later convenience, the expressions we shall derive are collected in Table 20.1.

The Helmholtz energy

The Helmholtz energy A is given by $A = U - TS$. This relation implies that $A(0) = U(0)$, so substitution for U and S using eqn 1 leads to the very simple expression

$$A - A(0) = -kT \ln Q \tag{2}$$

The pressure

Once we know the Helmholtz function, we can derive an expression for the pressure. It follows from classical thermodynamics (by analogy with eqn 5.7) that

$$p = -\left(\frac{\partial A}{\partial V}\right)_T$$

Therefore, the pressure p of a system is related to its partition function by

$$p = kT\left(\frac{\partial \ln Q}{\partial V}\right)_T \tag{3}$$

This relation is entirely general, and may be used for any type of substance, including perfect gases, real gases, and liquids (see Section 20.5).

The enthalpy

Because U and p have now both been related to Q, the enthalpy H can also be related to it through $H = U + pV$:

$$H - H(0) = -\left(\frac{\partial \ln Q}{\partial \beta}\right)_V + kTV\left(\frac{\partial \ln Q}{\partial V}\right)_T \tag{4}$$

Example 20.1 *Calculating the enthalpy*

Calculate the enthalpy of a perfect monatomic gas at a temperature T.

Method. A direct approach would be to substitute $Q = q^N/N!$ into eqn 4. However, there is a short cut, because most of the work has already been done for this calculation. Thus, in Section 19.3 we showed that

$$U - U(0) = \tfrac{3}{2}nRT$$

and we know that $pV = nRT$.

Answer. It follows from these two relations that

$$H - H(0) = U - U(0) + pV = \tfrac{3}{2}nRT + nRT = \tfrac{5}{2}nRT$$

Exercise E20.1. Repeat the calculation by explicit use of eqn 4.

The Gibbs energy

One of the most important thermodynamic functions for chemistry is the Gibbs energy, $G = A + pV$. We can now express this function in terms of the partition function because we know the expressions for A and p:

$$G - G(0) = -kT \ln Q + kTV\left(\frac{\partial \ln Q}{\partial V}\right)_T \tag{5}$$

In this expression, $G(0)$ is the Gibbs energy at $T = 0$.

The expression for G takes a simple form for a perfect gas because pV can be replaced by nRT:

$$G - G(0) = -kT \ln Q + nRT$$

Furthermore, because $Q = q^N/N!$, it follows that

$$G - G(0) = -NkT \ln q + kT \ln N! + nRT$$
$$= -nRT \ln q + kT(N \ln N - N) + nRT$$
$$= -nRT \ln\left(\frac{q}{N}\right) \tag{6°}$$

with $N = nN_A$. It will turn out to be convenient to define the **molar partition function**, q_m (with units mol^{-1}), as

$$q_m = \frac{q}{n} \tag{7}$$

Then

$$G - G(0) = -nRT \ln\left(\frac{q_m}{N_A}\right) \tag{8°}$$

20.2 The molecular partition function

The energy of a molecule is the sum of contributions from its different modes of motion:

$$\varepsilon_j = \varepsilon_j^T + \varepsilon_j^R + \varepsilon_j^V + \varepsilon_j^E$$

where T denotes translation, R rotation, V vibration, and E the electronic contribution. This separation is only approximate (except for translation) because the modes are not completely independent, but in most cases it is satisfactory. The separation of the electronic and vibrational motions, for example, is justified by the Born–Oppenheimer approximation (Chapter 14), and the separation of the vibrational and rotational modes is valid to the extent that the molecule can be treated as a rigid rotor.

Given that the energy separates in this way, the partition function factorizes into contributions from each mode:

$$q = \sum_i e^{-\beta \varepsilon_i} = \sum_i e^{-\beta \varepsilon_i^T - \beta \varepsilon_i^R - \beta \varepsilon_i^V - \beta \varepsilon_i^E}$$

$$= \left(\sum_i e^{-\beta \varepsilon_i^T} \right) \left(\sum_i e^{-\beta \varepsilon_i^R} \right) \left(\sum_i e^{-\beta \varepsilon_i^V} \right) \left(\sum_i e^{-\beta \varepsilon_i^E} \right)$$

$$= q^T q^R q^V q^E$$

This factorization means that we can investigate each contribution separately.

The translational contribution

The translational partition function of a molecule of mass m in a container of volume V was derived in Section 19.2:

$$q^T = \frac{V}{\Lambda^3}, \qquad \Lambda = h \left(\frac{\beta}{2\pi m} \right)^{\frac{1}{2}} = \frac{h}{(2\pi m k T)^{\frac{1}{2}}} \tag{9}°$$

Notice that $q^T \to \infty$ as $T \to \infty$ because an infinite number of states become accessible as the temperature is raised. Even at room temperature $q^T \approx 2 \times 10^{28}$ for an O_2 molecule in a vessel of volume 100 cm^3.

The thermal wavelength Λ lets us judge whether the approximations that led to the expression for q^T are valid. The approximations are valid if many states are occupied, which requires V/Λ^3 to be large. That will be so if Λ is small compared with the linear dimensions of the container. For H_2 at 25°C, $\Lambda = 71 \text{ pm}$, which is far smaller than any conventional container is likely to be (but comparable to pores in zeolites or cavities in clathrates). For O_2, a heavier molecule, $\Lambda = 18 \text{ pm}$.

The rotational contribution

As we saw in Example 19.2, the partition function of a non-symmetrical (AB) linear rotor is

$$q^R = \sum_J (2J + 1) e^{-\beta h c B J (J+1)} \tag{10}$$

The direct method of calculating q^R is to substitute the experimental values of the rotational energy levels into this expression and to sum the series numerically.

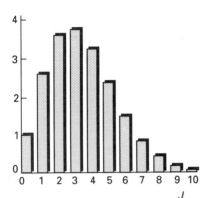

20.1 The contributions to the rotational partition function of an HCl molecule at 25°C. The vertical axis is the value of $(2J+1)e^{-\beta hcBJ(J+1)}$. Successive terms (which are proportional to the populations of the levels) pass through a maximum because the population of individual states decreases exponentially, but the degeneracy of the levels increases with J.

Example 20.2 *Evaluating the rotational partition function explicitly*

Evaluate the rotational partition function of $^1H^{35}Cl$ at 25°C, given that $B = 10.591 \text{ cm}^{-1}$.

Method. We use eqn 10, and evaluate it term by term. A useful relation is that $kT/hc = 207.22 \text{ cm}^{-1}$ at 298.15 K.

Answer. We draw up the following table on the basis that $hcB/kT = 0.05111$.

J	0	1	2	3	4	...	10
$J(J+1)$	0	2	6	12	20	...	110
$x = 2J+1$	1	3	5	7	9	...	21
$y = e^{-0.05111J(J+1)}$	1	0.903	0.736	0.542	0.360	...	0.004
xy	1	2.71	3.68	3.79	3.24	...	0.08

The contributions to the sum are illustrated in Fig. 20.1. The sum required by eqn 10 (the sum of the numbers in the last row of the table) is 19.9; hence $q^R = 19.9$ at this temperature. Taking J up to 16 gives $q^R = 19.902$.

Comment. Notice that about 10 J-levels are significantly populated but the number of populated *states* is larger on account of the $(2J+1)$-fold degeneracy of each level. We shall shortly encounter the approximation that $q^R \approx kT/hcB$, which in the present case gives $q^R = 19.6$, in good agreement with the exact value, and with much less work. Calculations based on eqn 10 are easily programmed for a microcomputer.

Exercise E20.2. Evaluate the rotational partition function for HCl at 0°C.
[18.26]

At room temperature $kT/hc \approx 200 \text{ cm}^{-1}$. The rotational constants of many molecules are close to 1 cm^{-1} and often smaller:

	CO$_2$	I$_2$	HI	HCl
B/cm^{-1}	0.39	0.037	6.5	10.6

(The very light H$_2$ molecule, for which $B = 60.9 \text{ cm}^{-1}$, is an exception.) It follows that at normal temperatures, many rotational levels are populated. When this is the case, the partition function may be approximated as follows

$$\text{Linear rotors:} \quad q^R = \frac{kT}{hcB}$$

$$\text{Non-linear rotors:} \quad q^R = \left(\frac{kT}{hc}\right)^{\frac{3}{2}}\left(\frac{\pi}{ABC}\right)^{\frac{1}{2}}$$

where A, B, and C are the rotational constants of the molecule. However, before using these expressions, read on (to eqns 12 and 13).

JUSTIFICATION

When many rotational states are occupied and kT is much larger than the separation between neighbouring states, the sum in the partition function can be approximated by an integral:

$$q^R = \int_0^\infty (2J + 1)e^{-\beta hcBJ(J+1)}\,dJ$$

Although this integral looks complicated, it can be evaluated without much effort by noticing that it can also be written as

$$q^R = -\frac{1}{\beta hcB}\int_0^\infty \left(\frac{d}{dJ}e^{-\beta hcBJ(J+1)}\right)dJ$$

Then, because the integral of a derivative of a function is the function itself,

$$q^R = -\frac{1}{\beta hcB}e^{-\beta hcBJ(J+1)}\Big|_0^\infty = \frac{1}{\beta hcB}$$

The calculation for a non-linear molecule is along the same lines, but slightly trickier (see *Further reading*).

A useful way of expressing the temperature above which the approximation is valid is to introduce the **rotational temperature** θ_R, which is defined as

$$k\theta_R = hcB \qquad\qquad (11)$$

Then 'high temperature' means $T \gg \theta_R$. Some typical values are

	I_2	CO_2	HI	HCl	CH_4	H_2
θ_R/K	0.053	0.56	7.5	9.4	15	88

The value for H_2 is abnormally high and we must be careful with the approximation for this molecule.

The general conclusion at this stage is that molecules with large moments of inertia (and hence small rotational constants and low rotational temperatures) have large rotational partition functions. The large value of q^R reflects the closeness in energy (compared with kT) of the rotational levels in large, heavy molecules, and the large number of them that are populated at normal temperatures.

We must take care, however, not to include too many rotational states in the sum. For a homonuclear diatomic molecule or a symmetrical linear ($D_{\infty h}$) molecule (such as CO_2 or $HC\equiv CH$), a rotation through $180°$ results in an indistinguishable state of the molecule. Hence, the number of thermally accessible states is only half the number that can be occupied by a heteronuclear diatomic molecule, where rotation through $180°$ does result in a distinguishable state. Therefore, for a symmetrical linear molecule,

$$q^R = \frac{kT}{2hcB}$$

The equations for symmetrical and non-symmetrical molecules can be combined into a single expression by introducing the **symmetry number** σ, which is the number of indistinguishable orientations of the molecule. Then

$$q^R = \frac{kT}{\sigma hcB} \tag{12}$$

For a heteronuclear diatomic molecule $\sigma = 1$; for a homonuclear diatomic molecule or a symmetrical ($D_{\infty h}$) linear molecule, $\sigma = 2$.

JUSTIFICATION

The quantum mechanical origin of the symmetry factor is the Pauli principle, which forbids the occupation of certain states. We saw in Section 16.7, for example, that H_2 may occupy only rotational states with even J if its nuclear spins are paired (*para*-hydrogen), and only odd J states if its nuclear spins are parallel (*ortho*-hydrogen). There are three states of *ortho*-H_2 to each value of J (because there are three parallel spin states of the nuclei).

To set up the rotational partition function we note that 'ordinary' molecular hydrogen is a mixture of one part *para*-H_2 (with only its even-J rotational states occupied) and three parts *ortho*-H_2 (with only its odd-J rotational states occupied). Therefore, the average partition function per molecule is

$$q^R = \frac{1}{4}\left\{\sum_{\text{even } J} (2J+1)e^{-\beta hcBJ(J+1)} + 3\sum_{\text{odd } J} (2J+1)e^{-\beta hcBJ(J+1)}\right\}$$

The odd-J states are more heavily weighted than the even-J states (Fig. 20.2). From the illustration we see that we would obtain approximately the same answer for the partition function (the sum of all the populations) if each J term contributed half its normal value to the sum. That is, the last equation can be approximated as

$$q^R = \tfrac{1}{2}\sum_{J} (2J+1)e^{-\beta hcBJ(J+1)}$$

and this approximation is very good when many terms contribute (at high temperatures).

The same type of argument may be used for linear symmetrical molecules in which identical bosons are interchanged by rotation (such as CO_2). As pointed out in Section 16.7, if the nuclear spin of the bosons is 0, then only even-J states are admissible. Because only half the rotational states are occupied, the rotational partition function is only half the value of the sum obtained by allowing all values of J to contribute (Fig. 20.3).

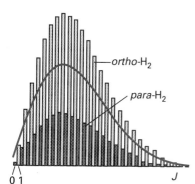

20.2 The values of the individual terms $(2J+1)e^{-\beta hcBJ(J+1)}$ contributing to the mean partition function of a 3:1 mixture of *ortho*- and *para*-H_2. The partition function is the sum of all these terms. At high temperatures the sum is approximately equal to the sum of the terms over all values of J, each with a weight of $\tfrac{1}{2}$. This is the sum of the contributions indicated by the curve.

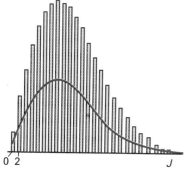

20.3 The relative populations of the rotational energy levels of CO_2. Only states with even J values are occupied. The full line shows the smoothed, averaged population of levels.

The same care must be exercised for other types of symmetrical molecule, and for a non-linear molecule we write

$$q^R = \frac{1}{\sigma}\left(\frac{kT}{hc}\right)^{\frac{3}{2}}\left(\frac{\pi}{ABC}\right)^{\frac{1}{2}} \tag{13}$$

Some typical values of the symmetry numbers required are

	H_2O	NH_3	CH_4	C_6H_6
σ:	2	3	12	12

The value $\sigma(H_2O) = 2$ reflects the fact that a 180° rotation about its C_2 axis interchanges two indistinguishable atoms. In NH_3, there are three indistinguishable orientations around its C_3 axis. For CH_4, any of three 120° rotations about any of its four C–H bonds leaves the molecule in an indistinguishable state, so the symmetry number is $3 \times 4 = 12$. For benzene, any of six orientations around its C_6 axis leaves it apparently unchanged, as does a rotation of 180° around any of six C_2 axes in the plane of the molecule.

A more formal way of arriving at the value of the symmetry number is to note that σ is the order (the number of elements) of the rotational subgroup of the molecules, the point group of the molecule with all but the identity and the rotations removed. The rotational subgroup of H_2O is $\{E, C_2\}$ so $\sigma = 2$. The rotational subgroup of NH_3 is $\{E, 2C_3\}$, so $\sigma = 3$. This recipe makes it easy to find the symmetry numbers for more complicated molecules. The rotational subgroup of CH_4 is obtained from the T character table (Data section) as $\{E, 8C_3, 3C_2\}$, so $\sigma = 12$. For benzene, the rotational subgroup of D_{6h} is $\{E, 2C_6, 2C_3, C_2, 3C_2', 3C_2''\}$, so $\sigma = 12$.

Example 20.3 *Estimating a rotational partition function*

Estimate the rotational partition function of ethene at 25°C given that $A = 4.828$ cm^{-1}, $B = 1.0012$ cm^{-1}, and $C = 0.828$ cm^{-1}.

Method. We use eqn 13 with $kT/hc = 207.223$ cm^{-1}. Next, identify the molecular point group (for example, by using the chart in Fig. 15.14 or the shapes in Fig. 15.15). The symmetry number is obtained by deciding on the rotational subgroup of the molecular point group, and counting the number of elements in the group.

Answer. The point group of the molecule is D_{2h}. From that group's character table, the rotational subgroup of the molecule consists of the elements $(E, C_{2x}, C_{2y}, C_{2z})$. The order of this subgroup is 4; therefore $\sigma = 4$. Then, because $ABC = 4.0033$ cm^{-3}, it follows that $q^R = 661$.

Comment. Ethene is quite a big molecule. The energy levels are close together (compared with kT at room temperature), and many are significantly populated at room temperature.

Exercise E20.3. Evaluate the rotational partition function of pyridine, C_5H_5N, at room temperature ($A = 0.2014$ cm^{-1}, $B = 0.1936$ cm^{-1}, $C = 0.0987$ cm^{-1}).   [4.3×10^4]

The vibrational contribution

The vibrational partition function of a molecule is calculated by substituting the measured vibrational energy levels into the exponentials appearing in the definition of q^V, and summing then numerically. In a polyatomic molecule each normal mode (Section 16.13) has its own partition function (so long as the anharmonicities are so small that the modes are independent). The overall vibrational partition function is the product of the individual partition functions:

$$q^V = q^V(1)q^V(2) \ldots$$

where $q^V(K)$ is the partition function for the Kth normal mode and is calculated by direct summation of the observed spectroscopic levels.

Example 20.4 *Estimating the vibrational partition function*

Given that a typical value of the vibrational partition function of one normal mode is about 1.1, estimate the overall vibrational partition function of a non-linear molecule containing 10 atoms.

Method. According to the remarks above, the overall vibrational partition function will be of the order of 1.1^n where n is the number of normal modes. Therefore, we need to calculate the number of normal modes of a non-linear molecule. That calculation was described in Section 16.13.

Answer. We saw in Section 16.13 that the number of normal modes is $3N - 6$ for a non-linear molecule containing N atoms. Therefore, in the present case $n = 24$. The overall vibrational partition function is therefore approximately

$$q^V \approx (1.1)^{24} = 9.9$$

Comment. Even though vibrational modes are not appreciably excited, there may be so many modes in a molecule that overall their excitation is significant.

Exercise E20.4. Repeat the calculation for a linear molecule containing 10 atoms.   [10.8]

If the vibrational excitation is not too great, the harmonic approximation may be made, and the vibrational energy levels written as

$$E_v = (v + \tfrac{1}{2})hc\tilde{v}, \qquad v = 0, 1, 2, \ldots$$

If we measure energies from the zero-point level (our general rule), then the permitted values are $\varepsilon_v = vhc\tilde{v}$ and the partition function is

$$q^V = \sum_v e^{-\beta vhc\tilde{v}} = \sum_v (e^{-\beta hc\tilde{v}})^v$$

We met this sum in Section 19.2 (which is no accident: the ladder-like array of levels in Fig. 19.4 is exactly the same as that of a harmonic oscillator). The series can be summed in the same way, and gives

$$q^V = \frac{1}{1 - e^{-\beta hc\tilde{v}}} \qquad (14)$$

In a polyatomic molecule, each normal mode gives rise to a partition function of this form (Fig. 20.4).

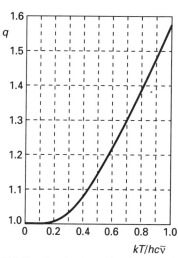

20.4 The vibrational partition function of a molecule in the harmonic approximation. Note that the partition function is linearly proportional to the temperature when the temperature is high ($T \gg \theta_V$).

Example 20.5 *Calculating a vibrational partition function*

The wavenumbers of the three normal modes of H_2O are 3656.7 cm^{-1}, 1594.8 cm^{-1}, and 3755.8 cm^{-1}. Evaluate the vibrational partition function at 1500 K.

Method. We use eqn 14 for each mode, and then form the product of the three contributions. At 1500 K, $kT/hc = 1042.5 \text{ cm}^{-1}$.

Mode:	1	2	3
$\tilde{\nu}/\text{cm}^{-1}$	3656.7	1594.8	3755.8
$hc\tilde{\nu}/kT$	3.508	1.530	3.603
q^V	1.031	1.276	1.028

The overall vibrational partition function is therefore

$$q^V = 1.031 \times 1.276 \times 1.028 = 1.353$$

Comment. The vibrations of H_2O are at such high wavenumber that even at 1500 K most of the molecules are in their vibrational ground state.

Exercise E20.5. Repeat the calculation for CO_2, where the vibrational wavenumbers are 1388 cm^{-1}, 667.4 cm^{-1}, and 2349 cm^{-1}, the second being the doubly degenerate bending mode.   [6.71]

In many molecules the vibrational wavenumbers are so great that $\beta hc\tilde{\nu} > 1$. For example, the lowest vibrational wavenumber of CH_4 is 1306 cm^{-1}, so $\beta hc\tilde{\nu} = 6$ at room temperature. C—H stretches normally lie in the range 2850 to 2960 cm^{-1}, so for them $\beta hc\tilde{\nu} = 14$. In these cases $e^{-\beta hc\tilde{\nu}}$ may be neglected in the denominator of q^V (for example, $e^{-6} = 0.002$), and the vibrational partition function for a single mode is simply $q^V \approx 1$ (implying that only the zero-point level is significantly occupied).

Now consider the case of bonds so weak that $\beta hc\tilde{\nu} \ll 1$. When this condition is satisfied, the partition function may be approximated by expanding the exponential:

$$q^V = \frac{1}{1 - (1 - \beta hc\tilde{\nu} + \ldots)} \approx \frac{1}{\beta hc\tilde{\nu}}$$

That is, for weak bonds at high temperatures,

$$q^V = \frac{kT}{hc\tilde{\nu}} \tag{15}$$

The temperatures for which this expansion is valid can be expressed in terms of the **vibrational temperature** θ_V, which is defined through $k\theta_V = hc\tilde{\nu}$. In terms of the vibrational temperature, 'high temperature' means $T \gg \theta_V$. Some typical values are

	I_2	F_2	HCl	H_2
$\tilde{\nu}/\text{cm}^{-1}$	215	892	2990	4400
θ_V/K	309	1280	4300	6330

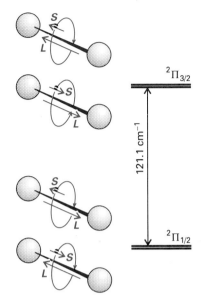

20.5 The doubly degenerate ground electronic level of NO (with the spin and orbital angular momentum around the axis in opposite directions) and the doubly degenerate first excited level (with the spin and orbital momenta parallel). The upper level is thermally accessible at room temperature.

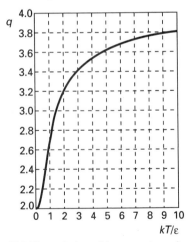

20.6 The variation with temperature of the electronic partition function of an NO molecule. Note that the curve resembles that for a two-level system (Fig. 19.3), but rises from 2 (the degeneracy of the lower level) and approaches 4 (the total number of states) at high temperatures.

The value for H_2 is abnormally high because the atoms are so light and the vibrational frequency consequently high.

The electronic contribution

Electronic energy separations from the ground state are usually very large, so for most cases $q^E = 1$. An important exception arises in the case of atoms and molecules having electronically degenerate ground states, in which case

$$q^E = g^E$$

where g^E is the degeneracy of the electronic ground state. The alkali metal atoms, for example, have doubly degenerate ground states (corresponding to the two orientations of their electron spin), so $q^E = 2$.

Some atoms and molecules have low-lying electronically excited states (at high enough temperatures, all atoms and molecules have thermally accessible excited states). One case is NO, which has the configuration ... π^1 (the molecule has one electron more than N_2). The orbital angular momentum may take two orientations with respect to the molecular axis, and the spin angular momentum may also take two, giving four states in all (Fig. 20.5). The energy of the two states in which the orbital and spin momenta are parallel (giving the $^2\Pi_{\frac{3}{2}}$ level) is slightly greater than that of the two other states in which they are antiparallel (giving the $^2\Pi_{\frac{1}{2}}$ level). The separation, which arises from spin–orbit coupling (Section 13.7), is only 121 cm^{-1}. Hence, at normal temperatures, all four states are thermally accessible. If we denote the energies of the two levels as $E_{\frac{1}{2}} = 0$ and $E_{\frac{3}{2}} = \varepsilon$, the partition function is

$$q^E = \sum_{\text{levels } j} g_j e^{-\beta \varepsilon_j} = 2 + 2e^{-\beta \varepsilon}$$

The variation of this partition function with temperature is shown in Fig. 20.6. At $T = 0$, $q^E = 2$, because only the doubly degenerate ground state is accessible. At very high temperatures, q^E approaches 4 because all four states are accessible. At 25°C, $q^E = 2.8$.

The overall partition function

The partition functions for each mode of motion of a molecule are collected in Table 20.2. The overall partition function is the product of each contribution. For a diatomic molecule with no low-lying electronically excited states

$$q = g^E \times \frac{V}{\Lambda^3} \times \frac{kT}{\sigma hcB} \times \frac{1}{1 - e^{-\beta hc\tilde{v}}} \qquad (16)$$

Overall partition functions obtained in this way are approximate because they assume that the rotational levels are very close and that the vibrational levels are harmonic. These approximations are avoided by evaluating the sums explicitly using the energy levels identified spectroscopically.

Table 20.2 Contributions to the molecular partition function*

Translation

$$q = \frac{V}{\Lambda^3} \qquad \Lambda = \left(\frac{h^2\beta}{2\pi m}\right)^{\frac{1}{2}}$$

$$\Lambda/\text{pm} = \frac{1749}{(T/\text{K})^{\frac{1}{2}}(M/\text{g mol}^{-1})^{\frac{1}{2}}}$$

$$\frac{q_m^{\ominus}}{N_A} = \frac{kT}{p^{\ominus}\Lambda^3} = 2.561 \times 10^{-2}(T/\text{K})^{\frac{5}{2}}(M/\text{g mol}^{-1})^{\frac{3}{2}}$$

Rotation
  (a) *Linear molecules*

$$q = \frac{1}{\sigma hcB\beta} = \frac{0.6950}{\sigma} \times \frac{T/\text{K}}{(B/\text{cm}^{-1})}$$

  (b) *Non-linear molecules*

$$q = \frac{1}{\sigma}\left(\frac{1}{hc\beta}\right)^{\frac{3}{2}}\left(\frac{\pi}{ABC}\right)^{\frac{1}{2}} = \frac{1.0270}{\sigma} \times \frac{(T/\text{K})^{\frac{3}{2}}}{(ABC/\text{cm}^{-3})^{\frac{1}{2}}}$$

Vibration

$$q = \frac{1}{1-e^{-hc\tilde{v}\beta}} = \frac{1}{1-e^{-a}} \qquad a = \frac{1.4388(\tilde{v}/\text{cm}^{-1})}{T/\text{K}}$$

Electronic

$$q = g$$

where g is the degeneracy of the electronic ground state (when that is the only accessible level); at high temperatures, evaluate q explicitly.

---

* $\beta = 1/kT$. It is often useful to note that

$$\frac{hc}{k} = 1.438\ 79\ \text{cm K}$$

See also inside front cover for further information.

Example 20.6 *Calculating a thermodynamic function from spectroscopic data*

Calculate the value of $G_m^{\ominus} - G_m^{\ominus}(0)$ for $H_2O(g)$ at 1500 K given that $A = 27.8778\ \text{cm}^{-1}$, $B = 14.5092\ \text{cm}^{-1}$, and $C = 9.2869\ \text{cm}^{-1}$ and using the information in Example 20.5.

Method. The starting point is eqn 8:

$$G_m - G_m(0) = -RT\ln\left(\frac{q_m}{N_A}\right)$$

For the standard value, we evaluate the translational partition function at $p^{\ominus}$. The vibrational partition function was calculated in Example 20.5. Use the expressions in Table 20.2 for the other contributions.

Answer. For this C_{2v} molecule, $\sigma = 2$. Since $m = 18.015\ \text{u}$, $q_m^{T\ominus} = 1.706 \times 10^8\ \text{mol}^{-1}$. For the vibrational contribution we have already found that $q^V = 1.353$. For the rotational contribution, $q^R = 486.7$. The overall molar partition function is therefore $q_m^{\ominus}/N_A = 1.123 \times 10^{11}$, so

$$G_m^{\ominus} - G_m^{\ominus}(0) = -(8.3145 \text{ J K}^{-1} \text{ mol}^{-1}) \times (1500 \text{ K}) \times \ln(1.123 \times 10^{11})$$
$$= -317.3 \text{ kJ mol}^{-1}$$

Exercise E20.6. Repeat the calculation for CO_2. The vibrational data are given in Exercise E20.5; $B = 0.3902 \text{ cm}^{-1}$. $[-366.6 \text{ kJ mol}^{-1}]$

USING STATISTICAL THERMODYNAMICS

Any thermodynamic quantity can now be calculated from a knowledge of the energy levels of molecules: we have merged thermodynamics and spectroscopy. In this section, we indicate how to do the calculations for four important properties.

20.3 Mean energies

It is often useful to know the **mean energy** $\langle \varepsilon \rangle$ of various modes of motion. When the molecular partition function can be factorized into contributions from each mode, the mean energy of each mode is

$$\langle \varepsilon^M \rangle = -\frac{1}{q^M}\left(\frac{\partial q^M}{\partial \beta}\right)_V, \qquad \text{with } M = T, R, V, \text{ or } E \tag{17}$$

The mean translational energy

To see a pattern emerging, we consider first a one-dimensional system of length X, for which $q^T = X/\Lambda$ with $\Lambda = h(\beta/2\pi m)^{\frac{1}{2}}$. Then, if we note that q^T is a constant times $\sqrt{\beta}$,

$$\langle \varepsilon^T \rangle = -\frac{\Lambda}{X}\left(\frac{\partial}{\partial \beta}\frac{X}{\Lambda}\right)_V = -\beta^{\frac{1}{2}}\frac{d}{d\beta}\left(\frac{1}{\beta^{\frac{1}{2}}}\right) = \frac{1}{2\beta} = \tfrac{1}{2}kT$$

For a molecule free to move in three dimensions, the analogous calculation leads to

$$\langle \varepsilon^T \rangle = \tfrac{3}{2}kT \tag{18}°$$

Both conclusions are in agreement with the classical equipartition theorem (see the Introduction), that the mean energy of each quadratic contribution to the energy is $\tfrac{1}{2}kT$. Furthermore, the fact that the mean energy is independent of the size of the container is consistent with the thermodynamic result that the internal energy of a perfect gas is independent of its volume (Section 5.1).

The mean rotational energy

The mean rotational energy of a linear molecule is obtained from the partition function given in eqn 10:

$$q^R = \sum_J (2J + 1)e^{-\beta \varepsilon J(J+1)}, \qquad \text{with } \varepsilon = hcB$$

When the temperature is low ($T < \theta_R$), the series must be summed term

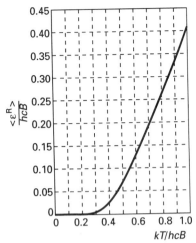

20.7 The mean rotational energy of a non-symmetrical linear rotor as a function of temperature. At high temperatures $(T \gg \theta_R)$, the energy is linearly proportional to the temperature, in accord with the equipartition theorem.

by term, which gives

$$q^R = 1 + 3e^{-2\beta\varepsilon} + 5e^{-6\beta\varepsilon} + \ldots$$

Hence

$$\langle \varepsilon^R \rangle = \frac{\varepsilon(6e^{-2\beta\varepsilon} + 30e^{-6\beta\varepsilon} + \ldots)}{1 + 3e^{-2\beta\varepsilon} + 5e^{-6\beta\varepsilon} + \ldots}$$

This function is plotted in Fig. 20.7. At high temperatures $(T \gg \theta_R)$, q^R is given by eqn 12, $q = 1/\sigma hc\beta B$, and

$$\langle \varepsilon^R \rangle = -\frac{1}{q^R} \left(\frac{\partial q^R}{\partial \beta} \right)_V = \frac{1}{\beta} = kT \tag{19}$$

The high-temperature result is also in agreement with the equipartition theorem, for the classical expression for the energy of a linear rotor (Section 16.4) is

$$E_K = \tfrac{1}{2}I_a\omega_a^2 + \tfrac{1}{2}I_b\omega_b^2$$

(There is no rotation around the line of atoms.) It follows from the equipartition theorem that the mean rotational energy is $2 \times \tfrac{1}{2}kT = kT$.

The mean vibrational energy

The vibrational partition function is given in eqn 14. Because q^V is independent of the volume, it follows that

$$\frac{dq^V}{d\beta} = \frac{d}{d\beta} \left(\frac{1}{1 - e^{-\beta hc\tilde{\nu}}} \right) = \frac{hc\tilde{\nu}e^{-\beta hc\tilde{\nu}}}{(1 - e^{-\beta hc\tilde{\nu}})^2}$$

and hence that

$$\langle \varepsilon^V \rangle = \frac{hc\tilde{\nu}}{e^{\beta hc\tilde{\nu}} - 1} \tag{20}$$

This formula is exact, apart from the zero-point energy of $\tfrac{1}{2}hc\tilde{\nu}$, which can be added to the right-hand side if required. The variation of the mean energy with temperature is illustrated in Fig. 20.8.

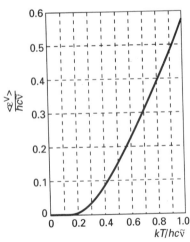

20.8 The mean vibrational energy of a molecule in the harmonic approximation as a function of temperature. At high temperatures $(T \gg \theta_V)$, the energy is linearly proportional to the temperature, in accord with the equipartition theorem.

Example 20.7 *Deriving the high-temperature form of an expression*

Derive an expression for the mean energy of a harmonic oscillator when the temperature is high.

Method. By 'high temperature' in this vibrational context is meant $T \gg \theta_V$, or $\beta hc\tilde{\nu} \ll 1$. The exponential functions can therefore be expanded and all but the leading terms discarded.

Answer. It follows from eqn 20 that

$$\langle \varepsilon^V \rangle = \frac{hc\tilde{\nu}}{(1 + \beta hc\tilde{\nu} + \ldots) - 1} \approx \frac{1}{\beta} = kT$$

Comment. A mean energy of kT is in agreement with the value predicted by the classical equipartition theorem, because the energy of a one-dimensional oscillator is

$$E = \tfrac{1}{2}mv_x^2 + \tfrac{1}{2}kx^2$$

and the mean energy of each quadratic term is $\tfrac{1}{2}kT$. Note also how the full expression, which is plotted in Fig. 20.8, tends to linearity at high temperatures.

Exercise E20.7. Suppose that we falsely believe that $h = 0$ (as in classical physics). Calculate the mean energy of a harmonic oscillator at a temperature T in the limit as $h \to 0$. $\qquad$ [kT]

20.4 Heat capacities

The constant-volume heat capacity is defined as follows:

$$C_V = \left(\frac{\partial U}{\partial T}\right)_V \tag{21a}$$

The derivative with respect to T can be converted into a derivative with respect to β by using

$$\frac{\mathrm{d}}{\mathrm{d}T} = \frac{\mathrm{d}\beta}{\mathrm{d}T} \times \frac{\mathrm{d}}{\mathrm{d}\beta} = -\frac{1}{kT^2}\frac{\mathrm{d}}{\mathrm{d}\beta} = -k\beta^2\frac{\mathrm{d}}{\mathrm{d}\beta}$$

Therefore, an equivalent definition is

$$C_V = -k\beta^2\left(\frac{\partial U}{\partial \beta}\right)_V \tag{21b}$$

Because the internal energy of a perfect gas is a sum of contributions, the heat capacity is also a sum of contributions from each mode, and the contribution of the mode M is

$$C_V^{\mathrm{M}} = N\left(\frac{\partial \langle \varepsilon^{\mathrm{M}} \rangle}{\partial T}\right)_V = -Nk\beta^2\left(\frac{\partial \langle \varepsilon^{\mathrm{M}} \rangle}{\partial \beta}\right)_V \tag{21c}$$

The individual contributions

The temperature is always high enough (so long as the gas is above its condensation temperature) for the mean translational energy to be $\tfrac{3}{2}kT$, the equipartition value. Therefore, the molar constant-volume heat capacity is

$$C_{V,\mathrm{m}}^{\mathrm{T}} = N_{\mathrm{A}}\frac{\mathrm{d}(\tfrac{3}{2}kT)}{\mathrm{d}T} = \tfrac{3}{2}R \tag{22}°$$

Translation is the only mode of motion for a monatomic gas, so for such a gas

$$C_{V,\mathrm{m}}^{\mathrm{T}} = \tfrac{3}{2}R = 12.47\,\mathrm{J\,K^{-1}\,mol^{-1}}$$

This is a very reliable result: helium, for example, has this value over a range of 2000 K. We saw in Section 3.3 that the molar constant-pressure heat capacity of a perfect gas is related to $C_{V,\mathrm{m}}$ by

$$C_{p,\mathrm{m}} = C_{V,\mathrm{m}} + R$$

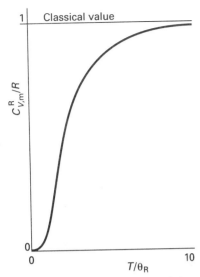

20.9 The temperature dependence of the rotational contribution to the heat capacity of a linear molecule.

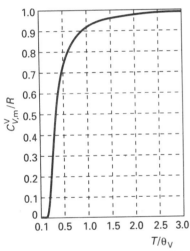

20.10 The temperature dependence of the vibrational heat capacity of a molecule in the harmonic approximation calculated by using eqn 24. Note that the heat capacity is within 10 per cent of its classical value for temperatures greater than θ_V.

We can therefore also conclude at once that, for a monatomic perfect gas,

$$C_{p,m} = \tfrac{5}{2}R = 20.78 \text{ J K}^{-1} \text{ mol}^{-1}$$

and that

$$\gamma = \frac{C_p}{C_V} = \tfrac{5}{3} \qquad (23)°$$

When the temperature is high enough for the rotations of the molecules to be highly excited (when $T \gg \theta_R$), we can use the equipartition value $\langle \varepsilon^R \rangle = kT$ (for a linear rotor) to obtain

$$C_{V,m}^R = R$$

For non-linear molecules, the mean rotational energy rises to $\tfrac{3}{2}kT$, so the molar rotational heat capacity rises to $\tfrac{3}{2}R$ when $T \gg \theta_R$. When the temperature is very low, only the lowest rotational state is occupied, and the rotations do not contribute to the heat capacity ($C_V^R = 0$). At intermediate temperatures, we can calculate the rotational heat capacity by differentiating the equation for $\langle \varepsilon^R \rangle$ obtained earlier. The resulting (untidy) expression, which is plotted in Fig. 20.9, shows that the contribution rises from zero (when $T = 0$) to the equipartition value (when $T \gg \theta_R$). Because the translational contribution is always present, we can expect the molar heat capacity of a gas of diatomic molecules ($C_{V,m}^T + C_{V,m}^R$) to rise from $12.5 \text{ J K}^{-1} \text{ mol}^{-1}$ as the temperature is increased above θ_R.

Molecular vibrations contribute to the heat capacity, but only when the temperature is high enough for them to be significantly excited. The equipartition mean energy is kT per mode, so the maximum contribution to the molar heat capacity is R. However, it is very unusual for the vibrations to be so highly excited that equipartition is valid, and it is more appropriate to use the full expression for the vibrational heat capacity, which is obtained by differentiating eqn 20:

$$C_{V,m}^V = Rf^2, \qquad \text{where } f = \frac{\theta_V}{T}\left\{\frac{e^{-\theta_V/2T}}{1 - e^{-\theta_V/T}}\right\} \qquad (24)$$

In this expression $\theta_V = hc\tilde{\nu}/k$, the vibrational temperature. The curve in Fig. 20.10 shows how the vibrational heat capacity depends on temperature. Note that, even when the temperature is only slightly above the vibrational temperature, the heat capacity is close to its equipartition value.

The overall heat capacity

The total heat capacity of a system is the sum of each contribution (Fig. 20.11). When equipartition is valid (when the temperature is well above the characteristic temperature of the mode, $T \gg \theta_M$) we can estimate the heat capacity by counting the numbers of modes that are active.

In gases, all three translational modes are always active, and contribute $\tfrac{3}{2}R$ to the molar heat capacity. If we denote the number of active

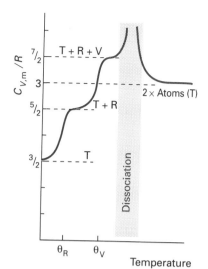

20.11 The general temperature dependence of the heat capacity of diatomic molecules is as shown here. Each mode becomes active when its characteristic temperature is exceeded. The heat capacity becomes very large when the molecule dissociates because the energy is used to cause dissociation and not to raise the temperature. Then it falls back to the translation-only value of the atoms.

rotational modes by v_R^* (so that for most molecules at normal temperatures $v_R^* = 2$ for linear molecules, and $v_R^* = 3$ for non-linear molecules), then the rotational contribution is $\frac{1}{2}v_R^* R$. If the temperature is high enough for v_V^* vibrational modes to be active, the vibrational contribution to the molar heat capacity is $v_V^* R$. In most cases $v_V^* \approx 0$. Therefore, the total molar heat capacity is

$$C_{V,m} = \tfrac{1}{2}(3 + v_R^* + 2v_V^*)R \qquad (25)$$

Example 20.8 *Estimating the molar heat capacity of a gas*

Estimate the molar constant-volume heat capacity of water vapour at 100°C. Vibrational wavenumbers are given in Example 20.5; the rotational constants of the molecule are 27.9, 14.5, and 9.3 cm^{-1}.

Method. We need to assess whether the rotational and vibrational modes are active by computing their characteristic temperatures from the data (to do so, we use $hc/k = 1.439$ cm K).

Answer. The characteristic temperatures (in round numbers) of the vibrations are 5300 K, 2300 K, and 5400 K and the vibrations are therefore inactive at 373 K. The three rotational modes have characteristic temperatures 40 K, 21 K, and 13 K, and so are fully active, like the three translational modes. The translational contribution is $\frac{3}{2}R = 12.5$ J K^{-1} mol^{-1}. Fully active rotations contribute a further 12.5 J K^{-1} mol^{-1}. Therefore, a value close to 25 J K^{-1} mol^{-1} is predicted.

Comment. The experimental value is 26.1 J K^{-1} mol^{-1}. The discrepancy is probably due to deviations from perfect gas behaviour.

Exercise E20.8. Estimate the molar constant-volume heat capacity of gaseous I_2 at 25°C ($B = 0.037$ cm^{-1}). [21 J K^{-1} mol^{-1}]

20.5 Equations of state

The canonical partition function Q is a function of the volume and the temperature of the system and the number of molecules it contains. Therefore, eqn 3 for the pressure in terms of the partition function has the form

$$p = f(n, V, T)$$

That is, eqn 3 is an equation of state. The relation between p and Q is therefore a very important route to the equations of state of real gases in terms of intermolecular forces, for the latter can be built into Q.

Example 20.9 *Deriving an equation of state*

Derive the perfect-gas law from the canonical partition function.

Method. Equation 3 has the form of an expression for p in terms of the variables on which Q depends. Therefore, we need to substitute the explicit formula for Q for a gas of independent, indistinguishable molecules (see Table 19.1).

Answer. For a perfect gas of independent molecules, we use $Q = q^N/N!$ with $q = V/\Lambda^3$:

$$p = kT\left(\frac{\partial \ln Q}{\partial V}\right)_T = \frac{kT}{Q}\left(\frac{\partial Q}{\partial V}\right)_T = \frac{NkT}{q}\left(\frac{\partial q}{\partial V}\right)_T$$

$$= nRT \times \frac{\Lambda^3}{V} \times \frac{1}{\Lambda^3} = \frac{nRT}{V}$$

To derive this relation, we have made use of

$$\left(\frac{\partial q}{\partial V}\right)_T = \frac{1}{\Lambda^3}$$

and $NkT = nN_AkT = nRT$.

Comment. This calculation can be regarded as yet another way of deducing that $\beta = 1/kT$.

Exercise E20.9. Derive the equation of state of a sample for which $Q = q^N f/N!$, where f depends on the volume.

$$[p = nRT/V + kT(\partial \ln f/\partial V)_T]$$

Real gases differ from perfect gases in their equations of state (Section 1.3), the existence of a non-zero Joule–Thomson coefficient (Section 3.2), and in various transport properties (which we leave until Chapter 24). The meeting place of theory and experiment is the virial equation of state

$$\frac{pV_m}{RT} = 1 + \frac{B}{V_m} + \frac{C}{V_m^2} + \dots$$

where B is the second virial coefficient and C is the third virial coefficient. Virial coefficients vary with temperature. In the following paragraphs we show in outline how B is related to the canonical partition function.

The total kinetic energy of a gas of molecules is the sum of the kinetic energies of the individual molecules even in the presence of interactions between them. Therefore, in a real gas the canonical partition function factorizes into a part arising from the kinetic energy, which is the same as for the perfect gas, and a factor called the **configuration integral** Z which depends on the intermolecular potentials. We shall therefore write

$$Q = \frac{Z}{\Lambda^{3N}} \tag{26}$$

For a perfect gas

$$Z = \frac{V^N}{N!} \tag{27}°$$

because Q is then the partition function for a collection of non-interacting particles. For a real gas, Z is related to the total potential energy V_N of interaction of all the particles by

$$Z = \frac{1}{N!}\int e^{-\beta V_N}\, d\boldsymbol{r}_1\, d\boldsymbol{r}_2 \dots d\boldsymbol{r}_N \tag{28}$$

JUSTIFICATION

We shall not justify the form of eqn 28 in detail, but instead will verify that this expression gives the perfect-gas partition function when there is no interaction, when

$$V_N = 0 \qquad \text{and } e^{-\beta V_N} = 1$$

It then follows that

$$Z = \frac{1}{N!} \int d\mathbf{r}_1 \, d\mathbf{r}_2 \ldots d\mathbf{r}_N = \frac{V^N}{N!}$$

because $\int d\mathbf{r} = V$, where V is the volume of the container.

When we consider only interactions between pairs of particles the configuration integral simplifies to

$$Z_2 = \tfrac{1}{2} \int e^{-V_2/kT} \, d\mathbf{r}_1 \, d\mathbf{r}_2 \tag{29}$$

The second virial coefficient then turns out to be

$$B = -\frac{N_A}{2V} \int f \, d\mathbf{r}_1 \, d\mathbf{r}_2, \qquad \text{where } f = e^{-V_2/kT} - 1 \tag{30a}$$

The quantity f is the **Mayer f-function**: it goes to zero when the two particles are so far apart that $V_2 = 0$. When the intermolecular interaction depends only on the separation r of the particles and not on their relative orientation, as in the interaction of closed-shell atoms and tetrahedral and octahedral molecules, eqn 30a simplifies to

$$B = -2\pi N_A \int_0^\infty f r^2 \, dr \tag{30b}$$

The integral can be evaluated (usually numerically) by substituting an expression for the intermolecular potential energy.

Intermolecular potential energies are discussed in more detail in Chapter 22, where several expressions are developed for them. At this stage we can illustrate how eqn 30 is used by considering the **hard-sphere potential**, which is infinite when the separations of the two molecules r is less than or equal to a certain value σ, and is zero for greater separations. Then

$$e^{-\beta V_2} = 0 \qquad f = -1 \qquad \text{when } r \leq \sigma \ (\text{and } V_2 = \infty)$$

$$e^{-\beta V_2} = 1 \qquad f = 0 \qquad \text{when } r > \sigma \ (\text{and } V_2 = 0)$$

It follows from eqn 30b that the second virial coefficient is

$$B = 2\pi N_A \int_0^\sigma r^2 \, dr = \tfrac{2}{3}\pi N_A \sigma^3$$

In Section 1.5, we saw that the part of the van der Waals equation of state that represented the hard-sphere repulsive part of the intermolecular potential (i.e. b) contributed to B in exactly this way.

The calculation of B just described raises the question as to whether a potential can be found which, when the virial coefficients are evaluated, gives the full van der Waals equation of state. Such a potential can be found: it consists of a hard-sphere repulsive core and a long-range, shallow attractive region. A further point is that once a second virial coefficient has been calculated for a given intermolecular potential, it is possible to calculate other thermodynamic properties that depend on the form of the potential. For example, it is possible to calculate the isothermal Joule–Thomson coefficient, μ_T (Section 3.2), from the thermodynamic relation

$$\lim_{p \to 0} \mu_T = B - T\frac{dB}{dT}$$

and from the result calculate the Joule–Thomson coefficient itself using eqn 13 of Section 3.2.

20.6 Residual entropies

Entropies may be calculated from spectroscopic data; they may also be measured experimentally (Section 4.4). In many cases there is good agreement, but in some the experimental entropy is less than the calculated value. One possibility is that the experimental determination failed to take a phase transition into account. Another is that some disorder is present in the solid even at $T = 0$. The entropy at $T = 0$ is then greater than zero, and is called the **residual entropy**.

The origin and magnitude of the residual entropy can be explained by considering a crystal composed of AB molecules, where A and B are similar atoms (such as CO, with its very small electric dipole moment). There may be so little energy difference between ...AB AB AB AB..., ...AB BA BA AB..., and other random arrangements that the molecules adopt either orientation at random in the solid. We can readily calculate the entropy arising from residual disorder by using the Boltzmann formula $S = k \ln W$. To do so, we suppose that two orientations are equally probable, and that the sample consists of N molecules. Because the same energy can be achieved in 2^N different way (because each molecule can take either of two orientations), the total number of ways of achieving the same energy is $W = 2^N$. It follows that

$$S = k \ln 2^N = Nk \ln 2 = nR \ln 2$$

We can therefore expect a residual molar entropy of $R \ln 2 = 5.8\,\mathrm{J\,K^{-1}}$ $\mathrm{mol^{-1}}$ for solids composed of molecules that can adopt either of two orientations at $T = 0$. If s orientations are possible, then the residual molar entropy will be

$$S_m = R \ln s \tag{31}$$

An $FClO_3$ molecule, for example, can adopt four orientations with about the same energy, and the calculated residual entropy of $R \ln 4 = 11.5\,\mathrm{J\,K^{-1}\,mol^{-1}}$ is in good agreement with the experimental value $(10.1\,\mathrm{J\,K^{-1}\,mol^{-1}})$. For CO, the measured residual entropy is $5\,\mathrm{J\,K^{-1}\,mol^{-1}}$, which is close to $R \ln 2$, the value expected for a random structure of the form ...CO CO OC CO OC OC... .

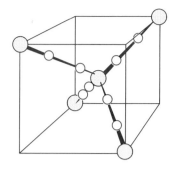

20.12 The possible locations of H atoms around a central O atom in an ice crystal are shown by the spheres. Only one of the locations on each bond may be occupied by an atom, and two H atoms must be close to the O atom and two H atoms must be distant from it.

20.13 The six possible arrangements of H atoms in the locations identified in Fig. 20.12.

The residual entropy of ice is $3.4\,\mathrm{J\,K^{-1}\,mol^{-1}}$. This value can be explained in terms of the hydrogen-bonded structure of the solid. Each O atom is surrounded tetrahedrally by four H atoms, two of which are attached by short σ bonds, the other two being attached by long hydrogen bonds (Fig. 20.12). The randomness lies in which two of the four bonds are short, and an approximate analysis leads to $S_m(0) \approx R \ln \frac{3}{2} = 3.4\,\mathrm{J\,K^{-1}\,mol^{-1}}$, in good agreement with the experimental value.

JUSTIFICATION

Consider a sample of ice that consists of N H_2O molecules. Each of the $2N$ H atoms can be in one of two positions: either close to or far from an O atom (as shown in Fig. 20.12). There are therefore 2^{2N} possible arrangements. However, not all these arrangements are acceptable. Indeed, of the $2^4 = 16$ ways of arranging four H atoms around one O atom, only six have two short and two long OH distances and hence are acceptable (Fig. 20.13). Therefore, the number of permitted arrangements is

$$W = 2^{2N} \times \left(\frac{6}{16}\right)^N = \left(\frac{3}{2}\right)^N$$

20.7 Equilibrium constants

The Gibbs energy of a system of independent molecules is given by eqn 8 as

$$G - G(0) = -nRT \ln\left(\frac{q_m}{N_A}\right)$$

where q_m is the molar partition function, q/n. The equilibrium constant K of a reaction is related to the standard Gibbs energy of reaction by

$$\Delta_r G^{\ominus} = -RT \ln K$$

To calculate the equilibrium constant, we must combine these two equations. We shall consider gas phase reactions in which the equilibrium constant is expressed in terms of the partial pressures of the reactants and products.

The relation between K and the partition function

To find an expression for the standard reaction Gibbs energy we need expressions for the standard molar Gibbs energies, $G^{\ominus}/n$, of each species. For these expressions, we need the value of the molar partition function when $p = p^{\ominus}$ (where $p^{\ominus} = 1$ bar): we denote this the **standard molar partition function** $q_m^{\ominus}$. Because only the translational component depends on the pressure, we can find $q_m^{\ominus}$ by evaluating the partition function with V replaced by $V_m^{\ominus}$, with $V_m^{\ominus} = RT/p^{\ominus}$. For a component J It follows that

$$G_{J,m}^{\ominus} = G_{J,m}^{\ominus}(0) - RT \ln\left(\frac{q_{J,m}^{\ominus}}{N_A}\right) \tag{32}°$$

where $q_{J,m}^{\ominus}$ is the standard molar partition function of species J. By combining these expressions (as shown in the *Justification* below), it turns out that the equilibrium constant for a general reaction is given by the expression

$$K = \prod_J \left(\frac{q_J^{\ominus}}{N_A}\right)^{\nu_J} \times e^{-\Delta E_0/RT} \tag{33}°$$

where $\prod$ means form the product of the term that follows it[1] and ΔE_0 is the difference in energy between the ground states of the products and reactants (this term is defined more precisely in the *Justification*).

JUSTIFICATION

We shall consider the general reaction

$$0 = \sum_J \nu_J J$$

The standard molar reaction Gibbs energy for this reaction is

$$\Delta_r G^{\ominus} = \sum_J \nu_J G_{J,m}^{\ominus} = \sum_J \nu_J G_{J,m}^{\ominus}(0) - RT \sum_J \nu_J \ln\left(\frac{q_{J,m}^{\ominus}}{N_A}\right)$$

For a perfect gas $G(0) = U(0)$,[2] and the first term on the right is

$$\sum_J \nu_J G_{J,m}^{\ominus}(0) = \Delta_r G^{\ominus}(0) = \Delta_r U^{\ominus}(0)$$

The quantity $\Delta_r U^{\ominus}(0)$ is the difference in (molar) energy between the zero-point levels of the reactants and products (weighted by the stoichiometric numbers), and is normally written ΔE_0:

$$\Delta E_0 = \sum_J \nu_J U_J^{\ominus}(0)$$

An example of the application of this formula follows immediately after this *Justification*.

For the last part of the calculation we write

$$\Delta_r G^{\ominus} = \Delta E_0 - RT \sum_J \ln\left(\frac{q_{J,m}^{\ominus}}{N_A}\right)^{\nu_J} = -RT\left\{-\frac{\Delta E_0}{RT} + \ln \prod_J \left(\frac{q_{J,m}^{\ominus}}{N_A}\right)^{\nu_J}\right\}$$

For the second equality, we have used the fact that the sum of logarithms is the logarithm of products.[3] At this stage we can pick out an expression for K by comparing this equation with $\Delta_r G^{\ominus} = -RT \ln K$, which gives

$$\ln K = -\frac{\Delta E_0}{RT} + \ln \prod_J \left(\frac{q_{J,m}^{\ominus}}{N_A}\right)^{\nu_J}$$

1 That is,
$$\prod_J x_J = x_1 \times x_2 \times x_3 \times \ldots$$

2 This follows from $G = H - TS = U + pV - TS = U + nRT - TS = U$ at $T = 0$.
3 That is,
$$\ln x_1 + \ln x_2 + \ln x_3 + \ldots = \ln(x_1 x_2 x_3 \ldots)$$

This expression is easily rearranged into eqn 33 by taking antilogarithms of both sides.

Example 20.10 *Calculating the reaction energy difference*

Calculate ΔE_0 for the reaction

$$N_2(g) + 3H_2(g) \rightarrow 2NH_3(g)$$

Method. First, write the reaction in the form $0 = \text{sum of species}$, and identify the stoichiometric numbers. Then use those numbers in the expression

$$\Delta E_0 = \sum_J v_J U_J^{\ominus}(0)$$

The internal energies can then be expressed relative to the internal energies (at $T = 0$) of the free atoms, and hence in terms of the dissociation energies D_0 of the species.

Answer. The chemical equation for the reaction may be written

$$0 = 2NH_3(g) - N_2(g) - 3H_2(g)$$

It follows that

$$\Delta E_0 = 2U(NH_3) - U(N_2) - 3U(H_2)$$
$$= 2 \times 3 \times D_0(N\text{---}H) - D_0(N\text{≡}N) - 3 \times D_0(H\text{---}H)$$

Exercise E20.10. Express ΔE_0 for the reaction $2H_2(g) + O_2(g) \rightarrow 2H_2O(g)$ in terms of bond dissociation energies.

$$[\Delta E_0 = 2D_0(O\text{---}H) - 2D_0(H\text{---}H) - D_0(O\text{=}O)]$$

A dissociation equilibrium

We shall illustrate the application of eqn 33 to an equilibrium in which a diatomic molecule X_2 dissociates into its atoms: $X_2(g) \rightleftharpoons 2X(g)$.

The chemical equation is

$$0 = 2X(g) - X_2(g) \qquad v(X) = 2, \qquad v(X_2) = -1$$

Therefore, the equilibrium constant is

$$K = \prod_J a_J^{v_J} = \prod_J \left(\frac{p_J}{p^{\ominus}}\right)^{v_J} = \left(\frac{p_X}{p^{\ominus}}\right)^2 \left(\frac{p_{X_2}}{p^{\ominus}}\right)^{-1} = \frac{p_X^2}{p_{X_2}p^{\ominus}}$$

Equation 33 gives the relation of this equilibrium constant to the partition functions of the species involved in the reaction:

$$K = \left(\frac{q_{X,m}^{\ominus}}{N_A}\right)^2 \left(\frac{q_{X_2,m}^{\ominus}}{N_A}\right)^{-1} e^{-\Delta E_0/RT} = \left(\frac{q_{X,m}^{\ominus 2}}{q_{X_2,m}^{\ominus} N_A}\right) e^{-\Delta E_0/RT} \qquad (34)^{\circ}$$

In this expression, ΔE_0 is the dissociation energy D_0 of the molecule

$$\Delta E_0 = 2U(X) - U(X_2) = D_0(X_2)$$

The molar partition functions of the atoms X relate only to their translational motion and any electronic degeneracy:

$$q_{X,m}^{\ominus} = g_X \frac{V_m^{\ominus}}{\Lambda_X^3} \qquad \text{with } V_m^{\ominus} = \frac{RT}{p^{\ominus}}$$

The diatomic molecule X_2 also has rotational and vibrational degrees of freedom, so its standard molar partition function is

$$q_{X_2,m}^{\ominus} = g_{X_2} q_{X_2}^R q_{X_2}^V \frac{V_m^{\ominus}}{\Lambda_{X_2}^3}$$

It follows that the equilibrium constant is

$$K = \frac{g_X^2}{g_{X_2}} \times \frac{\Lambda_{X_2}^3 V_m^{\ominus}}{\Lambda_X^6 N_A} \times \frac{1}{q_{X_2}^R q_{X_2}^V} \times e^{-D_0/RT}$$

All the quantities in this expression can be calculated from spectroscopic data.

Example 20.11 *Evaluating an equilibrium constant*

Evaluate the equilibrium constant for the dissociation $Na_2(g) \rightleftharpoons 2Na(g)$ at 1000 K from the following data: $B = 0.1547 \text{ cm}^{-1}$, $\tilde{v} = 159.2 \text{ cm}^{-1}$, $D_0 = 70.4 \text{ kJ mol}^{-1}$ (0.73 eV). The Na atoms have doublet ground terms.

Method. The partition functions required are specified in eqn 34. They can be evaluated by using expressions in Table 20.2.

Answer. The partition functions and other quantities required are as follows:

$$\Lambda(Na_2) = 8.14 \text{ pm} \qquad \Lambda(Na) = 11.5 \text{ pm}$$

$$q^R(Na_2) = 2246 \qquad q^V(Na_2) = 4.885$$

$$g(Na) = 2 \qquad g(Na_2) = 1$$

Then, with $V_m^{\ominus} = 8.206 \times 10^{-2} \text{ m}^3 \text{ mol}^{-1}$, eqn 34 evaluates to

$$K = 4 \times \frac{(8.14 \times 10^{-12} \text{ m})^3 \times (8.206 \times 10^{-2} \text{ m}^3 \text{ mol}^{-1})}{(1.15 \times 10^{-11} \text{ m})^6 \times (6.022 \times 10^{23} \text{ mol}^{-1})}$$

$$\times \frac{1}{2246 \times 4.885} \times e^{-8.47}$$

$$= 2.42$$

Comment. For conversion to an equilibrium constant in terms of molar concentrations, use $[J] = p_J/RT$.

Exercise E20.11. Evaluate K at 1500 K.   [52]

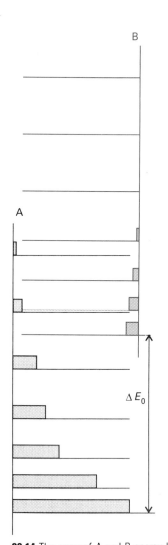

20.14 The array of A and B energy levels. At equilibrium all are accessible, and the equilibrium composition of the system reflects the overall Boltzmann distribution of populations. As ΔE_0 increases, A becomes dominant.

Contributions to the equilibrium constant

We are now in a position to appreciate the physical basis of equilibrium constants. To show what is involved, we consider a simple $A \rightleftharpoons B$ gas-phase equilibrium.

Figure 20.14 shows two sets of energy levels; one set of states belongs

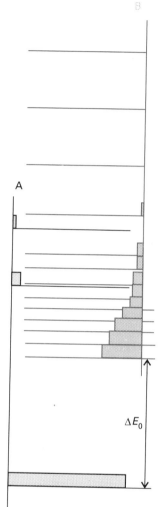

20.15 It is important to take into account the densities of states of the molecules. Even though B might lie well above A in energy (i.e. ΔE_0 is large and positive), B might have so many states that its total population dominates in the mixture. In classical thermodynamic terms, we have to take entropies into account as well as enthalpies when considering equilibria.

to A, and the other belongs to B. The populations of the states are given by the Boltzmann distribution, and are independent of whether any given state happens to belong to A or to B. We can therefore imagine a single Boltzmann distribution spreading, without distinction, over the two sets of states. If the spacings of A and B are similar (as in Fig. 20.14), and B lies above A, then the diagram indicates that A will dominate in the equilibrium mixture. However, if B has a high density of states (a large number of states in a given energy range, as in Fig. 20.15), then, even though its zero-point energy lies above that of A, the species B might still dominate at equilibrium.

It is quite easy to show (see the *Justification* that follows) that the ratio of numbers of A and B molecules at equilibrium is given by

$$\frac{N_B}{N_A} = \frac{q_B}{q_A} e^{-\Delta E_0/RT} \tag{35a}$$

and therefore that the equilibrium constant for the reaction is

$$K = \frac{q_B}{q_A} e^{-\Delta E_0/RT} \tag{35b}$$

just as would be obtained from eqn 33.[4]

JUSTIFICATION

The population in a state i of the composite (A + B) system is

$$n_i = N\frac{e^{-\beta \varepsilon_i}}{q}$$

where N is the total number of molecules. The total number of A molecules is the sum of these populations taken over the states belonging to A; these states we label a with energies ε_a. The total number of B molecules is the sum over the states belonging to B; these states we label b with energies ε_b' (we explain the prime on ε_b' in a moment):

$$N_A = \sum_a n_a = \frac{N}{q} \sum_a e^{-\beta \varepsilon_a} \qquad N_B = \sum_b n_b = \frac{N}{q} \sum_b e^{-\beta \varepsilon_b'}$$

The sum over the states of A is its partition function q_A, so

$$N_A = N \times \frac{q_A}{q}$$

The sum over the states of B is also a partition function, but the energies are measured from the ground state of the total system, which is the ground state of A. However, because $\varepsilon_b' = \varepsilon_b + \Delta \varepsilon_0$, where $\Delta \varepsilon_0$ is the separation of zero-point energies (Fig. 20.14),

$$N_B = \frac{N}{q} \sum_b e^{-\beta(\varepsilon_b + \Delta \varepsilon_0)} = \frac{N}{q} \left(\sum_b e^{-\beta \varepsilon_b} \right) e^{-\beta \Delta \varepsilon_0} = N \times \frac{q_B}{q} e^{-\Delta E_0/RT}$$

---

4 For an A $\rightleftharpoons$ B equilibrium, the V factors in the partition functions cancel, so the appearance of q in place of $q^{\ominus}$ has no effect. In the case of a more general reaction, the conversion from q to $q^{\ominus}$ comes about at the stage of converting the pressures that occur in K to numbers of molecules.

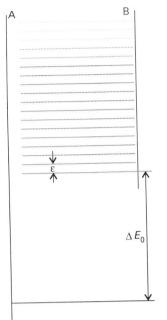

20.16 The model used in the text for exploring the effects of energy separations and densities of states on equilibria. B can dominate so long as ΔE_0 is not too large.

The switch from $\Delta\varepsilon_0/k$ to $\Delta E_0/R$ in the last step is the conversion of molecular energies to molar energies.

The equilibrium constant of the $A \rightleftharpoons B$ reaction is proportional to the ratio of the numbers of the two types of molecule. Therefore,

$$K = \frac{N_B}{N_A} = \frac{q_B}{q_A} e^{-\Delta E_0/RT}$$

The content of eqn 35 can be seen most clearly by exaggerating the molecular features that contribute to it. We shall suppose that A has only a single accessible level, which implies that $q_A = 1$, and that B has a large number of evenly, closely spaced levels (Fig. 20.16). The partition function of B is then

$$q_B = \frac{kT}{\varepsilon}$$

In this model system, the equilibrium constant is

$$K = \frac{kT}{\varepsilon} e^{-\Delta E_0/RT} \tag{36}$$

When ΔE_0 is very large, the exponential term dominates and $K \approx 0$, which implies that very little B is present at equilibrium. When ΔE_0 is small but still positive, K can exceed 1 because the factor kT/ε may be large enough to overcome the small size of the exponential term. The size of K then reflects the predominance of B at equilibrium on account of its high density of states. At low temperatures $K \approx 0$, and the system consists entirely of A. At high temperatures the exponential approaches 1 and the pre-exponential factor is large. Hence B becomes dominant. We see that in this endothermic reaction (endothermic because B lies above A), a rise in temperature favours B, because its states become accessible. This behaviour is what we saw, from the outside, in Chapter 9.

The model also shows why the Gibbs energy G, and not just the enthalpy, determines the position of equilibrium. It shows that the density of states (and hence the entropy) of each species as well as their relative energies control the distribution of populations and hence the value of the equilibrium constant.

CHECK LIST OF KEY IDEAS

1 The calculation of the **Helmholtz energy** and the **pressure** from the partition function (eqns 2 and 3).

2 The calculation of the **enthalpy** and the **Gibbs energy** from the partition function (eqns 4 and 5).

3 The definition of the **molar partition function** (eqn 7).

4 The calculation of the **translational partition function** of a gas-phase molecule (eqn 9).

5 The significance of the **rotational temperature** of a molecule and its use for judging when the high-temperature form of the rotational partition function is valid (eqn 11).

6 The calculation of the **rotational partition function** of a gas-phase molecule (eqns 12 and 13).

7 The significance and calculation of the **symmetry number** of a molecule (Section 20.2).

8 The calculation of the **vibrational partition function** of a molecule (eqn 14) and its high-temperature form (eqn 15).

9 The significance of the **vibrational temperature** of a molecule and its use for judging when the high-temperature form of the vibrational function is valid (following eqn 15).

10 The **electronic contribution** to the partition function and the **overall partition function** (eqn 16).

11 The calculation of a Gibbs energy from spectroscopic data (Example 20.6).

12 The calculation of the **mean energy** of each mode of motion of a molecule (Section 20.3) and the justification of the **equipartition theorem**.

13 The calculation of the contribution of different modes of motion to the heat capacity of a sample from the partition function (eqn 21c) and its estimation (eqn 25).

14 The derivation of the perfect-gas **equation of state** from a partition function (Example 20.9) and the relation between the **second virial coefficient** and the intermolecular potential energy (eqn 30).

15 The definition and interpretation of the **residual entropy** of substances that are disorderly at $T = 0$ (eqn 31).

16 The calculation of the **equilibrium constant** of a gas-phase reaction in terms of the partition functions of the reactants and products (eqn 33).

17 The interpretation of the relation between the equilibrium constant and the standard reaction Gibbs energy in terms of the distribution of energy levels of the reactants and the products (Section 20.7).

EXERCISES

20.1. Use the equipartition theorem to estimate the constant volume molar heat capacity of (a) I_2, (b) CH_4, (c) C_6H_6 in the gas phase at 25°C.

20.2. Estimate $\gamma = C_p/C_V$ for gaseous ammonia and methane. Do this calculation with and without the vibrational contribution to the energy. Which is closer to the expected experimental value at 25°C?

20.3. Estimate the rotational partition function of HCl at (a) 25°C and (b) 250°C.

20.4. Give the symmetry number for each of the following molecules: (a) CO, (b) O_2, (c) H_2S, (d) SiH_4, and (e) $CHCl_3$.

20.5. Calculate the rotational partition function of H_2O at 298 K from its rotational constants 27.878 cm^{-1}, 14.509 cm^{-1}, and 9.287 cm^{-1}. Above what temperature is the high-temperature approximation valid?

20.6. From the results of Exercise 20.5 calculate the rotational contribution to the molar entropy of gaseous water at 25°C.

20.7. Calculate the rotational partition function of CH_4 (a) by direct summation of the energy levels at 298 K and 500 K and (b) by the high-temperature approximation.

20.8. The bond length of O_2 is 120.75 pm. Use the high-temperature approximation to calculate the rotational partition function of the molecule at 300 K.

20.9. The NOF molecule is a nonlinear molecule with rotational constants 3.1752 cm^{-1}, 0.3951 cm^{-1}, and 0.3505 cm^{-1}. Calculate the rotational partition function of the molecule at (a) 25°C and (b) 100°C.

20.10. Plot the molar heat capacity of a collection of harmonic oscillators as a function of T/θ_V and predict the vibrational heat capacity of acetylene at (a) 298 K, (b) 500 K. The normal modes (and their degeneracies in parentheses) occur at wavenumbers 612(2), 729(2), 1974, 3287, and 3374 cm^{-1}.

20.11. The CO_2 molecule is linear, and its vibrational wavenumbers are 1388.2 cm^{-1}, 667.4 cm^{-1}, and 2349.2 cm^{-1}, the last being doubly degenerate and the others non-degenerate. The rotational constant of the molecule is 0.3902 cm^{-1}. Calculate the rotational and vibrational contributions to the molar Gibbs energy at 298 K.

20.12. The ground level of Cl is $^2P_{\frac{3}{2}}$ and a $^2P_{\frac{1}{2}}$ level lies 881 cm^{-1} above it. Calculate the electronic contribution to the molar heat capacity of Cl atoms at (a) 500 K and (b) 900 K.

20.13. The first electronically excited state of O_2 is $^1\Delta_g$ and lies $7918.1\ cm^{-1}$ above the ground state, which is $^3\Sigma_g^-$. Calculate the electronic contribution to the molar Gibbs energy of O_2 at 400 K.

20.14. The ground state of the Co^{2+} ion in $CoSO_4 \cdot 7H_2O$ may be regarded as $^4T_{\frac{9}{2}}$. The entropy of the solid at temperatures below 1 K is derived almost entirely from the electron spin. Estimate the molar entropy of the solid at these temperatures.

20.15 Calculate the residual molar entropy of a solid in which the molecules can adopt (a) three, (b) five, (c) six orientations of equal energy at $T = 0$.

20.16. Suppose that the hexagonal molecule $C_6H_nF_{6-n}$ has a residual entropy on account of the similarity of the H and F atoms. Calculate it for each value of n.

20.17. An average human DNA molecule has 5×10^8 dinucleotides (rungs on the DNA ladder) of four different kinds. If each rung were a random choice of one of these four possibilities, what would be the residual entropy associated with this typical DNA molecule?

20.18. Calculate the standard molar entropy of $N_2(g)$ at 298 K from its rotational constant $B = 1.9987\ cm^{-1}$ and its vibrational wavenumber $\tilde{v} = 2358\ cm^{-1}$. The thermochemical value is $192.1\ J\,K^{-1}\,mol^{-1}$. What does this suggest about the solid at $T = 0$?

20.19. Calculate the equilibrium constant of the reaction $I_2(g) \rightleftharpoons 2I(g)$ at 1000 K from the following data for I_2: $\tilde{v} = 214.36\ cm^{-1}$, $B = 0.0373\ cm^{-1}$, $D_e = 1.5422\ eV$. The ground state of the I atoms is $^2P_{\frac{3}{2}}$, implying four-fold degeneracy.

20.20 Calculate the value of K at 298 K for the gas-phase isotopic exchange reaction $2\,^{79}Br^{81}Br \rightleftharpoons\ ^{79}Br^{79}Br +\ ^{81}Br^{81}Br$. The Br_2 molecule has a non-degenerate ground state, with no other electronic states nearby. Base the calculation on the wavenumber of the vibration of $^{79}Br^{81}Br$, which is $323.33\ cm^{-1}$.

PROBLEMS
Numerical problems

20.1. The NO molecule has a doubly degenerate electronic ground state and a doubly degenerate excited state at $121.1\ cm^{-1}$. Calculate the electronic contribution to the molar heat capacity of the molecule at (a) 50 K, (b) 298 K, and (c) 500 K.

20.2. Explore whether a magnetic field can influence the heat capacity of a paramagnetic molecule by calculating the electronic contribution to the heat capacity of the NO_2 molecule in a magnetic field. Estimate the total constant volume heat capacity using equipartition, and calculate the percentage change in heat capacity brought about by a 5.0-T magnetic field at (a) 50 K and (b) 298 K.

20.3. The energy levels of a CH_3 group attached to a larger fragment are given by the expression for a particle on a ring so long as it is rotating freely. What is the high-temperature contribution to the heat capacity and the entropy of such a freely rotating group at 25°C? The moment of inertia of CH_3 about its C_3 axis is $5.341 \times 10^{-47}\ kg\,m^2$.

20.4. Calculate the temperature dependence of the heat capacity of para-H_2 (in which only rotational states with even values of J are populated) at low temperatures on the basis that its rotational levels $J = 0$ and $J = 2$ constitute a system that resembles a two-level system except for the degeneracy of the upper level. Use $B = 60.864\ cm^{-1}$ and sketch the heat capacity curve. The experimental heat capacity of para-H_2 does in fact show a peak at low temperatures.

20.5. The pure rotational microwave spectrum of HCl has absorption lines at the following wavenumbers (in cm^{-1}): 21.19, 42.37, 63.56, 84.75, 105.93, 127.12, 148.31, 169.49, 190.68, 211.87, 233.06, 254.24, 275.43, 296.62, 317.80, 338.99, 360.18, 381.36, 402.55, 423.74, 444.92, 466.11, 487.30, and 508.48. Calculate the rotational partition function at 25°C by direct summation.

20.6. Calculate and plot as a function of temperature, in the range 300–1000 K, the equilibrium constant for the reaction $CD_4(g) + HCl(g) \rightleftharpoons CHD_3(g) + DCl(g)$ using the following data (numbers in parentheses are degeneracies); $\tilde{v}(CHD_3)/cm^{-1} = 2993(1)$, 2142(1), 1003(3), 1291(2), 1036(2); $\tilde{v}(CD_4)/cm^{-1} = 2109(1)$, 1092(2), 2259(3), 996(3); $\tilde{v}(HCl)/cm^{-1} = 2991$; $\tilde{v}(DCl)/cm^{-1} = 2145$; $B(HCl)/cm^{-1} = 10.59$; $B(DCl)/cm^{-1} = 5.445$; $B(CHD_3)/cm^{-1} = 3.28$; $A(CHD_3)/cm^{-1} = 2.63\ cm^{-1}$; $B(CD_4)/cm^{-1} = 2.63$.

20.7. The exchange of deuterium between acid and water is an important type of equilibrium, and we can examine it using spectroscopic data on the molecules. Calculate the equilibrium constant at (a) 298 K and (b) 800 K for the gas-phase exchange reaction $H_2O + DCl \rightleftharpoons HCl + HDO$ from the following data: $\tilde{v}(H_2O)/cm^{-1} = 3656.7$, 1594.8, 3755.8; $\tilde{v}(HDO)/cm^{-1} = 2726.7$, 1402.2, 3707.5; $A(H_2O)/cm^{-1} = 27.88$, $B(H_2O)/cm^{-1} = 14.51$; $C(H_2O)/cm^{-1} = 9.29$; $A(HDO)/cm^{-1} = 23.38$; $B(HDO)/cm^{-1} = 9.102$; $C(HDO)/cm^{-1} = 6.417$; $B(HCl)/cm^{-1} = 10.59$

cm^{-1}; $B(\text{DCl})/\text{cm}^{-1} = 5.449$ cm^{-1}; $\tilde{\nu}(\text{HCl})/\text{cm}^{-1} = 2991$; $\tilde{\nu}(\text{DCl})/\text{cm}^{-1} = 2145$.

Theoretical problems

20.8. Derive the Sackur–Tetrode equation for a monatomic gas confined to a two-dimensional surface, and hence derive an expression for the standard molar entropy of condensation to form a mobile surface film.

20.9. Derive expressions for the internal energy, heat capacity, entropy, Helmholtz energy, and Gibbs energy of a harmonic oscillator. Express the results in terms of the vibrational temperature θ_V and plot graphs of each property against T/θ_V.

20.10. Although expressions like $\varepsilon = -\text{d} \ln q/\text{d}\beta$ are useful for formal manipulations in statistical thermodynamics, and for expressing thermodynamic functions in neat formulas, they are sometimes more trouble than they are worth in practical applications. When presented with a table of energy levels, it is often much more convenient to evaluate the following sums directly:

$$q = \sum_j e^{-\beta \varepsilon_j} \qquad \dot{q} = \sum_j (\beta \varepsilon_j) e^{-\beta \varepsilon_j} \qquad \ddot{q} = \sum_j (\beta \varepsilon_j)^2 e^{-\beta \varepsilon_j}$$

(a) Derive expressions for the internal energy, heat capacity, and entropy in terms of these three functions. (b) Apply the technique to the calculation of the electronic contribution to the constant volume molar heat capacity of magnesium vapour at 5000 K using the following data:

Term:	Degeneracy:	$\tilde{\nu}/\text{cm}^{-1}$
^1S	1	0
$^3\text{P}_0$	1	21 850
$^3\text{P}_1$	3	21 870
$^3\text{P}_2$	5	21 911
$^1\text{P}_1$	3	35 051
^3S	3	41 197

20.11 Determine whether a magnetic field can influence the value of an equilibrium constant. Consider the equilibrium $I_2(g) \rightleftharpoons 2I(g)$ at 1000 K, and calculate the ratio of equilibrium constants $K(B)/K$, where $K(B)$ is the equilibrium constant when a magnetic field B is present and removes the degeneracy of the four states of the $^2\text{P}_{\frac{3}{2}}$ level. Data on the species are given in Exercise 20.19. The electronic g value of the atoms is $\frac{4}{3}$. Calculate the field required to change the equilibrium constant by 1 per cent.

20.12. The heat capacity ratio of a gas determines the spread of sound in it through the formula

$$c_s = \left(\frac{\gamma R T}{M}\right)^{\frac{1}{2}}$$

where $\gamma = C_p/C_V$ and M is the molar mass of the gas. Deduce an expression for the speed of sound in a perfect gas of (a) diatomic, (b) linear triatomic, (c) non-linear triatomic molecules at high temperatures (with translation and rotation active). Estimate the speed of sound in air at 25°C.

21 Diffraction techniques

In this chapter we return to the techniques that are used to determine structure, but now the emphasis is on the geometrical arrangement of atoms and the distribution of electrons rather than the energy levels of molecules. All the techniques described in this chapter make use of the property of diffraction of waves by objects with dimensions similar to the wavelength of the waves. The most important technique of this kind is X-ray diffraction, and the bulk of the chapter deals with it.

First, we see how to describe the regular arrangement of atoms in crystals. We pay particular attention to the symmetry of the arrangement, for that helps to simplify the discussion, and to the separations of planes of atoms, which are important in the discussion of diffraction. Then we consider the basic principles of X-ray diffraction, and show that the dimensions of unit cells and their symmetries may be inferred from relatively straightforward X-ray diffraction experiments on powdered samples. After that, we turn to the most valuable technique, single-crystal X-ray diffraction, and show how the intensities of the diffraction pattern can be interpreted in terms of the distribution of electron density in a unit cell. This technique allows us to identify the locations of atoms and to measure bond lengths and bond angles.

We then explore some of the principles that govern the crystal structures that X-ray diffraction reveals. In many cases, a crystal structure may be understood in terms of the problem of packing together hard spheres of uniform size, of differing size, and perhaps of opposite electric charge.

In the concluding sections of the chapter we see how neutron diffraction and electron diffraction bear close similarities to X-ray diffraction, but provide complementary information.

A characteristic property of waves is that they interfere with each other, giving a greater displacement where peaks or troughs coincide and a smaller displacement where peaks coincide with troughs (Fig. 21.1).

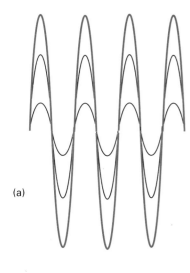

(a)

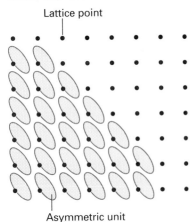

(b)

21.1 When two waves are in the same region of space they interfere. Depending on their relative phase, they may interfere (a) constructively, to give an enhanced amplitude, or (b) destructively, to give a smaller amplitude.

Lattice point

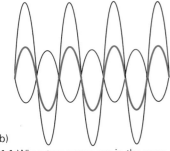

Asymmetric unit

21.2 Each lattice point specifies the location of an asymmetric unit. The crystal lattice is the array of lattice points; the crystal structure is the collection of asymmetric units arranged according to the lattice.

According to classical electromagnetic theory, the intensity of electromagnetic radiation is proportional to the square of the amplitude of the waves. Therefore, the regions of constructive and destructive interference show up as regions of enhanced and diminished intensities. The phenomenon of **diffraction** is the interference that is caused by an object in the path of waves, and the pattern of varying intensity that results is called the **diffraction pattern**. Diffraction occurs when the dimensions of the diffracting object are comparable to the wavelength of the radiation.

The diffraction of waves by atoms and molecules is utilized in several powerful techniques for the determination of the structures of molecules and solids. X-rays have wavelengths comparable to bond lengths in molecules and the spacing of atoms in crystals (about 100 pm), so the rays are diffracted by them. By analysing the diffraction pattern it is possible to draw up a detailed picture of the location of atoms even of such complex molecules as proteins. Electrons moving at about $20\,000\ km\ s^{-1}$ (after acceleration through about 4 kV) have wavelengths of 40 pm, and may also be diffracted by molecules, surfaces, and thin slices of solids. Neutrons generated in a nuclear reactor, and then slowed to thermal velocities, have similar wavelengths and may also be used for diffraction studies.

CRYSTAL STRUCTURE

Early in the history of modern science it was suggested that the regular external form of crystals implied an internal regularity of their constituents. In this section we shall see how to describe the arrangement of atoms inside crystals.

21.1 Lattices and unit cells

We need to distinguish the individual entities from which a crystal is built from the patterns these entities form. First, there is the **asymmetric unit**, the atom, ions, or molecule from which the crystal is built. Then there is the **space lattice**, the pattern formed by points representing the locations of the asymmetric units (Fig. 21.2). The space lattice is, in effect, an abstract scaffolding for the crystal structure.

More formally, the space lattice is a three-dimensional, infinite array of points, each of which is surrounded in an identical way by its neighbours, and which defines the basic structure of the crystal. In some cases there may be an asymmetric unit centred on each lattice point, but that is not necessary. The crystal structure itself is obtained by associating with each lattice point an identical asymmetric unit. These asymmetric units might be as small as an atom or as large as a virus. The crystal **structure** is the array of atoms obtained by associating an asymmetric unit with each lattice point.

The **unit cell** is an imaginary parallelepiped (parallel-sided figure) that contains one unit of the translationally repeating pattern (Fig. 21.3). A unit cell can be thought of as the fundamental unit from which the entire crystal may be constructed by purely translational displacements (like

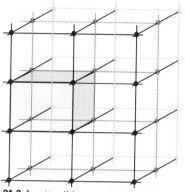

21.3 A unit cell is a parallel-sided (but not necessarily rectangular) figure from which the entire crystal structure can be constructed by using only translations (not reflections, rotations, or inversions).

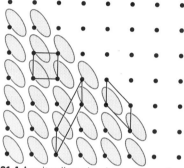

21.4 A unit cell can be chosen in a variety of ways, as shown here. It is conventional to choose the cell that represents the full symmetry of the lattice. In this rectangular lattice, the rectangular unit cell would normally be adopted.

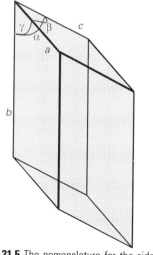

21.5 The nomenclature for the sides and angles of a unit cell.

bricks in a wall). A unit cell is formed by joining neighbouring lattice points by straight lines (Fig. 21.4), so that each unit cell has one lattice point at its origin. Such unit cells are called **primitive**. It is sometimes convenient (but not necessary), to draw **non-primitive unit cells** that have lattice points at their centres or on pairs of faces. An infinite number of different unit cells can describe the same lattice, but we normally choose the one with sides that have the shortest lengths and that are most nearly perpendicular to each other. The lengths of the sides of a unit cell are denoted a, b, and c, and the angles between them are denoted α, β, and γ (Fig. 21.5).

Unit cells are classified into seven **crystal systems** in terms of the rotational symmetry elements they possess. The unit cell of a cubic lattice, for example, has four threefold axes in a tetrahedral array (Fig. 21.6a). The unit cell of a monoclinic lattice has one twofold axis (Fig. 21.6b). The unit cell of a triclinic lattice has no rotational symmetry, and typically all three sides and all three angles are different (Fig. 21.6c). The **essential symmetries**, the elements that must be present for the unit cell to belong to a particular system, are listed in Table 21.1.

There are only 14 distinct crystal lattices in three dimensions. These **Bravais lattices** are illustrated in Fig. 21.7. It is conventional to portray these lattices by primitive unit cells in some cases and non-primitive unit cells in others. Primitive unit cells (those with lattice points only at the corners) are denoted P. A **body-centred unit cell** (I) also has a lattice point at its centre. A **face-centred unit cell** (F) has lattice points at its corners and also at the centres of its six faces. A **side-centred unit cell** (A, B, or C) has lattice points at its corners and at the centres of two opposite faces. As remarked above, there is no need for an asymmetric unit actually to lie at one of these lattice points, but in some cases, particularly for simple structures, it is both possible and convenient to regard them as the location of an atom, ion, or molecule.

21.2 The identification of lattice planes

The spacing of the lattice points in a crystal is an important quantitative aspect of its structure and its investigation by diffraction techniques. However, there are many different sets of planes (Fig. 21.8), and we need to be able to label them. Since two-dimensional lattices are easier to visualize than three-dimensional lattices, we shall introduce the concepts involved by referring to two dimensions initially, and then extend the conclusions by analogy to three dimensions.

The Miller indices

Consider a two-dimensional rectangular lattice formed from a unit cell of sides a, b (as in Fig. 21.8). Each plane in the illustration can be distinguished by the distances at which it intersects the a and b axes. One way of labelling each *set* of parallel planes would therefore be to quote the smallest intersection distances. For example, we could denote the four sets in the illustration as

$$(1a, 1b) \qquad (\tfrac{1}{2}a, \tfrac{1}{3}b) \qquad (-1a, 1b) \qquad (\infty a, 1b)$$

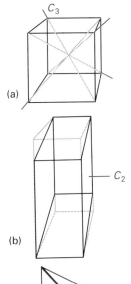

21.6 (a) A unit cell belonging to the cubic system has four threefold axes arranged tetrahedrally. (b) A unit cell belonging to the monoclinic system has a twofold axis. (c) A triclinic unit cell has no axes of rotational symmetry.

Table 21.1 The seven crystal systems

System	Essential symmetries
Triclinic	None
Monoclinic	One C_2 axis
Orthorhombic	Three perpendicular C_2 axes
Rhombohedral	One C_3 axis
Tetragonal	One C_4 axis
Hexagonal	One C_6 axis
Cubic	Four C_3 axes in a tetra-hedral arrangement

21.7 The fourteen Bravais lattices. The points are lattice points, and are not necessarily occupied by atoms. P denotes a primitive unit cell (R is used for a trigonal lattice), I a body-centred unit cell, F a face-centred unit cell, and C (or A or B) a cell with lattice points on two opposite faces.

If we agree to quote distances along the axes as multiples of the lengths of the unit cell, then we can label the planes more simply as

$$(1, 1) \qquad (\tfrac{1}{2}, \tfrac{1}{3}) \qquad (-1, 1) \qquad (\infty, 1)$$

Suppose now that the lattice in Fig. 21.8 is the top view of a three-dimensional orthorhombic lattice in which the unit cell has a length c in the z direction. All four sets of planes intersect the z-axis at infinity, so the full labels are

$$(1, 1, \infty) \qquad (\tfrac{1}{2}, \tfrac{1}{3}, \infty) \qquad (-1, 1, \infty) \qquad (\infty, 1, \infty)$$

The presence of fractions and ∞ in the labels is inconvenient. They can be eliminated by taking the reciprocals of the labels. As we shall see, taking reciprocals turns out to have further advantages. The **Miller indices** are the reciprocals of intersection distances (with fractions cleared, if taking the reciprocal results in a fraction). For example, the $(1, 1, \infty)$ planes in Fig. 21.8a are the (110) planes in the Miller notation.

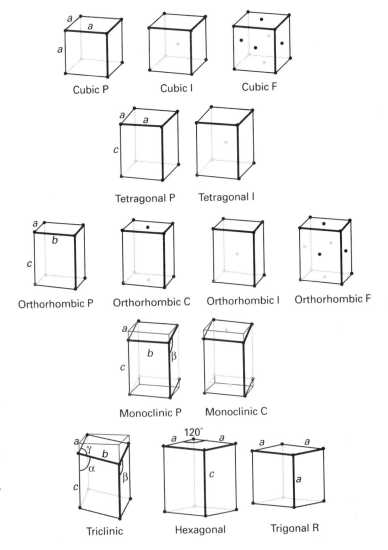

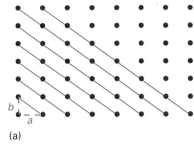

(a)

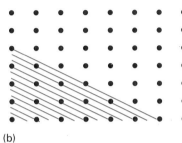

(b)

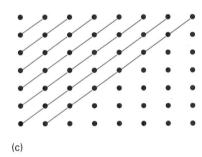

(c)

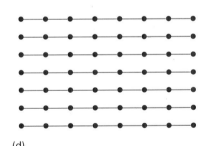

(d)

21.8 Some of the planes that can be drawn through the points of the space lattice and their corresponding Miller indices (*hkl*): (a) (110), (b) (230), (c) ($\bar{1}$10), (d) (010).

Similarly, the $(\frac{1}{2}, \frac{1}{3}, \infty)$ planes in Fig. 21.8b become the (230) planes in the Miller system. Negative indices are written with a bar over the number, and Fig. 21.8c shows the ($\bar{1}$10) planes. The Miller indices for the four sets of planes in Fig. 21.8 are therefore

$$(110) \quad (230) \quad (\bar{1}10) \quad (010)$$

Figure 21.9 shows some planes in three dimensions, including an example of a lattice with non-orthogonal axes.

It is helpful to keep in mind the fact that the (*hkl*) planes divide *a* into *h* equal parts, *b* into *k* parts, and *c* into *l* parts. Moreover, the smaller the value of *h* in (*hkl*), the more nearly parallel the plane is to the *a*-axis. The same is true of *k* and the *b*-axis and *l* and the *c*-axis. When $h = 0$, the planes intersect the *a*-axis at infinity, so the (0*kl*) planes are parallel to the *a*-axis. Similarly, the (*h*0*l*) planes are parallel to *b* and the (*hk*0) planes are parallel to *c*.

The separation of planes

The Miller indices are very useful for expressing the separation of planes. The separation of the (*hk*0) planes in the square lattice shown in Fig. 21.10 is given by

$$\frac{1}{d_{hk0}^2} = \frac{h^2 + k^2}{a^2}, \qquad \text{or } d_{hk0} = \frac{a}{(h^2 + k^2)^{\frac{1}{2}}}$$

By extension to three dimensions, the separation of the (*hkl*) planes of a cubic lattice is given by

$$\frac{1}{d_{hkl}^2} = \frac{h^2 + k^2 + l^2}{a^2}, \qquad \text{or } d_{hkl} = \frac{a}{(h^2 + k^2 + l^2)^{\frac{1}{2}}} \tag{1}$$

The corresponding expression for a general orthorhombic lattice is the generalization of the last equation:

$$\frac{1}{d_{hkl}^2} = \frac{h^2}{a^2} + \frac{k^2}{b^2} + \frac{l^2}{c^2} \tag{2}$$

Example 21.1 *Using the Miller indices*

Calculate the separation of (a) the (123) planes and (b) the (246) planes of an orthorhombic cell with $a = 0.82$ nm, $b = 0.94$ nm, and $c = 0.75$ nm.

Method. For the first part, we simply substitute the information into eqn 2. For the second part, instead of repeating the calculation, we note that, if all three Miller indices are multiplied by a factor *n*, then their separation is reduced by that factor:

$$\frac{1}{d_{nh,nk,nl}^2} = \frac{(nh)^2}{a^2} + \frac{(nk)^2}{b^2} + \frac{(nl)^2}{c^2} = n^2\left(\frac{h^2}{a^2} + \frac{k^2}{b^2} + \frac{l^2}{c^2}\right) = \frac{n^2}{d_{hkl}^2}$$

which implies that

$$d_{nh,nk,nl} = \frac{d_{hkl}}{n}$$

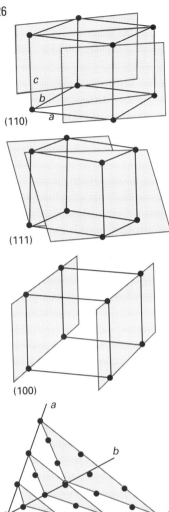

(110)

(111)

(100)

(111)

21.9 Some representative planes in three dimensions and their Miller indices. Note that a 0 indicates that a plane is parallel to the corresponding axis, and that the indexing may also be used for unit cells with non-orthogonal axes.

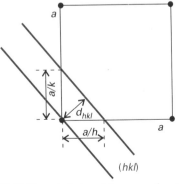

21.10 The calculation of the separation of the planes (*hkl*) in terms of the Miller indices for a square lattice.

Answer. Substituting the indices into eqn 2 gives

$$\frac{1}{d_{123}^2} = \frac{1^2}{(0.82\,\text{nm})^2} + \frac{2^2}{(0.94\,\text{nm})^2} + \frac{3^2}{(0.75\,\text{nm})^2} = 22\,\text{nm}^{-2}$$

Hence, $d_{123} = 0.21$ nm. It then follows immediately that d_{246} is half this value, or 0.11 nm (as illustrated for d_{220} and d_{110} in Fig. 21.11).

Exercise E21.1. Calculate the separation of (a) the (133) planes and (b) the (399) planes in the same lattice.   [0.19 nm, 0.063 nm]

X-RAY DIFFRACTION

X-Rays, which are electromagnetic radiation with wavelengths of about 100 pm, may be produced by bombarding a metal with high-energy electrons (Fig. 21.12). The electrons decelerate as they plunge into the metal and generate radiation with a continuous range of wavelengths called **Bremsstrahlung** (*Bremse* is German for brake, *Strahlung* for ray). Superimposed on the continuum are a few high-intensity, sharp peaks (Fig. 21.13). These peaks arise from collisions of the incoming electrons with the electrons in the inner shells of the atoms. A collision expels an electron from an inner shell, and an electron of higher energy drops into the vacancy, emitting the excess energy as an X-ray photon (Fig. 21.14).

21.3 The Bragg law

Wilhelm Röntgen discovered X-rays in 1895. Seventeen years later Max von Laue suggested that they might be diffracted when passed through a crystal, for by then he had realized that their wavelengths are comparable to the separation of lattice planes. Laue's suggestion was confirmed almost immediately by Walter Friedrich and Paul Knipping, and has grown since then into a technique of extraordinary power.

The earliest approach to the analysis of diffraction patterns produced by crystals was to regard a lattice plane as a mirror, and to model a crystal as stacks of reflecting lattice planes of separation *d* (Fig. 21.15). The model makes it easy to calculate the angle the crystal must make to the incoming beam of X-rays for constructive interference to occur. It has also given rise to the name **reflection** to denote an intense spot arising from constructive interference.

The path-length difference of the two rays shown in Fig. 21.15 is

$$\text{AB} + \text{BC} = 2d \sin \theta$$

where θ is the **glancing angle**. For many glancing angles the path-length difference is not an integral number of wavelengths, and the waves interfere destructively. However, when the path-length difference is an integral number of wavelengths (AB + BC = $n\lambda$), the reflected waves are in phase and interfere constructively. It follows that a bright reflection should be observed when the glancing angle satisfies the **Bragg law**

$$n\lambda = 2d \sin \theta$$

Reflections with $n = 2, 3, \ldots$ are called **second-order**, **third-order**, and

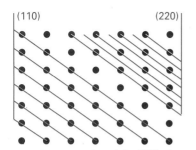

(110)   (220)

21.11 The separation of the (220) planes is half that of the (110) planes. In general, (the separation of the planes (nh, nk, nl) is a factor of n times smaller than the separation of the (hkl) planes.

so on; they correspond to pathlength differences of 2, 3, . . . wavelengths. In modern work it is normal to absorb the n into d, to write the Bragg law as

$$\lambda = 2d \sin \theta \tag{3}$$

and to regard the nth order reflection as arising from the (nh, nk, nl) planes (see Example 21.1).

The primary use of the Bragg law is in the determination of the spacing between the layers in the lattice, for once the angle θ corresponding to a reflection has been determined, d may readily be calculated.

Example 21.2 *Using the Bragg law*

A reflection from the (111) planes of a cubic crystal was observed at a glancing angle of 11.2° when $CuK\alpha$ X-rays of wavelength 154 pm were used. What is the length of the side of the unit cell?

Method. The separation of the planes can be determined from the Bragg law. Because the crystal is cubic, the separation is related to the length of the side of the unit cell a by eqn 1, which may therefore be solved for a.

Answer. According to eqn 3, (111) planes responsible for the diffraction have separation

$$d_{111} = \frac{\lambda}{2 \sin \theta} = \frac{154 \text{ pm}}{2 \times \sin 11.2°} = 396 \text{ pm}$$

The separation of the (111) planes of a cubic lattice of side a is given by eqn 1 as

$$d_{111} = \frac{a}{\sqrt{3}}$$

Therefore,

$$a = \sqrt{3} \times d_{111} = 687 \text{ pm}$$

Exercise E21.2. Calculate the angle at which the same lattice will give a reflection from the (123) planes. [24.8°]

21.4 The powder method

Laue's original method consisted of passing a broad-band beam of X-rays into a single crystal, and recording the diffraction pattern photographically. The idea behind the approach was that a crystal might not be suitably orientated to act as a diffraction grating for a single wavelength, but, whatever its orientation, the Bragg law would be satisfied for at least one of the wavelengths if a range of wavelengths was used. There is currently a resurgence of interest in this approach because synchrotron radiation (Section 16.1) spans a range of X-ray wavelengths.

The Debye–Scherrer method

An alternative technique to Laue's was developed by Peter Debye and Paul Scherrer and independently by Albert Hull, who used monochromatic radiation and a powdered sample. When the sample is a

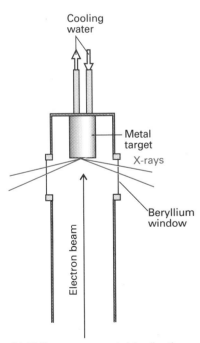

Cooling water

Metal target

X-rays

Beryllium window

Electron beam

21.12 X-rays are generated by directing an electron beam on to a cooled metal target. Beryllium is transparent to X-rays (on account of the small number of electrons in each atom) and is used for the windows.

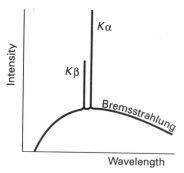

21.13 The X-ray emission from a metal consists of a broad, featureless Bremsstrahlung background, with sharp transitions superimposed on it. The label K indicates that the radiation comes from a transition in which an electron falls into a vacancy in the K shell of the atom (the shell with $n = 1$).

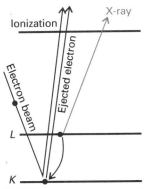

21.14 The processes that contribute to the generation of X-rays. An incoming electron collides with an electron (in the K shell), and ejects it. Another electron (from the L shell in this illustration) falls into the vacancy and emits its excess energy as an X-ray photon.

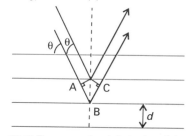

21.15 The conventional derivation of the Bragg law treats each lattice plane as reflecting the incident radiation. The path lengths differ by AB + BC, which depends on the glancing angle θ. Constructive interference (a 'reflection') occurs when AB + BC is equal to an integral number of wavelengths.

powder, at least some of the crystallites will be orientated so as to satisfy the Bragg condition for each set of planes (hkl). For example, some of the crystallites will be oriented so that their (111) planes, of spacing d_{111}, give rise to diffracted intensity at the glancing angle θ (Fig. 21.16). The crystallites with this glancing angle will lie at all possible angles around the incoming beam, so the diffracted beams lie on a cone around the incident beam of half-angle 2θ. Other crystallites will be oriented with different planes satisfying the Bragg law. They give rise to a cone of diffracted intensity with a different half-angle. In principle, each set of (hkl) planes gives rise to a diffraction cone, because some of the randomly orientated crystallites will have the correct angle to diffract the incident beam.

The original Debye–Scherrer method is illustrated in Fig. 21.17. The sample is in a capillary tube, which is rotated to ensure that the crystallites are randomly orientated. The diffraction cones are photographed as arcs of circles where they cut the strip of film, and two typical patterns are shown in Fig. 21.18. In modern diffractometers the sample is spread on a flat plate and the diffraction pattern, together with the intensities of the reflections, is monitored electronically (Figs 21.19 and 21.20).

Powder diffraction techniques are used to identify a sample of a solid substance by comparison of the positions of the diffraction lines and their intensities with a large data bank (*The powder diffraction file*, which is maintained by the Joint Committee on Powder Diffraction Standards, JCPDS, and contains information on over 30 000 substances). Powder diffraction data are also used to determine phase diagrams, for different solid phases result in different diffraction patterns, and to determine the relative amounts of each phase present in a mixture. The technique is also used for the initial determination of the dimensions and symmetries of unit cells, as the following section explains.

Indexing the reflections

The angle θ is measured from the location of the diffraction intensity maxima and from its value the separation of the planes responsible for the diffraction can be calculated by using the Bragg law. If the values of h, k, and l for the planes responsible for that reflection are known, then the dimensions of the unit cell can be deduced. The crux of the technique is therefore the **indexing** of the reflection, or ascribing the indexes hkl to it.

Some types of unit cell give characteristic and easily recognizable patterns of lines. For example, in a cubic lattice of unit cell dimension a the spacing is given by eqn 1, so the angles at which the (hkl) planes give reflections are given by

$$\sin \theta_{hkl} = (h^2 + k^2 + l^2)^{\frac{1}{2}} \times \frac{\lambda}{2a} \tag{4}$$

The reflections are then predicted by substituting the values of h, k, and l:

(hkl):	(100)	(110)	(111)	(200)	(210)	(211)	(220)	(300)	(221)	(310)
$h^2 + k^2 + l^2$	1	2	3	4	5	6	8	9	9	10

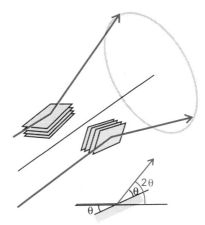

21.16 The same set of planes in two microcrystallites with different orientations around the direction of the incident beam give diffracted rays that lie on a cone. The full powder diffraction pattern is formed by cones corresponding to reflections from all the sets of (*hkl*) planes that satisfy the Bragg law. (A reflection at a glancing angle θ gives rise to a reflection at an angle 2θ to the direction of the incident beam; see inset.)

etc. Notice that 7 (and 15, . . .) is missing because the sum of the squares of three integers cannot equal 7 (or 15, . . .). Therefore the pattern has omissions that are characteristic of the cubic P lattice.

Example 21.3 *Identifying the unit cell*

A powder diffraction photograph of the element polonium gave lines at the following distances (in mm) from the centre spot when 71.0-pm Mo X-rays were used in a camera of radius 5.73 cm: 12.1, 17.1, 21.0, 24.3, 27.2, 29.9, 34.7, 36.9, 38.9, 40.9, 42.8. Identify the unit cell and determine its dimensions.

Method. From eqn 4 we write

$$\sin^2\theta = A \times (h^2 + k^2 + l^2), \qquad A = \left(\frac{\lambda}{2a}\right)^2$$

We need to convert the line distances to $\sin^2\theta$, determine the common factor A, and then find $h^2 + k^2 + l^2$. The angle θ in radians is related to the distances D of the reflection line from the centre of the pattern by $\theta = D/2R$ **(1)**. We can convert the angle in radians to an angle in degrees by multiplication by $360°/2\pi$. Because $R = 57.3$ mm, we have

$$\theta/° = \frac{\frac{1}{2}D}{57.3 \text{ mm}} \times \frac{360°}{2\pi} = \tfrac{1}{2}(D/\text{mm})$$

so the conversion is very simple.

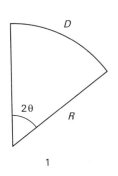

Answer. We draw up the following table:

D/mm	12.1	17.1	21.0	24.3	27.2	29.9	34.7	36.9	38.9	40.9	42.8
$\theta/°$	6.05	8.55	10.5	12.2	13.6	15.0	17.4	18.5	19.5	20.5	21.4
$100\sin^2\theta$	1.11	2.21	3.32	4.47	5.53	6.70	8.94	10.1	11.1	12.3	13.3

The common divisor is 1.11/100. Divide through to identify $h^2 + k^2 + l^2$:

$h^2 + k^2 + l^2$	1	2	3	4	5	6	8	9	10	11	12

The corresponding indexes are

(100) (110) (111) (200) (210) (211) (220) (300) (310) (311) (222)
(221)

We have now indexed the lines. Note the absence of $h^2 + k^2 + l^2 = 7$, which indicates a primitive cubic (cubic P) cell. From $(\lambda/2a)^2 = 0.0111$, we find $a = 337$ pm.

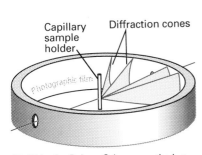

21.17 In the Debye–Scherrer method, a monochromatic X-ray beam is diffracted by a powder sample. The crystallites give rise to cones of intensity, which are detected by a photographic film wrapped round the circumference of the camera.

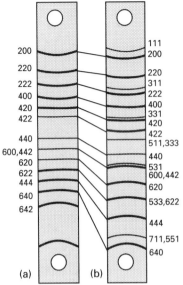

21.18 X-ray powder photographs of (a) KCl, (b) NaCl and the indexed reflections. The smaller number of lines in (a) is a consequence of the similarity of the K⁺ and Cl⁻ scattering factors, as discussed later in the chapter. Note that the film covers one-half of the circumference of the powder camera, the lines at the top corresponding to the largest diffraction angles.

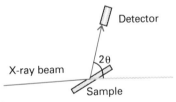

21.19 In a modern X-ray powder diffractometer the sample is spread on a flat plate and the diffraction intensity is detected electronically.

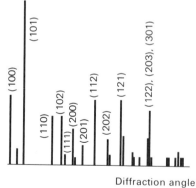

21.20 A typical X-ray powder diffraction pattern recorded electronically (schematic).

Comment. Later we shall see that additional information comes from the intensities of the lines. Note that a cunning choice of R saves a lot of work.

Exercise E21.3. In the same camera, another cubic crystal gave reflections at the following distances (in mm): 2.3, 3.2, 4.0, 4.6, 5.1, 5.6. Identify the unit cell (it will be helpful to refer to Fig. 21.26 on p. 734) and its dimensions.   [Cubic I; 550 pm]

Systematic absences

The X-ray powder diffraction patterns for NaCl and KCl shown in Fig. 21.18 are remarkably different for two such similar structures. Both crystals consist of two mutually interpenetrating face-centred cubic lattices, one of Na⁺ ions or K⁺ ions and the other of Cl⁻ ions (Fig. 21.21). The solution to the problem is found in the scattering strengths of the ions and the interference between waves scattered by the cations and anions. Thus, some reflections from the Na⁺ ions are in phase with the Cl⁻ reflections, and the two reflections augment one another to give more intense maxima. For other orientations, the two sets of reflections may be out of phase, and tend to cancel, but, as the scattering strengths of the two ions are different, the cancellation is incomplete. For KCl, however, the scattering strengths of K⁺ and Cl⁻, which have the same numbers of electrons, are very similar, and cancellation is complete. The ions in KCl therefore all look very similar (Fig. 21.22) and, instead of appearing to be face-centred cubic, the powder diffraction pattern is that typical of a lattice with a primitive cubic unit cell.

The general form of the diffraction pattern of a crystal composed of atoms and ions with different scattering strengths can be predicted by considering a crystal composed of A and B atoms with scattering strengths measured by their **scattering factors**, f_A and f_B. If the scattering factor is large, then the atoms contribute strongly to the scattering of X-rays that result in the diffraction pattern. The scattering factor of an atom is related to the electron density distribution in the atom, $\rho(r)$, by

$$f = 4\pi \int_0^\infty \rho(r)\, \frac{\sin kr}{kr}\, r^2\, dr, \qquad k = \frac{4\pi}{\lambda}\sin\theta \qquad (5)$$

The value of f is greatest in the forward direction and smaller for directions away from the forward direction (Fig. 21.23). In the forward direction (for $\theta = 0$), f is equal to the total number of electrons in the atom. Any detailed analysis of the intensities of reflections must take this dependence on direction into account.

JUSTIFICATION

For forward scattering, $\theta = 0$ so $k = 0$. However, because k occurs in the numerator and denominator in the integrand in eqn 5, we must take the limit $kr \to 0$. We first note that $\sin x = x + \dots$, so

$$\frac{\sin x}{x} \to \frac{x}{x} = 1 \qquad \text{as } x \to 0$$

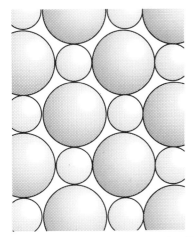

21.21 A fragment of the structure of NaCl and KCl: showing one plane of ions. Each cation (small spheres) has anions (large spheres) as its nearest neighbours. (For the three-dimensional structure, see Fig. 21.40.)

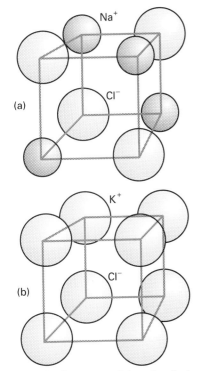

21.22 (a) One octant of the unit cell of NaCl; the Na^+ ions and Cl^- ions have different numbers of electrons and hence have different scattering factors. (b) One octant of the unit cell of KCl; the K^+ ions and Cl^- ions have the same numbers of electrons and hence have similar scattering factors. The diffraction pattern in this case is that of a primitive cubic lattice.

The factor $(\sin kr)/kr$ is therefore equal to 1 for forward scattering. It follows that in the forward direction

$$f = 4\pi \int_0^\infty \rho(r) r^2 \, dr$$

The integral over the electron density ρ (the number of electrons per unit volume) multiplied by the volume element $4\pi r^2 \, dr$ is the total number of electrons N_e in the atom. Hence, in the forward direction,

$$f = N_e$$

For example, the scattering factors of Na^+, K^+, and Cl^- are 10, 18, and 18, respectively.

The scattering factor is smaller in non-forward directions because $(\sin kr)/kr < 1$ for $\theta > 0$, so the integral is smaller than the value calculated above.

We shall now calculate the diffraction pattern to expect when two kinds of atom are present in the unit cell. Apart from the amplitude of the scattered waves, the most important feature is their relative phase, for that determines whether the waves add or subtract from one another (Fig. 21.24). We begin by showing that, if in the unit cell there is an A atom at the origin and a B atom at the coordinates (xa, yb, zc), then the phase difference between the (hkl) reflections of the A and B atoms is

$$\phi = 2\pi(hx + ky + lz) \tag{6}$$

JUSTIFICATION

Consider the crystal shown schematically in Fig. 21.25 with an A atom at the corner of the unit cell. The (100) reflection corresponds to two waves from adjacent A planes that have a phase difference of 2π. If there is a B atom at a fraction x of the distance between the two A planes, then it gives rise to a wave with a phase difference $2\pi x$ relative to an A reflection (Fig. 21.25a). To see this conclusion, note that, if $x = 0$, there is no phase difference; if $x = 0.5$, the phase difference is π; if $x = 1$, the B atom lies where the lower atom is and the phase difference is 2π. Now consider a (200) reflection (Fig. 21.25b). There is now a 4π difference between the waves from the two A layers, and if B were to lie at $x = 0.5$ it would give rise to a wave that differed in phase by 2π from the wave from the upper A layer. Thus, for a general fractional position x, the phase difference for a (200) reflection is $4\pi x$, or more specifically $2 \times 2\pi x$. For a general $(h00)$ reflection, the phase difference is therefore $h \times 2\pi x$. For three dimensions, this result generalizes to eqn 6.

The A and B reflections interfere destructively with each other when the phase difference is π (180°). If the atoms have the same scattering power, the total intensity is then zero. For example, if the unit cells are

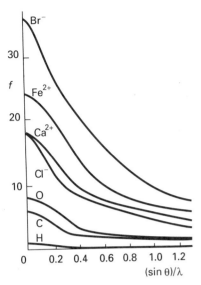

21.23 The variation of the scattering factor of atoms and ions with atomic number and angle. The scattering factor in the forward direction ($\theta = 0$) is equal to the number of electrons present in the species.

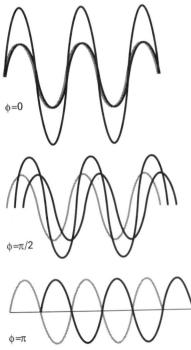

21.24 The interference between waves depends on their relative phase ϕ. For $\phi = 0$, there is no phase difference and the waves interfere constructively; destructive interference is complete when $\phi = \pi$ (out-of-phase).

cubic I with a B atom at $x = y = z = \frac{1}{2}$, then the A, B phase difference is $\pi(h + k + l)$. Therefore, all reflections for odd values of $h + k + l$ vanish because the waves are displaced in phase by 180°. Hence the diffraction pattern for a cubic I lattice can be constructed from that for the cubic P lattice (a cubic lattice without points at the centre of its unit cells) by striking out all reflections with odd values of $h + k + l$. Recognition of these **systematic absences** in a powder spectrum (Fig. 21.26) immediately indicates a cubic I lattice.

If the amplitude of the waves scattered from A is f_A at the detector, that of the waves scattered from B is $f_B e^{i\phi}$, with ϕ the phase difference given in eqn 6. The total amplitude at the detector is therefore

$$F = f_A + f_B e^{i\phi}$$

Because the intensity is proportional to the square modulus of the amplitude of the wave, the intensity I at the detector is

$$I_{hkl} \propto F^*F = (f_A + f_B e^{-i\phi})(f_A + f_B e^{i\phi})$$

This expression expands to

$$I_{hkl} \propto f_A^2 + f_B^2 + f_A f_B(e^{i\phi} + e^{-i\phi}) = f_A^2 + f_B^2 + 2f_A f_B \cos \phi$$

The cosine term either adds to or subtracts from $f_A^2 + f_B^2$ depending on the value of ϕ, which in turn depends on $h, k,$ and l (through eqn 6). Hence, there is a variation in the intensities of the lines with different hkl, which is exactly what is observed for NaCl.

21.5 Single-crystal X-ray diffraction

The method developed by the Braggs (William and his son Lawrence, who later jointly won the Nobel prize) is the foundation of almost all modern work in X-ray crystallography. They used a single crystal and a monochromatic beam of X-rays, and rotated the crystal until a reflection was detected. There are many different sets of planes in a crystal, so there are many angles at which a reflection occurs. The complete set of data consists of the list of angles at which reflections are observed and their intensities.

The technique

The preliminary investigation of a crystal, to identify the symmetry of its unit cell and its dimensions, makes use of a photographic technique. One method is to use an **oscillation camera**, in which the crystal (which typically might be of side 0.1 mm) is set on a mount called a 'goniometer head' and oscillated through about 10°. The diffraction pattern is recorded on a cylindrical film surrounding the sample. The problem of overlapping reflections is overcome by the **Weissenberg technique** in which a screen is placed in front of the film so that only one set of reflections is exposed, and gearing the oscillating goniometer head to the film so that the latter moves parallel to the axis of oscillation of the crystal. Although the Weissenberg technique greatly simplifies the

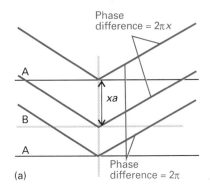

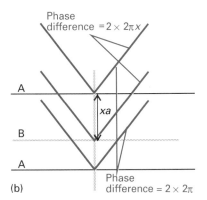

21.25 Diffraction from a crystal containing two kinds of atom. (a) For a first-order reflection from the A planes there is a phase difference of 2π between waves reflected by neighbouring planes, but (b) for a second-order reflection the phase difference is 4π. The reflection from a B plane at a fractional distance x from an A plane has a phase that is x times these phase differences.

indexing problem, the photographs are severely distorted. A refinement that overcomes the difficulty is to have a different coupling between the motions of the crystal and the screened film, and in the **precession camera technique** the pattern is undistorted and can be indexed quite readily.

Computing techniques are now available that lead not only to automatic indexing but also to the automated determination of the shape, symmetry, and size of the unit cell. A variety of techniques are also available for sampling large amounts of data, including area detectors and image plates. The traditional technique uses a **four-circle diffractometer** (Fig. 21.27). The crystal is set in an arbitrary orientation in the goniometer head. The computer linked to the diffractometer determines the unit cell dimensions and the settings of the diffractometer's four angles that are needed to observe any particular (hkl) reflection. The computer controls the settings, and moves the diffractometer to each one in turn. The diffraction intensity is measured at each setting (using some kind of crystal detector or photomultiplier) and background intensities are assessed by making measurements at slightly different settings.

Structure factors

The problem we now address is how to interpret the data from a diffractometer in terms of the structure of the crystal. To do so, we must go beyond the simple Bragg law.

If the unit cell contains several atoms with scattering factors f_i and coordinates (x_ia, y_ib, z_ic), then the overall amplitude of a wave diffracted by the (hkl) planes is a generalization of the expression $F = f_A + f_B e^{i\phi}$ obtained earlier:

$$F_{hkl} = \sum_i f_i e^{i\phi_i} \qquad \phi_i = 2\pi(hx_i + ky_i + lz_i) \qquad (7)$$

The sum is over all the atoms in the unit cell. The quantity F_{hkl} is called the **structure factor**. The intensity of the (hkl) reflection is proportional to $|F_{hkl}|^2$.

Example 21.4 *Calculating a structure factor*

Calculate the structure factors for the unit cell in Fig. 21.28.

Method. The structure factor is given by eqn 7. To use this equation, we consider the ions at the locations specified in Fig. 21.28. We write f^+ for the Na^+ scattering factor and f^- for the Cl^- scattering factor. Note that ions in the body of the cell contribute to the scattering with a strength f. However, ions on faces are shared between two cells (use $\frac{1}{2}f$), those on edges by four cells (use $\frac{1}{4}f$), and those at corners by eight cells (use $\frac{1}{8}f$). Two useful relations are

$$e^{i\pi} = -1 \qquad \cos\phi = \tfrac{1}{2}(e^{i\phi} + e^{-i\phi})$$

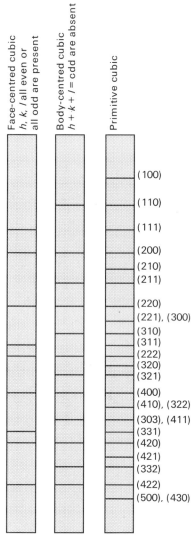

21.26 The powder photographs and the systematic absences of the cubic I and cubic F unit cells. Comparison of the observed photograph with patterns like these enable the unit cell to be identified. The locations of the lines give the cell dimensions.

Answer. From eqn 7, and summing over the coordinates of all 27 atoms in the illustration:

$$F_{hkl} = f^+\left(\frac{1}{8} + \frac{1}{8}e^{2\pi il} + \ldots + \frac{1}{2}e^{2\pi i(\frac{1}{2}h + \frac{1}{2}k + l)}\right)$$
$$+ f^-\left(e^{2\pi i(\frac{1}{2}h + \frac{1}{2}k + \frac{1}{2}l)} + \frac{1}{4}e^{2\pi i(\frac{1}{2}h)} + \ldots + \frac{1}{4}e^{2\pi i(\frac{1}{2}h + l)}\right)$$

To simplify this 27-term expression, we use

$$e^{2\pi ih} = e^{2\pi ik} = e^{2\pi il} = 1$$

because h, k, and l are all integers:

$$F_{hkl} = f^+\{1 + \cos(h + k)\pi + \cos(h + l)\pi + \cos(k + l)\pi\}$$
$$+ f^-\{(-1)^{h+k+l} + \cos h\pi + \cos l\pi + \cos k\pi\}$$

Then, since $\cos h\pi = (-1)^h$,

$$F_{hkl} = f^+\{1 + (-1)^{h+k} + (-1)^{h+l} + (-1)^{k+l}\}$$
$$+ f^-\{(-1)^{h+k+l} + (-1)^h + (-1)^k + (-1)^l\}$$

Now note that:

if h, k, and l are all even, $F_{hkl} = f^+\{1 + 1 + 1 + 1\} + f^-\{1 + 1 + 1 + 1\}$
$$= 4(f^+ + f^-)$$

if h, k, l are all odd, $F_{hkl} = 4(f^+ - f^-)$

if one index is odd and two are even, or vice versa, $F_{hkl} = 0$

Comment. Note that the hkl all-odd reflections are less intense than the hkl all-even. For $f^+ = f^-$, which is the case for identical atoms in a simple cubic arrangement, the hkl all-odd have zero intensity, corresponding to the systematic absences of simple cubic unit cells.

Exercise E21.4. Deduce the rule for the systematic absences of a cubic I lattice.   [$h + k + l$ odd]

The electron density

The details of the electron distribution inside a unit cell are contained in the structure factor. As we can see from eqn 7, F depends on the atoms present (through f_i) and on their locations (through $hx_i + ky_i + lz_i$). Given that we know the value of F for all the reflections (from their intensities), we can construct the electron density $\rho(r)$ in the unit cell by using the relation

$$\rho(r) = \frac{1}{V}\sum_{hkl} F_{hkl}e^{-2\pi i(hx + ky + lz)} \tag{8}$$

where V is the volume of the unit cell. Equation 8 is called a **Fourier synthesis** of the electron density.

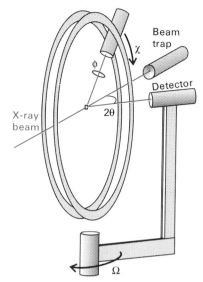

21.27 A four-circle diffractometer. The settings of the orientations (ϕ, χ, θ, and Ω) of the components is controlled by computer; each (hkl) reflection is monitored in turn, and their intensities are recorded.

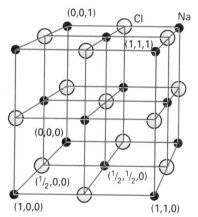

21.28 The location of the atoms for the structure factor calculation in Example 21.4. The tinted circles are Na^+, the open circles are Cl^-.

Example 21.5 *Calculating an electron density by Fourier synthesis*

Consider the ($h00$) planes of a crystal extending indefinitely in the x direction. In an X-ray analysis the structure factors were found as follows:

h:	0	1	2	3	4	5	6	7	8	9	10	11	12	13	14	15
F_h	16	−10	2	−1	7	−10	8	−3	2	−3	6	−5	3	−2	2	−3

(and $F_{-h} = F_h$). Construct a plot of the electron density along the x-axis of the unit cell.

Method. Because $F_{-h} = F_h$, it follows from eqn 8 that

$$V\rho(\boldsymbol{r}) = \sum_{h=-\infty}^{+\infty} F_h e^{-2\pi ihx} = F_0 + \sum_{h=1}^{+\infty} (F_h e^{-2\pi ihx} + F_{-h} e^{2\pi ihx})$$

$$= F_0 + \sum_{h=1}^{+\infty} F_h (e^{-2\pi ihx} + e^{2\pi ihx}) = F_0 + 2 \sum_{h=1}^{+\infty} F_h \cos(2\pi hx)$$

and we evaluate the sum (truncated at $h = 15$) for $x = 0$, 0.1, 0.2, ..., 1.0 (or more finely if a computer is available).

Answer. The results are plotted in Fig. 21.29 (the full line).

Comment. The positions of three atoms can be discerned very readily. The more terms there are included, the more accurate the density plot. Terms corresponding to high values of h (short-wavelength cosine terms in the sum) account for the finer details of the electron density; low values of h account for the broad features.

Exercise E21.5. Write a short computer program to evaluate Fourier sums, and then experiment with different structure factors (including changing signs as well as amplitudes).

The phase problem

From the measured intensities I_{hkl} we get the structure factors F_{hkl}, and then evaluate eqn 8 to find the electron density ρ. Unfortunately, I_{hkl} is proportional to the square modulus $|F_{hkl}|^2$, so we cannot say whether we should use $+|F_{hkl}|$ or $-|F_{hkl}|$ in the sum. In fact, the difficulty is more severe for non-centrosymmetric unit cells, because, if we write F_{hkl} as the complex number $|F_{hkl}| e^{i\alpha}$, where α is the phase of F_{hkl} and $|F_{hkl}|$ is its magnitude, then the intensity lets us determine $|F_{hkl}|$ but tells us nothing of its phase, which may lie anywhere from 0 to 360°. This ambiguity is called the **phase problem**. Some way must be found to assign phases to the structure factors, for otherwise the sum for ρ could not be evaluated and the method would be useless.

Example 21.6 *Illustrating the phase problem*

Repeat Example 21.5, but suppose that all the structure factors after $h = 6$ have the same positive phase.

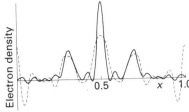

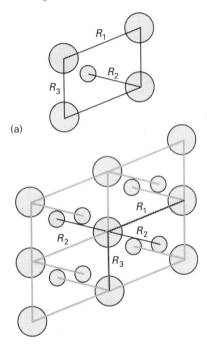

21.29 The plot of the electron density calculated in Examples 21.5 (full line) and 21.6 (broken line). Note how a different choice of phases for the structure factors leads to a markedly different structure (and, in the second case, for an unacceptable negative electron density in two regions).

21.30 The Patterson synthesis corresponding to the pattern in (a) is the pattern in (b). The distance and orientation of each spot from the origin gives the orientation and separation of one atom–atom separation in (a).

Method. Repeat the evaluation of the 16-term sum, but with positive values of F for $h > 6$.

Answer. The resulting electron density is plotted in Fig. 21.29 with a broken line.

Comment. This example illustrates the importance of the phase of the structure factor: different phase choices give quite different electron densities.

Exercise E21.6. Calculate the electron density using structure factors with signs that alternate as h increases.

The phase problem can be overcome to some extent by a variety of methods. One procedure that is widely used for inorganic materials with a reasonably small number of atoms in a unit cell is the **Patterson synthesis**. Instead of the structure factors F, the values of $|F|^2$, which can be obtained without ambiguity from the intensities, are used in an expression that resembles eqn 8:

$$P(\mathbf{r}) = \frac{1}{V} \sum_{hkl} |F_{hkl}|^2\, e^{-2\pi i(hx + ky + lz)} \tag{9}$$

The outcome of a Patterson synthesis is a map of the vector separations of the atoms (the distances and directions between atoms) in the unit cell. Thus, if atom A is at the coordinates (x_A, y_A, z_A) and atom B is at (x_B, y_B, z_B), then there will be a peak at $(x_A - x_B, y_A - y_B, z_A - z_B)$ in the Patterson map. There will also be a peak at the negative of these coordinates, because there is a vector from B to A as well as a vector from A to B. The height of the peak in the map is proportional to the product of the atomic numbers of the two atoms, $Z_A Z_B$. For example, if the unit cell has the structure shown in Fig. 21.30a, then the Patterson synthesis would be the map shown in Fig. 21.30b, where the location of each spot relative to the origin gives the orientations and separations of all the pairs of atoms in the original structure.

If some atoms are heavy, they dominate the scattering (because their scattering factors are large, of the order of their atomic number) and their locations may be deduced quite readily. The sign of F can now be calculated from the locations of the heavy atoms in the unit cell, and to a high probability the phase calculated for them will be the same as the phase for the entire unit cell. To see why this is so, we have to note that a structure factor of a centrosymmetric cell has the form

$$F = (\pm)f_{heavy} + (\pm)f_{light} + (\pm)f_{light} + \cdots$$

where f_{heavy} is the scattering factor of the heavy atom and f_{light} the scattering factors of the light atoms. (An expression of this form, but for atoms of similar atomic number, was derived in Example 21.4.) The f_{light} are all much smaller than f_{heavy}, and their phases are more or less random if the atoms are distributed throughout the unit cell. Therefore, the net effect of the f_{light} is to change F only slightly from f_{heavy}, and we can be reasonably confident that F will have the same sign as that calculated from the location of the heavy atom. This phase can then be combined with the observed $|F|$ (from the reflection intensity) to perform a Fourier synthesis of the full electron density in the unit cell, and hence to locate the light atoms as well as the heavy atoms.

In the technique known as **isomorphous replacement**, heavy atoms are introduced artificially into a complex molecule without affecting its structure significantly. Then the principal features of its structure are established by a Patterson synthesis before the locations of the lighter atoms are explored.

Modern structural analyses also make extensive use of the **direct method**. The direct method is based on statistical procedures and depends on the possibility of treating the atoms in a unit cell as being virtually randomly distributed (from the radiation's point of view), and then to use statistical techniques to compute the probabilities that the phases have a particular value. It is possible to deduce relations between some structure factors and sums (and sums of squares) of others, which have the effect of constraining the phases to particular values (with high probability, so long as the structure factors are large). For example, the **Sayre probability relation** has the form

$$\text{sign of } F_{h+h',\,k+k',\,l+l'} \text{ is probably equal to } (\text{sign of } F_{hkl})$$
$$\times (\text{sign of } F_{h'k'l'}) \quad \textbf{(10)}$$

For example, if F_{122} and F_{232} are both negative, then it is highly likely that F_{354} will be positive.

Because statistical relations can be built into computer programs, the determination of a crystal structure from a collection of observed intensities has become almost completely automated. Unfortunately, the reliability of the direct approach decreases as $1/N^{1/2}$, where N is the number of atoms per unit cell. Nevertheless, current techniques allow direct techniques to be used for crystals with $N < 100$.

The overall procedure

The X-ray technique depends on the availability of a single crystal of the sample. Only small crystals are needed (of side 0.1 to 0.5 mm, and even smaller if synchrotron sources are used), but they need to be of high quality. This need has led to a great deal of effort to obtain single crystals of large, biologically important molecules, such as proteins. Once a single crystal has been obtained, the indexing and determination of the unit cell characteristics are carried out in a diffractometer. Next, the intensities are measured and interpreted as structure factors with signs assigned by the direct method or by a preliminary Patterson synthesis. The electron density can then be calculated, and drawn out as a contour diagram, such as that in Fig. 21.31. Compilations of data on cell symmetries and dimensions and the locations of atoms in organic compounds are available in the *Cambridge crystallographic data file*, and similar information for inorganic compounds is published in the *Inorganic crystal structure data base*. We shall take up the subject again in Chapter 23 when we consider the structures of macromolecules.

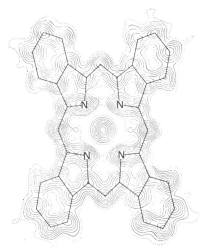

0.0 0.1 0.2 0.3 0.4 0.5
nm

21.31 The electron density contours of nickel phthalocyanine computed by Fourier synthesis from the observed structure factors (J. M. Robertson, *Organic crystals and molecules*, Cornell University Press (1953)). The conventional structure has been superimposed. Notice that in this early example of the technique the hydrogen atoms do not show.

Various correction techniques may be employed. For example, the atoms vibrate about their mean positions, which blurs their locations. The magnitude of the blurring can be calculated, and it can be eliminated from the electron density map. Mathematical **refinement** techniques are then used to improve the approximate atom coordinates obtained from imperfectly phased Fourier syntheses.

The remarkable and striking results obtained from X-ray diffraction suggest that it is an ideal technique for the determination of structures. This would be true were it not for a number of limitations. The first is its restriction to the solid state. We need to remember that in their natural environment biological molecules might unwind significantly, and that in the solid the packing of molecules might impose constraints on their stereochemistry. This problem is probably not as severe as it might seem, for samples of biochemical macromolecules often contain a high proportion of water, and the immediate environment of the macromolecule is quite similar to its environment in a living cell. When it is wished to study fluid samples, nuclear magnetic resonance can be used, and the development of special resonance techniques such as the nuclear Overhauser effect (Section 18.6) has enabled enzymes to be studied in their natural environment. Another limitation (which makes NMR complementary) is the poor response of X-ray diffraction to the presence of hydrogen atoms (because they have so few scattering electrons). However, the sensitivity of modern instruments is such that the detection of hydrogen atoms has now become largely routine.

INFORMATION FROM X-RAY ANALYSIS

The bonding within a solid may be of various kinds. Simplest of all (in principle) are metals, where electrons are delocalized over arrays of identical cations and bind the whole together into a rigid but malleable structure. In many cases their crystal structures are a result of the manner in which spherical metal cations can pack together into an orderly array.

In covalent solids, covalent bonds in a definite spatial orientation link the atoms in a network extending through the crystal. The demands of directional bonding, which have only a small effect on the structures of many metals, now override the geometrical problem of packing spheres together, and elaborate and extensive structures may be formed. A famous example of a covalent solid is diamond (Fig. 21.32), in which each sp^3-hybridized C atom is bonded tetrahedrally to its four neighbours. Covalent solids are often hard and unreactive.

Molecular solids, which are the subject of the overwhelming majority of modern structural determinations, are bonded together by van der Waals interactions (which we treat in detail in Chapter 22). The observed crystal structure is nature's solution to the problem of condensing objects of various shapes into an aggregate of minimum energy (actually, for temperatures above zero, of minimum Gibbs energy). The prediction of the structure is a very difficult task and rarely possible. The problem is made more complicated by the role of hydrogen bonds, which in some cases dominate the crystal structure, as in ice (Fig. 21.33), but

21.32 A fragment of the structure of diamond. Each C atom is tetrahedrally bonded to four neighbours. This framework-like structure results in a rigid crystal with a high thermal conductivity.

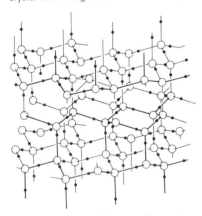

21.33 A fragment of the crystal structure of ice (ice-I). Each O atom is at the centre of a tetrahedron of four O atoms at a distance of 276 pm. The central O atom is attached by two short O–H bonds to two H atoms and by two long hydrogen bonds to the H atoms of two of the neighbouring molecules. Overall, the structure consists of planes of hexagonal puckered rings of H_2O molecules (like the chair form of cyclohexane).

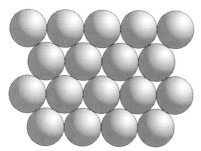

21.34 The first layer of close-packed spheres used to build a three-dimensional close-packed structure.

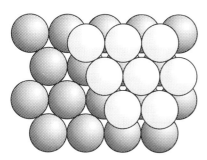

21.35 The second layer of close-packed spheres occupies the dips of the first layer. The two layers are the AB component of the close-packed structure.

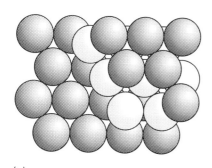

(a)

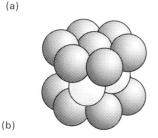

(b)

21.36 (a) The third layer of close-packed spheres might occupy the dips lying directly above the spheres in the first layer, resulting in an ABA structure that corresponds to hexagonal close-packing. (b) A fragment of the structure showing its hexagonal symmetry.

in others (e.g. phenol) distort a structure that is determined largely by the van der Waals interactions.

21.6 The packing of identical spheres: metal crystals

Most metallic elements crystallize in one of three simple forms, two of which can be explained in terms of organizing spheres into the closest possible packing.

Close packing

A **close-packed layer** of identical spheres, one with maximum utilization of space, can be formed as shown in Fig. 21.34. A close-packed three-dimensional structure is formed by stacking such close packed layers on top of each other. However, this stacking can be done in different ways and can result in close-packed **polytypes**, or structures that are identical in two dimensions (the close-packed layers) but differ in the third dimension.

A second close-packed layer can be formed by placing spheres in the depressions of the first layer (Fig. 21.35). The third layer may be added in either of two ways. In one, the spheres are placed so that they reproduce the first layer (Fig. 21.36a), to give an ABA pattern of layers. Alternatively, the spheres may be placed over the gaps in the first layer (Fig. 21.37a), so giving an ABC pattern.

Two polytypes are formed if the two stacking patterns are repeated in the vertical direction. If the ABA pattern is repeated, to give the sequence of layers ABABAB..., then the spheres are **hexagonally close-packed** (hcp). Alternatively, if the ABC pattern is repeated, to give the sequence ABCABC..., then the spheres are **cubic close-packed** (ccp). The origins of these names can be seen by referring to Figs 21.36b and 21.37b, respectively. The ccp structure gives rise to face-centred unit cells, and so may also be denoted cubic F (or fcc, for face-centred cubic). It is also possible to have ABCABAB... structures and even random sequences; however, the hcp and ccp polytypes are the most important, and we shall deal only with them. Some of the elements possessing these structures are listed in Table 21.2.

The compactness of the ccp and hcp structures is indicated by their **coordination number**, the number of atoms immediately surrounding any selected atom, which is 12 in both cases. Another measure of their compactness is the **packing fraction**, the fraction of space occupied by the spheres, which is 0.740. That is, in a close-packed solid of identical spheres, 26.0 per cent of the volume is empty space. The fact that many metals are closed-packed accounts for one of their common characteristics, their high densities.

Example 21.7 *Calculating the packing fraction*

Calculate the packing fraction of an hcp lattice.

Method. The strategy to adopt is first to calculate the volume of a unit cell, and then to calculate the total volume of the spheres that fully or partially occupy it. The first part of the calculation is a straightforward

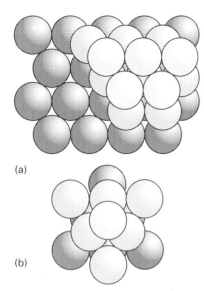

(a)

(b)

21.37 (a) Alternatively, the third layer might lie in the dips that are not above the spheres in the first layer, resulting in an ABC structure that corresponds to cubic close-packing. (b) A fragment of the structure showing its cubic symmetry.

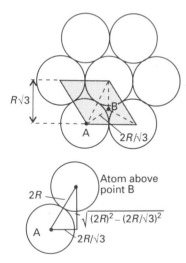

21.38 The calculation of the packing fraction of an hcp unit cell.

exercise in geometry, using the fact that the volume of a rectangular prism is the area of the base times the height. The second part involves counting the fraction of spheres that occupy the cell.

Answer. Refer to Fig. 21.38. The area of the base of the unit cell (the tinted parallelogram) is $\sqrt{3}R \times 2R$. The next layer of atoms lies above the point marked B, which is at a distance $R/\cos 30° = 2R/\sqrt{3}$ from the centre of the adjacent ion. The overlaying ion has a centre a distance $2R$ from the ions just mentioned, and so is at a height $\{(2R)^2 - (2R/\sqrt{3})^2\}^{\frac{1}{2}} = 2(\frac{2}{3})^{\frac{1}{2}}R$. The height of the unit cell is twice this value, or $4(\frac{2}{3})^{\frac{1}{2}}R$. The volume of the unit cell is (area) × (height) = $(2 \times \sqrt{3})R^2 \times 4(\frac{2}{3})^{\frac{1}{2}}R = 8 \times \sqrt{2}R^3$. There are the equivalent of two complete ions per unit cell, so the volume occupied per ion is $4 \times \sqrt{2}R$. The volume of a spherical ion is $\frac{4}{3}\pi R^3$. Therefore, the packing fraction, the fraction of 'full' space is $(\frac{4}{3}\pi R^3)/(4 \times \sqrt{2}R^3) = \pi/(3 \times \sqrt{2}) = 0.740$.

Comment. A ccp structure has the same packing fraction. The packing fraction of a cubic I (bcc) structure is 0.68 and that of a cubic P structure is 0.52, so both are less closely packed than the hcp and ccp structures.

Exercise E21.7. Calculate the packing fraction of a cubic I solid composed of identical spheres.   [0.68]

Less closely packed structures

As shown in Table 21.2, a number of common metals adopt structures that are less than close-packed. The departure from close packing suggests that specific covalent bonding between neighbouring atoms is beginning to influence the structure and impose a specific geometrical arrangement. One such arrangement results in a cubic I (bcc, for body-centred cubic) structure, with one sphere at the centre of a cube formed by eight others. The coordination number of a bcc structure is only 8, but there are six more atoms not much further away than the eight nearest neighbours. The packing fraction is only 0.68, showing that only about two-thirds of the available space is actually occupied.

Table 21.2 The crystal structures of some elements

Structure	Element
hcp*	Be, Cd, Co, He, Mg, Sc, Ti, Zn
fcc* (ccp, cubic F)	Ag, Al, Ar, Au, Ca, Cu, Kr, Ne, Ni, Pd, Pb, Pt, Rh, Rn, Sr, Xe
bcc (cubic I)	Ba, Cs, Cr, Fe, K, Li, Mn, Mo, Rb, Na, Ta, W, V
cubic P	Po

* Close-packed structures.

21.7 Ionic crystals

When crystals of compounds of monatomic ions are modelled by stacks of spheres it is essential to allow for the different ionic radii (generally

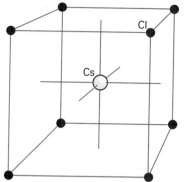

21.39 The caesium-chloride structure consists of two interpenetrating simple cubic lattices, one of cations and the other of anions, so that each cube of ions of one kind has a counter-ion at its centre.

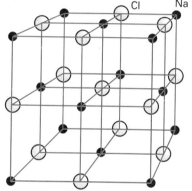

21.40 The rock-salt (NaCl) structure consists of two mutually interpenetrating slightly expanded face-centred cubic lattices. The entire assembly shown here is the unit cell.

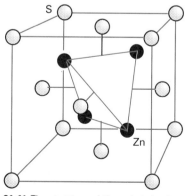

21.41 The structure of the sphalerite form of ZnS showing the location of the Zn atoms in the tetrahedral holes formed by the array of S atoms. (There is an S atom at the centre of the cube inside the tetrahedron of Zn atoms.)

with the cations smaller than the anions) and different charges. The coordination number of an ion is the number of nearest neighbours of opposite charge. Cations and anions may have different environments in the same crystal.

Even if, by chance, the ions have the same size, the problems of ensuring that the unit cells are electrically neutral makes it impossible to achieve 12-coordinate close-packed structures. As a result, ionic solids are generally less dense than metals. The best packing that can be achieved is the eight-coordination of the **caesium-chloride structure** (Fig. 21.39) in which each cation is surrounded by eight anions and each anion is surrounded by eight cations. In the caesium-chloride structure, an ion of one charge occupies the centre of a cubic unit cell with eight counterions at its corners. The structure is adopted by CsCl itself and by CsBr, CsI, CaS, CsCN (with some distortion), and CuZn.

When the radii of the ions differ more than in CsCl, even eight-coordinate packing cannot be achieved. One common structure adopted is the six-coordinate **rock-salt structure** (Fig. 21.40) typified by NaCl. In this structure, each cation is surrounded by six anions and each anion is surrounded by six cations. The rock-salt structure can be interpreted as the interpenetration of two slightly expanded cubic F (fcc) lattices, one of cations and the other of anions. It is the structure of NaCl itself and of many other MX compounds, including KBr, AgCl, MgO, and ScN.

The switch from the caesium-chloride to the rock-salt structure occurs (sometimes) in accord with the **radius-ratio rule**, which is based on the value of the radius ratio

$$\gamma = \frac{r_{smaller}}{r_{larger}} \tag{11}$$

The two radii are those of the larger and smaller ions in the crystal. The radius-ratio rule is derived by considering the geometrical problem of packing the maximum number of hard spheres of one radius around a hard sphere of a different radius. The rule states that the caesium-chloride structure should be expected when

$$\gamma > \sqrt{3} - 1 = 0.732$$

and that the rock-salt structure should be expected when

$$\sqrt{2} - 1 = 0.414 < \gamma < 0.732$$

For $\gamma < 0.414$, the most efficient packing leads to four-coordination of the type exhibited by the sphalerite (or zinc blende) form of ZnS (Fig. 21.41). The deviation of a structure from the prediction is often taken to be an indication of a shift from ionic towards covalent bonding; however, a major source of unreliability is the arbitrariness of ionic radii and their variation with coordination number.

Ionic radii are derived from the internuclear distance between adjacent ions in a crystal. However, we need to apportion the total distance between the two ions by defining the radius of one ion and then inferring the radius of the other ion. One scale that is widely used is based on the

Table 21.3* Ionic radii, R/pm

Na$^+$	102(6†), 116(8)
K$^+$	138(6), 151(8)
F$^-$	128(2), 131(4)
Cl$^-$	181 (Close packing)

* More values are given in the Data section at the end of this volume.
† Coordination number.

2 D(+)–Tartaric acid

3 L(−)–Tartaric acid

value 140 pm for the radius of the O^{2-} ion (Table 21.3); however, other scales are also available (such as one based on F$^-$ for discussing halides), and it is essential not to mix values from different scales. Because ionic radii are so arbitrary, predictions based on them (such as with the radius-ratio rule) must be viewed cautiously.

21.8 Absolute configurations

Although it has long been possible to separate enantiomers (mirror-image chiral isomers, Section 15.3), it was not until X-ray diffraction techniques were developed that the absolute stereochemical configuration of an isomer could be determined. It is now possible to state, for example, that D-tartaric acid (**2**) is the isomer responsible for rotating light clockwise (i.e. it is the (+) isomer), and that L-tartaric acid (**3**) is the (−) isomer. The X-ray method is not trivial, because enantiomers give almost identical diffraction patterns. The information about the absolute configuration is contained in the phase of the diffracted radiation, and its extraction is based on a technique developed by J. M. Bijvoet.

Consider first the diagrams in Fig. 21.42a and a′, which represent an idealized crystal and its mirror image. This model resembles the arrangement of Zn and S atoms in zinc blende, which was the first absolute configuration to be determined (by Koster, Kroll, and Prins in 1930). The technique we are about to describe was used to show that the shiny (111) faces of the crystal (Fig. 21.43) have S atoms on the surface whereas the dull (111) faces have Zn atoms on the surface. Each plane of atoms gives rise to a scattered wave, and the superpositions are shown in Fig. 21.42. Note that the two superpositions have the same amplitude, but differ in phase. The diffraction pattern therefore has the same intensity for each enantiomer and, at this stage, cannot be used to distinguish them.

The essence of the method is to use X-rays that are close to an absorption frequency of one species of atom in the sample. Koster, Kroll, and Prins, for example, used gold $L\alpha$ radiation (127.6 pm) which is close to the beginning of a zinc absorption band (which commences at 128.3 pm). In Bijvoet's development of their approach for a study of tartaric acid, a Rb atom is incorporated into the compound (he used sodium rubidium tartrate) with X-rays from a zirconium target. Atoms with absorptions close to the X-ray frequency introduce an extra phase shift in the scattered X-ray. A simple way of picturing the additional phase shift is to imagine the X-rays as exciting the atom, and being delayed in the process. The effect is called **anomalous scattering**.

If the layer marked A in the crystal contains the anomalous scatterers, then the scattered waves are as shown in Fig. 21.42b and b′. The essential point is that the superpositions now differ slightly in amplitude, not just phase, so the diffracted intensities are slightly different in each case. Therefore, the enantiomers can in fact be distinguished because the scattering intensities differ.

Modern diffractometers are so sensitive that the incorporation of a heavy atom is no longer strictly necessary. It is now possible to detect the

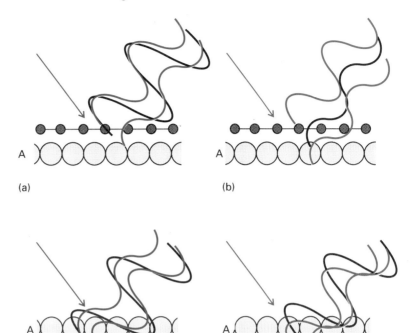

21.42 The two versions of the two layers of atoms represent enantiomers. The interference between their scattered waves results in composite waves that differ in phase (a and a'), but the absolute phase cannot be determined, and the intensities of the reflections are identical. If, however, A modifies the phase of the waves it scatters, then the resultant superpositions differ in amplitude as well as phase (b and b'), the reflections have different intensities, and the absolute configuration can be established.

(a)   (b)   (a')   (b')

small intensity variations arising from the light atoms normally present. However, the procedure is much easier and more reliable if some moderately heavy atoms (such as S or Cl) are present. Anomalous scattering depends strongly on the wavelength of the X-radiation. Thus, atoms lighter than S and Cl give little effect for Mo $K\alpha$ radiation, and Cu $K\alpha$ radiation must then be used.

NEUTRON AND ELECTRON DIFFRACTION

A neutron generated in a reactor and slowed to thermal velocities by repeated collisions with a moderator (such as graphite) until it is travelling at about 4 km s^{-1} has a wavelength of about 100 pm. Because 100 pm is comparable to X-ray wavelengths, similar diffraction phenomena can be expected. In practice, a range of wavelengths occurs in a neutron beam, but a monochromatic beam can be selected by diffraction from a germanium crystal.

Example 21.8 *Calculating the typical wavelength of thermal neutrons*

Calculate the typical wavelength of neutrons that have reached thermal equilibrium with their surroundings at 100 °C.

Method. We need to relate the wavelength to the temperature. There are two linking steps. First, the de Broglie relation expresses the wavelength in terms of the linear momentum. Then the linear momentum can be

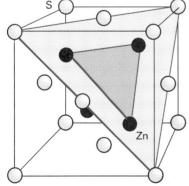

21.43 The (111) faces of the sphalerite crystal have either S atoms above Zn atoms or, as shown here, Zn atoms above S atoms.

expressed in terms of the kinetic energy, the mean value of which is given in terms of the temperature by the equipartition theorem.

Answer. The de Broglie relation states that $\lambda = h/p$. Then, from the equipartition theorem we know that the mean translational kinetic energy of a neutron at a temperature T travelling in the x-direction is $E_K = \frac{1}{2}kT$. The kinetic energy is also equal to $p^2/2m$, where p is the momentum of the neutron and m is its mass. Hence,

$$p = (mkT)^{\frac{1}{2}}$$

It follows that the neutron's wavelength is

$$\lambda = \frac{h}{(mkT)^{\frac{1}{2}}}$$

Therefore, at 100 °C,

$$\lambda = \frac{6.626 \times 10^{-34}\,\text{J s}}{\{(1.675 \times 10^{-27}\,\text{kg}) \times (1.381 \times 10^{-23}\,\text{J K}^{-1}) \times (373\,\text{K})\}^{\frac{1}{2}}}$$

$$= 226\,\text{pm}$$

Exercise E21.8. Calculate the temperature needed for the average wavelength of the neutrons to be 100 pm. $[1.6 \times 10^3\,°C]$

Electrons can be accelerated to precisely controlled energies by a known potential difference. When accelerated through 10 keV they acquire a wavelength of 12 pm, which makes them suitable for structural studies too.

21.9 Neutron diffraction

The scattering of X-rays is caused by the oscillations an incoming electromagnetic wave generates in the electrons of atoms. The scattering of neutrons is a nuclear phenomenon. Neutrons pass through the electronic structures of atoms and interact with the nuclei through the strong nuclear force that is responsible for binding nucleons together. As a result, the intensity with which neutrons are scattered is independent of the number of electrons. Whereas X-ray scattering factors increase strongly with atomic number, neutron scattering factors vary much less strongly; nor do they vary with angle. As a result, in contrast to X-rays, neutron diffraction is not dominated by the heavy atoms present in a molecule. Neutron diffraction therefore shows up the positions of hydrogen nuclei much more clearly than X-rays. Similarly, although neighbouring elements in the periodic table have almost identical X-ray scattering factors and hence are almost indistinguishable in X-ray diffraction, their neutron scattering strengths may be significantly different. As a result, it is possible to distinguish atoms of elements such as Ni and Co that are present in the same compound and to study order–disorder phase transitions in FeCo.

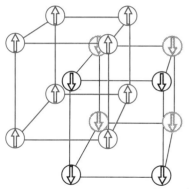

21.44 If the spins of atoms at lattice points are orderly, as in this antiferromagnetic material, where the spins of one set of atoms are aligned antiparallel to those of the other set, then neutron diffraction detects two interpenetrating simple cubic lattices on account of the magnetic interaction of the neutron with the atoms, but X-ray diffraction would see only a single bcc lattice.

The difference in sensitivity to hydrogen nuclei can have a pronounced effect on the measurement of C—H bond lengths. Because X-rays respond to accumulations of electrons, the weak peaks in an X-ray diffraction map represent the locations of the bulk of the electron density in the bonds, and this density may be shifted towards the C atom. For example, X-ray measurements on sucrose give $R(C—H) = 96$ pm; neutron measurements, which respond to the location of the nuclei, give $R(C—H) = 109.5$ pm. The O—H bond lengths in sucrose show similar differences, being 79 pm by X-rays but 97 pm by neutrons.

Another property of neutrons that distinguishes them from X-ray photons is their possession of a magnetic moment due to their spin. This magnetic moment can couple to the magnetic fields of ions in a crystal (if they have unpaired electrons) and modify the diffraction pattern. A simple example of **magnetic scattering** is provided by metallic chromium. The lattice is cubic I (bcc), and the diffraction pattern using X-rays has systematic absences. These absences are not observed when neutrons are used because the structure is such that atoms at the body-centre location have magnetic moments opposite to those at the corners, and the structure is better regarded as consisting of two interpenetrating lattices (Fig. 21.44) of magnetically different Cr atoms. Therefore, although the atoms are identical as far as X-rays are concerned, they are different from the viewpoint of neutrons, and diffraction intensity is observed at the predicted systematic X-ray absences. Neutron diffraction is especially important for investigating these magnetically ordered lattices.

21.10 Electron diffraction

Electrons are scattered strongly by their interaction with the charges of electrons and nuclei, and so cannot be used to study the interiors of solid samples. However, they can be used to study molecules in the gas phase, on surfaces, and in thin films.

A typical electron diffraction apparatus is illustrated in Fig. 21.45. Electrons are emitted from the hot filament on the left and accelerated through a potential gradient. They then pass through the stream of gas, and on to a fluorescent screen. (The modifications used when studying surfaces are described in Section 28.9.) The gaseous sample presents all possible orientations of atom–atom separations to the electron beam, and the resulting diffraction pattern is like an X-ray powder photograph. The pattern consists of a series of concentric undulations on a background with an intensity that decreases steadily with increasing scattering angle (Fig. 21.46a). The undulations are due to the sharply defined scattering from the nuclear positions, and the background is due to scattering from the less well-defined continuous electron density distribution in the molecule. One way of eliminating the unwanted background is to insert a rotating heart-shaped disk in front of the screen. The disk exposes the outer parts more than the inner, and helps to emphasize the undulations (Fig. 21.46b).

The scattering from a pair of nuclei separated by a distance R_{ij} and

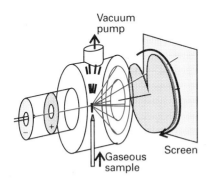

21.45 The layout of an electron diffraction apparatus. The diffraction pattern is photographed from the fluorescent screen. A rotating heart-shaped sector emphasizes the scattering from the nuclear positions and suppresses the smoothly varying background due to scattering from the continuous electron distribution in the molecules.

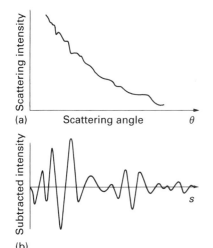

(a)

Scattering intensity

Scattering angle   θ

(b)

Subtracted intensity

s

21.46 (a) The scattering intensity consists of a smoothly varying background with undulations superimposed. (b) The undulations are emphasized if a sector is rotated in front of the screen, and then the densitometer trace taken from the photograph plotted against $s = (4\pi/\lambda)\sin\frac{1}{2}\theta$.

orientated at a definite angle to the incident beam can be calculated. The overall diffraction pattern is then calculated by allowing for all possible orientations of this pair of atoms. In other words, we integrate over orientations. When the molecule consists of a number of atoms, we sum over the contribution from all pairs, and the total intensity has an angular variation given by the **Wierl equation**:

$$I(\theta) = \sum_{i,j} f_i f_j \frac{\sin sR_{ij}}{sR_{ij}} \qquad s = \frac{4\pi}{\lambda}\sin\frac{1}{2}\theta \qquad (12)$$

where λ is the wavelength of the electrons in the beam and θ is the scattering angle. The electronic scattering factors f are a measure of the intensity of the electron scattering powers of the atoms.

The electron diffraction pattern gives the distances between all possible pairs of atoms in the molecule (not just those bonded together). When there are only a few atoms, the peaks can be analysed reasonably quickly, and the analysis proceeds by assuming a geometry and calculating the intensity pattern using the Wierl equation. The best fit is then taken as the actual molecular geometry.

CHECK LIST OF KEY IDEAS

1 The significance of an **asymmetric unit** and a **lattice** in the description of the structures of crystals (Section 21.1).

2 The **unit cell** of a crystal lattice and the classification of lattices into **crystal systems** on the basis of their **essential symmetries** (Section 21.1).

3 The **Bravais lattices** and their classification into different types (Fig. 21.7).

4 The description of lattice planes using **Miller indices** (Section 21.2) and the calculation of their separation (eqn 2).

5 The principles of **X-ray diffraction**, and the interpretation of a **reflection** using the **Bragg law** (eqn 3).

6 The **Debye–Scherrer method** using a powdered sample (Section 21.4).

7 **Indexing** the reflections in the Debye–Scherrer method and identifying the symmetry of the unit cell (Example 21.3).

8 The origin of **systematic absences** (Section 21.4) and the intensity of scattering in terms of the **scattering factors** of atoms (eqn 5).

9 The **Bragg technique** of X-ray diffraction using a single crystal, and a **four-circle diffractometer** (Section 21.5).

10 The interpretation of X-ray diffraction in terms of the **structure factor** (eqn 7 and Example 21.4) and the use of structure factors in the **Fourier synthesis** of the electron density (eqn 8 and Example 21.5).

11 The **phase problem**, the **Patterson synthesis**, and the **direct method** for the solution of the phase problem (Section 21.5).

12 **Close-packed structures** of identical spheres, their possible **polytypes**, and the **coordination number** and **packing fraction** of the lattice (Section 21.6).

13 The **caesium-chloride structure** (Fig. 21.39) and the **rock-salt structure** (Fig. 21.40) encountered for some ionic solids.

14 The **radius-ratio rule** for predicting structures and the conventional definition of **ionic radius** (Section 21.7).

15 The determination of **absolute configurations** using **anomalous scattering** (Section 21.8).

16 The principles of **neutron diffraction**, particularly for the detection of the location of hydrogen atoms and the role of **magnetic scattering** (Section 21.9).

17 The principles of **electron diffraction** (Section 21.10) and the interpretation of the scattering in terms of the **Wierl equation** (eqn 12).

EXERCISES

21.1. Equivalent lattice points within the unit cell of a Bravais lattice have identical surroundings. What points within a face-centred cubic unit cell are equivalent to the point $(\frac{1}{2}, 0, 0)$?

21.2. Find the Miller indices of the planes that intersect the crystallographic axes at the distances $(2a, 3b, 2c)$ and $(2a, 2b, \infty c)$.

21.3. Calculate the separations of the planes (111), (211), and (100) in a crystal in which the cubic unit cell has side 432 pm.

21.4. The glancing angle of a Bragg reflection from a set of crystal planes separated by 99.3 pm is 20.85°. Calculate the wavelength of the X-rays.

21.5. What are the values of 2θ of the first three diffraction lines of bcc iron when the X-ray wavelength is 58 pm? The unit cell dimension is 286.64 pm.

21.6. Copper $K\alpha$ radiation consists of two components of wavelengths 154.433 and 154.051 pm. Calculate the separation of the diffraction lines arising from the two components in a powder diffraction pattern recorded in a camera of radius 5.74 cm from planes of separation 77.8 pm.

21.7. The compound Rb_3TlF_6 has a tetragonal unit cell with dimensions $a = 651$ pm and $c = 934$ pm. Calculate the volume of the unit cell.

21.8. The orthorhombic unit cell of $NiSO_4$ has the dimensions $a = 634$ pm, $b = 784$ pm, and $c = 516$ pm, and the density of the solid is estimated as 3.9 g cm^{-3}. Determine the number of formula units per unit cell and calculate a more precise value of the density.

21.9. The unit cells of $SbCl_3$ are orthorhombic with dimensions $a = 812$ pm, $b = 947$ pm, and $c = 637$ pm. Calculate the spacing of the (411) planes.

21.10. A substance known to have a cubic unit cell gives reflections with Cu $K\alpha$ radiation (wavelength 154 pm) at the glancing angles 19.4°, 22.5°, 32.6°, and 39.4°. The reflection at 32.6° is known to be due to the (220) planes. Index the other reflections.

21.11. Potassium nitrate crystals have orthorhombic unit cells of dimensions $a = 542$ pm, $b = 917$ pm, and $c = 645$ pm. Calculate the glancing angles for the (100), (010), (111) reflections using Cu $K\alpha$ radiation (154 pm).

21.12. Copper(I) chloride forms cubic crystals with four formula units per unit cell. The only reflections present in a powder photograph are those with either all-even indices or all-odd indices. What is the symmetry of the unit cell?

21.13. A powder diffraction photograph from tungsten shows lines which index as (110), (200), (211), (220), (310), (222), (321), (400), ... Identify the symmetry of the unit cell.

21.14. The coordinates, in units of a, of the atoms in a simple cubic lattice are $(0, 0, 0)$, $(0, 1, 0)$, $(0, 0, 1)$, $(0, 1, 1)$, $(1, 0, 0)$, $(1, 1, 0)$, $(1, 0, 1)$, and $(1, 1, 1)$. Calculate the structure factors F_{hkl} when all the atoms are identical.

21.15. The coordinates of the four I atoms in the unit cell of KIO_4 are $(0, 0, 0)$, $(0, \frac{1}{2}, \frac{1}{2})$, $(\frac{1}{2}, \frac{1}{2}, \frac{1}{2})$, $(\frac{1}{2}, 0, \frac{3}{4})$. By calculating the phase of the I reflection in the structure factor, show that the I atoms contribute no net intensity to the (114) reflection.

21.16. Calculate the packing fraction for close-packed cylinders.

21.17. Verify that the radius ratios for six- and eightfold coordination are 0.414 and 0.732 as stated in the text.

21.18. From the data in Table 21.3 determine the radius of the smallest cation that can have (a) sixfold and (b) eightfold coordination with the O^{2-} ion.

21.19. Calculate the atomic packing factor for diamond.

21.20. The carbon–carbon bond length in diamond is 154.45 pm. If diamond were considered to be a close-packed structure of hard spheres with radii equal to half the bond length, what would be its expected density? The diamond lattice is face-centred cubic and its actual density is 3.516 g cm^{-3}. Can you explain the discrepancy?

21.21. Although the crystallization of large biological molecules may not be as readily accomplished as that of small molecules, their crystal lattices are no different. Tobacco seed globulin forms face-centred cubic crystals with a unit cell dimension of 12.3 nm and a density of 1.287 g cm^{-3}. Determine its molar mass.

21.22. Is there an expansion or a contraction as titanium transforms from hexagonal close-packed to body-centred cubic? The atomic radius of titanium is 145.8 pm in hcp but 142.5 pm in bcc.

21.23. In a Patterson synthesis, the spots correspond to the lengths and directions of the vectors joining the atoms in a unit cell. Sketch the pattern that would be obtained for (a) a planar, regular isolated BF_3 molecule, (b) the C atoms in an isolated benzene molecule.

21.24. The unit cell dimensions of NaCl, KCl, NaBr, and KBr, all of which crystallize in face-centred cubic lattices, are 562.8, 627.7, 596.2, and 658.6 pm, respectively. In each case, anions and cations are in contact along an edge of the unit cell. Does this information support the contention that ionic radii are constants independent of the counterion?

21.25. What velocity should neutrons have if they are to have wavelength 50 pm?

21.26. Calculate the wavelength of neutrons that have reached thermal equilibrium by collision with a moderator at 300 K.

21.27. What accelerating potential difference must be applied to electrons to generate a beam with wavelength 18 pm?

21.28. Calculate the wavelengths of electrons that have been accelerated through (a) 1.0 kV, (b) 10 kV, (c) 40 kV.

21.29. Predict from the Wierl equation the positions of the first maximum and first minimum in the neutron and electron diffraction patterns of the $^{35}Cl_2$ molecule obtained with neutrons of 80 pm wavelength and electrons of 4.0 pm wavelength. $R(Cl—Cl) = 198.75$ pm.

PROBLEMS
Numerical problems

21.1. In the early days of X-ray crystallography there was an urgent need to know the wavelengths of X-rays. One technique was to measure the diffraction angle from a mechanically ruled grating. Another method was to estimate the separation of lattice planes from the measured density of a crystal. The density of NaCl is 2.17 g cm^{-3} and the (100) reflection using Pb $K\alpha$ radiation occurred at 6.0°. Calculate the wavelength of the X-rays.

21.2. The element polonium crystallizes in a cubic system. Bragg first-order reflections, with X-rays of wavelength 154 pm, occur at $\sin \theta = 0.225$, 0.316, and 0.388 from the (100), (110), and (111) sets of planes. The separation between the sixth and seventh lines in the powder spectrum is larger than between the fifth and sixth lines. Is the unit cell simple, body-centred, or face-centred? Calculate the unit cell dimension.

21.3. The powder diffraction patterns of (a) tungsten, (b) copper obtained in a camera of radius 28.7 mm are shown in Fig. 21.47. Both were obtained with 154 pm X-rays and the scales are marked. Identify the unit cell in each case, and calculate the lattice spacing. Estimate the metallic radii of W and Cu.

21.4. Elemental silver reflects X-rays of wavelengths 154.18 pm at angles of 19.076°, 22.171°, and 32.256°. However, there are no other reflections at angles of less than 33°. Assuming a cubic unit cell, determine its type and dimension. Calculate the density of silver.

21.5. Genuine pearls consist of concentric layers of calcite ($CaCO_3$) crystals in which the trigonal axes are oriented along the radii. The nucleus of a cultured pearl is a piece of mother-of-pearl that has been worked into a sphere on a lathe. The oyster then deposits concentric layers of calcite on the central seed. Suggest an X-ray method for distinguishing between real and cultured pearls.

21.6. In their book *X-rays and crystal structures* (which begins 'It is now two years since Dr. Laue conceived the idea ...') the Braggs give a number of simple examples of X-ray analysis. For instance, they report that the reflection from (100) planes in KCl occurs at 5°23′, but for NaCl it occurs at 6°0′ for X-rays of the same wavelength. If the side of the NaCl unit cell is 564 pm, what is the side

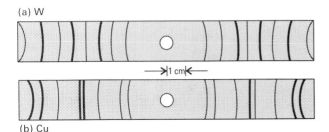

(a) W

|→ 1 cm |←

(b) Cu

21.47 The powder diffraction patterns of (a) tungsten, (b) copper.

of the KCl unit cell? The densities of KCl and NaCl are 1.99 and 2.17 g cm^{-3}, respectively. Do these values support the X-ray analysis?

21.7. The volume of a monoclinic unit cell is $abc \sin \beta$. Naphthalene has a monoclinic unit cell with two molecules per cell and sides in the ratio 1.377:1:1.436. The angle β is 122°49′ and the density of the solid is 1.152 g cm^{-3}. Calculate the dimensions of the cell.

21.8. Calculate the coefficient of thermal expansion of diamond given that the (111) reflection shifts from 22°2′25″ to 21°57′59″ on heating a crystal from 100 K to 300 K and that 154.0562 pm X-rays are used.

21.9. Use the Wierl equation to predict the appearance of the electron diffraction pattern of CCl_4 with an (as yet) undetermined C—Cl bond length but of known tetrahedral symmetry. Take $f_{Cl} = 17f$ and $f_C = 6f$ and note that $R(Cl—Cl) = (\frac{8}{3})^{\frac{1}{2}}R(C—Cl)$. Plot I/f^2 against $x = sR(C, Cl)$. In an actual experiment using 10.0 keV electrons the positions of the maxima occurred at 3°10′, 5°22′, and 7°54′ and minima occurred at 1°46′, 4°6′, 6°40′, and 9°10′. What is the C—Cl bond length in CCl_4?

Theoretical problems

21.10. Show that the separation of the (*hkl*) planes in an orthorhombic crystal with sides a, b, and c is given by eqn 2.

21.11. Show that the volume of a triclinic unit cell of sides a, b, and c and angles α, β, and γ is

$$V = abc(1 - \cos^2 \alpha - \cos^2 \beta - \cos^2 \gamma + 2 \cos \alpha \cos \beta \cos \gamma)^{\frac{1}{2}}$$

Use this expression to derive expressions for monoclinic and orthorhombic unit cells. For the derivation, it may be helpful to use the result from vector analysis that $V = \boldsymbol{a} \cdot \boldsymbol{b} \times \boldsymbol{c}$ and to calculate V^2 initially.

21.12. Calculate the packing fractions of (a) a primitive cubic lattice, (b) a bcc unit cell, (c) an fcc unit cell.

21.13. The coordinates, in units of a, of the A atoms, with scattering factor f_A, in a cubic lattice are (0, 0, 0), (0, 1, 0), (0, 0, 1), (0, 1, 1), (1, 0, 0), (1, 1, 0), (1, 0, 1), and (1, 1, 1). There is also a B atom, with scattering factor f_B, at $(\frac{1}{2}, \frac{1}{2}, \frac{1}{2})$. Calculate the structure factors F_{hkl} and predict the form of the powder diffraction pattern when (a) $f_A = f$, $f_B = 0$, (b) $f_B = \frac{1}{2}f_A$, and (c) $f_A = f_B = f$.

22

The electric and magnetic properties of molecules

In this chapter we examine some of the electric and magnetic properties of molecules and interpret them in terms of electronic structure. The properties we consider include the electric dipole moments and polarizabilities of molecules and some related properties that include refractive index, optical activity, and intermolecular forces. All these properties reflect the degree to which the nuclei of atoms exert control over the electrons in a molecule, either by causing electrons to accumulate in particular regions, or by permitting them to respond more or less strongly to the effects of external fields. We also discuss the analogous magnetic properties, particularly the magnetizability and the magnetic susceptibilities of molecules, and see the origins of the distinction between paramagnetic and diamagnetic substances.

The electric properties, and to a smaller extent the magnetic properties, of molecules are responsible for many of the properties of bulk matter. The small imbalances of charge distributions in molecules allow them to interact with one another and with externally applied fields. One result of this interaction is the weak cohesion of molecules to form the bulk phases of matter. Thus, we need to understand the material in this chapter if we are to understand gas imperfections and the formation of liquids. These interactions are also important for understanding the shapes adopted by biological and synthetic macromolecules, as we shall see in the next chapter. The failure of nuclear charges to control the surrounding electrons totally means that those electrons can respond to external fields. One particularly important example is their response to an electromagnetic field other than by absorption and emission: the resulting properites include refractive index and optical activity.

1

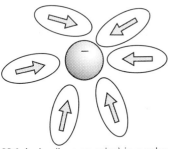

22.1 An ion (here an anion) in a polar solvent is solvated as a result of its interaction with the dipoles of the solvent molecules.

ELECTRIC PROPERTIES

An **electric dipole** consists of two electric charges q and $-q$ separated by a distance l, and is represented by a vector $\boldsymbol{\mu}$ that points from the negative charge to the positive charge. The magnitude of this vector (**1**) is ql and is called the **electric dipole moment**, μ.[1] Dipole moments are generally reported in the unit **debye**, D, where[2]

$$1\,\mathrm{D} = 3.336 \times 10^{-30}\,\mathrm{C\,m}$$

The debye is named after Peter Debye, a pioneer in the study of dipole moments of molecules. The dipole moment of a pair of charges e and $-e$ separated by 100 pm is $1.6 \times 10^{-29}\,\mathrm{C\,m}$, corresponding to 4.8 D. Dipole moments of small molecules are typically about 1 D.

22.1 Permanent and induced electric dipole moments

A **polar molecule** is a molecule with a permanent electric dipole moment. The permanent dipole moment arises from the partial charges on the atoms in the molecule that arise from differences in electronegativity or other features of bonding. Non-polar molecules may acquire an **induced dipole moment** in an electric field on account of the distortion the field causes in their electronic distributions and nuclear positions. Similarly, polar molecules may have their existing dipole moments modified by the applied field.

Permanent and induced dipole moments are important in chemistry through their contribution to intermolecular forces (as we describe later) and in their contribution to the ability of a substance to act as a solvent for ionic solids. The latter ability stems from the fact that one end of a dipole is Coulombically attracted to an ion of opposite charge and hence contributes an exothermic term to the enthalpy of solution (Fig. 22.1).

The average electric dipole moment per unit volume of a sample is called its **polarization** P. The polarization of an isotropic fluid sample is zero in the absence of an applied field because the molecules adopt random orientations and the average dipole moment is zero. In the presence of a field, the dipoles become partially aligned because some orientations have lower energies than others. As a result, the average dipole moment is non-zero. Moreover, as we shall see, there is an additional contribution from the dipole moment induced by the field.

Polar molecules

It was explained in Section 16.4 how the Stark effect is used to measure the electric dipole moments of molecules for which a rotational spectrum can be observed. When microwave spectroscopy cannot be used (because the sample is not volatile, because it decomposes, or because it consists

---

1 In elementary chemistry, an electric dipole is represented by the arrow $\longmapsto$ added to the Lewis structure for the molecule, with the + marking the positive end, as in $\overset{+\longmapsto}{\mathrm{HCl}}$. Note that the direction of $\longmapsto$ is opposite to that of the vector $\boldsymbol{\mu}$.
2 The strange conversion factor stems from the original definition of the debye in terms of c.g.s. units: 1 D is the dipole moment of two equal and opposite charges of magnitude 1 e.s.u. separated by 1 Å.

Table 22.1* Dipole moments (μ) and polarizability volumes (α')

	μ/D	$\alpha'/(10^{-30}\,m^3)$
CCl_4	0	10.5
H_2	0	0.819
H_2O	1.85	1.48
HCl	1.08	2.63
HI	0.42	5.45

* More values are given in the Data section at the end of this volume.

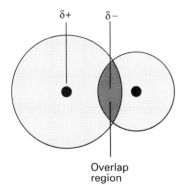

Overlap region

22.2 One of the contributions to the dipole moment of a molecule is the imbalance of charge arising from the overlap of orbitals of different radii. This diagram shows how the charge accumulation leads to a region of negative charge closer to the smaller atom.

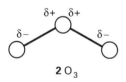

2 O_3

of molecules so complex that their rotational spectra cannot be interpreted), the dipole moment may be obtained by measurements on a liquid or solid bulk sample using a method explained later. In the following pages we refer to the sample as a **dielectric**, by which we mean a polarizable, non-conducting medium.

All heteronuclear diatomic molecules are polar to some extent, and typical values include 1.08 D for HCl and 0.42 D for HI (Table 22.1). A *very* approximate relation between the dipole moment and the difference in electronegativities of the two atoms, $\Delta\chi$, is

$$\mu/D = \Delta\chi \tag{1}$$

The more electronegative atom is normally the negative end of the dipole, but there are exceptions, particularly when antibonding orbitals are occupied.[3] Thus the dipole moment of CO is very small (0.12 D), but the negative end of the dipole is on the C atom even though oxygen is more electronegative than carbon. The interpretation of electric dipole moments is made even more complicated by the fact that a difference in atomic radii can result in an imbalance of electron density because the enhanced charge density associated with the overlap region lies closer to the nucleus of the smaller atom (Fig. 22.2). Such a **homopolar contribution** to the total dipole moment can arise even in the absence of a difference in electronegativity between the two atoms.

A polyatomic molecule is non-polar if it fulfils certain symmetry criteria. We saw in Section 15.3 that a molecule is non-polar if it belongs to a D point group or to one of the cubic or icosahedral point groups. We also saw that the dipole moment of a polar molecule with a symmetry axis cannot lie perpendicular to that axis (the dipole moment of NH_3, for instance, lies parallel to its C_3 axis). The symmetry criterion is more important than the question of whether or not the atoms in the molecule are the same. Thus the homonuclear triatomic molecule O_3 (which is angular, with C_{2v} symmetry) is allowed to be polar by symmetry considerations and, in fact, is polar because the electron density on the central O atom is different from that on the two outer O atoms. The dipole moment of the molecule lies parallel to the C_2 axis of the molecule (2). The heteronuclear triatomic molecule CO_2 (which is linear, with $D_{\infty h}$ symmetry) is strictly non-polar by symmetry even though the two O atoms have different electronegativities from the C atom. In CO_2, the dipole moments associated with each CO bond point in opposite directions and cancel.

To a first approximation it is possible to resolve the dipole moment of a polyatomic molecule into contributions of various components (Fig. 22.3). Thus, 1, 4-dichlorobenzene is non-polar on account of the cancellation of the two equal but opposing moments associated with the presence of Cl atoms on opposite sides of the ring (the molecule has D_{2h} symmetry, so it is necessarily non-polar). Its isomer 1,2-dichlorobenzene

3 We remarked in Section 14.7 that the major contribution to an antibonding orbital is made by the atomic orbitals of the less electronegative atom. Hence, if that orbital is occupied there may be so much electron density on the less electronegative atom that that atom has a partial negative charge.

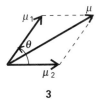

3

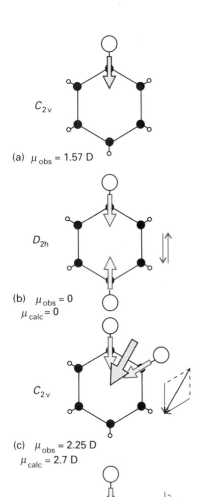

(a) μ_{obs} = 1.57 D

(b) μ_{obs} = 0
    μ_{calc} = 0

(c) μ_{obs} = 2.25 D
    μ_{calc} = 2.7 D

(d) μ_{obs} = 1.48 D
    μ_{calc} = 1.6 D

22.3 The resultant dipole moments (grey) of the dichlorobenzene isomers can be obtained approximately by vectorial addition of two chlorobenzene dipole moments (1.57 D).

(which has C_{2v} symmetry with the C_2 axis lying along the bisector of the angle between the two CCl bonds) has a dipole moment that is approximately the resultant of two monochlorobenzene dipole moments arranged at 60°. The technique of vector addition can be applied with fair success to other series of related molecules, and the resultant of two dipole moments that make an angle θ to each other (3) is obtained from

$$\mu^2 = \mu_1^2 + \mu_2^2 + 2\mu_1\mu_2 \cos \theta \tag{2a}$$

When the two dipole moments are equal, this equation simplifies to

$$\mu = 2\mu_1 \cos \tfrac{1}{2}\theta \tag{2b}$$

Induced dipole moments

In the presence of an applied electric field, the induced dipole moment μ^* is proportional to the field strength $\mathscr{E}$, and we write

$$\mu^* = \alpha \mathscr{E} \tag{3a}$$

The constant of proportionality α is the **polarizability** of the molecule. The greater the polarizability, the larger the induced dipole moment for a given applied field. When the applied field is very strong (as in laser beams), the induced moment is not strictly linear in the strength of the field, and we write

$$\mu^* = \alpha \mathscr{E} + \tfrac{1}{2}\beta \mathscr{E}^2 + \ldots \tag{3b}$$

The coefficient β is called the **hyperpolarizability** of the molecule.

The polarizability has the units (coulomb-metre)^2 per joule, $C^2 \, m^2 \, J^{-1}$, when the field strength is expressed in volts per metre ($V \, m^{-1}$) and the dipole moment is expressed in coulomb-metres ($C \, m$). That collection of units is awkward, so α is usually converted to a **polarizability volume** α' by using the relation

$$\alpha' = \frac{\alpha}{4\pi\varepsilon_0} \tag{4}$$

where ε_0 is the vacuum permittivity. Because the units of $4\pi\varepsilon_0$ are coulomb-squared per joule per metre ($C^2 \, J^{-1} \, m^{-1}$), it follows that α' has the dimensions of volume (hence its name).[4] Polarizability volumes are similar in magnitude to actual molecular volumes (of the order of $10^{-30} \, m^3$). The larger the polarizability volume, the larger the polarizability of the molecule.

Some experimental polarizability volumes of molecules are given in Table 22.1. The values reflect the strengths with which the nuclear charges control the electron distributions and prevent their distortion by the applied field. If the molecule has few electrons, they are tightly controlled by the nuclear charges and the polarizability of the molecule is low. If the molecule contains large atoms with electrons some distance from the nucleus, the nuclear control is less, the electron distribution is flabbier, and the polarizability is greater.

4 When using older compilations of data, it is useful to note that polarizability volumes in cm^3 have the same numerical values as the 'polarizabilities' reported using c.g.s. electrical units, so the tabulated values previously called 'polarizabilities' can be used directly.

JUSTIFICATION

The quantum mechanical expression for the mean polarizability is

$$\alpha = \tfrac{2}{3} \sum_n \frac{|\mu_{0n}|^2}{E_n - E_0}$$

where μ_{0n} is the magnitude of the integral

$$\boldsymbol{\mu}_{0n} = \int \psi_0^* \boldsymbol{\mu} \psi_n \, d\tau$$

with $\boldsymbol{\mu}$ the electric dipole moment operator. The sum is over the excited states, the wavefunctions of which are ψ_n with energies E_n. The general content of the expression for the polarizability can be exposed by approximating the excitation energies by a mean value ΔE and supposing that the transition dipole moments are approximately equal to the charge of an electron multiplied by the radius R of the molecule. Then

$$\alpha \approx \frac{2e^2 R^2}{3 \, \Delta E}$$

This expression shows that α increases with the size of the molecule and with the ease with which it can be excited (the smaller the value of ΔE).

In the same vein, it is possible to demonstrate why polarizability volumes are similar in magnitude to molecular volumes. If the excitation energy is approximated by the energy needed to remove an electron to infinity from a distance R for a single positive charge, we can write $\Delta E \approx e^2/4\pi\varepsilon_0 R$. When this expression is substituted into the equation above and both sides are divided by $4\pi\varepsilon_0$, we obtain $\alpha' \approx \tfrac{2}{3}R^3$, which is of the same order of magnitude as the molecular volume.

For all molecules other than those belonging to one of the cubic or icosahedral groups, the polarizability depends on the orientation of the molecule with respect to the field. We saw in Section 16.6 that the anisotropy of the polarizability of a molecule was essential to the observation of a rotational Raman spectrum. The values reported in Table 22.1 are mean values. The polarizability volume of benzene when the field is applied perpendicular to the ring is $12.3 \times 10^{-30} \, m^3$ and is $6.7 \times 10^{-30} \, m^3$ when the field is applied in the plane of the ring.

Polarization at high frequencies

When the applied field changes direction periodically, the permanent dipole moments reorientate and follow the field. However, when the frequency of the field is high, the molecules cannot change direction fast enough to follow it and they then make no contribution to the polarization of the sample. Because a molecule takes about 1 ps to turn through about 1 radian in a fluid, the loss of this contribution to the polarization occurs when measurements are made at frequencies greater than about 10^{11} Hz (in the microwave region). We say that the **orientation polarization**, the polarization arising from the permanent dipole moments, is lost at such high frequencies.

The next contribution to the polarization to be lost as the frequency is taken higher is the **distortion polarization**, the polarization that arises from the distortion of the positions of the nuclei by the applied field. The molecule is bent and stretched by the applied field and its molecular dipole changes accordingly. The time it takes for a molecule to bend is approximately the inverse of its vibrational frequency, so the distortion polarization disappears when the frequency of the radiation is increased through the infrared. The disappearance occurs in stages: as shown in the following *Justification*, each successive stage occurs as the incident frequency rises above the frequency of a particular mode of motion.

JUSTIFICATION

In this *Justification* we show that the contribution to the polarizability associated with a certain mode of motion is lost once the frequency has been raised beyond a typical absorption frequency for that type of motion. The expression for the polarizability of a molecule in the presence of a field that is oscillating at a frequency ω is

$$\alpha(\omega) = \frac{2}{3\hbar} \sum_n \frac{\omega_{n0} |\mu_{n0}|^2}{\omega_{n0}^2 - \omega^2}$$

The quantities in this expression (which is valid so long as ω is not close to ω_{n0}) are the same as those in the previous *Justification*, with $\hbar\omega_{n0} = E_n - E_0$. The two limits of this expression are worth noting. If $\omega = 0$, then the equation reduces to the expression for the static polarizability given earlier. On the other hand, if ω is very large (and much larger than any excitation frequency of the molecule), then the polarizability becomes

$$\alpha(\omega) = -\frac{2}{3\hbar\omega^2} \sum_n \omega_{n0} |\mu_{n0}|^2 \to 0 \quad \text{as} \quad \omega \to \infty$$

That is, when the incident frequency is higher than any excitation frequency, the polarizability becomes zero. The argument applies to each type of excitation, vibrational as well as electronic, and accounts for the successive decrease in polarizability as the frequency is increased.

At even higher frequencies, in the visible region, only the electrons are mobile enough to respond to the rapidly changing direction of the applied field. The polarization that remains is now due entirely to the distortion of the electron distribution, and the surviving contribution to the molecular polarizability is called the **electronic polarizability**.

The relative permittivity

When two charges q_1 and q_2 are separated by a distance r in a vacuum, the potential energy of their interaction is

$$V = \frac{q_1 q_2}{4\pi\varepsilon_0 r} \tag{5a}$$

When the same two charges are immersed in a medium (such as air or a liquid), their potential energy is reduced to

$$V = \frac{q_1 q_2}{4\pi\varepsilon r} \tag{5b}$$

where ε is the **permittivity** of the medium. The permittivity is normally expressed in terms of the dimensionless **relative permittivity** ε_r (which is also called the dielectric constant) of the medium:

$$\varepsilon_r = \frac{\varepsilon}{\varepsilon_0} \tag{6}$$

It follows from elementary electrostatics that the relative permittivity of a substance is measured by comparing the capacitance of a capacitor with and without the sample present (C and C_0, respectively) and using

$$\varepsilon_r = \frac{C}{C_0} \tag{7}$$

For example, if the capacitance of an empty cell is 5.01 pF (where F denotes the farad, $1\,\text{F} = 1\,\text{C V}^{-1}$, which is used to report capacitances, and $1\,\text{pF} = 10^{-12}\,\text{F}$), but is 57.1 pF when filled with camphor, then the relative permittivity of camphor is

$$\varepsilon_r = \frac{57.1\,\text{pF}}{5.01\,\text{pF}} = 11.4$$

The relative permittivity can have a very significant effect on the strength of the interactions between ions in solution. For instance, water has $\varepsilon_r = 78$ at 25°C, and the interionic Coulombic interaction energy is reduced by nearly two orders of magnitude from its vacuum value. We examined some of the consequences of this reduction in Chapter 10.

The relative permittivity of a substance is large if its molecules are polar and highly polarizable. The quantitative relation between the relative permittivity and the electric properties of the molecules is expressed by the **Debye equation**:

$$\frac{\varepsilon_r - 1}{\varepsilon_r + 2} = \frac{\mathscr{N}}{3\varepsilon_0}\left(\alpha + \frac{\mu^2}{3kT}\right) \tag{8a}$$

where $\mathscr{N}$ is the number density of molecules, the number per unit volume. As shown in the *Justification* that follows, the term $\mu^2/3kT$ stems from the thermal averaging of the electric dipole moment in the presence of the applied field. The same expression, but without the permanent dipole moment contribution, is called the **Clausius–Mossotti equation**:

$$\frac{\varepsilon_r - 1}{\varepsilon_r + 2} = \frac{\mathscr{N}\alpha}{3\varepsilon_0} \tag{8b}$$

The Clausius–Mossotti equation is used when there is no contribution from permanent electric dipole moments to the polarization, either because the molecules are non-polar or because the frequency of the applied field is so high that the molecules cannot orientate quickly enough to follow it.

JUSTIFICATION

When a field $\mathscr{E}$ is present, the energy of a molecule that makes an angle θ to the field is

$$E = -\mu \mathscr{E} \cos \theta$$

Some molecular orientations have lower energy than others, and the mean dipole moment $\langle \mu \rangle$ (and therefore the polarization) depends on a competition between the aligning influence of the field and the randomizing influence of thermal motion. The net dipole moment of a fluid sample, parallel to the applied field, is evaluated using the Boltzmann distribution (Section 19.1) for a sample at a temperature T, and is

$$\langle \mu \rangle = \mu \mathscr{L} \qquad \mathscr{L} = \frac{e^x + e^{-x}}{e^x - e^{-x}} - \frac{1}{x}, \qquad \text{with } x = \frac{\mu \mathscr{E}}{kT}$$

The function $\mathscr{L}$ is called the **Langevin function**.

Under most circumstances x is very small (e.g. if $\mu = 1\,\mathrm{D}$ and $T = 300\,\mathrm{K}$, then x exceeds 0.01 only if the field strength exceeds $100\,\mathrm{kV\,cm^{-1}}$, and most measurements are done at much lower strengths). When $x \ll 1$, the exponentials in the Langevin function can be expanded, and the largest term that survives is

$$\mathscr{L} = \tfrac{1}{3}x + \ldots$$

Therefore, the average molecular dipole moment is

$$\langle \mu \rangle = \frac{\mu^2 \mathscr{E}}{3kT}$$

If the number density of molecules (the number per unit volume) is $\mathscr{N}$, then the mean dipole moment per unit volume in the presence of the electric field is

$$P = \mathscr{N}\langle \mu \rangle = \frac{\mathscr{N}\mu^2 \mathscr{E}}{3kT}$$

as used in the text.

The Langevin function $\mathscr{L}$ is derived by considering a molecule of dipole moment μ that makes an angle θ to the z-axis (the direction of the field). This molecule gives a contribution $\mu \cos \theta$ to the average dipole moment of the sample. The probability $\mathscr{P}$ that a dipole has the orientation θ is given by the expression

$$\mathscr{P} = \frac{e^{x \cos \theta}}{\displaystyle\int_0^\pi e^{x \cos \theta} \sin \theta \, d\theta} \qquad x = \frac{\mu \mathscr{E}}{kT}$$

where the integral in the denominator ensures that $\mathcal{P}$ is normalized to 1. It follows that the average value of the component of the dipole moment parallel to the applied electric field is

$$\langle \mu \rangle = \int_0^\pi \mu \cos \theta \, \mathcal{P} \sin \theta \, d\theta = \frac{\mu \int_0^\pi e^{x \cos \theta} \cos \theta \sin \theta \, d\theta}{\int_0^\pi e^{x \cos \theta} \sin \theta \, d\theta}$$

The integral takes on a simpler appearance when we write $y = \cos \theta$ and note that $dy = -\sin \theta \, d\theta$:

$$\langle \mu \rangle = \frac{\mu \int_{-1}^1 y e^{xy} \, dy}{\int_{-1}^1 e^{xy} \, dy}$$

Then we use

$$\int_{-1}^1 e^{xy} \, dy = \frac{e^x - e^{-x}}{x}$$

$$\int_{-1}^1 y e^{xy} \, dy = \frac{e^x + e^{-x}}{x} - \frac{e^x - e^{-x}}{x^2}$$

It is now straightforward algebra to combine these two results and to obtain the Langevin function.

Equations 8a and 8b are normally written in terms of the mass density ρ of the sample by using $\mathcal{N} = \rho N_A / M$, where M is the molar mass of the molecules, and the **molar polarization** P_m, which is calculated from the expression

$$P_m = \frac{N_A}{3\varepsilon_0} \left(\alpha + \frac{\mu^2}{3kT} \right) \tag{9a}$$

Then

$$\frac{\varepsilon_r - 1}{\varepsilon_r + 2} = \frac{\rho P_m}{M} \tag{9b}$$

We see from eqn 9 that the polarizability and permanent dipole moment of the molecules in a sample can be determined by measuring ε_r at a series of temperatures, calculating P_m, and plotting it against $1/T$. The slope of the graph is $N_A \mu^2 / 9\varepsilon_0 k$ and its intercept at $1/T = 0$ is $N_A \alpha / 3\varepsilon_0$.

Example 22.1 *Determining dipole moment and polarizability*

The capacitance of a cell containing camphor was measured at a series of temperatures; when empty the cell has a capacitance of 5.01 pF. Use the following data to find the dipole moment and the polarizability volume of the molecule.

$\theta/°C$	$\rho/(g\,cm^{-3})$	C/pF	$\theta/°C$	$\rho/(g\,cm^{-3})$	C/pF
0	0.99	62.6	120	0.97	40.6
20	0.99	57.1	140	0.96	38.1
40	0.99	54.1	160	0.95	35.6
60	0.99	50.1	200	0.91	31.1
80	0.99	47.6			
100	0.99	44.6			

Method. According to eqn 9, we need to calculate ε_r at each temperature, form $(\varepsilon_r - 1)/(\varepsilon_r + 2)$, and then multiply by M/ρ to form P_m. Next, we should plot P_m against $1/T$ and expect a straight line. The intercept at $1/T = 0$ is $N_A\alpha/3\varepsilon_0 = (4\pi N_A/3)\alpha'$ and the slope is $N_A\mu^2/9\varepsilon_0 k$.

Answer. For camphor, $M = 152.23\ g\,mol^{-1}$. We can therefore use the data to draw up the following table:

$\theta/°C$	$10^3/(T/K)$	ε_r	$(\varepsilon_r - 1)/(\varepsilon_r + 2)$	$P_m/(cm^3\,mol^{-1})$
0	3.66	12.5	0.793	122
20	3.41	11.4	0.776	119
40	3.19	10.8	0.766	118
60	3.00	10.0	0.750	115
80	2.83	9.50	0.739	114
100	2.68	8.90	0.725	111
120	2.54	8.10	0.703	110
140	2.42	7.60	0.688	109
160	2.31	7.11	0.670	107
200	2.11	6.21	0.634	106

The points are plotted in Fig. 22.4. The intercept lies at 82.7, so $\alpha' = 3.3 \times 10^{-23}\ cm^3$. The slope is 10.9, so $\mu = 4.46 \times 10^{-30}\ C\,m$, corresponding to 1.34 D.

22.4 The plot of $P_m/(cm^3\,mol^{-1})$ against $(10^3\,K)/T$ used in Example 22.1 for the determination of the polarizability and dipole moment of camphor.

Comment. Because the Debye equation describes molecules that are free to rotate, the data show that camphor, which does not melt until 175°C, is rotating even in the solid. It is an approximately spherical molecule (4).

Exercise E22.1. The relative permittivity of chlorobenzene is 5.71 at 20°C and 5.62 at 25°C. Assuming a constant density (1.11 g cm^{-3}), estimate its polarizability volume and dipole moment.   [$1.4 \times 10^{-23}\ cm^3$, 1.2 D]

4 Camphor

22.2 Refractive index

The **refractive index** n_r, of a medium is the ratio of the speed of light in a vacuum c to its speed c' in the medium:

$$n_r = \frac{c}{c'} \tag{10}$$

It follows from the Maxwell equations, which describe the properties of electromagnetic radiation, that the refractive index at a (visible or ultraviolet) specified frequency is related to the relative permittivity at

Table 22.2* Refractive indices (at different wavelengths of light) relative to air at 20°C

	434 nm	589 nm	656 nm
$C_6H_6(l)$	1.524	1.501	1.497
$CS_2(l)$	1.675	1.628	1.618
$H_2O(l)$	1.340	1.333	1.331
$KI(s)$	1.704	1.666	1.658

* More values are given in the Data section.

that frequency by

$$n_r = \varepsilon_r^{\frac{1}{2}} \tag{11}$$

The molar polarization P_m and hence the molecular polarizability α can therefore be measured at frequencies typical of visible light (about 10^{15} to 10^{16} Hz) by measuring the refractive index of the sample (Table 22.2) and using the Clausius–Mossotti equation.

The refractive index is related to the molecular polarizability because the propagation of light through a dielectric can be imagined to occur by the incident light inducing an oscillating dipole moment, which then radiates light of the same frequency. The newly generated radiation is delayed in phase relative to the incident radiation, so it propagates more slowly through the medium than through a vacuum. Since photons of high-frequency light carry more energy than those of low-frequency light, they can distort the electronic distributions of the molecules in their path more effectively. Therefore, after allowing for the loss of contributions from low-frequency modes of motion, we can expect the electronic polarizabilities of molecules, and hence the refractive index, to increase as the incident frequency rises towards an absorption frequency. This dependence on frequency is the origin of the dispersion of white light by a prism: the refractive index is greater for blue light than for red, and therefore the blue rays are bent more than the red. The term **dispersion** is a name carried over from this phenomenon to mean the variation of the refractive index, or of any property, with frequency. Fig. 22.5 shows the typical dispersion of the polarization of a sample.

The concept of refractive index is closely related to the property of optical activity. An **optically active** substance is a substance that rotates the plane of polarization of plane-polarized light. As shown in the

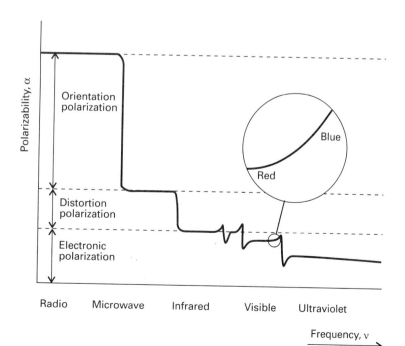

22.5 The general form of the variation of the polarizability with the frequency of the applied field. Note the considerable reduction in polarizability when the field is reversing direction so rapidly that the polar molecules cannot reorientate quickly enough to follow it. The inset shows the variation of the electronic polarizability in the visible region near an electronic excitation of the molecule.

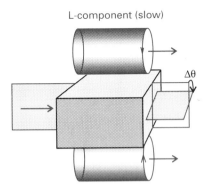

L-component (slow)

R-component (fast)

22.6 Linearly polarized light entering a sample (from the left) can be regarded as the superposition of two counter-rotating circularly polarized components (represented by the two cylindrical objects, which are actually superimposed inside the sample) with a definite phase relation. If one component propagates more rapidly than the other in the medium, when they emerge the phase relation is changed, and the resultant is plane-polarized light rotated through an angle $\Delta\theta$ to its original orientation.

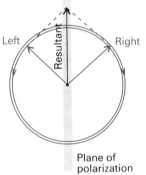

Left   Right

Resultant

Plane of polarization

22.7 The superposition shown in Fig. 22.6 as viewed by an observer facing the oncoming beam.

following *Justification*, the effect arises from the difference of the refractive indices for right and left circularly polarized light, n_R and n_L, respectively.[5] A sample in which these two refractive indices are different is said to be **circularly birefringent**.

JUSTIFICATION

That the angle of optical rotation is proportional to the difference between the refractive indices for left- and right-circularly polarized light may be seen by reference to Fig. 22.6.

Before entering the medium, the beam is plane-polarized at an angle $\theta = 0$. This plane-polarized light may be regarded as a superposition of two oppositely rotating circularly polarized components (Fig. 22.7). On entering the medium, one component propagates faster than the other if their refractive indices are different and, if the sample is of length l, then the difference in the times of passage is

$$\Delta t = \frac{l}{c_R} - \frac{l}{c_L}$$

where c_R and c_L are the speeds of the two components. In terms of the refractive indices the difference is

$$\Delta t = (n_R - n_L) \times \frac{l}{c}$$

The phase difference between the two components when they emerge from the sample is therefore

$$\Delta\theta = c\,\Delta t \times \frac{2\pi}{\lambda} = (n_R - n_L) \times \frac{2\pi l}{\lambda}$$

where λ is the wavelength of the light. The two rotating electric vectors have a different phase when they leave the sample from the value they had initially, so their superposition gives rise to a plane-polarized beam rotated through an angle $\Delta\theta$ relative to the plane of the incoming beam. Hence, the angle of optical rotation is proportional to the difference in refractive index, $n_R - n_L$.

To explain why the refractive indices depend on the handedness of the light, we must examine why the polarizabilities depend on the handedness. The full theory of optical rotation is complicated, but one interpretation is that, if a molecule has a helical structure (including, if the molecule is small, a structure that can be regarded as being a fragment of a helix), then its polarizability depends on whether or not the electric field of the incident radiation rotates in the same sense as the helix. Molecules having a helical structure are chiral, which is the criterion for optical activity discussed in Section 15.3.

---

5 We use the normal convention that in right-handed circularly polarized light the electric vector rotates clockwise as seen by an observer facing the oncoming beam (Fig. 22.6).

The angle of optical rotation varies with the frequency of light. This variation is called **optical rotatory dispersion** (ORD). It arises from the individual dispersions of the polarizabilities (and refractive indices) for left- and right-circularly polarized light, and can be used to investigate the stereochemistry of molecules.

Associated with the differences in the two refractive indices (the circular birefringence of the medium) is a difference in absorption intensities I_R and I_L for right- and left-circularly polarized light. This difference is known as **circular dichroism** (CD). The **CD spectrum** of a sample is a plot of the variation of $I_L - I_R$ with frequency of the radiation. Circular dichroism is particularly useful for determining the absolute configurations of d-metal complexes, because complexes with similar geometries give CD spectra with similar features.

INTERMOLECULAR FORCES

Van der Waals forces are the attractive interactions between closed-shell molecules. They include the net attractive interactions between the partial charges of polar molecules. There are also repulsive interactions that are responsible for the prevention of the complete collapse of matter to nuclear densities. The repulsive interactions arise from Coulombic repulsions and, indirectly, from the Pauli principle and the exclusion of electrons from regions of space where the orbitals of closed-shell species overlap. In this section we consider the attractive forces between molecules, and see how they are related to the electrical properties treated in Section 22.1.

22.3 Interactions between dipoles

Most of the work in this section will be based in one way or the other on the Coulombic potential energy of interaction between two charges (eqn 5a). It is easy to adapt this expression to give the interaction between a charge and a dipole and to extend it to the interaction between two dipoles.

The potential energy of interaction

It is shown in the *Justification* that follows that the potential energy of interaction between a point dipole $\mu_1 = q_1 l$ and the point charge q_2 in the arrangement shown in Fig. 22.8 is

$$V = -\frac{\mu_1 q_2}{4\pi\varepsilon_0} \times \frac{1}{r^2} \qquad (12)$$

(A *point* dipole is a dipole in which the separation between the charges is much smaller than the distance at which the dipole is being observed.) This expression should be multiplied by $\cos\theta$ when the point charge lies at an angle θ to the axis of the dipole. The potential energy rises towards zero (the value at infinite separation of the charge and the dipole) more rapidly than that between two point charges because, from the viewpoint of the point charge, the partial charges of the dipole seem to merge and cancel as the distance r increases (Fig. 22.9).

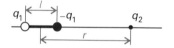

22.8 The potential energy of interaction between a dipole and a point charge is the sum of the repulsion of like charges and the, attraction of opposite charges. For a point dipole, $l \ll r$.

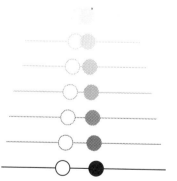

22.9 There are two contributions to the diminishing field of an electric dipole with distance (here seen from the side). The potentials of the charges decrease (shown here by a fading intensity) and the two charges appear to merge, so their combined effect approaches zero more rapidly than by the distance effect alone.

JUSTIFICATION

The sum of the potential energies of repulsion between like charges and attraction between opposite charges in the orientation shown in Fig. 22.8 is

$$V = \frac{1}{4\pi\varepsilon_0}\left(-\frac{q_1q_2}{r - \frac{1}{2}l} + \frac{q_1q_2}{r + \frac{1}{2}l}\right)$$

Because $l \ll r$ for a point dipole, this expression can be simplified by writing

$$V = \frac{q_1q_2}{4\pi\varepsilon_0 r}\left(-\frac{1}{1 - x} + \frac{1}{1 + x}\right)$$

where $x = l/2r$, and then expanding the terms in x and retaining only the leading term:

$$V = \frac{q_1q_2}{4\pi\varepsilon_0 r}\{-(1 + x + \ldots) + (1 - x + \ldots)\}$$

$$= -\frac{q_1q_2}{4\pi\varepsilon_0 r} \times 2x = -\frac{q_1q_2 l}{4\pi\varepsilon_0 r^2}$$

With $\mu_1 = q_1 l$, this expression becomes eqn 12.

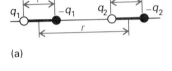

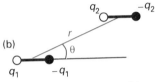

22.10 The potential energy of interaction between two dipoles is the sum of the repulsions of like charges and the attractions of opposite charges. (a) A collinear arrangement of dipoles and (b) a more general arrangement, but still with the dipoles parallel.

Example 22.2 *Calculating the interaction energy of two dipoles*

Calculate the potential energy of interaction of two dipoles in the arrangements shown in Fig. 22.10a when their separation is r.

Method. We proceed in exactly the same way as in the *Justification*, but now the total interaction energy is the sum of four pairwise terms, two attractions between opposite charges, which contribute negative terms to the potential energy, and two repulsions between like charges, which contribute positive terms.

Answer. The sum of the four contributions is

$$V = \frac{1}{4\pi\varepsilon_0}\left(-\frac{q_1q_2}{r + l} + \frac{q_1q_2}{r} + \frac{q_1q_2}{r} - \frac{q_1q_2}{r - l}\right)$$

$$= -\frac{q_1q_2}{4\pi\varepsilon_0}\left(\frac{1}{1 + x} - 2 + \frac{1}{1 - x}\right)$$

with $x = l/r$. As before, we expand the two terms in x using

$$\frac{1}{1 + x} = 1 - x + x^2 - \ldots \qquad \frac{1}{1 - x} = 1 + x + x^2 + \ldots$$

and retain only the first surviving term, which is equal to $2x^2$. This step results in the expression

$$V = -\frac{q_1q_2}{4\pi\varepsilon_0 r} \times 2x^2$$

Therefore, because $\mu_1 = q_1 l$ and $\mu_2 = q_2 l$, the potential energy of interaction in the alignment shown in Fig. 22.10a is

$$V = -\frac{\mu_1\mu_2}{4\pi\varepsilon_0} \times \frac{2}{r^3}$$

Comment. Notice that the interaction energy approaches zero more rapidly than for the previous case: now both interacting entities seem neutral to each other at large separations.

Exercise E22.2. Derive an expression for the potential energy when the dipoles are in the arrangement shown in Fig. 22.10b.

$$[V = (\mu_1\mu_2/4\pi\varepsilon_0 r^3)(1 - 3\cos^2\theta)]$$

The various expressions for the interaction of charges and dipoles are summarized in Table 22.3. It is quite easy to extend the formulae given there to obtain expressions for the energy of interaction of higher **multipoles**, or arrays of point charges (Fig. 22.11). Specifically, an **n-pole** is an array of n point charges with an n-pole moment but no lower moment. Thus, a **monopole** is a point charge, and the monopole moment is what we normally call the overall charge. A **dipole**, as we have seen, is an array of two charges that has no monopole moment (no net charge). A **quadrupole** consists of an array of four point charges that has neither net charge nor dipole moment (as for CO_2 molecules, **5**). An **octupole** consists of an array of eight point charges that sum to zero and which has neither a dipole moment nor a quadrupole moment (as for CH_4 molecules, **6**). The feature to remember is that the interaction energy falls off more rapidly the higher the order of the multipole. For the interaction of an 2^n-pole with an 2^m-pole, the potential energy varies with distance as

$$V \propto \frac{1}{r^{n+m-1}} \tag{13}$$

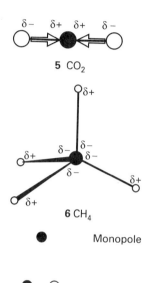

5 CO_2

6 CH_4

 Monopole

 Dipole

Quadrupole

Quadrupole

 Octupole

 Octupole

22.11 Typical charge arrays corresponding to electric multipole moments. The field arising from an arbitrary finite charge distribution can be expressed as the superposition of the fields arising from a superposition of multipoles.

Table 22.3 Multipole interaction potential energies

Interaction type	Distance dependence of potential energy	Typical energy/(kJ mol^{-1})	Comment
Ion–ion	$1/r$	250	Only between ions
Ion–dipole	$1/r^2$	15	
Dipole–dipole	$1/r^3$	2	Between stationary polar molecules
	$1/r^6$	0.3	Between rotating polar molecules
London (dispersion)	$1/r^6$	2	Between all types of molecules

The energy of a hydrogen bond A—H $\cdots$ B is typically 20 kJ mol^{-1} and occurs on contact for A, B = N, O, or F.

The reason for the even steeper decrease with distance is the same as before: the array of charges seems to blend together into neutrality more rapidly with distance the higher the number of individual charges that contribute to the multipole.

The electric field

The same kind of argument can be used to establish the distance dependence of the strength of the electric field generated by a dipole. We shall need this expression when we calculate the dipole moment induced in one molecule by another.

The starting point for the calculation is the strength of the electric field[6] generated by a point electric charge:

$$\mathscr{E} = \frac{q}{4\pi\varepsilon_0 r^2} \tag{14}$$

The field generated by a dipole is the sum of the fields generated by each partial charge. For the point-dipole arrangement shown in Fig. 22.12, the same procedure that was used to derive the potential energy gives

$$\mathscr{E} = \frac{\mu}{4\pi\varepsilon_0} \times \frac{2}{r^3} \tag{15}$$

The electric field of a multipole (in this case a dipole) decreases more rapidly with distance than a monopole (a point charge).

Dipole–dipole interactions

The potential energy of interaction between two polar molecules is a complicated function of the angle between them. However, when the two dipoles are parallel (as in Fig. 22.10b), the energy is simply

$$V = \frac{\mu_1\mu_2}{4\pi\varepsilon_0} \times \frac{f}{r^3}, \qquad \text{with } f = 1 - 3\cos^2\theta \tag{16}$$

This expression applies to polar molecules in a fixed orientation in a solid.

In a fluid of freely rotating molecules, the interaction between dipoles averages to zero because f changes sign as θ changes, and its average value, $\langle f \rangle$, is zero; see the following *Justification*. Physically, the like partial charges of two freely rotating molecules are close together as much as the two opposite charges, and the repulsion of the former is cancelled by the attraction of the latter.

JUSTIFICATION

The average value of f is the sum of the values it has in each infinitesimal region on the surface of a sphere (Fig. 22.13) divided by the surface area

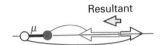

22.12 The electric field of a dipole is the sum of the opposing fields from the positive and negative charges, each of which is proportional to $1/r^2$. The difference is proportional to $1/r^3$.

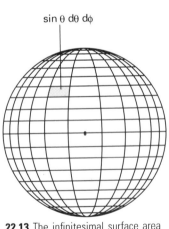

$\sin\theta\,d\theta\,d\phi$

22.13 The infinitesimal surface area element used for evaluating average values of functions over a sphere.

---

6 The electric field is actually a vector, and we cannot simply add and subtract magnitudes without taking into account the directions of the fields. In the cases we consider this will not be a complication because the two charges of the dipoles will be collinear and give rise to fields in the same direction. Be careful, though, with more general arrangements of charges.

of the sphere (4π, if the radius is 1). We note that the infinitesimal area element in polar coordinates is $\sin\theta\, d\theta\, d\phi$. Therefore,

$$\langle f\rangle = \frac{1}{4\pi}\int_0^{2\pi} d\phi\int_0^{\pi}(1-3\cos^2\theta)\sin\theta\, d\theta$$

$$= \frac{1}{2}\int_0^{\pi}(1-3\cos^2\theta)\sin\theta\, d\theta$$

To evaluate the remaining integral, we make the substitution $x = \cos\theta$, which implies that $dx = -\sin\theta\, d\theta$. Then

$$\langle f\rangle = \frac{1}{2}\int_{-1}^{1}(1-3x^2)\, dx = 0$$

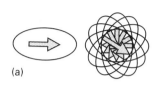

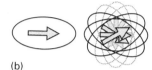

22.14 (a) When a pair of molecules can adopt all relative orientations with equal probability, the favourable and unfavourable orientations cancel, and the average interaction is zero. (b) In an actual fluid the favourable interactions slightly predominate (in accord with the Boltzmann distribution for states of different energies) and there is a net attractive interaction.

The interaction energy of two freely rotating dipoles is zero. However, because their mutual potential energy depends on their relative orientation, the molecules do not in fact rotate completely freely even in a gas. In fact, the lower-energy orientations are marginally favoured, so there is a non-zero average interaction between polar molecules (Fig. 22.14). We show in the following *Justification* that the average potential energy of two rotating molecules that are separated by a distance r is

$$\langle V\rangle = -\frac{C}{r^6}, \qquad \text{where } C = \frac{2\mu_1^2\mu_2^2}{3(4\pi\varepsilon_0)^2 kT} \tag{17}$$

JUSTIFICATION

The detailed calculation of the interaction energy is quite complicated, but the form of the final answer can be constructed quite simply. First, we note that the average interaction energy of two polar molecules rotating at a fixed separation r is given by

$$\langle V\rangle = \frac{\mu_1\mu_2}{4\pi\varepsilon_0 r^3}\langle f\mathcal{P}\rangle$$

where $\mathcal{P}$ is a weighting factor in the averaging, and is equal to the probability that a particular orientation θ will be adopted. This probability is given by the Boltzmann distribution $\mathcal{P}\propto e^{-E/kT}$ with E interpreted as the potential energy of interaction of the two dipoles in that orientation. That is,

$$\mathcal{P}\propto e^{-V/kT}, \qquad \text{where } V = \frac{\mu_1\mu_2 f}{4\pi\varepsilon_0 r^3}$$

When the potential energy of interaction of the two dipoles is very small compared with the energy of thermal motion, we can make use of the fact that $V\ll kT$, expand the exponential function in $\mathcal{P}$, and retain only the first two terms:

$$\mathcal{P}\propto 1-\frac{V}{kT}+\dots$$

The average interaction energy is therefore

$$\langle V^* \rangle = \frac{\mu_1 \mu_2}{4\pi\varepsilon_0 r^3} \left(\langle f \rangle - \frac{\mu_1 \mu_2}{4\pi\varepsilon_0 k T r^3} \langle f^2 \rangle + \cdots \right)$$

The average value of f is zero, so the first term vanishes. However, the average value of f^2 is non-zero because f^2 is positive at all orientations, so we can write

$$\langle V \rangle \propto -\frac{\mu_1^2 \mu_2^2}{(4\pi\varepsilon_0)^2 k T r^6} \langle f^2 \rangle$$

The average value $\langle f^2 \rangle$ is a number that we can expect to be close to 1 (because f^2 ranges from 0 to 4) and in fact turns out to be $\frac{2}{3}$ when the calculation is carried through in detail. Therefore, the final result is that

$$\langle V \rangle = -\frac{C}{r^6}, \qquad \text{where } C = \frac{2\mu_1^2 \mu_2^2}{3(4\pi\varepsilon_0)^2 k T}$$

as quoted in eqn 17.

The important features of eqn 17 are its negative sign (the average interaction is attractive), the dependence of the average interaction energy on the inverse sixth power of the separation, and its inverse dependence on the temperature. The last feature reflects the way that the greater thermal motion overcomes the mutual orientating effects of the dipoles at higher temperatures. The inverse sixth power arises from the inverse third power of the interaction potential energy that is weighted by the energy in the Boltzmann term, which is also proportional to the inverse cube of the separation.

At 25°C the average interaction energy for pairs of molecules with $\mu = 1\,\text{D}$ is about $-0.07\,\text{kJ mol}^{-1}$ when the separation is $0.5\,\text{nm}$. This energy should be compared with the average molar kinetic energy of $\frac{3}{2}RT = 3.7\,\text{kJ mol}^{-1}$ at the same temperature. The interaction energy is much smaller than the energies involved in the making and breaking of chemical bonds.

Dipole–induced-dipole interactions

A polar molecule with dipole moment μ_1 can induce a dipole μ_2^* in a polarizable molecule. The induced dipole interacts with the permanent dipole of the first molecule, and the two are attracted together. It is shown in the following *Justification* that the average interaction energy when the separation of the molecules is r is

$$V = -\frac{C}{r^6}, \qquad \text{where } C = \frac{\mu_1^2 \alpha_2'}{\pi\varepsilon_0} \tag{18}$$

where α_2' is the polarizability volume of molecule 2 and μ_1 is the permanent dipole moment of molecule 1.

JUSTIFICATION

The general form of the potential energy of interaction can be justified in a fairly straightforward calculation that shows the essentials of the derivation. Thus, if we assume that the two dipoles (the permanent and the induced dipole, μ_1 and μ_2^*, respectively) already exist, then their energy of interaction is given by eqn 16:

$$V = -\frac{2\mu_1\mu_2^*}{4\pi\varepsilon_0 r^3}$$

However, the induced dipole moment depends on the field generated by the polar molecule, and hence on the separation of the two molecules. Because we can write

$$\mu_2^* = \alpha_2 \mathscr{E}$$

where α_2 is the polarizability of molecule 2 and $\mathscr{E}$ is the field generated by molecule 1 (the polar molecule), the potential energy is:

$$V = -\frac{2\mu_1\alpha_2\mathscr{E}}{4\pi\varepsilon_0 r^3}$$

The electric field generated by the polar molecule is given by eqn 15, so:

$$V = -\frac{2\mu_1\alpha_2}{4\pi\varepsilon_0 r^3} \times \frac{2\mu_1}{4\pi\varepsilon_0 r^3}$$

As the induced dipole follows the direction of the inducing dipole (Fig. 22.15), we do not need to take account of the effects of thermal motion: both dipoles remain aligned however fast the molecules tumble.

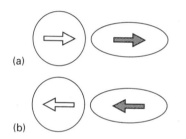

22.15 (a) A polar molecule (dark arrow) can induce a dipole (light arrow) in a non-polar molecule, and (b) the latter's orientation follows the former's, so that the interaction does not average to zero.

The dipole–induced-dipole interaction energy is independent of the temperature because thermal motion has no effect on the averaging process. Moreover, like the dipole–dipole interaction, the potential energy depends on $1/r^6$: this distance dependence stems from the $1/r^3$ dependence of the field (and hence the magnitude of the induced dipole) and the $1/r^3$ dependence of the potential energy of interaction between the permanent and induced dipoles. For a molecule with $\mu = 1\ \mathrm{D}$ (such as HCl) near a molecule of polarizability volume $\alpha' = 10 \times 10^{-30}\ \mathrm{m}^3$ (such as benzene, Table 22.1) the average interaction energy is about $-0.8\ \mathrm{kJ\ mol}^{-1}$ when the separation is 0.3 nm.

Induced-dipole–induced-dipole interactions

Non-polar molecules (including closed-shell atoms, such as Ar) attract one another even though neither has a permanent dipole moment. The evidence for the existence of interactions between them is the formation of condensed phases of non-polar substances, such as the condensation of hydrogen or argon to a liquid at low temperatures and the fact that benzene is a liquid at normal temperatures.

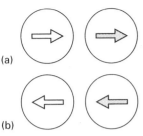

(a)

(b)

22.16 (a) In the dispersion interaction, an instantaneous dipole (shaded arrow) on one molecule induces a dipole (light arrow) on another molecule, and the two dipoles then interact to lower the energy. (b) The two instantaneous dipoles are correlated and, although they occur in different orientations at different instants, the interaction does not average to zero.

The interaction between non-polar molecules arises from the instantaneous transient dipoles that all molecules possess as a result of fluctuations in the instantaneous positions of electrons. To see the origin of the interaction, suppose that the electrons in one molecule flicker into an arrangement that gives it an instantaneous dipole moment μ_1^*. This dipole generates an electric field that polarizes the other molecule, and induces in it an instantaneous dipole moment μ_2^*. The two dipoles attract each other and the potential energy of the pair is lowered. Although the first molecule will go on to change the size and direction of its instantaneous dipole, the electron distribution of the second molecule will follow it; that is, the two dipoles are correlated in direction (Fig. 22.16). Because of this correlation, the attraction between the two instantaneous dipoles does not average to zero, and gives rise to an induced-dipole–induced-dipole interaction. This interaction is called either the **dispersion interaction** or the **London interaction** (for Fritz London, who first described it).

Polar molecules also interact by a dispersion interaction: such molecules also possess instantaneous dipoles, the only difference being that the time average of each fluctuating dipole does not vanish, but corresponds to the permanent dipole.

The strength of the dispersion interaction depends on the polarizability of the first molecule because the instantaneous dipole moment μ_1^* depends on the looseness of the nuclear charge's control over the outer electrons. The strength of the interaction also depends on the polarizability of the second molecule, for that polarizability determines how readily a dipole can be induced by another molecule. The actual calculation of the dispersion interaction is quite involved, but a reasonable approximation to the interaction energy is given by the **London formula**:

$$V = -\frac{C}{r^6}, \qquad \text{where } C = \tfrac{2}{3}\alpha_1'\alpha_2'\frac{I_1 I_2}{I_1 + I_2} \tag{19}$$

where I_1 and I_2 are the ionization energies of the two molecules (Table 2.6). Once again, the interaction turns out to be proportional to the inverse sixth power of the separation.

In the case of two CH_4 molecules we can substitute $\alpha' = 2.6 \times 10^{-30}\,\text{m}^3$ and $I \approx 7\,\text{eV}$ to obtain $V = -5\,\text{kJ mol}^{-1}$ for $r = 0.3\,\text{nm}$. A very rough check on this figure is the enthalpy of vaporization of methane, which is $8.2\,\text{kJ mol}^{-1}$. (The comparison is insecure partly because the enthalpy of vaporization is a many-body quantity and partly because the long-distance assumption breaks down.) The dispersion interaction generally dominates all the interactions between molecules other than hydrogen bonds.

Hydrogen bonding

The interactions we have described so far are universal in the sense that they are possessed by all molecules independent of their specific identity. However, there is a type of interaction that is possessed by molecules that have a particular constitution. A **hydrogen bond** is an attractive

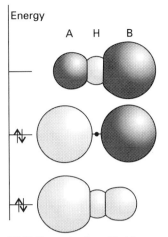

Energy

22.17 The molecular orbital interpretation of the formation of an A—H · · · B hydrogen bond. From the three A, H, and B orbitals, three molecular orbitals can be formed (their relative contributions are represented by the sizes of the spheres). Only the two lower energy orbitals are occupied, and there may therefore be a net lowering of energy compared with the separate AH and B species.

interaction between two closed-shell species that arises from a link of the form A—H · · · B, where A and B are highly electronegative elements and B possesses a lone pair of electrons. Hydrogen bonding is conventionally regarded as being limited to N, O, and F but, if B is an anionic species (such as Cl^-), then it may also participate in hydrogen bonding. There is no strict cut-off for an ability to participate in hydrogen bonding, but N, O, and F participate most effectively.

The formation of a hydrogen bond may be expressed in a variety of ways. It can be regarded as a particular example of delocalized molecular orbital formation in which A, H, and B each supply one atomic orbital from which three molecular orbitals are constructed (Fig. 22.17). The A and H1s orbitals are those used to form the A—H bond in the AH molecule and the B orbital accommodates the lone pair on B. In the combined species, there are four electrons to accommodate (two from the A—H bond, two from the lone pair of B), and they occupy the two lowest molecular orbitals of the AHB fragment. Because the uppermost (most antibonding) orbital is vacant, it is feasible for the net effect to be a lowering of energy, and hence the formation of a hydrogen bond.

In practice, the strength of the bond is found to be about $20 \, kJ \, mol^{-1}$. Because the bonding depends on orbital overlap, it is virtually a contact-like interaction that is turned on when AH touches B and is zero as soon as the contact is broken. If hydrogen bonding is present, it dominates the van der Waals interactions we have described. The properties of liquid and solid water, for example, are dominated by the hydrogen bonding between H_2O molecules.

The total attractive interaction

We shall consider molecules that are unable to participate in hydrogen bond formation. The total attractive interaction energy between rotating molecules is then the sum of the three van der Waals contributions discussed above. (Only the dispersion interaction contributes if both molecules are non-polar.) All three vary as the inverse sixth power of the separation, so we may write

$$V = -\frac{C_6}{r^6} \tag{20}$$

where C_6 is a coefficient that depends on the identity of the molecules.

Although attractive interactions between molecules are often expressed in the form of eqn 20, we must remember that this equation has only limited validity. First, we have taken into account only dipolar interactions, for they have the longest range and are dominant if the average separation of the molecules is large. However, in a complete treatment we should also consider quadrupolar and higher-order multipole interactions, particularly if the molecules do not have dipole moments. Second, the expressions have been derived by assuming that the molecules can rotate reasonably freely. That is not the case in most solids, and then the dipole–dipole interaction is proportional to $1/r^3$ because the Boltzmann averaging procedure is irrelevant when the molecules are trapped into a fixed orientation.

A different kind of limitation is that eqn 20 relates to the interactions of *pairs* of molecules, and there is no reason to suppose that the energy of interaction of three (or more) molecules is the sum of the pairwise interaction energies alone. The total dispersion energy of three closed-shell atoms, for instance, is given approximately by the **Axilrod–Teller formula**:

$$V = -\frac{C_6}{r_{AB}^6} - \frac{C_6}{r_{BC}^6} - \frac{C_6}{r_{CA}^6} + \frac{C'}{(r_{AB}r_{BC}r_{CA})^3} \tag{21}$$

where

$$C' = a(3\cos\theta_A \cos\theta_B \cos\theta_C + 1)$$

The parameter a is approximately equal to $\frac{3}{4}\alpha'C_6$; the angles θ are the internal angles of the triangle formed by the three atoms (**7**). The term in C' (which represents the non-additivity of the pairwise interactions) is negative for a linear arrangement of atoms (so that arrangement is stabilized) and positive for an equilateral triangular cluster. It is found that the three-body term contributes about 10 per cent of the total interaction energy in liquid argon.

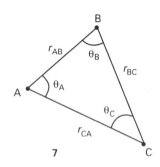

7

22.4 Repulsive and total interactions

When molecules are squeezed together, the nuclear and electronic repulsions and the rising electronic kinetic energy begin to dominate the attractive forces. The repulsions increase steeply with decreasing separation (Fig. 22.18) in a way that can be deduced only by very extensive, complicated molecular structure calculations of the kind described in Chapter 14.

In many cases, however, progress can be made by using a greatly simplified representation of the potential energy, where the details are ignored and the general features expressed by a few adjustable parameters. One such approximation is the **hard-sphere potential**, in which it is assumed that the potential energy rises abruptly to infinity as soon as the particles come within a separation d, which is called the **collision diameter**:

$$V = \infty \text{ for } r \le d \qquad V = 0 \text{ for } r > d \tag{22}$$

This very simple potential is surprisingly useful for assessing a number of properties, as we shall see.

Another widely used approximation is to suppose that the potential energy of the short-range repulsive interaction is inversely proportional to a high power of r.

$$V = \frac{C_n}{r^n}$$

So long as $n > 6$, this positive contribution to the potential energy dominates the $1/r^6$ potential energy arising from the attractive interaction at short separations because C_n/r^n is then larger than C_6/r^6. The sum of this repulsive interaction and the attractive interaction given by eqn 20 is

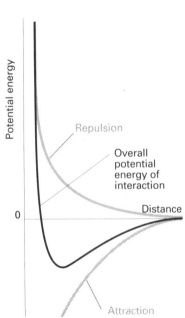

22.18 The general form of an intermolecular potential energy curve. At long range the interaction is attractive, but at close range the repulsions dominate.

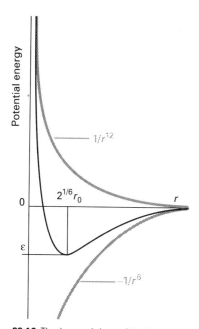

22.19 The Lennard-Jones (12, 6)-potential and the relation of the parameters to the features of the curve. The grey lines are the two contributions.

Table 22.4* Lennard-Jones (12, 6) parameters†

	$(\varepsilon/k)/K$	r_0/pm
Ar	124	342
CCl$_4$	327	588
He	10.22	258
N$_2$	91.5	368

* More values are given in the Data section.

† ε is expressed as an effective temperature by division by Boltzmann's constant k.

called the **Lennard-Jones (n, 6)-potential**:

$$V = \frac{C_n}{r^n} - \frac{C_6}{r^6} \tag{23a}$$

For mathematical reasons it is convenient to select $n = 12$, and the resulting **(12,6)-potential** is often written in the form

$$V = 4\varepsilon\left\{\left(\frac{r_0}{r}\right)^{12} - \left(\frac{r_0}{r}\right)^{6}\right\} \tag{23b}$$

The distance dependence of the potential energy is drawn in Fig. 22.19. The two parameters are now ε, the depth of the well, and r_0, the separation at which $V = 0$. The well minimum occurs at $r_e = 2^{\frac{1}{6}}r_0$. Some typical values are listed in Table 22.4. Although the (12, 6)-potential has been used in many calculations, there is plenty of evidence to show that $1/r^{12}$ is a very poor representation of the repulsive potential, and that an exponential form e^{-r/r_0} is greatly superior. An exponential function is more faithful to the exponential decay of atomic wavefunctions at large distances, and hence to the overlap that is responsible for repulsion. The potential with an exponential repulsive term and a $1/r^6$ attractive term is known as an **exponential-6 potential**.

22.5 Molecular interactions in beams

A notable advance in the experimental study of intermolecular forces has come from the development of **molecular beams**, which consist of a narrow beam of molecules travelling though an evacuated vessel. The beam is directed towards other molecules, and the scattering that occurs on impact is related to the intermolecular interactions.

The basic principles

The basic arrangement for a molecular beam experiment is shown in Fig. 22.20. The slotted disks make up the **velocity selector**. They rotate in the path of the beam and allow only those molecules having a certain speed to pass. There are also more sophisticated devices for generating molecules with a desired velocity. Among the most important are

22.20 The basic arrangement of a molecular beam apparatus. The atoms or molecules emerge from a heated source, and pass through the velocity sector, a train of rotating disks. The scattering occurs from the target gas (which might take the form of another beam), and the flux of particles entering the detector set at some angle is recorded.

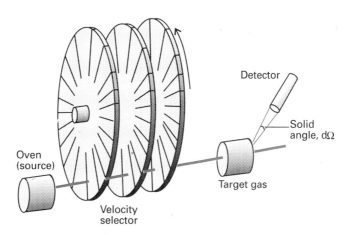

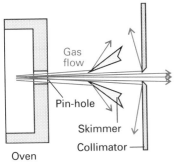

22.21. A supersonic nozzle skims off some of the molecules of the beam and leads to a beam with well-defined velocity.

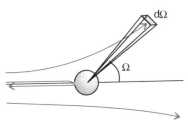

22.22 The definition of the solid angle for scattering.

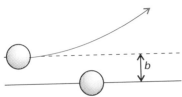

22.23 The definition of the impact parameter b as the perpendicular separation of the initial paths of the particles.

supersonic nozzles, in which the molecules stream out of the source with a very narrow velocity distribution and are then stripped down to a beam by passing through a skimmer shaped like an inverted cone (Fig. 22.21). Other types of selection are also possible: for example, electric fields may be used to deflect polar molecules and to obtain a beam of aligned molecules. The target gas may be either a bulk sample or another molecular beam. The latter **crossed beam technique** gives a lot of information because the states of both the target and projectile molecules may be controlled. The intensity of the incident beam is measured by the **incident beam flux** I which is the number of particles per unit area per unit time.

The detectors may consist of a chamber fitted with a sensitive pressure gauge or an ionization detector, in which the incoming molecule is first ionized and then detected electronically. The state of the scattered molecules may also be determined spectroscopically, and is of interest when the collisions change their vibrational or rotational states.

The experimental observations

The primary experimental information from a molecular beam experiment is the fraction of the molecules in the incident beam that are scattered into a particular direction. The fraction is normally expressed in terms of dI, the number of molecules that are scattered per unit time into a cone that represents the area covered by the 'eye' of the detector (Fig. 22.22). This number is reported as the **differential scattering cross-section** σ, the constant of proportionality between the value of dI and the intensity of the incident beam, the number density of target molecules, and the infinitesimal path length dx through the sample:

$$dI = \sigma I \mathcal{N} \, dx \tag{24}$$

The value of σ (which has the dimensions of area) depends on the **impact parameter** b, the initial perpendicular separation of the paths of the colliding molecules (Fig. 22.23), and the details of the intermolecular potential. The role of the impact parameter is most easily seen by considering the impact of two hard spheres (Fig. 22.24). If $b = 0$, the lighter projectile is on a trajectory that leads to a head-on collision, so that the only scattering intensity is detected when the detector is at $\theta = 180°$. If the impact parameter is so great that the spheres do not make contact $(b > R_A + R_B)$, there is no scattering and the scattering cross-section is zero at all angles except $\theta = 0$. Glancing blows, with $0 < b \leq R_A + R_B$, lead to scattering intensity in cones around the initial line of flight direction (Fig. 22.24c).

Scattering effects

The scattering pattern of real molecules, which are not hard spheres, depends on the details of the intermolecular potential, including the anisotropy that is present when the molecules are non-spherical. The scattering also depends on the relative speed of approach of the two

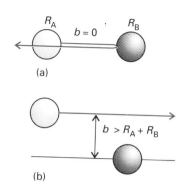

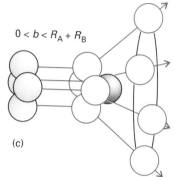

22.24 Three typical cases for the collisions of two hard spheres: (a) $b = 0$, giving backward scattering; (b) $b > R_A + R_B$, giving forward scattering; (c) $0 < b < R_A + R_B$, leading to scattering into one direction on a ring of possibilities. (The target molecule is taken to be so heavy that it remains virtually stationary.)

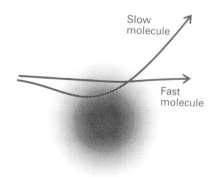

22.25 The extent of scattering may depend on the relative speed of approach as well as the impact parameter. The dark zone represents the repulsive core; the fuzzy outer zone represents the long-range attractive potential.

particles: a very fast particle might pass through the interaction region without much deflection, whereas a slower one on the same path might be temporarily captured and undergo considerable deflection (Fig. 22.25). The variation of the scattering cross-section with the relative speed of approach should therefore give information about the strength and range of the intermolecular potential.

A further point is that the outcome of collisions is determined by quantum, not classical, mechanics. The wave nature of the particles can be taken into account, at least to some extent, by drawing all classical trajectories that take the projectile particle from source to detector, and then considering the effects of interference between them.

Two quantum mechanical effects are of great importance. A particle with a certain impact parameter might approach the attractive region of the potential in such a way that it is deflected towards the repulsive core (Fig. 22.26), which then repels it out through the attractive region to continue its flight in the forward direction. Some molecules, however, also travel in the forward direction because they have impact parameters so large that they are undeflected. The wavefunctions of the particles that take the two types of path interfere, and the intensity in the forward direction is modified. The effect is called **glory scattering**. The same phenomenon accounts for the optical glory effect, in which a bright halo can sometimes be seen surrounding an illuminated object. (The coloured rings around the shadow of an aircraft cast on clouds by the sun, and often seen in flight, is an example of an optical glory.)

The second quantum effect we need consider is the observation of a strongly enhanced scattering in a non-forward direction. This effect is called **rainbow scattering** because the same mechanism accounts for the appearance of an optical rainbow. The origin of the phenomenon is illustrated in Fig. 22.27. As the impact parameter decreases, there comes a stage at which the scattering angle passes through a maximum and the interference between the paths results in a strongly scattered beam. The rainbow angle is the angle for which $d\theta/db = 0$ and the scattering is strong.

The detailed analysis of scattering data can be very complicated, but the main outcome should be clear: the intensity distribution of scattered particles can be related to the intermolecular potential, and a detailed picture can be built up of its radial and angular variation. One outcome is the value of C_6 in the van der Waals interaction and the testing of the Lennard-Jones potential (or some more elaborate version). It is found, for example, that although $1/r^6$ is quite a good representation of the attractive component of the potential, as remarked before the repulsive component is not at all well described by $1/r^{12}$.

MAGNETIC PROPERTIES

The magnetic and electric properties of molecules are analogous. For instance, some molecules possess permanent magnetic dipole moments, and an applied magnetic field can induce a magnetic moment.

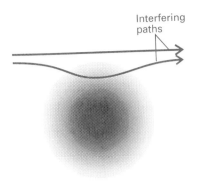

22.26 Two paths leading to the same destination will interfere quantum mechanically; in this case they give rise to glory scattering in the forward direction.

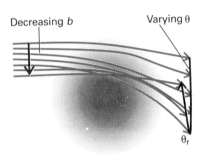

22.27 The interference of paths leading to rainbow scattering. The rainbow angle θ_r is the maximum scattering angle reached as b is decreased. Interference between the numerous paths at that angle modifies the scattering intensity markedly.

22.6 Magnetic susceptibility

The analogue of the electric polarization P is the **magnetization** M, the magnetic dipole moment per unit volume. The magnetization induced by a field of strength H is proportional to H, and we write

$$M = \chi H \tag{25}$$

where χ is the dimensionless **volume magnetic susceptibility**. A closely related quantity is the **molar magnetic susceptibility** χ_m:

$$\chi_m = \chi V_m \tag{26}$$

where V_m is the molar volume of the substance (we shall soon see why it is sensible to introduce this quantity). The **magnetic flux density** B is related to the applied field strength and the magnetization by

$$B = \mu_0(H + M) = \mu_0(1 + \chi)H \tag{27}$$

where μ_0 is the fundamental constant known as the **vacuum permeability**:

$$\mu_0 = 4\pi \times 10^{-7} \, \mathrm{J\,C^{-2}\,m^{-1}\,s^2}$$

The magnetic flux density can be thought of as the density of magnetic lines of force permeating the medium. This density is increased if M adds to H (when $\chi > 0$), but the density is decreased if M opposes H (when $\chi < 0$). Materials for which χ is positive are called **paramagnetic**. Those for which χ is negative are called **diamagnetic**.

Just as polar molecules contribute a term proportional to $\mu^2/3kT$ to the electric polarization of a medium, so molecules with a permanent magnetic dipole moment of magnitude m contribute to the magnetization an amount proportional to $m^2/3kT$. An applied field can also induce a magnetic moment to an extent determined by the **magnetizability** ξ of the molecules, and the magnetic analogue of eqn 9a is

$$\chi = \mathcal{N}\mu_0\left(\xi + \frac{m^2}{3kT}\right) \tag{28a}$$

We can now see why it is convenient to introduce χ_m, for the product of the number density $\mathcal{N}$ and the molar volume is Avogadro's constant N_A:

$$\mathcal{N} \times V_m = \frac{N}{V} \times V_m = \frac{nN_A}{nV_m} \times V_m = N_A$$

Hence

$$\chi_m = N_A\mu_0\left(\xi + \frac{m^2}{3kT}\right) \tag{28b}$$

and the density dependence of the susceptibility has been eliminated. The expression for χ_m is in agreement with the empirical **Curie law**:

$$\chi_m = A + \frac{C}{T} \tag{29}$$

with $A = N_A\mu_0\xi$ and $C = N_A\mu_0 m^2/3k$.

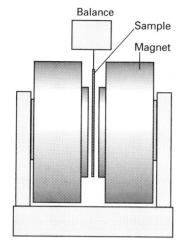

22.28 The arrangement of the Gouy balance for measuring magnetic susceptibilities. A paramagnetic sample appears to weigh more and a diamagnetic sample appears to weigh less when the magnetic field is on.

Table 22.5* Magnetic susceptibilities at 298 K

	$\chi/10^{-6}$	$\chi_m/(10^{-5}\ cm^3\ mol^{-1})$
$H_2O(l)$	-90	-160
$NaCl(s)$	-13.9	-38
$Cu(s)$	-9.6	-6.8
$CuSO_4\cdot$ $5H_2O(s)$	$+176$	$+1920$

* More values are given in the Data section.

The magnetic susceptibility is traditionally measured with a **Gouy balance**. This instrument consists of a sensitive balance from which the sample hangs in the form of a narrow cylinder (Fig. 22.28) and lies between the poles of a magnet. If the sample is paramagnetic, it is drawn into the field, and its apparent weight is greater than when the field is off. A diamagnetic sample tends to be expelled from the field and appears to weigh less when the field is turned on. The balance is normally calibrated against a sample of known susceptibility. A modern version of the determination makes use of a SQUID, a superconducting quantum interference device.

Some experimental values are listed in Table 22.5; a typical paramagnetic volume susceptibility is about 10^{-3}, and a typical diamagnetic volume susceptibility is about $(-)10^{-5}$. The permanent magnetic moment can be extracted from susceptibility measurements by plotting χ against $1/T$.

22.7 The permanent magnetic moment

The permanent magnetic moment of a molecule arises from any unpaired electron spins in the molecule. We saw in Section 13.9 that the magnetic moment of an electron is proportional to its spin, and the magnitude of the magnetic moment is proportional to the magnitude of the angular momentum, $\{s(s+1)\}^{\frac{1}{2}}\hbar$:

$$m = g_e\{s(s+1)\}^{\frac{1}{2}}\mu_B \tag{30}$$

The factor g_e is the electron's g-factor ($g_e = 2.0023$) and the Bohr magneton μ_B is:

$$\mu_B = \frac{e\hbar}{2m_e} \tag{31}$$

If there are several electron spins in each molecule, they combine to a total spin S, and then $s(s+1)$ should be replaced by $S(S+1)$. It follows that the spin contribution to the molar magnetic susceptibility is

$$\chi_m = \frac{N_A g_e^2 \mu_0 \mu_B^2 S(S+1)}{3kT} \tag{32}$$

This expression shows that the susceptibility is positive, and so that the spin magnetic moments contribute to the paramagnetic susceptibilities of materials. The contribution decreases with increasing temperature because the thermal motion randomizes the spin orientations. In practice, a contribution to the paramagnetism also arises from the orbital angular momenta of electrons: we have discussed the spin-only contribution.

Example 22.3 *Calculating the magnetic susceptibility*

Calculate the volume and molar paramagnetic susceptibilities of a sample of a complex salt with three unpaired electrons per complex cation at 298 K, given that its mass density is $3.24\ g\ cm^{-3}$ and its molar mass is $200\ g\ mol^{-1}$.

Method. We use eqn 32. A convenient quantity in calculations like this is

$$\frac{N_A g_e^2 \mu_0 \mu_B^2}{3k} = 6.3001 \text{ cm}^3 \text{ K mol}^{-1}$$

Consequently,

$$\chi_m = 6.3001 \times \frac{S(S + 1)}{T/K} \text{ cm}^3 \text{ mol}^{-1}$$

To obtain the volume magnetic susceptibility, the molar susceptibility is divided by the molar volume (eqn 26), which is calculated from $V_m = M/\rho$.

Answer. Substitution of the data with $S = \frac{3}{2}$ gives

$$\chi_m = 6.3001 \times \frac{15/4}{298} \text{ cm}^3 \text{ mol}^{-1} = 7.9 \times 10^{-2} \text{ cm}^3 \text{ mol}^{-1}$$

The molar volume of the substance is $61.7 \text{ cm}^3 \text{ mol}^{-1}$, so the volume susceptibility is 1.3×10^{-3}.

Comment. Note that χ is dimensionless and (in this case) positive, indicating paramagnetism. The molar susceptibility has the same dimensions as molar volume. Note also that the density is not needed for the calculation of χ_m.

Exercise E22.3. Repeat the calculation for a complex with five unpaired spins, a density of 2.87 g cm^{-3}, and a molar mass of 322.4 g mol^{-1} at 273 K. $[\chi_m = 0.20 \text{ cm}^3 \text{ mol}^{-1}; \chi = 1.79 \times 10^{-3}]$

(a)

(b)

22.29 In (a) a ferromagnetic material the electron spins are locked into a parallel alignment over large domains; in (b) an antiferromagnetic material the electron spins are locked into an antiparallel arrangement.

At low temperatures some paramagnetic solids make a phase transition to a state in which large domains of spins align with parallel orientations. This co-operative alignment gives rise to a very strong magnetization and is called **ferromagnetism** (Fig. 22.29a). In other cases the co-operative effect leads to alternating spin orientations, so they are locked into a low-magnetization arrangement (Fig. 22.29b) to give an **antiferromagnetic** phase. The ferromagnetic phase has a non-zero magnetization in the absence of an applied field, but the antiferromagnetic phase has a zero magnetization because the spin magnetic moments cancel. The ferromagnetic transition occurs at the **Curie temperature**, and the antiferromagnetic transition occurs at the **Néel temperature**.

22.8 Induced magnetic moments

Whereas an electric field polarizes a molecule by stretching it, a magnetic field magnetizes a molecule by twisting it. More precisely, an applied magnetic field induces the circulation of electronic currents. These currents give rise to a magnetic field which usually opposes the applied field, so the substance is diamagnetic. In a few cases the induced field augments the applied field, and the substance is then paramagnetic.

The great majority of molecules with no unpaired electron spins are diamagnetic. This is because the diamagnetic flow occurs within the orbitals of the molecule that are occupied in its ground state, whereas an orbital contribution to the paramagnetic susceptibility depends on currents being stimulated by forcing the electrons to pass through unoccupied, higher-energy orbitals. When orbital paramagnetism does occur it can be distinguished from spin paramagnetism by the fact that it is temperature independent: this is why it is called **temperature-independent paramagnetism** (TIP).

We can summarize these remarks as follows. All molecules have a diamagnetic component to their susceptibility, but it is dominated by spin paramagnetism if the molecules have unpaired electrons. In a few cases (where there are low-lying excited states) TIP is strong enough to make the molecules paramagnetic even though all their electrons are paired.

CHECK LIST OF KEY IDEAS

1 The definition of the **electric dipole moment** and the approximate additivity of dipole moments in polyatomic molecules (Section 22.1).

2 The **polarizability** and **polarizability volume** of a molecule and its contribution to the **polarization** of a medium (Section 22.1).

3 The **polarization** of a medium as the mean dipole moment per unit volume (Section 22.1).

4 The variation of the polarizability with frequency and the **orientation, distortion,** and **electronic contributions** (Section 22.1).

5 The **Debye equation** for the contribution of the permanent dipole moment and the polarizability to the polarization (eqn 8a) and the **Clausius-Mossotti equation** for the polarization when the permanent dipole moment does not contribute (eqn 8b).

6 The **refractive index** of a substance and its **dispersion** (Section 22.2).

7 **Circular birefringence,** the origin of **optical activity,** and **optical rotatory dispersion** (Section 22.2).

8 The **circular dichroism** of chiral molecules (Section 22.2).

9 The **potential energy** of interaction between a charge and an electric dipole (eqn 12) and between two **multipoles** (eqn 13).

10 The **electric field** developed by an electric dipole (eqn 15).

11 The origin of **van der Waals forces,** the attractive forces, between molecules and the origin of repulsive forces (Section 22.3).

12 The average **dipole-dipole interaction** of rotating polar molecules (eqn 17).

13 The **dipole-induced-dipole interaction** between a polar molecule and a polarizable molecule (eqn 18).

14 The **induced-dipole-induced-dipole interaction** between molecules and the **London formula** (eqn 19) for the **dispersion interaction.**

15 The formation of a **hydrogen bond** (Section 22.3).

16 The contribution of **three-body interactions** and the **Axilrod-Teller formula** (eqn 21).

17 The **hard-sphere potential** (eqn 22) and the **Lennard-Jones (n, 6)-potential** (eqn 23).

18 The study of molecular interactions in **molecular beams** and the significance of the **differential scattering cross-section** and the **impact parameter** (Section 22.5).

19 The phenomena of **glory scattering** and **rainbow scattering** (Section 22.5).

20 The **magnetic susceptibility** of a substance and its classification as **paramagnetic** and **diamagnetic** (Section 22.6).

21 The **Curie law** of magnetism (eqn 29).

22 The contribution of electronic spin to the magnetic susceptibility and the transition to **ferromagnetic** and **antiferromagnetic** phases (Section 22.7).

23 The contribution of **induced magnetic moments** to the magnetic susceptibility and the origin of **temperature-independent paramagnetism** (Section 22.8).

EXERCISES

22.1. Calculate the capacitance of a condenser when the space between the plates is filled with a substance of relative permittivity 35.5. When the space is a vacuum, the capacitance is 6.2 pF.

22.2. The molar polarization of fluorobenzene vapour is proportional to T^{-1}, and is $70.62 \, cm^3 \, mol^{-1}$ at 351.0 K and $62.47 \, cm^3 \, mol^{-1}$ at 423.2 K. Calculate the polarizability and dipole moment of the molecule.

22.3. At 0°C, the molar polarization of liquid chlorine trifluoride is $27.18 \, cm^3 \, mol^{-1}$ and its density is $1.89 \, g \, cm^{-3}$. Calculate the relative permittivity of the liquid.

22.4. The ClF_3 molecule has five electron pairs around the central Cl atom, and so may have either three equatorial F atoms or two axial and one equatorial F atoms. Since the molecule is polar, which structure does it have?

22.5. The refractive index of CH_2I_2 is 1.732 for 656 nm light. Its density at 20°C is $3.32 \, g \, cm^{-3}$. Calculate the polarizability of the molecule at this wavelength.

22.6. The dipole moments of the bonds C—F, C—O, and C=O are 1.4, 1.2, and 2.7 D, respectively. The bond lengths are 141, 143, and 122 pm, respectively. Estimate the percentage ionic character of the bonds. How well do the results correlate with the electronegativity differences of the atoms in the bonds?

22.7. The electric dipole moment of toluene (methylbenzene) is 0.4 D. Estimate the dipole moments of the three xylenes (dimethylbenzene). Which answer can you be sure about?

22.8. Calculate the resultant of two dipole moments of magnitude 1.5 D and 0.80 D that make an angle 109.5° to each other.

22.9. Calculate the magnitude and direction of the dipole moment of the following arrangement of charges in the xy-plane: $3e$ at $(0, 0)$, $-e$ at $(0.32 \, nm, 0)$, and $-2e$ at an angle of 20° from the x-axis and a distance of 0.23 nm from the origin.

22.10. The polarizability volume of H_2O is $1.48 \times 10^{-24} \, cm^3$; calculate the dipole moment of the molecule (in addition to the permanent dipole moment) induced by an applied electric field of strength $1.0 \, kV \, cm^{-1}$.

22.11. The polarizability volume of H_2O at optical frequencies is $1.5 \times 10^{-24} \, cm^3$: estimate the refractive index of water. The experimental value is 1.33; what may be the origin of the discrepancy?

22.12. The dipole moment of chlorobenzene is 1.57 D and its polarizability volume is $1.23 \times 10^{-23} \, cm^3$. Estimate its relative permittivity at 25°C, when its density is $1.173 \, g \, cm^{-3}$.

22.13. A solution of an optically active substance shows an optical rotation of 250° in a cell of 10 cm length at 500 nm. What is the difference of the refractive indices of left and right circularly polarized light through this substance?

22.14. The Lennard-Jones $(12, 6)$-potential gives the potential energy of interaction between molecules. Given that the force is the negative slope of the potential, calculate the distance dependence of the force acting between the molecules. What is the separation at which the force is zero?

22.15. The magnetic moment of $CrCl_3$ is $3.81 \mu_B$. How many unpaired electrons does the Cr atom possess?

22.16. Calculate the molar susceptibility of benzene given that its volume susceptibility is -7.2×10^{-7} and its density $0.879 \, g \, cm^{-3}$ at 25°C.

22.17. According to Lewis theory, an O_2 molecule should be diamagnetic. However, experimentally it is found that $\chi_m/(m^3 \, mol^{-1}) = (1.22 \times 10^{-5} \, K)/T$. Determine the number of unpaired spins in O_2. How is the problem of the Lewis structure resolved?

22.18. Data on a single crystal of MnF_2 give $\chi_m = 0.1463 \, cm^3 \, mol^{-1}$ at 294.53 K. Determine the effective number of Bohr magnetons in this compound and compare your result with the theoretical value.

22.19. Estimate the spin-only molar susceptibility of $CuSO_4 \cdot 5H_2O$ at 25°C.

22.20. Approximately how large must the magnetic induction B be for the orientational energy of a $S = 1$ system to be comparable to kT at 298 K?

PROBLEMS

Numerical problems

22.1. Suppose an H_2O molecule ($\mu = 1.85$ D) approaches an anion. What is the favourable orientation of the molecule? Calculate the electric field in $(V\,m^{-1})$ experienced by the anion when the water dipole is (a) 1.0 nm, (b) 0.3 nm, (c) 30 nm from the ion.

22.2. An H_2O molecule is aligned by an external electric field of strength $1.0\,kV\,m^{-1}$ and an Ar atom ($\alpha' = 1.66 \times 10^{-24}\,cm^3$) is brought up slowly from one side. At what separation is it energetically favourable for the H_2O molecule to flip over and point towards the approaching Ar atom?

22.3. The relative permittivity of chloroform was measured over a range of temperatures with the following results:

$\theta/°C$	-80	-70	-60	-40	-20	0	20
ε_r	3.1	3.1	7.0	6.5	6.0	5.5	5.0
$\rho/(g\,cm^{-3})$		1.65	1.64	1.64	1.61	1.57	1.53	1.50

The freezing point of chloroform is $-64°C$. Account for these results and calculate the dipole moment and polarizability volume of the molecule.

22.4. The relative permittivities of methanol (melting point $-95°C$) corrected for density variation are given below. What molecular information can be deduced from these values? Take $\rho = 0.791\,g\,cm^{-3}$ at $20°C$.

$\theta/°C$	-185	-170	-150	-140	-110	-80	-50	-20	0	20
ε_r	3.2	3.6	4.0	5.1	67	57	49	42	38	34

22.5. In his classic book *Polar molecules*, Debye reports some early measurements of the polarizability of ammonia. From the selection below, determine the dipole moment and the polarizability volume of the molecule.

T/K	292.2	309.0	333.0	387.0	413.0	446.0
$P_m/(cm^3\,mol^{-1})$	57.57	55.01	51.22	44.99	42.51	39.59

The refractive index of ammonia at 273 K and 100 kPa is 1.000379 (for yellow sodium light). Calculate the molar polarizability of the gas at this temperature and at 293.2 K. Combine the value calculated with the static molar polarizability at 292.2 K and deduce from this information alone the molecular dipole moment.

22.6. Values of the molar polarization of gaseous water at 100 kPa as determined from capacitance measurements are given below as a function of temperature.

T/K	384.3	420.1	444.7	484.1	522.0
$P_m/(cm^3\,mol^{-1})$	57.4	53.5	50.1	46.8	43.1

Calculate the dipole moment of H_2O and its polarizability volume.

Theoretical problems

22.7 Calculate the potential energy of the interaction between two linear quadrupoles when they are (a) collinear, (b) parallel and separated by a distance r.

22.8 Show that in a gas (for which the refractive index is close to 1) the refractive index depends on the pressure as $n_r = 1 + const \times p$, and find the constant of proportionality. Go on to show how to deduce the polarizability volume of a molecule from measurements of the refractive index of a gaseous sample.

22.9. The refractive index of benzene is constant (at 1.51) from 0.4 up to 0.55 GHz (in the microwave region of the spectrum), but then shows a series of oscillations between 1.47 and 1.54. Throughout the same frequency range, methylbenzene shows a higher refractive index (about 1.55), the same oscillations as in benzene, and additional oscillations between 1.52 and 1.56 near 0.4 GHz. Account for these observations.

22.10. Acetic acid vapour contains a proportion of planar, hydrogen-bonded dimers. The relative permittivity of pure liquid acetic acid is 7.14 at 290 K and increases with increasing temperature. Suggest an interpretation of the latter observation. What effect should isothermal dilution have on the relative permittivity of solutions of acetic acid in benzene?

22.11. Show that the mean interaction energy of N atoms of diameter d interacting with a potential energy of the form C_6/R^6 is given by $U = -2\pi N^2 C_6/3Vd^3$, where V is the volume in which the molecules are confined and all effects of clustering are ignored. Hence, find a connection between the van der Waals parameter a and C_6 from $n^2a/V^2 = (\partial U/\partial V)_T$.

22.12. Suppose that the repulsive term in a Lennard-Jones (12, 6)-potential is replaced by an exponential function of the form $\exp(-r/d)$. Sketch the form of the potential energy and locate the distance at which it is a minimum.

22.13. The 'cohesive energy density' is defined as U/V, where U is the mean potential energy of attraction within the sample and V its volume. Show that the cohesive energy density is equal to $-\frac{1}{2}\mathcal{N}^2 \int V(R)\,d\tau$, where $\mathcal{N}$ is the number density of the molecules, $V(R)$ is their attractive potential energy, and where the integration ranges from d to infinity and over all angles. Go on to show that the cohesive energy density of a uniform distribution of molecules that interact by a van der Waals attraction of the form $-C_6/R^6$ is equal to $(2\pi/3)(N_A^2/d^3M^2)\rho^2C_6$, where ρ is the mass density of the solid sample and M is the molar mass of the molecules.

22.14. Consider the collision between a hard-sphere molecule of radius R_1 and mass m_1 and an infinitely massive impenetrable sphere of radius R_2. Plot the scattering angle θ as a function of the impact parameter b.

Carry out the calculation using simple geometrical considerations.

22.15. The dependence of the scattering characteristics of atoms on the energy of the collision can be modelled as follows. We suppose that the two colliding atoms behave as impenetrable spheres, as in Problem 22.14, but that the effective radius of the heavy atom depends on the speed v of the light atom. Suppose its effective radius depends on v as $R_2 \exp(-v/v^*)$ where v^* is a constant. Take $R_1 = \frac{1}{2}R_2$ for simplicity and an impact parameter $b = \frac{1}{2}R_2$, and plot the scattering angle as a function of (a) speed, (b) kinetic energy of approach.

22.16. The magnetizability ξ and the volume and molar magnetic susceptibilities can be calculated from the wavefunctions of molecules. For instance, the magnetizability of a hydrogenic atom is given by the expression $\xi = -(e^2/6m_e)\langle r^2 \rangle$, where $\langle r^2 \rangle$ is the mean value of r^2 in the atom. Calculate ξ and χ_m for the ground state of a hydrogenic atom.

22.17. An NO molecule has thermally accessible electronically excited states. It also has an unpaired electron, and so may be expected to be paramagnetic. However, its ground state is not paramagnetic because the magnetic moment of the orbital motion of the unpaired electron almost exactly cancels the spin magnetic moment. The first excited state (at $121\ cm^{-1}$) is paramagnetic because the orbital magnetic moment adds to, rather than cancels, the spin magnetic moment. The upper state has a magnetic moment of $2\mu_B$. Because the upper state is thermally accessible, the paramagnetic susceptibility of NO shows a pronounced temperature dependence even near room temperature. Calculate the molar paramagnetic susceptibility of NO and plot it as a function of temperature.

23 Macromolecules

Macromolecules exhibit a range of properties and problems that illustrate a wide variety of physical chemical principles. They need to be characterized in terms of their molar mass and their geometrical size and shape. However, as their molar masses are so great, special techniques need to be used to determine them. This chapter describes some of the approaches that are used. As the molecules are so large, the solutions they form depart strongly from ideality, so techniques for accommodating these departures need to be developed. Although the shapes of large biomolecules can be determined by X-ray diffraction, synthetic polymers have less regular shapes in solution, and only their general shape can be inferred: this chapter describes some of the techniques that are used. Another major problem concerns the influences that determine the shapes of the molecules. We consider a range of influences in this chapter, beginning with the structureless random coil and ending with the strictly constraining forces that operate in polypeptides.

There are macromolecules everywhere, inside us and outside us. Some are natural: they include polysaccharides such as cellulose, polypeptides such as enzymes, and nucleic acids such as DNA. Others are synthetic: they include **polymers** such as nylon and polystyrene that are manufactured by stringing together and (in some cases) cross-linking smaller units known as **monomers**. Life in all its forms, from its intrinsic nature to its technological interaction with its environment, is the chemistry of macromolecules.

Although the concepts of physical chemistry apply equally to macromolecules as well as to small molecules, macromolecules do give rise to special questions and problems. These problems include the determination of their sizes, the shapes and the lengths of polymer chains, and the large deviations from ideality of their solutions. We concentrate on these special characteristics here, and treat the relation of the rate of formation

of synthetic polymers to their physical properties in Sections 26.4 and
26.5.

SIZE AND SHAPE

X-ray diffraction (Chapter 21) can reveal the position of almost every
atom, even in highly complex molecules. However, there are several
reasons why other techniques must also be used. In the first place, the
sample might be a mixture of polymers with different chain lengths and
extents of cross-linking, in which case sharp X-ray images are unob-
tainable. Even if all the molecules in the sample are identical, it might
prove impossible to obtain a single crystal. Furthermore, although the
work on enzymes, proteins, and DNA has shown how immensely
stimulating the data can be, the information is incomplete. For instance,
what can be said about the shape of the molecule in its natural
environment, a biological cell? What can be said about the response of its
shape to changes in its environment? Shape and function go hand in
hand, and it is essential to know how the shapes of biological macromolecules, which often carry both acidic and basic groups, respond to the
pH of the medium. It is also useful to be able to follow the collapse of a
macromolecule into a less orderly form: this denaturation is often
accompanied by loss of function, but when it happpens in a controlled
way it is sometimes an essential step in the fulfilment of function, as in
the replication of DNA.

23.1 Mean molecular masses

A complication that we need to address at the outset is the fact that
samples of synthetic polymers and many biomacromolecules consist of
molecules covering a range of molar masses. A pure protein is
monodisperse, meaning that it has a single, definite molar mass. (There
may be small variations, such as one amino acid replacing another,
depending on the source of the sample.) A synthetic polymer is
polydisperse, in the sense that a sample is a mixture of molecules with
various chain lengths and molar masses. The various techniques that are
used to measure molar masses result in different types of mean value. For
example, the mean obtained from the determination of molar mass by
osmometry gives the **number-average molar mass** $\bar{M}_n$ which is the
mean molar mass obtained by weighting each molar mass by the number
of molecules of that molar mass present in the sample:

$$\bar{M}_n = \frac{1}{N} \sum_i N_i M_i \qquad (1)$$

In this definition N_i is the number of molecules with molar mass M_i and
there are N molecules in all. (The number average is also used for the
mean score in a test or height of a population.)

   Other experiments give a different average. For example, we shall see
that viscosity measurements give the **viscosity-average molar mass**
$\bar{M}_v$, light-scattering experiments give the **weight-average molar mass**
$\bar{M}_w$, and sedimentation experiments can be used to obtain the **Z-average**

molar mass $\bar{M}_Z$. Although such averages are often best left as empirical quantites, some may be interpreted in terms of the composition of the sample. Thus, the weight-average molar mass (which would be better called the mass-average molar mass) is the average calculated by weighting the molar masses of the molecules by the mass of each one present in the sample:

$$\bar{M}_w = \frac{1}{m} \sum_i m_i M_i \tag{2a}$$

In this expression, m_i is the total mass of molecules of molar mass M_i and m is the total mass of the sample. Because $m_i = N_i M_i / N_A$, we can also express the mass average as

$$\bar{M}_w = \frac{\sum\limits_i N_i M_i^2}{\sum\limits_i N_i M_i} \tag{2b}$$

Hence, the mass-average molar mass is proportional to the mean square molar mass. Similarly, the Z-average molar mass can be interpreted in terms of the mean cubic molar mass:

$$\bar{M}_Z = \frac{\sum\limits_i N_i M_i^3}{\sum\limits_i N_i M_i^2} \tag{2c}$$

Example 23.1 *Calculating number and mass averages*

Determine the number-average and the weight-average molar masses for a sample of poly(vinyl chloride) from the data shown in the margin.

Method. The relevant equations are eqns 1 and 2. The two averages are obtained by weighting the molar mass within each interval by the number and mass, respectively, of the molecules in each interval. The amounts in each interval are obtained by dividing the mass of the sample in each interval by the average molar mass for that interval. Because number of molecules is proportional to amount of substance (the number of moles), the number-weighted average can be obtained directly from the amounts of substance in each interval.

Molar mass interval/ (kg mol^{-1})	Average molar mass within interval/ (kg mol^{-1})	Mass of sample within interval/g
5–10	7.5	9.6
10–15	12.5	8.7
15–20	17.5	8.9
20–25	22.5	5.6
25–30	27.5	3.1
30–35	32.5	1.7

Answer. The amounts of substance in each interval are as follows:

Interval	5–10	10–15	15–20	20–25	25–30	30–35	
Molar mass/(kg mol^{-1})	7.5	12.5	17.5	22.5	27.5	32.5	
Amount/mmol	1.3	0.70	0.51	0.25	0.11	0.052	Total: 2.9

The number-average molar mass is therefore

$$\bar{M}_n/(\text{kg mol}^{-1}) = \frac{1}{2.9}(1.3 \times 7.5 + 0.70 \times 12.5 + 0.51 \times 17.5$$

$$+ 0.25 \times 22.5 + 0.11 \times 27.5 + 0.052 \times 32.5)$$

$$= 13$$

The weight-average molar mass can be calculated directly from the data after noting that the total mass of the sample is 37.6 g:

$$\bar{M}_w/(\text{kg mol}^{-1}) = \frac{1}{37.6}(9.6 \times 7.5 + 8.7 \times 12.5 + 8.9 \times 17.5$$
$$+ 5.6 \times 22.5 + 3.1 \times 27.5 + 1.7 \times 32.5)$$
$$= 16$$

Comment. Note the different values of the two averages. In this instance, $\bar{M}_w/\bar{M}_n = 1.2$.

Exercise E23.1. Evaluate the Z-average molar mass of the sample.

[19 kg mol^{-1}]

Whereas at first sight it might appear troublesome to have several types of average, the observation that they have different values gives additional information about the range of molar masses in the sample. In the determination of protein molar masses we expect the various averages to be the same because the sample is monodisperse (unless there has been degradation). In samples of synthetic polymers there is normally a range of molar masses and the different averages are expected to yield different values. Typical synthetic materials have $\bar{M}_w/\bar{M}_n \approx 3$. The term 'monodisperse' is conventionally applied to synthetic polymers in which this ratio is less than 1.1. One consequence of a narrow molar mass distribution for synthetic polymers is often a higher crystallinity, and therefore density and melting point. The spread of values is controlled by the choice of catalyst and reaction conditions (Sections 26.4 and 26.5).

23.2 Colligative properties

The classical methods of determining molar mass utilize colligative properties (Section 7.5). For macromolecules, where the number of molecules in solution may be very small even though the mass of the solute may be appreciable, only osmometry is sufficiently sensitive (Fig. 23.1). We shall see that molar masses are often determined more by empirical comparisons than by absolute measurements.

Osmometry

The van't Hoff equation (eqn 7.22) for the osmotic pressure of an ideal solution resembles the perfect gas equation of state:

$$\Pi = [P]RT \tag{3}$$

In this expression, [P] is the molar concentration of the macromolecule P, and Π is the osmotic pressure. The extension of the equation to non-ideal solutions is written like the virial equation for real gases

$$\Pi = [P]RT(1 + B[P] + \ldots) \tag{4}$$

and B is an **osmotic virial coefficient**.

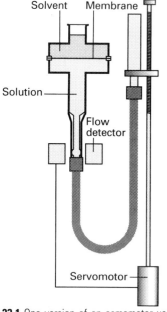

23.1 One version of an osmometer used to determine the molar masses of macromolecules. The pressure on the solution is adjusted until there is no flow through the semipermeable membrane: its value is the osmotic pressure.

The most straightforward thermodynamic justification of the osmotic virial expansion, and an interpretation of B, is obtained from the expression for the osmotic pressure in terms of the chemical potential. We saw in Section 7.5 that

$$-RT \ln x_A = \int_p^{p+\Pi} V_m \, dp$$

where V_m is the molar volume of the pure solvent and x_A is its mole fraction. The more general form of this expression for non-ideal solutions has the solvent activity a_A in place of x_A:

$$-RT \ln a_A = \int_p^{p+\Pi} V_m \, dp$$

Because the integral is equal to ΠV_m if the solvent is incompressible, the osmotic pressure is given by

$$\Pi V_m = -RT \ln a_A$$

If the solution is ideal

$$\ln a_A = \ln x_A = \ln(1 - x_P) \approx -x_P$$

where x_P is the mole fraction of the solute macromolecule and we have supposed that $x_P \ll 1$. This approximation leads to the van't Hoff equation. If the solution is non-ideal, we suppose that $-x_P$ is the first term in a series,[1] and write

$$\ln a_A = -x_P(1 + B'x_P + \ldots)$$

On substitution of this expansion into the expression for Π, eqn 4 is obtained.

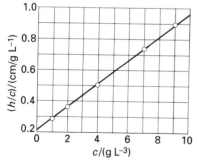

23.2 The plot of h/c against c used for the determination of molar mass (from the intercept) and the osmotic virial coefficient (from the slope).

Because the molar concentration $[P]$ is related to the mass concentration c by $[P] = c/\bar{M}_n$, where $\bar{M}_n$ is the number-average molar mass of P, another version of eqn 4 is

$$\frac{\Pi}{c} = \frac{RT}{\bar{M}_n}\left(1 + \frac{B}{\bar{M}_n}c + \ldots\right) \tag{5}$$

Therefore, by plotting Π/c (or h/c, because h is proportional to Π) against c and extrapolating the data to $c = 0$, the value of $\bar{M}_n$ can be obtained from the intercept at $c = 0$ and B can be obtained from the slope (Fig. 23.2). The procedure was illustrated in Example 7.7.

---

1 The virial expansion in the following equation is not a trivial result, but it is confirmed by the McMillan–Mayer theory of solutions of non-electrolytes. That is, it is not obvious that the expansion of $\ln a$ should have this form. A counterexample that we have already met is in the Debye–Hückel theory (Section 10.2), where we saw that for solutions of electrolytes the first term in the deviation of the activity coefficient from 1 is proportional to the square root of the concentration of the ions.

Macromolecules given strongly non-ideal solutions. This is partly because, being so large, they displace a large quantity of solvent instead of replacing individual solvent molecules with negligible disturbance. In thermodynamic terms, the displacement of solvent molecules implies that the entropy change is especially important when a macromolecule dissolves.[2] Furthermore, its great bulk means that a macromolecule is unable to move freely through the solution as it is excluded from the regions occupied by others. There are also significant contributions to the Gibbs energy from the enthalpy of solution, largely because solvent–solvent interactions are more favorable than the macromolecule–solvent interactions that replace them.

The coefficient B arises largely from the effect of excluded volume. If we imagine a solution of a macromolecule being built by the successive addition of macromolecules to the solvent, each one being excluded by the ones that preceded it, then the value of B turns out to be

$$B = \tfrac{1}{2}N_A v_P \tag{6}$$

where v_P is the excluded volume due to a single molecule.

Example 23.2 *Estimating the volume of polymer molecules*

Use the information in Example 7.7 to estimate the volume of the polymer molecules regarded as impenetrable spheres.

Method. The excluded volume of spherical molecules of volume v is $v_P = 8v$ because the minimum separation of the centres of two spheres is the sum of their radii. We can estimate v_P from the osmotic virial coefficient B by using eqn 6, and can find B from the slope of the graph plotted in Fig. 7.21. To do so we use (as in Example 7.7):

$$\frac{h}{c} = \frac{RT}{\rho g \bar{M}_n}\left(1 + \frac{B}{\bar{M}_n}c\right)$$

Answer. The intercept $RT/\rho g\bar{M}_n$ was found in Example 7.7 to be $0.21\ \text{cm}/(\text{g L}^{-1})$. The slope of the straight line in Fig. 7.21, which is equal to $(RT/\rho g\bar{M}_n) \times B/\bar{M}_n$, is $0.073\ (\text{cm g}^{-1}\text{L})/(\text{g L}^{-1})$. It follows that

$$\frac{\text{slope}}{\text{intercept}} = \frac{B}{\bar{M}_n} = \frac{0.073(\text{cm g}^{-1}\text{L})/(\text{g L}^{-1})}{0.21\ \text{cm g}^{-1}\text{L}} = \frac{0.35}{\text{g L}^{-1}}$$

Therefore

$$B = 0.35/(\text{g L}^{-1}) \times (123 \times 10^3\ \text{g mol}^{-1}) = 43 \times 10^3\ \text{L mol}^{-1}$$

Equation 6 then implies that

$$v_P = \frac{2B}{N_A} = 1.4 \times 10^{-22}\ \text{m}^3$$

---

2 A good starting point for a discussion of the entropy of mixing is a generalization of the expression given in Section 7.2 for the ideal entropy of mixing: instead of using eqn 7.8, which depends on the mole fractions of the components, it turns out to be better to use

$$\Delta_{mix}S = -nR(x_A \ln v_A + x_B \ln v_B)$$

where v_J is the 'volume fraction' of component J. This equation reduces to the mole fraction expression when the molecular volumes of the two components are the same, but is a better starting point when they are markedly different.

From this value of v_P it follows that the molecular volume is approximately 1.8×10^4 nm^3.

Comment. The radius of the molecule is approximately 16 nm.

Exercise E23.2. Another sample in the same solvent resulted in the following heights of solution at the same temperature: 0.22, 0.53, 1.39, 3.32, 5.02 cm. Calculate the molar mass and the molecular volume of the solute. [14 kg mol^{-1}, 9×10^{-23} m^3]

In broad terms, the excluded volume contributes to the excess entropy of solution (the entropy change in excess of the ideal value, Section 7.4), and the attractions and repulsions between macromolecules contribute to the excess enthalpy. For most solute/solvent systems there is a unique temperature (which is not always experimentally attainable) at which these effects cancel and the solution is virtually ideal. This temperature (the analogue of the Boyle temperature for real gases) is called the **Flory theta temperature** θ. At the Flory theta temperature, the osmotic virial coefficient B is zero. As an example, for polystyrene in cyclohexane $\theta \approx 306$ K, the exact value depending on the average molar mass of the polymer. A solution at its Flory theta temperature is called a **θ solution**. Because a θ solution behaves nearly ideally, its thermodynamic and structural properties are easier to describe even though the concentration is not low.

Vapour-phase osmometry

The technique of **vapour-phase osmometry** (Fig. 23.3) is used to study polymers that have molar masses too low to be measured by membrane osmometry.

In the technique, a droplet of solution is placed on one thermistor (a temperature probe) and a droplet of pure solvent is placed on another thermistor. The two droplets are surrounded by an atmosphere of solvent vapour. Because the vapour pressure of a solvent in a solution is lower than when the solvent is pure, the net rate of condensation of solvent on to a droplet of solution is greater than the rate of condensation on to a droplet of pure solvent. It follows that more heat is liberated in the solution droplet, and the rise in temperature is greater there than in the droplet of pure solvent. The difference in temperature is measured for a series of concentrations and extrapolated to zero concentration. After calibration by using samples of known molar mass, the molar mass of the sample can be inferred from the temperature difference between the two thermistors.

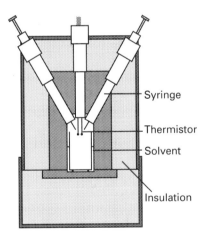

Syringe

Thermistor

Solvent

Insulation

23.3 A vapour-phase osmometer. The syringes introduce droplets of solvent and solution on to the thermistors, and the differences in temperature (arising from the different rates of condensation) are noted.

Polyelectrolytes and dialysis

Some polymers are strings of acid groups, as in poly(acrylic acid) $-(CH_2CHCOOH)_n-$, or strings of bases, as in nylon $-[NH(CH_2)_6-NHCO(CH_2)_4CO]_n-$; proteins have both acid and base groups. Macromolecules may therefore be **polyelectrolytes** and, depending on their state of ionization, **polyanions** or **polycations**. A macromolecule with mixed cation and anion character is known as a **polyampholyte**.

One consequence of dealing with polyelectrolytes is that it is necessary to know the extent of ionization before osmotic data can be interpreted. For example, suppose the sodium salt of a polyelectrolyte is present in solution as νNa^+ ions and a single polyanion $P^{\nu-}$, then it gives rise to $\nu + 1$ particles for each formula unit of salt that dissolves. If we guess that $\nu = 1$ when in fact $\nu = 10$, then the estimate of the molar mass will be wrong by an order of magnitude. We can find a way out of this difficulty by considering another feature of charged macromolecules.

Suppose the solution of the polyelectrolyte $Na_\nu P$ also contains added NaCl, and that it is in contact through a semipermeable membrane (such as a cell wall) with another salt solution. Furthermore, suppose the membrane is permeable to the solvent and to the salt ions, but not to the polyanion. This arrangement is one that actually occurs in living systems, where osmosis is an important feature of cell operation. The presence of the salt affects the osmotic pressure because the anions and cations cannot migrate through the membrane to an arbitrary extent. Apart from small imbalances of charge close to the membrane that give rise to transmembrane potentials, electrical neutrality must be preserved in the bulk on both sides of the membrane: if an anion migrates, a cation must accompany it.

The presence of a high concentration of added salt to each side of a semipermeable membrane ensures that the effective difference in concentrations is due solely to the presence of the polyanion P on one side of the membrane, for the number of cations the polymer provides is insignificant in comparison with the number supplied by the additional salt. Hence, under such circumstances we can expect the osmotic pressure to be given by

$$\Pi = RT[P]$$

a result independent of the value of ν. Therefore, if we measure the osmotic pressure in the presence of high concentrations of salt, the molar mass may be obtained unambiguously.

JUSTIFICATION

Suppose that $Na_\nu P$ is at a molar concentration [P] on one side of the membrane, and that NaCl is added to each side. On the left (L) there are $P^{\nu-}$, Na^+, and Cl^- ions. On the right (R) there are Na^+ and Cl^- ions. The condition for equilibrium is that the chemical potential of NaCl should be the same on both sides of the membrane, so a net flow of Na^+ and Cl^- ions occurs until $\mu_L(NaCl) = \mu_R(NaCl)$. This equality occurs when

$$\mu^{\ominus}(NaCl) + RT \ln a_L(Na^+) + RT \ln a_L(Cl^-)$$
$$= \mu^{\ominus}(NaCl) + RT \ln a_R(Na^+) + RT \ln a_R(Cl^-)$$

or

$$RT \ln a_L(Na^+)a_L(Cl^-) = RT \ln a_R(Na^+)a_R(Cl^-)$$

If we ignore activity coefficients, the two expressions are equal when

$$[Na^+]_L[Cl^-]_L = [Na^+]_R[Cl^-]_R$$

As the Na^+ ions are supplied by the polyelectrolyte as well as the added salt, the conditions for bulk electrical neutrality are

$$[Na^+]_L = [Cl^-]_L + v[P]$$
$$[Na^+]_R = [Cl^-]_R$$

We can now combine these three conditions to obtain expressions for the differences in ion concentrations across the membrane:

$$[Na^+]_L - [Na^+]_R = \frac{v[P][Na^+]_L}{[Na^+]_L + [Na^+]_R} = \frac{v[P][Na^+]_L}{2[Cl^-] + v[P]}$$

$$[Cl^-]_L - [Cl^-]_R = -\frac{v[P][Cl^-]_L}{[Cl^-]_L + [Cl^-]_R} = -\frac{v[P][Cl^-]_L}{2[Cl^-]}$$

where

$$[Cl^-] = \tfrac{1}{2}([Cl^-]_L + [Cl^-]_R)$$

The quantity $[Cl^-]$ is the average concentration of Cl^- ions on each side of the membrane.

The final step is to note that the osmotic pressure depends on the difference in the numbers of solute particles on each side of the membrane. That being so, the van't Hoff equation

$$\Pi = RT[\text{Solute}]$$

becomes

$$\Pi = RT\{([P] + [Na^+]_L + [Cl^-]_L) - ([Na^+]_R + [Cl^-]_R)\}$$

$$= RT[P](1 + B[P]), \quad \text{with } B = \frac{v^2[Cl^-]_R}{4[Cl^-]^2 + 2v[Cl^-][P]}$$

When the concentration of added salt is so great that $[Cl^-]_L$ and $[Cl^-]_R$ are both much larger than $[P]$, it follows that $B[P] \ll 1$ and this expression reduces to

$$\Pi = RT[P]$$

as given in the text.

A second point arises from the effect of added salt. There is often interest in the extent to which ions are bound to macromolecules, especially when a membrane (such as a cell wall) separates two regions. The equations

$$[Na^+]_L - [Na^+]_R = \frac{v[P][Na^+]_L}{2[Cl^-] + v[P]}$$

$$[Cl^-]_L - [Cl^-]_R = -\frac{v[P][Cl^-]_L}{2[Cl^-]} \tag{7}$$

which are derived in the preceding *Justification* show that cations will dominate the anions in the compartment containing the polyanion (because the concentration difference is positive for Na^+ and negative for Cl^-) as a result of the equilibrium and electroneutrality conditions. The equilibrium distribution of ions in two compartments connected by a semipermeable membrane, in one of which there is a polyelectrolyte, is called a **Donnan equilibrium**.

Example 23.3 *Analysing a Donnan equilibrium*

Two equal volumes of 0.200 M NaCl(aq) are separated by a membrane. A macromolecule of molar mass 55 kg mol^{-1}, which cannot pass through the membrane, is added as its sodium salt Na_6P in a concentration of 50 g L^{-1} to the left-hand compartment. What are the equilibrium concentrations of Na^+ and Cl^- in each compartment?

Method. We use eqn 7 to calculate the concentration differences and calculate their sum as

$$[Na^+]_L + [Na^+]_R = [Cl^-]_L + [Cl^-]_R + v[P] = 2[Cl^-] + v[P]$$

Then use $[Cl^-] = 0.200 \text{ mol L}^{-1}$.

Answer. Because $[P] = 9.1 \times 10^{-4} \text{ mol L}^{-1}$, eqn 7 gives

$$[Na^+]_L - [Na^+]_R$$
$$= \frac{6 \times (9.1 \times 10^{-4} \text{ mol L}^{-1}) \times [Na^+]_L}{2 \times (0.200 \text{ mol L}^{-1}) + 6 \times (9.1 \times 10^{-4} \text{ mol L}^{-1})}$$

$$[Na^+]_L + [Na^+]_R = 2 \times (0.200 \text{ mol L}^{-1})$$
$$+ 6 \times (9.1 \times 10^{-4} \text{ mol L}^{-1})$$
$$= 0.405 \text{ mol L}^{-1}$$

Solving these equations gives

$$[Na^+]_L = 0.204 \text{ mol L}^{-1} \qquad [Na^+]_R = 0.201 \text{ mol L}^{-1}$$

Then

$$[Cl^-]_R = [Na^+]_R = 0.201 \text{ mol L}^{-1}$$
$$[Cl^-]_L = [Na^+]_L - 6[P] = 0.199 \text{ mol L}^{-1}$$

Comment. Note how the Na^+ accumulates slightly in the compartment containing the macromolecule.

Exercise E23.3. Repeat the calculation for 0.300 M NaCl(aq), a polyelectrolyte $Na_{10}P$ of molar mass 33 kg mol^{-1} at a mass concentration of 50.0 g L^{-1}.
[Na^+: 0.311 mol L^{-1}, 0.304 mol L^{-1}]

23.3 Sedimentation

In a gravitational field, heavy particles settle towards the foot of a column of solution by the process called **sedimentation**. The rate of sedimentation depends on the strength of the field and on the masses and shapes of

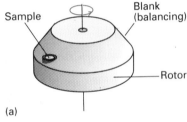

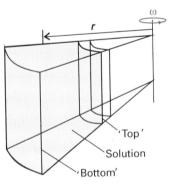

23.4 (a) An ultracentrifuge head. The sample on one side is balanced by a blank diametrically opposite. (b) Detail of the sample cavity: the 'top' surface is the inner surface, and the centrifugal force causes sedimentation towards the outer surface; a particle at a radius r experiences a force of magnitude $mr\omega^2$

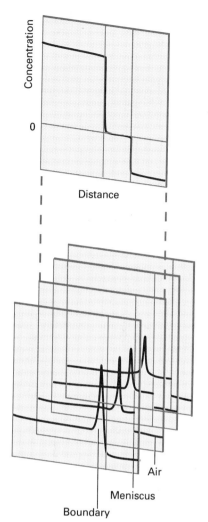

23.5 A Schlieren photograph indicates regions where the refractive index is changing. This set of diagrams, corresponding to a series of times, shows the sedimentation of the solute; one has been interpreted in terms of the concentration profile through the cell.

the particles. Spherical molecules (and compact molecules in general) sediment faster than rod-like or extended molecules. For example, DNA helices sediment much faster when they are denatured to a random coil, and so sedimentation rates can be used to study denaturation. When the sample is at equilibrium, the particles are dispersed over a range of heights in accord with the Boltzmann distribution (because the gravitational field competes with the stirring effect of thermal motion). The spread of heights depends on the masses of the molecules, so the equilibrium distribution is another way of determining molar mass.

Sedimentation is normally very slow, but it can be accelerated by replacing the gravitational field by a centrifugal field. The latter can be achieved in an **ultracentrifuge,** which is essentially a cylinder that can be rotated at high speed about its axis with a sample in a cell near its periphery (Fig. 23.4). Modern ultracentrifuges can produce accelerations equivalent to about 10^5 that of gravity ('$10^5 g$'). Initially the sample is uniform, but the 'top' (innermost) boundary of the solute moves outwards as sedimentation proceeds. The rate at which the boundary moves can be monitored by making use of the effect of the concentration on the refractive index of the sample. Since there is a sharp change of refractive index between the solution and the solvent left behind by the receding solute, the sample behaves like a prism and bends the light passing through it. The **Schlieren optical system** (which is also used to study air flow in wind tunnels and shock tubes) turns a refractive index gradient into an intensity difference (Fig. 23.5). Alternatively, in the **interference technique,** the concentration profile is monitored through the effect of refractive index on the interference on two beams of light, one coming through the sample and the other through a blank (Fig. 23.6).

The rate of sedimentation

A solute particle of mass m has an effective mass $m_{eff} = bm$ on account of the buoyancy of the medium, with

$$b = 1 - \rho v_s \tag{8}$$

where ρ is the solution density, v_s is the solute's specific volume (its volume per unit mass), and $\rho m v_s$ is the mass of solvent displaced by the solute. (More precisely, v_s is the partial specific volume, in the sense described in Section 7.1.) The solute particles at a distance r from the axis of a rotor spinning at an angular velocity ω experience a centrifugal force of magnitude $m_{eff} r \omega^2$. The acceleration outwards is countered by a frictional force proportional to the speed s of the particles through the medium. This force is written fs, where f is the **frictional coefficient.** The particles therefore adopt a **drift speed,** a steady speed through the medium, which is found by equating the two forces $m_{eff} r \omega^2$ and fs. The forces are equal when

$$s = \frac{m_{eff} r \omega^2}{f} = \frac{bmr\omega^2}{f}$$

The drift speed depends on the angular velocity and the radius, and it is

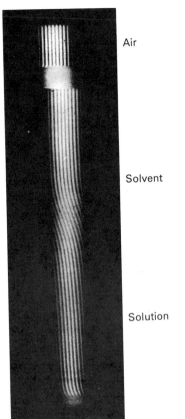

23.6 An interference photograph also shows the concentration profile, but in a different manner. The regions of air, solvent, and solution have been marked. (Provided by Professor D. Freifelder, from his *Physical biochemistry*, W. H. Freeman & Co. (1982).)

convenient to focus on the **sedimentation constant** S, which is defined as

$$S = \frac{s}{r\omega^2} \qquad \text{(9a)}$$

Then, because the average mass of an individual molecule is related to the average molar mass $\bar{M}_n$ through $m = \bar{M}_n/N_A$,

$$S = \frac{b\bar{M}_n}{fN_A} \qquad \text{(9b)}$$

Example 23.4 *Determining a sedimentation constant*

The sedimentation of bovine serum albumin (BSA) was monitored at 25°C. The initial radius of the solute surface was 5.50 cm and during centrifugation at 56 850 r.p.m. it receded as follows:

t/s	0	500	1000	2000	3000	4000	5000
r/cm	5.50	5.55	5.60	5.70	5.80	5.91	6.01

Calculate the sedimentation coefficient.

Method. Equation 9a can be interpreted as a differential equation for $s = dr/dt$ in terms of r, so we need to integrate it to obtain a formula for r in terms of t. The integrated expression will be an expression for r as a function of t, and will suggest how to plot the data and obtain from it the sedimentation constant.

Answer. Equation 9a may be written

$$\frac{dr}{dt} = r\omega^2 S$$

This equation integrates to

$$\ln \frac{r}{r_0} = \omega^2 St$$

It follows that a plot of $\ln(r/r_0)$ against t should be a straight line of slope $\omega^2 S$. Use $\omega = 2\pi\nu$, where ν is in cycles/s, and draw up the following table:

t/s	0	500	1000	2000	3000	4000	5000
$100 \ln(r/r_0)$	0	0.900	1.80	3.57	5.31	7.19	8.87

The straight line graph has slope 1.79×10^{-5}, so $\omega^2 S = 1.79 \times 10^{-5}\,s^{-1}$. Because $\omega = 2\pi \times (56\,850/60)\,s^{-1} = 5.95 \times 10^3\,s^{-1}$, it follows that $S = 5.1 \times 10^{-13}\,s$.

Comment. We develop this result below. The unit $10^{-13}\,s$ is sometimes called a 'svedberg' and denoted Sv; in this case $S = 5.1$ Sv. Accurate results are obtained by extrapolating to zero concentration.

Exercise E23.4 Calculate the sedimentation constant given the following data (the other conditions being the same as above):

t/s	0	500	1000	2000	3000	4000	5000
r/cm	5.65	5.68	5.71	5.77	5.84	5.90	5.97

[3.08 Sv]

Table 23.1* Frictional coefficients and molecular geometry†

a/b	Prolate	Oblate
2	1.04	1.04
4	1.18	1.17
6	1.31	1.28
8	1.43	1.37
10	1.54	1.46

* More values and analytical expressions are given in the Data section at the end of this volume.

† Entries are the ratio f/f_0 where $f_0 = 6\pi\eta c$ with $c = (ab^2)^{\frac{1}{3}}$ for prolate ellipsoids and $c = (a^2b)^{\frac{1}{3}}$ for oblate ellipsoids; $2a$ is the major axis and $2b$ is the minor axis.

To make progress we need to know the frictional constant f. For a spherical particle of radius a in a solvent of viscosity η and, for solute molecules that are not small compared with the solvent molecules, f is given by **Stokes' relation**

$$f = 6\pi a\eta \tag{10}$$

Therefore, for spherical molecules,

$$S = \frac{b\bar{M}_n}{6\pi a\eta N_A} \tag{11}$$

and S may be used to determine either $\bar{M}_n$ or a. If the molecules are not spherical we use the appropriate value of f given in Table 23.1. As always when dealing with macromolecules, the measurements must be carried out at a series of concentrations and then extrapolated to zero concentration in order to avoid the complications that arise from the interference between bulky molecules.

At this stage it appears that we need to know the molecular radius a (and in general the frictional coefficient f) to obtain the molar mass from the value of S. Fortunately, this requirement can be avoided by drawing on the **Stokes–Einstein relation** between f and the **diffusion coefficient** D:

$$f = \frac{kT}{D} \tag{12}$$

The diffusion coefficient is a measure of the rate at which molecules spread down a concentration gradient (it is treated in detail in Sections 24.10 and 24.11). It can be measured by observing the rate at which a concentration boundary spreads on the rate at which a more concentrated solution diffuses into a less concentrated one. Some typical values are given in Table 23.2. The diffusion coefficient may also be measured using light scattering, as we shall see later. Then, it follows from eqns 9b and 12 that

$$\bar{M}_n = \frac{SRT}{bD} \tag{13}$$

This result is independent of the shape of the solute molecules. It follows that we can find the molar mass by combining measurements of sedimentation and diffusion rates (for S and D, respectively).

Table 23.2* Diffusion coefficients in water at 20°C

	$M/(\text{kg mol}^{-1})$	$D/(\text{m}^2\,\text{s}^{-1})$
Sucrose	0.342	4.59×10^{-10}
Lysozyme	14.1	1.04×10^{-10}
Haemoglobin	68	6.9×10^{-11}
Collagen	345	6.9×10^{-12}

* More values are given in the Data section.

Example 23.5 *Interpreting a sedimentation experiment*

Use the result from Example 23.4 in combination with the data below to calculate the molar mass of BSA. Estimate its axial ratio on the basis that it is a prolate ellipsoid. Take $D = 6.97 \times 10^{-11}\,\text{m}^2\,\text{s}^{-1}$, $\rho = 1.0024\,\text{g cm}^{-3}$, $v_s = 0.734\,\text{cm}^3\,\text{g}^{-1}$, $\eta \times 10^{-3} = 0.890\,\text{kg m}^{-1}\,\text{s}^{-1}$, and the temperature as 25°C.

Method. First, we use eqn 13 (with $b = 1 - \rho v_s$) to find $\bar{M}_n$. Then, to find the axial ratio, we must find f/f_0 and use Table 23.1; f is given by eqn 12.

The quantity $f_0 = 6\pi c\eta$ is based on the assumption that the molecule is a sphere of radius c; that radius is obtained from v_s by using $v = (4\pi/3)c^3$, with v related to v_s by $v = v_s\bar{M}_n/N_A$.

Answer. Substitution of the data into eqn 13 with $b = 1 - \rho v_s$ gives $\bar{M}_n = 69$ kg mol^{-1}. Then, using $f = kT/D = 5.91 \times 10^{-11}$ kg s^{-1},

$$v = v_s \times \frac{\bar{M}_n}{N_A} = 0.734 \text{ cm}^3 \text{ g}^{-1} \times \frac{69 \times 10^3 \text{ g mol}^{-1}}{6.022 \times 10^{23} \text{ mol}^{-1}}$$

$$= 8.4 \times 10^{-20} \text{ cm}^3, \quad \text{or } 8.4 \times 10^{-26} \text{ m}^3$$

$$c = \left(\frac{3v}{4\pi}\right)^{\frac{1}{3}} = 2.7 \times 10^{-9} \text{ m}$$

$$f_0 = 6\pi c\eta = 4.5 \times 10^{-11} \text{ kg s}^{-1}$$

Therefore, $f/f_0 = 1.3$. Reference to Table 23.1 shows that this ratio corresponds to an axial ratio of about 6.

Comment. The ellipsoid is like a small cigar, six times longer than it is broad. Extrapolation to zero concentration gives an axial ratio of 4.4.

Exercise E23.5. Use the result of Exercise E23.4 to find the molar mass and the axial ratio of that macromolecule. Use the following additional data: $D = 5.89 \times 10^{-11}$ m^2 s^{-1}, $\rho = 1.0024$ g cm^{-3}, $v_s = 0.728$ cm^3 g^{-1}, $\eta = 0.890 \times 10^{-3}$ kg m^{-1} s^{-1}, and the same temperature.

[48 kg mol^{-1}, $f/f_0 = 1.7$, corresponding to an axial ratio of more than 10]

Sedimentation equilibria

The difficulty with using sedimentation rates to measure molar masses lies in the inaccuracies inherent in the determination of diffusion coefficients, such as the blurring of the boundary by convection currents. This problem can be avoided by allowing the system to reach equilibrium, for the transport property D is then no longer relevant.

Because the number of solute molecules with any given potential energy E is proportional to $e^{-E/kT}$, the ratio of the concentrations at different heights (or radii in a centrifuge) can be used to determine their masses. The kinetic energy of a molecule of mass m_{eff} arising from its circulation around the axis of the centrifuge is

$$E = \tfrac{1}{2}m_{eff}r^2\omega^2$$

when it is travelling in a circle of radius r with angular velocity ω. Therefore, the ratio of the concentrations at radii r_1 and r_2 is

$$\frac{c_1}{c_2} = \frac{N_1}{N_2} = \frac{e^{-E_1/kT}}{e^{-E_2/kT}} = e^{-bm\omega^2(r_2^2 - r_1^2)/2kT}$$

so

$$\bar{M}_w = \frac{2RT}{(r_2^2 - r_1^2)b\omega^2}\ln\frac{c_2}{c_1} \tag{14}$$

This expression acknowledges that a detailed analysis shows that the weight-averaged molar mass is given by this technique; an alternative treatment of the data leads to the *Z*-average molar mass. The centrifuge is run more slowly in this technique than in the sedimentation rate method in order to avoid having all the solute pressed in a thin film against the bottom of the cell. At these slower speeds, several days may be needed for equilibrium to be reached.

Electrophoresis

Many macromolecules are charged and move in an electric field: this motion is called **electrophoresis**. In **gel electrophoresis** the migration takes place through a cross-linked polyacrylamide gel. The mobilities of macromolecules depend on their masses and their shapes, and a constant drift speed is reached when the driving force $ez\mathscr{E}$ (where z is the charge number and $\mathscr{E}$ is the field strength) is matched by the viscous retarding force fs.

One way of avoiding the problem of knowing neither the hydrodynamic shape of the molecules nor their charge is to denature them in a controlled way. Sodium dodecylsulfate has been found to be very useful in this respect. It denatures proteins into rods by forming a complex with them, so all proteins, whatever their initial shapes, are made rod-like. Moreover, most proteins have been found to bind a constant amount of the anion per unit mass, so the charge per protein molecule is well regulated. The molar mass of the protein is determined by comparing its mobility in its rod-like complexed form with standard samples.

The charge on a protein depends on the pH, and hence the rate of migration varies with pH. This apparent difficulty can be used to distinguish proteins. For example, at a given pH the rate of migration of haemoglobin from people with sickle-cell anaemia is different from that in a sample taken from people without the disease. This difference is an indication that there are different charges on the protein molecule, which in turn is ascribed to the presence of a different amino acid residue in the polypeptide chain.

Gel permeation chromatography

All the techniques discussed so far have certain drawbacks, including the time needed to obtain data, and the often awkward interpretation of that data. Much of this difficulty has been swept away by a technique that makes use of beads of porous polymeric material about 0.1 mm in diameter that capture molecules selectively, according to their size. In the technique of **gel permeation chromatography** (GPC), which is now the most widely used technique for molar mass determinations of polymers, a solution of the polymer sample is filtered through a column. The small molecules, which can permeate into the porous structure of the gel, require a long **elution time**, or time to pass through a particular length of column, whereas the larger ones, which are not captured, pass through rapidly. The average molar mass of a macromolecule may therefore be determined by observing its elution time in a column calibrated against standard samples.

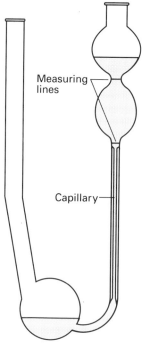

23.7 An Ostwald viscometer. The viscosity is measured by noting the time required for the liquid to drain between the two marks.

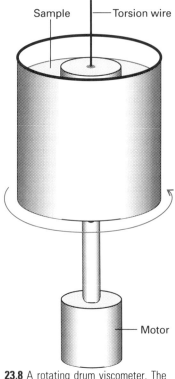

23.8 A rotating drum viscometer. The torque on the inner drum is observed when the outer container is rotated.

The range of molar masses that can be determined by GPC can be altered by selecting columns made from polymers with different degrees of cross-linking and of different materials. The elution time depends on shape in a complicated way and the technique works best if the molecules are spherical. Polystyrene gels are used for investigations of non-polar polymers in non-polar solvents, and porous glass gels are used for more polar systems. Because elution is performed under pressure, molar mass determinations may be completed within a few minutes, in striking contrast to the time required for more classical techniques. Moreover, only a few milligrams of material are needed for highly reliable measurements.

23.4 Viscosity

The presence of a macromolecular solute increases the viscosity of a solution. The effect is large even at low concentrations, because big molecules affect the fluid flow over a long range. At low concentrations the viscosity of the solution η is related to the viscosity of the pure solvent η^*, by

$$\eta = \eta^*(1 + [\eta]c + \ldots) \tag{15}$$

The **intrinsic viscosity** $[\eta]$ is the analogue of a virial coefficient (and has the dimensions of 1/concentration). It follows from eqn 15 that

$$[\eta] = \lim_{c \to 0} \left(\frac{\eta/\eta^* - 1}{c} \right) \tag{16}$$

Viscosities are measured in several ways. In the **Ostwald viscometer** shown in Fig. 23.7, the time taken for the solution to flow through the capillary is noted, and compared with a standard sample. The method is well suited to the determination of $[\eta]$ because the ratio of the viscosities of the solution and the pure solvent is proportional to the drainage times t and t^* after correcting for different densities ρ and ρ^*:

$$\frac{\eta}{\eta^*} = \frac{t}{t^*} \times \frac{\rho}{\rho^*}$$

(In practice, the two densities are only rarely significantly different.) This ratio can be used directly in eqn 16. Viscometers in the form of rotating concentric cylinders are also used (Fig. 23.8), and the torque on the inner cylinder is monitored while the outer one is rotated. Such **rotating drum viscometers** have the advantage over the Ostwald type that the shear gradient between the cylinders is simpler than in the capillary, and effects of the kind we shall shortly describe can be studied more easily.

There are many complications in the interpretation of viscosity measurements. Much (but not all) the work is based on empirical observations, and the determination of molar mass is usually based on comparisons with standard nearly monodisperse samples. Some regularities are observed that help in the determination. For example, it is found that θ solutions of macromolecules often fit the expression

$$[\eta] = K\bar{M}_v^a \tag{17}$$

Table 23.3* Intrinsic viscosity

Macromolecule	Solvent	$\theta/°C$	$K/(cm^3\,g^{-1})$	a
Polystyrene	Benzene	25	9.5×10^{-3}	0.74
Polyisobutylene	Benzene	23	8.3×10^{-2}	0.50
Various proteins	Guanidine hydrochloride $+HSCH_2CH_2OH$		7.2×10^{-3}	0.66

* More values are given in the Data section.

where K and a are constants that depend on the solvent and type of macromolecule (Table 23.3); the viscosity-average molar mass appears in this expression. As an example, solutions of poly(γ-benzyl-L-glutamate) in its rod-like form have an intrinsic viscosity four times greater than when it is denatured and the rods collapse into random coils. Conversely, solutions of natural ribonuclease are less viscous than the denatured form: this observation suggests that the natural protein is more compact than when it is denatured.

Example 23.6 *Using intrinsic viscosity to measure molar mass*

The viscosities of a series of solutions of polystyrene in toluene were measured at 25°C with the following results:

$c/(g\,L^{-1})$	0	2.0	4.0	6.0	8.0	10.0
$\eta/(10^{-4}\,kg\,m^{-1}\,s^{-1})$	5.58	6.15	7.74	6.35	7.98	8.64

Calculate the intrinsic viscosity and estimate the molar mass of the polymer by using eqn 17 with $K = 3.80 \times 10^{-5}\,L\,g^{-1}$ and $a = 0.63$.

Method. The intrinsic viscosity is defined in eqn 16; therefore, form this ratio at the series of data points and extrapolate to $c = 0$.

Answer. We draw up the following table:

$c/(g\,L^{-1})$	0	2.0	4.0	6.0	8.0	10.0
η/η^*	1	1.102	1.208	1.317	1.430	1.549
$100[(\eta/\eta^*) - 1]/(c/g\,L^{-1})$		5.11	5.20	5.28	5.38	5.49

The points are plotted in Fig. 23.9. The extrapolated intercept at $c = 0$ is 0.0504, so $[\eta] = 0.0504\,L\,g^{-1}$. Therefore,

$$\bar{M}_v = \left(\frac{[\eta]}{K}\right)^{1/a} = 90 \times 10^3\,g\,mol^{-1}$$

Comment. When $\eta \approx \eta^*$,

$$\ln\frac{\eta}{\eta^*} = \ln\left(1 + \frac{\eta - \eta^*}{\eta^*}\right) \approx \frac{\eta - \eta^*}{\eta^*} = \frac{\eta}{\eta^*} - 1$$

This relation is exact in the limit that η coincides with η^*, which is true when $c = 0$. Hence, $[\eta]$ can also be defined as the limit of $(1/c)\ln(\eta/\eta^*)$ as $c \to 0$. The intercept is identified more precisely by plotting both functions.

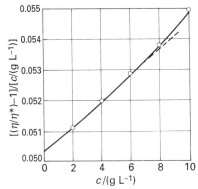

23.9 The plot used for the determination of intrinsic viscosity, which is taken from the intercept at $c = 0$; see Example 23.6.

Exercise E23.6. Evaluate the viscosity-average molar mass by using the second plotting technique.   [90 kg mol^{-1}]

One complication in viscosity measurements is that in some cases it is found that the fluid is non-Newtonian in the sense that its viscosity changes as the rate of flow increases. A decrease in viscosity with increasing rate of flow indicates the presence of long rod-like molecules that are orientated by the flow and hence slide past each other more freely. In some somewhat rare cases the stresses set up by the flow are so great that long molecules are broken up, with further consequences for the viscosity.

23.5 Light scattering

When electromagnetic radiation falls on an object, it forces the electron distribution in the object to oscillate and hence to radiate. If the medium is perfectly homogeneous (for example, a perfect crystal or a completely random collection of molecules that is homogeneous on the scale of the wavelength of the radiation, like a sample of water), all the secondary waves interfere destructively except in the original propagation direction. Therefore, an observer sees the beam only when looking towards the source along the initial direction. If the medium is inhomogeneous (an imperfect crystal or a solution containing foreign bodies, such as macromolecules in a solvent or smoke in air), radiation is scattered into other directions too. A familiar example is light scattered by specks of dust in a sunbeam (and in advertisers' photographs of laser beams).

Scattering by particles with diameters much smaller than the wavelength of the incident radiation is called **Rayleigh scattering**. The intensity of Rayleigh scattered radiation depends on $1/\lambda^4$, and shorter wavelength radiation is scattered more intensely than longer. The blue of the sky arises from the more intense scattering of the blue component of white sunlight by the molecules of the atmosphere. The intensity also depends on the scattering angle θ, and is proportional to $1 + \cos^2\theta$ when the light is unpolarized and to $\cos^2\theta$ when it is polarized (Fig. 23.10). In practice it turns out to be easier to make observations in a non-forward direction. The intensity also depends on the strength of the interaction of

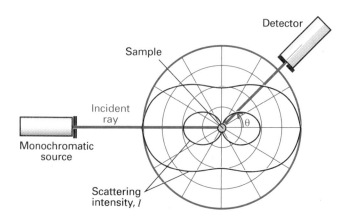

23.10 Rayleigh scattering from a sample of point-like particles follows a $1 + \cos^2\theta$ dependence (outer trace on the polar plot) when unpolarized light is used, but a $\cos^2\theta$ dependence (inner trace) when plane polarized light is used (e.g. when the source is a laser).

the light with the molecules, the interaction being large when their polarizability is large.

When these remarks are combined into a quantitative theory, it turns out that the scattering intensity I at the angle θ is

$$I = AI_0\bar{M}_w g[\text{P}] \qquad g = \begin{cases} 1 + \cos^2\theta & \text{for unpolarized light} \\ \cos^2\theta & \text{for polarized light} \end{cases} \qquad (18)$$

In this expression, I_0 is the incident intensity, $[\text{P}]$ is the molar concentration of the solute, $\bar{M}_w$ its weight-average molar mass, and A is a constant that depends on the refractive index of the solution, the wavelength, and the distance of the detector from the sample. Equation 18 is an 'ideal' result in the sense that it ignores the complications that arise from the interactions between solute particles, and in an actual experiment it is important to extrapolate to zero concentration.

Turbidity

Because a solute scatters light away from the forward direction, the transmitted intensity is reduced. The intensity in the forward direction is given by a type of Beer–Lambert law (Section 16.2): if the incident intensity is I_0, then the intensity that survives after passing through a solution of length l is

$$I = I_0 e^{-\tau l} \qquad (19)$$

where τ is the **turbidity**. For macromolecules of moderate size the turbidity is related to the mass concentration c by

$$\frac{Hc}{\tau} = \frac{1}{\bar{M}_w}(1 + 2Bc + \ldots), \qquad \text{where } H \propto \frac{\mathrm{d}n}{\mathrm{d}c}$$

and n is the refractive index of the solution. Therefore, by plotting Hc/τ against c, the intercept gives the weight-average molar mass. Typical orders of magnitude of τ are $10^{-5}\,\text{cm}^{-1}$ for pure transparent liquids, $10^{-3}\,\text{cm}^{-1}$ for polymers at 1 per cent concentration, and $10\,\text{cm}^{-1}$ for milk. For the first of these values, the sample would need to be 1 km long before the intensity drops to $1/e$ of its initial value as a result of scattering alone (absorption in fact dominates). The last value, loosely interpreted, means that we can see about 1 mm into a glass of milk.

Large-particle scattering

When the wavelength of the incident radiation is comparable to the size of the scattering particles, scattering may occur at different sites of the same molecule (Fig. 23.11) and the interference between different rays is important.[3] As a result, the scattering intensity is distorted from the form

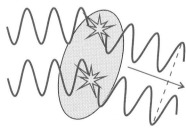

23.11 When light scatters from different regions of the same molecule the emergent waves interfere and modify the Rayleigh scattering intensity distribution. The modification can be interpreted in terms of the shapes of the macromolecules.

3 The effect accounts for the appearance of clouds, which, although we see them by scattered light, look white, not blue like the sky. The reason is that the water molecules group together into droplets of a size comparable to the wavelength of light, and scatter cooperatively. Although blue light scatters more strongly, more molecules can contribute cooperatively when the wavelength is longer (as for red light), so the net result is uniform scattering for all wavelengths: white light scatters as white light. This paper looks white for the same reason. Cigarette smoke is blue before it is inhaled, but brownish after it is exhaled because the particles aggregate in the lungs.

characteristic of small-particle, Rayleigh scattering of light given in eqn 18. A measure of the distortion is the ratio

$$P = \frac{I_{\text{observed}}}{I_{\text{Rayleigh}}}$$

measured at several angles, where I_{observed} is the observed intensity at each angle and I_{Rayleigh} is the intensity predicted for Rayleigh scattering at that angle.

If the molecule is regarded as composed of a number of atoms i at distances R_i from a convenient point, then interference occurs between the radiation scattered by each pair. The scattering from all the particles is then calculated by allowing for contributions from all possible orientations of each pair of atoms in each molecule. This description is very much like the one used in the discussion of electron diffraction (Section 21.10), so we can expect the intensity pattern to be described by a kind of Wierl equation. This turns out to be so and, if there are N atoms in the macromolecule and if all are assumed to have the same scattering power, then

$$P = \frac{1}{N^2} \sum_i \sum_j \frac{\sin sR_{ij}}{sR_{ij}}, \qquad \text{where } s = \frac{4\pi}{\lambda} \sin \tfrac{1}{2}\theta \qquad (20)$$

(Compare eqn 21.12.) In this expression, R_{ij} is the separation of atoms i and j, and λ is the wavelength of the incident radiation. The observed intensity is equal to PI_{Raleigh}, with I_{Rayleigh} given by eqn 18.

Small-particle scattering

When the molecule is much smaller than the wavelength of the incident radiation in the sense that $sR_{ij} \ll 1$ (for example, if $R = 5$ nm, and $\lambda = 500$ nm, then all the sR_{ij} are about 0.1), then we show in the *Justification* that follows that the deviation from Rayleigh scattering is proportional to the square of the **radius of gyration** R_g, of the molecule:

$$P - 1 \propto R_g^2$$

The radius of gyration is the radius of a thin hollow spherical shell of the same mass and moment of inertia as the molecule (Fig. 23.12), and is calculated formally from the expression:[4]

$$R_g = \frac{1}{N} \left(\tfrac{1}{2} \sum_{i,j} R_{ij}^2 \right)^{\frac{1}{2}} \qquad (21)$$

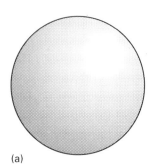

(a)

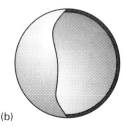

(b)

23.12 (a) A spherical molecule and (b) the hollow spherical shell that has the same rotational characteristics. The radius of the hollow shell is the radius of gyration of the molecule. The radius of gyration of a solid sphere of radius R is $0.77R$.

JUSTIFICATION

When $sR_{ij} \ll 1$ we can use the expansion $\sin x = x - \tfrac{1}{6}x^3 + \ldots$ to write

$$\sin sR_{ij} = sR_{ij} - \tfrac{1}{6}(sR_{ij})^3 + \ldots$$

---

4 In Problem 23.25 this definition is shown to be equivalent to another and more easily visualized one in the case of a chain of identical atoms: the radius of gyration is the root mean square distance of the atoms from the centre of mass.

It then follows that

$$P = \frac{1}{N^2} \sum_i \sum_j \left\{ 1 - \frac{1}{6}(sR_{ij})^2 + \ldots \right\} = 1 - \frac{s^2}{6N^2} \sum_{i,j} R_{ij}^2 + \ldots$$

The sum over the squares of the separations gives the radius of gyration of the molecule. Therefore

$$P \approx 1 - \frac{1}{3}s^2 R_g^2$$

which shows that $P - 1$ is proportional to R_g^2.

Table 23.4* Radius of gyration

	$M/(\text{kg mol}^{-1})$	R_g/nm
Serum albumin	66	2.98
Polystyrene	3.2×10^3	50†
DNA	4×10^3	117

* More values are given in the Data section.
† In a poor solvent.

Because the deviation from Rayleigh scattering is determined by R_g, an analysis of the scattering intensity should give the value of R_g for the molecule in solution. This quantity in turn can be interpreted in terms of the size of the molecule. For example, a solid sphere of radius R has $R_g = (3/5)^{\frac{1}{2}}R$, and a long thin rod of length l has $R_g = l/(2\sqrt{3})$. Once again, it must be emphasized that the analysis must be performed on data obtained by extrapolation to zero concentration. Some experimental values are listed in Table 23.4.

The use of laser light has led to further refinements in the application and interpretation of light scattering. There has been a shift of emphasis towards the investigation of the time dependence of the positions of atoms and the orientation of macromolecules in solution. The properties can be studied by measuring the shift of frequency that occurs when monochromatic light is scattered by a moving target in the technique called **dynamic light scattering**. In particular, laser light scattering can be used for the direct determination of the diffusional characteristics of macromolecules, and provides a fast, direct, and reliable method for the measurement of diffusion coefficients, even of macromolecules of low stability.

CONFORMATION AND CONFIGURATION

The **primary structure** of a macromolecule is the sequence of small molecular residues making up the chain (or network if there is cross-linking). In the case of a synthetic polymer, virtually all the residues are identical, and it is sufficient to name the monomer used in the synthesis. Thus, the repeating unit of polyethylene is $-CH_2CH_2-$, and the primary structure of the chain is specified by denoting it as $-(CH_2CH_2)_n-$.

The concept of primary structure ceases to be trivial in the case of synthetic copolymers and biological macromolecules, for, in general, these substances are chains formed from different molecules. Proteins, for example, are **polypeptides**, the name signifying chains formed from numbers of different amino acids (about 20 occur naturally) strung together by the **peptide link**, $-CO-NH-$. The determination of the primary structure is then a highly complex problem of chemical analysis called **sequencing**. The **degradation** of a polymer is a disruption of its primary structure, when the chain breaks into shorter components.

The **secondary structure** of macromolecules refers to the (often local) spatial, well-characterized arrangement of the basic structural units. The secondary structure of an isolated molecule of polyethylene is a random coil, whereas that of a protein is a highly organized arrangement determined largely by hydrogen bonds, and taking the form of helices or sheets in various segments of the molecule. The loss of secondary structure is called **denaturation**. When the hydrogen bonds in a protein are destroyed (for instance, by heating, as when cooking an egg) the structure denatures into a random coil.

The difference between primary and secondary structure is closely related to the difference between the configuration and the conformation of a chain. The term **configuration** refers to the structural features that can be changed only by breaking chemical bonds and forming new ones. Thus, the chains –A–B–C– and –A–C–B– have different configurations. The term **conformation** refers to the spatial arrangement of the different parts of a chain, and one conformation can be changed into another by rotating one part of a chain round the bond joining it to another.

The term **tertiary structure** refers to the overall three-dimensional structure of the molecule. For instance, many proteins have a helical secondary structure, but in many proteins the helix is so bent and distorted that the molecule has a globular tertiary structure. The term **quaternary structure** refers to the manner in which some molecules are formed by the aggregation of others. Haemoglobin is a famous example: each molecule consists of four subunits of two types (the α and the β chains).

23.6 Random coils

As the first step in unravelling the various aspects of structure, we consider the most likely conformation of a chain of identical units that are incapable of forming hydrogen bonds or any other type of specific bond. Polyethylene is a simple example, but the general idea applies to a denatured protein too. The simplest model is a **freely jointed chain**, in which any bond is free to make any angle with respect to the preceding one (Fig. 23.13a); the residues are assumed to occupy zero volume so that different parts of the chain can occupy the same region of space. The model is obviously an oversimplification, because a bond is actually constrained to a cone of angles around a direction defined by its neighbour (Fig. 23.13b).

The random coil is the least structured conformation of a polymer chain, and thus corresponds to the state of maximum conformational entropy. Any stretching of the coil introduces order and reduces the entropy. Conversely, the formation of a random coil from a more extended form is a spontaneous process (so long as enthalpy contributions do not interfere). The elasticity of a perfect elastomer, a flexible polymer in which the internal energy is independent of the extension, may be discussed in these terms (see *Further information 15*). The random coil model is also a helpful starting point for estimating the orders of magnitude of the hydrodynamic properties (such as sedimentation rates) of polymers and denatured proteins in solution.

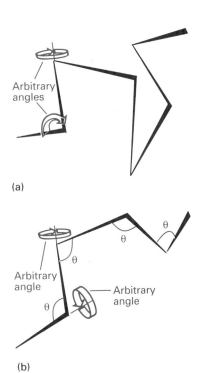

(a)

(b)

23.13 (a) A freely jointed chain is like a three-dimensional random walk, each step being in an arbitrary direction but of the same length. (b) A better description is obtained by fixing the bond angle (e.g. at the tetrahedral angle) and allowing free rotation about a bond direction.

The radial distribution

A random coil is a version of the three-dimensional random walk (Section 24.13), in which each bond represents a step taken in a random direction. For our present purposes all we need to know is that the probability of the ends of the chain lying in the range R to $R + dR$ is $f\,dR$, where

$$f = 4\pi\left(\frac{a}{\pi^{\frac{1}{2}}}\right)^{3} R^{2}e^{-a^{2}R^{2}}, \qquad \text{where } a = \left(\frac{3}{2Nl^{2}}\right)^{\frac{1}{2}} \tag{22}$$

In this expression, N is the number of bonds and l is the bond length.[5] Equation 22 shows that in some coils (the proportion being given by the value of f with R large) the ends may be far apart, whereas in others their separation is small. An alternative interpretation is to regard each coil as writhing continually from one conformation to another; then $f\,dR$ is the probability that at any instant the chain will be found with the separation of its ends between R and $R + dR$.

Measures of size

There are several measures of the geometrical size of a random coil. The **contour length** R_{c} is the length of the macromolecule measured along its backbone from atom to atom (the maximum distance that the random walker could walk). For a polymer of N monomer units each of length l, the contour length is

$$R_{c} = Nl \tag{23}$$

The **root mean square separation** R_{rms}, is a measure of the average separation of the ends of a random coil. It is the square root of the mean value of R^{2}, calculated by weighting each possible value of R^{2} with the probability that R occurs:

$$R_{rms} = \left(\int_{0}^{\infty} R^{2}f\,dR\right)^{\frac{1}{2}} = N^{\frac{1}{2}} \times l \tag{24}$$

We see that, as the number of monomer units increases, the root mean square separation of its end increases as $N^{\frac{1}{2}}$ (and, consequently, its volume increases as $N^{\frac{3}{2}}$). Similarly, the radius of gyration of the coil is

$$R_{g} = \left(\frac{N}{6}\right)^{\frac{1}{2}} \times l \tag{25}$$

The radius of gyration also increases as $N^{\frac{1}{2}}$.

Example 23.7 *Calculating the dimensions of a random coil*

Calculate the mean separation of the ends of a freely jointed polymer chain of N bonds of length l.

---

5 Here and elsewhere we are ignoring the fact that the chain cannot be longer than Nl. Although eqn 22 gives a non-zero probability for $R > Nl$, the values are so small that the errors in pretending that R can range up to infinity are negligible.

Method. The general expression for the mean nth power of the end-to-end separation is

$$\langle R^n \rangle = \int_0^\infty R^n f \, dR$$

which should be used with $n = 1$ and f from eqn 22.

Answer. The mean separation is

$$\langle R \rangle = 4\pi\left(\frac{a}{\pi^{\frac{1}{2}}}\right)^3 \int_0^\infty R^3 e^{-a^2 R^2} \, dR = \frac{2}{a\pi^{\frac{1}{2}}} = \left(\frac{8}{3\pi}\right)^{\frac{1}{2}} N^{\frac{1}{2}} l$$

The standard integral we have used is

$$\int_0^\infty x^3 e^{-a^2 x^2} \, dx = \frac{1}{2a^4}$$

Comment. The result must be multiplied by a factor when the chain is not freely jointed: see below.

Exercise E23.7. Evaluate the root mean square separation of the ends of the chain.
$$[N^{\frac{1}{2}} l]$$

Constrained chains

Before making use of these conclusions we must remove the freedom for bond angles to take any value. This adjustment is simple for long chains, for we can take groups of neighbouring bonds and consider the direction of their resultant. Although the individual bonds are constrained to a single cone of angle θ, the resultant of several lies in a random direction. By concentrating on such groups rather than individuals, it turns out that for long chains the average values given above should be multiplied by

$$F = \left(\frac{1 - \cos\theta}{1 + \cos\theta}\right)^{\frac{1}{2}}$$

For tetrahedral bonds, for which $\cos\theta = -\frac{1}{3}$ (that is, $\theta = 109.5°$), $F = \sqrt{2}$. Therefore:

$$R_{rms} = (2N)^{\frac{1}{2}} l \qquad R_g = \left(\frac{N}{3}\right)^{\frac{1}{2}} l \qquad (26)$$

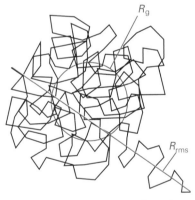

23.14 A random coil in three dimensions. This one contains about 200 units. The root mean square distance between the ends (R_{rms}) and the radius of gyration (R_g) are indicated.

For example, in the case of a polyethylene chain with $M = 2.8 \text{ kg mol}^{-1}$, corresponding to $N = 200$, because $l = 154$ pm for a C—C bond, we find $R_{rms} = 3.1$ nm and $R_g = 1.3$ nm (Fig. 23.14). This value of R_g means that, on average, the coils rotate like hollow spheres of radius 1.3 nm and mass equal to the molecular mass.

The model of a randomly coiled molecule is still an approximation even after the bond angles have been restricted because it does not take into account the impossibility of two or more atoms occupying the same place. Such self-avoidance tends to swell the coil, so it is better to regard R_{rms} and R_g as lower bounds to the actual values. The model also ignores the role of the solvent: a poor solvent will tend to cause the coil to tighten so that solute–solvent contacts are minimized; a good solvent does the opposite.

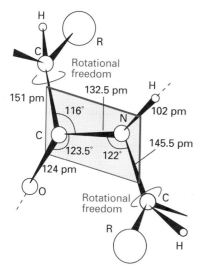

23.15 The dimensions that characterize the peptide link. The C—CO—NH—C atoms define a plane (the C—N bond has partial double-bond character), but there is rotational freedom around the C—CO and N—C bonds.

23.7 Helices and sheets

Natural macromolecules need a precisely maintained conformation to function. The achievement of a specific conformation is the major remaining problem in protein synthesis, for, although primary structures can be built, the product is inactive because the secondary structure cannot yet be produced.

The Corey–Pauling rules

The origin of the secondary structures of proteins is found in the rules formulated by Linus Pauling and Robert Corey. The essential feature is the stabilization of structures by hydrogen bonds involving the peptide link. The latter can act both as a donor of the H atom (the NH part of the link) and as an acceptor (the CO part). The **Corey–Pauling rules** are as follows:

1. The atoms of the peptide link lie in a plane (Fig. 23.15).
2. The N, H, and O atoms of a hydrogen bond lie in a straight line (with displacements of H tolerated up to not more than 30° from the N—O vector).
3. All NH and CO groups are engaged in hydrogen bonding.

The rules are satisfied by two structures. One, in which hydrogen bonding occurs between peptide links of the same chain, is the **α helix**. The other, in which hydrogen bonding links different chains, is the **β-pleated sheet**; this form is the secondary structure of the protein fibroin, the constituent of silk.

The α helix is illustrated in Fig. 23.16. Each turn of the helix contains 3.6 amino acid residues, so the period of the helix corresponds to 5 turns (18 residues). The pitch of a single turn (the distance betwen points separated by 360°) is 544 pm. The N—H $\cdots$ O bonds lie parallel to the axis and link every fifth group (so residue i is linked to residues $i-4$ and $i+4$). There is freedom for the helix to be arranged as either a right- or a left-handed screw, but the overwhelming majority of natural polypeptides are right-handed on account of the preponderance of the L-configuration of the naturally occurring amino acids, as we explain below. The reason for their preponderance is uncertain, but it may be related to the symmetries of fundamental particles and the non-conservation of parity (the fact that this universe behaves differently from its hypothetical mirror image).

Conformational energy

The stabilities of different polypeptide geometries can be investigated by calculating the total potential energy of all the interactions between nonbonded atoms, and looking for a minimum. It turns out, in agreement with experience, that a right-handed α helix of L-amino acids is marginally more stable than a left-handed helix of the same acids.

The geometry of the chain can be specified by two angles, ϕ (the torsional angle for the N—C bond) and ψ (the torsional angle for the

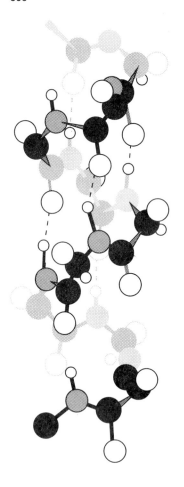

H   C   N   O

23.16 The polypeptide α helix. There are 3.6 residues per turn, and a translation along the helix of 151 pm per residue, giving a pitch of 544 pm. The diameter (ignoring side chains) is about 600 pm.

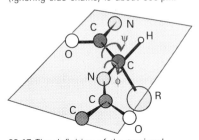

23.17 The definition of the torsional angles ψ and ϕ between two peptide units. In this case (an α-L-polypeptide) the chain has been drawn in its all-trans form, with $\psi = \phi = 180°$.

C—C bond). The illustration in Fig. 23.17 defines these angles,[6] and shows the all-*trans* form of the chain, in which all ϕ and ψ are 180°. A helix is obtained when all the ϕ are equal and when all the ψ are equal. For a right-handed α helix, all $\phi = -57°$ and all $\psi = -47°$. For a left-handed α helix, both angles are positive. Because only two angles are needed to specify the conformation of a helix, and they range from $-180°$ to $+180°$, the potential energy of the entire molecule can be represented by a point on a plane in which one axis represents ϕ and the other represents ψ.

The potential energy of a given conformation (ϕ, ψ) can be calculated by using the expressions developed in Sections 22.3 and 22.4. For example, the interaction energy of two atoms separated by a distance R (which we know once ϕ and ψ are specified) can be given the Lennard-Jones (12, 6) form. If the partial charges on the atoms (arising from ionic character in the bonds) are known, a Coulombic contribution of the form $1/R$ can be included. This is sometimes done by ascribing charges $-0.28e$ and $+0.28e$ to N and H, respectively, and $-0.39e$ and $+0.39e$ to O and C, respectively. There is also a torsional contribution arising from the barrier to internal rotation of one bond relative to another (just like the barrier to internal rotation in ethane), which is normally expressed as

$$V = A(1 + \cos 3\phi) + B(1 + \cos 3\psi)$$

in which A and B are constants of the order of 1 kJ mol^{-1}.

The potential energy contours for the helical form of polypeptide chains formed from the nonchiral amino acid glycine (R = H) and the chiral amino acid L-alanine are shown in Figs 23.18a and 23.18b, respectively. They were computed by summing all the contributions described above for each choice of angles, and then plotting contours of equal potential energy. The glycine map is symmetrical, with minima of equal depth near $\phi = -80°$, $\psi = +90°$ and at $\phi = +80°$, $\psi = -90°$. In contrast, the map for L-alanine is unsymmetrical, and there are three distinct low-energy conformations (marked I, II, III). The minima of regions I and II lie close to the angles typical of right- and left-handed α helices, but the former has a lower minimum, which is consistent with the formation of right-handed helices from the naturally occurring L-amino acids.

23.8 Higher-order structure

Helical polypeptide chains are folded into a tertiary structure if there are other bonding influences between the residues of the chain that are strong enough to overcome the interactions responsible for the secondary structure. The folding influences include disulfide (–S–S–) links, ionic interactions (which depend on the pH), and strong hydrogen bonds (such a O—H $\cdots$ O—), and are illustrated by the structure of myoglobin (Fig. 23.19), the full structure (of 2600 atoms) having been determined by

6 The sign convention is that a positive angle means that the front atom must be rotated clockwise to bring it into an eclipsed position relative to the rear atom.

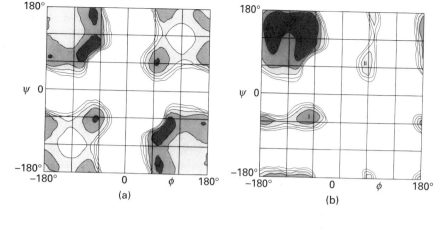

23.18 Energy contour diagrams for (a) a glycyl residue of a polypeptide chain and (b) an alanyl residue. The darker the shading the lower the potential energy. The glycyl diagram is symmetrical, but regions I and II in the alanine diagram correspond to right- and left-handed helices, are unsymmetrical, and the minimum in region I lies lower than that in region II. (After D. A. Brant and P. J. Flory, *J. Mol. Biol.* **23,** 47 (1967).)

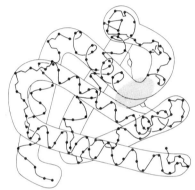

23.19 The structure of myoglobin. Only the α-carbon atom positions are shown. The haem group, the oxygen binding group, is shown as a shaded region. (Based on M. F. Perutz, copyright *Scientific American*, 1964; with permission.)

X-ray diffraction. The folding of the basic α helix caused by disulfide links can be seen in the structure: about 77 per cent of the structure is α helix, the rest being involved in the folds.

Proteins with $M > 50$ kg mol^{-1} are often found to be aggregates of two or more polypeptide chains. The possibility of such quaternary structure often confuses the determination of their molar masses, for different techniques might give values differing by factors of 2 or more. Haemoglobin, which consists of four myoglobin-like chains, is an example.

Protein denaturation can be caused by several means, and different aspects of structure may be affected. The permanent waving of hair, for example, is reorganization at the quaternary level. Hair is a form of the protein keratin, and its quaternary structure is thought to be a multiple helix, with the α helices bound together by disulfide links and hydrogen bonds (although there is some dispute about its precise structure). The process of permanent waving consists of disrupting these links, unravelling the keratin quaternary structure, and then reforming it into a more fashionable disposition. The 'permanence' is only temporary, however, because the structure of the newly formed hair is genetically controlled. Incidentally, normal hair grows at a rate that requires at least 10 twists of the keratin helix to be produced each second, so very close inspection of the human scalp would show it to be literally writhing with activity.

Denaturation at the secondary level is brought about by agents that destroy hydrogen bonds. Thermal motion may be sufficient, in which case denaturation is a kind of intramolecular melting. When eggs are cooked, the albumin is denatured irreversibly, and the protein collapses into a structure resembling a random coil. This **helix-coil transition** is sharp, like ordinary melting, because it is a cooperative process: when one hydrogen bond has been broken it is easier to break its neighbours, and then even easier to break theirs, and so on. The disruption cascades down the helix, and the transition occurs sharply. Denaturation may also be brought about chemically. For instance, a solvent that forms stronger hydrogen bonds than those within the helix will compete successfully for the NH and CO groups. Acids and bases can cause denaturation by protonation or deprotonation of various groups.

CHECK LIST OF KEY IDEAS

1 The definition of **number-average molar mass** (eqn 1), the **weight-average molar mass** (eqn 2), and the term **viscosity-average molar mass**.

2 The use of **osmometry** for the determination of molar mass (Section 23.2).

3 The determination and significance of the **osmotic virial coefficient** (Section 23.2 and eqn 4).

4 The significance of the **Flory theta temperature** and a **θ solution** (Section 23.2).

5 The technique of **vapour-phase osmometry** (Section 23.2).

6 The role of **polyelectrolytes** in osmosis and the **Donnan equilibrium** established when a polyelectrolyte affects the distribution of ion concentrations (eqn 7).

7 The rate of **sedimentation** and the use of the **sedimentation constant** in the determination of molar mass (Section 23.3).

8 The **sedimentation equilibrium** established in a gravitational field or in a centrifuge (Section 23.3) and its use for the determination of molar mass (eqn 14).

9 The determination of molar mass by **electrophoresis** and **gel permeation chromatography** (Section 23.3).

10 The definition of **intrinsic viscosity** (eqn 16) and its correlation with the viscosity-average molar mass of polymers (eqn 17).

11 The use of **light scattering** to determine the weight-average molar mass of a macromolecule (Section 23.5).

12 The significance of the **turbidity** of a solution (eqn 19).

13 The differences between light scattering by large particles and small particles (Section 23.5).

14 The **primary**, **secondary**, **tertiary**, and **quaternary structures** of polypeptides.

15 The properties of **random coils** (Section 23.6), including the **random walk** for calculating the **radial distribution** (eqn 22) and the dimensions of a random coil (Example 23.7).

16 The **Corey–Pauling rules** for the secondary structure of polypeptides (Section 23.7) and the formation of the **α helix**.

17 The **conformational energy analysis** of polypeptides (Section 23.7).

EXERCISES

23.1. Calculate the number-average molar mass and the mass-average molar mass of a mixture of equal amounts of two polymers, one having $M = 62$ kg mol^{-1} and the other $M = 78$ kg mol^{-1}.

23.2. A polymer chain consists of 700 segments, each 0.90 nm long. If the chain were ideally flexible, what would be the r.m.s. separation of the ends of the chain?

23.3. The radius of gyration of a long-chain molecule is found to be 7.3 nm. The chain consists of tetrahedral C—C links. Assume the chain is randomly coiled and estimate the number of links in the chain.

23.4. Calculate the contour length (the length of the extended chain) and the root mean square separation (the end-to-end distance) for polyethylene with a molar mass of 280 kg mol^{-1}. Use $R(\text{C—C}) = 154$ pm.

23.5. What is the relative rate of sedimentation for two spherical particles of the same density, but which differ in radius by a factor of 10?

23.6. Find the drift speed of a particle of radius 20 μm and density 1750 kg m^{-3} which is settling from suspension in water (density 1000 kg m^{-3}) under the influence of gravity alone. The viscosity of water is 8.9×10^{-4} kg m^{-1} s^{-1}.

23.7. At 20°C human haemoglobin has a specific volume of 0.749×10^{-3} m^3 kg^{-1}, a sedimentation constant of 4.48 Sv, and a diffusion coefficient of 6.9×10^{-11} m^2 s^{-1}. Determine its molar mass from this information. Take $\rho \approx \rho(\text{H}_2\text{O}) = 0.9982$ g cm^{-3}.

23.8. At 20°C the diffusion coefficient of a macromolecule is found to be 8.3×10^{-11} m^2 s^{-1}. Its sedimentation constant is 3.2 Sv in a solution of density

1.06 g cm^{-3}. The specific volume of the macromolecule is 0.656 cm^3 g^{-1}. Determine the molar mass of the macromolecule.

23.9. A solution consists of solvent, and a solute which is 30 per cent by mass of a dimer with $M = 30$ kg mol^{-1} and 70 per cent of its monomer. What average molar mass would be obtained from measurement of: (a) osmotic pressure, (b) light scattering?

23.10. A polyelectrolyte Na$_{20}$P with $M = 100$ kg mol^{-1} at a concentration 1.00 g/(100 cm^3) was equilibrated in the presence of 0.0010 M NaCl(aq) (i.e. $[Na^+]_R = 0.0010$ mol L^{-1}). What is the value of $[Na^+]_L$ at equilibrium?

23.11. At the start of a membrane equilibrium experiment, the first compartment contains 1.00 L of solution with an NaX concentration of 0.100 mol L^{-1}, where X$^-$ cannot pass through the membrane. The second compartment has 2.00 L of 0.030 M NaCl(aq). Find the concentration of Cl$^-$ ion in the first compartment after equilibrium is established.

23.12. The data from a sedimentation equilibrium experiment performed at 300 K on a macromolecular solute in aqueous solution show that a graph of $\ln c$ against r^2 is a straight line with a slope of 729 cm^{-2}. The rotational speed of the centrifuge was 50 000 r.p.m. The specific volume of the solute is 0.61 cm^3 g^{-1}. Calculate the molar mass of the solute.

23.13. Calculate the radial acceleration (as so many g) in a cell placed at 6.0 cm from the centre of rotation in an ultracentrifuge operating at 80 000 r.p.m.

23.14. Cotton consists of the polymer cellulose, which is a linear chain of glucose molecules. The chains are held together by hydrogen bonding. When a cotton shirt is ironed, it is first moistened, then heated under pressure. Explain this process.

PROBLEMS
Numerical problems

23.1. The concentration dependence of the osmotic pressure of solutions of a macromolecule at 20°C was found to be as follows:

$c/(g\,L^{-1})$	1.21	2.72	5.08	6.60
Π/Pa	134	321	655	898

Determine the molar mass of the macromolecule and the osmotic virial coefficient.

23.2. The osmotic pressure of a fraction of poly(vinyl chloride) in a ketone solvent was measured at 25°C. The density of the solvent (which is virtually equal to the density of the solution) was 0.798 g cm^{-3}. Calculate the molar mass and the osmotic virial coefficient B of the fraction from the following data:

$c/(g/100\,cm^3)$	0.200	0.400	0.600	0.800	1.000
h/cm	0.48	1.2	1.86	2.76	3.88

23.3. The concentration dependence of the viscosity of a polymer solution is found to be as follows:

$c/(g\,L^{-1})$	1.32	2.89	5.73	9.17
$\eta/(g\,m^{-1}\,s^{-1})$	1.08	1.20	1.42	1.73

The viscosity of the solvent is 0.985 g m^{-1} s^{-1}. What is the intrinsic viscosity of the polymer?

23.4. In a sedimentation experiment the position of the boundary as a function of time was found to be as follows:

t/min	15.5	29.1	36.4	58.2
r/cm	5.05	5.09	5.12	5.19

The rotational speed of the centrifuge was 45 000 r.p.m. Calculate the sedimentation constant of the solute.

23.5. In an ultracentrifuge experiment at 20°C on bovine serum albumin the following data were obtained: $\rho = 1.001$ g cm^{-3}, $v_s = 1.112$ cm^3 g^{-1}, $\omega/2\pi = 322$ Hz,

r/cm	5.0	5.1	5.2	5.3	5.4
$c/(mg\,cm^{-3})$	0.536	0.284	0.148	0.077	0.039

Evaluate the molar mass of the sample.

23.6. Calculate the speed of operation (in r.p.m.) of a centrifuge needed to obtain a readily measurable concentration gradient in a sedimentation equilibrium experiment. Take that gradient to be a concentration at the bottom of the cell about five times greater that at the top. Use $r_{top} = 5.0$ cm, $r_{bottom} = 7.0$ cm, $M \approx 10^5$ g mol^{-1}, $\rho v_s \approx 0.75$, $T = 298$ K.

23.7. At the start of a Donnan equilibrium experiment, the first compartment contains 2.00 L of solution which is 0.015 M in the polyelectrolyte Na$_2$P(aq) and 0.010 M in NaCl(aq). The second compartment has 2.00 L of solution which is 0.0050 M in NaCl(aq). What is the potential difference across the membrane arising from the Na$^+$ ion concentration difference at 300 K?

23.8. Investigation of the composition of the solutions used to study the osmotic pressure due to a polyelectrolyte with $v = 20$ showed that at equilibrium the concentrations corresponded to $[Cl^-] \approx 0.020$ mol L^{-1}. Calculate the osmotic virial coefficient for $v = 20$. Does it dominate the effect of excluded volume?

23.9. Sedimentation studies on haemoglobin in water gave a sedimentation constant $S = 4.5$ Sv at 20°C. The diffusion coefficient is 6.3×10^{-11} m^2 s^{-1} at the same temperature. Calculate the molar mass of haemoglobin using $v_s = 0.75$ cm^3 g^{-1} for its partial specific volume and $\rho = 0.998$ g cm^{-3} for the density of the solution. Estimate the effective radius of the haemoglobin molecule given that the viscosity of the solution is 1.00×10^{-3} kg m^{-1} s^{-1}.

23.10. The times of flow of dilute solutions of polystyrene in benzene through a viscometer at 25°C are given in the table below. From these data, calculate the molar mass of the polystyrene samples. Since the solutions are dilute, assume that the densities of the solutions are the same as pure benzene. η(benzene) $= 0.601 \times 10^{-3}$ kg m^{-1} s^{-1} (0.601 cP) at 25°C.

$c/(\text{g L}^{-1})$	0.00	2.22	5.00	8.00	10.00
t/s	208.2	248.1	303.4	371.8	421.3

23.11. The table below gives light scattering data obtained with light of wavelength 546.1 nm on aqueous solutions of sucrose at 25°C. Calculate the apparent molar mass of sucrose and compare to the known value.

$c/(\text{g L}^{-1})$	35.2	61.4	106	163
$(1000 Hc/\tau)/(\text{mol kg}^{-1})$	2.84	2.91	3.08	3.32

23.12. The diffusion coefficient for bovine serum albumin, a prolate ellipsoid, is 6.97×10^{-11} m^2 s^{-1} at 20°C, its partial specific volume is 0.734 cm^3 g^{-1}, and its sedimentation constant is 5.01 Sv in a solution of density 1.0023 g cm^{-3} and viscosity 1.00×10^{-3} kg m^{-1} s^{-1}. Estimate its dimensions.

23.13. The rate of sedimentation of a recently isolated protein was monitored at 20°C and with a rotor speed of 50 000 r.p.m. The boundary receded as follows:

t/s	0	300	600	900	1200	1500	1800
r/cm	6.127	6.153	6.179	6.206	6.232	6.258	6.284

Calculate the sedimentation constant and the molar mass of the protein on the basis that is partial specific volume is 0.728 cm^3 g^{-1} and its diffusion coefficient is 7.62×10^{-11} m^2 s^{-1} at 20°C, the density of the solution then being 0.9981 g cm^{-3}. Suggest a shape for the protein given that the viscosity of the solution is 1.00×10^{-3} kg m^{-1} s^{-1} at 20°C.

23.14. The viscosities of solutions of polyisobutylene in benzene were measured at 24°C (the θ temperature for the system) with the following results:

$c/(\text{g}/100\ \text{cm}^3)$	0	0.2	0.4	0.6	0.8	1.0
$\eta/(10^{-3}\ \text{kg m}^{-1}\text{s}^{-1})$	0.647	0.690	0.733	0.777	0.821	0.865

Use the information in Table 23.3 to deduce the molar mass of the polymer.

23.15. Evaluate the radius of gyration of (a) a solid sphere of radius a, (b) a long straight rod of radius a and length l.

Show that in the case of a solid sphere of specific volume v_s,

$$R_g/\text{nm} \approx 0.056902 \times \{(v_s/\text{cm}^3\,\text{g}^{-1})(M/\text{g mol}^{-1})\}^{\frac{1}{3}}$$

Evaluate R_g for a species with $M = 100$ kg mol^{-1}, $v_s = 0.750$ cm^3 g^{-1}, and, in the case of the rod, of radius 0.50 nm.

23.16. Use the information below and the expression for R_g of a solid sphere quoted in Problem 23.15 to classify the species below as globular or rod-like.

	Measured		
	$M/(\text{g mol}^{-1})$	$v_s/(\text{cm}^3\,\text{g}^{-1})$	R_g/nm
Serum albumin	66×10^3	0.752	2.98
Bushy stunt virus	10.6×10^6	0.741	12.0
DNA	4×10^6	0.556	117.0

23.17. In formamide as solvent, poly(γ-benzyl-L-glutamate) is found by light scattering experiments to have a radius of gyration proportional to M; in contrast, polystyrene in butanone has R_g proportional to $M^{1/2}$. Present arguments to show that the first polymer is a rigid rod, while the second is a random coil.

23.18. The structures of crystalline macromolecules may be determined by X-ray diffraction techniques by methods similar to those for smaller molecules. Fully crystalline polyethylene has its chains aligned in an orthorhombic unit cell of dimensions 740 pm $\times$ 493 pm $\times$ 253 pm. There are two repeating CH_2CH_2 units per unit cell. Calculate the theoretical density of fully crystalline polyethylene. The actual density ranges from 0.92 to 0.95 g cm^{-3}.

Theoretical problems

23.19. A polymerization process produced a Gaussian distribution of polymers in the sense that the proportion of molecules having a molar mass in the range M to $M + dM$ was proportional to $\exp\{-(M - \bar{M})^2/2\Gamma\}$ dM. What is the number-average molar mass when the distribution is narrow?

23.20. Calculate the excluded volume in terms of the molecular volume on the basis that the molecules are spheres of radius a. Evaluate the osmotic virial coefficient in the case of bushy stunt virus, $a = 14.0$ nm, and haemoglobin, $a = 3.2$ nm. Evaluate the percentage deviation of the osmotic pressures of 1.00 g/(100 cm^3) solutions of bushy stunt virus ($M = 1.07 \times 10^4$ kg mol^{-1}) and haemoglobin ($M = 66.5$ kg mol^{-1}) from the ideal solution values.

23.21. The effective radius of a random coil a is related to its radius of gyration R_g by $a = \gamma R_g$, with $\gamma = 0.85$. Deduce an expression for the osmotic virial coefficient B

in terms of the number of chain units for (a) a freely jointed chain, (b) a chain with tetrahedral bond angles. Evaluate B for $l = 154$ pm and $N = 4000$. Estimate the osmotic virial coefficient B for randomly coiled polyethylene chain of arbitrary M and evaluate it for $M = 56$ kg mol^{-1}.

23.22. Show that the ratio $[Na^+]_L/[Na^+]_R$ in the Donnan equilibrium is equal to $x + (1 + x^2)$, where $x = v[P]/2[Na^+]_R$ and sketch the ratio as a function of the poly-electrolyte concentration.

23.23. Consider the thermodynamic description of stretching rubber. The observables are the tension t and length l (like p and V for gases). Since $dw = t \, dl$, the basic equation is $dU = T \, dS + t \, dl$. ($p \, dV$ terms are supposed negligible throughout.) If $G = U - TS - tl$, find expressions for dG and dA and deduce the Maxwell relations

$$\left(\frac{\partial S}{\partial l}\right)_T = -\left(\frac{\partial t}{\partial T}\right)_l, \qquad \left(\frac{\partial S}{\partial t}\right)_T = \left(\frac{\partial l}{\partial T}\right)_t.$$

Go on to deduce the equation of state for rubber,

$$\left(\frac{\partial U}{\partial l}\right)_T = t - T\left(\frac{\partial t}{\partial T}\right)_l.$$

23.24. On the assumption that the tension required to keep a sample at a constant length is proportional to the temperature ($t = aT$, the analogue of $p \propto T$), show that the tension can be ascribed to the dependence of the entropy on the length of the sample. Account for this result in terms of the molecular nature of the sample.

23.25. The radius of gyration is defined in eqn 21. Show that an equivalent definition is that R_g is the average root mean square distance of the atoms or groups (all assumed to be of the same mass); that is, that $R_g^2 = (1/N) \sum_j R_j^2$, where R_j is the distance of atom j from the centre of mass.

23.26. Use eqn 22 to deduce expressions for (a) the root mean square separation of the ends of the chain, (b) the mean separation of the ends, and (c) their most probable separation. Evaluate these three quantities for an $N = 4000$, $l = 154$ pm fully flexible chain.

23.27. Construct a two-dimensional random walk by using a random number generating program of a computer or calculator. Construct a walk of 50 and 100 steps. If there are many people working on the problem, investigate the mean and most probable separations in the plots by direct measurement. Do they vary as $N^{\frac{1}{2}}$?

23.28. Evaluate eqn 20 for P for a long rigid rod of N identical scattering units with separation l, and assume that the units are so closely spaced that sums may be replaced by integrals. Plot P as a function of θ on polar graph paper in the case $l \approx \lambda$. The integral over $(\sin x/x) \, dx$ between 0 and z is the 'sine-integral' Si(z), and is tabulated in Abramowitz and Stegun, *Handbook of mathematical functions*.

23.29. Sections of the space shuttle *Challenger*'s solid fuel rocket boosters were sealed together with O-ring rubber seals of 11 m circumference. These seals failed at 0°C, a temperature well above the crystallization temperature of the rubber. Speculate on why the failure occurred.

PART THREE

Change

In Part 3 of the text we consider the processes responsible for chemical change. We prepare the ground for a discussion of the rates of chemical reactions by considering the motion of molecules in gases and in liquids. Then we establish the precise meaning of reaction rate, and see how the overall rate, and the complex behaviour of some reactions, may be expressed in terms of elementary steps and the atomic events that take place when molecules meet. Special physical and chemical events take place at surfaces, including catalysis, and we see how to describe them. A special type of surface is that of an electrode, and we shall see how to describe and understand the rate at which electrons are transferred between an electrode and a species in solution.

24

Molecules in motion

One of the simplest types of molecular motion to describe is the purely chaotic motion of molecules of a perfect gas. We see that the kinetic theory can be used to account for the rates at which molecules and energy migrate through gases and that simple expressions for the rates can be derived. Molecular mobility is particularly important in liquids, and we shall see a little of the structure of this phase and the motion of molecules in it. Another simple kind of motion is the largely uniform motion of ions in solution in the presence of a potential difference. The migration of ions can be discussed reasonably simply in terms of a simple model. Molecular and ionic motion have common features and, by considering them from a more general viewpoint, we find expressions that govern the migration of properties in all kinds of matter. One of the most useful consequences of this general approach is the formulation of the diffusion equation, which is an equation that shows how matter and energy spread through media of various kinds. Finally, we build a simple model for all types of molecular motion, in which the molecules migrate in a series of small steps, and see that it accounts for many of the properties of migrating molecules in both gases and condensed phases.

The general approach we describe in this chapter provides techniques for discussing the motion of all kinds of particles in all kinds of fluids. We shall set the scene by considering a simple type of motion, that of molecules in a perfect gas, and then we shall see that molecular motion in liquids shows a number of similarities.

MOLECULAR MOTION IN GASES

Little more need be said about the structure of perfect gases than has already been said in Section 1.3. There we saw that the equilibrium properties of a gas can be understood in terms of the kinetic theory, which is based on a model of a gas in which the molecules are in

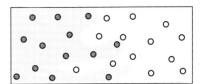

24.1 In the process of diffusion, molecules of one species mingle with molecules of another species.

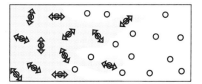

24.2 In the process of thermal conduction, molecules with different energies of thermal motion (represented by the arrows) spread into each others' regions.

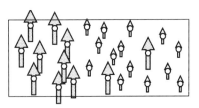

24.3 When an ion migrates under the influence of an electric field (represented by the tinting), it contributes to electrical conduction. Positive and negative ions migrate in opposite directions.

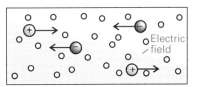

24.4 When molecules with different linear momenta (represented by the lengths of the arrows) migrate into each others' regions they contribute to the viscosity of the fluid.

ceaseless, random motion. Here we shall develop the kinetic theory to deal with gases that are not at internal equilibrium. In particular, we shall concentrate on transport properties. A **transport property** of a substance is its ability to transfer matter, energy, or some other specified property from one place to another. Four examples of transport properties are

> **Diffusion**, the migration of matter down a concentration gradient (Fig. 24.1).
> **Thermal conduction**, the migration of energy down a temperature gradient (Fig. 24.2).
> **Electrical conduction**, the migration of electric charge down a potential gradient (Fig. 24.3).
> **Viscosity**, the migration of linear momentum down a velocity gradient (Fig. 24.4).

We shall find it convenient to include **effusion**, the emergence of a gas from a container through a small hole, in the discussion.

For the purposes of this discussion, it will be useful to have in mind the following two expressions, which were derived in Chapter 1. One is the mean free path λ of molecules in a gas:

$$\lambda = \frac{kT}{\sqrt{2}\sigma p} \tag{1}°$$

where σ is the collision cross-section (this is eqn 1.19). In a container of constant volume, the mean free path is independent of temperature because p is itself proportional to the temperature and its variation cancels the T in the numerator. The second property is the mean speed $\bar{c}$ of the molecules:

$$\bar{c} = \left(\frac{8kT}{\pi m}\right)^{\frac{1}{2}} = \left(\frac{8RT}{\pi M}\right)^{\frac{1}{2}} \tag{2}°$$

(This expression was derived in Example 1.7.) The features to note are that the mean speed is proportional to $T^{\frac{1}{2}}$ and inversely proportional to $M^{\frac{1}{2}}$, where M is the molar mass of the molecules.

24.1 Collisions with walls and surfaces

The key to accounting for transport in the gas phase is the rate at which molecules strike an area (which may be an imaginary area embedded in the gas, or part of a real wall). As we shall soon see, we shall need to know that the number of collisions per unit area per unit time, Z_W, is related to the pressure and temperature of the gas by

$$Z_W = \frac{p}{(2\pi mkT)^{\frac{1}{2}}} \tag{3}°$$

When $p = 100\ \text{kPa}$ (1 bar) and $T = 300\ \text{K}$, there are about 3×10^{23} collisions per second per square centimetre.

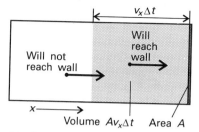

24.5 Only molecules within a distance $v_x\,\Delta t$ with $v_x > 0$ can reach the wall on the right in an interval Δt.

JUSTIFICATION

Consider a wall of area A perpendicular to the x-axis (Fig. 24.5). If a molecule has $v_x > 0$, then it will strike the wall within an interval Δt if it lies within a distance $v_x\,\Delta t$ of it. (If $v_x < 0$, the molecule is moving away from the wall.) Therefore, all molecules in the volume $A v_x\,\Delta t$, and with positive velocities, will strike the wall in the interval Δt. The total number of collisions in this interval is therefore the average of this quantity multiplied by the number density $\mathcal{N}$ of molecules:

$$\text{Number of collisions} = \mathcal{N} A\,\Delta t \int_0^\infty v_x f(v_x)\,\mathrm{d}v_x$$

(Notice that the integration is over only positive velocities, for we count only molecules travelling to the right.) We can evaluate the integral easily by using the velocity distribution (Section 1.3 and *Further information 1*):

$$\int_0^\infty v_x f(v_x)\,\mathrm{d}v_x = \left(\frac{m}{2\pi kT}\right)^{\frac{1}{2}} \int_0^\infty v_x \mathrm{e}^{-mv_x^2/2kT}\,\mathrm{d}v_x = \left(\frac{kT}{2\pi m}\right)^{\frac{1}{2}}$$

Therefore, the number of collisions per unit time per unit area is

$$Z_\mathrm{W} = \left(\frac{kT}{2\pi m}\right)^{\frac{1}{2}} \mathcal{N} = \tfrac{1}{4}\bar{c}\mathcal{N}$$

Then, since $\mathcal{N} = nN_\mathrm{A}/V = p/kT$,

$$Z_\mathrm{W} = \frac{p\bar{c}}{4kT} = \frac{p}{(2\pi mkT)^{\frac{1}{2}}}$$

which is eqn 3.

24.2 The rate of effusion

The empirical observations on effusion that we need to note and explain are summarized by **Graham's law of effusion**, that the rate of effusion is inversely proportional to the square root of the molar mass. The essential basis of this result is that, as remarked above, the mean speed of molecules is inversely proportional to $M^{\frac{1}{2}}$, so the rate at which they strike the area of the hole is similarly inversely proportional to $M^{\frac{1}{2}}$. However, by using the expression for the rate of collisions, we can obtain a more detailed expression for the rate of effusion and hence use effusion data more effectively.

We need to note that, when a gas at a pressure p and temperature T is separated from a vacuum by a very small hole, the rate of escape of its molecules is equal to the rate at which they strike the area of the hole (which is given by eqn 3). Therefore, if the area of the hole is A_0, the number of molecules that escape per unit time is

$$\text{Rate of effusion} = Z_\mathrm{W} A_0 = \frac{pA_0}{(2\pi mkT)^{\frac{1}{2}}} = \frac{pA_0 N_\mathrm{A}}{(2\pi RT)^{\frac{1}{2}}} \times \frac{1}{M^{\frac{1}{2}}} \qquad (4)°$$

(In the last step we have used $R = N_A k$ and $M = m N_A$). We see that the rate is inversely proportional to $M^{\frac{1}{2}}$, in accord with Graham's law.

Example 24.1 *Deducing the time dependence of the pressure of an effusing gas*

Derive an expression that shows how the pressure of an effusing gas varies with time if the container is not replenished as the gas escapes (as in a punctured vessel).

Method. The rate of effusion is proportional to the pressure of the gas in the container; so, as gas effuses and the pressure falls, the rate of effusion will decrease. To find the explicit expression, we need to set up a differential equation relating dp/dt to p, and then integrate it. The rate of effusion, as given by eqn 4, is the *number* of molecules per unit time that leave the container. The first step is to relate the rate of change of pressure to the rate of change of the number of molecules by using the perfect gas law in the form $pV = NkT$.

Answer. The rate of change of pressure of a gas in a container at constant pressure and temperature is related to the rate of change of the number of molecules present by

$$\frac{dp}{dt} = \frac{kT}{V}\frac{dN}{dt}$$

The rate of change of the number of molecules is equal to the collision frequency multiplied by the area of the hole:

$$\frac{dN}{dt} = -Z_W A_0 = -\frac{pA_0}{(2\pi m kT)^{\frac{1}{2}}}$$

Substitution of this expression into the one above gives

$$\frac{dp}{dt} = -\left(\frac{kT}{2\pi m}\right)^{\frac{1}{2}}\frac{A_0}{V} \times p$$

This expression integrates to

$$p = p_0 e^{-t/\tau}, \qquad \text{where } \tau = \left(\frac{2\pi m}{kT}\right)^{\frac{1}{2}}\frac{V}{A_0}$$

Comment. The pressure decreases exponentially towards zero; the decrease is faster the higher the temperature, the bigger the hole, the lighter the molecules, and the smaller the volume of the container.

Exercise E24.1. Show that $t_{\frac{1}{2}}$, the time required for the pressure to decrease to half its initial value, is independent of the initial pressure.

$$[t_{\frac{1}{2}} = \tau \ln 2]$$

Equation 4 is the basis of the **Knudsen method** for the determination of the vapour pressures of liquids and solids, particularly of substances with very low vapour pressures. Thus, if the vapour pressure of a sample is p and the sample is enclosed in a cavity with a small hole, then the rate of loss of mass from the container is proportional to p.

Example 24.2 *Calculating the vapour pressure from a mass loss*

Some caesium (melting point 29°C; boiling point 686°C) was introduced into a container and heated to 500°C. When a 0.50-mm hole was opened in the container, a mass loss of 385 mg was measured in the course of 100 s. Calculate the vapour pressure of liquid caesium at 500°C.

Method. The pressure of vapour is constant inside the container despite the effusion of molecules because the hot liquid replenishes the vapour (the vapour pressure is constant at a given temperature). The rate of effusion is therefore constant, and given by eqn 4. To express the rate in terms of mass, the number of molecules that escape should be multiplied by the mass of each molecule.

Answer. The mass loss Δw in an interval Δt is related to the collision frequency by

$$\Delta w = Z_W A_0 m \, \Delta t$$

where A_0 is the area of the hole and m the molecular mass. It follows that

$$Z_W = \frac{\Delta w}{A_0 m \, \Delta t}$$

Because Z_W is related to the pressure by eqn 3, we can write

$$p = (2\pi m k T)^{\frac{1}{2}} Z_W = \left(\frac{2\pi k T}{m}\right)^{\frac{1}{2}} \frac{\Delta w}{A_0 \, \Delta t}$$

In terms of the molar mass M, this expression is

$$p = \left(\frac{2\pi R T}{M}\right)^{\frac{1}{2}} \frac{\Delta w}{A_0 \, \Delta t}$$

Because $M = 132.9 \text{ g mol}^{-1}$, substitution of the data gives $p = 11$ kPa (using $1 \text{ Pa} = 1 \text{ N m}^{-2} = 1 \text{ J m}^{-3}$), or 83 Torr.

Exercise E24.2. How long would it take 1.0 g of Cs atoms to effuse out of the oven?

[225 s]

24.3 Migration down gradients

The description of effusion is very simple because a gas at a uniform pressure is allowed to escape into a vacuum, and the molecules that have escaped from the container do not return into it. The other four transport properties we shall consider are more complicated, because the sample is not uniform (for example, there may be a temperature gradient), and molecules that travel in one direction are free to return to their initial region (Fig. 24.6). Nevertheless, considerable progress can be made in showing how the rate of migration in any medium depends on the non-uniformity as expressed by the steepness of the slope of the

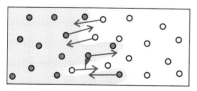

24.6 When considering transport properties other than effusion, the migration of molecules in both directions must be considered, and the *net* flux of a property calculated.

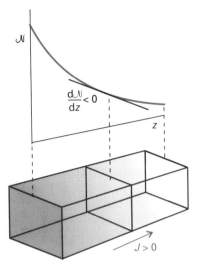

24.7 The flux of particles down a concentration gradient. Fick's first law states that the flux of matter (the number of particles per unit area per unit time) is proportional to the density gradient at that point.

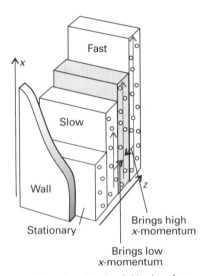

24.8 The viscosity of a fluid arises from the transport of linear momentum. In this illustration the fluid is undergoing laminar flow, and particles bring their initial momentum when they enter a new layer. If they arrive with high x-component of momentum they accelerate the layer; if they arrive with low x-component of momentum they retard the layer.

concentration with distance. Moreover, the kinetic theory can be used to derive some useful expressions for transport in gases.

The rate of migration of a property is measured by its **flux** J, the quantity of that property passing through unit area per unit time. If matter is flowing (as in diffusion), then we speak of a *matter flux* of so many molecules per square metre per second; if the property is energy (as in thermal conduction), then we speak of the *energy flux* and express it in joules per square metre per second, and so on.

Experimental observations on transport properties show that the flux of a property is usually proportional to the slope of the variation of a related property with distance. For example, the flux of matter diffusing parallel to the z-axis of a container is found to be proportional to the slope of the variation of the concentration in that direction:

$$J(\text{matter}) \propto \frac{\mathrm{d}\mathscr{N}}{\mathrm{d}z}$$

In this expression, $\mathscr{N}$ is the number density of particles, the number of particles per unit volume. The proportionality of the flux of matter to the concentration gradient is sometimes called **Fick's first law of diffusion**. Similarly, the rate of thermal conduction (the flux of the energy associated with thermal motion) is found to be proportional to the temperature gradient:

$$J(\text{energy}) \propto \frac{\mathrm{d}T}{\mathrm{d}z}$$

The sign of the flux is significant. If $J > 0$, then the flux is towards positive z; if $J < 0$, then the flux is towards negative z. Matter flow occurs down a concentration gradient, from high concentration to low concentration, and so, if $\mathrm{d}\mathscr{N}/\mathrm{d}z < 0$ (implying that the concentration decreases to the right, Fig. 24.7), then J is positive (flow to the right). Therefore, the coefficient of proportionality in the matter flux expression must be negative, and we write it $-D$. The constant D is then called the **diffusion coefficient** of the species in the medium (for example, of sucrose in water). Hence

$$J(\text{matter}) = -D \frac{\mathrm{d}\mathscr{N}}{\mathrm{d}z} \tag{5}$$

The energy of thermal motion migrates down a temperature gradient, and the same reasoning leads to

$$J(\text{energy}) = -\kappa \frac{\mathrm{d}T}{\mathrm{d}z} \tag{6}$$

where κ is the **coefficient of thermal conductivity**.

To see the connection between the flux of momentum and the viscosity, consider a fluid in a state of **Newtonian flow**, which can be imagined as occurring by a series of layers moving past one another (Fig. 24.8). The layer next to the wall of the vessel is stationary, and the velocity of successive layers varies linearly with the distance z from the wall. Molecules continuously move between the layers and bring with them the x-component of linear momentum that they possessed in their original layer. A layer is retarded by molecules arriving from a more slowly moving layer because they have a lower momentum in the x direction. A layer is accelerated by molecules arriving from a more rapidly moving layer. Because fast layers are retarded and slow layers are accelerated by the arriving molecules, the layers tend towards a uniform velocity. We interpret the retarding effect of the slow layers on the fast layers as the fluid's viscosity.

Because the retarding effect depends on the transfer of the x-component of linear momentum into the layer of interest, the viscosity depends on the flux of this x-component in the z-direction (as shown in Fig. 24.8). The flux of the x-component of momentum is proportional to dv_x/dz because there is no net flux when all the layers move at the same velocity. We can therefore write

$$J(x\text{-component of momentum}) \propto \frac{dv_x}{dz}$$

and hence that

$$J(x\text{-component of momentum}) = -\eta \frac{dv_x}{dz} \qquad (7)$$

The constant η is the **coefficient of viscosity** (or simply 'the viscosity').

Example 24.3 *Deducing the units of a transport coefficient*

Deduce the SI unit of viscosity.

Method. The units of η must be such as to make the units on the right of eqn 7 equal to those on the left. So, we write down the dimensions of the known physical quantities, and select units for η that satisfy the equality.

Answer. The flux of momentum has the dimensions of [momentum] [area]$^{-1}$[time]$^{-1}$, which in SI units is

$$[\text{flux}] = (\text{kg m s}^{-1}) \times \text{m}^{-2} \times \text{s}^{-1} = \text{kg m}^{-1}\,\text{s}^{-2}$$

The dimensions of the velocity gradient are [velocity] $\times$ [length]$^{-1}$, which in SI units is

$$[\text{velocity gradient}] = (\text{m s}^{-1}) \times \text{m}^{-1} = \text{s}^{-1}$$

Therefore, to satisfy

$$\text{kg m}^{-1}\,\text{s}^{-2} = [\text{viscosity}] \times \text{s}^{-1}$$

we need

$$[\text{viscosity}] = \text{kg m}^{-1}\,\text{s}^{-1}$$

Comment. Viscosities are often reported in poise (P), which is defined as $1\,\text{P} = 10^{-1}\,\text{kg m}^{-1}\,\text{s}^{-1}$. The units centipoise $(1\,\text{cP} = 10^{-2}\,\text{P})$ and milli-poise $(1\,\text{mP} = 10^{-3}\,\text{P})$ are also often commonly encountered.

Exercise E24.3. What is the SI unit of thermal conductivity?

$$[\text{J K}^{-1}\,\text{m}^{-1}\,\text{s}^{-1}]$$

Table 24.1 Transport properties of perfect gases

Property	Transported quantity	Simple kinetic theory	Units
Diffusion	Matter	$D = \frac{1}{3}\lambda\bar{c}$	$m^2\,s^{-1}$
Thermal conductivity	Energy	$\kappa = \frac{1}{3}\lambda\bar{c}C_{V,m}[X]$	$J\,K^{-1}\,m^{-1}\,s^{-1}$
		$= \dfrac{\bar{c}C_{V,m}}{(3\sqrt{2})\sigma N_a}$	
Viscosity	Momentum	$\eta = \frac{1}{3}\lambda\bar{c}m\mathcal{N}$	$kg\,m^{-1}\,s^{-1}$
		$= \dfrac{m\bar{c}}{(3\sqrt{2})\sigma}$	

24.4 Transport properties of a perfect gas

We shall now see how the kinetic theory can be used to justify Fick's law and deduce the value of the transport coefficients for a perfect gas. With these expressions established, we shall be able to see how transport properties vary with the conditions.

Diffusion

As shown in the *Justification* that follows and summarized in Table 24.1, the kinetic theory leads to the result that, for a perfect gas,

$$D = \tfrac{1}{3}\lambda\bar{c} \tag{8}°$$

The mean free path λ decreases as the pressure is increased, so D decreases with increasing pressure and, as a result, the gas molecules diffuse more slowly. The mean speed $\bar{c}$ increases with the temperature, so D also increases with temperature. As a result, molecules in a hot sample diffuse more quickly than those in a cool sample (for a given concentration gradient). Because the mean free path decreases with the collision cross-section of the molecules, the diffusion coefficient is greater for small molecules than for large molecules.

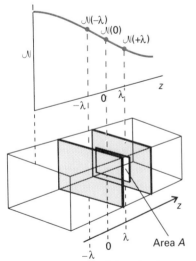

24.9 The calculation of the rate of diffusion of a gas considers the net flux of molecules through a plane of area A as a result of arrivals from, on average, a distance λ away, where λ is the mean free path.

JUSTIFICATION

Consider the arrangement depicted in Fig. 24.9. On average, the molecules passing through the area A at $z = 0$ have travelled about one mean free path λ since their last collision. Therefore, the number density where they originated is $\mathcal{N}(z)$ evaluated at $z = -\lambda$. This number density is approximately

$$\mathcal{N}(-\lambda) = \mathcal{N}(0) - \lambda\left(\frac{d\mathcal{N}}{dz}\right)_0$$

where the subscript 0 indicates that the slope should be evaluated at $z = 0$. Because the average number of impacts on the imaginary window of area A_0 from the left during an interval Δt is $Z_W A_0\,\Delta t$, and we saw in the previous *Justification* that $Z_W = \frac{1}{4}\mathcal{N}\bar{c}$, the flux from left to right,

$J(L \rightarrow R)$, arising from the supply of molecules on the left is

$$J(L \rightarrow R) = \frac{\frac{1}{4}A_0 \mathcal{N}(-\lambda)\bar{c}\,\Delta t}{A_0\,\Delta t} = \frac{1}{4}\mathcal{N}(-\lambda)\bar{c}$$

There is also a flux of molecules from right to left. On average, the molecules making the journey have originated from $z = +\lambda$ where the number density is $\mathcal{N}(\lambda)$. Therefore,

$$J(L \leftarrow R) = \frac{1}{4}\mathcal{N}(\lambda)\bar{c}$$

The average number density at $z = +\lambda$ is approximately

$$\mathcal{N}(\lambda) = \mathcal{N}(0) + \lambda\left(\frac{d\mathcal{N}}{dz}\right)_0$$

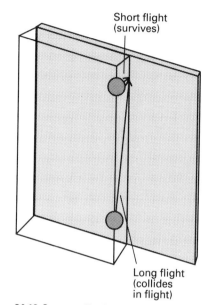

Short flight (survives)

Long flight (collides in flight)

24.10 One complication ignored in the simple treatment is that some particles might make a long flight to the plane even though they are only a short perpendicular distance away, and therefore they have a higher chance of colliding during their journey.

The flow from the more concentrated region on the left dominates the backwash, and the net flux is

$$J = J(L \rightarrow R) - J(L \leftarrow R)$$

$$= \frac{1}{4}\bar{c}\left\{\left[\mathcal{N}(0) - \lambda\left(\frac{d\mathcal{N}}{dz}\right)_0\right] - \left[\mathcal{N}(0) + \lambda\left(\frac{d\mathcal{N}}{dz}\right)_0\right]\right\}$$

$$= -\frac{1}{2}\lambda\bar{c}\left(\frac{d\mathcal{N}}{dz}\right)_0$$

This equation shows that the flux is proportional to the gradient of concentration, in agreement with Fick's law.

At this stage it looks as though we can pick out a value of the diffusion coefficient by comparing the equation just derived with eqn 5, and write $D = \frac{1}{2}\lambda\bar{c}$. It must be remembered, however, that the calculation is quite crude, and is little more than an assessment of the order of magnitude of D. One aspect that has not been taken into account is illustrated in Fig. 24.10, which shows that, although a molecule may have begun its journey very close to the window, it could have a long flight before it gets there. Because the path is long, the molecule is likely to collide before reaching the window, so it ought to be added to the graveyard of other molecules that have collided. Taking this effect into account involves a lot of work, but the end result is the appearance of a factor of $\frac{2}{3}$ representing the lower flux. The modification results in

$$D = \frac{1}{3}\lambda\bar{c}$$

as in eqn 8.

Table 24.2* Transport properties of gases at 1 atm

	$\kappa/(\text{mJ cm}^{-2}\,\text{s}^{-1}$ $(\text{K cm}^{-1})^{-1})$	$\eta/\mu\text{P}\dagger$	
	273 K	273 K	293 K
Ar	0.163	210	223
CO_2	0.145	136	147
He	1.442	187	196
N_2	0.240	166	176

* More values are given in the Data section at the end of this volume.
† $1\,\mu\text{P} = 10^{-7}\,\text{kg m}^{-1}\,\text{s}^{-1}$.

Thermal conduction

To explain thermal conduction we need to account for the proportionality of the flux of the energy of thermal motion to the temperature gradient, and then to find an expression for the coefficient of thermal conductivity, κ. Some experimental values are given in Table 24.2. Thermal conductivities are made use of in the 'Pirani gauge', by which pressure is measured by monitoring the temperature of a heated wire, and in 'katharometer detectors' in gas–liquid chromatography (GLC), where changes of composition are detected similarly.

According to the kinetic theory of gases, and as shown in the following *Justification*, the thermal conductivity of a perfect gas A of molar concentration [A] is given by the expression

$$\kappa = \tfrac{1}{3}\lambda\bar{c}C_{V,\mathrm{m}}[\mathrm{A}] \tag{9}°$$

where $C_{V,\mathrm{m}}$ is the molar heat capacity at constant volume.[1]

JUSTIFICATION

Suppose each molecule carries an average energy $\varepsilon = vkT$, where v is a number of the order of 1 and can be obtained from the equipartition theorem (see the *Introduction*). For monatomic particles, $v = \tfrac{3}{2}$. When one molecule passes through the imaginary window, it transports that energy on average. We shall suppose that the number density is uniform (so there is no diffusion of matter) but that the temperature is not. On average, molecules arrive from the left after travelling a mean free path from their last collision in a hotter region and therefore with a higher energy. They arrive from the right after travelling a mean fee path from a cooler region. The energy fluxes in the two directions are therefore

$$J(\mathrm{L}\rightarrow\mathrm{R}) = \tfrac{1}{4}\bar{c}\mathcal{N}\varepsilon(-\lambda) \qquad \varepsilon(-\lambda) = vk\left\{T - \lambda\left(\frac{\mathrm{d}T}{\mathrm{d}z}\right)_0\right\}$$

$$J(\mathrm{L}\leftarrow\mathrm{R}) = \tfrac{1}{4}\bar{c}\mathcal{N}\varepsilon(\lambda) \qquad \varepsilon(\lambda) = vk\left\{T + \lambda\left(\frac{\mathrm{d}T}{\mathrm{d}z}\right)_0\right\}$$

and the net energy flux is

$$J = J(\mathrm{L}\rightarrow\mathrm{R}) - J(\mathrm{L}\leftarrow\mathrm{R}) = -\tfrac{1}{2}v\lambda\bar{c}k\mathcal{N}\left(\frac{\mathrm{d}T}{\mathrm{d}z}\right)_0$$

As before, we multiply by $\tfrac{2}{3}$ to take long flight paths into account, and so arrive at

$$J = -\tfrac{1}{3}v\lambda\bar{c}k\mathcal{N}\left(\frac{\mathrm{d}T}{\mathrm{d}z}\right)_0$$

The energy flux is proportional to the temperature gradient, as we wanted to show. Comparison of this equation with eqn 6 shows that

$$\kappa = \tfrac{1}{3}v\lambda\bar{c}k\mathcal{N}$$

For a perfect gas $C_{V,\mathrm{m}} = vkN_{\mathrm{A}}$, so we can express this relation as

$$\kappa = \tfrac{1}{3}\lambda\bar{c}C_{V,\mathrm{m}}[\mathrm{A}]$$

where [A] is the molar concentration of A.[2]

---

1 To use eqn 9 with SI units, the mean free path should be in metres, the mean speed in metres per second ($\mathrm{m\,s}^{-1}$), the molar heat capacity in joules per kelvin per mole ($\mathrm{J\,K}^{-1}\,\mathrm{mol}^{-1}$), and the molar concentration in moles per cubic metre ($\mathrm{mol\,m}^{-3}$). Note that $1\ \mathrm{mol\,L}^{-1} = 10^3\ \mathrm{mol\,m}^{-3}$.

2 We have used

$$\mathcal{N} = N/V = nN_{\mathrm{A}}/V = N_{\mathrm{A}}[\mathrm{A}]$$

where N_{A} is Avogadro's constant and $[\mathrm{A}] = n/V$.

Because λ is inversely proportional to the pressure, and hence inversely proportional to the molar concentration of the gas, it follows that $\kappa \propto \bar{c} C_{V,m}$ and the thermal conductivity is independent of the pressure. The physical reason for this independence is that the thermal conductivity can be expected to be large when many molecules are available to transport the energy, but the presence of so many molecules limits their mean free path and they cannot carry the energy over a great distance. These two effects balance.

The thermal conductivity is found experimentally to be independent of the pressure, except when the pressure is very low. At low pressures $\kappa \propto p$ because λ is then greater than the dimensions of the apparatus and the distance over which the energy is transported is determined by the size of the container and not by the other molecules present. The flux is still proportional to the number of carriers, but the length of the journey no longer depends on λ, so $\kappa \propto [A] \propto p$.

Example 24.4 *Estimating the thermal conductivity of a gas*

Estimate the thermal conductivity of air at room temprature.

Method. We can use eqn 9, noting that the constant-volume molar heat capacity of a gas of diatomic molecules is $\frac{5}{2}R$ (Section 20.4) and that the molar mass of air is about 29 g mol^{-1}. Take $\sigma = 0.42$ nm^2 for O_2 and N_2 from Table 1.2 and estimate $\bar{c}$ from eqn 2. It is sensible to proceed algebraically as far as possible, and then to insert numerical values after cancellation of terms.

Answer. The algebraic expression for κ is

$$\kappa = \frac{1}{3} \times \frac{kT}{\sqrt{2}\,\sigma p} \times \left(\frac{8RT}{\pi M}\right)^{\frac{1}{2}} \times \frac{5}{2} R \times \frac{p}{RT} = \frac{5}{3\pi^{\frac{1}{2}}} \frac{k}{\sigma} \left(\frac{RT}{M}\right)^{\frac{1}{2}}$$

Substitution of the data then gives

$$\kappa = \frac{5}{3\pi^{\frac{1}{2}}} \times \frac{1.381 \times 10^{-23}\,\text{J K}^{-1}}{0.42 \times 10^{-18}\,\text{m}^2}$$
$$\times \left(\frac{(8.3145\,\text{J K}^{-1}\,\text{mol}^{-1}) \times (298\,\text{K})}{29 \times 10^{-3}\,\text{kg mol}^{-1}}\right)^{\frac{1}{2}}$$
$$= 9.0 \times 10^{-3}\,\text{J K}^{-1}\,\text{m}^{-1}\,\text{s}^{-1}$$

Comment. A more convenient but unwieldy form of the result is $\kappa = 9.0 \times 10^{-2}$ mJ cm^{-2} s^{-1}/(K cm^{-1}). Then, in a temperature gradient of 1 K cm^{-1} the heat flux would be about 0.1 mJ cm^{-2} s^{-1}.

Exercise E24.4. Would carbon dioxide be a better insulator under the same conditions? $[\kappa = 6 \times 10^{-2}$ mJ cm^{-2} s^{-1} (K cm^{-1})$^{-1}$; yes]

The viscosity of a perfect gas

We have seen that viscosity is related to the flux of momentum. Once again, as we show in the following *Justification*, the kinetic theory of gases can be used to construct an expression for the viscosity coefficient. It is found that

$$\eta = \tfrac{1}{3} M \lambda \bar{c} [A] \qquad\qquad (10)^\circ$$

where [A] is the molar concentration of the gas molecules and M their molar mass.

JUSTIFICATION

Molecules travelling from the right in Fig. 24.11 (from a faster layer to a slower one) transport a momentum $m v_x(\lambda)$ to their new layer at $z = 0$, and those travelling from the left transport $m v_x(-\lambda)$ to it. If it is assumed that the density is uniform (an approximation), then the number of impacts per unit area per unit time on the imaginary window is $\tfrac{1}{4} \mathcal{N} \bar{c}$. Those from the right on average carry a momentum

$$m v_x(\lambda) = m v_x(0) + m\lambda \left(\frac{\mathrm{d}v_x}{\mathrm{d}z}\right)_0$$

Those from the left bring a momentum

$$m v_x(-\lambda) = m v_x(0) - m\lambda \left(\frac{\mathrm{d}v_x}{\mathrm{d}z}\right)_0$$

The net flux of x-momentum in the z-direction is therefore

$$J = \tfrac{1}{4} \mathcal{N} \bar{c} \left\{ \left[m v_x(0) - m\lambda \left(\frac{\mathrm{d}v_x}{\mathrm{d}z}\right)_0 \right] - \left[m v_x(0) + m\lambda \left(\frac{\mathrm{d}v_x}{\mathrm{d}z}\right)_0 \right] \right\}$$

$$= -\tfrac{1}{2} \mathcal{N} m \lambda \bar{c} \left(\frac{\mathrm{d}v_x}{\mathrm{d}z}\right)_0$$

We see that the flux is proportional to the velocity gradient, as we wished to show. By comparing this expression with eqn 7, and multiplying by $\tfrac{2}{3}$ in the normal way, we can identify the viscosity coefficient as

$$\eta = \tfrac{1}{3} \mathcal{N} m \lambda \bar{c}$$

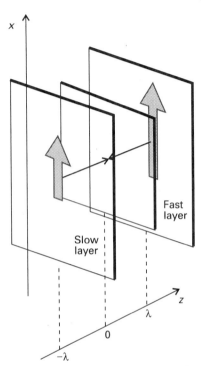

24.11 The calculation of the viscosity of a gas examines the net x-component of momentum brought to a plane from faster and slower layers on average a mean free path away.

As is the case also for thermal conductivity, the viscosity is independent of the pressure. Thus $\lambda \propto 1/p$ and $[A] \propto p$, implying that $\eta \propto \bar{c}$, independent of p. The physical reason is also the same: more molecules are available to transport the momentum, but they carry it less far on account of the shorter mean free path. An unexpected result is that, because $\bar{c} \propto T^{\frac{1}{2}}$, the viscosity coefficient is proportional to $T^{\frac{1}{2}}$. That is, the viscosity of a gas *increases* with temperature. This conclusion is explained when we remember that at high temperatures the molecules travel more quickly, so the flux of momentum is greater.[3]

There are two main techniques for measuring viscosities of gases. One depends on the rate of damping of the torsional oscillations of a disk hanging in the gas. The half-life of the decay of the oscillation depends on the viscosity and the design of the apparatus, and it needs to be

3 As we shall see in Section 24.6, the viscosity of a liquid *decreases* with temperature because intermolecular interactions must be overcome.

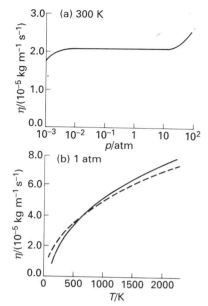

24.12 The experimental results for (a) the pressure dependence of the viscosity of argon, and (b) its temperature dependence. The broken line in the latter is the calculated value. Fitting the observed and calculated curves is one way of determining the collision cross-section.

calibrated. The other method is based on **Poiseuille's formula** for the rate of flow of a fluid through a tube of radius r:

$$\frac{\mathrm{d}V}{\mathrm{d}t} = \frac{(p_1^2 - p_2^2)\pi r^4}{16l\eta p_0} \tag{11}$$

where V is the volume flowing, p_1 and p_2 are the pressures at each end of the tube of length l, and p_0 is the pressure at which the volume is measured.

Such measurements confirm that the viscosities of gases are independent of pressure over a wide range. For instance, the results for argon from 10^{-3} atm to 10^2 atm are shown in Fig. 24.12, and we see that η is constant from about 0.01 atm to 20 atm. The measurements also confirm (to a lesser extent) the $T^{\frac{1}{2}}$ dependence. The dotted line in the illustration shows the calculated values using $\sigma = 22 \times 10^{-20}\,\mathrm{m}^2$, implying a collision diameter of 260 pm, in contrast to the van der Waals diameter of 335 pm obtained from the density of the solid. The agreement is not too bad, considering the simplicity of the model and the neglect of intermolecular forces.

Example 24.5 *Measuring the viscosity using Poiseuille's method*

A Poiseuille flow experiment was carried out to measure the viscosity of air at 298 K. The sample was allowed to flow through a 100-cm tube of internal diameter 1.00 mm. The high-pressure end was at 765 Torr and the low-pressure end was at 760 Torr. The volume was measured at the latter pressure. In 100 s a volume of 90.2 cm³ passed through the tube. What is the viscosity of air at this temperature?

Method. We use eqn 11, after converting pressures to pascals by using 1 Torr = 133.3 Pa.

Answer. The rate of flow is

$$\frac{\mathrm{d}V}{\mathrm{d}t} = \frac{90.2\,\mathrm{cm}^3}{100\,\mathrm{s}} = 9.02 \times 10^{-7}\,\mathrm{m}^3\,\mathrm{s}^{-1}$$

Then, since

$$p_1^2 - p_2^2 = 1.355 \times 10^8\,\mathrm{Pa}^2$$

and

$$\frac{\pi r^4}{16lp_0} = 1.21 \times 10^{-19}\,\mathrm{N}^{-1}\,\mathrm{m}^5$$

it follows from eqn 11 that $\eta = 1.8 \times 10^{-5}\,\mathrm{kg\,m}^{-1}\,\mathrm{s}^{-1}$.

Comment. The kinetic theory expression (eqn 10) gives $\eta = 1.4 \times 10^{-5}\,\mathrm{kg\,m}^{-1}\,\mathrm{s}^{-1}$, so the agreement is quite good. Viscosities are often expressed in centipoise (cP) or (for gases) micropoise (μP), the conversion being 1 cP = $10^{-3}\,\mathrm{kg\,m}^{-1}\,\mathrm{s}^{-1}$: the viscosity of air at 20°C is 180 μP.

Exercise E24.5. What volume would be collected if the pressure gradient were doubled, other conditions remaining constant?   [180 cm³]

THE MOTION OF MOLECULES AND IONS IN LIQUIDS

As a first step in dealing with the much more difficult problem of motion in liquids, we shall outline what is currently known about the structure of a simple liquid. Then we shall turn attention to a particularly simple type of motion through a liquid, that of an ion, and see that the information that motion provides can be used to infer the behaviour of uncharged species too.

24.5 The structure of liquids

The starting point for the discussion of solids is the well-ordered structures of perfect crystals. The starting point for the discussion of gases is the totally chaotic distribution of the molecules of a perfect gas. Liquids lie between these two extremes: there is some short-range order but little long-range order.

The radial distribution function

The particles of a liquid are held together by intermolecular forces, but their kinetic energies are comparable to their potential energies. As a result, the whole structure is very mobile. The mathematical description of the average locations of the particles is expressed in terms of the **radial distribution function** g; this function is defined so that $gr^2\,dr$ is the probability that a molecule will be found in the range dr at a distance r from another.

In a perfect crystal, g is a periodic array of sharp spikes, representing the certainty (in the absence of defects and thermal motion) that particles lie at definite locations. This regularity continues out to the edges of the crystal, so we say that crystals have **long-range order**. When the crystal melts, the long-range order is lost and, wherever we look at long distances from a given particle, there is equal probability of finding a second particle. Close to the first particle, though, the nearest neighbours might still adopt approximately their original relative positions and, even if they are displaced by newcomers, the new particles might adopt their vacated positions. It is still possible to detect a sphere of nearest neighbours at a distance r_1, and perhaps beyond them a sphere of next-nearest neighbours at r_2. The existence of this **short-range order** means that the radial distribution function can be expected to oscillate at short distances, with a peak at r_1, a smaller peak at r_2, and perhaps some more structure beyond that.

The shape of the radial distribution function can be determined by X-ray diffraction, for g can be extracted from the diffuse diffraction pattern characteristic of liquid samples in much the same way as a crystal structure is obtained from X-ray diffraction of crystals. The shells of local structure shown in the example in Fig. 24.13 (for water) are unmistakable. Closer analysis shows that any given H_2O molecule is surrounded by other molecules at the corners of a tetrahedron, similar to the arrangement in ice (Fig. 21.33). The form of g at 100°C shows that the intermolecular forces (in this case, largely hydrogen bonds) are strong enough to affect the local structure right up to the boiling point.

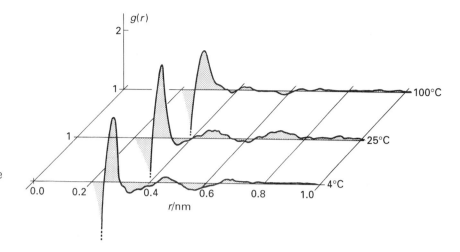

24.13 The radial distribution function of the oxygen atoms in liquid water at three temperatures. Note the expansion as the temperature is raised. (A. H. Narten, M. D. Danford, and H. A. Levy, *Discuss. Faraday Soc.* **43**, 97 (1967).)

The calculation of g

Because the radial distribution function can be calculated by making assumptions about the intermolecular forces, it can be used to test theories of liquid structure. However, even a fluid of hard spheres without attractive interactions (a collection of ball-bearings in a container) gives a function that oscillates near the origin (Fig. 24.14), and one of the factors influencing, and sometimes dominating, the structure of a liquid is the geometrical problem of stacking together reasonably hard spheres. Indeed, a liquid of hard spheres shows a more pronounced oscillatory behaviour at a given temperature than any other type of liquid. The attractive part of the potential modifies this basic structure but sometimes only quite weakly. One of the reasons behind the difficulty of describing liquids theoretically is the importance of both the attractive and repulsive (hard core) components of the potential.

There are several ways of building the intermolecular potential into the calculation of g. Numerical methods take a box of 10^3 particles (the number increases as computers grow more powerful), and the rest of the liquid is simulated by surrounding the box with replications of the original box (Fig. 24.15). Then, whenever a particle leaves the box through one of its faces, its image arrives through the opposite face so that the total number remains constant.

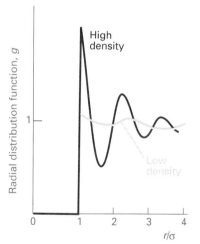

24.14 The radial distribution function for a simulation of a liquid using impenetrable hard spheres (ball bearings).

Monte Carlo methods

In the **Monte Carlo method**, the particles in the box are moved (usually one at a time) through small but otherwise random distances, and the total potential energy is calculated using one of the intermolecular potentials discussed in Chapter 22. Whether or not this new configuration is accepted is then judged from the following rules:

(1) If the potential energy is not greater than before the change, then the configuration is accepted.

(2) If the potential energy is greater than before the change, then it is accepted or rejected with the probability of acceptance in

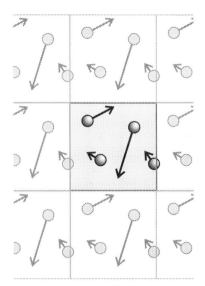

proportion to the value of $e^{-\Delta V_N/kT}$, where ΔV_N is the change in total potential energy of the N particles in the box.

It can be shown that this procedure ensures that at equilibrium the probability of occurrence of any configuration is proportional to the Boltzmann factor $e^{-V_N/kT}$. The configurations generated in this way can then be analyzed for the value of g simply by counting the number of pairs of particles with a separation r and averaging the result over the whole collection of configurations.

Molecular dynamics

In the **molecular dynamics** approach, the history of an initial arrangement is followed by calculating the trajectories of all the particles under the influence of the intermolecular potentials. Newton's laws are used to predict where each particle will be after a short time interval (about 10^{-15} s, which is shorter than the average time between collisions), and then the calculation is repeated for tens of thousands of such steps. The procedure gives a series of snapshots of the liquid, and g can be calculated as before.

Once g is known it can be used to calculate the thermodynamic properties of liquids. For example, the contribution of the pairwise additive intermolecular potential to the internal energy is given by the integral,

$$U = \frac{2\pi N^2}{V} \int_0^\infty g V_2 r^2 \, dr \tag{12}$$

That is, it is the average two-particle potential energy weighted by $g(r)r^2 \, dr$, which is the probability that the pair of particles have a separation between r and $r + dr$. Likewise, the contribution to the pressure of pairwise interactions is given by

$$\frac{pV}{nRT} = 1 - \frac{2\pi N}{kTV} \int_0^\infty g\left\{r\left(\frac{dV_2}{dr}\right)\right\} r^2 \, dr \tag{13}$$

The quantity $r(dV_2/dr)$ is called the **virial** (hence the term 'virial equation of state').

Liquid crystals

A feature that makes calculations even more difficult is the possibility that molecules have strongly anisotropic interactions. An important extreme case of anisotropy gives rise to a **mesophase**, a phase intermediate between solid and liquid.

A mesophase may arise when molecules have highly anisotropic shapes, such as being long and thin, as in (**1**), or disk-like. When the solid melts, some aspects of the long-range order characteristic of the solid may be retained, and the new phase may be a **liquid crystal**, a substance having liquid-like imperfect long-range order in at least one direction in space but positional or orientational order in at least one other direction (Fig. 24.16). One type of retained long-range order gives rise to a **smectic phase** (from the Greek word for soapy), in which the

24.15 In a two-dimensional molecular dynamics simulation that uses periodic boundary conditions, when one particle leaves the cell its mirror image enters through the opposite face.

CN

1

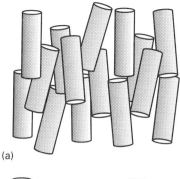

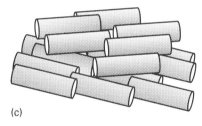

24.16 The arrangement of molecules in (a) the nematic phase, (b) the smectic phase, and (c) the cholesteric phase of liquid crystals. In the cholesteric phase, the stacking of layers continues to give a helical arrangement of molecules.

molecules align themselves in layers. Other materials, and some smectic liquid crystals at higher temperatures, lack the layered structure but retain a parallel alignment: this mesophase is called a **nematic phase** (from the Greek for thread, which refers to the observed defect structure of the phase). In the **cholesteric phase** (from the Greek for bile solid) the molecules lie in sheets at angles that change slightly between each sheet. That is, they form helical structures with a pitch that depends on the temperature. As a result, cholesteric liquid crystals diffract light and have colours that depend on the temperature. The strongly anisotropic optical properties of nematic liquid crystals, and their response to electric fields, is the basis of their use as data displays in calculators and watches.

Although there are many liquid crystalline materials, some difficulty is often experienced in achieving a technologically useful temperature range for the existence of the mesophase. To overcome this difficulty, mixtures can be used. An example of the type of phase diagram that is then obtained is shown in Fig. 24.17. As can be seen, the mesophase exists over a wider range of temperatures than either liquid crystalline material alone.

24.6 Molecular motion in liquids

The motion of molecules in liquids can be studied experimentally by a variety of methods. Relaxation time measurements in NMR and ESR (Section 18.5) can be interpreted in terms of the mobilities of the particles, and have been used to show that some types of molecule rotate in a series of small (about 5°) steps, but that others jump through about 1 radian (57°) in each step. Another important technique is **inelastic neutron scattering**, in which the energy neutrons collect or discard as they pass through a sample is interpreted in terms of the motion of its particles, and very detailed information can be obtained.

More mundane than these experiments are viscosity measurements (some values are given in Table 24.3). Unlike in a gas, if a molecule is to move in a liquid, it must escape from its neighbours. Consequently, a molecule moves only if it acquires at least a minimum energy. The probability that a molecule has at least an energy E_a is proportional to $e^{-E_a/RT}$, so the mobility of the molecules in the liquid should follow this type of temperature dependence. Because the coefficient of viscosity η is inversely proportional to the mobility of the particles, we should expect that

$$\eta \propto e^{E_a/RT} \qquad (14)$$

This expression implies that the viscosity should decrease exponentially with increasing temperature. Such a variation is found experimentally, at least over reasonably small temperature ranges (Fig. 24.18).

One problem with the interpretation of viscosity measurements is that the change in density of the liquid as it is heated makes a pronounced contribution to the temperature variation of the viscosity. Thus, the temperature dependence of viscosity at constant volume, when the density is constant, is much less than at constant pressure. The intermolecular forces between the molecules of the liquid govern the

Table 24.3* Viscosities of liquids at 298 K, $\eta/(10^{-3}\,\text{kg m}^{-1}\,\text{s}^{-1})$

Benzene	0.601
Mercury	1.53
Pentane	0.224
Water†	0.891

* More values are given in the Data section.

† The viscosity of water corresponds to 0.891 cP.

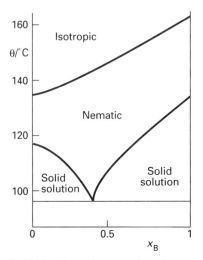

24.17 The phase diagram at 1 atm for a binary system of two liquid crystalline materials, 4,4′-dimethoxyazoxybenzene (A) and 4,4′-diethoxyazoxybenzene (B).

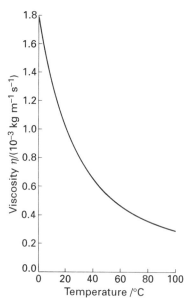

24.18 The experimental temperature dependence of the viscosity of water. As the temperature is increased, more molecules are able to escape from the potential wells provided by their neighbours, and so the liquid becomes more fluid. A plot of $\ln \eta$ against $1/T$ is a straight line with positive slope.

magnitude of E_a, but the problem of calculating it is immensely difficult and still largely unsolved.

Further insight into the nature of molecular motion can be obtained by studying the motion of ions in solution, for they can be dragged through the solution by the application of a potential difference between two electrodes immersed in the sample. By studying the transport of charge through electrolyte solutions it is possible to build up a picture of the events that occur in them and, in some cases, to extrapolate the conclusions to species that are neutral molecules.

The motion of ions in the presence of an electric field is simple (as long as we do not enquire too closely into the details) because the ions drift steadily through the solvent towards one electrode or the other. We shall see that the electric force acting on the ions can be regarded as a special case of a generalized force. The formulation of this generalized force will enable us to discuss the motion of ions and neutral molecules in greater detail and so describe their diffusion through solutions even in the absence of electric fields. We shall then be in a position to discuss (in the following chapters) the rates of chemical reactions, which depend on the migration of particles through media of various kinds.

24.7 The conductivity of electrolyte solutions

The fundamental measurement used to study the motion of ions is that of the electrical resistance of the solution. The standard technique is to incorporate a **conductivity cell** (Fig. 24.19) into one arm of a resistance bridge and to search for the balance point, as explained in standard texts on electricity. The main complication is that alternating current must be used because a direct current would lead to electrolysis and to **polarization**, the modification of the composition of the layers of solution in contact with the electrodes. The use of alternating current (with a frequency of about 1 kHz) may avoid polarization because the charging that occurs on one-half of the cycle is undone during the second half (if the reverse reaction is kinetically feasible).

Conductance and conductivity

The **conductance** G of a solution is the inverse of its **resistance** R: the lower the resistance of a solution, the greater its conductance. Resistance is expressed in ohms, Ω, so the conductance of a sample is expressed in Ω^{-1}. The reciprocal ohm used to be called the mho, but its official designation is now the **siemens**, S, and $1\,S = 1\,\Omega^{-1}$.

The resistance of a sample increases with its length l and decreases with its cross-sectional area A. We therefore write

$$R = \rho \times \frac{l}{A} \tag{15}$$

The constant of proportionality ρ is called the **resistivity** of the sample. The **conductivity** κ is the inverse of the resistivity, so

$$R = \frac{1}{\kappa} \times \frac{l}{A}, \qquad \text{or } \kappa = \frac{l}{RA} \tag{16}$$

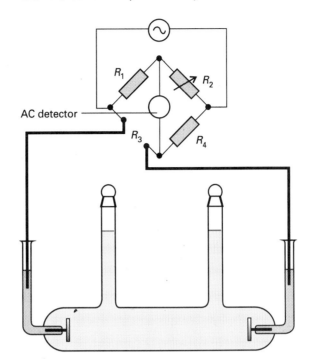

24.19 The conductivity of an electrolyte solution is measured by making the cell one arm of a resistance bridge. The balance point is obtained when the resistances satisfy $R_1/R_2 = R_3/R_4$ (or, since the current is alternating, a similar relation for the impedances).

With the resistance in ohms and the dimensions in metres, it follows that the units of κ are siemens per metre (S m^{-1}; the units S cm^{-1} are sometimes more convenient).

It is unreliable to calculate the conductivity direct from the resistance of the sample and the cell dimensions l and A because the current distribution is complicated. In practice, the cell is calibrated using a sample (typically an aqueous solution of potassium chloride) of known conductivity, κ^*, and a **cell constant** C is determined from

$$\kappa^* = \frac{C}{R^*}$$

where R^* is the resistance of the standard. If the sample was a resistance R in the same cell, then its conductivity is

$$\kappa = \frac{C}{R}$$

The conductivity of a solution depends on the number of ions present, and it is normal to introduce the **molar conductivity** Λ_m, which is defined as

$$\Lambda_m = \frac{\kappa}{c} \tag{17}$$

where c is the molar concentration of the added electrolyte. The molar conductivity is normally expressed[4] in siemens centimetre-squared per mole (S cm^2 mol^{-1}).

---

4 The conductivity is normally available in S cm^{-1} and the molar concentration in mol L^{-1}, so the practical relation is

$$\Lambda_m = 1000 \times \frac{\kappa/(\text{S cm}^{-1})}{c/(\text{mol L}^{-1})} \text{S cm}^2\,\text{mol}^{-1}$$

Example 24.6 *Calculating the molar conductivity of a solution*

The molar conductivity of 0.100 M KCl(aq) at 298 K is 129 S cm² mol⁻¹. The measured resistance in a conductivity cell was 28.44 Ω. The resistance was 31.60 Ω when the same cell contained 0.0500 M NaOH(aq). Calculate the molar conductivity of NaOH(aq) at that concentration.

Method. First, we need to determine the cell constant, and then use this constant and the given resistance to calculate the conductivity of the test solution. The molar conductivity is then found by dividing the conductivity by the molar concentration.

Answer. The conductivity of a 0.100 M KCl(aq) solution, which contains 1.00×10^{-1} mol L⁻¹ of KCl (corresponding to 1.00×10^{-4} mol cm⁻³) is

$$\kappa = \Lambda_m c = 129 \text{ S cm}^2 \text{ mol}^{-1} \times (1.00 \times 10^{-4} \text{ mol cm}^{-3})$$
$$= 1.29 \times 10^{-2} \text{ S cm}^{-1}$$

It follows that the cell constant is

$$C = 1.29 \times 10^{-2} \text{ S cm}^{-1} \times 28.44 \ \Omega = 0.367 \text{ cm}^{-1}$$

Therefore, for NaOH(aq),

$$\kappa = \frac{C}{R} = \frac{0.367 \text{ cm}^{-1}}{31.60 \ \Omega} = 1.16 \times 10^{-2} \text{ S cm}^{-1}$$

The molar conductivity of NaOH(aq) is therefore

$$\Lambda_m = \frac{1.16 \times 10^{-2} \text{ S cm}^{-1}}{5.00 \times 10^{-5} \text{ mol cm}^{-3}} = 232 \text{ S cm}^2 \text{ mol}^{-1}$$

Comment. It is good practice to calibrate cells with solutions having conductances close to the conductance of the sample of interest.

Exercise E24.6. The same cell was used to study 0.100 M NH₄Cl(aq), and a resistance of 28.50 Ω was measured. Calculate the molar conductivity of NH₄Cl(aq) at this concentration.   [129 S cm² mol⁻¹]

The molar conductivity of an electrolyte would be independent of concentration if κ were proportional to the concentration of the electrolyte. However, in practice, the molar conductivity is found to vary with the concentration. One reason for this variation is that the number of ions in the solution might not be proportional to the concentration of the electrolyte. For instance, the concentration of ions in a solution of a weak acid depends on the concentration of the acid in a complicated way, and doubling the nominal concentration of the acid does not double the number of ions. Second, because ions interact with each other strongly, the conductivity of a solution is not exactly proportional to the number of ions present.

Measurement of the concentration dependence of molar conductivities shows that there are two classes of electrolyte. The characteristic of a

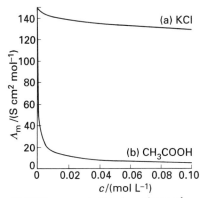

24.20 The concentration dependence of the molar conductivities of (a) a typical strong electrolyte and (b) a typical weak electrolyte.

strong electrolyte is that its molar conductivity decreases only slightly as its concentration is increased (Fig. 24.20). The characteristic of a weak electrolyte is that its molar conductivity is normal at concentrations close to zero, but falls sharply to low values as the concentration increases. The classification depends on the solvent employed as well as the solute: lithium chloride, for example, is a strong electrolyte in water but a weak electrolyte in propanone.

Strong electrolytes

Strong electrolytes are substances that are fully ionized in solution, and include ionic solids and strong acids. As a result of their complete ionization, the concentration of ions in solution is proportional to the concentration of electrolyte added.

In an extensive series of measurements during the nineteenth century, Friedrich Kohlrausch showed that at low concentrations the molar conductivities of strong electrolytes vary as the square root of the concentration:

$$\Lambda_m = \Lambda_m^\circ - \mathcal{K}c^{\frac{1}{2}} \tag{18}$$

This variation is called **Kohlrausch's law.** The constant Λ_m° is the **limiting molar conductivity,** the molar conductivity in the limit of zero concentration (when the ions do not interact with each other). The constant $\mathcal{K}$ is a coefficient which is found to depend more on the stoichiometry of the electrolyte (i.e., whether it is of the form MA, or M_2A, etc.) than on its specific identity.

Kohlrausch was also able to confirm that Λ_m° can be expressed as the sum of contributions from its individual ions. If the limiting molar conductivity of the cations is denoted λ_+ and that of the anions λ_-, then his **law of the independent migration of ions** is that

$$\Lambda_m^\circ = \nu_+\lambda_+ + \nu_-\lambda_- \tag{19}°$$

where ν_+ and ν_- are the numbers of cations and anions per formula unit of electrolyte (e.g. $\nu_+ = \nu_- = 1$ for HCl, NaCl, and $CuSO_4$, but $\nu_+ = 1$, $\nu_- = 2$ for $MgCl_2$). This simple result, which can be understood on the grounds that the ions behave independently in the limit of zero concentration, lets us predict the limiting molar conductivity of any strong electrolyte from the data in Table 24.4. For example, the limiting molar conductivity of $BaCl_2$ in water at 298 K is

$$\Lambda_m^\circ = (127.2 + 2 \times 76.3)\ S\ cm^2\ mol^{-1} = 279.8\ S\ cm^2\ mol^{-1}$$

Table 24.4* Limiting ionic conductivities in water at 298 K, $\lambda/(S\ cm^2\ mol^{-1})$

H^+	349.6	OH^-	199.2
Na^+	50.1	Cl^-	76.3
K^+	73.5	Br^-	78.1
Zn^{2+}	105.6	SO_4^{2-}	160.0

* More values are given in the Data section.

Weak electrolytes

Weak electrolytes are substances that are not fully ionized in solution. They include weak Brønsted acids and bases such as CH_3COOH and NH_3. The marked concentration dependence of their conductivities arises from the displacement of the equilibrium

$$HA(aq) + H_2O(l) \rightleftharpoons H_3O^+(aq) + A^-(aq) \qquad K_a = \frac{a(H_3O^+)a(A^-)}{a(HA)}$$

towards the right at low molar concentrations.

The conductivity depends on the number of ions in the solution, and therefore on the **degree of ionization** α of the electrolyte. The degree of ionization is defined so that, for the acid HA at a nominal concentration c, at equilibrium

$$[H_3O^+] = \alpha c \qquad [A^-] = \alpha c \qquad [HA] = (1 - \alpha)c$$

If we ignore activity coefficients, the acidity constant is approximately

$$K_a = \frac{\alpha^2 c}{1 - \alpha} \tag{20}$$

The degree of ionization when the concentration is c is obtained by rearranging this expression into a quadratic equation for α and solving it to

$$\alpha = \frac{K_a}{2c}\left\{\left(1 + \frac{4c}{K_a}\right)^{\frac{1}{2}} - 1\right\} \tag{21}$$

If the molar conductivity of the hypothetical fully ionized electrolyte is Λ_m', then since only a fraction α is actually present as ions in the actual solution, the measured molar conductivity Λ_m is given by

$$\Lambda_m = \alpha \Lambda_m'$$

with α given by eqn 21. When the concentration of ions in solution is very low, we can approximate Λ_m' by its limiting value, and write

$$\Lambda_m = \alpha \Lambda_m^\circ \tag{22}$$

Example 24.7 *Using conductivity measurements to determine* pK_a

The resistance of a 0.0100 M CH_3COOH(aq) solution was measured at 298 K in the same cell as in Example 24.6, and found to be 2220 Ω. Find the degree of ionization of the acid at this concentration and its pK_a.

Method. To calculate K_a we need the value of α, which can be obtained from the ratio of Λ_m (obtained by using the cell constant determined in Example 24.6) and the value of Λ_m° (obtained from the data in Table 24.4).

Answer. Because the cell constant is $C = 0.367 \text{ cm}^{-1}$, we find $\Lambda_m = 16.5 \text{ S cm}^2 \text{ mol}^{-1}$. From the data in Table 24.4 we find $\Lambda_m^\circ = 390.5 \text{ S cm}^2 \text{ mol}^{-1}$. Therefore, $\alpha = 0.0423$. The acidity constant is then obtained by substituting this value of α into eqn 20, and is 1.9×10^{-5}, implying that p$K_a = 4.72$.

Comment. The thermodynamic value of pK_a is obtained by repeating the determination with different concentrations and extrapolating to zero concentration.

Exercise E24.7. The resistance of 0.0250 M HCOOH(aq) was measured as 444 Ω in the same cell; evaluate the pK_a of the acid.   [3.74]

Once we know K_a, we can use eqn 21 to predict the concentration dependence of the molar conductivity. The result agrees quite well with

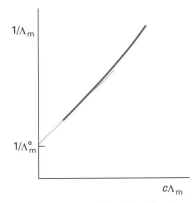

24.21 The graph used to determine the limiting value of the molar conductivity of a solution by extrapolation to zero concentration.

the experimental curve in Fig. 24.20. Alternatively, we can use the concentration dependence of Λ_m in measurements of the limiting molar conductance. It is quite easy to manipulate the expression for K_a into the form

$$\frac{1}{\alpha} = 1 + \frac{\alpha c}{K_a}$$

Then, using eqn 21, we obtain **Ostwald's dilution law**:

$$\frac{1}{\Lambda_m} = \frac{1}{\Lambda_m^\circ} + \frac{\Lambda_m c}{K_a(\Lambda_m^\circ)^2} \tag{23}$$

This equation suggests a procedure for determining the limiting molar conductivity of a solution, for it implies that, if $1/\Lambda_m$ is plotted against $c\Lambda_m$, then the intercept at $c = 0$ will be $1/\Lambda_m^\circ$ (Fig. 24.21).

24.8 The mobilities of ions

To interpret conductivity measurements we need to know why ions move at different rates, why they have different molar conductivities, and why the molar conductivities of strong electrolytes decrease with the square root of the concentration. The central concept we shall build on is that the greater the mobility of an ion in solution, the greater its contribution to the conductivity.

The drift speed

When the potential difference between two electrodes a distance l apart is $\Delta\phi$, the ions in the solution between them experience a uniform electric field of magnitude

$$\mathscr{E} = \frac{\Delta\phi}{l}$$

In such a field an ion of charge[5] ze experiences a force of magnitude

$$\mathscr{F} = ze\mathscr{E} = \frac{ze\,\Delta\phi}{l}$$

A cation responds by accelerating towards the negative electrode and an anion responds by accelerating towards the positive electrode. However, as an ion moves through the solvent it experiences a frictional retarding force $\mathscr{F}'$ that is proportional to its speed. If we assume that the Stokes formula (eqn 23.10) for a sphere of radius a and speed s applies even on a microscopic scale (and independent evidence from magnetic resonance suggests that it often gives at least the right order of magnitude), then we can write this retarding force as

$$\mathscr{F}' = fs \qquad f = 6\pi\eta a$$

The two forces act in opposite directions, and the ions reach a terminal

5 In this chapter we disregard the sign of the charge number and so avoid notational complications.

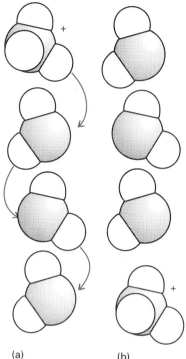

(a)   (b)

24.22 The mechanism of conduction by water. Because protons can be transferred between molecules, the current is transported very rapidly down a chain by an effective, not an actual, migration of a proton as each H_2O molecule passes one proton to its neighbour. Proton transfer between neighbouring molecules occurs when one molecule rotates into such a position that an $O—H \cdots O$ hydrogen bond can flip into being an $O \cdots H—O$ hydrogen bond.

speed, the **drift speed**, when the accelerating force $\mathscr{F}$ is balanced by the viscous drag $\mathscr{F}'$. The net force is zero ($\mathscr{F} = \mathscr{F}'$) when

$$s = \frac{ze\mathscr{E}}{f} \tag{24}$$

Because the drift speed governs the rate at which charge is transported, we might expect the conductivity to decrease with increasing solution viscosity and ion size. Experiments confirm these predictions for bulky ions (such as R_4N^+ and RCO_2^-) but not for small ions. For example, the molar conductivities of the alkali metal ions increase from Li^+ to Cs^+ (Table 24.4) even though the ionic radii increase. The paradox is resolved when we realize that the radius a in the Stokes formula is the **hydrodynamic radius** of the ion, its effective radius in the solution taking into account all the H_2O molecules it carries in its hydration sphere. Small ions give rise to stronger electric fields than large ones,[6] so small ions are more extensively solvated than big ions. Thus, an ion of small ionic radius may have a large hydrodynamic radius because it drags many solvent molecules through the solution as it migrates. The hydrating H_2O molecules are often very labile, however, and NMR and isotope studies have shown that the exchange between the co-ordination sphere of the ion and the bulk solvent is very rapid.

The proton, although it is very small, has a very high molar conductivity (Table 24.4)! The explanation is that the proton conducts by a mechanism that does not involve its actual motion through the solution. According to the **Grotthus mechanism**, or **chain mechanism**, of proton migration, instead of a single, highly solvated proton moving through the solution, there is an effective motion of a proton that involves the rearrangement of bonds in a group of water molecules (Fig. 24.22). The conductivity is governed by the rates at which the water molecules can rotate into orientations in which they can accept or donate protons and the rate at which the protons tunnel from one end of a hydrogen bond to another (from $O—H \cdots O$ to $O \cdots H—O$).

Ion mobilities

According to eqn 24, the drift speed of an ion is proportional to the strength of the applied field. We write

$$s = u\mathscr{E} \tag{25}$$

where the coefficient of proportionality u is called the **mobility** of the ion. Comparison of eqns 24 and 25 shows that

$$u = \frac{ze}{f} = \frac{ze}{6\pi\eta a} \tag{26}$$

For an order of magnitude estimate we can take $z = 1$ and a the radius of an ion such as Cs^+ (which might be typical of a smaller ion plus its hydration sphere), which is 170 pm. For the viscosity, we use $\eta = 1.0$ cP

---

6 The electric field at the surface of a sphere of radius r is proportional to ze/r^2, and so the smaller the radius the stronger the field.

Table 24.5* Ionic mobilities in water at 298 K, $u/(10^{-8}\,m^2\,s^{-1}\,V^{-1})$

H^+	36.23	OH^-	20.64
Na^+	5.19	Cl^-	7.91
K^+	7.62	Br^-	8.09
Zn^{2+}	5.47	SO_4^{2-}	8.29

* More values are given in the Data section.

$(1.0 \times 10^{-3}\,kg\,m^{-1}\,s^{-1}$, Table 24.3). Then u works out as about $5 \times 10^{-8}\,m^2\,V^{-1}\,s^{-1}$. This value means that when there is a potential difference of 1 V across a solution of length 1 cm (so $\mathscr{E} = 100\,V\,m^{-1}$), the drift speed is typically about $5\,\mu m\,s^{-1}$. That speed might seem slow, but not when expressed on a molecular scale, for it corresponds to an ion passing about 10^4 solvent molecules per second.

Some ionic mobilities are given in Table 24.5. We show below how they are related to conductivities and how they may be measured.

Mobility and conductivity

The usefulness of ionic mobilities is that they provide a link between measurable and theoretical quantities. As a first step we establish in the *Justification* below the following relation between an ion's mobility and its molar conductivity:

$$\lambda = zuF \tag{27}°$$

where F is Faraday's constant ($F = N_A e$).

JUSTIFICATION

To keep the calculation simple, we ignore signs in the following, and concentrate on the magnitudes of quantities: the direction of ion flux can always be decided by common sense.

Consider a solution of a strong electrolyte at a molar concentration c. Let each formula unit give rise to v_+ cations of charge z_+e and v_- anions of charge z_-e. The concentration of each type of ion is therefore vc (with $v = v_+$ or v_-), and the number density of each type is vcN_A. The number of ions of one kind that pass through an imaginary window of area A (Fig. 24.23) during an interval Δt is equal to the number within the distance $s\,\Delta t$, and therefore to the number in the volume $s\,\Delta tA$. The number of ions of that kind in this volume is equal to $s\,\Delta tA \times vcN_A$. The flux through the window (the number of this type of ion passing through the window per unit area per unit time) is therefore

$$J(\text{ions}) = \frac{s\,\Delta tA \times vcN_A}{A\,\Delta t} = svcN_A$$

Each ion carries a charge ze, and so the flux of charge is

$$J(\text{charge}) = zsvceN_A = zsvcF$$

Because $s = u\mathscr{E}$, the flux is

$$J(\text{charge}) = zuvcF\mathscr{E}$$

The current I through the window due to the ions we are considering is the charge flux times the area:

$$I = JA = zuvcFA\mathscr{E}$$

Because the electric field is the potential gradient, $\Delta\phi/l$, we can write

$$I = \frac{zuvcFA\,\Delta\phi}{l}$$

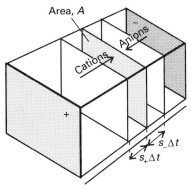

24.23 In the calculation of the current, all the cations within a distance $s_+\Delta t$ (i.e. those in the volume $As_+\Delta t$) will pass through the area A. The anions in the corresponding volume the other side of the window will also contribute to the current similarly.

Current and potential difference are related by Ohm's law, that

$$\text{Current} = \frac{\text{potential difference}}{\text{resistance}} = \frac{\Delta\phi}{R} = \frac{\kappa A\, \Delta\phi}{l}$$

where we have used eqn 16 (that $\kappa = l/RA$). On comparing the last two expressions we find that $\kappa = zuvcF$. Division by the molar concentration of ions, vc, then results in eqn 27.

Equation 27 applies to the cations and to the anions. Therefore, for the solution itself in the limit of zero concentration (when there are no interionic interactions),

$$\Lambda_m^\circ = (z_+ u_+ v_+ + z_- u_- v_-)F \tag{28a}^\circ$$

For a symmetrical $z:z$ electrolyte (e.g. $CuSO_4$), this equation simplifies to

$$\Lambda_m^\circ = z(u_+ + u_-)F \tag{28b}^\circ$$

We have already estimated the typical ionic mobility as $5 \times 10^{-8}\,\mathrm{m^2\,V^{-1}\,s^{-1}}$ so, with $z \approx 1$ for both the cation and anion, we can anticipate that a typical limiting molar conductivity should be about $10^2\,\mathrm{S\,cm^2\,mol^{-1}}$, in accord with experiment. The experimental value for KCl, for instance, is $150\,\mathrm{S\,cm^2\,mol^{-1}}$.

Transport numbers

The **transport number** t is defined as the fraction of total current carried by the ions of a specified type. For a solution of two kinds of ion, the cation transport number is

$$t_+ = \frac{I_+}{I} \tag{29}$$

where I_+ is the current carried by the cation and I is the total current through the solution. The anion transport number, t_-, is defined analogously in terms of I_-, the anion current. Because the total current I is the sum of the cation and anion currents, it follows that

$$t_+ + t_- = 1$$

The **limiting transport number** t° is defined in the same way but for the limit of zero concentration of the electrolyte solution. We shall consider only these limiting values from now on, for that avoids the problem of ionic interactions.

The conductivity, and hence the current that can be ascribed to each type of ion is related to the mobility of the ion by eqn 27. Hence the relation between t° and u is

$$t^\circ = \frac{zuv}{z_+ u_+ v_+ + z_- u_- v_-} \tag{30a}^\circ$$

For a symmetrical electrolyte (in which the charge numbers and hence the v_i are the same for both ions), eqn 30a simplifies to

$$t° = \frac{u}{u_+ + u_-} \qquad (30b)°$$

Moreover, because the ionic conductivities are related to the mobilities by eqn 27, it follows from eqn 30a that

$$t° = \frac{v\lambda}{v_+\lambda_+ + v_-\lambda_-} = \frac{v\lambda}{\Lambda_m°}$$

and hence, for each type of ion,

$$v\lambda = t°\Lambda_m° \qquad (31)°$$

Consequently, because there are independent ways of measuring transport numbers of ions, we can determine the individual ionic conductivities and (through eqn 27) the ionic mobilities.

The measurement of transport numbers

One of the most accurate methods for measuring transport numbers is the **moving boundary method**, in which the motion of a boundary between two ionic solutions having a common ion is observed as a current flows.

Let MX be the salt of interest and NX a salt giving a denser solution. The solution of NX is called the **indicator solution**; it occupies the lower part of a vertical tube (Fig. 24.24). The MX solution, which is called the **leading solution**, occupies the upper part of the tube. There is a sharp boundary between the two solutions. The indicator solution must be denser than the leading solution, and the mobility of the M ions must be greater than that of the N ions.[7] Thus, if any M ions diffuse into the lower solution, they will be pulled upwards more rapidly than the N ions around them, and the boundary will reform. The interpretation of the experiment makes use of the relation (see the *Justification* that follows) between the distance moved by the boundary in the time Δt for which a current I is passed:

$$t = \frac{zcVF}{I\,\Delta t}$$

Hence, by measuring V from the distance moved, the transport number and hence the conductivity and mobility of the ions can be determined.

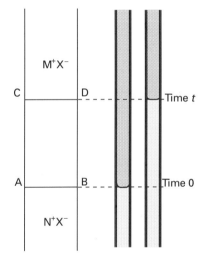

24.24 In the moving boundary method for the measurement of transport numbers the distance moved by the boundary is observed as a current is passed. All the M ions in the volume between AB and CD must have passed through CD if the boundary moves from AB to CD.

JUSTIFICATION

When a current I is passed for a time Δt, the boundary moves from AB to CD, and so all the M ions in the volume between AB and CD must have passed through CD. That number is cVN_A, so the charge they transfer through the plane is $zcVeN_A$. However, the total charge transferred when

---

7 One procedure is to add bromothymol blue indicator to a slightly alkaline solution of the ion of interest and to use a cadmium electrode at the lower end of the vertical tube. The electrode produces Cd^{2+} ions, which are slow moving and slightly acidic (the hydrated ion is a Brønsted acid), and the boundary is revealed by the colour change of the indicator.

a current I flows for an interval Δt is $I\,\Delta t$. Therefore, the fraction due to the motion of the M ions, which is their transport number, is

$$t = \frac{zcVF}{I\,\Delta t}$$

In the **Hittorf method**, an electrolytic cell is divided into three compartments and an amount $I\,\Delta t$ of electricity is passed. An amount $I\,\Delta t / z_+ F$ of cations is discharged at the cathode but an amount $t_+ (I\,\Delta t / z_+ F)$ of cations migrates into the cathode compartment. The net change in the amount of cations in that compartment is therefore

$$\text{Net change} = (t_+ - 1)\frac{I\,\Delta t}{z_+ F} = -t_- \frac{I\,\Delta t}{z_+ F}$$

Hence, by measuring the change of composition in the cathode compartment, the anion transport number t_- can be deduced. Likewise, the change in composition of the anode compartment is $-t_+(I\,\Delta t / z_- F)$, which gives the cation transport number t_+.

Transport numbers may also be measured using galvanic cells. In particular, the measurement is made on a **cell with transference**, which is a galvanic cell with a liquid boundary across which ions may pass from one electrode compartment to the other. The zero-current cell potential of a cell with transference E_t having an electrode reversible with respect to anions, for example

$$\text{Ag} \,|\, \text{AgCl(s)} \,|\, \text{HCl}(m_1) \,|\, \text{HCl}(m_2) \,|\, \text{AgCl(s)} \,|\, \text{Ag}$$

is related to the potential of a cell with the same overall reaction but without transference E, such as

$$\text{Ag} \,|\, \text{AgCl(s)} \,|\, \text{HCl}(m_1) \,|\, \text{H}_2(\text{g}) \,|\, \text{Pt} \,|\, \text{H}_2(\text{g}) \,|\, \text{HCl}(m_2) \,|\, \text{AgCl(s)} \,|\, \text{Ag}$$

by

$$E_t = t_+ E$$

Therefore, comparison of the two cell potentials gives the transport number, in this case of H^+.

JUSTIFICATION

The argument is similar to that used to analyse the Hittorf method. Consider the consequences of passing 1 mol of electrons through the cell

$$\text{Ag} \,|\, \text{AgCl(s)} \,|\, \text{HCl}(m_1) \,|\, \text{HCl}(m_2) \,|\, \text{AgCl(s)} \,|\, \text{Ag}$$

In the right-hand electrode compartment, 1 mol Cl^- is formed but t_- mol Cl^- migrate out of it across the junction, giving a net change of $(1 - t_-)$ mol $= t_+$ mol. In the left-hand electrode compartment 1 mol Cl^- is removed from the solution (to form 1 mol AgCl), but t_- mol Cl^- flows in across the junction. The net change is therefore $(-1 + t_-)$ mol $= -t_+$ mol. The reaction Gibbs energy when 1 mol of electrons is passed is

therefore

$$\Delta_r G = t_+\{\mu(Cl^-, m_2) - \mu(Cl^-, m_1)\} = t_+ RT \ln \frac{a_2}{a_1}$$

Because $\Delta_r G = -FE$, it follows that

$$E_t = -\frac{t_+ RT}{F} \ln \frac{a_2}{a_1}$$

For the same cell without transference, the Nernst equation gives

$$E = -\frac{RT}{F} \ln \frac{a_2}{a_1}$$

and the ratio of the two cell potentials is the transport number t_+.

24.9 Conductivities and ion–ion interactions

The remaining problem is to account for the $c^{\frac{1}{2}}$ dependence of the Kohlrausch law (eqn 18). In Section 10.2 we saw something similar: the activity coefficients of ions also depend on $c^{\frac{1}{2}}$ and depend on their charge type rather than their specific identities. That $c^{\frac{1}{2}}$ dependence was explained in terms of the properties of the ionic atmosphere around each ion, and we can suspect that the same explanation applies here too.

We need to modify the picture of an ionic atmosphere as a spherical haze of charge when an electric field is applied to a solution of ions and all the ions drift in a definite direction. In the first place, the ions forming the atmosphere do not adjust to the moving ion infinitely quickly, and the atmosphere is incompletely formed in front of the moving ion and is incompletely decayed in its wake (Fig. 24.25). The overall effect is the displacement of the centre of charge of the atmosphere a short distance behind the moving ion. Because the two charges are opposite, the result is a retardation of the moving ion. This reduction of the ions' mobility is called the **relaxation effect**. A confirmation of the picture is obtained by observing the conductivities of ions at high frequencies, which are greater than at low frequencies, because the atmosphere does not have time to follow the rapidly changing direction of motion of the ion, and its effect averages to zero.

The ionic atmosphere also introduces another effect on the motion of the ions. We have seen that the moving ion experiences a viscous drag. When the ionic atmosphere is present this drag is enhanced because the ionic atmosphere moves in an opposite direction to the central ion. The enhanced viscous drag, which is called the **electrophoretic effect**, reduces the mobility of the ions, and hence also reduces their conductivities.

The quantitative formulation of these effects is far from simple, but the **Debye–Hückel–Onsager theory** is an attempt to obtain quantitative expressions at about the same level of sophistication as the Debye–Hückel theory itself. The theory leads to a Kohlrausch-like expression in

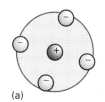

(a)

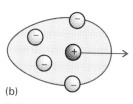

(b)

24.25 (a) In the absence of an applied field, the ionic atmosphere is spherically symmetric, but (b) when a field is present it is distorted and the centres of negative and positive charge no longer coincide. The attraction between the opposite charges retards the motion of the central ion.

Table 24.6* Debye–Hückel–Onsager coefficients for (1,1)-electrolytes at 298 K

Solvent	$A/(\text{S cm}^2 \text{ mol}^{-1}/(\text{mol L}^{-1})^{\frac{1}{2}})$	$B/(\text{mol L}^{-1})^{-\frac{1}{2}}$
Methanol	156.1	0.923
Propanone	32.8	1.63
Water	60.2	0.229

* More values are given in the Data section.

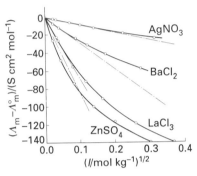

24.26 The dependence of molar conductivities on the square root of the ionic strength, and comparison (broken lines) with the dependence predicted by the Debye–Hückel–Onsager theory.

which

$$\mathcal{K} = A + B\Lambda_m^\circ \tag{32}$$

with

$$A = \frac{z^2 e F^2}{3\pi\eta}\left(\frac{2}{\varepsilon RT}\right)^{\frac{1}{2}} \qquad B = \frac{qz^3 e F}{24\pi\varepsilon RT}\left(\frac{2}{\varepsilon RT}\right)^{\frac{1}{2}}$$

where ε is the electric permittivity of the solvent and $q = 0.586$ for a 1,1-electrolyte. The numerical values of A and B for 1,1-electrolytes in several solvents are listed in Table 24.6. The slopes of the conductivity curves are predicted to depend on the charge type of the electrolyte, as the Kohlrausch measurements require, and some comparisons between theory and experiment are shown in Fig. 24.26. The agreement is quite good at very low concentrations (less than about 10^{-3} M depending on the charge type).

DIFFUSION

We are now in a position to generalize the discussion of ionic motion to cover the migration of neutral molecules and of ions in the absence of an applied electric field. We shall do this by expressing ion motion in a more general way, and will then discover that the same equations apply even when the charge on the particles is zero.

24.10 The thermodynamic view

We saw in Part 1 that, at constant temperature and pressure, the maximum non-expansion work that can be done per mole when a substance moves from a location where its chemical potential is μ to a location where it is $\mu + d\mu$ is $dw = d\mu$. In a system in which the chemical potential depends on the position x,

$$dw = d\mu = \left(\frac{\partial\mu}{\partial x}\right)_{p,T} dx$$

We also saw in Chapter 2 (Table 2.1) that in general work can always be expressed in terms of an opposing force (which here we shall write $\mathcal{F}$), and that

$$dw = -\mathcal{F}\,dx$$

By comparing these expressions, we see that the slope of the chemical potential can be interpreted as an effective force per mole of molecules. We shall write this **thermodynamic force** as

$$\mathcal{F} = -\left(\frac{\partial\mu}{\partial x}\right)_{p,T} \tag{33}$$

There is not necessarily a real force pushing the particles down the slope of the chemical potential. As we shall see, the force may represent the spontaneous tendency of the molecules to disperse as a consequence of the Second Law and the hunt for maximum entropy.

The thermodynamic force of a concentration gradient

In a solution where the activity of the particles is a, the chemical potential is

$$\mu = \mu^{\ominus} + RT \ln a$$

If the solution is not uniform the activity depends on the position and we can write

$$\mathscr{F} = -RT \left(\frac{\partial \ln a}{\partial x} \right)_{p,T}$$

If the solution is ideal, a may be replaced by the concentration c, and then

$$\mathscr{F} = -\frac{RT}{c} \left(\frac{\partial c}{\partial x} \right)_{p,T} \tag{34}°$$

because $(\mathrm{d}/\mathrm{d}x) \ln c = (1/c)\, \mathrm{d}c/\mathrm{d}x$.

Example 24.8 *Calculating the thermodynamic force*

Suppose that the concentration of a solute decays exponentially along the length of a container. Calculate the thermodynamic force on the solute at 25°C when the concentration decreases to half its value in every 10 cm.

Method. The thermodynamic force is calculated by differentiating the concentration with respect to distance, as in eqn 34; therefore, we need to write an expression for the variation of the concentration with distance, and then differentiate it.

Answer. The concentration varies with position as

$$c = c_0 e^{-x/\lambda}$$

where λ is the decay constant. It is stated that the concentration falls to $\frac{1}{2}c_0$ when $x = 10$ cm, so we can find λ from

$$\tfrac{1}{2} = e^{-(10\ \mathrm{cm})/\lambda}$$

That is,

$$\lambda = \frac{10\ \mathrm{cm}}{\ln 2}$$

From the exponential form of the concentration it follows that

$$\frac{\mathrm{d}c}{\mathrm{d}x} = -\frac{c}{\lambda}$$

and therefore, from eqn 34, that

$$\mathscr{F} = \frac{RT}{\lambda}$$

Therefore,

$$\mathscr{F} = \frac{(8.3145\ \mathrm{J\,K^{-1}\,mol^{-1}}) \times (298\ \mathrm{K}) \times \ln 2}{1.0 \times 10^{-1}\ \mathrm{m}} = 17\ \mathrm{kN\,mol^{-1}}$$

Comment. We have used $1\,J = 1\,N\,m$. Note that the thermodynamic force is a molar quantity.

Exercise E24.8. Calculate the thermodynamic force on the molecules of molar mass M in a vertical tube in a gravitational field on the surface of the Earth and evaluate it for molecules of molar mass $100\,g\,mol^{-1}$. Comment on its magnitude relative to that just calculated.

> $[\mathscr{F} = -Mg, -0.98\,N\,mol^{-1}$; the force arising from the concentration gradient greatly dominates that arising from the gravitational gradient.]

Fick's first law of diffusion

In Section 24.4 it was shown that Fick's first law of diffusion (that the particle flux is proportional to the concentration gradient) could be deduced from the kinetic theory of gases. We shall now show that it can be deduced more generally and that it applies to the diffusion of species in condensed phases too.

We suppose that the flux of diffusing particles is motion in response to a thermodynamic force arising from a concentration gradient. The particles reach a steady drift speed s when the thermodynamic force $\mathscr{F}$ is matched by the viscous drag. This drift speed is proportional to the thermodynamic force, and we write $s \propto \mathscr{F}$. However, the particle flux J is proportional to the drift speed, and the thermodynamic force is proportional to the concentration gradient dc/dx. The chain of proportionalities ($J \propto s$, $s \propto \mathscr{F}$, and $\mathscr{F} \propto dc/dx$) implies that J is proportional to dc/dx, which is the content of Fick's law.

The Einstein relation

Fick's law for the particle flux in amount of molecules per unit area per unit time is

$$J = -D\frac{dc}{dx}$$

In this expression, D is the diffusion coefficient and dc/dx is the slope of the molar concentration.[8] The flux is related to the drift speed by

$$J = sc$$

This relation follows from the argument that we have used several times before. Thus, all particles within a distance $s\,\Delta t$, and therefore in a volume $s\,\Delta tA$, can pass through a window of area A in an interval Δt. Hence, the amount of substance that can pass through the window in that interval is $s\,\Delta tA \times c$; so

$$sc = -D\frac{dc}{dx}$$

---

8 This relation is derived from eqn 5 by dividing both sides by Avogadro's constant, which converts numbers into amounts (numbers of moles).

If now we express dc/dx in terms of $\mathscr{F}$ by using eqn 34, we find

$$s = -\frac{D}{c}\frac{dc}{dx} = \frac{D}{RT} \times \mathscr{F}$$

Therefore, once we know the effective force and the diffusion coefficient D, we can calculate the drift speed of the particles (and vice versa) whatever the origin of the force.

There is one case where we already know the drift speed and the effective force acting on a particle: an ion in solution has a drift speed $s = u\mathscr{E}$ when it experiences a force $ez\mathscr{E}$ from an electric field of strength $\mathscr{E}$. Therefore, substituting these known values into the equation above gives

$$u\mathscr{E} = \frac{D}{RT} \times zF\mathscr{E} \qquad \text{or} \quad u = \frac{zFD}{RT}$$

This equation rearranges into the very important result known as the **Einstein relation** between the diffusion coefficient and the ionic mobility:

$$D = \frac{uRT}{zF} \tag{35}°$$

On inserting the typical value $u = 5 \times 10^{-8}\,\text{m}^2\,\text{s}^{-1}\,\text{V}^{-1}$, we find $D \approx 1 \times 10^{-9}\,\text{m}^2\,\text{s}^{-1}$ at 25°C as a typical value of the diffusion coefficient of an ion in water.

The Nernst–Einstein equation

We can develop the Einstein relation in two ways. In the first place, it can be used to find the relation between the molar conductivity of an electrolyte and the diffusion coefficients of its ions. First, using eqns 27 and 35, we write

$$\lambda = zuF = \frac{z^2 D F^2}{RT} \tag{36}°$$

for each ion. Then, from $\Lambda_m° = \nu_+\lambda_+ + \nu_-\lambda_-$, the limiting molar conductivity is

$$\Lambda_m° = \frac{F^2}{RT}(\nu_+ z_+^2 D_+ + \nu_- z_-^2 D_-) \tag{37}°$$

which is the **Nernst–Einstein equation**. One of its applications is to the determination of ionic diffusion coefficients from conductivity measurements; another is to the prediction of conductivities using models of ionic diffusion (see below).

The Stokes–Einstein equation

Equations 26 ($u = ez/f$) and 35 relate the mobility of an ion to the frictional force and to the diffusion coefficient, respectively. The two expressions can be combined to give

$$D = \frac{kT}{f} \tag{38}$$

This expression is called the **Stokes–Einstein equation**. If the frictional force is described by Stokes's law, then we also obtain a relation between the diffusion coefficient and the viscosity of the medium:

$$D = \frac{kT}{6\pi\eta a} \tag{39}$$

An important feature of eqn 38 (and of its special case, eqn 39) is that it is independent of the charge of the diffusing species. Therefore, it also applies in the limit of vanishingly small charge; that is, it also applies to neutral molecules. Consequently, we may use viscosity measurements to estimate the diffusion coefficients for molecules in solution. (This relation was in fact used in Section 23.3 in the discussion of macromolecules.) It must not be forgotten, however, that both equations depend on the assumption that the viscous drag is proportional to the speed. Some diffusion coefficients are listed in Table 24.7.

Table 24.7* Diffusion coefficients at 298 K, $D/(10^{-9}\,\text{m}^2\,\text{s}^{-1})$

H^+ in water	9.31
I_2 in hexane	4.05
Na^+ in water	1.33
Sucrose in water	0.521

* More values are given in the Data section.

Example 24.9 *Interpreting the mobility of an ion*

Use the experimental value of the mobility to evaluate the diffusion coefficient, limiting molar conductivity, and effective hydrodynamic radius of SO_4^{2-} in water at 298 K.

Method. The starting point is the mobility of the ion, which we can take from Table 24.5. The diffusion coefficient can then be determined from the Einstein relation (eqn 35). The ionic conductivity is related to the mobility by eqn 27. To obtain the hydrodynamic radius of the ion, we can use the Stokes–Einstein equation to find f and the Stokes law force to relate f to the hydrodynamic radius a.

Answer. According to Table 24.5, the mobility of the ion is $8.29 \times 10^{-8}\,\text{m}^2\,\text{s}^{-1}\,\text{V}^{-1}$. Then, from eqn 35,

$$D = \frac{uRT}{zF} = 1.1 \times 10^{-9}\,\text{m}^2\,\text{s}^{-1}$$

From eqn 27 for the ionic conductivity:

$$\lambda = zuF = 1.6 \times 10^{-2}\,\text{S}\,\text{m}^2\,\text{mol}^{-1}$$

Finally, from $f = 6\pi\eta a$ using 1.00 cP (or $1.00 \times 10^{-3}\,\text{kg}\,\text{m}^{-1}\,\text{s}^{-1}$) for the viscosity of water (Table 24.3):

$$a = \frac{kT}{6\pi\eta D} = 2.0 \times 10^{-10}\,\text{m} = 200\,\text{pm}$$

Comment. The bond length in SO_4^{2-} is 144 pm, so the radius calculated here is plausible and consistent with a small degree of solvation.

Exercise E24.9. Repeat the calculation for the NH_4^+ ion.

$$[1.96 \times 10^{-9}\,\text{m}^2\,\text{s}^{-1}, 7.4\,\text{mS}\,\text{m}^2\,\text{mol}^{-1}, 110\,\text{pm}]$$

Experimental support for the relations derived above comes from conductivity measurements. In particular, **Walden's rule** is the empirical observation that the product $\eta\Lambda_m$ is very approximately constant for the same ions in different solvents. Since $\Lambda_m \propto D$, and we have just seen that $D \propto 1/\eta$, we do indeed predict that $\Lambda_m \propto 1/\eta$ as Walden's rule requires. The usefulness of the rule, however, is muddied by the role of solvation: different solvents solvate the same ions to different extents, and so both the hydrodynamic radius and the viscosity change with the solvent.

24.11 The diffusion equation

We now turn to the discussion of time-dependent diffusion processes, where we are interested in the spreading of inhomogeneities with time. One example is the temperature of a metal bar that has been heated at one end: if the source of heat is removed, then the bar gradually settles down into a state of uniform temperature. If the source of heat is maintained and the bar can radiate, it settles down into a steady state of non-uniform temperature. Another example (and one more relevant to chemistry) is the concentration distribution in a solvent to which a solute is added. We shall focus on the description of the diffusion of particles, but similar arguments apply to the diffusion of physical properties, such as temperature. Our aim is to obtain an equation for the rate of change of the concentration of particles in an inhomogeneous region.

The central equation of this section is the **diffusion equation**, which relates the rate of change of concentration at a point to the spatial variation of the concentration at that point:

$$\frac{\partial c}{\partial t} = D\frac{\partial^2 c}{\partial x^2} \tag{40}$$

(This equation used to be called Fick's second law of diffusion.)

JUSTIFICATION

Consider a thin slab of cross-sectional area A that extends from x to $x + l$ (Fig. 24.27). Let the concentration at x be c at the time t. The amount of particles that enter the slab per unit time is JA, so the rate of increase in molar concentration inside the slab (which has volume Al) on account of the flux from the left is

$$\frac{dc}{dt} = \frac{JA}{Al} = \frac{J}{l}$$

There is also an outflow through the right-hand window. The flux through that window is J', and the rate of change of concentration that results is

$$\frac{dc}{dt} = -\frac{J'A}{Al} = -\frac{J'}{l}$$

The net rate of change of concentration is therefore

$$\frac{dc}{dt} = \frac{J - J'}{l}$$

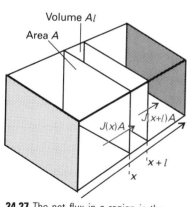

24.27 The net flux in a region is the difference between the flux entering from the region of high concentration (on the left) and the flux leaving to the region of low concentration (on the right).

Each flux is proportional to the concentration gradient at the window. So, by using Fick's first law we can write

$$J - J' = -D\frac{dc}{dx} + D\frac{dc'}{dx}$$

$$= -D\frac{dc}{dx} + D\frac{d}{dx}\left\{c + \left(\frac{dc}{dx}\right)l\right\}$$

$$= Dl\frac{d^2c}{dx^2}$$

When this relation is substituted into the expression for the rate of change of concentration in the slab, we get eqn 40.

The significance of the diffusion equation

The diffusion equation shows that the rate of change of concentration is proportional to the curvature (more precisely, to the second derivative) of the concentration with respect to distance. If the concentration changes sharply from point to point (if the distribution is highly wrinkled) then the concentration changes rapidly with time. If the curvature is zero, then the concentration is constant in time. If the concentration decreases linearly with distance, then the concentration at any point is constant because the inflow of particles is exactly balanced by the outflow.

The diffusion equation can be regarded as a mathematical formulation of the intuitive notion that there is a natural tendency for the wrinkles in a distribution to disappear. More succinctly: Nature abhors a wrinkle.

Diffusion with convection

The transport of particles arising from the motion of a streaming fluid is called **convection**. If for the moment we ignore diffusion, then the flux of particles through an area A in an interval Δt when the fluid is flowing at a velocity v can be calculated in the way we have used several times before (by counting the particles within a distance $v\,\Delta t$), and is

$$J = \frac{cAv\,\Delta t}{A\,\Delta t} = cv \tag{41}$$

This J is called the **convective flux**. The rate of change of concentration in a slab of thickness l and area A is, by the same argument as before,

$$\frac{dc}{dt} = \frac{J - J'}{l} = \left\{c - \left[c + \left(\frac{\partial c}{\partial x}\right)l\right]\right\}\frac{v}{l} = -v\frac{\partial c}{\partial x}$$

(We have assumed that the velocity does not depend on the position.)

When both diffusion and convection are of similar importance, the total change of concentration in a region is the sum of the two effects, and the **generalized diffusion equation** is

$$\frac{\partial c}{\partial t} = D\frac{\partial^2 c}{\partial x^2} - v\frac{\partial c}{\partial x} \tag{42}$$

A further refinement, which is important in chemistry, is the possibility that the concentrations of particles may change as a result of reaction. When reactions are included in eqn 42 (as we describe in Section 27.2), we get a powerful differential equation for discussing the properties of reacting, diffusing, convecting systems, which is the basis of reactor design in the chemical industry and of the utilization of resources in living cells.

Solutions of the equation

The diffusion equation, eqn 40, is a second-order differential equation with respect to space and a first-order differential equation with respect to time. Therefore, we must specify two boundary conditions for the spatial dependence and a single initial condition for the time dependence.

As an illustration, we consider the example of a solvent in which the solute is initially coated on one surface of the container (e.g. a layer of sugar on the bottom of a deep beaker of water). The single initial condition is that at $t = 0$ all N_0 particles are concentrated on the yz-plane (of area A) at $x = 0$. The two boundary conditions are derived from the requirements (1) that the concentration must everywhere be finite and (2) that the total amount (number of moles) of particles present is n_0 at all times. These requirements imply that the flux of particles is zero at the top and bottom surfaces of the system. Under these conditions it is found that

$$c = \frac{n_0}{A(\pi Dt)^{\frac{1}{2}}} e^{-x^2/4Dt} \tag{43}$$

as may be verified by direct substitution. Figure 24.28 shows the shape of the concentration distribution at various times, and it is clear that the concentration spreads and tends to uniformity.

Another useful result is for a localized concentration of solute in a three-dimensional solvent (a sugar lump suspended in a large flask of water). The concentration of diffused solute is spherically symmetrical, and at a radius r is

$$c = \frac{n_0}{8(\pi Dt)^{\frac{3}{2}}} e^{-r^2/4Dt} \tag{44}$$

Other chemically (and physically) interesting arrangements can be treated, but the solutions are more cumbersome.

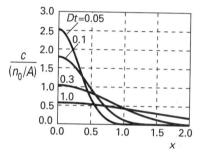

24.28 The concentration profiles above a plane from which a solute is diffusing. The curves are plots of eqn 43. The units of Dt and x are arbitrary, but are related so that Dt/x^2 is dimensionless. For example, if x is in metres, Dt would be in metres^2, and so for $D = 10^{-9}$ m^2 s^{-1}, '$Dt = 0.1$' corresponds to $t = 10^4$ s.

The measurement of diffusion coefficients

The solutions of the diffusion equation are useful for experimental determinations of diffusion coefficients. In the **capillary technique**, a capillary tube, open at one end and containing a solution, is immersed in a well stirred larger quantity of solvent, and the change of concentration in the tube is measured at a series of times. The solute diffuses from the open end of the capillary at a rate that can be calculated by solving the diffusion equation with the appropriate boundary conditions, so D may be determined. In the **diaphragm technique**, the diffusion occurs through the capillary pores of a sintered glass diaphragm separating the well-stirred solution and solvent. The concentrations are monitored and then related to the solutions of the diffusion equation corresponding to this arrangement.

24.12 Diffusion probabilities

The solutions of the diffusion equation can be used to predict the concentration of particles (or the value of some other physical quantity, such as the temperature in a non-uniform system) at any location. We can also use them to calculate the net distance through which the particles diffuse in a given time.

Example 24.10 *Calculating the net distance of diffusion*

Calculate the net distance travelled on average by particles in a time t if they are diffusing in a medium with diffusion coefficient D.

Method. We need to calculate the probability that a particle will be found at a certain distance from the origin, and then calculate the average by weighting each distance by that probability.

Answer. The number of particles in a slab of thickness dx and area A at x, where the molar concentration is c, is $cAN_A\,dx$. The probability that any of the $N_0 = n_0 N_A$ particles is in the slab is therefore $cAN_A\,dx/N_0$. If the particle is in the slab, it has travelled a distance x from the origin. Therefore, the mean distance travelled by all the particles is the sum of each x weighted by the probability of its occurrence:

$$\langle x \rangle = \int_0^\infty \frac{xcAN_A}{N_0}\,dx = \frac{1}{(\pi Dt)^{\frac{1}{2}}} \int_0^\infty x e^{-x^2/4Dt}\,dx$$
$$= 2\left(\frac{Dt}{\pi}\right)^{\frac{1}{2}}$$

Comment. The average distance of diffusion varies as the square root of the lapsed time. If we use the Stokes–Einstein equation for the diffusion coefficient, the mean distance travelled by particles of radius a in a solvent of viscosity η is

$$\langle x \rangle = \left(\frac{2kTt}{3\pi^2 \eta a}\right)^{\frac{1}{2}}$$

Exercise E24.10. Derive an expression for the root mean square distance travelled by diffusing particles in a time t. $[\langle x^2 \rangle^{\frac{1}{2}} = (2Dt)^{\frac{1}{2}}]$

As shown in Example 24.10, the average distance travelled by a diffusing particle in a time t (in an arrangement like that illustrated in Fig. 24.28) is

$$\langle x \rangle = 2\left(\frac{Dt}{\pi}\right)^{\frac{1}{2}} \tag{45a}$$

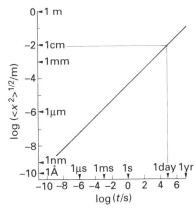

24.29 The root mean square distance covered by particles with $D = 5 \times 10^{-10}$ m^2 s^{-1}. Note the great slowness of diffusion.

and the mean square distance travelled in the same time is

$$\langle x^2 \rangle = 2Dt \tag{45b}$$

The latter is a valuable measure of the spread of particles when they can diffuse in both directions from the origin (for then $\langle x \rangle = 0$ at all times). The root mean square distance travelled by particles with a typical diffusion coefficient ($D = 5 \times 10^{-10}$ m^2 s^{-1}) is illustrated in Fig. 24.29, which shows how long it takes for diffusion to increase the net distance travelled on average to about 1 cm in an unstirred solution. The graph shows that diffusion is a very slow process (which is why solutions are stirred, to encourage mixing by convection).

24.13 The statistical view

An intuitive picture of diffusion is of the particles moving in a series of small steps and gradually migrating from their original positions. We shall explore this idea using a model in which the particles can jump through a distance λ in a time τ. The total distance travelled by a particle in a time t is therefore $t\lambda/\tau$. However, the particle will not necessarily be found at that distance from the origin. The direction of each step may be different, and the net distance travelled must take the changing directions into account. If we simplify the discussion by allowing the particles to travel only along a straight line (the x-axis), and for each step (to the left or the right) to be through the same distance λ, then we obtain the one-dimensional random walk.

The random walk is described in *Further information 16*. We show there that the probability of being at a distance x from the origin after a time t is

$$P = \left(\frac{2\tau}{\pi t}\right)^{\frac{1}{2}} e^{-x^2\tau/2t\lambda^2} \tag{46}$$

This equation has precisely the same form as eqn 43, the differences of detail arising from the fact that in the present calculation the particles can migrate in either direction from the origin. Moreover, they can be found only at discrete points separated by λ instead of being anywhere on a continuous line. The fact that the two expressions are so similar suggests that diffusion can indeed be interpreted as the outcome of a large number of steps in random directions.

We can now relate the coefficient D to the step length λ and the rate at which the jumps occur. Thus, by comparing the two exponents in eqn 43 and eqn 46 we can immediately write down the **Einstein–Smoluchowski equation**:

$$D = \frac{\lambda^2}{2\tau} \tag{47}$$

For example, suppose that a SO_4^{2-} ion jumps through its own diameter each time it makes a move in an aqueous solution: then, because $D = 1.1 \times 10^{-9}$ m^2 s^{-1} and $a = 210$ pm, it follows from $\lambda = 2a$ that $\tau = 80$ ps. Because τ is the time for one jump, the ion makes 1×10^{10} jumps per second.

The Einstein–Smoluchowski equation is the central link between the microscopic details of particle motion and the macroscopic parameters relating to diffusion (e.g. the diffusion coefficient and, through the Stokes–Einstein equation, the viscosity). It also brings us back full circle to the properties of the perfect gas. For if we interpret λ/τ as $\bar{c}$, the mean speed of the molecules, and interpret λ as a mean free path, then we can recognize in the Einstein–Smoluchowski equation exactly the same expression as we obtained from the kinetic theory of gases in eqn 8. That is, the diffusion of a perfect gas is a random walk with an average step size equal to the mean free path.

CHECK LIST OF KEY IDEAS

1 The calculation of the **frequency of collisions** with the walls of a container (eqn 3).

2 The **rate of effusion** of a gas through a hole in a container (eqn 4).

3 The definition of a **transport coefficient** and the **flux** of a property, and the statement of **Fick's first law of diffusion** (eqn 5).

4 The definition of the **diffusion coefficient** (eqn 5), the **coefficient of thermal conductivity** (eqn 6), and the **coefficient of viscosity** (eqn 7).

5 The calculation of the **diffusion coefficient** from the kinetic theory of gases (eqn 8) and its variation with pressure and temperature.

6 The calculation of the **coefficient of thermal conductivity** from the kinetic theory of gases (eqn 9) and its variation with pressure and temperature.

7 The calculation and measurement of the **coefficient of viscosity** of a perfect gas (eqn 10) and its variation with pressure and temperature.

8 The description of the structure of a liquid in terms of the **radial distribution function** (Section 24.5).

9 The calculation of the radial distribution function by using **Monte Carlo methods** or **molecular dynamics** (Section 24.5).

10 The structure of **liquid crystals** and their classification (Section 24.5).

11 The variation of the **viscosity of a liquid** with temperature (eqn 14)

12 The measurement of the **conductance** of an electrolyte solution and the definition of the **conductivity** (eqn 16) and the **molar conductivity** (eqn 17).

13 The classification of electrolytes as **strong** or **weak** (Section 24.7) and their conductivity characteristics (Fig. 24.20).

14 **Kohlrausch's law** for the molar conductivities of strong electrolytes (eqn 18) and the **limiting molar conductivity**.

15 The **law of the independent migration of ions** (eqn 19).

16 The **degree of ionization** of a weak electrolyte (eqn 21) and **Ostwald's dilution law** for the molar conductivity of the solution (eqn 23).

17 The **drift speed** of an ion in an electric field (eqn 24) and the variation of ionic conductivities with their effective hydrodynamic radii.

18 The **Grotthus mechanism** for conduction by protons (Fig. 24.22).

19 The definition of the **mobility** of an ion (eqn 25) and its relation to the ionic conductivity (eqn 27).

20 The **transport number** of an ion (eqn 29) and its measurement by the **moving boundary** and **Hittorf** methods (Section 24.8).

21 The effect of **ion–ion interactions** on the conductivities of ions and the **Debye–Hückel–Onsager theory** (eqn 32).

22 The definition of **thermodynamic force** (eqn 33), its calculation (Example 24.8), and its use in the derivation of Fick's first law of diffusion.

23 The **Einstein relation** between the diffusion coefficient and the ionic mobility (eqn 35).

24 The **Nernst–Einstein equation** for the molar conductivity in terms of the diffusion coefficient (eqn 37).

25 The **Stokes–Einstein equation** between the

diffusion coefficient and the frictional constant (eqn 38).

26 The derivation of the **diffusion equation** (eqn 40) and the inclusion of **convection** (eqn 42).

27 The measurement of diffusion coefficients (Section 24.11).

28 The **net distance** that a molecule diffuses (eqn 45).

29 The interpretation of diffusion as a **random walk** (Section 24.13) and the derivation of the **Einstein–Smoluchowski equation** for the diffusion coefficient in terms of the jump length and time of a random walk (eqn 47).

EXERCISES

24.1. A solid surface with dimensions 2.5 mm × 3.0 mm is exposed to argon gas at 90 Pa and 500 K. How many collisions do the Ar atoms make with this surface in 15 s?

24.2. An effusion cell has a circular hole, which is 2.50 mm in diameter. If the molar mass of the solid in the cell is 260 g mol^{-1} and its vapour pressure is 0.835 Pa at 400 K, by how much will the mass of the solid decrease in a period 2.00 h?

24.3. Calculate the flux of energy arising from a temperature gradient of 2.5 K m^{-1} in a sample of argon in which the mean temperature is 273 K.

24.4. Use the experimental value of the thermal conductivity of neon at 273 K (Table 24.2) to estimate the collision cross-section of Ne atoms.

24.5. In a double-glazed window, the panes of glass are separated by 5.0 cm. What is the rate of transfer of heat by conduction from the warm room (25°C) to the cold exterior (−10°C) through a window of area 1.0 m^2? What power of heater is required to make good the loss of heat?

24.6. A manometer was connected to a bulb containing carbon dioxide under slight pressure. The gas was allowed to escape through a small pinhole, and the time for the manometer reading to drop from 75 cm to 50 cm was 52 s. When the experiment was repeated using nitrogen (for which $M = 28.02$ g mol^{-1}) the same fall took place in 42 s. Calculate the molar mass of carbon dioxide.

24.7. A space vehicle of internal volume 3.0 m^3 is struck by a meteor and a hole of 0.10 mm radius is formed. If the oxygen pressure within the vehicle is initially 80 kPa and its temperature 298 K, how long will it take to fall to 70 kPa?

24.8. Use the experimental value of the coefficient of viscosity for neon (Table 24.2) to estimate the collision cross-section of Ne atoms at 273 K.

24.9. Calculate the inlet pressure required to maintain a flow rate of 9.5×10^5 L h^{-1} of nitrogen at 293 K flowing through a pipe of length 8.50 m and diameter 1.00 cm. The pressure of gas as it leaves the tube is 1.00 bar. The volume of the gas is measured at that pressure.

24.10. Calculate the viscosity of air at (a) 273 K, (b) 298 K, (c) 1000 K. Take $\sigma \approx 0.40$ nm^2. (The experimental values are 173 μP at 273 K, 182 μP at 20°C, and 394 μP at 600°C.)

24.11. Calculate the thermal conductivities of (a) argon, (b) helium at 300 K and 1.0 mbar. Each gas is confined in a cubic vessel of 10 cm side, one wall being at 310 K and the one opposite at 295 K. What is the rate of flow of energy as heat from one wall to the other in each case?

24.12. The viscosity of carbon dioxide was measured by comparing its rate of flow through a long narrow tube (using Poiseuille's formula) with that of argon. For the same pressure differential, the same volume of carbon dioxide passed through the tube in 55 s as that of argon in 83 s. The viscosity of argon at 25°C is 208 μP; what is the viscosity of carbon dioxide? Estimate the molecular diameter of carbon dioxide.

24.13. Calculate the thermal conductivity of air ($C_{V,m} = 21.0$ J K^{-1} mol^{-1}, $\sigma = 0.40$ nm^2) at room temperature.

24.14. Calculate the diffusion constant of argon at 25°C and (a) 1 Pa, (b) 100 kPa, (c) 10 MPa. If a pressure gradient of 0.10 atm cm^{-1} is established in a pipe, what is the flow of gas due to diffusion?

24.15. A $0.0200 \, mol \, L^{-1}$ aqueous KCl solution has a molar conductivity of $138.3 \, S \, cm^2 \, mol^{-1}$ at 25°C, and in a cell its resistance was found to be $74.58 \, \Omega$. Calculate the cell constant.

24.16. The molar conductivity of a strong electrolyte in water at 25°C was found to be $109.9 \, S \, cm^2 \, mol^{-1}$ for a concentration of $6.2 \times 10^{-3} \, mol \, L^{-1}$ and $106.1 \, S \, cm^2 \, mol^{-1}$ for a concentration of $1.50 \times 10^{-2} \, mol \, L^{-1}$. Estimate the limiting molar conductivity of the electrolyte.

24.17. The mobility of the negative ion in a $1:1$ electrolyte in aqueous solution at 25°C is measured as $6.85 \times 10^{-8} \, m^2 \, s^{-1} \, V^{-1}$. Calculate the molar ionic conductivity.

24.18. The mobility of the Rb^+ ion in aqueous solution is $7.92 \times 10^{-8} \, m^2 \, s^{-1} \, V^{-1}$ at 25°C. The potential difference between two electrodes placed in the solution is 35.0 V. If the electrodes are 8.00 mm apart, what is the drift speed of the Rb^+ ion?

24.19. What fraction of the total current is carried by Li^+ when current flows through an aqueous solution of LiBr at 25°C?

24.20. The limiting molar conductivities of KCl, KNO_3, and $AgNO_3$ are 149.9, 145.0, and $133.4 \, S \, cm^2 \, mol^{-1}$, respectively (all at 25°C). What is the limiting molar conductivity of AgCl at this temperature?

24.21. At 25°C the molar ionic conductivities of Li^+, Na^+, and K^+ are 38.7, 50.1, and $73.5 \, S \, cm^2 \, mol^{-1}$, respectively. What are their mobilities?

24.22. The mobility of the NO_3^- ion in aqueous solution at 25°C is $7.40 \times 10^{-8} \, m^2 \, s^{-1} \, V^{-1}$. Calculate its diffusion coefficient in water at 25°C.

24.23. The diffusion coefficient of CCl_4 in heptane at 25°C is $3.17 \times 10^{-9} \, m^2 \, s^{-1}$. Estimate the time required for a CCl_4 molecule to have a root mean square displacement of 5.0 mm. (Hint: treat as a three-dimensional problem.)

24.24. Estimate the effective radius of a sugar (sucrose) molecule in water at 25°C given that its diffusion coefficient is $5.2 \times 10^{-10} \, m^2 \, s^{-1}$ and that the viscosity of water is 1.00 cP.

24.25. The diffusion coefficient for molecular iodine in benzene is $2.13 \times 10^{-9} \, m^2 \, s^{-1}$. How long does a molecule take to jump through about one molecular diameter (approximately the fundamental jump length for translational motion)?

24.26. What is the root mean square distance travelled in three dimensions by (a) an iodine molecule in benzene, (b) a sucrose molecule in water at 25°C in 1.0 s?

24.27. About how long, on average, does it take for the molecules in Exercise 24.26 to drift to a point (a) 1 mm, (b) 1 cm from their starting points in three dimensions?

24.28. The diffusion coefficient of a particular kind of t-RNA molecule is $D = 1.0 \times 10^{-11} \, m^2 \, s^{-1}$ in the medium of a cell interior. How long does it take molecules produced in the cell nucleus to reach the walls of the cell at a distance 1.0 μm, corresponding to the radius of the cell?

PROBLEMS
Numerical problems

24.1. Enrico Fermi, the great Italian scientist, was a master at making good approximate calculations based on little or no actual data. Hence, such calculations are often called 'Fermi calculations'. Do a Fermi calculation on how long it would take for a gaseous air-borne cold virus of molar mass $1 \times 10^4 \, kg \, mol^{-1}$ to travel the distance between two conversing people 1 m apart by diffusion in still air.

24.2. Calculate the ratio of the thermal conductivities of gaseous hydrogen at 300 K to gaseous hydrogen at 10 K. Be circumspect.

24.3. In the Knudsen method for the determination of vapour pressure, a weighed amount of a sample is heated inside a container, in the wall of which there is a small hole. The mass loss over a period of time can be related to the vapour pressure at the temperature of the experiment. If Δw is the mass lost in an interval Δt through a circular hole of radius R, find an expression relating the vapour pressure p to Δw and Δt. A Knudsen cell was used to determine the vapour pressure of germanium at 1000°C. During an interval of 7200 s the mass loss through a hole of radius 0.50 mm amounted to $4.3 \times 10^{-2} \, mg$. What is the vapour pressue of germanium at 1000°C? Assume the gas to be monatomic.

24.4. In a study of the catalytic properties of a titanium surface it was necessary to maintain the surface free from contamination. Calculate the collision frequency per square centimetre of surface made by O_2 molecules at (a) 100 kPa, (b) 1.000 Pa and 300 K. Estimate the number of collisions made with a single surface atom in each second. The conclusions underline the importance of working at very low pressures (much lower than 1 Pa, in fact) in order to study the properties of uncontaminated surfaces. Take the nearest-neighbour distance as 291 pm.

24.5. The nuclide ^{244}Bk (berkelium) decays by producing α particles, which capture electrons and form He atoms. Its half-life is 4.4 h. A 1.0-mg sample was placed in a 1.0 cm^3 container that was impermeable to α radiation, but there was also a hole of radius 2.0 μm in the wall. What is the pressure of helium at 298 K, inside the container after (a) 1.0 h, (b) 10 h?

24.6. An atomic beam is designed to function with (a) cadmium, (b) mercury. The source is an oven maintained at 380 K, there being a small slit of dimensions 1.0 cm long by 1.0×10^{-3} cm wide. The vapour pressure of cadmium is 0.13 Pa and that of mercury is 152 Pa at this temperature. What is the atomic current (the number of atoms per unit time) in the beams?

24.7. Conductivities are often measured by comparing the resistance of a cell filled with the sample to its resistance when filled with some standard solution, such as aqueous potassium chloride. The conductivity of water is 7.6×10^{-4} S cm^{-1} at 25°C and the conductivity of 0.100 mol L^{-1} aqueous KCl is 1.1639×10^{-2} S cm^{-1}. A cell had a resistance of 33.21 Ω when filled with 0.100 mol L^{-1} KCl solution and 300.0 Ω when filled with 0.100 mol L^{-1} acetic acid. What is the molar conductivity of acetic acid at that concentration and temperature?

24.8. The resistances of a series of aqueous NaCl solutions, formed by successive dilution of a sample, were measured in a cell with cell constant 0.2063 cm^{-1}. The following resistances were found:

$c/$(mol L^{-1})	0.0005	0.001	0.005	0.010	0.020	0.050
R/Ω	3314	1669	342.1	174.1	89.08	37.14

Verify that the molar conductivity follows the Kohlrausch law and find the limiting molar conductivity. Determine the coefficient $\mathcal{K}$. Use the value of $\mathcal{K}$ (which should depend only on the nature, not the identity of the ions) and the information that $\lambda(\text{Na}^+) = 50.1$ S cm^2 mol^{-1} and $\lambda(\text{I}^-) = 76.8$ S cm^2 mol^{-1} to predict (a) the molar conductivity, (b) the conductivity, (c) the resistance it would show in the cell, of an 0.010 mol L^{-1} aqueous solution of NaI at 25°C.

24.9. After correction for the water conductivity, the conductivity of a saturated solution of AgCl in water at 25°C was found to be 1.887×10^{-6} S cm^{-1}. What is the solubility of silver chloride at this temperature?

24.10. The resistances of aqueous acetic acid solutions were measured at 25°C in a cell with cell constant 0.2063 cm^{-1} with the following results:

$c/$(mol L^{-1})	0.00049	0.00099	0.00198	0.01581	0.06323	0.2529
R/Ω	6146	4210	2927	1004	497	253

Draw the appropriate graph to obtain values of pK_a.

24.11. What are the drift speeds of Li$^+$, Na$^+$, and K$^+$ in water when a potential difference of 10 V is applied across a 1.00 cm conductivity cell? How long would it take an ion to move from one electrode to the other? In conductivity measurements it is normal to use alternating current: what are the displacements of the ions in (a) cm, (b) solvent diameters, about 300 pm, during a half cycle of 1 kHz applied potential?

24.12. The mobilities of H$^+$ and Cl$^-$ at 22°C in water are 3.623×10^{-3} and 7.91×10^{-4} cm^2 s^{-1} V^{-1}, respectively. What proportion of the current is carried by the protons in 1.0×10^{-3}M hydrochloric acid? What fraction do they carry when the NaCl is added to the acid so that it is 1.0 mol L^{-1} in the salt? Note how concentration as well as mobility governs the transport of current.

24.13. In a moving boundary experiment on KCl the apparatus consisted of a tube of internal diameter 4.146 mm, and it contained aqueous KCl at a concentration of 0.021 mol L^{-1}. A steady current of 18.2 mA was passed, and the boundary advanced as follows:

$\Delta t/$s	200	400	600	800	1000
$x/$mm	64	128	192	254	318

Find the transport number of K$^+$, its mobility, and its ionic conductivity.

24.14. The proton possesses abnormal mobility in water, but does it behave normally in liquid ammonia? To investigate this question, a moving-boundary technique was used to determine the transport number of NH$_4^+$ in liquid ammonia (the analogue of H$_3$O$^+$ in liquid water) at -40°C (J. Baldwin, J. Evans, and J. B. Gill, *J. Chem. Soc. A*, 3389 (1971)). A steady current of 5.000 mA was passed for 2500 s, during which time the boundary formed between mercury(II) iodide and ammonium iodide ammoniacal solutions moved 286.9 mm in a 0.01365 mol kg^{-1} solution and 92.03 mm in a 0.4255 mol kg^{-1} solution. Calculate the transport number of NH$_4^+$ at these concentrations, and comment on the mobility of the proton in liquid ammonia. The bore of the tube is 4.146 mm and the density of liquid ammonia is 0.682 g cm^{-3}.

24.15. A dilute solution of potassium permanganate in water at 25°C was prepared. The solution was in a horizontal 10 cm tube, and at first there was a linear gradation of intensity of the purple solution from the left (where the concentration was 0.100 mol L^{-1}) to the right (where the concentration was 0.050 mol L^{-1}). What is the magnitude and sign of the thermodynamic force acting on the solute (a) close to the left face of the container, (b) in the middle, (c) close to the right face? Give the force per mole and force per molecule in each case.

24.16. Estimate the diffusion coefficients and the effective hydrodynamic radii of the alkali metal cations in water from their mobilities at 25°C. Estimate the approximate number of water molecules that are dragged along by the cations. Ionic radii are given in Table 21.3.

24.17. Nuclear magnetic resonance can be used to determine the mobility of molecules in liquids. A set of measurements on methane in carbon tetrachloride showed that its diffusion coefficient is $2.05 \times 10^{-9}\,m^2\,s^{-1}$ at $0°C$ and $2.89 \times 10^{-9}\,m^2\,s^{-1}$ at $25°C$. Deduce what information you can about the mobility of methane in carbon tetrachloride.

24.18. A concentrated sucrose solution is poured into a 5.0 cm diameter cylinder. Take it to be 10 g of sugar in 5.0 cm^3 of water. A further 1.0 L of water is then poured very carefully on top of the layer, without disturbing it. Ignore gravitational effects, and pay attention only to diffusional processes. Find the concentration at 5.0 cm above the lower layer after a lapse of (a) 10 s, (b) 1 year.

24.19. In a series of observations on the displacement of rubber latex spheres of radius $0.212\,\mu m$ the mean square displacements after selected time intervals were on average as follows:

t/s	30	60	90	120
$10^{12}\langle x^2 \rangle /m^2$	88.2	113.4	128	144

These results were originally used to find the value of Avogadro's constant, but there are now better ways of determining it, so the data can be used to find another quantity. Find the effective viscosity of water at the temperature of this experiment (25°C).

Theoretical problems

24.20. The rate of growth of droplets of lead from a vapour has been studied in the laboratories of an oil company (J. B. Homer and A. Prothero, *J. Chem. Soc. Faraday Trans.* I, **69**, 673 (1973)). Virtually all the lead condensed within 0.5 ms of the initiation of the run, and the concentration in the gas phase was no more than about 3×10^{15} atoms/cm^3. Find an expression for the rate of growth of the radius of the spherical particles. Take $T = 935\,K$ and assume that every atom sticks to the growing surface when it collides with it.

24.21. Show how the ratio of two transport numbers t', t'' for two cations in a mixture depends on their concentrations c', c'', and their mobilities u', u''.

24.22. Confirm that eqn 43 is a solution of the diffusion equation with the correct initial value.

24.23. The diffusion equation is valid when many elementary steps are taken in the time interval of interest; but the random walk calculation lets us discuss distributions for short times as well as for long. Use eqn 2 in *Further information 16* to calculate the probability of being six paces from the origin (that is, at $x = 6\lambda$) after (a) four, (b) six, (c) twelve steps.

24.24. Write a program for calculating P in a one-dimensional random walk, and evaluate the probability of being at $x = 6\lambda$ for $n = 6, 10, 20, \ldots, 60$. Compare the numerical value with the analytical value in the limit of a large number of steps. At what value of n is the discrepancy no more than 0.1 per cent?

The rates of chemical reactions

This chapter is the first of a sequence that explores the rates of chemical reactions. The chapter begins with a discussion of the definition of reaction rate and outlines the techniques for its measurement. The results of such measurements show that reaction rates depend on the concentration of reactants (and products) in characteristic ways that can be expressed in terms of differential equations known as rate laws. The solutions of these equations are then used to predict the concentrations of species at any time after the start of the reaction. The form of the rate law also provides insight into the series of elementary steps by which a reaction takes place. The key task in this connection is to be able to derive a rate law from a proposed mechanism (and then compare it with experiment). We see that simple elementary steps have simple rate laws and that these simple rate laws can be combined together by invoking one or more approximations. These approximations include the concept of the rate-determining step of a reaction, the steady-state concentration of a reaction intermediate, and the existence of a pre-equilibrium. These concepts and techniques are then applied to two simple but important cases: the description of enzyme kinetics and the elucidation of the mechanism of simple gas-phase reactions.

One reason for studying the rates of reactions is the practical importance of being able to predict how quickly a reaction mixture approaches equilibrium. The rate might depend on variables under our control, such as the pressure, the temperature, and the presence of a catalyst, and we may be able to optimize the rate by the appropriate choice of conditions. Another reason is that the study of reaction rates leads to an understanding of the **mechanisms** of reactions, their analysis into a sequence of elementary steps. For example, it might be discovered that the reaction of hydrogen and bromine to form hydrogen bromide proceeds by the dissociation of Br_2, the attack of a Br atom on H_2, and several subsequent steps, and not by a single event in which an H_2 molecule collides with a Br_2 molecule and the atoms exchange partners to form two HBr molecules.

This chapter introduces the principles of chemical kinetics by showing how reaction rates may be measured and interpreted. The remaining chapters of this part of the text then develop this material in more detail and apply it to more complicated or more specialized cases.

EMPIRICAL CHEMICAL KINETICS

The first step in the kinetic analysis of reactions is to establish the stoichiometry of the reaction and identify any side reactions. The basic data of chemical kinetics are then the concentrations of the reactants and products at different times after a reaction has been initiated. The rates of most chemical reactions are sensitive to the temperature (for reasons we explore later), so the temperature of the reaction mixture must be held constant throughout the course of the reaction. This requirement puts severe demands on the design of an experiment. Gas-phase reactions, for instance, are often carried out in a vessel held in contact with a substantial block of metal. Liquid-phase reactions, including flow reactions, must be carried out in an efficient thermostat.

25.1 Experimental techniques

The method used to monitor the concentrations depends on the species involved and the rapidity with which their concentrations change. Many reactions reach equilibrium over periods of minutes or hours, and several techniques may then be used to follow the changing concentrations.

Monitoring the progress of a reaction

A reaction in which at least one component is a gas might result in an overall change in pressure in a system of constant volume, so its progress may be followed by recording the variation of pressure with time. An example is the decomposition of nitrogen(V) oxide,

$$2N_2O_5(g) \rightarrow 4NO_2(g) + O_2(g)$$

For each mole of N_2O_5 molecules destroyed, $\frac{5}{2}$ moles of gas molecules are formed, so the total pressure increases as the reaction proceeds (if the volume is constant). A disadvantage of this method is that it is not specific: all the gas-phase molecules contribute to the pressure.

Example 25.1 *Monitoring the variation in pressure*

Predict how the total pressure varies during the gas-phase decomposition of N_2O_5.

Method. The total pressure (at constant volume and temperature and assuming perfect gas behaviour) is proportional to the number of gas-phase molecules. Therefore, we need to determine the number. To do so, we express the progress of the reaction in terms of the fraction α of N_2O_5 molecules that have reacted.

Answer. Let the initial pressure be p_0 and the initial amount of N_2O_5 molecules present be n. When a fraction α of the N_2O_5 molecules has decomposed, the amounts of the components in the reaction mixture are:

	N_2O_5	NO_2	O_2	Total
Amount:	$n(1-\alpha)$	$2\alpha n$	$\frac{1}{2}\alpha n$	$n(1+\frac{3}{2}\alpha)$

When $\alpha = 0$ the pressure is p_0, so at any stage the total pressure is

$$p = (1 + \tfrac{3}{2}\alpha)p_0$$

When the reaction is complete, the pressure will have risen to $\frac{5}{2}$ times its initial value.

Exercise E25.1. Repeat the calculation for the decomposition $2NOBr(g) \rightarrow 2NO(g) + Br_2(g)$. $\qquad [p = (1 + \tfrac{1}{2}\alpha)p_0]$

Spectrophotometry, the measurement of the intensity of absorption in a particular spectral region, is widely applicable and is especially useful when one substance (and only one) in the reaction mixture has a strong characteristic absorption in a conveniently accessible region of the spectrum. For example, the reaction

$$H_2(g) + Br_2(g) \rightarrow 2HBr(g)$$

can be followed by measuring the absorption of visible light by bromine. If the reaction changes the number or type of ions present in a solution, then it may be followed by monitoring the conductivity of the solution. If hydrogen ions are produced or consumed, then the reaction may be followed by monitoring the pH of the solution.

Other methods of determining composition include titration, mass spectrometry, gas chromatography, and magnetic resonance. Polarimetry, the observation of the optical activity of a reaction mixture, is occasionally applicable.

Application of the techniques

In a **real-time analysis** the composition of the system is analysed while the reaction is in progress. Either a small sample is withdrawn or the bulk solution is monitored. In the **quenching method** the reaction is stopped after it has been allowed to proceed for a certain time, and the composition is analysed at leisure. The quenching (of the entire mixture or of a sample drawn from it) can be achieved either by cooling it suddenly, by adding the mixture to a large volume of solvent, or by rapid neutralization of an acid reagent (or by some analogous reaction). This method is suitable only for reactions that are slow enough for there to be little reaction during the time it takes to quench the mixture. In recent years notable advances have been made in the study of fast reactions, which we shall take to be reactions complete in less than about 1 s (and often very much less), and the present thrust of chemical kinetics is to ever shorter time scales. With special laser techniques it is now possible to observe processes occurring in a few tens of femtoseconds (1 fs = 10^{-15} s).

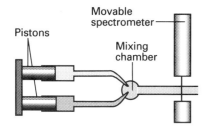

25.1 The arrangement used in the flow technique for studying reaction rates. The reactants are injected into the mixing chamber at a steady rate. The location of the spectrometer corresponds to different times after initiation.

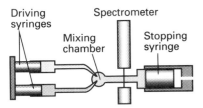

25.2 In the stopped-flow technique the reagents are driven quickly into the mixing chamber by the driving syringes and then the time dependence of the concentrations is monitored.

In the **flow method** the reactants are mixed as they flow together in a chamber (Fig. 25.1). The reaction continues as the thoroughly mixed solutions flow through the outlet tube, and observation of the composition at different positions along the tube (e.g. spectrophotometrically) is equivalent to the observation of the reaction mixture at different times after mixing.

The disadvantage of conventional flow techniques is that a large volume of reactant solution is necessary. This disadvantage is particularly important for fast reactions, because to spread the reaction over a length of tube the flow must be rapid. The **stopped-flow technique** (Fig. 25.2) avoids this disadvantage. The two solutions are mixed very rapidly by injecting them into a tangential mixing chamber designed to ensure that the flow is turbulent and that complete mixing occurs. Beyond the mixing chamber there is an observation cell fitted with a stopping syringe, which moves back as the liquids flood in but comes up against a stop when a required volume (typically about 1 mL) has been injected. The reaction then continues in the thoroughly mixed solution and is monitored. The suitability of the stopped-flow technique for the study of small samples means that it is appropriate for biochemical reactions, and it has been widely used to study the kinetics of enzyme action.

In **flash photolysis** the gaseous or liquid sample is exposed to a brief photolytic flash of light and then the contents of the reaction chamber are monitored. Although discharge lamps can be used for flashes of about 10^{-5} s, most work is now done with lasers with flashes of about 10^{-9} s duration. Mode locking (Section 17.5) has made picosecond pulses available (1 ps $= 10^{-12}$ s), and many studies are now carried out on that time scale. The world record for pulse shortness at the time of writing stands at about 4 fs. Both emission and absorption spectroscopy may be used to monitor the reaction, and the spectra are observed electronically or photographically at a series of times following the flash.

25.2 The rates of reactions

The general outcome of such experiments is the observation that reaction rates depend on the composition and the temperature of the reaction mixture. The next few sections look at these observations in more detail.

The definition of rate

Consider a reaction of the form

$$A + B \rightarrow C$$

in which at some instant the concentrations of the participants are [A], [B], and [C]. Measures of the rate of the reaction are the **rate of formation** of a product (C) and the **rate of consumption** of one of the reactants (A or B). The rate of consumption of the reactant A is

$$v_A = -\frac{d[A]}{dt}$$

and the rate of formation of the product C is

$$v_C = \frac{d[C]}{dt}$$

Both rates are positive. In the present case, the reaction stoichiometry implies that the rate of formation of C is equal to the rate of consumption of either A or B, because, whenever a C molecule is formed, one A molecule and one B molecule are destroyed. For a reaction with a more complicated stoichiometry, such as

$$A + 2B \rightarrow 3C + D$$

the relation between the various rates of formation and consumption is more complicated. In this case

Rate of formation of C = 3 × rate of consumption of A

More specifically and completely,

$$\frac{d[D]}{dt} = \frac{1}{3}\frac{d[C]}{dt} = -\frac{d[A]}{dt} = -\frac{1}{2}\frac{d[B]}{dt}$$

The ambiguity in the definition of rate is avoided if we define the **rate of reaction** v as

$$v = \frac{1}{v_J}\frac{d[J]}{dt} \tag{1}$$

where v_J is the stoichiometric coefficient of substance J (with v_J negative for reactants and positive for products, Section 2.5). Now there is a single rate for the entire equation.

Example 25.2 *Reporting rates of reaction*

The rate of formation of NO in the reaction

$$2NOBr(g) \rightarrow 2NO(g) + Br_2(g)$$

was reported as $1.6 \times 10^{-4}\,mol\,L^{-1}\,s^{-1}$. What is the rate of reaction, and the rate of consumption of NOBr?

Method. Write the chemical equation in the form $0 = (products) - (reactants)$ and identify the stoichiometric numbers. Then use eqn 1 with $d[NO]/dt = 1.6 \times 10^{-4}\,mol\,L^{-1}\,s^{-1}$.

Answer. The reaction is formally

$$0 = 2NO(g) + Br_2(g) - 2NOBr(g)$$

so $v_{NO} = +2$. It follows that

$$v = \frac{1}{2}\frac{d[NO]}{dt} = \frac{1}{2} \times (1.6 \times 10^{-4}\,mol\,L^{-1}\,s^{-1})$$
$$= 8.0 \times 10^{-5}\,mol\,L^{-1}\,s^{-1}$$

Then, because $v_{NOBr} = -2$, the rate at which the concentration of NOBr changes is

$$\frac{d[NOBr]}{dt} = \nu_{NOBr} \times v = -2 \times (8.0 \times 10^{-5}\,mol\,L^{-1}\,s^{-1})$$

$$= -1.6 \times 10^{-4}\,mol\,L^{-1}\,s^{-1}$$

The rate of consumption of NOBr is therefore $1.6 \times 10^{-4}\,mol\,L^{-1}\,s^{-1}$.

Comment. The rate of formation of a substance is the negative of the rate of consumption. Note that rates of reaction are normally expressed in moles per litre per second ($mol\,L^{-1}\,s^{-1}$).

Exercise E25.2. The rate of change of molar concentration of CH_3 radicals in the reaction $2CH_3(g) \rightarrow CH_3CH_3(g)$ was reported as $d[CH_3]/dt = -1.2\,mol\,L^{-1}\,s^{-1}$ under particular conditions. What is (a) the rate of reaction and (b) the rate of formation of CH_3CH_3?

[(a) $0.60\,mol\,L^{-1}\,s^{-1}$,   (b) $0.60\,mol\,L^{-1}\,s^{-1}$]

Rate laws and rate constants

It is often found that the rate of reaction is proportional to the concentrations of the reactants raised to a power. For example, it may be found that the rate is proportional to the concentrations of two reactants A and B, and that

$$v = k[A][B] \tag{2}$$

where each concentration is raised to the first power. The coefficient k is called the **rate constant** for the reaction. The rate constant is independent of the concentrations but depends on the temperature. An experimentally determined equation of this kind is called the **rate law** of the reaction. More formally, a rate law is an equation that expresses the rate of reaction as a function of the concentrations of all the species present in the overall chemical equation for the reaction.

A practical application of a rate law is that, once we know it and the value of the rate constant, we can predict the rate of reaction from the composition of the mixture. Moreover, as we shall see later, by knowing the rate law we can go on to predict the composition of the reaction mixture at a later stage of the reaction. The theoretical usefulness of a rate law is that it is a guide to the mechanism of the reaction, for any proposed mechanism must be consistent with the observed rate law.

Reaction order

The power to which the concentration of a species is raised in a rate law is the **order** of the reaction with respect to that species. A reaction with the rate law in eqn 2 is first-order in A and first-order in B. The **overall order** of a reaction is the sum of the orders of all the components. The rate law in eqn 2 is therefore second-order overall.

A reaction need not have an integral order, and many gas-phase reactions do not. For example, if a reaction is found to have the rate law

$$v = k[A]^{\frac{1}{2}}[B]$$

then it is half-order in A, first-order in B, and three-halves order overall. When a rate law is not of the form $[A]^x[B]^y[C]^z \ldots$, the reaction does not have an order. Thus, the experimentally determined rate law for the gas-phase reaction $H_2 + Br_2 \rightarrow 2HBr$ is

$$v = \frac{k[H_2][Br_2]^{\frac{3}{2}}}{[Br_2] + k'[HBr]} \tag{3}$$

Although the reaction is first order in H_2, it has an indefinite order with respect to both Br_2 and HBr and overall (except under certain simplifying conditions, such as $[Br_2] \gg k'[HBr]$).

The rate law of a reaction is determined experimentally and cannot, in general, be inferred from the reaction equation. The reaction of hydrogen and bromine, for example, has a very simple stoichiometry, but its rate law (eqn 3) is quite complicated. The thermal decomposition of nitrogen(V) oxide has the following rate law:

$$2N_2O_5(g) \rightarrow 4NO_2(g) + O_2(g) \qquad v = k[N_2O_5]$$

The reaction is first-order. In some cases, however, the rate law does happen to reflect the reaction stoichiometry. This is the case with the oxidation of nitrogen(II) oxide, which under certain conditions has a third-order rate law:

$$2NO(g) + O_2(g) \rightarrow 2NO_2(g) \qquad v = k[NO]^2[O_2]$$

Some reactions obey a zero-order rate law, and therefore have a rate that is independent of the concentration of the reactant (so long as some is present). Thus, the catalytic decomposition of phosphine (PH_3) on hot tungsten at high pressures has the rate law

$$v = k$$

The PH_3 decomposes at a constant rate until it has almost entirely disappeared. Only heterogeneous reactions can have rate laws that are zero-order overall.

These remarks point to three problems. First, we must see how to identify the rate law and obtain the rate constant from the experimental data. We shall concentrate on this aspect in this chapter. Second, we must see how to construct reaction mechanisms that are consistent with the rate law. We shall introduce the techniques of doing so in this chapter and develop them further in Chapter 26. Third, we must account for the values of the rate constants and explain their temperature dependence. We shall see a little of what is involved in this chapter, but leave the details until Chapter 27.

The determination of the rate law

The first step in the experimental investigation of a reaction rate is to identify all the products, and to investigate whether any transient intermediates and side reactions are involved.

The determination of rate laws is simplified by the **isolation method** in which the concentrations of all the reactants except one are in large excess. If B is in large excess, for example, then it is a good approximation to take its concentration as constant throughout the reaction. Although the true rate law might be

$$v = k[A][B]$$

we can approximate $[B]$ by $[B]_0$ and write

$$v = k'[A], \quad \text{where } k' = k[B]_0$$

which has the form of a first-order rate law. Because the true rate law has been forced into first-order form by assuming that the concentration of B is constant, it is called a **pseudofirst-order rate law**. Had the rate law been more complicated, such as

$$v = \frac{k_1[A]^2[B]^{\frac{1}{2}}}{k_2 + k_3[B]}$$

then the isolation technique with B in excess would result in

$$v = k[A]^2, \quad \text{where } k = \frac{k_1[B]_0^{\frac{1}{2}}}{k_2 + k_3[B]_0}$$

This is a **pseudosecond-order rate law** and it is much easier to analyse and identify than the complete law. The dependence of the rate on the concentration of each of the reactants may be found by isolating them in turn (by having all the other substances present in large excess), and so constructing a picture of the overall rate law.

In the method of **initial rates**, which is often used in conjunction with the isolation method, the rate is measured at the beginning of the reaction for several different initial concentrations of reactants. We shall suppose that the rate law for a reaction with A isolated is

$$v = k[A]^a$$

Then its initial rate v_0 is given by the initial value of the concentration of A:

$$v_0 = k[A]_0^a$$

Taking logarithms:

$$\log v_0 = \log k + a \log [A]_0 \tag{4}$$

For a series of initial concentrations, a plot of the logarithms of the initial rates against the logarithms of the initial concentrations of A should be a straight line with slope a.

Example 25.3 *Using the method of initial rates*

The recombination of iodine atoms in the gas phase in the presence of argon was investigated and the order of the reaction was determined by the method of initial rates. The initial rates of reaction of $2I(g) + Ar(g) \rightarrow I_2(g) + Ar(g)$ were as follows:

$[I]_0/(10^{-5}\,\text{mol L}^{-1})$		1.0	2.0	4.0	6.0
$v_0/(\text{mol L}^{-1}\,\text{s}^{-1})$	(a)	8.70×10^{-4}	3.48×10^{-3}	1.39×10^{-2}	3.13×10^{-2}
	(b)	4.35×10^{-3}	1.74×10^{-2}	6.96×10^{-2}	1.57×10^{-1}
	(c)	8.69×10^{-3}	3.47×10^{-2}	1.38×10^{-1}	3.13×10^{-1}

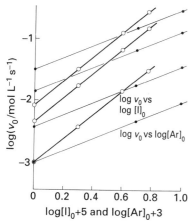

25.3 The plot of $\log v_0$ against $\log [I]_0$ for a given $[Ar]_0$ and against $\log[Ar]_0$ for a given $[I]_0$. The horizontal scales have been adjusted so as to fit all the data on to a single graph, for only the slopes are significant.

The Ar concentrations are (a) $1.0 \times 10^{-3} \,\text{mol}\,L^{-1}$, (b) $5.0 \times 10^{-3}\,\text{mol}\,L^{-1}$, and (c) $1.0 \times 10^{-2}\,\text{mol}\,L^{-1}$. Determine the orders of reaction with respect to the I and Ar atom concentrations and the rate constant.

Method. We need to plot the logarithm of the initial rate, $\log v_0$, against $\log [I]_0$ for a given concentration of Ar, and then against $\log [Ar]_0$ for a given concentration of I. The slopes of the two lines are the orders of reaction with respect to I and Ar, respectively. The intercepts with the vertical axis give $\log k$; alternatively, evaluate $k = v/[A]^a[B]^b$.

Answer. The plots are shown in Fig. 25.3. The slopes are 2 and 1, respectively, so the (initial) rate law is

$$v_0 = k[I]_0^2[Ar]_0$$

This rate law signifies that the reaction is second-order in [I], first-order in [Ar], and third-order overall. The intercept corresponds to $\log(k/\text{mol}^{-2}\,L^2\,s^{-1}) = 9.9$.

Comment. The units of k come automatically from the calculation, and are always such as to convert the product of concentrations to concentration per unit time (e.g. $\text{mol}\,L^{-1}\,s^{-1}$).

Exercise E25.3. The initial rate of a reaction depended on concentration of a substance J as follows:

$[J]_0/(10^{-3}\,\text{mol}\,L^{-1})$	5.0	8.2	17	30
$v_0/(10^{-7}\,\text{mol}\,L^{-1}\,s^{-1})$	3.6	9.6	41	130

Determine the order of the reaction with respect to J and calculate the rate constant.
$$[2, 1.4 \times 10^{-2}\,L\,\text{mol}^{-1}\,s^{-1}]$$

25.3 Integrated rate laws

The method of initial rates might not reveal the full rate law, for in a complex reaction the products may participate in the reaction and affect the rate once they have been formed. For example, products participate in the synthesis of HBr, because eqn 3 shows that the full rate law depends on the concentration of HBr. To avoid this difficulty the rate law should be fitted to the data throughout the reaction. The fitting may be done, in simple cases at least, by using a proposed rate law to predict the concentration of any component at any time and comparing it with the data. A law should also be tested by observing whether the addition of products or, for gas-phase reactions, a change in the surface-to-volume ratio in the reaction chamber affect the rate.

Rate laws are differential equations. Therefore, we must integrate them if we want to find the concentrations as a function of time. Now that computers are so widely available, even the most complex rate laws may be integrated numerically. However, in a number of simple cases analytical solutions are easily obtained, and prove to be very useful. We shall examine a few of these simple cases here, and illustrate the computational approach in Chapter 26.

First-order reactions

As shown in the *Justification* that follows, the first-order rate law for the consumption of a reactant A

$$\frac{d[A]}{dt} = -k[A] \tag{5a}$$

has the solution

$$\ln\left(\frac{[A]}{[A]_0}\right) = -kt \tag{5b}$$

$$[A] = [A]_0 e^{-kt} \tag{5c}$$

These two equations are versions of an **integrated rate law**, the integrated form of the rate equation.

JUSTIFICATION

Equation 5a rearranges to

$$\frac{d[A]}{[A]} = -k\,dt$$

which can be integrated directly. Initially (at $t = 0$), the concentration of A is $[A]_0$, and at a later time t it is $[A]$, so we write

$$\int_{[A]_0}^{[A]} \frac{d[A]}{[A]} = -\int_0^t k\,dt$$

Because the integral of $1/x$ is $\ln x$, eqns 5b and 5c are obtained immediately.

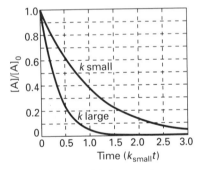

25.4 The exponential decay of the reactant in a first-order reaction. The larger the rate constant, the more rapid the decay. For this graph, $k_{large} = 3k_{small}$.

Equation 5b shows that, if $\ln([A]/[A]_0)$, is plotted against t, then a first-order reaction will give a straight line of slope $-k$. Some rate constants determined in this way are given in Table 25.1. Equation 5c shows that in a first-order reaction the reactant concentration decreases exponentially with time with a rate determined by k (Fig. 25.4).

Table 25.1* Kinetic data for first-order reactions

Reaction	Phase	$\theta/°C$	k/s^{-1}	$t_{\frac{1}{2}}$
$2N_2O_5 \rightarrow 4NO_2 + O_2$	g	25	3.38×10^{-5}	2.85 h
$2N_2O_5 \rightarrow 4NO_2 + O_2$	$Br_2(l)$	25	4.27×10^{-5}	2.25 h
$C_2H_6 \rightarrow 2CH_3$	g	700	5.46×10^{-4}	21.2 min

* More values are given in the Data section at the end of this volume.

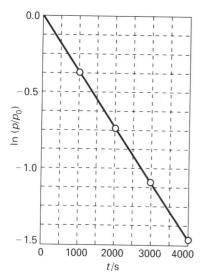

25.5 The determination of the rate constant of a first-order reaction: a straight line is obtained when ln [A] (or ln p) is plotted against t; the slope gives k.

Example 25.4 *Analysing a first-order reaction*

The variation in the partial pressure of azomethane with time was followed at 600 K, with the results given below. Confirm that the decomposition

$$CH_3N_2CH_3(g) \rightarrow CH_3CH_3(g) + N_2(g)$$

is first-order in azomethane, and find the rate constant at 600 K.

t/s	0	1000	2000	3000	4000
$p/(10^{-2}\,Torr)$	8.20	5.72	3.99	2.78	1.94

Method. As indicated in the text, to confirm that a reaction is first-order, we plot $\ln([A]/[A]_0)$ against time and expect a straight line. Because the partial pressure of a gas is proportional to its concentration, it is equivalent to plot $\ln(p/p_0)$ against t. If a straight line is obtained, then its slope can be identified with $-k$.

Answer. Figure 25.5 shows the plot of $\ln(p/p_0)$ against t. The plot is straight, confirming a first-order reaction, and its slope is -3.6×10^{-4}. Therefore, $k = 3.6 \times 10^{-4}\,s^{-1}$.

Exercise E25.4. In a particular experiment, it was found that the concentration of N_2O_5 in liquid bromine varied with time as follows:

t/s	0	200	400	600	1000
$[N_2O_5]/mol\,L^{-1}$	0.110	0.073	0.048	0.032	0.014

Confirm the reaction order and determine the rate constant.

$$[1, 2.1 \times 10^{-3}\,s^{-1}]$$

Half-lives

A useful indication of the rate of a first-order chemical reaction is the **half-life** $t_{\frac{1}{2}}$, of a substance, the time it takes for its concentration to fall to half the initial value. The time for [A] to decrease from $[A]_0$ to $\frac{1}{2}[A]_0$ in a first-order reaction is given by eqn 5 as

$$kt_{\frac{1}{2}} = -\ln\left(\frac{\frac{1}{2}[A]_0}{[A]_0}\right) = -\ln\frac{1}{2} = \ln 2$$

Hence

$$t_{\frac{1}{2}} = \frac{\ln 2}{k} \tag{6}$$

(It is sometimes useful to remember that $\ln 2 = 0.693$.) The main point to note about this result is that, *for a first-order reaction, the half-life of a reactant is independent of its initial concentration.* Hence, if the concentration of A at some arbitrary stage of the reaction is [A], then it will have fallen to $\frac{1}{2}[A]$ after a further interval of $(\ln 2)/k$. Some half-lives are given in Table 25.1.

Second-order reactions

We show in the *Justification* that follows that the integrated form of the second-order rate law

$$\frac{d[A]}{dt} = -k[A]^2 \tag{7a}$$

is

$$\frac{1}{[A]} - \frac{1}{[A]_0} = kt \tag{7b}$$

This expression rearranges to

$$[A] = \frac{[A]_0}{1 + kt[A]_0} \tag{7c}$$

JUSTIFICATION

Equation 7a is integrated by rearranging it to

$$-\frac{d[A]}{[A]^2} = k\,dt$$

The concentration of A is $[A]_0$ at $t = 0$ and $[A]$ at a time t later. Therefore, this expression integrates as follows:

$$-\int_{[A]_0}^{[A]} \frac{d[A]}{[A]^2} = \int_0^t k\,dt$$

Because the integral of $1/x^2$ is $-1/x$, we obtain eqn 7b by substitution of the limits.

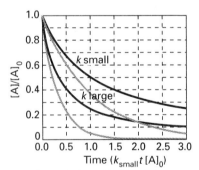

25.6 The variation with time of the concentration of a reactant in a second-order reaction. The pale line is the corresponding decay in a first-order reaction with the same initial rate. In each case, $k_{large} = 3k_{small}$.

Equation 7b shows that to test for a second-order reaction we should plot $1/[A]$ against t and expect a straight line. The slope of the graph is k. Some rate constants determined in this way are given in Table 25.2. Equation 7c lets us predict the concentration of A at any time after the start of the reaction. It shows that the concentration of A approaches zero more slowly than in a first-order reaction with the same initial rate (Fig. 25.6).

It is easy to show from eqn 7b by substituting $t = t_{\frac{1}{2}}$ and $[A] = \frac{1}{2}[A]_0$

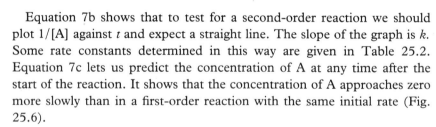

Table 25.2 *Kinetic data for second-order reactions

Reaction	Phase	$\theta/°C$	$k/(\text{L mol}^{-1}\,\text{s}^{-1})$
$2NOBr \rightarrow 2NO + Br_2$	g	10	0.80
$2I \rightarrow I_2$	g	23	7×10^9
$CH_3Cl + CH_3O^-$	$CH_3OH(l)$	20	2.29×10^{-6}

* More values are given in the Data section.

that the half-life of a species A that is consumed in a second-order reaction is proportional to $1/[A]_0$. Therefore, unlike in a first-order reaction, the half-life varies with the initial concentration. This dependence is indicative of second-order kinetics, for there are few examples of reactions with order higher than 2. A practical consequence is that species that decay by second-order reactions (which includes some environmentally harmful substances) may persist in low concentrations for long periods because their half-lives are long when their concentrations are low.

Another type of second-order reaction is one that is first-order in each of two reactants A and B:

$$\frac{d[A]}{dt} = -k[A][B] \tag{8a}$$

Such a rate law cannot be integrated until we know how the concentration of B is related to that of A. For example, if the reaction has the stoichiometry

$$A + B \rightarrow Products$$

and the initial concentrations are $[A]_0$ and $[B]_0$, then it is shown in the following *Justification* that, at a time t after the start of the reaction, the concentrations satisfy the relation

$$kt = \frac{1}{[B]_0 - [A]_0} \ln\left(\frac{[B]/[B]_0}{[A]/[A]_0}\right) \tag{8b}$$

JUSTIFICATION

It follows from the reaction stoichiometry that, when the concentration of A has fallen to $[A]_0 - x$, the concentration of B will have fallen to $[B]_0 - x$ (because each A that disappears entails the disappearance of one B). It follows that

$$\frac{d[A]}{dt} = -k([A]_0 - x)([B]_0 - x)$$

Then, because $d[A]/dt = -dx/dt$, the rate law is

$$\frac{dx}{dt} = k([A]_0 - x)([B]_0 - x)$$

The initial condition is that $x = 0$ when $t = 0$; so the integration required is

$$\int_0^x \frac{dx}{([A]_0 - x)([B]_0 - x)} = \int_0^t k \, dt$$

The integral on the right is simply kt. It follows that

$$kt = \int_0^x \frac{dx}{([A]_0 - x)([B]_0 - x)}$$

$$= \frac{1}{[B]_0 - [A]_0} \int_0^x \left\{ \frac{1}{[A]_0 - x} - \frac{1}{[B]_0 - x} \right\} dx$$

$$= \frac{1}{[B]_0 - [A]_0} \left\{ \ln\left(\frac{[A]_0}{[A]_0 - x}\right) - \ln\left(\frac{[B]_0}{[B]_0 - x}\right) \right\}$$

This expression can be simplified and rearranged into eqn 8b by combining the two logarithms and noting that $[A] = [A]_0 - x$ and $[B] = [B]_0 - x$.

Similar calculations may be carried out to find the integrated rate laws for other orders, and some are listed in Table 25.3. However, because all the laws considered so far disregard the possibility that the reverse reaction is important, none of them describes the overall rate when the reaction is close to equilibrium. At that stage the products may be so abundant that the reverse reaction must be taken into account. In practice, however, most kinetic studies are made on reactions that are far from equilibrium, and the reverse reactions are unimportant.

25.4 Reactions approaching equilibrium

We can explore the variation of the composition with time close to equilibrium by considering the reaction in which A forms B and both forward and reverse reactions are first-order

$$A \rightarrow B \qquad v = k[A]$$
$$B \rightarrow A \qquad v = k'[B]$$

The concentration of A is reduced by the forward reaction (at a rate $k[A]$) but it is increased by the reverse reaction (at a rate $k'[B]$). The net rate of change is therefore

$$\frac{d[A]}{dt} = -k[A] + k'[B]$$

If the initial concentration of A is $[A]_0$ and there is no B present initially, then at all times $[A] + [B] = [A]_0$. Therefore

$$\frac{d[A]}{dt} = -k[A] + k'([A]_0 - [A]) = -(k + k')[A] + k'[A]_0$$

The solution of this first-order differential equation (as may be checked by differentiation) is

$$[A] = \frac{k' + ke^{-(k+k')t}}{k + k'} \times [A]_0 \tag{9}$$

The time dependence predicted by this equation is drawn in Fig. 25.7. As $t \rightarrow \infty$, the concentrations reach their equilibrium values, which are

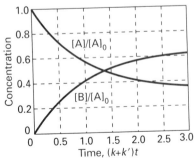

25.7 The approach of concentrations to their equilibrium values is predicted by eqn 9 for a reaction $A \rightleftharpoons B$ that is first-order in each direction, and for which $k = 2k'$.

Table 25.3 Integrated rate laws

Order	Reaction	Rate law*	$t_{\frac{1}{2}}$
0	$A \rightarrow P$	$v = k$	$\dfrac{[A]_0}{2k}$
		$kt = x$ for $0 \leq x \leq [A]_0$	
1	$A \rightarrow P$	$v = k[A]$	$\dfrac{\ln 2}{k}$
		$kt = \ln \dfrac{[A]_0}{[A]_0 - x}$	
2	$A \rightarrow P$	$v = k[A]^2$	$\dfrac{1}{k[A]_0}$
		$kt = \dfrac{x}{[A]_0([A]_0 - x)}$	
	$A + B \rightarrow P$	$v = k[A][B]$	
		$kt = \dfrac{1}{[B]_0 - [A]_0} \ln \dfrac{[A]_0([B]_0 - x)}{([A]_0 - x)[B]_0}$	
	$A + 2B \rightarrow P$	$v = k[A][B]$	
		$kt = \dfrac{1}{[B]_0 - 2[A]_0} \ln \dfrac{[A]_0([B]_0 - 2x)}{([A]_0 - x)[B]_0}$	
	$A \rightarrow P$ with autocatalysis		
		$v = k[A][P]$	
		$kt = \dfrac{1}{[A]_0 + [P]_0} \ln \dfrac{[A]_0([P]_0 + x)}{([A]_0 - x)[P]_0}$	
3	$A + 2B \rightarrow P$	$v = k[A][B]^2$	
		$kt = \dfrac{2x}{(2[A]_0 - [B]_0)([B]_0 - 2x)[B]_0}$	
		$\quad + \dfrac{1}{(2[A]_0 - [B]_0)^2} \ln \dfrac{[A]_0([B]_0 - 2x)}{([A]_0 - x)[B]_0}$	
$n \geq 2$	$A \rightarrow P$	$v = k[A]^n$	$\dfrac{2^{n-1} - 1}{(n-1)k[A]_0^{n-1}}$
		$kt = \dfrac{1}{n-1} \left\{ \dfrac{1}{([A]_0 - x)^{n-1}} - \dfrac{1}{[A]_0^{n-1}} \right\}$	

* $x = [P]$, and $v = dx/dt$.

given by the equation as:

$$[A]_\infty = \frac{k'}{k + k'} \times [A]_0$$

$$[B]_\infty = [A]_0 - [A]_\infty = \frac{k}{k + k'} \times [A]_0$$

It follows that the equilibrium constant of the reaction is

$$K = \frac{[B]_\infty}{[A]_\infty} = \frac{k}{k'} \tag{10}$$

Equation 10 is a very important result, because it relates the thermodynamic quantity, the equilibrium constant, to quantities relating to rates. The practical importance of eqn 10 is that, if one of the rate constants can be measured, then the other may be obtained if the equilibrium constant is known.

Example 25.5 *Relating the equilibrium constant to the rate constants*

Show that, for a reaction that takes place in a sequence of steps, the overall equilibrium constant is a product of ratios of the rate constants for each step.

Method. It is sufficient to consider a reasonably general but simple two-step reaction sequence, such as

$A + B \rightleftarrows C + D$ (second-order in each direction;

k_a forward, k_a' reverse)

$C \rightleftarrows E + F$ (first-order forward, second-order reverse;

k_b forward, k_b' reverse)

Overall: $A + B \rightleftarrows D + E + F$

(Note the use of $\rightleftarrows$ to signify a reaction that may run in either direction; the symbol $\rightleftharpoons$ is reserved for reactions at equilibrium.) We write the expressions for the net rate of formation of one species from each step (A and C, for instance). Then we set these net rates equal to zero, corresponding to the attainment of equilibrium. That results in relations between the concentrations. Finally, we write the equilibrium constant for the overall reaction, insert the relations just established, and express the result in terms of the rate constants.

Answer. The net rates of change of the concentrations of A and C are

$$\frac{d[A]}{dt} = -k_a[A][B] + k_a'[C][D]$$

$$\frac{d[C]}{dt} = -k_b[C] + k_b'[E][F]$$

At equilibrium, all the reactions are individually at equilibrium, and setting the net rates each equal to zero gives

$$\frac{[C][D]}{[A][B]} = \frac{k_a}{k_a'} \qquad \frac{[E][F]}{[C]} = \frac{k_b}{k_b'}$$

The equilibrium constant of the overall reaction is therefore

$$K = \frac{[D][E][F]}{[A][B]} = \frac{[C][D][E][F]}{[A][B][C]}$$

$$= \frac{[C][D]}{[A][B]} \times \frac{[E][F]}{[C]} = \frac{k_a}{k_a'} \times \frac{k_b}{k_b'}$$

Comment. When an overall reaction is the sum of several steps, the equilibrium constant is a product of all the ratios:

$$K = \frac{k_a}{k_a'} \times \frac{k_b}{k_b'} \ldots$$

where the ks are the rate constants for the individual steps and the k's are those for the corresponding reverse reaction steps.

Exercise E25.5. Confirm the general equation for the reaction sequence $A + B \rightarrow 2C$, $C + D \rightarrow 2P$, $C + B \rightarrow D$ at equilibrium.

25.5 The temperature dependence of reaction rates

It is found that the rates of most reactions increase as the temperature is raised. Many reactions fall somewhere in the range spanned by the hydrolysis of methyl ethanoate (where the rate constant at 35°C is 1.82 times that at 25°C) and hydrolysis of sucrose (where the factor is 4.13).

The Arrhenius parameters

An empirical observation is that many reactions have rate constants that follow the **Arrhenius equation**

$$\ln k = \ln A - \frac{E_a}{RT} \tag{11}$$

That is, for many reactions it is found that a plot of $\ln k$ against $1/T$ gives a straight line. The Arrhenius equation is often written as

$$k = A e^{-E_a/RT} \tag{12a}$$

The factor A is called the **pre-exponential factor** or the **frequency factor**; E_a is called the **activation energy**. Collectively, the two quantities are called the **Arrhenius parameters** of the reaction, and some experimental values are given in Table 25.4. Equation 12a is

Table 25.4* Arrhenius parameters

(1) First-order reactions	A/s^{-1}	$E_a/(kJ\ mol^{-1})$
$CH_3NC \rightarrow CH_3CN$	3.98×10^{13}	160
$2N_2O_5 \rightarrow 4NO_2 + O_2$	4.94×10^{13}	103.4
(2) Second-order reactions	$A/(L\ mol^{-1}\ s^{-1})$	$E_a/(kJ\ mol^{-1})$
$OH + H_2 \rightarrow H_2O + H$	8.0×10^{10}	42
$NaC_2H_5O + CH_3I$ in ethanol	2.42×10^{11}	81.6

* More values are given in the Data section.

sometimes written in an alternative form that combines the two parameters:

$$k = e^{-\Delta^{\ddagger}G/RT}, \qquad \text{or } -RT \ln k = \Delta^{\ddagger}G \qquad (12b)$$

The quantity $\Delta^{\ddagger}G$ is called the **activation Gibbs energy**. In this form, the expression for the rate constant strongly resembles the formula for the equilibrium constant in terms of the standard reaction Gibbs energy (eqn 9.7a). We shall see the origin of this formal analogy in Chapter 27.

For the present chapter we shall regard the Arrhenius parameters as purely empirical quantities that enable us to discuss the variation of rate constants with temperature. However, it is worth anticipating the interpretation that we shall put on E_a in Section 27.1. There we shall see that *the activation energy is the minimum energy that reactants must have in order to form products.* For example, in a gas-phase reaction there are numerous collisions each second, but only a tiny proportion of them are sufficiently energetic to lead to reaction. The fraction of collisions with a kinetic energy in excess of an energy E_a is given by the Boltzmann distribution as $e^{-E_a/RT}$, exactly as in eqn 12a. Hence, the exponential factor in eqn 12a can be interpreted as the fraction of collisions that have enough energy to lead to reaction.

The analogous interpretation of the pre-exponential factor is that it is a measure of the rate at which collisions occur irrespective of their energy. Hence the product of A and the exponential factor in eqn 12a gives the rate of successful collisions. We shall develop these remarks in Chapter 27 and see that they have their analogues for reactions that take place in liquids.

Example 25.6 *Determining the Arrhenius parameters*

The rate of the second-order decomposition of acetaldehyde (ethanal, CH_3CHO) was measured over the temperature range 700–1000 K, and the rate constants are reported below. Find the activation energy and the pre-exponential factor.

T/K	700	730	760	790	810	840	910	1000
$k/(\text{L mol}^{-1}\,\text{s}^{-1})$	0.011	0.035	0.105	0.343	0.789	2.17	20.0	145

Method. According to eqn 11, the data can be analysed by plotting $\ln (k/\text{L mol}^{-1}\,\text{s}^{-1})$ against $1/(T/\text{K})$ and getting a straight line. The slope of this line is $(-E_a/R)/\text{K}$ and the intercept at $1/T = 0$ is $\ln A$.

Answer. We plot $\ln k$ against $1/T$ in Fig. 25.8. The least-squares best fit of the line is with slope -2.21×10^4 and intercept 27.0. Therefore,

$$E_a = (2.21 \times 10^4\,\text{K}) \times (8.3145\,\text{J K}^{-1}\,\text{mol}^{-1}) = 184\,\text{kJ mol}^{-1}$$
$$A = e^{27.0}\,\text{L mol}^{-1}\,\text{s}^{-1} = 5.3 \times 10^{11}\,\text{L mol}^{-1}\,\text{s}^{-1}$$

Comment. Note that A has the same units as k. The slopes and intercepts of graphs are always dimensionless, and care must be taken to relate the numerical value to the physical quantity by noting how the data have been plotted.

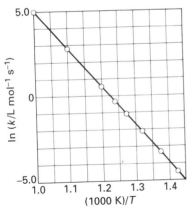

25.8 The Arrhenius plot of $\ln k$ against $1/T$ for the decomposition of CH_3CHO, and the best straight line. The slope gives $-E_a/R$ and the intercept at $1/T = 0$ gives $\ln A$.

Exercise E25.6. Determine A and E_a from the following data:

T/K	300	350	400	450	500
$k/(\text{L mol}^{-1}\,\text{s}^{-1})$	7.9×10^6	3.0×10^7	7.9×10^7	1.7×10^8	3.2×10^8

$$[8 \times 10^{10}\,\text{L mol}^{-1}\,\text{s}^{-1}, 23\,\text{kJ mol}^{-1}]$$

The temperature dependence of some reactions is not Arrhenius-like. However, it is still possible to express the strength of the dependence by defining an activation energy as

$$E_a = RT^2 \frac{\text{d} \ln k}{\text{d}T} \tag{13a}$$

This definition reduces to the earlier one (as the slope of an Arrhenius plot) for a temperature-independent activation energy. Thus, by using $\text{d}(1/T) = -\text{d}T/T^2$ we can rearrange eqn 13a into

$$E_a = -R \frac{\text{d} \ln k}{\text{d}(1/T)} \tag{13b}$$

which integrates to eqn 11 if E_a is independent of temperature. However, the definition in eqn 13 is more general than eqn 11, because it allows E_a to be obtained from the slope (at the temperature of interest) of a plot of $\ln k$ against $1/T$ even if the Arrhenius plot is not a straight line.

Equation 13a shows that the higher the activation energy, the stronger the temperature dependence of the rate constant. That is, a high activation energy signifies that the rate constant changes rapidly with temperature.

Temperature-jump relaxation methods

The term **relaxation** denotes the return of a system to equilibrium. It is used in chemical kinetics to indicate that an externally applied influence has shifted the equilibrium position of a reaction, normally suddenly, and that the reaction is adjusting to the equilibrium composition characteristic of the new conditions (Fig. 25.9). We shall consider the response of reaction rates to a **temperature jump**, a sudden change in temperature. Studies of relaxation times are of considerable importance for the investigation of fast reactions.

When a sudden temperature increase is applied to a simple $A \rightleftharpoons B$ equilibrium that is first-order in each direction, the composition relaxes to the new equilibrium composition exponentially:

$$x = x_0 e^{-t/\tau}, \qquad \text{with } \frac{1}{\tau} = k_a + k_b \tag{14}$$

where x is the departure from equilibrium at the new temperature and x_0 is the departure from equilibrium immediately after the temperature jump.

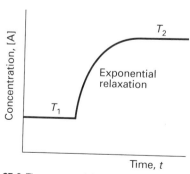

25.9 The exponential relaxation of an equilibrium concentration to a new equilibrium when a sample is subjected to a temperature jump from T_1 to T_2.

JUSTIFICATION

At the initial temperature, the rate of change of [A] is

$$\frac{d[A]}{dt} = -k'_a[A] + k'_b[B]$$

At equilibrium under these conditions, $d[A]/dt = 0$, and the concentrations are $[A]'_{eq}$ and $[B]'_{eq}$. Therefore,

$$k'_a[A]'_{eq} = k'_b[B]'_{eq}$$

When the temperature is increased suddenly, the rate constants change to k_a and k_b but the concentrations of A and B remain for an instant at their old equilibrium values. As the system is no longer at equilibrium, it readjusts to the new equilibrium concentrations, which are now given by

$$k_a[A]_{eq} = k_b[B]_{eq}$$

and it does so at a rate that depends on the new rate constants.

We write the deviation of [A] from its new equilibrium value as x, so $[A] = x + [A]_{eq}$ and $[B] = [B]_{eq} - x$. The concentration of A then changes as follows:

$$\frac{d[A]}{dt} = -k_a(x + [A]_{eq}) + k_b(-x + [B]_{eq}) = -(k_a + k_b)x$$

because the two terms involving the equilibrium concentrations cancel. Because $d[A]/dt = dx/dt$, this equation is a first-order differential equation with the solution given in eqn 14.

Equation 14 shows that the concentration of A (and of B) relaxes into the new equilibrium at a rate determined by the sum of the two new rate constants. Because the equilibrium constant under the new conditions is $K = k_a/k_b$, its value may be combined with the relaxation time measurement to find the individual k_a and k_b.

One way of achieving a temperature jump is to discharge an electric current through a sample made conducting by the addition of ions, but laser discharges can also be used. Temperature jumps of between 5 and 10 K can be achieved in about 10^{-6} s. An important application of the technique has been to the determination of the rate of the proton transfer reaction

$$H_3O^+(aq) + OH^-(aq) \rightarrow 2H_2O(l)$$

by monitoring the conductivity of the sample. By using a burst of microwave power to heat the sample, a relaxation time of about 40 μs was measured at room temperature. This relaxation time corresponds to $k = 1.4 \times 10^{11}$ L mol^{-1} s^{-1}, making it one of the fastest liquid-phase reactions known (but the reaction is faster in ice, where $k = 8.6 \times 10^{12}$ L mol^{-1} s^{-1}).

Example 25.7 *Analysing a temperature-jump experiment*

The $H_2O(l) \rightleftharpoons H^+(aq) + OH^-(aq)$ equilibrium relaxes in 37 μs at 298 K and $pH \approx 7$, and $pK_w = 14.01$. Calculate the rate constants for the forward and reverse reactions.

Method. We need to derive an expression for the relaxation time τ in terms of k_1 (forward, first-order reaction) and k_2 (reverse, second-order reaction). We can proceed as above, but it will be necessary to make the assumption that the deviation from equilibrium (x) is so small that terms in x^2 can be neglected. Relate k_1 and k_2 through the equilibrium constant, but be careful with units because K_w is dimensionless.

Answer. The forward rate is $k_1[H_2O]$ and the reverse rate is $k_2[H^+][OH^-]$. The same procedure as before leads to

$$\frac{1}{\tau} = k_1 + k_2([H^+] + [OH^-])$$

The equilibrium condition is

$$k_1[H_2O]_e = k_2[H^+]_e[OH^-]_e$$

From this expression it follows that

$$\frac{k_1}{k_2} = \frac{[H^+][OH^-]}{[H_2O]} = \frac{K_w(mol\,L^{-1})^2}{[H_2O]} = \frac{K_w}{55.6}\,mol\,L^{-1}$$

because the molar concentration of pure water is $55.6\,mol\,L^{-1}$. If we write $K = K_w/55.6 = 1.8 \times 10^{-16}$, we obtain

$$\frac{1}{\tau} = k_2(K\,mol\,L^{-1} + [H^+] + [OH^-])$$

$$= k_2(K + K_w^{\frac{1}{2}} + K_w^{\frac{1}{2}})\,mol\,L^{-1} = (2.0 \times 10^{-7}) \times k_2\,mol\,L^{-1}$$

Hence,

$$k_2 = \frac{1}{(37 \times 10^{-6}\,s) \times (2.0 \times 10^{-7}\,mol\,L^{-1})} = 1.4 \times 10^{11}\,L\,mol^{-1}\,s^{-1}$$

It follows that

$$k_1 = k_2 K\,mol\,L^{-1} = 2.4 \times 10^{-5}\,s^{-1}$$

Comment. Notice how we keep track of units: K and K_w are dimensionless; k_2 is expressed in $L\,mol^{-1}\,s^{-1}$ and k_1 is expressed in s^{-1}.

Exercise E25.7. Derive an expression for the relaxation time of a concentration when the reaction $A + B \rightleftharpoons C + D$ is second-order in both directions.
$$[1/\tau = k([A] + [B])_e + k'([C] + [D])_e]$$

The equilibrium constant of a reaction varies with temperature if the reaction enthalpy is non-zero (Section 9.4), so the temperature-jump method is widely applicable. The equilibrium composition also depends on pressure in some cases, and **pressure-jump techniques** may then be used.

ACCOUNTING FOR THE RATE LAWS

We now move on to the second stage of the analysis of kinetic data, their explanation in terms of a postulated reaction mechanism.

25.6 Elementary reactions

Most reactions occur in a sequence of steps called **elementary reactions**, each of which involves only one or two molecules or ions. A typical elementary reaction is

$$H + Br_2 \rightarrow HBr + Br$$

(We do not specify the phase of the species when giving the chemical equation for an elementary reaction.) This equation signifies that an H atom attacks a Br_2 molecule to produce an HBr molecule and a Br atom. The **molecularity** of an elementary reaction is the number of molecules coming together to react. In a **unimolecular reaction**, a single molecule shakes itself apart or its atoms into a new arrangement, as in the isomerization of cyclopropane (**1**) to propene:

1 C_3H_6, cyclopropane

$$cyclo\text{-}C_3H_6 \rightarrow CH_3CH{=}CH_2$$

In a **bimolecular reaction**, a pair of molecules collide and exchange energy, atoms, or groups of atoms, or undergo some other kind of change, as in the reaction between H and Br_2 specified above. It is most important to distinguish molecularity from order:

> Reaction order is an empirical quantity, and obtained from the experimental rate law. The molecularity refers to an elementary reaction proposed as an individual step in a mechanism.

In contrast to reactions in general, the rate law of an elementary reaction can be written down from its chemical equation. Thus, the rate law of a unimolecular elementary reaction is first-order in the reactant:

$$A \rightarrow Products \qquad \frac{d[A]}{dt} = -k[A] \qquad (15)$$

A unimolecular reaction is first-order because the number of A molecules that decay in a short interval is proportional to the number available to decay. (Ten times as many decay in the same interval when there are initially 1000 A molecules than when there are only 100 present.) Therefore, the rate of decomposition of A is proportional to its molar concentration.

An elementary bimolecular reaction has a second-order rate law:

$$A + B \rightarrow Products \qquad \frac{d[A]}{dt} = -k[A][B] \qquad (16)$$

A bimolecular reaction is second-order because its rate is proportional to the rate at which the reactant species meet, which is proportional to their concentrations. Therefore, if we believe (or simply postulate) that a reaction is a single-step, bimolecular process, then we can write down the rate law (and then go on to test it). Bimolecular elementary reactions are believed to account for many homogeneous reactions, such as the dimerizations of alkenes and dienes and reactions such as

$$CH_3I(alc) + CH_3CH_2O^-(alc) \rightarrow CH_3OCH_2CH_3(alc) + I^-(alc)$$

(where 'alc' signifies alcohol solution). The mechanism of the last reaction is believed to be the single elementary step

$$CH_3I + CH_3CH_2O^- \rightarrow CH_3OCH_2CH_3 + I^-$$

The mechanism is consistent with the observed rate law

$$v = k[CH_3I][CH_3CH_2O^-]$$

The interpretation of a rate law is full of pitfalls, partly because a second-order rate law, for instance, can also result from a complex reaction scheme. We shall see below how to string simple steps together into a mechanism and how to arrive at the corresponding rate law. For the present we emphasize that *if the reaction is an elementary bimolecular process, then it has second-order kinetics, but if the kinetics are second-order, then the reaction might be complex*. The postulated mechanism can be explored only by detailed detective work on the system, and by investigating whether side products or intermediates appear during the course of the reaction. Detailed analysis of this kind was one of the ways, for example, in which the reaction $H_2(g) + I_2(g) \rightarrow 2HI(g)$ was shown to proceed by a complex reaction after many years during which it had been accepted on good, but insufficiently meticulous evidence, that it was a fine example of a simple bimolecular reaction in which atoms exchanged partners during a collision.

25.7 Consecutive elementary reactions

Some reactions proceed through the formation of an intermediate, as in the consecutive unimolecular reactions

$$A \xrightarrow{k_a} B \xrightarrow{k_b} C$$

An example is the decay of a radioactive family, such as

$$^{239}U \xrightarrow{23.5\,min} {}^{239}Np \xrightarrow{2.35\,days} {}^{239}Pu$$

(The times are half-lives.) We can discover the characteristics of this type of reaction by setting up the rate laws for the net rate of change of the concentration of each substance.

The variation of concentrations with time

The rate of unimolecular decomposition of A is

$$\frac{d[A]}{dt} = -k_a[A] \tag{17a}$$

and A is not replenished. The intermediate B is formed from A (at a rate $k_a[A]$) but decays to C (at a rate $k_b[B]$). Its net rate of formation is

therefore

$$\frac{d[B]}{dt} = k_a[A] - k_b[B] \tag{17b}$$

The product C is formed by the unimolecular decay of B:

$$\frac{d[C]}{dt} = k_b[B] \tag{17c}$$

We suppose that initially only A is present, and that its concentration is $[A]_0$.

The first of the rate laws is an ordinary first-order decay, so

$$[A] = [A]_0 e^{-k_a t} \tag{18a}$$

When this equation is substituted in eqn 17b, and we set $[B]_0 = 0$, the solution is

$$[B] = \frac{k_a}{k_b - k_a} \times (e^{-k_a t} - e^{-k_b t})[A]_0 \tag{18b}$$

At all times

$$[A] + [B] + [C] = [A]_0$$

so it follows that

$$[C] = \left\{ 1 + \frac{k_a e^{-k_b t} - k_b e^{-k_a t}}{k_b - k_a} \right\}[A]_0 \tag{18c}$$

We see that the intermediate's concentration rises to a maximum, and then falls to zero (Fig. 25.10). The concentration of the product C rises from zero and reaches $[A]_0$.

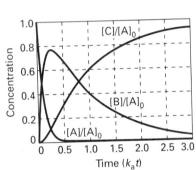

25.10 The concentrations of A, B, and C in the consecutive reaction scheme A → B → C. The curves are plots of eqn 18 with $k_a = 10k_b$. If B is the desired product, it is important to be able to predict when its concentration is greatest; see Example 25.8.

Example 25.8 *Analysing consecutive reactions*

Suppose that in an industrial batch process a substance A produces the desired product B which goes on to decay to a worthless product C, each stage of the reaction being first-order. At what time will product B be present in greatest concentration?

Method. The time dependence of the concentration of B is given by eqn 18b, We can find by differentiation the time at which it passes through a maximum by calculating d[B]/dt and setting the resulting rate equal to zero.

Answer. It follows from eqn 18b that

$$\frac{d[B]}{dt} = \frac{-k_a[A]_0(k_a e^{-k_a t} - k_b e^{-k_b t})}{k_b - k_a}$$

This rate is equal to zero when

$$k_a e^{-k_a t} = k_b e^{-k_b t}$$

Therefore, the time at which B has its maximum concentration is

$$t = \frac{1}{k_a - k_b} \ln \frac{k_a}{k_b}$$

Comment. For a given value of $k_a > k_b$, as k_b increases both the time at which [B] is a maximum and the yield of B increase.

Exercise E25.8. Calculate the maximum concentration of B and justify the last remark.

$$\left[\frac{[B]_{max}}{[A]_0} = \left(\frac{k_a}{k_b}\right)^c, \text{ with } c = \frac{k_b}{k_a - k_b} \right]$$

The rate-determining step

Suppose now that $k_b \gg k_a$; then whenever a B molecule is formed it decays rapidly into C. Because

$$e^{-k_b t} \ll e^{-k_a t} \qquad \text{and} \qquad k_b - k_a \approx k_b$$

eqn 18c reduces to

$$[C] = (1 - e^{-k_a t})[A]_0$$

which shows that the formation of C depends on only the *smaller* of the two rate constants. That is, the rate of formation of C depends on the rate at which B is formed, not on the rate at which B changes into C. For this reason, the step $A \rightarrow B$ is called the **rate-determining step** of the reaction. Its existence has been likened to building a six-lane highway up to a single-lane bridge: the traffic flow is governed by the rate of crossing the bridge. Similar remarks apply to more complicated reaction mechanisms and, in general, the rate-determining step is the one with the smallest rate constant.

The steady-state approximation

One feature of the calculation so far has probably not gone unnoticed: there is a considerable increase in mathematical complexity as soon as the reaction mechanism has more than a couple of steps. A mechanism involving many steps will be unsolvable analytically, and we shall need alternative methods of solution. One approach is to integrate the rate laws numerically. An alternative, which continues to be widely used because it leads to convenient expressions and more readily digestible results, is to make an approximation.

The **steady-state approximation** assumes that, during the major part of the reaction, the rates of change of concentrations of all reaction intermediates are negligibly small (Fig. 25.11):

$$\frac{d[\text{Intermediate}]}{dt} \approx 0$$

This approximation greatly simplifies the discussion of reaction schemes. For example, when we apply it to the consecutive first-order mechanism,

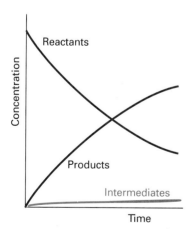

25.11 The basis of the steady-state approximation. It is supposed that the concentrations of intermediates remain small and hardly change during most of the course of the reaction.

we set $d[B]/dt = 0$ in eqn 17b, which then becomes

$$k_a[A] - k_b[B] = 0$$

Then

$$[B] = \frac{k_a}{k_b} \times [A]$$

On substituting this value of [B] into eqn 17c, that equation becomes

$$\frac{d[C]}{dt} = k_b[B] = k_a[A]$$

and we see that C is formed by a first-order decay of A, with a rate constant k_a, the rate constant of the slower, rate-determining, step. We can write down the solution of this equation at once by substituting the solution for [A], eqn 18a, and integrating:

$$[C] = k_a[A]_0 \int_0^t e^{-k_a t}\, dt = (1 - e^{-k_a t})[A]_0$$

This is the same (approximate) result as before, but obtained much more quickly.

Example 25.9 *Using the steady-state approximation*

Account for the rate law for the decomposition of N_2O_5

$$2N_2O_5(g) \rightarrow 4NO_2(g) + O_2(g) \qquad v = k[N_2O_5]$$

on the basis of the following mechanism:

$$\begin{aligned}
&N_2O_5 \rightarrow NO_2 + NO_3 &\qquad& k_a \\
&NO_2 + NO_3 \rightarrow N_2O_5 &\qquad& k_a' \\
&NO_2 + NO_3 \rightarrow NO_2 + O_2 + NO &\qquad& k_b \\
&NO + N_2O_5 \rightarrow 3NO_2 &\qquad& k_c
\end{aligned}$$

Method. We first identify the intermediates (the species that occur in the reaction steps but do not appear in the overall reaction) and write expressions for their net rates of formation. Then, all net rates of change of the concentrations of intermediates are set equal to zero and the resulting equations are solved algebraically.

Answer. The intermediates are NO and NO_3; the net rates of change of their concentrations are

$$\frac{d[NO]}{dt} = k_b[NO_2][NO_3] - k_c[NO][N_2O_5]$$

$$\frac{d[NO_3]}{dt} = k_a[N_2O_5] - k_a'[NO_2][NO_3] - k_b[NO_2][NO_3]$$

According to the steady-state approximation, we set both rates equal to zero:

$$k_b[NO_2][NO_3] - k_c[NO][N_2O_5] = 0$$
$$k_a[N_2O_5] - k_a'[NO_2][NO_3] - k_b[NO_2][NO_3] = 0$$

The net rate of change of concentration of N_2O_5 is

$$\frac{d[N_2O_5]}{dt} = -k_a[N_2O_5] + k_a'[NO_2][NO_3] - k_c[NO][N_2O_5]$$

and replacing the concentrations of the intermediates by using the equations above gives

$$\frac{d[N_2O_5]}{dt} = -\frac{2k_a k_b[N_2O_5]}{k_a' + k_b}$$

Because $v(N_2O_5) = -2$, it follows that the reaction rate is

$$v = k[N_2O_5], \qquad \text{where } k = \frac{k_a k_b}{k_a' + k_b}$$

in accord with observation.

Comment. The N_2O_5 decomposition is a problematic reaction, since its rate decreases more quickly than expected at low pressures. It is believed that this is due to changes in the values of the rate constants themselves (particularly k_a').

Exercise E25.9. Derive the rate law for the decomposition of ozone in the reaction $2O_3(g) \rightarrow 3O_2(g)$ on the basis of the (incomplete) mechanism

$$
\begin{array}{ll}
O_3 \rightarrow O_2 + O & k_a \\
O_2 + O \rightarrow O_3 & k_a' \\
O + O_3 \rightarrow 2O_2 & k_b
\end{array}
\qquad
\left[v = \frac{k_a k_b[O_3]^2}{k_a'[O_2] + k_b[O_3]} \right]
$$

Pre-equilibria

From a simple sequence of consecutive reactions we now turn to a slightly more complicated mechanism:

$$A + B \underset{k_a'}{\overset{k_a}{\rightleftharpoons}} C \overset{k_b}{\longrightarrow} P \tag{19}$$

where C denotes the intermediate. This scheme involves a **pre-equilibrium**, in which an intermediate is in equilibrium with the reactants. A pre-equilibrium arises when the rates of formation of the intermediate and its decay back into reactants are much faster than its rate of formation of products; thus, the condition is possible when $k_a' \gg k_b$ but not when $k_b \gg k_a'$. Because we assume that A, B, and C are in equilibrium, we can write

$$\frac{[C]}{[A][B]} = K, \qquad \text{with } K = \frac{k_a}{k_a'}$$

In writing these equations, we are presuming that the rate of reaction of

C to form P is too slow to affect the maintenance of the pre-equilibrium (see Example 25.10). The rate of formation of P may now be written:

$$\frac{d[P]}{dt} = k_b[C] = k_b K[A][B]$$

This rate law has the form of a second-order rate law with a composite rate constant:

$$\frac{d[P]}{dt} = k[A][B], \qquad \text{with } k = k_b K = \frac{k_a k_b}{k'_a} \qquad (20)$$

Example 25.10 *Analysing a pre-equilibrium*

Repeat the pre-equilibrium calculation but without ignoring the fact that C is slowly leaking away as it forms P.

Method. Begin by writing the net rates of change of the concentrations of the substances and then invoke the steady-state approximation for the intermediate C. Use the resulting expression to obtain the rate of change of the concentration of P.

Answer. The net rates of change of P and C are

$$\frac{d[P]}{dt} = k_b[C]$$

$$\frac{d[C]}{dt} = k_a[A][B] - k'_a[C] - k_b[C] \approx 0$$

The second equation solves to

$$[C] = \frac{k_a[A][B]}{k'_a + k_b}$$

When we substitute this into the expression for the rate of formation of P, we obtain

$$\frac{d[P]}{dt} = k[A][B], \qquad \text{with } k = \frac{k_a k_b}{k'_a + k_b}$$

Comment. This expression reduces to that in eqn 20 when the rate constant for the decay of C into products is much smaller than that for its decay into reactants, $k_b \ll k'_a$.

Exercise E25.10. Show that the pre-equilibrium mechanism

$$2A \underset{}{\overset{K}{\rightleftharpoons}} B \text{ followed by } B + C \xrightarrow{k_b} P$$

results in an overall third-order reaction. $[d[P]/dt = k_b K[A]^2[B]]$

Third-order reactions

An example will help to show how a pre-equilibrium assumption helps to elucidate a mechanism. The oxidation of nitrogen(II) oxide is found to be third-order overall:

$$2NO(g) + O_2(g) \rightarrow 2NO_2(g) \qquad \frac{d[NO_2]}{dt} = k[NO]^2[O_2] \qquad (21)$$

One explanation might be that the reaction is a single termolecular simple step. However, such a step would require the simultaneous collision of three particles, and ternary collisions are infrequent. Furthermore, it is observed that the reaction rate decreases as the temperature is raised. This observation points to a complex reaction mechanism, because simple reactions almost always go faster at higher temperatures.

A mechanism that accounts for the rate law and the temperature dependence is a pre-equilibrium

$$\text{(a)} \quad 2NO \rightleftharpoons N_2O_2 \qquad K = \frac{[N_2O_2]}{[NO]^2}$$

followed by a simple bimolecular reaction

$$\text{(b)} \quad N_2O_2 + O_2 \rightarrow 2NO_2 \qquad \frac{d[NO_2]}{dt} = k_b[N_2O_2][O_2]$$

The reaction rate is obtained by combining the two equations into

$$\frac{d[NO_2]}{dt} = k_b K[NO]^2[O_2]$$

which has the observed overall third-order form.

The mechanism is consistent with the anomalous temperature dependence because K decreases with temperature: the dimerization of NO is exothermic, and from thermodynamics we know that K decreases as the temperature is raised (Section 9.4). Therefore, provided k_b does not increase more sharply, the rate constant $k = k_b K$ decreases as the temperature is increased and the effective activation energy of the reaction is negative.

Example 25.11 *Calculating an apparent activation energy*

Calculate the activation energy of a reaction with a pre-equilibrium step and show how it might be negative.

Method. The formal procedure for calculating an activation energy is specified in eqn 13a. As we have just seen, the overall rate constant for a reaction with pre-equilibrium has the form $k = k_a k_b / k_a'$, so insert this expression into eqn 13a. Suppose that each individual step has an Arrhenius form with activation energies ε_a, ε_a', and ε_b, respectively.

Answer. From eqn 13a we obtain

$$E_a = RT^2 \frac{d \ln k}{dT} = RT^2 \frac{d}{dT}\left(\ln \frac{k_b k_a}{k_a'} \right)$$

$$= RT^2\left(\frac{d \ln k_b}{dT} + \frac{d \ln k_a}{dT} - \frac{d \ln k_a'}{dT} \right)$$

$$= \varepsilon_b + \varepsilon_a - \varepsilon_a'$$

If the activation energy ε_a' is larger than the sum of the other two activation energies, then E_a will be negative, and the reaction will form products more slowly at higher temperatures than at low.

Comment. A reaction with a high activation energy is more responsive to temperature than one with a low activation energy. Therefore, a way of understanding the conclusion is to say that the rate of decay of AB into reactants increases more quickly with temperature than either of the forward reactions.

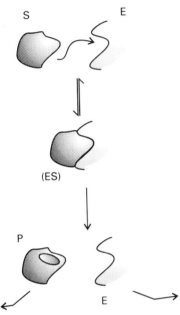

25.12 The basis of the Michaelis–Menten mechanism of enzyme action. Only a fragment of the large enzyme molecule E is shown.

The Michaelis–Menten mechanism

Another example of a reaction in which an intermediate is formed is the **Michaelis–Menten mechanism** of enzyme action. The rate of an enzyme-catalysed reaction in which a substrate S is converted into products P is found to depend on the concentration of the enzyme E even though the enzyme undergoes no net change. The proposed mechanism (Fig. 25.12) is

$$E + S \underset{k_a'}{\overset{k_a}{\rightleftharpoons}} ES \xrightarrow{k_b} P + E \tag{22}$$

In this mechanism, ES denotes a bound state of the enzyme and its substrate. This mechanism has the same form as that treated in Example 25.10, so we can conclude at once that

$$[ES] = \frac{k_a[E][S]}{k_a' + k_b}$$

[E] and [S] are the concentrations of the free enzyme and free substrate and, if $[E]_0$ is the total concentration of enzyme, then

$$[E] + [ES] = [E]_0$$

Because only a little enzyme is added, the free substrate concentration is almost the same as the total substrate concentration, and we can ignore the fact that [S] differs slightly from $[S]_{total}$. Therefore,

$$[ES] = \frac{k_a([E]_0 - [ES])[S]}{k_a' + k_b}$$

which rearranges to

$$[ES] = \frac{k_a[E]_0[S]}{k_a' + k_b + k_a[S]}$$

It follows that the rate of formation of product is

$$\frac{d[P]}{dt} = k[E]_0, \qquad \text{where } k = \frac{k_b[S]}{K_M + [S]} \tag{23a}$$

The **Michaelis constant** K_M is

$$K_M = \frac{k_b + k_a'}{k_a} \tag{23b}$$

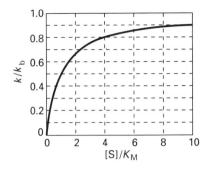

25.13 The variation of the effective rate constant k with substrate concentration according to the Michaelis–Menten mechanism.

According to eqn 23a, the rate of enzymolysis varies linearly with the enzyme concentration, but in a more complicated manner with the concentration of substrate (Fig. 25.13). Thus, when $[S] \gg K_M$, the rate law in eqn 23a reduces to

$$\frac{d[P]}{dt} = k_b[E]_0$$

and is zero-order in S. This result means that the rate is constant: there is so much S present that it saturates the enzyme, and product formation is limited by the rate at which ES falls apart. Moreover, the rate of formation of products is a maximum, and $k_b[E]_0$ is called the **maximum velocity** of the enzymolysis: k_b itself is called the **maximum turnover number**. When so little S is present that $[S] \ll K_M$, then the rate of formation of products is

$$\frac{d[P]}{dt} = \frac{k_b}{K_M}[E]_0[S]$$

Now the rate is proportional to $[S]$ as well as to $[E]_0$.

It follows from eqn 23a that

$$\frac{1}{k} = \frac{1}{k_b} + \frac{K_M}{k_b[S]} \tag{24}$$

Hence, a plot of $1/k$ against $1/[S]$ will give the value of k_b (from the intercept at $1/[S] = 0$) and K_M (from the slope, K_M/k_b). However, the plot cannot give the values of the individual rate coefficients k_a and k_a' that appear in K_M. The stopped-flow technique can give the additional data needed, because the rate of formation of the enzyme–substrate complex can be found by monitoring its concentration after mixing enzyme and substrate. This procedure gives k_a, and k_a' can then be found by combining this result with the value of K_M.

25.8 Unimolecular reactions

A number of gas-phase reactions follow first-order kinetics, as in the isomerization of cyclopropane mentioned earlier:

$$cyclo\text{-}C_3H_6 \rightarrow CH_3CH{=}CH_2 \qquad v = k[cyclo\text{-}C_3H_6]$$

The problem with first-order rate laws is that presumably a molecule acquires enough energy to react as a result of its collisions with other molecules. However, collisions are simple *bi*molecular events, so how can they result in a first-order rate law? First-order gas-phase reactions are widely called 'unimolecular reactions' because they also involve an elementary unimolecular step in which the reactant molecule changes into the product. This term must be used with caution, though, because the overall mechanism has bimolecular as well as unimolecular steps.

The Lindemann–Hinshelwood mechanism

The first successful explanation of unimolecular reactions was provided by Frederick Lindemann in 1921 and then elaborated by Cyril Hin-

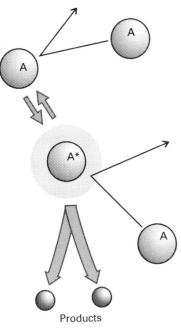

25.14 A representation of the Lindemann–Hinshelwood mechanism of unimolecular reactions. The species A is excited by collision with A, and the excited A molecule (A*) may either be deactivated by a collision with A or go on to decay by a unimolecular process to form products.

shelwood. In the **Lindemann–Hinshelwood mechanism** it is supposed that a reactant molecule A becomes energetically excited by collision with another A molecule (Fig. 25.14):

$$A + A \rightarrow A^* + A \qquad \frac{d[A^*]}{dt} = k_a[A]^2 \tag{25a}$$

The energized molecule might lose its excess energy by collision with another molecule:

$$A + A^* \rightarrow A + A \qquad \frac{d[A^*]}{dt} = -k_a'[A^*][A] \tag{25b}$$

Alternatively, the excited molecule might shake itself apart and form products P. That is, it might undergo the unimolecular decay

$$A^* \rightarrow P \qquad \frac{d[A^*]}{dt} = -k_b[A^*] \tag{25c}$$

If the unimolecular step is slow enough to be the rate-determining step, the overall reaction will have first-order kinetics, as observed. We can demonstrate this explicitly by applying the steady-state approximation to the net rate of formation of A*:

$$\frac{d[A^*]}{dt} = k_a[A]^2 - k_a'[A^*][A] - k_b[A^*] \approx 0$$

This equation solves to

$$[A^*] = \frac{k_a[A]^2}{k_b + k_a'[A]}$$

so the rate law for the formation of P is

$$\frac{d[P]}{dt} = k_b[A^*] = \frac{k_a k_b[A]^2}{k_b + k_a'[A]} \tag{26}$$

At this stage the rate law is not first-order. However, if the rate of deactivation by (A^*, A) collisions is much greater than the rate of unimolecular decay, in the sense that

$$k_a'[A^*][A] \gg k_b[A^*], \qquad \text{or } k_a'[A] \gg k_b \tag{27}$$

then we can neglect k_b in the denominator and obtain

$$\frac{d[P]}{dt} = k[A], \qquad \text{where } k = \frac{k_a k_b}{k_a'} \tag{28}$$

Equation 28 is a first-order rate law, as we set out to show.

The Lindemann–Hinshelwood mechanism can be tested because it predicts that as the concentration (and therefore the partial pressure) of A is reduced the reaction should switch to overall second-order kinetics. Thus, when $k_a'[A] \ll k_b$, the rate law in eqn 26 is approximately

$$\frac{d[P]}{dt} = k_a[A]^2$$

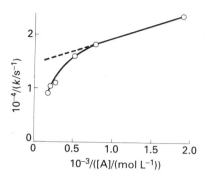

25.15 The pressure dependence of the unimolecular isomerization of trans-CHD=CHD showing a pronounced departure from the straight line predicted by eqn 30 based on the Lindemann–Hinshelwood mechanism.

The physical reason for the change of order is that at low pressures the rate-determining step is the bimolecular formation of A^*. If we write the full rate law in eqn 26 as

$$\frac{d[P]}{dt} = k[A], \qquad \text{with } k = \frac{k_a k_b [A]}{k_b + k_a'[A]} \qquad (29)$$

then the expression for the effective rate-constant k can be rearranged to

$$\frac{1}{k} = \frac{k_a'}{k_a k_b} + \frac{1}{k_a[A]} \qquad (30)$$

Hence, a test of the theory is to plot $1/k$ against $1/[A]$ and to expect a straight line.

Whereas the Lindemann–Hinshelwood mechanism agrees in general with the switch in order of unimolecular reactions, it does not agree in detail. A typical graph of $1/k$ against $1/[A]$ is shown in Fig. 25.15. The graph has a pronounced curvature, corresponding to a larger value of k (a smaller value of $1/k$) at high pressures (low $1/[A]$).

Improvements to the theory

One of the reasons for the discrepancy is that the Lindemann–Hinshelwood mechanism fails to recognize that specific excitation of the molecule may be required before reaction occurs. For example, in the isomerization of cyclopropane the crucial step is the breaking of one C—C bond, and this occurs when that bond is highly vibrationally excited. In the collision leading to excitation, however, the excess energy is shared among all the bonds and the rotation of the molecule, so the isomerization occurs only after the energy accumulates in the critical bond. The existence of this delay suggests that we should distinguish between the **energized molecule** A^*, where the excess energy is dispersed over many modes, and the **activated state** $A^\dagger$, where the excitation is specific and the molecule is poised for reaction. The unimolecular part of the mechanism therefore ought to be modified to

$$A^* \rightarrow A^\dagger \rightarrow P$$

with a rate constant for each step. The values of the various rate constants are related to the numbers and frequencies of the vibrational modes available by the **Rice–Ramsperger–Kassel theory** (RRK theory) of unimolecular reactions, and by the more sophisticated **RRKM theory** (where M is Marcus), which also takes into account the effects of molecular rotation. Descriptions of these theories will be found in the books mentioned in *Further reading*.

CHECK LIST OF KEY IDEAS

1 The **techniques** for following the concentrations of reactants and products (Section 25.1).

2 The definition of the **rate of reaction** (eqn 1).

3 The significance of a **rate law** and the **rate constant** of a reaction (Section 25.2).

4 The definition of **reaction order** and the **overall order** of a reaction (Section 25.2).

5 The **isolation method** for simplifying the determination of reaction order, the significance of a **pseudofirst-order rate law,** and the method of **initial rates** (eqn 4).

6 The **integrated rate laws** for first- and second-order reactions (eqns 5 and 7).

7 The definition of the **half-life** of a species in a reaction and its independence of the concentration for first-order reactions (eqn 6).

8 The time dependence of **reactions approaching equilibrium** when the reverse reaction is significant (eqn 9) and the relation between the **equilibrium constant** and the rate constants of forward and reverse reactions (eqn 10 and Example 25.5).

9 The **Arrhenius equation** (eqn 11) for the temperature dependence of the rate constant and the **Arrhenius parameters** of a reaction.

10 The use of **relaxation methods** and particularly **temperature-jump relaxation** to measure rate constants (Section 25.5).

11 The **reaction mechanism** in terms of **elementary reactions** and the classification of elementary reactions according to their **molecularity** (Section 25.6).

12 The rate laws for elementary **unimolecular** and **bimolecular reactions** (eqns 15 and 16).

13 The overall rate law stemming from **consecutive reactions** (Section 25.7) and the **rate-determining step** in a mechanism.

14 The **steady-state approximation** for simplifying the analysis of a kinetic scheme (Section 25.7 and Example 25.9).

15 The assumption that a **pre-equilibrium** has been established and the approximate analysis of a kinetic scheme (Section 25.7).

16 The **Michaelis–Menten mechanism** of enzyme action (eqn 22).

17 The **Lindemann–Hinshelwood mechanism** of **unimolecular reactions** (Section 25.8) and improvements to the theory.

EXERCISES

25.1. The rate of the reaction $A + 2B \rightarrow 3C + D$ was reported as $1.0 \, \text{mol} \, \text{L}^{-1} \, \text{s}^{-1}$. State the rates of formation and consumption of the participants.

25.2. The rate of formation of C in the reaction $2A + B \rightarrow 2C + 3D$ is $1.0 \, \text{mol} \, \text{L}^{-1} \, \text{s}^{-1}$. State the reaction rate, and the rates of formation or consumption of A, B, and D.

25.3. The rate law for the reaction in Exercise 25.1 was found to be as $v = k[A][B]$. What are the units of k? Express the rate law in terms of the rates of formation and consumption of (a) A, (b) C.

25.4. The rate law for the reaction in Exercise 25.2 was reported as $d[C]/dt = k[A][B][C]$. Express it in terms of the reaction rate; what are the units for k in each case?

25.5. At 518°C, the rate of decomposition of a sample of gaseous acetaldehyde, initially at a pressure of 363 Torr, was $1.07 \, \text{Torr} \, \text{s}^{-1}$ when 5.0 per cent had reacted and $0.76 \, \text{Torr} \, \text{s}^{-1}$ when 20.0 per cent had reacted. Determine the order of the reaction.

25.6. At 518°C, the half-life for the decomposition of a sample of gaseous acetaldehyde initially at 363 Torr was

410 s. When the pressure was 169 Torr the half-life was 880 s. Determine the order of the reaction.

25.7. The rate constant for the first-order decomposition of N_2O_5 in the reaction $2N_2O_5(g) \rightarrow 4NO_2(g) + O_2(g)$ is $k = 3.38 \times 10^{-5} \, \text{s}^{-1}$ at 25°C. What is the half life of N_2O_5? What will be the pressure, initially 500 Torr, after (a) 10 s, (b) 10 min after initiation of the reaction?

25.8. A second-order reaction of the type $A + B \rightarrow P$ was carried out in a solution that was initially $0.050 \, \text{mol} \, \text{L}^{-1}$ in A and $0.080 \, \text{mol} \, \text{L}^{-1}$ in B. After 1.0 h the concentration of A had fallen to $0.020 \, \text{mol} \, \text{L}^{-1}$. (a) Calculate the rate constant. (b) What is the half-life of the reactants?

25.9. If the rate laws are expressed with (a) concentrations in moles per litre, (b) pressures in kilopascals, what are the units of the second-order and third-order rate constants?

25.10. The half-life for the (first-order) radioactive decay of ^{14}C is 5730 years (it emits β rays with an energy of 0.16 MeV). An archaeological sample contained wood that had only 72 per cent of the ^{14}C found in living trees. What is its age?

25.11. One of the hazards of nuclear explosions is the generation of ^{90}Sr and its subsequent incorporation in place of calcium in bones. This nuclide emits β rays of energy 0.55 MeV, and has a half-life of 28.1 years. Suppose 1.00 µg was absorbed by a newly born child. How much will remain after (a) 18 years (b) 70 years, if none is lost metabolically?

25.12. The second-order rate constant for the reaction

$$CH_3COOC_2H_5(aq) + OH^-(aq)$$
$$\rightarrow CH_3CO_2^-(aq) + CH_3CH_2OH(aq)$$

is $0.11 \, L \, mol^{-1} \, s^{-1}$. What is the concentration of ester after (a) 10 s, (b) 10 min when ethyl acetate is added to sodium hydroxide so that the initial concentrations are $[NaOH] = 0.050 \, mol \, L^{-1}$ and $[CH_2COOC_2H_5] = 0.100 \, mol \, L^{-1}$?

25.13. A reaction $2A \rightarrow P$ has a second-order rate law with $k = 3.50 \times 10^{-4} \, L \, mol^{-1} \, s^{-1}$. Calculate the time required for the concentration of A to change from $0.260 \, mol \, L^{-1}$ to $0.011 \, mol \, L^{-1}$.

25.14. The rate constant for the decomposition of a certain substance is $2.80 \times 10^{-3} \, L \, mol^{-1} \, s^{-1}$ at 30°C and $1.38 \times 10^{-2} \, L \, mol^{-1} \, s^{-1}$ at 50°C. Evaluate the Arrhenius parameters of the reaction.

25.15. The reaction $2H_2O_2(aq) \rightarrow 2H_2O(l) + O_2(g)$ is catalysed by Br^- ions. If the mechanism is:

$$H_2O_2(aq) + Br^-(aq) \rightarrow H_2O(l) + BrO^-(aq) \quad \text{(slow)}$$
$$BrO^-(aq) + H_2O_2(aq) \rightarrow H_2O(l) + O_2(g) + Br^-(aq) \quad \text{(fast)}$$

give the order of the reaction with respect to the various participants.

25.16. The reaction mechanism

$$A_2 \rightleftharpoons 2A \quad \text{(fast)}$$
$$A + B \rightarrow P \quad \text{(slow)}$$

involves an intermediate A. Deduce the rate law for the reaction.

25.17. Consider the following mechanism for renaturation of a double helix from its strands A and B:

$$A + B \rightleftharpoons \text{unstable helix} \quad \text{(fast)}$$
$$\text{Unstable helix} \rightarrow \text{stable double helix} \quad \text{(slow)}$$

Derive the rate equation for the formation of the double helix and express the rate constant of the renaturation reaction in terms of the rate constants of the individual steps.

25.18. Show that $t_{\frac{1}{2}} \propto 1/[A]^{n-1}$ for a reaction that is nth-order in A.

25.19. The enzyme-catalysed conversion of a substrate at 25°C has a Michaelis constant of $0.035 \, mol \, L^{-1}$. The rate of the reaction is $1.15 \times 10^{-3} \, mol \, L^{-1} \, s^{-1}$ when the substrate concentration is $0.110 \, mol \, L^{-1}$. What is the maximum velocity of this enzymolysis?

25.20. Find the condition for which the reaction rate of an enzymolysis that follows Michaelis–Menten kinetics is half its maximum value.

25.21. The effective rate constant for a gaseous reaction that has a Lindemann–Hinshelwood mechanism is $2.50 \times 10^{-4} \, s^{-1}$ at 1.30 kPa and $2.10 \times 10^{-5} \, s^{-1}$ at 12 Pa. Calculate the rate constant for the activation step in the mechanism.

25.22. The pK_a of NH_4^+ is 9.25 at 25°C. The rate constant at 25°C for the reaction of NH_4^+ and OH^- to form aqueous NH_3 is $4.0 \times 10^{10} \, L \, mol^{-1} \, s^{-1}$. Calculate the rate constant for proton transfer to NH_3. What relaxation time would be observed if a temperature jump were applied to a solution of $0.15 \, mol \, L^{-1} \, NH_3(aq)$ to bring it to 25°C?

25.23. The equilibrium $A \rightleftharpoons B + C$ is subjected to a temperature jump to 25°C that slightly increases the concentrations of B and C. The measured relaxation time is 3.0 µs. The equilibrium constant for the system is 2.0×10^{-16} at 25°C, and the equilibrium concentrations of B and C at 25°C are both $2.0 \times 10^{-4} \, mol \, L^{-1}$. Calculate the rate constants for the forward and reverse reactions.

25.24. Show that the reaction $A \rightleftharpoons B + C$, first-order forwards, second-order reverse, relaxes exponentially for small displacements from equilibrium. Find an expression for the relaxation time in terms of k_1 and k_2.

PROBLEMS

25.1. The data below apply to the formation of urea from ammonium cyanate, $NH_4CNO \rightarrow NH_2CONH_2$. Initially 22.9 g of ammonium cyanate was dissolved in enough water to prepare 1.00 L of solution. Determine the order of the reaction, the rate constant, and the mass of ammonium cyanate left after 300 min.

t/min	0	20.0	50.0	65.0	150
$m(\text{urea})/\text{g}$	0	7.0	12.1	13.8	17.7

25.2. The data below apply to the reaction, $(CH_3)_3CBr +$

$H_2O \rightarrow (CH_3)_3COH + HBr$. Determine the order of the reaction, the rate constant, and the molar concentration of $(CH_3)_3CBr$ after 43.8 h.

t/h	0	3.15	6.20	10.00	18.30	30.80
$[(CH_3)_3CBr]/(10^{-2} \text{ mol L}^{-1})$	10.39	8.96	7.76	6.39	3.53	2.07

25.3. The thermal decomposition of an organic nitrile produced the following data:

$t/(10^3 \text{ s})$	0	2.00	4.00	6.00	8.00	10.00	12.00	∞
[nitrile]/(mol L^{-1})	1.10	0.86	0.67	0.52	0.41	0.32	0.25	0

Determine the order of the reaction and the rate constant.

25.4. The following data have been obtained for the decomposition of $N_2O_5(g)$ at 67°C according to the reaction $2N_2O_5(g) \rightarrow 4N_2(g) + O_2(g)$. Determine the order of the reaction, the rate constant, and the half-life. It is not necessary to obtain the result graphically, you may do a calculation using estimates of the rates of change of concentration.

t/min	0	1.00	2.00	3.00	4.00	5.00
[N_2O_5]/(mol L^{-1})	1.000	0.705	0.497	0.349	0.246	0.173

25.5. A first-order decomposition reaction is observed to have the following rate constants at the indicated temperatures. Estimate the activation energy.

$k/(10^{-3} \text{ s}^{-1})$	2.46	45.1	576
$\theta/°C$	0	20.0	40.0

25.6. The gas-phase decomposition of acetic acid at 1189 K proceeds by way of two parallel first-order reactions:

(1) $CH_3COOH \rightarrow CH_4 + CO_2$ $k_1 = 3.74 \text{ s}^{-1}$

(2) $CH_3COOH \rightarrow CH_2=C=O + H_2O$ $k_2 = 4.65 \text{ s}^{-1}$

What is the maximum percentage yield of the ketene CH_2CO obtainable at this temperature?

25.7. The composition of a liquid phase reaction $2A \rightarrow B$ was followed by a spectrophotometric method with the following results:

t/min	0	10	20	30	40	∞
[B]/(mol L^{-1})	0	0.089	0.153	0.200	0.230	0.312

Determine the order of the reaction and its rate constant.

25.8. Sucrose is readily hydrolysed to glucose and fructose in acidic solution. The hydrolysis can be monitored by measuring the angle of rotation of plane-polarized light passing through the solution. From the angle of rotation the concentration of sucrose can be determined. An experiment on the hydrolysis of sucrose in 0.50 M HCl(aq) produced the following data:

t/min	0	14	39	60	80	110	140	170	210
[sucrose]/ (mol L^{-1})	0.316	0.300	0.274	0.256	0.238	0.211	0.190	0.170	0.146

Determine the rate constant of the reaction and the average lifetime of a sucrose molecule.

25.9. The ClO radical decays rapidly by way of the reaction, $2ClO \rightarrow Cl_2 + O_2$. The following data have been obtained:

$t/(10^{-3} \text{ s})$	0.12	0.62	0.96	1.60	3.20	4.00	5.75
[ClO]/(10^{-6} mol L^{-1})	8.49	8.09	7.10	5.79	5.20	4.77	3.95

Determine the rate constant of the reaction.

25.10. Cyclopropane isomerizes into propene when heated to 500°C in the gas phase. The extent of conversion for various initial pressures has been followed by gas chromatography by allowing the reaction to proceed for a time with various initial pressures:

p_0/Torr	200	200	400	400	600	600
t/s	100	200	100	200	100	200
p/Torr	186	173	373	347	559	520

where p_0 is the initial pressure and p is the final pressure of cyclopropane. What is the order and rate constant for the reaction under these conditions?

25.11. The addition of hydrogen halides to alkenes has played a fundamental role in the investigation of organic reaction mechanisms. In one study (M. J. Haugh and D. R. Dalton, *J. Am. Chem. Soc.* **97**, 5674 (1975)), high pressures of hydrogen chloride (up to 25 atm) and propene (up to 5 atm) were examined over a range of temperatures and the amount of 2-chloropropane formed was determined by NMR. Show that if the reaction $A + B \rightarrow P$ proceeds for a short time Δt, the concentration of product follows $[P]/[A] = k[A]^{m-1}[B]^n \Delta t$ if the reaction is mth-order in A and nth-order in B. In a series of runs the ratio of [chloropropane] to [propene] was independent of [propene] but the ratio of [chloropropane] to [HCl] for constant amounts of propene depended on [HCl]. For $\Delta t \approx 100$ h (which is short on the time scale of the reaction) the latter ratio rose from zero to 0.05, 0.03, 0.01 for $p(HCl) = 10$ atm, 7.5 atm, 5.0 atm. What are the orders of the reaction with respect to each reactant?

25.12. Show that the following mechanism can account for the rate law of the reaction in Problem 25.11:

$2HCl \rightleftharpoons (HCl)_2$ K_1

$HCl + CH_3CH=CH_2 \rightleftharpoons$ complex K_2

$(HCl)_2 + $ complex $\rightarrow CH_3CHClCH_3 + 2HCl$ k, slow

What further tests could you apply to verify this mechanism?

25.13. In the experiments described in Problems 25.11 and 25.12 an inverse temperature dependence of the reaction rate was observed, the overall rate of reaction at 70°C being roughly one-third that at 19°C. Estimate the apparent activation energy and the activation energy of the rate-determining step given that the enthalpies of the two equilibria are both of the order of $-14\ kJ\ mol^{-1}$.

25.14. The second-order rate constants for the reaction of oxygen atoms with aromatic hydrocarbons have been measured (R. Atkinson and J. N. Pitts, *J. Phys. Chem.* **79**, 295 (1975)). In the reaction with benzene the rate constants are 1.44×10^7 at 300.3 K, 3.03×10^7 at 341.2 K, and 6.9×10^7 at 392.2 K, all in $L\ mol^{-1}\ s^{-1}$. Find the pre-exponential factor and activation energy of the reaction.

25.15. In Problem 25.10 the isomerization of cyclopropane over a limited pressure range was examined. If the Lindemann mechanism of first-order reactions is to be tested we also need data at low pressures. These have been obtained (H. O. Pritchard, R. G. Sowden, and A. F. Trotman-Dickenson, *Proc. R. Soc.* **A217**, 563 (1953)):

$p/Torr$	84.1	11.0	2.89	0.569	0.120	0.067
$10^4 k_{eff}/s^{-1}$	2.98	2.23	1.59	0.857	0.392	0.303

Test the Lindemann theory with these data.

25.16. The initial rate of O_2 production by the action of an enzyme on a substrate was measured for a range of substrate concentrations; the data are below. Evaluate the Michaelis constant for the reaction.

$[S]/(mol\ L^{-1})$	0.050	0.017	0.010	0.005	0.003
rate/($mm^3\ min^{-1}$)	16.6	12.4	10.1	6.6	3.3

Theoretical problems

25.17. The equilibrium $A \rightleftharpoons B$ is first-order in both directions. Derive an expression for the concentration of A as a function of time when the initial molar concentrations of A and B are $[A]_0$ and $[B]_0$. What is the final composition of the system?

25.18. Derive an integrated expression for a second-order rate law $v = k[A][B]$ for a reaction of stoichiometry $2A + 3B \rightarrow P$.

25.19. Derive the integrated form of a third-order rate law $v = k[A]^2[B]$ in which the stoichiometry is $2A + B \rightarrow P$ and the reactants are initially present in (a) their stoichiometric proportions, (b) with B present initially in twice the amount.

25.20. Set up the rate equations for the reaction mechanism:

$$A \underset{k_a'}{\overset{k_a}{\rightleftharpoons}} B \underset{k_b'}{\overset{k_b}{\rightleftharpoons}} C$$

Show that the mechanism is equivalent to

$$A \underset{k_{eff}'}{\overset{k_{eff}}{\rightleftharpoons}} C$$

under specified circumstances.

25.21. Show that the ratio $t_{\frac{1}{2}}/t_{\frac{3}{4}}$ where $t_{\frac{1}{2}}$ is the half-life and $t_{\frac{3}{4}}$ is the time for the concentration of A to decrease to $\frac{3}{4}$ of its initial value (implying $t_{\frac{3}{4}} < t_{\frac{1}{2}}$) can be written as a function of n alone, so it can be used as a rapid assessment of the order of a reaction.

25.22. Many enzyme-catalysed reactions are consistent with a modified version of the Michaelis–Menten mechanism in which the second step is also reversible. For this mechanism obtain an expression for the rate of formation of product and find its limiting behaviour for large and small concentrations of substrate.

25.23. Derive an equation for the steady-state rate of the sequence of reactions $A \rightleftharpoons B \rightleftharpoons C \rightleftharpoons D$ with [A] maintained at a fixed value and the product D removed as soon as it is formed.

26

The kinetics of complex reactions

This chapter extends the material introduced in Chapter 25 by showing how more complex reaction mechanisms are treated. In particular, we deal with chain reactions and see that, although the mechanisms can be complicated, under certain circumstances very simple rate laws are obtained. On the other hand, we shall also see that chain mechanisms can lead to complex rate laws. Under certain circumstances, a chain reaction can become violently explosive, and we see some of the reasons for this behaviour. An important application of these more complicated techniques is to the kinetics of polymerization reactions. Here we shall see that there are two major classes of polymerization process, and the average molar mass of the product varies with time in characteristic ways. It follows that the time for which a polymerization reaction runs can influence the physical properties of the resulting polymer. Finally, we shall consider the most complex types of reaction in which the concentrations of the intermediates and products oscillate in time. We shall see that, although the rate laws may be well known, under certain circumstances the composition of a system may be unpredictable. The physical reason for such behaviour is the complex web of relations between the concentrations of different species. The mathematical reason (which reflects the underlying chemical phenomena) is the non-linearity of the coupled differential equations that describe the kinetics of the reaction.

Many reactions take place by mechanisms that involve several elementary steps, and some take place at a useful rate only if a catalyst is present. Certain other reactions take place by mechanisms in which there is positive or negative feedback, in which the products of the reaction influence the rate at which more products are produced. In this chapter we see how to develop the ideas introduced in Chapter 25 to deal with these special kinds of reactions.

CHAIN REACTIONS

Many gas-phase reactions and liquid-phase polymerization reactions are **chain reactions**. In a chain reaction an intermediate produced in one step generates an intermediate in a subsequent step, then that intermediate generates another intermediate, and so on.

26.1 The structure of chain reactions

The intermediates responsible for the propagation of a chain reaction are called **chain carriers**. In a **radical chain reaction** the chain carriers are radicals (species with unpaired electrons). Ions may also propagate chains. In nuclear fission the chain carriers are neutrons.

The classification of reaction steps

The first chain carriers are formed in the **initiation step** of the reaction. For example, Cl atoms are formed by the dissociation of Cl_2 molecules either as a result of vigorous intermolecular collisions in a **thermolysis** (a reaction initiated by heat) or as a result of absorption of a photon in a **photolysis** (a reaction stimulated by light). The chain carriers produced in the initiation step attack other reactant molecules in the **propagation steps**, and each attack gives rise to a new chain carrier. An example is the attack of a methyl radical on ethane:

$$\cdot CH_3 + CH_3CH_3 \rightarrow CH_4 + \cdot CH_2CH_3$$

(The dot signifies the unpaired electron and marks the radical.) In some cases the attack results in the production of more than one chain carrier. An example of such a **branching step** is

$$\cdot O\cdot + H_2O \rightarrow HO\cdot + HO\cdot$$

where the attack of one O atom on an H_2O molecule forms two $\cdot OH$ radicals (an O atom has the configuration $[He]2s^2 2p^4$, with two unpaired electrons).

The chain carrier might attack a product molecule formed earlier in the reaction. Because this attack reduces the net rate of formation of product, it is called a **retardation step**. For example, in a photochemical reaction in which HBr is formed from H_2 and Br_2, an H atom might attack an HBr molecule, leading to H_2 and Br:

$$\cdot H + HBr \rightarrow H_2 + \cdot Br$$

Retardation does not end the chain, because one radical ($\cdot H$) gives rise to another ($\cdot Br$), but it does deplete the concentration of the product. Elementary reactions in which radicals combine and end the chain are called **termination steps**, as in

$$CH_3CH_2\cdot + \cdot CH_2CH_3 \rightarrow CH_3CH_2CH_2CH_3$$

In an **inhibition step** radicals are removed other than by chain termination, such as by reaction with the walls of the vessel or with foreign radicals:

$$CH_3CH_2\cdot + \cdot R \rightarrow CH_3CH_2R$$

The NO molecule has an unpaired electron and is a very efficient chain inhibitor. The observation that a reaction is quenched when NO is introduced is a good indication that a radical chain mechanism is in operation.

The rate laws of chain reactions

A chain reaction can lead to a simple rate law. As a first example, consider the **pyrolysis** (thermal decomposition in the absence of air) of acetaldehyde, CH_3CHO, which is found to be of three-halves order in CH_3CHO:

$$CH_3CHO(g) \xrightarrow{\Delta} CH_4(g) + CO(g) \qquad \frac{d[CH_4]}{dt} = k[CH_3CHO]^{\frac{3}{2}} \qquad (1)$$

Some ethane is also detected. (The Δ signifies an elevated temperature.)

In 1934, F. O. Rice and K. F. Herzfeld showed that simple rate laws can follow from quite complex chain mechanisms. They proposed a number of specific schemes, which are now known as **Rice–Herzfeld mechanisms**. The Rice–Herzfeld mechanism for the pyrolysis of acetaldehyde is as follows:

(a) Initiation:
$$CH_3CHO \rightarrow \cdot CH_3 + \cdot CHO \qquad v = k_a[CH_3CHO]$$
(b) Propagation:
$$CH_3CHO + \cdot CH_3 \rightarrow CH_4 + CH_3CO\cdot \qquad v = k_b[CH_3CHO][\cdot CH_3]$$
(c) Propagation:
$$CH_3CO\cdot \rightarrow \cdot CH_3 + CO \qquad v = k_c[CH_3CO\cdot]$$
(d) Termination:
$$\cdot CH_3 + \cdot CH_3 \rightarrow CH_3CH_3 \qquad v = k_d[\cdot CH_3]^2$$

As we shall see, this mechanism captures the principal features of the reaction but does not accommodate the formation of various by-products, such as propanone (acetone), CH_3COCH_3, and propanal, CH_3CH_2CHO.

To test the proposed mechanism we need to show that it leads to the observed rate law. To do so, we need to make the steady-state approximation to simplify the work. A more sophisticated procedure is solve the set of differential equations numerically, and then to compare the predictions with observation.

According to the steady-state approximation (Section 25.7), the net rate of change of the intermediates ($\cdot CH_3$ and $CH_3CO\cdot$) may be set equal to zero:

$$\frac{d[\cdot CH_3]}{dt} = k_a[CH_3CHO] - k_b[CH_3CHO][\cdot CH_3]$$

$$+ k_c[CH_3CO\cdot] - k_d[\cdot CH_3]^2 = 0$$

$$\frac{d[CH_3CO\cdot]}{dt} = k_b[CH_3CHO][\cdot CH_3] - k_c[CH_3CO\cdot] = 0$$

The sum of the two equations is

$$k_a[CH_3CHO] - k_d[\cdot CH_3]^2 = 0$$

which implies that the steady-state concentration of $\cdot CH_3$ radicals is

$$[\cdot CH_3] = \left(\frac{k_a}{k_d}\right)^{\frac{1}{2}} [CH_3CHO]^{\frac{1}{2}}$$

It follows that the rate of formation of CH_4 is

$$\frac{d[CH_4]}{dt} = k_b[CH_3CHO][\cdot CH_3] = k_b\left(\frac{k_a}{k_d}\right)^{\frac{1}{2}} [CH_3CHO]^{\frac{3}{2}} \qquad (2)$$

which is in agreement with the three-halves order observed experimentally (eqn 1). However, as we have already indicated, the true mechanism must be more complicated than the Rice–Herzfeld mechanism because other products are formed in significant quantities. A more complex scheme has been proposed, in which the $\cdot CHO$ radical participates in further reactions.

In many cases, a chain reaction leads to a complicated rate law. An example is the reaction between H_2 and Br_2:

$$H_2(g) + Br_2(g) \rightarrow 2HBr(g) \qquad \frac{d[HBr]}{dt} = \frac{k[H_2][Br_2]^{\frac{3}{2}}}{[Br_2] + k'[HBr]} \qquad (3)$$

The following radical chain mechanism has been proposed to account for this law:

(a) Initiation: $Br_2 \rightarrow 2Br\cdot$ $\qquad\qquad\qquad v = k_a[Br_2]$

(At low pressures this elementary reaction is bimolecular and second order in Br_2.)

(b) Propagation: $Br\cdot + H_2 \rightarrow HBr + H\cdot$ $\qquad v = k_b[Br\cdot][H_2]$
$\qquad\qquad\qquad H\cdot + Br_2 \rightarrow HBr + Br\cdot$ $\qquad v = k_b'[H\cdot][Br_2]$
(c) Retardation: $H\cdot + HBr \rightarrow H_2 + Br\cdot$ $\qquad v = k_c[H\cdot][HBr]$
(d) Termination: $Br\cdot + \cdot Br + M \rightarrow Br_2 + M$ $\qquad v = k_d[Br\cdot]^2$

(The third body M removes the energy of recombination; the constant concentration of M has been absorbed into the rate constant.) Other possible termination steps include the recombination of H atoms to form H_2 and the combination of H and Br atoms; however, it turns out that only Br atom recombination is important. The net rate of formation of the product HBr is

$$\frac{d[HBr]}{dt} = k_b[Br\cdot][H_2] + k_b'[H\cdot][Br_2] - k_c[H\cdot][HBr]$$

As for the acetaldehyde reaction, we can now either analyse the rate equations numerically or look for approximate solutions and see if they agree with the empirical rate law. The latter approach is carried out in the following example.

Example 26.1 *Deriving the rate equation of a chain reaction*

Derive the rate law for the formation of HBr according to the mechanism given above.

Method. We make the steady-state approximation for the concentrations of any intermediates ($\cdot$H and $\cdot$Br in the present case) and set the net rates of change of their concentrations equal to zero. Begin by writing down the net rates of formation of the intermediates. Set these equal to zero, solve the resulting algebraic equations for the concentrations of the intermediates, and then use the resulting expressions in the equation for the net rate of formation of HBr.

Answer. The net rates of formation of the two intermediates are

$$\frac{d[H\cdot]}{dt} = k_b[Br\cdot][H_2] - k_b'[H\cdot][Br_2] - k_c[H\cdot][HBr] = 0$$

$$\frac{d[Br\cdot]}{dt} = 2k_a[Br_2] - k_b[Br\cdot][H_2] + k_b'[H\cdot][Br_2]$$
$$+ k_c[H\cdot][HBr] - 2k_d[Br\cdot]^2 = 0$$

The steady-state concentrations of the intermediates are therefore

$$[Br\cdot] = \left(\frac{k_a}{k_d}\right)^{\frac{1}{2}} [Br_2]^{\frac{1}{2}}$$

$$[H\cdot] = \frac{k_b(k_a/k_d)^{\frac{1}{2}}[H_2][Br_2]^{\frac{1}{2}}}{k_b'[Br_2] + k_c[HBr]}$$

When we substitute these concentrations into the expression for $d[HBr]/dt$, we obtain

$$\frac{d[HBr]}{dt} = \frac{2k_b(k_a/k_d)^{\frac{1}{2}}[H_2][Br_2]^{\frac{3}{2}}}{[Br_2] + (k_c/k_b')[HBr]}$$

This equation has the same form as the empirical rate law (eqn 3), and the two empirical rate constants can be identified as

$$k = 2k_b\left(\frac{k_a}{k_d}\right)^{\frac{1}{2}} \qquad k' = \frac{k_c}{k_b'}$$

Exercise E26.1. Deduce the rate law when the initiation step is bimolecular in Br_2. In practice, the recombination would be termolecular, but ignore this feature here.

$$\left[\frac{d[HBr]}{dt} = \frac{2k_b(k_a/k_d)^{\frac{1}{3}}[H_2][Br_2]^2}{[Br_2] + (k_c/k_b')[HBr]}\right]$$

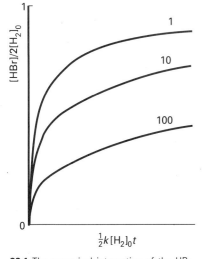

26.1 The numerical integration of the HBr rate law, Example 26.1, can be used to explore how the concentration of HBr changes with time. These runs began with stoichiometric proportions of H_2 and Br_2; the curves are labelled with the value of $2k' - 1$.

In the examples we have considered, the observed rate laws are reproduced by the mechanisms. That used to be essentially the end of the calculation (but not of the experimental investigation). Now, though, we can make use of computers to integrate the approximate rate law numerically, and hence predict the time dependence of the HBr concentration. Some typical results are plotted in Fig. 26.1: they show how the rate of formation of HBr is slow when the inhibition step is important.

26.2 Explosions

A **thermal explosion** is due to the rapid increase of reaction rate with temperature. If the energy of an exothermic reaction cannot escape, the temperature of the reaction system rises and the reaction goes faster. The acceleration of the rate results in a faster rise of temperature, and so the reaction goes even faster . . . catastrophically fast. A **chain-branching explosion** may occur when there are chain-branching steps in a reaction, for then the number of chain carriers grows exponentially and the rate of reaction may cascade into an explosion.

An example of both types of explosion is provided by the reaction between hydrogen and oxygen:

$$2H_2(g) + O_2(g) \rightarrow 2H_2O(g)$$

Although the net reaction is very simple, the mechanism is very complex and has not yet been fully elucidated. It is known that a chain reaction is involved, and that the chain carriers include $\cdot H$, $\cdot O \cdot$, $\cdot OH$, and (under certain conditions) $\cdot O_2H$. Some steps are:

Initiation: $H_2 + \cdot(O_2)\cdot \rightarrow 2 \cdot OH$

Propagation: $H_2 + \cdot OH \rightarrow \cdot H + H_2O$

             $\cdot(O_2)\cdot + \cdot H \rightarrow \cdot O\cdot + \cdot OH$ (branching)

             $\cdot O\cdot + H_2 \rightarrow \cdot OH + \cdot H$ (branching)

The last two branching steps can lead to a chain-branching explosion.

The occurrence of an explosion depends on the temperature and pressure of the system, and the explosion regions for the reaction are shown in Fig. 26.2. At very low pressures the system is outside the explosion region and the mixture reacts smoothly. At these pressures the chain carriers produced in the branching steps can reach the walls of the container where they combine (with an efficiency that depends on the composition of the walls). Increasing the pressure (along the vertical line in the illustration) takes the system through the **first explosion limit** (if the temperature is greater than about 730 K). The mixture then explodes because the chain carriers react before reaching the walls and the branching reactions are explosively efficient. The reaction is smooth when the pressure is above the **second explosion limit**. The concentration of molecules in the gas is then so great that the radicals produced in the branching reaction combine in the body of the gas, and reactions such as $O_2 + \cdot H \rightarrow \cdot O_2H$ can occur. Recombination reactions like this are facilitated by three-body collisions because the third body (M) can remove the excess energy:

$$O_2 + \cdot H + M \rightarrow \cdot O_2H + M^*$$

At low pressures three-particle collisions are unimportant and recombination is much slower. At higher pressures, when three-particle collisions are important, the explosive propagation of the chain by the radicals is partially quenched because the branching steps are diverted into simple propagation steps (see the following example). If the pressure

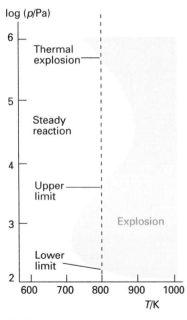

26.2 The explosion limits of the $H_2 + O_2$ reaction. In the explosive regions the reaction proceeds explosively when heated homogeneously.

is increased to above the **third explosion limit,** the reaction rate increases so much that a thermal explosion occurs.

Example 26.2 *Examining the explosion behaviour of a chain reaction*

Consider the following mechanism for the reaction of hydrogen and oxygen in the fuel-rich regime:

Initiation:	$H_2 \rightarrow 2H\cdot$	$v = \text{constant } (v^{init})$
Propagation:	$H_2 + \cdot OH \rightarrow \cdot H + H_2O$	$v = k_1[\cdot OH][H_2]$
Branching:	$\cdot(O_2)\cdot + \cdot H \rightarrow \cdot O\cdot + \cdot OH$	$v = k_2[H\cdot][O_2]$
	$\cdot O\cdot + H_2 \rightarrow \cdot OH + \cdot H$	$v = k_3[\cdot O\cdot][H_2]$
Termination:	$H\cdot + \text{wall} \rightarrow \frac{1}{2}H_2$	$v = k_4[H\cdot]$
	$H\cdot + \cdot(O_2)\cdot + M \rightarrow HO_2\cdot + M$	$v = k_5[H\cdot][O_2]$

Show that an explosion occurs when the rate of chain branching exceeds that of chain termination.

Method. We shall identify the onset of explosion with the rapid increase in the concentration of radicals, and for simplicity will identify that concentration with the concentration of H· atoms, which probably outnumber the highly reactive ·OH and ·O· radicals. When confronted by a reaction mechanism, it is sensible to set up the corresponding rate laws for the reaction intermediates and then to apply the steady-state approximation. Note that, in this simplified scheme, $HO_2\cdot$ radicals play no propagation role.

Answer. The rate of formation of radicals v^{rad} is identified with $d[H\cdot]/dt$; therefore we need

$$v^{rad} = v^{init} + k_1[\cdot OH][H_2] - k_2[H\cdot][O_2]$$
$$+ k_3[\cdot O\cdot][H_2] - k_4[H\cdot] - k_5[H\cdot][O_2]$$

The rates of formation of the intermediates are

$$\frac{d[\cdot OH]}{dt} = -k_1[\cdot OH][H_2] + k_2[H\cdot][O_2] + k_3[\cdot O\cdot][H_2] \approx 0$$

$$\frac{d[\cdot O\cdot]}{dt} = k_2[H\cdot][O_2] - k_3[\cdot O\cdot][H_2] \approx 0$$

These two equations solve to

$$[\cdot O\cdot] = \frac{k_2[H\cdot][O_2]}{k_3[H_2]} \qquad [\cdot OH] = \frac{2k_2[H\cdot][O_2]}{k_1[H_2]}$$

The rate of formation of radicals is therefore

$$v^{rad} = v^{init} + (2k_2[O_2] - k_4 - k_5[O_2])[H\cdot]$$

If then we write $k_{branch} = 2k_2[O_2]$, a measure of the rate of the more important chain-branching step, and $k_{term} = k_4 + k_5[O_2]$, a measure of the rate of chain termination, then this differential equation is

$$\frac{d[H\cdot]}{dt} = v^{init} + (k_{branch} - k_{term})[H\cdot]$$

There are two solutions. If termination dominates branching in the sense that $k_{term} > k_{branch}$, which it will at low oxygen concentrations, then

$$[radicals] \approx [H\cdot] = \frac{v^{init}\{1 - e^{-(k_{term} - k_{branch})t}\}}{k_{term} - k_{branch}}$$

As can be seen from Fig. 26.3a, in this regime, there is steady combustion of hydrogen. However, if branching dominates termination in the sense that $k_{branch} > k_{term}$, which it will at high oxygen concentrations, then

$$[radicals] \approx [H\cdot] = \frac{v^{init}\{e^{(k_{branch} - k_{term})t} - 1\}}{k_{branch} - k_{term}}$$

As can be seen from Fig. 26.3b, there is now an explosive increase in the concentration of radicals.

Comment. Although the steady-state approximation is highly questionable under explosive conditions, the calculation at least gives an indication of the basis for the transition from smooth combustion to explosive reaction conditions.

Exercise E26.2. What is the variation in radical composition when the rates of branching and termination are equal?   [[H·] = $v^{init}t$]

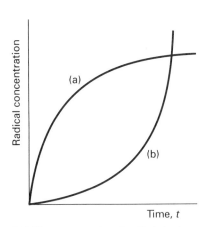

26.3 The concentration of radicals in the fuel-rich regime of the hydrogen–oxygen reaction (a) under steady combustion conditions, (b) in the explosive region.

26.3 Photochemical reactions

Many reactions can be initiated by the absorption of light by one of the mechanisms described in Section 17.2. The most important of all are the photochemical processes that capture the radiant energy of the Sun. Some of these reactions lead to the heating of the atmosphere during the daytime by absorption of ultraviolet radiation (Fig. 26.4). Others include the absorption of red and blue light by chlorophyll and the subsequent use of the energy to bring about the synthesis of carbohydrates from carbon dioxide and water. Without photochemical processes, the world would be simply a warm, sterile rock. A summary of the processes that can occur following photochemical excitation is given in Table 26.1.

Quantum yield

A molecule acquires energy to react by absorbing photons. The **Stark–Einstein law** states that one photon is absorbed by each molecule responsible for the primary photochemical process. However, the law is not valid under all conditions, because in a laser beam the intensity is so great that a single molecule may absorb several photons.

The Stark–Einstein law is valid under normal conditions. However, it leaves open the possibility that, even though a reactant molecule absorbs

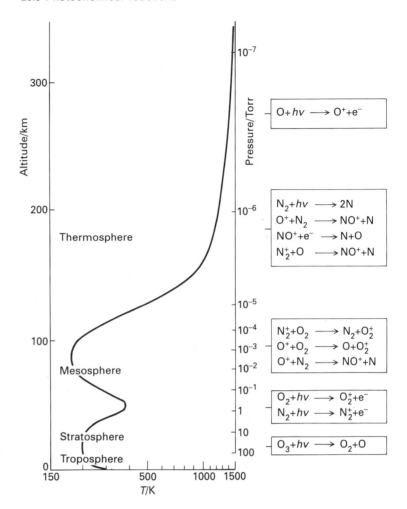

26.4 The temperature profile through the atmosphere and some of the reactions that occur. The temperature peak at about 50 km is due to the absorption of solar radiation by the O_2 and N_2 ionization reactions.

one photon, the excited molecule does not form products: there are many ways in which the excitation may be lost other than by dissociation or ionization. We therefore speak of the **primary quantum yield** ϕ, the number of reactant molecules producing specified primary products (atoms or ions, for instance) for each photon absorbed.

The primary product of photon absorption—a radical, a photoexcited molecule, or an ion—may be successful in initiating a process that leads to products. As a result of one successful initiation, many reactant molecules might be consumed. The **overall quantum yield** Φ is the number of reactant molecules that react (or product molecules that are formed) for each photon absorbed. In the photolysis of HI, for example, the processes are

$$HI + h\nu \rightarrow H\cdot + I\cdot$$
$$H\cdot + HI \rightarrow H_2 + I\cdot$$
$$2I\cdot \rightarrow I_2$$

The overall quantum yield is 2 because the absorption of one photon leads to the destruction of two HI molecules. In a chain reaction Φ may

Table 26.1 Photochemical processes†

Primary absorption $S \rightarrow S^*$
followed by vibrational and rotational relaxation

Physical processes

Fluorescence: $S^* \rightarrow S + h\nu$

Collision-induced emission:
$S^* + M \rightarrow S + M + h\nu$

Stimulated emission:
$S^* + h\nu \rightarrow S + 2h\nu$

Intersystem crossing (ISC): $S^* \rightarrow T^*$

Phosphorescence: $T^* \rightarrow S + h\nu$

Internal conversion (IC): $S^* \rightarrow S'$

Singlet electronic energy transfer:
$S^* + S \rightarrow S + S^*$

Energy pooling: $S^* + S^* \rightarrow S^{**} + S$

Triplet electronic energy transfer:
$T^* + S \rightarrow S + T^*$

Triplet–triplet absorption: $T^* + h\nu \rightarrow T^{**}$

Ionization

Penning ionization:
$A^* + B \rightarrow A + B^+ + e^-$

Dissociative ionization:
$A^* + B-C \rightarrow A-B^+ + C + e^-$

Collisional ionization:
$A^* + B \rightarrow A^+ + B + e^- \text{(or } B^-)$

Associative ionization:
$A^* + B \rightarrow AB^+ + e^-$

Chemical processes

Dissociation: $A-B^* \rightarrow A + B$

Addition or insertion: $A^* + B \rightarrow AB$

Abstraction or fragmentation:
$A^* + B \rightarrow C + D$

Isomerization:
$A^* \rightarrow A'$

Dissociative excitation:
$A^* + C-D \rightarrow A + C^* + D$

† S denotes a singlet state and T a triplet state; A, B, and M are arbitrary.

be very large, and values of about 10^4 are common. In such cases the chain acts as a chemical amplifier of the initial absorption step.

Example 26.3 *Using the quantum yield*

The overall quantum yield for the formation of ethene from 4-heptanone with 313-nm light is 0.21. How many molecules of the ketone per second, and what amount per second, are destroyed when the sample is irradiated with a 50-W, 313-nm source under conditions of total absorption?

Method. Calculate the number of photons emitted by the lamp per second; all are absorbed (by assertion); the number of molecules destroyed per second is the number of photons absorbed multiplied by the quantum yield Φ.

Answer. The energy of a 313-nm photon is

$$\frac{hc}{\lambda} = 6.35 \times 10^{-19}\,\text{J}$$

A 50-W source therefore generates photons at a rate

$$\frac{50\,\text{J s}^{-1}}{6.35 \times 10^{-19}\,\text{J}} = 7.9 \times 10^{19}\,\text{s}^{-1}$$

The number of 4-heptanone molecules destroyed per second is therefore 0.21 times this quantity, or $1.7 \times 10^{19}\,\text{s}^{-1}$, which corresponds to $2.8 \times 10^{-5}\,\text{mol s}^{-1}$.

Comment. The quantum yield depends on the wavelength of radiation used. One mole of photons is called an *einstein*.

Exercise E26.3. The overall quantum yield for another reaction at 290 nm is 0.30. For what length of time must irradiation with a 100-W source continue in order to destroy 1.0 mol of molecules?   [3.8 h]

Photochemical rate laws

As an example of how to incorporate the photochemical activation step into a mechanism, consider the photochemical activation of the reaction

$$H_2(g) + Br_2(g) \rightarrow 2HBr(g)$$

In place of the first step in the thermal reaction we have

$$Br_2 \xrightarrow{h\nu} 2Br \cdot \qquad v = I_{abs}$$

where I_{abs} is the number of photons of the appropriate frequency absorbed per unit time per unit volume. It follows that I_{abs} should take the place of $k_a[Br_2]$ in the thermal reaction scheme (if we assume that $\phi_{Br} = 2$), so from Example 26.1 we can write

$$\frac{d[HBr]}{dt} = \frac{2k_b(1/k_d)^{\frac{1}{2}}[H_2][Br_2]I_{abs}^{\frac{1}{2}}}{[Br_2] + (k_c/k_b')[HBr]} \qquad (4)$$

Equation 4 predicts that the reaction rate should depend on the square root of the absorbed light intensity, which is confirmed experimentally.

Photosensitization

The reactions of a molecule that does not absorb directly can be stimulated if another absorbing molecule is present because the latter may be able to transfer its energy during a collision. An example of this **photosensitization** process is the reaction often used to generate atomic hydrogen, the irradiation of hydrogen gas containing a trace of mercury using 254-nm radiation from a mercury discharge lamp. The Hg atoms are excited by resonant absorption of the radiation, and then collide with H_2 molecules. Two reactions then take place:

$$Hg^* + H_2 \rightarrow Hg + 2H\cdot$$
$$Hg^* + H_2 \rightarrow HgH + H\cdot$$

The latter reaction, which is monitored by detecting the HgH spectrophotometrically, accounts for 67 per cent of the process (but 76 per cent when deuterium is used in place of hydrogen). It is the initiation step for other mercury-photosensitized reactions, such as the synthesis of formaldehyde from carbon monoxide and hydrogen:

$$H\cdot + CO \rightarrow HCO\cdot$$
$$HCO\cdot + H_2 \rightarrow HCHO + H\cdot$$
$$2HCO\cdot \rightarrow HCHO + CO$$

Note that the last step is termination by disproportionation rather than by combination. Photosensitization also plays an important role in solution kinetics, and molecules containing the carbonyl chromophore (such as benzophenone, $C_6H_5COC_6H_5$) are often used to trap the incident light and to transfer it to some potentially reactive species.

POLYMERIZATION KINETICS

In **chain polymerization** an activated monomer M attacks another monomer, links to it, then that unit attacks another monomer, and so on. The monomer is used up slowly through the reaction by linking to the growing chains (Fig. 26.5). High polymers are formed rapidly and, as we shall see in detail later, only the yield and not the molar mass of the polymer is increased by allowing long reaction times. In **stepwise polymerization** any two monomers present in the reaction mixture can link together at any time (Fig. 26.6), and growth is not confined to chains that are already forming. A consequence of the process is that monomers are removed early in the reaction and (as we shall see) the average molar mass of the product grows with time.

26.4 Chain polymerization

Chain polymerization results in the rapid growth of an individual polymer chain for each activated monomer. It commonly occurs by addition, often by a radical chain process. Examples include the addition polymerizations of ethene, methyl methacrylate, and styrene. The central feature

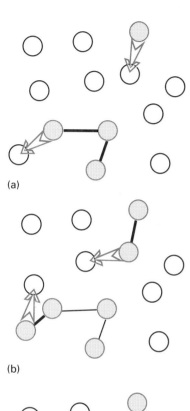

(a)

(b)

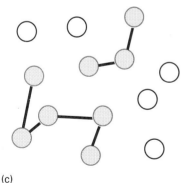

(c)

26.5 The process of chain polymerization. Chains grow as each chain acquires additional monomers.

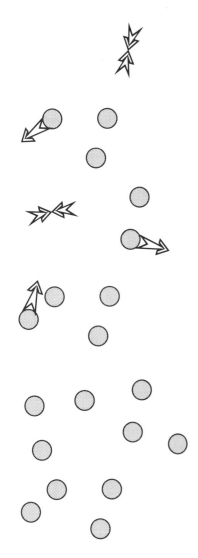

26.6 In stepwise polymerization, growth can start at any pair of monomers, and so new chains begin to form throughout the reaction.

of the kinetic analysis (which is summarized in the following *Justification*) is that the rate of polymerization is proportional to the square root of the initiator concentration:

$$v = k[I]^{\frac{1}{2}}[M] \tag{5}$$

JUSTIFICATION

There are three basic reaction steps in a chain polymerization process:

(a) Initiation:

$$I \to 2R\cdot \qquad v = k_i[I]$$
$$M + \cdot R \to \cdot M_1 \qquad \text{(fast)}$$

where I is the initiator and $\cdot M_1$ is a monomeric radical. We have shown a reaction in which a radical is produced, but in some polymerizations the initiation step leads to the formation of a cation or an anion chain carrier. The rate-determining step is the formation of the radicals $R\cdot$ by homolysis of the initiator, so the rate of initiation is equal to the v given above.

(b) Propagation:

$$M + \cdot M_1 \to \cdot M_2$$
$$M + \cdot M_2 \to \cdot M_3$$
$$\cdots \cdots \cdots \cdots \cdots$$
$$M + \cdot M_{n-1} \to \cdot M_n \qquad v = k_p[M][\cdot M]$$

Because this chain of reactions propagates quickly, the rate at which the total concentration of radicals grows is equal to the rate of the rate-determining initiation step. It follows that we can write

$$\frac{d[\cdot M]}{dt} = 2\phi k_i[I]$$

where ϕ is the yield of the initiation step, the fraction of radicals $R\cdot$ that successfully initiate a chain.

(c) Termination:

$$\cdot M_n + \cdot M_m \to M_{n+m}$$

If we suppose that the rate of termination is independent of the length of the chain, then its rate law is

$$v = k_t[\cdot M]^2$$

and the rate of change of radical concentration by this process is

$$\frac{d[\cdot M]}{dt} = -2k_t[\cdot M]^2$$

In practice, other termination steps may intervene. Side reactions may also occur, such as **chain transfer**, in which the reaction

$$M + \cdot M_n \rightarrow \cdot M + M_n$$

initiates a new chain at the expense of the one currently growing.

The total radical concentration is approximately constant throughout the main part of the polymerization because the rate at which radicals are formed by initiation is approximately the same as the rate at which they are removed by termination. Hence, we can use the steady-state approximation to write

$$\frac{d[\cdot M]}{dt} = 2\phi k_i[I] - 2k_t[\cdot M]^2 = 0$$

The steady-state concentration of radical chains is therefore

$$[\cdot M] = \left(\frac{\phi k_i}{k_t}\right)^{\frac{1}{2}}[I]^{\frac{1}{2}}$$

Because the rate of propagation of the chains (the rate at which the monomer is consumed) is

$$\frac{d[M]}{dt} = -k_p[\cdot M][M]$$

it follows that

$$\frac{d[M]}{dt} = -k_p\left(\frac{\phi k_i}{k_t}\right)^{\frac{1}{2}}[I]^{\frac{1}{2}}[M]$$

which has the form of eqn 5.

The **kinetic chain length** v is a measure of the efficiency of the chain propagation mechanism. It is defined as the ratio of the number of monomer units consumed per active centre produced in the initiation step:

$$v = \frac{\text{number of monomer units consumed}}{\text{number of active centres produced}} \tag{6}$$

The kinetic chain length is therefore equal to the ratio of the propagation and initiation rates:

$$v = \frac{\text{propagation rate}}{\text{initiation rate}}$$

Because the initiation rate is equal to the termination rate, we can write this expression (using the rate constants introduced in the *Justification*) as

$$v = \frac{k_p[\cdot M][M]}{2k_t[\cdot M]^2} = \frac{k_p[M]}{2k_t[\cdot M]}$$

When the steady-state expression is substituted for the radical concentration, we obtain

$$v = k[M][I]^{-\frac{1}{2}}, \qquad \text{with } k = \frac{k_p}{2(\phi k_i k_t)^{-\frac{1}{2}}} \tag{7}$$

We see that the slower the initiation of the chain (the smaller the initiator concentration and the smaller the initiation rate constant), the greater the kinetic chain length.

Example 26.4 *Using the kinetic chain length*

Estimate the average number of units in a polymer produced by a chain mechanism in which termination occurs by combination of radicals.

Method. Because termination occurs by the combination of two radicals, the number of monomers in a polymer molecule $\langle n \rangle$ produced by the reaction is the sum of the numbers in the two combining polymer chains. The latter are both, on average, equal to v (by eqn 6 each initiation leads to a polymer of v monomer units).

Answer. The average number of monomers in a product molecule is

$$\langle n \rangle = 2v = 2k[M][I]^{-\frac{1}{2}}$$

with k given in eqn 7.

Comment. We see that the slower the initiation (as expressed by the rate constant k_i and the initiator concentration $[I]$), the longer the average chain length, and therefore the higher the molar mass of the polymer. Some of the consequences of molar mass for polymers were explored in Chapter 23: now we see how we can exercise kinetic control over them.

Exercise E26.4. Another termination mechanism is a disproportionation reaction of the form $M\cdot + \cdot M \rightarrow M + :M$. Calculate the average polymer length.

$$[\langle n \rangle = k[M][I]^{-\frac{1}{2}}]$$

26.5 Stepwise polymerization

Stepwise polymerization commonly proceeds by a **condensation reaction,** in which a small molecule (typically H_2O) is eliminated in each step. Stepwise polymerization is the mechanism of production of polyamides, as in the formation of nylon-66:

$$H_2N-(CH_2)_6-NH_2 + HOOC-(CH_2)_4-COOH$$
$$\rightarrow H_2N-(CH_2)_6-NH-CO-(CH_2)_4-COOH + H_2O$$
$$\rightarrow H[NH-(CH_2)_6-NH-CO-(CH_2)_4-CO]_n OH + nH_2O$$

Polyesters and polyurethanes are formed similarly (the latter without elimination). A polyester, for example, can be regarded as the outcome of the stepwise condensation of a hydroxyacid $HO-M-COOH$. We shall consider the formation of a polyester from such a monomer, and measure its progress in terms of the concentration of the $-COOH$ groups in the sample (which we shall denote A), since these gradually disappear as the condensation proceeds. Because the condensation reaction can occur between molecules containing any number of monomer units, chains of many different lengths can grow in the reaction mixture. We shall show how to use the reaction scheme to predict molar mass distributions.

The rate law of stepwise polymerization

The condensation can be expected to be overall second-order in the concentration of the –OH and –COOH (or A) groups:

$$\frac{d[A]}{dt} = -k[-OH][A]$$

However, because there is one –OH group for each –COOH group, this equation is the same as

$$\frac{d[A]}{dt} = -k[A]^2$$

If it is assumed that the rate constant for the condensation is independent of the chain length, then k remains constant throughout the reaction. The solution of this rate law is given by eqn 25.7, and is

$$[A] = \frac{[A]_0}{1 + kt[A]_0} \tag{8}$$

The fraction p of –COOH groups that have condensed at time t is

$$p = \frac{[A]_0 - [A]}{[A]_0} \tag{9}$$

so

$$p = \frac{kt[A]_0}{1 + kt[A]_0} \tag{10}$$

Example 26.5 *Calculating the degree of polymerization*

Find an expression for the growth with time of the degree of polymerization of a polymer formed by a stepwise process.

Method. The number of monomers per polymer molecule is the ratio of the initial concentration of A, $[A]_0$, to the number of end groups, $[A]$, at the time of interest, because there is one A group per polymer molecule. For example, if there were initially 1000 A groups and there are now only 10, each polymer must be 100 units long on average. The value of $[A]$ can be expressed in terms of p, and hence in terms of the time by using eqn 10.

Answer. The average number of monomers per polymer molecule is

$$\langle n \rangle = \frac{[A]_0}{[A]} = \frac{1}{1 - p}$$

When we substitute the value for p given in eqn 10, we find

$$\frac{1}{1 - p} = 1 + kt[A]_0$$

Therefore, the degree of polymerization grows as

$$\langle n \rangle = 1 + kt[A]_0$$

Comment. The length grows in proportion to the time, so the longer stepwise polymerization proceeds, the higher the average molar mass of the product.

Exercise E26.5. Derive the expression for the time-dependence of p for a stepwise polymerization in which the reaction is acid-catalysed by the $-COOH$ acid functional group. The rate law is $d[A]/dt = -k[A]^2[-OH]$.

$$[p = 1 - \alpha^{-\frac{1}{2}}, \alpha = 1 + 2kt[A]_0^2]$$

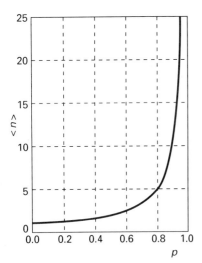

26.7 The average chain length of a polymer as a function of the fraction of reacted monomers p. Note that p must be very close to 1 for the chains to be long.

The statistics of polymerization

We can arrive at the same conclusion about the degree of polymerization as in Example 26.5 by considering the probability p that a group (e.g. $-COOH$) has formed a link to another molecule. At the start of the reaction $p = 0$; at the end $p = 1$.

The average number of monomers linked together in a polymer that has been formed by stepwise polymerization increases as links are formed. That is, the average number increases as the probability that a monomer has not formed a link decreases. Because the latter probability is equal to $1 - p$, we can conclude that $\langle n \rangle$ is inversely proportional to $1 - p$. The constant of proportionality can be obtained quite simply by noting that, if $p = 0$, then the number in each chain is 1 (because no polymerization has occurred). It follows that[1]

$$\langle n \rangle = \frac{1}{1 - p} \tag{11}$$

which is the same expression as that obtained from kinetics in Example 26.5 (Fig. 26.7). It is clear from the illustration that we need p very close to 1 in order to obtain polymers with high molar masses.

Because p grows with time according to eqn 10, the degree of polymerization also grows with time:

$$\langle n \rangle = 1 + kt[A]_0 \tag{12}$$

Therefore the number-average molar mass increases with time as

$$\bar{M}_n = \langle n \rangle M_{mon} = (1 + kt[A]_0)M_{mon} \tag{13}$$

Equation 13 is the origin of the remark made earlier that, in stepwise polymerization (in contrast to chain polymerization), the molar mass of the product increases the longer the reaction proceeds.

CATALYSIS AND OSCILLATION

If the activation energy of a reaction is high, at normal temperatures only a small proportion of molecular encounters result in reaction. A catalyst lowers the activation energy of the reaction by providing an alternative

1 A more quantitative argument is given in the fourth edition of this text (1990), p. 835.

path that avoids the slow, rate-determining step of the uncatalysed reaction, and results in a higher reaction rate at the same temperature. Catalysts can be very effective; for instance, the activation energy for the decomposition of hydrogen peroxide in solution is $76 \, \mathrm{kJ \, mol^{-1}}$, and the reaction is slow at room temperature. When a little iodide is added, the activation energy falls to $57 \, \mathrm{kJ \, mol^{-1}}$, and the rate increases by a factor of 2000. Enzymes, which are biological catalysts, are very specific and can have a dramatic effect on the reactions they control. The activation energy for the acid hydrolysis of sucrose is $107 \, \mathrm{kJ \, mol^{-1}}$, but the enzyme saccharase reduces it to $36 \, \mathrm{kJ \, mol^{-1}}$, corresponding to an acceleration of the reaction by a factor of 10^{12} at blood temperature (310 K).

A **homogeneous catalyst** is a catalyst that is in the same phase as the reaction mixture (e.g. an acid added to an aqueous solution). A **heterogeneous catalyst** is in a different phase (e.g. a solid catalyst for a gas-phase reaction). We examine heterogeneous catalysis in Chapter 28 and consider only homogeneous catalysis here.

26.6 Homogeneous catalysis

Some idea of the mode of action of homogeneous catalysts can be obtained by examining the kinetics of the bromide-catalysed decomposition of hydrogen peroxide:

$$2H_2O_2(aq) \rightarrow 2H_2O(aq) + O_2(g)$$

The reaction is believed to proceed through the following pre-equilibrium:

$$H_3O^+ + HOOH \rightleftharpoons HOOH_2^+ + H_2O \qquad K = \frac{[HOOH_2^+]}{[HOOH][H_3O^+]}$$

$$HOOH_2^+ + Br^- \rightarrow HOBr + H_2O \qquad v = k[HOOH_2^+][Br^-]$$

$$HOBr + HOOH \rightarrow H_3O^+ + O_2 + Br^- \qquad \text{(fast)}$$

(In the equilibrium constant, the activity of H_2O has been set equal to 1.) Because the second step is rate-determining, the rate law of the overall reaction is predicted to be

$$\frac{d[O_2]}{dt} = kK[HOOH][H_3O^+][Br^-]$$

in agreement with the observed dependence of the rate on the Br^- concentration and the pH of the solution. The observed activation energy is that of the effective rate coefficient kK. In the absence of Br^- ions the reaction cannot proceed through the path set out above, and a different and much higher activation energy is observed.

Two important types of homogeneous catalysis are **acid catalysis** and **base catalysis**, and many organic reactions involve one or the other (and sometimes both). Brønsted acid catalysis is the transfer of a proton to the substrate:

$$X + HA \rightarrow HX^+ + A^- \qquad HX^+ \text{ then reacts}$$

It is the primary process in the solvolysis of esters, keto-enol tautomerism, and the inversion of sucrose, Brønsted base catalysis is the transfer of a hydrogen ion from the substrate to a base:

$$XH + B \rightarrow X^- + BH^+ \qquad X^- \text{ then reacts}$$

It is the primary step in the isomerization and halogenation of organic compounds, and of the Claisen and aldol reactions.

26.7 Autocatalysis

The phenomenon of **autocatalysis** is the acceleration of a reaction by the products. For example, in a reaction $A \rightarrow P$ it may be found that the rate law is

$$v = k[A][P] \qquad\qquad (14)$$

so the reaction rate increases as products are formed. (The reaction gets started because there are usually other reaction routes for the formation of some P initially, which then takes part in the autocatalytic reaction proper.) An example of autocatalysis is provided by two steps in the Belousov–Zhabotinskii reaction (BZ reaction) that will figure in discussions later in the section:

$$BrO_3^- + HBrO_2 + H_3O^+ \rightarrow 2BrO_2 + 2H_2O$$
$$2BrO_2 + 2Ce(III) + 2H_3O^+ \rightarrow 2HBrO_2 + Ce(IV) + 2H_2O$$

The product $HBrO_2$ is a reactant in the first step.

Example 26.6 *Calculating concentrations in an autocatalytic reaction*

Set up and solve the rate equation for the $A \rightarrow P$ autocatalytic reaction described by eqn 14.

Method. It is convenient to write $[A] = [A]_0 - x$, $[P] = [P]_0 + x$ and then to write down the expression for the rate of change of either species in terms of x.

Answer. The rate law is

$$\frac{dx}{dt} = k([A]_0 - x)([P]_0 + x)$$

Integration by partial fractions, using

$$\frac{1}{([A]_0 - x)([P]_0 + x)} = \frac{1}{[A]_0 + [P]_0} \times \left(\frac{1}{[A]_0 - x} + \frac{1}{[P]_0 + x} \right)$$

gives

$$\frac{1}{[A]_0 + [P]_0} \ln \left(\frac{([P]_0 + x)[A]_0}{[P]_0([A]_0 - x)} \right) = kt$$

This expression can be rearranged into

$$\frac{x}{[P]_0} = \frac{e^{at} - 1}{1 + be^{at}}$$

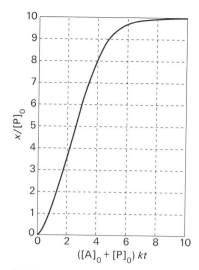

26.8 The concentration of product during the autocatalysed A → P reaction discussed in Example 26.6 (using $b = 0.1$).

with

$$a = ([A]_0 + [P]_0)k \qquad b = \frac{[P]_0}{[A]_0}$$

Comment. The solution is plotted in Fig. 26.8. The rate of reaction is slow initially (little P present), then fast (when P and A are both present), and finally slow again (when A has disappeared).

Exercise E26.6. At what time is the reaction rate a maximum?

$$[t_{\max} = -(1/ka) \ln b]$$

The industrial importance of autocatalysis (which occurs in a number of reactions, such as oxidations) is that the rate of the reaction can be maximized by ensuring that the optimum concentrations of reactant and product are always present.

26.8 Oscillating reactions

One consequence of autocatalysis is the possibility that the concentrations of reactants, intermediates, and products will vary periodically either in space or in time (Fig. 26.9). Chemical oscillation is the

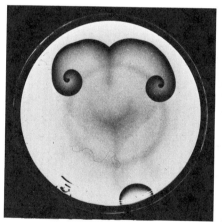
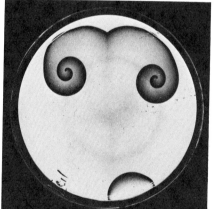

26.9 Some reactions show oscillations in time; some show spatially periodic variations. This sequence of photographs shows the emergence of a spatial pattern.

analogue of electrical oscillation, with autocatalysis playing the role of positive feedback. Oscillating reactions are much more than a laboratory curiosity. While they are known to occur in only a few cases in industrial processes, there are many examples in biochemical systems where a cell plays the role of a chemical reactor. Oscillating reactions, for example, maintain the rhythm of the heartbeat. They are also known to occur in the glycolytic cycle, in which one molecule of glucose is used to produce (through enzyme-catalysed reactions involving ATP) two molecules of ATP. All the metabolites in the chain oscillate under some conditions, and do so with the same period but with different phases.

The Lotka–Volterra mechanism

We shall use an autocatalytic reaction of a particularly simple form that illustrates how these oscillations may occur. The actual chemical examples that have been discovered so far have a different mechanism, as we shall see. The **Lotka–Volterra mechanism** is as follows:

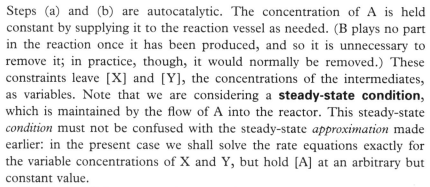

$$\text{(a)} \quad A + X \rightarrow 2X \qquad \frac{d[A]}{dt} = -k_a[A][X]$$

$$\text{(b)} \quad X + Y \rightarrow 2Y \qquad \frac{d[X]}{dt} = -k_b[X][Y] \qquad \textbf{(15)}$$

$$\text{(c)} \quad Y \rightarrow B \qquad \frac{d[B]}{dt} = k_c[Y]$$

Steps (a) and (b) are autocatalytic. The concentration of A is held constant by supplying it to the reaction vessel as needed. (B plays no part in the reaction once it has been produced, and so it is unnecessary to remove it; in practice, though, it would normally be removed.) These constraints leave [X] and [Y], the concentrations of the intermediates, as variables. Note that we are considering a **steady-state condition**, which is maintained by the flow of A into the reactor. This steady-state *condition* must not be confused with the steady-state *approximation* made earlier: in the present case we shall solve the rate equations exactly for the variable concentrations of X and Y, but hold [A] at an arbitrary but constant value.

The Lotka–Volterra equations can be solved numerically, and the results can be depicted in two ways. One way is to plot [X] and [Y] against time (Fig. 26.10). The same information can be displayed more succinctly by plotting one concentration against the other: the representation gives the series of closed curves shown in Fig. 26.11.

The periodic variation of the concentrations of the intermediates can be explained as follows. At some stage there may be only a little X present, but reaction (a) provides more, and the production of X autocatalyses the production of even more X. There is therefore a surge of X. However, as X is formed, reaction (b) can begin. It occurs slowly initially because [Y] is small, but autocatalysis leads to a surge of Y. This surge, though, removes X, so reaction (a) slows, and less X is produced.

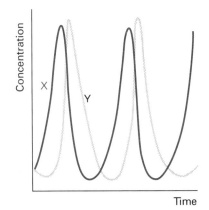

26.10 The periodic variation of the concentrations of the intermediates X and Y in a Lotka–Volterra reaction. The system is in a steady-state, but not at equilibrium.

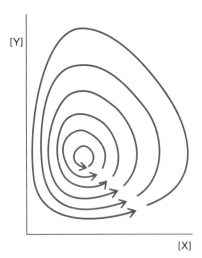

26.11 An alternative representation of the periodic variation of [X] and [Y] is to plot one against the other. Then the system describes closed orbits, different ones being obtained with different starting conditions.

Because less X is now available, reaction (b) slows. As less Y becomes available to remove X, X has a chance to surge forward again, and so on.

The brusselator

Another very interesting set of equations is called the **brusselator** (because it is an oscillator that originated with Ilya Prigogine's group in Brussels):

(a) $A \rightarrow X$ $\qquad \dfrac{d[X]}{dt} = k_a[A]$

(b) $2X + Y \rightarrow 3Y$ $\qquad \dfrac{d[Y]}{dt} = k_b[X]^2[Y]$

(c) $B + X \rightarrow Y + C$ $\qquad \dfrac{d[Y]}{dt} = k_c[B][X]$

(d) $X \rightarrow D$ $\qquad \dfrac{d[X]}{dt} = -k_d[X]$

(16)

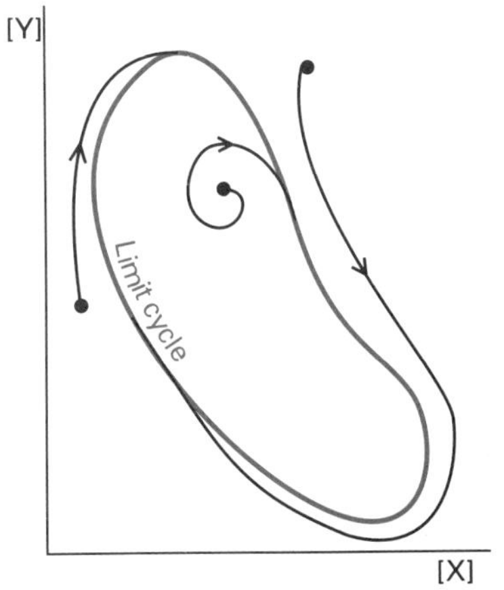

26.12 Some oscillating reactions approach a closed trajectory whatever their starting conditions. The closed trajectory is called a 'limit cycle'.

Because the reactants (A and B) are maintained at constant concentration, the two variables are the concentrations of X and Y. These two concentrations may be calculated by solving the rate equations numerically, and the results are plotted in Fig. 26.12. The interesting feature is that, whatever the initial concentrations of X and Y, the system settles down into the same periodic variation of concentrations. The common trajectory to which the system migrates is called a **limit cycle**, and its period depends on the values of the rate constants. A limit cycle is an example of a structure called by mathematicians an **attractor** because it appears to attract to itself trajectories in its vicinity. In conventional chemistry, the equilibrium state is an attractor. In systems maintained far from equilibrium, a limit cycle may act as an attractor.

The oregonator

One of the first oscillating reactions to be reported and studied systematically was the BZ reaction mentioned earlier, which takes place in a mixture of potassium bromate, malonic acid, and a cerium(IV) salt in an acidic solution. The mechanism has been elucidated by Richard Noyes, Richard Field, and Endre Körös, and involves 18 elementary steps and 21 different chemical species. The main features of this awesomely complex mechanism can be reproduced by the following **oregonator** (so called because Noyes and his group work in Oregon), where X stands for $HBrO_2$, Y for Br^-, Z for Ce^{4+}, A for BrO_3^-, and C for HBrO:

(a) $A + Y \rightarrow X$

(b) $X + Y \rightarrow C$

(c) $A + X \rightarrow 2X + Z$ $\qquad$ (17)

(d) $2X \rightarrow D$

(e) $Z \rightarrow Y$

A, B (which denotes malonic acid or bromomalonate, and does not occur in the mechanism), C, and D are held constant (in the model). The oscillations arise in a similar way to the brusselator, and can be traced to the autocatalysis in step c and the linkage between the reactions provided by the other steps.

Bistability

Attempts have been made to discover the underlying causes of oscillation more deeply than simply recognizing the role of autocatalysis. It appears that in one class of systems three conditions must be fulfilled in order to obtain oscillations:

(1) The reactions must be far from equilibrium.
(2) The reactions must have autocatalytic steps.
(3) The system must be able to exist in two steady states.

The last criterion is called **bistability**, and is a property that takes us well beyond what is familiar from the equilibrium properties of systems. The property of bistability is a chemical version of supercooling, in which the temperature of a liquid may be lowered beneath its freezing point without solidifying.

Consider a reaction in which there are two intermediates X and Y. If the concentration of Y is at some high value in a reactor, and X is added, then the concentration of Y might decrease as shown by the upper line in Fig. 26.13. If X is at some high value, then as Y is added the reaction might result in the slow increase of Y as shown by the lower line. However, in each case, a concentration may be reached (A or B) at which the concentration will jump from one curve to the other (just as a supercooled liquid might suddenly solidify). The two curves represent the two stable states of the bistable system. Neither state is an equilibrium state in the thermodynamic sense: they occur in states that are well removed from equilibrium, and the concentrations of X and Y represent the consequences of reactants continuously flowing into and of products flowing out of the reactor.

Now consider what happens when a third type of intermediate Z is present. Suppose Z reacts with both X and Y. In the absence of Z the flows of material might correspond to a state on the upper curve of Fig. 26.14. However, as Z reacts with Y to produce X, the state of the system moves along the curve (to the right, as Y decreases and X increases) until at A a sudden transition occurs to the lower curve. Then Z reacts with X and produces Y, which means that the composition moves to the left along the lower curve. There comes a point (B), however, when the concentration of X has been reduced so much, and that of Y has risen so much, that there is a sudden transition to the upper curve, when the process begins again. It is the leaping from one stable state to another that we see as the sudden surge or depletion of the concentration of a species (Fig. 26.15).

By studying the regions of concentrations and the rate constants for the individual steps of a reaction it is now becoming possible to predict the

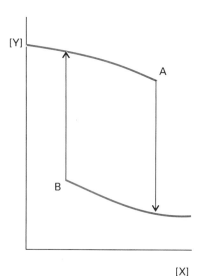

26.13 A system showing bistability. As the concentration of X is increased (by adding it to a reactor) the concentration of Y decreases along the upper curve but at A it drops sharply to a low value. If X is then decreased, the concentration of Y increases along the lower curve, but rises sharply to a high value at B.

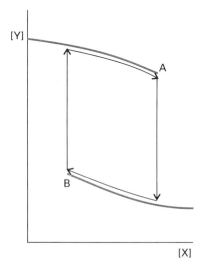

26.14 Chemical oscillation in a bistable system occurs as a result of the effect of a third substance Z which can react with X to produce Y and with Y to produce X. As a result, the system switches periodically between the upper and lower curves.

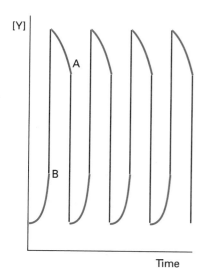

26.15 The leaping from one branch to another of a bistable system appears to the observer as a periodic surge and depletion of concentration.

occurrence of bistable chemical systems and to anticipate the occurrence of oscillations (as in the behaviour of cool flames). We are still far, however, from being able to use these ideas to account for gene expression and the patterns on tigers and butterflies, in both of which it is thought that these processes play a part.

26.9 Chemical chaos

A remarkable development of chemical kinetics in recent years has been the discovery of rate laws for complex reactions that have innately chaotic solutions. Specifically, it is found that the solutions of certain sets of rate laws, although fully determined by the structure of the equations, cannot be used to predict the composition of the system from its initial composition. The kinetic equations that we have been considering are of such richness that it should be hardly surprising that they, or minor elaborations of them, can display such **chaotic** solutions. Instead of a reaction showing periodic oscillatory behaviour, the concentrations burst into chaotic oscillation, and the concentrations of the intermediates show unpredictable amplitudes or unpredictable frequencies (Fig. 26.16). Such behaviour can be literally a matter of life and death, for should the heartbeat become chaotic, it may result in death. On a larger scale, whole economies (which may also be described by differential equations for the flow of goods and money) may collapse into revolutionary chaos.[2]

There are several ways in which a reaction can be steered into a chaotic regime. For example, it may be the case in certain systems that, as a parameter (such as a flow rate through the reaction vessel or the rate of stirring) is changed, the period of the limit cycle successively doubles (Fig. 26.17). When this **period-doubling** occurs, the system must circulate twice round the cycle before an initial pair of concentrations is restored. Then, after a further modification of the parameter, the period doubles again, and the limit cycle circulates four times before repeating itself. As each period doubles, the periodicity of the motion becomes less apparent, and finally appears like random fluctuations. The trajectory of

26.16 The onset of chaos as a result of period-doubling. (a) Steady-state oscillation, (b) the oscillation after one period-doubling step, (c) the chaotic regime after many period-doubling steps.

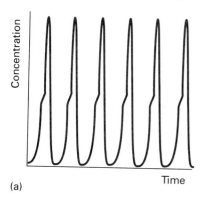

(a)

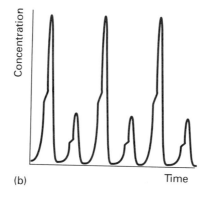

(b)

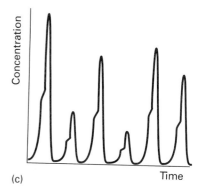

(c)

2 The type of chaos that stems from the structure of well-defined (and often quite simple) non-linear differential equations is called *deterministic chaos,* for the behaviour of the solutions is predictable but infinitely sensitive to the initial conditions. It is not clear that revolutionary chaos is deterministic, but some simple economic models may be.

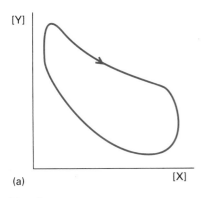

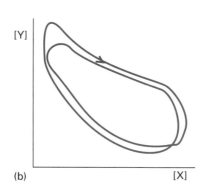

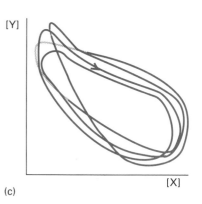

26.17 Successive trajectories in concentration space (a) before period doubling, (b) after one period-doubling step, (c) after several period-doubling steps.

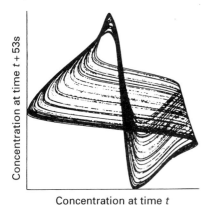

26.18 A representation of a strange attractor for a system showing chemical chaos.

the composition of the system is now highly complex, and may take a path that never retraces its path or crosses itself. Such a path is called a **strange attractor** (Fig. 26.18).

It should be understood, however, that the term 'chaos' is in certain respects misleading. First, the conditions under which certain systems of differential equations display period doubling can be specified exactly, and the progression through period-doubling cycles shows considerable regularity that is common to many classes of systems. In so far as the composition can be specified *exactly*, then a later composition can be predicted. Our inability to predict the composition of a system that is in a chaotic regime stems from our inability *experimentally* to know the initial conditions exactly or make a later determination of the composition at an *exact* later instant. The unpredictability of a chaotic system lies not in the formulation or solution of the differential equations that describe the rates of processes, but in our ability to relate those solutions to the practical system of interest given the inherent imprecision of experimental observations.

CHECK LIST OF KEY IDEAS

1 The structure of **chain reactions** in terms of **initiation, propagation, branching, retardation,** and **termination** (Section 26.1).

2 The **rate laws** of chain reactions using the steady-state approximation (Section 26.1).

3 The origin of **thermal explosions** and of **chain-branching explosions** (Section 26.2).

4 The processes that may follow **photoexcitation** (Table 26.1).

5 The **primary quantum yield** and the **overall**

quantum yield of photochemical reactions (Section 26.3).

6 The formulation of **photochemical rate laws** (as in eqn 4).

7 The principles of **photosensitization** (Section 26.3).

8 The rate law governing **chain polymerization** and the characteristics of the molar mass distribution to which it leads (Section 26.4).

9 The **kinetic chain length** in a polymerization

process (eqn 6) and the average molar mass (Example 26.4).

10 The rate law governing **stepwise polymerization** and the molar mass to which it leads (Section 26.5).

11 The calculation of the **degree of polymerization** (Example 26.5) and the growth of the **number-average molar mass** as polymerization proceeds (eqn 13).

12 The effect of **homogeneous catalysis** on the rate of reaction (Section 26.6).

13 The role of **autocatalysis** in chemical kinetics

(Section 26.7 and Example 26.6).

14 The **Lotka–Volterra mechanism** of chemical oscillation (Section 26.8) and the concepts of **attractor** and **limit cycle** for systems that are far from equilibrium.

15 The **brusselator** and the **oregonator** as examples of chemical oscillation, and the property of **bistability** (Section 26.8).

16 Chemical reactions may enter a **chaotic regime** when the composition cannot, in practice, be predicted from the initial conditions (Section 26.9).

EXERCISES

In the following exercises and problems, it is recommended that rate constants are labelled with the number of the step in the proposed reaction mechanism and that any reverse steps are labelled similarly but with a prime.

26.1. Derive the rate law for the decomposition of ozone in the reaction $2O_3(g) \rightarrow 3O_2(g)$ on the basis of the following proposed mechanism:

(1) $O_3 \rightleftarrows O_2 + O$ (forward k_1, reverse k_1')
(2) $O + O_3 \rightarrow 2O_2$

26.2. On the basis of the following proposed mechanism, account for the experimental fact that the rate law for the decomposition $2N_2O_5(g) \rightarrow 4NO_2(g) + O_2(g)$ is $v = k[N_2O_5]$.

(1) $N_2O_5 \rightleftarrows NO_2 + NO_3$ (forward k_1, reverse k_1')
(2) $NO_2 + NO_3 \rightarrow NO_2 + O_2 + NO$
(3) $NO + N_2O_5 \rightarrow 3NO_2$

26.3. A slightly different mechanism for the decomposition of N_2O_5 from that in Exercise 26.2 has also been proposed. It differs only in the last step, which is replaced by

(3) $NO + NO_3 \rightarrow 2NO_2$

Show that this mechanism leads to the same overall rate law.

26.4. Photolysis of $Cr(CO)_6$ in the presence of certain molecules M can give rise to the following reaction sequence:

(1) $Cr(CO)_6 + h\nu \rightarrow Cr(CO)_5 + CO$
(2) $Cr(CO)_5 + CO \rightarrow Cr(CO)_6$
(3) $Cr(CO)_5 + M \rightarrow Cr(CO)_5M$
(4) $Cr(CO)_5M \rightarrow Cr(CO)_5 + M$

Suppose that the absorbed light intensity is so weak that $I \ll k_4[Cr(CO)_5M]$. Find the factor f in the equation $d[Cr(CO)_5M]/dt = -f[Cr(CO)_5M]$. Show that a graph of $1/f$ against [M] should be a straight line.

26.5. Consider the following mechanism for the thermal decomposition of R_2

(1) $R_2 \rightarrow 2R$
(2) $R + R_2 \rightarrow P_B + R'$
(3) $R' \rightarrow P_A + R$
(4) $2R \rightarrow P_A + P_B$

where R_2, P_A, P_B are stable hydrocarbons and R and R' are radicals. Find the dependence of the rate of decomposition of R_2 on the concentration of R_2.

26.6. Refer to Fig. 26.2 and determine the pressure range for branching-chain explosion in the hydrogen–oxygen reaction at 700 K, 800 K, and 900 K.

26.7. In a photochemical reaction $A \rightarrow 2B + C$, the quantum efficiency with 500 nm light is 2.1×10^2 mol einstein^{-1}. After exposure of 300 mmol of A to the light, 2.28 mmol of B is formed. How many photons were absorbed by A?

26.8. In an experiment to measure the quantum efficiency of a photochemical reaction, the absorbing substance was exposed to 490 nm light from a 100 W source for 45 minutes. The intensity of the transmitted light was 40 per cent of the intensity of the incident light. As a result of irradiation, 0.344 mol of the absorbing substance decomposed. Determine the quantum efficiency.

26.9. The condensation reaction of acetone $(CH_3)_2CO$ in aqueous solution is catalysed by bases B, which react reversibly with acetone to form the carbanion $C_3H_5O^-$. The carbanion then reacts with a molecule of acetone to give the product. A simplified version of the mechanism is

(1) $AH + B \rightarrow BH^+ + A^-$
(2) $A^- + BH^+ \rightarrow AH + B$
(3) $A^- + AH \rightarrow product$

where AH stands for acetone and A^- denotes its carbanion. Use the steady-state approximation to find the concentration of the carbanion and derive the rate equation for the formation of the product.

26.10. Consider the acid-catalysed reaction

(1) $HA + H^+ \rightleftharpoons HAH^+$
    (k_1 forward, k_1' reverse; both fast)

(2) $HAH^+ + B \rightarrow BH^+ + AH$ (k_2, slow)

Deduce the rate law and show that it can be made independent of the specific term $[H^+]$.

26.11. Consider the following chain mechanism:

(1) $AH \rightarrow A\cdot + H\cdot$
(2) $A\cdot \rightarrow B\cdot + C$
(3) $AH + B\cdot \rightarrow A\cdot + D$
(4) $A\cdot + B\cdot \rightarrow P$

Identify the initiation, propagation, and termination steps, and use the steady-state approximation to deduce that the decomposition of AH is first-order in AH.

PROBLEMS

Numerical problems

26.1. The number of photons falling on a sample can be determined by a variety of methods, of which the classical one is chemical actinometry. The decomposition of oxalic acid $(COOH)_2$ in the presence of uranyl sulfate $(UO_2)SO_4$ proceeds according to the sequence

(1) $UO^{2+} + h\nu \rightarrow (UO^{2+})^*$
(2) $(UO^{2+})^* + (COOH)_2 \rightarrow UO^{2+} + H_2O + CO_2 + CO$

with a quantum efficiency of 0.53 at the wavelength used. The amount of oxalic acid remaining after exposure can be determined by titration (with $KMnO_4$) and the extent of decomposition used to find the number of incident photons. In a particular experiment, the actionometry solution consisted of 5.232 g anhydrous oxalic acid, 25.0 cm^3 water (together with the uranyl salt). After exposure for 300 s the remaining solution was titrated with 0.212 M $KMnO_4$(aq), and 17.0 cm^3 were required for complete oxidation of the remaining oxalic acid. What is the rate of incidence of photons at the wavelength of the experiment? Express the answer in photon/second and einstein/second.

26.2. When benzophenone is illuminated with ultraviolet light it is excited into a singlet state. This singlet changes rapidly into a triplet, which phosphoresces. Triethylamine acts as a quencher for the triplet. In an experiment in methanol as solvent, the phosphorescence intensity varied with amine concentration as shown below. A flash photolysis experiment had also shown that the half-life of the fluorescence in the absence of quencher is 29 μs. What is the value of k_q?

$[Q]/(mol\ L^{-1})$	0.0010	0.0050	0.0100
I_f/(arbitrary units)	0.41	0.25	0.16

26.3. Studies of combustion reactions depend on knowing the concentrations of H atoms and OH radicals. Measurements on a flow system using ESR for the detection of radicals gave information on the reactions

(1) $H + NO_2 \rightarrow OH + NO$ $k_1 = 2.9 \times 10^{10}\ L\ mol^{-1}\ s^{-1}$
(2) $OH + OH \rightarrow H_2O + O$ $k_2 = 1.55 \times 10^9\ L\ mol^{-1}\ s^{-1}$
(3) $O + OH \rightarrow O_2 + H$ $k_3 = 1.1 \times 10^{10}\ L\ mol^{-1}\ s^{-1}$

(J. N. Bradley, W. Hack, K. Hoyermann, and H. G. Wagner, *J. Chem. Soc. Faraday Trans.* I, 1889 (1973)). Using initial H atom and NO_2 concentrations of $4.5 \times 10^{-10}\ mol\ cm^{-3}$ and $5.6 \times 10^{-10}\ mol\ cm^{-3}$, respectively, compute and plot curves showing the O, O_2, and OH concentrations as a function of time in the range 0–10 ns.

26.4. In a flow study of the reaction between O atoms and Cl_2 (J. N. Bradley, D. A. Whytock, and T. A. Zaleski, *J. Chem. Soc. Faraday Trans.* I, 1251 (1973)) at high chlorine pressures, plots of $\ln [O]_0/[O]$ against distances l along the flow tube, where $[O]_0$ is the oxygen concentration at zero chlorine pressure, gave straight lines. Given the flow velocity as $6.66\ m\ s^{-1}$ and the data below, find the rate coefficient for the reaction $O + Cl_2 \rightarrow ClO + Cl$.

l/cm	$\ln([O]_0/[O])$	l/cm	$\ln([O]_0/[O])$
0	0.27	10	0.46
2	0.31	12	0.50
4	0.34	14	0.55
6	0.38	16	0.56
8	0.45	18	0.60

$[O]_0 = 3.3\ 10^{-8}\ mol\ L^{-1}$, $[Cl_2] = 2.54 \times 10^{-7}\ mol\ L^{-1}$, $p = 1.70$ Torr.

26.5. Models of population growth are analogous to chemical reaction rate equations. In the model due to Malthus (1798) the rate of change of the population N of the planet is assumed to be given by $dN/dt = \text{births} - \text{deaths}$. The numbers of births and deaths are proportional to the population, with proportionality constants b and d. Obtain the integrated rate law. How well does it fit the (very approximate) data below on the population of the planet as a function of time?

Year	$N/10^9$	Year	$N/10^9$
1750	0.5	1960	3
1825	1	1974	4
1922	2	1987	5

Theoretical problems

26.6. An autocatalytic reaction $A \rightarrow P$ is observed to have the rate law $d[P]/dt = k[A]^2[P]$. Solve the rate law for initial concentrations $[A]_0$ and $[P]_0$. Calculate the time at which the rate reaches a maximum.

26.7. Another reaction with the stoichiometry $A \rightarrow P$ has the rate law $d[P]/dt = k[A][P]^2$; integrate the rate law for initial concentrations $[A]_0$ and $[P]_0$. Calculate the time at which the rate reaches a maximum.

26.8. The Rice–Herzfeld mechanism for the dehydrogenation of ethane is

$$CH_3CH_3 \rightarrow 2CH_3\cdot, \quad k_a$$
$$CH_3\cdot + CH_3CH_3 \rightarrow CH_4 + CH_3CH_2\cdot, \quad k_b$$
$$CH_3CH_2\cdot \rightarrow CH_2{=}CH_2 + H\cdot, \quad k_c$$
$$H\cdot + CH_3CH_3 \rightarrow H_2 + CH_3CH_2\cdot, \quad k_d$$
$$H\cdot + CH_3CH_2\cdot \rightarrow CH_3CH_3, \quad k_e$$

Show that first-order kinetics are obtained if k_a is small. How may the conditions be changed so that the reaction shows different orders?

26.9. Express the root mean square deviation of the molar mass of a condensation polymer in terms of p, where $\langle M^2 \rangle^{\frac{1}{2}} = \{\langle M^2 \rangle - \langle \bar{M} \rangle^2\}^{\frac{1}{2}}$, and deduce its time dependence.

26.10. Calculate the ratio of the mean cube molar mass to the mean square molar mass in terms of (a) the fraction p, (b) the chain length.

26.11. Derive an expression for the rate of disappearance of a species A in a photochemical reaction for which the mechanism is:

(1) initiation with light of intensity I, $A \rightarrow 2R\cdot$
(2) propagation, $A + R\cdot \rightarrow R\cdot + B$
(3) termination, $R\cdot + R\cdot \rightarrow R_2$

Hence, show that rate measurements will give only a combination of k_2 and k_3 if a steady state is reached, but that both may be obtained if a steady state is not reached. The latter is the basis of the rotating sector method of studying photochemical reactions (see *Further reading*).

26.12. The photochemical chlorination of chloroform in the gas phase has been found to follow the rate law $d[CCl_4]/dt = k[Cl_2]^{\frac{1}{2}}I_a^{\frac{1}{2}}$. Devise a mechanism that leads to this rate law when the chlorine pressure is high.

26.13. Conventional equilibrium considerations do not apply when a reaction is being driven by light absorption. Thus the steady-state concentration of products and reactants might differ significantly from equilibrium values. For instance, suppose the reaction $A \rightarrow B$ is driven by light absorption, and that its rate is I_a, but that the reverse reaction $B \rightarrow A$ is bimolecular and second-order with a rate $k[B]^2$. What is the stationary-state concentration of B? Why does this 'photostationary state' differ from the equilibrium state?

26.14. Write a program for the integration of the Lotka–Volterra equations and arrange for it to plot the concentration of Y against that of X. Explore the consequences of varying the starting concentrations for the integration. Identify the conditions (the concentrations of X and Y) corresponding to the steady state of the Lotka–Volterra equations (the point at the centre of the orbits).

26.15. Set up the overall rate equations for the concentrations of X and Y in terms of the oregonator and explore the periodic properties of the solutions.

26.16. Many biological and biochemical processes involve autocatalytic steps. In the SIR model of the spread and decline of infectious diseases the population is divided into three classes; the susceptibles S, who can catch the disease, the infectives I, who have the disease and can transmit it, and the removed class R, who have either had the disease and recovered, are dead, or are immune or isolated. The model mechanism for this process implies the following rate laws:

$$\frac{dS}{dt} = -rSI \qquad \frac{dI}{dt} = rSI - aI \qquad \frac{dR}{dt} = aI$$

What are the autocatalytic steps of this mechanism? Find the conditions on the ratio a/r that decide whether the disease will spread (an epidemic) or die out. Show that a constant population is built into this system, namely that $S + I + R = N$, meaning that the time-scales of births, deaths by other causes, and migration are assumed large compared to that of the spread of the disease.

27

Molecular reaction dynamics

In this chapter we examine the details of reactions and see how rate constants can be calculated or interpreted in terms of the processes that occur when reactant molecules meet. There are three levels of discussion. The simplest quantitative account of reaction rates is in terms of collision theory. We shall see that collision theory can be used only for the discussion of reactions between simple species in the gas phase and that it fails badly when the reactants are more complex than atoms. Reactions in solution can be classified into those that have rates that depend on the speed at which molecules diffuse through the solution and those that have rates that depend on the rate at which energy accumulates in a pair of molecules that have encountered one another. The former can be expressed quantitatively in terms of the diffusion equation.

The next level of discussion is in terms of activated complex theory, where it is assumed that the reactant molecules form a complex that can be discussed in terms of the population of its energy levels. Activated complex theory inspires a thermodynamic approach to reaction rates, in which the rate constant is expressed in terms of thermodynamic parameters. This approach is quite useful for parametrizing the rates of reactions in solution.

The highest level of sophistication is in terms of potential energy surfaces and the motion of molecules over these surfaces. As we shall see, such an approach gives a very intimate picture of the events that occur when reactions occur, and is open to experimental study through molecular beam techniques.

Now we are at the heart of chemistry. Here we examine the details of what happens to molecules at the climax of reactions. Extensive changes of structure are taking place and energies the size of dissociation energies are being redistributed among bonds: old bonds are being ripped apart and new bonds are being formed.

As may be imagined, the calculation of the rates of such processes from first principles is very difficult. Nevertheless, like so many intricate problems, the broad features can be established quite simply, and only when we enquire more deeply do the complications emerge. In this chapter we look at three levels of approach to the calculation of a rate constant for an elementary bimolecular reaction. Although a great deal of information can be obtained from gas-phase reactions, most reactions of interest take place in solution, and we shall also see to what extent their rates can be predicted.

REACTIVE ENCOUNTERS

In this section we consider two elementary approaches to the calculation of reaction rates, one relating to gas-phase reactions and the other to reactions in solution. Both approaches are based on the view that reactant molecules must meet, and that reaction takes place only if they have a certain minimum energy. In the collision theory of bimolecular gas-phase reactions, which we mentioned briefly in Section 25.5, products are formed only if the collision is sufficiently energetic. If the collision is insufficiently energetic, then the colliding reactant molecules separate again. In solution, on the other hand, the reactant molecules may diffuse together and *then* acquire the necessary energy from their immediate surroundings while they are in contact.

27.1 Collision theory

We shall consider the bimolecular elementary reaction

$$A + B \rightarrow P \qquad v = k_2[A][B]$$

where P denotes products, and aim to calculate the second-order rate constant k_2.

The basic calculation

The **collision density** Z_{AB} is the number of (A, B) collision per unit volume per unit time. The frequency of collisions of a *single* molecule in a gas was calculated in Section 1.3. As shown in the following *Justification*, that result can be adapted to derive the following expression for the collision density:

$$Z_{AB} = \sigma \left(\frac{8kT}{\pi\mu} \right)^{\frac{1}{2}} N_A^2 [A][B] \tag{1}$$

In this expression σ is the collision cross-section (Fig. 27.1)

$$\sigma = \pi d^2, \qquad d = \tfrac{1}{2}(d_A + d_B)$$

and μ is the reduced mass of A and B:

$$\frac{1}{\mu} = \frac{1}{m_A} + \frac{1}{m_B}, \qquad \text{so } \mu = \frac{m_A m_B}{m_A + m_B}$$

Collision densities may be very large. For example, in nitrogen at room temperature and pressure, with $d = 280$ pm, $Z = 5 \times 10^{34}\,\text{s}^{-1}\,\text{m}^{-3}$.

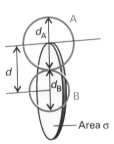

27.1 The collision cross-section for two molecules can be considered to be the area within which the projectile molecule (A) must enter around the target molecule (B) in order for a collision to occur. If the diameters of the two molecules are d_A and d_B, the radius of the target area is $d = \tfrac{1}{2}(d_A + d_B)$ and the cross-section is πd^2.

JUSTIFICATION

The collision frequency z for a single A molecule with the B molecules present is given by eqn 1.17a modified to the case of two dissimilar molecules:

$$z = \sigma \bar{c} N_A [B]$$

The number of B molecules per unit volume has been written as $N_A[B]$, where [B] is their molar concentration and N_A is Avogadro's constant. The mean *relative* speed of A and B molecules is

$$\bar{c}_{rel} = \left(\frac{8kT}{\pi\mu}\right)^{\frac{1}{2}}, \qquad \text{where } \frac{1}{\mu} = \frac{1}{m_A} + \frac{1}{m_B}$$

This expression reduces to eqn 1.17a when A and B are identical. The rate of collisions of all the A molecules in the gas with the B molecules is z multiplied by the number of A molecules present. The collision density is then obtained by dividing the resulting expression by the volume. Because the number of A molecules per unit volume can be expressed as $N_A[A]$, where [A] is the molar concentration of A and N_A is Avogadro's constant, we obtain

$$Z_{AB} = z N_A [A] = \sigma \bar{c} N_A^2 [A][B]$$

Substitution of the expression above for the mean relative speed gives eqn 1.

According to collision theory, the rate of change in the molar concentration of A molecules per unit time is the product of the collision density and the probability that a collision occurs with sufficient energy:

$$\frac{d[A]}{dt} = -\frac{Z_{AB} \times f}{N_A}$$

Avogadro's constant has been introduced to convert the rate of change in the number of A molecules per unit volume to the rate of change of molar concentration of A. In this expression f is the fraction of collisions that occur with a kinetic energy along the line of approach in excess of a certain minimum value E_a, the activation energy of the reaction. This fraction is given by the Boltzmann distribution as

$$f = e^{-E_a/RT} \tag{2}$$

JUSTIFICATION

The Boltzmann distribution over a continuous range of energies is

$$f(E) = \frac{1}{kT} e^{-E/kT}$$

where the factor $1/kT$ ensures that the distribution is normalized to 1.

It follows that the fraction with an energy in excess of E_a is

$$f = \int_{E_a}^{\infty} f(E) \, dE = \frac{1}{kT} \int_{E_a}^{\infty} e^{-E/kT} \, dE = e^{-E_a/kT}$$

Then, with E_a interpreted as a molar quantity, which requires k to be replaced by R, we obtain eqn 2.

At this stage we can combine eqns 1 and 2 and write

$$\frac{d[A]}{dt} = -\sigma \left(\frac{8kT}{\pi \mu} \right)^{\frac{1}{2}} N_A e^{-E_a/RT} [A][B]$$

It follows that

$$k_2 = \sigma \left(\frac{8kT}{\pi \mu} \right)^{\frac{1}{2}} N_A e^{-E_a/RT} \tag{3a}$$

Equation 3a has the Arrhenius form

$$k_2 = A e^{-E_a/RT} \tag{3b}$$

so long as the exponential temperature dependence dominates the weak square-root temperature dependence of the pre-exponential factor.[1] It follows that the activation energy E_a can be identified with the minimum kinetic energy along the line of approach that is needed for reaction, and that the pre-exponential factor is a measure of the rate at which collisions occur in the gas.

Comparison with experiment

The simplest procedure for calculating k_2 is to use for σ the values obtained for non-reactive collisions (e.g. typically those obtained from viscosity measurements) or from tables of molecular radii. Table 27.1

Table 27.1* Arrhenius parameters for gas-phase reactions

	$A/(\text{L mol}^{-1}\,\text{s}^{-1})$ Experiment	Theory	$E_a/(\text{kJ mol}^{-1})$	P
$2NOCl \rightarrow 2NO + 2Cl$	9.4×10^9	5.9×10^{10}	102.0	0.16
$2ClO \rightarrow Cl_2 + O_2$	6.3×10^7	2.5×10^{10}	0.0	2.5×10^{-3}
$H_2 + C_2H_4 \rightarrow C_2H_6$	1.24×10^6	7.3×10^{11}	180.0	1.7×10^{-6}
$K + Br_2 \rightarrow KBr + Br$	1.0×10^{12}	2.1×10^{11}	0.0	4.8

* More values are given in the Data section at the end of this volume.

1 From the general definition of activation energy in eqn 25.13a:

$$RT^2 \frac{d \ln k_a}{dT} = E_a + \tfrac{1}{2} RT$$

so the activation energy is weakly temperature-dependent. Generally, $E_a \gg \tfrac{1}{2} RT$, and the $\tfrac{1}{2} RT$ can be neglected.

compares some values of the pre-exponential factor A calculated in this way with values obtained from Arrhenius plots of $\ln k$ against $1/T$ (Section 25.5). One of the reactions shows fair agreement between theory and experiment, but for others there are major discrepancies. In some cases the experimental values are orders of magnitude smaller than those calculated, which suggests that the collision energy is not the only criterion for reaction and that some other feature, such as the relative orientation of the colliding species, is important. Moreover, one reaction in the table has a pre-exponential factor larger than theory, which seems to indicate that the reaction occurs more quickly than the particles collide!

The steric requirement

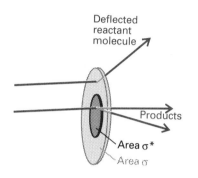

Deflected
reactant
molecule

Products

Area σ^*

Area σ

27.2 The collision cross-section is the target area that results in simple deflection of the projectile molecule; the reaction cross-section is the corresponding area for chemical change to occur on collision.

We can express the disagreement between experiment and theory by replacing σ in eqn 1 by the **reactive cross-section** σ^*. The reactive cross-section can be regarded as the target area for reaction, whereas σ itself is the target area of a simple non-reactive collision (Fig. 27.2). Sometimes it is convenient to express σ^* as a multiple of σ by introducing a **steric factor** P and writing

$$\sigma^* = P\sigma \tag{4}$$

Then the rate constant becomes

$$k_2 = P\sigma\left(\frac{8kT}{\pi\mu}\right)^{\frac{1}{2}}N_A e^{-E_a/RT} \tag{5}$$

The steric factor is normally found to be several orders of magnitude smaller than 1.

Example 27.1 *Estimating a steric factor (1)*

Estimate the steric factor for the reaction $H_2 + C_2H_4 \rightarrow C_2H_6$ at 628 K given that the pre-exponential factor is $1.24 \times 10^6 \, \text{L mol}^{-1}\,\text{s}^{-1}$.

Method. To calculate P, we need to calculate the pre-exponential factor A by using eqn 5 and then compare the answer with experiment: the ratio is P. Collision cross-sections for non-reactive encounters are listed in Table 1.2. The best way to estimate the collision cross-section for dissimilar spherical species is to calculate the collision diameter for each one (from $\sigma = \pi d^2$), to calculate the mean of the two diameters, and then to calculate the cross-section for that mean diameter. However, as neither species is spherical, a simpler but more approximate procedure is just to take the average of the two collision cross-sections.

Answer. The reduced mass of the colliding pair is

$$\mu = \frac{m_1 m_2}{m_1 + m_2} = 3.12 \times 10^{-27} \, \text{kg}$$

because $M_1 = 2.016 \, \text{g mol}^{-1}$ for H_2 and $M_2 = 28.05 \, \text{g mol}^{-1}$ for C_2H_4.

Hence

$$\left(\frac{8kT}{\pi\mu}\right)^{\frac{1}{2}} = 2.66 \times 10^3 \, \text{m s}^{-1}$$

From Table 1.2, $\sigma(H_2) = 0.27 \, \text{nm}^2$ and $\sigma(C_2H_4) = 0.64 \, \text{nm}^2$, giving a mean collision cross-section of $\sigma = 0.46 \, \text{nm}^2$. Therefore,

$$A = P\sigma\left(\frac{8kT}{\pi\mu}\right)^{\frac{1}{2}} N_A = P \times 7.37 \times 10^{11} \, \text{L mol}^{-1} \, \text{s}^{-1}$$

Experimentally $A = 1.24 \times 10^6 \, \text{L mol}^{-1} \, \text{s}^{-1}$, so it follows that $P = 1.7 \times 10^{-6}$.

Comment. The very small value of P is one reason why catalysts are needed to bring this reaction about at a reasonable rate. As a general guide, the more complex the molecules, the smaller the value of P.

Exercise E27.1. It is found for the reaction $NO + Cl_2 \rightarrow NOCl + Cl$ that $A = 4.0 \times 10^9 \, \text{L mol}^{-1} \, \text{s}^{-1}$ at 298 K. Use $\sigma(NO) = 0.42 \, \text{nm}^2$ and $\sigma(Cl_2) = 0.93 \, \text{nm}^2$ to estimate the P factor for the reaction.   [0.02]

If we use eqn 5 to write the rate as

$$k_2[A][B] = P \times Z_{AB} \times f$$

then we see the three quantities that govern its magnitude. The factor f is the energy criterion. The factor Z_{AB} is the transport property, which governs how often the particles come together. The factor P takes care of the local properties of the reaction, such as the orientations necessary for reaction and the details of how close they must come.

An example of a reaction for which it is possible to estimate the steric factor is

$$K + Br_2 \rightarrow KBr + Br$$

for which the experimental value of P is 4.8. In this reaction, the distance of approach at which reaction occurs appears to be considerably larger than the distance needed for deflection of the path of the approaching molecules in a non-reactive collision. It has been proposed that the reaction proceeds by a **harpoon mechanism**. This brilliant name is based on a model of the reaction which pictures the K atom as approaching a Br_2 molecule; when the two are close enough an electron (the harpoon) flips across from K to Br_2. In place of two neutral particles there are now two ions, so there is a Coulombic attraction between them: this attraction is the line on the harpoon. Under its influence the ions move together (the line is wound in), the reaction takes place, and $KBr + Br$ emerge. The harpoon extends the cross-section for the reactive encounter, and the reaction rate is greatly underestimated by taking for the collision cross-section the value for simple mechanical contact between $K + Br_2$.

Example 27.2 *Estimating a steric factor (2)*

Estimate the value of P for the harpoon mechanism by calculating the distance at which it is energetically favourable for the electron to leap from K to Br_2.

Method. We should begin by identifying all the contributions to the energy of interaction between the colliding species. There are three contributions to the energy of the process $K + Br_2 \rightarrow K^+ + Br_2^-$. The first is the ionization energy I of K. The second is the electron affinity E_{ea} of Br_2. The third is the Coulombic interaction energy between the ions when they have been formed: when their separation is R this energy is $-e^2/4\pi\varepsilon_0 R$. The electron flips across when the sum of these three contributions changes from positive to negative (that is, when the sum is zero).

Answer. The net change in energy when the transfer occurs at a separation R is

$$E = I - E_{ea} - \frac{e^2}{4\pi\varepsilon_0 R}$$

The ionization energy I is larger than E_{ea}, so E becomes negative only when R has decreased to less than some critical value $R*$ given by

$$\frac{e^2}{4\pi\varepsilon_0 R*} = I - E_{ea}$$

When the particles are at this separation, the harpoon shoots across, so we can identify the reactive cross-section as

$$\sigma* = \pi R*^2$$

This value of $\sigma*$ implies that the steric factor is

$$P = \frac{\sigma*}{\sigma} = \frac{R*^2}{d^2} = \left(\frac{e^2}{4\pi\varepsilon_0 d(I - E_{ea})} \right)^2$$

where $d = R(K) + R(Br_2)$. With $I = 420 \text{ kJ mol}^{-1}$, $E_{ea} \approx 250 \text{ kJ mol}^{-1}$, and $d = 400$ pm, we find $P \approx 12$, which is consistent with the experimental value.

Exercise E27.2. Estimate the value of P for the harpoon reaction between Na and Cl_2 for which $d \approx 350$ pm; take $E_{ea} \approx 230 \text{ kJ mol}^{-1}$.   [6]

Example 27.2 illustrates two points about steric factors. First, the concept of a steric factor is not wholly useless because in some cases its numerical value can be estimated. Second (and more pessimistically) most reactions are much more complex than $K + Br_2$, and we cannot expect to obtain P so easily. What we need is a more powerful theory that lets us calculate, and not merely guess, its value. We go some way to setting up that theory in Section 27.4.

27.2 Diffusion-controlled reactions

Encounters between reactants in solution occur in a very different manner from encounters in gases. Particles have to jostle their way through the solvent, so their encounter frequency is considerably less than in a gas. However, because a particle also migrates only slowly away from a location, two particles encountering each other stay near each other for much longer than in a gas, and such an **encounter pair** may accumulate enough energy to react even though it does not have enough energy to do so when it is first formed. The activation energy of a reaction is a much more complicated quantity in solution than in a gas because the encounter pair is surrounded by solvent and the energy of the entire local assembly of reactant and solvent molecules must be considered.

Classes of reaction

The complicated overall process can be divided into simpler parts by setting up a simple kinetic scheme. We suppose that the rate of formation of an encounter pair AB is second-order in the reactants A and B:

$$A + B \rightarrow AB \qquad v = k_d[A][B]$$

As we shall see, k_d (where the d signifies diffusion) is determined by the diffusional characteristics of A and B. The encounter pair can break up without reaction or it can go on to form products P. If we suppose that both processes are pseudofirst-order reactions (with the solvent perhaps playing a role), then we can write

$$AB \rightarrow A + B \qquad v = k_d'[AB]$$
$$AB \rightarrow P \qquad v = k_a[AB]$$

The steady-state concentration of AB can now be found from the equation for the net rate of change of concentration of AB:

$$\frac{d[AB]}{dt} = k_d[A][B] - k_d'[AB] - k_a[AB] = 0$$

This expression solves to

$$[AB] = \frac{k_d[A][B]}{k_a + k_d'}$$

The overall rate law for the formation of products is therefore

$$\frac{d[P]}{dt} = k_a[AB] = k_2[A][B], \qquad \text{where } k_2 = \frac{k_a k_d}{k_a + k_d'} \qquad (6)$$

Two limits can now be distinguished. If the rate of separation of the unreacted encounter pair is much slower than the rate at which it forms products, then $k_d' \ll k_a$ and the effective rate constant is

$$k_2 \approx \frac{k_a k_d}{k_a} = k_d \qquad (7a)$$

In this **diffusion-controlled limit** the rate of reaction is governed by the rate at which the reactant particles diffuse through the medium. An indication that a reaction is diffusion-controlled is that its rate constant is of the order of $10^9 \, L \, mol^{-1} \, s^{-1}$ or greater. Because the combination of radicals involves very little activation energy, radical and atom recombination reactions are often diffusion-controlled.

An **activation-controlled reaction** arises when a substantial activation energy is involved in the reaction of AB. Then $k_a \ll k_d'$ and

$$k_2 \approx \frac{k_a k_d}{k_d'} = k_a K \tag{7b}$$

where K is the equilibrium constant for $A + B \rightleftharpoons AB$. In this limit, the reaction proceeds at the rate at which energy accumulates in the encounter pair from the surrounding solvent. Some experimental data are given in Table 27.2.

Table 27.2* Arrhenius parameters for reactions in solution

	Solvent	$A/(L \, mol^{-1} \, s^{-1})$	$E_a/(kJ \, mol^{-1})$
$(CH_3)_3CCl$ solvolysis	Water	7.1×10^{16}	100
	Ethanol	3.0×10^{13}	112
	Chloroform	1.4×10^4	45
$CH_3CH_2Br + OH^-$	Ethanol	4.3×10^{11}	89.5

* More values are given in the Data section.

Diffusion and reaction

The rate of a diffusion-controlled reaction is calculated by considering the rate at which the reactants diffuse together. As shown in the following *Justification*, the rate constant for a diffusion-controlled reaction in which the two reactant molecules react if they come within a distance R^* of one another is

$$k_d = 4\pi R^* D N_A \tag{8}$$

where D is the sum of the diffusion coefficients of the two reactant species in the solution.

JUSTIFICATION

According to the diffusion equation $(\partial[B]/\partial t = D_B \nabla^2[B])$, the concentration of B when the system has reached a steady state $(\partial[B]/\partial t = 0)$ satisfies $\nabla^2[B]_r = 0$, where the subscript r signifies a quantity that varies with the distance r, subject to the boundary conditions that $[B]_r$ has its bulk value $[B]$ as $r \to \infty$ and is zero at $r = R^*$, the distance at which reaction occurs. For a spherically symmetrical system, ∇^2 can be replaced by radial derivatives alone (see Table 11.1), so the equation satisfied by

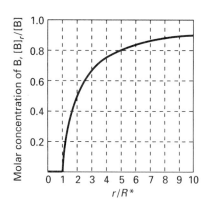

27.3 The concentration profile for reaction in solution when a molecule B diffuses towards another reactant molecule and reacts if it reaches R^*.

$[B]_r$ is

$$\frac{d^2[B]_r}{dr^2} + \frac{2}{r}\frac{d[B]_r}{dr} = 0$$

The general solution of this equation is

$$[B]_r = a + \frac{b}{r}$$

The boundary conditions then enable us to identify the constants as $a = [B]$, the bulk concentration of B, and $b = -R^*[B]$, and hence

$$[B]_r = \left(1 - \frac{R^*}{r}\right)[B]$$

The variation of concentration expressed by this equation is illustrated in Fig. 27.3.

The rate of reaction is the (molar) flux, J, of the reactant B towards A multiplied by the area of the spherical surface of radius R^*:

$$\text{rate of reaction} = 4\pi R^{*2}J$$

From Fick's first law (eqn 24.5), the flux towards A is proportional to the concentration gradient at a radius R^*:

$$J = D_B\left(\frac{d[B]_r}{dr}\right)_{\text{at } r=R^*} = D_B\frac{[B]}{R^*}$$

(A sign change has been introduced because we are interested in the flux towards decreasing values of r.) When this condition is substituted into the previous equation we obtain

$$\text{rate of reaction} = 4\pi R^* D_B[B]$$

The rate of the diffusion-controlled reaction is equal to the average flow of B molecules to all the A molecules in the sample. If the bulk concentration of A is $[A]$, then the number of A molecules in the sample of volume V is $N_A[A]V$; the global flow of all B to all A is therefore $4\pi R^* D_B N_A[A][B]V$. It is unrealistic to suppose that all A are stationary; this feature is removed by replacing D_B by the sum of the diffusion coefficients of the two species and writing $D = D_A + D_B$. Then the rate of change of concentration of AB is

$$\frac{d[AB]}{dt} = 4\pi R^* D N_A[A][B]$$

Hence, the diffusion-controlled rate constant is

$$k_d = 4\pi R^* D N_A$$

which is eqn 8.

Equation 8 can be taken further by incorporating the Stokes–Einstein equation (Section 24.10) relating the diffusion constant and the hydrodynamic radius R_A and R_B of each molecule in a medium of viscosity η:

$$D_A = \frac{kT}{6\pi\eta R_A} \qquad D_B = \frac{kT}{6\pi\eta R_B}$$

As these relations are approximate, little extra error is introduced if we write $R_A = R_B = \frac{1}{2}R^*$, which leads to

$$k_d = \frac{8RT}{3\eta} \tag{9}$$

(The R in this equation is the gas constant.) The radii have cancelled because, although the diffusion constants are smaller when the radii are large, the reactive collision radius is larger and the particles need travel a shorter distance to meet. In this approximation, the rate constant is independent of the identities of the reactants, and depends only on the temperature and the viscosity of the solvent. For example, the rate constant for the recombination of I atoms in hexane at 298 K, when the viscosity of the solvent is 0.326 cP (with $1\,\text{cP} = 10^{-3}\,\text{kg m}^{-1}\,\text{s}^{-1}$) is

$$k_d = \frac{8 \times (8.3145\,\text{J K}^{-1}\,\text{mol}^{-1}) \times (298\,\text{K})}{3 \times (0.326 \times 10^{-3}\,\text{kg m}^{-1}\,\text{s}^{-1})}$$
$$= 2.0 \times 10^7\,\text{m}^3\,\text{mol}^{-1}\,\text{s}^{-1}$$

Because $1\,\text{m}^3 = 10^3\,\text{L}$, this result corresponds to $2.0 \times 10^{10}\,\text{L mol}^{-1}\,\text{s}^{-1}$. The experimental value is $1.3 \times 10^{10}\,\text{L mol}^{-1}\,\text{s}^{-1}$, so the agreement is very good considering the approximations involved.

27.3 The material-balance equation

The diffusion of reactants plays an important role in many chemical processes, such as the diffusion of O_2 molecules into red blood corpuscles and the diffusion of a gas towards a catalyst. We can have a glimpse of the kinds of calculations involved by considering the generalized diffusion equation, Section 24.11, but generalized even more to take into account the possibility that the diffusing, convecting molecules are also reacting.

The formulation of the equation

Consider a small volume element in a chemical reactor (or a biological cell). During a small time interval Δt, the total change of the number of molecules of a species J is

Net change of number of J in the volume element

= Number entering − Number leaving

+ Number formed by reaction

− Number consumed by reaction

The rate of change of the concentration is this expression divided by the volume and the time interval. We shall now show how to express the relation as a differential equation.

The net rate at which J molecules enter the region by diffusion and convection is given by eqn 24.42:

$$\frac{\partial [J]}{\partial t} = D\frac{\partial^2 [J]}{\partial x^2} - v\frac{\partial [J]}{\partial x}$$

The net rate of change due to chemical reaction is

$$\frac{\partial [J]}{\partial t} = -k[J]$$

if we suppose that J disappears by a pseudofirst-order reaction. Therefore, the equation becomes

$$\frac{\partial [J]}{\partial t} = D\frac{\partial^2 [J]}{\partial x^2} - v\frac{\partial [J]}{\partial x} - k[J] \tag{10}$$

A similar equation, with different reaction terms, applies to each species present. Equation 10 is called the **material-balance equation**.

Solutions of the equation

The material-balance equation is a second-order partial differential equation, and it is far from easy to solve in general. Some idea of how it is solved can be obtained by considering the special case in which there is no convective motion (as in an unstirred reaction vessel):

$$\frac{\partial [J]}{\partial t} = D\frac{\partial^2 [J]}{\partial x^2} - k[J] \tag{11}$$

If the solution in the absence of reaction (i.e. $k = 0$ in this equation) is [J], then it is easy to show that the solution in the presence of reaction ($k > 0$) is

$$[J]^* = k\int_0^t [J]e^{-kt}\,dt + [J]e^{-kt} \tag{12}$$

We have already met one solution of the diffusion equation in the absence of reaction: eqn 24.43 is the solution for a system in which initially a layer of $n_0 N_A$ molecules are spread over a plane of area A:

$$[J] = \frac{n_0 e^{-x^2/4Dt}}{A(\pi Dt)^{\frac{1}{2}}}$$

When this expression is substituted into eqn 12 and the integral is evaluated, we obtain the concentration of J as it diffuses away from its initial surface layer and undergoes reaction in the solution above (Fig. 27.4).

Even this relatively simple example has led to a difficult equation to compute, and only in some special cases can the full material-balance equation be solved analytically. Most modern work on reactor design and cell kinetics now uses numerical methods to solve the equation, and detailed solutions for realistic environments can be obtained reasonably easily. Some of the interesting applications include the exploration of the spatial periodicity of the reactions mentioned in Section 26.8.

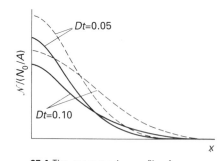

27.4 The concentration profiles for a diffusing, reacting system (e.g. a column of solution) in which one reactant is initially in a layer at $x = 0$. In the absence of reaction (broken lines) the concentration profiles are the same as in Fig. 24.28. (Arbitrary values for D and k have been taken.)

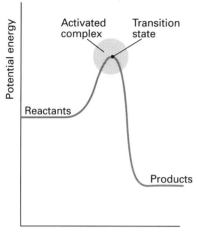

27.5 A reaction profile. The horizontal axis is the reaction coordinate, and the vertical axis is potential energy. The activated complex is the region near the potential maximum, and the transition state corresponds to the maximum itself.

ACTIVATED COMPLEX THEORY

We shall now consider a more detailed calculation of rate constants using the concepts of statistical thermodynamics developed in Chapter 20. The approach we describe, which is called **activated complex theory** (ACT), has the advantage that a quantity corresponding to the steric factor appears automatically, and P does not need to be grafted on to an equation as an afterthought. Activated complex theory is an attempt to identify the principal features governing the size of a rate constant in terms of a model of the events that take place during the reaction.

27.4 The reaction coordinate and the transition state

The general features of how the potential energy of the reactants A and B change in the course of a bimolecular elementary reaction are shown in Fig. 27.5. The horizontal axis of the diagram represents the course of the individual reaction event (a bimolecular collision in a gas-phase reaction) and is called the **reaction coordinate**.

Initially, only the reactants A and B are present. As the reaction event proceeds, A and B come into contact, distort, and begin to exchange or discard atoms. The potential energy rises to a maximum, and the cluster of atoms that corresponds to the region close to the maximum is called the **activated complex**. After the maximum, the potential energy falls as the atoms rearrange in the cluster, and it reaches a value characteristic of the products. The climax of the reaction is at the peak of the potential energy. Here two reactant molecules have come to such a degree of closeness and distortion that a small further distortion will send them in the direction of products. This crucial configuration is called the **transition state** of the reaction. Although some molecules entering the transition state might revert to reactants, if they pass through this configuration then it is inevitable that products will emerge from the encounter.

As an example, consider the approach of an H atom to an F_2 molecule. For simplicity, we imagine the approach as occurring along the F—F bond direction. At great distances apart, the potential energy is the sum of the potential energies of H and F_2. When H and F_2 are so close that their orbitals start to overlap, the F—F bond begins to stretch and a bond begins to form between H and the nearer F atom. The H atom comes closer, the F—F bond lengthens, the H—F bond shortens and strengthens, and the atoms enter the range of locations characteristic of the activated complex. There comes a stage when the activated complex has maximum potential energy and is poised at the transition state. An infinitesimal compression of the H—F bond and a stretch of F—F take the complex through the transition state. Points further along the reaction coordinate represent stages at which the H—F bond forms more fully and the F—F bond breaks. Motion along the reaction coordinate from left to right therefore represents the progress of H and F_2 through these configurations. Whether or not a colliding pair actually crosses the potential barrier depends on the kinetic energy the molecules have initially, for they must be able to climb the barrier and attain the transition state.

In an actual reaction, H atoms approach F_2 molecules from all angles and the specification of the reaction coordinate is a subtle problem. We shall therefore regard the coordinate simply as an indication of the distortions in the reactant molecules as the activated complex is formed, the critical transition state is reached, and the product molecules emerge. At the transition state, motion along the reaction coordinate corresponds to a complicated collective vibration-like motion of all the atoms in the complex. An order-of-magnitude estimate of the lifetime of a transition state is that it is about the period of vibration of the breaking and forming bonds.

27.5 The Eyring equation

Activated complex theory pictures a reaction between A and B as proceeding through the formation of an activated complex $C^{\ddagger}$ that falls apart by unimolecular decay into products P with a rate constant $k^{\ddagger}$:

$$C^{\ddagger} \to P \qquad v = k^{\ddagger}[C^{\ddagger}]$$

The concentration of the activated complex is likely to be proportional to the concentrations of the reactants, and later we show explicitly that

$$[C^{\ddagger}] = K^{\ddagger}[A][B]$$

where $K^{\ddagger}$ is a proportionality constant (with dimensions 1/concentration). It follows that

$$v = k_2[A][B], \qquad \text{where } k_2 = k^{\ddagger}K^{\ddagger}$$

Our task is to calculate the unimolecular rate constant $k^{\ddagger}$ and the proportionality constant $K^{\ddagger}$.

The rate of decay of the activated complex

The activated complex can form products if it passes through the transition state. If its vibration-like motion along the reaction coordinate occurs with a frequency ν, then the frequency with which the cluster of atoms forming the complex approaches the transition state is also ν. However, it is possible that not every oscillation along the reaction coordinate takes the complex through the transition state. For instance, the centrifugal effect of rotations might also be an important contribution to the break-up of the complex, and in some cases the complex might be rotating too slowly or rotating rapidly but about the wrong axis. Therefore, we suppose that the rate of passage of the complex through the transition state is proportional to the vibrational frequency along the reaction coordinate, and write

$$k^{\ddagger} = \kappa \nu \tag{13}$$

where κ is the **transmission coefficient**. In many cases $\kappa \approx 1$.

The concentration of the activated complex

The simplest procedure for estimating the concentration of the activated complex is to assume that there is a pre-equilibrium between the reactants and the complex,[2] and to write

$$A + B \rightleftharpoons C^{\ddagger} \qquad K = \frac{(p_C/p^{\ominus})}{(p_A/p^{\ominus})(p_B/p^{\ominus})} = \frac{p_C p^{\ominus}}{p_A p_B}$$

The partial pressures p_J can be expressed in terms of the concentrations by using $p_J = RT[J]$, which gives

$$[C^{\ddagger}] = K \times \frac{RT}{p^{\ominus}}[A][B]$$

from which it follows that

$$K^{\ddagger} = \frac{RT}{p^{\ominus}} \times K$$

The remaining task is to calculate the equilibrium constant K.

We saw in Section 20.7 how to calculate equilibrium constants from structural data. Equation 20.34 of that section can be used directly, which in this case gives

$$K = \frac{N_A q_C^{\ominus}}{q_A^{\ominus} q_B^{\ominus}} \times e^{-\Delta E_0/RT}$$

where ΔE_0 is the molar energy separation between the zero-point levels of $C^{\ddagger}$ and $A + B$:

$$\Delta E_0 = E_0(C^{\ddagger}) - E_0(A) - E_0(B)$$

and the $q_J^{\ominus}$ are the standard molar partition functions as defined in Section 20.1. Note that the units of N_A and q_J are both mol^{-1}, so K is dimensionless.

In the final step of this part of the calculation, we focus attention on the partition function of the activated complex. We have already assumed that a vibration of the activated complex $C^{\ddagger}$ along the reaction coordinate tips it through the transition state. The partition function for this vibration is

$$q = \frac{1}{1 - e^{-h\nu/kT}}$$

where ν is its frequency (the same frequency that determines $k^{\ddagger}$). This frequency is much lower than for an ordinary molecular vibration because the oscillation corresponds to the complex falling apart (and so

---

2 In earlier editions of this text a different line of argument was used: we recognized that almost nothing is known about the populations of the levels of the activated complex. Since nothing is known, the most honest approach is to assume that the populations depend on the energy and not on the identity of the level (i.e. whether it belongs to A, B, or $C^{\ddagger}$). This 'least-prejudiced' approach leads to the same results as here, but has the advantage that it avoids the presumption of equilibrium between the reactants and their activated complex. Its disadvantage is that it is slightly less direct. The second edition should be consulted for details.

the force constant is very low). Therefore, because $h\nu/kT \ll 1$, the exponential may be expanded and the partition function reduces to

$$q = \frac{1}{1 - \left(1 - \dfrac{h\nu}{kT} + \ldots\right)} \approx \frac{kT}{h\nu}$$

We can therefore write

$$q_C = \frac{kT}{h\nu} \times \bar{q}_C$$

where $\bar{q}$ denotes the partition function for all the other modes in the complex. The constant $K^{\ddagger}$ is therefore

$$K^{\ddagger} = \frac{kT}{h\nu} \times \bar{K} \tag{14a}$$

where

$$\bar{K} = \frac{RT}{p^{\ominus}} \times \frac{N_A \bar{q}_C^{\ominus}}{q_A^{\ominus} q_B^{\ominus}} \times e^{-\Delta E_0/RT} \tag{14b}$$

with $\bar{K}$ a kind of equilibrium constant, but with one vibrational mode of $C^{\ddagger}$ discarded.

The rate constant

We can now combine all the parts of the calculation into

$$k_2 = k^{\ddagger} K^{\ddagger} = \kappa \nu \times \frac{kT}{h\nu} \times \bar{K}$$

At this stage the unknown frequencies ν cancel and we obtain the **Eyring equation**:

$$k_2 = \kappa \frac{kT}{h} \times \bar{K} \tag{15}$$

The factor $\bar{K}$ is given by eqn 14 in terms of the partition functions of A, B, and $C^{\ddagger}$, so, in principle, we now have an explicit expression for calculating the second-order rate constant for a bimolecular reaction in terms of the molecular parameters for the reactants and the activated complex and the quantity κ (which is often set equal to 1).

The partition functions for the reactants can normally be calculated quite readily, using either spectroscopic information about their energy levels or the approximate expressions set out in Table 20.2. The difficulty with the Eyring equation, however, lies in the calculation of the partition function of the activated complex, for $C^{\ddagger}$ cannot normally be investigated spectroscopically and in general we need to make assumptions about its size, shape, and structure. We shall illustrate what is involved for two simple cases.

The collision of structureless particles

As a first example, consider the case of two structureless particles A and B colliding to give an activated complex that resembles a diatomic molecule. Because the reactants J = A, B are structureless 'atoms' the only contributions to their partition functions are the translational terms:

$$q_J^{\ominus} = \frac{V_m^{\ominus}}{\Lambda_J^3}, \quad \text{with } \Lambda_J = \frac{h}{(2\pi m_J kT)^{\frac{1}{2}}}, \quad V_m^{\ominus} = \frac{RT}{p^{\ominus}}$$

The activated complex is a diatomic cluster of mass $m_C = m_A + m_B$ and moment of inertia I. It has one vibrational mode, but that mode corresponds to motion along the reaction coordinate and therefore does not appear in $\bar{q}_C$. It follows that the molar partition function of the activated complex is

$$\bar{q}_C^{\ominus} = \frac{2IkT}{\hbar^2} \times \frac{V_m^{\ominus}}{\Lambda_C^3}$$

The moment of inertia of a diatomic molecule of bond length r is μr^2, where μ is the effective mass, so the expression for the rate constant is

$$
\begin{aligned}
k_2 &= \kappa \times \frac{kT}{h} \times \frac{RT}{p^{\ominus}} \times \frac{N_A \Lambda_A^3 \Lambda_B^3}{\Lambda_C^3 V_m^{\ominus}} \times \frac{2IkT}{\hbar^2} \times e^{-\Delta E_0/RT} \\
&= \kappa \frac{kT}{h} \times N_A \left(\frac{\Lambda_A \Lambda_B}{\Lambda_C}\right)^3 \left(\frac{2IkT}{\hbar^2}\right) \times e^{-\Delta E_0/RT} \\
&= \kappa N_A \left(\frac{8kT}{\pi\mu}\right)^{\frac{1}{2}} \pi r^2 e^{-\Delta E_0/RT}
\end{aligned}
$$

Finally, by identifying the reactive cross-section σ^* as $\kappa\pi r^2$, we arrive at precisely the same expression as that obtained from simple collision theory (eqn 5).

Example 27.3 *Estimating the steric factor from the Eyring equation*

Obtain an order-of-magnitude estimate of the P factor for the reaction between two non-linear molecules.

Method. We need to use the Eyring equation twice, first for two structureless molecules and then for two molecules with internal structure (and which can rotate and vibrate). For the order-of magnitude estimate, assume that all the translational partition functions are the same. Likewise, also assume that all rotational and all vibrational partition functions (q_r and q_v, respectively) are the same. An N-atomic non-linear molecule has three translational, three rotational, and $3N - 6$ vibrational modes.

Answer. For no internal modes of the colliding molecules,

$$q_A^{\ominus} = q_B^{\ominus} = \frac{V_m^{\ominus}}{\Lambda^3} \qquad \bar{q}_C^{\ominus} = q_r^2 \times \frac{V_m^{\ominus}}{\Lambda^3}$$

Therefore, from eqn 15:

$$k_2 = \kappa \frac{kT}{h} \times \frac{RT}{p^{\ominus}} \times \frac{N_A q_r^2 \Lambda^6}{\Lambda^3 V_m^{\ominus}} \times e^{-\Delta E_0/RT} = \kappa \frac{RT}{h} \times q_r^2 \Lambda^3 \times e^{-\Delta E_0/RT}$$

Now allow for internal structure. The partition functions then become

$$q_A^{\ominus} = q_r^3 q_v^{3N-6} \times \frac{V_m^{\ominus}}{\Lambda^3}$$

$$q_B^{\ominus} = q_r^3 q_v^{3N'-6} \times \frac{V_m^{\ominus}}{\Lambda^3}$$

$$\bar{q}_C^{\ominus} = q_r^3 q_v^{3(N+N')-7} \times \frac{V_m^{\ominus}}{\Lambda^3}$$

One vibrational mode of C has been discarded, as required by the Eyring equation. It then follows that

$$k_2 = \kappa \frac{kT}{h} \times \frac{RT}{p^{\ominus}} \times \frac{N_A q_r^3 q_v^{3(N+N')-7} \Lambda^6}{q_r^6 q_v^{3(N+N')-12} \Lambda^3 V_m^{\ominus}} \times e^{-\Delta E_0/RT}$$

$$= \kappa \frac{RT}{h} \times \frac{q_v^5 \Lambda^3}{q_r^3} \times e^{-\Delta E_0/RT}$$

Comparison of the two expressions for k_2 gives

$$P = \left(\frac{q_v}{q_r}\right)^5$$

Because $q_v/q_r \approx 1/50$, $P \approx 3 \times 10^{-9}$.

Comment. The calculation suggests that, for reactions with similar activation energies, the rate constants for reactions between complex molecules in the gas phase should be much slower than reactions between simple molecules.

Exercise E27.3. Estimate the steric factor for two linear molecules that form a non-linear activated complex. $[P = q_v/q_r \approx 1/50]$

The kinetic isotope effect

As a second example, consider the effect of deuteration on a reaction in which the rate-determining step is the scission of a C—H bond. The reaction coordinate corresponds to the stretching of the C—H bond, and the potential energy profile is shown in Fig. 27.6. On deuteration, the dominant change is the reduction of the zero-point energy of the bond (because the deuterium atom is heavier). The whole reaction profile is not lowered, however, because the relevant vibration in the activated complex has a very low force constant, so there is little zero-point energy associated with the reaction coordinate in either the proton or the deuteron forms of the complex.

We assume that the deuteration affects only the reaction coordinate, and hence that the partition functions for all the other internal modes remain unchanged. The translational partition functions are changed by

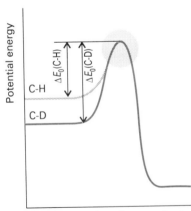

27.6 Changes in the reaction profile when a bond undergoing cleavage is deuterated. The only significant change is to the zero-point energy of the reactants, which is lower for C–D than for C–H. As a result, the activation energy is greater for C–D than for C–H.

deuteration, but the mass of the rest of the molecule is normally so great that the change is insignificant. The value of ΔE_0 changes on account of the change of zero-point energy, and

$$\Delta E_0(C\text{—}D) - \Delta E_0(C\text{—}H) = N_A\{\tfrac{1}{2}\hbar\omega(C\text{—}D) - \tfrac{1}{2}\hbar\omega(C\text{—}H)\}$$

$$= \tfrac{1}{2}N_A\hbar k_f^{\frac{1}{2}}\left(\frac{1}{\mu_{CH}^{\frac{1}{2}}} - \frac{1}{\mu_{CD}^{\frac{1}{2}}}\right)$$

where k_f is the force constant of the bond and μ the effective mass. Because all the partition functions are the same (by assumption), the rate constants for the two species should be in the ratio

$$\frac{k(C\text{—}D)}{k(C\text{—}H)} = e^{-\lambda} \qquad \text{where } \lambda = \frac{\hbar k_f^{\frac{1}{2}}}{2kT}\left(\frac{1}{\mu_{CH}^{\frac{1}{2}}} - \frac{1}{\mu_{CD}^{\frac{1}{2}}}\right) \tag{16}$$

Note that $\lambda > 0$ because $\mu_{CD} > \mu_{CH}$. This equation predicts that at room temperature C—H cleavage should be about seven times faster than C—D cleavage, other conditions being equal. The details, however, should not conceal the principal reason for the difference, which is that C—H bond scission has a higher activation energy on account of its greater zero-point energy.

The experimental observation of the activated complex

Until very recently there were no direct spectroscopic observations on activated complexes, for they have a very fleeting existence and often survive for only a picosecond or so. However, the development of femtosecond-pulsed lasers (1 fs = 10^{-15} s) and their application to chemistry in the form of **femtochemistry** have made it possible to make observations on species that have such short lifetimes that in a number of respects they resemble an activated complex.

In a typical experiment, a femtosecond pulse is used to excite a molecule to a dissociative state, and then a second femtosecond pulse is fired at a series of intervals after the dissociating pulse. The frequency of the second pulse is set at an absorption of one of the free fragmentation products, so its absorption is a measure of the abundance of the dissociation product. For example, when ICN is dissociated by the first pulse, the emergence of CN from the photoactivated state can be monitored by watching the growth of the free CN absorption (or, more commonly, its laser-induced fluorescence). In this way it has been found that the CN signal remains zero until the fragments have separated by about 600 pm, which takes about 205 fs.

Some sense of the progress that has been made in the study of the intimate mechanism of chemical reactions can be obtained by considering an analogue of the harpoon reaction introduced in Section 27.1. The decay of the ion pair Na^+X^-, where X is a halogen, has been studied by exciting it with a femtosecond pulse to an excited state that corresponds to a covalently bonded NaX molecule. The second probe pulse examines the system at an absorption frequency either of the free Na atom or at a frequency at which the atom absorbs when it is a part of the complex.

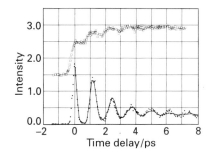

27.7 Femtosecond spectroscopic results for the reaction in which NaI‡ separates into Na + I. The full circles are the absorption of the complex and the open circles the absorption of the free Na atoms (A. H. Zewail, *Science* **242**, 1645 (1988)).

The latter frequency depends on the Na–X distance, so an absorption (in practice, a laser-induced fluorescence) is obtained each time the vibration of the complex returns it to that separation.

A typical set of results for NaI is shown in Fig. 27.7. The bound Na absorption intensity shows up as a series of pulses that recur in about 1 ps, showing that the complex vibrates with about that period. The decline in intensity shows the rate at which the complex can dissociate as the two atoms swing away from each other. The complex does not dissociate on every outward-going swing because there is a chance that the I atom can be harpooned again, in which case it fails to make good its escape. The free Na absorption also grows in an oscillating manner, showing the periodicity of the vibration of the complex that gives it a chance to dissociate each picosecond. The precise period of the oscillation in NaI is 1.25 ps, corresponding to a vibrational wavenumber of 27 cm^{-1} (recall that the activated complex theory assumes that such a vibration has a very low frequency). The complex survives for about 10 oscillations. In contrast, although the oscillation frequency of NaBr is similar, it barely survives one oscillation.

Femtosecond spectroscopy has also been used to examine analogues of the activated complex involved in bimolecular reactions. Thus, a molecular beam can be used to produce a **van der Waals molecule** in which two species are loosely bonded together by intermolecular forces. An example is the van der Waals molecule IH$\cdots$OCO between HI and CO_2, where the interaction is by hydrogen bonding. The HI bond can be dissociated by a femtosecond pulse, and the H atom is ejected towards the O atom of the neighbouring CO_2 molecule to form HOCO. Hence, the van der Waals molecule is a source of a species that resembles the activated complex of the reaction

$$H + CO_2 \rightarrow [HOCO]^\ddagger \rightarrow HO + CO$$

The probe pulse is tuned to the OH radical, which enables the evolution of $[HOCO]^\ddagger$ to be studied in real time.

27.6 Thermodynamic aspects

The statistical version of activated complex theory rapidly runs into difficulties because only rarely is anything known about the structure of the activated complex. However, the concepts that it introduces, principally that of an equilibrium between the reactants and the activated complex, have motivated a more general, empirical approach in which the activation process is expressed in terms of thermodynamic functions.

Activation parameters

If we accept that $\bar{K}$ is an equilibrium constant (despite one mode of C‡ having been discarded), we can express it in terms of a (molar) **Gibbs energy of activation** $\Delta^\ddagger G$ through

$$\Delta^\ddagger G = -RT \ln \bar{K} \tag{17a}$$

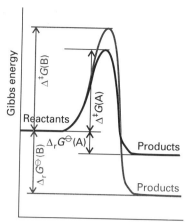

27.8 For a related series of reactions, as the magnitude of the standard reaction Gibbs energy increases (from A to B), so the activation barrier decreases. The approximate correlation between $\Delta^{\ddagger}G$ and $\Delta_r G^{\ominus}$ is the origin of linear free energy relations.

Then the rate constant becomes

$$k_2 = \kappa \frac{kT}{h} \times \frac{RT}{p^{\ominus}} e^{-\Delta^{\ddagger}G/RT} \tag{17b}$$

Because $G = H - TS$, the Gibbs energy of activation can be divided into an **entropy of activation** $\Delta^{\ddagger}S$ and an **enthalpy of activation** $\Delta^{\ddagger}H$, by writing

$$\Delta^{\ddagger}G = \Delta^{\ddagger}H - T\Delta^{\ddagger}S \tag{18}$$

Gibbs free energies, enthalpies, and entropies of activation are widely used to report experimental reaction rates, especially for organic reactions in solution. They are encountered when relationships between equilibrium constants and rates of reaction are explored using **correlation analysis**, in which $\ln K$ (which is equal to $-\Delta_r G^{\ominus}/RT$) is plotted against $\ln k$ (which is proportional to $-\Delta^{\ddagger}G/RT$). In many cases the correlation is linear, signifying that, as the reaction becomes thermodynamically more favourable, its rate constant increases (Fig. 27.8). This linear correlation is the origin of the alternative name **linear free energy relation** (LFER; see *Further reading*).

When eqn 18 is used in eqn 17b and κ is absorbed into the entropy term, we obtain

$$k_2 = B e^{\Delta^{\ddagger}S/R} e^{-\Delta^{\ddagger}H/RT}, \qquad \text{where } B = \frac{kT}{h} \times \frac{RT}{p^{\ominus}} \tag{19}$$

This equation can be expressed in Arrhenius form by using the definition of E_a in eqn 25.13a. This procedure gives

$$\Delta^{\ddagger}H = E_a - 2RT \tag{20a}$$

for a bimolecular gas-phase reaction, whereas for a reaction in solution (and for a unimolecular reaction in either phase)

$$\Delta^{\ddagger}H = E_a - RT \tag{20b}$$

It follows from eqn 20 that k_2 can be written in the Arrhenius form $k_2 = A e^{-E_a/RT}$ if we write (for a bimolecular gas-phase reaction)

$$A = e^2 B \times e^{\Delta^{\ddagger}S/R} \tag{21}$$

Hence, to calculate the entropy of activation from the pre-exponential factor in a bimolecular gas phase reaction, we use

$$\Delta^{\ddagger}S = R\left(\ln \frac{A}{B} - 2\right) \tag{22}$$

Example 27.4 *Calculating activation parameters*

Calculate the activation entropy, enthalpy, and Gibbs energy of the second-order hydrogenation of ethene at 355°C using information from Table 27.1.

Method. The enthalpy of activation is obtained from the activation energy by using eqn 20a. The entropy of activation can be obtained from eqn 22; the practical form of this equation is obtained by writing

$$\frac{A}{B} = \frac{(5.7723 \times 10^{-10}) \times (A/\text{L mol}^{-1}\,\text{s}^{-1})}{(T/\text{K})^2}$$

At this point, the two parameters can be combined by using eqn 18.

Answer. Since $E_a = 180 \text{ kJ mol}^{-1}$, from eqn 20a

$$\Delta^{\ddagger}H = E_a - 2RT = (180 - 2 \times 5.2) \text{ kJ mol}^{-1} = +170 \text{ kJ mol}^{-1}$$

From the table, $A = 1.24 \times 10^6 \text{ L mol}^{-1}\,\text{s}^{-1}$; hence

$$\Delta^{\ddagger}S = (8.3145 \text{ J K}^{-1}\,\text{mol}^{-1})$$
$$\times \left\{ \ln\left(\frac{(5.7723 \times 10^{-10}) \times (1.24 \times 10^6)}{(628)^2} \right) - 2 \right\}$$
$$= -184 \text{ J K}^{-1}\,\text{mol}^{-1}$$

Finally, from eqn 18,

$$\Delta^{\ddagger}G = \Delta^{\ddagger}H - T\Delta^{\ddagger}S = +286 \text{ kJ mol}^{-1}$$

Comment. Notice the large negative entropy of activation. The simple collision theory magnitude of A calculated in Example 27.1 corresponds to a much less negative value ($-73 \text{ J K}^{-1}\,\text{mol}^{-1}$).

Exercise E27.4. Repeat the calculation for the reaction $\text{K} + \text{Br}_2 \rightarrow \text{KBr} + \text{Br}$ at 298 K.

$$[-2.5 \text{ kJ mol}^{-1},\ -58 \text{ J K}^{-1}\,\text{mol}^{-1},\ +15 \text{ kJ mol}^{-1}]$$

We can now make contact with the collision theory expression for the rate constant. In that theory, the pre-exponential factor A is determined by the frequency of collisions in the gas. But a collision event corresponds to a decrease in entropy (because two molecules come together and therefore reduce the disorder of the gas). Hence, the negative value of $\Delta^{\ddagger}S$ reflects the occurrence of collisions. Furthermore, collisions with well-defined relative orientations correspond to an even greater reduction in entropy than is brought about by collisions in general, so the entropy of activation should then be even more negative. This reduction in entropy reduces the value of A, a feature taken into account in collision theory by the steric factor P.

Reactions between ions

The thermodynamic version of activated complex theory makes it relatively simple to discuss reactions in solution. The statistical theory is very complicated to apply because the solvent plays a role in the activated complex. In the thermodynamic approach we combine the rate law

$$\frac{\text{d}[P]}{\text{d}t} = k^{\ddagger}[C^{\ddagger}]$$

with the thermodynamic equilibrium constant

$$K = \frac{a_C}{a_A a_B} = K_\gamma \times \frac{[C^\ddagger]}{[A][B]}$$

where K_γ is the ratio of activity coefficients,

$$K_\gamma = \frac{\gamma_C}{\gamma_A \gamma_B}$$

Then

$$\frac{d[P]}{dt} = k_2[A][B] \qquad \text{where } k_2 = \frac{k^\ddagger K}{K_\gamma}$$

If k_2° is the rate constant when the activity coefficients are 1 (that is, $k_2^\circ = k^\ddagger K$), we can write

$$k_2 = \frac{k_2^\circ}{K_\gamma} \tag{23}$$

At low concentrations the activity coefficients can be expressed in terms of the ionic strength, I, of the solution by using the Debye–Hückel limiting law (Section 10.2, particularly eqn 10.6) in the form

$$\log \gamma_J = -A z_J^2 I^{\frac{1}{2}}$$

with I understood to be $I/m^\ominus$. Then

$$\log k_2 = \log k_2^\circ - A\{z_A^2 + z_B^2 - (z_A + z_B)^2\}I^{\frac{1}{2}} = \log k_2^\circ + 2A z_A z_B I^{\frac{1}{2}} \tag{24}$$

(The charges of A and B are z_A and z_B, so the charge of the activated complex is $z_A + z_B$; the z_J are positive for cations and negative for anions.)

Equation 24 expresses the **kinetic salt effect**, the variation of the rate constant of a reaction between ions with the ionic strength of the solution (Fig. 27.9). If the reactant ions have the same sign (as in a reaction between cations or between anions), then increasing the ionic strength by the addition of inert ions increases the rate constant. The formation of a single, highly charged ionic complex from two less highly charged ions is favoured by a high ionic strength because the new ion has a denser ionic atmosphere and interacts with it more strongly. Conversely, ions of opposite charge react more slowly in solutions of high ionic strength. Now the charges cancel and the complex has a less favourable interaction with its atmosphere than the separated ions.

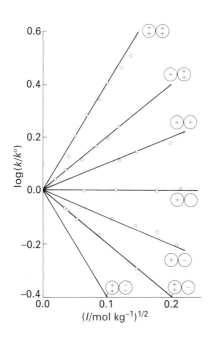

27.9 Experimental tests of the kinetic salt effect for reactions in water at 298 K. The ion types are shown as spheres, and the slopes of the lines are those given by the Debye–Hückel limiting law and eqn 24.

Example 27.5 *Analysing the kinetic salt effect*

The rate constant for the base hydrolysis of $[CoBr(NH_3)_5]^{2+}$ varies with ionic strength as tabulated below. What can be deduced about the charge of the activated complex in the rate-determining stage?

$I/(\text{mol kg}^{-1})$	0.005	0.010	0.015	0.020	0.025	0.030
k/k°	0.718	0.631	0.562	0.515	0.475	0.447

Method. According to eqn 24, we need to plot $\log(k/k^\circ)$ against $I^{\frac{1}{2}}$, when

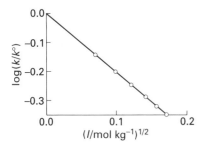

27.10 The experimental ionic strength dependence of the rate constant of a hydrolysis reaction: the slope gives information about the charge types involved in the activated complex of the rate determining step. See Example 27.5.

the slope will give $1.02z_A z_B$, from which we can infer the charges of the ions involved in the formation of the activated complex.

Answer. Form the following table:

$I/(\text{mol kg}^{-1})$	0.005	0.010	0.015	0.020	0.025	0.030
$I^{\frac{1}{2}}/(\text{mol kg}^{-1})^{\frac{1}{2}}$	0.071	0.100	0.122	0.141	0.158	0.173
$\log(k/k^\circ)$	-0.14	-0.20	-0.25	-0.29	-0.32	-0.35

These points are plotted in Fig. 27.10. The slope of the (least squares) straight line is -2.1, indicating that $z_A z_B = -2$. Because $z_A = -1$ for the OH$^-$ ion, if that ion is involved in the formation of the activated complex, then the charge number of the second ion is $+2$. This suggests that the pentaamminebromocobalt(III) cation participates in the formation of the activated complex.

Exercise E27.5. An ion of charge number $+1$ is known to be involved in the activated complex of a reaction. Deduce the charge number of the other ion from the following data:

$I/(\text{mol kg}^{-1})$	0.005	0.010	0.015	0.020	0.025	0.030
k/k°	0.98	0.97	0.97	0.96	0.96	0.95

[-1]

THE DYNAMICS OF MOLECULAR COLLISIONS

We now come to the third and most detailed level of our examination of the factors that govern the rates of reactions. Molecular beams allow us to study collisions between molecules in preselected states, and can be used to determine the states of the products of a reactive collision. Information of this kind is essential if a full picture of the reaction is to be built, because the rate constant is an average over events in which reactants in different initial states evolve into products in their final states.

27.7 Reactive collisions

Detailed experimental information comes from molecular beams, especially **crossed molecular beams** (Fig. 27.11). The detector for the products of the collision of two beams can be moved to different angles, so the angular distribution of the products can be determined. The detector can also distinguish between different energy states of the products. Therefore, because the molecules in the incoming beams can be prepared with different energies (e.g. with different translational energies by using rotating sectors and supersonic nozzles, and with different vibrational energies by using selective excitation with lasers) and with different orientations (by using electric fields), it is possible to study the dependence of the success of collisions on these variables and to study how they affect the properties of the product molecules.

One method for examining the energy distribution in the products is **infrared chemiluminescence**, in which vibrationally excited molecules emit infrared radiation as they return to their ground states. By studying

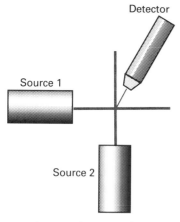

27.11 In a cross-beam experiment, state-selected molecules are generated in two separate sources, and are directed perpendicular to one another. The detector responds to molecules (which may be product molecules if chemical reaction occurs) scattered into a chosen direction.

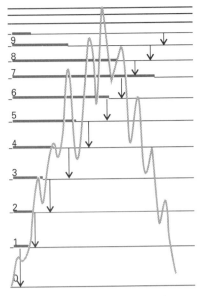

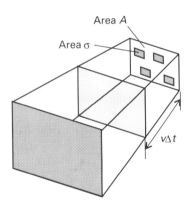

27.12 Infrared chemiluminescence from CO produced in the reaction $O + CS \rightarrow CO + S$ arises from the non-equilibrium populations of the vibrational states of CO and the radiative relaxation to equilibrium.

27.13 The basis of the relation between the rate constant and the reaction cross-section. Only molecules within a distance $v\,\Delta t$ can reach the target zone in an interval Δt.

the intensities of the infrared emission spectrum, the populations of the vibrational states may be determined (Fig. 27.12). Another method makes use of **laser-induced fluorescence**. In this technique, a laser is used to excite a product molecule from a specific vibration–rotation level and the intensity of the fluorescence from the upper state is monitored and interpreted in terms of the population of the initial vibration–rotation state.

The concept of collision cross-section was introduced in connection with collision theory in Section 27.1. Molecular beam studies provide a more sophisticated version of this quantity, for they provide the **state-to-state cross-section** $\sigma_{nn'}(v)$. This quantity, which depends on the speed v of approach of the particles, is the target area presented by one species to another for scattering from one state n to another state n'. The **state-to-state rate constant** $k_{nn'}$ is obtained by averaging the product of the speed and the state-to-state cross-section over a Maxwell distribution of speeds (Section 1.3) at the temperature of interest:

$$k_{nn'}(T) = \langle v\sigma_{nn'}(v)\rangle_T \tag{25}$$

JUSTIFICATION

Consider a beam of molecules incident on a wall of area A (Fig. 27.13). In an interval Δt, the number of molecules reaching the wall is the number in the region of length $v\,\Delta t$ and hence of area $vA\,\Delta t$, which is $\mathcal{N}vA\,\Delta t$, where $\mathcal{N}$ is the number density of particles. The proportion of these molecules that change their state is the ratio of the state-to-state cross-sectional area to the area of the wall (Fig. 27.14), which is σ/A. Therefore, the number of product molecules produced per unit time is

$$\frac{(\mathcal{N}vA\,\Delta t) \times \left(\dfrac{\sigma}{A}\right)}{\Delta t} = \mathcal{N}v\sigma$$

averaged over all speeds. On setting this rate of production equal to $k\mathcal{N}$, we obtain eqn 25.

The rate constant for the reaction itself is the sum of the state-to-state rate constants over all final states (because a reaction is successful whatever the final state of the products) and over a Boltzmann-weighted sum of initial states (because the reactants are initially present with a characteristic distribution of populations at a temperature T):

$$k_2 = \sum_{n,n'} k_{nn'}(T)f_n(T) \tag{26}$$

where $f_n(T)$ is the Boltzmann factor at a temperature T.

It follows that, if we can determine or calculate the state-to-state cross-sections for a wide range of approach speeds and initial and final states, then we have a route to the calculation of the rate constant for the reaction.

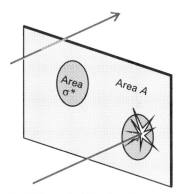

27.14 The probability of undergoing reaction is proportional to the ratio of the reaction cross-section to the area of the rectangle A.

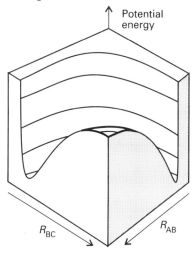

27.15 The potential energy surface for the H + H_2 reaction when the atoms are constrained to be collinear.

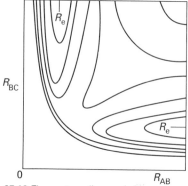

27.16 The contour diagram (with contours of equal potential energy) corresponding to the surface in Fig. 27.15. R_e marks the equilibrium bond length of an H_2 molecule (strictly, it relates to the arrangement when the third atom is at infinity).

27.8 Potential energy surfaces

One of the most important concepts for discussing beam results and calculating the state-to-state collision cross-section is the **potential energy surface** of a reaction, the potential energy as a function of the relative positions of all the atoms taking part in the reaction. For a collision between an H atom and an H_2 molecule, for instance, the potential energy surface is the plot of the potential energy for all relative locations of the three hydrogen nuclei. Detailed calculations show that the approach of an atom along the H–H axis requires less energy for reaction than any other approach, so initially we confine our attention to a collinear approach. Two parameters are required to define the nuclear separations: one is the H_A–H_B separation R_{AB}, and the other is the H_B–H_C separation R_{BC}.

At the start of the encounter R_{AB} is infinite and R_{BC} is the H_2 equilibrium bond length. At the end of a successful reactive encounter R_{AB} is equal to the equilibrium bond length and R_{BC} is infinite. The total energy of the three-atom system depends on their relative separations, and can be found by doing a molecular structure calculation using the Born–Oppenheimer approximation (p. 462) in which all atom locations are regarded as frozen in their instantaneous positions. The plot of the total energy of the system against each value of R_{AB} and R_{BC} gives the potential energy surface of this collinear reaction (Fig. 27.15). This surface is normally depicted as a contour diagram (Fig. 27.16).

When R_{AB} is very large, the variation in potential energy represented by the surface as R_{BC} changes are those of an isolated H_2 molecule as its bond length is altered. A section through the surface at $R_{AB} = \infty$, for example, is the same as the H_2 potential energy curve drawn in Fig. 14.3. At the edge of the diagram where R_{BC} is very large, a section through the surface is the molecular potential energy curve of an isolated H_AH_B molecule.

The actual path of the atoms in the course of the encounter depends on their total energy, the sum of their kinetic and potential energies. However, we can obtain an initial idea of the paths available to the system by considering the potential energy surface alone, and looking for paths that correspond to least potential energy. For example, consider the changes in potential energy as H_A approaches H_BH_C. If the H_B–H_C bond length is constant during the initial approach of H_A, then the potential energy of the H_3 cluster would rise along the path marked A in Fig. 27.17. We see that the potential energy rises to a high value as H_A is pushed into the molecule, and then decreases sharply as H_C breaks off and separates a great distance. An alternative reaction path can be imagined (B) in which the H_B–H_C bond length increases while H_A is still far away. It is clear that both paths, although feasible if the molecules have sufficient initial kinetic energy, take the three atoms to regions of high potential energy.

The path of least potential energy is the one marked C. It corresponds to R_{BC} lengthening as H_A approaches and begins to form a bond with H_B. The H_B–H_C bond relaxes at the demand of the incoming atom and,

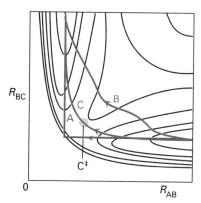

27.17 Various trajectories through the potential energy surface shown in Fig. 27.16. Path A corresponds to a route in which R_{BC} is held constant as H_A approaches; path B corresponds to a route in which R_{BC} lengthens at an early stage during the approach of H_A, and path C is the route along the floor of the potential valley.

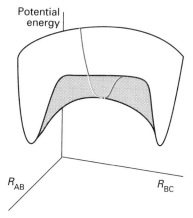

27.18 The transition state is a set of configurations (here, marked by the line across the saddle point) through which successful reactive trajectories must pass.

although the potential energy rises, it climbs only as far as the saddle-shaped region of the surface, to the **saddle point** marked $C^{\ddagger}$. The encounter of least potential energy is one in which the atoms take route C up the floor of the valley, through the saddle point, and down the floor of the other valley as H_C recedes and the new H_A–H_B bond achieves its equilibrium length. This path is the reaction coordinate we met in Section 27.4.

We can now make contact with the activated complex theory of reaction rates. In terms of trajectories on potential surfaces, the transition state can be identified with a critical geometry such that every trajectory that goes through this geometry goes on to react (Fig. 27.18).

27.9 Some results from experiments and calculations

In this section we see how some of the questions just raised are answered, and how the study of collisions and the calculation of potential energy surfaces gives some insight into the course of reactions.

The type of question explored by molecular beam studies can be introduced by considering how the molecules move across their potential energy surface when they also possess kinetic energy. To travel successfully from reactants to products the incoming molecules must possess enough kinetic energy to be able to climb to the saddle point of the potential surface. Therefore, the shape of the surface can be explored experimentally by changing the relative speed of approach (by selecting the beam velocity) and the degree of vibrational excitation and observing whether reaction occurs and whether the products emerge in a vibrationally excited state (Fig. 27.19). For example, one question that can be answered is whether it is better to smash the reactants together with a lot of translational kinetic energy or to ensure instead that they approach in highly excited vibrational states. Thus, is trajectory C_2^* in Fig. 27.19b, where the $H_B H_C$ molecule is initially vibrationally excited, more efficient at leading to reaction than the trajectory C_1^* in Fig. 27.19a in which the total energy is the same but has a high translational kinetic energy? We see below something of the answers to questions like this.

The direction of attack and separation

Figure 27.20 shows the results of a calculation of the potential energy as an H atom approaches an H_2 molecule from different angles, the H_2 bond being allowed to relax to the optimum length in each case. The potential barrier is least for collinear attack, as we assumed earlier. (But we must be aware that other lines of attack are feasible and contribute to the overall rate.) In contrast, Fig. 27.21 shows the potential energy changes that occur as a Cl atom approaches an HI molecule. The lowest barrier occurs for approaches within a cone of half-angle 30° surrounding the H atom. The relevance of this result to the calculation of the steric factor of collision theory should be noted: not every collision is successful, because not every one lies within the reactive cone.

If the collision is sticky, so that when the reactants collide they orbit around each other, then the products can be expected to emerge in

27.19 Some successful (*) and unsuccessful encounters. (a) C_1^* corresponds to the path along the foot of the valley; (b) C_2^* corresponds to an approach of A to a vibrating BC molecule, and the formation of a vibrating AB molecule as C departs; (c) C_3 corresponds to A approaching a non-vibrating BC molecule, but with insufficient translational kinetic energy; (d) C_4 corresponds to A approaching a vibrating BC molecule, but still the energy, and the phase of the vibration, is insufficient for reaction.

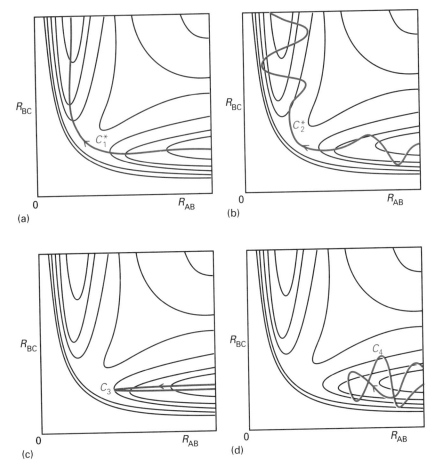

(a) (b)

(c) (d)

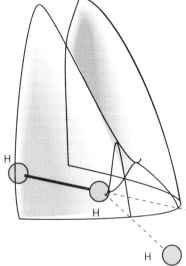

27.20 An indication of the anisotropy of the potential energy changes as H approaches H_2 with different angles of attack. The collinear attack has the lowest potential barrier to reaction. The surface indicates the potential energy profile along the reaction coordinate for each configuration (two cross-sections through the surface are indicated).

random directions because all memory of the approach direction has been lost. A rotation takes about 1 ps, so if the collision is over in less than that time the complex will not have had time to rotate and the products will be thrown off in a specific direction. In the collision of K and I_2, for example, most of the products are thrown off in the forward direction.[3] This product distribution is consistent with the harpoon mechanism (Section 27.1) because the transition takes place at long range. In contrast, the collision of K with CH_3I leads to reaction only if the molecules approach each other very closely. This mechanism is like K bumping into a brick wall, and the KI product bouncing out in the backward direction. The detection of this anisotropy in the angular distribution of products gives an indication of the distance and orientation of approach needed for reaction, as well as showing that the event is complete in less than 1 ps.

---

3 There is subtlety here. In molecular beam work the remarks normally refer to directions in a centre-of-mass coordinate system. The origin of the coordinates is the centre of mass of the colliding reactants, and the collision takes place when the molecules are at the origin. The way in which centre-of-mass coordinates are constructed and the events in them interpreted involves too much detail for our present purposes, but we should bear in mind that 'forward' and 'backward' have unconventional meanings. The details are explained in the books in *Further reading*.

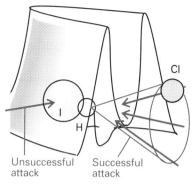

27.21 The potential energy barrier for the approach of Cl to HI. In this case successful encounters occur only when Cl approaches within a cone surrounding the H atom.

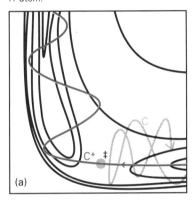

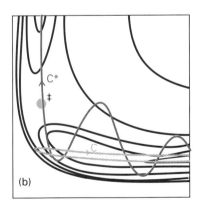

27.22 (a) An attractive potential energy surface. A successful encounter (C*) involves high translational kinetic energy and results in a vibrationally excited product. (b) A repulsive potential energy surface. A successful encounter (C*) involves initial vibrational excitation and the products have high translational kinetic energy. A reaction that is attractive in one direction is repulsive in the reverse direction.

Attractive and repulsive surfaces

Some reactions are very sensitive to whether the energy has been predigested into a vibrational mode or left as the relative translational kinetic energy of the colliding molecules. For example, if two HI molecules are hurled together with more than twice the activation energy of the reaction, then no reaction occurs if all the energy is translational. For $F + HCl \rightarrow Cl + HF$, for example, it has been found that the reaction is about five times as efficient when the HCl is in its first vibrational excited state than when, although it has the same total energy, it is in its vibrational ground state.

The origin of these requirements can be found by examining the potential energy surface. Figure 27.22a shows an **attractive surface** in which the saddle point occurs early in the reaction coordinate. Figure 27.22b is a **repulsive surface** in which the saddle point occurs late. A surface that is attractive in one direction is repulsive in the reverse direction.

Consider first the attractive surface. If the original molecule is vibrationally excited, then a collision with an incoming molecule takes the system along C. This path is bottled up in the region of the reactants, and does not take the system to the saddle point. If, however, the same amount of energy is present solely as translational kinetic energy, then the system moves along C* and travels smoothly over the saddle point into products. We can therefore conclude that reactions with attractive potential energy surfaces proceed more efficiently if the energy is in relative translational motion. Moreover, the potential surface shows that once past the saddle point the trajectory runs up the steep wall of the product valley, and then rolls from side to side as it falls to the foot of the valley as the products separate. In other words, the products emerge in a vibrationally excited state.

Now consider the repulsive surface (Fig. 27.22b). On trajectory C the collisional energy is largely in translation. As the reactants approach, the potential energy rises. Their path takes them up the opposing face of the valley, and they are reflected back into the reactant region. This path corresponds to an unsuccessful encounter, even though the energy is sufficient for reaction. On C* some of the energy is in the vibration of the reactant molecule and the motion causes the trajectory to weave from side to side up the valley as it approaches the saddle point. This motion may be sufficient to tip the system round the corner to the saddle point and then on to products. In this case, the product molecule is expected to be in an unexcited vibrational state. It follows that reactions with repulsive potential surfaces can be expected to proceed more efficiently if the excess energy is present as vibrations. This is the case with the $H + Cl_2 \rightarrow HCl + Cl$ reaction, for instance.

Classical trajectories

A clear picture of the reaction event can be obtained by using classical mechanics to calculate the trajectories of the atoms taking place in a reaction. Figure 27.23 shows the result of such a calculation of the

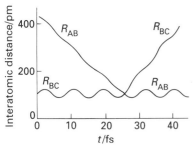

27.23 The calculated trajectories for a reactive encounter between A and a vibrating BC molecule leading to the formation of a vibrating AB molecule. This direct-mode reaction is between H and H_2. (M. Karplus, R. N. Porter, and R. D. Sharma, *J. Chem. Phys.* **43,** 3258 (1965).)

27.24 An example of the trajectories calculated for a complex-mode reaction, $KCl + NaBr \rightarrow KBr + NaCl$, in which the collision cluster has a long lifetime. (P. Brumer and M. Karplus, *Faraday Disc. Chem. Soc.* **55,** 80 (1973).)

positions of the three atoms in the reaction $H + H_2 \rightarrow H_2 + H$, the horizontal coordinate now being time and the vertical coordinate the separations. This illustration shows clearly the vibration of the original molecule and the approach of the attacking atom. The reaction itself, the switch of partners, takes place very rapidly and is an example of a **direct mode** process. Then the new molecule shakes, but quickly settles down to steady, harmonic vibration as the expelled atom departs. In contrast, Fig. 27.24 shows an example of a **complex mode** process, in which the activated complex survives for an extended period. The reaction in the illustration is the exchange reaction

$$KCl + NaBr \rightarrow KBr + NaCl$$

The tetraatomic activated complex survives for about 5 ps, during which time the atoms make about 15 oscillations before dissociating into products.

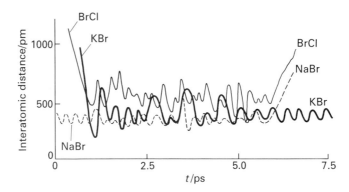

Although this kind of calculation gives a good sense of what happens during a reaction, its limitations must be kept in mind. In the first place, a real gas-phase reaction occurs with a wide variety of different speeds and angles of attack. In the second place, the motion of the atoms, electrons, and nuclei is governed by quantum mechanics. The concept of trajectory then fades and is replaced by the unfolding of a wavefunction that represents initially the reactants and finally products. Nevertheless, recognition of these limitations should not be allowed to obscure the fact that recent advances in molecular reaction dynamics have given us a first glimpse of the processes going on at the core of reactions.

CHECK LIST OF KEY IDEAS

1 The **collision theory** of bimolecular gas phase reaction rates and the significance of the **activation energy** and **pre-exponential factor** (Section 27.1).

2 The **reactive cross-section** and the **steric factor** (Examples 27.1 and 27.2).

3 **Diffusion-controlled** and **activation-controlled** reactions in solution (Section 27.2).

4 The **diffusion-controlled rate constant** in terms of the solvent viscosity (eqn 9).

5 The formulation of the **material-balance equa-**

tion (eqn 10) and some of its solutions (Section 27.3).

6 The formulation of **activated complex theory** in terms of the **reaction coordinate** and the **transition state** (Section 27.4).

7 The derivation and application of the **Eyring equation** (eqn 15).

8 The **kinetic isotope effect** and its analysis in terms of activated complex theory (eqn 16).

9 The use of **ultrashort pulses** in **femto-chemistry** and the direct observation of the activated complex (Section 27.5).

10 The description of rate constants in terms of

thermodynamic functions of activation and the origin of **linear free energy relations** (Section 27.6).

11 The interpretation of collision theory in terms of the **entropy of activation** (Section 27.6).

12 The rate constant for reactions in solution and the **kinetic salt effect** (eqn 24).

13 The interpretation of reaction rates in terms of **potential energy surfaces** (Section 27.8).

14 The trajectories of reactants and products on **attractive** and **repulsive** potential surfaces and the distinction between **direct mode** and **complex mode** reactions (Section 27.9).

EXERCISES

27.1. Calculate the collision frequency z and the collision density Z in (a) ammonia, $R = 190$ pm, (b) carbon monoxide, $R = 180$ pm at 25°C and 100 kPa. What is the percentage increase when the temperature is raised by 10 K at constant volume?

27.2. Collision theory depends on knowing the fraction of molecular collisions having at least the kinetic energy E_a along the line of flight. What is this fraction when (a) $E_a = 10$ kJ mol^{-1}, (b) $E_a = 100$ kJ mol^{-1} at (1) 300 K and (2) 100 K?

27.3. Calculate the percentage increase in the fractions in Exercise 27.2 when the temperature is raised by 10 K.

27.4. Use the collision theory of gas-phase reactions to calculate the theoretical value of the second-order rate constant for the reaction $H_2(g) + I_2(g) \rightarrow 2HI(g)$ at 650 K, assuming that it is elementary bimolecular. The collision cross-section is 0.38 nm^2, the reduced mass is 3.32×10^{-27} kg, and the activation energy is 171 kJ mol^{-1}.

27.5. A typical diffusion constant for small molecules in aqueous solution at 25°C is 5×10^{-9} m^2 s^{-1}. If the critical reaction distance is 0.4 nm, what value is expected for the second-order rate constant for a diffusion-controlled reaction?

27.6. Calculate the magnitude of the diffusion-controlled rate constant at 298 K for species in' (a) pentane, (b) decylbenzene. The viscosities are 2.2×10^{-4} kg m^{-1} s^{-1} and 3.36×10^{-3} kg m^{-1} s^{-1} (0.22 cP and 3.36 cP), respectively.

27.7. Calculate the magnitude of the diffusion-controlled rate constant at 298 K for the recombination of two atoms

in water, for which $\eta = 0.89$ cP. Assuming the concentration of the reacting species is 1.0×10^{-3} mol L^{-1} initially, how long does it take for the concentration of the atoms to fall to half that value? Assume the reaction is elementary.

27.8. For the gaseous reaction $A + B \rightarrow P$, the reactive cross-section obtained from the experimental value of the pre-exponential factor is 9.2×10^{-22} m^2. The collision cross-sections of A and B estimated from the transport properties are 0.95 and 0.65 nm^2, respectively. Calculate the P-factor for the reaction.

27.9. A and B, which are neutral species with diameters 588 pm and 1650 pm respectively, undergo the diffusion-controlled elementary reaction $A + B \rightarrow P$ in a solvent of viscosity 2.37×10^{-3} kg m^{-1} s^{-1} at 40°C. Calculate the initial rate d[P]/dt if the initial concentrations of A and B are 0.150 and 0.330 mol L^{-1}, respectively.

27.10. The reaction of propylxanthate ion in acetic acid buffer solutions has the mechanism $A^- + H^+ \rightarrow P$. Near 30°C the rate constant is given by the empirical expression $k_2 = 2.05 \times 10^{13} e^{(-8681\ K)/T}$ L mol^{-1} s^{-1}. Evaluate the energy and entropy of activation at 30°C.

27.11. When the reaction in Exercise 27.10 occurs in a dioxane–water mixture that is 30 per cent dioxane by mass, the rate constant fits $k_2 = 7.78 \times 10^{14} e^{-(9134\ K)/T}$ L mol^{-1} s^{-1} near 30°C. Calculate $\Delta^{\ddagger}G$ for the reaction at 30°C.

27.12. The gas-phase association reaction between F_2 and IF_5 is first-order in each of the reactants. The energy of activation for the reaction is 58.6 kJ mol^{-1}. At 65°C the rate constant is $7.84 \times 10^{-3} \, \text{kPa}^{-1} \, \text{s}^{-1}$. Calculate the entropy of activation at 65°C.

27.13. Calculate the molar entropy of activation for a collision between two essentially structureless particles at 300 K, taking $M = 50 \, \text{g mol}^{-1}$ and $\sigma = 0.4 \, \text{nm}^2$.

27.14. The pre-exponential factor for the gas-phase decomposition of ozone at low pressures is 4.6×10^{12} $\text{L mol}^{-1} \, \text{s}^{-1}$ and its activation energy is $10.0 \, \text{kJ mol}^{-1}$. What is (a) the entropy of activation, (b) the enthalpy of activation, (c) the Gibbs energy of activation at 298 K?

27.15. The base-catalysed bromination of nitromethane-d_3 in water at room temperature (298 K) proceeds 4.3 times more slowly than the bromination of the non-deuterated material. Account for this difference. Use $k_f(\text{C–H}) = 450 \, \text{N m}^{-1}$.

27.16. Predict the order of magnitude of the isotope effect on the relative rates of displacement of (a) ^1H and ^3H, (b) ^{16}O and ^{18}O. Will raising the temperature enhance the difference? Take $k_f(\text{CH}) = 450 \, \text{N m}^{-1}$, $k_f(\text{CO}) = 1750 \, \text{N m}^{-1}$.

27.17. The rate constant of the reaction

$$\text{H}_2\text{O}_2(\text{aq}) + \text{I}^-(\text{aq}) + \text{H}^+(\text{aq}) \rightarrow \text{H}_2\text{O}(\text{l}) + \text{HIO}(\text{aq})$$

is sensitive to the ionic strength of the aqueous solution in which the reaction occurs. At 25°C and at an ionic strength of $0.0525 \, \text{mol kg}^{-1}$, $k = 12.2 \, \text{L}^2 \, \text{mol}^{-2} \, \text{min}^{-1}$.

Use the Debye–Hückel limiting law to estimate the rate constant at zero ionic strength.

27.18. The rate constant of the reaction

$$\text{I}^-(\text{aq}) + \text{H}_2\text{O}_2(\text{aq}) \rightarrow \text{H}_2\text{O}(\text{l}) + \text{IO}^-(\text{aq})$$

varies slowly with ionic strength, even though the Debye–Hückel limiting law predicts no effect. Use the following data from 25°C to find the dependence of $\log k$ on the ionic strength:

$I/(\text{mol kg}^{-1})$	0.0207	0.0525	0.0925	0.1575
$k/(\text{L mol}^{-1} \, \text{min}^{-1})$	0.663	0.670	0.679	0.694

Evaluate the limiting value of k at zero ionic strength. What does the result suggest for the dependence of $\log \gamma$ on ionic strength for a neutral molecule in an electrolyte solution even though eqn 24 predicts no effect?

27.19. Use the Debye–Hückel limiting law to show that changes in ionic strength can affect the rate of reaction catalysed by H^+ from the ionization of a weak acid. Consider the mechanism $\text{H}^+(\text{aq}) + \text{B}(\text{aq}) \rightarrow \text{P}$, where H^+ comes from the ionization of the weak acid, HA. The weak acid has a fixed concentration. First show that log $[\text{H}^+]$, derived from the ionization of HA, depends on the activity coefficients of ions and thus depends on the ionic strength. Then find the relationship between log (rate) and log $[\text{H}^+]$ to show that the rate also depends on the ionic strength.

PROBLEMS
Numerical problems

27.1. In the dimerization of methyl radicals at 25°C the experimental pre-exponential factor is 2.4×10^{10} $\text{L mol}^{-1} \, \text{s}^{-1}$. What is (a) the reactive cross-section, (b) the P factor for the reaction if the C—H bond length is 154 pm?

27.2. Nitrogen dioxide reacts bimolecularly in the gas phase to give $2\text{NO} + \text{O}_2$. The temperature dependence of the second-order rate constant for the rate law $d[\text{P}]/dt = k[\text{NO}_2]^2$ is given below. What is the P-factor and the reactive cross-section for the reaction?

T/K	600	700	800	1000
$k/(\text{cm}^3 \, \text{mol}^{-1} \, \text{s}^{-1})$	4.6×10^2	9.7×10^3	1.3×10^5	3.1×10^6

Take $\sigma = 0.60 \, \text{nm}^2$.

27.3. The diameter of the methyl radical is about 308 pm. What is the maximum rate constant in the expression $d[\text{C}_2\text{H}_6]/dt = k[\text{CH}_3]^2$ for second-order recombination of radicals at room temperature? 10 per cent of a 1.0-L sample of ethane at 298 K and 100 kPa is dissociated into methyl radicals. What is the minimum time for 90 per cent recombination?

27.4. The rates of thermolysis of a variety of *cis*- and *trans*-azoalkanes have been measured over a range of temperatures in order to settle a controversy concerning the mechanism of the reaction. In ethanol an unstable *cis*-azoalkane decomposed at a rate that was followed by observing the N_2 evolution, and this led to the rate constants listed below (P. S. Engel and D. J. Bishop, *J. Am. Chem. Soc.* **97**, 6754 (1975)). Calculate the enthalpy, entropy, energy, and Gibbs energy of activation at -20°C.

$\theta/\text{°C}$	-24.82	-20.73	-17.02	-13.00	-8.95
$10^4 \times k/\text{s}^{-1}$	1.22	2.31	4.39	8.50	14.3

27.5. In an experimental study of a bimolecular reaction in aqueous solution, the second-order rate constant was measured at 25°C and at a variety of ionic strengths and the results are tabulated below. It is known that a singly charged ion is involved in the rate-determining step. What is the charge on the other ion involved?

$I/(\text{mol kg}^{-1})$	0.0025	0.0037	0.0045	0.0065	0.0085
$k_2/(\text{L mol}^{-1} \, \text{s}^{-1})$	1.05	1.12	1.16	1.18	1.26

27.6. The total cross-sections for reactions between alkali metal atoms and halogen molecules are given in the table below (R. D. Levine and R. B. Bernstein, *Molecular reaction dynamics*, Clarendon Press, Oxford, 72 (1974)). Assess the data in terms of the harpoon mechanism.

σ^*/nm^2	Cl_2	Br_2	I_2
Na	1.24	1.16	0.97
K	1.54	1.51	1.27
Rb	1.90	1.97	1.67
Cs	1.96	2.04	1.95

Electron affinities are approximately 1.3 eV (Cl_2), 1.2 eV (Br_2), and 1.7 eV (I_2), and ionization energies are 5.1 eV (Na), 4.3 eV (K), 4.2 eV (Rb), and 3.9 eV (Cs).

Theoretical problems

27.7. Confirm that eqn 12 is a solution of eqn 11, where [J] is a solution of the same equation but with $k = 0$ and for the same initial conditions.

27.8. Evaluate [J]* numerically by writing a program for the numerical integration of eqn 12, and explore the effect of increasing reaction rate constant on the spatial distribution of J.

27.9. Estimate the orders of magnitude of the partition functions involved in a rate expression. State the order of magnitude of q_m^T/N_A, q^R, q^V, q^E for typical molecules. Check that in the collision of two structureless molecules the order of magnitude of the pre-exponential factor is of the same order as that predicted by collision theory. Go on to estimate the P-factor for a reaction in which $A + B \rightarrow P$, and A and B are non-linear triatomic molecules.

27.10. The major difficulty in applying activated complex theory (and, it must be admitted, in devising straightforward problems to illustrate it) is to decide on the structure of the activated complex and to ascribe appropriate bond strengths and lengths to it. The following exercise gives some familiarity with the difficulties involved, yet leads to a numerical result for a reaction of some interest. Consider the attack of H on D_2, which is one step in the $H_2 + D_2$ reaction. Suppose that the H approaches D_2 from the side and forms a complex in the form of an isosceles triangle. Take the H–D distance as 30 per cent greater than in H_2 (74 pm) and the D–D distance as 20 per cent greater than in H_2. Let the critical coordinate be the antisymmetric stretching vibration in which one H—D bond stretches as the other shortens. Let all the vibrations be at about 1000 cm^{-1}. Estimate k_2 for this reaction at

400 K using the experimental activation energy of about 35 kJ mol^{-1}.

27.11. Now change the model of the activated complex in Problem 27.10 and make it linear. Use the same estimated molecular bond lengths and vibrational frequencies to calculate k_2 for this choice of model.

27.12. Clearly, there is much scope for modifying the parameters of the models of the activated complex in the last pair of problems. Write and run a program that allows you to vary the structure of the complex and the parameters in a plausible way, and look for a model (or more than one model) that gives a value of k close to the experimental value, 4×10^5 L mol^{-1} s^{-1}.

27.13. The Eyring equation can also be applied to physical processes. As an example, consider the rate of diffusion of an atom stuck to the surface of a solid. Suppose that in order to move from one site to another it has to reach the top of the barrier where it can vibrate classically in the vertical direction and in one horizontal direction, but vibration along the other horizontal direction takes it into the neighbouring site. Find an expression for the rate of diffusion, and evaluate it for W atoms on a tungsten surface ($E_a = 60$ kJ mol^{-1}). Suppose that the vibration frequencies at the transition state are (a) the same as, (b) one-half the value for the adsorbed atom. What is the value of the diffusion coefficient D at 500 K? (Take the site separation as 316 pm and $v = 1 \times 10^{11}$ Hz.)

27.14. Suppose now that the adsorbed, migrating species treated in Problem 27.13 is a spherical molecule, and that it can rotate clasically as well as vibrate at the top of the barrier, but that at the adsorption site itself it can only vibrate. What effect does this have on the diffusion constant? Take the molecule to be methane, for which $B = 5.24$ cm^{-1}.

27.15. Show that the intensities of a molecular beam before and after passing through a chamber of length l containing inert scattering atoms are related by $I = I_0 e^{-\mathcal{N}\sigma l}$, where σ is the collision cross-section and $\mathcal{N}$ the number density of scattering atoms.

27.16. In a molecular beam experiment to measure collision cross-sections it was found that the intensity of a CsCl beam was reduced to 60 per cent of its initial intensity on passage through CH_2F_2 at 10 μTorr, but that when the target was Ar at the same pressure the intensity was reduced by only 10 per cent. What are the relative cross-sections of the two types of collision? Why is one much larger than the other?

28

The properties of surfaces

Much change occurs at the surfaces of liquids and solids. The surface of a liquid is where vaporization and condensation occur, and we shall see a little of the events there. However, the surface of a liquid has interesting properties of its own, including surface tension, and we shall see how the shape of a surface affects the behaviour of the liquid. These properties are modified if a solute is present, particularly if that solute is a surface-active agent, and the role of a solute can be described thermodynamically and explored experimentally. Surface effects also profoundly affect the properties of colloids, particularly their stabilities.

The second part of the chapter deals with the events that take place at solid surfaces, particularly the surfaces of metals. We shall see how solids grow at their surfaces and how the details of the structure and composition of solid surfaces can be determined experimentally. A major part of the material concerns the extent to which a surface is covered and its variation with the pressure and temperature. All this material is put to use in the closing section of the chapter, where we examine how surfaces affect the rate and course of chemical change by acting as the sites of catalysis.

Processes at surfaces govern most aspects of daily life, including life itself. Single layers of molecules attached to solid surfaces reduce friction and mechanical wear and inhibit corrosion. Layers of molecules on liquid surfaces are used to reduce the rate of evaporation of water in arid regions and to stabilize foams, and their properties are important when considering the properties and treatment of pollutants, as in marine oil spills.

Processes at solid surfaces govern the viability of industry both constructively, as in catalysis, and destructively, as in corrosion. Processes at surface layers one or two molecules thick govern the viability of living things, for biological cells have walls that consist of monolayers or bilayers of organic molecules that protect and contain the contents of the cell and allow access to it. Chemical reactions at surfaces may differ sharply from reactions in the bulk. This is particularly true of reactions at solid surfaces, where reaction pathways of much lower activation energy may be provided, and hence result in catalysis. Reactions are also different at liquid surfaces, where there may be significantly different ion concentrations from those in the bulk. The special case of electron transfer reactions at electrode surfaces is treated in the next chapter.

THE PROPERTIES OF LIQUID SURFACES

In this section we concentrate on the liquid–vapour interface, which is interesting because it is so mobile and can be prepared free from local irregularities and defects.

28.1 Surface tension

Liquids tend to adopt shapes that minimize their surface area, for then the maximum number of molecules are in the bulk and hence surrounded by neighbours. Droplets of liquids therefore tend to be spherical, because a sphere is the shape with the smallest surface-to-volume ratio. However, there may be other forces present that compete against the tendency to form this ideal shape and, in particular, gravity may flatten spheres into puddles or oceans.

Surface effects may be expressed in the language of Helmholtz and Gibbs energies. The link between these quantities and the surface area is the work needed to change the area by a given amount, and the fact that dA and dG are equal (under different conditions) to the work done in changing the energy of a system. The work needed to change the surface area σ of a sample by an infinitesimal amount $d\sigma$ is proportional to $d\sigma$, and we write

$$dw = \gamma \, d\sigma \tag{1}$$

The coefficient γ is called the **surface tension**; its dimensions are energy/area ($J\,m^{-2}$). However, as in Table 28.1, values of γ are usually reported in newtons per metre ($N\,m^{-1}$, because $1\,J = 1\,N\,m$). At constant volume and temperature the work of surface formation can be identified with the change in the Helmholtz energy, and we can write

$$dA = \gamma \, d\sigma \tag{2}$$

Because the Helmholtz energy decreases ($dA < 0$) if the surface area decreases ($d\sigma < 0$), surfaces have a natural tendency to contract. This is a more formal way of expressing what we have already described.

Example 28.1 *Using the surface tension*

Calculate the work needed to raise a wire of length l and to stretch the surface of a liquid through a height h in the arrangement shown in Fig. 28.1. Disregard gravitational potential energy.

Method. According to eqn 1, the work required to create a surface area, given that the surface tension does not vary as the surface is formed, is

Table 28.1* Surface tensions of liquids at 293 K

	$\gamma/(mN\,m^{-1})$†
Benzene	28.86
Mercury	472
Methanol	22.6
Water	72.75

* More values are given in the Data section at the end of this volume.
† Note that $1\,N\,m^{-1} = 1\,J\,m^{-2}$.

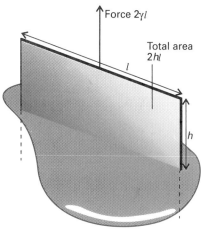

28.1 The model used for calculating the work of forming a liquid film when a wire of length l is raised and pulls the surface with it through a height h.

$w = \gamma\sigma$. Therefore, all we need do is to calculate the surface area of the two-sided rectangle formed as the frame is withdrawn from the liquid.

Answer. When the wire of length l is raised through a height h it increases the area of the liquid by twice the area of the rectangle (because there is a surface on each side). The total increase is therefore $2lh$. The work done is therefore $2\gamma lh$.

Comment. The work can be expressed as a force × distance by writing it as $\gamma l \times 2h$, and identifying γl as the opposing force on the wire of length l. This is why γ is called a tension and why its units are often chosen to be newtons per metre ($N\,m^{-1}$, so γl is a force in newtons).

Exercise E28.1. Calculate the work of creating a spherical cavity of radius r in a liquid of surface tension γ. $[4\pi r^2\gamma]$

28.2 Curved surfaces

A liquid surface is not in general flat. We have seen, for example, that the effect of surface tension is to minimize the surface area, and that minimization of area may result in the formation of a curved surface, as in a bubble. We shall now see that there are two consequences of curvature, and hence of the surface tension, that are relevant to the properties of liquids. One is that the vapour pressure of a liquid depends on the curvature of its surface. The other is the capillary rise (or fall) of liquids in narrow tubes.

Bubbles, cavities, and droplets

By a **bubble** we mean a region in which vapour (and possibly air too) is trapped by a thin film; by a **cavity** we mean a vapour-filled hole in a liquid. What are widely called 'bubbles' in liquids are therefore strictly cavities: true bubbles have two surfaces (one on each side of the film); cavities have only one. The treatments of both are similar, but a factor of 2 is required for bubbles so as to take into account the doubled surface area. By a **droplet** is meant a small volume of liquid at equilibrium surrounded by its vapour (and possibly also air).

The most important conclusion for us is that the pressure on the concave side of an interface is always greater than the pressure on the convex side. This relation is expressed by the **Laplace equation**, which is derived in the following *Justification*:

$$p_{in} = p_{out} + \frac{2\gamma}{r} \tag{3}$$

The Laplace equation also shows that the difference in pressure decreases to zero as the radius of curvature becomes infinite (when the surface is flat). Small cavities have small radii of curvature, and so the pressure difference across their surface is quite large. For instance, a 0.10-mm radius 'bubble' (actually, a cavity) in champagne implies a pressure difference of 1.5 kPa, which is enough to sustain a 15-cm column of water.

JUSTIFICATION

The cavities in a liquid are at equilibrium when the tendency for their surface area to decrease is balanced by the rise of internal pressure which would then result. If the pressure inside a cavity is p_{in} and its radius is r, then the outward force is pressure $\times$ area $= 4\pi r^2 p_{in}$. The force inwards arises from the sum of the external pressure p_{out} and the surface tension. The change in surface area when the radius of a sphere changes from r to $r + dr$ is

$$d\sigma = 4\pi(r + dr)^2 - 4\pi r^2 = 8\pi r \, dr$$

(The second-order infinitesimal $(dr)^2$ is ignored.) The work done when the surface is stretched by this amount is therefore

$$dw = 8\pi\gamma r \, dr$$

As force $\times$ distance is work, the force opposing stretching through a distance dr when the radius is r is

$$F = 8\pi\gamma r$$

When the outward and inward forces are balanced,

$$4\pi r^2 p_{in} = 4\pi r^2 p_{out} + 8\pi\gamma r$$

which rearranges into eqn 3.

We saw in Section 6.3 that the vapour pressure of a liquid depends on the pressure applied to the liquid. Because curving a surface gives rise to a pressure differential of $\Delta p = 2\gamma/r$, we can expect the vapour pressure above a curved surface to be different from that above a flat surface. By substituting this value of Δp into eqn 6.3 we obtain the **Kelvin equation** for the vapour pressure of a liquid when it is dispersed as droplets of radius r:

$$p = p^* e^{2\gamma V_m/rRT} \tag{4a}$$

The analogous expression for the vapour pressure inside a cavity can be written at once. The pressure of liquid outside the cavity is less than the pressure inside, so the only change is in the sign of the exponent in the last expression:

$$p = p^* e^{-2\gamma V_m/rRT} \tag{4b}$$

Nucleation

For droplets of water of radius 10^{-3} and 10^{-6} mm the ratios p/p^* at 25°C are about 1.001 and 3.0, respectively. The second figure, although quite large, is unreliable because at that radius the droplet is less than about 10 molecules in diameter and the basis of the calculation is suspect. The first figure shows that the effect is usually small: nevertheless it may have important consequences. Consider, for example, the formation of a cloud.

Warm, moist air rises into the cooler regions higher in the atmosphere. At some altitude the temperature is so low that the vapour becomes thermodynamically unstable with respect to the liquid and we expect it to condense into a cloud of liquid droplets. The initial step can be imagined as a swarm of water molecules congregating into a microscopic droplet. Because the initial droplet is so small it has an enhanced vapour pressure. Therefore, instead of growing it evaporates. This process effectively stabilizes the vapour because an initial tendency to condense is overcome by a heightened tendency to evaporate. The vapour phase is then said to be **supersaturated**. It is thermodynamically unstable with respect to the liquid but not unstable with respect to the small droplets that need to form before the bulk liquid phase can appear, so the formation of the latter by a simple, direct mechanism is hindered.

Clouds do form, so there must be a mechanism. Two processes are responsible. The first is that a sufficiently large number of molecules might congregate into a droplet so big that the enhanced evaporative effect is unimportant. The chance of one of these **spontaneous nucleation centres** forming is low, and in rain formation it is not a dominant mechanism. The more important process depends on the presence of minute dust particles or other kinds of foreign matter. These **nucleate** the condensation (that is, provide centres at which it can occur) by providing surfaces to which the water molecules can attach. A cloud chamber, a device once used to track the paths of elementary particles, works on the same principle. In a very clean environment a supersaturated mixture of water vapour and air does not condense, but when an ionizing, swiftly moving elementary particle passes through, the ions formed in its path nucleate the condensation and the trajectory is mapped as a streak of condensed water.

Liquids may be **superheated** above their boiling temperatures and **supercooled** below their freezing temperatures. In each case the thermodynamically stable phase is not achieved on account of the kinetic stabilization that occurs in the absence of nucleation centres. For example, superheating occurs because the vapour pressure inside a cavity is artifically low, so any cavity that does form tends to collapse. This instability is encountered when an unstirred beaker of water is heated, for its temperature may be raised above its boiling point. Violent bumping often ensues as spontaneous nucleation leads to bubbles big enough to survive. To ensure smooth boiling at the true boiling temperature, nucleation centres, such as small pieces of sharp-edged glass or bubbles (cavities) of air, should be introduced. The bubble chamber, another device for tracking elementary particles, works on a similar principle, but depends on the nucleation of the evaporation of superheated liquid hydrogen by ionizing radiation.

28.3 Capillary action

The tendency of liquids to rise up capillary tubes (tubes of narrow bore), which is called **capillary action**, is a consequence of surface tension. Consider what happens when a glass capillary tube is first immersed in

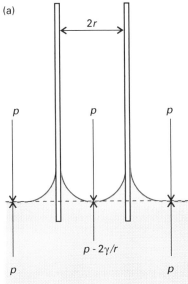

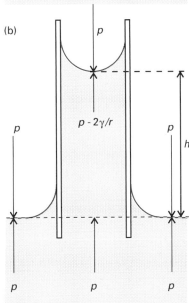

28.2 (a) When a capillary tube is first stood in a liquid, the latter climbs up the walls so curving the surface. The pressure just under the meniscus is less than that arising from the atmosphere by $2\gamma/r$. (b) Equality of pressure at equal heights throughout the liquid is achieved if the liquid rises in the tube. The condition of equilibrium is when the hydrostatic pressure at the foot of the column (which is equal to ρgh) cancels the pressure difference arising from the curved surface.

water or any liquid that has a tendency to adhere to the walls. The energy is lowest when a thin film covers as much of the glass as possible. As this film creeps up the inside wall it has the effect of curving the surface of the liquid inside the tube (Fig. 28.2a). This curvature implies that the pressure just beneath the curving meniscus is less than the atmospheric pressure by approximately $2\gamma/r$, where r is the radius of the tube and we assume a hemispherical surface. The pressure immediately under the flat surface outside the tube is p, the atmospheric pressure, but inside the tube under the curved surface it is only $p - 2\gamma/r$. The excess external pressure presses the liquid up the tube until hydrostatic equilibrium (equal pressures at equal depths) has been reached (Fig. 28.2b).

Capillary rise

The pressure exerted by a column of liquid of density ρ and height h is

$$p = \rho gh \tag{5}$$

This hydrostatic pressure matches the pressure difference $2\gamma/r$ at equilibrium. Therefore, the height of the column at equilibrium is obtained by equating $2\gamma/r$ and ρgh, which gives

$$h = \frac{2\gamma}{\rho gr} \tag{6}$$

This simple expression provides a reasonably accurate way of measuring the surface tension of liquids. For example, if at 25°C water rises through 7.36 cm in a capillary of radius 0.20 mm, then its surface tension at that temperature is

$$\gamma = \tfrac{1}{2}\rho ghr$$
$$= \tfrac{1}{2} \times (997.1\ \text{kg m}^{-3}) \times (9.81\ \text{m s}^{-2})$$
$$\qquad \times (7.36 \times 10^{-2}\ \text{m}) \times (0.20 \times 10^{-3}\ \text{m})$$
$$= 72\ \text{mN m}^{-1}$$

Surface tension decreases with increasing temperature (Fig. 28.3).

When the adhesive forces between the liquid and the material of the capillary wall are weaker than the cohesive forces within the liquid (as for mercury in glass), the liquid in the tube retracts from the walls. This retraction curves the surface with the concave, high-pressure side downwards. To equalize the pressure at the same depth throughout the liquid the surface must fall to compensate for the heightened pressure arising from its curvature. This process results in a capillary depression.

The contact angle

In many cases there is a non-zero angle between the edge of the meniscus and the wall. If this **contact angle** is θ_c, then eqn 6 is modified by multiplying the right-hand side by $\cos\theta_c$.

The origin of the contact angle can be traced to the balance of forces at the line of contact between the liquid and the solid (Fig. 28.4). If the solid/gas, solid/liquid, and liquid/gas surface tensions (essentially the

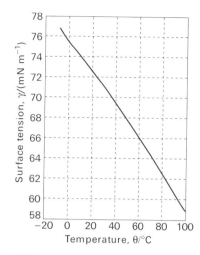

28.3 The variation of the surface tension of water with temperature.

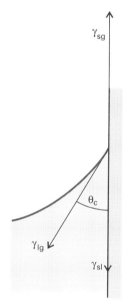

28.4 The balance of forces that results in a contact angle θ_c.

energy needed to create unit area of each of the interfaces) are denoted γ_{sg}, γ_{sl} and γ_{lg}, respectively, then the forces are in balance if

$$\gamma_{sg} = \gamma_{sl} + \gamma_{lg} \cos \theta_c$$

This expression solves to

$$\cos \theta_c = \frac{\gamma_{sg} - \gamma_{sl}}{\gamma_{lg}} \tag{7}$$

If $0 < \theta_c < 90°$ (which occurs when $\gamma_{sl} < \gamma_{sg}$), then the liquid 'wets' (spreads over) the surface fully. In this case, no work is needed to create the solid/liquid interface. If $\theta_c \approx 180°$ (which occurs when $\gamma_{sg} = 0$, and no work is needed to prepare the gas/solid interface), a drop of liquid on the surface of a solid remains separated from it by a thin film of vapour. For mercury in contact with glass, $\theta_c = 140°$, which requires $\gamma_{sl} > \gamma_{sg}$. This condition arises when less energy is needed to create a solid/gas interface than a solid/liquid interface, and is attributed to the strong cohesive forces within the liquid.

SURFACTANTS

From pure liquid surfaces we now turn to the properties of a liquid surface that has a composition differing from that of the bulk liquid. In particular, we consider the role of surfactants at surfaces. A **surfactant** or **surface-active agent** is a species that is active at the interface between two phases, such as at the interface between hydrophilic and hydrophobic phases. A surfactant accumulates at the interface, and modifies its surface tension.

28.4 The surface excess

We begin by establishing the relation between the concentration of surfactant at a surface and the change in surface tension it brings about. To do so, we consider two phases α and β in contact and suppose that the system consists of several components J, each one present in an overall amount n_J. If the components were distributed uniformly through the two phases right up to the interface, which is taken to be a plane of surface area σ, then the total Gibbs energy G would be the sum of the Gibbs energies of both phases:

$$G = G(\alpha) + G(\beta)$$

However, the components are not uniformly distributed because one may accumulate at the interface. As a result, the sum of the two Gibbs energies differs from G by an amount called the **surface Gibbs energy** $G(\sigma)$:

$$G(\sigma) = G - \{G(\alpha) + G(\beta)\}$$

Similarly, if it is supposed that the concentration of a species J is uniform right up to the interface, then from its volume we would conclude that it contains an amount $n_J(\alpha)$ of J in phase α and an amount $n_J(\beta)$ in phase β. However, because a species may accumulate at the interface, the total amount of J differs from the sum of these two amounts by

$$n_J(\sigma) = n_J - \{n_J(\alpha) + n_J(\beta)\}$$

This excess can be expressed as an amount per unit area of the surface by introducing the **surface excess** Γ_J:

$$\Gamma_J = \frac{n_J(\sigma)}{\sigma} \tag{8}$$

$n_J(\sigma)$ and Γ_J may be either positive (an accumulation of J at the interface) or negative (a deficiency there).

The relation between the change in surface tension and the composition of a surface (as expressed by the surface excess) was derived by Gibbs. The first step in the derivation is to obtain the following relation, which is called the **Gibbs surface-tension equation**, between the changes in the chemical potentials of the substances present in the interface and the change in surface tension:

$$d\gamma = -\sum_J \Gamma_J \, d\mu_J \tag{9}$$

JUSTIFICATION

A general change in G is brought about by changes in T, p, and the n_J:

$$dG = -S \, dT + V \, dp + \gamma \, d\sigma + \sum_J \mu_J \, dn_J$$

When this relation is applied to G, $G(\alpha)$, and $G(\beta)$ we find

$$dG(\sigma) = -S(\sigma) \, dT + \gamma \, d\sigma + \sum_J \mu_J \, dn_J(\sigma)$$

because at equilibrium the chemical potential of each component is the same in every phase, $\mu_J(\alpha) = \mu_J(\beta) = \mu_J(\sigma)$. Just as in the discussion of partial molar quantites (Section 7.1), the last equation integrates at constant temperature to

$$G(\sigma) = \gamma\sigma + \sum_J \mu_J n_J(\sigma)$$

We are seeking a connection between the change of surface tension $d\gamma$ and the change of composition at the interface. Therefore, we use the argument which in Section 7.1 led to the Gibbs–Duhem equation (eqn 7.6), but this time we compare the expression

At constant temperature: $dG(\sigma) = \gamma \, d\sigma + \sum_J \mu_J \, dn_J(\sigma)$

with the expression for the same quantity but derived from the preceding equation:

At constant temperature:

$$dG(\sigma) = \gamma \, d\sigma + \sigma \, d\gamma + \sum_J \mu_J \, dn_J(\sigma) + \sum_J n_J(\sigma) \, d\mu_J$$

The comparison implies that

At constant temperature: $\quad \sigma\,d\gamma + \sum_J n_J(\sigma)\,d\mu_J = 0$

Division by σ then gives eqn 9.

Now consider a simplified model of the interface in which the 'oil' and 'water' phases are separated by a geometrically flat surface. This approximation implies that only the surfactant, S, accumulates at the surface, and hence that Γ_{Oil} and Γ_{Water} are both zero. Then the Gibbs equation becomes

$$d\gamma = -\Gamma_S\,d\mu_S$$

For dilute solutions,

$$d\mu_S = RT\,d\ln c$$

where c is the concentration of the surfactant. It follows that

$$d\gamma = -RT\Gamma_S\frac{dc}{c}$$

at constant temperature, or

$$\left(\frac{\partial\gamma}{\partial c}\right)_T = -\frac{RT\Gamma_S}{c} \tag{10}°$$

If the surfactant accumulates at the interface, its surface excess is positive and eqn 10 implies that $(\partial\gamma/\partial c)_T$ is then negative. That is, *the surface tension decreases when a solute accumulates at a surface.* Conversely, if the concentration dependence of γ is known, then the surface excess may be predicted and used to infer the area occupied by each surfactant molecule on the surface.

28.5 The experimental study of surface films

The compositions of surface layers have been investigated by the simple (but technically elegant) procedure of slicing thin layers off the surfaces of solutions and analysing their compositions. The physical properties of surface films have also been investigated. Surface films one molecule thick are called **monolayers** or **Langmuir–Blodgett films** after two of the people who developed experimental techniques for studying them. The primary goal of such investigations is to establish the relation between the surface tension and the surface concentration of a monolayer of surfactant molecules, and hence the proximity of the surfactant molecules to one another on the surface.

The principal apparatus used for the study of surface monolayers is a **surface film balance** (Fig. 28.5). This device consists of a shallow trough and a barrier that can be moved along the surface of the liquid in the trough, and hence compress any monolayer on the surface. The **surface pressure** π, the difference between the surface tension of the pure solvent and the solution ($\pi = \gamma^* - \gamma$), is measured by using a

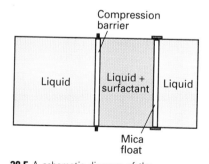

Compression barrier

Liquid | Liquid + surfactant | Liquid

Mica float

28.5 A schematic diagram of the apparatus used to measure the surface pressure and other characteristics of a surface film. The surfactant is spread on the surface of the liquid in the trough, and then compressed horizontally by moving the compression barrier towards the mica float. The latter is connected to a torsion wire, so the difference in force on either side of the float can be monitored.

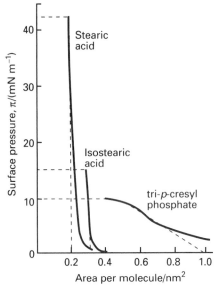

28.6 The variation of surface pressure with the area occupied by each surfactant molecule. The collapse pressures are indicated by the horizontal dashed lines.

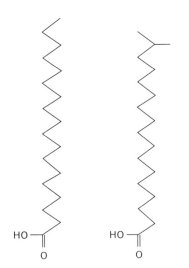

1 Stearic acid   **2** Isostearic acid

3 tri–*p*–cresyl phosphate

torsion wire attached to a strip of mica that rests on the surface and against which one edge of the monolayer is pressed. The parts of the apparatus that are in touch with the liquids are coated in polytetrafluoroethylene to eliminate effects arising from the liquid–solid interface. In an actual experiment, a small amount (about 0.01 mg) of the surfactant under investigation is dissolved in a volatile solvent and then poured on to the surface of the water; the compression barrier is then moved across the surface and the surface pressure exerted on the mica bar is monitored.

Some typical results are shown in Fig. 28.6. One parameter obtained from the isotherms is the area occupied by the molecules when the monolayer is closely packed. This quantity is obtained from the extrapolation of the steepest part of the isotherm to the horizontal axis. As can be seen from the illustration, even though stearic acid (**1**) and isostearic acid (**2**) are chemically very similar (they differ only in the location of a methyl group at the end of a long hydrocarbon chain), they occupy significantly different areas in the monolayer. Neither, though, occupies as much area as the tri-*p*-cresyl phosphate molecule (**3**), which is like a wide bush rather than a lanky tree.

The second feature to note from Fig. 28.6 is that the tri-*p*-cresyl phosphate isotherm is much less steep than the stearic acid isotherms. This difference indicates that the tri-*p*-cresyl phosphate film is more compressible than the stearic acid film, which is consistent with their different molecular structures.

A third feature of the isotherms is the **collapse pressure**, the highest surface pressure. When the monolayer is compressed beyond the point represented by the collapse pressure, the monolayer buckles and collapses into a film several molecules thick. As can be seen from the isotherms in Fig. 28.6, stearic acid has a high collapse pressure, but that of tri-*p*-cresyl phosphate is significantly smaller, indicating a much weaker film.

COLLOIDAL SYSTEMS

Now we turn from largely flat liquid surfaces to surfaces that are highly curved, and consider colloids. A **colloid** is a dispersion of small particles of one material in another. 'Small' means something less than about 500 nm in diameter (about the wavelength of visible light). In general, colloidal particles are aggregates of numerous atoms or molecules, but are too small to be seen with an ordinary optical microscope. They pass through most filter papers, but can be detected by light-scattering, sedimentation, and osmosis.

Colloids are relevant to a discussion of surfaces because the ratio of their surface area to their volume is so large that their properties are dominated by events at their surfaces. For example, a cube of material of side 1 cm has a surface area of 6 cm^2, but when it is dispersed as 10^{18} little cubes of side 10 nm the total surface area is 6×10^6 cm^2 (about the size of a tennis court).

28.6 Classification and preparation

The name given to the colloid depends on the two phases involved. A **sol** is a dispersion of a solid in a liquid (such as clusters of gold atoms in water) or of a solid in a solid (such as ruby glass, which is a gold-in-glass sol, and achieves its colour by scattering). An **aerosol** is a dispersion of a liquid in a gas (like fog and many sprays) or a solid in a gas (such as smoke): the particles are often large enough to be seen with a microscope. An **emulsion** is a dispersion of a liquid in a liquid (such as milk).

A further classification of colloids is as **lyophilic** (solvent-attracting) and **lyophobic** (solvent-repelling). If the sovent is water, the terms **hydrophilic** and **hydrophobic**, respectively, are used instead. Lyophobic colloids include the metal sols. Lyophilic colloids generally have some chemical similarity to the solvent, such as −OH groups able to form hydrogen bonds. A **gel** is a semirigid mass of a lyophilic sol in which all the dispersion medium has penetrated into the sol particles.

The preparation of aerosols can be as simple as sneezing (which produces an imperfect aerosol). Laboratory and commercial methods make use of several techniques. Material (e.g. quartz) may be ground in the presence of the dispersion medium. Passing a heavy electric current through a cell may lead to the sputtering (crumbling) of an electrode into colloidal particles. Arcing between electrodes immersed in the support medium also produces a colloid. Chemical precipitation sometimes results in a colloid. A precipitate (e.g. silver iodide) already formed may be dispersed by the addition of a **peptizing agent** (e.g. potassium iodide). Clays may be peptized by alkalis, the OH$^-$ ion being the active agent.

Emulsions are normally prepared by shaking the two components together vigorously, although some kind of **emulsifying agent** usually has to be added to stabilize the product. This emulsifying agent may be a soap (a long-chain carboxylic acid), a surfactant, or a lyophilic sol that forms a protective film around the dispersed phase. In milk, which is an emulsion of fats in water, the emulsifying agent is casein, a protein containing phosphate groups. It is clear from the formation of cream on the surface of milk that casein is not completely successful in stabilizing milk: the dispersed fats coalesce into oily droplets that float to the surface. This coagulation may be prevented by ensuring that the emulsion is dispersed very finely initially: intense agitation with ultrasonics brings this about, the product being 'homogenized' milk.

One way to form an aerosol is to tear apart a spray of liquid with a jet of gas. The dispersal is aided if a charge is applied to the liquid, for then electrostatic repulsions help to blast it apart into droplets. This procedure may also be used to produce emulsions, for the charged liquid phase may be directed into another liquid.

Colloids are often purified by **dialysis**. The aim is to remove much (but not all, for reasons explained later) of the ionic material that may have accompanied their formation. As in the discussion of the Donnan effect in Section 23.2, a membrane (e.g. cellulose) is selected that is permeable to solvent and ions, but not to the colloid particles. Dialysis is very slow, and is normally accelerated by applying an electric field and making use of the charges carried by many colloid particles; the technique is then called **electrodialysis**.

28.7 Surface, structure, and stability

The existence of colloids gives rise to a number of problems concerning their ability to persist for long periods.

The stability of colloids

As a result of their great surface area, colloids are thermodynamically unstable with respect to the bulk. This instability can be expressed thermodynamically by noting that because $dG = \gamma \, d\sigma$, where γ is the surface tension and $d\sigma$ the change in surface area, it follows that dG is negative for a decrease in area. The survival of colloids must therefore be a consequence of the kinetics of collapse: colloids are thermodynamically unstable but kinetically non-labile (rather like a mixture of hydrogen and oxygen gases at room temperature).

At first sight, even the kinetic argument seems to fail: colloidal particles attract each other over large distances, so there is a long-range force that tends to condense them into a single blob. The reasoning behind this remark is as follows. The energy of attraction between two individual atoms, one in each colloidal particle, varies as their separation as $1/R_{ij}^6$ (Section 22.3). The sum of all these pairwise interactions, however, decreases only as approximately $1/R^2$ (the precise variation depending on the shape of the particles and their closeness), where R is the separation of the centres of the particles. The sum has a much longer range than the $1/R^6$ dependence characteristic of individual particles and small molecules.

Several factors oppose the long-range dispersion attraction. For example, there may be a protective film at the surface of the colloid particles that stabilizes the interface and cannot be penetrated when two particles touch. Thus the surface atoms of a platinum sol in water react chemically and are turned into $-Pt(OH)_3H_3$, and this layer encases the particle like a shell. A fat can be emulsified by a soap because the long hydrocarbon tails penetrate the oil droplet but the carboxylate head groups (or other hydrophilic groups in synthetic detergents) surround the surface, form hydrogen bonds with water, and give rise to a shell of negative charge that repels a possible approach from another similarly charged particle.

Micelle formation and the hydrophobic interaction

Soap molecules can cluster together as **micelles**, which are colloid-sized clusters of molecules, for their hydrophobic tails tend to congregate, and their hydrophilic heads provide protection (Fig. 28.7). Micelles form only above the **critical micelle concentration** (CMC) and above the **Krafft temperature**. The CMC is detected by noting a pronounced discontinuity in physical properties of the solution, particularly the molar

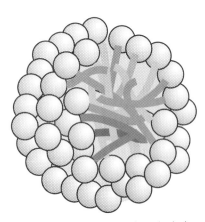

28.7 A schematic version of a spherical micelle. The hydrophilic groups are represented by spheres and the hydrophobic hydrocarbon chains are represented by the stalks; these stalks are mobile.

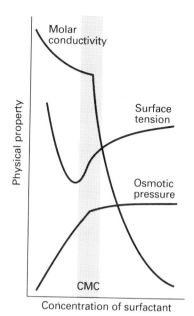

28.8 The typical variation of some physical properties of an aqueous solution of sodium dodecylsulfate close to the critical micelle concentration (CMC).

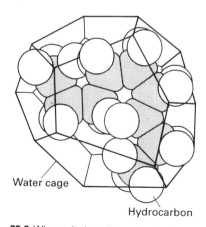

28.9 When a hydrocarbon molecule is surrounded by water, the H_2O molecules form a clathrate cage. As a result of this acquisition of structure, the entropy of the water decreases, so the dispersal of the hydrocarbon into the water is entropy-opposed; its coalescence is entropy-favoured.

conductivity (Fig. 28.8). The hydrocarbon interior of a micelle is like a droplet of oil. Nuclear magnetic resonance shows that the hydrocarbon tails are mobile, but slightly more restricted than in the bulk. Micelles are important in industry and biology on account of their solubilizing function: matter can be transported by water after it has been dissolved in their hydrocarbon interiors. For this reason, micellar systems are used as detergents and drug carriers, and for organic synthesis, froth flotation, and petroleum recovery.

Non-ionic surfactant molecules may cluster together in clumps of 1000 or more, but ionic species tend to be disrupted by the electrostatic repulsions between head groups and are normally limited to groups of between 10 and 100. The micelle population is often polydisperse (that is, spread over a range of numbers of molecules in each micelle), and the shapes of the individual micelles vary with concentration. Although spherical micelles do occur, micelles are more commonly flattened spheres close to the CMC. Some micelles at concentrations well above the CMC form extended parallel sheets, called **lamellar micelles**, two molecules thick. The individual molecules lie perpendicular to the sheets, with hydrophilic groups on the outside in aqueous solution and on the inside in non-polar media. Such lamellar micelles show a close resemblance to biological membranes, and are often a useful model on which to base investigations of biological structures. In concentrated solutions micelles formed from surfactant molecules may take the form of long cylinders and stack together in reasonably close-packed (hexagonal) arrays. These orderly arrangements of micelles are called **lyotropic mesomorphs** and, more colloquially, **liquid crystalline phases**.

The thermodynamics of micelle formation shows that the enthalpy of formation in aqueous systems is probably positive (that is, that they are endothermic) with $\Delta H \approx 1-2$ kJ per mole of surfactant. That micelles do form above the CMC indicates that the entropy change accompanying their formation must then be positive, and measurements suggest a value of about $+140 \, \text{J K}^{-1} \, \text{mol}^{-1}$ at room temperature. The fact that the entropy change is positive even though the molecules are clustering together shows that there must be a contribution to the entropy from the solvent and that solvent molecules must be more free to move once the solute molecules have herded into small clusters. This interpretation is plausible, because each individual solute molecule is held in an organized solvent cage (Fig. 28.9), but once the micelle has formed the solvent molecules need form only a single (admittedly larger) cage. The increase in energy when hydrophobic groups cluster together and reduce their structural demands on the solvent is the origin of the **hydrophobic interaction** that tends to stabilize groupings of hydrophobic groups in biological macromolecules. The hydrophobic interaction is an example of an ordering process that is stabilized by a tendency toward greater disorder of the solvent.

The electric double layer

A major source of kinetic stability of colloids is the existence of an electric charge on the surfaces of the particles. On the account of this charge, ions of opposite charge tend to cluster nearby, and an ionic atmosphere is formed, just as was described for ions in Section 10.2.

Two regions of charge must be distinguished. First, there is a fairly immobile layer of ions that adhere tightly to the surface of the colloidal particle, and which may include water molecules (if that is the support medium). The radius of the sphere that captures this rigid layer is called the **radius of shear**, and is the major factor determining the mobility of the particles. The electric potential at the radius of shear relative to its value in the distant, bulk medium is called the **zeta potential** ζ or the **electrokinetic potential**. Second, the charged unit atrracts an oppositely charged atmosphere of mobile ions. The inner shell of charge and the outer ionic atmosphere is called the **electric double layer**.

The theory of the stability of lyophobic dispersions was developed by D. Derjaguin and L. Landau and independently by E. Verwey and J. T. G. Overbeek, and is known as the **DLVO theory**. It assumes that there is a balance between the repulsive interaction between the charges of the electric double layers on neighbouring particles and the attractive interactions arising from van der Waals interactions between the molecules in the particles. The potential energy arising from the repulsion of double layers on particles of radius a has the form

$$V_{\text{Repulsion}} = +\frac{Aa^2\zeta^2}{R}e^{-s/r_D} \tag{11a}$$

where A is a constant, ζ is the zeta potential,[1] R is the separation of centres, s is the separation of the surfaces of the two particles ($s = R - 2a$ for spherical particles of radius a), and r_D is the thickness of the double layer. This expression is valid for small particles with a thick double layer ($a \ll r_D$). When the double layer is thin ($a \gg r_D$), the expression is replaced by

$$V_{\text{Repulsion}} = +\frac{Aa\zeta^2}{2}\ln(1 + e^{-s/r_D}) \tag{11b}$$

In each case the thickness of the double layer can be estimated from an expression like that derived for the thickness of the ionic atmosphere in the Debye–Hückel theory (eqn 5 of *Further information 4*):

$$r_D = \left(\frac{\varepsilon RT}{2\rho F^2 I}\right)^{\frac{1}{2}} \tag{12}$$

In this expression, I is the ionic strength of the solution. The potential energy arising from the attractive interaction has the form

$$V_{\text{Attraction}} = -\frac{B}{s} \tag{13}$$

where B is another constant. The variation of the total potential energy with separation is shown in Fig. 28.10.

At high ionic strengths the ionic atmosphere is dense and the potential shows a secondary minimum at large separations. Aggregation of the

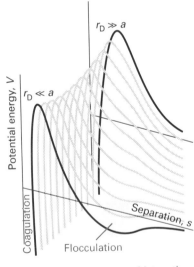

28.10 The potential energy of interaction as a function of the separation of the centres of the two particles and its variation with the ratio of the particle size a to the thickness of the electric double layer r_D. The regions labelled coagulation and flocculation show the dips in the potential energy curves where these processes occur.

---

1 The actual potential is that of the surface of the particles; there is some danger in identifying it with the zeta potential. See the references in *Further reading*.

particles arising from the stabilizing effect of this secondary minimum is called **flocculation**. The flocculated material can often be redispersed by agitation because the well is so shallow, and full **coagulation**, the irreversible blending together of distinct particles into large particles, has not occurred. The latter occurs when the separation of the particles is so small that they enter the primary minimum of the potential energy curve and van der Waals forces are dominant.

The ionic strength is increased by the addition of ions, particularly those of high charge type, so such ions act as flocculating agents. This increase is the basis of the empirical **Schulze–Hardy rule**, that hydrophobic colloids are flocculated most efficiently by ions of opposite charge type and high charge number. The Al^{3+} ions in alum are very effective, and are used to induce the congealing of blood. When river water containing colloidal clay flows into the sea, the salt water induces flocculation and coagulation, and is a major cause of silting in estuaries.

Metal oxide sols tend to be positively charged whereas sulfur and the noble metals tend to be negatively charged. Naturally occurring macromolecules also acquire a charge when dispersed in water, and an important feature of proteins and other natural macromolecules is that their overall charge depends on the pH of the medium. For instance, in acidic environments protons attach to basic groups, and the net charge of the macromolecule is positive; in basic media the net charge is negative as a result of proton loss. At the **isoelectric point** the pH is such that there is no net charge on the macromolecule.

Example 28.2 *Determining the isoelectric point*

The drift speed of bovine serum albumin (BSA) under the influence of an electric field in aqueous solution was monitored at several values of pH, and the data are listed below (opposite signs indicate opposite directions of travel). What is the isoelectric point of the protein?

pH	4.20	4.56	5.20	5.65	6.30	7.00
Speed/($\mu m\ s^{-1}$)	+0.50	+0.18	−0.25	−0.65	−0.90	−1.25

Method. The macromolecule has zero electrophoretic mobility when it is uncharged. Therefore, the isoelectric point is the pH at which it does not migrate in an electric field. We should therefore plot speed against pH and find by interpolation the pH of zero mobility.

Answer. The data are plotted in Fig. 28.11. The drift speed is zero at pH = 4.8; hence pH = 4.8 is the isoelectric point.

Comment. For some species, the isoelectric point must be obtained by extrapolation because the macromolecule might not be stable over the whole pH range.

Exercise E28.2. The following data were obtained for another protein:

pH	4.5	5.0	5.5	6.0
Speed/($\mu m\ s^{-1}$)	−0.10	−0.20	−0.30	−0.35

Estimate the pH of the isoelectric point.

[4.3]

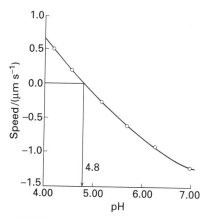

28.11 The plot of drift speed against pH by which the isoelectric point of a macromolecule can be determined: it corresponds to the pH at which the drift speed in the presence of an electric field is zero.

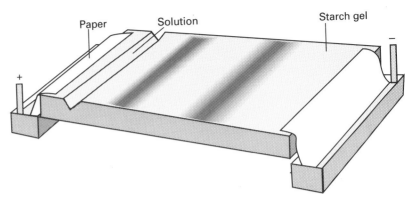

28.12 The layout of a simple electrophoresis apparatus. The sample is introduced into the trough in the gel, and the different components form separate bands as they migrate under the influence of the electric field.

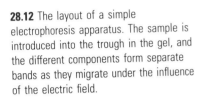

The primary role of the electric double layer is to confer kinetic stability. Colliding colloidal particles break through the double layer and coalesce only if the collision is sufficiently energetic to disrupt the layers of ions and solvating molecules, or if thermal motion has stirred away the surface accumulation of charge. This disruption may occur at high temperatures, which is one reason why sols precipitate when they are heated. The protective role of the double layer is the reason why it is important not to remove all the ions when a colloid is being purified by dialysis, and why proteins coagulate most readily at their isoelectric point.

The presence of charge on colloidal particles and natural macromolecules also permits us to control their motion, such as in dialysis and electrophoresis. Apart from its application to the determination of molar mass (Section 23.3), electrophoresis has several analytical and technological applications. One analytical application is to the separation of different macromolecules, and a typical apparatus is illustrated in Fig. 28.12. Technical applications include the painting of objects by airborne charged paint droplets, and electrophoretic rubber forming by deposition of charged rubber molecules on anodes formed into the shape of the desired product (e.g. surgical gloves).

THE GROWTH AND STRUCTURE OF SOLID SURFACES

Now we turn to a consideration of solid surfaces. In this section we see how surfaces are extended and crystals grow, and begin to picture the structures of solid surfaces that are responsible for catalysis. The attachment of particles to a surface is called **adsorption**. The substance that adsorbs is the **adsorbate** and the underlying material is the **adsorbent** or **substrate**. The reverse of adsorption is **desorption**.

28.8 Surface growth

A simple picture of a perfect crystal surface is as a tray of oranges in a grocery store (Fig. 28.13). A gas molecule that collides with the surface can be imagined as a ping-pong ball bouncing erratically over the oranges. The molecule loses energy as it bounces, but it is likely to escape from the surface before it has lost enough kinetic energy to be

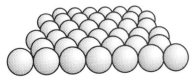

28.13 A schematic diagram of the flat surface of a solid. This primitive model is largely supported by scanning-tunnelling microscope images (see Fig. 28.33).

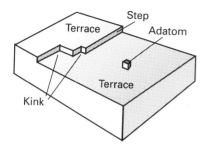

28.14 Some of the kinds of defects that may occur on otherwise perfect terraces. Defects play an important role in surface growth and catalysis.

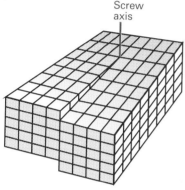

28.15 A screw dislocation occurs where one region of the crystal is pushed up through one or more unit cells relative to another region. The cut extends to the screw axis. As atoms lie along the step the dislocation rotates round the screw axis, and is not annihilated.

28.16 The spiral growth arising from the propagation of a screw axis is clearly visible in this photograph of an alkane crystal. (B. R. Jennings and V. J. Morris, *Atoms in contact*, Clarendon Press, Oxford (1974); courtesy of Dr A. J. Forty.)

trapped. The same is true, to some extent, of an ionic crystal in contact with a solution. There is little energy advantage for an ion in solution to discard some of its solvating molecules and stick at an exposed position on the surface. However, the actual changes in energy and entropy (collectively, the Gibbs energy) are difficult to assess because an ion interacts strongly with other ions below the surface of the solid and discards its hydrating molecules as it attaches to the surface.

The role of defects

The picture changes when the surface has defects, for then there are ridges of incomplete layers of atoms or ions. A typical type of surface defect is a **step** between two otherwise flat layers of atoms called **terraces** (Fig. 28.14). A step defect might itself have defects for it might have **kinks**. When an atom settles on a terrace it bounces across it under the influence of the intermolecular potential, and might come to a step or a corner formed by a kink. Instead of interacting with a single terrace atom it now interacts with several, and the interaction may be strong enough to trap it. Likewise, when ions deposit from solution, the loss of the solvation interaction is offset by a strong Coulombic interaction between the arriving ions and several ions at the surface defect.

Dislocations

Not all kinds of defect result in sustained surface growth. As the process of settling into ledges and kinks continues, there comes a stage when an entire lower terrace has been covered. At this stage the surface defects have been eliminated, and growth will cease. For continuing growth, a surface defect is needed that propagates as the crystal grows. We can see what form of defect this must be by considering the types of **dislocations**, or discontinuities in the regularity of the lattice, that exist in the bulk of a crystal. One reason for their formation may be that the crystal grows so quickly that its particles do not have time to settle into states of lowest potential energy before being trapped in position by the deposition of the next layer.

A special kind of dislocation is the **screw dislocation** shown in Fig. 28.15. Imagine a cut in the crystal, with the particles to the left of the cut pushed up through the distance of one unit cell. The unit cells now form a continuous spiral around the end of the cut, which is called the **screw axis**. A path encircling the screw axis spirals up to the top of the crystal, and where the dislocation breaks through to the surface it takes the form of a spiral ramp.

The surface defect formed by a screw dislocation is a step, possibly with kinks, where growth can occur. The incoming particles lie in ranks on the ramp, and successive ranks reform the step at an angle to its initial position. As deposition continues the step rotates around the screw axis, and is not eliminated. Growth may therefore continue indefinitely. Several layers of deposition may occur, and the edges of the spirals might be cliffs several atoms high (Fig. 28.16).

Propagating spiral edges can also give rise to flat terraces (Fig. 28.17).

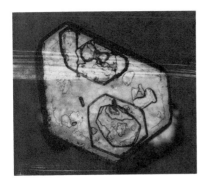

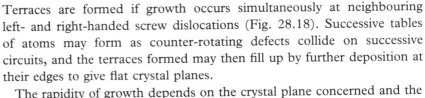

Terraces are formed if growth occurs simultaneously at neighbouring left- and right-handed screw dislocations (Fig. 28.18). Successive tables of atoms may form as counter-rotating defects collide on successive circuits, and the terraces formed may then fill up by further deposition at their edges to give flat crystal planes.

The rapidity of growth depends on the crystal plane concerned and the *slowest* growing faces dominate the appearance of the crystal. This is explained in Fig. 28.19, where we see that, although the horizontal face grows forward most rapidly, it grows itself out of existence, and the slower-growing faces survive.

28.17 The spiral growth pattern is sometimes concealed because the terraces are subsequently completed by further deposition. This accounts for the appearance of this cadmium iodide crystal. (H. M. Rosenberg, *The solid state*, Clarendon Press, Oxford (1978).)

28.9 Surface composition

The first step in understanding how surfaces catalyse reactions is to characterize the clean adsorbent surface. In this context 'clean' means much more than scrubbing the sample and handling it with care. Under normal conditions a surface is constantly bombarded with gas particles and a freshly prepared surface is covered very quickly. Just how quickly can be estimated using the kinetic theory of gases and the expression it gives (eqn 24.3) for the number of collisions per unit area per unit time when the pressure is p:

$$Z_W = \frac{p}{(2\pi m k T)^{\frac{1}{2}}} \tag{14a}$$

A practical form of this equation is

$$Z_W = (2.63 \times 10^{24}\,\mathrm{m}^{-2}\,\mathrm{s}^{-1}) \times \frac{p/\mathrm{Pa}}{\{(T/\mathrm{K}) \times (M/\mathrm{g\,mol}^{-1})\}^{\frac{1}{2}}} \tag{14b}$$

where M is the molar mass of the gas. For air ($M \approx 29\,\mathrm{g\,mol}^{-1}$) at 1 atm and 25°C the collision frequency is $3 \times 10^{27}\,\mathrm{m}^{-2}\,\mathrm{s}^{-1}$. Because $1\,\mathrm{m}^2$ of metal surface consists of about 10^{19} atoms, each atom is struck about 10^8 times each second. Even if only a few collisions leave a molecule adsorbed to the surface, the time for which a freshly prepared surface remains clean is very short.

High-vacuum techniques

The obvious way to retain cleanliness is to reduce the pressure. When it is reduced to $100\,\mu\mathrm{Pa}$ (as in a simple vacuum system) the collision frequency falls to about $10^{18}\,\mathrm{m}^{-2}\,\mathrm{s}^{-1}$, corresponding to one hit per surface atom in each 0.1 s. Even that is too brief in most experiments, and in **ultra-high vacuum** (UHV) techniques pressures as low as $0.1\,\mu\mathrm{Pa}$ ($10^{-9}\,\mathrm{Torr}$, when $Z_W = 10^{15}\,\mathrm{m}^{-2}\,\mathrm{s}^{-1}$) are reached on a routine basis and $1\,\mathrm{nPa}$ ($10^{-11}\,\mathrm{Torr}$, when $Z_W = 10^{13}\,\mathrm{m}^{-2}\,\mathrm{s}^{-1}$) are reached with special care. These collision frequencies correspond to each surface atom being hit once every 10^4 to 10^6 s, or about once a day.

The layout of a typical UHV apparatus is such that the whole of the evacuated part can be heated to 200–300°C for several hours to drive gas molecules from the walls. All the taps and seals are of metal in order to avoid contamination from greases. The sample is usually in the form of a

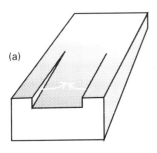

(a)

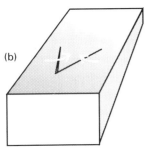

(b)

28.18 Counter-rotating screw dislocations on the same surface lead to the formation of terraces. Four stages of one cycle of growth are shown here. Subsequent deposition can complete each terrace. (The illustration is continued on the next page.)

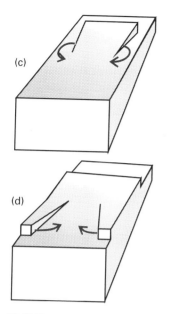

(c)

(d)

28.18 (Continued)

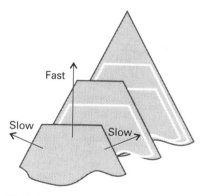

28.19 The slower-growing faces of a crystal dominate its final external appearance. Three successive stages of the growth are shown.

thin foil, a filament, or a sharp point. Where there is interest in the role of specific crystal planes the sample is a single crystal with a freshly cleaved face. Initial surface cleaning is achieved either by heating it electrically or by bombarding it with accelerated gaseous ions. The latter procedure demands care because ion bombardment can shatter the surface structure and leave it an amorphous jumble of atoms. High-temperature annealing is then required to return the surface to an ordered state.

Ionization techniques

Surface composition can be determined by a variety of ionization techniques. The same techniques can be used to detect any remaining contamination after cleaning and to detect layers of material adsorbed later in the experiment. Their common feature is that the **escape depth** of the electrons, the maximum depth from which ejected electrons come, is in the range 0.1–1.0 nm, which ensures that only surface species contribute.

One technique that may be used is photoelectron spectroscopy (Section 17.8), which in surface studies is normally called **photoemission spectroscopy**. X-Rays or hard ultraviolet ionizing radiation may be used, but X-PES (which examines inner-shell binding energies) seems to be better than UV-PES for the analysis of composition because it is able to fingerprint the materials present (Fig. 28.20). UV-PES, which examines electrons ejected from valence shells, is more suited to establishing the bonding characteristics and the details of valence shell electronic structures of substances on the surface. Its usefulness is its ability to reveal which orbitals of the adsorbate are involved in the bond to the substrate. For instance, the principal difference between the PES results on free benzene and benzene adsorbed on palladium is in the energies of the π electrons. This difference is interpreted as meaning that the C_6H_6 molecules lie parallel to the surface and are attached to it through their π orbitals. In contrast, pyridine is known to stand more or less perpendicular to the surface, and is attached to it by means of a σ bond formed by the nitrogen lone pair. The ejected electrons may be focused and an image built up of the surface from which they have been ejected. This is the technique of **photoelectron spectromicroscopy** (PESM), and a typical image is shown in Fig. 28.21.

Electron energy-loss spectroscopy

Several kinds of vibrational spectroscopy have been developed to study adsorbates and to show whether dissociation has occurred. Infrared and Raman spectroscopy have been greatly improved by the development of laser and Fourier transform techniques, and have largely overcome the difficulties arising from low intensities on account of the low surface coverages normally encountered under laboratory conditions.

A hybrid version of PES and vibrational spectroscopy is **electron energy-loss spectroscopy** (EELS) in which the energy loss suffered by

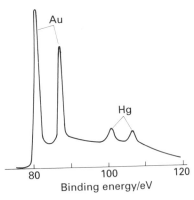

28.20 The X-ray photoelectron emission spectrum of a sample of gold contaminated with a surface layer of mercury. (M. W. Roberts and C. S. McKee, *Chemistry of the metal–gas interface*, Oxford (1978).)

4

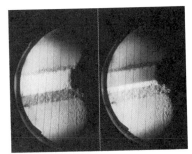

28.21 Photoelectron spectromicroscopy images of a sectional silicon diode (left) unbiased and (right) with −30 V reverse bias. (Photograph provided by Professor D. W. Turner.) The technique allows the detection of regions of a sample with different compositions and can be used, as here, to monitor the changes to those regions that occur when the sample is modified.

28.22 The electron energy loss spectrum of CO adsorbed on Pt(111). The results for three different pressures are shown, and the growth of the additional peak at about 200 meV (1600 cm^{-1}) should be noted. (Spectra provided by Professor H. Ibach.)

a beam of electrons is monitored when they are reflected from a surface. As in optical Raman spectroscopy, the spectrum of energy loss can be interpreted in terms of the vibrational spectrum of the adsorbate. High resolution and sensitivity are attainable, and the technique is sensitive to light elements (to which X-ray techniques are insensitive). Very tiny amounts of adsorbate can be detected, and one report estimates that about 48 atoms of phosphorus were detected in one sample. As an example, Fig. 28.22 shows the EELS result for CO on the (111) face of a platinum crystal as the extent of surface coverage increases. The main peak above 100 meV arises from CO attached perpendicular to the surface by a single Pt atom. As the coverage increases the smaller peak next to it increases in intensity. This peak is due to CO at a bridge site, where it is attached to two Pt atoms, as in (**4**).

Auger electron spectroscopy

A very important technique, which is widely used in the microelectronics industry, is **Auger electron spectroscopy**. The **Auger effect** is the emission of a second electron after high-energy radiation has expelled another. The first electron to depart leaves a hole in a low-lying orbital, and an upper electron falls into it. The energy this releases may result either in the generation of radiation, which is called **X-ray fluorescence** (Fig. 28.23a) or in the ejection of another electron (Fig. 28.23b). The latter is the secondary electron of the Auger effect. The energies of the secondary electrons are characteristic of the material present, so the Auger effect effectively takes a fingerprint of the sample (Fig. 28.24). In practice, the Auger spectrum is normally obtained by irradiating the sample with an electron beam rather than electromagnetic radiation.

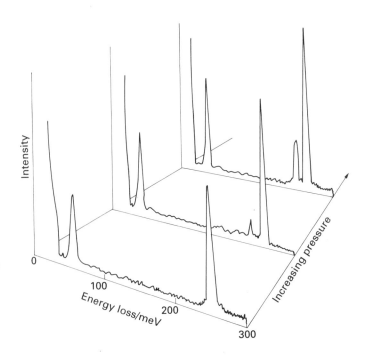

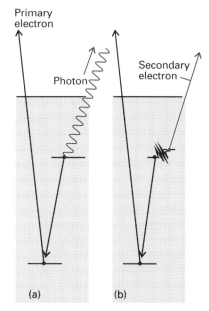

28.23 When an electron is expelled from a solid (a) an electron of higher energy may fall into the vacated orbital and emit an X-ray photon to produce X-ray fluorescence. Alternatively (b) the electron falling into the orbital may give up its energy to another electron, which is ejected in the Auger effect.

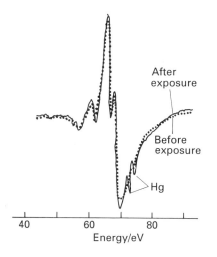

28.24 An Auger spectrum of the same sample used for Fig. 28.20 taken before and after deposition of mercury. (M. W. Roberts and C. S. McKee, *Chemistry of the metal—gas interface*, Oxford (1978).)

The technique known as **surface-extended X-ray absorption fine-structure spectroscopy** (SEXAFS) makes use of the intense X-radiation from synchrotron sources (Section 16.1). The technique makes use of the oscillations in X-ray absorbance that are observed on the high-frequency side of the absorption edge (the start of an X-ray absorption band) of a substance. It arises from a quantum mechanical interference between the wavefunction of a photoejected electron and parts of that electron's wavefunction that are scattered by neighbouring atoms. If the waves interfere destructively, then the photoelectron appears with lower probability and the X-ray absorption is correspondingly less. If the waves interfere constructively, then the photoelectron amplitude is higher, the electron has a higher probability of emerging, and, correspondingly, the X-ray absorption is greater. The oscillations therefore contain information about the number and distances of the neighbouring atoms. Such studies show that a solid's surface is much more plastic than had previously been thought, and that it undergoes reconstruction in response to adsorbates that are present. An example is given in Fig. 28.25, which shows how a palladium surface is reconstructed in different ways and to different extents at different temperatures and with varying surface coverage of CO.

Low-energy electron diffraction

One of the most informative techniques for determining the arrangement of the atoms close to the surface is **low-energy electron diffraction** (LEED). This technique is essentially electron diffraction (Section 21.10), but the sample is now the surface of a solid. The use of low-energy electrons ensures that the diffraction is caused only by atoms on and close to the surface. The experimental arrangement is shown in Fig. 28.26, and typical LEED patterns, obtained by photographing the fluorescent screen through the viewing port, are shown in Fig. 28.27.

The LEED pattern portrays the two-dimensional structure of the surface. By studying how the diffraction intensities depend on the energy of the electron beam it is also possible to infer some details about the vertical location of the atoms and to measure the thickness of the surface layer, but the interpretation of LEED data is much more complicated than the interpretation of bulk X-ray data. The pattern is sharp if the surface is well-ordered for distances long compared with the wavelength of the incident electrons. In practice, sharp patterns are obtained for surfaces ordered to depths of about 20 nm and more. Diffuse patterns indicate either a poorly ordered surface or the presence of impurities. If the LEED pattern does not correspond to the pattern expected by extrapolation of the bulk surface to the surface, then either a reconstruction of the surface has occurred or there is order in the arrangement of an adsorbed layer.

LEED experiments show that the surface of a crystal rarely has exactly the same form as a slice through the bulk. As a general rule, it is found that metal surfaces are simply truncations of the bulk lattice, but the distance between the top layer of atoms and the one below is contracted

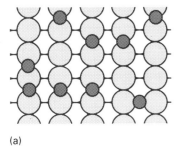

(a)

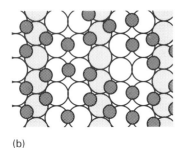

(b)

28.25 The restructuring of a surface that sometimes accompanies chemisorption. The diagrams illustrate the changes that occur when CO adsorbs on the (110) face of palladium. (a) Up to 0.3 monolayers, the surface structure is unchanged. (b) A metastable structure produced at low temperature when the surface is saturated.

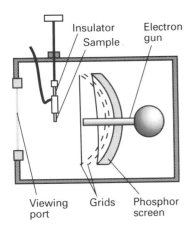

28.26 A schematic diagram of the apparatus used for a LEED experiment. The electrons diffracted by the surface layers are detected by the fluorescence they cause on the phosphor screen.

by around 5 per cent. Semiconductors generally have surfaces reconstructed to a depth of several layers. Reconstruction occurs in ionic solids. For example, in lithium fluoride the Li^+ and F^- ions close to the surface apparently lie on slightly different planes. An actual example of the detail that can now be obtained from refined LEED techniques (specifically ATLEED, where the prefix AT stands for 'automated tensor' and denotes a specific manner of handling the data computationally), is shown in Fig. 28.28 for CH_3C- (ethylidyne) adsorbed on a (111) plane of rhodium.

The presence of terraces, steps, and kinks in a surface shows up in LEED patterns, and their densities (the number of defects per unit area) can be estimated. The importance of this will emerge later. Three examples of how steps and kinks affect the pattern are shown in Fig. 28.29. The samples used were obtained by cleaving a crystal at different angles to a plane of atoms. Only terraces are produced when the cut is parallel to the plane, and the density of steps increases as the angle of the cut increases. The observation of additional structure in the LEED patterns, not merely blurring, shows that the steps are arrayed regularly.

Field emission and ionization microscopy

Until recently, the most spectacular portrayals of surface structure were obtained from two closely related techniques.

In **field emission microscopy** (FEM) the sample is in the form of a filament etched to a sharp tip and enclosed within a chamber fitted with a fluorescent screen. When a large potential difference is applied between the sample and the screen, electrons are stripped out towards the screen and give a flash of light where they strike it. The ease with which the electrons can escape from the metal depends on the variation of the work function (Section 11.1) with the composition and structure of the surface, and the screen shows a corresponding variation of intensity (Fig. 28.30). The change in the FEM pattern when material has been deposited can be used to detect the places where atoms are most likely to stick.

The technique of **field ionization microscopy** (FIM) is a development of FEM. The apparatus is virtually the same, but the potential difference is reversed with the fluorescent screen made negative relative to the tip. In the experiment a small quantity of gas (typically helium) is admitted. An atom strikes the tip and bounces over its terraced, cobbled surface until it strikes a protruding atom, such as one on the rim of a ledge (Fig. 28.31). Protruding atoms are able to ionize the gas atom and, immediately the positive ion (He^+) is formed, the potential difference plucks it off towards the screen, where its collision generates fluorescence.

The spatial resolution that can be achieved with FIM depends on the transverse motion of the gas ions. Such motion can be reduced by cooling the tip to about 20 K, when the resolution is of the order of atomic dimensions and the positions of individual atoms can be discerned (Fig. 28.32). This remarkable picture, in which the small bright

(a)

(b)

28.27 LEED photographs of (a) a clean platinum surface and (b) after its exposure to propyne, $CH_3C\equiv CH$. (Photographs provided by Professor G. A. Samorjai.)

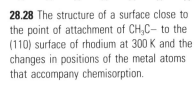

28.28 The structure of a surface close to the point of attachment of CH_3C- to the (110) surface of rhodium at 300 K and the changes in positions of the metal atoms that accompany chemisorption.

28.29 LEED patterns may be used to assess the defect density of a surface. The photographs correspond to a platinum surface with (a) low defect density, (b) regular steps separated by about six atoms, and (c) regular steps with kinks. (Photographs provided by Professor G. A. Samorjai.)

spots are caused by individual terrace atoms, shows the power of the technique. However, we should not forget its limitations. First, ionization occurs unequally at different atoms, and many of the atoms on the surface and even on the edges are insufficiently exposed to cause ionization and so remain invisible. Second, the sample must be in the form of a tip, and be made of material strong enough to withstand the very high electric fields required. Despite these limitations, FIM is remarkable for its ability to portray the positions of individual atoms.

An elegant refinement of FIM identifies individual atoms on an otherwise clean tip. **Atom-probe FIM** is the ultimate in surgery. The FIM image of an adsorbed atom is brought into coincidence with a hole in the fluorescent screen. The imaging gas is removed, and a pulse of potential difference plucks off the atom (as an ion). It moves in the same direction as did the gas ions, and passes through the hole in the screen. Behind the screen is a mass spectrometer, and so the atom can be identified.

Apart from being the ultimate analytical technique (because as well as knowing what the atom is we also know exactly where it was in the sample) events can be observed that on an atomic scale are really dramatic. For example, the analysing pulse lasts for about 2 ns, and during that time the evaporation of about 10 atomic layers is sometimes observed. This corresponds to a rate of evapoaration equivalent to the surface receding at about 1 m s^{-1}.

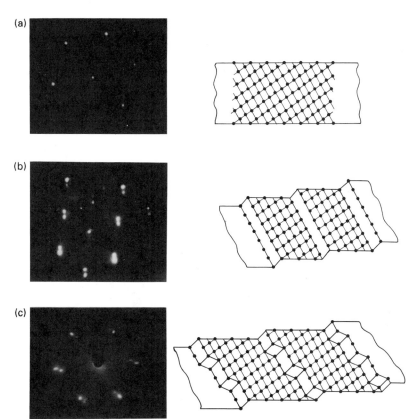

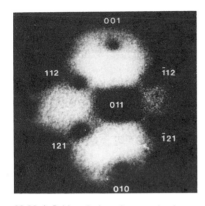

28.30 A field emission photograph of a tungsten tip of radius 210 nm and the assignment to the exposed crystal faces. (M. Prutton, *Surface physics*, Clarendon Press, Oxford (1975).)

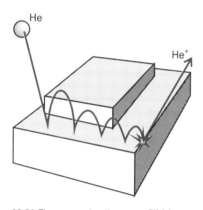

28.31 The events leading to an FIM image of a surface. The He atom migrates across the surface until it is ionized at an exposed atom, when it is pulled off by the externally applied potential. (The bouncing motion is due to the intermolecular potential, not gravity!)

Scanning-tunnelling microscopy

The **scanning-tunnelling microscope** (STM), which was developed during the early 1980s, has made possible the most striking life-like images of surfaces and of the adsorbates on them.

The central component of the microscope is a platinum–rhodium needle, which is scanned across the sample. The needle is attached to a cylinder of piezoelectric ceramic that expands and contracts lengthwise in response to an electric current. When the tip of the needle is brought very close to the surface, electrons tunnel across the space and modify the potential difference there. The monitoring electronics are designed to maintain a constant potential difference across the gap between the tip and the sample by passing a current through the ceramic that keeps the size of the gap constant. The equipment monitors the current needed, and interprets it to give the height of the tip above the surface. Because the tunnelling probability is very sensitive to the size of the gap (Section 12.3), the microscope can detect tiny, atom-scale variations in the height of the surface. An example of the kind of image obtained with a clean surface is shown in Fig. 28.33, where the cliff is only one atom high. A spectacular demonstration of the power of the technique for displaying the shape of adsorbed species is shown in Fig. 28.34, which shows a single molecule adsorbed on a surface.

Molecular beam techniques

Whereas many important studies have been carried out simply by exposing a surface to a gas, modern work is increasingly making use of

28.32 An FIM image of an iridium tip showing the details of the exposed crystal faces. (Photograph provided by Professor E. W. Müller.)

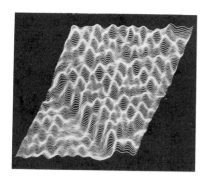

28.33 The type of image that can be obtained with scanning-tunnelling microscopy. The sample is of a silicon surface and the cliff is one atom high. (Sang-il Park and C. F. Quaite.)

molecular beam techniques. One advantage is that the activity of specific crystal planes can be investigated by directing the beam on to an orientated surface with known step and kink densities (as measured by LEED). Furthermore, if the adsorbate reacts at the surface the products (and their angular distributions) can be analysed as they are ejected from the surface and pass into a mass spectrometer. Another advantage is that the time of flight of a particle may be measured and interpreted in terms of its residence time on the surface. In this way a very detailed picture can be constructed of the events taking place during reactions at surfaces.

THE EXTENT OF ADSORPTION

The extent of surface coverage is normally expressed as the fractional coverage θ:

$$\theta = \frac{\text{Number of adsorption sites occupied}}{\text{Number of adsorption sites available}} \tag{15}$$

The fractional coverage is often expressed in terms of the volume of adsorbate adsorbed by $\theta = V/V_\infty$, where V_∞ is the volume of adsorbate corresponding to complete monolayer coverage. The rate of adsorption, $d\theta/dt$, is the rate of change of surface coverage, and can be determined by observing the change of fractional coverage with time.

Among the principal techniques for measuring $d\theta/dt$ are flow methods, in which the sample itself acts as a pump because adsorption removes particles from the gas. One commonly used technique is therefore to monitor the rates of flow of gas into and out of the system: the difference is the rate of gas uptake by the sample. Integration of this rate then gives the fractional coverage at any stage. In **flash desorption** the sample is suddenly heated (electrically) and the rise of pressure this leads to is interpreted in terms of the amount originally on the sample. The interpretation may be confused by the desorption of a compound (e.g. WO_3 from oxygen on tungsten). **Gravimetry**, in which the sample is

28.34 A scanning-tunnelling microscope image of a liquid crystal molecule (5-nonyl-2-nonoxylphenylpyrimidine) adsorbed on a graphite surface. (J. S. Foster *et al.*, *Nature*, **338**, 137 (1988).)

weighed on a microbalance during the experiment, can also be used, as can radioactive tracers. In the latter case, the radioactivity of the sample is measured after exposure to an isotopically labelled gas.

28.10 Physisorption and chemisorption

Molecules and atoms can attach to surfaces in two ways. In **physisorption** (an abbreviation of 'physical adsorption'), there is a van der Waals interaction (for example, a dispersion or a dipolar interaction) between the adsorbate and the substrate. Van der Waals interactions have a long range but are weak, and the energy released when a particle is physisorbed is of the same order of magnitude as the enthalpy of condensation. Such small amounts of energy can be absorbed as vibrations of the lattice and dissipated as thermal motion, and a molecule bouncing across the surface will gradually lose its energy and finally adsorb to it in the process called **accommodation**. The enthalpy of physisorption can be measured by monitoring the rise in temperature of a sample of known heat capacity, and typical values are in the region of $20 \, kJ \, mol^{-1}$ (Table 28.2). This small enthalpy change is insufficient to lead to bond breaking, so a physisorbed molecule retains its identity, although it might be distorted by the presence of the surface.

In **chemisorption** (an abbreviation of 'chemical adsorption'), the particles stick to the surface by forming a chemical (usually covalent) bond, and tend to find sites that maximize their coordination number with the substrate. The enthalpy of chemisorption is very much greater than that for physisorption, and typical values are in the region of $200 \, kJ \, mol^{-1}$ (Table 28.3). A chemisorbed molecule may be torn apart at the demand of the unsatisfied valencies of the surface atoms, and the existence of molecular fragments on the surface as a result of chemisorption is one reason why solid surfaces catalyse reactions.

Except in special cases, chemisorption must be exothermic. A spontaneous process requires a negative ΔG. Because the translational freedom of the adsorbate is reduced when it is adsorbed, ΔS is negative. Therefore, in order for $\Delta G = \Delta H - T \Delta S$ to be negative, ΔH must be negative (and the process exothermic). Exceptions may occur if the adsorbate dissociates and has high translational mobility on the surface. For example, H_2 adsorbs endothermically on glass because there is a large increase of translational entropy accompanying the dissociation of the molecules into atoms that move quite freely over the surface. In its case, the entropy change in the process $H_2(g) \rightarrow 2H(glass)$ is sufficiently positive to overcome the small positive enthalpy change.

The principal test for distinguishing chemisorption from physisorption used to be the enthalpy of adsorption. Values less negative than $-25 \, kJ \, mol^{-1}$ were taken to signify physisorption, and values more negative than about $-40 \, kJ \, mol^{-1}$ were taken to signify chemisorption. However, this criterion is by no means foolproof and spectroscopic techniques that identify the adsorbed species are now available.

The enthalpy of adsorption depends on the extent of surface coverage, mainly because the adsorbate particles interact. If the particles repel each

Table 28.2* Maximum observed enthalpies of physisorption, $\Delta_{ad}H^{\ominus}/(kJ \, mol^{-1})$

CH_4	-21
H_2	-84
H_2O	-59
N_2	-21

* More values are given in the Data section.

Table 28.3* Enthalpies of chemisorption, $\Delta_{ad}H^{\ominus}/(kJ \, mol^{-1})$

Adsorbate	Adsorbent (substrate)		
	Cr	Fe	Ni
C_2H_4	-427	-285	-243
CO		-192	
H_2	-188	-134	
NH_3		-188	-155

* More values are given in the Data section.

other (as for CO on palladium) the enthalpy of adsorption becomes less exothermic (less negative) as coverage increases. Moreover, LEED studies show that such species settle on the surface in a disordered way until packing requirements demand order. If the adsorbate particles attract each other (as for O_2 on tungsten), then they tend to cluster together in islands, and growth occurs at the borders. These adsorbates also show order–disorder transitions when they are heated enough for thermal motion to overcome the particle–particle interactions, but not so much that they are desorbed.

28.11 Adsorption isotherms

The free gas and the adsorbed gas are in dynamic equilibrium, and the fractional coverage of the surface depends on the pressure of the overlying gas. The variation of θ with pressure at a chosen temperature is called the **adsorption isotherm**.

The Langmuir isotherm

The simplest isotherm is based on three assumptions:

(1) Adsorption cannot proceed beyond monolayer coverage.

(2) All sites are equivalent and the surface is uniform (that is, the surface is perfectly flat on a microscopic scale).

(3) The ability of a molecule to adsorb at a given site is independent of the occupation of neighbouring sites.

The dynamic equilibrium is

$$A(g) + M(surface) \rightleftharpoons AM(surface)$$

with rate constants k_a for adsorption and k_d for desorption. The rate of change of surface coverage due to adsorption is proportional to the partial pressure p of A and the number of vacant sites $N(1 - \theta)$, where N is the total number of sites:

$$\frac{d\theta}{dt} = k_a pN(1 - \theta)$$

The rate of change of θ due to desorption is proportional to the number of adsorbed species, $N\theta$:

$$\frac{d\theta}{dt} = -k_d N\theta$$

At equilibrium the net rate of adsorption is zero, and solving for θ gives the **Langmuir isotherm**:

$$\theta = \frac{Kp}{1 + Kp} \qquad \text{where } K = \frac{k_a}{k_d} \tag{16}$$

Example 28.3 *Using the Langmuir isotherm*

The data given below are for the adsorption of CO on charcoal at 273 K.

Confirm that they fit the Langmuir isotherm, and find the constant K and the volume corresponding to complete coverage. In each case V has been corrected to 1 atm.

p/Torr	100	200	300	400	500	600	700
V/cm^3	10.2	18.6	25.5	28.4	36.9	41.6	46.1

Method. From eqn 16,

$$Kp\theta + \theta = Kp$$

With $\theta = V/V_\infty$, where V_∞ is the volume corresponding to complete coverage, this expression can be rearranged into

$$\frac{p}{V} = \frac{p}{V_\infty} + \frac{1}{KV_\infty}$$

Hence, a plot of p/V against p should give a straight line of slope $1/V_\infty$ and intercept $1/KV_\infty$.

Answer. The data for the plot are as follows:

p/Torr	100	200	300	400	500	600	700
$(p$/Torr$)/(V$/cm$^3)$	9.80	10.8	11.8	12.7	13.6	14.4	15.2

The points are plotted in Fig. 28.35. The slope is 0.0090, so $V_\infty = 110$ cm^3. The intercept at $p = 0$ is 9.0, so

$$K = \frac{1}{(110 \text{ cm}^3) \times (9.0 \text{ Torr cm}^{-3})} = 1.0 \times 10^{-3} \text{ Torr}^{-1}$$

Comment. At high surface coverage, the data would deviate from a straight line. The dimensions of K are 1/pressure.

Exercise E28.3. Repeat the calculation for the following data:

p/Torr	100	200	300	400	500	600	700
V/cm^3	10.3	19.3	27.3	34.1	40.0	45.5	48.0

$$[150 \text{ cm}^3, 7.4 \times 10^{-3} \text{ Torr}^{-1}]$$

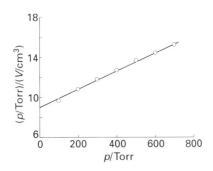

28.35 The Langmuir isotherm predicts that a straight line should be obtained when p/V is plotted against p. This is a test of the isotherm using the information in Example 28.3.

For adsorption with dissociation, the rate of adsorption is proportional to the pressure and to the probability that *both* atoms will find sites, which is proportional to the *square* of the number of vacant sites,

$$\frac{d\theta}{dt} = k_a p\{N(1 - \theta)\}^2$$

The rate of desorption is proportional to the frequency of encounters of atoms on the surface, and is therefore second-order in the number of atoms present:

$$\frac{d\theta}{dt} = -k_d(N\theta)^2$$

The condition for the net rate of adsorption to be zero leads to the

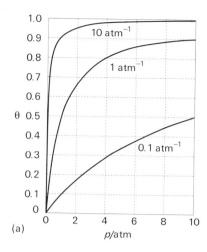

(a)

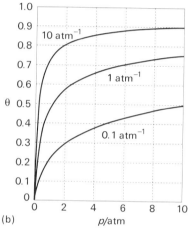

(b)

28.36 The Langmuir isotherm for (a) non-dissociative and (b) dissociative adsorption for different values of K.

isotherm

$$\theta = \frac{(Kp)^{\frac{1}{2}}}{1 + (Kp)^{\frac{1}{2}}} \tag{17}$$

The surface coverage now depends more weakly on pressure than for non-dissociative adsorption.

The shapes of the Langmuir isotherms with and without dissociation are shown in Fig. 28.36. The fractional coverage increases with increasing pressure, and approaches 1 only at very high pressure, when the gas is forced on to every available site of the surface. Different curves (and therefore values of K) are obtained at different temperatures, and the temperature dependence of K can be used to determine the **isosteric enthalpy of adsorption**, $\Delta_{ad}H^{\ominus}$, the standard enthalpy of adsorption at a fixed surface coverage. To determine this quantity we recognize that K is essentially an equilibrium constant, and then use the van't Hoff equation (eqn 9.12) to write

$$\left(\frac{\partial \ln K}{\partial T}\right)_{\theta} = \frac{\Delta_{ad}H^{\ominus}}{RT^2} \tag{18}$$

Example 28.4 *Measuring the isosteric enthalpy of adsorption*

The data below show the pressures of CO needed for the volume of adsorption (corrected to 1 atm and 273 K) to be 10.0 cm^3 using the same sample as in Example 28.3. Calculate the adsorption enthalpy at this surface coverage.

T/K	200	210	220	230	240	250
$p/Torr$	30.0	37.1	45.2	54.0	63.5	73.9

Method. The Langmuir isotherm can be rearranged to

$$Kp = \frac{\theta}{1 - \theta}$$

Therefore, when θ is constant,

$$\ln K + \ln p = \text{constant}$$

It follows from eqn 18 that

$$\left(\frac{\partial \ln p}{\partial T}\right)_{\theta} = -\left(\frac{\partial \ln K}{\partial T}\right)_{\theta} = -\frac{\Delta_{ad}H^{\ominus}}{RT^2}$$

With $d(1/T) \, dT = -1/T^2$ this expression rearranges to

$$\left(\frac{\partial \ln p}{\partial (1/T)}\right)_{\theta} = \frac{\Delta_{ad}H^{\ominus}}{R}$$

Therefore, a plot of $\ln p$ against $1/T$ should be a straight line of slope $\Delta_{ad}H^{\ominus}/R$.

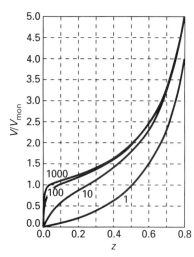

28.37 The isosteric enthalpy of adsorption can be obtained from the slope of the plot of $\ln p$ against $1/T$, where p is the pressure needed to achieve the specified surface coverage. The data used are from Example 28.4.

Answer. We draw up the following table:

T/K	200	210	220	230	240	250
$10^3/(T/K)$	5.00	4.76	4.55	4.35	4.17	4.00
$\ln(p/\text{Torr})$	3.40	3.61	3.81	3.99	4.15	4.30

The points are plotted in Fig. 28.37. The slope (of the least squares fitted line) is -0.90, so

$$\Delta_{\text{ad}}H^{\ominus} = -(0.90 \times 10^3 \, \text{K}) \times R = -7.5 \, \text{kJ mol}^{-1}$$

Comment. The value of K can be used to obtain a value of $\Delta_{\text{ad}}G^{\ominus}$, and then that value combined with $\Delta_{\text{ad}}H^{\ominus}$ to obtain the standard entropy of adsorption.

Exercise E28.4. Repeat the calculation using the following data:

T/K	200	210	220	230	240	250
p/Torr	32.4	41.9	53.0	66.0	80.0	96.0

$$[-9.0 \, \text{kJ mol}^{-1}]$$

The BET isotherm

If the initial overlayer may act as a substrate for further (e.g. physical) adsorption then, instead of the isotherm levelling off to some saturated value at high pressures, it can be expected to rise indefinitely. The most widely used isotherm dealing with multilayer adsorption was derived by Stephen Brunauer, Paul Emmett, and Edward Teller, and is called the **BET isotherm**:

$$\frac{V}{V_{\text{mon}}} = \frac{cz}{(1-z)\{1-(1-c)z\}} \quad \text{where } z = \frac{p}{p^*} \qquad (19)$$

In this expression, p^* is the vapour pressure above a layer of adsorbate that is more than one molecule thick and which resembles a pure bulk liquid, V_{mon} is the volume corresponding to monolayer coverage, and c is a constant which is large when the enthalpy of desorption from a monolayer is large compared with the enthalpy of vaporization of the liquid adsorbate:

$$c \approx e^{(\Delta_{\text{des}}H^{\ominus} - \Delta_{\text{vap}}H^{\ominus})/RT} \qquad (20)$$

The shapes of BET isotherms are illustrated in Fig. 28.38. They rise indefinitely as the pressure is increased because there is no limit to the amount of material that may condense when multilayer coverage may occur.

28.38 Plots of the BET isotherm for different values of c. The value of V/V_{mon} rises indefinitely because the adsorbate may condense on the covered substrate surface.

Example 28.5 *Using the BET isotherm*

The data below relate to the adsorption of N_2 on rutile (TiO_2) at 75 K. Confirm that they fit a BET isotherm in the range of pressures reported, and determine V_{mon} and c.

p/Torr	1.20	14.0	45.8	87.5	127.7	164.4	204.7
V/mm^3	601	720	822	935	1046	1146	1254

At 75 K, $p^* = 570$ Torr. The volumes have been corrected to 1 atm and 273 K and refer to 1.0 g of substrate.

Method. Equation 19 can be reorganized into

$$\frac{z}{(1-z)V} = \frac{1}{cV_{mon}} + \frac{(c-1)z}{cV_{mon}}$$

It follows that $(c-1)/cV_{mon}$ can be obtained from the slope of a plot of the expression on the left against z, and cV_{mon} can be found from the intercept at $z = 0$. The results can then be combined to give c and V_{mon}.

Answer. We draw up the following table:

p/Torr	1.20	14.0	45.8	87.5	127.7	164.4	204.7
$10^3 z$	2.11	24.6	80.4	154	224	288	359
$10^4 z/\{(1-z)(V/\text{mm}^3)\}$	0.035	0.350	1.06	1.95	2.76	3.53	4.47

These points are plotted in Fig. 28.39. The least-squares best line has an intercept at 0.034, so

$$\frac{1}{cV_{mon}} = 0.034 \times 10^{-4}\,\text{mm}^{-3} = 3.4 \times 10^{-6}\,\text{mm}^{-3}$$

The slope of the line is 1.23×10^{-2}, so

$$\frac{c-1}{cV_{mon}} = (1.23 \times 10^{-2}) \times 10^3 \times 10^{-4}\,\text{mm}^{-3} = 1.23 \times 10^{-3}\,\text{mm}^{-3}$$

Solving these equations gives $c - 1 = 362$, or $c = 363$, and $V_{mon} = 810\,\text{mm}^3$.

Comment. At 1 atm and 273 K, $810\,\text{mm}^3$ corresponds to $3.6 \times 10^{-5}\,\text{mol}$, or 2.2×10^{19} atoms. Because each atom occupies an area of about $0.16\,\text{nm}^2$, the surface area of the sample is about $3.5\,\text{m}^2$.

Exercise E28.5. Repeat the calculation for the following data:

p/Torr	1.20	14.0	45.8	87.5	127.7	164.4	204.7
V/cm^3	235	559	649	719	790	860	950

[290, 620 cm^3]

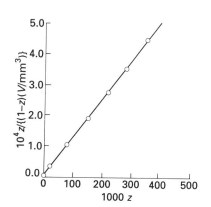

28.39 The BET isotherm can be tested, and the parameters determined, by plotting $z/(1-z)V$ against $z = p/p^*$. The data are from Example 28.5.

When the coefficient c is large ($c \gg 1$), the BET isotherm takes the simpler form

$$\frac{V}{V_{mon}} = \frac{1}{1-z} \tag{21}$$

This expression is applicable to unreactive gases on polar surfaces, for which $c \approx 10^2$ because $\Delta_{des}H^\ominus$ is then significantly greater than $\Delta_{vap}H^\ominus$ (eqn 20). The BET isotherm fits experimental observations moderately well over restricted pressure ranges, but it errs by underestimating the extent of adsorption at low pressures and by overestimating it at high pressures.

Other isotherms

An assumption of the Langmuir isotherm is the independence and equivalence of the adsorption sites. Deviations from the isotherm can often be traced to the failure of these assumptions. For example, the enthalpy of adsorption often becomes less negative as θ increases, which suggests that the energetically most favourable sites are occupied first. Various attempts have been made to take these variations into account. The **Temkin isotherm**,

$$\theta = c_1 \ln (c_2 p) \tag{22}$$

where c_1 and c_2 are constants, corresponds to supposing that the adsorption enthalpy changes linearly with pressure. The **Freundlich isotherm**,

$$\theta = c_1 p^{1/c_2} \tag{23}$$

corresponds to a logarithmic change. Different isotherms agree with experiment more or less well over restricted ranges of pressure, but they remain largely empirical. Empirical, however, does not mean useless, for if the parameters of a reasonably reliable isotherm are known, reasonably reliable results can be obtained for the extent of surface coverage under various conditions. This kind of information is essential for any discussion of heterogeneous catalysis.

Example 28.6 *Assessing the Freundlich and Langmuir isotherms*

Decide whether the Freundlich isotherm is a better representation than the Langmuir isotherm for the data in Example 28.3.

Method. If we write $\theta = V/V_{mon}$, and rearrange eqn 23 into

$$\ln V = \ln (c_1 V_{mon}) + \frac{1}{c_2} \ln p$$

then we can plot $\ln V$ against $\ln p$ and compare the least-squares fit of the points with the plot in Example 28.3. The best way of doing this is to quote the correlation coefficient of the least-squares lines, a measure of the fit of the data to a straight line (see the Appendix of the *Solutions manual*).

Answer. Draw up the following table.

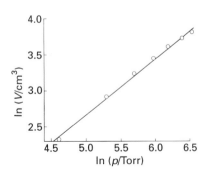

28.40 The Freundlich isotherm may be tested by plotting $\ln V$ against $\ln p$ because it predicts that a straight line should be obtained. The data are from Example 28.6.

p/Torr	100	200	300	400	500	600	700
$\ln (p/\text{Torr})$	4.61	5.30	5.70	5.99	6.21	6.40	6.55
$\ln (V/\text{cm}^3)$	2.32	2.92	3.24	3.45	3.61	3.73	3.83

The points are plotted in Fig. 28.40. The correlation coefficient of the least-squares best line is 0.9968. The value for the Langmuir plot is 0.9979. Hence the latter is marginally better.

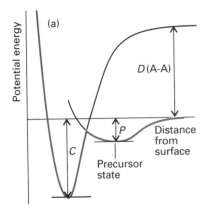

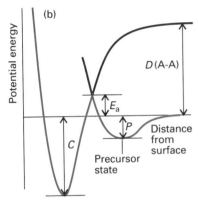

28.41 The potential energy profiles for the dissociative chemisorption of an A–A molecule. In each case P is the enthalpy of (non-dissociative) physisorption and C that for chemisorption (at $T = 0$). The relative locations of the curves determines whether the chemisorption is (a) not activated or (b) activated.

Comment. The Freundlich isotherm is often used in discussions of adsorption from liquid solutions, when it is written

$$w = c_1 \times c^{1/c_2}$$

where w is the mass fraction adsorbed (the mass of solute adsorbed per unit mass of adsorbent) and c is the solution's concentration.

Exercise E28.6. Examine the suitability of the Temkin isotherm to fit the data. $[V \propto \ln p, 0.9731]$

28.12 The rates of surface processes

Figure 28.41 shows how the potential energy of a molecule varies with its distance from the substrate surface. As the particle approaches the surface its energy falls as it becomes physisorbed into the **precursor state** for chemisorption. Dissociation into fragments often takes place as a molecule moves into its chemisorbed state, and after an initial increase of energy as the bonds stretch there is a sharp decrease as the adsorbate–substrate bonds reach their full strength. Even if the molecule does not fragment, there is likely to be an initial increase of potential energy as it adjusts its bonds as it approaches the surface.

In most cases, therefore, we can expect there to be a potential energy barrier separating the precursor and chemisorbed states. This barrier, though, might be low, and might not rise above the energy of a distant, stationary particle (as in Fig. 28.41a). In this case, chemisorption is not an activated process and can be expected to be rapid. Many gas adsorptions on clean metals appear to be non-activated. In some cases the barrier rises above the zero axis (as in Fig. 28.41b); such chemisorptions are activated and slower than the non-activated kind. An example is H_2 on copper, which has an activation energy somewhere in the region of $20–40 \, \text{kJ} \, \text{mol}^{-1}$.

One point that emerges from this discussion is that rates are not good criteria for distinguishing between physisorption and chemisorption. Chemisorption can be fast if the activation energy is small or zero, but it may be slow if the activation energy is large. Physisorption is usually fast, but it can appear to be slow if adsorption is taking place on a porous medium.

The rate of adsorption

The rate at which a surface is covered by adsorbate depends on the ability of the substrate to dissipate the energy of the incoming particle as thermal motion as it crashes on to the surface. If the energy is not dissipated quickly, the particle migrates over the surface until a vibration expels it into the overlying gas or it reaches an edge. The proportion of collisions with the surface that successfully lead to adsorption is called the **sticking probability** s:

$$s = \frac{\text{Rate of adsorption of particles by the surface}}{\text{Rate of collision of particles with the surface}} \tag{24}$$

The denominator can be calculated from kinetic theory, and the numerator can be measured by observing the rate of change of pressure.

Values of s vary widely. For example, at room temperature CO has s in the range 0.1–1.0 for several d-metal surfaces, but for N_2 on rhenium $s < 10^{-2}$, indicating that more than 100 collisions are needed before one molecule sticks successfully. Beam studies on specific crystal planes show a pronounced specificity: for N_2 on tungsten, s ranges from 0.74 on the (320) faces down to less than 0.01 on the (110) faces at room temperature.

Example 28.7 *Calculating the time needed for surface coverage*

Calculate the time for 10 per cent of the sites on a (100) surface of a certain substrate to be covered with nitrogen at 298 K when the pressure is $0.30\,\mu Pa$ and the sticking probability is 0.55. Take the substrate to have a bcc structure with lattice constant 316 pm.

Method. The number of atoms on the surface should be calculated first. The number of collisions required to cover the whole surface is the number of atoms divided by the sticking probability; but we need to note that each N_2 molecule occupies two sites. The time required for that number of collisions is the number divided by the number of collisions per unit time, which can be obtained from Z_W (eqn 14).

Answer. The distance between lattice sites on the (100) face is 316 pm; therefore, each atom accounts for $(316\,pm)^2$ of the surface. The number of atoms in $1.00\,m^2$ is therefore $(1.00\,m^2)/(316\,pm)^2 = 1.00 \times 10^{19}$, so the surface density of sites is $1.00 \times 10^{19}\,m^{-2}$. From eqn 14, $Z_W = 8.6 \times 10^{15}\,m^{-2}\,s^{-1}$. Therefore, the time required for 10 per cent coverage is

$$t = \frac{0.10 \times (1.00 \times 10^{19}\,m^{-2})}{2 \times 0.55 \times Z_W} = 1.1 \times 10^2\,s$$

Comment. Covering the entire surface will take longer than 10 times 110 s because the sticking probability decreases with increasing coverage.

Exercise E28.7. Calculate the time for the same sample to be covered to the extent of 20 per cent.   [2.2×10^2 s]

The sticking probability decreases as the surface coverage increases (Fig. 28.42). A simple assumption is that s is proportional to $1 - \theta$, the fraction uncovered, and it is common to write

$$s = (1 - \theta)s_0$$

where s_0 is the sticking probability on a perfectly clean surface. The results in the illustration do not fit this expression because they show that s remains close to s_0 until the coverage has risen to about 6×10^{13} molecules cm^{-2}, and then falls steeply. The explanation is probably that the colliding molecule does not enter the chemisorbed state at once, but moves over the surface until it encounters an empty site.

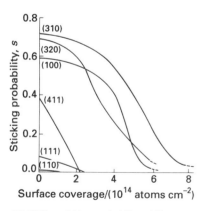

28.42 The sticking probability of N_2 on various faces of a tungsten crystal and its dependence on surface coverage. Note the very low sticking probability for the (110) and (111) faces. (Data provided by Professor D. A. King.)

The rate of desorption

Desorption is always activated because the particles have to be lifted from the foot of a potential well. A physisorbed particle vibrates in its shallow potential well, and might shake itself off the surface after a short time. The temperature dependence of the first-order rate of departure can be expected to be Arrhenius-like, with an activation energy comparable to the enthalpy of physisorption:

$$k_d = A e^{-E_d/RT}$$

where E_d is the activation energy for desorption. Therefore, the half-life for remaining on the surface has a temperature dependence

$$t_{\frac{1}{2}} = \frac{\ln 2}{k_d} = \tau_0 e^{E_d/RT} \qquad \text{where } \tau_0 = \frac{\ln 2}{A} \tag{25}$$

If we suppose that $1/\tau_0$ is approximately the same as the vibrational frequency of the weak particle–surface bond (about 10^{12} Hz) and $E_d \approx 25$ kJ mol^{-1}, then residence half-lives of around 10^{-8} s are predicted at room temperature. Lifetimes close to 1 s are obtained only by lowering the temperature to about 100 K. For chemisorption, with $E_d = 100$ kJ mol^{-1} and guessing that $\tau_0 = 10^{-14}$ s (because the adsorbate–substrate bond is quite stiff), we expect a residence half-life of about 3×10^3 s (about an hour) at room temperature, decreasing to 1 s at about 350 K.

The desorption activation energy can be measured in several ways. However, we must be guarded in its interpretation because it often depends on the fractional coverage and so may change as desorption proceeds. Moreover, the transfer of concepts such as 'reaction order' and 'rate constant' from bulk studies to surfaces is hazardous, and there are few examples of strictly first-order or second-order desorption kinetics (just as there are few integral-order reactions in the gas phase too).

If we disregard these complications, then one way of measuring the desorption activation energy is to monitor the rate of increase in pressure when the sample is maintained at a series of temperatures, and to attempt to make an Arrhenius plot. A more sophisticated technique is the determination of the **flash desorption spectrum** of the sample. The basic observation is that in a pumped vessel there is a pressure surge when the temperature is raised to the value at which the activation energy is overcome and desorption occurs rapidly; but once the desorption has occurred there is no more to escape from the surface, so the pressure falls again as the temperature continues to rise and pumping continues. The flash desorption spectrum, the plot of the pressure against temperature, therefore consists of a peak, the location of which depends on the desorption activation energy. In the example shown in Fig. 28.43 three peaks are shown, indicating the presence of three sites with different activation energies.

In many cases only a single activation energy (and a single peak in the flash desorption spectrum) is observed. When several peaks are observed

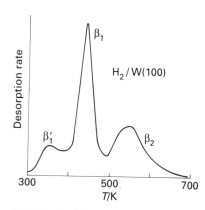

28.43 The flash desorption spectrum of H_2 on the (100) face of tungsten. The three peaks indicate the presence of three sites with different adsorption enthalpies and therefore different desorption activation energies. (P. W. Tamm and L. D. Schmidt, *J. Chem. Phys.* **51**, 5352 (1969).)

they might correspond to adsorption on different crystal planes or to multilayer adsorption. For instance, Cd atoms on tungsten show two activation energies, one of 18 and the other of $90\,kJ\,mol^{-1}$. The explanation is that the more tightly bound Cd atoms are attached directly to the substrate, and the less strongly bound are in a layer (or layers) above the primary overlayer. Chemisorption cannot normally exceed monolayer coverage because a hydrocarbon gas, for instance, cannot chemisorb on to a surface already covered with its fragments; a metal, on the other hand, as in this case, can provide a surface capable of further chemisorption.

Another example of a system showing two desorption activation energies is CO on tungsten, the values being 120 and $300\,kJ\,mol^{-1}$. The explanation is believed to be the existence of two types of metal–adsorbate binding site, one involving a simple M—CO bond, the other adsorption with dissociation into individually adsorbed C and O atoms. In some cases two desorption peaks may occur even though there is only one type of site. This complication arises when there are significant interactions between the adsorbate particles so that at low surface coverages the enthalpy of adsorption is significantly different from its value at high coverages.

Mobility on surfaces

A further aspect of the strength of the interactions between adsorbate and substrate is the former's mobility. Mobility is often a vital feature of a catalyst's activity, because it might be impotent if the reactant molecules adsorb so strongly that they cannot migrate. The activation energy for diffusion over a surface need not be the same as for desorption because the particles may be able to move through valleys between potential peaks without leaving the surface completely. In general, it turns out that the activation energy for migration is about 10–20 per cent of the energy of the surface–adsorbate bond, but that the actual value depends on the extent of coverage. The defect structure of the sample (which depends on the temperature) may also play a dominant role because the adsorbed particles might find it easier to skip across a terrace than to roll along the foot of a step, and they might become trapped in vacancies in an otherwise flat terrace. Diffusion may also be easier across one crystal face than another, so the surface mobility depends on which lattice planes are exposed.

There are two very elegant FIM-based methods for determining the diffusion characteristics of an adsorbate. One involves depositing an adsorbate on a surface plane at low temperature, taking an FIM image before and after the temperature is raised, and monitoring the migration of the boundary as the adsorbate floods across the crystal faces at different rates. In a modification of the technique, an individual atom is imaged, the temperature is raised, and then lowered after a definite interval. A new image is then taken, and the new position of the atom measured (Fig. 28.44). A sequence of pictures shows that the atom makes a random walk across the surface, and the diffusion coefficient D

28.44 FIM micrographs showing the migration of Re atoms on rhenium during 3 s intervals at 375 K. (Photographs provided by Professor G. Ehrlich.)

can be inferred from the mean distance d travelled in an interval τ using the two-dimensional random walk expression $d = (D\tau)^{\frac{1}{2}}$. The value of D for different crystal planes at different temperatures can be determined directly in this way, and the activation energy for migration over each plane obtained from the Arrhenius-like expression

$$D = D_0 e^{-E_D/RT} \tag{26}$$

where E_D is the activation energy for diffusion. Typical values for W atoms on tungsten have E_D in the range 57–$87\ \text{kJ mol}^{-1}$ and $D_0 \approx 3.8 \times 10^{-11}\ \text{m}^2\,\text{s}^{-1}$. For CO on tungsten, the activation energy falls from $144\ \text{kJ mol}^{-1}$ at low surface coverage to $88\ \text{kJ mol}^{-1}$ when the coverage is high.

CATALYTIC ACTIVITY AT SURFACES

A catalyst acts by providing an alternative reaction path with a lower activation energy (Table 28.4). It does not disturb the final equilibrium composition of the system, only the rate at which that equilibrium is approached. In this section we shall consider **heterogeneous catalysis**, in which the catalyst and the reagents are in different phases. For simplicity, we shall consider only gas/solid systems and the solids we consider will be primarily metals. In practice, industry makes use of a wide range of complex solid catalysts, including oxides and zeolites.

Table 28.4* Activation energies of catalysed reactions

Reaction	Catalyst	$E_a/(\text{kJ mol}^{-1})$
$2HI \rightarrow H_2 + I_2$	None	184
	Au	105
	Pt	59
$2NH_3 \rightarrow N_2 + 3H_2$	None	350
	W	162

* More values are given in the Data section.

28.13 Adsorption and catalysis

Heterogeneous catalysis normally depends on at least one reactant being adsorbed (usually chemisorbed) and modified to a form in which it readily undergoes reaction. Often this modification takes the form of a fragmentation of the reactant molecules.

The Eley–Rideal mechanism

In the **Eley–Rideal mechanism** of a surface-catalysed reaction, a gas-phase molecule collides with another molecule adsorbed on the surface. The rate of formation of product is expected to be proportional to the partial pressure p_B of the non-adsorbed gas B and the extent of surface coverage θ_A of the adsorbed gas A. It follows that the rate law should be

$$A + B \rightarrow P \qquad v = k p_B \theta_A$$

The rate constant k might be much larger than for the uncatalysed gas-phase reaction because the reaction on the surface has a low activation energy and the adsorption itself is often not activated.

If we know the adsorption isotherm for A, we can express the rate law in terms of its partial pressure p_A. For example, if the adsorption of A follows a Langmuir isotherm in the pressure range of interest, then the rate law would be

$$v = \frac{kKp_Ap_B}{1 + Kp_A} \tag{27}$$

If A were a diatomic molecule that adsorbed as atoms, we would substitute the isotherm in eqn 17 instead.

According to eqn 27, when the partial pressure of A is high (in the sense $Kp_A \gg 1$) there is almost complete surface coverage, and the rate is equal to kp_B. Now the rate-determining step is the collision of B with the adsorbed fragments. When the pressure of A is low ($Kp_A \ll 1$), perhaps because of its reaction, the rate is equal to kKp_Ap_B; now the extent of surface coverage is important in the determination of the rate.

The Langmuir–Hinshelwood mechanism

In the **Langmuir–Hinshelwood mechanism** of surface-catalysed reactions, the reaction takes place by encounters between molecular fragments and atoms adsorbed on the surface. We therefore expect the rate law to be second-order in the extent of surface coverage:

$$A + B \rightarrow P \qquad v = k\theta_A\theta_B$$

Insertion of the appropriate isotherms for A and B then gives the reaction rate in terms of the partial pressures of the reactants. For example, if A and B follow Langmuir isotherms, and adsorb without dissociation, so that

$$\theta_A = \frac{K_A p_A}{1 + K_A p_A + K_B p_B} \qquad \theta_B = \frac{K_B p_B}{1 + K_A p_A + K_B p_B} \tag{28}$$

then it follows that the rate law is

$$v = \frac{kK_AK_Bp_Ap_B}{(1 + K_A p_A + K_B p_B)^2} \tag{29}$$

The parameters K in the isotherms and the rate constant k are all temperature-dependent, so the overall temperature dependence of the rate may be strongly non-Arrhenius.

Example 28.8 *Interpreting the kinetics of a catalysed reaction*

The decomposition of phosphine (PH_3) on tungsten is first-order at low pressures and zero-order at high pressures. Account for these observations.

Method. Write down a plausible rate law in terms of an adsorption isotherm and explore its form in the limits of high and low pressure.

Answer. If the rate is supposed to be proportional to the surface coverage, then we can write

$$v = k\theta = \frac{kKp}{1 + Kp}$$

where p is the pressure of phosphine. When the pressure is so low that $Kp \ll 1$,

$$v = kKp$$

and the decomposition is first-order. When $Kp \gg 1$,

$$v = k$$

and the decomposition is zero-order.

Comment. Many heterogeneous reactions are first-order, which indicates that the rate-determining stage is the adsorption process.

Exercise E28.8. Suggest the form of the rate law for the deuteration of NH_3 in which D_2 adsorbs dissociatively and extensively (that is, $Kp \gg 1$, with p the partial pressure of D_2), and NH_3 (with partial pressure p') adsorbs at different sites.
$$[v = k(Kp)^{\frac{1}{2}}K'p'/(1 + K'p')]$$

Molecular beam studies

Molecular beam studies are able to give detailed information about catalysed reactions. It has become possible to investigate how the catalytic activity of a surface depends on its structure as well as its composition. For instance, the cleavage of C—H and H—H bonds appears to depend on the presence of steps and kinks, and a terrace often has only minimal catalytic activity. The reaction

$$H_2 + D_2 \rightarrow 2HD$$

has been studied in detail, and it is found that terrace sites are inactive but one molecule in ten reacts when it strikes a step. Although the step itself might be the important feature, it may be that the presence of the step merely exposes a more reactive crystal face (the step face itself). Likewise, the dehydrogenation of hexane to hexene depends strongly on the kink density, and it appears that kinks are needed to cleave C—C bonds. These observations suggest a reason why even small amounts of impurities may poison a catalyst: they are likely to attach to step and kink sites, and so impair the activity of the catalyst entirely. A constructive outcome is that the extent of dehydrogenation may be controlled relative to other types of reactions by seeking impurities that adsorb at kinks and act as specific poisons.

28.14 Examples of catalysis

Almost the whole of modern chemical industry depends on the development, selection, and application of catalysts (Table 28.5). All we can hope to do is this section is to give a brief indication of some of the problems involved. Other than the ones we consider, these include the

Table 28.5 Properties of catalysts

Catalyst	Function	Examples
Metals	Hydrogenation Dehydrogenation	Fe, Ni, Pt, Ag
Semiconducting oxides and sulfides	Oxidation Desulfurization	NiO, ZnO, MgO, Bi_2O_3/MoO_3
Insulating oxides	Dehydration	Al_2O_3, SiO_2, MgO
Acids	Polymerization Isomerization Cracking Alkylation	H_3PO_4, H_2SO_4, SiO_2/Al_2O_3

danger of the catalyst being poisoned by by-products or impurities and economic considerations relating to cost and lifetime.

The activity of a catalyst depends on the strength of chemisorption as indicated by the 'volcano' curve in Fig. 28.45 (it is so-called on account of its general shape). In order to be active, the catalyst should be extensively covered by adsorbate, which is the case if chemisorption is strong. On the other hand, if the strength of the substrate–adsorbate bond becomes too great, the activity declines either because the other reactant molecules cannot react with the adsorbate or because the adsorbate molecules are immobilized on the surface. This suggests that the activity of a catalyst should initially increase with strength of adsorption (as measured, for instance, by the enthalpy of adsorption) and then decline, and that the most active catalysts should be those lying near the summit of the volcano. The most active metals are those lying close to the middle of the d block.

Many metals are suitable for adsorbing gases, and the general order of adsorption strengths decreases along the series O_2, C_2H_2, C_2H_4, CO, H_2, CO_2, N_2. Some of these molecules adsorb dissociatively (e.g. H_2). Elements from the d block, such as iron, vanadium, and chromium, show a strong activity towards all these gases, but manganese and copper are unable to adsorb N_2 and CO_2. Metals towards the left of the periodic table (e.g. magnesium and lithium) can adsorb (and, in fact, react with) only the most active gas (O_2). These trends are summarized in Table 28.6.

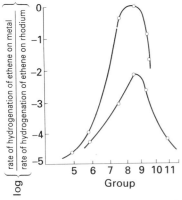

28.45 A volcano curve of catalytic activity arises because, although the reactants must adsorb reasonably strongly, they must not adsorb so strongly that they are immobilized. The lower curve refers to the first series of d-block metals, the upper curve to the second and third series d-block metals. The group numbers relate to the periodic table inside the back cover.

Table 28.6 Chemisorption abilities

	O_2	C_2H_2	C_2H_4	CO	H_2	CO_2	N_2
Ti, Cr, Mo, Fe	+	+	+	+	+	+	+
Ni, Co	+	+	+	+	+	+	−
Pd, Pt	+	+	+	+	+	−	−
Mn, Cu	+	+	+	+	±	−	−
Al, Au	+	+	+	+	−	−	−
Li, Na, K	+	+	−	−	−	−	−
Mg, Ag, Zn, Pb	+	−	−	−	−	−	−

+, Strong chemisorption; ±, chemisorption; −, no chemisorption. See G. C. Bond, *Heterogeneous catalysis*, Oxford University Press (1986) for further information.

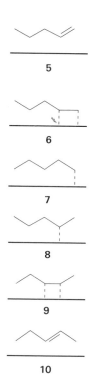

5

6

7

8

9

10

Hydrogenation

An example of catalytic action is found in the hydrogenation of alkenes. The alkene (5) adsorbs by forming two bonds with the surface (6), and on the same surface there may be adsorbed H atoms. When an encounter occurs, one of the alkene–surface bonds is broken (6 → 7 or 8) and later an encounter with a second H atom releases the fully hydrogenated hydrocarbon, which is the thermodynamically more stable species.

The evidence for a two-stage reaction is the appearance of different isomeric alkenes in the mixture. The formation of isomers comes about because while the hydrocarbon chain is waving about over the surface of the metal, it might chemisorb again (8 → 9) and desorb to 10, an isomer of the original 5. The new alkene would not be formed if the two hydrogen atoms attached simultaneously.

A major industrial application of catalytic hydrogenation is to the formation of edible fats from vegetable and animal oils. Raw oils obtained from sources such as the soya bean have the structure $CH_2(O_2CR)CH(O_2CR')CH_2(O_2CR'')$, where R, R', and R'' are long-chain hydrocarbons with several double bonds. One disadvantage of the presence of many double bonds is that the oils are susceptible to atmospheric oxidation, and therefore are liable to become rancid. The geometrical configuration of the chains is responsible for the liquid nature of the oil, and in many applications a solid fat is at least much better and often necessary. Controlled partial hydrogenation of an oil with a catalyst carefully selected so that hydrogenation is incomplete and so that the chains do not isomerize (nickel, in fact), is used on a wide scale to produce edible fats. The process, and the industry, is not made any easier by the seasonal variation of the number of double bonds in the oils.

Oxidation

Catalytic oxidation is also widely used in industry and in pollution control. Although in some cases it is desirable to achieve complete oxidation (as in the production of nitric acid from ammonia); in others partial oxidation is the aim. For example, the complete oxidation of propene to carbon dioxide and water is wasteful, but its partial oxidation to propenal (acrolein, $CH_2{=}CHCHO$) is the start of important industrial processes. Likewise, the controlled oxidations of ethene to ethanol, acetaldehyde, and (in the presence of acetic acid or chlorine) to vinyl acetate or vinyl chloride are the initial stages of very important chemical industries.

Some of these reactions are catalysed by d-metal oxides of various kinds. The physical chemistry of oxide surfaces is very complex, as can be appreciated by considering what happens during the oxidation of propene to acrolein on bismuth molybdate. The first stage is the adsorption of the propene molecule with loss of a hydrogen to form the allyl radical, $CH_2{=}CHCH_2\cdot$. An O atom in the surface can now transfer to this radical, leading to the formation of acrolein and its desorption from the surface. The H atom also escapes with a surface O atom, and goes on to form H_2O, which leaves the surface. The surface is left with vacancies and metal ions in lower oxidation states. These vacancies are attacked by O_2 molecules in the overlying gas, which then chemisorb as O^{2-} ions, so reforming the catalyst. This sequence of events involves great upheavals of the surface, and some materials break up under the stress.

Cracking and reforming

Many of the small organic molecules used in the preparation of all kinds of chemical products come from oil. These small building blocks of polymers, perfumes, and petrochemicals in general, are usually cut from the long-chain hydrocarbons drawn from the Earth as petroleum. The catalytically induced fragmentation of the long-chain hydrocarbons is called **cracking**, and is often brought about on silica–alumina catalysts. These catalysts act by forming unstable carbocations, which dissociate and rearrange to more highly branched isomers. These branched isomers burn more smoothly and efficiently in internal combustion engines, and are used to produce higher octane fuels.

Catalytic **reforming** uses a dual-function catalyst, such as a dispersion of platinum and acidic alumina. The platinum provides the metal function, and brings about dehydrogenation and hydrogenation. The alumina provides the acidic function, being able to form carbocations from alkenes. The sequence of events in catalytic reforming shows up very clearly the complications that must be unravelled if a reaction as important as this is to be understood and improved. The first step is the attachment of the long-chain hydrocarbon by chemisorption to the platinum. In this process first one and then a second H atom is lost, and an alkene is formed. The alkene migrates to a Brønsted acid site, where it accepts a proton and attaches to the surface as a carbocation. This carbocation can undergo several different reactions. It can break into two, isomerize into a more highly branched form, or undergo varieties of ring-closure. Then it loses a proton, escapes from the surface, and migrates (possibly through the gas) as an alkene to a metal part of the catalyst where it is hydrogenated. We end up with a rich selection of smaller molecules that can be withdrawn, fractionated, and then used as raw materials for other products.

CHECK LIST OF KEY IDEAS

1 The concept of **surface tension** (Section 28.1).

2 The **Laplace equation** (eqn 3) for the difference in pressure across a curved surface and the **Kelvin equation** (eqn 4) for the effect of curvature on the vapour pressure of a substance.

3 The role of **nucleation** in the formation of condensed phases (Section 28.2).

4 The phenomenon of **capillary action** (Section

28.3) and the height to which a liquid can rise in a capillary tube (eqn 6), including the role of the **contact angle** (eqn 7).

5 The definition of **surface excess** (eqn 8) and the **Gibbs surface-tension equation** (eqn 9) between the change in surface tension and the surface excess.

6 The experimental study of surface films in terms

of **Langmuir–Blodgett films** and the use of a **surface film balance** (Section 28.5) to determine **surface pressure**.

7 The classification and preparation of **colloids** (Section 28.6) and the origin of their stabilities in terms of the **electric double layer** (Section 28.7).

8 The formation of **micelles** (Section 28.7).

9 The **DVLO theory** of the stability of lyophobic dispersions (Section 28.7, eqns 11–13).

10 The role of **defects** in the growth of surfaces and the self-propagating property of a **screw dislocation** (Section 28.8).

11 The importance of **high-vacuum techniques** when studying surfaces (Section 28.9).

12 The study of surface composition using **photoemission spectroscopy, electron energy-loss spectroscopy,** and **Auger electron spectroscopy** (Section 28.9).

13 The composition of surfaces from **low-energy electron diffraction, field emission microscopy,** and **field ionization microscopy** (Section 28.9).

14 The use of **scanning-tunnelling microscopy** to show details of surface structure (Section 28.9).

15 The definition of **fractional coverage** (eqn 15) and its determination.

16 The processes of **physisorption** and **chemisorption** and the experimental distinctions between them (Section 28.10).

17 The derivation of the **Langmuir isotherm** (eqn 16) and its modification when the adsorbate dissociates (eqn 17).

18 The determination of the **isosteric enthalpy of adsorption** from the adsorption isotherm (eqn 18 and Example 28.4).

19 The **Brunauer–Emmett–Teller isotherm** when multilayer adsorption may occur (eqn 19).

20 The **Temkin isotherm** (eqn 22) and the **Freundlich isotherm** (eqn 23).

21 The **rate of adsorption** in terms of the **sticking probability** (Section 28.12 and eqn 24).

22 The **rate of desorption** in terms of the **desorption activation energy** (eqn 25) and the technique of **flash desorption spectroscopy** (Section 28.12).

23 The **mobility** of adsorbates on surfaces (Section 28.12).

24 The **Eley–Rideal mechanism** and the **Langmuir–Hinshelwood mechanism** of heterogeneous catalysis (Section 28.13).

25 The significance of a **volcano curve** and examples of catalysis (Section 28.14).

EXERCISES

28.1. The surface tensions of aqueous salt solutions are normally greater than that of water itself. Does the salt accumulate at the surface?

28.2. How many molecules of cetanol (of cross-sectional area 2.58×10^{-19} m^2) can be adsorbed on the surface of a spherical drop of dodecane of radius 17.8 nm?

28.3. Calculate the vapour pressure of a spherical droplet of water of radius 10 nm at 20°C. The vapour pressure of bulk water at that temperature is 2.3 kPa and its density is 0.9982 g cm^{-3}.

28.4. The contact angle for water on clean glass is close to zero. Calculate the surface tension of water at 20°C given that at that temperature water climbs to a height of 4.96 cm in a clean glass capillary tube of internal radius 0.300 mm. The density of water at 20°C is 998.2 kg m^{-3}.

28.5. Calculate the pressure differential of water across the surface of a spherical droplet of radius 200 nm at 20°C.

28.6. The values of γ_{lg} and γ_{lw} are 27.5 and 8.5 mN m^{-1}, respectively, for octanol (l) and water (w) at 20°C; g denotes air. Will octanol spread on a water surface?

28.7. Calculate the surface excess of l-aminobutanoic acid in a 0.10 M aqueous solution at 20°C given that $d\gamma/d(\ln c) = -40 \,\mu$N m^{-1}. Convert the answer to the number of molecules per square metre, and calculate the area occupied by a molecule.

28.8. Calculate the change in Gibbs energy when a

spherical droplet of mercury of diameter 1.0 mm is distorted to an oblate spheroid of double the surface area at 20°C.

28.9. Calculate the frequency of molecular collision per square centimetre of surface in a vessel containing (a) hydrogen, (b) propane at 25°C when the pressure is (i) 100 Pa, (ii) 0.10 μTorr.

28.10. What pressure of argon gas is required to produce a collision rate of $4.5 \times 10^{20}\,s^{-1}$ at 425 K on a circular surface of diameter 1.5 mm?

28.11. Calculate the average rate at which He atoms strike a Cu atom in a surface formed by exposing a (100) plane in metallic copper to helium gas at 80 K and a pressure of 35 Pa. Crystals of copper are face-centred cubic with a cell edge of 361 pm.

28.12. A monolayer of N_2 molecules is adsorbed on the surface of 1.00 g of an Fe/Al_2O_3 catalyst at 77 K, the boiling point of liquid nitrogen. Upon warming, the nitrogen occupies $2.86\,cm^3$ at 0°C and 760 Torr. What is the surface area of the catalyst? The effective area of an N_2 molecule is $0.167\,nm^2$.

28.13. The volume of oxygen gas at 0°C and 101 kPa adsorbed on the surface of 1.00 g of a sample of silica at 0°C was $0.284\,cm^3$ at 142.4 Torr and $1.430\,cm^3$ at 760 Torr O_2. What is the value of V_{mon}?

28.14. The enthalpy of adsorption of CO on a surface is found to be $-120\,kJ\,mol^{-1}$. Is the adsorption physisorption or chemisorption? Estimate the mean lifetime of a CO molecule on the surface at 400 K.

28.15. The half-life for which an oxygen atom remains adsorbed to a tungsten surface is 0.36 s at 2548 K and 3.49 s at 2362 K. Find the activation energy for desorption. What is the pre-exponential factor for these tightly chemisorbed atoms?

28.16. The chemisorption of hydrogen on manganese is activated, but only weakly so. Careful measurements have shown that it proceeds 35 per cent faster at 1000 K than at 600 K. What is the activation energy for chemisorption?

28.17. The adsorption of a gas is described by the Langmuir isotherm with $K = 0.85\,kPa^{-1}$ at 25°C. Calculate the pressure at which the fractional surface coverage is (a) 0.15, (b) 0.95.

28.18. A certain solid sample adsorbs 0.44 mg of CO when the pressure of the gas is 26.0 kPa and the temperature is 300 K. The mass of gas adsorbed when the pressure is 3.0 kPa and the temperature is 300 K is 0.19 mg. The Langmuir isotherm is known to describe the adsorption. Find the fractional coverage of the surface at the two pressures.

28.19. What half-life would an H atom have on a surface at 298 K if its desorption activation energy were (a) $15\,kJ\,mol^{-1}$, (b) $150\,kJ\,mol^{-1}$? Take $\tau_0 = 0.10\,ps$. For how long on average would the same atoms remain at 1000 K?

28.20. A solid in contact with a gas at 12 kPa and 25°C adsorbs 2.5 mg of the gas and obeys the Langmuir isotherm. The enthalpy change when 1.00 mmol of the adsorbed gas is desorbed is +10.2 J. What is the equilibrium pressure for the adsorption of 2.5 mg of gas at 40°C?

28.21. Hydrogen iodide is very strongly adsorbed on gold but only slightly adsorbed on platinum. Assume the adsorption follows the Langmuir isotherm and predict the order of the HI decomposition reaction $2HI \rightarrow H_2 + I_2$ on each of the two metal surfaces.

28.22. Suppose it is known that ozone adsorbs on a particular surface in accord with a Langmuir isotherm. How could you use the pressure dependence of the fractional coverage to distinguish between adsorption (a) without dissociation, (b) with dissociation into $O + O_2$, (c) with dissociation into $O + O + O$?

28.23. Nitrogen gas adsorbed on charcoal to the extent of $0.921\,cm^3\,g^{-1}$ at 490 kPa and 190 K, but at 250 K the same amount of adsorption was achieved only when the pressure was increased to 3.2 MPa. What is the molar enthalpy of adsorption of nitrogen on charcoal?

28.24. In an experiment on the adsorption of oxygen on tungsten it was found that the same volume of oxygen was desorbed in 27 min at 1856 K, 2 min at 1978 K, and 0.3 min at 2070 K. What is the activation energy of desorption? How long would it take for the same amount to desorb at (a) 298 K, (b) 3000 K?

28.25. Ammonia was introduced into a bulb at a pressure of 27 kPa. At 856°C it was found that a tungsten catalyst brought about a presssure change of 8 kPa in 500 s, and 15 kPa in 1000 s. What is the order of the catalysed decomposition? Account for the result.

PROBLEMS

Numerical problems

28.1. The surface pressure can be expressed as $\pi\sigma = n(\sigma)RT$, where σ is the area of the surface. The surface tensions of a series of aqueous solutions of a surfactant were measured at 20°C, with the following results:

$[A]/(mol\,L^{-1})$	0	0.10	0.20	0.30	0.40	0.50
$\gamma/(mN\,m^{-1})$	72.8	70.2	67.7	65.1	62.8	59.8

Calculate the surface excess concentration and the surface pressure exerted by the surfactant, and investigate whether the equation quoted is satisfied.

28.2. The surface tensions of solutions of salts in water at molar concentration c can be expressed in the form $\gamma = \gamma^* + (c/\text{mol L}^{-1})\,\Delta\gamma$. The values of $\Delta\gamma$ at 20°C and near $c = 1\,\text{mol L}^{-1}$ are as follows: $\Delta\gamma/(\text{mN m}^{-1}) = 1.4$ (KCl), 1.64(NaCl), 2.7(Na$_2$CO$_3$). Calculate the surface excess concentrations when the bulk concentrations are $1.0\,\text{mol L}^{-1}$.

28.3. The concentration and corresponding surface tensions of aqueous solution of butanol were measured at 20°C with the following results:

$c/(\text{mol L}^{-1})$	0.0264	0.0536	0.1050	0.2110	0.4330
$\gamma/(\text{mN m}^{-1})$	68.00	63.14	56.31	48.08	38.87

Determine the area occupied per molecule.

28.4. The surface tensions of aqueous solution of NH$_4$NO$_3$ have been measured at 20°C and have been found to fit the equation $\gamma/(\text{mN m}^{-1}) = 72.75 + c/(\text{mol L}^{-1})$. Calculate the surface excess for a solution of concentration $1.00\,\text{mol L}^{-1}$

28.5. Nickel is face-centred cubic with a unit cell of side 352 pm. What is the number of atoms per square centimetre exposed on a surface formed by (a) (100), (b) (110), (c) (111) planes? Calculate the frequency of molecular collisions per surface atom in a vessel containing (1) hydrogen, (2) propane at 25°C when the pressure is (i) 100 Pa, (ii) 0.10 μTorr.

28.6. The data below are for the chemisorption of hydrogen on copper powder at 25°C. Confirm that they fit the Langmuir isotherm at low coverages. Then find the value of K for the adsorption equilibrium and the adsorption volume corresponding to complete coverage.

p/Torr	0.19	0.97	1.90	4.05	7.50	11.95
V/cm^3	0.042	0.163	0.221	0.321	0.411	0.471

28.7. The data for the adsorption of ammonia on barium fluoride are reported below. Confirm that they fit a BET isotherm and find values of c and V_{mon}.
(a) $\theta = 0$°C, $p^* = 3222$ Torr:

p/Torr	105	282	492	594	620	755	798
V/cm^3	11.1	13.5	14.9	16.0	15.5	17.3	16.5

(b) $\theta = 18.6$°C, $p^* = 6148$ Torr:

p/Torr	39.5	62.7	108	219	466	555	601	765
V/cm^3	9.2	9.8	10.3	11.3	12.9	13.1	13.4	14.1

28.8. Carbon monoxide adsorbs on mica, and the data for 90 K are given below. Decide whether the Langmuir or the Freundlich isotherm is a better representation of the system. What is the value of K? Given that the total sample area is $6.2 \times 10^3\,\text{cm}^2$, calculate the area occupied by each adsorbed molecule.

p/Torr	100	200	300	400	500	600
V/cm^3	0.130	0.150	0.162	0.166	0.175	0.180

What volume of carbon monoxide would be adsorbed by the mica 90 K when the pressure is 1.00 atm?

28.9. The following data have been obtained for the adsorption of H$_2$ on the surface of 1.00 g of copper at 0°C. The volume of H$_2$ below is the volume that the gas would occupy at STP (0°C and 1 atm).

p/atm	0.050	0.100	0.150	0.200	0.250
V/mL	1.22	1.33	1.31	1.36	1.40

Determine the volume of H$_2$ necessary to form a monolayer and estimate the surface area of the copper sample. The density of liquid hydrogen is $0.0708\,\text{g cm}^{-3}$.

28.10. The designers of a new industrial plant wanted to use a catalyst code-named CR-1 in a step involving the fluorination of butadiene. As a first step in the investigation they determined the form of the adsorption isotherm. The volume of butadiene adsorbed per gram of CR-1 at 15°C varied with pressure as given below. Is the Langmuir isotherm suitable over this pressure range?

p/Torr	100	200	300	400	500	600
V/cm^3	17.9	33.0	47.0	60.8	75.3	91.3

Investigate whether the BET isotherm gives a better description of the adsorption of butadiene on CR-1. At 15°C, p^* (butadiene) = 200 kPa. Find V_{mon} and c.

28.11. The adsorption of solutes on solids from liquids often follows a Freundlich isotherm. Check the applicability of this isotherm to the following data for the adsorption of acetic acid on charcoal at 25°C and find the value of the parameters c_1 and c_2.

$[\text{acid}]/(\text{mol L}^{-1})$	0.05	0.10	0.50	1.0	1.5
w_a/g	0.04	0.06	0.12	0.16	0.19

w_a is the mass adsorbed per unit mass of charcoal.

28.12. In some catalytic reactions the products may adsorb more strongly than the reacting gas. This is the case, for instance, in the catalytic decomposition of ammonia on platinum at 1000°C. As a first step in examining the kinetics of this type of process, show that the rate of ammonia decomposition should follow

$$\frac{dp(\text{NH}_3)}{dt} = -k_c \frac{p(\text{NH}_3)}{p(\text{H}_2)}$$

in the limit of very strong adsorption of hydrogen. Start by showing that when a gas J adsorbs very strongly, and its pressure is $p(\text{J})$, that the fraction of uncovered sites is approximately $1/Kp(\text{J})$. Solve the rate equation for the catalytic decomposition of NH_3 on platinum and show that a plot of $F(t) = (1/t) \ln (p/p_0)$ against $G(t) = (p - p_0)/t$, where p is the pressure of ammonia, should give a straight line from which k_c can be determined. Check the rate law on the basis of the data below, and find k_c for the reaction.

t/s	0	30	60	100	160	200	250
$p/$Torr	100	88	84	80	77	74	72

Theoretical problems

28.13. The deposition of atoms and ions on a surface depends on their ability to stick, and therefore on the energy changes that occur. As an illustration, consider a two-dimensional square lattice of univalent positive and negative ions separated by 200 pm, and consider a cation approaching the upper terrace of this array from the top of the page. Calculate, by direct summation, its Coulombic interaction when it is in an empty lattice point directly above an anion. Now consider a high cliff-like step in the same lattice, and let the approaching ion go into the corner formed by the step and the terrace. Calculate the Coulombic energy for this position, and decide on the likely settling point for a deposited cation.

28.14. Although the attractive van der Waals interaction between individual molecules varies as R^{-6}, the interaction of a molecule with a nearby solid (a homogeneous collection of molcules) varies as R^{-3}, where R is its vertical distance above the surface. Confirm this assertion.

Calculate the interaction energy between an Ar atom and the surface of solid argon on the basis of a Lennard-Jones (6, 12)-potential. Estimate the equilibrium distance of an atom above the surface.

28.15. Show from the Gibbs surface tension equation that, if the surface excess is proportional to the concentration, then the surface phase satisfies the equation of state of a two-dimensional perfect gas.

28.16. Use the Gibbs adsorption isotherm (another name for eqn 9), to show that the volume adsorbed per unit area of solid, V_a/σ, is related to the pressure of the gas by $V_a = -(\sigma/RT)(d\mu/d \ln p)$, where μ is the chemical potential of the adsorbed gas.

28.17. If the dependence of the chemical potential of the gas on the extent of surface coverage is known, the Gibbs adsorption isotherm, eqn 9, can be integrated to give a relation between V_a and p, as in a normal adsorption isotherm. For instance, suppose that the change in the chemical potential of a gas when it adsorbs is of the form $d\mu = -c_2(RT/\sigma) dV_a$, where c_2 is a constant of proportionality: show that the Gibbs isotherm leads to the Freundlich isotherm in this case.

28.18. Finally we come full circle and return to the Langmuir isotherm. Find the form of $d\mu$ that, when inserted in the Gibbs adsorption isotherm, leads to the Langmuir isotherm.

29 Dynamic electrochemistry

In this final chapter of the text we examine one more example of chemical change, that of the transfer of electrons at electrodes. The approach we adopt is largely phenomenological and draws on the thermodynamic language inspired by activated complex theory. First, a model of a solution–electrode interface is constructed, and that model is used to derive a relation, the Butler–Volmer equation, between the current density at the electrode and the overpotential. The latter is the difference between the electrode potential when a current is flowing and when it is not. We shall see that there is a characteristic relation between the current and the overpotential at an electrode, which can be used to identify the species in solution. Moreover, the variation of current with overpotential can be used to infer details of the electron-transfer mechanism responsible for the redox process at the electrode. The Butler–Volmer equation can also be used to analyse the behaviour of working cells and to demonstrate how their potentials differ from the zero-current value when they are in use. Finally, the same approach can be used to analyse the kinetics of reactions that are responsible for corrosion, and point the way to methods of decreasing its rate.

The economic consequences of electrochemistry are almost incalculable. Most of the modern methods of generating electricity are inefficient, and the development of fuel cells could revolutionize our production and deployment of energy. Today we produce energy inefficiently to produce goods that then decay by corrosion. Each step of this wasteful sequence could be improved by discovering more about the kinetics of electrochemical processes. Similarly, the emerging techniques of organic and inorganic electrosynthesis, where an electrode is an active component of an industrial process, depend on a detailed knowledge of the factors affecting their rates. One example is from nylon production, in which adiponitrile is synthesized commercially by the electrolytic reductive coupling (hydrodimerization) of acrylonitrile.

Much of electrochemistry depends on processes that occur at the interface of an electrode and an ionic solution, and the kinetic problem examined in this chapter is the rate at which ions can be discharged at electrodes. A measure of this rate is the **current density** j, the electric current per unit area (the charge flux). Most of the discussion will be concerned with the properties that control the size of the current density. In an electrolytic cell (an electrochemical cell in which a non-spontaneous chemical reaction is driven by an external supply of electricity) deposition and gas evolution occur significantly only when the applied potential exceeds the zero-current cell potential to an extent called the **overpotential** η. We shall establish a relation between the overpotential and the current, and see what determines their values. The potential of a galvanic cell (an electrochemical cell in which a spontaneous reaction produces electricity) is smaller when it is producing current than when it is not producing current. We shall develop relations that enable us to estimate the potential of a working cell.

PROCESSES AT ELECTRODES

When only equilibrium properties are of interest there is no need to know the details of the charge separation responsible for the potential difference at an interface (just as we do not need to propose a reaction mechanism when discussing equilibria). However, a description of the interface is essential when we are interested in the *rate* of charge transfer.

29.1 The electrical double layer

An electrode becomes positively charged relative to the solution nearby if electrons leave the electrode and decrease the local cation concentration. The most primitive model of the interface is an **electrical double layer**, which consists of a sheet of positive charge at the surface of the electrode and a sheet of negative charge next to it in the solution (or vice versa).

A more detailed picture of the interface can be constructed by speculating about the arrangement of ions and electric dipoles in the solution. In the **Helmholtz model** of the double layer it is supposed that the solvated ions range themselves along the surface of the electrode but are held away from it by the molecules in their hydration spheres (Fig. 29.1). The location of the sheet of ionic charge, which is called the **outer Helmholtz plane,** is identified as the plane running through the solvated ions. The Helmholtz model ignores the disrupting effect of thermal motion, which tends to break up and disperse the rigid wall of charge. In the **Gouy–Chapman model** of the **diffuse double layer**, the disordering effect of thermal motion is taken into account in much the same way as in the Debye–Hückel model of the ionic atmosphere of an ion (Section 10.2) with the latter's single central ion replaced by an infinite, plane electrode.

Figure 29.2 shows how the local concentrations of cations and anions differ in the Gouy–Chapman model from their bulk concentrations. As expected, ions of opposite charge cluster close to the electrode, and ions of the same charge are repelled from it. The modification of the local concentrations near an electrode implies that it might be misleading to

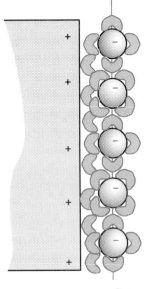

Outer
Helmholtz
plane

29.1 A simple model of the electric double layer treats it as two rigid planes of charge: one plane, the outer Helmholtz plane being due to the ions in solution, the other, the inner Helmholtz plane, being due to the charge on the electrode.

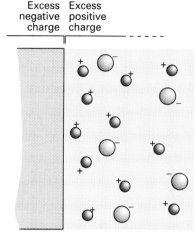

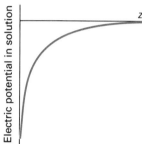

Electric potential in solution

z

29.2 The Gouy–Chapman model of the electric double layer treats it as an atmosphere of counter-charge, similar to the Debye–Hückel theory of ionic atmospheres.

Excess negative charge

Excess positive charge

use activity coefficients characteristic of the bulk to discuss the thermo-dynamic properties of ions near the interface. This is one of the reasons why measurements in dynamic electrochemistry are almost always done using a large excess of supporting electrolyte (e.g. a 1 M solution of a salt, an acid, or a base). Under such conditions the activity coefficients are almost constant because the inert ions dominate the effects of local changes caused by any reactions taking place.

Neither the Helmholtz nor the Gouy–Chapman model is a very good representation of the structure of the double layer. The former over-emphasizes the rigidity of the local solution; the latter underemphasizes its structure. The two are combined in the **Stern model**, in which the ions closest to the electrode are constrained into a rigid Helmholtz plane while outside that plane the ions are dispersed as in the Gouy–Chapman model.

The most important property of the double layer is the effect it has on the electric potential near the electrode. The potential at the interface can be analysed by imagining the separation of the electrode from the solution, but with the charges of the metal and the solution frozen in position.

A positive test charge at great distances from the isolated electrode experiences a Coulomb potential that varies inversely with distance (Fig. 29.3). As the test charge approaches the electrode it enters a region where the potential varies more slowly. This change in behaviour can be traced (using electrostatics) to the fact that the surface charge is not point-like but is spread over an area. At about 100 nm from the surface the potential varies only slightly with distance, and its value in this region is called the **Volta potential**, or the **outer potential**, ψ. As the test charge is taken through the skin of electrons on the surface of the electrode, the potential it experiences changes until it reaches the inner, bulk metal environment. This extra potential is called the **surface potential** χ. The total potential inside the electrode is called the **Galvani potential** ϕ.

A similar sequence of changes of potential is observed as a positive test charge is brought up to and through the solution surface. The potential changes to its Volta value as the charge approaches the charged medium, then to its Galvani value as it is taken into the bulk.

When the electrode and solution are reassembled (without any change of charge distribution) the potential difference between points in the bulk metal and the bulk solution is the **Galvani potential difference** $\Delta\phi$. Apart from a constant, this Galvani potential difference is the electrode potential that was at the centre of our discussion in Chapter 10. We shall ignore the constant, and identify changes in $\Delta\phi$ with changes in electrode potential.

JUSTIFICATION

To demonstrate the relation between $\Delta\phi$ and E we consider the cell

$$\text{Pt} \mid \text{H}_2(\text{g}) \mid \text{H}^+(\text{aq}) \mid\mid \text{M}^+(\text{aq}) \mid \text{M}(\text{s})$$

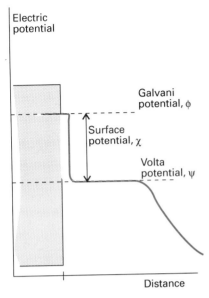

29.3 The variation of potential with distance from an electrode that has been separated from the electrolyte solution without there being an adjustment of charge. A similar diagram applies to the separated solution.

and the half-reactions

$$M^+(aq) + e^- \rightarrow M(s)$$

$$H^+(aq) + e^- \rightarrow \tfrac{1}{2}H_2(g)$$

The Gibbs energies of these two half-reactions can be expressed in terms of the chemical potentials of all the species. However, in doing so we must take into account the fact that the species are present in phases with different electric potentials. Thus, a cation in a region of positive potential has a higher chemical potential (is chemically more active in a thermodynamic sense) than in a region of zero potential.

The contribution of an electrical potential to the chemical potential is calculated by noting that the electrical work of adding a charge ze to a region where the potential is ϕ is

$$w_e = ze \times \phi$$

and therefore that the work per mole is

$$w_e = zF \times \phi$$

where F is Faraday's constant. Because at constant temperature and pressure the maximum electrical work can be identified with the change in Gibbs energy (Section 10.4), the difference in chemical potential of an ion with and without the electrical potential present is

$$\bar{\mu} - \mu = zF\phi$$

the chemical potential of an ion in the presence of an electric potential is called its **electrochemical potential** $\bar{\mu}$. It follows from the last equation that

$$\bar{\mu} = \mu + zF\phi$$

When $z = 0$ (a neutral species), the electrochemical potential is equal to the chemical potential.

The Gibbs energy for the two half-reactions at the electrode is now written in terms of the electrochemical potentials of the species. To do so, we note that the cations M^+ are in the solution where the Galvani potential is ϕ_S and the electrons are in the electrode where the Galvani potential is ϕ_M:

$$\begin{aligned}\Delta_r G_R &= \bar{\mu}(M) - \{\bar{\mu}(M^+) + \bar{\mu}(e^-)\} \\ &= \mu(M) - \{\mu(M^+) + F\phi_S + \mu(e^-) - F\phi_M\} \\ &= \mu(M) - \mu(M^+) - \mu(e^-) + F\,\Delta\phi_R\end{aligned}$$

$\Delta\phi_R$ is the Galvani potential difference at the right-hand electrode. Likewise, in the hydrogen half-reaction, the electrons are in the metal electrode at a potential ϕ_M and the H^+ ions are in the solution where the potential is ϕ_S:

$$\begin{aligned}\Delta_r G_L &= \tfrac{1}{2}\bar{\mu}(H_2) - \{\bar{\mu}(H^+) + \bar{\mu}(e^-)\} \\ &= \tfrac{1}{2}\mu(H_2) - \mu(H^+) - \mu(e^-) + F\,\Delta\phi_L\end{aligned}$$

$\Delta\phi_L$ is the Galvani potential difference at the left-hand electrode.

The overall change in Gibbs energy is

$$\Delta_r G_R - \Delta_r G_L = \mu(M) + \mu(H^+) - \mu(M^+) - \tfrac{1}{2}\mu(H_2) + F(\Delta\phi_R - \Delta\phi_L)$$
$$= \Delta_r G + F(\Delta\phi_R - \Delta\phi_L)$$

where $\Delta_r G$ is the Gibbs energy of the cell reaction. When the cell is balanced against an external source of potential the entire system is at equilibrium. The overall reaction Gibbs energy is then zero[1] and the last equation becomes

$$0 = \Delta_r G + F(\Delta\phi_R - \Delta\phi_L)$$

which rearranges to

$$\Delta_r G = -F(\Delta\phi_R - \Delta\phi_L)$$

If we compare this result with the result established in Section 10.4 that

$$\Delta_r G = -FE \qquad E = E_R - E_L$$

we can conclude that

$$E_R - E_L = \Delta\phi_R - \Delta\phi_L$$

This is the result we wanted to show, for it implies that the Galvani potential difference at each electrode can differ from the electrode potential by a constant at most (which cancels when the difference is taken).

29.2 The rate of charge transfer

Because an electrode reaction is heterogeneous, it is natural to express its rate as the amount of material produced per unit area of electrode per unit time (that is, as a flux of product). A first-order heterogeneous rate law therefore has the form

Amount produced per unit area per unit time = k[species]

where [species] is the molar concentration of the relevant species in solution. The term on the left is expressed in amount per unit area per unit time, and [species] is an amount per unit volume, so k has dimensions of length per unit time (e.g. $cm\,s^{-1}$). If the molar concentration of the oxidized and reduced materials outside the double layer are [Ox] and [Red], respectively, the rate of reduction of Ox is

Rate of reduction of Ox = k_c[Ox]

and the rate of oxidation of Red is

Rate of oxidation of Red = k_a[Red]

(The notation k_c and k_a is justified below.)

Now consider a reaction at the electrode in which an ion is reduced by

---

1 The cell reaction itself is not necessarily at equilibrium: its tendency to change is balanced against the external source and *overall* there is stalemate.

(a)   (b)

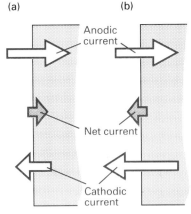

29.4 The net current is defined as the difference $j_a - j_c$. (a) When $j_a > j_c$ the net current is anodic, and there is a net oxidation of the species in solution. (b) When $j_c > j_a$ the net current is cathodic, and the net process is reduction.

the transfer of a single electron in the rate-determining step.[2] Then the net current density at the electrode is the difference of the current densities arising from the reduction of Ox and the oxidation of Red. Because the redox processes at the electrode involve the transfer of one electron per reaction event, the current densities j arising from the redox processes are the rates (as expressed above) multiplied by the charge transferred per mole of reaction, which is given by Faraday's constant. Therefore, there is a **cathodic current density** of magnitude

$$j_c = Fk_c[Ox]$$

arising from the reduction. There is also an opposing **anodic current density** of magnitude

$$j_a = Fk_a[Red]$$

arising from the oxidation.[3] The net current density at the electrode (Fig. 29.4) is the difference

$$j = j_a - j_c = Fk_a[Red] - Fk_c[Ox] \tag{1}$$

The activation Gibbs energy

If an ion (or neutral molecule) is to participate in reduction or oxidation at an electrode, it must discard any solvating molecules, migrate through the electrical double layer, and adjust its hydration sphere as it receives or discards electrons. Likewise, an ion or molecule already at the inner plane must be detached and migrate into the bulk. Because both processes are activated, we can expect to write their rate constants in the form suggested by activated complex theory (Section 27.6) as

$$k = Be^{-\Delta^{\ddagger}G/RT} \tag{2}$$

In this expression $\Delta^{\ddagger}G$ is the activation Gibbs energy and B is a constant with the same dimensions as k.

When eqn 2 is inserted into eqn 1 we obtain

$$j = FB_a[Red]e^{-\Delta^{\ddagger}G_a/RT} - FB_c[Ox]e^{-\Delta^{\ddagger}G_c/RT} \tag{3}$$

This expression allows the activation Gibbs energies to be different for the cathodic and anodic processes. That they are different is the central feature of the remaining discussion. Note that when $j_a > j_c$, so that $j > 0$, the current is anodic (Fig. 29.4a); when $j_c > j_a$, so that $j < 0$, the current is cathodic (Fig. 29.4b).

The Butler–Volmer equation

Now we relate j to the Galvani potential difference, which varies across the double layer as shown schematicaly in Fig. 29.5.

Consider the reduction reaction. An electron is transferred from the

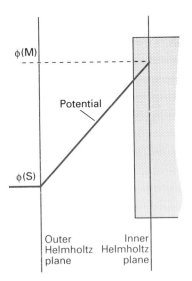

29.5 The electric potential varies linearly between two plane parallel sheets of charge, and its effect on the Gibbs energy of the transition state depends on the extent to which the latter resembles the species at the inner or outer planes.

2 The last phrase is important: in the deposition of cadmium, for instance, only one electron is transferred in the rate-determining step even though overall the deposition involves the transfer of two electrons.

3 Recall from Chapter 10 that the cathode is the site of reduction; hence the cathodic current is the current arising from reduction. The anode is the site of oxidation, and the anodic current is the current arising from oxidation.

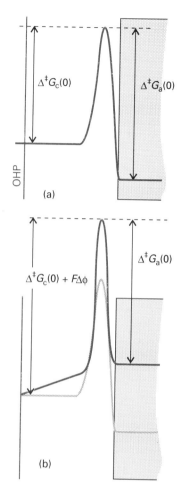

OHP

(a)

(b)

29.6 When the transition state resembles a species that has undergone reduction (and which is represented here by a peak in the reaction profile that lies very close to the inner plane) the activation Gibbs energy for the anodic current is almost unchanged, but the full effect applies to the cathodic current. OHP is the outer Helmholtz plane.

electrode where the potential is ϕ_M to the solution where it is ϕ_S. There is therefore an electrical contribution to the work of magnitude $e\,\Delta\phi$. If the transition state of the activated complex is product-like (as represented by the peak of the reaction profile being close to the electrode in Fig. 29.6), the activation Gibbs energy is changed from $\Delta^{\ddagger}G_c(0)$, the value it has in the absence of a potential difference across the double layer, to

$$\Delta^{\ddagger}G_c = \Delta^{\ddagger}G_c(0) + F\,\Delta\phi$$

Thus, if the electrode is more positive than the solution, $\Delta\phi > 0$, then more work has to be done to form an activated complex from Ox; in this case the activation Gibbs energy is increased. If the transition state is reactant-like (represented by the peak of the reaction profile being close to the outer plane of the double-layer in Fig. 29.7), then $\Delta^{\ddagger}G_c$ is independent of $\Delta\phi$. In a real system the transition state has an intermediate resemblance to these extremes (Fig. 29.8) and the activation Gibbs energy for reduction may be written as

$$\Delta^{\ddagger}G_c = \Delta^{\ddagger}G_c(0) + \alpha F\,\Delta\phi \tag{4}$$

The parameter α is called the **transfer coefficient**, and lies in the range 0 to 1. Experimentally, α is often found to be about $\frac{1}{2}$.

Now consider the oxidation of Red. In this case Red discards an electron to the electrode, and so the extra work is zero if the transition state is reactant-like (represented by a peak close to the electrode in Fig. 29.6) and the full $-F\,\Delta\phi$ if it resembles the product (the peak close to the outer plane Fig. 29.7). In general (Fig. 29.8), the activation Gibbs energy for this anodic process is

$$\Delta^{\ddagger}G_a = \Delta^{\ddagger}G_a(0) - (1-\alpha)F\,\Delta\phi \tag{5}$$

The two activation Gibbs energies can now be inserted in place of the values used in eqn 3 with the result that

$$\begin{aligned}
j &= FB_a[\text{Red}] \times e^{-\Delta^{\ddagger}G_a(0)/RT} \times e^{(1-\alpha)F\,\Delta\phi/RT} \\
&\quad - FB_c[\text{Ox}] \times e^{-\Delta^{\ddagger}G_c(0)/RT} \times e^{-\alpha F\,\Delta\phi/RT}
\end{aligned} \tag{6}$$

This is an explicit, if complicated, expression for the net current density in terms of the potential difference.

The appearance of eqn 6 can be simplified. First, in a purely cosmetic step we write

$$f = \frac{F}{RT}$$

Next, we identify the individual cathodic and anodic current densities:

$$\left.\begin{aligned}
j_a &= FB_a[\text{Red}] \times e^{-\Delta^{\ddagger}G_a(0)/RT} \times e^{(1-\alpha)f\,\Delta\phi} \\
j_c &= FB_c[\text{Ox}] \times e^{-\Delta^{\ddagger}G_c(0)/RT} \times e^{-\alpha f\,\Delta\phi}
\end{aligned}\right\} j = j_a - j_c \tag{7}$$

Example 29.1 *Calculating the current density*

Calculate the change in cathodic current density at an electrode when the potential difference changes from 1.0 V to 2.0 V at 25°C.

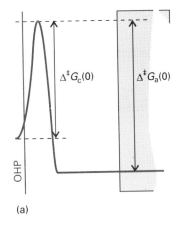

(a)

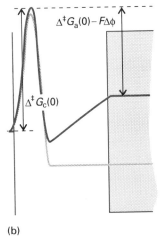

(b)

29.7 When the transition state resembles a species that has undergone oxidation (and which is represented here by a peak in the reaction profile that lies very close to the outer Helmholtz plane) the cathodic current activation Gibbs energy is almost unchanged but the anodic current activation Gibbs energy is strongly affected.

Method. Equation 7 can be used to express the ratio of cathodic current densities j'_c and j_c when the potential differences are $\Delta\phi'$ and $\Delta\phi$. Then we can insert a typical value of α (namely, $\frac{1}{2}$) and the data.

Answer. The ratio is

$$\frac{j'_c}{j_c} = e^{-\alpha f(\Delta\phi' - \Delta\phi)}$$

So, with $\alpha = \frac{1}{2}$ and $\Delta\phi' - \Delta\phi = 1.0\text{ V}$,

$$\alpha f \times (\Delta\phi' - \Delta\phi) = \frac{\frac{1}{2} \times (9.6485 \times 10^4\text{ C mol}^{-1}) \times (1.0\text{ V})}{(8.3145\text{ J K}^{-1}\text{ mol}^{-1}) \times (298\text{ K})} = 19$$

Hence,

$$\frac{j'_c}{j_c} = e^{-19} = 6 \times 10^{-9}$$

Comment. This huge change in current density occurs for a very mild and easily applied change of conditions. We can appreciate why the change is so great by realizing that a change of potential difference by 1 V changes the activation Gibbs energy by $(1\text{ V}) \times F$, or about 50 kJ mol^{-1}, which has an enormous effect on the rates.

Exercise E29.1. Calculate the change in anodic current density under the same circumstances. $[j'_a/j_a = 2 \times 10^8]$

If the cell is balanced against an external source, the difference in Galvani potential $\Delta\phi$ can be identified as the (zero-current) electrode potential E, and we can write[4]

$$j_a = FB_a[\text{Red}] \times e^{-\Delta^{\ddagger}G_a(0)/RT} \times e^{(1-\alpha)fE}$$
$$j_c = FB_c[\text{Ox}] \times e^{-\Delta^{\ddagger}G_c(0)/RT} \times e^{-\alpha fE}$$

When these equations apply, there is no net current at the electrode (as the cell is balanced), so the two current densities must be equal. From now on we denote them both as j_0, which is called the **exchange current density**.

When the cell is producing current the electrode potential changes from its zero-current value E to its working value E', and the difference is the **overpotential** η:

$$\eta = E' - E \tag{8}$$

Hence $\Delta\phi$ changes to

$$\Delta\phi = E + \eta$$

It follows that the two current densities become

$$j_a = j_0 e^{(1-\alpha)f\eta} \qquad j_c = j_0 e^{-\alpha f\eta} \tag{9}$$

---

4 We are assuming that we can identify the Galvani potential and the zero-current electrode potential. As explained earlier, they differ by a constant amount, which may be regarded as absorbed into the constant B.

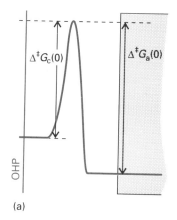

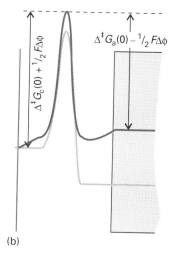

(a)

(b)

29.8 When the transition state is intermediate in its resemblance to reduced and oxidized species, as represented here by a peak located at an intermediate position as measured by α (with $0 < \alpha < 1$), both activation Gibbs energies are affected.

Then from eqn 7 we obtain the **Butler–Volmer equation**:

$$j = j_0\{e^{(1-\alpha)f\eta} - e^{-\alpha f\eta}\} \tag{10}$$

The Butler–Volmer equation is the basis of all that follows.

The low overpotential limit

When the overpotential is so small that $f\eta \ll 1$ (in practice, η less than about 0.01 V) the exponentials in eqn 10 can be expanded using $e^x = 1 + x + \ldots$ to give

$$j = j_0\{1 + (1-\alpha)f\eta + \ldots - (1 - \alpha f\eta + \ldots)\} \approx j_0 f\eta \tag{11}$$

This equation shows that the current density is proportional to the overpotential, so at low overpotentials the interface behaves like a conductor that obeys Ohm's law. When there is a small positive overpotential the current is anodic ($j > 0$ when $\eta > 0$), and when the overpotential is small and negative the current is cathodic ($j < 0$ when $\eta < 0$). The relation can also be reversed to calculate the potential difference that must exist if a current density j has been established by some external circuit:

$$\eta = \frac{RTj}{Fj_0} \tag{12}$$

The importance of this interpretation will become clear below.

Example 29.2 *Calculating the current at an electrode*

The exchange current density of a Pt $|$ H$_2$(g) $|$ H$^+$(aq) electrode at 298 K is 0.79 mA cm^{-2}. What current flows through a standard electrode of total area 5.0 cm^2 when the overpotential is $+5.0$ mV?

Method. This example is a straightforward application of eqn 12. It is useful to note that, at 298 K, $f = F/RT = 1/(25.69 \text{ mV})$. The current flowing through an electrode is the product of the current density and the area of the electrode.

Answer. The current density is

$$j = j_0 f\eta = \frac{(0.79 \text{ mA cm}^{-2}) \times (5.0 \text{ mV})}{25.69 \text{ mV}} = 0.15 \text{ mA cm}^{-2}$$

The current through the electrode is therefore 0.75 mA.

Comment. The current density is anodic, which means that the oxidation reaction dominates when the overpotential is positive. Because $f\eta = 0.2$, the linear approximation is reasonably valid.

Exercise E29.2. What would be the current at pH = 2.0, the other conditions being the same?
$$[-17 \text{ mA}]$$

The high overpotential limit

When the overpotential is large and positive, corresponding to the electrode being the anode in electrolysis, the second exponential in eqn 10 is much smaller than the first, and may be neglected. Then

$$j = j_0 e^{(1-\alpha)f\eta}$$

so

$$\ln j = \ln j_0 + (1 - \alpha)f\eta \tag{13a}$$

When the overpotential is large but negative (corresponding to the cathode in electrolysis), the first exponential in eqn 10 may be neglected. Then

$$j = -j_0 e^{-\alpha f\eta}$$

so

$$\ln (-j) = \ln j_0 - \alpha f\eta \tag{13b}$$

The plot of the logarithm of the current density against the overpotential is called a **Tafel plot**. The slope gives the value of α and the intercept at $\eta = 0$ gives the exchange current density.

The experimental arrangement used for a Tafel plot is shown in Fig. 29.9. The electrode of interest is called the **working electrode**, and the current flowing through it is controlled externally. If its area is A and the current is I, then the current density across its surface is I/A. The potential difference across the interface cannot be measured directly, but the potential of the working electrode relative to a third electrode, the **reference electrode**, can be measured with a high-impedance voltmeter, and no current flows in that half of the circuit. The reference electrode is in contact with the solution close to the working electrode through a 'Luggin capillary', which helps to eliminate any ohmic potential difference that might arise accidentally. Changing the current flowing through the working circuit causes a change of potential of the working electrode, and that change is measured with the voltmeter. The overpotential is then obtained by taking the difference between the potentials measured with and without a flow of current through the working circuit.

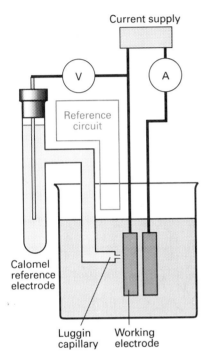

29.9 The general arrangement for electrochemical rate measurements. The external source establishes a current between the electrodes, and its effect on the potential difference of either of them relative to the reference electrode is observed. No current flows in the reference circuit.

(Labels in figure: Current supply; V; A; Reference circuit; Calomel reference electrode; Luggin capillary; Working electrode)

Example 29.3 *Interpreting a Tafel plot*

The data below refer to the anodic current through a $2.0\ \text{cm}^2$ platinum electrode in contact with a Fe^{3+},Fe^{2+} aqueous solution at $298\ \text{K}$. Calculate the exchange current density and the transfer coefficient for the electrode process.

η/mV	50	100	150	200	250
I/mA	8.8	25.0	58.0	131	298

Method. The anodic process is the oxidation $Fe^{2+}(aq) \rightarrow Fe^{3+}(aq) + e^-$. To analyse the data, we make a Tafel plot (of $\ln j$ against η) using the

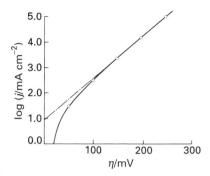

29.10 A Tafel plot is used to measure the exchange current density (given by the extrapolated intercept at $\eta = 0$) and the transfer coefficient (from the slope). The data are from Example 29.3.

anodic form (eqn 13a). The intercept at $\eta = 0$ is $\ln j_0$ and the slope is $(1 - \alpha)f$.

Answer. Draw up the following table:

η/mV	50	100	150	200	250
j/(mA cm^{-2})	4.4	12.5	29.0	56.6	149
$\ln (j$/mA cm^{-2})	1.50	2.53	3.37	4.18	5.00

The points are plotted in Fig. 29.10;. The high overpotential region gives a straight line of intercept 0.92 and slope 0.0163. From the former it follows that $\ln (j_0/\text{mA cm}^{-2}) = 0.92$, so $j_0 = 2.5\,\text{mA cm}^{-2}$. From the latter.

$$(1 - \alpha) \times \frac{F}{RT} = 0.0163\,\text{mV}^{-1}$$

so $\alpha = 0.58$.

Comment. Note that the Tafel plot is non-linear for $\eta < 150\,\text{mV}$; in this region $\alpha f = 6$ and the approximation that $\alpha f \gg 1$ is starting to fail.

Exercise E29.3. Repeat the analysis using the following cathodic current data:

η/mV	−50	−100	−150	−200	−250	−300
I/mA	−0.3	−1.5	−6.4	−27.6	−118.6	−510

$[\alpha = 0.75, j_0 = 0.040\,\text{mA cm}^{-2}]$

Table 29.1* Exchange current densities and transfer coefficients at 298 K

Reaction	Electrode	j_0/(A cm^{-2})	α
$2H^+ + 2e^- \rightarrow H_2$	Pt	7.9×10^{-4}	
	Ni	6.3×10^{-6}	0.58
	Pb	5.0×10^{-12}	
$Fe^{3+} + e^- \rightarrow Fe^{2+}$	Pt	2.5×10^{-3}	0.58

* More values are given in the Data section at the end of this volume.

Some experimental values for the Butler–Volmer parameters are given in Table 29.1. From them we can see that exchange current densities vary over a very wide range. For example, the N_2, N_3^- couple on platinum has $j_0 = 10^{-76}\,\text{A cm}^{-2}$ whereas the H^+, H_2, couple on platinum has $j_0 = 8 \times 10^{-4}\,\text{A cm}^{-2}$, a difference of 73 orders of magnitude. Exchange currents are generally large when the redox process involves no bond-breaking (as in the $[Fe(CN)_6]^{3-}, [Fe(CN)_6]^{4-}$ couple) or if only weak bonds are broken (as in Cl_2, Cl^-). They are generally small when more than one electron needs to be transferred, or when multiple or strong bonds are broken, as in the N_2, N_3^- couple and in redox reactions of organic compounds.

29.3 Polarization

Electrodes with potentials that change only slightly when a current passes through them are classified as **non-polarizable**. Those with strongly current-dependent potentials are classified as **polarizable**. From the linearized equation (eqn 11) it is clear that the criterion for low polarizability is high exchange current density (so that η may be small even though j is large). The calomel and H_2/Pt electrodes are both highly non-polarizable, which is one reason why they are so extensively used in equilibrium electrochemistry measurements.

Concentration polarization

One of the assumptions in the derivation of the Butler–Volmer equation is the uniformity of concentration near the electrode. This assumption fails at high current densities because migration of ions towards the electrode from the bulk is slow and may become rate-determining. A larger overpotential is then needed to produce a given current. This effect is called **concentration polarization** and its contribution to the total overpotential is called the **polarization overpotential** η^c.

For simplicity, we consider a case for which the concentration polarization dominates all the rate processes. We shall consider a redox couple of the type M^{z+}, M. Under zero-current conditions, when the net current density is zero, the electrode potential is related to the activity a of the ions in the solution by the Nernst equation (eqn 10.13):

$$E = E^\ominus + \frac{RT}{zF} \ln a$$

We have remarked that electrode kinetics are normally studied using a large excess of support electrolyte so as to keep the mean activity coefficients approximately constant. Therefore, the constant activity coefficient in $a = \gamma c$ may be absorbed into E, and we write the **formal potential** of the electrode as

$$E^\circ = E^\ominus + \frac{RT}{zF} \ln \gamma \tag{14}$$

Then the electrode potential is

$$E = E^\circ + \frac{RT}{zF} \ln c$$

When the cell is producing current, the active ion concentration just outside the double layer changes to c' and the electrode potential changes to

$$E' = E^\circ + \frac{RT}{zF} \ln c'$$

The concentration overpotential is therefore

$$\eta^c = E' - E = \frac{RT}{zF} \ln\left(\frac{c'}{c}\right) \tag{15}$$

We now suppose that the solution has its bulk concentration up to a distance δ from the outer Helmholtz plane, and then falls linearly to c' at the plane itself. This **Nernst diffusion layer** is illustrated in Fig. 29.11. The thickness of the Nernst layer (which is typically 0.1 mm, and strongly dependent on the condition of hydrodynamic flow) is quite different from that of the electric double layer (which is typically less than 1 nm, and unaffected by stirring). The concentration gradient through the Nernst layer is

$$\frac{dc}{dx} = \frac{c' - c}{\delta}$$

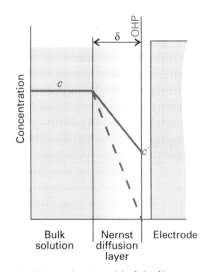

29.11 In a simple model of the Nernst diffusion layer there is a linear variation in concentration between the bulk and the outer Helmholtz plane (OHP); the thickness of the layer depends strongly on the state of flow of the fluid.

This gradient gives rise to a flux of ions towards the electrode, which replenishes the cations as they are reduced. The (molar) flux J is proportional to the concentration gradient and, according to Fick's law (Section 24.10),

$$J = -D \left(\frac{\partial c}{\partial x} \right)$$

Therefore, the flux towards the electrode is

$$J = D \times \frac{c - c'}{\delta}$$

The current density towards the electrode is the product of the particle flux and the charge per mole of ions, zF:

$$j = zFJ = zFD \times \frac{c - c'}{\delta}$$

The maximum rate of diffusion across the Nernst layer occurs when the gradient is steepest, which is when $c' = 0$. This concentration occurs when an electron from an ion that diffuses across the layer is snapped over the activation barrier, through the double layer, and on to the electrode. No flow of current can exceed the **limiting current density** j_{lim}, which is given by

$$j_{lim} = zFJ_{lim} = \frac{zFDc}{\delta} \tag{16}$$

Example 29.4 *Estimating the limiting current density*

Estimate the limiting current density at 298 K for an electrode in a 0.10 M Cu^{2+}(aq) unstirred solution in which the thickness of the diffusion layer is about 0.3 mm.

Method. We estimate D from the ionic conductivity $\lambda = 107$ S cm^2 mol^{-1} (Table 24.4) and the Nernst–Einstein equation (Section 24.10):

$$\lambda = \frac{z^2 F^2 D}{RT}$$

Answer. Equation 16 becomes

$$j_{lim} = \frac{cRT\lambda}{zF\delta}$$

Therefore, with $\delta = 0.3$ mm, $c = 0.10$ mol L^{-1}, $z = 2$, and $T = 298$ K, this expression gives $j_{lim} = 5$ mA cm^{-2}.

Comment. The result implies that the current towards a 1 cm^2 electrode cannot exceed 5 mA in this solution. If the solution is stirred, or if the electrode surface is moving (as in polarography; see below) the Nernst layer is much thinner.

Exercise E29.4. Evaluate the limiting current density for an Ag(s) |

$Ag^+(aq)$ electrode in a 0.010 M $Ag^+(aq)$ solution at 298 K. Take $\delta \approx 0.03 \text{ mm}$. [5 mA cm^{-2}]

The concentration c' is related to the current density at the double layer by

$$c' = c - \frac{j\delta}{zFD} \tag{17}$$

Hence, as the current density is increased, the concentration falls below the bulk value. However, this decline in concentration is small when the diffusion coefficient is large, for then the ions are very mobile and can quickly replenish any ions that have been removed.

Finally, we substitute eqn 17 into eqn 15 and obtain the following expressions for the overpotential in terms of the current density and vice versa:

$$\eta^c = \frac{RT}{zF} \ln\left(1 - \frac{j\delta}{zcFD}\right) \tag{18a}$$

$$j = \frac{zcFD}{\delta}(1 - e^{zf\eta^c}) \tag{18b}$$

Polarography and voltammetry

In the analytical technique of **polarography**, the current passing through a solution is measured as the potential difference between two electrodes is changed. The current climbs with the potential difference until the limiting value is reached and, as the limiting current is characteristic of the ion present, we may be able to identify them from their current–voltage characteristics. When several different ions are present, each one reaches its limiting current density at a characteristic value, and the current changes in a series of waves. Each ion is identified by measuring the **half-wave potential**, the potential mid-way between the initial and final potentials of the wave (Fig. 29.12 and Table 29.2). The limiting current density itself can be interpreted in terms of the concentrations of the ions.

One of the problems with the procedure is the contamination of the electrode surface. This problem is avoided by using a **dropping mercury electrode** (Fig. 29.13) in which the surface is continuously renewed as drops form at the end of the capillary and then fall off. The growth of the drop, and therefore of the surface area of the electrode, accounts for the oscillations in the polarogram.

The kinetics of electrode processes are also commonly studied by **voltammetry**, in which the current is monitored as the potential of the electrode is changed, and by **chronopotentiometry**, in which the potential is monitored as the current flow is changed.

The kind of output from a **linear-sweep voltammetry experiment** is illustrated in Fig. 29.14. Initially the potential is low, and the cathodic current is due to the migration of ions in the solution. However, as the potential approaches the reduction potential of the reducible solute, the

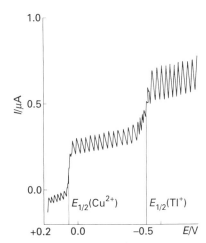

29.12 A polarogram is obtained when the current flowing through a cell is plotted against the applied potential. This polarogram shows the presence of Cu^{2+} and Tl^+ ions by their characteristic half-wave potentials $E_{\frac{1}{2}}$.

Table 29.2* Half-wave potentials at 298 K, $E_{\frac{1}{2}}$/V†

Cd^{2+}	-0.60
Cu^{2+}	$+0.04$
Fe^{3+}	0.0
Zn^{2+}	-1.00

* More values are given in the Data section.
† Values refer to ions in 0.1 M KCl(aq), relative to the calomel electrode.

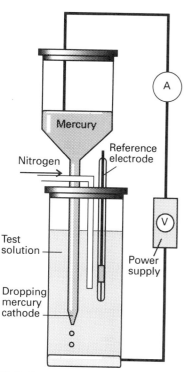

29.13 The dropping mercury electrode ensures that a renewed, clean surface is provided continuously. The growth of the droplets accounts for the periodic variation of the current in Fig. 29.12.

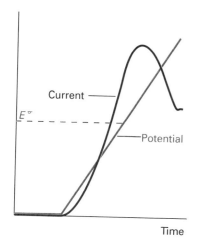

29.14 The change of potential with time and the resulting current/potential curve in a linear-sweep voltammetry experiment.

cathodic current grows. Soon after the potential exceeds the reduction potential, the current declines on account of the concentration polarization of the electrode, because now there is a shortage of reducible solute molecules close to the electrode. A modification of the technique is **cyclic voltammetry** in which the potential is returned cyclically to its initial value. A typical cyclic voltammogram is shown in Fig. 29.15. The shape of the curve is initially like that of a linear sweep experiment, but after the potential begins to fall there is a rapid change in current on account of the high concentration of oxidizable species close to the electrode. When the potential is close to the potential required to oxidize the reduced species, there is a substantial anodic current until all the oxidation is complete, and the current returns to zero. The overall shape of the curve gives details of the kinetics of the electrode process.

Kinetic processes are also studied by using a **rotating-disk electrode** (Fig. 29.16). The electrode is a small flat disk set in a vertical rotating axle. The rotation of its surface sets up a steady hydrodynamic flow which circulates the solution over its face. The flow pattern can be calculated, and the limiting current related to the speed of rotation.

Various modifications of the rotating-disk electrode have been developed. One modification is to pulse the potential difference and then observe the growth of current towards its limiting value. Another is the **ring-disk electrode**, in which the central spinning disk is surrounded by an electrode in the form of a narrow ring. The ring can be set at a different potential from that of the disk and hence may be used to reduce or oxidize the products that have been formed at the disk and then swept into its vicinity by the hydrodynamic flow. Analysis of the currents at both electrodes, the concentration profiles (Fig. 29.17), and knowledge of the time it takes for the products formed at the disk to flow to the ring gives very detailed information about electrode kinetics. This information includes the identities of the reactants and the rates of electron transfer.

ELECTROCHEMICAL PROCESSES

To induce current to flow through an electrolytic cell and bring about a non-spontaneous cell reaction, the applied potential difference must exceed the zero-current potential by at least the **cell overpotential**. The cell overpotential is the sum of the overpotentials at the two electrodes and the ohmic drop ($I \times R$) due to the current through the electrolyte. The additional potential needed to achieve a detectable rate of reaction may need to be large when the exchange current density at the electrodes is small. For similar reasons, a working galvanic cell generates a smaller potential than under zero-current conditions. In this section we see how to cope with both aspects of the overpotential.

29.4 Electrolysis

The rate of gas evolution or metal deposition during electrolysis can be estimated from the Butler–Volmer equation and tables of exchange current densities. The exchange current density depends strongly on the nature of the electrode surface, and changes in the course of the

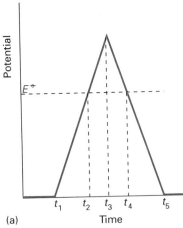

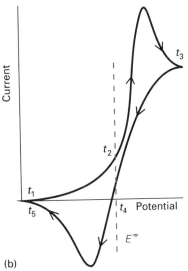

29.15 (a) The change of potential with time and (b) the resulting current/potential curve in a cyclic voltammetry experiment.

electrodeposition of one metal on another. A very crude criterion is that significant evolution or deposition occurs only if the overpotential exceeds about 0.6 V.

Example 29.5 *Estimating the relative rates of electrolysis*

Derive an expression for the relative rates of electrodeposition and hydrogen evolution in a solution in which both may occur.

Method. We can calculate the ratio of the cathodic currents by using eqn 13. For simplicity, assume equal transfer coefficients.

Answer. From eqn 13,

$$\frac{j'}{j} = \frac{j_0'}{j_0} \times e^{(\eta - \eta')\alpha f}$$

Comment. This equation shows that metal deposition is favoured by a large exchange-current density and relatively high hydrogen evolution overvoltage (so that $\eta - \eta'$ is positive and large).

Exercise E29.5. Deduce an expression for the ratio when the hydrogen evolution is limited by transport across a diffusion layer.

$$[j'/j = (\delta c j_0'/FD)e^{-\alpha \eta' f}]$$

A glance at Table 29.1 shows the wide range of exchange current densities for a metal/hydrogen electrode. The most sluggish exchange currents occur for lead and mercury, and the value of $10^{-12}\,\text{A cm}^{-2}$ corresponds to a monolayer of atoms being replaced in about 5 years. For such systems, a high overpotential is needed to induce significant hydrogen evolution. In contrast, the value for platinum ($10^{-3}\,\text{A cm}^{-2}$) corresponds to a monolayer being replaced in 0.1 s, so gas evolution occurs for a much lower overpotential.

The exchange current density also depends on the crystal face exposed. For the deposition of copper on copper, the (100) face has $j_0 = 1\,\text{mA cm}^{-2}$, so for the same overpotential the (100) face grows at 2.5 times the rate of the (111) face, for which $j_0 = 0.4\,\text{mA cm}^{-2}$.

29.5 The characteristics of working cells

We expect the cell potential to decrease as current is generated because it is then no longer working reversibly and can therefore do less than maximum work.

The potentials of working cells

We shall consider the cell

$$M\,|M^+(aq)|\,|M'^+(aq)|\,M'$$

and ignore all the complications arising from liquid junctions. The working potential of the cell is

$$E' = \Delta\phi_R - \Delta\phi_L$$

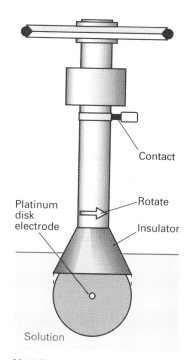

29.16 The dropping mercury electrode has been largely replaced by the rotating-disk electrode, because the hydrodynamic flow over its face can be calculated more reliably.

Because the working potential differences differ from their zero-current values by overpotentials, we can write

$$\Delta\phi_X = E_X + \eta_X$$

where X is L or R for the left or right electrode, respectively. The working cell potential is therefore

$$E' = E + \eta_R - \eta_L$$

with E the zero-current cell potential. We should subtract from this expression the ohmic potential difference IR_s, where R_s is the cell's internal resistance:

$$E' = E + \eta_R - \eta_L - IR_s \tag{19}$$

The ohmic term is a contribution to the cell's irreversibility—it is a thermal dissipation term—so the sign of IR_s is always such as to reduce the potential in the direction of zero.

The overpotentials in eqn 19 can be calculated from the Butler–Volmer equation for a given current I being drawn. We shall simplify the equations by supposing that the areas A of the electrodes are the same, that only one electron is transferred in the rate-determining steps at the electrodes, that the transfer coefficients are both $\frac{1}{2}$, and that the high-overpotential limit of the Butler–Volmer equation may be used. Then from eqns 13 and 19 we find

$$E' = E - \frac{4RT}{F}\ln\left(\frac{I}{A\bar{j}}\right) - IR_s, \qquad \text{where } \bar{j} = (j_{L0}j_{R0})^{\frac{1}{2}} \tag{20}$$

The concentration overpotential also reduces the cell potential. If we use the Nernst diffusion layer model for each electrode, then the total change of potential arising from concentration polarization is given by eqn 18 as

$$E' = E - \frac{RT}{zF}\ln\left\{\left(1 - \frac{I}{Aj_{\text{lim,L}}}\right)\left(1 - \frac{I}{Aj_{\text{lim,R}}}\right)\right\} \tag{21}$$

This contribution can be added to the one in eqn 20 to obtain a full (but still very approximate) expression for the cell potential when a current I is being drawn:

$$E' = E - IR_s + \frac{2RT}{zF}\ln g, \quad \text{where } g = (I/A\bar{j})^{2z}\left\{\left(1 - \frac{I}{Aj_{\text{lim,L}}}\right)\left(1 - \frac{I}{Aj_{\text{lim,R}}}\right)\right\}^{\frac{1}{2}} \tag{22}$$

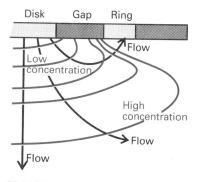

29.17 Schematic concentration profiles of a species in the vicinity of a ring-disk electrode. Note that the concentration changes in the gap region and is constant over the surfaces of the two electrodes.

This equation depends on a lot of parameters, but an example of its general form is given in Fig. 29.18. Notice the very steep decline of working potential when the current is high and close to the limiting value for one of the electrodes.

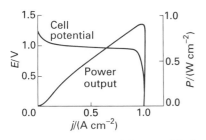

29.18 The dependence of the potential of a working cell on the current being drawn and the corresponding power output calculated using eqns 22 and 23, respectively. Notice the sharp decline in power just after the maximum.

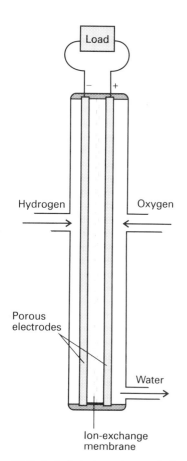

29.19 A single hydrogen/oxygen fuel cell. In practice, a battery of many cells is used.

The power output of working cells

Because the power P supplied by a working cell is $I \times E'$, from eqn 22 we can write

$$P = IE - I^2 R_s + \frac{2IRT}{zF} \ln g \tag{23}$$

The first term on the right is the power that would be produced if the cell retained its zero-current potential when delivering current. The second term is the power generated uselessly as heat as a result of the resistance of the electrolyte. The third term is the reduction of the potential at the electrodes as a result of drawing current.

The general dependence of power output on the current drawn is illustrated in Fig. 29.18. Notice how maximum power is achieved just before the concentration polarization quenches the cell's performance. Information of this kind is essential if the optimum conditions for operating electrochemical devices are to be found and their performance improved.

29.6 Fuel cells and secondary cells

We considered the thermodynamic properties of galvanic cells in Chapter 10. Here we shall mention some of their kinetic aspects in the light of the concepts introduced in this chapter.

Fuel cells

A fuel cell operates like a conventional galvanic cell with the exception that the reactants are supplied from outside rather than forming an integral part of its construction. A fundamental and important example of a fuel cell is the hydrogen/oxygen cell (Fig. 29.19). The electrolyte is concentrated aqueous potassium hydroxide maintained at 200°C and 20–40 atm; the electrodes are porous nickel in the form of sheets of compressed powder. The cathode reaction is the reduction

$$O_2(g) + 2H_2O(l) + 4e^- \rightarrow 4OH^-(aq) \qquad E^\ominus = +0.40\,V$$

and the anode reaction is the oxidation

$$H_2(g) + 2OH^-(aq) \rightarrow 2H_2O(l) + 2e^- \qquad E^\ominus = -0.83\,V$$

Because the overall reaction

$$2H_2(g) + O_2(g) \rightarrow 2H_2O(l) \qquad E^\ominus = +1.23\,V$$

is exothermic as well as spontaneous, it is less favourable at 200°C than at 25°C, so the cell potential is lower. However, the increased pressure compensates for the increased temperature, and at 200°C and 40 atm $E \approx +1.2\,V$.

Example 29.6 *Calculating the potential of a fuel cell*

Calculate the potential of a reversible propane/oxygen fuel cell operating under standard conditions at 298 K.

Method. The route to the potential is through a calculation of the standard reaction Gibbs energy from tables of standard Gibbs energies of formation (Tables 2.11 and 2.12 in the Data section). The standard Gibbs energy of reaction is then converted to a standard potential by using $\Delta_r G^{\ominus} = -vFE^{\ominus}$. Begin by writing the cell reaction and calculating its standard Gibbs energy. Then identify the value of v by expressing the overall reaction in terms of half-reactions.

Answer. The cell reaction is

$$C_3H_8(g) + 5O_2(g) \rightarrow 3CO_2(g) + 4H_2O(l)$$

The standard Gibbs energy of the combustion is

$$\Delta_r G^{\ominus} = \{3(-394.4) + 4(-237.2) - (-23.5)\}\,kJ\,mol^{-1}$$
$$= -2108\,kJ\,mol^{-1}$$

The cell reaction is the combination of the following two half-reactions:

$$C_3H_8(g) + 6H_2O(l) \rightarrow 3CO_2(g) + 20H^+(aq) + 20e^-$$
$$5O_2(g) + 20H^+(aq) + 20e^- \rightarrow 10H_2O(l)$$

Therefore, $v = 20$. It follows that

$$E^{\ominus} = -\frac{(-2108\,kJ\,mol^{-1})}{20F} = +1.09\,V$$

Comment. The standard enthalpy of combustion at 298 K is $-2220\,kJ\,mol^{-1}$ and so, in a cell working with perfect efficiency, $112\,kJ\,mol^{-1}$ of energy is released as heat.

Exercise E29.6. Repeat the calculation for a methanol-powered fuel cell under the same conditions.
[1.21 V]

One advantage of the hydrogen/oxygen system is the large exchange current density of the hydrogen reaction. Unfortunately, the oxygen reaction has an exchange current density of only about $10^{-10}\,A\,cm^{-2}$, which limits the current available from the cell. One way round the difficulty is to use a catalytic surface (to increase j_0) with a large surface area. One type of highly developed fuel cell has phosphoric acid as the electrolyte and operates with hydrogen and air at about 200°C; the hydrogen is obtained from a reforming reaction on natural gas. The power output of batteries of such cells has reached the order of 10 MW. Cells with molten carbonate electrolytes at about 600°C can make use of natural gas directly. Solid-state electrolytes are also used. They include one version in which the electrolyte is a solid polymeric ionic conductor at about 100°C, but in current versions it requires very pure hydrogen to operate successfully. Solid ionic conducting oxide cells operate at about 1000°C and can use hydrocarbons directly as fuel.

The lead-acid battery

Electric storage cells operate as galvanic cells while they are producing electricity but as electrolytic cells while they are being charged by an external supply.

The **lead-acid battery** is an old device, but one well suited to the job of starting cars (and the only one available). During charging the cathode reaction is the reduction of Pb^{2+} and its deposition as lead on the lead electrode. Deposition occurs instead of the reduction of the acid to hydrogen because the latter has a low exchange current density on lead. The anode reaction during charging is the oxidation of Pb(II) to Pb(IV), which is deposited as the oxide PbO_2. On discharge the two reactions run in reverse. Because they have such high exchange-current densities the discharge can occur rapidly, which is why the lead-acid battery can produce large currents on demand.

CORROSION

A thermodynamic warning of the likelihood of corrosion is obtained by comparing the standard potentials of the metal reduction, such as

$$Fe^{2+}(aq) + 2e^- \rightarrow Fe(s) \qquad E^\ominus = -0.44 \text{ V}$$

with the values for one of the following half-reactions:

In acidic solution:

(a) $2H^+(aq) + 2e^- \rightarrow H_2(g)$ $\qquad\qquad E^\ominus = 0$

(b) $4H^+(aq) + O_2(g) + 4e^- \rightarrow 2H_2O(l)$ $\qquad E^\ominus = +1.23 \text{ V}$

In basic solution:

(c) $2H_2O(l) + O_2(g) + 4e^- \rightarrow 4OH^-(aq)$ $\qquad E^\ominus = +0.40 \text{ V}$

Because all three redox couples have standard potentials more positive than $E^\ominus(Fe^{2+}/Fe)$, all three can drive its oxidation. The electrode potentials we have quoted are standard values, and they change with the pH of the medium. For the first two:

$$E(a) = E^\ominus(a) + \frac{RT}{F} \ln a(H^+) = -0.059 \text{ V} \times pH$$

$$E(b) = E^\ominus(b) + \frac{RT}{F} \ln a(H^+) = 1.23 \text{ V} - 0.059 \text{ V} \times pH$$

These expressions let us judge at what pH the iron will have a tendency to oxidize (see Chapter 10). A thermodynamic discussion of corrosion, however, only indicates whether a tendency to corrode exists. If there is a thermodynamic tendency, we must examine the kinetics of the processes involved to see whether the process occurs at a significant rate.

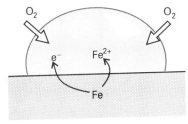

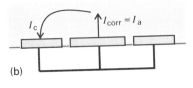

29.20 (a) A simple version of the corrosion process is that of a droplet of water, which is oxygen-rich near its boundary with air. The oxidation of the iron takes place in the region away from the oxygen because the electrons are transported through the metal. (b) The process may be modelled as a short-circuited electrochemical cell.

29.7 The rate of corrosion

A model of a corrosion system is shown in Fig. 29.20a. It can be taken to be a drop of slightly acidic (or basic) water containing some dissolved oxygen in contact with the metal. The oxygen near the edges of the droplet is reduced by electrons donated by the iron over an area A. Those electrons are replaced by others released elsewhere as $Fe \rightarrow Fe^{2+} + 2e^-$. This oxidative release occurs over an area A' under the oxygen-deficient inner region of the droplet. The droplet acts as a short-circuited galvanic cell (Fig. 29.20b).

The rate of corrosion is measured by the current of metal ions leaving the metal surface in the anodic region. This flux of ions gives rise to the **corrosion current** I_{corr}, which can be identified with the anodic current I_a. The corrosion current is related to the cell potential of the corrosion couple by

$$I_{corr} = \bar{A}\bar{j}_0 e^{\frac{1}{4}fE}, \qquad \text{where } \bar{j}_0 = (j_0 j_0')^{\frac{1}{2}} \text{ and } \bar{A} = (AA')^{\frac{1}{2}} \tag{24}$$

JUSTIFICATION

Because any current emerging from the anodic region must find its way to the cathodic region, the cathodic current I_c must also be equal to the corrosion current. In terms of the current densities at the oxidation and reduction sites, j and j', respectively, we can write

$$I_{corr} = jA = j'A' = (jj'AA')^{\frac{1}{2}} = \bar{j}\bar{A}, \qquad \text{where } \bar{j} = (jj')^{\frac{1}{2}} \text{ and } \bar{A} = (AA')^{\frac{1}{2}}$$

The Butler–Volmer equation is now used to express the current densities in terms of overpotentials. For simplicity we assume that the overpotentials are large enough for the high-overpotential limit (eqn 13) to apply, that polarization overpotential can be neglected, that the rate-determining step is the transfer of a single electron, and that the transfer coefficients are $\frac{1}{2}$. We also assume that, since the droplet is so small, there is negligible potential difference between the cathode and anode regions of the solution. Moreover, since it is short-circuited by the metal, the potential of the metal is the same in both regions, so the potential difference between the metal and the solution is the same in both regions too. This common potential difference is called the **corrosion potential difference** $\Delta\phi_{corr}$. The overpotentials in the two regions are therefore

$$\eta = \Delta\phi_{corr} - \Delta\phi_e \qquad \text{and} \qquad \eta' = \Delta\phi_{corr} - \Delta\phi_e'$$

so the current densities are

$$j = j_0 e^{\frac{1}{2}\eta f} = j_0 e^{\frac{1}{2}f\Delta\phi_{corr}} e^{-\frac{1}{2}f\Delta\phi_e}$$
$$j' = j_0' e^{-\frac{1}{2}\eta'f} = j_0' e^{-\frac{1}{2}f\Delta\phi_{corr}} e^{\frac{1}{2}f\Delta\phi_e'}$$

These expressions can be substituted into the expression for I_{corr} and $\Delta\phi' - \Delta\phi$ replaced by the difference of electrode potentials E to give eqn 24.

Several conclusions can be drawn from eqn 24. First, the rate of corrosion depends on the surfaces exposed: if either A or A' is zero, then the corrosion current is zero. This interpretation points to a trivial, yet often effective, method of slowing corrosion: cover the surface with a coating, such as paint. (Paint also increases the effective solution resistance between the cathode and anode patches on the surface.) Second, for corrosion reactions with similar exchange current densities, the rate of corrosion is high when E is large. That is, rapid corrosion can be expected when the oxidizing and reducing couples have widely differing electrode potentials.

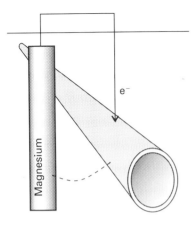

(a)

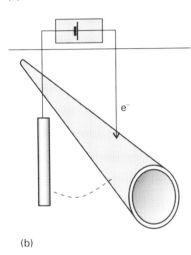

(b)

29.21 (a) In cathodic protection an anode of a more strongly reducing metal is sacrificed to maintain the integrity of the protected object (e.g. a pipeline, bridge, or boat). (b) In impressed-current cathodic protection electrons are supplied from an external cell so that the object itself is not oxidized. The broken lines depict the completed circuit through the soil.

The effect of the exchange current density on the corrosion rate can be seen by considering the specific case of iron in contact with acidified water. Thermodynamically, either oxygen reduction reaction (a) or (b) on p. 1026 is effective. However, the exchange-current density of reaction (b) on iron is only about 10^{-14} A cm^{-2}, while for (a) it is 10^{-6} A cm^{-2}. The latter therefore dominates kinetically, and iron corrodes by hydrogen evolution in acidic solution.

29.8 The inhibition of corrosion

Several techniques for inhibiting corrosion are available. Coating the surface with some impermeable layer, such as paint, may prevent the access of damp air. Unfortunately, this protection fails disastrously if the paint becomes porous. The oxygen then has access to the exposed metal and corrosion continues beneath the paintwork. Another form of surface coating is provided by **galvanizing**, the coating of an iron object with zinc. Because the latter's standard potential is -0.76 V, which is more negative than that of the iron couple, the corrosion of zinc is thermodynamically favoured and the iron survives (the zinc survives because it is protected by a hydrated oxide layer). In contrast, tin plating leads to a very rapid corrosion of the iron once its surface is scratched and the iron exposed because the tin couple ($E^{\ominus} = -0.14$ V) oxidizes the iron couple ($E^{\ominus} = -0.44$ V). Some oxides are inert kinetically in the sense that they adhere to the metal surface and form an impermeable layer over a fairly wide pH range. This **passivation**, or kinetic protection, can be seen as a way of decreasing the exchange currents by sealing the surface. Thus, aluminium is inert in air even though its standard potential is strongly negative (-1.66 V).

Another method of protection is to change the electric potential of the object by pumping in electrons that can be used to satisfy the demands of the oxygen reduction without involving the oxidation of the metal. In **cathodic protection**, the object is connected to a metal with a more negative standard potential (such as magnesium, -2.36 V). The magnesium acts as a sacrificial anode, supplying its own electrons to the iron and becoming oxidized to Mg^{2+} in the process (Fig. 29.21a). A block of magnesium replaced occasionally is much cheaper than the ship, building, or pipeline for which it is being sacrificed. In **impressed-current cathodic protection** (Fig. 29.21b) an external cell supplies the electrons and eliminates the need for iron to transfer its own.

CHECK LIST OF KEY IDEAS

1 The **electric double layer** and approximate models of its structure (Section 29.1).

2 The **Volta potential** and the **Galvani potential** and the latter's relation to the electrode potential, (Section 29.1).

3 The **cathodic** and **anodic current densities** at the electrodes and the net current density (eqn 1).

4 The definitions of **transfer coefficient** (eqn 4),

the **exchange current density**, and the **overpotential** (eqn 8).

5 The derivation of the **Butler–Volmer equation** (eqn 10).

6 The **low overpotential limit** (eqn 12) and **high overpotential limit** (eqn 13) of the Butler–Volmer equation and the **Tafel plot** for determining the properties of a working electrode (Example 29.3).

7 The classification of electrodes as **polarizable** and **non-polarizable** (Section 29.3).

8 The origin of **concentration polarization** and its description in terms of the **Nernst diffusion layer** and the **limiting current density** (eqn 16).

9 The technique of **polarography** and the study of electrode kinetics using a **rotating-disk electrode**, **voltammetry**, and **cyclic voltammetry** (Section 29.3).

10 The considerations that govern the kinetic feasibility of **electrolysis** (Section 29.4).

11 The potential of **working cells** (eqn 22) and their **power output** (eqn 23).

12 The operation of **fuel cells** and of **storage cells** (Section 29.6) and the kinetic considerations that determine their characteristics.

13 The electrochemical considerations relating to the rate of **corrosion** (Section 29.7) and its inhibition (Section 29.8).

EXERCISES

29.1. The Helmholtz model of the electric double layer is equivalent to a parallel plate capacitor. Hence the potential difference across the double layer is given by $\Delta\phi = \sigma d/\varepsilon$, where d is the distance between the plates and σ is the surface charge density. Assuming that this model holds for concentrated salt solutions, calculate the magnitude of the electric field on the surface of silica in a 5.0 M NaCl(aq) solution, for which the relative permittivity at 21°C is 48, if the surface charge density is 0.10 C m^{-2}.

29.2. The transfer coefficient of a certain electrode in contact with M^{3+} and M^{4+} in aqueous solution at 25°C is 0.39. The current density is found to be 55.0 mA cm^{-2} when the overvoltage is 125 mV. What is the overvoltage required for a current density of 75 mA cm^{-2}?

29.3. Determine the exchange current density from the information given in Exercise 29.2.

29.4. To a first approximation, significant evolution or deposition occurs in electrolysis only if the overpotential exceeds about 0.6 V. To illustrate this criterion determine the effect that increasing the overpotential from 0.40 V to 0.60 V has on the current density in the electrolysis of 1.0 M NaOH(aq), which is 1.0 mA cm^{-2} at 0.4 V and 25°C. Take $\alpha = 0.5$.

29.5. Use the data in Table 29.1 for the exchange current density and transfer coefficient for the reaction $2H^+ + 2e^- \rightarrow H_2$ on nickel at 25°C to determine what current density would be needed to obtain an overpotential of 0.20 V as calculated from (a) the Butler–Volmer equation, and (b) the Tafel equation. Is the validity of the Tafel approximation affected at lower overpotentials (of 0.1 V and less)?

29.6. Estimate the limiting current density at an electrode in which the concentration of Ag^+ ions is 2.5 mmol L^{-1} at 25°C. The thickness of the Nernst diffusion layer is 0.40 mm. The ionic conductivity of Ag^+ at infinite dilution and 25°C is 61.9 S cm^2 mol^{-1}.

29.7. A 0.10 M CdSO$_4$(aq) solution is electrolysed between a cadmium cathode and a platinum anode at 25°C with a current density of 1.00 mA cm^{-2}. The hydrogen overpotential is 0.60 V. What will be the concentration of Cd^{2+} ions when evolution of H_2 just begins at the cathode? Assume all activity coefficients are unity.

29.8. Take $\alpha = 0.5$ and plot j/j_0 as a function of the overpotential η at 298 K.

29.9. A typical exchange current density, that for H^+ discharge at platinum, is 0.79 mA cm^{-2} at 25°C. What is the current density at an electrode when its overpotential is (a) 10 mV, (b) 100 mV, (c) -5.0 V? Take $\alpha = 0.5$.

29.10. The exchange current density for a Pt | Fe^{3+}, Fe^{2+} electrode is 2.5 mA cm^{-2}. The standard potential of the electrode is $+0.77$ V. Calculate the current flowing through an electrode of surface area 1.0 cm^2 as a function of the potential of the electrode. Take unit activity for both ions.

29.11. Suppose that the electrode potential is set at 1.00 V. Calculate the current that flows for the ratio of activities $a(Fe^{2+})/a(Fe^{3+})$ in the range 0.1 to 10.0 and at 25°C.

29.12. What overpotential is needed to sustain a 20 mA current at the $Pt\,|\,Fe^{3+},Fe^{2+}$ electrode in which both ions are at a mean activity $a = 0.1$?

29.13. How many electrons or protons are transported through the double layer in each second when the Pt, $H_2\,|\,H^+$, $Pt\,|\,Fe^{3+}$, Fe^{2+}, and Pb, $H_2\,|\,H^+$ electrodes are at equilibrium at 25°C? Take the area as $1.0\ cm^2$ in each case. Estimate the number of times each second a single atom on the surface takes part in a electron transfer event, assuming an electrode atom occupies about $(280\ pm)^2$ of the surface.

29.14. What is the effective resistance at 25°C of an electrode interface when the overpotential is small? Evaluate it for $1.0\ cm^2$ (a) $Pt,H_2\,|\,H^+$, (b) $Hg,H_2\,|\,H^+$ electrodes.

29.15. State what happens when a platinum electrode in an aqueous solution containing both Cu^{2+} and Zn^{2+} ions at unit activity is made the cathode of an electrolysis cell.

29.16. What are the conditions that allow a metal to be deposited from aqueous acidic solution before hydrogen evolution occurs significantly? Why may silver be deposited from aqueous silver nitrate? Why may cadmium be deposited from aqueous cadmium sulfate? (The overpotential for hydrogen evolution on cadmium is about 1 V at current densities of $1\ mA\ cm^{-2}$.)

29.17. The exchange current density for H^+ discharge at zinc is about $50\ pA\ cm^{-2}$. The overpotential for H^+ discharge is about 0.3 V. Can zinc be deposited from a unit activity aqueous solution of a zinc salt?

29.18. The standard potential of the $Zn^{2+}\,|\,Zn$ electrode is -0.76 V at 25°C. The exchange current density for H^+ discharge at platinum is $0.79\ mA\ cm^{-2}$. Can zinc be plated on to platinum at the temperature? (Take unit activities.)

29.19. Can magnesium be deposited on a zinc electrode from a unit activity acid solution at 25°C?

29.20. The limiting current density for the reaction $I_3^- + 2e^- \rightarrow 3I^-$ at a platinum electrode is $28.9\ \mu A\ cm^{-2}$ when the concentration of KI is $6.6 \times 10^{-4}\ mol\ L^{-1}$ and the temperature 25°C. The diffusion coefficient of I_3^- is $1.14 \times 10^{-9}\ m^2\ s^{-1}$. What is the thickness of the diffusion layer?

29.21. Calculate the maximum (zero-current) potential difference of a nickel–cadmium cell, and the maximum possible power output when 100 mA is drawn at 25°C.

29.22. Calculate the thermodynamic limit to the zero-current potential of fuel cells operating on (a) hydrogen and oxygen, (b) methane and air. Use the Gibbs energy information in the Data section, and take the species to be in their standard states at 25°C.

29.23. Which of the following metals has a thermodynamic tendency to corrode in moist air at $pH = 7$: Fe, Cu, Pb, Al, Ag, Cr, Co? Take as a criterion of corrosion a metal ion concentration of at least $10^{-6}\ mol\ L^{-1}$.

29.24. The corrosion current density j_{corr} at an iron anode is $1.0\ A\ m^{-2}$. What is the corrosion rate in millimetres per year? Assume uniform corrosion.

PROBLEMS
Numerical problems

29.1. In an experiment on the $Pt\,|\,H_2\,|\,H^+$ electrode in dilute H_2SO_4 the following current densities were observed at 25°C. Evaluate α and j_0 for the electrode.

η/mV	50	100	150	200	250
$j/(mA\ cm^{-2})$	2.66	8.91	29.9	100	335

How would the current density at this electrode depend on the overpotential of the same set of magnitudes but of opposite sign?

29.2. The standard electrode potentials of lead and tin are -126 and -136 mV, respectively, at 25°C, and the overvoltages for their deposition are close to zero. What should their relative activities be in order to ensure simultaneous deposition from a mixture?

29.3. The maximum theoretical efficiency of a fuel cell may be defined as $\varepsilon = |\Delta_r G/\Delta_r H| = |\nu FE/\Delta_r H|$. However,

as indicated in eqn 22, the potential of a working cell is reduced from the zero-current potential. In a hydrogen/oxygen fuel cell with platinum electrodes operating at 373 K, the exchange current densities are $j_a = 100\ mA\ m^{-2}$ and $j_c = 3.00\ mA\ m^{-2}$. Calculate the efficiency of the cell when operated at a curent density of $300\ mA\ m^{-2}$ with internal resistance equivalent to $0.500\ \Omega\ m^2$. Assume $\alpha \approx 0.5$ and $j/j_{lim} \approx 0.5$ for both electrodes and assume $E = E^{\ominus}$. Compare the resulting value to the theoretical maximum efficiency of the cell and to the theoretical maximum efficiency of a heat engine that uses the combustion of hydrogen and oxygen and operates between 373 and 673 K.

29.4. Estimating the power output and potential of a cell under operating conditions is very difficult, but eqn 22 summarizes, in an approximate way, some of the parameters involved. As a first step in manipulating this expression, identify all the quantities that depend on the ionic concentrations. Express E in terms of the concentration and conductivities of the ions present in the cell. Estimate the parameters for $Zn|ZnSO_4(aq)||CuSO_4(aq)|Cu$. Take electrodes of area $5\,cm^2$ separated by $5\,cm$. Ignore both potential differences and resistance of the liquid junction. Take the concentration as $1\,mol\,L^{-1}$, the temperature $25°C$, and neglect activity coefficients. Plot E as a function of the current drawn. On the same graph, plot the power output of the cell. What current corresponds to maximum power?

29.5. Consider a cell in which the current is activation controlled. Show that the current for maximum power can be estimated by plotting $\log(I/I_0)$ and $c_1 - c_2 I$ against I (where $I_0 = A^2 j_0 j_0'$ and c_1 and c_2 are constants), and looking for the point of intersection of the curves. Carry through this analysis for the cell in Problem 29.4 ignoring all concentration overpotentials.

29.6. Estimate the magnitude of the corrosion current for a patch of zinc of area $0.25\,cm^2$ in contact with a similar area of iron in an aqueous environment at $25°C$. Take the exchange current densities as $1\,\mu A\,cm^{-2}$ and the local ion concentrations at $1\,\mu mol\,L^{-1}$.

29.7. The corrosion potential of iron immersed in a deaerated acidic solution of $pH = 3$ is $-0.720\,V$ as measured at $25°C$ relative to the standard calomel electrode with potential $0.2802\,V$. A Tafel plot of cathodic current density against overpotential yields a slope of $18\,V^{-1}$ and

the hydrogen ion exchange current density $j_0 = 0.10\,\mu A\,cm^{-2}$. Calculate the corrosion rate in milligrams of iron per square centimetre per day $(mg\,cm^{-2}\,d^{-1})$.

Theoretical problems

29.8. If $\alpha = \frac{1}{2}$, an electrode interface is unable to rectify alternating current because the current density curve is symmetrical about $\eta = 0$. When $\alpha \neq \frac{1}{2}$, the magnitude of the current density depends on the sign of the overpotential, so some degree of 'Faradaic rectification' may be obtained. Suppose that the overpotential varies as $\eta = \eta_0 \cos \omega t$. Derive an expression for the mean flow of current (averaged over a cycle) for general α, and confirm that the mean current is zero when $\alpha = \frac{1}{2}$. In each case work in the limit of small η_0 but to second-order in $\eta_0 F/RT$. Calculate the mean direct current at $25°C$ for a $1.0\,cm^2$ hydrogen-platinum electrode with $\alpha = 0.38$ when the overpotential varies between $\pm 10\,mV$ at $50\,Hz$.

29.9. Now suppose that the overpotential is in the high-overpotential region at all times even though it is oscillating. What waveform will the current across the interface show if it varies linearly and periodically (as a sawtooth waveform) between η_- and η_+ around η_0? Take $\alpha = \frac{1}{2}$.

29.10. Derive an expression for the current density at an electrode where the rate process is diffusion-controlled and η^c is known. Sketch the form of j/j_L as a function of η^c. What changes occur if anion currents are involved?

Further information

FURTHER INFORMATION 1: The Maxwell–Boltzmann distribution

The mean value of a property X that can have a continuous range of values (like a molecular speed) is

$$\langle X \rangle = \int Xf(X)\, dX$$

The function $f(X)$ is the distribution of the property X. It gives the probability that the property lies in the range X to $X + dX$ (which from now on we write $(X, X + dX)$). For example, $f(s)$ is the distribution of the speed s and $f(s)\, ds$ is the probability that the speed lies in the range $(s, s + ds)$. If the probability that one property X lies in the range $(X, X + dX)$ is $f(X)\, dX$ and the probability that an *independent* property Y lies in the range $(Y, Y + dY)$ is $f(Y)\, dY$, then the probability of X and Y lying simultaneously in these ranges is the product of the individual probabilities:

$$f(X, Y)\, dX\, dY = f(X)f(Y)\, dX\, dY$$

For example, the probabilities of occurrence of components of velocity v_x and v_y of molecules in a gas are independent of one another, and the probability that a molecule has a velocity x-component in the range $(v_x, v_x + dv_x)$ and simultaneously a y-component in the range $(v_y, v_y + dv_y)$ is the product of the two probabilities. In such cases the joint distribution $f(X, Y)$ is the product of the individual distributions:

$$f(X, Y) = f(X)f(Y)$$

The preceding remarks provide enough background to establish the Maxwell–Boltzmann distribution of speed in a perfect gas. The three velocity components are independent of one another, so the joint distribution is the product of the three individual distributions:

$$f(v_x, v_y, v_z) = f(v_x)f(v_y)f(v_z)$$

Because the probability of finding a molecule with the velocity $+|v_x|$ is the same as the probability of finding it with the velocity $-|v_x|$, and likewise for the other components, the individual distributions must depend on the *square* of the components, not on the components themselves. We therefore write the last relation as

$$f(v_x, v_y, v_z) = f(v_x^2)f(v_y^2)f(v_z^2)$$

We next assume that the probability of a molecule having a particular range of velocity components is independent of its direction of flight. That is, we assume that f depends on the speed (s, the magnitude of the velocity), where $s^2 = v_x^2 + v_y^2 + v_z^2$, but not on the individual components. For example, the probability that a molecule has a velocity with components in an infinitesimal range at the velocity $(1.0\,\text{km s}^{-1}, 2.0\,\text{km s}^{-1}, 3.0\,\text{km s}^{-1})$, and speed $3.7\,\text{km s}^{-1}$, is the same as the probability that it has a velocity in the same infinitesimal range at the velocity $(2.0\,\text{km s}^{-1}, 1.0\,\text{km s}^{-1}, 3.0\,\text{km s}^{-1})$, and hence the same speed of $3.7\,\text{km s}^{-1}$, or any other set of components that correspond to a speed of $3.7\,\text{km s}^{-1}$. It follows that f depends only on $v_x^2 + v_y^2 + v_z^2$, so the last equation becomes

$$f(v_x^2 + v_y^2 + v_z^2) = f(v_x^2)f(v_y^2)f(v_z^2)$$

Only an exponential function satisfies such a relation (because $e^a e^b e^c = e^{a+b+c}$). Consequently, because $f(v_x) = f(v_x^2)$,

$$f(v_x) = K e^{\pm \zeta v_x^2}$$

where K and ζ are constants. The two constants are the same for $f(v_y)$ and $f(v_z)$, because the distributions are the same in each direction. Because the probability of very high velocities is very small, we can discard the solution with $+$ in the exponent.

To determine the constant K, we note that there is unit probability of finding a velocity in the range $-\infty < v_x < \infty$. Therefore, we can write the total probability as

$$\int_{-\infty}^{\infty} f(v_x)\,\mathrm{d}v_x = 1$$

Substitution of the expression given above then leads to

$$1 = \int_{-\infty}^{\infty} f(v_x)\,\mathrm{d}v_x = K \int_{-\infty}^{\infty} e^{-\zeta v_x^2}\,\mathrm{d}v_x = K\left(\frac{\pi}{\zeta}\right)^{1/2}$$

Therefore $K = (\zeta/\pi)^{1/2}$. To determine ζ, we calculate the mean value of a property we already know. The mean value of v_x^2 is

$$\langle v_x^2 \rangle = \int_{-\infty}^{\infty} v_x^2 f(v_x)\,\mathrm{d}v_x = (\zeta/\pi)^{1/2} \int_{-\infty}^{\infty} v_x^2 e^{-\zeta v_x^2}\,\mathrm{d}v_x$$

The integral is standard, and is equal to $\frac{1}{2}(\pi/\zeta^3)^{1/2}$. It follows that

$$\langle v_x^2 \rangle = \left(\frac{\zeta}{\pi}\right)^{1/2} \times \frac{1}{2}\left(\frac{\pi}{\zeta^3}\right)^{1/2} = \frac{1}{2\zeta}$$

The mean square speed is therefore $3/(2\zeta)$. However, in the text we deduced that the mean square speed is given by eqn 1.15; hence we can conclude that

$$\zeta = \frac{m}{2kT}$$

Therefore, the complete form of the velocity distribution is

$$f(v_x) = \left(\frac{m}{2\pi kT}\right)^{1/2} e^{-mv_x^2/2kT}$$

This expression is the Maxwell–Boltzmann distribution of molecular velocities.

Finally, we can derive the distribution of speeds of molecules irrespective of their direction of motion. The probability that a molecule has velocity components in the range $(v_x, v_x + dv_x)$, $(v_y, v_y + dv_y)$, $(v_z, v_z + dv_z)$ is

$$f(v_x, v_y, v_z)\, dv_x\, dv_y\, dv_z = f(v_x)f(v_y)f(v_z)\, dv_x\, dv_y\, dv_z$$

$$= \left(\frac{m}{2\pi kT}\right)^{\frac{3}{2}} e^{-mv^2/2kT}\, dv_x\, dv_y\, dv_z$$

the probability $f(s)\, ds$ that the molecules have a speed in the range $(s, s + ds)$ is the sum of the probabilities that it lies in any of the volume elements $dv_x\, dv_y\, dv_z$ in the spherical shell of radius s (Fig. 1). The sum of the volume elements on the right-hand side of the last expression is the volume of this shell, which is $4\pi s^2\, ds$. Therefore,

$$f(s) = 4\pi \left(\frac{m}{2\pi kT}\right)^{\frac{3}{2}} s^2 e^{-ms^2/2kT}$$

which is the Maxwell distribution of speeds, eqn 1.16.

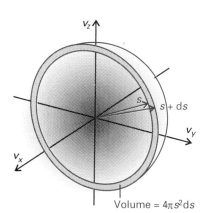

Volume = $4\pi s^2 ds$

1 To calculate the probability that molecules have a speed in the range s to $s + ds$, we calculate the probability that their velocity vector lies somewhere between the walls of a shell of radii s and $s + ds$.

FURTHER INFORMATION 2: Relations between partial derivatives

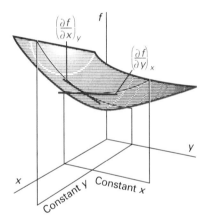

the local slopes with x and y held constant.

A partial derivative of a function of more than one variable, such as $f(x, y)$, is the slope of the function with respect to *one* of the variables, all the other variables being held constant (Fig. 1). Although a partial derivative shows how a function changes when one variable changes, it may be used to determine how the function changes when more than one variable changes by an infinitesimal amount. Thus, if f is a function of x and y, then, when x and y change by dx and dy, respectively, f changes by

$$df = \left(\frac{\partial f}{\partial x}\right)_y dx + \left(\frac{\partial f}{\partial y}\right)_x dy$$

For example, if $f = ax^3y + by^2$,

$$\left(\frac{\partial f}{\partial x}\right)_y = 3ax^2y \qquad \left(\frac{\partial f}{\partial y}\right)_x = ax^3 + 2by$$

Then, when x and y undergo infinitesimal changes, f changes by

$$df = 3ax^2y\,dx + (ax^3 + 2by)\,dy$$

Partial derivatives may be taken in any order:

$$\frac{\partial^2 f}{\partial x\,\partial y} = \frac{\partial^2 f}{\partial y\,\partial x}$$

For the function f given above, it is easy to verify that

$$\left(\frac{\partial}{\partial y}\left(\frac{\partial f}{\partial x}\right)_y\right)_x = 3ax^2 \qquad \left(\frac{\partial}{\partial x}\left(\frac{\partial f}{\partial y}\right)_x\right)_y = 3ax^2$$

In the following, z is a variable on which x and y depend (for example, x, y, and z might correspond to p, V, and T).

Relation no. 1. When x is changed at constant z:

$$\left(\frac{\partial f}{\partial x}\right)_z = \left(\frac{\partial f}{\partial x}\right)_y + \left(\frac{\partial f}{\partial y}\right)_x \left(\frac{\partial y}{\partial x}\right)_z$$

Relation no. 2 (the inverter).

$$\left(\frac{\partial x}{\partial y}\right)_z = \frac{1}{\left(\dfrac{\partial y}{\partial x}\right)_z}$$

Relation no. 3 (the permuter).

$$\left(\frac{\partial x}{\partial y}\right)_z = -\left(\frac{\partial x}{\partial z}\right)_y \left(\frac{\partial z}{\partial y}\right)_x$$

By combining this relation and relation no. 2 we obtain **Euler's chain relation**:

$$\left(\frac{\partial x}{\partial y}\right)_z \left(\frac{\partial y}{\partial z}\right)_x \left(\frac{\partial z}{\partial x}\right)_y = -1$$

Relation no. 4. This relation establishes whether or not df is an exact differential.

$$df = g(x, y)\, dx + h(x, y)\, dy \text{ is exact if } \left(\frac{\partial g}{\partial y}\right)_x = \left(\frac{\partial h}{\partial x}\right)_y$$

If df is exact, its integral between specified limits is independent of the path.

FURTHER INFORMATION 3:
Electrical quantities

The fundamental expression in **electrostatics**, the interactions of stationary electric charges, is the **Coulomb potential energy** of one charge of magnitude q at a distance r from another charge q':

$$V = \frac{1}{4\pi\varepsilon_0} \times \frac{qq'}{r}$$

That is, the potential energy is inversely proportional to the separation of the charges. The fundamental constant ε_0 is the **vacuum permittivity**; its value is $\varepsilon_0 = 8.854 \times 10^{-12} \, \mathrm{J}^{-1} \, \mathrm{C}^2 \, \mathrm{m}^{-1}$. (Note that with r in metres, m, and the charges in coulombs, C, the potential energy is in joules, J.) The potential energy is equal to the work that must be done to bring up a charge q from infinity to a distance r from a charge q'. In a medium other than a vacuum, the potential energy of interaction between two charges is reduced, and the vacuum permittivity is replaced by the **permittivity**, ε, of the medium. It is common to express the permittivity as a multiple of the vacuum permittivity, and to write $\varepsilon = \varepsilon_r \varepsilon_0$, where ε_r is the **relative permittivity** (or dielectric constant) of the medium. For water at 25°C, $\varepsilon_r = 78.54$.

The potential energy of a charge q in the presence of another charge q' can be expressed in terms of the **Coulomb potential**, ϕ:

$$V = q \times \phi, \quad \text{where} \quad \phi = \frac{1}{4\pi\varepsilon_0} \times \frac{q'}{r}$$

The units of potential are joules per coulomb, $\mathrm{J} \, \mathrm{C}^{-1}$, so that, when ϕ is multiplied by a charge in coulombs, the result is in joules. The combination joules per coulomb occurs widely in electrostatics, and is called a *volt*, V:

$$1 \, \mathrm{V} = 1 \, \mathrm{J} \, \mathrm{C}^{-1}$$

(which implies that $1\,V\,C = 1\,J$). If there are several charges $q_1, q_2, \ldots$ present in the system, then the total potential experienced by the charge q is the sum of the potential generated by each charge:

$$\phi = \phi_1 + \phi_2 + \ldots$$

The motion of charge gives rise to an electric **current** I. Electric current is measured in amperes, A, where

$$1\,A = 1\,C\,s^{-1}$$

If the electric charge is that of electrons (as it is through metals and semiconductors), then a current of $1\,A$ represents the flow of 6×10^{18} electrons per second. If the current flows from a region of potential ϕ_i to ϕ_f, through a **potential difference** $\Delta\phi = \phi_f - \phi_i$, then the rate of doing work is the current (the rate of transfer of charge) multiplied by the potential difference, $I \times \Delta\phi$. The rate of doing work is called **power**, P, so

$$P = I \times \Delta\phi$$

With current in amperes and the potential difference in volts, the power works out in joules per second, or watts, W:

$$1\,W = 1\,J\,s^{-1}$$

The total energy supplied in a time t is the power (the energy per second) multiplied by the time:

$$E = P \times t = I\,\Delta\phi \times t$$

The energy is obtained in joules with the current in amperes, the potential difference in volts, and the time in seconds.

FURTHER INFORMATION 4:
The Debye–Hückel theory

We imagine a solution in which all the ions have their actual positions, but in which their Coulombic interactions have been turned off. The difference in chemical potential between the ideal and real solutions is then identified with w_e, the electrical work of charging the system. Therefore, for a salt M_pX_q, it follows from eqn 10.5 that

$$\ln \gamma_{\pm} = \frac{w_e}{sRT} \quad \text{where} \quad s = p + q \tag{1}$$

It follows that we must (a) find the final distribution of the ions and (b) the work of charging them in that distribution.

THE DISTRIBUTION OF IONS

The Coulomb potential at a distance r from an isolated ion of charge $z_i e$ in a medium of permittivity ε and relative permittivity ε_r is

$$\phi_i = \frac{Z_i}{r}, \quad \text{with} \quad Z_i = \frac{z_i e}{4\pi\varepsilon}$$

The effect of the ionic atmosphere is to cause the potential to decay with distance more sharply than this expression implies, and for the **shielded Coulomb potential** we write

$$\phi_i = \frac{Z_i}{r} e^{-r/r_D} \tag{2}$$

where r_D is called the **Debye length**. When r_D is large, the shielded potential is virtually the same as the unshielded potential. When r_D is small, the shielded potential is much smaller than the unshielded potential, even for short distances.

To calculate r_D we need to know how the charge density ρ_i of the ionic atmosphere, the charge per unit volume, varies with distance from the ion. Charge density and electrostatic potential are related by **Poisson's equation**, which for a spherically symmetrical charge distribution is

$$\frac{1}{r^2}\frac{\mathrm{d}}{\mathrm{d}r}\left(r^2\frac{\mathrm{d}\phi_i}{\mathrm{d}r}\right) = -\frac{\rho_i}{\varepsilon}$$

Substitution of the shielded potential results in

$$r_D^2 = -\frac{\varepsilon\phi_i}{\rho_i} \tag{3}$$

To solve this equation we need to relate ρ_i and ϕ_i.

The energy of an ion of charge $z_j e$ at a distance where it experiences the potential ϕ_i of the central ion i relative to its energy where it is far away in the bulk solution is

$$E = z_j e\phi_i$$

Therefore, according to the Boltzmann distribution (see the Introduction and Section 19.1) the ratio of the concentration c_j of ions at a distance r and the concentration in the bulk c_j° is:

$$\frac{c_j}{c_j^\circ} = \mathrm{e}^{-E/kT}$$

The charge density ρ_i at a distance r from the ion i is the concentration of each type of ion multiplied by their charges:

$$\rho_i = c_+ z_+ F + c_- z_- F = c_+^\circ F z_+ \mathrm{e}^{-z_+ e\phi_i/kT} + c_-^\circ F z_- \mathrm{e}^{-z_- e\phi_i/kT}$$

Since the average electrostatic interaction energy is small compared with kT we may write the last equation as

$$\rho_i = (c_+^\circ z_+ + c_-^\circ z_-)F - (c_+^\circ z_+^2 + c_-^\circ z_-^2)\left(\frac{F^2\phi_i}{RT}\right) + \dots$$

because $F = eN_A$ and $R = kN_A$. The first term in the expansion is zero because it is the charge density in the bulk, uniform solution, and the solution is electrically neutral. The unwritten terms are assumed to be too small to be significant. Expressed in terms of the ionic strength, eqn 10.7a, the one remaining term is

$$\rho_i = -\frac{2\rho F^2 I\phi_i}{RT} \tag{4}$$

where ρ is the density of the solution. We can now solve eqn 3 for r_D:

$$r_D = \left(\frac{\varepsilon RT}{2\rho F^2 I}\right)^{\frac{1}{2}} \tag{5}$$

THE WORK OF CHARGING

To calculate the activity coefficient we need to find the electrical work of charging the central ion when it is surrounded by its atmosphere. To do so, we need to know the potential at the ion due to its atmosphere, ϕ_{atmos}. This potential is the difference between the total potential, given by eqn 4, and the potential due to the central ion itself:

$$\phi_{atmos} = \phi - \phi_{central\,ion} = Z_i\left(\frac{e^{-r/r_D}}{r} - \frac{1}{r}\right)$$

The potential at the ion (at $r = 0$) is obtained by taking the limit of this expression as $r \to 0$, and is

$$\phi_{atmos}(0) = -\frac{Z_i}{r_D}$$

If the charge of the central ion were q and not $z_i e$, then the potential due to its atmosphere would be

$$\phi_{atmos}(0) = -\frac{q}{4\pi\varepsilon} \times \frac{1}{r_D}$$

The work of adding a charge dq to a region where the electrical potential is $\phi_{atmos}(0)$ is

$$dw_e = \phi_{atmos}(0)\,dq$$

Therefore, the total work of fully charging (per mole of ions) is

$$w_e = N_A \int_0^{z_i} \phi_{atmos}(0)\,dq = -\frac{N_A}{4\pi\varepsilon r_D} \int_0^{z_i} q\,dq$$

$$= -\frac{z_i^2 e^2 N_A}{8\pi\varepsilon r_D} = -\frac{z_i^2 F^2}{8\pi\varepsilon N_A r_D} \tag{6}$$

It follows from eqn 1 that the mean activity coefficient of the ions is

$$\ln \gamma_{\pm} = \frac{pw_{e,+} + qw_{e,-}}{sRT} = -\frac{(pz_+^2 + qz_-^2)F^2}{8\pi\varepsilon s N_A RT r_D}$$

However, for neutrality $pz_+ + qz_- = 0$, so

$$\ln \gamma_{\pm} = -\frac{|z_+ z_-|F^2}{8\pi\varepsilon N_A RT r_D}$$

When we replace r_D using eqn 5, and convert to common logarithms, this expression becomes

$$\log \gamma_{\pm} = -|z_+ z_-| A(I/m^{\ominus})^{\frac{1}{2}}$$

where

$$A = \frac{F^3 \ln 10}{4\pi N_A} \left(\frac{\rho m^{\ominus}}{2\varepsilon^2 R^3 T^3}\right)^{\frac{1}{2}}$$

This is the limiting law. For water at 298 K, when $\rho = 0.997\,\mathrm{g\,cm^{-3}}$ and $\varepsilon_r = 78.54$, $A = 0.509$.

FURTHER INFORMATION 5:
Classical mechanics

We shall see how classical mechanics describes the behaviour of objects in terms of two equations. One equation expresses the fact that the total energy is constant in the absence of external forces. The other equation expresses the response of particles to the forces acting on them.

THE TRAJECTORY IN TERMS OF THE ENERGY

The total energy of a particle is the sum of its **kinetic energy** E_K, the energy arising from its motion, and its **potential energy** $V(x)$, the energy arising from its position in a field of force:

$$E = E_K + V(x) \tag{1}$$

The **force** F is related to the potential energy by

$$F = -\frac{dV}{dx} \tag{2}$$

According to this expression, the direction of the force is towards decreasing potential energy (Fig. 1). The kinetic energy of a particle of mass m travelling with a speed v is

$$E_K = \tfrac{1}{2}mv^2 \tag{3}$$

It is often convenient to express kinetic energy in terms of the **linear momentum** p. The linear momentum is a vector quantity (that is, it has direction as well as magnitude, like the velocity v). Its magnitude p is related to the speed v of the particle by

$$p = mv \tag{4}$$

A heavy particle moving slowly can have a higher momentum than a light particle moving rapidly. The linear momentum vector points in the

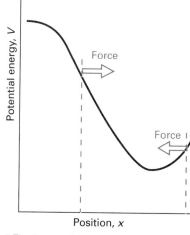

1 The force acting on a particle is determined by the slope of the potential energy at its location. The force points in the direction of decreasing potential energy.

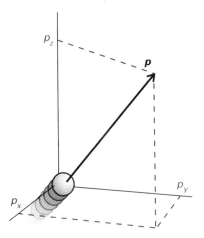

2 The linear momentum of a particle is a vector property, and points in the direction of motion.

direction of travel of the particle (Fig. 2). In terms of the linear momentum, the total energy of a particle is

$$E = \frac{p^2}{2m} + V(x) \tag{5}$$

These equations can be used in a number of ways. For example, it is easy to show that they predict that a particle will have a definite **trajectory**, or definite position and momentum at each instant. For example, consider a particle free to move in one direction (along the x-axis) in a region where $V = 0$ (so the energy is independent of position and there is no force acting). Because speed is the rate of change of position, $v = dx/dt$, it follows from eqn 3 that

$$\frac{dx}{dt} = \left(\frac{2E_K}{m}\right)^{\frac{1}{2}}$$

A solution of this equation is

$$x(t) = x(0) + \left(\frac{2E_K}{m}\right)^{\frac{1}{2}} t$$

The linear momentum is a constant:

$$p(t) = mv = m\frac{dx}{dt} = (2mE_K)^{\frac{1}{2}}$$

Hence, if we know the initial position and momentum, we can predict all later positions and momenta exactly.

NEWTON'S SECOND LAW

The second basic equation of classical mechanics is **Newton's second law of motion**, that the rate of change of momentum is equal to the force acting on the particle. In one dimension:

$$\frac{dp}{dt} = F \tag{6a}$$

Because $p = m\, dx/dt$, in one dimension, it is sometimes more convenient to write this equation as

$$m\frac{d^2x}{dt^2} = F \tag{6b}$$

The second derivative d^2x/dt^2 is the **acceleration** of the particle, its rate of change of velocity (along the x-axis). Newton's second law therefore states that the acceleration of a particle is proportional to the force it experiences. It follows that, if we know the force acting everywhere and at all times, then solving eqn 6b will also give the trajectory. This calculation is equivalent to the one based on E but is more suitable in some applications. For example, it can be used to show that, if a particle of mass m is initially stationary and is subjected to a constant force F for a time τ, then its kinetic energy increases from zero to

$$E_K = \frac{F^2 \tau^2}{2m} \tag{7}$$

and then remains at that energy after the force ceases to act. Because the applied force F and the time τ for which it acts may be varied at will, the energy of the particle may be increased to any value.

ROTATIONAL MOTION

The same type of calculation may be applied to free rotational motion, a form of motion that plays an important role in the electronic structure of atoms (where electrons are free to travel round the central nucleus) and in the spectroscopy of molecules (because free molecules rotate). Whereas the *translational* motion of a particle is expressed in terms of its *linear* momentum, the *rotational* motion of a particle about a central point is described by its **angular momentum J**. The angular momentum has direction as well as magnitude: its magnitude gives the rate at which a particle circulates and its direction indicates the axis of rotation (Fig. 3). The magnitude of the angular momentum J is given by the expression

$$J = I\omega \tag{8}$$

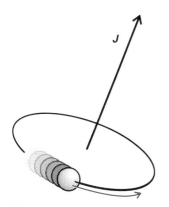

3 The angular momentum of a particle can be represented by a vector along the axis of rotation and perpendicular to the plane of rotation. The length of the vector denotes the magnitude of the angular momentum. The direction of motion is clockwise for an observer looking in the direction of the vector.

where ω is its **angular velocity**, its rate of change of angular position (in radians per second), and I is the **moment of inertia**.[1] For a point particle of mass m moving in a circle of radius r, the moment of inertia is given by the expression

$$I = mr^2$$

The angular momentum of a particle is large if it has a large moment of inertia (which is the case if it is heavy and moving in a circle of large radius) and its angular velocity is high (so that it is travelling rapidly in its circular path).

To accelerate a rotation it is necessary to apply a **torque T**, a twisting force, and Newton's equation is

$$\frac{dJ}{dt} = T$$

If a constant torque is applied for a time τ, the rotational energy of an initially stationary body is increased to

$$E_K = \frac{T^2 \tau^2}{2I} \tag{9}$$

(The reasoning behind this expression is exactly the same as that leading to eqn 7.) The implication of this equation is that an appropriate torque and period for which it is applied can excite the rotation to an arbitrary energy.

1 The analogous roles of m and I, of v and ω, and of p and J in the translational and rotational cases, respectively, should be remembered, because they provide a ready way of constructing and recalling equations.

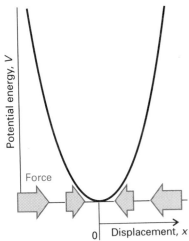

Potential energy, V

Force

0 | Displacement, x

4 The force acting on a particle that undergoes harmonic motion. The force is in the opposite direction to the displacement and proportional to the displacement. The corresponding potential energy is a parabola.

THE HARMONIC OSCILLATOR

A third fundamental type of motion is oscillatory motion, the kind of motion that atoms undergo in a vibrating molecule. A **harmonic oscillator** consists of a particle that experiences a restoring force proportional to its displacement from its equilibrium position:

$$F = -kx \tag{10}$$

An example is a particle joined to a rigid support by a spring. The constant of proportionality k is called the **force constant**, and the stiffer the spring the greater the force constant. The negative sign in F signifies that the direction of the force is opposite to that of the displacement (force, like momentum, velocity, and angular momentum, has direction as well as magnitude). Thus, when x is positive (displacement to the right), the force is negative (pushing towards the left), and vice versa (Fig. 4). A typical force constant of a chemical bond is about $500 \, \text{N m}^{-1}$.

The motion of a particle that undergoes harmonic motion is found by substituting the expression for the force, eqn 10, into Newton's equation, eqn 6b. The resulting equation is

$$m \frac{d^2x}{dt^2} = -kx \tag{11a}$$

and a solution is

$$x = A \sin \omega t \qquad p = m\omega A \cos \omega t, \qquad \text{where } \omega = \left(\frac{k}{m}\right)^{\frac{1}{2}} \tag{11b}$$

(To verify the solution for x, substitute it into the differential equation; the expression for the momentum is obtained from $p = m \, dx/dt$.) These expressions show that the position of the particle varies **harmonically** (that is, as $\sin \omega t$) with a frequency $\nu = \omega/2\pi$. They also show that the particle is stationary ($p = 0$) when the displacement x has its maximum value A, which is called the **amplitude** of the motion.

The total energy of a classical harmonic oscillator is proportional to the square of the amplitude of its motion. To confirm this remark we note that the kinetic energy is

$$E_K = \frac{p^2}{2m} = \frac{(m\omega A \cos \omega t)^2}{2m} = \tfrac{1}{2}m\omega^2 A^2 \cos^2 \omega t$$

Because $\omega = (k/m)^{\frac{1}{2}}$, this expression may be written

$$E_K = \tfrac{1}{2}kA^2 \cos^2 \omega t$$

The force on the oscillator is $F = -kx$, so it follows from the relation $F = -dV/dx$ that the potential energy of a harmonic oscillator is

$$V = \tfrac{1}{2}kx^2 \tag{12}$$

and hence that

$$V = \tfrac{1}{2}kA^2 \sin^2 \omega t$$

The total energy is therefore

$$E = \tfrac{1}{2}kA^2 \cos^2 \omega t + \tfrac{1}{2}kA^2 \sin^2 \omega t = \tfrac{1}{2}kA^2 \tag{13}$$

(We have used $\cos^2 \omega t + \sin^2 \omega t = 1$.) That is, the energy of the oscillator is constant and, for a given force constant, is determined by its maximum displacement. It follows that the energy of an oscillating particle can be raised to any value by stretching the spring to any desired amplitude A. It is important to note that the frequency of the motion depends only on the inherent properties of the oscillator (as represented by k and m) and is independent of the energy; the amplitude governs the energy, through $E = \tfrac{1}{2}kA^2$, and is independent of the frequency. In other words, the particle will oscillate at the same frequency regardless of the amplitude of its motion.

FURTHER INFORMATION 6:
Quantum mechanics

Quantum mechanics can be formulated in a variety of different ways. A reasonably elementary way is to set out a series of basic postulates and then to derive the entire edifice of the theory from them. These postulates are fundamental, and there is no deeper justification of them other than that their implications are found to agree with experiment. One set of postulates that can be used is as follows:

(1) The state of a system is fully described by the wavefunction ψ, which is a function of the coordinates of all the particles and of the time.

(2) Observables Ω are represented by operators $\hat{\Omega}$ chosen to satisfy a particular commutation relation (see below).

(3) When a system is described by a wavefunction ψ, the mean value of the observable Ω is equal to the expectation value of the corresponding operator.

(4) When ψ is an eigenfunction of the operator $\hat{\Omega}$ corresponding to the observable of interest, the determination of the value of Ω always leads to one result, the corresponding eigenvalue of $\hat{\Omega}$. When ψ is not an eigenfunction of the operator $\hat{\Omega}$, a single measurement of Ω yields a single result which is *one* of the eigenvalues of $\hat{\Omega}$, and the probability that a particular eigenvalue ω_n is measured is equal to $|c_n|^2$, where c_n is the coefficient of the eigenfunction ψ_n in the expansion of ψ in terms of these eigenfunctions.

In postulate 1, by 'describe' is meant that the wavefunction contains all the information open to experimental determination. In general, the wavefunction varies with time, which corresponds to the state of the system evolving with time (for instance, under the influence of applied electric and magnetic fields). The evolution of the wavefunction with time is given by the time-dependent Schrödinger equation

$$H\psi = i\hbar \frac{\partial \psi}{\partial t} \tag{1}$$

where H is the Hamiltonian operator for the system, the operator corresponding to the total energy.

Postulate 2 is incomplete as it stands, for we need to know the fundamental commutation rule that operators must satisfy. A **commutation rule** is a statement about the value of the **commutator**,

$$[\hat{\Omega}_1, \hat{\Omega}_2] = \hat{\Omega}_1\hat{\Omega}_2 - \hat{\Omega}_2\hat{\Omega}_1 \tag{2}$$

of two operators. Postulate 2 goes on to say that all operators in quantum mechanics must be selected so that the operators for position and linear momentum satisfy

$$[\hat{x}, \hat{p}] = i\hbar \tag{3}$$

This is a fundamental postulate and cannot be derived from deeper principles. To see that the choices

$$\hat{x} = x \times \qquad \hat{p} = \frac{\hbar}{i}\frac{d}{dx}$$

are consistent with the postulate, we consider the following calculation:

$$[\hat{x}, \hat{p}]\psi = (\hat{x}\hat{p} - \hat{p}\hat{x})\psi = \left(x \times \frac{\hbar}{i}\frac{d}{dx} - \frac{\hbar}{i}\frac{d}{dx}x \times\right)\psi$$

The second term in parentheses is the derivative of a product of functions (x and ψ), and hence may be written

$$\frac{\hbar}{i}\frac{d}{dx}x\psi = \frac{\hbar}{i}\left(\psi + x\frac{d\psi}{dx}\right)$$

On substitution of this expression into the one above, we find

$$[\hat{x}, \hat{p}]\psi = -\frac{\hbar}{i}\psi = i\hbar\psi$$

which agrees with the requirement about the value of the commutator.

Postulate 3 has already been considered in the text (Chapter 11). The expectation value is the average value of a quantity, such as the position or the momentum. To calculate the average value of a position, for example, the value of x is weighted by the probability that the particle will be found in a small region dx at x, which is $\psi^*\psi\,dx$ for a wavefunction normalized to 1, and then summed over all such positions:

$$\langle x \rangle = \int \psi^*x\psi\,dx$$

A similar expression applies to other observables. The operator lies between the two functions for reasons explained in the text.

Postulate 4 lies at the heart of the probabilistic interpretation of quantum mechanics. It means that, when the system is described by a wavefunction that is not an eigenfunction of the operator of interest (such as the linear momentum, for instance), the outcome is unpredictable. We know that we will measure *one* of the eigenvalues of the operator, but we cannot predict which one. The *average* value of all such measurements, however, is well defined, and can be predicted from the wavefunction by calculating the expectation value of the operator.

FURTHER INFORMATION 7:
Differential equations

A differential equation is a relation between derivatives of a function and the function itself, as in

$$a\frac{\mathrm{d}^2 y}{\mathrm{d}x^2} + b\frac{\mathrm{d}y}{\mathrm{d}x} + cy = 0$$

The coefficients a, b, etc need not be constants but may also be functions of x. The **order** of the equation is the order of the highest derivative that occurs in it, so the differential equation shown above is a second-order equation. Only rarely in science are differential equations of order higher than 2 encountered. A solution of a differential equation is an expression for y in terms of x and the coefficients (here written a, b, and c) that occur in it. The process of solving an equation is commonly termed integration, and in simple cases simple integration can be employed. A **general solution** of a differential equation is the most general solution of the equation and is expressed in terms of a number of constants. When the constants are chosen to accord with certain specified **initial conditions** (if one variable is the time) or certain **boundary conditions** (to fulfil certain spatial restrictions on the solutions), then we obtain the **particular solution** of the equation. A first-order differential equation requires the specification of *one* boundary (or initial) condition; a second-order differential equation requires the specification of *two* such conditions, and so on.

First-order differential equations may often be solved by direct integration. For example, the equation

$$\frac{\mathrm{d}y}{\mathrm{d}x} = axy$$

with a a constant may be rearranged into

$$\frac{\mathrm{d}y}{y} = ax\,\mathrm{d}x$$

and then integrated to

$$\ln y = \tfrac{1}{2}ax^2 + A$$

where A is a constant. If we know that $y = y_0$ when $x = 0$ (for instance), then it follows that $A = \ln y_0$, and hence the particular solution of the equation is

$$\ln y = \tfrac{1}{2}ax^2 + \ln y_0$$

This expression rearranges to

$$y = y_0 \mathrm{e}^{\tfrac{1}{2}ax^2}$$

First-order differential equations of a more complex form can often be solved by the appropriate substitution. For example, it is sensible to try the substitution $y = sx$, and to change the variables from x and y to x and s. An alternative transformation is $x = u + a$, $y = v + b$, and then to select a and b to simplify the form of the resulting expression.

Second-order differential equations are in general much more difficult to solve than first-order equations. The general solutions of many such equations are best found by referring to tables: the *Handbook of mathematical functions*, M. Abramowitz and I. A. Stegun, Dover, New York (1965), is a particularly helpful source of such information. One powerful approach (which may be applied to the Schrödinger equation for the hydrogen atom and the harmonic oscillator) is to express the solution as a power series:

$$y = \sum_{n=0}^{\infty} c_n x^n$$

and then to use the differential equation to find a relation between the coefficients. This approach results, for example, in the Hermite polynomials that form part of the solution of the harmonic oscillator Schrödinger equation. All the second-order differential equations that occur in this text can be found tabulated in compilations of solutions, and the specialized techniques that are needed to establish the form of the solutions may be found in mathematical texts.

A **partial differential equation** is a differential equation in more than one variable. An example is

$$\frac{\partial^2 y}{\partial t^2} = a\frac{\partial^2 y}{\partial x^2}$$

with y a function of the two variables x and t. In certain cases, partial differential equations may be separated into ordinary differential equations. Thus, the Schrödinger equation for the particle in a two-dimensional square well (eqn 12.11) may be separated by writing the wavefunction $\psi(x, y)$ as the product $X(x)Y(y)$, which results in the separation of the second-order partial differential equation into two second-order ordinary differential equations in the variables x and y. A good guide to the likely success of such a separation of variables procedure is the symmetry of the system, but some rather clever changes of variable are often needed. Once again, in elementary work, solutions of partial differential equations are best found by referring to tabulated sources.

A common approach to the solution of awkward differential equations that appear to have no analytical solutions (particularly the equations that arise in chemical kinetics and the Schrödinger equation for potentials of an awkward form) is to adopt numerical procedures. Software packages (such as Mathematica and MathCad) are now widely available that can be used to solve almost any equation numerically. The general form of such programs to solve an equation of the form $df/dx = g(x)$ first approximates the infinitesimal $df = g(x)\, dx$ by the small finite quantity $\delta f = g(x)\, \delta x$, so that

$$f(x + \delta x) = f(x) + g(x)\, \delta x$$

and then proceeds numerically with a program of the form shown in the margin.

There are, however, several problems with simple integration procedures. One is that it may be inaccurate to use the value of $g(x)$ at the start of the interval for which we are calculating $f(x + \delta x)$. This source of error is avoided by the use of **predictor–corrector methods** that make predictions of the later value of g on the basis of its most recent value. Another problem is that one term in the equation may vary slowly but another may vary rapidly. Thus, while it may be tempting to use a large step size for the former, the latter demands that the step size must be small. Differential equations with widely varying rates of change are termed **stiff** and are not easy to integrate without special routines.

Input step size
Input initial and final values of x
Evaluate f(initial x)
Loop from initial x to final x
    Evaluate $f(x + \text{step})$
        $= f(x) + g(x)*\text{step}$
    Store $f(x)$
Repeat loop

FURTHER INFORMATION 8:
The harmonic oscillator

The Schrödinger equation for a harmonic oscillator is

$$-\frac{\hbar^2}{2m}\frac{d^2\psi}{dx^2} + \tfrac{1}{2}kx^2\psi = E\psi$$

The appearance of this equation is simplified by introducing the following definitions:

$$y = \frac{x}{\alpha} \qquad \varepsilon = \frac{E}{\tfrac{1}{2}\hbar\omega} \qquad \alpha = \left(\frac{\hbar^2}{mk}\right)^{\frac{1}{4}} \qquad \omega = \left(\frac{k}{m}\right)^{\frac{1}{2}}$$

The equation then becomes

$$\frac{d^2\psi}{dy^2} + (\varepsilon - y^2)\psi = 0$$

One approach to solving this equation is first to find the **asymptotic solutions**, the solutions that show how the wavefunctions behave as $x \to \pm\infty$. When x (and therefore y) is very large, the term ε in parentheses may be neglected relative to y^2, and the equation simplifies to

$$\frac{d^2\psi}{dy^2} - y^2\psi = 0$$

This equation can be solved by straightforward integration, and one solution is

$$\psi \sim e^{y^2/2}$$

because

$$\frac{d^2\psi}{dy^2} = \frac{d^2}{dy^2}e^{y^2/2} = e^{y^2/2} + y^2e^{y^2/2} \sim y^2e^{y^2/2} = y^2\psi$$

(The sign $\sim$ indicates 'asymptotically equal', which means that two quantities become equal in a certain limit, in this case, as y approaches infinity.) However, this solution is physically unacceptable because it increases rapidly with the displacement, and approaches infinity. The other solution is $e^{-\frac{1}{2}y^2}$, which is well-behaved for large positive and negative displacements, and so is acceptable. Therefore, at this stage, we know that, at large displacements, all harmonic oscillator wavefunctions approach zero like the wings of a Gaussian bell-shaped curve.

The next step is to suppose that the exact wavefunction has the form

$$\psi = f e^{-\frac{1}{2}y^2}$$

where f is a function that does not increase to infinity more rapidly than $e^{-\frac{1}{2}y^2}$ decays towards zero: then $f e^{-\frac{1}{2}y^2}$ remains acceptable at large displacements. We substitute this trial solution into the exact equation and find

$$f'' - 2yf' + (\varepsilon - 1)f = 0$$

where $f' = \mathrm{d}f/\mathrm{d}y$ and $f'' = \mathrm{d}^2f/\mathrm{d}y^2$. This differential equation, which is called 'Hermite's equation' is one that has been thoroughly studied by mathematicians, and its solutions are known.

Acceptable solutions of Hermite's equation (those not going to infinity more quickly than the factor $e^{-\frac{1}{2}y^2}$ falls to zero, so that the product is never infinite) exist only for ε equal to a positive, odd integer. Therefore we write $\varepsilon = 2v + 1$, with $v = 0, 1, 2, \ldots$ It follows from the relation between E and ε that the energy levels of the oscillator are

$$E_v = (2v + 1) \times \tfrac{1}{2}\hbar\omega$$

The shapes of the wavefunctions can be obtained once we know the solutions of Hermite's equation. The solutions for v a positive integer are the Hermite polynomials H_v, which are polynomials that are generated by differentiating e^{-y^2} v times:

$$H_v = (-1)^v e^{y^2} \left(\frac{\mathrm{d}}{\mathrm{d}y}\right)^v e^{-y^2}$$

The explicit forms of the first few polynomials[1] can be found in Table 12.1 along with some of their more useful properties.

---

1 A polynomial in x has the form $a_0 + a_1 x + \ldots + a_n x^n$ running to a finite number of terms. It is unlike $\cos x$, for example, which runs to an infinite number of terms when expressed in the same way.

FURTHER INFORMATION 9: Rotational motion

The Schrödinger equation for rotation in three dimensions is

$$-\frac{\hbar^2}{2m}\nabla^2\psi + V\psi = E\psi \tag{1}$$

The effect of the operator ∇^2 on the wavefunction in spherical coordinates (the natural coordinates to use for this rotational problem) is

$$\nabla^2\psi = \frac{1}{r}\frac{\partial^2}{\partial r^2}r\psi + \frac{1}{r^2}\Lambda^2\psi \tag{2a}$$

where

$$\Lambda^2\psi = \frac{1}{\sin^2\theta}\left(\frac{\partial^2\psi}{\partial\phi^2}\right) + \frac{1}{\sin\theta}\frac{\partial}{\partial\theta}\left(\sin\theta\frac{\partial\psi}{\partial\theta}\right) \tag{2b}$$

These equations look fearsome, but they can be simplified quite readily. Thus, because the particle is moving on the surface of a sphere, the radius is not a variable and the differentiations with respect to r can be ignored. Moreover, on the surface of the sphere, V is a constant and may be set equal to zero. Therefore, eqn 1 simplifies to

$$-\frac{\hbar^2}{2mr^2}\Lambda^2\psi = E\psi$$

which rearranges to

$$\Lambda^2\psi = -\frac{2IE}{\hbar^2}\psi \tag{3}$$

Notice that the moment of inertia $I = mr^2$ has appeared automatically.

Equation 3 is a partial differential equation in the two angular variables θ and ϕ, but it may be separated into two ordinary differential equations by the separation of variables procedure. Thus, we write

$$\psi = \Theta\Phi$$

where Θ is a function of θ alone and Φ is a function of ϕ alone. With this trial factorization we obtain

$$\Lambda^2\psi = \frac{\Theta}{\sin^2\theta}\frac{d^2\Phi}{d\phi^2} + \frac{\Phi}{\sin\theta}\frac{d}{d\theta}\sin\theta\frac{d\Theta}{d\theta} = -\frac{2IE}{\hbar^2}\Theta\Phi$$

Division through by $\Theta\Phi$ gives

$$\frac{1}{\sin^2\theta}\frac{\Phi''}{\Phi} + \frac{1}{\Theta}\frac{1}{\sin\theta}\frac{d}{d\theta}(\Theta'\sin\theta) = -\frac{2IE}{\hbar^2}$$

where $\Theta' = d\Theta/d\theta$ and $\Phi'' = d^2\Phi/d\phi^2$. We ensure that each term depends on only one variable by multiplying through by $\sin^2\theta$, which gives:

$$\frac{\Phi''}{\Phi} + \frac{\sin\theta}{\Theta}\frac{d}{d\theta}(\Theta'\sin\theta) = -\frac{2IE}{\hbar^2}\sin^2\theta$$

The first term depends only on ϕ, and the rest depend only on θ. Therefore, by the same argument as in Section 12.2, the first term must be equal to a constant, which for future convenience we write $-m_l^2$. The separation is therefore successful and, after some rearrangement, the two ordinary differential equations we must solve are found to be

$$\Phi'' = -m_l^2\Phi$$

$$\sin\theta\frac{d}{d\theta}(\Theta'\sin\theta) + \left(\frac{2IE}{\hbar^2}\sin^2\theta - m_l^2\right)\Theta = 0$$

The first of this pair of equations is the same as for a particle on a ring. The cyclic boundary conditions are also the same, so the acceptable solutions are the same (eqn 12.26b). Hence, we can write

$$\Phi_{m_l} = \frac{1}{\sqrt{2\pi}}e^{im_l\phi} \qquad m_l = 0, \pm1, \pm2, \ldots$$

However, because m_l also occurs in the equation for Θ, we must be alert to the possibility that the boundary conditions on Θ may also limit the values that m_l may take.

The equation for Θ has been studied extensively by mathematicians. We turn it into a standard form by making the substitutions

$$\zeta = \cos\theta \qquad l(l+1) = \frac{2IE}{\hbar^2}$$

where l is a dimensionless number (which will shortly turn out to be to a quantum number). Since

$$\frac{d}{d\theta} = \frac{d\zeta}{d\theta}\frac{d}{d\zeta} = -\sin\theta\frac{d}{d\zeta}$$

$$\sin\theta = (1 - \cos^2\theta)^{\frac{1}{2}} = (1 - \zeta^2)^{\frac{1}{2}}$$

the equation transforms into

$$(1 - \zeta^2)\frac{d^2\Theta}{d\zeta^2} - 2\zeta\frac{d\Theta}{d\zeta} + \left(l(l+1) - \frac{m_l^2}{1 - \zeta^2}\right)\Theta = 0$$

This differential equation is known as the **associated Legendre equation** (because it is associated with another equation known by Legendre's name). It has acceptable solutions (i.e. solutions that are single-valued and do not become infinite anywhere) so long as two conditions are fulfilled:

$$\text{(a)}\ l = 0, 1, 2, \ldots \qquad \text{(b)}\ |m_l| \leqslant l$$

That is, acceptable solutions exist only for non-negative integral l, and the absolute value of the number m_l must not exceed l (this is the second constraint on m_l mentioned above). When these conditions are satisfied, the functions Θ, which now must be denoted $\Theta_{l,\,|m_l|}$, are the **associated Legendre functions**. They can be written in terms of sine and cosine functions, and the first few are listed in Table 12.3 as components of the spherical harmonics $Y = \Theta\Phi$.

FURTHER INFORMATION 10: Centre-of-mass coordinates

The Schrödinger equation for a hydrogen atom is

$$H_{\text{total}}\Psi = E\Psi$$

where

$$H_{\text{total}} = -\frac{\hbar^2}{2m_{\text{e}}}\nabla_{\text{e}}^2 - \frac{\hbar^2}{2m_{\text{N}}}\nabla_{\text{N}}^2 + V$$

with m_{e} the mass of the electron, m_{N} the mass of the nucleus, and V the Coulomb potential energy. The Laplacians differentiate with respect to the coordinates of the electron and the nucleus, so ∇_{e}^2, for instance, includes terms such as $\partial^2/\partial x_{\text{e}}^2$, where x_{e} is the x-coordinate of the electron. We wish to show that the equation can be expressed in terms of the separation of the particles $\boldsymbol{r}$ and the location of the centre of mass $\boldsymbol{R}$. We shall do the transformation of the derivatives with respect to the x-coordinate: the others follow in the same way.

The location of the centre of mass is at

$$X = \frac{m_{\text{e}}}{m}x_{\text{e}} + \frac{m_{\text{N}}}{m}x_{\text{N}}$$

with m the total mass. The separation of the particles is

$$x = x_{\text{e}} - x_{\text{N}}$$

We can now write

$$\frac{\partial}{\partial x_{\text{e}}} = \frac{\partial X}{\partial x_{\text{e}}} \times \frac{\partial}{\partial X} + \frac{\partial x}{\partial x_{\text{e}}} \times \frac{\partial}{\partial x} = \frac{m_{\text{e}}}{m} \times \frac{\partial}{\partial X} + \frac{\partial}{\partial x}$$

$$\frac{\partial}{\partial x_{\text{N}}} = \frac{m_{\text{N}}}{m} \times \frac{\partial}{\partial X} - \frac{\partial}{\partial x}$$

The x-part of the sum of the two Laplacians becomes

$$\frac{1}{m_e}\frac{\partial^2}{\partial x_e^2} + \frac{1}{m_N}\frac{\partial^2}{\partial x_N^2} = \frac{1}{m}\frac{\partial^2}{\partial X^2} + \left(\frac{1}{m_e} + \frac{1}{m_N}\right)\frac{\partial^2}{\partial x^2}$$

The y- and z-components can be dealt with similarly, so

$$\frac{1}{m_e}\nabla_e^2 + \frac{1}{m_N}\nabla_N^2 = \frac{1}{m}\nabla_{cm}^2 + \frac{1}{\mu}\nabla^2$$

with

$$\frac{1}{\mu} = \frac{1}{m_e} + \frac{1}{m_N}$$

Now write the overall wavefunction as

$$\Psi = \chi(\boldsymbol{R})\psi(\boldsymbol{r})$$

and the Hamiltonian as

$$H_{total} = H_{cm} + H$$

where

$$H_{cm} = -\frac{\hbar^2}{2m}\nabla_{cm}^2$$

$$H = -\frac{\hbar^2}{2\mu}\nabla^2 + V$$

and carry through the separation of variables procedure. The total equation is

$$\{H_{cm}\chi(\boldsymbol{R})\}\psi(\boldsymbol{r}) + \chi(\boldsymbol{R})\{H\psi(\boldsymbol{r})\} = E\chi(\boldsymbol{R})\psi(\boldsymbol{r})$$

and division by $\chi(\boldsymbol{R})\psi(\boldsymbol{r})$ gives

$$\frac{H_{cm}\chi(\boldsymbol{R})}{\chi(\boldsymbol{R})} + \frac{H\psi(\boldsymbol{r})}{\psi(\boldsymbol{r})} = E$$

The first term is independent of $\boldsymbol{r}$ and the second is independent of $\boldsymbol{R}$; hence the equation is separable.

FURTHER INFORMATION 11: The separation of the internal motion

The Schrödinger equation for a hydrogenic atom is

$$\frac{-\hbar^2}{2\mu} \nabla^2 \psi + V\psi = E\psi$$

In spherical polar coordinates the Laplacian operator separates into the Legendrian operator (which acts only on the angular variables) and derivatives with respect to r (see *Further information* 9), and the equation becomes

$$-\frac{\hbar^2}{2\mu} \left(\frac{1}{r} \frac{\partial^2(r\psi)}{\partial r^2} + \frac{1}{r^2} \Lambda^2 \psi \right) + V\psi = E\psi$$

We now write $\psi = RY$ and obtain

$$-\frac{\hbar^2}{2\mu} \left(\frac{Y}{r} \frac{\partial^2(rR)}{\partial r^2} + \frac{R}{r^2} \Lambda^2 Y \right) + VRY = ERY$$

Then division by RY and reorganization of terms gives

$$\left(-\frac{\hbar^2 r^2}{2\mu(rR)} \frac{\partial^2(rR)}{\partial r^2} + Vr^2 - Er^2 \right) - \frac{\hbar^2}{2\mu Y} \Lambda^2 Y = 0$$

The term in parentheses depends only on the radius and the remaining term depends only on the angles; hence each one is equal to a constant and the equation is separable. Because the spherical harmonics satisfy

$$\Lambda^2 Y = -l(l+1)Y$$

it follows that the last equation becomes

$$\left(-\frac{\hbar^2 r^2}{2\mu(rR)}\frac{\partial^2(rR)}{\partial r^2} + Vr^2 - Er^2\right) + \frac{\hbar^2 l(l+1)}{2\mu} = 0$$

which rearranges to

$$-\frac{\hbar^2}{2\mu}\frac{\partial^2(rR)}{\partial r^2} + \left(V + \frac{\hbar^2 l(l+1)}{2\mu r^2}\right)rR = ErR$$

which, with $\Pi = rR$, is eqn 13.8a.

FURTHER INFORMATION 12: The Pauli principle

The Pauli exclusion principle is a special case of a general principle called the **Pauli principle**:

> When the labels of any two identical fermions are exchanged, the total wavefunction changes sign. When the labels of any two identical bosons are exchanged, the total wavefunction retains the same sign.

Fermions are particles with half-integral spins, and include electrons, protons, neutrons, and some atomic nuclei. Bosons include photons ($s = 1$) and a number of atomic nuclei and atoms. By 'total wavefunction' we mean the entire wavefunction, including the spin of the particles.

Consider the wavefunction for two electrons $\Psi(1, 2)$. The Pauli principle implies that it is a fact of nature (which has its roots in the theory of relativity) that the wavefunction must change sign if we interchange the labels 1 and 2 wherever they occur in the function:

$$\Psi(1, 2) = -\Psi(2, 1)$$

The connection of this general form of the principle with the exclusion principle can be illustrated by the following argument, which has three stages.

An electron we might label 1 in a hydrogenic atom has a wavefunction that is the solution of the Schrödinger equation

$$H_1 \psi(1) = E_1 \psi(1)$$

An electron we might label 2 in a hydrogenic atom has a wavefunction that is a solution of

$$H_2 \psi(2) = E_2 \psi(2)$$

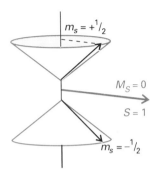

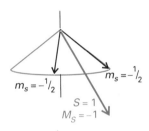

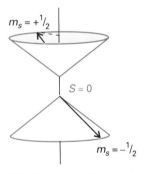

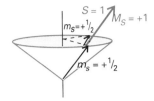

1 The four total spin states that can arise from two electrons. Three have a resultant corresponding to $S = 1$, and one has a zero resultant, corresponding to $S = 0$. This illustration is a summary of Figs 13.18 and 13.24.

When both electrons are present in the same atom, the Hamiltonian is

$$H = H_1 + H_2 + V_{12}$$

V_{12} is the potential energy of repulsion between the electrons. If this interaction is ignored, the joint wavefunction of the two electrons is $\psi(1)\psi(2)$ (see p. 439).

To apply the Pauli principle, we must deal with the total wavefunction, the wavefunction including spin. There are several possibilities for two spins:

both α, denoted $\alpha(1)\alpha(2)$

both β, denoted $\beta(1)\beta(2)$

one α, the other β, denoted $\alpha(1)\beta(2)$ or $\beta(1)\alpha(2)$

Because we cannot tell which electron is α and which is β, in the last case it is appropriate to express the spin states as the linear combinations

$$\sigma_+(1, 2) = \alpha(1)\beta(2) + \beta(1)\alpha(2)$$

$$\sigma_-(1, 2) = \alpha(1)\beta(2) - \beta(1)\alpha(2)$$

because these allow one spin to be α and the other β with equal probability (Fig. 1). The total wavefunction of the system is therefore the product of the orbital part $\psi(1)\psi(2)$ and one of the four spin states:

$$\psi(1)\psi(2)\alpha(1)\alpha(2) \qquad \psi(1)\psi(2)\beta(1)\beta(2)$$

$$\psi(1)\psi(2)\sigma_+(1, 2) \qquad \psi(1)\psi(2)\sigma_-(1, 2)$$

Now we take the Pauli principle into account. It says that if a wavefunction is to be acceptable (for electrons), it must change sign when the electrons are exchanged. In each case, exchanging the labels 1 and 2 converts the factor $\psi(1)\psi(2)$ into $\psi(2)\psi(1)$, which is the same, because the order of multiplying the functions does not change the value of the product. The same is true of $\alpha(1)\alpha(2)$ and $\beta(1)\beta(2)$. Therefore, the first two overall products are not allowed, because they do not change sign. The combination $\sigma_+(1, 2)$ changes to

$$\sigma_+(2, 1) = \alpha(2)\beta(1) + \beta(2)\alpha(1) = \sigma_+(1, 2)$$

because it is simply the original written in a different order. The third overall product is therefore also disallowed. Finally, consider $\sigma_-(1, 2)$:

$$\sigma_-(2, 1) = \alpha(2)\beta(1) - \beta(2)\alpha(1) = -\{\alpha(1)\beta(2) - \beta(1)\alpha(2)\}$$
$$= -\sigma_-(1, 2)$$

and so it does change sign (it is 'antisymmetric'). Therefore the product $\psi(1)\psi(2)\sigma_-(1, 2)$ also changes sign under particle exchange, and therefore it is acceptable.

Now we see that only one of the four possible states is allowed by the Pauli principle, and the one that survives has one α and one β spin. This is the content of the Pauli exclusion principle. The exclusion principle is irrelevant when the orbitals occupied by the electrons are different, and both electrons may then have (but need not have) the same spin state. Nevertheless, even then the overall wavefunction must still be antisymmetric overall, and must still satisfy the Pauli principle itself.

FURTHER INFORMATION 13:
Groups

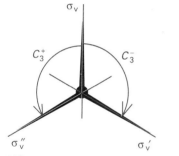

1 The symmetry operations of the group C_{3v}.

The mathematical theory of symmetry is called **group theory** because the symmetry operations of an object form what mathematicians call a 'group'. We shall explain this term by considering NH_3, which belongs to C_{3v} and is symmetrical under the operations E, C_3^+, C_3^-, σ_v, σ_v', and σ_v'' (Fig. 1).

The operation C_3^+ is a counterclockwise 120° rotation as seen from above and C_3^- is the corresponding clockwise rotation. It should be clear that the operation C_3^+ followed by C_3^- is the identity E. We can express this relation symbolically by writing

$$C_3^- C_3^+ = E$$

Similarly, two successive counterclockwise rotations by 120° (Fig. 2a) are equivalent to one clockwise rotation:

$$C_3^+ C_3^+ = C_3^-$$

We can see from Fig. 2b that C_3^+ followed by σ_v is equivalent to σ_v'', so we can write

$$\sigma_v C_3^+ = \sigma_v''$$

Note that in working out these relations, all the operations refer to some fixed arrangement of symmetry elements. That is, the planes and axes remain where they were first drawn on the page, and are unaffected by the performance of an operation. Note, too, that the second operation is written to the left of the first, so in the last example σ_v is carried out after C_3^+.

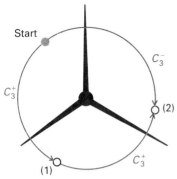

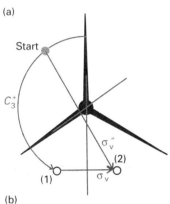

2 (a) The effect of two successive applications of the operation C_3^+ is equivalent to one application of the operation C_3^-, that is, $C_3^+ C_3^+ = C_3^-$. (b) The application of first C_3^+ and then σ_v is equivalent to a single application of the operation σ_v'', that is, $\sigma_v C_3^+ = \sigma_v''$.

The table of all such combinations is called the **group multiplication table**, and for C_{3v} is as follows:

Second operation	First operation					
	E	C_3^+	C_3^-	σ_v	σ_v'	σ_v''
E	E	C_3^+	C_3^-	σ_v	σ_v'	σ_v''
C_3^+	C_3^+	C_3^-	E	σ_v'	σ_v''	σ_v
C_3^-	C_3^-	E	C_3^+	σ_v''	σ_v	σ_v'
σ_v	σ_v	σ_v''	σ_v'	E	C_3^-	C_3^+
σ_v'	σ_v'	σ_v	σ_v''	C_3^+	E	C_3^-
σ_v''	σ_v''	σ_v'	σ_v	C_3^-	C_3^+	E

A glance at the group multiplication table shows that *the outcome of successive symmetry operations is always equivalent to a single symmetry operation of the group.* This behaviour is called the **group property**. The group property is the main feature of the structure of groups: a set of h operations $g_1, g_2, \ldots, g_h$ form a group if they satisfy the group property together with the following mild conditions:

(1) They include the identity E, an element for which $Eg_i = g_iE = g_i$ for all elements g_i.

(2) They include the inverse (g_i^{-1}) of each element, the element for which $g_ig_i^{-1} = g_i^{-1}g_i = E$.

(3) The rule of combination is associative, in the sense that $g_i(g_jg_k) = (g_ig_j)g_k$.

All symmetry operations on molecules satisfy these conditions.

Expressions like $C_3^+ C_3^- = E$ are symbolic ways of writing what happens when various physical operations are carried out in succession. However, it is possible to give them an actual algebraic significance by identifying the matrices that correspond to each operation. The matrix representation of a group was introduced in Chapter 15. It is easy to verify that the matrices given there multiply together in a manner that reproduces the group multiplication table.

FURTHER INFORMATION 14:
Undetermined multipliers

Suppose we need to find the maximum (or minimum) value of some function f that depends on several variables $x_1, x_2, \ldots, x_n$. When the variables undergo a small change from x_i to $x_i + \delta x_i$ the function changes from f to $f + \delta f$, where

$$\delta f = \sum_i \left(\frac{\partial f}{\partial x_i} \right) \delta x_i$$

At a minimum or maximum, $\delta f = 0$, so then

$$\sum_i \left(\frac{\partial f}{\partial x_i} \right) \delta x_i = 0 \tag{1}$$

If the x_i were all independent, all the δx_i would be arbitrary, and this equation could be solved by setting each $(\partial f / \partial x_i) = 0$ individually. When the x_i are not all independent, the δx_i are not all independent, and the simple solution is no longer valid. We proceed as follows.

Let the constraint connecting the variables be an equation of the form $g = 0$. For example, in Chapter 19, one constraint was $n_0 + n_1 + \ldots = N$, which can be written

$$g = 0 \qquad \text{with } g = (n_0 + n_1 + \ldots) - N$$

The constraint $g = 0$ is always valid, so g remains unchanged when the x_i are varied:

$$\delta g = \sum_i \left(\frac{\partial g}{\partial x_i} \right) \delta x_i = 0$$

Since δg is zero, we can multiply it by a parameter λ and add it to eqn 1:

$$\sum_{i=1}^{n} \left\{ \left(\frac{\partial f}{\partial x_i} \right) + \lambda \left(\frac{\partial g}{\partial x_i} \right) \right\} \delta x_i = 0 \tag{2}$$

Equation 2 can be solved for one of the δx, δx_n for instance, in terms of all the other δx_i. All those other δx_i ($i = 1, 2, \ldots, n - 1$) are independent, because there is only one constraint on the system. But here is the trick: λ is arbitrary; therefore we can choose it so that the coefficient of δx_n in eqn 2 is zero. That is, we choose λ so that

$$\left(\frac{\partial f}{\partial x_n}\right) + \lambda\left(\frac{\partial g}{\partial x_n}\right) = 0 \tag{3}$$

Then eqn 2 becomes

$$\sum_{i=1}^{n-1} \left\{\left(\frac{\partial f}{\partial x_i}\right) + \lambda\left(\frac{\partial g}{\partial x_i}\right)\right\} \delta x_i = 0$$

Now the $n - 1$ variations δx_i are independent, so the solution of this equation is

$$\left(\frac{\partial f}{\partial x_i}\right) + \lambda\left(\frac{\partial g}{\partial x_i}\right) = 0 \qquad i = 1, 2, \ldots, n - 1$$

But eqn 3 has exactly the same form as this equation, and so the maximum or minimum of f can be found by solving

$$\left(\frac{\partial f}{\partial x_i}\right) + \lambda\left(\frac{\partial g}{\partial x_i}\right) = 0 \qquad i = 1, 2, \ldots, n$$

The use of this approach was illustrated in Chapter 19 for two constraints and therefore two undetermined multipliers λ_1 and λ_2 (α and $-\beta$).

The multipliers λ cannot always remain undetermined. One approach is to solve eqn 3 instead of incorporating it into the minimization scheme. In Chapter 19 we used the alternative procedure of keeping λ undetermined until a property was calculated for which the value was already known. Thus, we found that $\beta = 1/kT$ by calculating the internal energy of a perfect gas.

FURTHER INFORMATION 15:
The elasticity of rubber

Consider a one-dimensional freely jointed polymer. The conformation can be expressed in terms of the number of bonds pointing to the right (N_R) and the number pointing to the left (N_L). The distance between the ends of the chain is $(N_R - N_L)l$, where l is the length of an individual bond. We write $n = N_R - N_L$ and the total number of bonds as $N = N_R + N_L$.

The number of ways of forming a chain with a given end-to-end distance nl is the number of ways of having N_R right-pointing and N_L left-pointing bonds, and is given by the binomial coefficient

$$W = \frac{N!}{N_L! \, N_R!} = \frac{N!}{\{\frac{1}{2}(N + n)\}! \, \{\frac{1}{2}(N - n)\}!}$$

The conformational entropy of the chain, $S = k \ln W$, is therefore

$$S/k = \ln N! - \ln N_R! - \ln N_L!$$

Since the factorials are large (except for large extensions), we can use Stirling's approximation in the form

$$\ln x! \approx \ln (2\pi)^{\frac{1}{2}} + (x + \tfrac{1}{2}) \ln x - x$$

to obtain

$$S/k = -\ln (2\pi)^{\frac{1}{2}} + (N + 1) \ln 2 + (N + \tfrac{1}{2}) \ln N \\ - \tfrac{1}{2} \ln \{(N + n)^{N+n+1}(N - n)^{N-n+1}\}$$

The most probable conformation of the chain is the one with the ends close together ($n = 0$), as may be confirmed by differentiation. Therefore, the maximum entropy is

$$S_{max}/k = -\ln (2\pi)^{\frac{1}{2}} + (N + 1) \ln 2 - \tfrac{1}{2} \ln N$$

The change in entropy when the chain is stretched from its most

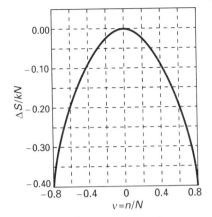

1 The change in molar entropy of a perfect rubber as its extension changes. $v = 1$ corresponds to complete extension; $v = 0$, the conformation of highest entropy, corresponds to the random coil.

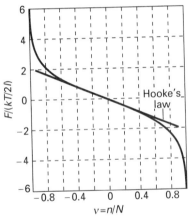

2 The restoring force F of a one-dimensional perfect rubber. For small deflections, F is linearly proportional to extension, corresponding to Hooke's law.

probable conformation until the distance between its ends is nl is therefore

$$\Delta S = S - S_{max} = -\tfrac{1}{2} kN \ln\{(1 + v)^{1+v}(1 - v)^{1-v}\} \tag{1}$$

with $v = n/N$. Equation 1 is plotted in Fig. 1. Note that it is negative for all extensions, and so we conclude that adiabatic contraction of the chain to its fully coiled state is spontaneous.

The work done on a piece of rubber when it is extended through a distance dx is $F\,dx$, where F is the restoring force. The First Law is therefore

$$dU = T\,dS - p\,dV + F\,dx$$

It follows that

$$\left(\frac{\partial U}{\partial x}\right)_{T,V} = T\left(\frac{\partial S}{\partial x}\right)_{T,V} + F$$

In a perfect rubber, as in a perfect gas, the internal energy is independent of the dimensions (at constant temperature), so $(\partial U/\partial x)_{T,V} = 0$. The restoring force is therefore

$$F = -T\left(\frac{\partial S}{\partial x}\right)_{T,V}$$

If the statistical expression for the entropy is introduced into this equation (and we evade problems arising from the constant volume constraint by assuming that the sample contracts laterally when it is stretched), we obtain

$$F = -\frac{T}{l}\left(\frac{\partial S}{\partial n}\right)_{T,V} = -\frac{T}{Nl}\left(\frac{\partial S}{\partial v}\right)_{T,V} = \frac{kT}{2l}\ln\left(\frac{1 + v}{1 - v}\right)$$

which is plotted in Fig. 2. At low extensions ($v \ll 1$),

$$F \approx \frac{vkT}{l} = (N_R - N_L)\frac{kT}{Nl}$$

and the sample obeys Hooke's law (that the restoring force is proportional to the displacement), but it departs from it at higher extensions.

FURTHER INFORMATION 16:
The random walk

We picture the one-dimensional random walk as taking place in steps of length λ, each of duration τ. Our task is to calculate the probability that a particle will be found at a distance x from the origin after a time t. During that time it will have taken N steps, with $N = t/\tau$. If N_R of these are steps to the right, and N_L are steps to the left (with $N_L + N_R = N$), then the net distance travelled is $x = (N_R - N_L)\lambda$. That is, to arrive at x, we must ensure that

$$N_R = \tfrac{1}{2}(N + s) \qquad \text{and} \quad N_L = \tfrac{1}{2}(N - s)$$

with $s = x/\lambda$. The probability of being at x after N steps of length λ is therefore the probability that, when N random steps are taken, the numbers to the right and the left are as given above.

The total number of different journeys for a walk of N steps is equal to $2N$, because each step can be in either of two directions. The number of journeys in which exactly N_R steps are taken to the right is equal to the number of ways of choosing N_R objects from N possibilities irrespective of the order:

$$W = \frac{N!}{N_R! \, (N - N_R)!} \tag{1}$$

As a check, consider all journeys of four steps: there are 16 possible journeys (1). There are six ways of taking two steps to the right and two to the left, which tallies with expression $4!/2!2! = 6$. The probability that the particle is back at the origin after four steps is therefore 6/16. The probability that it is out at $x = +4\lambda$ is only 1/16 because, in order to be there, all four steps must be to the right, and since $4!/4!0! = 1$ the chance of that occurring is 1/16.

LLLL LLLR LLRR LRRR RRRR

LLRL LRLR RLRR

LRLL LRRL RRLR

RLLL RLRL RRRL

RLLR

RRLL

1

Returning now to the general case, the probability of being at x after N steps is

$$P = \frac{\text{Number of journeys with } N_R \text{ steps to the right}}{\text{Total number of journeys}}$$

$$= \left\{ \frac{N!}{N_R! \, (N - N_R)!} \right\} \bigg/ 2^N = \left\{ \frac{N!}{[\frac{1}{2}(N + s)]! \, [\frac{1}{2}(N - s)]!} \right\} \bigg/ 2^N \qquad (2)$$

At this stage the expression for the probability does not seem to resemble the Gaussian function (eqn 24.46) obtained by solving the diffusion equation. However, consider the case when so much time has elapsed that the particles have taken many steps. Then the factorials may be expressed using Stirling's approximation

$$\ln N! \approx (N + \tfrac{1}{2}) \ln N - N + \ln (2\pi)^{\frac{1}{2}}$$

with similar expressions for the other factorials in eqn 2. Taking logarithms of eqn 2 then leads (after quite a lot of algebra) to

$$\ln P = \ln N! - \ln \{\tfrac{1}{2}(N + s)\}! - \ln \{\tfrac{1}{2}(N - s)\}! - N \ln 2$$

$$= \ln \left(\frac{2}{\pi N} \right)^{\frac{1}{2}} - \tfrac{1}{2}(N + s + 1) \ln \left(1 + \frac{s}{N} \right) - \tfrac{1}{2}(N - s + 1) \ln \left(1 - \frac{s}{N} \right)$$

So long as $s/N \ll 1$ (which is equivalent to x not being at great distances from the origin) we can use the approximation $\ln (1 + z) \approx z - \tfrac{1}{2}z^2$, and obtain

$$\ln P = \ln \left(\frac{2}{\pi N} \right)^{\frac{1}{2}} - \frac{s^2}{2N}$$

which rearranges to

$$P = \left(\frac{2}{\pi N} \right)^{\frac{1}{2}} e^{-s^2/2N}$$

and now looks very familiar. Finally, we replace s by x/λ and N by t/τ, and obtain eqn 24.46.

Further reading

INTRODUCTION
Articles of general interest
I. M. Mills, The choice of names and symbols for quantities in chemistry. *J. Chem. Educ.* **66,** 887 (1989).

G. B. Kauffman, The centenary of physical chemistry. *Educ. in Chem.* **24,** 168 (1987).

Texts and sources of data and information
J. W. Servos, *Physical chemistry from Ostwald to Pauling.* Princeton University Press (1990).

K. J. Laidler, *The world of physical chemistry.* Oxford University Press (1993).

I. M. Mills (ed.), *Quantities, units, and symbols in physical chemistry.* Blackwell Scientific, Oxford (1988).

V. Gold, K. L. Loening, A. D. McNaught, and P. Sehmi, *Compendium of chemical terminology (IUPAC recommendations).* Blackwell Scientific, Oxford (1987).

1
Articles of general interest
D. B. Clark, The ideal gas law at the center of the Sun. *J. Chem. Educ.* **66,** 826 (1989).

J. G. Eberhardt, The many faces of van der Waals's equation of state. *J. Chem. Educ.* **66,** 906 (1989).

J. M. Alvariño, J. Veguillas, and S. Valasco, Equations of state, collisional energy transfer, and chemical equilibrium in gases. *J. Chem. Educ.* **66,** 157 (1989).

G. Rhodes, Does a one-molecule gas obey Boyle's law? *J. Chem. Educ.* **69,** 16 (1992).

J. G. Eberhart, A least-squares technique for determining the van der Waals parameters from the critical constants. *J. Chem. Educ.* **69,** 220 (1992).

J. B. Ott, J. R. Coates, and H. T. Hall, Comparison of equations of state. *J. Chem. Educ.* **48,** 515 (1971).

Texts and sources of data and information
C. E. Hecht, *Statistical thermodynamics and kinetic theory*. W. H. Freeman & Co, New York (1990).

W. Kauzmann, *Kinetic theory of gases*. Addison-Wesley, Reading (1966).

J. O. Hirschfelder, C. F. Curtiss, and R. B. Bird, *The molecular theory of gases and liquids*. Wiley, New York (1954).

D. Tabor, *Gases, liquids, and solids*. Cambridge University Press (1979).

A. J. Walton, *Three phases of matter*. Oxford University Press (1983).

J. H. Dymond and E. B. Smith, *The virial coefficients of pure gases and mixtures*. Oxford University Press (1980).

M. Ross, Equations of state. In *Encyclopedia of applied physics* (ed. G. L. Trigg), **6,** 291. VCH, New York (1993).

2
Articles of general interest
E. A. Gislason and N. C. Craig, General definitions of work and heat in thermodynamic processes. *J. Chem. Educ.* **64,** 670 (1987).

J. B. Holbrook, R. Salry-Grant, B. C. Smith, and T. V. Tandel, Lattice enthalpies of ionic halides, hydrides, oxides, and sulfides: second-electron affinities of atomic oxygen and sulfur. *J. Chem. Educ.* **67,** 304 (1990).

E. R. Boyko and J. F. Belliveau, Simplification of some thermochemical equations. *J. Chem. Educ.* **67,** 743 (1990).

T. Solomon, Standard enthalpies of formation of ions in solution. *J. Chem. Educ.* **68,** 41 (1991).

R. D. Freeman, Conversion of standard (1 atm) thermodynamic data to the new standard-state pressure, 1 bar (10^5 Pa). *Bull. Chem. Thermodynamics*, **25,** 523 (1982).

Texts and sources of data and information
E. B. Smith, *Basic chemical thermodynamics*. Oxford University Press (1990).

P. A. Rock, *Chemical thermodynamics*. University Science Books, Mill Valley (1983).

M. L. McGlashan, *Chemical thermodynamics*. Academic Press, London (1979).

W. E. Dasent, *Inorganic energetics*. Cambridge University Press (1982).

D. A. Johnson, *Some thermodynamic aspects of inorganic chemistry*. Cambridge University Press (1982).

I. H. Segel and L. D. Segel, Energetics of biological processes. In *Encyclopedia of applied physics* (ed. G. L. Trigg), **6,** 207. VCH, New York (1993).

J. D. Cox and G. Pilcher, *Thermochemistry of organic and organometallic compounds*. Academic Press, New York (1970).

D. R. Stull, E. F. Westrum, and G. C. Sinke, *The chemical thermodynamics of organic compounds*. Wiley-Interscience, New York (1969).

D. B. Wagman, W. H. Evans, V. B. Parker, R. H. Schumm, I. Halow, S. M. Bailey, K. L. Churney, and R. L. Nuttall, *The NBS tables of chemical thermodynamic properties*. Published as *J. Phys. Chem. Reference Data*, **11** (Supplement 2), 1982.

M. W. Chase, Jr., C. A. Davies, J. R. Downey, Jr., D. J. Frurip, R. A. McDonald, and A. N. Syverud, *JANAF thermochemical tables. The NBS tables of chemical thermodynamic properties*. Published as *J. Phys. Chem. Reference Data*, **14** (Supplement 1), 1985.

J. B. Pedley, R. D. Naylor, and S. P. Kirby, *Thermochemical data of organic compounds*. Chapman and Hall, London (1986).

J. A. Dean (ed.), *Handbook of organic chemistry*. McGraw-Hill, New York (1987).

D. R. Lide (ed.), *Handbook of chemistry and physics*, Vol. 74. CRC Press, Boca Raton (1993).

A. M. James and M. P. Lord (ed.), *Macmillan's chemical and physical data*. Macmillan, London (1992).

J. Emsley, *The elements*. Clarendon Press, Oxford (1991).

3

Articles of general interest

S. M. Blinder, Mathematical methods in elementary thermodynamics. *J. Chem. Educ.* **43,** 85 (1966).

E. W. Anacker, S. E. Anacker, and W. J. Swartz, Some comments on partial derivatives in thermodynamics. *J. Chem. Educ.* **64,** 674 (1987).

G. A. Estévez, K. Yang, and B. B. Dasgupta, Thermodynamic partial derivatives and experimentally measurable quantities. *J. Chem. Educ.* **66,** 890 (1989).

Texts and sources of data and information

P. A. Rock, *Chemical thermodynamics*. University Science Books, Mill Valley (1983).

M. L. McGlashan, *Chemical thermodynamics*. Academic Press, London (1979).

K. E. Bett, J. S. Rowlinson, and G. Saville, *Thermodynamics for chemical engineers*. Athlone Press, London, and MIT Press, Cambridge (1975).

D. M. Hirst, *Mathematics for chemists*. Macmillan, London (1983).

4

Articles of general interest

K. Seidman and T. R. Michalik, The efficiency of reversible heat engines: The possible misinterpretation of a corollary to Carnot's theorem. *J. Chem. Educ.* **68,** 208 (1991).

H. B. Hollinger and M. J. Zenzen, Thermodynamic irreversibility: 1. What is it? *J. Chem. Educ.* **68,** 31 (1991).

E. F. Meyer, The Carnot cycle revisited. *J. Chem. Educ.* **65,** 873 (1988).

N. C. Craig, Entropy analyses of four familiar processes. *J. Chem. Educ.* **65,** 760 (1988).

P. G. Nelson, Derivation of the Second Law of thermodynamics from Boltzmann's distribution law. *J. Chem. Educ.* **65,** 390 (1988).

J. Waser and V. Schomaker, A note on thermodynamic inequalities. *J. Chem. Educ.* **65,** 393 (1988).

P. Djurdjević and I. Cutman, A simple method for showing that entropy is a function of state. *J. Chem. Educ.* **65,** 399 (1988).

W. H. Cropper, Walther Nernst and the last law. *J. Chem. Educ.* **64,** 3 (1987).

L. Glasser, Order, chaos, and all that! *J. Chem. Educ.* **66,** 997 (1989).

M. Barón, With Clausius from energy to entropy. *J. Chem. Educ.* **66,** 1001 (1989).

D. F. R. Gilson, Order and disorder and entropies of fusion. *J. Chem. Educ.* **69,** 23 (1992).

Texts and sources of data and information

P. W. Atkins, *The second law.* Scientific American Books, New York (1984).

J. B. Fenn, *Engines, energy, and entropy.* W. H. Freeman and Co, New York (1982).

J. M. Smith and H. C. Van Ness, *Introduction to chemical engineering thermodynamics.* McGraw-Hill, New York (1987).

K. E. Bett, J. S. Rowlinson, and G. Saville, *Thermodynamics for chemical engineers.* Athlone Press, London and MIT Press, Cambridge (1975).

P. A. Rock, *Chemical thermodynamics.* University Science Books, Mill Valley (1983).

N. C. Craig, *Entropy analysis.* VCH, New York (1992).

5
Articles of general interest

J. S. Winn, The fugacity of van der Waals gas. *J. Chem. Educ.* **65,** 772 (1988).

R. M. Noyes, Thermodynamics of a process in a rigid container. *J. Chem. Educ.* **69,** 470 (1992).

L. L. Combs, An alternative view of fugacity. *J. Chem. Educ.* **69,** 218 (1992).

L. D'Alessio, On the fugacity of a van der Waals gas: An approximate expression that separates attractive and repulsive forces. *J. Chem. Educ.* **70,** 96 (1993).

Texts and sources of data and information

J. M. Smith and H. C. Van Ness, *Introduction to chemical engineering thermodynamics.* McGraw-Hill, New York (1987).

K. E. Bett, J. S. Rowlinson, and G. Saville, *Thermodynamics for chemical engineers.* MIT Press, Cambridge (1975).

P. A. Rock, *Chemical thermodynamics.* University Science Books, Mill Valley (1983).

B. D. Wood, *Applications of thermodynamics*. Addison-Wesley, New York (1982).

G. N. Lewis and M. Randall, *Thermodynamics* (Revised by K. S. Pitzer and L. Brewer). McGraw-Hill, New York (1961).

6
Articles of general interest

M. E. Cardinali and C. Giomini, Boiling temperature vs. composition: an almost-exact explicit equation for a binary mixture following Raoult's law. *J. Chem. Educ.* **66**, 549 (1989).

K. M. Scholsky, Supercritical phase transitions at very high pressure. *J. Chem. Educ.* **66**, 989 (1989).

B. L. Earl, The direct relation between altitude and boiling point. *J. Chem. Educ.* **67**, 45 (1990).

Texts and sources of data and information

S. I. Sandler, *Chemical and engineering thermodynamics*. Wiley, New York (1989).

H. E. Stanley, *Introduction to phase transitions and critical phenomena*. Clarendon Press, Oxford (1971).

E. L. Skau and J. C. Arthur, Determination of melting and freezing temperatures. In *Techniques of chemistry* (ed. A. Weissberger and B. W. Rossiter) **5**, 105. Wiley, New York (1971).

J. R. Anderson, Determination of boiling and condensation temperatures. In *Techniques of chemistry* (ed. A. Weissberger and B. W. Rossiter) **5**, 199. Wiley, New York (1971).

7
Articles of general interest

E. F. Meyer, Thermodynamics of "mixing" of ideal gases: a persistent pitfall. *J. Chem. Educ.* **64**, 676 (1987).

M. J. Clugston, A mathematical verification of the Second law of thermodynamics from the entropy of mixing. *J. Chem. Educ.* **67**, 203 (1990).

J. J. Carroll, Henry's law: A historical view. *J. Chem. Educ.* **70**, 91 (1993).

H. F. Franzen, The freezing point depression law in physical chemistry. *J. Chem. Educ.* **65**, 1077 (1988).

R. L. Scott, Models for phase equilibria in fluid mixtures. *Acc. Chem. Res.* **20**, 97 (1987).

A. S. Kertes and C. J. King, Extraction chemistry of low molecular weight aliphatic alcohols. *Chem. Rev.* **87**, 687 (1987).

Texts and sources of data and information

S. I. Sandler, *Chemical and engineering thermodynamics*. Wiley, New York (1989).

J. S. Rowlinson and F. L. Swinton, *Liquids and liquid mixtures*. Butterworths, London (1982).

J. N. Murrell and E. A. Boucher, *Properties of liquids and solutions*. Wiley-Interscience, New York (1982).

J. R. Overton, Determination of osmotic pressure. In *Techniques of chemistry* (ed. A. Weissberger and B. W. Rossiter) **5**, 309. Wiley, New York (1971).

W. J. Mader and L. T. Brady, Determination of solubility. In *Techniques of chemistry* (ed. A. Weissberger and B. W. Rossiter) **5**, 257. Wiley, New York (1971).

M. R. J. Dack, Solutions and solubilities. In *Techniques of chemistry* (ed. A. Weissberger and B. W. Rossiter) **8**, Wiley, New York (1975).

8
Articles of general interest

M. Zhao, Z. Wang, and L. Xiao, Determining the number of independent components by Brinkley's method. *J. Chem. Educ.* **69**, 539 (1992).

R. J. Stead and K. Stead, Phase diagrams for ternary liquid systems. *J. Chem. Educ.* **67**, 385 (1990).

J. Kochansky, Liquid systems with more than two immiscible phases. *J. Chem. Educ.* **68**, 655 (1991).

R. Battino, The critical point and the number of degrees of freedom. *J. Chem. Educ.* **68**, 276 (1991).

A. Jayaraman, The diamond-anvil high-pressure cell. *Scientific American,* **250** (4), 42 (1984).

Texts and sources of data and information

S. I. Sandler, *Chemical and engineering thermodynamics.* Wiley, New York (1989).

A. Alper, *Phase diagrams,* Vols 1, 2, and 3. Academic Press, New York (1970).

A. Reisman, *Phase equilibria.* Academic Press, New York (1970).

J. Wisniak, *Phase diagrams*: *a literature source book.* Elsevier, Amsterdam (1981).

9
Articles of general interest

N. C. Craig, The chemists' delta. *J. Chem. Educ.* **64**, 668 (1987).

J. J. MacDonald, Equilibria and $\Delta G^{\ominus}$. *J. Chem. Educ.* **67**, 745 (1990).

H. R. Kemp, The effect of temperature and pressure on equilibria: a derivation of the van't Hoff rules. *J. Chem. Educ.* **64**, 482 (1987).

M. Deumié, B. Boulil, and O. Henri-Rousseau, On the minimum of the Gibbs free energy involved in chemical equilibrium: some further comments on the stability of the system. *J. Chem. Educ.* **64**, 201 (1987).

J. J. MacDonald, Equilibrium, free energy, and entropy: rates and differences. *J. Chem. Educ.* **67**, 380 (1990).

H. Xijun and Y. Xiuping, Influences of temperature and pressure on chemical equilibrium in nonideal systems. *J. Chem. Educ.* **68**, 295 (1991).

A. Dumon, A. Lichanot, and E. Poquet, Describing chemical transformation: from the extent of reaction ξ to the reaction advancement ratio χ. *J. Chem. Educ.* **70**, 29 (1993).

R. deLevie, Explicit expression of the general form of the titration curve in terms of concentration: writing a single closed-form expression for the titration curve for a variety of titrations without using approximations or segmentation. *J. Chem. Educ.* **70**, 209 (1993).

J. Gold and V. Gold, Le Chatelier's principle and the laws of van't Hoff. *Educ. in Chem.* **22**, 82 (1985).

R. T. Allsop and N. H. George, Le Chatelier—a redundant principle? *Educ. in Chem.* **21**, 54 (1984).

S. R. Logan, Entropy of mixing and homogeneous equilibria. *Educ. in Chem.* **25**, 44 (1988).

Texts and sources of data and information

M. J. Blandamer, *Chemical equilibria in solution.* Ellis Horwood/Prentice Hall, Hemel Hempstead (1992).

K. G. Denbigh, *The principles of chemical equilibrium.* Cambridge University Press (1971).

R. Holub and P. Vónka, *The chemical equilibrium of gaseous systems.* Reidel, Boston (1976).

W. R. Smith and R. W. Missen, *Chemical reaction equilibrium analysis.* Wiley, New York (1982).

J. T. Edsall and H. Guttfreund, *Biothermodynamics.* Wiley, New York (1983).

P. A. Rock, *Chemical thermodynamics.* University Science Books, Mill Valley (1983).

M. L. McGlashan, *Chemical thermodynamics.* Academic Press, London (1979).

W. E. Dasent, *Inorganic energetics.* Cambridge University Press (1982).

D. A. Johnson, *Some thermodynamic aspects of inorganic chemistry.* Cambridge University Press (1982).

F. M. Harold, *The vital force: a study of bioenergetics.* W. H. Freeman and Co, New York (1986).

N. C. Craig, *Entropy analysis.* VCH, New York (1992).

10

Articles of general interest

P. J. Morgan and E. Gileadi, Alleviating the common confusion caused by polarity in electrochemistry. *J. Chem. Educ.* **66**, 912 (1989).

J. J. MacDonald, Cathodes, terminals, and signs. *Educ. in Chem.* **25**, 52 (1988).

Y. Marcus, Ionic radii in aqueous solutions. *Chem. Rev.* **88**, 1475 (1988).

A. K. Covington, R. G. Bates, and D. A. Durst, Definition of pH scales, standard reference values, and related terminology. *Pure. Appl. Chem.* **57**, 531 (1985).

Texts and sources of data and information

J. Koryta, *Ions, electrodes, and membranes.* Wiley, New York (1991).

A. J. Bard and L. R. Faulkner, *Electrochemical methods.* Wiley, New York (1980).

D. B. Hibbert, *Introduction to electrochemistry.* Macmillan, Basingstoke (1993).

A. J. Bard, R. Parsons, and J. Jordan, *Standard potentials in aqueous solution*. Marcel Dekker, New York (1985).

J. Goodisman, *Electrochemistry: theoretical foundations*. Wiley, New York (1987).

D. Pletcher and F. C. Walsh, *Industrial electrochemistry*. Chapman and Hall, London (1989).

P. Reiger, *Electrochemistry*. Prentice Hall, Englewood Cliffs (1987).

J. O'M. Bockris, B. E. Conway, *et al.* (ed.), *Comprehensive treatise on electrochemistry*, Vols 1–22. Plenum, New York (1980).

R. R. Adžič and E. B. Yeager, Electrochemistry. In *Encyclopedia of applied physics* (ed. G. L. Trigg), **5**, 223. VCH, New York (1993).

11

Texts and sources of data and information

P. W. Atkins, *Quanta: a handbook of concepts*. Oxford University Press (1991).

P. W. Atkins, *Molecular quantum mechanics*. Oxford University Press (1983).

B. Cagnac and J. C. Pebay-Peyroula, *Modern atomic physics: fundamental principles*. Macmillan, London (1975).

T. S. Kuhn, *Black-body theory and the quantum discontinuity 1894–1912*. Oxford University Press (1978).

M. Jammer, *The conceptual development of quantum mechanics*. McGraw-Hill, New York (1966).

F. Hund, *The history of quantum theory*. Harrap & Co, London (1974).

D. C. Cassidy, *Uncertainty: The life and science of Werner Heisenberg*. W. H. Freeman & Co, New York (1992).

A. Pais, *Niels Bohr's times: in physics, philosophy, and polity*. Clarendon Press, Oxford (1991).

W. J. Moore, *Schrödinger: life and thought*. Cambridge University Press (1989).

12

Articles of general interest

G. L. Breneman, The two-dimensional particle in a box. *J. Chem. Educ.* **67**, 866 (1990).

H. F. Blanck, Introduction to a quantum mechanical harmonic oscillator using a modified particle-in-a-box problem. *J. Chem. Educ.* **69**, 98 (1992).

K. Volkamer and M. W. Lerom, More about the particle-in-a-box system: the confinement of matter and the wave-particle dualism. *J. Chem. Educ.* **69**, 100 (1992).

C. A. Hollingsworth, Accidental degeneracies of the particle in a box. *J. Chem. Educ.* **67**, 999 (1990).

W.-K. Li and S. M. Blinder, Particle in an equilateral triangle: exact solution of a nonseparable problem. *J. Chem. Educ.* **64**, 130 (1987).

T. McDermott and G. Henderson, Spherical harmonics in a cartesian frame. *J. Chem. Educ.* **67**, 915 (1990).

K. V. Mikkelsen and M. A. Ratner, Electron tunneling in solid-state electron-transfer reactions. *Chem. Rev.* **87**, 113 (1987).

Texts and sources of data and information

P. W. Atkins, *Quanta: a handbook of concepts.* Oxford University Press (1991).

P. W. Atkins, *Molecular quantum mechnics.* Oxford University Press (1983).

P. J. E. Peebles, *Quantum mechanics.* Princeton University Press, Princeton (1992).

G. C. Schatz and M. A. Ratner, *Quantum mechanics in chemistry,* Ellis Horwood/Prentice Hall, Hemel Hempstead (1993).

R. E. Christofferson, *Basic principles and techniques of molecular quantum mechanics.* Springer, New York (1989).

D. A. McQuarrie, *Quantum chemistry.* University Science Books, Mill Valley (1983).

R. N. Zare, *Angular momentum.* Wiley, New York (1988).

L. Pauling and E. B. Wilson, *Introduction to quantum mechanics.* McGraw-Hill, New York (1935).

13
Articles of general interest

R. D. Allendoerfer, Teaching the shapes of the hydrogenlike and hybrid atomic orbitals. *J. Chem. Educ.* **67**, 37 (1990).

G. L. Breneman, Order out of chaos: shapes of hydrogen orbitals. *J. Chem. Educ.* **65**, 31 (1988).

J. Mason, Periodic contractions among the elements: or, on being the right size. *J. Chem. Educ.* **65**, 17 (1988).

N. Agmon, Ionization potentials for isoelectronic series. *J. Chem. Educ.* **65**, 42 (1988).

E. M. R. Kiremire, A numerical algorithm technique for deriving Russell–Saunders (R–S) terms. *J. Chem. Educ.* **64**, 951 (1987).

L. Guofan and M. L. Elizey Jr., Finding the terms of configurations of equivalent electrons by partitioning total spins. *J. Chem. Educ.* **64**, 771 (1987).

E. R. Scerri, Transition metal configurations and limitations of the orbital approximation. *J. Chem. Educ.* **66**, 481 (1989).

R. T. Myers, The periodicity of electron affinity. *J. Chem. Educ.* **67**, 307 (1990).

N. Shenkuan, The physical basis of Hund's rule: orbital contraction effects. *J. Chem. Educ.* **69**, 800 (1992).

K. D. Sen, M. Slamet, and V. Sahni, Atomic shell structure in Hartree–Fock theory. *Chem. Phys. Lett.,* **205**, 313 (1993).

N. C. Pyper and M. Berry, Ionization energies revisited. *Educ. in Chem.* **27**, 135 (1990).

P. G. Nelson, Relative energies of 3*d* and 4*s* orbitals. *Educ. in Chem.* **29**, 84 (1992).

K. D. Sen, Electronegativity. In *Structure and bonding,* Vol. 6. Springer, New York (1987).

Texts and sources of data and information

P. W. Atkins, *Quanta: a handbook of concepts*. Oxford University Press (1991).

P. W. Atkins, *Molecular quantum mechanics*. Oxford University Press (1983).

T. P. Softley, *Atomic spectra* (Oxford Primer Series). Oxford University Press (1994).

C. F. Fischer, *The Hartree–Fock method for atoms*. Wiley, New York (1977).

M. Karplus and R. N. Porter, *Atoms and molecules*. Benjamin, New York (1970).

E. U. Condon and H. Odabaşi, *Atomic structure*. Cambridge University Press (1980).

S. Bashkin and J. O. Stonor, Jr, *Atomic energy levels and Grotrian diagrams*. North-Holland, Amsterdam (1975 *et seq.*).

K. Bonin and W. Happer, Atomic spectroscopy. In *Encyclopedia of applied physics* (ed. G. L. Trigg), **2**, 245 VCH, New York (1991).

B. Bederson, Atoms. In *Encyclopedia of applied physics* (ed. G. L. Trigg), **2**, 285. VCH, New York (1991).

J. C. Morrison, A. W. Weiss, K. Kirby, and D. Cooper, Electronic structure of atoms and molecules. In *Encyclopedia of applied physics* (ed. G. L. Trigg), **6**, 45. VCH, New York (1993).

14

Articles of general interest

G. A. Gallup, The Lewis electron-pair model, spectroscopy, and the role of the orbital picture in describing the electronic structure of molecules. *J. Chem. Educ.* **65,** 671 (1988).

A. B. Sannigrahi and T. Kir, Molecular orbital theory of bond order and valency. *J. Chem. Educ.* **65,** 647 (1988).

S. Nordholm, Delocalization—the key concept of covalent bonding. *J. Chem. Educ.* **65,** 581 (1988).

A. A. Woolf, Oxidation numbers and their limitations. *J. Chem. Educ.* **65,** 45 (1988).

A. D. Buckingham and T. W. Rowlands, Can addition of a bonding electron weaken a bond? *J. Chem. Educ.* **68,** 282 (1991).

E. I. von Nagy-Felsobuki, Hückel theory and photoelectron spectroscopy. *J. Chem. Educ.* **66,** 821 (1989).

J. R. Dias, A facile Hückel molecular orbital solution of buckminsterfullerene using chemical graph theory. *J. Chem. Educ.* **66,** 1012 (1989).

D. J. Klein and N. Trinajstić, Valence-bond theory and chemical structure. *J. Chem. Educ.* **67,** 633 (1990).

A. Pisanty, The electronic structure of graphite: a chemist's introduction to band theory. *J. Chem. Educ.* **68,** 804 (1991).

R. Parson, Visualizing the variation principle: an intuitive approach to interpreting the theorem in geometric terms. *J. Chem. Educ.* **70,** 115 (1993).

Texts and sources of data and information

M. J. Winter, *Chemical bonding* (Oxford Primer Series). Oxford University Press (1993).

J. N. Murrell, S. F. A. Kettle, and J. M. Tedder, *The chemical bond*. Wiley, New York (1985).

M. F. C. Ladd, *Chemical bonding in solids and fluids*. Ellis Horwood/Prentice Hall, Hemel Hempstead (1993).

C. A. Coulson, *The shape and structure of molecules* (revised by R. McWeeny). Oxford University Press (1982).

R. McWeeny, *Coulson's valence*. Oxford University Press (1979).

D. F. Shriver, P. W. Atkins, and C. H. Langford, *Inorganic chemistry*. Oxford University Press and W. H. Freeman & Co. (1994).

L. Pauling, *The nature of the chemical bond*. Cornell University Press, Ithaca (1960).

A. Hinchcliffe, *Computational quantum chemistry*. Wiley, New York (1988).

B. Webster, *Chemical bonding theory*. Blackwell Scientific, Oxford (1990).

A. Zewail (ed.). *The chemical bond: structure and dynamics*. Academic Press, San Diego (1992).

P. A. Cox, *The electronic structure and chemistry of solids*. Oxford University Press (1987).

J. C. Morrison, A. W. Weiss, K. Kirby, and D. Cooper, Electronic structure of atoms and molecules. In *Encyclopedia of applied physics* (ed. G. L. Trigg), **6**, 45. VCH, New York (1993).

N. H. March and J. F. Mucci, *Chemical physics of free molecules*. Plenum, New York (1993).

15

Articles of general interest

G. L. Breneman, Crystallographic symmetry point group notation flow chart. *J. Chem. Educ.* **64**, 216 (1987).

C. Contreras-Ortega, L. Vera, and E. Quiroz-Reyes, How great is the Great Orthogonality Theorem? *J. Chem. Educ.* **68**, 200 (1991).

Texts and sources of data and information

F. A. Cotton, *Chemical applications of group theory*. Wiley, New York (1990).

S. F. A. Kettle, *Symmetry and structure*. Wiley, New York (1985).

P. W. Atkins, *Molecular quantum mechanics*. Oxford University Press (1983).

D. C. Harris and M. D. Bertolucci, *Symmetry and spectroscopy*. Oxford University Press, New York (1978).

P. W. Atkins, M. S. Child, and C. S. G. Phillips, *Tables for group theory*. Oxford University Press (1970).

B. E. Douglas and C. Hollingsworth, *Symmetry in bonding and structure*. Academic Press, New York (1985).

16
Articles of general interest

N. C. Thomas, The early history of spectroscopy. *J. Chem. Educ.* **68**, 631 (1991).

R. Woods and G. Henderson, FTIR rotational spectroscopy. *J. Chem. Educ.* **64**, 921 (1987).

L. Glasser. Fourier transforms for chemists. Part I. Introduction to the Fourier transform. *J. Chem. Educ.* **64**, A228 (1987). Part II. Fourier transforms in chemistry and spectroscopy. *J. Chem. Educ.* **64**, A260 (1987).

W. D. Perkins, Fourier transform infrared spectroscopy: Part II. Advantages of FT-IR. *J. Chem. Educ.* **64**, A269 (1987).

P. L. Goodfriend, Diatomic vibrations revisited. *J. Chem. Educ.* **64**, 753 (1987).

A. R. Lacey, A student introduction to molecular vibrations. *J. Chem. Educ.* **64**, 756 (1987).

J. G. Verkade, A novel pictorial approach to teaching molecular motions in polyatomic molecules. *J. Chem. Educ.* **64**, 411 (1987).

F. A. Miller and G. B. Kauffman, C. V. Raman and the discovery of the Raman effect. *J. Chem. Educ.* **66**, 795 (1989).

M.-K. Ahn, A comparison of FTNMR and FTIR techniques. *J. Chem. Educ.* **66**, 802 (1989).

D. P. Strommen, Specific values of the depolarization ratio in Raman spectroscopy: their origins and significance. *J. Chem. Educ.* **69**, 803 (1992).

B. J. Bozlee, J. H. Luther, and M. Buraczewski, The infrared overtone intensity of a simple diatomic molecule: nitric oxide. *J. Chem. Educ.* **69**, 370 (1992).

Texts and sources of data and information

J. M. Hollas, *Modern spectroscopy*. Wiley, New York (1991).

J. M. Hollas, *High resolution spectroscopy*. Butterworth, London (1982).

W. Gordy and R. L. Cook, *Microwave molecular spectra*. Wiley, New York (1984).

S. Wilson (ed.), *Methods in computational chemistry*, Vol 4. *Molecular vibrations*. Plenum, New York (1992).

K. P. Huber and G. Herzberg, *Molecular spectra and molecular structure IV. Constants of diatomic molecules*. Van Nostrand-Reinhold, New York (1979).

L. Bellamy, *The infrared spectra of complex molecules*. Chapman and Hall, London (1980).

E. A. V. Ebsworth, D. W. H. Rankin, and S. Cradock, *Structural methods of inorganic chemistry*. Blackwell Scientific, Oxford (1992).

R. Drago, *Physical methods for chemists*. Saunders, Philadelphia (1992).

G. Herzberg, *Infrared and Raman spectra of polyatomic molecules*. Van Nostrand, New York (1945).

E. B. Wilson, J. C. Decius, and P. C. Cross, *Molecular vibrations*. McGraw-Hill, New York (1955).

17
Articles of general interest

R. B. Snadden, The iodine spectrum revisited. *J. Chem. Educ.* **64,** 919 (1987).

M. Allan, Electron spectroscopic techniques in teaching. *J. Chem. Educ.* **64,** 418 (1987).

F. Ahmed, A good example of the Franck–Condon principle. *J. Chem. Educ.* **64,** 427 (1987).

M. G. D. Baumann, J. C. Wright, A. B. Ellis, T. Kuech, and G. C. Lisesnky, Diode lasers. *J. Chem. Educ.* **69,** 89 (1992).

Texts and sources of data and information

J. M. Hollas, *Modern spectroscopy.* Wiley, New York (1991).

J. M. Hollas, *High resolution spectroscopy.* Butterworth, London (1982).

E. A. V. Ebsworth, D. W. H. Rankin, and S. Cradock, *Structural methods of inorganic chemistry.* Blackwell Scientific, Oxford (1992).

R. Drago, *Physical methods for chemists.* Saunders, Philadelphia (1992).

G. Herzberg, *Spectra of diatomic molecules.* Van Nostrand, New York (1950).

G. Herzberg, *Electronic spectra and electronic structure of polyatomic molecules.* Van Nostrand, New York (1966).

A. G. Gaydon, *Dissociation energies.* Chapman and Hall, London (1968).

R. P. Wayne, *Principles and applications of photochemistry.* Oxford University Press (1988).

D. L. Andrews, *Lasers in chemistry.* Springer-Verlag, Berlin (1990).

A. E. Siegman, *Lasers.* University Science Books, Mill Valley (1988).

W. Demtröder, *Laser spectroscopy.* Springer, Berlin (1988).

G. R. Fleming, *Chemical applications of ultrafast spectroscopy.* Oxford University Press (1986).

J. H. D. Eland, *Photoelectron spectroscopy.* Butterworth–Heinemann, London (1983).

C. R. Brundle and A. D. Baker (ed.), *Electron spectroscopy: theory, techniques, and applications,* Vols 1 and 2. Academic Press, London (1977).

18
Articles of general interest

R. H. Orcutt, A. straightforward way to determine relative intensities of spin–spin splitting lines of equivalent nuclei in NMR spectra. *J. Chem. Educ.* **64,** 763 (1987).

T. D. Lash and S. S. Lash, The use of Pascal-like triangles in describing first-order coupling patterns. *J. Chem. Educ.* **64,** 315 (1987).

T. A. Shaler, Generalization of Pascal's triangle to nuclei of any spin. *J. Chem. Educ.* **68,** 853 (1991).

L. J. Schwartz, A step-by-step picture of pulsed (time-domain) NMR. *J. Chem. Educ.* **65,** 959 and 752 (1988).

M.-K. Ahn, A comparison of FTNMR and FTIR techniques. *J. Chem. Educ.* **66,** 802 (1989).

D. J. Wink, Spin–lattice relaxation times in ^{1}H NMR spectroscopy. *J. Chem. Educ.* **66,** 810 (1989).

R. W. King and K. R. Williams, The Fourier transform in chemistry. Part 1. Nuclear magnetic resonance: introduction. *J. Chem. Educ.* **66,** A213 (1989). Part 2. Nuclear magnetic resonance: The single pulse experiment. *Ibid.* A243. Part 3. Multiple-pulse experiments. *J. Chem. Educ.* **67,** A93 (1990). Part 4. NMR: Two-dimensional methods. *Ibid.* A125. A glossary of NMR terms. *Ibid.* A100.

R. Freeman, Selective excitation in high-resolution NMR. *Chem. Rev.* **91,** 1397 (1991).

L. R. Dalton, A. Bain, and C. K. Westbrook, Recent advances in electron paramagnetic resonance. *Ann. Rev. Phys. Chem.* **41,** 389 (1990).

O. Haworth (ed.), Special issue on NMR spectroscopy. *Chem. in Britain,* **29,** 589 (1993).

Texts and sources of data and information

R. J. Abraham, J. Fisher, and P. Lofthus, *Introduction to NMR spectroscopy.* Wiley, New York (1991).

R. K. Harris, *Nuclear magnetic resonance spectroscopy.* Longman, London (1986).

A. E. Derome, *Modern NMR techniques for chemistry research.* Pergamon, Oxford (1987).

J. K. M. Sanders and B. K. Hunter, *Modern NMR spectroscopy.* Oxford University Press (1987).

R. Freeman, *A handbook of nuclear magnetic resonance spectroscopy.* Longman, London (1987).

E. A. V. Ebsworth, D. W. H. Rankin, and S. Cradock, *Structural methods of inorganic chemistry.* Blackwell Scientific, Oxford (1992).

R. Drago, *Physical methods for chemists.* Saunders, Philadelphia (1992).

N. M. Atherton, *Principles of electron spin resonance.* Ellis Horwood/Prentice Hall, Hemel Hempstead (1993).

R. G. Barnes, Electron paramagnetic resonance. In *Encyclopedia of applied physics* (ed. G. L. Trigg), **5,** 475. VCH, New York (1993).

19
Article of general interest

C. W. David, On the Legendre transformation and the Sackur–Tetrode equation. *J. Chem. Educ.* **65,** 876 (1988).

Texts and sources of data and information

P. W. Atkins, *The second law.* Scientific American Books, New York (1984).

R. P. H. Gasser and W. G. Richards, *Entropy and energy levels.* Oxford University Press (1988).

T. L. Hill, *An introduction to statistical thermodynamics.* Dover, New York (1986).

N. Davidson, *Statistical thermodynamics*. McGraw-Hill, New York (1962).

D. Chandler, *Introduction to statistical mechanics*. Oxford University Press (1987).

C. E. Hecht, *Statistical thermodynamics and kinetic theory*. W. H. Freeman & Co, New York (1990).

20
Articles of general interest

C. W. David, A tractable model for studying solution thermodynamics. *J. Chem. Educ.* **64,** 484 (1987).

R. A. Alberty, The effect of a catalyst on the thermodynamic properties and partition functions of a group of isomers. *J. Chem. Educ.* **65,** 409 (1988).

Texts and sources of data and information

A. Ben-Naim, *Statistical thermodynamics for chemists and biologists*. Plenum, New York (1992).

T. L. Hill, *An introduction to statistical thermodynamics*. Dover, New York (1986).

N. Davidson, *Statistical thermodynamics*. McGraw-Hill, New York (1962).

D. Chandler, *Introduction to statistical mechanics*. Oxford University Press (1987).

C. E. Hecht, *Statistical thermodynamics and kinetic theory*. W. H. Freeman & Co, New York (1990).

M. Ross, Equations of state. In *Encyclopedia of applied physics* (ed. G. L. Trigg), **6,** 291. VCH, New York (1993).

21
Articles of general interest

J. P. Glusker, Teaching crystallography to noncrystallographers. *J. Chem. Educ.* **65,** 474 (1988).

J. H. Enemark, Introducing chemists to X-ray structure determination. *J. Chem. Educ.* **65,** 491 (1988). Crystallographic resource list. *J. Chem. Educ.* **65,** 512 (1988).

J. P. Chesick, Fourier analysis and structure determination. Part 1. Fourier transform. *J. Chem. Educ.* **66,** 128 (1989). Part II. Pulse NMR and NMR imaging. *Ibid.* 283. Part III. X-ray crystal structure analysis. *Ibid.* 413.

T. Li and J. H. Worrell, Construction of the seven basic crystallographic units. *J. Chem. Educ.* **66,** 73 (1989).

J. P. Birk and P. R. Coffman, Finding the face-centered cube in the cubic close-packed structure. *J. Chem. Educ.* **69,** 953 (1992).

Texts and sources of data and information

M. F. C. Ladd and R. A. Palmer, *Structure determination by X-ray crystallography*. Plenum, New York (1985).

J. P. Glusker and K. N. Trueblood, *Crystal structure analysis*. Oxford University Press (1985).

E. A. V. Ebsworth, D. W. H. Rankin, and S. Cradock, *Structural methods of inorganic chemistry*. Blackwell Scientific, Oxford (1992).

R. Drago, *Physical methods for chemists*. Saunders, Philadelphia (1992).

C. Giacovazzo (ed.), *Fundamentals of crystallography*. Oxford University Press (1992).

W. B. Pearson and C. Chieh, Crystallography. In *Encyclopedia of applied physics* (ed. G. L. Trigg), **4**, 385. VCH, New York (1992).

A. F. Wells, *Structural inorganic chemistry*. Clarendon Press, Oxford (1984).

R. W. G. Wycoff, *Crystal structure* (five sections, and supplements). Wiley-Interscience, New York (1959).

22
Articles of general interest

C. E. Dykstra, Electrical polarization in diatomic molecules. *J. Chem. Educ.* **65,** 198 (1988).

P. Hobza and R. Zahradník, Intermolecular interactions between medium-sized systems. Nonempirical and empirical calculations of interaction energies: successes and failures. *Chem. Rev.* **88,** 871 (1988).

G. Chałasiński and M. Gutowski, Weak interactions between small systems. Models for studying the nature of intermolecular forces and challenging problems for *ab initio* calculations. *Chem. Rev.* **88,** 943 (1988).

A. D. Buckingham, P. W. Fowler, and J. M. Hutson, Theoretical studies of van der Waals molecules and intermolecular forces. *Chem. Rev.* **88,** 963 (1988).

H. Guerin, Influence of the well width on the third virial coefficient of the square-well intermolecular potential. *J. Chem. Educ.* **69,** 203 (1992).

J.-L. Barrat and M. L. Klein, Molecular dynamics simulations of supercooled liquids near the glass transition. *Ann. Rev. Phys. Chem.* **42,** 23 (1991).

L. N. Mulay and I. L. Mulay, Static magnetic techniques and applications. In *Physical methods of chemistry* (ed. B. W. Rossiter and J. E. Hamilton), **IIIB,** 133 (1989).

W. E. Hatfield, Magnetic measurements. In *Solid-state chemistry: techniques* (ed. A. K. Cheetham and P. Day). Clarendon Press, Oxford (1987).

C. E. Dykstra, Intermolecular electrical interaction: A key ingredient in hydrogen bonding. *Acc. Chem. Res.* **21,** 355 (1988).

J. Israelachvili, Solvation forces and liquid structure, as probed by direct force measurements. *Acc. Chem. Res.* **20,** 415 (1987).

Texts and sources of data and information

J. Israelachvili, *Intermolecular and surface forces*. Academic Press, New York (1985).

C. P. Smyth, Determination of dipole moments. In *Techniques of chemistry* (ed. A. Weissberger and B. W. Rossiter), **4,** 397. Wiley-Interscience, New York (1972).

M. Rigby, E. B. Smith, W. A. Wakeham, and G. C. Maitland, *The forces between molecules*. Oxford University Press (1986).

G. C. Maitland, M. Rigby, E. B. Smith, and W. A. Wakeham, *Intermolecular forces: their origin and determination*. Clarendon Press, Oxford (1981).

M. A. D. Fluendy and K. P. Lawley, *Molecular beams in chemistry*. Chapman and Hall, London (1973).

E. A. V. Ebsworth, D. W. H. Rankin, and S. Cradock, *Structural methods of inorganic chemistry*. Blackwell Scientific, Oxford (1992).

R. Drago, *Physical methods for chemists*. Saunders, Philadelphia (1992).

23
Article of general interest
J. P. Queslel and J. E. Mark, Advances in rubber elasticity and characterization of elastomeric networks. *J. Chem. Educ.* **64,** 491 (1987).

Texts and sources of data and information
F. W. Billmeyer, *Textbook of polymer science*. Wiley, New York (1984).

I. M. Ward and D. W. Hedley, *Mechanical properties of solid polymers*. Wiley, New York (1993).

H. R. Allcock and F. W. Lampe, *Contemporary polymer chemistry*. Prentice-Hall, Englewood Cliffs (1981).

E. G. Richards, *An introduction to physical properties of large molecules in solution*. Cambridge University Press (1980).

L. H. Sperling, *Physical polymer science*. Wiley-Interscience, New York (1986).

D. Freifelder, *Physical biochemistry*. W. H. Freeman & Co, New York (1978).

P. Flory, *Principles of polymer chemistry*. Cornell University Press, Ithaca (1953).

24
Articles of general interest
H. G. Herz, B. M. Braun, K. J. Müller, and R. Mauer, What is the physical significance of the pictures representing the Grotthuss H^+ conductance mechanism? *J. Chem. Educ.* **64,** 777 (1987).

B. L. Earl, Confusion in the expressions for transport coefficients. *J. Chem. Educ.* **66,** 147 (1989).

T. Kenney, Graham's law: defining gas velocities. *J. Chem. Educ.* **67,** 871 (1990).

Texts and sources of data and information
P. J. Collings, *Liquid crystals: Nature's delicate phase of matter*. Wiley, New York (1991).

G. A. Krestov, *et al.*, *Ionic solvation*. Ellis Horwood/Prentice Hall, Hemel Hempstead (1993).

W. Kauzmann, *Kinetic theory of gases*. Addison-Wesley, Reading (1966).

J. O. Hirschfelder, C. F. Curtiss, and R. B. Bird, *The molecular theory of gases and liquids*. Wiley, New York (1954).

D. Tabor, *Gases, liquids, and solids*. Cambridge University Press (1979).

A. J. Walton, *Three phases of matter*. Oxford University Press (1983).

R. B. Bird, W. E. Stewart, and E. N. Lightfoot, *Transport phenomena*. Wiley, New York (1960).

M. Spiro, Determination of transference numbers. In *Techniques of chemistry* (ed. A. Weissberger and B. W. Rossiter), **2A**, 205. Wiley-Interscience, New York (1971).

A. J. Bard and L. R. Faulkner, *Electrochemical methods: Fundamentals and applications*. Wiley, New York (1980).

P. J. Dunlop, B. J. Steel, and J. E. Lane, Experimental methods for studying diffusion in liquids, gases, and solids. In *Techniques of chemistry* (ed. A. Weissberger and B. W. Rossiter), **4**, 205. Wiley-Interscience, New York (1972).

J. Crank, *The mathematics of diffusion*. Clarendon Press, Oxford (1975).

J. S. Rowlinson and F. L. Swinton, *Liquids and liquid mixtures*. Butterworths, London (1982).

J. N. Murrell and E. A. Boucher, *Properties of liquids and solutions*. Wiley-Interscience, New York (1982).

M. P. Allen and D. Tildesley, *Computer simulation of liquids*. Clarendon Press, Oxford (1986).

D. G. Leaist, Diffusion and ionic conduction in liquids. In *Encyclopedia of applied physics* (ed. G. L. Trigg), **5**, 661. VCH, New York (1993).

25
Articles of general interest

E. Levin and J. G. Eberhart, Simplified rate-law integration for reactions that are first-order in each of the two reactants. *J. Chem. Educ.* **66**, 705 (1989).

M. N. Berberan-Santos and J. M. G. Martinho, Integration of kinetic rate equations by matrix methods. *J. Chem. Educ.* **67**, 375 (1990).

J. C. Reeve, Some provocative opinions on the terminology of chemical kinetics. *J. Chem. Educ.* **68**, 728 (1991).

H. Maskill, The extent of reaction and chemical kinetics. *Educ. in Chem.* **21**, 122 (1984).

S. R. Logan, The meaning and significance of "the activation energy" of a chemical reaction. *Educ. in Chem.* **23**, 148 (1986).

Texts and sources of data and information

J. I. Steinfeld, J. S. Francisco, and W. L. Hase, *Chemical kinetics and dynamics*. Prentice Hall, Englewood Cliffs (1989).

K. A. Connors, *Chemical kinetics*. VCH, Weinheim (1990).

K. J. Laidler, *Chemical kinetics*. Harper and Row, New York (1987).

C. H. Bamford and C. F. Tipper (ed.), *Comprehensive chemical kinetics*. Elsevier, Amsterdam (1969 *et seq.*).

B. B. Chance, Rapid flow methods. In *Techniques of chemistry* (ed. G. G. Hammes), **6B**, 5. Wiley-Interscience, New York (1974).

G. G. Hammes, Temperature-jump methods. In *Techniques of chemistry* (ed. G. G. Hammes), **6B**, 147. Wiley-Interscience, New York (1974).

W. Knoche, Pressure-jump methods. In *Techniques of chemistry* (ed. G. G. Hammes), **6B**, 187. Wiley-Interscience, New York (1974).

26
Articles of general interest

K. J. Laidler, Rate-controlling step: a necessary or useful concept? *J. Chem. Educ.* **65,** 250 (1988).

A. F. Shaaban, The integrated rate equation of the nitric oxide–oxygen reaction. *J. Chem. Educ.* **67,** 869 (1990).

J. N. Spencer, Competitive and coupled reactions. *J. Chem. Educ.* **69,** 281 (1992).

R. T. Raines and D. E. Hansen, An intuitive approach to steady-state kinetics. *J. Chem. Educ.* **65,** 757 (1988).

G. F. Swiegers, Applying the principles of chemical kinetics to population growth problems. *J. Chem. Educ.* **70,** 364 (1993).

F. Mata-Perez and J. F. Perez-Benito, The kinetic rate law for autocatalytic reactions. *J. Chem. Educ.* **64,** 925 (1987).

L. J. Soltzberg, Self-organization in chemistry. *J. Chem. Educ.* **66,** 187 (1989).

R. J. Field, The language of dynamics. *J. Chem. Educ.* **66,** 188 (1989).

R. M. Noyes, Some models of chemical oscillators. *J. Chem. Educ.* **66,** 190 (1989).

I. R. Epstein, The role of flow systems in far-from-equilibrium dynamics. *J. Chem. Educ.* **66,** 191 (1989).

R. J. Field and F. W. Schneider, Oscillating chemical reactions and nonlinear dynamics. *J. Chem. Educ.* **66,** 195 (1989).

E. Hughes, Jr., Solving differential equations in kinetics by using power series. *J. Chem. Educ.* **66,** 46 (1989).

W. Jahnke and A. R. Winfree, Recipes for Belousov–Zhabotinsky reagents. *J. Chem. Educ.* **68,** 320 (1991).

S. K. Scott, Oscillations in simple models of chemical systems. *Acc. Chem. Res.* **20,** 186 (1987).

P. Brumer and M. Shapiro, Coherence chemistry: controlling chemical reactions with lasers. *Acc. Chem. Res.* **22,** 407 (1989).

Texts and sources of data and information

K. R. Westererp, W. P. M. van Swaaij, and A. A. C. M. Beenackers, *Chemical reactor design and operation.* Wiley, New York (1984).

J. H. Espenson, *Chemical kinetics and reaction mechanics.* McGraw-Hill, New York (1981).

F. W. Billmeyer, *Textbook of polymer science.* Wiley-Interscience, New York (1984).

R. P. Wayne, *Principles and applications of photochemistry.* Oxford University Press (1988).

R. P. Wayne, *Chemistry of atmospheres.* Clarendon Press, Oxford (1991).

S. K. Scott, *Oscillations, waves, and chaos in chemical kinetics* (Oxford Primer Series). Oxford University Press (1994).

P. Gray and S. K. Scott, *Chemical oscillations and instabilities*. Clarendon Press, Oxford (1990).

27
Articles of general interest

G. M. Fernández, J. A. Sordo, and T. L. Sordo, Analysis of potential energy surfaces. *J. Chem. Educ.* **65,** 665 (1988).

K. J. Laidler, Just what is a transition state? *J. Chem. Educ.* **65,** 540 (1988).

M. A. Smith, The nature of distribution functions for colliding systems: calculation of averaged properties. *J. Chem. Educ.* **70,** 218 (1993).

H. Maskill, The Arrhenius equation. *Educ. in Chem.* **27,** 111 (1990).

S. R. Logan, The meaning and significance of 'the activation energy' of a chemical reaction. *Educ. in Chem.* **23,** 148 (1986).

I. Powis, Energy redistribution in unimolecular ion dissociation. *Acc. Chem. Res.* **20,** 179 (1987).

C. E. Klots, The reaction coordinate and its limitations: an experimental perspective. *Acc. Chem. Res.* **21,** 16 (1988).

I. W. M. Smith, Vibrational adiabaticity in chemical reactions. *Acc. Chem. Res.* **23,** 101 (1990).

P. R. Brooks, Spectroscopy of transition region species. *Chem. Rev.* **88,** 407 (1988).

J. Keizer, Diffusion effects on rapid bimolecular chemical reactions. *Chem. Rev.* **87,** 167 (1987).

D. W. Lupo and M. Quack, IR-laser photochemistry. *Chem. Rev.* **87,** 181 (1987).

D. G. Trular, R. Steckler, and M. S. Gordon, Potential energy surfaces for polyatomic reaction dynamics. *Chem. Rev.* **87,** 181 (1987).

A. H. Zewail, Laser femtochemistry. *Science,* **242,** 1645 (1988).

A. H. Zewail, Femtosecond transition-state dynamics. *Faraday Discuss. Chem. Soc.* **91,** 1 (1991).

M. S. Child, Molecular reaction dynamics. *Sci. Prog.* **70,** 73 (1986).

Texts and sources of data and information

K. J. Laidler, *Chemical kinetics*. Harper and Row, New York (1987).

J. Simons, *Energetic principles of chemical reactions*. Jones and Bartlett, Portola Valley (1983).

R. A. Marcus, Activated complex theory: current status, extensions, and applications. In *Techniques of chemistry* (ed. E. S. Lewis), **6A,** 13. Wiley-Interscience, New York (1974).

M. J. Blandamer, *Chemical equilibria in solutions*. Ellis Horwood/Prentice Hall, Hemel Hempstead (1992).

P. C. Jordan, *Chemical kinetics and transport*. Plenum, New York (1979).

I. W. M. Smith, *Kinetics and dynamics of elementary gas reactions*. Butterworths, London (1980).

J. I. Steinfeld, J. S. Francisco, and W. L. Hase, *Chemical kinetics and dynamics.* Prentice Hall, Englewood Cliffs (1989).

R. G. Gilbert and S. C. Smith, *Theory of unimolecular and recombination reactions.* Blackwell Scientific, Oxford (1990).

R. B. Bernstein, *Chemical dynamics via molecular beam and laser techniques.* Clarendon Press, Oxford (1982).

D. M. Hirst, *Potential energy surfaces.* Taylor and Francis, London (1985).

R. D. Levine and R. B. Bernstein, *Molecular reaction dynamics and chemical reactivity.* Clarendon Press, Oxford (1987).

R. van Eldik, T. Asano, and W. J. Le Noble, Activation and reaction volumes in solution. 2. *Chem. Rev.* **89,** 549 (1989).

R. W. Carr, Chemical kinetics. In *Encyclopedia of applied physics* (ed. G. L. Trigg), **3,** 345. VCH, New York (1992).

28
Articles of general interest

S. T. Oyama and G. A. Samorjai, Homogeneous, heterogeneous, and enzymatic catalysis. *J. Chem. Educ.* **65,** 765 (1988).

N.-F. Zhou, The availability of a simple form of Gibbs adsorption equation for mixed surfactants. *J. Chem. Educ.* **66,** 137 (1989).

T. E. Mallouk and H. Lee, Designer solids and surfaces. *J. Chem. Educ.* **67,** 829 (1990).

L. Glasser, The BET isotherm in 3D. *Educ. in Chem.* **25,** 178 (1988).

D. Langevin, Microemulsions. *Acc. Chem. Res.* **21,** 255 (1988).

D. Langevin, Micelles and microemulsions. *Ann. Rev. Phys. Chem.,* **43,** 341 (1992).

E. Shustorovich, Chemisorption theory: in search of the elephant. *Acc. Chem. Res.* **21,** 183 (1988).

S. L. Bernasek, State-resolved dynamics of chemical reactions at surfaces. *Chem. Rev.* **87,** 91 (1987).

B. B. Laird and A. D. J. Haymet, The crystal/liquid interface: structure and properties from computer simulation. *Chem. Rev.* **92,** 1819 (1992).

R. Parsons, Electrical double layer: recent experimental and theoretical developments. *Chem. Rev.* **90,** 813 (1990).

C. L. Perrin and T. J. Dwyer, Application of two-dimensional NMR to kinetics of chemical exchange. *Chem. Rev.* **90,** 935 (1990).

Texts and sources of data and information

B. C. Gates, *Catalytic chemistry.* Wiley, New York (1992).

Y. Moroi, *Micelles.* Plenum, New York (1992).

K. L. Mittal and D. O. Shah (ed.), *Surfactants in solution,* Vol. 11. Plenum, New York (1992).

K. L. Mittal (ed.), *Particles on surfaces,* Vol. 3. Plenum, New York (1991).

Further reading

V. P. Zhdanov, *Elementary physicochemical processes on solid surfaces.* Plenum, New York (1991).

S. R. Morrison, *The chemical physics of surfaces.* Plenum, New York (1990).

J. Israelachvili, *Intermolecular and surface forces.* Academic Press, New York (1985).

R. J. Hunter, *Foundations of colloid science*, Vols 1 and 2. Clarendon Press, Oxford (1987, 1989).

B. Dobiáš (ed.), *Coagulation and flocculation.* Marcel Dekker, New York (1993).

K. Tamaru, *Dynamic heterogeneous catalysis*, Academic Press, New York (1978).

A. Zangwill, *Physics at surfaces.* Cambridge University Press (1988).

R. Aveyard and D. A. Haydon, *An introduction to the principles of surface chemistry.* Cambridge University Press (1973).

A. W. Anderson, *Physical chemistry of surfaces.* Wiley-Interscience, New York (1990).

R. P. H. Gasser, *An introduction to chemisorption and catalysis by metals.* Clarendon Press, Oxford (1985).

M. W. Roberts and C. S. McKee, *Chemistry of the metal–gas interface.* Clarendon Press, Oxford (1978).

G. Somorjai, *Chemistry in two dimensions: surfaces.* Cornell University Press, Ithaca (1981).

G. C. Bond, *Heterogeneous catalysis: principles and applications.* Clarendon Press, Oxford (1986).

M. Boudart and G. Djéga-Mariadassou, *Kinetics of heterogeneous catalysis reactions.* Princeton University Press (1984).

29

Articles of general interest

R. M. Wightman and D. O. Wipf, High-speed cyclic voltammetry. *Acc. Chem. Res.* **23,** 64 (1990).

A. T. Hubbard, Electrochemistry at well-characterized surfaces. *Chem. Rev.* **88,** 633 (1988).

R. Parsons, Electrical double layer: recent experimental and theoretical developments. *Chem. Rev.* **90,** 813 (1990).

Texts and sources of data and information

J. Koryta, *Ions, electrodes, and membranes.* Wiley, New York (1991).

D. B. Hibbert, *Introduction to electrochemistry.* Macmillan, Basingstoke (1993).

J. O'M. Bockris, B. E. Conway, and R. E. White (ed.), *Modern aspects of electrochemistry*, Vol. 22. Plenum, New York (1992).

R. R. Adžić and E. B. Yeager, Electrochemistry. In *Encyclopedia of applied physics* (ed. G. L. Trigg), **5,** 223. VCH, New York (1993).

C. D. S. Tuck, *Modern battery construction.* Ellis Horwood, New York (1991).

D. Linden (ed.), *Handbook of batteries and cells.* McGraw-Hill, New York (1984).

D. R. Crow, *Principles and applications of electrochemistry*. Chapman and Hall, London (1988).

A. J. Bard and L. R. Faulkner, *Electrochemical techniques: fundamentals and applications*. Wiley-Interscience, New York (1979).

M. G. Fontanna and R. W. Staehle (ed.), *Advances in corrosion science and technology*. Plenum, New York (1980).

Data section

The following tables reproduce and expand the data given in the short tables in the text, and follow their numbering. Standard states refer to a pressure of $p^{\ominus} = 1$ bar. The general references are as follows:

AIP: D. E. Gray (ed.), *American Institute of Physics handbook*. McGraw-Hill, New York (1972).

AS: M. Abramowitz and I. A. Stegun (ed.), *Handbook of mathematical functions*. Dover, New York (1963).

E: J. Emsley, *The elements*. Clarendon Press, Oxford (1991).

HCP: D. R. Lide (ed.), *Handbook of chemistry and physics*. CRC Press, Boca Raton (1993).

JL: A. M. James and M. P. Lord (ed.), *Macmillan's chemical and physical data*. Macmillan, London (1992).

KL: G. W. C. Kaye and T. H. Laby (ed.), *Tables of physical and chemical constants*. Longman, London (1973).

LR: G. N. Lewis and M. Randall, revised by K. S. Pitzer and L. Brewer, *Thermodynamics*. McGraw-Hill, New York (1961).

NBS: *NBS Tables of chemical thermodynamic properties*, published as *J. Phys. and Chem. Reference Data*, **11**, Supplement 2 (1982).

RS: R. A. Robinson and R. H. Stokes, *Electrolyte solutions*. Butterworth, London (1959).

TDOC: J. B. Pedley, R. D. Naylor, and S. P. Kirby, *Thermochemical data of organic compounds*. Chapman and Hall, London (1986).

CONTENTS

The following list is a directory of all tables in the text; those included in this Data section are marked with an asterisk (*).

Physical properties of selected materials

	$\rho/(\text{g cm}^{-3})$ at 293 K†	T_f/K	T_b/K		$\rho/(\text{g cm}^{-3})$ at 293 K†	T_f/K	T_b/K
Elements							
Aluminium(s)	2.698	933.5	2740	HBr(g)	2.77	184.3	206.4
Argon(g)	1.381	83.8	87.3	HCl(g)	1.187	159.0	191.1
Boron(s)	2.340	2573	3931	HI(g)	2.85	222.4	237.8
Bromine(l)	3.123	265.9	331.9	H_2O(l)	0.997	273.2	373.2
Carbon(d, gr)	2.260	3700s		D_2O(l)	1.104	277.0	374.6
Carbon(s, d)	3.513			NH_3(g)	0.817	195.4	238.8
Chlorine(g)	1.507	172.2	239.2	KBr(s)	2.750	1003	1708
Copper(s)	8.960	1357	2840	KCl(s)	1.984	1049	1773s
Fluorine(g)	1.108	53.5	85.0	NaCl(s)	2.165	1074	1686
Gold(s)	19.320	1338	3080	H_2SO_4(l)	1.841	283.5	611.2
Helium(g)	0.125		4.22				
Hydrogen(g)	0.071	14.0	20.3	**Organic compounds**			
Iodine(s)	4.930	386.7	457.5	Acetaldehyde, CH_3CHO(l, g)	0.788	152	293
Iron(s)	7.874	1808	3023	Acetic acid, CH_3COOH(l)	1.049	289.8	391
Krypton(g)	2.413	116.6	120.8	Acetone, $(CH_3)_2CO$(l)	0.787	178	329
Lead(s)	11.350	600.6	2013	Aniline, $C_6H_5NH_2$(l)	1.026	267	457
Lithium(s)	0.534	453.7	1620	Anthracene, $C_{14}H_{10}$(s)	1.243	490	615
Magnesium(s)	1.738	922.0	1363	Benzene, C_6H_6(l)	0.879	278.6	353.2
Mercury(l)	13.546	234.3	629.7	Carbon tetrachloride, CCl_4(l)	1.63	250	349.9
Neon(g)	1.207	24.5	27.1	Chloroform, $CHCl_3$(l)	1.499	209.6	334
Nitrogen(g)	0.880	63.3	77.4	Ethanol, C_2H_5OH(l)	0.789	156	351.4
Oxygen(g)	1.140	54.8	90.2	Formaldehyde, HCHO(g)		181	254.0
Phosphorus(s, wh)	1.820	317.3	553	Glucose, $C_6H_{12}O_6$(s)	1.544	415	
Potassium(s)	0.862	336.8	1047	Methane, CH_4(g)		90.6	111.6
Silver(s)	10.500	1235	2485	Methanol, CH_3OH(l)	0.791	179.2	337.6
Sodium(s)	0.971	371.0	1156	Naphthalene, $C_{10}H_8$(s)	1.145	353.4	491
Sulfur(s, α)	2.070	386.0	717.8	Octane, C_8H_{18}(l)	0.703	216.4	398.8
Uranium(s)	18.950	1406	4018	Phenol, C_6H_5OH(s)	1.073	314.1	455.0
Xenon(g)	2.939	161.3	166.1	Sucrose, $C_{12}H_{22}O_{11}$(s)	1.588	457d	
Zinc(s)	7.133	692.7	1180				
Inorganic compounds							
$CaCO_3$(s, calcite)	2.71	1612	1171d				
$CuSO_4 \cdot 5H_2O$(s)	2.284	$383(-H_2O)$	$423(-5H_2O)$				

d: decomposes
s: sublimes
Data: AIP, E, HCP, KL
†For gases, at their boiling points.

Table 1.2 Collision cross-sections, σ/nm^2

Ar	0.36
C_2H_4	0.64
C_6H_6	0.88
CH_4	0.46
Cl_2	0.93
CO_2	0.52
H_2	0.27
He	0.21
N_2	0.43
Ne	0.24
O_2	0.40
SO_2	0.58

Data: KL

Table 1.3 Second virial coefficients, $B/(\text{cm}^3\,\text{mol}^{-1})$

	100 K	273 K	373 K	600 K
Air	−167.3	−13.5	3.4	19.0
Ar	−187.0	−21.7	−4.2	11.9
CH_4		−53.6	−21.2	8.1
CO_2		−142	−72.2	−12.4
H_2	−2.0	13.7	15.6	
He	11.4	12.0	11.3	10.4
Kr		−62.9	−28.7	1.7
N_2	−160.0	−10.5	6.2	21.7
Ne	−6.0	10.4	12.3	13.8
O_2	−197.5	−22.0	−3.7	12.9
Xe		−153.7	−81.7	−19.6

Data: AIP, JL. The values relate to the expansion in eqn 21b of Section 1.3; convert to eqn 21a using $B' = B/RT$. For Ar at 273 K, $C = 1200\,\text{cm}^6\,\text{mol}^{-1}$

Table 1.4 Critical constants

	p_c/atm	$V_c/(\text{cm}^3\,\text{mol}^{-1})$	T_c/K	Z_c	T_B/K		p_c/atm	$V_c/(\text{cm}^3\,\text{mol}^{-1})$	T_c/K	Z_c	T_B/K
Ar	48.00	75.25	150.72	0.292	411.5	HCl	81.5	81.0	324.7	0.248	
Br_2	102	135	584	0.287		He	2.26	57.76	5.21	0.305	22.64
C_2H_4	50.50	124	283.1	0.270		HI	80.8	423.2			
C_2H_6	48.20	148	305.4	0.285		Kr	54.27	92.24	209.39	0.291	575.0
C_6H_6	48.6	260	562.7	0.274		N_2	33.54	90.10	126.3	0.292	327.2
CH_4	45.6	98.7	190.6	0.288	510.0	Ne	26.86	41.74	44.44	0.307	122.1
Cl_2	76.1	124	417.2	0.276		NH_3	111.3	72.5	405.5	0.242	
CO_2	72.85	94.0	304.2	0.274	714.8	O_2	50.14	78.0	154.8	0.308	405.9
F_2	55	144				Xe	58.0	118.8	289.75	0.290	768.0
H_2	12.8	64.99	33.23	0.305	110.0						
H_2O	218.3	55.3	647.4	0.227							
HBr	84.0	363.0									

Data: AIP, KL

Table 1.5 Van der Waals constants

	$a/(\text{L}^2\,\text{atm}\,\text{mol}^{-1})$	$b/(10^{-2}\,\text{L}\,\text{mol}^{-1})$		$a/(\text{L}^2\,\text{atm}\,\text{mol}^{-1})$	$b/(10^{-2}\,\text{L}\,\text{mol}^{-1})$
Ar	1.363	3.219	He	0.03457	2.370
C_2H_4	4.530	5.714	Kr	2.349	3.978
C_2H_6	5.562	6.380	N_2	1.408	3.913
C_6H_6	18.24	11.54	Ne	0.2135	1.709
CH_4	2.283	4.278	NH_3	4.225	3.707
Cl_2	6.579	5.622	O_2	1.378	3.183
CO	1.505	3.985	SO_2	6.803	5.636
CO_2	3.640	4.267	Xe	4.250	5.105
H_2	0.2476	2.661			
H_2O	5.536	3.049			
H_2S	4.490	4.287			

Data: HCP, JL

Table 2.2 Temperature variation of molar heat capacities†

	a	$b/(10^{-3}\,K)$	$c/(10^5\,K)$
Monoatomic gases			
	20.78	0	0
Other gases			
Br_2	37.32	0.50	-1.26
Cl_2	37.03	0.67	-2.85
CO_2	44.22	8.79	-8.62
F_2	34.56	2.51	-3.51
H_2	27.28	3.26	0.50
I_2	37.40	0.59	-0.71
N_2	28.58	3.77	-0.50
NH_3	29.75	25.1	-1.55
O_2	29.96	4.18	-1.67
Liquids (from melting to boiling)			
$C_{10}H_8$, naphthalene	79.5	0.4075	0
I_2	80.33	0	0
H_2O	75.29	0	0
Solids			
Al	20.67	12.38	0
C (graphite)	16.86	4.77	-8.54
$C_{10}H_8$, naphthalene	-115.9	3.920×10^3	0
Cu	22.64	6.28	0
I_2	40.12	49.79	0
NaCl	45.94	16.32	0
Pb	22.13	11.72	0.96

†For $C_{p,m}/(J\,K^{-1}\,mol^{-1}) = a + bT + c/T^2$
Source: LR.

Table 2.3 Standard enthalpies of fusion and vaporization at the transition temperature, $\Delta_{trs}H^{\ominus}/(kJ\,mol^{-1})$

	T_f/K	Fusion	T_b/K	Vaporization		T_f/K	Fusion	T_b/K	Vaporization
Elements									
Ag	1234	11.30	2436	250.6	CS_2	161.2	4.39	319.4	26.74
Ar	83.81	1.188	87.29	6.506	H_2O	273.15	6.008	373.15	40.656
Br_2	265.9	10.57	332.4	29.45					44.016 at 298 K
Cl_2	172.1	6.41	239.1	20.41					
F_2	53.6	0.26	85.0	3.16	H_2S	187.6	2.377	212.8	18.67
H_2	13.96	0.117	20.38	0.916	H_2SO_4	283.5	2.56		
He	3.5	0.021	4.22	0.084	NH_3	195.4	5.652	239.7	23.35
Hg_2	234.3	2.292	629.7	59.30					
I_2	386.8	15.52	458.4	41.80	**Organic compounds**				
N_2	63.15	0.719	77.35	5.586	CH_4	90.68	0.941	111.7	8.18
Na	371.0	2.601	1156	98.01	CCl_4	250.3	2.5	350	30.0
O_2	54.36	0.444	90.18	6.820	C_2H_6	89.85	2.86	184.6	14.7
Xe	161	2.30	165	12.6	C_6H_6	278.61	10.59	353.2	30.8
					CH_3OH	175.2	3.16	337.2	35.27
Inorganic compounds									37.99 at 298 K
CCl_4	250.3	2.47	349.9	30.00	C_2H_5OH	156	4.60	352	43.5
CO_2	217.0	8.33	194.6	25.23 s					

Data: AIP   s denotes sublimation

Table 2.4 Limiting enthalpies of solution at 298 K, $\Delta_{sol}H^{\ominus}/(\text{kJ mol}^{-1})$

	F$^-$	Cl$^-$	Br$^-$	I$^-$	OH$^-$	CO$_3^{2-}$	NO$_3^-$	SO$_4^{2-}$
Li$^+$	+4.9	−37.0	−48.8	−63.3	−23.6	−18.2	−2.7	−29.8
Na$^+$	+1.90	+3.89	−0.6	−7.5	−44.5	−26.7	+20.4	−2.4
K$^+$	−17.74	+17.22	+19.9	+20.3	−57.1	−30.9	+34.9	+23.8
NH$_4^+$	−1.2	+14.8	+16.0	+13.7			+25.69	+6.6
Ag$^+$	−22.5	+65.5	+84.4	+112.2		+41.8	+22.6	+17.8
Mg^{2+}	−17.7	−160.0	−185.6	−213.2	+2.3	−25.3	−90.9	−91.2
Ca^{2+}	+11.5	−81.3	−103.1	−119.7	−16.7	−13.1	−19.2	−18.0
Al^{3+}	−27	−329	−368	−385				−350

The entry for X$^+$Y$^-$ is the limiting enthalpy of the process XY(s) → X$^+$(aq) + Y$^-$(aq).
Source: NBS

Table 2.6 Ionization energies, $E_i/(\text{kJ mol}^{-1})$

H							He
1312.0							2372.3
							5250.4

Li	Be	B	C	N	O	F	Ne
513.3	899.4	800.6	1086.2	1402.3	1313.9	1681	2080.6
7298.0	1757.1	2427	2352	2856.1	3388.2	3374	3952.2

Na	Mg	Al	Si	P	S	Cl	Ar
495.8	737.7	577.4	786.5	1011.7	999.6	1251.1	1520.4
4562.4	1450.7	1816.6	1577.1	1903.2	2251	2297	2665.2
		2744.6		2912			

K	Ca	Ga	Ge	As	Se	Br	Kr
418.8	589.7	578.8	762.1	947.0	940.9	1139.9	1350.7
3051.4	1145	1979	1537	1798	2044	2104	2350
		2963	2735				

Rb	Sr	In	Sn	Sb	Te	I	Xe
403.0	549.5	558.3	708.6	833.7	869.2	1008.4	1170.4
2632	1064.2	1820.6	1411.8	1794	1795	1845.9	2046
		2704	2943.0	2443			

Cs	Ba	Tl	Pb	Bi	Po	At	Rn
375.5	502.8	589.3	715.5	703.2	812	930	1037
2420	965.1	1971.0	1450.4	1610			
		2878	3081.5	2466			

Data: E

Table 2.7 Electron affinities, $E_{ea}/(\text{kJ mol}^{-1})$

H							He
72.8							−21

Li	Be	B	C	N	O	F	Ne
59.8	−18	23	122.5	−7	141	322	−29
					−844		

Na	Mg	Al	Si	P	S	Cl	Ar
52.9	−21	44	133.6	71.7	200.4	348.7	−35
					−532		

K	Ca	Ga	Ge	As	Se	Br	Kr
48.3	−186	36	116	77	195.0	324.5	−39

Rb	Sr	In	Sn	Sb	Te	I	Xe
46.9	−146	34	121	101	190.2	295.3	−41

Cs	Ba	Tl	Pb	Bi	Po	At	Rn
45.5	−46	30	35.2	101	186	270	−41

Data: E

Table 2.8 Bond dissociation enthalpies, $\Delta H^{\ominus}(A-B)/(kJ\,mol^{-1})$ at 298 K

Diatomic molecules

H—H	436	F—F	155	Cl—Cl	242	Br—Br	193	I—I	151
O=O	497	C=O	1076	N≡N	945				
H—O	428	H—F	565	H—Cl	431	H—Br	366	H—I	299

Polyatomic molecules

H—CH_3	435	H—NH_2	460	H—OH	492	H—C_6H_5	469
H_3C—CH_3	368	H_2C=CH_2	720	HC≡CH	962		
HO—CH_3	377	Cl—CH_3	352	Br—CH_3	293	I—CH_3	237
O=CO	531	HO—OH	213	O_2N—NO_2	54		

Data: HCP, KL

Table 2.9 Mean bond enthalpies, $B(A-B)/(kJ\,mol^{-1})$

	H	C	N	O	F	Cl	Br	I	S	P	Si
H	436										
C	412	348(i)									
		612(ii)									
		838(iii)									
		518(a)									
N	388	305(i)	163(i)								
		613(ii)	409(ii)								
		890(iii)	946(iii)								
O	463	360(i)	157	146(i)							
		743(ii)		497(ii)							
F	565	484	270	185	155						
Cl	431	338	200	203	254	242					
Br	366	276				219	193				
I	299	238				210	178	151			
S	338	259			496	250	212		264		
P	322									201	
Si	318		374	466							226

(i) Single bond, (ii) double bond, (iii) triple bond, (a) aromatic.
Data: HCP and L. Pauling, *The nature of the chemical bond.* Cornell University Press (1960).

Table 2.10 Standard enthalpies of atomization. See Table 2.12.

Table 2.11 Thermodynamic data for organic compounds (all values relate to 298 K)

	$M/$ g mol^{-1}	$\Delta_f H^{\ominus}/$ kJ mol^{-1}	$\Delta_f G^{\ominus}/$ kJ mol^{-1}	$S^{\ominus}/$ J K^{-1} mol^{-1}	$C_{p,m}/$ J K^{-1} mol^{-1}	$\Delta_c H^{\ominus}/$ kJ mol^{-1}
C(s) (graphite)	12.011	0	0	5.740	8.527	−393.51
C(s) (diamond)	12.011	+1.895	+2.900	2.377	6.113	−395.40
CO_2(g)	44.010	−393.51	−394.36	213.74	37.11	
Hydrocarbons						
CH_4(g), methane	16.04	−74.81	−50.72	186.26	35.31	−890
CH_3(g), methyl	15.04	+145.69	+147.92	194.2	38.70	
C_2H_2(g), ethyne	26.04	+226.73	+209.20	200.94	43.93	−1300
C_2H_4(g), ethene	28.05	+52.26	+68.15	219.56	43.56	−1411
C_2H_6(g), ethane	30.07	−84.68	−32.82	229.60	52.63	−1560
C_3H_6(g), propene	42.08	+20.42	+62.78	267.05	63.89	−2058
C_3H_6(g), cyclopropane	42.08	+53.30	+104.45	237.55	55.94	−2091
C_3H_8(g), propane	44.10	−103.85	−23.49	269.91	73.5	−2220
C_4H_8(g), 1-butene	56.11	−0.13	+71.39	305.71	85.65	−2717
C_4H_8(g), cis-2-butene	56.11	−6.99	+65.95	300.94	78.91	−2710
C_4H_8(g), trans-2-butene	56.11	−11.17	+63.06	296.59	87.82	−2707
C_4H_{10}(g), butane	58.13	−126.15	−17.03	310.23	97.45	−2878
C_5H_{12}(g), pentane	72.15	−146.44	−8.20	348.40	120.2	−3537
C_5H_{12}(l)	72.15	−173.1				
C_6H_6(l), benzene	78.12	+49.0	+124.3	173.3	136.1	−3268
C_6H_6(g)	78.12	+82.93	+129.72	269.31	81.67	−3302
C_6H_{12}(l), cyclohexane	84.16	−156	+26.8		156.5	−3902
C_6H_{14}(l), hexane	86.18	−198.7		204.3		−4163
$C_6H_5CH_3$(g), methylbenzene (toluene)	92.14	+50.0	+122.0	320.7	103.6	−3953
C_7H_{16}(l), heptane	100.21	−224.4	+1.0	328.6	224.3	
C_8H_{18}(l), octane	114.23	−249.9	+6.4	361.1		−5471
C_8H_{18}(l), iso-octane	114.23	−255.1				−5461
$C_{10}H_8$(s), naphthalene	128.18	+78.53				−5157
Alcohols and phenols						
CH_3OH(l), methanol	32.04	−238.66	−166.27	126.8	81.6	−726
CH_3OH(g)	32.04	−200.66	−161.96	239.81	43.89	−764
C_2H_5OH(l), ethanol	46.07	−277.69	−174.78	160.7	111.46	−1368
C_2H_5OH(g)	46.07	−235.10	−168.49	282.70	65.44	−1409
C_6H_5OH(s), phenol	94.12	−165.0	−50.9	146.0		−3054
Carboxylic acids, hydroxy acids and esters						
HCOOH(l), formic	46.03	−424.72	−361.35	128.95	99.04	−255
CH_3COOH(l), acetic	60.05	−484.5	−389.9	159.8	124.3	−875
CH_3COOH(aq)	60.05	−485.76	−396.46	178.7		
$CH_3CO_2^-$(aq)	59.05	−486.01	−369.31	86.6	−6.3	
$(COOH)_2$(s), oxalic	90.04	−827.2			117	−254
C_6H_5COOH(s), benzoic	122.13	−385.1	−245.3	167.6	146.8	−3227
$CH_3CH(OH)COOH$(s), lactic	90.08	−694.0				−1344
$CH_3COOC_2H_5$(l), ethyl acetate	88.11	−479.0	−332.7	259.4	170.1	−2231
Alkanals and alkanones						
HCHO(g), methanal	30.03	−108.57	−102.53	218.77	35.40	−571
CH_3CHO(l), ethanal	44.05	−192.30	−128.12	160.2		−1166
CH_3CHO(g)	44.05	−166.19	−128.86	250.3	57.3	−1192
CH_3COCH_3(l), propanone	58.08	−248.1	−155.4	200.4	124.7	−1790

Table 2.11 (Continued)

	$M/$ g mol^{-1}	$\Delta_f H^{\ominus}/$ kJ mol^{-1}	$\Delta_f G^{\ominus}/$ kJ mol^{-1}	$S^{\ominus}/$ J K^{-1} mol^{-1}	$C_{p,m}/$ J K^{-1} mol^{-1}	$\Delta_c H^{\ominus}/$ kJ mol^{-1}
Sugars						
$C_6H_{12}O_6$(s), α-D-glucose	180.16	−1274				−2808
$C_6H_{12}O_6$(s), β-D-glucose	180.16	−1268	−910	212		
$C_6H_{12}O_6$(s), β-D-fructose	180.16	−1266				−2810
$C_{12}H_{22}O_{11}$(s), sucrose	342.30	−2222	−1543	360.2		−5645
Nitrogen compounds						
$CO(NH_2)_2$(s), urea	60.06	−333.51	−197.33	104.60	93.14	−632
CH_3NH_2(g), methylamine	31.06	−22.97	+32.16	243.41	53.1	−1085
$C_6H_5NH_2$(l), aniline	93.13	+31.1				−3393
$CH_2(NH_2)COOH$(s), glycine	75.07	−532.9	−373.4	103.5	99.2	−969

Data: NBS, TDOC

Table 2.12 Thermodynamic data (all values relate to 298 K)

	$M/$ g mol^{-1}	$\Delta_f H^{\ominus}/$ kJ mol^{-1}	$\Delta_f G^{\ominus}/$ kJ mol^{-1}	$S^{\ominus}/$ J K^{-1} mol^{-1}	$C_{p,m}/$ J K^{-1} mol^{-1}
Aluminium (aluminum)					
Al(s)	26.98	0	0	28.33	24.35
Al(l)	26.98	+10.56	+7.20	39.55	24.21
Al(g)	26.98	+326.4	+285.7	164.54	21.38
Al^{3+}(g)	26.98	+5483.17			
Al^{3+}(aq)	26.98	−531	−485	−321.7	
Al$_2$O$_3$(s, α)	101.96	−1675.7	−1582.3	50.92	79.04
AlCl$_3$(s)	133.24	−704.2	−628.8	110.67	91.84
Argon					
Ar(g)	39.95	0	0	154.84	20.786
Antimony					
Sb(s)	121.75	0	0	45.69	25.23
SbH$_3$(g)	153.24	+145.11	+147.75	232.78	41.05
Arsenic					
As(s, α)	74.92	0	0	35.1	24.64
As(g)	74.92	+302.5	+261.0	174.21	20.79
As$_4$(g)	299.69	+143.9	+92.4	314	
AsH$_3$(g)	77.95	+66.44	+68.93	222.78	38.07
Barium					
Ba(s)	137.34	0	0	62.8	28.07
Ba(g)	137.34	+180	+146	170.24	20.79
Ba^{2+}(aq)	137.34	−537.64	−560.77	9.6	
BaO(s)	153.34	−553.5	−525.1	70.43	47.78
BaCl$_2$(s)	208.25	−858.6	−810.4	123.68	75.14
Beryllium					
Be(s)	9.01	0	0	9.50	16.44
Be(g)	9.01	+324.3	+286.6	136.27	20.79

Data section

Table 2.12 (Continued)

	M / g mol^{-1}	$\Delta_f H^{\ominus}$ / kJ mol^{-1}	$\Delta_f G^{\ominus}$ / kJ mol^{-1}	$S^{\ominus}$ / J K^{-1} mol^{-1}	$C_{p,m}$ / J K^{-1} mol^{-1}
Bismuth					
Bi(s)	208.98	0	0	56.74	25.52
Bi(g)	208.98	+207.1	+168.2	187.00	20.79
Bromine					
Br$_2$(l)	159.82	0	0	152.23	75.689
Br$_2$(g)	159.82	+30.907	+3.110	245.46	36.02
Br(g)	79.91	+111.88	+82.396	175.02	20.786
Br$^-$(g)	79.91	−219.07			
Br$^-$(aq)	79.91	−121.55	−103.96	82.4	−141.8
HBr(g)	90.92	−36.40	−53.45	198.70	29.142
Cadium					
Cd(s, γ)	112.40	0	0	51.76	25.98
Cd(g)	112.40	+112.01	+77.41	167.75	20.79
Cd^{2+}(aq)	112.40	−75.90	−77.612	−73.2	
CdO(s)	128.40	−258.2	−228.4	54.8	43.43
CdCO$_3$(s)	172.41	−750.6	−669.4	92.5	
Caesium (cesium)					
Cs(s)	132.91	0	0	85.23	32.17
Cs(g)	132.91	+76.06	+49.12	175.60	20.79
Cs$^+$(aq)	132.91	−258.28	−292.02	133.05	−10.5
Calcium					
Ca(s)	40.08	0	0	41.42	25.31
Ca(g)	40.08	+178.2	+144.3	154.88	20.786
Ca^{2+}(aq)	40.08	−542.83	−553.58	−53.1	
CaO(s)	56.08	−635.09	−604.03	39.75	42.80
CaCO$_3$(s) (calcite)	100.09	−1206.9	−1128.8	92.9	81.88
CaCO$_3$(s) (aragonite)	100.09	−1207.1	−1127.8	88.7	81.25
CaF$_2$(s)	78.08	−1219.6	−1167.3	68.87	67.03
CaCl$_2$(s)	110.99	−795.8	−748.1	104.6	72.59
CaBr$_2$(s)	199.90	−682.8	−663.6	130	
Carbon (for 'organic' compounds of carbon, see Table 2.11)					
C(s) (graphite)	12.011	0	0	5.740	8.527
C(s) (diamond)	12.011	+1.895	+2.900	2.377	6.113
C(g)	12.011	+716.68	+671.26	158.10	20.838
C$_2$(g)	24.022	+831.90	+775.89	199.42	43.21
CO(g)	28.011	−110.53	−137.17	197.67	29.14
CO$_2$(g)	44.010	−393.51	−394.36	213.74	37.11
CO$_2$(aq)	44.010	−413.80	−385.98	117.6	
H$_2$CO$_3$(aq)	62.03	−699.65	−623.08	187.4	
HCO$_3^-$(aq)	61.02	−691.99	−586.77	91.2	
CO$_3^{2-}$(aq)	60.01	−677.14	−527.81	−56.9	
CCl$_4$(l)	153.82	−135.44	−65.21	216.40	131.75
CS$_2$(l)	76.14	+89.70	+65.27	151.34	75.7
HCN(g)	27.03	+135.1	+124.7	201.78	35.86
HCN(l)	27.03	+108.87	+124.97	112.84	70.63
CN$^-$(aq)	26.02	+150.6	+172.4	94.1	

Table 2.12 (Continued)

	$M/$ g mol^{-1}	$\Delta_f H^{\ominus}/$ kJ mol^{-1}	$\Delta_f G^{\ominus}/$ kJ mol^{-1}	$S^{\ominus}/$ $\text{J K}^{-1}\text{mol}^{-1}$	$C_{p,m}/$ $\text{J K}^{-1}\text{mol}^{-1}$
Chlorine					
$Cl_2(g)$	70.91	0	0	223.07	33.91
$Cl(g)$	35.45	+121.68	+105.68	165.20	21.840
$Cl^-(g)$	35.45	−233.13			
$Cl^-(aq)$	35.45	−167.16	−131.23	56.5	−136.4
$HCl(g)$	36.46	−92.31	−95.30	186.91	29.12
$HCl(aq)$	36.46	−167.16	−131.23	56.5	−136.4
Chromium					
$Cr(s)$	52.00	0	0	23.77	23.35
$Cr(g)$	52.00	+396.6	+351.8	174.50	20.79
$CrO_4^{2-}(aq)$	115.99	−881.15	−727.75	50.21	
$Cr_2O_7^{2-}(aq)$	215.99	−1490.3	−1301.1	261.9	
Copper					
$Cu(s)$	63.54	0	0	33.150	24.44
$Cu(g)$	63.54	+338.32	+298.58	166.38	20.79
$Cu^+(aq)$	63.54	+71.67	+49.98	40.6	
$Cu^{2+}(aq)$	63.54	+64.77	+65.49	−99.6	
$Cu_2O(s)$	143.08	−168.6	−146.0	93.14	63.64
$CuO(s)$	79.54	−157.3	−129.7	42.63	42.30
$CuSO_4(s)$	159.60	−771.36	−661.8	109	100.0
$CuSO_4{\cdot}H_2O(s)$	177.62	−1085.8	−918.11	146.0	134
$CuSO_4{\cdot}5H_2O(s)$	249.68	−2279.7	−1879.7	300.4	280
Deuterium					
$D_2(g)$	4.028	0	0	144.96	29.20
$HD(g)$	3.022	+0.318	−1.464	143.80	29.196
$D_2O(g)$	20.028	−249.20	−234.54	198.34	34.27
$D_2O(l)$	20.028	−294.60	−243.44	75.94	84.35
$HDO(g)$	19.022	−245.30	−233.11	199.51	33.81
$HDO(l)$	19.022	−289.89	−241.86	79.29	
Fluorine					
$F_2(g)$	38.00	0	0	202.78	31.30
$F(g)$	19.00	+78.99	+61.91	158.75	22.74
$F^-(aq)$	19.00	−332.63	−278.79	−13.8	−106.7
$HF(g)$	20.01	−271.1	−273.2	173.78	29.13
Gold					
$Au(s)$	196.97	0	0	47.40	25.42
$Au(g)$	196.97	+366.1	+326.3	180.50	20.79
Helium					
$He(g)$	4.003	0	0	126.15	20.786
Hydrogen (see also deuterium)					
$H_2(g)$	2.016	0	0	130.684	28.824
$H(g)$	1.008	+217.97	+203.25	114.71	20.784
$H^+(aq)$	1.008	0	0	0	0

Table 2.12 (Continued)

	$M/$ g mol^{-1}	$\Delta_f H^{\ominus}/$ kJ mol^{-1}	$\Delta_f G^{\ominus}/$ kJ mol^{-1}	$S^{\ominus}/$ $\text{J K}^{-1}\text{mol}^{-1}$	$C_{p,m}/$ $\text{J K}^{-1}\text{mol}^{-1}$
Hydrogen (continued)					
$H_2O(l)$	18.015	−285.83	−237.13	69.91	75.291
$H_2O(g)$	18.015	−241.82	−228.57	188.83	33.58
$H_2O_2(l)$	34.015	−187.78	−120.35	109.6	89.1
Iodine					
$I_2(s)$	253.81	0	0	116.135	54.44
$I_2(g)$	253.81	+62.44	+19.33	260.69	36.90
$I(g)$	126.90	+106.84	+70.25	180.79	20.786
$I^-(aq)$	126.90	−55.19	−51.57	111.3	−142.3
$HI(g)$	127.91	+26.48	+1.70	206.59	29.158
Iron					
$Fe(s)$	55.85	0	0	27.28	25.10
$Fe(g)$	55.85	+416.3	+370.7	180.49	25.68
$Fe^{2+}(aq)$	55.85	−89.1	−78.90	−137.7	
$Fe^{3+}(aq)$	55.85	−48.5	−4.7	−315.9	
$Fe_3O_4(s)$ (magnetite)	231.54	−1118.4	−1015.4	146.4	143.43
$Fe_2O_3(s)$ (haematite)	159.69	−824.2	−742.2	87.40	103.85
$FeS(s, \alpha)$	87.91	−100.0	−100.4	60.29	50.54
$FeS_2(s)$	119.98	−178.2	−166.9	52.93	62.17
Krypton					
$Kr(g)$	83.80	0	0	164.08	20.786
Lead					
$Pb(s)$	207.19	0	0	64.81	26.44
$Pb(g)$	207.19	+195.0	+161.9	175.37	20.79
$Pb^{2+}(aq)$	207.19	−1.7	−24.43	10.5	
$PbO(s, \text{yellow})$	223.19	−217.32	−187.89	68.70	45.77
$PbO(s, \text{red})$	223.19	−218.99	−188.93	66.5	45.81
$PbO_2(s)$	239.19	−277.4	−217.33	68.6	64.64
Lithium					
$Li(s)$	6.94	0	0	29.12	24.77
$Li(g)$	6.94	+159.37	+126.66	138.77	20.79
$Li^+(aq)$	6.94	−278.49	−293.31	13.4	68.6
Magnesium					
$Mg(s)$	24.31	0	0	32.68	24.89
$Mg(g)$	24.31	+147.70	+113.10	148.65	20.786
$Mg^{2+}(aq)$	24.31	−466.85	−454.8	−138.1	
$MgO(s)$	40.31	−601.70	−569.43	26.94	37.15
$MgCO_3(s)$	84.32	−1095.8	−1012.1	65.7	75.52
$MgCl_2(s)$	95.22	−641.32	−591.79	89.62	71.38
Mercury					
$Hg(l)$	200.59	0	0	76.02	27.983
$Hg(g)$	200.59	+61.32	+31.82	174.96	20.786
$Hg^{2+}(aq)$	200.59	+171.1	+164.40	−32.2	
$Hg_2^{2+}(aq)$	401.18	+172.4	+153.52	84.5	
$HgO(s)$	216.59	−90.83	−58.54	70.29	44.06

Table 2.12 (Continued)

	$M/$ g mol^{-1}	$\Delta_f H^{\ominus}/$ kJ mol^{-1}	$\Delta_f G^{\ominus}/$ kJ mol^{-1}	$S^{\ominus}/$ J K^{-1} mol^{-1}	$C_{p,m}/$ J K^{-1} mol^{-1}
Mercury (Continued)					
$Hg_2Cl_2(s)$	472.09	−265.22	−210.75	192.5	102
$HgCl_2(s)$	271.50	−224.3	−178.6	146.0	
$HgS(s, black)$	232.65	−53.6	−47.7	88.3	
Neon					
$Ne(g)$	20.18	0	0	146.33	20.786
Nitrogen					
$N_2(g)$	28.013	0	0	191.61	29.125
$N(g)$	14.007	+472.70	+455.56	153.30	20.786
$NO(g)$	30.01	+90.25	+86.55	210.76	29.844
$N_2O(g)$	44.01	+82.05	+104.20	219.85	38.45
$NO_2(g)$	46.01	+33.18	+51.31	240.06	37.20
$N_2O_4(g)$	92.01	+9.16	+97.89	304.29	77.28
$N_2O_5(s)$	108.01	−43.1	+113.9	178.2	143.1
$N_2O_5(g)$	108.01	+11.3	+115.1	355.7	84.5
$HNO_3(l)$	63.01	−174.10	−80.71	155.60	109.87
$HNO_3(aq)$	63.01	−207.36	−111.25	146.4	−86.6
$NO_3^-(aq)$	62.01	−205.0	−108.74	146.4	−86.6
$NH_3(g)$	17.03	−46.11	−16.45	192.45	35.06
$NH_3(aq)$	17.03	−80.29	−26.50	111.3	
$NH_4^+(aq)$	18.04	−132.51	−79.31	113.4	79.9
$NH_2OH(s)$	33.03	−114.2			
$HN_3(l)$	43.03	+264.0	+327.3	140.6	43.68
$HN_3(g)$	43.03	+294.1	+328.1	238.97	98.87
$N_2H_4(l)$	32.05	+50.63	+149.43	121.21	139.3
$NH_4NO_3(s)$	80.04	−365.56	−183.87	151.08	84.1
$NH_4Cl(s)$	53.49	−314.43	−202.87	94.6	
Oxygen					
$O_2(g)$	31.999	0	0	205.138	29.355
$O(g)$	15.999	+249.17	+231.73	161.06	21.912
$O_3(g)$	47.998	+142.7	+163.2	238.93	39.20
$OH^-(aq)$	17.007	−229.99	−157.24	−10.75	−148.5
Phosphorus					
$P(s, wh)$	30.97	0	0	41.09	23.840
$P(g)$	30.97	+314.64	+278.25	163.19	20.786
$P_2(g)$	61.95	+144.3	+103.7	218.13	32.05
$P_4(g)$	123.90	+58.91	+24.44	279.98	67.15
$PH_3(g)$	34.00	+5.4	+13.4	210.23	37.11
$PCl_3(g)$	137.33	−287.0	−267.8	311.78	71.84
$PCl_3(l)$	137.33	−319.7	−272.3	217.1	
$PCl_5(g)$	208.24	−374.9	−305.0	364.6	112.8
$PCl_5(s)$	208.24	−443.5			
$H_3PO_3(s)$	82.00	−964.4			
$H_3PO_3(aq)$	82.00	−964.8			
$H_3PO_4(s)$	94.97	−1279.0	−1119.1	110.50	106.06
$H_3PO_4(l)$	94.97	−1266.9			
$H_3PO_4(aq)$	94.97	−1277.4	−1018.7	−222	
$PO_4^{3-}(aq)$	94.97	−1277.4	−1018.7	−222	
$P_4O_{10}(s)$	283.89	−2984.0	−2697.0	228.86	211.71
$P_4O_6(s)$	219.89	−1640.1			

Data section

Table 2.12 (Continued)

	$M/$ g mol^{-1}	$\Delta_f H^{\ominus}/$ kJ mol^{-1}	$\Delta_f G^{\ominus}/$ kJ mol^{-1}	$S^{\ominus}/$ J K^{-1} mol^{-1}	$C_{p,m}/$ J K^{-1} mol^{-1}
Potassium					
K(s)	39.10	0	0	64.18	29.58
K(g)	39.10	+89.24	+60.59	160.336	20.786
K$^+$(g)	39.10	+514.26			
K$^+$(aq)	39.10	−252.38	−283.27	102.5	21.8
KOH(s)	56.11	−424.76	−379.08	78.9	64.9
KF(s)	58.10	−576.27	−537.75	66.57	49.04
KCl(s)	74.56	−436.75	−409.14	82.59	51.30
KBr(s)	119.01	−393.80	−380.66	95.90	52.30
KI(s)	166.01	−327.90	−324.89	106.32	52.93
Silicon					
Si(s)	28.09	0	0	18.83	20.00
Si(g)	28.09	+455.6	+411.3	167.97	22.25
SiO$_2$(s, α)	60.09	−910.94	−856.64	41.84	44.43
Silver					
Ag(s)	107.87	0	0	42.55	25.351
Ag(g)	107.87	+284.55	+245.65	173.00	20.79
Ag$^+$(aq)	107.87	+105.58	+77.11	72.68	21.8
AgBr(s)	187.78	−100.37	−96.90	107.1	52.38
AgCl(s)	143.32	−127.07	−109.79	96.2	50.79
Ag$_2$O(s)	231.74	−31.05	−11.20	121.3	65.86
AgNO$_3$(s)	169.88	−124.39	−33.41	140.92	93.05
Sodium					
Na(s)	22.99	0	0	51.21	28.24
Na(g)	22.99	+107.32	+76.76	153.71	20.79
Na$^+$(aq)	22.99	−240.12	−261.91	59.0	46.4
NaOH(s)	40.00	−425.61	−379.49	64.46	59.54
NaCl(s)	58.44	−411.15	−384.14	72.13	50.50
NaBr(s)	102.90	−361.06	−348.98	86.82	51.38
NaI(s)	149.89	−287.78	−286.06	98.53	52.09
Sulfur					
S(s, α) (rhombic)	32.06	0	0	31.80	22.64
S(s, β) (monoclinic)	32.06	+0.33	+0.1	32.6	23.6
S(g)	32.06	+278.81	+238.25	167.82	23.673
S$_2$(g)	64.13	+128.37	+79.30	228.18	32.47
S^{2-}(aq)	32.06	+33.1	+85.8	−14.6	
SO$_2$(g)	64.06	−296.83	−300.19	248.22	39.87
SO$_3$(g)	80.06	−395.72	−371.06	256.76	50.67
H$_2$SO$_4$(l)	98.08	−813.99	−690.00	156.90	138.9
H$_2$SO$_4$(aq)	98.08	−909.27	−744.53	20.1	−293
SO$_4^{2-}$(aq)	96.06	−909.27	−744.53	20.1	−293
HSO$_4^-$(aq)	97.07	−887.34	−755.91	131.8	−84
H$_2$S(g)	34.08	−20.63	−33.56	205.79	34.23
H$_2$S(aq)	34.08	−39.7	−27.83	121	
HS$^-$(aq)	33.072	−17.6	+12.08	62.08	
SF$_6$(g)	146.05	−1209	−1105.3	291.82	97.28
Tin					
Sn(s, β)	118.69	0	0	51.55	26.99
Sn(g)	118.69	+302.1	+267.3	168.49	20.26

Table 2.12 (Continued)

	$M/$ g mol^{-1}	$\Delta_f H^\ominus/$ kJ mol^{-1}	$\Delta_f G^\ominus/$ kJ mol^{-1}	$S^\ominus/$ J K^{-1} mol^{-1}	$C_{p,m}/$ J K^{-1} mol^{-1}
Tin (Continued)					
$Sn^{2+}(aq)$	118.69	-8.8	-27.2	-17	
SnO(s)	134.69	-285.8	-256.9	56.5	44.31
SnO_2(s)	150.69	-580.7	-519.6	52.3	52.59
Xenon					
Xe(g)	131.30	0	0	169.68	20.786
Zinc					
Zn(s)	65.37	0	0	41.63	25.40
Zn(g)	65.37	$+130.73$	$+95.14$	160.98	20.79
$Zn^{2+}(aq)$	65.37	-153.89	-147.06	-112.1	46
ZnO(s)	81.37	-348.28	-318.30	43.64	40.25

Source: NBS

Table 2.13 Lattice enthalpies, $\Delta H_L^\ominus/(\text{kJ mol}^{-1})$

Halides

	F	Cl	Br	I
Li	1037	852	815	761
Na	926	787	752	705
K	821	717	689	649
Rb	789	695	668	632
Cs	750	676	654	620
Ag	969	912	900	886
Be		3017		
Mg		2524		
Ca		2255		
Sr		2153		

Oxides

MgO 3850   CaO 3461   SrO 3283   BaO 3114

Sulfides

MgS 3406   CaS 3119   SrS 2974   BaS 2832

Data: Principally D. Cubicciotti, *J. chem. Phys.*, **31**, 1646 (1959).

Table 2.14 Standard molar enthalpies of hydration at infinite dilution, $\Delta_{hyd}H^\ominus/(\text{kJ mol}^{-1})$

	Li^+	Na^+	K^+	Rb^+	Cs^+
F^-	-1026	-911	-828	-806	-782
Cl^-	-884	-783	-685	-664	-640
Br^-	-856	-742	-658	-637	-613
I^-	-815	-701	-617	-596	-572

Entries refer to $X^+(g) + Y^-(g) \rightarrow X^+(aq) + Y^-(aq)$.
Data: Principally J. O'M. Bockris and A. K. N. Reddy, *Modern electrochemistry*, Vol. 1. Plenum Press, New York (1970).

Table 2.15 Limiting enthalpies of formation of ions in aqueous solution, $\Delta_f H^\ominus/(\text{kJ mol}^{-1})$. See Table 2.12.

Table 2.16 Standard molar ion hydration enthalpies, $\Delta_{hyd}H^{\ominus}/(kJ\ mol^{-1})$ at 298 K

Cations

H^+	(-1090)	Ag^+	-464	Mg^{2+}	-1920
Li^+	-520	NH_4^+	-301	Ca^{2+}	-1650
Na^+	-405			Sr^{2+}	-1480
K^+	-321			Ba^{2+}	-1360
Rb^+	-300			Fe^{2+}	-1950
Cs^+	-277			Cu^{2+}	-2100
				Zn^{2+}	-2050
				Al^{3+}	-4690
				Fe^{3+}	-4430

Anions

OH^-	-460						
F^-	-506	Cl^-	-364	Br^-	-337	I^-	-296

Entries refer to $X^{\pm}(g) \rightarrow X^{\pm}(aq)$ based on $H^+(g) \rightarrow H^+(aq)$ $\Delta H^{\ominus} = -1090\ kJ\ mol^{-1}$.

Data: Principally J. O'M. Bockris and A. K. N. Reddy, *Modern electrochemistry*, Vol. 1. Plenum Press, New York (1970).

Table 3.1 Expansion coefficients α and isothermal compressibilities κ_T

	$\alpha/(10^{-4}\ K^{-1})$	$\kappa_T/(10^{-6}\ atm^{-1})$
Liquids		
Benzene	12.4	92.1
Carbon tetrachloride	12.4	90.5
Ethanol	11.2	76.8
Mercury	1.82	38.7
Water	2.1	49.6
Solids		
Copper	0.501	0.735
Diamond	0.030	0.187
Iron	0.354	0.597
Lead	0.861	2.21

The values refer to 20°C
Data: AIP(α), KL(κ_T).

Table 3.2 Inversion temperatures, normal freezing and boiling points, and Joule–Thomson coefficients at 1 atm and 298 K

	T_i/K	T_f/K	T_b/K	$\mu_{JT}/(K\ atm^{-1})$
Air	603			0.189 at 50°C
Argon	723	83.8	87.3	
Carbon dioxide	1500	194.7s		1.11 at 300 K
Helium	40		4.22	-0.062
Hydrogen	202	14.0	20.3	-0.03
Krypton	1090	116.6	120.8	
Methane	968	90.6	111.6	
Neon	231	24.5	27.1	
Nitrogen	621	63.3	77.4	0.27
Oxygen	764	54.8	90.2	0.31

s: sublimes
Data: AIP, JL, and M. W. Zemansky, *Heat and thermodynamics*. McGraw-Hill, New York (1957).

Table 4.1 Entropies (and temperatures) of phase transitions at 1 atm pressure, $\Delta_{trs}S/(J\,K^{-1}\,mol^{-1})$

	Fusion (at T_f)	Vaporization (at T_b)
Ar	14.17 (at 83.8 K)	74.53 (at 87.3 K)
Br_2	39.76 (at 265.9 K)	88.61 (at 332.4 K)
C_6H_6	38.00 (at 278.6 K)	87.19 (at 353.2 K)
CH_3COOH	40.4 (at 289.8 K)	61.9 (at 391.4 K)
CH_3OH	18.03 (at 175.2 K)	104.6 (at 337.2 K)
Cl_2	37.22 (at 172.1 K)	85.38 (at 239.0 K)
H_2	8.38 (at 14.0 K)	44.96 (at 20.38 K)
H_2O	22.00 (at 273.2 K)	109.0 (at 373.2 K)
H_2S	12.67 (at 187.6 K)	87.75 (at 212.0 K)
He	4.8 (at 1.8 K and 30 bar)	19.9 (at 4.22 K)
N_2	11.39 (at 63.2 K)	75.22 (at 77.4 K)
NH_3	28.93 (at 195.4 K)	97.41 (at 239.73 K)
O_2	8.17 (at 54.4 K)	75.63 (at 90.2 K)

Data: AIP

Table 4.2 Standard molar entropies of vaporization of liquids

	$\Delta_{vap}H^{\ominus}/(kJ\,mol^{-1})$	$\theta_b/°C$	$\Delta_{vap}S^{\ominus}/(J\,K^{-1}\,mol^{-1})$
Benzene	+30.8	80.1	+87.2
Carbon tetrachloride	+30.00	76.7	+85.8
Cyclohexane	+30.1	80.7	+85.1
Hydrogen sulfide	+18.7	−60.4	+87.9
Methane	+8.18	−161.5	+73.2
Water	+40.7	100.0	+109.1

Table 4.3 Standard Third-Law entropies at 298 K: see Tables 2.11 and 2.12

Table 4.4 Standard Gibbs energies of formation at 298 K: see Tables 2.11 and 2.12

Table 5.2 The fugacity coefficient of nitrogen at 273 K

p/atm	ϕ	p/atm	ϕ
1	0.99955	300	1.0055
10	0.9956	400	1.062
50	0.9812	600	1.239
100	0.9703	800	1.495
150	0.9672	1000	1.839
200	0.9721		

Data: LR

Table 7.1 Henry's law constants for gases at 298 K, K/Torr

	Water	Benzene
CH_4	3.14×10^5	4.27×10^5
CO_2	1.25×10^6	8.57×10^4
H_2	5.34×10^7	2.75×10^6
N_2	6.51×10^7	1.79×10^6
O_2	3.30×10^7	

Data: F. Daniels and R. A. Alberty, *Physical chemistry*. Wiley, New York (1980).

Table 7.2 Cryoscopic and ebullioscopic constants

	$K_f/(K\,kg\,mol^{-1})$	$K_b/(K\,kg\,mol^{-1})$
Acetic acid	3.90	3.07
Benzene	5.12	2.53
Camphor	40	
Carbon disulfide	3.8	2.37
Carbon tetrachloride	30	4.95
Naphthalene	6.94	5.8
Phenol	7.27	3.04
Water	1.86	0.51

Data: KL

Table 9.1 Acidity constants for aqueous solution at 298 K. (a) In order of acid strength

Acid	HA	A^-	K_a	pK_a
Hydroiodic	HI	I^-	10^{11}	-11
Perchloric	$HClO_4$	ClO_4^-	10^{10}	-10
Hydrobromic	HBr	Br^-	10^9	-9
Hydrochloric	HCl	Cl^-	10^7	-7
Sulfuric	H_2SO_4	HSO_4^-	10^2	-2
Hydronium ion	H_3O^+	H_2O	1	0.0
Oxalic	$(COOH)_2$	$HOOCCO_2^-$	5.9×10^{-2}	1.23
Sulfurous	H_2SO_3	HSO_3^-	1.5×10^{-2}	1.81
Hydrogensulfate ion	HSO_4^-	SO_4^{2-}	1.2×10^{-2}	1.92
Phosphoric	H_3PO_4	$H_2PO_4^-$	7.5×10^{-3}	2.12
Hydrofluoric	HF	F^-	3.5×10^{-4}	3.45
Formic	HCOOH	HCO_2^-	1.8×10^{-4}	3.75
Lactic	$CH_3CH(OH)COOH$	$CH_3CH(OH)CO_2^-$	1.4×10^{-4}	3.86
Hydrogenoxalate ion	$HOOCCO_2^-$	$(CO_2)_2^{2-}$	6.5×10^{-5}	4.19
Anilinium ion	$C_6H_5NH_3^+$	$C_6H_5NH_2$	2.3×10^{-5}	4.63
Acetic (ethanoic)	CH_3COOH	$CH_3CO_2^-$	1.8×10^{-5}	4.75
Butanoic	C_3H_7COOH	$C_3H_7CO_2^-$	1.5×10^{-5}	4.82
Propanoic	C_2H_5COOH	$C_2H_5CO_2^-$	1.4×10^{-5}	4.87
Pyridinium ion	$HC_5H_5N^+$	C_5H_5N	5.6×10^{-6}	5.25
Carbonic	H_2CO_3	HCO_3^-	4.3×10^{-7}	6.37
Hydrogen sulfide	H_2S	HS^-	9.1×10^{-8}	7.04
Dihydrogenphosphate ion	$H_2PO_4^-$	HPO_4^{2-}	6.2×10^{-8}	7.21
Hypochlorous	HClO	ClO^-	3.0×10^{-8}	7.53
Hydrazinium ion	$NH_2NH_3^+$	NH_2NH_2	5.9×10^{-9}	8.23
Hypobromous	HBrO	BrO^-	2.0×10^{-9}	8.69
Boric	$B(OH)_3$	$B(OH)_4^-$	7.2×10^{-10}	9.14
Ammonium ion	NH_4^+	NH_3	5.6×10^{-10}	9.25
Hydrogen cyanide	HCN	CN^-	4.9×10^{-10}	9.31
Glycinium ion	NH_2CH_2COOH	$NH_2CH_2CO_2^-$	1.7×10^{-10}	9.78
Trimethylammonium ion	$(CH_3)_3NH^+$	$(CH_3)_3N$	1.6×10^{-10}	9.81
Phenol	C_6H_5OH	$C_6H_5O^-$	1.3×10^{-10}	9.89
Hydrogencarbonate ion	HCO_3^-	CO_3^{2-}	5.6×10^{-11}	10.25
Hypoiodous	HIO	IO^-	2.3×10^{-11}	10.64
Methylammonium ion	$CH_3NH_3^+$	CH_3NH_2	2.2×10^{-11}	10.66
Dimethylammonium ion	$(CH_3)_2NH_2^+$	$(CH_3)_2NH$	1.9×10^{-11}	10.73
Triethylammonium ion	$(C_2H_5)_3NH^+$	$(C_2H_5)_3N$	1.7×10^{-11}	10.76
Ethylammonium ion	$C_2H_5NH_3^+$	$C_2H_5NH_2$	1.6×10^{-11}	10.81
Diethylammonium ion	$(C_2H_5)_2NH_2^+$	$(C_2H_5)_2NH$	1.0×10^{-11}	10.99
Hydrogenarsenate ion	$HAsO_4^{2-}$	AsO_4^{3-}	3.0×10^{-12}	11.53
Hydrogensulfide ion	HS^-	S^{2-}	1.1×10^{-12}	11.96
Hydrogenphosphate ion	HPO_4^{2-}	PO_4^{3-}	2.2×10^{-13}	12.67

Data: Principally HCP

Table 9.1 Acidity constants for aqueous solution at 298 K. (b) In alphabetical order of acid

Acid	HA	A^-	K_a	pK_a
Acetic	CH_3COOH	$CH_3CO_2^-$	1.8×10^{-5}	4.75
Ammonium ion	NH_4^+	NH_3	5.6×10^{-10}	9.25
Anilinium ion	$C_6H_5NH_3^+$	$C_6H_5NH_2$	2.3×10^{-5}	4.63
Boric	$B(OH)_3$	$B(OH)_4^-$	7.2×10^{-10}	9.14
Butanoic	C_3H_7COOH	$C_3H_7CO_2^-$	1.5×10^{-5}	4.82
Carbonic	H_2CO_3	HCO_3^-	4.3×10^{-7}	6.37
Diethylammonium ion	$(C_2H_5)_2NH_2^+$	$(C_2H_5)_2NH$	1.0×10^{-11}	10.99
Dihydrogenphosphate ion	$H_2PO_4^-$	HPO_4^{2-}	6.2×10^{-8}	7.21
Dimethylammonium ion	$(CH_3)_2NH_2^+$	$(CH_3)_2NH$	1.9×10^{-11}	10.73
Ethylammonium ion	$C_2H_5NH_3^+$	$C_2H_5NH_2$	1.6×10^{-11}	10.81
Formic	HCOOH	HCO_2^-	1.8×10^{-4}	3.75
Glycinium ion	NH_2CH_2COOH	$NH_2CH_2CO_2^-$	1.7×10^{-10}	9.78
Hydrazinium ion	$NH_2NH_3^+$	NH_2NH_2	5.9×10^{-9}	8.23

Table 9.1 (Continued)

Acid	HA	A^-	K_a		pK_a
Hydroiodic	HI	I^-	10^{11}	-11	
Hydrobromic	HBr	Br^-	10^9	-9	
Hydrochloric	HCl	Cl^-	10^7	-7	
Hydrofluoric	HF	F^-	3.5×10^{-4}	3.45	
Hydrogenarsenate ion	$HAsO_4^{2-}$	AsO_4^{3-}	3.0×10^{-12}	11.53	
Hydrogencarbonate ion	HCO_3^-	CO_3^{2-}	4.8×10^{-11}	10.32	
Hydrogen cyanide	HCN	CN^-	4.9×10^{-10}	9.31	
Hydrogenoxalate ion	$HOOCCO_2^-$	$(CO_2)_2^{2-}$	6.5×10^{-5}	4.19	
Hydrogenphosphate ion	HPO_4^{2-}	PO_4^{3-}	2.2×10^{-13}	12.67	
Hydrogensulfate ion	HSO_4^-	SO_4^{2-}	1.2×10^{-2}	1.92	
Hydrogen sulfide	H_2S	HS^-	9.1×10^{-8}	7.04	
Hydrogensulfide ion	HS^-	S^{2-}	1.1×10^{-12}	11.96	
Hydronium ion	H_3O^+	H_2O	1	0.0	
Hypobromous	HBrO	BrO^-	2.0×10^{-9}	8.69	
Hypochlorous	HClO	ClO^-	3.0×10^{-8}	7.53	
Hypoiodous	HIO	IO^-	2.3×10^{-11}	10.64	
Lactic	$CH_3CH(OH)COOH$	$CH_3CH(OH)CO_2^-$	1.4×10^{-4}	3.86	
Methylammonium ion	$CH_3NH_3^+$	CH_3NH_2	2.2×10^{-11}	10.66	
Oxalic	$(COOH)_2$	$HOOCCO_2^-$	5.9×10^{-2}	1.23	
Perchloric	$HClO_4$	ClO_4^-	10^{10}	-10	
Phenol	C_6H_5OH	$C_6H_5O^-$	1.3×10^{-10}	9.89	
Phosphoric	H_3PO_4	$H_2PO_4^-$	7.5×10^{-3}	2.12	
Propanoic	C_2H_5COOH	$C_2H_5CO_2^-$	1.4×10^{-5}	4.87	
Pyridinium ion	$HC_5H_5N^+$	C_5H_5N	5.6×10^{-6}	5.25	
Sulfuric	H_2SO_4	HSO_4^-	10^2	-2	
Sulfurous	H_2SO_3	HSO_3^-	1.5×10^{-2}	1.81	
Triethylammonium ion	$(C_2H_5)_3NH^+$	$(C_2H_5)_3N$	1.7×10^{-11}	10.76	
Trimethylammonium ion	$(CH_3)_3NH^+$	$(CH_3)_3N$	1.6×10^{-10}	9.81	

Table 10.1 Standard thermodynamic functions of ions in solution at 298 K. See Table 2.12

Table 10.2 Relative permittivities (dielectric constants) at 298 〈

Non-polar molecules		Polar molecules	
Methane	1.70	Water	78.54
(at $-173°C$)			80.37 at 20°C
Carbon tetrachloride	2.228	Ammonia	16.9
			22.4 at $-33°C$
Cyclohexane	2.015	Hydrogen sulfide	9.26 at $-85°C$
Benzene	2.274	Methanol	32.63
		Ethanol	24.30
		Nitrobenzene	34.82

Data: HCP

Table 10.4 Mean activity coefficients in water at 298 K

$m/m^\ominus$	HCl	KCl	$CaCl_2$	H_2SO_4	$LaCl_3$	$In_2(SO_4)_3$
0.001	0.966	0.966	0.888	0.830	0.790	
0.005	0.929	0.927	0.789	0.639	0.636	0.16
0.01	0.905	0.902	0.732	0.544	0.560	0.11
0.05	0.830	0.816	0.584	0.340	0.388	0.035
0.10	0.798	0.770	0.524	0.266	0.356	0.025
0.50	0.769	0.652	0.510	0.155	0.303	0.014
1.00	0.811	0.607	0.725	0.131	0.387	
2.00	1.011	0.577	1.554	0.125	0.954	

Data: RS, HCP, and S. Glasstone, *Introduction to electrochemistry*. Van Nostrand (1942).

Table 10.5 Standard potentials at 298 K. (a) In electrochemical order

Reduction half-reaction	$E^{\ominus}/V$	Reduction half-reaction	$E^{\ominus}/V$
Strongly oxidizing		$Bi^{3+} + 3e^- \rightarrow Bi$	+0.20
$H_4XeO_6 + 2H^+ + 2e^- \rightarrow XeO_3 + 3H_2O$	+3.0	$Cu^{2+} + e^- \rightarrow Cu^+$	+0.16
$F_2 + 2e^- \rightarrow 2F^-$	+2.87	$Sn^{4+} + 2e^- \rightarrow Sn^{2+}$	+0.15
$O_3 + 2H^+ + 2e^- \rightarrow O_2 + H_2O$	+2.07	$AgBr + e^- \rightarrow Ag + Br^-$	+0.07
$S_2O_8^{2-} + 2e^- \rightarrow 2SO_4^{2-}$	+2.05	$Ti^{4+} + e^- \rightarrow Ti^{3+}$	0.00
$Ag^{2+} + e^- \rightarrow Ag^+$	+1.98	$2H^+ + 2e^- \rightarrow H_2$	0, by definition
$Co^{3+} + e^- \rightarrow Co^{2+}$	+1.81	$Fe^{3+} + 3e^- \rightarrow Fe$	−0.04
$H_2O_2 + 2H^+ + 2e^- \rightarrow 2H_2O$	+1.78	$O_2 + H_2O + 2e^- \rightarrow HO_2^- + OH^-$	−0.08
$Au^+ + e^- \rightarrow Au$	+1.69	$Pb^{2+} + 2e^- \rightarrow Pb$	−0.13
$Pb^{4+} + 2e^- \rightarrow Pb^{2+}$	+1.67	$In^+ + e^- \rightarrow In$	−0.14
$2HClO + 2H^+ + 2e^- \rightarrow Cl_2 + 2H_2O$	+1.63	$Sn^{2+} + 2e^- \rightarrow Sn$	−0.14
$Cd^{2+} + 2e^- \rightarrow Cd$	−0.40	$AgI + e^- \rightarrow Ag + I^-$	−0.15
$Ce^{4+} + e^- \rightarrow Ce^{3+}$	+1.61	$Ni^{2+} + 2e^- \rightarrow Ni$	−0.23
$2HBrO + 2H^+ + 2e^- \rightarrow Br_2 + 2H_2O$	+1.60	$Co^{2+} + 2e^- \rightarrow Co$	−0.28
$MnO_4^- + 8H^+ + 5e^- \rightarrow Mn^{2+} + 4H_2O$	+1.51	$In^{3+} + 3e^- \rightarrow In$	−0.34
$Mn^{3+} + e^- \rightarrow Mn^{2+}$	+1.51	$Tl^+ + e^- \rightarrow Tl$	−0.34
$Au^{3+} + 3e^- \rightarrow Au$	+1.40	$PbSO_4 + 2e^- \rightarrow Pb + SO_4^{2-}$	−0.36
$Cl_2 + 2e^- \rightarrow 2Cl^-$	+1.36	$Ti^{3+} + e^- \rightarrow Ti^{2+}$	−0.37
$Cr_2O_7^{2-} + 14H^+ + 6e^- \rightarrow 2Cr^{3+} + 7H_2O$	+1.33	$In^{2+} + e^- \rightarrow In^+$	−0.40
$O_3 + H_2O + 2e^- \rightarrow O_2 + 2OH^-$	+1.24	$Cr^{3+} + e^- \rightarrow Cr^{2+}$	−0.41
$O_2 + 4H^+ + 4e^- \rightarrow 2H_2O$	+1.23	$Fe^{2+} + 2e^- \rightarrow Fe$	−0.44
$ClO_4^- + 2H^+ + 2e^- \rightarrow ClO_3^- + H_2O$	+1.23	$In^{3+} + 2e^- \rightarrow In^+$	−0.44
$MnO_2 + 4H^+ + 2e^- \rightarrow Mn^{2+} + 2H_2O$	+1.23	$S + 2e^- \rightarrow S^{2-}$	−0.48
$Br_2 + 2e^- \rightarrow 2Br^-$	+1.09	$In^{3+} + e^- \rightarrow In^{2+}$	−0.49
$Pu^{4+} + e^- \rightarrow Pu^{3+}$	+0.97	$U^{4+} + e^- \rightarrow U^{3+}$	−0.61
$NO_3^- + 4H^+ + 3e^- \rightarrow NO + 2H_2O$	+0.96	$Cr^{3+} + 3e^- \rightarrow Cr$	−0.74
$2Hg^{2+} + 2e^- \rightarrow Hg_2^{2+}$	+0.92	$Zn^{2+} + 2e^- \rightarrow Zn$	−0.76
$ClO^- + H_2O + 2e^- \rightarrow Cl^- + 2OH^-$	+0.89	$Cd(OH)_2 + 2e^- \rightarrow Cd + 2OH^-$	−0.81
$Hg^{2+} + 2e^- \rightarrow Hg$	+0.86	$2H_2O + 2e^- \rightarrow H_2 + 2OH^-$	−0.83
$NO_3^- + 2H^+ + e^- \rightarrow NO_2 + H_2O$	+0.80	$Cr^{2+} + 2e^- \rightarrow Cr$	−0.91
$Ag^+ + e^- \rightarrow Ag$	+0.80	$Mn^{2+} + 2e^- \rightarrow Mn$	−1.18
$Hg_2^{2+} + 2e^- \rightarrow 2Hg$	+0.79	$V^{2+} + 2e^- \rightarrow V$	−1.19
$Fe^{3+} + e^- \rightarrow Fe^{2+}$	+0.77	$Ti^{2+} + 2e^- \rightarrow Ti$	−1.63
$BrO^- + H_2O + 2e^- \rightarrow Br^- + 2OH^-$	+0.76	$Al^{3+} + 3e^- \rightarrow Al$	−1.66
$Hg_2SO_4 + 2e^- \rightarrow 2Hg + SO_4^{2-}$	+0.62	$U^{3+} + 3e^- \rightarrow U$	−1.79
$MnO_4^{2-} + 2H_2O + 2e^- \rightarrow MnO_2 + 4OH^-$	+0.60	$Mg^{2+} + 2e^- \rightarrow Mg$	−2.36
$MnO_4^- + e^- \rightarrow MnO_4^{2-}$	+0.56	$Ce^{3+} + 3e^- \rightarrow Ce$	−2.48
$I_2 + 2e^- \rightarrow 2I^-$	+0.54	$La^{3+} + 3e^- \rightarrow La$	−2.52
$Cu^+ + e^- \rightarrow Cu$	+0.52	$Na^+ + e^- \rightarrow Na$	−2.71
$I_3^- + 2e^- \rightarrow 3I^-$	+0.53	$Ca^{2+} + 2e^- \rightarrow Ca$	−2.87
$NiOOH + H_2O + e^- \rightarrow Ni(OH)_2 + OH^-$	+0.49	$Sr^{2+} + 2e^- \rightarrow Sr$	−2.89
$Ag_2CrO_4 + 2e^- \rightarrow 2Ag + CrO_4^{2-}$	+0.45	$Ba^{2+} + 2e^- \rightarrow Ba$	−2.91
$O_2 + 2H_2O + 4e^- \rightarrow 4OH^-$	+0.40	$Ra^{2+} + 2e^- \rightarrow Ra$	−2.92
$ClO_4^- + H_2O + 2e^- \rightarrow ClO_3^- + 2OH^-$	+0.36	$Cs^+ + e^- \rightarrow Cs$	−2.92
$[Fe(CN)_6]^{3-} + e^- \rightarrow [Fe(CN)_6]^{4-}$	+0.36	$Rb^+ + e^- \rightarrow Rb$	−2.93
$Cu^{2+} + 2e^- \rightarrow Cu$	+0.34	$K^+ + e^- \rightarrow K$	−2.93
$Hg_2Cl_2 + 2e^- \rightarrow 2Hg + 2Cl^-$	+0.27	$Li^+ + e^- \rightarrow Li$	−3.05
$AgCl + e^- \rightarrow Ag + Cl^-$	+0.22	**Strongly reducing**	

Table 10.5 Standard electrode potentials at 298 K. (b) In alphabetical order

Reduction half-reaction	$E^{\ominus}/V$	Reduction half-reaction	$E^{\ominus}/V$
$Ag^+ + e^- \rightarrow Ag$	+0.80	$I_2 + 2e^- \rightarrow 2I^-$	+0.54
$Ag^{2+} + e^- \rightarrow Ag^+$	+1.98	$I_3^- + 2e^- \rightarrow 3I^-$	+0.53
$AgBr + e^- \rightarrow Ag + Br^-$	+0.0713	$In^+ + e^- \rightarrow In$	−0.14
$AgCl + e^- \rightarrow Ag + Cl^-$	+0.22	$In^{2+} + e^- \rightarrow In^+$	−0.40
$Ag_2CrO_4 + 2e^- \rightarrow 2Ag + CrO_4^{2-}$	+0.45	$In^{3+} + 2e^- \rightarrow In^+$	−0.44
$AgF + e^- \rightarrow Ag + F^-$	+0.78	$In^{3+} + 3e^- \rightarrow In$	−0.34
$AgI + e^- \rightarrow Ag + I^-$	−0.15	$In^{3+} + e^- \rightarrow In^{2+}$	−0.49
$Al^{3+} + 3e^- \rightarrow Al$	−1.66	$K^+ + e^- \rightarrow K$	−2.93
$Au^+ + e^- \rightarrow Au$	+1.69	$La^{3+} + 3e^- \rightarrow La$	−2.52
$Au^{3+} + 3e^- \rightarrow Au$	+1.40	$Li^+ + e^- \rightarrow Li$	−3.05
$Ba^{2+} + 2e^- \rightarrow Ba$	−2.91	$Mg^{2+} + 2e^- \rightarrow Mg$	−2.36
$Be^{2+} + 2e^- \rightarrow Be$	−1.85	$Mn^{2+} + 2e^- \rightarrow Mn$	−1.18
$Bi^{3+} + 3e^- \rightarrow Bi$	+0.20	$Mn^{3+} + e^- \rightarrow Mn^{2+}$	+1.51
$Br_2 + 2e^- \rightarrow 2Br^-$	+1.09	$MnO_2 + 4H^+ + 2e^- \rightarrow Mn^{2+} + 2H_2O$	+1.23
$BrO^- + H_2O + 2e^- \rightarrow Br^- + 2OH^-$	+0.76	$MnO_4^- + 8H^+ + 5e^- \rightarrow Mn^{2+} + 4H_2O$	+1.51
$Ca^{2+} + 2e^- \rightarrow Ca$	−2.87	$MnO_4^- + e^- \rightarrow MnO_4^{2-}$	+0.56
$Cd(OH)_2 + 2e^- \rightarrow Cd + 2OH^-$	−0.81	$MnO_4^{2-} + 2H_2O + 2e^- \rightarrow MnO_2 + 4OH^-$	+0.60
$Cd^{2+} + 2e^- \rightarrow Cd$	−0.40	$Na^+ + e^- \rightarrow Na$	−2.71
$Ce^{3+} + 3e^- \rightarrow Ce$	−2.48	$Ni^{2+} + 2e^- \rightarrow Ni$	−0.23
$Ce^{4+} + e^- \rightarrow Ce^{3+}$	+1.61	$NiOOH + H_2O + e^- \rightarrow Ni(OH)_2 + OH^-$	+0.49
$Cl_2 + 2e^- \rightarrow 2Cl^-$	+1.36	$NO_3^- + 2H^+ + e^- \rightarrow NO_2 + H_2O$	+0.80
$ClO^- + H_2O + 2e^- \rightarrow Cl^- + 2OH^-$	+0.89	$NO_3^- + 4H^+ + 3e^- \rightarrow NO + 2H_2O$	+0.96
$ClO_4^- + 2H^+ + 2e^- \rightarrow ClO_3^- + H_2O$	+1.23	$NO_3^- + H_2O + 2e^- \rightarrow NO_2^- + 2OH^-$	+0.10
$ClO_4^- + H_2O + 2e^- \rightarrow ClO_3^- + 2OH^-$	+0.36	$O_2 + 2H_2O + 4e^- \rightarrow 4OH^-$	+0.40
$Co^{2+} + 2e^- \rightarrow Co$	−0.28	$O_2 + 4H^+ + 4e^- \rightarrow 2H_2O$	+1.23
$Co^{3+} + e^- \rightarrow Co^{2+}$	+1.81	$O_2 + e^- \rightarrow O_2^-$	−0.56
$Cr^{2+} + 2e^- \rightarrow Cr$	−0.91	$O_2 + H_2O + 2e^- \rightarrow HO_2^- + OH^-$	−0.08
$Cr_2O_7^{2-} + 14H^+ + 6e^- \rightarrow 2Cr^{3+} + 7H_2O$	+1.33	$O_3 + 2H^+ + 2e^- \rightarrow O_2 + H_2O$	+2.07
$Cr^{3+} + 3e^- \rightarrow Cr$	−0.74	$O_3 + H_2O + 2e^- \rightarrow O_2 + 2OH^-$	+1.24
$Cr^{3+} + e^- \rightarrow Cr^{2+}$	−0.41	$Pb^{2+} + 2e^- \rightarrow Pb$	−0.13
$Cs^+ + e^- \rightarrow Cs$	−2.92	$Pb^{4+} + 2e^- \rightarrow Pb^{2+}$	+1.67
$Cu^+ + e^- \rightarrow Cu$	+0.52	$PbSO_4 + 2e^- \rightarrow Pb + SO_4^{2-}$	−0.36
$Cu^{2+} + 2e^- \rightarrow Cu$	+0.34	$Pt^{2+} + 2e^- \rightarrow Pt$	+1.20
$Cu^{2+} + e^- \rightarrow Cu^+$	+0.16	$Pu^{4+} + e^- \rightarrow Pu^{3+}$	+0.97
$F_2 + 2e^- \rightarrow 2F^-$	+2.87	$Ra^{2+} + 2e^- \rightarrow Ra$	−2.92
$Fe^{2+} + 2e^- \rightarrow Fe$	−0.44	$Rb^+ + e^- \rightarrow Rb$	−2.93
$Fe^{3+} + 3e^- \rightarrow Fe$	−0.04	$S + 2e^- \rightarrow S^{2-}$	−0.48
$Fe^{3+} + e^- \rightarrow Fe^{2+}$	+0.77	$S_2O_8^{2-} + 2e^- \rightarrow 2SO_4^{2-}$	+2.05
$[Fe(CN)_6]^{3-} + e^- \rightarrow [Fe(CN)_6]^{4-}$	+0.36	$Sn^{2+} + 2e^- \rightarrow Sn$	−0.14
$2H^+ + 2e^- \rightarrow H_2$	0, by definition	$Sn^{4+} + 2e^- \rightarrow Sn^{2+}$	+0.15
$2H_2O + 2e^- \rightarrow H_2 + 2OH^-$	−0.83	$Sr^{2+} + 2e^- \rightarrow Sr$	−2.89
$2HBrO + 2H^+ + 2e^- \rightarrow Br_2 + 2H_2O$	+1.60	$Ti^{2+} + 2e^- \rightarrow Ti$	−1.63
$2HClO + 2H^+ + 2e^- \rightarrow Cl_2 + 2H_2O$	+1.63	$Ti^{3+} + e^- \rightarrow Ti^{2+}$	−0.37
$H_2O_2 + 2H^+ + 2e^- \rightarrow 2H_2O$	+1.78	$Ti^{4+} + e^- \rightarrow Ti^{3+}$	0.00
$H_4XeO_6 + 2H^+ + 2e^- \rightarrow XeO_3 + 3H_2O$	+3.0	$Tl^+ + e^- \rightarrow Tl$	−0.34
$Hg_2^{2+} + 2e^- \rightarrow 2Hg$	+0.79	$U^{3+} + 3e^- \rightarrow U$	−1.79
$Hg_2Cl_2 + 2e^- \rightarrow 2Hg + 2Cl^-$	+0.27	$U^{4+} + e^- \rightarrow U^{3+}$	−0.61
$Hg^{2+} + 2e^- \rightarrow Hg$	+0.86	$V^{2+} + 2e^- \rightarrow V$	−1.19
$2Hg^{2+} + 2e^- \rightarrow Hg_2^{2+}$	+0.92	$V^{3+} + e^- \rightarrow V^{2+}$	−0.26
$Hg_2SO_4 + 2e^- \rightarrow 2Hg + SO_4^{2-}$	+0.62	$Zn^{2+} + 2e^- \rightarrow Zn$	−0.76

Table 12.2 The error function

z	erf z	z	erf z
0	0	0.50	0.520 50
0.01	0.011 28	0.55	0.563 32
0.02	0.022 56	0.60	0.603 86
0.03	0.033 84	0.65	0.642 03
0.04	0.045 11	0.70	0.677 80
0.05	0.056 37	0.75	0.711 16
0.06	0.067 62	0.80	0.742 10
0.07	0.078 86	0.85	0.770 67
0.08	0.090 08	0.90	0.796 91
0.09	0.101 28	0.95	0.820 89
0.10	0.112 46	1.00	0.842 70
0.15	0.168 00	1.20	0.910 31
0.20	0.222 70	1.40	0.952 28
0.25	0.276 32	1.60	0.976 35
0.30	0.328 63	1.80	0.989 09
0.35	0.379 38	2.00	0.995 32
0.40	0.428 39		
0.45	0.475 48	Data: AS	

Table 13.3 Effective atomic numbers

	H							He
Z	1							2
$1s$:	1							1.69

	Li	Be	B	C	N	O	F	Ne
Z	3	4	5	6	7	8	9	10
$1s$:	2.69	3.68	4.68	5.67	6.66	7.66	8.65	9.64
$2s$:	1.28	1.91	2.58	3.22	3.85	4.49	5.13	5.76
$2p$:			2.42	3.14	3.83	4.45	5.10	5.76

	Na	Mg	Al	Si	P	S	Cl	Ar
	11	12	13	14	15	16	17	18
$1s$:	10.63	11.61	12.59	13.57	14.56	15.54	16.52	17.51
$2s$:	6.57	7.39	8.21	9.02	9.82	10.63	11.43	12.23
$2p$:	6.80	7.83	8.96	9.94	10.96	11.98	12.99	14.01
$3s$:	2.51	3.31	4.12	4.90	5.64	6.37	7.07	7.76
$3d$:			4.07	4.29	4.89	5.48	6.12	6.76

Data: E. Clementi and D. L. Raimondi, *Atomic screening constants from SCF functions.* IBM Research Note NJ-27 (1963).

Table 14.1 Some hybridization schemes

Coordination number	Arrangement	Composition
2	Linear	sp, pd, sd
	Angular	sd
3	Trigonal planar	sp^2, p^2d
	Unsymmetrical planar	spd
	Trigonal pyramidal	pd^2
4	Tetrahedral	sp^3, sd^3
	Irregular tetrahedral	spd^2, p^3d, pd^3
	Square planar	p^2d^2, sp^2d
5	Trigonal bipyramidal	sp^3d, spd^3
	Tetragonal pyramidal	$sp^2d^2, sd^4, pd^4, p^3d^2$
	Pentagonal planar	p^2d^3
6	Octahedral	sp^3d^2
	Trigonal prismatic	spd^4, pd^5
	Trigonal antiprismatic	p^3d^3

Source: H. Eyring, J. Walter, and G. E. Kimball, *Quantum chemistry.* Wiley (1944).

Table 16.2 Properties of diatomic molecules

	$\bar{\nu}_0/\text{cm}^{-1}$	B/cm^{-1}	r/pm	$k/(\text{N m}^{-1})$	$D/(\text{kJ mol}^{-1})$
$^1\text{H}_2^+$	2321.8	29.8	106	160.0	255.8
$^1\text{H}_2$	4400.39	60.864	74.138	574.9	432.1
$^2\text{H}_2$	3118.46	30.442	74.154	577.0	439.6
$^1\text{H}^{19}\text{F}$	4138.32	20.956	91.680	965.7	564.4
$^1\text{H}^{35}\text{Cl}$	2990.95	10.593	127.45	516.3	427.7
$^1\text{H}^{81}\text{Br}$	2648.98	8.465	141.44	411.5	362.7
$^1\text{H}^{127}\text{I}$	2308.09	6.511	160.92	313.8	294.9
$^{14}\text{N}_2$	2358.07	1.9987	109.76	2293.8	941.7
$^{16}\text{O}_2$	1580.36	1.4457	120.75	1176.8	493.5
$^{19}\text{F}_2$	891.8	0.8828	141.78	445.1	154.4
$^{35}\text{Cl}_2$	559.71	0.2441	198.75	322.7	239.3

Data: AIP

Table 16.3 Typical vibration wavenumbers, $\bar{\nu}/\text{cm}^{-1}$

C—H stretch	2850–2960	C—Cl stretch	600–800
C—H bend	1340–1465	C—Br stretch	500–600
C—C stretch, bend	700–1250	C—I stretch	500
C=C stretch	1620–1680	CO_3^{2-}	1410–1450
C≡C stretch	2100–2260	NO_3^-	1350–1420
O—H stretch	3590–3650	NO_2^-	1230–1250
H-bonds	3200–3570	SO_4^{2-}	1080–1130
C=O stretch	1640–1780	Silicates	900–1100
C≡N stretch	2215–2275		
N—H stretch	3200–3500		
C—F stretch	1000–1400		

Data: L. J. Bellamy, *The infrared spectra of complex molecules* and *Advances in infrared group frequencies*. Chapman and Hall.

Table 17.1 Colour, frequency, and energy of light

Colour	λ/nm	$\nu/(10^{14}\ \text{Hz})$	$\bar{\nu}/(10^4\ \text{cm}^{-1})$	E/eV	$E/(\text{kJ mol}^{-1})$
Infrared	>1000	< 3.00	<1.00	<1.24	<120
Red	700	4.28	1.43	1.77	171
Orange	620	4.84	1.61	2.00	193
Yellow	580	5.17	1.72	2.14	206
Green	530	5.66	1.89	2.34	226
Blue	470	6.38	2.13	2.64	254
Violet	420	7.14	2.38	2.95	285
Near ultraviolet	300	10.0	3.33	4.15	400
Far ultraviolet	< 200	>15.0	>5.00	>6.20	>598

Data: J. G. Calvert and J. N. Pitts, *Photochemistry*. Wiley, New York (1966).

Table 17.2 Absorption characteristics of some groups and molecules

Group	$\bar{\nu}_{max}/(10^4\ \text{cm}^{-1})$	λ_{max}/nm	$\varepsilon_{max}/(\text{L mol}^{-1}\ \text{cm}^{-1})$
C=C($\pi^* \leftarrow \pi$)	6.10	163	1.5×10^4
	5.73	174	5.5×10^3
C=O($\pi^* \leftarrow n$)	3.7–3.5	270–290	10–20
—N=N—	2.9	350	15
	>3.9	<260	Strong
—NO$_2$	3.6	280	10
	4.8	210	1.0×10^4
C$_6$H$_5$—	3.9	255	200
	5.0	200	6.3×10^3
	5.5	180	1.0×10^5
[Cu(OH$_2$)$_6$]$^{2+}$(aq)	1.2	810	10
[Cu(NH$_3$)$_4$]$^{2+}$(aq)	1.7	600	50
H$_2$O($\pi^* \leftarrow n$)	6.0	167	7.0×10^3

Data section

Table 18.1 Nuclear spin properties

Nuclide	Natural abundance %	Spin I	Magnetic moment μ/μ_N	g-value	NMR frequency at 1 T, ν/MHz
^{1}n*		$\frac{1}{2}$	−1.9130	−3.8260	29.167
^{1}H	99.9844	$\frac{1}{2}$	2.792 85	5.5857	42.576
^{2}H	0.0156	1	0.857 45	0.857 45	6.536
^{3}H*		$\frac{1}{2}$	−2.127 65	−4.2553	32.434
^{10}B	19.6	3	1.8005	0.6002	4.574
^{11}B	80.4	$\frac{3}{2}$	2.6884	1.7923	13.660
^{13}C	1.108	$\frac{1}{2}$	0.7023	1.4046	10.705
^{14}N	99.635	1	0.403 56	0.403 56	3.076
^{17}O	0.037	$\frac{5}{2}$	−1.893	−0.7572	5.772
^{19}F	100	$\frac{1}{2}$	2.628 35	5.2567	40.054
^{31}P	100	$\frac{1}{2}$	1.1317	2.2634	17.238
^{33}S	0.74	$\frac{3}{2}$	0.6434	0.4289	3.266
^{35}Cl	75.4	$\frac{3}{2}$	0.8218	0.5479	4.171
^{37}Cl	24.6	$\frac{3}{2}$	0.6841	0.4561	3.472

* Radioactive.
μ is the magnetic moment of the spin state with the largest value of m_I: $\mu = g_I\mu_N I$ and μ_N is the nuclear magneton (see inside front cover).
Data: KL

Table 18.2 Hyperfine coupling constants for atoms, a/mT

Nuclide	Spin	Isotropic coupling	Anisotropic coupling
^{1}H	$\frac{1}{2}$	50.8(1s)	
^{2}H	1	7.8(1s)	
^{13}C	$\frac{1}{2}$	113.0(2s)	6.6(2p)
^{14}N	1	55.2(2s)	4.8(2p)
^{19}F	$\frac{1}{2}$	1720(2s)	108.4(2p)
^{31}P	$\frac{1}{2}$	364(3s)	20.6(3p)
^{35}Cl	$\frac{3}{2}$	168(3s)	10.0(3p)
^{37}Cl	$\frac{3}{2}$	140(3s)	8.4(3p)

Data: P. W. Atkins and M. C. R. Symons, *The structure of inorganic radicals.* Elsevier, Amsterdam (1967).

Table 21.3 Ionic radii†

Li$^+$(4)	Be^{2+}(4)	B^{3+}(4)	N^{3-}	O^{2-}(6)	F$^-$(6)
59	27	12	171	140	133
Na$^+$(6)	Mg^{2+}(6)	Al^{3+}(6)	P^{3-}	S^{2-}(6)	Cl$^-$(6)
102	72	53	212	184	181
K$^+$(6)	Ca^{2+}(6)	Ga^{3+}(6)	As^{3-}	Se^{2-}(6)	Br$^-$(6)
138	100	62	222	198	196
Rb$^+$(6)	Sr^{2+}(6)	In^{3+}(6)		Te^{2-}(6)	I$^-$(6)
149	116	79		221	220
Cs$^+$(6)	Ba^{2+}(6)	Tl^{3+}(6)			
167	136	88			

d-block elements (high-spin ions)

Sc^{3+}(6)	Ti^{4+}(6)	Cr^{3+}(6)	Mn^{3+}(6)	Fe^{2+}(6)	Co^{3+}(6)	Cu^{2+}(6)	Zn^{2+}(6)
73	60	61	65	63	61	73	75

† Numbers in parentheses are the coordination numbers of the ions. Values for ions without a coordination number stated are estimates.
Data: R. D. Shannon and C. T. Prewitt, *Acta Cryst.*, **B25**, 925 (1969).

Table 22.1 Dipole moments, polarizabilities, and polarizability volumes

	$\mu/(10^{-30}\,\text{C m})$	μ/D	$(\alpha'/10^{-30}\,\text{m}^3)$	$\alpha/(10^{-40}\,\text{J}^{-1}\,\text{C}^2\,\text{m}^2)$
Ar	0	0	1.66	1.85
C_2H_5OH	5.64	1.69		
$C_6H_5CH_3$	1.20	0.36		
C_6H_6	0	0	10.4	11.6
CCl_4	0	0	10.5	11.7
CH_2Cl_2	5.24	1.57	6.80	7.57
CH_3Cl	6.24	1.87	4.53	5.04
CH_3OH	5.70	1.71	3.23	3.59
CH_4	0	0	2.60	2.89
$CHCl_3$	3.37	1.01	8.50	9.46
CO	0.390	0.117	1.98	2.20
CO_2	0	0	2.63	2.93
H_2	0	0	0.819	0.911
H_2O	6.17	1.85	1.48	1.65
HBr	2.67	0.80	3.61	4.01
HCl	3.60	1.08	2.63	2.93
He	0	0	0.20	0.22
HF	6.37	1.91	0.51	0.57
HI	1.40	0.42	5.45	6.06
N_2	0	0	1.77	1.97
NH_3	4.90	1.47	2.22	2.47
$1,2\text{-}C_6H_4(CH_3)_2$	2.07	0.62		

Data: HCP and C. J. F. Böttcher and P. Bordewijk, *Theory of electric polarization*. Elsevier, Amsterdam (1978).

Table 22.2 Refractive indices relative to air at 20°C

	434 nm	589 nm	656 nm
Benzene	1.5236	1.5012	1.4965
Carbon tetrachloride	1.4729	1.4676	1.4579
Carbon disulfide	1.6748	1.6276	1.6182
Ethanol	1.3700	1.3618	1.3605
KCl(s)	1.5050	1.4904	1.4973
KI(s)	1.7035	1.6664	1.6581
Methanol	1.3362	1.3290	1.3277
Methylbenzene	1.5170	1.4955	1.4911
Water	1.3404	1.3330	1.3312

Data: AIP

Table 22.4 Lennard-Jones (12,6)-potential parameters

	$(\varepsilon/k)/\text{K}$	σ/pm
Ar	124	342
Br_2	520	427
C_2H_4	205	423
C_6H_6	440	527
CCl_4	327	588
CH_4	137	382
Cl_2	357	412
CO_2	190	400
H_2	33.3	297
He	10.22	258
N_2	91.5	368
Ne	35.7	279
O_2	113	343
Xe	229	406

Data: J. O. Hirschfelder, C. F. Curtiss, and R. B. Bird, *Molecular theory of gases and liquids*. Wiley, New York (1954).

Table 22.5 Magnetic susceptibilities at 298 K

	$\chi/10^{-6}$	$\chi_m/(10^{-4}\,\text{cm}^3\,\text{mol}^{-1})$
Water	−90	−16.0
Benzene	−7.2	−6.4
Cyclohexane	−7.9	−8.5
Carbon tetrachloride	−8.9	−8.4
NaCl(s)	−13.9	−3.75
Cu(s)	−96	−6.8
S(s)	−12.9	−2.0
Hg(l)	−28.5	−4.2
$CuSO_4 \cdot 5H_2O$(s)	+176	+192
$MnSO_4 \cdot 4H_2O$(s)	+2640	$+2.79 \times 10^3$
$NiSO_4 \cdot 7H_2O$(s)	+416	+600
$FeSO_4(NH_4)_2SO_4 \cdot 6H_2O$(s)	+755	$+1.51 \times 10^3$
Al(s)	+22	+2.2
Pt(s)	+262	+22.8
Na(s)	+7.3	+1.7
K(s)	+5.6	+2.5

Data: KL and $\chi_m = \chi M/\rho$.

Table 23.1 Frictional coefficients and molecular geometry

Major axis Minor axis	Prolate	Oblate
2	1.04	1.04
3	1.11	1.10
4	1.18	1.17
5	1.25	1.22
6	1.31	1.28
7	1.38	1.33
8	1.43	1.37
9	1.49	1.42
10	1.54	1.46
50	2.95	2.38
100	4.07	2.97

Data: K. E. Van Holde, *Physical biochemistry*. Prentice Hall, Englewood Cliffs (1971).

Sphere; radius a, $c = a$ f_0

Prolate ellipsoid; major axis $2a$, minor axis $2b$, $c = (ab^2)^{1/3}$

$$\left\{ \frac{(1 - b^2/a^2)^{1/2}}{(b/a)^{2/3} \ln\{[1 + (1 - b^2/a^2)^{1/2}]/(b/a)\}} \right\} f_0$$

Oblate ellipsoid; major axis $2a$, minor axis $2b$, $c = (a^2 b)^{1/3}$

$$\left\{ \frac{(a^2/b^2 - 1)^{1/2}}{(a/b)^{2/3} \arctan[(a^2/b^2 - 1)^{1/2}]} \right\} f_0$$

Long rod; length l, radius a, $c = (3a^2 l/4)^{1/3}$

$$\left\{ \frac{(l/2a)^{2/3}}{(3/2)^{1/3}\{2\ln(l/a) - 0.11\}} \right\} f_0$$

In each case $f_0 = 6\pi\eta c$ with the appropriate value of c.

Table 23.2 Diffusion coefficients of macromolecules in water at 20°C

	$M/(\text{kg mol}^{-1})$	$D/(10^{-10}\ \text{m}^2\ \text{s}^{-1})$
Sucrose	0.342	4.586
Ribonuclease	13.7	1.19
Lysozyme	14.1	1.04
Serum albumin	65	0.594
Haemoglobin	68	0.69
Urease	480	0.346
Collagen	345	0.069
Myosin	493	0.116

Data: C. Tanford, *Physical chemistry of macromolecules*. Wiley, New York (1961).

Table 23.3 Intrinsic viscosity

Macromolecule	Solvent	$\theta/°C$	$K/(10^{-3}\,cm^3\,g^{-1})$	a
Polystyrene	Benzene	25	9.5	0.74
	Cyclohexane	34†	81	0.50
Polyisobutylene	Benzene	23†	83	0.50
	Cyclohexane	30	26	0.70
Amylose	0.33 M KCl(aq)	25†	113	0.50
Various protein‡	Guanidine hydrochloride + $HSCH_2CH_2OH$		7.16	0.66

† The θ temperature.
‡ Use $[\eta] = KN^a$, N the number of amino acid residues.
Data: K. E. Van Holde, *Physical biochemistry.* Prentice Hall, Englewood Cliffs (1971).

Table 23.4 Radius of gyration of some macromolecules

	$M/(kg\,mol^{-1})$	R_g/nm
Serum albumin	66	2.98
Myosin	493	46.8
Polystyrene	3.2×10^3	50 (in poor solvent)
DNA	4×10^3	117.0
Tobacco mosaic virus	3.9×10^4	92.4

Data: C. Tanford, *Physical chemistry of macromolecules.* Wiley, New York (1961).

Table 24.2 Transport properties of gases at 1 atm

	$\kappa/(mJ\,cm^{-2}\,s^{-1}\,(K\,cm^{-1})^{-1})$ 273 K	$\eta/\mu P$ 273 K	$\eta/\mu P$ 293 K
Air	0.241	173	182
Ar	0.163	210	223
C_2H_4	0.164	97	103
CH_4	0.302	103	110
Cl_2	0.79	123	132
CO_2	0.145	136	147
H_2	1.682	84	88
He	1.442	187	196
Kr	0.087	234	250
N_2	0.240	166	176
Ne	0.465	298	313
O_2	0.245	195	204
Xe	0.052	212	228

Data: KL

Table 24.3 Viscosities of liquids at 298 K, $\eta/(10^{-3}\,kg\,m^{-1}\,s^{-1})$

Benzene	0.601
Carbon tetrachloride	0.880
Ethanol	1.06
Mercury	1.55
Methanol	0.553
Pentane	0.224
Sulfuric acid	27
Water†	0.891

† The viscosity of water over its entire liquid range is represented with less than 1 per cent error by the expression

$$\log(\eta_{20}/\eta) = A/B,$$

$$A = 1.37023(t-20) + 8.36 \times 10^{-4}(t-20)^2$$

$$B = 109 + t \qquad t = \theta/°C$$

Convert $kg\,m^{-1}\,s^{-1}$ to centipoise (cP) by multiplying by 10^3 (so that $\eta \approx 1\,cP$ for water).
Data: AIP, KL.

Table 24.4 Limiting ionic conductivities in water at 298 K, $\lambda/(\text{S cm}^2\,\text{mol}^{-1})$

Cations		Anions	
Ba^{2+}	127.2	Br^-	78.1
Ca^{2+}	119.0	$CH_3CO_2^-$	40.9
Cs^+	77.2	Cl^-	76.35
Cu^{2+}	107.2	ClO_4^-	67.3
H^+	349.6	CO_3^{2-}	138.6
K^+	73.50	$(CO_2)_2^{2-}$	148.2
Li^+	38.7	F^-	55.4
Mg^{2+}	106.0	$[Fe(CN)_6]^{3-}$	302.7
Na^+	50.10	$[Fe(CN)_6]^{4-}$	442.0
$[N(C_2H_5)_4]^+$	32.6	I^-	76.8
$[N(CH_3)_4]^+$	44.9	NO_3^-	71.46
NH_4^+	73.5	OH^-	199.1
Rb^+	77.8	SO_4^{2-}	160.0
Sr^{2+}	118.9		
Zn^{2+}	105.6		

Data: KL, RS.

Table 24.5 Ionic mobilities in water at 298 K, $u/(10^{-8}\,\text{m}^2\,\text{s}^{-1}\,\text{V}^{-1})$

Cations		Anions	
Ag^+	6.42	Br^-	8.09
Ca^{2+}	6.17	$CH_3CO_2^-$	4.24
Cu^{2+}	5.56	Cl^-	7.91
H^+	36.23	CO_3^{2-}	7.46
K^+	7.62	F^-	5.70
Li^+	4.01	$[Fe(CN)_6]^{3-}$	10.5
Na^+	5.19	$[Fe(CN)_6]^{4-}$	11.4
NH_4^+	7.63	I^-	7.96
$[N(CH_3)_4]^+$	4.65	NO_3^-	7.40
Rb^+	7.92	OH^-	20.64
Zn^{2+}	5.47	SO_4^{2-}	8.29

Data: Principally Table 24.4 and $u = \lambda/zF$.

Table 24.6 Debye–Hückel–Onsager coefficients for (1,1)-electrolytes at 25°C

Solvent	$A/(\text{S cm}^2\,\text{mol}^{-1}/(\text{mol L}^{-1})^{\frac{1}{2}})$	$B/(\text{mol L}^{-1})^{-\frac{1}{2}}$
Acetone (propanone)	32.8	1.63
Acetonitrile	22.9	0.716
Ethanol	89.7	1.83
Methanol	156.1	0.923
Nitrobenzene	44.2	0.776
Nitromethane	125.1	0.708
Water	60.20	0.229

Data: J. O'M. Bockris and A. K. N. Reddy, *Modern electrochemistry*. Plenum, New York (1970).

Table 24.7 Diffusion coefficients at 25°C, $D/(10^{-9}\,\text{m}^2\,\text{s}^{-1})$

Molecules in liquids				Ions in water			
I_2 in hexane	4.05	H_2 in $CCl_4(l)$	9.75	K^+	1.96	Br^-	2.08
in benzene	2.13	N_2 in $CCl_4(l)$	3.42	H^+	9.31	Cl^-	2.03
CCl_4 in heptane	3.17	O_2 in $CCl_4(l)$	3.82	Li^+	1.03	F^-	1.46
Glycine in water	1.055	Ar in $CCl_4(l)$	3.63	Na^+	1.33	I^-	2.05
Dextrose in water	0.673	CH_4 in $CCl_4(l)$	2.89			OH^-	5.30
Sucrose in water	0.5216	H_2O in $CCl_4(l)$	2.26				
		CH_3OH in water	1.58				
		C_2H_5OH in water	1.24				

Data: AIP and (for the ions) $\lambda = zuF$ in conjunction with Table 24.5.

Table 25.1 Kinetic data for first-order reactions

	Phase	$\theta/°C$	k/s^{-1}	$t_{\frac{1}{2}}$
$2N_2O_5 \rightarrow 4NO_2 + O_2$	g	25	3.38×10^{-5}	5.70 h
	$HNO_3(l)$	25	1.47×10^{-6}	131 h
	$Br_2(l)$	25	4.27×10^{-5}	4.51 h
$C_2H_6 \rightarrow 2CH_3$	g	700	5.36×10^{-4}	21.6 min
Cyclopropane $\rightarrow$ propene	g	500	6.71×10^{-4}	17.2 min
$CH_3N_2CH_3 \rightarrow C_2H_6 + N_2$	g	327	3.4×10^{-4}	34 min
Sucrose $\rightarrow$ glycose + fructose	$aq(H^+)$	25	6.0×10^{-5}	3.2 h

g: High pressure gas-phase limit.
Data: Principally K. J. Laidler, *Chemical kinetics.* Harper & Row, New York (1987); M. J. Pilling, *Reaction kinetics.* Clarendon Press, Oxford (1974); J. Nicholas, *Chemical kinetics.* Harper & Row, New York (1976). See also JL.

Table 25.2 Kinetic data for second-order reactions

	Phase	$\theta/°C$	$k/(L\,mol^{-1}\,s^{-1})$
$2NOBr \rightarrow 2NO + Br_2$	g	10	0.80
$2NO_2 \rightarrow 2NO + O_2$	g	300	0.54
$H_2 + I_2 \rightarrow 2HI$	g	400	2.42×10^{-2}
$D_2 + HCl \rightarrow DH + DCl$	g	600	0.141
$2I \rightarrow I_2$	g	23	7×10^9
	hexane	50	1.8×10^{10}
$CH_3Cl + CH_3O^-$	methanol	20	2.29×10^{-6}
$CH_3Br + CH_3O^-$	methanol	20	9.23×10^{-6}
$H^+ + OH^- \rightarrow H_2O$	water	25	1.35×10^{11}

Data: Principally K. J. Laidler, *Chemical kinetics.* Harper & Row, New York (1987); M. J. Pilling, *Reaction kinetics.* Clarendon Press, Oxford (1974); J. Nicholas, *Chemical kinetics.* Harper & Row, New York (1976).

Table 25.4 Arrhenius parameters

First-order reactions	A/s^{-1}	$E_a/(kJ\,mol^{-1})$
Cyclopropane $\rightarrow$ propene	1.58×10^{15}	272
$CH_3NC \rightarrow CH_3CN$	3.98×10^{13}	160
cis-CHD=CHD $\rightarrow$ *trans*-CHD=CHD	3.16×10^{12}	256
Cyclobutane $\rightarrow 2C_2H_4$	3.98×10^{15}	261
$C_2H_5I \rightarrow C_2H_4 + HI$	2.51×10^{13}	209
$C_2H_6 \rightarrow 2CH_3$	2.51×10^{17}	384
$2N_2O_5 \rightarrow 4NO_2 + O_2$	4.94×10^{13}	103
$N_2O \rightarrow N_2 + O$	7.94×10^{11}	250
$C_2H_5 \rightarrow C_2H_4 + H$	1.0×10^{13}	167

Second-order, gas phase	$A/(L\,mol^{-1}\,s^{-1})$	$E_a/(kJ\,mol^{-1})$
$O + N_2 \rightarrow NO + N$	1×10^{11}	315
$OH + H_2 \rightarrow H_2O + H$	8×10^{10}	42
$Cl + H_2 \rightarrow HCl + H$	8×10^{10}	23
$2CH_3 \rightarrow C_2H_6$	2×10^{10}	ca. 0
$NO + Cl_2 \rightarrow NOCl + Cl$	4.0×10^9	85
$SO + O_2 \rightarrow SO_2 + O$	3×10^8	27
$CH_3 + C_2H_6 \rightarrow CH_4 + C_2H_5$	2×10^8	44
$C_6H_5 + H_2 \rightarrow C_6H_6 + H$	1×10^8	ca. 25

25.4 (Continued)

Second-order, solution	$A/(\text{L mol}^{-1}\,\text{s}^{-1})$	$E_a/(\text{kJ mol}^{-1})$
$C_2H_5ONa + CH_3I$ in ethanol	2.42×10^{11}	81.6
$C_2H_5Br + OH^-$ in water	4.30×10^{11}	89.5
$C_2H_5I + C_2H_5O^-$ in ethanol	1.49×10^{11}	86.6
$CH_3I + C_2H_5O^-$ in ethanol	2.42×10^{11}	81.6
$C_2H_5Br + OH^-$ in ethanol	4.30×10^{11}	89.5
$CO_2 + OH^-$ in water	1.5×10^{10}	38
$CH_3I + S_2O_3^{2-}$ in water	2.19×10^{12}	78.7
Sucrose + H_2O in acidic water	1.50×10^{15}	107.9
$(CH_3)_3CCl$ solvolysis		
in water	7.1×10^{16}	100
in methanol	2.3×10^{13}	107
in ethanol	3.0×10^{13}	112
in acetic acid	4.3×10^{13}	111
in chloroform	1.4×10^4	45
$C_6H_5NH_2 + C_6H_5COCH_2Br$		
in benzene	91	34

Data: Principally J. Nicholas, *Chemical kinetics.* Harper and Row, New York (1976) and A. A. Frost and R. G. Pearson, *Kinetics and mechanism.* Wiley, New York (1961).

Table 27.1 Arrhenius parameters for gas-phase reactions

	$A/(\text{L mol}^{-1}\,\text{s}^{-1})$		$E_a/(\text{kJ mol}^{-1})$	P
	Experiment	Theory		
$2NOCl \rightarrow 2NO + Cl_2$	9.4×10^9	5.9×10^{10}	102.0	0.16
$2NO_2 \rightarrow 2NO + O_2$	2.0×10^9	4.0×10^{10}	111.0	5.0×10^{-2}
$2ClO \rightarrow Cl_2 + O_2$	6.3×10^7	2.5×10^{10}	0.0	2.5×10^{-3}
$H_2 + C_2H_4 \rightarrow C_2H_6$	1.24×10^6	7.3×10^{11}	180	1.7×10^{-6}
$K + Br_2 \rightarrow KBr + Br$	1.0×10^{12}	2.1×10^{11}	0.0	4.8

Data: Principally M. J. Pilling, *Reaction kinetics.* Clarendon Press, Oxford (1974).

Table 27.2 Arrhenius parameters for reactions in solution. See Table 25.4

Table 28.1 Surface tensions of liquids at 293 K

	$\gamma/(\text{mN m}^{-1})$
Benzene	28.88
Carbon tetrachloride	27.0
Ethanol	22.8
Hexane	18.4
Mercury	472
Methanol	22.6
Water	72.75
	72.0 at 25°C
	58.0 at 100°C

Data: KL

Table 28.2 Maximum observed enthalpies of physisorption, $\Delta_{ad}H^{\ominus}/(\text{kJ mol}^{-1})$

C_2H_2	-38
C_2H_4	-34
CH_4	-21
Cl_2	-36
CO	-25
CO_2	-25
H_2	-84
H_2O	-59
N_2	-21
NH_3	-38
O_2	-21

Data: D. O. Haywood and B. M. W. Trapnell, *Chemisorption.* Butterworth (1964).

Table 28.3 Enthalpies of chemisorption, $\Delta_{ad}H^{\ominus}/(kJ\ mol^{-1})$

Adsorbate	Adsorbent (substrate)											
	Ti	Ta	Nb	W	Cr	Mo	Mn	Fe	Co	Ni	Rh	Pt
H_2		−188			−188	−167	−71	−134			−117	
N_2		−586						−293				
O_2						−720					−494	−293
CO	−640							−192	−176			
CO_2	−682	−703	−552	−456	−339	−372	−222	−225	−146	−184		
NH_3				−301				−188		−155		
C_2H_4		−577		−427	−427			−285		−243	−209	

Data: D. O. Haywood and B. M. W. Trapnell, *Chemisorption*. Butterworth (1964).

Table 28.4 Activation energies of catalysed reactions

	Catalyst	$E_a/(kJ\ mol^{-1})$
$2HI \rightarrow H_2 + I_2$	None	184
	Au(s)	105
	Pt(s)	59
$2NH_3 \rightarrow N_2 + 3H_2$	None	350
	W(s)	162
$2N_2O \rightarrow 2N_2 + O_2$	None	245
	Au(s)	121
	Pt(s)	134
$(C_2H_5)_2O$ pyrolysis	None	224
	$I_2(g)$	144

Data: G. C. Bond, *Heterogeneous catalysis*. Clarendon Press, Oxford (1986).

Table 29.1 Exchange current densities and transfer coefficients at 298 K

Reaction	Electrode	$j_0/(A\ cm^{-2})$	α
$2H^+ + 2e^- \rightarrow H_2$	Pt	7.9×10^{-4}	
	Cu	1×10^{-6}	
	Ni	6.3×10^{-6}	0.58
	Hg	7.9×10^{-13}	0.50
	Pb	5.0×10^{-12}	
$Fe^{3+} + e^- \rightarrow Fe^{2+}$	Pt	2.5×10^{-3}	0.58
$Ce^{4+} + e^- \rightarrow Ce^{3+}$	Pt	4.0×10^{-5}	0.75

Data: Principally J. O'M. Bockris and A. K. N. Reddy, *Modern electrochemistry*. Plenum, New York (1970).

Table 29.2 Half-wave potentials at 298 K, $E_{\frac{1}{2}}/V$

Cd^{2+}	−0.60
Co^{2+}	−1.4
Cr^{3+}	−0.91
Cu^{2+}	+0.4
Fe^{2+}	−1.3
Fe^{3+}	0.0
Tl^+	−0.46
Zn^{2+}	−1.0

Data: LJ. Values refer to ions in 0.1 M KCl(aq), relative to the calomel electrode.

Character tables

The groups C_1, C_s, C_i

C_1 (1)	E	$h = 1$
A	1	

$C_s = C_h$ (m)	E	σ_h	$h = 2$	
A'	1	1	x, y, R_z	x^2, y^2, z^2, xy
A''	1	-1	z, R_x, R_y	yz, xz

$C_i = S_2$ (1̄)	E	i	$h = 2$	
A_g	1	1	R_x, R_y, R_z	$x^2, y^2, z^2, xy, xz, yz$
A_u	1	-1	x, y, z	

The groups C_n

C_2 (2)	E	C_2	$h = 2$	
A	1	1	z, R_z	x^2, y^2, z^2, xy
B	1	-1	x, y, R_x, R_y	yz, xz

C_3 (3)	E	C_3	C_3^2	$\varepsilon = \exp(2\pi i/3) \quad h = 3$	
A	1	1	1	z, R_z	$x^2 + y^2, z^2$
E	$\begin{Bmatrix} 1 \\ 1 \end{Bmatrix}$	$\begin{matrix} \varepsilon \\ \varepsilon^* \end{matrix}$	$\begin{matrix} \varepsilon^* \\ \varepsilon \end{matrix}$	$(x, y)(R_x, R_y)$	$(x^2 - y^2, xy)(yz, xz)$

C_4 (4)	E	C_4	C_2	C_4^3	$h = 4$	
A	1	1	1	1	z, R_z	$x^2 + y^2, z^2$
B	1	-1	1	-1		$x^2 - y^2, xy$
E	$\begin{Bmatrix} 1 \\ 1 \end{Bmatrix}$	$\begin{matrix} i \\ -i \end{matrix}$	$\begin{matrix} -1 \\ -1 \end{matrix}$	$\begin{matrix} -i \\ i \end{matrix}$	$(x, y)(R_x, R_y)$	(yz, xz)

The groups C_{nv}

C_{2v} (2mm)	E	C_2	$\sigma_v(xz)$	$\sigma_v'(yz)$	$h=4$	
A_1	1	1	1	1	z	x^2, y^2, z^2
A_2	1	1	-1	-1	R_z	xy
B_1	1	-1	1	-1	x, R_y	xz
B_2	1	-1	-1	1	y, R_x	yz

C_{3v} (3m)	E	$2C_3$	$3\sigma_v$	$h=6$	
A_1	1	1	1	z	x^2+y^2, z^2
A_2	1	1	-1	R_z	
E	2	-1	0	$(x, y)(R_x, R_y)$	$(x^2-y^2, xy)(xz, yz)$

C_{4v} (4mm)	E	$2C_4$	C_2	$2\sigma_v$	$2\sigma_d$	$h=8$	
A_1	1	1	1	1	1	z	x^2+y^2, z^2
A_2	1	1	1	-1	-1	R_z	
B_1	1	-1	1	1	-1		x^2-y^2
B_2	1	-1	1	-1	1		xy
E	2	0	-2	0	0	$(x, y)(R_x, R_y)$	(xz, yz)

C_{5v}	E	$2C_5$	$2C_5^2$	$5\sigma_v$	$h=10, \alpha=72°$	
A_1	1	1	1	1	z	x^2+y^2, z^2
A_2	1	1	1	-1	R_z	
E_1	2	$2\cos\alpha$	$2\cos 2\alpha$	0	$(x, y)(R_x, R_y)$	(xz, yz)
E_2	2	$2\cos 2\alpha$	$2\cos\alpha$	0		(x^2-y^2, xy)

C_{6v} (6mm)	E	$2C_6$	$2C_3$	C_2	$3\sigma_v$	$3\sigma_d$	$h=12$	
A_1	1	1	1	1	1	1	z	x^2+y^2, z^2
A_2	1	1	1	1	-1	-1	R_z	
B_1	1	-1	1	-1	1	-1		
B_2	1	-1	1	-1	-1	1		
E_1	2	1	-1	-2	0	0	$(x, y)(R_x, R_y)$	(xz, yz)
E_2	2	-1	-1	2	0	0		(x^2-y^2, xy)

$C_{\infty v}$	E	C_2	$2C_\phi$	$\infty\sigma_v$	$h=\infty$	
$A_1(\Sigma^+)$	1	1	1	1	z	z^2, x^2+y^2
$A_2(\Sigma^-)$	1	1	1	-1	R_z	
$E_1(\Pi)$	2	-2	$2\cos\phi$	0	$(x, y), (R_x, R_y)$	(xz, yz)
$E_2(\Delta)$	2	2	$2\cos 2\phi$	0		(xy, x^2-y^2)
$\vdots$	$\vdots$	$\vdots$	$\vdots$	$\vdots$		

The groups D_n

D_2 (222)	E	$C_2(z)$	$C_2(y)$	$C_2(x)$	$h = 4$	
A	1	1	1	1		x^2, y^2, z^2
B_1	1	1	-1	-1	z, R_z	xy
B_2	1	-1	1	-1	y, R_y	xz
B_3	1	-1	-1	1	x, R_x	yz

D_3 (32)	E	$2C_3$	$3C_2$	$h = 6$	
A_1	1	1	1		$x^2 + y^2, z^2$
A_2	1	1	-1	z, R_z	
E	2	-1	0	$(x, y)(R_x, R_y)$	$(x^2 - y^2, xy)(xz, yz)$

The groups D_{nh}

D_{2h} (mmm)	E	$C_2(z)$	$C_2(y)$	$C_2(x)$	i	$\sigma(xy)$	$\sigma(xz)$	$\sigma(yz)$	$h = 8$	
A_g	1	1	1	1	1	1	1	1		x^2, y^2, z^2
B_{1g}	1	1	-1	-1	1	1	-1	-1	R_z	xy
B_{2g}	1	-1	1	-1	1	-1	1	-1	R_y	xz
B_{3g}	1	-1	-1	1	1	-1	-1	1	R_x	yz
A_u	1	1	1	1	-1	-1	-1	-1		
B_{1u}	1	1	-1	-1	-1	-1	1	1	z	
B_{2u}	1	-1	1	-1	-1	1	-1	1	y	
B_{3u}	1	-1	-1	1	-1	1	1	-1	x	

D_{3h} (6m2)	E	$2C_3$	$3C_2$	σ_h	$2S_3$	$3\sigma_v$	$h = 12$	
A_1'	1	1	1	1	1	1		$x^2 + y^2, z^2$
A_2'	1	1	-1	1	1	-1	R_z	
E'	2	-1	0	2	-1	0	(x, y)	$(x^2 - y^2, xy)$
A_1''	1	1	1	-1	-1	-1		
A_2''	1	1	-1	-1	-1	1	z	
E''	2	-1	0	-2	1	0	(R_x, R_y)	(xz, yz)

The groups D_{nh} (continued)

D_{4h} $(4/mmm)$	E	$2C_4$	C_2	$2C_2'$	$2C_2''$	i	$2S_4$	σ_h	$2\sigma_v$	$2\sigma_d$	$h=16$	
A_{1g}	1	1	1	1	1	1	1	1	1	1		$x^2+y^2,\ z^2$
A_{2g}	1	1	1	−1	−1	1	1	1	−1	−1	R_z	
B_{1g}	1	−1	1	1	−1	1	−1	1	1	−1		x^2-y^2
B_{2g}	1	−1	1	−1	1	1	−1	1	−1	1		xy
E_g	2	0	−2	0	0	2	0	−2	0	0	(R_x, R_y)	(xz, yz)
A_{1u}	1	1	1	1	1	−1	−1	−1	−1	−1		
A_{2u}	1	1	1	−1	−1	−1	−1	−1	1	1	z	
B_{1u}	1	−1	1	1	−1	−1	1	−1	−1	1		
B_{2u}	1	−1	1	−1	1	−1	1	−1	1	−1		
E_u	2	0	−2	0	0	−2	0	2	0	0	(x, y)	

D_{5h}	E	$2C_5$	$2C_5^2$	$5C_2$	σ_h	$2S_5$	$2S_5^3$	$5\sigma_v$	$h=20,\ \alpha=72°$	
A_1'	1	1	1	1	1	1	1	1		$x^2+y^2,\ z^2$
A_2'	1	1	1	−1	1	1	1	−1	R_z	
E_1'	2	$2\cos\alpha$	$2\cos 2\alpha$	0	2	$2\cos\alpha$	$2\cos 2\alpha$	0	(x, y)	
E_2'	2	$2\cos 2\alpha$	$2\cos\alpha$	0	2	$2\cos 2\alpha$	$2\cos\alpha$	0		$(x^2-y^2,\ xy)$
A_1''	1	1	1	1	−1	−1	−1	−1		
A_2''	1	1	1	−1	−1	−1	−1	1	z	
E_1''	2	$2\cos\alpha$	$2\cos 2\alpha$	0	−2	$-2\cos\alpha$	$-2\cos 2\alpha$	0	(R_x, R_y)	(xz, yz)
E_2''	2	$2\cos 2\alpha$	$2\cos\alpha$	0	−2	$-2\cos 2\alpha$	$-2\cos\alpha$	0		

D_{6h} $(6/mmm)$	E	$2C_6$	$2C_3$	C_2	$3C_2'$	$3C_2''$	i	$2S_3$	$2S_6$	σ_h	$3\sigma_d$	$3\sigma_v$	$h=24$	
A_{1g}	1	1	1	1	1	1	1	1	1	1	1	1		$x^2+y^2,\ z^2$
A_{2g}	1	1	1	1	−1	−1	1	1	1	1	−1	−1	R_z	
B_{1g}	1	−1	1	−1	1	−1	1	−1	1	−1	1	−1		
B_{2g}	1	−1	1	−1	−1	1	1	−1	1	−1	−1	1		
E_{1g}	2	1	−1	−2	0	0	2	1	−1	−2	0	0	(R_x, R_y)	(xz, yz)
E_{2g}	2	−1	−1	2	0	0	2	−1	−1	2	0	0		$(x^2-y^2,\ xy)$
A_{1u}	1	1	1	1	1	1	−1	−1	−1	−1	−1	−1		
A_{2u}	1	1	1	1	−1	−1	−1	−1	−1	−1	1	1	z	
B_{1u}	1	−1	1	−1	1	−1	−1	1	−1	1	−1	1		
B_{2u}	1	−1	1	−1	−1	1	−1	1	−1	1	1	−1		
E_{1u}	2	1	−1	−2	0	0	−2	−1	1	2	0	0	(x, y)	
E_{2u}	2	−1	−1	2	0	0	−2	1	1	−2	0	0		

$D_{\infty h}$	E	$\infty C_2'$	$2C_\phi$	i	$\infty\sigma_v$	$2S_\phi$	$h=\infty$	
$A_{1g}(\Sigma_g^+)$	1	1	1	1	1	1		$z^2,\ x^2+y^2$
$A_{1u}(\Sigma_u^+)$	1	−1	1	−1	1	−1	z	
$A_{2g}(\Sigma_g^-)$	1	−1	1	1	−1	1	R_z	
$A_{2u}(\Sigma_u^-)$	1	1	1	−1	−1	−1		
$E_{1g}(\Pi_g)$	2	0	$2\cos\phi$	2	0	$-2\cos\phi$	(R_x, R_y)	(xz, yz)
$E_{1u}(\Pi_u)$	2	0	$2\cos\phi$	−2	0	$2\cos\phi$	(x, y)	
$E_{2g}(\Delta_g)$	2	0	$2\cos 2\phi$	2	0	$2\cos 2\phi$		$(xy,\ x^2-y^2)$
$E_{2u}(\Delta_u)$	2	0	$2\cos 2\phi$	−2	0	$-2\cos 2\phi$		
$\vdots$	$\vdots$	$\vdots$	$\vdots$	$\vdots$	$\vdots$	$\vdots$		

The groups D_{nd}

$D_{2d} = V_d$ ($\bar{4}2m$)	E	$2S_4$	C_2	$2C_2'$	$2\sigma_d$	$h = 8$	
A_1	1	1	1	1	1		$x^2 + y^2, z^2$
A_2	1	1	1	-1	-1	R_z	
B_1	1	-1	1	1	-1		$x^2 - y^2$
B_2	1	-1	1	-1	1	z	xy
E	2	0	-2	0	0	(x, y) (R_x, R_y)	(xz, yz)

D_{3d} ($\bar{3}m$)	E	$2C_3$	$3C_2$	i	$2S_6$	$3\sigma_d$	$h = 12$	
A_{1g}	1	1	1	1	1	1		$x^2 + y^2, z^2$
A_{2g}	1	1	-1	1	1	-1	R_z	
E_g	2	-1	0	2	-1	0	(R_x, R_y)	$(x^2 - y^2, xy)(xz, yz)$
A_{1u}	1	1	1	-1	-1	-1		
A_{2u}	1	1	-1	-1	-1	1	z	
E_u	2	-1	0	-2	1	0	(x, y)	

D_{4d}	E	$2S_8$	$2C_4$	$2S_8^3$	C_2	$4C_2'$	$4\sigma_d$	$h = 16$	
A_1	1	1	1	1	1	1	1		$x^2 + y^2, z^2$
A_2	1	1	1	1	1	-1	-1	R_z	
B_1	1	-1	1	-1	1	1	-1		
B_2	1	-1	1	-1	1	-1	1	z	
E_1	2	$\sqrt{2}$	0	$-\sqrt{2}$	-2	0	0	(x, y)	
E_2	2	0	-2	0	2	0	0		$(x^2 - y^2, xy)$
E_3	2	$-\sqrt{2}$	0	$\sqrt{2}$	-2	0	0	(R_x, R_y)	(xz, yz)

The cubic groups

T_d ($\bar{4}3m$)	E	$8C_3$	$3C_2$	$6S_4$	$6\sigma_d$	$h = 24$	
A_1	1	1	1	1	1		$x^2 + y^2 + z^2$
A_2	1	1	1	-1	-1		
E	2	-1	2	0	0		$(2z^2 - x^2 - y^2, x^2 - y^2)$
T_1	3	0	-1	1	-1	(R_x, R_y, R_z)	
T_2	3	0	-1	-1	1	(x, y, z)	(xy, xz, yz)

The cubic groups (continued)

O_h (*m3m*)	E	$8C_3$	$6C_2$	$6C_4$	$3C_2$ $(=C_4^2)$	i	$6S_4$	$8S_6$	$3\sigma_h$	$6\sigma_d$	$h = 48$	
A_{1g}	1	1	1	1	1	1	1	1	1	1		$x^2 + y^2 + z^2$
A_{2g}	1	1	−1	−1	1	1	−1	1	1	−1		
E_g	2	−1	0	0	2	2	0	−1	2	0		$(2z^2 - x^2 - y^2, x^2 - y^2)$
T_{1g}	3	0	−1	1	−1	3	1	0	−1	−1	(R_x, R_y, R_z)	
T_{2g}	3	0	1	−1	−1	3	−1	0	−1	1		(xz, yz, xy)
A_{1u}	1	1	1	1	1	−1	−1	−1	−1	−1		
A_{2u}	1	1	−1	−1	1	−1	1	−1	−1	1		
E_u	2	−1	0	0	2	−2	0	1	−2	0		
T_{1u}	3	0	−1	1	−1	−3	−1	0	1	1	(x, y, z)	
T_{2u}	3	0	1	−1	−1	−3	1	0	1	−1		

The icosahedral group

I	E	$12C_5$	$12C_5^2$	$20C_3$	$15C_2$	$h = 60$	
A	1	1	1	1	1		$z^2 + y^2 + z^2$
T_1	3	$\frac{1}{2}(1 + \sqrt{5})$	$\frac{1}{2}(1 - \sqrt{5})$	0	−1	(x, y, z) (R_x, R_y, R_z)	
T_2	3	$\frac{1}{2}(1 - \sqrt{5})$	$\frac{1}{2}(1 + \sqrt{5})$	0	−1		
G	4	−1	−1	1	0		
H	5	0	0	−1	1		$(2z^2 - x^2 - y^2, x^2 - y^2, xy, yz, zx)$

Further information: P. W. Atkins, M. S. Child, and C. S. G. Phillips, *Tables for group theory*. Oxford University Press (1970).

Answers to exercises

1.1 10 atm.
1.2 (a) 24 atm; (b) 22 atm.
1.3 (a) 2.57 kTorr; (b) 3.38 atm.
1.4 30 K.
1.5 30 lb in^{-1}.
1.6 4.22×10^{-2}.
1.7 2.67 Mg.
1.8 $8.3147 \, \text{J} \, \text{K}^{-1} \, \text{mol}^{-1}$.
1.9 $0.082\,061\,5 \, \text{L atm} \, \text{K}^{-1} \, \text{mol}^{-1}$, $31.9987 \, \text{g mol}^{-1}$.
1.10 S_8.
1.11 6.2 kg.
1.12 (a) 0.758, 0.242, 561 Torr, 179 Torr; (b) 0.751, 0.239, 0.010, 556 Torr, 117 Torr, 7.4 Torr.
1.13 (a) 3.14 L; (b) 212 Torr.
1.14 $169 \, \text{g mol}^{-1}$.
1.15 $16.4 \, \text{g mol}^{-1}$.
1.16 97.8 kPa.
1.17 $-272°C$.
1.18 $-270°C$.
1.19 $-233°N$.
1.20 (a) 9.975; (b) 1.
1.21 (a) 72 K; (b) 0.95 km s^{-1}; (c) 72 K.
1.22 (a) 0.475 km s^{-1}; (b) 4×10^4 m; (c) 1×10^{-2} s^{-1}.
1.23 (a) 640, 1260, 2300 m s^{-1}; (b) 320, 630, 1150 m s^{-1}.
1.24 81 mPa.
1.25 24 MPa.
1.26 1×10^3 nm.
1.27 (a) 5×10^{10} s^{-1}; (b) 5×10^9 s^{-1}; (c) 5×10^3 s^{-1}.
1.28 4×10^8 s^{-1}.
1.29 (a) 6.7 nm; (b) 67 nm; (c) 6.7 cm.
1.30 9.06×10^{-3}.
1.31 Independent.
1.32 (a) 1.0 atm, 8.2×10^2 atm; (b) 1.0 atm, 1.7×10^3 atm.
1.33 $67.8 \, \text{mL mol}^{-1}$, 54.5 atm, 120 K.
1.34 (a) 0.88; (b) 1.2 L.
1.35 140 atm.
1.36 (a) $0.124 \, \text{L mol}^{-1}$; (b) $0.109 \, \text{L mol}^{-1}$.
1.37 $0.1353 \, \text{L mol}^{-1}$, 0.6957, 0.657.
1.38 (a) 50.7 atm; (b) 35.1 atm, 0.693.
1.39 (a) 8.7 mL; (b) $-0.15 \, \text{L mol}^{-1}$.
1.40 (a) 0.67, 0.33; (b) 2.0 atm, 1.0 atm; (c) 3.0 atm.
1.41 $32.9 \, \text{cm}^3 \, \text{mol}^{-1}$, $1.33 \, \text{L}^2 \, \text{atm mol}^{-2}$, 0.24 nm.

1.42 (a) 1.4 kK; (b) 0.28 nm.
1.43 (a) 3.64 kK, 8.7 atm; (b) 2.60 kK, 4.5 atm; (c) 46.7 K, 0.18 atm.
1.44 $4.6 \times 10^{-5} \, \text{m}^3 \, \text{mol}^{-1}$, 0.66.

2.1 (a) -98 J, -16 J.
2.2 -2.6 kJ.
2.3 -1.0×10^2 J.
2.4 (a) $\Delta U = 0$, $\Delta H = 0$, $q = +1.57$ kJ, $w = -1.57$ kJ; (b) $\Delta U = 0$, $\Delta H = 0$, $q = +1.13$ kJ, $w = -1.13$ kJ; (c) all 0.
2.5 1.33 atm, $\Delta U = +1.25$ kJ, $w = 0$, $q = +1.25$ kJ.
2.6 (a) -88 J, (b) -167 J.
2.7 $+124$ J.
2.8 $\Delta U = -37.55$ kJ, $\Delta H = -40.656$ kJ, $q = -40.656$ kJ, $w = -3.10$ kJ.
2.9 -1.5 kJ.
2.10 85.0 MJ.
2.11 (a) $\Delta U = +26.8$ kJ, $\Delta H = +28.3$ kJ, $q = +28.3$ kJ, $w = -1.45$ kJ; (b) $\Delta U = +26.8$ kJ, $\Delta H = +28.3$ kJ, $q = +26.8$ kJ, $w = 0$.
2.12 $-125 \, \text{kJ mol}^{-1}$.
2.13 $C_{p,m} = 30 \, \text{J} \, \text{K}^{-1} \, \text{mol}^{-1}$, $C_{V,m} = 22 \, \text{J} \, \text{K}^{-1} \, \text{mol}^{-1}$.
2.14 $80 \, \text{J} \, \text{K}^{-1}$.
2.15 $\Delta U = +1.6$ kJ, $\Delta H = +2.2$ kJ, $q = +2.2$ kJ.
2.16 $\Delta U = +12$ kJ, $\Delta H = +13$ kJ, $q = +13$ kJ, $w = -1.0$ kJ.
2.17 $-4564.7 \, \text{kJ mol}^{-1}$.
2.18 $-126 \, \text{kJ mol}^{-1}$.
2.19 $+53 \, \text{kJ mol}^{-1}$, $+33 \, \text{kJ mol}^{-1}$.
2.20 $-1152 \, \text{kJ mol}^{-1}$.
2.21 $-432 \, \text{kJ mol}^{-1}$.
2.22 $641 \, \text{J} \, \text{K}^{-1}$.
2.23 $1.58 \, \text{kJ} \, \text{K}^{-1}$, $+2.05$ K.
2.24 (a) $-2.80 \, \text{MJ mol}^{-1}$; (b) $-2.80 \, \text{MJ mol}^{-1}$; (c) $-1.28 \, \text{MJ mol}^{-1}$.
2.25 $+65.49 \, \text{kJ mol}^{-1}$.
2.26 $-383 \, \text{kJ mol}^{-1}$.
2.27 $+1.90 \, \text{kJ mol}^{-1}$.
2.28 9.8 m.
2.29 (a) $-2205 \, \text{kJ mol}^{-1}$; (b) $-2200 \, \text{kJ mol}^{-1}$.
2.30 (a) Exothermic; (b, c) endothermic.

2.31 (a) $\nu(CO_2) = +1$, $\nu(H_2O) = +2$, $\nu(CH_4) = -1$, $\nu(O_2) = -2$; (b) $\nu(C_2H_2) = +1$, $\nu(C) = -2$, $\nu(H_2) = -1$; (c) $\nu(Na^+) = +1$, $\nu(Cl^-) = +1$, $\nu(NaCl) = -1$.
2.32 (a) $-57.20 \, \text{kJ mol}^{-1}$; (b) $-176.01 \, \text{kJ mol}^{-1}$; (c) $-32.88 \, \text{kJ mol}^{-1}$; (d) $-55.84 \, \text{kJ mol}^{-1}$.
2.33 (a) $-144.40 \, \text{kJ mol}^{-1}$, $-111.92 \, \text{kJ mol}^{-1}$; (b) $-92.31 \, \text{kJ mol}^{-1}$, $-241.82 \, \text{kJ mol}^{-1}$.
2.34 $-1368 \, \text{kJ mol}^{-1}$.
2.35 (a) $-392.1 \, \text{kJ mol}^{-1}$; (b) $-946.6 \, \text{kJ mol}^{-1}$; (c) $+52.5 \, \text{kJ mol}^{-1}$.
2.36 $-56.98 \, \text{kJ mol}^{-1}$.
2.37 (a) $+131.29 \, \text{kJ mol}^{-1}$, $+128.81 \, \text{kJ mol}^{-1}$; (b) $+130.02 \, \text{kJ mol}^{-1}$, $+127.54 \, \text{kJ mol}^{-1}$.
2.38 $-175 \, \text{kJ mol}^{-1}$, $-173 \, \text{kJ mol}^{-1}$, $+174 \, \text{kJ mol}^{-1}$.
2.39 $-1892.2 \, \text{kJ mol}^{-1}$.

3.2 $dz = 2axy^3 \, dx + 3ax^2y^2 \, dy$.
3.3 (a) $dz = (2x - 2y + 2) \, dx + (4y - 2x - 4) \, dy$.
3.4 $dz = (y + 1/x) \, dx + (x - 1) \, dy$.
3.5 $(\partial C_V / \partial V)_T = (\partial (\partial U / \partial V)_T \, \partial T)_V$.
3.6 $(\partial H / \partial U)_p = 1 + p(\partial V / \partial U)_p$.
3.7 $dV = (\partial V / \partial p)_T \, dp + (\partial V / \partial T)_p \, dT$; $d \ln V = -\kappa_T \, dp + \alpha \, dT$.
3.8 0, 0.
3.9 (b) $\alpha = 1/T$, $\kappa_T = 1/p$.
3.10 $0.71 \, \text{K atm}^{-1}$.
3.11 $+135 \, \text{J mol}^{-1}$.
3.12 $1.31 \times 10^{-3} \, \text{K}^{-1}$.
3.13 1×10^3 atm.
3.14 $-7.2 \, \text{J atm}^{-1} \, \text{mol}^{-1}$, 8.1 kJ.
3.15 $w = -3.2$ kJ, $q = 0$, $\Delta T = -38$ K, $\Delta U = -3.2$ kJ, $\Delta H = -4.5$ kJ.
3.16 $w = +4.1$ kJ, $q = 0$, $\Delta U = +4.1$ kJ, $\Delta H = +5.4$ kJ, $p_f = 5.2$ atm, $V_f = 11.8$ L.
3.17 9.4 L, 288 K, -0.46 kJ.
3.18 (a) $+0.9 \, \text{mm}^3$; (b) $+0.02 \, \text{mm}^3$.
3.19 -4.2 atm.
3.20 $w = -1.6$ kJ, $q = 0$, $\Delta U = -1.6$ kJ, $\Delta H = -2.1$ kJ, $\Delta T = -28$ K.
3.21 (a) 226 K; (b) 238 K.

4.1 (a) $+92 \, \mathrm{J \, K^{-1}}$; (b) $+67 \, \mathrm{J \, K^{-1}}$.
4.2 $152.67 \, \mathrm{J \, K^{-1} \, mol^{-1}}$.
4.3 $+8.92 \, \mathrm{J \, K^{-1}}$.
4.4 $-22.1 \, \mathrm{J \, K^{-1}}$.
4.5 $w = +4.1 \, \mathrm{kJ}$, $q = 0$,
$\Delta U = +4.1 \, \mathrm{kJ}$, $\Delta H = +5.4 \, \mathrm{kJ}$,
$\Delta S = 0$.
4.6 $+12.9 \, \mathrm{J \, K^{-1}}$.
4.7 Not reversible.
4.8 $54.9 \, \mathrm{kJ}$, $-195 \, \mathrm{J \, K^{-1}}$.
4.9 $+26 \, \mathrm{J \, K^{-1}}$.
4.10 $6.6 \, \mathrm{L}$.
4.11 $+2.8 \, \mathrm{J \, K^{-1}}$.
4.12 ΔH(overall) $= 0$,
ΔH(individual) $= \pm 1.9 \times 10^{2} \, \mathrm{kJ}$,
ΔS(overall) $= +93.4 \, \mathrm{J \, K^{-1}}$.
4.13 (a) $q = 0$; (b) $w = -20 \, \mathrm{J}$;
(c) $\Delta U = -20 \, \mathrm{J}$;
(d) $\Delta T = -0.35 \, \mathrm{K}$;
(e) $\Delta S = +0.60 \, \mathrm{J \, K^{-1}}$.
4.14 $-87.8 \, \mathrm{J \, K^{-1} \, mol^{-1}}$.
4.15 (a) $-386.1 \, \mathrm{J \, K^{-1} \, mol^{-1}}$;
(b) $-49.0 \, \mathrm{J \, K^{-1} \, mol^{-1}}$;
(c) $-153.1 \, \mathrm{J \, K^{-1} \, mol^{-1}}$;
(d) $-21.0 \, \mathrm{J \, K^{-1} \, mol^{-1}}$;
(e) $+512.0 \, \mathrm{J \, K^{-1} \, mol^{-1}}$.
4.16 (a) $-521.5 \, \mathrm{kJ \, mol^{-1}}$;
(b) $+25.8 \, \mathrm{kJ \, mol^{-1}}$;
(c) $-178.7 \, \mathrm{kJ \, mol^{-1}}$;
(d) $-212.40 \, \mathrm{kJ \, mol^{-1}}$;
(e) $-5798 \, \mathrm{kJ \, mol^{-1}}$.
4.17 (a) $-522.1 \, \mathrm{kJ \, mol^{-1}}$;
(b) $+25.78 \, \mathrm{kJ \, mol^{-1}}$;
(c) $-178.6 \, \mathrm{kJ \, mol^{-1}}$;
(d) $-212.55 \, \mathrm{kJ \, mol^{-1}}$;
(e) $-5798 \, \mathrm{kJ \, mol^{-1}}$.
4.18 $-93.05 \, \mathrm{kJ \, mol^{-1}}$.
4.19 $-50 \, \mathrm{kJ \, mol^{-1}}$.
4.20 (a) $+2.9 \, \mathrm{J \, K^{-1}}$, $-2.9 \, \mathrm{J \, K^{-1}}$, 0;
(b) $+2.9 \, \mathrm{J \, K^{-1}}$, 0, $+2.9 \, \mathrm{J \, K^{-1}}$;
(c) 0, 0, 0.
4.21 $\Delta S = n(C_{V,\mathrm{m}} - R) \ln 2$.
4.22 $817.90 \, \mathrm{kJ \, mol^{-1}}$.
4.23 0.11, 0.38.
4.24 (a) 0.500; (b) $0.50 \, \mathrm{kJ}$, $0.5 \, \mathrm{kJ}$.
4.25 $+0.95 \, \mathrm{J \, K^{-1} \, mol^{-1}}$.
4.26 (a) 0; (b) $20 \, \mathrm{kJ}$.
4.27 $201 \, \mathrm{K}$.
4.28 $3.15 \, \mathrm{kJ}$.
4.29 (a) 14; (b) 8.8.
4.30 $6.11 \, \mathrm{kJ}$, $61.1 \, \mathrm{s}$.

5.1 $(\partial S/\partial V)_T = \alpha/\kappa_T$,
$(\partial S/\partial p)_T = -\alpha V$.
5.2 $-3.8 \, \mathrm{J}$.
5.3 $-36.5 \, \mathrm{J \, K^{-1}}$.
5.4 $0.89 \, \mathrm{g \, cm^{-3}}$.
5.5 $15.7 \, \mathrm{atm}$, $+8.25 \, \mathrm{kJ}$.
5.6 $+7.3 \, \mathrm{kJ \, mol^{-1}}$.
5.7 $-0.55 \, \mathrm{kJ \, mol^{-1}}$.
5.8 $-2.63 \times 10^{-8} \, \mathrm{Pa^{-1}}$, 0.88.
5.9 $+10 \, \mathrm{kJ}$.
5.10 $+11 \, \mathrm{kJ \, mol^{-1}}$.
5.11 $p = RT/(V_\mathrm{m} - b) - a/V_\mathrm{m}^2$.
5.12 $V = (RT/p)(1 + B'p/RT$
$+ C'p^2/RT + D'p^3/RT)$.
5.13 $(\partial S/\partial V)_T = R/(V_\mathrm{m} - b)$, ΔS
greater for van der Waals gas.

6.1 $303 \, \mathrm{K}$ ($30^{\circ}\mathrm{C}$).
6.2 $+16 \, \mathrm{kJ \, mol^{-1}}$,
$+45.2 \, \mathrm{J \, K^{-1} \, mol^{-1}}$.
6.3 $+20.80 \, \mathrm{kJ \, mol^{-1}}$.
6.4 (a) $+34.08 \, \mathrm{kJ \, mol^{-1}}$,
(b) $350.5 \, \mathrm{K}$.
6.5 $281.8 \, \mathrm{K}$ ($8.7^{\circ}\mathrm{C}$).
6.6 $25 \, \mathrm{g \, s^{-1}}$.
6.7 (a) $1.7 \, \mathrm{kg}$; (b) $31 \, \mathrm{kg}$; (c) $1.4 \, \mathrm{g}$.
6.8 Yes; $\geq 3 \, \mathrm{Torr}$.
6.12 (a) $+49 \, \mathrm{kJ \, mol^{-1}}$; (b) $216^{\circ}\mathrm{C}$;
(c) $+99 \, \mathrm{J \, K^{-1} \, mol^{-1}}$.
6.13 (a) $29 \, \mathrm{kJ \, mol^{-1}}$; (b) $168 \, \mathrm{Torr}$,
$576 \, \mathrm{Torr}$.
6.14 $272.80 \, \mathrm{K}$.
6.15 0.07630 (7.6 per cent).
6.16 $196.0 \, \mathrm{K}$, $11.1 \, \mathrm{Torr}$.

7.1 $886.8 \, \mathrm{cm^3}$.
7.2 $56.3 \, \mathrm{cm^3 \, mol^{-1}}$.
7.3 $6.4 \, \mathrm{MPa}$.
7.4 $0.13 \, \mathrm{MPa}$.
7.5 $K_\mathrm{f} = 32 \, \mathrm{K} \, (\mathrm{mol \, kg^{-1}})^{-1}$,
$K_\mathrm{b} = 5.22 \, \mathrm{K} \, (\mathrm{mol \, kg^{-1}})^{-1}$.
7.6 $82 \, \mathrm{g \, mol^{-1}}$.
7.7 $381 \, \mathrm{g \, mol^{-1}}$.
7.8 $-0.09^{\circ}\mathrm{C}$.
7.9 $+1.2 \, \mathrm{J \, K^{-1}}$, $-0.35 \, \mathrm{kJ}$.
7.10 $+4.7 \, \mathrm{J \, K^{-1} \, mol^{-1}}$.
7.11 $-18.5 \, \mathrm{kJ}$, $+61.9 \, \mathrm{J \, K^{-1}}$, 0.
7.12 (a) $1:1$; (b) 0.8600.
7.13 (a) $3.4 \, \mathrm{mmol \, kg^{-1}}$;
(b) $34 \, \mathrm{mmol \, kg^{-1}}$.
7.14 $0.51 \, \mathrm{mmol \, N_2 \, kg^{-1}}$;
$0.27 \, \mathrm{mmol \, O_2 \, kg^{-1}}$.
7.15 $0.17 \, \mathrm{mol \, L^{-1}}$.
7.16 $-0.16^{\circ}\mathrm{C}$.
7.17 $24 \, \mathrm{g \, kg^{-1}}$.
7.18 $11 \, \mathrm{kg \, kg^{-1}}$.
7.19 $87 \, \mathrm{kg \, mol^{-1}}$.
7.20 $14 \, \mathrm{kg \, mol^{-1}}$.
7.21 Raoult's law basis: $a_\mathrm{A} = 0.833$,
$\gamma_\mathrm{A} = 0.93$, $a_\mathrm{B} = 0.125$,
$\gamma_\mathrm{B} = 1.25$; Henry's law basis:
$a_\mathrm{B} = 2.8$, $\gamma_\mathrm{B} = 1.25$.
7.22 $a = 0.9701$, $\gamma = 0.980$.
7.23 $p(\mathrm{CCl_4}) = 32.2 \, \mathrm{Torr}$,
$p(\mathrm{Br_2}) = 6.2 \, \mathrm{Torr}$,
$p(\mathrm{Total}) = 38.4 \, \mathrm{Torr}$,
$y(\mathrm{CCl_4}) = 0.839$, $y(\mathrm{Br_2}) = 0.16$.
7.24 $-3.536 \, \mathrm{kJ \, mol^{-1}}$.
7.25 $a_\mathrm{A} = 0.499$, $a_\mathrm{M} = 0.668$,
$\gamma_\mathrm{A} = 1.25$, $\gamma_\mathrm{M} = 1.11$.

8.1 $x_\mathrm{A} = 0.920$, $y_\mathrm{A} = 0.968$.
8.2 $440 \, \mathrm{Torr}$, $x_\mathrm{A} = 0.268$.
8.3 (a) Yes; (b) $y_\mathrm{A} = 0.830$.
8.4 (a) $154 \, \mathrm{Torr}$; (b) $y_\mathrm{de} = 0.67$.
8.5 (a) $48 \, \mathrm{Torr}$, (b) $y_\mathrm{B} = 0.77$;
(c) $34 \, \mathrm{Torr}$.
8.6 (a) $y_\mathrm{T} = 0.36$; (b) $y_\mathrm{T} = 0.82$.
8.7 (a) 2; (b) 2; (c) 3.
8.8 2, 2.
8.9 (a) 1, 2; (b) 2, 2.
8.10 (a) 3, 2; (b) 1.
8.11 (a) 2, 2; (b) 2.
8.12 (a) No.
8.21 (a) 80 per cent Ag by mass.

8.22 (b) $620 \, \mathrm{Torr}$; (c) $490 \, \mathrm{Torr}$;
(d) $x_\mathrm{Hex} = 0.5$, $y_\mathrm{Hex} = 0.72$;
(e) $x_\mathrm{Hex} = 0.50$, $y_\mathrm{Hex} = 0.30$;
(f) $n_\mathrm{g} = 1.7 \, \mathrm{mol}$, $n_\mathrm{l} = 0.3 \, \mathrm{mol}$.
8.31 (a) 2; (b) 3; (c) 1; (d) 3.
8.32 (a) $19.5 \, \mathrm{mol \, kg^{-1}}$;
(b) $23.8 \, \mathrm{mol^{-1} \, kg^{-1}}$.

9.1 $-2.42 \, \mathrm{kJ \, mol^{-1}}$.
9.2 3.01.
9.3 (a) 2.85×10^{-6};
(b) $+240 \, \mathrm{kJ \, mol^{-1}}$; (c) 0.
9.4 (a) 0.1411; (b) $+4.855 \, \mathrm{kJ \, mol^{-1}}$;
(c) 0; (d) 14.556.
9.6 (a) $-68.36 \, \mathrm{kJ \, mol^{-1}}$, 9.6×10^{11};
(b) $-69.7 \, \mathrm{kJ \, mol^{-1}}$, 1.3×10^9.
9.7 (a) $-308.84 \, \mathrm{kJ \, mol^{-1}}$, 1.3×10^{54};
(b) $-306.52 \, \mathrm{kJ \, mol^{-1}}$,
3.5×10^{49}.
9.8 (a) 0.087 (A), 0.370 (B), 0.196
(C), 0.438 (D); (b) 0.33;
(c) 0.33; (d) $+2.8 \, \mathrm{kJ \, mol^{-1}}$.
9.9 $1.5 \, \mathrm{kK}$.
9.10 $+2.77 \, \mathrm{kJ \, mol^{-1}}$,
$-16.5 \, \mathrm{J \, K^{-1} \, mol^{-1}}$.
9.11 $+12.3 \, \mathrm{kJ \, mol^{-1}}$.
9.12 $-41.0 \, \mathrm{kJ \, mol^{-1}}$.
9.13 (a) 50 per cent; (b) no change.
9.14 0.904, 0.096.
9.15 (a, c, e).
9.16 (b, d).
9.17 (a) $+53 \, \mathrm{kJ \, mol^{-1}}$;
(b) $-53 \, \mathrm{kJ \, mol^{-1}}$.
9.18 $-14.38 \, \mathrm{kJ \, mol^{-1}}$, product
formation.
9.19 (a) 9.24; (b) $-12.9 \, \mathrm{kJ \, mol^{-1}}$;
(c) $+161 \, \mathrm{kJ \, mol^{-1}}$;
(d) $+248 \, \mathrm{J \, K^{-1} \, mol^{-1}}$.
9.20 $+56.1 \, \mathrm{kJ \, mol^{-1}}$.
9.21 (a) $1110 \, \mathrm{K}$; (b) $397 \, \mathrm{K}$.
9.22 5.40, 3.61.
9.23 (a) 5.13; (b) 8.88; (c) 2.88.
9.24 8.3.
9.26 (a) $\mathrm{Na_2HPO_4/H_3PO_4}$;
(b) $\mathrm{H_2PO_4^-/HPO_4^{2-}}$.

10.1 $-218.66 \, \mathrm{kJ \, mol^{-1}}$.
10.2 $1.25 \times 10^{-5} \, \mathrm{mol \, L^{-1}}$.
10.3 $-290 \, \mathrm{kJ \, mol^{-1}}$.
10.5 $0.90 \, \mathrm{mol \, kg^{-1}}$.
10.6 $0.320 \, \mathrm{mol \, kg^{-1}}$.
10.7 (a) $2.73 \, \mathrm{g}$; (b) $2.92 \, \mathrm{g}$.
10.8 $0.25 \, \mathrm{mol \, kg^{-1}}$.
10.9 $\gamma_{\pm} = (\gamma_+ \gamma_-^2)^{\frac{1}{3}}$.
10.10 0.56.
10.11 1×10^4 per cent.
10.12 -2.1
10.13 $-1108 \, \mathrm{kJ \, mol^{-1}}$.
10.14 $+34.2 \, \mathrm{mV}$.
10.15 $-1.18 \, \mathrm{V}$.
10.16 (a) $\mathrm{Ag^+(aq) + e^- \rightarrow Ag(s)}$,
$\mathrm{Zn^{2+}(aq) + 2e^- \rightarrow Zn(s)}$,
$\mathrm{2Ag^+(aq) + Zn(s) \rightarrow}$
$\mathrm{2Ag(s) + Zn^{2+}(aq)}$, $+1.56 \, \mathrm{V}$;
(b) $\mathrm{H^+(aq) + e^- \rightarrow \frac{1}{2}H_2(g)}$,
$\mathrm{Cd^{2+}(aq) + 2e^- \rightarrow Cd(s)}$,
$\mathrm{Cd(s) + 2H^+(aq) \rightarrow}$
$\mathrm{Cd^{2+}(aq) + H_2(g)}$, $+0.40 \, \mathrm{V}$;

(c) $Cr^{3+}(aq) + 3e^- \to Cr(s)$,
$[Fe(CN)_6]^{3-}(aq) + e^- \to$
$[Fe(CN)_6]^{4-}(aq)$,
$Cr^{3+}(aq) + 3[Fe(CN)_6]^{4-}(aq) \to$
$Cr(s) + 3[Fe(CN)_6]^{3-}(aq)$,
-1.10 V;
(d) $Ag_2CrO_4(s) + 2e^- \to$
$2Ag(s) + CrO_4^{2-}(aq)$,
$Cl_2(g) + 2e^- \to 2Cl^-(aq)$,
$Ag_2CrO_4(s) + 2Cl^-(aq) \to$
$2Ag(s) + CrO_4^{2-}(aq) + Cl_2(g)$,
-0.91 V;
(e) $Sn^{4+}(aq) + 2e^- \to Sn^{2+}(aq)$,
$Fe^{3+}(aq) + e^- \to Fe^{2+}(aq)$;
$Sn^{4+}(aq) + 2Fe^{2+}(aq) \to$
$Sn^{2+}(aq) + 2Fe^{3+}(aq)$,
-0.62 V;
(f) $MnO_2(s) + 4H^+(aq) + 2e^- \to$
$Mn^{2+}(aq) + 2H_2O(l)$,
$Cu^{2+}(aq) + 2e^- \to Cu(s)$,
$Cu(s) + MnO_2(s) + 4H^+(aq) \to$
$Cu^{2+}(aq) + Mn^{2+}(aq) + 2H_2O(l)$,
$+0.89$ V.

10.17 (a) $Zn(s) \mid ZnSO_4(aq) \mid\mid CuSO_4(aq) \mid$
$Cu(s)$, $+1.10$ V;
(b) Pt,
$H_2(g) \mid HCl(aq) \mid AgCl(s) \mid Ag(s)$,
$+0.22$ V; (c) Pt, $H_2(g) \mid H^+(aq)$,
$H_2O(l) \mid O_2(g)$, Pt, $+1.23$ V;
(d) $Na(s) \mid NaOH(aq) \mid H_2(g)$,
Pt, $+1.88$ V;
(e) Pt, $H_2(g) \mid HI(aq) \mid I_2(s)$,
Pt, $+0.54$ V.

10.18 See entries above.
10.19 (a) -1.20 V; (b) -1.18 V.
10.20 (a) -363 kJ mol^{-1};
(b) -405 kJ mol^{-1};
(c) -291 kJ mol^{-1};
(d) $+122$ kJ mol^{-1}.
10.21 (a) $+0.324$ V; (b) $+0.45$ V.
10.22 (a) -2.45 V; (b) $+1.627$ V.
10.23 $+1.92$ V.
10.24 (a) $E = E^\ominus - (2RT/F) \ln(\gamma_\pm m)$;
(b) -89.89 kJ mol^{-1};
(c) $+0.223$ V.
10.25 -0.62 V.
10.26 -1.24 V, $+223.1$ kJ mol^{-1},
$+300.3$ kJ mol^{-1},
$+221.8$ kJ mol^{-1} at 35°C.
10.27 (a) 1.6×10^{-8} mol kg^{-1};
(b) $+0.12$ V.
10.28 (a) 6.5×10^9; (b) 1.5×10^{12};
(c) 2.8×10^{-16}; (d) 1.7×10^{16};
(e) 8.2×10^{-7}.
10.29 $+0.49$ V, 4×10^{16}.
10.30 $+1.22$ V.
10.31 $1.80 \times 10^{-10} \to 1.78 \times 10^{-10}$,
$9.04 \times 10^{-7} \to 5.1 \times 10^{-7}$.
10.32 $E = E^\ominus - (RT/6F) \ln(a(Cr^{3+})^2/$
$a(Cr_2O_7^{2-})a(H^+)^{14})$.
10.33 0.86.
10.34 0.
10.35 (a) 1×10^{-8} mol kg^{-1};
(b) 1×10^{-16}.

11.1 1.7 MW.
11.2 5.2×10^{16} Hz.
11.3 262 nm.

11.4 2.42 cm s^{-1}.
11.5 3.32 pm.
11.6 1.6 Mm s^{-1}.
11.7 8.83×10^{-28} kg m s^{-1},
0.969 km s^{-1}.
11.8 50.6 nm.
11.9 0.70 nm.
11.10 $E/(10^{-19}$ J), $E/(kg\,mol^{-1})$:
(a) 3.31, 199; (b) 3.61, 218;
(c) 4.97, 299; (d) 9.93, 598;
(e) 1.32×10^{-15} J,
7.98×10^5 kJ mol^{-1};
(f) 1.99×10^{-23} J,
0.012 kJ mol^{-1}.
11.11 $v/(m\,s^{-1})$: (a) 0.66; (b) 0.72;
(c) 0.99; (d) 1.98; (e) 2640;
(f) 3.96×10^{-5}.
11.12 21 m s^{-1}.
11.13 (a) 2.8×10^{18}; (b) 2.8×10^{20}.
11.14 6 kK.
11.15 (a) No ejection;
(b) 3.19×10^{-19} J, 837 km s^{-1}.
11.16 (a) 7×10^{-19} J, 400 kJ mol^{-1};
(b) 7×10^{-20} J, 40 kJ mol^{-1};
(c) 7×10^{-34} J,
4×10^{-13} kJ mol^{-1}.
11.17 (a) 6.6×10^{-29} m;
(b) 6.6×10^{-36} m; (c) 99.7 pm.
11.18 (a) 123 pm; (b) 3.88 pm.
11.19 1.1×10^{-28} m s^{-1}, 1×10^{-27} m.
11.20 5×10^{-25} kg m s^{-1},
5×10^5 m s^{-1}.
11.21 1.12×10^{-15} J.

12.1 (a) 1.81×10^{-19} J, 110 kJ mol^{-1},
1.1 eV, 9.1×10^3 cm^{-1};
(b) 6.6×10^{-19} J, 400 kJ mol^{-1},
4.1 eV, 3.3×10^4 cm^{-1}.
12.2 (a) 0.04, (b) 0.
12.3 (a) $-(\hbar^2/2m)(d^2\psi/dx^2) = E\psi$;
(b) 0, $h^2/4L^2$.
12.4 $L/6$, $L/2$, $5L/6$.
12.5 3.
12.6 23 per cent.
12.7 7.26×10^{10}, 1.71×10^{-31} J,
27.5 pm.
12.8 4.30×10^{-21} J.
12.9 278 N m^{-1}.
12.10 2.63 μm.
12.11 $+1.09$ μm.
12.12 (a) 3.4×10^{-35} J;
(b) 3.3×10^{-33} J;
(c) 2.2×10^{-29} J;
(d) 3.14×10^{-20} J.
12.13 5.61×10^{-21} J.
12.14 $N = 1/(2\pi)^{\frac{1}{2}}$.
12.15 117 pm.
12.16 1.49×10^{-34} J s; 0,
$\pm 1.05 \times 10^{-34}$ J s.

13.1 14.0 eV.
13.2 $r = 4a_0$.
13.3 101 pm, 376 pm.
13.4 $N = 2/a_0^{\frac{3}{2}}$.
13.5 $\langle V \rangle = 2E(1s)$, $\langle T \rangle = -E(1s)$.
13.6 $r^* = 5.24a_0/Z$.
13.7 (Angular momentum/$\hbar$, angular
nodes, radial nodes) = (a) 0, 0, 0;

(b) 0, 0, 2; (c) $6\frac{1}{2}$, 2, 0;
(d) $2\frac{1}{2}$, 1, 0; (e) $2\frac{1}{2}$, 1, 1.
13.8 (a) $\frac{5}{2}$, 3, 2; (b) $\frac{7}{2}$, $\frac{5}{2}$.
13.9 2 or 1, 1 or 0.
13.10 8, 7, 6, 5, 4, 3, 2.
13.11 (a) 1; (b) 9; (c) 25.
13.12 $L = 2$, $S = 0$, $J = 2$.
13.13 $r = 0.35a_0$.
13.14 (a) 110 pm; (b) 86 pm.
13.15 (b, c, e).
13.16 (a) 2; (b) 6; (c) 10; (d) 18.
13.17 (a) $[Ar]3d^8$; (b) $S = 1$, $M_S = 0$,
± 1, $S = 0$, $M_S = 0$.
13.18 (a) 1 (3), 0 (1); (b) $\frac{3}{2}$ (4), $\frac{1}{2}$ (2), $\frac{1}{2}$
(2); (c) 2 (5), 1 (3), 1 (3), 0 (1),
1 (3), 0 (1).
13.19 $^3D_3$, $^3D_2$, $^3D_1$, $^1D_2$ with $^3D < {}^1D$.
13.20 (a) 0 (1); (b) $\frac{3}{2}$ (4), $\frac{1}{2}$ (2); (c) 3
(7), 2 (5), 1 (3); (d) $\frac{7}{2}$ (8), $\frac{5}{2}$ (6),
$\frac{3}{2}$ (4), $\frac{1}{2}$ (2).
13.21 (a) $^2S_{\frac{1}{2}}$; (b) $^2P_{\frac{3}{2}}$, $^2P_{\frac{1}{2}}$; (c) $^2D_{\frac{5}{2}}$,
$^2D_{\frac{3}{2}}$; (d) $^2P_{\frac{3}{2}}$, $^2P_{\frac{1}{2}}$.
13.22 $+2$.
13.23 2.1 T.

14.1 $\psi = H(1)O(2) + H(2)O(1)$.
14.2 (a) $\psi_{ion} =$
$c_1O(1)O(2) + c_2H(1)H(2)$;
(b) $\psi = A\psi_{ion} + B\psi_{cov}$.
14.3 (a) $\psi = \{H(1)O(2) + H(2)O(1)\}$
$\times \{H(3)O(4) + H(4)O(3)\}$;
(b) $\psi = \{H(1)h(2) + H(2)h(1)\}$
$\times \{H(3)h(4) + H(4)h(3)\}$.
14.6 $\psi =$
$\{N_{Az}(1)N_{Bz}(2) + N_{Az}(2)N_{Bz}(1)\}$
$\times \{N_{Ay}(3)N_{By}(4) + N_{Ay}(4)N_{By}(3)\}$
$\times \{N_{Ax}(5)N_{Bx}(6) + N_{Ax}(6)N_{Bx}(5)\}$.
14.7 3.7 per cent s character.
14.8 (a) $1\sigma^2(1)$; (b) $1\sigma^2 2\sigma^{*2}(0)$;
(c) $1\sigma^2 2\sigma^{*2} 1\pi^4(2)$.
14.9 (a) $1\sigma^2 2\sigma^{*1}$;
(b) $1\sigma^2 2\sigma^{*2} 1\pi^3 3\sigma^2$;
(c) $1\sigma^2 2\sigma^{*2} 3\sigma^2 1\pi^2 2\pi^2 \pi^{*2}$.
14.10 (a) $1\sigma^2 2\sigma^{*2} 1\pi^4 3\sigma^2$;
(b) $1\sigma^2 2\sigma^{*2} 1\pi^4 3\sigma^2 2\pi^{*1}$;
(c) $1\sigma^2 2\sigma^{*2} 1\pi^4 3\sigma^2$.
14.11 C_2.
14.12 (a) C_2, CN; (b) NO, O_2, F_2.
14.14 (a) g; (c) g; (d) u.
14.16 1.6.
14.17 $\psi = \frac{1}{2}\psi_A \pm 3\frac{1}{2}/2\psi_B$.
14.18 $\frac{1}{2}$, 0.
14.19 3, u.
14.21 $1\sigma_g^1 1\pi_u^1$.
14.22 $N = 1/(1 + 2\lambda + \lambda^2)^{\frac{1}{2}}$.
14.25 (c, g).
14.28 (a) $a_{2u}^2 e_{1g}^4$, $6\alpha + 8\beta$; (b) $a_{2u}^2 e_{1g}^3$,
$5\alpha + 7\beta$.
14.29 (a) Neutral; (b) anion;
(c) butadiene.

15.1 E, C_3, $3\sigma_v$.
15.2 (a, b, c).
15.3 Yes.
15.6 i, σ_h, i, S_4, respectively.
15.8 (a) R_3; (b) C_{2v}; (c) D_{3h}; (d) $D_{\infty h}$;
(e) $C_{\infty v}$; (f) D_3; (g) C_{4v}; (h) C_s.

15.9 (a) C_{2v}; (b) $C_{\infty v}$; (c) C_{3v}; (d) D_{2h}; (e) C_{2v}; (f) C_{2h}.

15.10 (a) D_{2h}; (b) D_{2h}; (c) C_{2v} (1, 2), C_{2v} (1, 3), D_{2h} (1, 4).

15.12 (a) $C_{\infty v}$; (b) D_{5h}; (c) C_{2v}; (d) D_{4h}; (e) O_h; (f) T_d.

15.13 (a) NO_2, N_2O, $CHCl_3$, 1,2-dichlorobenzene, 1,3-dichlorobenzene; (b) none.

15.14 d_{xy}.

15.15 $B_1(x)$, $B_2(y)$, $A_1(z)$.

15.16 A_2.

15.17 (a) E_{1u}, A_{2u}; (b) B_{3u}, B_{2u}, B_{1u}.

16.1 (a) 2.84×10^{-12}; (b) 8.99×10^{13}; (c) 1.31×10^{-26}; (d) 1.65×10^{-30}.

16.2 0.409 THz.

16.3 (a) 2.707×10^{-47} kg m^2; (b) 129.0 pm.

16.4 (a) 4.442×10^{-47} kg m^2; 165.9 pm.

16.5 0.16 kN m^{-1}.

16.6 232.1 pm.

16.7 106.5 pm, 115.6 pm.

16.8 116.1 pm, 155.9 pm.

16.9 20 475 cm^{-1}.

16.10 2699.77 cm^{-1}.

16.11 1.08 per cent.

16.12 328.7 N m^{-1}.

16.13 381.3 cm^{-1}, 378.3 cm^{-1}.

16.14 2.177 eV.

16.15 $4A_1 + A_2 + 2B_1 + 2B_2$.

16.16 b, d, e, f, g, h.

16.17 b, c, d, e, f, g.

16.18 a, b, d, e, f.

16.19 $0.999\,999\,925 \times 660$ nm, 6.36×10^7 m s^{-1}.

16.20 2.4×10^7 m s^{-1}, 8.4×10^5 K.

16.21 (a) 5×10 ps; (b) 5 ps; (c) 2 ns.

16.22 (a) 50 cm^{-1}; (b) 0.5 cm^{-1}.

16.23 (a) 0.067; (b) 0.20.

16.24 160.4 pm.

16.25 515.6 (HCl), 411.8 (HBr), 314.2 (HI).

16.26 2143.7 (DCl), 1885.9 (DBr), 1640.1 (DI).

16.27 1580.38 cm^{-1}, 7.644×10^{-3}.

16.28 5.15 eV.

16.29 198.9 pm.

16.30 (a) 3; (b) 6; (c) 12; (d) 30.

16.31 (a) All; (b) symmetric stretch: Raman, antisymmetric stretch and bends: IR.

16.32 (a) Raman active.

16.33 Linear. ν_1: 1400 cm^{-1}; ν_2 (bend): 540 cm^{-1}, ν_3: 2360 cm^{-1}.

17.1 80 per cent.

17.2 6.28×10^4 L mol^{-1} cm^{-1}.

17.3 7.9×10^5 cm^2 mol^{-1}.

17.4 1.5 mmol L^{-1}.

17.5 5.44×10^7 L mol^{-1} cm^{-2}.

17.6 No.

17.7 Diene: 243 nm, butene: 192 nm.

17.10 450 L mol^{-1} cm^{-1}.

17.11 159 L mol^{-1} cm^{-1}, 23 per cent.

17.12 (a) 0.9 m; (b) 3 m.

17.13 (a) 5×10^7 L mol^{-1} cm^{-1}; (b) 3×10^5 L mol^{-1} cm^{-1}.

18.1 (a) 0; (b) 3; (c) 1; (d) $\frac{1}{2}$; (e) $\frac{3}{2}$.

18.2 600 MHz.

18.3 $(-1.625 \times 10^{-26}$ J$) \times m_I$.

18.4 d.

18.5 6.116×10^{-26} J.

18.6 3.523 T.

18.7 (a) 5.87, 38.3, 23.4, 81.3, 6.24, 14.5; (b) 11.7, 76.6, 46.8, 163, 12.5, 29.0 T.

18.8 (a) 1×10^{-6}; (b) 5.1×10^{-6}; (c) 3.4×10^{-5}.

18.9 10; (a) 1; (b) 10.

18.10 (a) 11 μT; (b) 110 μT.

18.11 7.885 kHz.

18.12 (a) 1.90 kHz; (b) 3.80 kHz.

18.14 1.9×10^{-3} s^{-1}.

18.16 203 MHz.

18.18 b.

18.19 0.59 mT, 20 μs.

18.20 0.2 kT, 10 mT.

18.21 1 T.

18.22 28 THz.

18.23 2.0022.

18.24 2.3 mT, 2.003.

18.25 3330.2, 332.2, 332.8, 334.8 mT, 1:1:1:1.

18.26 (a) 1:3:3:1; (b) 1:3:6:7:6:3:1.

18.27 (a) 331.9 mT; (b) 1.201 T.

18.28 Resolved.

18.29 $\frac{3}{2}$.

19.1 1.

19.2 (a) 2.57×10^{27}; (b) 7.26×10^{27}.

19.3 (a) 15.9 pm, 5.04 pm; (b) 2.47×10^{26}, 7.82×10^{27}.

19.4 2.83.

19.5 3.156.

19.6 2.45 kJ mol^{-1}.

19.7 354 K.

19.9 (a) 0.72; (b) 0.996.

19.11 (a) 5×10^{-5}, 0.4, 0.904; (b) 1.4; (c) 22 J mol^{-1}; (d) 1.6 J K^{-1} mol^{-1}; (e) 4.8 J K^{-1} mol^{-1}.

19.12 4303 K.

19.13 (a) 138 J K^{-1} mol^{-1}; (b) 146 J K^{-1} mol^{-1}.

19.14 5.18 J K^{-1} mol^{-1}.

19.15 a, b, d.

19.17 -15.63 kJ mol^{-1}.

20.1 (a) $5R/2$; (b) $3R$; (c) $3R$.

20.2 NH_3: 1.33 and 1.11 (1.31); CH_4: 1.33 and 1.08 (1.31).

20.3 (a) 19.6; (b) 34.3.

20.4 (a) 1; (b) 2; (c) 2; (d) 12; (e) 3.

20.5 43.1, 40 K.

20.6 43.76 J K^{-1} mol^{-1}.

20.7 (a) 36.63, 79.17; (b) 36.3, 78.9.

20.8 71.2.

20.9 (a) 7.97×10^3; (b) 1.12×10^4.

20.10 (a) 14.93 J K^{-1} mol^{-1}; (b) 25.65 J K^{-1} mol^{-1}.

20.11 -13.8 kJ mol^{-1}, -0.20 kJ mol^{-1}.

20.12 (a) $0.236R$, (b) $0.193R$.

20.13 -3.65 kJ mol^{-1}.

20.14 11.5 J K^{-1} mol^{-1}.

20.15 (a) 9 J K^{-1} mol^{-1}; (b) 13 J K^{-1} mol^{-1}; (c) 15 J K^{-1} mol^{-1}.

20.16 $R \ln s$.

20.17 9.57×10^{-15} J K^{-1}.

20.18 191.4 J K^{-1} mol^{-1}.

20.19 3.70×10^{-3}.

20.20 0.25.

21.1 $(1, \frac{1}{2}, 0)$, $(1, 0, \frac{1}{2})$, $(\frac{1}{2}, \frac{1}{2}, \frac{1}{2})$.

21.2 (323), (110).

21.3 249, 176, 432 pm.

21.4 70.7 pm.

21.5 17°, 23°, 29°.

21.6 0.215.

21.7 3.96×10^{-28} m^3.

21.8 4, 4.01 g cm^{-3}.

21.9 190 pm.

21.10 (111), (200), (311).

21.11 8.17°, 4.82°, 11.75°.

21.12 fcc.

21.13 bcc.

21.14 $F_{hkl} = 1$.

21.16 0.9069.

21.18 (a) 58.0 pm; (b) 102 pm.

21.19 0.340.

21.20 7.654 g cm^{-3}.

21.21 361 kg mol^{-1}.

21.22 +1.6 per cent.

21.24 Yes.

21.25 7.9 km s^{-1}.

21.26 252 pm.

21.27 4.6 kV.

21.28 (a) 39 pm; (b) 12 pm; (c) 6.1 pm.

21.29 5.8°, 17°; 0.3°, 0.9°.

22.1 220 pF.

22.2 1.01×10^{-39} J^{-1} C^2 m^2 $(9.1 \times 10^{-24}$ cm^3), 1.7 D.

22.3 4.8.

22.4 The second (C_{2v}).

22.5 1.42×10^{-39} J^{-1} C^2 m^2 $(1.28 \times 10^{-23}$ cm^3).

22.7 (a) 0 (by symmetry); (b) 0.7 D; (c) 0.4 D.

22.8 1.4 D.

22.9 37 D at 11.7° to x-axis.

22.10 4.9 μD.

22.11 1.36.

22.12 18.

22.13 6.9×10^{-6}.

22.14 $2^{\frac{1}{6}}r_0$.

22.15 3.

22.16 -6.4×10^{-5} cm^3 mol^{-1}.

22.17 2.

22.18 5.237 (5.923).

22.19 +0.016 cm^3 mol^{-1}.

22.20 222 T.

23.1 71 kg mol^{-1}.

23.2 24 nm.

23.3 6.7×10^3.

23.4 1.54 μm, 15.4 nm.

23.5 100.

23.6 $0.73 \, \text{mm s}^{-1}$.
23.7 $63 \, \text{kg mol}^{-1}$.
23.8 $31 \, \text{kg mol}^{-1}$.
23.9 (a) $18 \, \text{kg mol}^{-1}$; (b) $20 \, \text{kg mol}^{-1}$.
23.10 $0.67 \, \text{mmol L}^{-1}$.
23.11 $6.7 \, \text{mmol L}^{-1}$.
23.12 $3.4 \, \text{Mg mol}^{-1}$.
23.13 $4.3 \times 10^5 \, g$.

24.1 1.9×10^{20}.
24.2 $104 \, \text{mg}$.
24.3 $4.1 \, \mu\text{J cm}^{-2} \, \text{s}^{-1}$.
24.4 $0.056 \, \text{nm}^2$.
24.5 $17 \, \text{W}$, $28 \, \text{W}$.
24.6 $43 \, \text{g mol}^{-1}$.
24.7 $30 \, \text{h}$.
24.8 $0.142 \, \text{nm}^2$.
24.9 $205 \, \text{kPa}$.
24.10 (a) $130 \, \mu\text{P}$; (b) $130 \, \mu\text{P}$;
 (c) $240 \, \mu\text{P}$.
24.11 (a) $5.4 \, \text{mJ K}^{-1} \, \text{m}^{-1} \, \text{s}^{-1}$, $8.1 \, \text{mW}$;
 (b) $29 \, \text{mJ K}^{-1} \, \text{m}^{-1} \, \text{s}^{-1}$, $44 \, \text{mW}$.
24.12 $390 \, \text{pm}$.
24.13 $5.4 \, \text{mJ K}^{-1} \, \text{m}^{-1} \, \text{s}^{-1}$,
 $9.6 \, \text{mJ K}^{-1} \, \text{m}^{-1} \, \text{s}^{-1}$.
24.14 (a) $11 \, \text{m}^2 \, \text{s}^{-1}$;
 (b) $1.1 \times 10^{-5} \, \text{m}^2 \, \text{s}^{-1}$;
 $1.1 \times 10^{-7} \, \text{m}^2 \, \text{s}^{-1}$.
24.15 $0.2063 \, \text{cm}^{-1}$.
24.16 $116.7 \, \text{S cm}^2 \, \text{mol}^{-1}$.
24.17 $66.1 \, \text{S cm}^2 \, \text{mol}^{-1}$.
24.18 $347 \, \mu\text{m s}^{-1}$.
24.19 0.331.
24.20 $138.3 \, \text{S cm}^2 \, \text{mol}^{-1}$.
24.21 $4.01 \times 10^{-8} \, \text{m}^2 \, \text{s}^{-1} \, \text{V}^{-1}$,
 $5.19 \times 10^{-8} \, \text{m}^2 \, \text{s}^{-1} \, \text{V}^{-1}$,
 $7.52 \times 10^{-8} \, \text{m}^2 \, \text{s}^{-1} \, \text{V}^{-1}$.
24.22 $1.90 \times 10^{-9} \, \text{m}^2 \, \text{s}^{-1}$.
24.23 $1.3 \, \text{ks}$.
24.24 $420 \, \text{pm}$.
24.25 $21 \, \text{ps}$.
24.26 (a) $113 \, \mu\text{m}$; (b) $56 \, \mu\text{m}$.
24.27 (a) $80 \, \text{s}$, $320 \, \text{s}$; (b) $0.80 \, \text{ks}$, $32 \, \text{ks}$.
24.28 $17 \, \text{ms}$.

25.1 C: $3.0 \, \text{mol L}^{-1} \, \text{s}^{-1}$,
 D: $1.0 \, \text{mol L}^{-1} \, \text{s}^{-1}$,
 A: $1.0 \, \text{mol L}^{-1} \, \text{s}^{-1}$,
 B: $2.0 \, \text{mol L}^{-1} \, \text{s}^{-1}$.
25.2 A: $0.50 \, \text{mol L}^{-1} \, \text{s}^{-1}$,
 D: $1.5 \, \text{mol L}^{-1} \, \text{s}^{-1}$,
 A: $1.0 \, \text{mol L}^{-1} \, \text{s}^{-1}$,
 B: $0.50 \, \text{mol L}^{-1} \, \text{s}^{-1}$.
25.3 $\text{L mol}^{-1} \, \text{s}^{-1}$; (a) $k[\text{A}][\text{B}]$;
 (b) $3k[\text{A}][\text{B}]$.
25.4 $v = \frac{1}{2}k[\text{A}][\text{B}][\text{C}]$; $\text{L}^2 \, \text{mol}^{-2} \, \text{s}^{-1}$.
25.5 2.

25.6 2.
25.7 $25.1 \, \text{ks}$; (a) $499 \, \text{Torr}$;
 (b) $424 \, \text{Torr}$.
25.8 (a) $4.1 \times 10^{-3} \, \text{L mol}^{-1} \, \text{s}^{-1}$;
 (b) A: $2.6 \, \text{ks}$, B: $7.4 \, \text{ks}$.
25.9 (a) $\text{L mol}^{-1} \, \text{s}^{-1}$, $\text{L}^2 \, \text{mol}^{-2} \, \text{s}^{-1}$;
 (b) $\text{kPa}^{-1} \, \text{s}^{-1}$, $\text{kPa}^{-2} \, \text{s}^{-1}$.
25.10 $2.7 \, \text{Ma}$.
25.11 (a) $0.64 \, \mu\text{g}$; (b) $0.18 \, \mu\text{g}$.
25.12 (a) 45, $95 \, \text{mmol L}^{-1}$;
 (b) 1, $51 \, \text{mmol L}^{-1}$.
25.13 $124 \, \text{ks}$.
25.14 $64.9 \, \text{kJ mol}^{-1}$, $4.32 \, \text{mol L}^{-1} \, \text{s}^{-1}$.
25.15 $1(\text{H}_2\text{O}_2)$, $1(\text{Br}^-)$.
25.16 $v = k_2 K^{\frac{1}{2}}[\text{A}_2]^{\frac{1}{2}}[\text{B}]$.
25.17 $v = (k_1 k_2 / k_2')[\text{A}][\text{B}]$.
25.19 $1.52 \, \text{mmol L}^{-1} \, \text{s}^{-1}$.
25.20 $[\text{S}] = K_{\text{M}}$.
25.21 $1.9 \, \text{MPa}^{-1} \, \text{s}^{-1}$.
25.22 $7.1 \times 10^5 \, \text{s}^{-1}$, $7.63 \, \text{ns}$.
25.23 $1.7 \times 10^{-7} \, \text{s}^{-1}$,
 $8.5 \times 10^8 \, \text{mol L}^{-1} \, \text{s}^{-1}$.
25.24 $1/\tau = k_1 + k_2([\text{B}]_e + [\text{C}]_e)$.

26.6 None, 0.16 to $4.0 \, \text{kPa}$,
 $> 0.11 \, \text{kPa}$.
26.7 3.3×10^{18}.
26.8 0.521.
26.11 Initiation: 1; propagation: 2, 3;
 termination: 4.

27.1 (a) $9.6 \times 10^9 \, \text{s}^{-1}$,
 $1.2 \times 10^{35} \, \text{m}^{-3} \, \text{s}^{-1}$;
 (b) $6.7 \times 10^9 \, \text{s}^{-1}$,
 $8.3 \times 10^{34} \, \text{m}^{-3} \, \text{s}^{-1}$; 1.7 per cent.
27.2 (a) 0.018, 0.30; (b) 3.9×10^{-18},
 6.0×10^{-6}.
27.3 (a) 13 per cent, 1.2 per cent;
 (b) 130 per cent, 12 per cent.
27.4 $1.8 \times 10^{-2} \, \text{L mol}^{-1} \, \text{s}^{-1}$.
27.5 $3 \times 10^{10} \, \text{L mol}^{-1} \, \text{s}^{-1}$.
27.6 (a) $3.0 \times 10^{10} \, \text{L mol}^{-1} \, \text{s}^{-1}$;
 (b) $2.0 \times 10^9 \, \text{L mol}^{-1} \, \text{s}^{-1}$.
27.7 $7.4 \times 10^9 \, \text{L mol}^{-1} \, \text{s}^{-1}$; $0.17 \, \mu\text{s}$.
27.8 1.2×10^{-3}.
27.9 $1.8 \times 10^8 \, \text{mol L}^{-1} \, \text{s}^{-1}$.
27.10 $74.7 \, \text{kJ mol}^{-1}$, $-25 \, \text{J K}^{-1} \, \text{mol}^{-1}$
27.11 $+74.4 \, \text{kJ mol}^{-1}$.
27.12 $-96.6 \, \text{J K}^{-1} \, \text{mol}^{-1}$.
27.13 $-76 \, \text{J K}^{-1} \, \text{mol}^{-1}$.
27.14 (a) $-45.8 \, \text{J K}^{-1} \, \text{mol}^{-1}$;
 (b) $+5.0 \, \text{kJ mol}^{-1}$;
 (c) $+18.7 \, \text{kJ mol}^{-1}$.
27.16 (a) 0.071; (b) 0.89; reduced.
27.17 $20.9 \, \text{L}^2 \, \text{mol}^{-2} \, \text{min}^{-1}$.
27.18 $0.658 \, \text{L mol}^{-1} \, \text{min}^{-1}$.

27.19 $\log v \propto I^{\frac{1}{2}}$.

28.1 No.
28.2 1.54×10^4.
28.3 $2.6 \, \text{kPa}$.
28.4 $72.8 \, \text{mN m}^{-1}$.
28.5 $728 \, \text{kPa}$.
28.6 Yes.
28.7 $1.6 \times 10^{-8} \, \text{mol m}^{-2}$,
 $9.9 \times 10^{15} \, \text{m}^{-2}$, $101 \, \text{nm}^2$.
28.8 $+1.5 \, \text{mJ}$.
28.9 (a) 1.1×10^{21}, 1.4×10^{14};
 (b) 2.3×10^{20}, 3.1×10^{13}.
28.10 $94 \, \text{Torr}$.
28.11 $3.4 \times 10^5 \, \text{s}^{-1}$.
28.12 $12.8 \, \text{m}^2$.
28.13 $20.5 \, \text{cm}^3$.
28.14 Chemisorption, $50 \, \text{s}$.
28.15 $610 \, \text{kJ mol}^{-1}$, $0.11 \, \text{ps}$.
28.16 $3.7 \, \text{kJ mol}^{-1}$.
28.17 (a) $0.21 \, \text{kPa}$; (b) $22 \, \text{kPa}$.
28.18 0.83, 0.36.
28.19 (a) $40 \, \text{ps}$, $0.6 \, \text{ps}$; (b) $20 \, \text{Ts}$, $7 \, \mu\text{s}$.
28.20 $14 \, \text{kPa}$.
28.21 0 on gold, 1 on platinum.
28.23 $-13 \, \text{kJ mol}^{-1}$.
28.24 $700 \, \text{kJ mol}^{-1}$; (s) $2 \times 10^{104} \, \text{min}$;
 (b) $50 \, \mu\text{s}$.
28.25 0.

29.1 $0.24 \, \text{GV m}^{-1}$.
29.2 $138 \, \text{mV}$.
29.3 $2.8 \, \text{mA cm}^{-2}$.
29.4 Increase $\times 50$.
29.5 (a) $0.17 \, \text{mA cm}^{-2}$;
 (b) $0.17 \, \text{mA cm}^{-2}$.
29.6 $0.99 \, \text{A m}^{-2}$.
29.7 $0.2 \, \mu\text{mol L}^{-1}$.
29.9 (a) $0.31 \, \text{mA cm}^{-2}$;
 (b) $5.41 \, \text{mA cm}^{-2}$;
 $8.3 \, \text{A cm}^{-2}$.
29.12 $108 \, \text{mV}$.
29.13 (a) $4.9 \times 10^{15} \, \text{cm}^{-2} \, \text{s}^{-1}$, $3.8 \, \text{s}^{-1}$;
 (b) $1.6 \times 10^{16} \, \text{cm}^{-2} \, \text{s}^{-1}$, $12 \, \text{s}^{-1}$;
 (c) $3.1 \times 10^7 \, \text{cm}^{-2} \, \text{s}^{-1}$,
 $2.4 \times 10^{-8} \, \text{s}^{-1}$.
29.14 (a) $33 \, \Omega$; (b) $33 \, \text{G}\Omega$.
29.17 Yes.
29.18 No.
29.19 No.
29.20 $0.25 \, \text{mm}$.
29.21 $-1.30 \, \text{V}$, $0.13 \, \text{W}$.
29.22 (a) $+1.23 \, \text{V}$;
 (b) $-817.9 \, \text{kJ mol}^{-1}$.
29.23 Fe, Al, Co, Cr if O_2 absent; all if O_2 present.
29.24 $1.2 \, \text{mm year}^{-1}$.

Answers to problems

1.1 $0.50 \, m^3$.

1.2 1.5 kPa.

1.3 3.2×10^{-2} atm.

1.4 $p = \rho RT/M$, 46.0 g mol^{-1}.

1.5 $-272.95°C$.

1.6 (a) 4.6 kmol; (b) 130 kg′ (c) 120 kg.

1.7 102 g mol^{-1}, CH_2FCF_3.

1.8 (a) 0.184 Torr; (b) 68.6 Torr; (c) 0.184 Torr.

1.9 0.33 atm (N_2), 0 (H_2), 1.33 atm (NH_3), 1.66 atm.

1.11 (a) 1.8 mph E; (b) 56 mph E; (c) 56 mph.

1.12 (a) $5'9\frac{1}{2}''$; (b) $5'9\frac{1}{2}''$.

1.13 $v = (2gR)^{\frac{1}{2}}$.

1.14 (a) 12.5 L mol^{-1}; (b) 12.3 L mol^{-1}.

1.15 0.927, 0.208 L.

1.16 (a) 0.941 L; (c) 5.11 L.

1.17 (a) 0.1353 L mol^{-1}; (b) 0.6957; (c) 0.70; (d) 0.72.

1.18 210 K, 0.28 nm.

1.19 5.649 L^2 atm mol^{-1}, 59.4 cm^3 mol^{-1}, 21 atm.

1.21 $c^* = (2kT/m)^{\frac{1}{2}}$.

1.22 $c_{mean} = (\pi kT/2m)^{\frac{1}{2}}$.

1.23 $v/v_{initial} = 0.47 v_{initial}$.

1.24 (a) 0.61, 0.39; (b) 0.53, 0.47.

1.25 3.02×10^{-3}, 4.9×10^{-6}.

1.27 $B = b - a/RT$, $C = b^2$, 1.26 L^2 atm mol^{-2}, 34.6 cm^3 mol^{-1}.

1.28 $V_c = 3C/B$, $T_c = B^2/3RC$, $p_c = B^3/27C^3$, $\frac{1}{3}$.

1.29 $B' = B/RT$, $C' = (C - B^2)/R^2T^2$.

1.30 -0.18 atm^{-1}, 2.1 L mol^{-1}.

1.31 $(dV_m/dT)_p = (RV_m + b)/(2pV_m + RT)$.

1.32 No.

1.33 1.11.

1.34 $(p - p_0)/p_0 =$ (a) 0.00; (b) 0.05.

2.1 2.6 MJ.

2.2 $+37$ K, 4.09 kg.

2.3 (a) -3.46 kJ; (b) 0; (c) -3.46 kJ; (d) $+24.0$ kJ; $+27.5$ kJ.

2.4 $T_2 = 546$ K, $T_3 = 273$ K; Step 1 → 2: $w = -2.27$ kJ, $q = +5.67$ kJ, $\Delta U = +3.40$ kJ, $\Delta H = +5.67$ kJ; Step 2 → 3: $w = 0$, $q = -3.40$ kJ, $\Delta U = -3.40$ kJ, $\Delta H = -5.67$ kJ;

Step 3 → 1: $w = +1.57$ kJ, $q = -1.57$ kJ, $\Delta U = 0$, $\Delta H = 0$; Cycle: $w = -0.70$ kJ, $q = +0.70$ kJ, $\Delta U = 0$, $\Delta H = 0$.

2.5 (a) -0.27 kJ; (b) -0.92 kJ.

2.6 -8.9 kJ, -8.9 kJ.

2.7 $w = 0$, $\Delta U = +2.35$ kJ, $\Delta H = +3.03$ kJ.

2.8 $w = -25$ J, $q = +109$ J, $\Delta H = +97$ J.

2.9 $w = -1.7$ kJ, $\Delta U = +20.5$ kJ, $\Delta H = +22.2$ kJ, $\Delta H_m = +40$ kJ mol^{-1}.

2.10 $+98.7$ kJ mol^{-1}, $+95.8$ kJ mol^{-1}.

2.11 36.5 L.

2.12 -87.33 kJ mol^{-1}.

2.13 (a) 2878, 3537, 5471 kJ mol^{-1}; (b) 49.51, 49.02, 47.89 kJ g^{-1}.

2.14 $\Delta_{trs}H - \Delta_{trs}U = -93.9$ kJ mol^{-1}.

2.15 -2.13 MJ mol^{-1}, -1.267 MJ mol^{-1}.

2.16 $+17.7$ kJ mol^{-1}, $+66.8$ kJ mol^{-1}, $+116.0$ kJ mol^{-1}.

2.17 More exothermic by 5376 kJ mol^{-1}.

2.18 (a) $-2Fa/\pi$; (b) 0.

2.19 $w = -nRT \ln(V_2/V_1) + n^2RTB(1/V_2 - 1/V_1) + \frac{1}{2}n^3RTC(1/V_2^2 - 1/V_1^2)$; (a) -1.5 kJ; (b) -1.6 kJ.

2.20 (a) 0.39 mol, 0.50 L, 0.50 L; (b) $+19$ kJ; (c) -3.0 kJ; (d) $\Delta U = 0$, $\Delta V = 0$.

2.21 $w = -nRT \ln\{(V_2 - nb)/(V_1 - nb)\} - n^2a(1/V_2 - 1/V_1)$.

2.22 (a) 60 J; (b) -70 J; (c) $+10$ J; (d) $+50$ J.

2.23 $w_r = 3bw/a$; $w_r = -\frac{8}{9}nT_r \ln\{(V_{r,2} - 1)/(V_{r,1} - 1)\} - n(1/V_{r,2} - 1/V_{r,1})$; $w_r/n = -\frac{8}{9}\ln\{\frac{1}{2}(3x - 1)\} - 1/x + 1$.

2.24 (a) p/T, $-p/V$; $(1 + na/RTV)p/T$, $\{na/RTV - V/(V - nb)\}p/V$.

2.25 $\Delta H(T_2) = \Delta H(T_1) + \Delta a(T_2 - T_1) + \frac{1}{2}\Delta b(T_2^2 - T_1^2) - \Delta c(1/T_2 - 1/T_1)$; error about 0.02 kJ mol^{-1}.

3.1 -0.23 cm^3, -2.8 cm^3.

3.2 (a) $+0.75$ kJ mol^{-1}; (b) $+0.75$ kJ mol^{-1}.

3.3 41.40 J K^{-1} mol^{-1}.

3.4 -30.5 J mol^{-1}.

3.5 $T_1 = 546.3$ K, $V_2 = 44.83$ L, $T_3 = 414.0$ K; Step 1 → 2: $w = -3148$ J, $q = +3148$ J, $\Delta U = 0$, $\Delta H = 0$; Step 2 → 3: $w = +1100$ J, $q = -2750$ J, $\Delta U = -1650$ J, $\Delta H = -2750$ J; Step 3 → 1: $w = -2750$ J, $q = 0$, $\Delta U = -2750$ J, $\Delta H = +2750$ J.

3.6 (a) $w = 0$, $q = +6.19$ kJ, $\Delta U = +6.19$ kJ, $\Delta H = +8.67$ kJ; (b) $w = -6.19$ kJ, $q = 0$, $\Delta U = -6.19$ kJ, $\Delta H = +8.67$ kJ; (c) $w = +4.29$ kJ, $q = -4.29$ kJ, $\Delta U = 0$, $\Delta H = 0$; cycle: $w = -1.90$ kJ, $q = +1.90$ kJ, $\Delta U = 0$, $\Delta H = 0$.

3.7 1.67.

3.8 Not exact.

3.9 Not exact; dq/T is exact.

3.10 $dw = (y + z) \, dx + (x + z) \, dy + (x + y) \, dz$.

3.12 $(\partial H/\partial p)_T = -\mu C_p$.

3.13 $C_{p,m} - C_{V,m} = nR$.

3.17 $dp = \{R/(V_m - b)\} \, dT + \{2a/V_m^3 - RT/(V_m - b)^2\} \, dV_m$.

3.18 $+3.73$ kJ.

3.19 $dp = \{a(V_m - 2b)/V_m^3 - p\} \, dV_m/(V_m - b) + (p + a/V_m^2) \, dT/T$.

3.20 $(\partial T/\partial p)_V = (V_m - b)/R$.

3.21 $\alpha = RT^2(V_m - b)/\{RTV_m^3 - 2a(V_m - b)^2\}$, $\kappa_T = V_m^2(V_m - b)^2/\{RTV_m^3 - 2a(V_m - b)^2\}$.

3.22 $\mu C_p = (1 - b\zeta/V_m)V/(\zeta - 1)$, $\zeta = RTV_m^3/2a(V_m - b)^2$, 1.46 K atm^{-1}, $T_I = (\frac{27}{4})T_c(1 - b/V_m)^2$, 2021 K.

3.24 9.2 J K^{-1} mol^{-1}.

3.25 $\Delta H = C_p(T_f - T_i)$.

3.26 322 m s^{-1}.

4.1 (a) -21.3 J K^{-1} mol^{-1}, $+21.7$ J K^{-1} mol^{-1}; (b) -111.2 J K^{-1} mol^{-1}, -1.5 J K^{-1} mol^{-1}.

4.2 $+11$ J K^{-1}.

4.3 (a) 57.0°C, -43.9 kJ, $+146$ J K^{-1}, $+28$ J K^{-1}; (b) 53.5°C.

4.4 (a) $+50.74$ J K^{-1}, -11.5 J K^{-1};

(b) +3.46 kJ; (c) +3.46 kJ;
(d) +39.2 J K^{-1}, −39.2 J K^{-1}.

4.5 Step 1 → 2: $w = -11.5$ kJ,
$q = +11.5$ kJ, $\Delta U = 0$, $\Delta H = 0$,
$\Delta S = +19.1$ J K^{-1},
$\Delta S_{sur} = -19.1$ J K^{-1}; Step 2 → 3:
$w = -3.74$ kJ, $q = 0$,
$\Delta U = -3.74$ kJ, $\Delta H = -6.23$,
$\Delta S = 0$, $\Delta S_{sur} = 0$; Step 3 → 4:
$w = +5.74$ kJ, $q = -5.74$ kJ,
$\Delta U = 0$, $\Delta H = 0$,
$\Delta S = +19.1$ J K^{-1},
$\Delta S_{sur} = -19.1$ J K^{-1}; Step 4 → 1:
$w = +3.74$ kJ, $q = 0$,
$\Delta U = +3.74$ kJ, $\Delta H = +6.23$ kJ,
$\Delta S = 0$, $\Delta S_{sur} = 0$; Cycle:
$w = -5.8$ kJ, $q = +5.8$ kJ,
$\Delta U = 0$, $\Delta H = 0$, $\Delta S = 0$;
$\Delta S_{sur} = 0$.

4.6 Step 1 → 2: $w = -2.27$ kJ,
$q = +5.67$ kJ, $\Delta U = +3.40$ kJ,
$\Delta H = +5.67$, $\Delta S = +14.4$ J K^{-1},
$\Delta S_{tot} \geq 0$, $\Delta G \leq 0$; Step 2 → 3:
$w = 0$, $q = -3.40$ kJ,
$\Delta U = -3.40$ kJ, $\Delta H = -5.67$ kJ,
$\Delta S = -8.64$ J K^{-1}, $\Delta S_{sur} \geq 0$,
$\Delta G \leq 0$; Step 3 → 1:
$w = +1.57$ kJ, $q = -1.57$ kJ,
$\Delta U = 0$, $\Delta H = 0$,
$\Delta S = -5.76$ J K^{-1},
$\Delta G = +1.57$ kJ; Cycle:
$w = -0.70$ kJ, $q = +0.70$ kJ,
$\Delta U = 0$, $\Delta H = 0$, $\Delta S = 0$,
$\Delta G = 0$.

4.7 Path 1: $w = -2.74$ kJ,
$q = +2.74$ kJ, $\Delta U = 0$, $\Delta H = 0$,
$\Delta S = +9.13$ J K^{-1},
$\Delta S_{sur} = -9.13$ J K^{-1}, $\Delta S_{tot} = 0$;
Path 2: $w = -1.66$ kJ,
$q = +1.66$ kJ, $\Delta U = 0$, $\Delta H = 0$,
$\Delta S = +9.13$ J K^{-1},
$\Delta S_{sur} = -5.53$ J K^{-1},
$\Delta S_{tot} = +3.60$ J K^{-1}.

4.8 Path 1: $T_f = 227$ K,
$w = -9.1 \times 10^2$ J, $\Delta H = -1.5$ kJ,
$\Delta S = 0$, $\Delta S_{sur} = 0$, $\Delta S_{tot} = 0$; Path
2: $T_f = 240$ K, $w = -7.5$ kJ,
$\Delta H = -1.2$ kJ, $\Delta S = +1.12$ J K^{-1},
$\Delta S_{sur} = 0$, $\Delta S_{tot} = +1.12$ J K^{-1}.

4.9 Process 1: $\Delta S = +5.8$ J K^{-1},
$\Delta S_{sur} = -5.8$ J K^{-1}, $\Delta H = 0$,
$\Delta T = 0$, $\Delta A = -1.7$ kJ,
$\Delta G = -1.7$ kJ; Process 2:
$\Delta S = +5.8$ J K^{-1},
$\Delta S_{sur} = -1.7$ J K^{-1}, $\Delta H = 0$,
$\Delta T = 0$, $\Delta A = -1.7$ kJ,
$\Delta G = -1.7$ kJ; Process 3:
$\Delta S = +3.9$ J K^{-1}, $\Delta S_{sur} = 0$,
$\Delta H = -0.84$ kJ, $\Delta T = -41$ K, ΔA
and ΔG indeterminate.

4.10 (a) 200.7 J K^{-1} mol^{-1};
(b) 232.0 J K^{-1} mol^{-1}.

4.11 +45.4 J K^{-1}, +51.2 J K^{-1}.

4.12 −160.07 kJ mol^{-1}.

4.13 (a) +17.0 J K^{-1}; (b) +36 J K^{-1}.

4.14 (a) 0.11 kJ mol^{-1};
(b) 0.11 kJ mol^{-1}.

4.15 (a) 63.88 J K^{-1} mol^{-1};

(b) 66.08 J K^{-1} mol^{-1}.

4.16 7.8 km.

4.17 At 298 K: +41.16 kJ mol^{-1},
+42.08 J K^{-1} mol^{-1}; at 398 K:
+40.84 kJ mol^{-1},
+41.05 J K^{-1} mol^{-1}.

4.18 +96.864 J K^{-1} mol^{-1},
+76.9 J K^{-1} mol^{-1}.

4.19 293.5 J K^{-1} mol^{-1} at 200 K.

4.20 +34.4 kJ mol^{-1}, 243 J K^{-1} mol^{-1}
at 298 K.

4.21 0.61 μJ, 6.7 μJ.

4.25 $\Delta S = nC_{p,m} \ln(T_f/T_h)$
$+ nC_{p,m} \ln(T_f/T_c)$,
$+22.6$ J K^{-1}.

4.26 −21 K.

4.27 Step 1: $\Delta S = 0$, $\Delta S_{sur} = 0$,
Step 2: $\Delta S = +33$ J K^{-1},
$\Delta S_{sur} = -33$ J K^{-1};
Step 3: $\Delta S = 0$, $\Delta S_{sur} = 0$;
Step 4: $\Delta S = -33$ J K^{-1},
$\Delta S_{sur} = +33$ J K^{-1}.

4.30 6.86 kJ.

5.1 −501 kJ mol^{-1}.

5.2 (a) +7 kJ mol^{-1},
(b) +107 kJ mol^{-1}.

5.3 −6 kJ mol^{-1}.

5.4 73 atm.

5.7 $(\partial p/\partial S)_V = \alpha T/\kappa_T C_V$.

5.9 (a) $(\partial H/\partial p)_T = 0$;
(b) $(\partial H/\partial p)_T =$
$\{nb - (2na/RT)\lambda^2\}/$
$\{1 - (2na/RTV)\lambda^2\}$,
$\lambda = 1 - b/V_m \approx -8.3$ J atm^{-1},
−8 J.

5.11 (a) 3.0×10^{-3} atm; (b) 0.30 atm.

5.12 $(\partial C_V/\partial V)_T$
$= (RT/V_m^2)(\partial^2(BT)/\partial T^2)$.

5.14 $\pi_T = ap/RTV_m$.

5.15 0.02 per cent.

5.16 (a) $\Delta_r G' = \tau \Delta_r G + (1 - \tau)\Delta_r H$;
(b) $\Delta_r G' =$
$\tau \Delta_r G + (1 - \tau)(\Delta_r H - T\Delta_r C_p)$
$- T'\Delta C_p \ln \tau$, $\tau = T'/T$.

5.18 $q_{rev} = nRT \ln\{(V_f - nb)/(V_i - nb)\}$.

5.19 −0.50 kJ.

5.20 $G' = G + p^*V_0(1 - e^{-p/p^*})$,
expansion.

5.21 $\ln \phi = Bp/RT + (C - B^2)p^2/$
$2R^2T^2 + \cdots$, 0.999 atm.

5.22 $\phi = 2e^{\lambda - 1}/(1 + \lambda^{\frac{1}{2}})$,
$\lambda = 1 + 4pB/R$.

6.1 9 atm.

6.2 +5.42 kPa K^{-1}, 2.5 per cent.

6.3 (a) −22.0 J K^{-1} mol^{-1};
(b) −109.0 J K^{-1} mol^{-1},
+110 J mol^{-1}.

6.4 (a) −1.63 cm^3 mol^{-1};
(b) +30.1 L mol^{-1},
+0.6 kJ mol^{-1}.

6.5 234.4 K.

6.6 22°C.

6.7 (a) 357 K (84°C);
(b) +37.8 kJ mol^{-1}.

6.8 (a) 227.5°C; (b) +55 kJ mol^{-1}.

6.11 9.8 Torr.

6.12 $1/T_h = 1/T_b + Mbh/T\Delta_{vap}H$,
363 K (90°C).

6.13 $(\partial^2\mu/\partial T^2)_p = C_{p,m}/T$.

7.1 $K_A = 15.58$ kPa, $K_B = 47.03$ kPa.

7.2 17.7 cm^3 mol^{-1} (NaCl),
18.07 cm^3 mol^{-1} (H$_2$O).

7.3 −1.4 cm^3 mol^{-1} (MgSO$_4$),
18.04 cm^3 mol^{-1} (H$_2$O).

7.4 57.9 mL ethanol, 45.8 mL water,
+0.96 cm^3.

7.5 36 to 62 g mol^{-1}.

7.6 4.

7.9 −4.6 kJ.

7.10 $\mu_A = \mu_A^* + RT \ln x_A + gRTx_B^2$.

8.2 (a) 2150°C; (b) x(MgO) = 0.35,
y(MgO) = 0.18, ratio 0.4;
(c) 2640°C.

8.10 2, 81 g.

9.1 (a) +4.48 kJ mol^{-1};
(b) 0.101 atm.

9.5 $\Delta_r G^\ominus(T)/(\text{kJ mol}^{-1})$
$= 78 - 0.161(T/\text{K})$.

9.6 1.69×10^{-5}.

9.7 5.71, −103 kJ mol^{-1}.

9.8 +14.7 kJ mol^{-1}, +18 g kJ mol^{-1}.

9.9 7 mmol H$_2$, 107 mmol I$_2$,
786 mmol HI.

9.10 −66.1 kJ mol^{-1},
−142 J K^{-1} mol^{-1}.

9.11 1.800×10^{-3} (at 973 K),
1.109×10^{-2} (at 1073 K),
4.848×10^{-2} (at 1173 K);
+158 kJ mol^{-1}.

9.12 Gas: −137 kJ mol^{-1},
−469 J K^{-1} mol^{-1}; liquid:
−102 kJ mol^{-1},
−312 J K^{-1} mol^{-1}.

9.14 $\xi = 1 - 1/(1 + ap/p^\ominus)^{\frac{1}{2}}$.

9.15 0.140.

9.16 $\Delta_r G' = \Delta_r G + (T - T')\Delta_r S$
$+ \alpha \Delta a + \beta \Delta b + \gamma \Delta c$,
$\alpha = T' - T - T' \ln(T'/T)$,
$\beta = \frac{1}{2}(T'^2 - T^2) - T'(T' - T)$,
$\gamma = 1/T - 1/T'$
$+ \frac{1}{2}T'(1/T'^2 - 1/T^2)$;
−225.31 kJ mol^{-1}.

10.1 Pb(s)|PbSO$_4$(s)|PbSO$_4$(aq)|
|Hg$_2$SO$_4$(aq)|Hg$_2$SO$_4$(s)|Hg(l),
+0.98 V.

10.2 (a) 4.0 mmol kg^{-1},
0.12 mmol kg^{-1}; (b) 0.74, 0.60;
(c) 5.9; (d) +1.102 V;
(e) +1.091 V.

10.3 2.0.

10.4 (a) +1.23 V; (b) +1.09 V.

10.5 0.637 bar, 22.3 bar, 28.6 bar.

10.6 (a) $E = E^\ominus - (38.54$ mV$)$
$\times \{\ln(4m) + \ln \gamma_\pm\}$;
(b) +1.0304 V; (c) 6.84×10^{34};
(d) 0.763; (e) 0.75;
(f) −77.6 J K^{-1} mol^{-1},
−259.9 kJ mol^{-1}.

10.7 +0.268 38 V.

10.8 14.23 at 20.0°C, $+74.9\,\text{kJ mol}^{-1}$, $+80.0\,\text{kJ mol}^{-1}$, $-17.1\,\text{J K}^{-1}\,\text{mol}^{-1}$.

10.9 0.533.

10.10 (a) $+0.2223\,\text{V}$; (b) 1.10, 0.796.

10.12 $-131.25\,\text{kJ mol}^{-1}$, $+56.7\,\text{J K}^{-1}\,\text{mol}^{-1}$, $-167.10\,\text{kJ mol}^{-1}$.

10.14 (a) $E = E^{\ominus} + (2.303RT/F)\text{pOH}$; (b) $E = E^{\ominus} + (2.303RT/F) \times (\text{p}K_w - \text{pH})$; (c) $-37.6\,\text{mV}$.

10.15 $-1.2\,\text{V}$.

11.1 (a) $1.6 \times 10^{-33}\,\text{J m}^{-3}$; (b) $2.5 \times 10^{-4}\,\text{m}^{-3}$.

11.2 $6.29 \times 10^{-34}\,\text{J s}$.

11.3 (a) $7.47 \times 10^{-29}\,\text{J m}^{-3}$; (b) $4.59 \times 10^{-14}\,\text{J m}^{-3}$; (c) $3.49 \times 10^{-11}\,\text{J m}^{-3}$; (a) $0.807\,\text{J m}^{-3}$; (b) $1.67\,\text{J m}^{-3}$; (c) $2.10\,\text{J m}^{-3}$.

11.4 (a) 2231 K, $0.031R$; (b) 343 K, $0.897R$.

11.5 (a) 0.020; (b) 0.007; (c) 6×10^{-6}; (d) 0.5; (e) 0.61.

11.6 9.0×10^{-6}, 1.2×10^{-6}.

11.7 $\lambda_{max} T = hc/5k$.

11.8 (a) $N = (2/L)^{\frac{1}{2}}$; (b) $1/c(2L)^{\frac{1}{2}}$; (c) $1/(\pi a^3)^{\frac{1}{2}}$; (d) $1/(32\pi a^5)^{\frac{1}{2}}$.

11.9 (a) $N = 1/(32\pi a_0^5)^{\frac{1}{2}}$; (b) $N = 1/(32\pi a_0^5)^{\frac{1}{2}}$.

11.10 (a) ik; (c) 0.

11.11 (a) -1; (b) -1.

11.12 (a) $-k^2$; (b) $-k^2$; (c) 0; (d) 0.

11.13 (a) $\cos^2 \chi$; (b) $\sin^2 \chi$; (c) $0.95 e^{ikx} \pm 0.32 e^{-ikx}$.

11.14 $\hbar^2 k^2/2m$.

11.15 (a) $k\hbar$; (b) 0; (c) 0.

11.16 (a) $6a_0$, $42a_0^2$; (b) $4a_0$, $30a_0^2$.

11.17 (a) $-e^2/4\pi\varepsilon_0 a_0$; (b) $\hbar^2/m_e a_0^2$.

11.19 (a) 1; (b) $2x$; (c) $\hbar$.

12.1 $1.24 \times 10^{-39}\,\text{J}$, 2.2×10^9, $1.8 \times 10^{-30}\,\text{J}$.

12.2 (a) $1.60 \times 10^{-19}\,\text{J}$; (b) $2.42 \times 10^{14}\,\text{Hz}$; (c) total number of electrons = 1.12212.

12.3 CO $(1900\,\text{N m}^{-1})$ $> \text{NO}\ (1600\,\text{N m}^{-1})$ $> \text{HCl}\ (516\,\text{N m}^{-1})$ $> \text{HBr}\ (412\,\text{N m}^{-1})$ $> \text{HI}\ (314\,\text{N m}^{-1})$.

12.4 $1.30 \times 10^{-27}\,\text{J}$, $\pm\hbar$.

12.5 $E/(10^{-22}\,\text{J}) = 0$, 2.62, 7.86, 15.72.

12.6 $E = (h^2/8m) \times (n_1^2/L_1^2 + n_2^2/L_2^2 + n_3^2/L_3^2)$.

12.7 (a) $N^2/2\kappa$; (b) $N^2/4\kappa^2$.

12.8 $g = \frac{1}{2}(mk/\hbar^2)^{\frac{1}{2}}$.

12.9 $\langle T \rangle = \frac{1}{2}(v + \frac{1}{2})\hbar\omega$.

12.10 0, $\frac{3}{4}(2v^2 + 2v + 1)\alpha^4$.

12.11 (a) $L/(12)^{\frac{1}{2}}$, $nh/2L$; (b) $\{(v + \frac{1}{2})\hbar/\omega m\}^{\frac{1}{2}}$, $\{(v + \frac{1}{2})\hbar\omega m\}^{\frac{1}{2}}$.

12.12 $\mu_{v+1,v} = \alpha\{(v + 1)/2\}^{\frac{1}{2}}$, $\mu_{v-1,v} = \alpha (v/2)^{\frac{1}{2}}$.

12.13 $\langle T \rangle = -\frac{1}{2}\langle V \rangle$.

12.14 (a) $+\hbar$, $\hbar^2/2I$; (b) $-2\hbar$, $2\hbar^2/I$; (c) 0, $\hbar^2/2I$; (d) $\hbar \cos 2\chi$, $\hbar^2/2I$.

12.15 (a) 0, 0; (b) $3\hbar^2/I$, $6\frac{1}{2}\hbar$; (c) $6\hbar^2/I$, $2(3\frac{1}{2})\hbar$.

12.17 $\cos \theta = m_l/\{l(l + 1)\}^{\frac{1}{2}}$, $54°44'$.

12.18 $-(a^2 + b^2 + c^2)$.

12.19 $l_x = (\hbar/i)(y\partial/\partial z - z\partial/\partial y)$ and cyclic permutations, $[l_x, l_y] = i\hbar l_z$.

13.1 $n_2 \to 6$, 7503, 5908, 5129, ..., 3908 nm.

13.2 397.13 nm, 3.40 eV.

13.3 $987\,663\,\text{cm}^{-1}$, $137\,175\,\text{cm}^{-1}$, $185\,187\,\text{cm}^{-1}$, 122.5 eV.

13.4 5.39 eV.

13.5 $A = 38.50\,\text{cm}^{-1}$.

13.6 $3.3429 \times 10^{-27}\,\text{kg}$, 1.000 272.

13.7 7621, 10 228, $11\,552\,\text{cm}^{-1}$, 6.80 eV.

13.8 0.420 pm.

13.9 $2s$.

13.10 106 pm.

13.12 $p_x \pm ip_y$.

13.14 $2.66a_0$.

13.15 $E = -(Z^2 e^4 m_e/32\pi^2\varepsilon_0^2\hbar^2)(1/n^2)$.

13.17 $a_{\text{Ps}} = 2a_0$, $E_{1,\text{Ps}} = \frac{1}{2}E_{1,\text{H}}$.

14.4 (a) 8.6×10^{-7}, 2.0×10^{-6}; (b) 8.6×10^{-7}, 2.0×10^{-6}; (c) 3.7×10^{-7}, 0; (d) 4.9×10^{-7}, 5.5×10^{-7}.

14.5 1.9 eV, 130 pm.

14.7 2.7 eV (460 nm).

14.13 (a) Nonplanar; (b) planar.

15.1 (a) D_{3d}; (b) D_{3d}, C_{2v}; (c) D_{2h}; (d) D_3; (e) D_{4d}.

15.2 trans-CHCl=CHCl.

15.3 $C_2 \sigma_h = i$.

15.6 $D(C_6) = +1$, -1, respectively.

15.8 $A_1 + T_2$, s and p, (d_{xy}, d_{yz}, d_{zx}) span T_2.

15.9 (a) All 5 d orbitals; (b) all except $d_{xy}(A_2)$.

15.10 (a) $2A_1 + A_2 + 2B_1 + 2B_2$; (b) $A_1 + E$; (c) $A_1 + T_1 + T_2$; (d) $A_{2u} + T_{1u} + T_{2u}$.

15.11 (a) Yes, (b) no; (c) yes.

15.13 $4A_1 + 2B_1 + 3B_2 + A_2$.

16.1 (a) 2.1×10^{-6} 1.3 MHz, $0.0063\,\text{cm}^{-1}$; (b) 9.7×10^{-7}, 6.6 kHz, $0.0004\,\text{cm}^{-1}$.

16.2 700 MHz, 1 Torr.

16.3 596 GHz, $19.9\,\text{cm}^{-1}$, 0.503 mm, $B = 9.941\,\text{cm}^{-1}$.

16.4 From 112.83 pm to 123.52 pm.

16.5 $R(\text{CC}) = 139.6$ pm, $R(\text{CH}) = 108.5$ pm.

16.6 $k = 93.8\,\text{N m}^{-1}$, $142.81\,\text{cm}^{-1}$, 3.36 eV.

16.7 (a) $2143.26\,\text{cm}^{-1}$; (b) $12.8195\,\text{kJ mol}^{-1}$; (c) $1.85563\,\text{kN m}^{-1}$; (d) $1.91\,\text{cm}^{-1}$; (e) 113 pm.

16.8 $2.728 \times 10^{-47}\,\text{kg m}^2$, 129.5 pm; DCl lines at 10.56, 21.11, 31.67, ... cm^{-1}.

16.9 HCl: 128.393 pm, DCl: 128.13 pm.

16.10 $R(\text{CuBr}) = 218$ pm.

16.11 $R(\text{CO}) = 116.28$ pm, $R(\text{CS}) = 155.97$ pm.

16.12 (a) 5.15 eV; (b) 5.20 eV.

16.15 $J_{max} = (kT/2hcB)^{\frac{1}{2}} - \frac{1}{2}$, 30, $J_{max} = (kT/hcB)^{\frac{1}{2}} - \frac{1}{2}$, 6.

16.16 349 pm, 401 pm, 458 pm.

17.1 $49\,354\,\text{cm}^{-1}$.

17.2 5.1147 eV.

17.3 $14\,660\,\text{cm}^{-1}$.

17.4 $4.8 \times 10^4\,\text{L mol}^{-1}\,\text{cm}^{-2}$.

17.5 $1.1 \times 10^6\,\text{L mol}^{-1}\,\text{cm}^{-2}$.

17.6 5.07 eV.

17.7 0.2 ms.

17.11 (a) Allowed; (b) forbidden.

17.15 $f = (RS/a_0)^{\frac{1}{2}}$.

17.16 Lengthen; to red.

17.17 (a) $3 + 1$, $3 + 3$; (b) $4 + 4$, $2 + 2$.

18.1 10.3 T, 2.42×10^{-5}, β.

18.2 $57\,\text{kJ mol}^{-1}$.

18.3 1·992, 2.002.

18.4 6.9 mT, 2.1 mT.

18.5 $1:2:3:2:1$ quintet of $1:4:6:4:1$ quintets.

18.6 0.10, 0.38; (a) 3.8; (b) 0.52; 3.8, 131°.

18.8 158 pm.

18.10 $-0.89\,\mu\text{T}$.

18.11 $I = \frac{1}{2}A\tau/\{1 + (\omega_0 - \omega)^2\tau^2\}$.

19.1 No.

19.2 3.5 fK, 7.41.

19.3 (a) 5.00, 6.25; (b) 1.00 at 298 K, 0.80 at 5000 K; 6.5×10^{-11} at 298 K and 0.12 at 5000 K; (c) $13.38\,\text{J K}^{-1}\,\text{mol}^{-1}$, $18.07\,\text{J K}^{-1}\,\text{mol}^{-1}$.

19.4 0.257, 0.336, 0.396, 0.011.

19.5 (a) 0.64, 0.36; (b) $0.52\,\text{kJ mol}^{-1}$.

19.6 (a) 1.049, $1.65\,\text{J K}^{-1}\,\text{mol}^{-1}$; (b) 1.56, $8.37\,\text{J K}^{-1}\,\text{mol}^{-1}$.

19.7 (a) 1; (b) {2, 2, 0, 1, 0, 0} and {2, 1, 2, 0, 0, 0}.

19.8 {4, 2, 2, 1, 0, 0, 0, 0, 0, 0}.

19.9 (a) 160 K.

19.10 (a) 2.104; (b) $1.547RT$, $13.08\,\text{J K}^{-1}\,\text{mol}^{-1}$, $19.27\,\text{J K}^{-1}$.

19.11 104 K; (b) $q = 1 + a$; (c) $2Nk \ln 2$.

20.1 (a) $0.354R$; (b) $0.079R$; (c) $0.029R$.

20.2 (a) 0.1 per cent; (b) 4×10^{-3} per cent.

20.3 4.2, $15\,\text{J K}^{-1}\,\text{mol}^{-1}$.

20.5 19.89.

20.7 (a) 3.89; (b) 2.41.

20.10 (b) $5.41\,\text{J K}^{-1}\,\text{mol}^{-1}$.

20.11 100 T.

20.12 $350\,\text{m s}^{-1}$.

21.1 117 pm.

21.2 P, 342 pm.

21.3 (a) bcc, 316 pm, 137 pm; (b) fcc, 361 pm, 128 pm.

21.4 $10.51\,\mathrm{g\,cm^{-3}}$.

21.6 628 pm, yes.

21.7 834, 606, 870 pm.

21.8 $\alpha_{\mathrm{Vol}} = 4.8 \times 10^{-5}\,\mathrm{K^{-1}}$, $\alpha_{\mathrm{Lin}} = 1.6 \times 10^{-5}\,\mathrm{K^{-1}}$.

21.9 177 pm.

21.12 (a) 0.5236; (b) 0.6802; (c) 0.7405.

21.13 (a) No absences; (b) alternation ($h + k + l$ odd or even); (c) $h + k + l$ odd missing.

22.1 (a) $0.11\,\mathrm{GV\,m^{-1}}$; (b) $4\,\mathrm{GV\,m^{-1}}$; (c) $4\,\mathrm{kV\,m^{-1}}$.

22.2 2.4 nm.

22.3 $1.2 \times 10^{-23}\,\mathrm{cm^3}$, 0.86 D.

22.4 $1.38 \times 10^{-23}\,\mathrm{cm^3}$, 0.35 D.

22.5 $2.24 \times 10^{-24}\,\mathrm{cm^3}$, 1.58 D.

22.6 1.85, $1.36 \times 10^{-24}\,\mathrm{cm^3}$.

22.7 (a) $6Q_1Q_2/\pi\varepsilon_0 r^5$; (b) $9Q_1Q_2/4\pi\varepsilon_0 r^5$.

22.8 $n_{\mathrm{r}} = 1 + (2\pi\alpha'/kT)$.

22.11 $a = 2\pi N_{\mathrm{A}}^2 C_6/3d^3$.

22.16 $\xi = -e^2 a_0^2/2m_{\mathrm{e}}$, $\chi_{\mathrm{m}} = -N_{\mathrm{A}}\mu_0 e^2 a_0^2/2m_{\mathrm{e}}$.

23.1 $23.1\,\mathrm{kg\,mol^{-1}}$, $1.02\,\mathrm{m^3\,mol^{-1}}$.

23.2 $155\,\mathrm{kg\,mol^{-1}}$, $13.7\,\mathrm{m^3\,mol^{-1}}$.

23.3 $0.0716\,\mathrm{L\,g^{-1}}$.

23.4 5.0 Sv.

23.5 $65.6\,\mathrm{kg\,mol^{-1}}$.

23.6 3500 rpm.

23.7 $-27\,\mathrm{mV}$.

23.8 $5\,\mathrm{m^3\,mol^{-1}}$.

23.9 3.4 nm.

23.10 $0.21\,\mathrm{Mg\,mol^{-1}}$.

23.11 $365\,\mathrm{g\,mol^{-1}}$ ($342\,\mathrm{g\,mol^{-1}}$).

23.12 $6.2\,\mathrm{nm} \times 1.8\,\mathrm{nm}$.

23.13 5.14 Sv, $60.1\,\mathrm{kg\,mol^{-1}}$.

23.14 $158\,\mathrm{kg\,mol^{-1}}$.

23.15 (a) $(\tfrac{5}{3})^{\frac{1}{2}}a$; (b) $l/2(3)^{\frac{1}{2}}$; 46 nm.

23.18 $1.01\,\mathrm{g\,cm^{-3}}$.

23.20 BSV: $28\,\mathrm{m^2\,mol^{-1}}$, 2.6 per cent; Hb: $0.33\,\mathrm{m^3\,mol^{-1}}$, 5 per cent.

23.21 (a) $0.39\,\mathrm{m^3\,mol^{-1}}$; (b) $1.1\,\mathrm{m^3\,mol^{-1}}$.

23.23 $\mathrm{d}G = -S\,\mathrm{d}T - l\,\mathrm{d}t$, $\mathrm{d}A = -S\,\mathrm{d}T + t\,\mathrm{d}l$.

23.26 (a) $lN^{\frac{1}{2}}$, 9.74 nm; (b) $(8N/3\pi)^{\frac{1}{2}}l$, 8.97 nm; (c) $(2N/3)^{\frac{1}{2}}l$, 7.95 nm.

24.1 100 Ms.

24.2 9.1.

24.3 7.3 mPa.

24.4 (a) $1.6 \times 10^8\,\mathrm{s^{-1}}$; (b) $1.6 \times 10^3\,\mathrm{s^{-1}}$.

24.5 (a) 100 Pa; (b) 24 Pa.

24.6 (a) $2 \times 10^{14}\,\mathrm{s^{-1}}$; (b) $1 \times 10^{20}\,\mathrm{s^{-1}}$.

24.7 $5.3\,\mathrm{S\,cm^2\,mol^{-1}}$.

24.8 (a) $119.2\,\mathrm{S\,cm^2\,mol^{-1}}$; (b) $1.192\,\mathrm{mS\,cm^{-1}}$; (c) $173.1\,\Omega$.

24.9 $1.36 \times 10^{-5}\,\mathrm{mol\,L^{-1}}$.

24.10 4.73.

24.11 $40\,\mu\mathrm{m\,s^{-1}}$, $5.2\,\mu\mathrm{m\,s^{-1}}$, $7.6\,\mu\mathrm{m\,s^{-1}}$; 250 s, 190 s, 130 s; 13 nm, 17 nm, 24 nm.

24.12 0.82, 0.0028.

24.13 0.48, $7.5 \times 10^{-8}\,\mathrm{m^2\,s^{-1}\,V^{-1}}$, $72\,\mathrm{S\,cm^2\,mol^{-1}}$.

24.14 0.278, 0.280.

24.15 (a) $12\,\mathrm{kN\,mol^{-1}}$, $2.1 \times 10^{-20}\,\mathrm{N\,molecule^{-1}}$; (b) $17\,\mathrm{kN\,mol^{-1}}$, $2.8 \times 10^{-20}\,\mathrm{N\,molecule^{-1}}$; (c) $25\,\mathrm{kN\,mol^{-1}}$, $4.1 \times 10^{-20}\,\mathrm{N\,molecule^{-1}}$.

24.17 $E_{\mathrm{a}} = 9.3\,\mathrm{kJ\,mol^{-1}}$.

24.18 (a) $c.0$; (b) $63\,\mathrm{mmol\,L^{-1}}$.

24.19 $1.2\,\mathrm{g\,m^{-1}\,s^{-1}}$.

24.21 $t'/t'' = c'u'z'/c''u''z''$.

24.23 (a) 0; (b) 0.016; (c) 0.054.

24.24 $N > 60$.

25.1 2, $59\,\mathrm{mL\,mol^{-1}\,min^{-1}}$, 2.94 g.

25.2 1, $1.51 \times 10^{-5}\,\mathrm{s^{-1}}$, $9.82\,\mathrm{mmol\,L^{-1}}$.

25.3 1, $0.12\,\mathrm{s^{-1}}$.

25.4 1, $5.84 \times 10^{-3}\,\mathrm{s^{-1}}$, 1.98 min.

25.5 $97.0\,\mathrm{kJ\,mol^{-1}}$.

25.6 55.4 per cent.

25.7 1, $2.8 \times 10^{-4}\,\mathrm{s^{-1}}$.

25.8 $3.65 \times 10^{-3}\,\mathrm{min^{-1}}$, 274 min.

25.9 $2.42 \times 10^7\,\mathrm{L\,mol^{-1}\,s^{-1}}$.

25.10 1, $7.2 \times 10^{-4}\,\mathrm{s^{-1}}$.

25.11 Propene: 1, HCl: 3.

25.13 $-20\,\mathrm{kJ\,mol^{-1}}$, $+8\,\mathrm{kJ\,mol^{-1}}$.

25.14 $1.14 \times 10^{10}\,\mathrm{L\,mol^{-1}\,s^{-1}}$, $16.7\,\mathrm{kJ\,mol^{-1}}$.

25.16 $10\,\mathrm{mmol\,L^{-1}}$.

26.1 $1.9 \times 10^{20}\,\mathrm{s^{-1}}$, 3.1×10^{-4} einstein $\mathrm{s^{-1}}$.

26.2 $5.1 \times 10^8\,\mathrm{L\,mol^{-1}\,s^{-1}}$.

26.4 $5.0 \times 10^7\,\mathrm{L\,mol^{-1}\,s^{-1}}$.

26.11 Ratio $= M(1 + 4p + p^2)/(1 - p^2)$.

26.15 $[\mathrm{X}] = k_{\mathrm{c}}/k_{\mathrm{b}}$, $[\mathrm{Y}] = k_{\mathrm{a}}[\mathrm{A}]/k_{\mathrm{b}}$.

26.17 Step 1 autocatalytic.

27.1 (a) $0.044\,\mathrm{nm^2}$; (b) 0.15.

27.2 0.007, $0.0040\,\mathrm{nm^2}$.

27.3 $1.7 \times 10^{11}\,\mathrm{L\,mol^{-1}\,s^{-1}}$, 3.6 ns.

27.4 $+83.8\,\mathrm{kJ\,mol^{-1}}$, $+19.1\,\mathrm{J\,K^{-1}\,mol^{-1}}$, $85.9\,\mathrm{kJ\,mol^{-1}}$, $+79.0\,\mathrm{kJ\,mol^{-1}}$.

27.5 $2-$.

27.9 $P \approx 5.2 \times 10^{-6}$.

27.10 $1.8 \times 10^6\,\mathrm{L\,mol^{-1}\,s^{-1}}$.

27.11 $1.5 \times 10^6\,\mathrm{L\,mol^{-1}\,s^{-1}}$.

27.13 (a) $2.7 \times 10^{-15}\,\mathrm{m^2\,s^{-1}}$; (b) $1.1 \times 10^{-14}\,\mathrm{m^2\,s^{-1}}$.

27.14 $8 \times 10^{-12}\,\mathrm{m^2\,s^{-1}}$.

27.16 5.

28.2 KCl: $-5.7 \times 10^{-11}\,\mathrm{mol\,cm^{-2}}$; Na: $-6.7 \times 10^{-11}\,\mathrm{mol\,cm^{-2}}$; $-11.1 \times 10^{-11}\,\mathrm{mol\,cm^{-2}}$.

28.3 $0.379\,\mathrm{nm^2}$.

28.4 $-4.10 \times 10^{-7}\,\mathrm{mol\,m^{-2}}$.

28.6 $0.52\,\mathrm{Torr^{-1}}$.

28.7 (a) 164, $13.1\,\mathrm{cm^3}$; (b) 264, $12.5\,\mathrm{cm^3}$.

28.8 Langmuir, $0.017\,\mathrm{Torr^{-1}}$, $0.13\,\mathrm{nm^2}$, $0.19\,\mathrm{cm^3}$.

28.9 1.4 mL, $5.9\,\mathrm{m^2}$.

28.10 BET better; $75.4\,\mathrm{cm^3}$, 3.98.

28.11 2.4, 0.16.

28.12 $0.02\,\mathrm{Torr^{-1}}$.

28.14 $U \propto C_6/R^3$, 294 pm.

28.18 $\mathrm{d}\mu' = (RTV_\infty/\sigma)\,\mathrm{d}\ln(1 - \theta)$.

29.1 0.38, $0.78\,\mathrm{mA\,cm^{-2}}$.

29.2 $a(\mathrm{Sn^{2+}}) \approx 2.2a(\mathrm{Pb^{2+}})$.

29.3 60 per cent, 45 per cent.

29.6 $6\,\mu\mathrm{A}$.

29.7 $0.28\,\mathrm{mg\,cm^{-2}\,d^{-1}}$.

29.8 $7.2\,\mu\mathrm{A}$.

29.10 $j/j_{\mathrm{L}} = 1 - \exp(F\eta^{\mathrm{c}}/RT)$.

Copyright acknowledgements

The permission of the respective copyright holders to reproduce tables of data and extracts from them is gratefully acknowledged as follows: Butterworths (28.2, 28.3 [© 1964]), Professor J. G. Calvert (17.1, 17.2 [© 1966], Dr J. Emsley (2.6, 2.7 [© 1989]), Chapman and Hall (16.3 [© 1975], 2.11, 9.1 [© 1986]), Cornell University Press (2.9 [© 1960]), CRC Press (1.3, 2.8, 9.1, 10.2, 10.5, 22.1 [© 1979]), Dover Publications Inc. (12.2 [© 1965]), Elsevier Scientific Publishing Company (22.1 [© 1978]), Professor A. A. Frost (26.3), Longman (1.2, 1.4, 2.2, 2.15, 3.1, 7.2, 9.1, 21.3, 22.5, 24.2, 24.4, 24.5, 29.2 [© 1973]), McGraw Hill Book Company (1.3, 1.4, 2.3, 2.6, 3.1, 3.2, 4.1, 16.2, 22.2, 24.3, 24.7 [© 1975], 5.2 [© 1961], 25.1, 25.2 [© 1965], 3.2 [© 1968]), National Bureau of Standards (2.11, 2.12 [© 1982]), Dr J. Nicholas (25.1, 25.2, 25.4 [© 1976]), Oxford University Press (25.1, 25.2, 25.4 [© 1975], 28.4, 28.5, 28.6 [© 1974]), Pergamon Press Ltd (22.3 [© 1961]), Prentice Hall Inc. (23.1, 23.3 [© 1971]), John Wiley and Sons Inc. (7.1 [© 1975], 22.5 [© 1954], 23.2, 23.4 [© 1961]). The sources of the material are quoted at the foot of each table.

The permission of the following individuals, institutions, and journals for reproduction of illustrations is gratefully acknowledged: Dr R. F. Barrow (11.9), Professor G. C. Bond (28.4), Professor G. Erlich (28.44), Dr A. J. Forty (28.16), Dr J. Foster (28.34), Professor D. Freifelder (23.6), Professor H. Ibach (28.22), Professor M. Karplus (27.23, 27.24), Dr W. Kiefer and Dr H. J. Bernstein (16.55), Professor D. A. King (28.42), Professor D. A. Long (16.53), Dr G. Morris (18.31), Professor E. W. Müller (28.32), Dr A. H. Narten (24.13), Open University Press (17.42 [© 1979]), Oxford University Press (28.16, 28.17), Dr M. Prutton (28.30), Professor C. F. Quate (28.33), Professor M. W. Roberts (28.20), Dr H. M. Rosenberg (28.17), Professor G. A. Somorjai (28.27, 29.29), *Scientific American* (23.19), Professor D. W. Turner (28.21), Professor A. H. Zewail (27.2).

Index

(T) after a page number refers to a Table in the text (and usually in the *Data section*).

You're going to have problems with physical chemistry. Here are the solutions.

SOLUTIONS MANUAL FOR PHYSICAL CHEMISTRY

FIFTH EDITION

Peter Atkins, Oxford University
Charles Trapp, University of Louisville

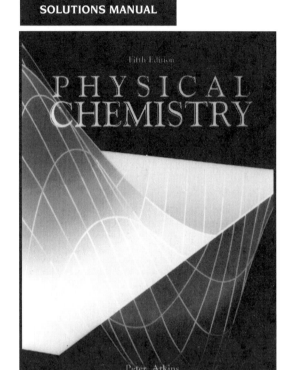

To succeed in physical chemistry, it is essential that you develop good problem-solving skills—skills that go beyond plugging numbers into formulas. This **SOLUTIONS MANUAL** offers invaluable assistance.

In the **Manual** you will find:

- **Fully-detailed solutions and answers** to all problems and exercises in **Physical Chemistry, Fifth Edition.**

- **Special strategy sections** in each solution, which will help you zero in on the essence of the problem and reason your way through it.

- **Unique compatibility with the main text.** This student resource was cowritten by the author of the textbook. It shares the same unique voice and vision of **PHYSICAL CHEMISTRY, Fifth Edition.**

Look for **SOLUTIONS MANUAL FOR PHYSICAL CHEMISTRY, 5/e,** at your college bookstore. If it is not available, ask the bookstore manager to order it.

ISBN 1-7167-2403-0

Published by ▪▪ W. H. FREEMAN AND COMPANY

The Periodic Table

Group →	1	2	3	4	5	6	7	8	9	10	11	12	13/III	14/IV	15/V	16/VI	17/VII	18/VIII
Period 1	1 H 1.008																	2 He 4.003
Period 2	3 Li 6.941	4 Be 9.012											5 B 10.81	6 C 12.01	7 N 14.01	8 O 16.00	9 F 19.00	10 Ne 20.18
Period 3	11 Na 22.99	12 Mg 24.30											13 Al 26.98	14 Si 28.09	15 P 30.97	16 S 32.07	17 Cl 35.45	18 Ar 39.95
Period 4	19 K 39.10	20 Ca 40.08	21 Sc 44.96	22 Ti 47.88	23 V 50.94	24 Cr 52.00	25 Mn 54.94	26 Fe 55.85	27 Co 58.93	28 Ni 58.69	29 Cu 63.55	30 Zn 65.39	31 Ga 69.72	32 Ge 72.61	33 As 74.92	34 Se 78.96	35 Br 79.90	36 Kr 83.80
Period 5	37 Rb 85.47	38 Sr 87.62	39 Y 88.91	40 Zr 91.22	41 Nb 92.91	42 Mo 95.94	43 Tc 98.91	44 Ru 101.1	45 Rh 102.9	46 Pd 106.4	47 Ag 107.9	48 Cd 112.4	49 In 114.8	50 Sn 118.7	51 Sb 121.8	52 Te 127.6	53 I 126.9	54 Xe 131.3
Period 6	55 Cs 132.9	56 Ba 137.3	La-Lu	72 Hf 178.5	73 Ta 180.9	74 W 183.8	75 Re 186.2	76 Os 190.2	77 Ir 192.2	78 Pt 195.1	79 Au 197.0	80 Hg 200.6	81 Tl 204.4	82 Pb 207.2	83 Bi 209.0	84 Po 210.0	85 At 210.0	86 Rn 222.0
Period 7	87 Fr 223.0	88 Ra 226.0	Ac-Lr	104 Unq	105 Unp	106 Unh	107 Uns	108 Uno	109 Une	......								

s block, d block, p block

Lanthanides

57 La 138.9	58 Ce 140.1	59 Pr 140.9	60 Nd 144.2	61 Pm 144.9	62 Sm 150.4	63 Eu 152.0	64 Gd 157.2	65 Tb 158.9	66 Dy 162.5	67 Ho 164.9	68 Er 167.3	69 Tm 168.9	70 Yb 173.0	71 Lu 175.0

Actinides

89 Ac 227.0	90 Th 232.0	91 Pa 231.0	92 U 238.0	93 Np 237.0	94 Pu 239.1	95 Am 243.1	96 Cm 247.1	97 Bk 247.1	98 Cf 252.1	99 Es 252.1	100 Fm 257.1	101 Md 256.1	102 No 259.1	103 Lr 260.1

f block